Progress in New Crops

Proceedings of the Third National Symposium

NEW CROPS

NEW OPPORTUNITIES, NEW TECHNOLOGIES

Indianapolis, Indiana, October 22–25, 1996

Under the Auspices of

The Indiana Center for New Crops and Plant Products, Purdue University

Indiana Business Modernization and Technology Corporation (BMT)

Association for the Advancement of Industrial Crops (AAIC)

Cosponsored by

United States Department of Agriculture (USDA)

Agricultural Research Service (ARS)

Cooperative State Research Education and Extension Service (CSREES)

Alternative Agricultural Research and Commercialization Center (AARC)

American Society for Horticultural Science

Crop Science Society of America

American Society of Agronomy

The Society for Economic Botany

Center for Alternative Plant and Animal Products, University of Minnesota

Industrial Sponsors

Procter & Gamble

Dow Elanco

Maple Leaf Farms

PROGRESS IN NEW CROPS

Edited by
Jules Janick
Purdue University

Jacket photo: The photograph of *Capsicums*
is courtesy of Paul Bosland.

ISBN 0-9615027-3-8
Printed in the United States of America

ASHS Press
600 Cameron Street
Alexandria, VA 22314-2562
telephone 703.836.4606 • fax 703.836.2024 • e-mail ashspres@ashs.org

CONTENTS

PART III: NEW CROPS, NEW TECHNOLOGIES

CEREALS AND PSEUDOCEREALS

GRAIN LEGUMES

FORAGES

OILSEEDS

INDUSTRIAL CROPS

VEGETABLES

FLORAL AND LANDSCAPE CROPS

MEDICINAL, AROMATIC, SPICE, AND BIOACTIVE CROPS

List of Contributors

Ozzie Abaye
Department of Crop and Soil Environmental Sciences
Virginia Tech
Blacksburg, Virginia 24061-0404

Olayiwola Adeola
Purdue University
Department of Animal Sciences
1151 Lilly Hall
West Lafayette, Indiana 47907-1151

Muhammad Afzal
Plant Genetic Resources Institute
National Agricultural Research Centre
Islamabad, Pakistan

Vernon L. Anderson
North Dakota State University
Carrington Research Extension Center
Box 219
Carrington, North Dakota 58421

C. Anderson
Department of Plant, Soil, and Insect Sciences
University of Wyoming
Laramie, Wyoming 82071

Robert G. Anderson
Department of Horticulture and Landscape Architecture
N-318 Ag Science Bldg N
University of Kentucky
Lexington, Kentucky 40546-0091

David J. Andrews
University of Nebraska
Agronomy Department
Lincoln, Nebraska 68583-0915

Paul Angers
Indiana Center for New Crops and Plant Products
Purdue University
1165 Horticulture Building
West Lafayette, Indiana 47907-1165

José Luis Angulo Sánchez
Centro de Investigación en Química Aplicada
Saltillo, Coahuila
México 25100

Robert Armstrong
USDA/AARC
14th & Independence SW, Rm 0156
Washington, DC 20250-0401

Ricardo Ayerza (h)
Office of Arid Land Studies
Bioresources Research Facility
The University of Arizona
250 E Valencia Rd
Tucson, Arizona 85706

Brian S. Baldwin
Mississippi State University
Department of Plant & Soil Sciences
Box 9555
Mississippi State, Mississippi 39762

Sharon Bale
Department of Horticulture and Landscape Architecture
N-318 Agricultural Science Bldg. N
University of Kentucky
Lexington, Kentucky 40546-0091

David D. Baltensperger
University of Nebraska
Panhandle Research & Extension Center
4502 Avenue I
Scottsbluff, Nebraska 69361-4939

Martha A. Belury
Department of Foods & Nutrition
Purdue University
West Lafayette, Indiana 47907

Marisol Berti
Facultad di Agronomia
Universidad de Concepción
Casilla 537
Chillán, Chile

Harbans L. Bhardwaj
Agricultural Research Station
Virginia State University
Box 9152
Petersburg, Virginia 23806

B. Bjornsen
KFX Tech, Inc.
Gillette, Wyoming 82716

Melvin G. Blase
Department of Agricultural Economics
University of Missouri
200 Mumford Hall
Columbia, Missouri 65211

Bruce Bordelon
Department of Horticulture
Purdue University
1165 Horticulture Building
West Lafayette, Indiana 47907-1165

Paul W. Bosland
Department of Agronomy and Horticulture
New Mexico State University
Box 30001, Dept 3Q
Las Cruces, New Mexico 88003-8003

Gerard Boswinkel
ATO-DLO
Haagsteeg 6/Postbus 17
Wageningen, 6700 AA
The Netherlands

Joe H. Bouton
Department of Crop & Soil Sciences
3111 Plant Sciences Building
University of Georgia
Athens, Georgia 30602-7272

Roberto Brevedan
CERZOS
Departmento de Agronomía
Universidad Nacional del Sur
8000 Bahía Blanca, Argentina

Andrew Broderick
University of Western Sydney–Hawkesbury
School of Horticulture
Locked Bag No. 1
Richard, New South Wales 2753
Australia

Theresa Browning
Indiana Wine Grape Council
5610 Crawfordsville Rd.
Suite 2004
Indianapolis, Indiana 46624

David L. Bubenheim
NASA Ames Research Center
Space Technology Division
Regenerative Life Support Branch
Moffettfield, California 94035

E.J. Callan
Agronomy Department
Lousiana State University
Agricultural Center
Baton Rouge, Lousiana 70893-2110

Richard J. Campbell
Fairchild Tropical Garden
11935 Old Cutler Rd.
Miami, Florida 33156

Marita Cantwell
University of California
Mann Laboratory
Department of Vegetable Crops
Davis, California 95616-8631

Anthony Capelle
Cebeco-Handelsraad
31 Blaak Rotterdam
P.O. Box 182
NL-3000 AD Rotterdam
The Netherlands

Frieda Caplan
Frieda's, Inc.
4465 Corporate Center Dr
Los Alamitos, California 90720

Kenneth D. Carlson
USDA/ARS/NCAUR
1815 North University St
Peoria, Illinois 61604-3999

Patrick M. Carr
North Dakota State University
Dickinson Research Extension Center
470 State Avenue, Suite 101
Dickinson, North Dakota 58601

Catherine G. Cavaletto
Department of Horticulture
University of Hawaii at Manoa
3190 Maile Way
Honolulu, Hawaii 96822

Ellsworth P. Christmas
Department of Agronomy
Purdue University
West Lafayette, Indiana 47907

Dario Cisneros
Integrated Ingredients
Spice Islands
1808 Berino Road
Berino, New Mexico 88024

Charles R. Clement
Instituto Nacional de Pesquisas da Amazônia
Cx. Postal 478
69011-920 Manaus
Amazônia, Brazil

Wayne Coates
Office of Arid Land Studies
Bioresources Research Facility
The University of Arizona
250 E Valencia Rd
Tucson, Arizona 85706

Terry A. Coffelt
USDA/ARS Water Conservation Lab
4331 E. Broadway Rd.
Phoenix, Arizona 85040

Marnix M. Cokelaere
Interdisciplinary Research Center
Katholieke Universiteit Leuven
Campus Kortrijk
E. Sabbelaan 53
B-8500 Kortrijk, Belgium

Victoria P. Collins
Department of Animal Science
University of Kentucky
Lexington, Kentucky 40564

John A. Considine
Faculty of Agriculture
The University of Western Australia
Nedlands, Western Australia 6907
Australia

Charles G. Cook
USDA/ARS
2413 East Hwy 83
Weslaco, Texas 78596

Katrina Cornish
USDA/ARS/WRRC/PBT
800 Buchanan St
Albany, California 94710

Gordon M. Cragg
Natural Products Branch
Fairview Center 206, NCI/FCRDC
P.O. Box B
Frederick, Maryland 21702-1201

F. Petrus Cuperus
ATO-DLO
Haagsteeg 6/Postbus 17
Wageningen, 6700 AA
The Netherlands

Paul Daenens
Laboratory of Toxicology
Katholieke Universiteit Leuven
E. Van Edenstraat 4
B-3000 Leuven, Belgium

Astrid Damboeck
Plant Breeding Department
University of Agriculture
Gregor Mendel-Str. 33
A-1180 Vienna, Austria

Lonni R. Davenport
Department of Horticulture
Purdue University
1165 Horticulture Building
West Lafayette, IN 47907-1165

H. Maelor Davies
University of Kentucky
Tobacco and Health Research Institutes
Cooper and Lexington Drives
Lexington, Kentucky 40546-0236

Eddy Decuypere
Laboratory of Physiology and Immunology of Domestic Animals
Faculty of Agronomy
Katholieke Universiteit Leuven
B-3001 Heverlee, Belgium

Joseph DeFrank
Department of Horticulture
University of Hawaii at Manoa
3190 Maile Way
Honolulu, Hawaii 96822

Johannes T.P. Derksen
ATO-DLO
Haagsteeg 6/Postbus 17
Wageningen, 6700 AA
The Netherlands

David A. Dierig
USDA/ARS Water Conservation Lab
4331 E. Broadway Rd.
Phoenix, Arizona 85040

Anna Eckl
Plant Breeding Department
University of Agriculture
Gregor Mendel-Str. 33
A-1180 Vienna, Austria

Steven Edwardson
Minn-Dak Growers Ltd.
P.O. Box 1307
393 28th Ave East
Dickinson, North Dakota 58602

Eric D. Eriksmoen
Hettinger Research Extension Center
P.O. Box 1377
Hettinger, North Dakota 58639

André Estragnat
CETIOM
174 Avenue Victor Hugo
75116 Paris, France

Peter Felker
Center for Semi-Arid Forest Resources
Texas A&M University–Kingsville
Kingsville, Texas 78363

Louise Ferguson
Department of Pomology
University of California
Davis, California 95616

Jorge F.S. Ferreira
USDA/ARS
Southern Weed Science Lab
P.O. Box 350
Stoneville, Mississippi 38776

Inés Figueroa
Facultad di Agronomia
Universidad de Concepción
Casilla 537
Chillán, Chile

Scott A. Fitterer
Department of Crop and Weed Science
North Dakota State University
166 Loftsgard Hall
Fargo, North Dakota 58105

Trish Flaster
Botanical Liaisons
1180 Crestmoor
Boulder, Colorado 80303

Kent Fleming
Department of Agricultural and Resource Economics
University of Hawaii
3050 Maile Way
Honolulu, Hawaii 96822

Gerda Flo
Interdisciplinary Research Center
Katholieke Universiteit Leuven
Campus Kortrijk
E. Sabbelaan 53
B-8500 Kortrijk, Belgium

K. Frank
Institute of Agriculture and Natural Resources
University of Nebraska
Lincoln, Nebraska 68583

John C. Gardner
North Dakota State University
Carrington Research Extension Center, Box 219
Carrington, North Dakota 58421

K.M. Gilbertson
Montana State University
Central Agricultural Research Center
HC90, Box 20
Moccasin, Montana 59462-9702

John Gordon
Paw Paw Foundation
1385 Campbell Blvd.
Amherst, New York 14228-1404

A. Gray
Department of Plant, Soil, and Insect Sciences
University of Wyoming
Laramie, Wyoming 82071

Fred A. Gray
University of Wyoming
Department of Plant, Soil, and Insect Sciences
P.O. Box 3354
Laramie, Wyoming 82071-3354

Robin W. Groose
University of Wyoming
Department of Plant, Soil, and Insect Sciences
P.O. Box 3354
Laramie, Wyoming 82071-3354

Tony Haigh
University of Western Sydney–Hawkesbury
School of Horticulture
Locked Bag No. 1
Richard 2753 New South Wales
Australia

Anwar A. Hamama
Agricultural Research Station
Virginia State University
Box 9152
Petersburg, Virginia 23806

Andy Hankins
Agricultural Research Station
Virginia State University
Petersburg, Virginia 23806

Wayne W. Hanna
USDA/ARS
Department of Crop & Soil Science
Costal Plain Experiment Station
Tifton, Georgia 31793

James J. Hanzel
North Dakota State University
Department of Plant Sciences
Fargo, North Dakota 58105

Michael Hermann
International Potato Center (CIP)
Casilla 17-21-1977
Quito, Ecuador

Amots Hetzroni
Institute of Agricultural Engineering
B.O. Box 6
Bet Dagan 50250, Israel

Felicitas Hevia
Facultad di Agronomia
Universidad de Concepción
Casilla 537
Chillán, Chile

Lysbeth Hof
CPRO-DLO
Post Bus 16
Wageningen 6700 AA
The Netherlands

Paul Holford
University of Western Sydney–Hawkesbury
School of Horticulture
Locked Bag No. 1
Richard, New South Wales 2753
Australia

Jules Janick
Indiana Center for New Crops and Plant Products
Purdue University
1165 Horticulture Building
West Lafayette, Indiana 47907-1165

Diana Jasso Cantú
Universidad Autónoma Agraria
Antonio Narro Buenavista
Saltillo Coahulia, México 25315

Johnson G. Jato
Faculty of Medicine and Biomedical Science
University of Yaounde
Yaounde, Cameroon

Wenwei Jia
Department of Horticulture and Landscape Architecture
N-318 Ag Science Bldg N
University of Kentucky
Lexington, Kentucky 40546-0091

Duane L. Johnson
Department of Soil & Crop Science
Colorado State University
Fort Collins, Colorado 80523

B.L. Johnson
Department of Crop and Weed Science
North Dakota State University
166 Loftsgard Hall
Fargo, North Dakota 58105

Holly A. Johnson
Department of Medicinal Chemistry and Pharmacognosy
Purdue University
403 Robert E. Heine Pharmacy Building
West Lafayette, Indiana 47907

Gary D. Jolliff
Department of Crop & Soil Science
Oregon State University
Corvallis, Oregon 97331-3002

Jean Paul Joublan
Facultad di Agronomia
Universidad de Concepción
Casilla 537
Chillán, Chile

James W. Julian
Department of Horticulture
Purdue University
1165 Horticulture Building
West Lafayette, IN 47907-1165

Makoto Kawase
Shikoku National Agricultural Experiment Station
Sen-yucho 1-3-1
Zentsuji, Kagawa 765, Japan

C.W. Kennedy
Agronomy Department
Lousiana State University
Agricultural Center
Baton Rouge, Lousiana 70893-2110

Dale King
Purdue University
Department of Animal Sciences
1151 Lilly Hall
West Lafayette, Indiana 47907-1151

David W. Koch
University of Wyoming
Department of Plant, Soil, and Insect Sciences
P.O. Box 3354
Laramie, Wyoming 82071-3354

P. Kolster
ATO-DLO
Haagsteeg 6/Postbus 17
Wageningen, 6700 AA
The Netherlands

E. Koppelman
KFX Tech, Inc.
Gillette, Wyoming 82716

James M. Krall
University of Wyoming
Research and Extension Center
Route 1, P.O. Box 374
Torrington, Wyoming 82240

Joseph O. Kuti
Texas A&M University–Kingville
Horticultural Crops Research Laboratory
John Howe Agriculture Laboratory Center
Kingville, Texas 78363

Françoise Labalette
CETIOM
174 Avenue Victor Hugo
75116 Paris, France

Bradley V. Lawrence
Purdue University
Department of Animal Sciences
1151 Lilly Hall
West Lafayette, Indiana 47907-1151

Roger H. Lawson
USDA/ARS
National Program Staff
Beltsville, Maryland 20705

Peter R. Lister
University of Western Sydney–Hawkesbury
School of Horticulture
Locked Bag No. 1
Richard, New South Wales 2753
Australia

Gilbert R. Lovell
USDA/ARS
Southern Plant Introduction Station
1109 Experiment Street
Griffin, Georgia 30223-1797

Gustavo Luayza
CERZOS
Departmento de Agronomía
Universidad Nacional del Sur
8000 Bahía Blanca, Argentina

Thomas A. Lumpkin
Department of Crop & Soil Sciences
Washington State University
Pullman, Washington 99164-6420

Richard M. Manshardt
Department of Horticulture
University of Hawaii at Manoa
3190 Maile Way
Honolulu, Hawaii 96822

Glenn B. Martin
North Dakota State University
Dickinson Research Extension Center
470 State Avenue, Suite 101
Dickinson, North Dakota 58601

E. Massoura
ATO-DLO
Haagsteeg 6/Postbus 17
Wageningen, 6700 AA
The Netherlands

Lawrence T. McHargue
Department of Biology
Southern California College
55 Fair Drive
Costa Mesa, California 92626

Steven P. McLaughlin
Office of Arid Lands Studies
The University of Arizona
Tucson, Arizona 85721

Jerry L. McLaughlin
Department of Medicinal Chemistry and Pharmacognosy
Purdue University
403 Robert E. Heine Pharmacy Bldg.
West Lafayette, Indiana 47907

Tadesse Mebrahtu
Agricultural Research Station
Virginia State University
Petersburg, Virginia 23806

Antoine Messéan
CETIOM
174 Avenue Victor Hugo
75116 Paris, France

Michael V. Mickelbart
Department of Horticulture
Purdue University
1165 Horticulture Building
West Lafayette, Indiana 47907-1165

Cary A. Mitchell
Department of Horticulture
Purdue University
1165 Horticulture Building
West Lafayette, Indiana 47907-1165

Y. Mizrahi
The Institutes for Applied Research
Ben Gurion University of the Negev
P.O. Box 1025
Beer Sheva 84110, Israel

Ali I. Mohamed
Agricultural Research Station
Virginia State University
Box 9152
Petersburg, Virginia 23806

C. Mondragon-Jacobo
Department of Horticulture
Purdue University
1165 Horticulture Building
West Lafayette, Indiana 47907-1165

John Mood, Jr.
59-515 Akanoho Place
Haleiwa, Hawaii 96712

Mario R. Morales
Indiana Center for New Crops and Plant Products
Purdue University
1165 Horticulture Building
West Lafayette, IN 47907-1165

David Morris
Institute for Local Self-Reliance
1313 5th St, SE, Suite 306
Minneapolis, Minnesota 55414

David A. Morrison
University of Technology–Sydney
Department of Environmental Biology and Horticulture
P.O. Box 123
Broadway, New South Wales 2007
Australia

Julia Morton
University of Miami
Morton Collectanea
Box 248204
Coral Gables, Florida 33124

Jimmy Mullins
Agricultural Research Station
Virginia State University
Petersburg, Virginia 23806

Robert L. Myers
Office of Sustainable Agricultural Programs
USDA, Ag Box 0910
Washington, DC 20250-0910

Natalie Y. Nagai
Department of Horticulture
University of Hawaii at Manoa
3190 Maile Way
Honolulu, Hawaii 96822

Hiroki Nakayama
National Institute of Agrobiological Resources
Kannondai 2-1-2
Tsukuba, Ibaraki 305, Japan

Francis S. Nakayama
USDA/ARS Water Conservation Lab
4331 E. Broadway Rd.
Phoenix, Arizona 85040

A. Nerd
The Institutes for Applied Research
Ben Gurion University of the Negev
P.O. Box 1025
Beer Sheva 84110, Israel

Francis H. Nicholls
John K King & Sons Ltd.
Coggleshall Colchester
Essex CO6 1TH
United Kingdom

Kwangok P. Nickel
Department of Foods & Nutrition
Purdue University
West Lafayette, Indiana 47907

Xunli Nie
University of California
Mann Laboratory
Department of Vegetable Crops
Davis, California 95616-8631

David C. Nielsen
USDA-ARS
Central Great Plains Research Station
P.O. Box 400
Akron, Colorado 80720

S. Suzanne Nielsen
Department of Food Science
Purdue University
West Lafayette, Indiana 47907

Bart C. O'Brien
Rancho Santa Ana Botanic Garden
1500 N. College Avenue
Claremont, California 91711

Kazutoshi Okuno
National Institute of Agrobiological Resources
Kannondai 2-1-2
Tsukuba, Ibaraki 305, Japan

N. Rick Olson
Hettinger Research Extension Center
P.O. Box 1377
Hettinger, North Dakota 58639

Rosana Palomo
CERZOS
Departmento de Agronomía
Universidad Nacional del Sur
8000 Bahía Blanca, Argentina

Jeffrey F. Pedersen
USDA\ARS 344 Keim Hall
University of Nebraska
P.O. Box 830937
Lincoln, Nebraska 68583

S. Perez-Gonzalez
INIFAP-CENGUA Program de Frutales
Prol. Zaragoza 408
Jardines de la Hacienda Qro.
Mexico 76180

Gunter Pfaff
Betsy's Tempeh
14780 Beardslee
Perry, Michigan 48872

Julie A. Plummer
Faculty of Agriculture
The University of Western Australia
Nedlands, Western Australia 6907
Australia

D. Morris Porter
USDA/ARS
NE Plant, Soil and Water Laboratory
University of Maine
Orono, Maine 04469

Jane Potter Gates
USDA/ARS
National Agricultural Library
10301 Baltimore Avenue, Room 304
Beltsville, Maryland 20705-2351

John T.A. Proctor
University of Guelph
Ontario Agricultural College
Department of Horticultural Science
Guelph, Ontario N1G 2W1 Canada

John F. Rajewski
Agronomy Department
University of Nebraska
Lincoln, Nebraska 68583-0915

Muddappa Rangappa
Agricultural Research Station
Virginia State University
Box 9152
Petersburg, Virginia 23806

J.E. Ranney
Montana State University
Yellowstone County Extension
Courthouse 202, P.O. Box 35,021
Billings, Montana 59107-5021

Dennis T. Ray
University of Arizona
Department of Plant Sciences
Tucson, Arizona 85721

Paul F. Reese, Jr.
Asgrow Seed Company
Kalamazoo, Michigan 49002

Nancy A. Reichert
Mississippi State University
Department of Plant & Soil Sciences
Box 9555
Mississippi State, Mississippi 39762

Charlie L. Rife
Kansas State University
Department of Agronomy
2004 Throckmorton Plant Science Center
Manhattan, Kansas 66506

Raúl Rodriguez García
Universidad Autónoma Agraria
Antonio Narro Buenavista
Saltillo Coahulia, México 25315

Mark S. Roh
USDA/ARS
U.S. National Arboretum
Floral and Nursery Plants Research Unit
Beltsville, Maryland 20705

Richard J. Roseberg
Oregon State University
Southern Oregon Research &
Extension Center
569 Hanley Road
Medford, Oregon 97502

Daniel J. Royse
316 Buckhout Laboratory
Department of Plant Pathology
Pennsylvania State University
University Park, Pennsylvania 16802

Peter Ruckenbauer
Plant Breeding Department
University of Agriculture
Gregor Mendel-Str. 33
A-1180 Vienna, Austria

J.P. Salgado
Kansas State University
Department of Agronomy
2004 Throckmorton Plant Science Center
Manhattan, Kansas 66506

D. Schafferman
Department of Field Crops and Natural Resources
Institute of Field and Garden Crops
ARO The Volcani Center
Bet-Dagan 50250, Israel

Greg Schlick
NASA Ames Research Center
Space Technology Division
Regenerative Lif e Support Branch
Moffettfield, California 94035

A.A. Schneiter
Department of Crop & Weed Science
North Dakota State University
166 Loftsgard Hall
Fargo, North Dakota 58105

Heinrich Schrems
Saatbau Linz
Zuchtstation Reichersberg
A-4981 Reichersberg, Austria

Thomas P. Schuman
University of Southern Mississippi
Department of Polymer Science
Box 10037
Hattiesburg, Mississippi 39406-0037

Humberto Serri
Facultad di Agronomia
Universidad de Concepción
Casilla 537
Chillán, Chile

I. Shamir
Department of Field Crops and Natural Resources
Institute of Field and Garden Crops
ARO The Volcani Center
Bet-Dagan 50250, Israel

Robert C. Shearman
Center for Grassland Studies
University of Nebraska
Keim Hall
Lincoln, Nebraska 68583-0937

Betsy Shipley
Betsy's Tempeh
14780 Beardslee
Perry, Michigan 48872

M. Siedell
Institute of Agriculture and Natural Resources
University of Nebraska
Lincoln, Nebraska 68583

Deborah J. Siler
USDA/ARS/WRRC/PBT
800 Buchanan St
Albany, California 94710

James E. Simon
Indiana Center for New Crops and Plant Products
Purdue University
1165 Horticulture Building
West Lafayette, Indiana 47907-1165

Bharat P. Singh
Fort Valley State College
Agricultural Research Station
Fort Valley, Georgia 31030-3298

Kenneth M. Snader
Natural Products Branch
Fairview Center 206, NCI/FCRDC
P.O. Box B
Frederick, Maryland 21702-1201

Jonathan Sobels
Seedco, LTD.
78 Burbridge Road
Hilton, South Australia 5033
Australia

Jacqueline Soule
Frederick Meijer Gardens
3411 Bradford NE
Grand Rapids, Michigan 49546

Z.E. (Ted) Spadek
Agricultural Chemistry Laboratory
Chemistry Centre of Western Australia
125 Hay Street
East Perth, Western Australia 6004
Australia

Gilbert F. Stallknecht
Montana State University
Central Agricultural Research Center
HC90, Box 20
Moccasin, Montana 59462-9702

David E. Starner
Northern Piedmont Agricultural Research & Extension Center
Virginia Tech
Box 448
Orange, Virginia 22960

Karen Stott
University of Western Sydney–Hawkesbury
School of Horticulture
Locked Bag No. 1
Richard 2753 New South Wales
Australia

John Strang
Department of Horticulture and Landscape Architecture
N-318 Ag Science Bldg N
University of Kentucky
Lexington, Kentucky 40546-0091

Glenn H. Sullivan
Department of Horticulture
Purdue University
1165 Horticulture Building
West Lafayette, IN 47907-1165

Shelby F. Thames
University of Southern Mississippi
Department of Polymer Science
Box 10037
Hattiesburg, Mississippi 39406-0037

Anson E. Thompson (deceased)
USDA/ARS Water Conservation Lab
4331 E. Broadway Rd.
Phoenix, Arizona 85040

Eliseo S. Torres
Texas A&M University–Kingville
Horticultural Crops Research Laboratory
John Howe Agriculture Laboratory Center
Kingville, Texas 78363

Joan Tous
Institut de Recerca i Tecnologia Agroalimentaries (IRTA)
Departament d'Arboricultura Mediterranea
Centre de Mas Bove, Apartat 415
43280 Reus, Spain

Maurits Van Boven
Laboratory of Toxicology
Katholieke Universiteit Leuven
E. Van Edenstraat 4
B-3000 Leuven, Belgium

J.M. Vereijken
ATO-DLO
Haagsteeg 6/Postbus 17
Wageningen, 6700 AA
The Netherlands

Sabien Vermaut
Laboratory of Physiology and Immunology of Domestic Animals
Faculty of Agronomy
Katholieke Universiteit Leuven
B-3001 Heverlee, Belgium

Noel Vietmeyer
National Research Council
BOSTID
2101 Constitution Avenue NW
Washington, DC 20418

Kenneth P. Vogel
USDA/ARS
University of Nebraska
Keim Hall
Lincoln, Nebraska 68583-0937

Johann Vollmann
Plant Breeding Department
University of Agriculture
Gregor Mendel-Str. 33
A-1180 Vienna, Austria

Cheri Wagner
Indiana Wine Grape Council
5610 Crawfordsville Rd.
Suite 2004
Indianapolis, Indiana 46624

Joanne M. Wann
Faculty of Agriculture
The University of Western Australia
Nedlands, Western Australia 6907
Australia

Charles L. Webber III
USDA, ARS, SCARL
P.O. Box 159
Lane, OK 74555

Gregory E. Welbaum
Department of Horticulture
Virginia Tech
Blacksburg, Virginia 24061-0327

George A. White
USDA/ARS
National Clonal Repository
University of California
Davis, California 95616

Wayne F. Whitehead
Fort Valley State College
Agricultural Research Station
Fort Valley, Georgia 31030-3298

Rosemarie Wilckens
Facultad di Agronomia
Universidad de Concepción
Casilla 537
Chillán, Chile

Mary Witt
Department of Horticulture and Landscape Architecture
N-318 Ag Science Bldg N
University of Kentucky
Lexington, Kentucky 40546-0091

Umedi L. Yadava
Agricultural Research Station
School of Agriculture
Fort Valley State College
Fort Valley, Georgia 31030-3298

Mas Yamaguchi
University of California
Department of Vegetable Crops
Davis, California 95616-8631

Zohara Yaniv
Department of Field Crops and Natural Resources
Institute of Field and Garden Crops
ARO The Volcani Center
Bet-Dagan 50250, Israel

Li Mei Yun
University of Wyoming
Department of Plant, Soil, and Insect Sciences
P.O. Box 3354
Laramie, Wyoming 82071-3354

Francis Zee
National Clonal Germplasm Repository
USDA/ARS
Hilo, Hawaii 96720

Ru Jing Zong
Transfresh Corporation
607 Brunken Ave.
Davis, California 95616

M. Zur
Department of Field Crops and Natural Resources
Institute of Field and Garden Crops
ARO The Volcani Center
Bet-Dagan 50250, Israel

Preface

Jules Janick

This volume, ***Progress in New Crops,*** is the Proceedings of the Third National Symposium, entitled **New Crops: New Opportunities, New Technologies,** held October 22–25, 1995, in Indianapolis, Indiana. The objectives of the symposium, as in the previous two symposia, were to provide a national forum for leading authorities from industry, government, experiment stations, and academia, as well as growers, to discuss the status and future of new crop development. Although the symposium was national in scope, a significant international presence indicated the global appeal of new and alternate crops.

As in the previous symposia, we encountered a wave of euphoria, enthusiasm, and good feelings about new crops and their potential for enriching the agriculture of the United States and other countries. Some questioned if we were preaching to the choir and proselytizing the converted. While one goal of these meetings was to strengthen the fervor of the faithful, it is clear that apocalyptic changes that increase the relevance of new crops are indeed on the horizon. These include declines in crop subsidies, increasing concerns over sustainable agriculture, rural discontent, and demand for diversity in diets.

Enthusiasm for new and alternate crops is manifested by increasing emphasis on farm diversification, by greater industry interest in plant resources, and by continual expansion of food offerings in supermarkets. Recent changes in the 1995 Farm Bill have created agricultural legislation decreasing crop subsidies that will encourage increasing diversification in US agriculture and move farmers to become more attuned to market forces.

Many persons have contributed to the success of this symposium and volume. I gratefully acknowledge the other members of the program committee (Robert Anderson, Robert Kleiman, Steve Knapp, Dan Kugler, Rob Myers, Dennis Ray, Henry Shands, and James Simon) and the sponsors and benefactors listed on the title page, who backed up their belief in new crops with tangible support necessary to make the symposium and thus, this volume, a reality. I especially acknowledge the expertise of Anna L. Whipkey, who was in charge of all computer operations for both the symposium and this volume and who transformed the edited manuscripts into computer files which were transferred directly to print.

This volume's contents include papers from invited speakers as well as from those who presented posters. Contributions cover a diverse assortment of crops, topics, and views. The book is divided into three parts. Part I (New Crops, New Opportunities) includes the two keynote addresses and papers on international and national aspects of new crops. Part II (New Crops Policy and Programs) includes three subsections: Policy, Marketing and Commercialization, and Information Explosion. The banquet address by Frieda Caplan describing the origins and development of Frieda's Inc. leads off the Commercialization subsection. Part III (New Crops, New Technologies) is subdivided, as in the other two volumes, by major crop groupings (Cereals and Pseudocereals, Grain Legumes, Forages, Oilseeds, Industrial Crops, Fiber Crops, Fruits and Nuts, Vegetables, Floral and Landscape Crops, Medicinal, Aromatic, Spice, and Bioactive Crops, and Crops in Space).

This volume and its two predecessors, *Advances in New Crops*, published in 1990 by Timber Press, Portland Oregon (Proceedings of the First National Symposium New Crops: Research, Development, Economics) and *New Crops*, published by Wiley, New York in 1993 (Proceeding of the Second National Symposium New Crops: Exploration, Research, Commercialization) provide an overview of new and underexploited crops including economics, exploration, commercialization, research, and development. Each volume has a detailed index, so that the three-volume set constitutes a veritable encyclopedia of new and underexploited crops.

Finally, it is my pleasure to dedicate this work to Frieda Caplan and Julia Morton*, two gracious women who have made an enormous impact on the field of new crops. Frieda Caplan, by innovative marketing and promotion, has, without a doubt, done more than any person to make new food crops known to the Amercian consumer. Julia Morton, a distinguished economic botanist, has compiled an enormous treasury of technical information through her many books and her devotion to the Morton Collectanea, a specialized reference and research department devoted to the acquisition, classification, physical collation, and maintenance of subject-filed data in the field of economic botany, with special emphasis on the tropics and subtropics. Their contributions to our field can only be repaid if we redouble our efforts to ensure that the enormous potential of new crops is realized.

Julia Morton

Frieda Caplan

*Julia Morton passed away on September 10, 1996, while this volume was in press. We will sorely miss her presence.

PART I
NEW CROPS, NEW OPPORTUNITIES

New Crops: Solutions for Global Problems

Noel Vietmeyer

During the period most of us were growing up, the greatest threat to mother earth was a nuclear holocaust. Thankfully humanity seems to have put that terrifying period behind. Now the greatest threats facing our planet are what are generally called the global problems: hunger, malnutrition, deforestation, desertification, soil erosion, the loss of soil fertility, and several others. The vital question today is: can we also put those behind? Presently most people are pessimistic. Indeed there is a widespread belief that the global problems are intractable, or at least so demanding of research that it will be next century before solutions come clear. However I do not believe that, and the reason for my optimism lies in the whole subject of new crops and in the skills of people you might call "new-croppers." Here's my thinking.

The so-called global problems are certainly global in their impact, but hunger, malnutrition, deforestation, desertification, and the rest originate mainly in the hot zone. For convenience I'll call that zone "the tropics," but I'm using the terminology to mean the region between the tropics of Capricorn and Cancer, and I'm including the arid lands as well as the humid ones. In all the debate over global problems, the point being missed is that the tropics happen to be the location of the planet's greatest collection of biodiversity. It is in Africa, Asia, and Latin America where the most species are to be found. Thus there is a convergence between problems and plants; the greatest global hazards originating in the region where is found the greatest array of botanical wealth.

To better understand what I'm driving at, consider the following. In Africa, where the hunger problem is concentrated, there are actually 2,000 species of native food plants. Both there and in the rest of the tropics, the region where malnutrition is severest, can be found 3,000 different fruits as well as over 1,000 vegetables. In the lands where deforestation is so destructive are to be found over 20,000 trees, some of them extremely fast growing. In the desertifying regions can be found many useful drought-tolerant plants—I don't know how many, but it runs into the hundreds. And as far as soil degradation is concerned, the legume family (Fabaceae, Mimosaceae, and Caesalpiniaceae) has 18,000 species, most of which are the nitrogen fixers nature uses as botanical shock troops to achieve a bridgehead on worn-out soils. Many of these 18,000 legumes—collectively comprising the plant kingdom's third largest family—are native to the tropics and adapted to the lands that are wearing out.

Despite the existence of all this herbaceous heritage, hardly any tropical plant is being employed to full advantage to relieve the pervasive problems. Of Africa's 2,000 food plants, only sorghum gets major research, and then mainly to improve its ability to feed cattle in our part of the world. Of the 3,000 fruits, only banana and pineapple get solid global support, and again much of the effort is aimed at helping the rich rather than the needy. Of the 20,000 tropical trees, only a handful are undergoing global advancement, and financial support for even those few is hardly stellar. (Included in the few partially supported species are leucaena, mangium, calliandra, and neem, which my little program helped lift out of obscurity during the last 20 years.)

That the bulk of useful tropical biodiversity is underexploited is one of the glaring oversights of this era. The creation of more crops and the support for new-croppers is of global importance. Specialists who can sort through underexploited plants and locate the few, the very few, that can become new tools for use against the earth's main long-term problems are among the most valuable of all scientists.

New-croppers may carry out their sorting processes in various ways. Books of the type Mark Dafforn and I have been producing in our National Research Council program exemplify one way. Some of you probably don't know these books, but residents of developing countries know them. Over the years we have produced about 40 titles and distributed more than 700,000 copies to Africa, Asia, and Latin America, mostly without charge.

Our publications have covered such things as new crops for food, industrial materials, firewood, forage, timber, and environmental protection. In each, we integrate the wisdom and experience of hundreds of specialists and build the case for the crop with a lively text supported by instructive pictures and intriguing sidebars. The goal is to create a combination that not only convinces and educates readers of various levels of sophistication and interest, but motivates them to act. Put another way, we describe promising new crops in a balanced but bold manner that stimulates people and governments to undertake their own initiatives. The process works: individuals from many walks of life have seized upon these plants and carried them forward. Even at this conference you can glimpse the results. We helped in the early initiation phases of guayule, jojoba, the "lost" crops of the Americas, prosopis, amaranth, quinoa, yacon, leucaena, and others you are going to hear about in the next few days. However it is in countries of the tropics that we've had our greatest impacts. Two prime examples are vetiver and mangium.

The King of Thailand has become perhaps the greatest supporter of vetiver, the species described in our 1993 book, *Vetiver Grass: A Thin Green Line Against Erosion*. In Thailand the agriculture department, forests department, roads department, and border police are all propagating and planting this formerly obscure grass whose hedges are so dense they block the passage of soil. Also dozens of nongovernmental organizations (NGOs) throughout the tropics are adopting vetiver, and a network of researchers and a home page on the Internet have sprung up. In fact, vetiver is now working as an erosion barrier in 106 countries. Starting in 1979 we helped bring international recognition to the Australian rainforest tree *Acacia mangium*, known as mangium. At that time this fast-growing legume's potential was known to only a handful of foresters in Borneo. Today this tree has risen to become one of the top four reforestation species in Asia. The Indonesian government, alone, is relying on it to reforest 4.4 million ha.

Despite such gratifying results, our activity is now essentially dead. In 1993 the prime sponsor—the Global Bureau of the Agency for International Development—refused to keep funding the program. Subsequent years of dedicated effort have failed to produce any substitute funding source adequate to the task of keeping it alive. This is sad because the hot zone has not only plants and problems, it has people with a hunger for knowledge and an eagerness to put new crops to work. Give them information about a nutritious fruit or fast-growing tree and—especially if the information is presented in a bright and motivational way with names and addresses of contacts for seeds or local advice—they will adopt it with a zest amounting almost to reverence. Each day requests from such people pour into my office from around the Third World. Most of the writers currently want *Neem; A Tree for Solving Global Problems*. In the last three years I've sent out more than 30,000 copies of that book along with almost 40,000 copies of the vetiver counterpart, which is one reason why neem has become tropical forestry's hottest tree and vetiver is being employed in 106 countries.

TROPICAL NEW CROPS

There's no shortage of new topics still to take up. Indeed our files are bulging with ideas on plants that seem to have as much promise as any we've ever publicized before. If I can get the project funded again, which I'm hopeful of still accomplishing, we will bring world recognition to potential new crops such as the following.

Peachpalm

This little-known Central American palm (*Guilielma gasipaes*, Arecaceae) produces fruits that look like tiny American footballs. Boiled, they have the texture of new potatoes and the taste of chestnuts. Of all foods, they are perhaps the most balanced for human nutrition. Having just the right proportions of carbohydrate, protein, fat, vitamins, and minerals they theoretically need no supplement. A few special types contain an oil very rich in vitamin A, a deficiency of which blinds about a million children each year. As a plantation crop, peachpalm is said to be capable of providing more protein and more carbohydrate per hectare than corn.

Vetiver

Beyond stopping soil erosion, this grass (*Vetiveria zizanioides*, Poaceae) has vital uses for which it is not yet being employed. For one thing, vetiver hedges block the rushing runoff from tropical storms, and thereby help water sink in where it fell on the slopes. In the right locations vetiver hedges can be operating literally a few weeks after planting, so the plant offers the promise of "instant" working watersheds that improve year-round water availability as well as perhaps mitigate disastrous floods and mudslides. Also vetiver is a potential tool for renovating polluted sites because it can survive in, and perhaps "mine," soils contaminated with heavy metals, alkali as well as some salt. Tolerant of soluble aluminum and severe acidity (almost to pH 3), it thrives in the so-called "laterite," the red, infertile, aluminum-rich, and very acid soil that keeps much of the tropics a wasteland. Vetiver, then, is potentially a tool for bringing millions of abandoned hectares back into some form of productive use while greatly benefiting the tropical environment.

Mangium

Mangium is also worth revisiting because of its newly emerging global potentials. For example, Indonesia plants this tree on such a giant scale because its foresters have found that, given a little help, mangium outcompetes the vigorous and tenacious imperata grass, the curse of the tropical landscape. Further, under the protective canopy of this nitrogen-fixing leguminous (Mimosaceae) tree an understory of shade-loving native species often springs up vigorously. Thus, mangium could become a major weapon for healing damaged tropical forests, reversing the loss of forest biodiversity and providing fuelwood, cash from lumber sales, and improved soil all in one.

Velvet Bean

In Central America the velvet bean (*Mucuna deeringiana*, Fabaceae) is creating a lot of enthusiasm among farmers and environmentalists. This weed-smothering, nitrogen-fixing herbaceous legume protects the land and helps crops yield well with few inputs. It seems likely to prove broadly applicable throughout the tropics, and may well provide a way to retain and even restore fertility on vast acreages of degraded farmland. In addition, several dozen legume genera—including *Mucuna*, *Pueraria*, *Lotus*, *Lotononis*, and *Vicia*—offer the likelihood of finding equally successful groundcovers, including some for restoring various extreme soils and seemingly impossible tropical sites.

Vetch

An innovative U.S. Department of Agriculture researcher at Beltsville, Maryland has already turned one groundcover species to such good use that he produces record yields of vegetables on land once so worn-out it was considered the worst on his experiment station. Among his several innovations, Aref Abdul-Baki uses hairy vetch (*Vicia villosa*, Fabaceae) as a winter mulch and a powerful flail mower instead of a plow. In the fall season he mows the remains of his tomatoes, beans, melons or corn and into the resultant "mash" of chopped residues he plants vetch seeds. The resulting vetch crop covers the land during the winter season, rebuilding fertility and tilth as it suppresses weeds and protects against erosion. In the spring season, Abdel-Baki mows the vetch and immediately plants vegetable seedlings into the new "mash" of chopped up plants. The thick pad of rotting vetch remains covering the soil through the vegetable-growing season, suppressing weeds, insulating the roots from the sun's heat and slowly releasing nutrients to the crop. In this way, Abdul-Baki needs little or no fertilizer or pesticide and (because the land is never exposed) experiences neither weed problems nor erosion. Thus high yields are achieved at little cost. In 1995, he achieved three times the commercial yield of tomatoes at less than one-tenth the cost (much of the savings coming because he needed none of the plastic film commercial tomato growers use as mulch). Through normal channels this knowledge can reach American vegetable growers, but my kind of activity could project it worldwide to probable great effect.

Tropical Fruits

We all know the kind of tropical fruits hotels serve in places such as Bangkok or Singapore—gorgeous sweet things with lively flavors. However, tropical fruits are much more than that—they are among the most diverse and amazing of foodstuffs: In the region where children are dying or losing their sight for lack of nutrients there are fruits surprisingly rich in vitamin A as well as others with many times the level of vitamin C of an orange. Certain tropical fruits aren't at all sweet. These can be thought of as "staple fruits," more like potato than peach, and they include breadfruit, jakfruit, and the above mentioned peachpalm. Tree species like these have a special importance because they offer not only food but reforestation, shade, shelter and the many other benefits of a permanent tree cover in the tropics.

Moringa

This extremely fast growing woody species (*Moringa oleifera*, Moringaceae) doesn't look like much, but it could open up a new category of crops: "vegetable trees." Moringa produces long pods with the appearance of giant green beans and the taste of asparagus. It also produces masses of very small leaflets that are boiled and eaten like spinach. Being so small, the leaflets sun dry in just a few hours and can then be put in a jar and stored for the off-season, a time when dietary minerals and vitamins are often scarce. In addition to providing these natural supplements, the moringa tree yields seeds that clarify turbid water. Compounds in its seeds make traces of silt and clay settle out as effectively as the alum our water departments use. In the rural tropics, moringa seeds could be employed to make water safer for drinking and cooking. Taken all round, this species could be a powerful new weapon against two great scourges, malnutrition and water-borne disease.

Lost Crops of Africa

Obviously, something must be done about all the neglected crops of Africa. Early in 1996 we at the NRC will publish a 400 page book showcasing the promising native cereals. Then we'd like to compile a volume dealing with the several dozen highly promising African fruits. Later, we hope to complete a volume covering the equal number of promising native vegetables. For all those forgotten African food plants there is presently no readily available promotional materials, let alone guides to such things as nutritional content, soil and climatic limits, varieties that yield the most nutritious parts, or the best ways to prevent pests. Given a little attention, however, these plants are potential wonder weapons against hunger. This is particularly because they are adapted to the challenging conditions and to the needs of Africans.

Quality-Protein Maize

The protein of this new and high-yielding form of maize has a nutritional quality almost the equal of milk protein. A little-known program directed by Nobel Laureate Norman Borlaug and former President Jimmy Carter is having success with QPM on a considerable scale in Ghana. Moreover the crop has caught on in Brazil, where 100,000 hectares are now grown annually. But impressive as all that is, the super-nutritious corn represents a negligible proportion of the world's third largest staple. A new NRC book could help take it to a new level of use and acceptance worldwide. The result might be millions of hogs and chickens raised more efficiently and malnutrition becoming much easier to stamp out in Africa and Latin America, where corn is the main sustenance for millions of the poor. Indeed, the influence could be so far reaching that I hazard a guess that within our lifetime all corn—even that in Iowa—would be switched to the quality-protein type.

Malted Barley

I've found that a surprising change occurs when you stir a little malted barley into any hot boiled staple: the whole mess of porridge, pabulum, mash, or mush turns as liquid as water, and almost instantly. In seconds anyone can produce liquid rice, liquid potato, liquid corn, and so forth. There's nothing magical about the process; malted barley is rich in amylase enzymes that digest starch. It comes in an inexpensive, food-grade flour for use in malted milkshakes, pizza crusts, and baked goods of a hundred and one kinds. (The sample I use was actually made for Grape-Nuts cereal.) Such a readily available and wholesome liquefier could be a major answer to malnutrition. Almost everywhere mothers have starchy staples; it is fluid food for their babies that they lack. Malted-barley, used as a catalyst, would allow millions who cannot obtain cow's milk or baby formula to transform part of their staples into palatable, safe, nutrient-dense liquids their babies can sip from a cup or suck from a bottle at weaning time.

Aquatic Plants

Here in the United States we treat our wastewaters in steel and concrete facilities; everything is monitored and measured, and the whole process is wonderfully successful. But in the developing world things are different. Poor countries can afford neither the equipment, the supplies nor the electricity, not to mention the maintenance. Therefore, most human waste in the tropics is just released onto the land or into the waterways. But the ingredients in human waste are the foods for plants. A crop of aquatic plants absorbs those nutrients while also providing the root and stem surfaces where bacteria and viruses—including those that cause cholera, hepatitis, typhoid, dysentery, and diarrhea—are destroyed by natural oxygenation or physical detention. Already several dozen small towns in the United States clean their wastewaters by trickling them through small marshes filled with bulrushes, reeds, duckweed, or other aquatic species. The time has come to adapt this technique for rendering wastewaters safe in the tropics. There is an especial need to include plants that both thrive in wastewaters and produce things useful to the poor, such as firewood, thatch, and handicraft materials. Possibilities abound.

Plants for Absorbing Carbon Dioxide

I don't know if the world is getting warmer, if the seas are rising or if the weather is getting more violent, but I do know that the level of carbon dioxide in the atmosphere is rising. Accordingly, it seems prudent that nations do whatever is practicable to reduce that increase. Of course they can cut their carbon dioxide emissions, which is going to be especially painful for industrialized and highly motorized countries like ours. But we new-croppers should also be highlighting species for absorbing the surpluses. The best are likely to be species for growing in the tropics, where temperatures remain high and plants absorb carbon dioxide year round. Choices can be made from the 20,000 trees I mentioned earlier, some of which are legumes that could turn the vast barren lands into massive sinks for greenhouse gas.

Making a noticeable change in the atmospheric imbalance is no pipe dream. Indonesia's mangium plantings involve some 5 billion trees, and (given the fact that mangium can grow 3 m in a year) each tree absorbs perhaps 20 kg of carbon dioxide annually. Back-of-the-envelope calculations suggest that Indonesia's mangiums will absorb about 2% of the surplus carbon dioxide that United States emits. Even if I'm wrong in the calculations (a likelihood verging on certainty), the amount of greenhouse gas removed will be substantial. Moreover, there's every reason to expect Indonesia's experience can be replicated many fold elsewhere.

Around this theme of using new crops to rebalance the atmosphere are numerous possible elaborations. For one thing, the process doesn't require trees at all. In fact, the growing roots of perennial grasses and other herbaceous species can sequester carbon dioxide underground. Thailand's anticipated billions of vetiver plants—a single experiment station near Chiang Rai expects by the year 2001 to have propagated 1.5 billion—will alone make a substantial contribution because the massive network of roots descend vertically 5 m and more. Using vetiver for all its above-ground benefits—erosion control, watershed restoration, pollution mining and the rest—might provide a great means for squirreling atmospheric carbon dioxide underground as an environmental by-product.

Support for such plantings might come from unexpected sources. For instance, the power company here in Indianapolis might sponsor the planting of millions of trees in, let's say, Belize or Burundi or Bangladesh. It could then boast of being innocent of contributing to global warming. Such efforts would also help reforest the tropics and provide shade and shelter for the world's most beleaguered peoples. Moreover, anyone who planted tropical food trees for carbon-dioxide sequestration would be helping feed the hungry while ensuring that the trees would not be cut down willy nilly the next time someone lazy needs a little firewood.

Crops for Political Stability

Certain new-crop introductions have been more influential on history than the politicians of the time (witness the development of soybean during the Harding and Coolidge era), and several geopolitically hot spots now need new crops more than they need new politics. These particular hot spots are concentrated in the semiarid lands where populations are increasing and arable soils are decreasing. Social tensions among already restive populations are being thereby exacerbated.

One such area is North Africa, a vast and desertifying area whose social collapse could take us from the Cold War to something like a reprise of the Crusades. One (hopefully fanciful) scenario involves Western troops facing hordes of villagers-turned-fanatics marching out of places such as Algeria and Libya and heading toward Israel, Mecca, and the world's main oil spigots around the Gulf. Of course things are unlikely to ever get that bad, but it seems to me new-croppers everywhere should be urgently funded to work on crops to stabilize North Africa's desertifying regions. Possible target species include jojoba, several species of prosopis, spineless cactus, tagasaste (a leguminous shrub from the Canary Islands), neem, leucaena, vetch, and jatropha (a plant that thrives in degraded lands and produces an oil that can be used for running diesel engines). These and other extremely drought-tolerant species can provide income, forage, and food as well as better living conditions and improved rural environments. Most of all, by bringing hope to the masses these new crops can help retain social stability, perhaps making one of the greatest contributions to long-term peace.

The other geopolitically vital desertifying area is South Africa. Here, too, the situation is both fragile and fraught with potential mass disillusionment and possible major disaster. The world is lucky to have Nelson Mandela, but he is not a young man and new-croppers should be helping him get South Africa quickly on its feet. Again the areas of greatest impoverishment are arid. New drought-tolerant plants (many of them the same as just listed) could turn out to be more significant than any of the diplomatic and political initiatives under consideration.

APPLIED BIODIVERSITY

Although I could describe other plants with globally significant potentials, I hope these few examples give a sense of the immensely important possibilities. The truth is that waiting among the unused tropical biodiversity is a cornucopia of powerful living tools. New-croppers could and should be raising them up and getting them into everyday use.

To best accomplish this and to most efficiently capture the power inherent in such species, we really need a new branch of science. You might call it "Applied Biodiversity," the science for applying the full spectrum of natural-resources to the world's (not to mention the nation's) problems. That new discipline would be more vital in a practical sense than applied physics or applied mathematics. Indeed, applied biodiversity could be the most valuable of all the sciences. Yet it doesn't exist.

In theory, building this new frontier of science shouldn't be difficult. The biodiversity is there, the need is there, the latent interest is there, and only modest financial support is required. But it will indeed be difficult because the barrier is one of the mind. Currently, conventional crops dominate the priorities of all decisionmakers and institutions. This is understandable, however the present attitudes are skewed so steadfastly against creating new funding initiatives that grand opportunities for progress are being lost.

The species and situations I've mentioned above exemplify what might be achieved for mother earth. But no matter how important such topics could be, no part of government seems capable of supporting efforts to check them out; not the National Science Foundation, the U.S. Department of Agriculture, the Agency for International Development, the Department of Energy, nor the Environmental Protection Agency. And in my experience, most foundations are just as resistant to the possibilities of applying biodiversity against the global problems.

It's up to us to change this. We must find ways to get the new-croppers' knowledge of plants and techniques of crop assessment into global play. In other words, the knowledge, visions, insights, spirit of endeavor, and capabilities represented at this conference is of historical significance. To my way of thinking, new-croppers are among the most needed people on earth. With our special skills we can provide the tools for solving the problems that threaten today's future the way the nuclear holocaust did during that agonizing period when most of us were growing up.

NEW CROP PUBLICATIONS IN THE BOSTID INNOVATION SERIES (NATIONAL ACADEMY OF SCIENCES)

More Water for Arid Lands. 1974. 149 p.
Underexploited Tropical Plants with Promising Economic Value. 1975. 184 p.
Making Aquatic Weeds Useful. 1976. 169 p.
Guayule: An Alternative Source of Natural Rubber. 1977. 77 p.
Tropical Legumes. 1979. 326 p.
Firewood Crops: Shrub and Tree Species for Energy Production. 1980. 236 p.
Water Buffalo: New Prospects for an Underutilized Animal. 1981. 111 p.
The Winged Bean: A High-Protein Crop for the Humid Tropics. 1981. 41 p.
Producer Gas: Another Fuel for Motor Transport. 1983. 95 p.
Mangium and Other Fast-Growing Acacias. 1983. 56 p.
Calliandra: A Versatile Tree for the Humid Tropics. 1983. 45 p.
Butterfly Farming in Papua New Guinea. 1983. 33 p.
Crocodiles as a Resource for the Tropics. 1983. 52 p.
Little-Known Asian Animals With Promising Economic Future. 1983. 125 p.
Firewood Crops: Shrub and Tree Species for Energy Production, Volume 2. 1983. 103 p.
Casuarinas: Nitrogen-Fixing Trees for Adverse Sites. 1984. 112 p.
Amaranth: Modern Prospects for an Ancient Crop. 1984. 74 p.
Leucaena: Promising Forage and Tree Crop (Second edition, 1984. 93 p.
Jojoba: A New Crop for Arid Lands. 1985. 100 p.
Quality-Protein Maize. 1988. 100 p.
Triticale: A Promising Addition to the World's Cereal Grains. 1989. 103 p.
Lost Crops of the Incas: Little-Known Plants of the Andes with Promise for Worldwide Cultivation. 1989. 415 p.
Microlivestock: Little-Known Small Animals with a Promising Economic Future. 1991. 450 p.
Neem: A Tree for Solving Global Problems. 1992. 149 p.
Vetiver: A Thin Green Line Against Erosion. 1993. 185 p.
Lost Crops of Africa: Volume 1—Grains. 1996. 383 p.
Lost Crops of Africa: Volume 2—Fruits. (In preparation).
Lost Crops of Africa: Volume 3—Vegetables. (In preparation).
Underexploited Tropical Fruits with Promising Economic Value. (In preparation).
Tropical Fruits for Solving Global Problems. (In preparation).

New Crops or New Uses for Old Crops: Where Should the Emphasis Be?

Shelby F. Thames and Thomas P. Schuman

Agriculture offers what other raw material sources cannot, i.e., renewability, efficient energy conversion, and beneficial environmental effects. Agricultural raw materials can alleviate petroleum demands, a raw material of finite supply. A number of environmental benefits are realized from plant growth including the absorption/conversion process of the sun's energy.

Current agriculture growth is limited to relatively few consumable crops though expanded use via industrial applications promises to broaden their market potential. Maize, soybean, and rapeseed are examples of consumable crops that are currently being considered as industrially significant fuels, solvents, coating resins, and/or lubricants. In addition, a wide variety of new, "alternative" crops are being examined as substitutes or adjuvants to existing sources of food and non-food raw materials. The use of agricultural materials as industrial raw materials is accomplished via accurate analysis of raw material composition(s) and structural identification of chemical components. Given this information, the value of the raw material source can be determined either as an unmodified source and/or as derivatized products. Finally the competitive use or application of the material and/or its derivatives must be determined by its incorporation into and testing of new industrial products.

The field of polymer application is immense and offers much potential for growth. Markets for polymeric industrial materials include plastics, coatings, adhesives, and elastomers. More than one half of all polymer personnel are employed in the coatings industry. In 1994, the coatings industry generated 1200 million gallons of coating shipments or an equivalent to 14 billion dollars in revenues. Moreover, the coatings industry is ever mindful of existing and impending environmental regulations and disposal issues that must be solved.

The question of new crops versus old crops is mute. The decision of which to use is entirely dependent upon raw material composition, processing, future risk (i.e., pollution cleanup), and product performance properties. An "old," over-produced crop can be instantly revitalized by the advent of innovative technologies. Future needs/demands will expand for renewable raw materials, for disposal control, and for pollution. Clearly, the science of biodegradation will receive much impetus in the years ahead. Indeed, a well-planned, consistent product design, synthesis, and market and investment protocol for future product development exist and will be the subject of this paper.

THE DEVELOPMENT OF AGRICULTURE

Agriculture has historically been associated with development of food sources through domestication of wild species discovered in nature. Hunting and gathering food dominated primordial behaviors, where a rich diversity of plants and animals were consumed depending on the location and season (Solbrig and Solbrig 1994). Thus, the move by humans to harness the potential of agriculture provided a perceived power over nature. The improved food supply created conditions allowing population growth, human concentration, and thus civilization. Agriculture fostered civilization, trade, sedentary life, work specialization, wealth/accumulation of wealth, and appreciation loss for the rural class while diets narrowed, social distinctions increased, and manufacturing was launched (Carrier 1923; Solbrig and Solbrig 1994).

Agriculture began in four independent centers: southwest Asia, southeast Asia, Mexico, and Peru (Solbrig and Solbrig 1994; Fig. 1), which influenced other geographical areas. Several experts believe these centers to be the ancestors of modern horticulture and animal husbandry (Vavilov 1926; Wilkinson 1972).

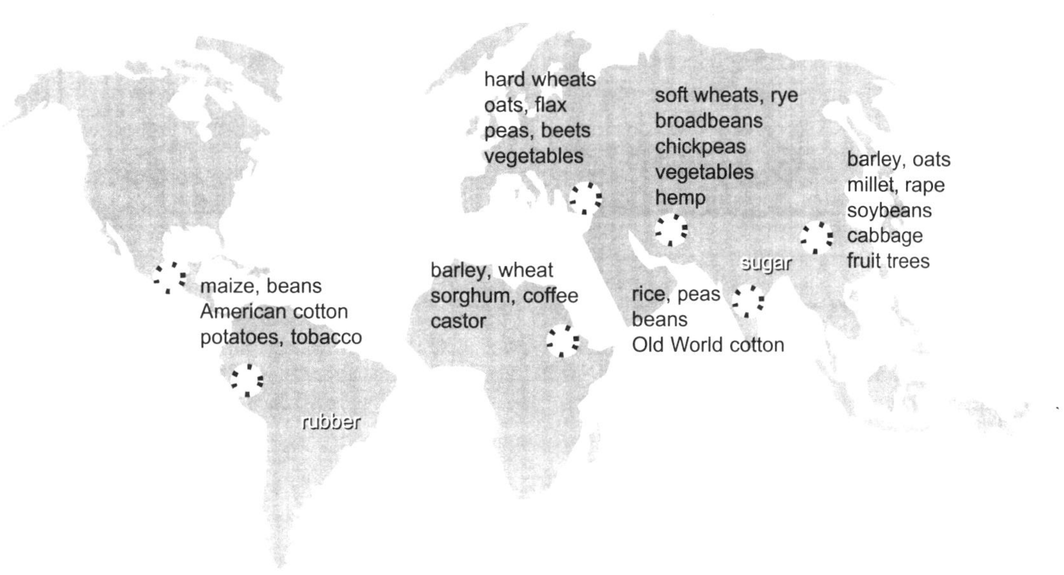

Fig. 1. Abbreviated map displaying the centers of origin for plant agriculture.

Southwest Asia, comprised of India, south Afghanistan, near Cashmere, Iran, Asia Minor, and the Caucasus, was responsible for soft wheats, rye, small-seeded flax, lentils, broadbeans, chickpeas, assorted vegetables, and Old World cotton. Southeast Asia, China, Tibet, Nepal, and surrounding areas domesticated barley, naked oats, millets, and soybeans. Southeast Asians also founded crucifers and many fruit trees. The Americas were responsible for maize, potatoes, beans, tobacco, Jerusalem artichokes, and American cotton. Vavilov ascribes the development of hard wheats, cultivated oats, large-seeded flax, peas, large-seeded lentils, beets, and other vegetables to the lands around the Mediterranean Sea. He credits an Ethiopian region as the nesting ground for barley, wheat, sorghum, and coffee. As people migrated from region to region, they carried crops with them, spreading agriculture from centers of origin into "noncenters" (Solbrig and Solbrig 1994). These Old World "noncenters" expanded to physical barriers, such as the oceans.

Once the physical barriers to the Old World were crossed, crop species were further exchanged. Sugar, a storage-stable confection of little nutritional value was the first world crop unrelated to food/feed survival. It was grown strictly for cash profits beginning around 600 AD. Tobacco brought the habit of smoking to Europe in the early 17th century, while coffee was smuggled to Martinique by a young French naval officer spawning a coffee industry in Brazil and Columbia (Solbrig and Solbrig 1994). Cotton fabric, the cloth of luxury, was popular when Eli Whitney invented the cotton gin in 1793 and revolutionized cotton fiber. Likewise, seedlings from the rubber tree, *Hevea*, were smuggled to England from South America by Henry Wickham and later transported to the British Malaysian colony where hevea rubber has become a thriving industry (Allen 1949). Coffee, sugar, wheat, and grapes are examples of crops introduced into the New World, while potatoes, tomatoes, maize, and tobacco were returned from the New World to the Old World.

AGRICULTURE FOR FOOD/FEED

Early agriculture was associated with food for survival. A condensed list of crops were chosen, either subliminally or purposefully, to increase efficiency. The reduced number of variables, in terms of proper nurturing, increased the chance of successful harvest; thus, assuring survival through narrow but plentiful sustenance. For example, the introduction of the potato as a staple food during the Industrial Revolution was likely responsible for the large population increase (Solbrig and Solbrig 1994). Since the Industrial Revolution, two important events precipitating the modern agricultural crises are worldwide population increase and the conversion to hydrocarbon fuels.

Rural communities, initially located near the food or agricultural source, expanded rapidly into highly populated areas of commerce. The fastest growing cities were historically located on the most fertile lands (Brown 1981; Solbrig and Solbrig 1994) reducing available land for cultivation. In 1980, for example, 2,000 acres (800 ha) of U.S. prime farmland were gobbled up by urban sprawl **each day** (Jackson 1980). W.H. Davis (1972) stated that a million acres (400,000 ha) of farmland were destroyed annually from urban expansion alone, and projected a decrease in agricultural land from 2.6 acres (1 ha) in 1972, to 1.2 acres (0.5 ha) per person by the year 2000. As population increases, a greater demand on agricultural production continues regardless of the reduced land area.

Probably the largest influence on modern society is fossil fuels, primarily petroleum. Fossil fuels are now a necessary evil to our present day society although chemical refineries, ocean tankers, and automobiles pollute land, water, and air. In addition to modern conveniences, energy is required to subsidize our fertilizers, pesticides, and harvesters permitting maximum crop yields on minimal acreage in order to meet today's demands for food, shelter, clothing, transportation, and heat. Petroleum products have not only allowed enhanced agricultural production but permeated raw material markets due to ebbing natural productivity compared to demand and scarcity in new cropland (Brown 1981). For example, world consumption of wood products for housing and paper production was at 0.62 m^3 per capita in 1976 (Brown 1981), converting to 2.5 billion m^3 of wood harvested. Wood capacity has since dropped off per capita (0.60 m^3 per person in 1980) from a high of 0.67 m^3 per person in 1964, as it was unable to keep pace with world population growth. Petroleum filled this production gap in the form of polymeric wood composites. Similarly, petroleum in the form of fertilizer is essential for producing more sustenance from less available land.

Plants extract nutrients from the soil during their growth cycle. If land is over planted, the crop yields are reduced over time. Crop rotation, allowing the ground to lie fallow or planting crops that promote nitrogen fixation, restores soil fertility as well as the application of natural or man-made fertilizers. Nutrients are being replenished by synthetic fertilization in ever intensifying amounts. In the U.S. during the period of 1945 to 1975, there was an increase from 10,000 to 400,000 t, producing only a twofold yield (Jackson 1980). World use of fertilizer increased from 15,000,000 to 114,000,000 t/year from 1950 to 1980, respectively. Herbicide and pesticide use has also escalated (Brown 1981; Solbrig and Solbrig 1994). Pesticides were applied at an accelerating dosage of 168% per unit production increase in the years 1950 to 1967 (Jackson 1980). It is important to remember that most fertilizers and pesticides are petroleum in origin.

Yet petroleum is ultimately limited in supply, therefore coal, the probable substitute for petroleum, will eventually compete for land and water use. Fig. 2 shows a disturbing overlap of coal reserves and agricultural regions in the maize and wheat production belts. Technologies beyond petroleum must include not only energy and chemical resources for agriculture, i.e., fertilizers, agricultural chemicals, and fuel for harvesters, but also industrial materials for civilian and critical/strategic military needs (Wheaton 1990).

Of greatest importance to society's future is that only agriculture can provide sustenance for our growing population on an otherwise shrinking planet. Agricultural production is limited only by the number and type of plants grown. Diverse planting avoids widespread dependence on a single plant, for example, the Irish potato famine of 1845 (Solbrig and Solbrig 1994). Recent trends in agriculture thus seek alternatives that diversify rather than concentrate on a few crops (New Farm and Forest Products Task Force 1987; Conrad 1993; Williams and Haq 1993; Babb 1990). Alternative crops can be either food or non-food related and are usually produced in smaller volumes, thus not competing with large, existing markets. However, alternative crops have the potential for spawning new industries such as soybean. Soybean was a North American alternative crop until just a few decades ago. Examples of alternative crops for food include canola, sunflower, rapeseed, high-bush blueberries, and broccoli. Non-food alternative crops include vernonia, lesquerella, tung, guayule, and kenaf as well as industrial strains of sunflower and rapeseed.

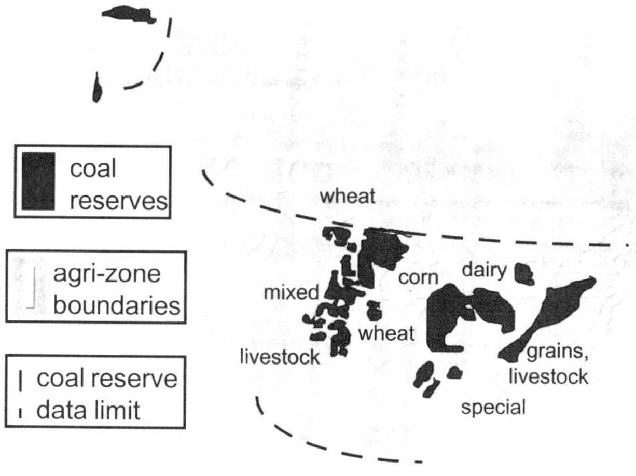

Coal reserves: U.S. Department of Energy. 1992. Supplement to the Annual Energy Outlook. United States Government Printing Office, Washington, D.C.
Agricultural zones: Gregor, 1970.

Fig. 2. Map of coal reserves (United States) and agricultural production zones (North America). Mixed = grains and livestock; maize = grains and livestock; wheat = grain farming; dairy = dairy livestock; special = primarily rice, cotton, and sugarcane.

INDUSTRIAL AGRICULTURE

A number of alternative crops target industrial raw material production. In fact, most modern industrial materials were originally of agricultural origin; linseed, tung, and castor oil coatings; rubber, casein, and soy adhesives; natural rubber elastomers; tallow and castor lubricants; cellulose and protein fibers. Petroleum displaced many of the agricultural material markets through post-war availability, convenience and performance, but primarily through cost advantages. Now a dwindling petroleum supply is shifting the focus from petroleum-based acrylic coatings, plastics, phenolic and formaldehyde adhesives, butyl-styrene elastomers, hydrocarbon greases, and synthetic fibers back to agriculture. Lesquerella, an alternative crop, can open new hydroxy oil markets for lubricants and waxes. As a source of natural rubber, guayule can contribute hypoallergenic latex to the medical product industry. Cotton, silk, and natural fibers are again in demand replacing petroleum-based nylons and polyesters. Kenaf is a fast growing plant with the capability to extend wood fibers for paper and corrugate manufacture alleviating timber demands.

Well established crops are also enjoying innovative successes. Maize is attempting to expand in the area of alternative fuels, i.e., ethanol. Newly constructed factories produce poly(lactide) from maize carbohydrates for biodegradable and medical/surgical polymers. Tung oil derivatives such as methyl eleostearate are useful as a reactive diluent in high solids, low volatile organic content (VOC) coatings. Soy protein is again showing promise in adhesives and high performance plastics, e.g., Environ of Phenix Composites. Soy oils are finding use in diesel fuel, letter-print inks, and industrial solvents.

Although a 12,000 year history in food/feed production exists, only recently has agriculture focused on production of industrial materials. Biomass currently enjoys use in textiles, paper and paperboard as fibers, sizing, and adhesives. However, use in plastics, coatings, resins, and composites is negligible due to the adoption of synthetics (Narayan 1994). Synthetics displaced agricultural materials, flooding the market because of their low costs. This implies low-to-no profit margin, i.e., strictly large volume farming or a high petroleum price, is required for agricultural materials to replace petroleum. Current strategies target high-priced, low-volume specialty markets to bridge the gap with low-price, high-volume commodity chemicals (USDA 1992). Examples of these markets are additives for foodstuffs, paints and coatings, lubricants, and fuels such as rheological aids, dispersants, adhesion promoters, and anti-knock or smog-reducing products.

NEW MARKETS FOR AGRICULTURE

Soybean

Soybean is a valuable raw material source for food and industrial applications. Lecithin, an oleophosphatide obtained from soybean oil processing (Fig. 3), is a well-known emulsifier and anti-oxidant for food, cosmetic, and pharmaceutical industries. Soybean oil is composed of mixed fatty acid triglycerides, primarily linoleic, oleic, and palmitic acids. While soybean oils were first used for manufacture of low molecular weight polyesters (alkyd resins) in the paint and coatings industry, synthetically-derived coatings have reduced soybean oil demand. Epoxidized soybean oil comprises 6.25% (100 billion lbs) of the plasticizer market, 87.5% of which is served by petroleum raw materials. Soybean proteins are very nutritious, having an amino acid composition against which all other plant meals are compared (Lusas and Rhee 1994; Wilson 1994). However, taste and flatulence factors inhibit most 100% soy protein foods; exceptions are Tofu, soy milk, and various fermented products. Currently soy proteins are used industrially as paper/textile sizings, fermentation media, adhesives, and recently as a building material (Johnson and Meyers 1994).

The polymer structure of protein lends well to chemical reaction and application in industrial materials (Fig. 3). Although the amino acid composition of proteins can vary widely, the polymer backbone structure remains the same. The polyamide feature of protein contributes both chemical and physical properties that can be altered by processing. Hydrolysis of protein yields lower molecular weight polyamides that can provide better mixing, reactivity, and uniformity.

Plastics and composites have been synthesized from soy proteins in combination with phenol and formaldehyde, filler, and plasticizer (Johnson and Meyers 1994). However, these plastics were susceptible to water absorption and longer cure times though somewhat improved by the addition of wood flour. Water sensitivity is partially due to the presence of carbohydrate impurities within the protein raw material (Lambuth 1989), which can be fixed to reduce water sensitivity via reactions with silicates (Johnson and Meyers 1994), formaldehyde (Brother and McKinney 1940), and perchlorate and periodate (Wurtzburg 1986). Environ, a protein product of Phenix Composites, is made from recycled newspaper, soy flour, and polymer binder. The material is innovative, utilizing agricultural raw materials and paper recycle material, thus reducing the quantity of solid waste. Paper, e.g., newspaper and paperboard, comprises 35%, or 58 million tons, of the landfilled solid wastes produced in the U.S. annually (Wolfe 1991). Other new uses for soybean proteins are expected in the areas of water resistant wood adhesives, biodegradable plastics and composites, textile fibers, and emulsifiers (United Soybean Board 1995).

Soybean oils are being applied as hydraulic fluids, form release for concrete, improved inks, and as a raw material to produce polyester and nylon monomers (United Soybean Board 1995). Saponification of the oil yields a glycerol and fatty acid mixture (Fig. 4). The profits obtained by recovery of glycerol are usually sufficient to cover process costs, thus free fatty acids, by volume, cost nearly the same as the whole oil. Esterification of the fatty acids with methanol affords fatty methyl esters, a class of materials useful for diesel fuels or environmentally friendly, industrial solvents. Methyl esters are less expensive than ethyl esters since ethanol not only costs more than methanol but the same esterification requires nearly twice the volume. Maize oils are also being examined for similar applications.

Maize

Maize (or corn in the U.S.) is a well-established crop grown in the northern central U.S. Kernels contain (poly)saccharides and oil but maize is primarily considered as a consumable grain and not usually noted for its oil producing capacity (Alexander 1989). A truly old crop for North America, the maize industry has been rejuvenated by new plant developments and new technologies. For instance, maize carbohydrates are used as raw material for microbial fermentations that yield citric, lactic, gluconic and ketogluconic acids, ethanol, and various polyols (Wit et al. 1993; Fig. 5). A direct benefit from the production of ethanol alone increased the price of a bushel of U.S. maize by 19.25 cents (Brownback 1992).

Cargill and Ecochem have recently installed polymer plants for production of poly(lactide) synthesized from maize-derived lactic acid, a novel biocompatible and biodegradable material for medical sutures and implants, drug delivery, and commercial packaging (Narayan 1994; Fig. 6). Poly(hydroxybutyrate) and poly(hydroxyvalerate) (Fig. 7), biodegradable thermoplastic polymers for personal care containers, medical uses, films, paper coatings, and even fishing nets (Office of Technology Assessment 1993), are also produced directly from fermentation processes (Narayan 1994; Shimamura et al. 1994; Eggink et al. 1993).

Fermentation processing is unique in that many processes are relatively independent of carbon nutrient structure. For example, fermentation of glucose, sucrose, starch, oleic acid, or 3-hydroxybutyrate (Nakamura et al. 1992) can each be used to produce poly(3-hydroxybutyrate). The variable for control of polymer structure is entirely dependent on proper selection of the microbe (bacteria) and its particular genome, or genetic code. Thus, bacteria can be genetically engineered to custom produce a raw material, either a monomer for

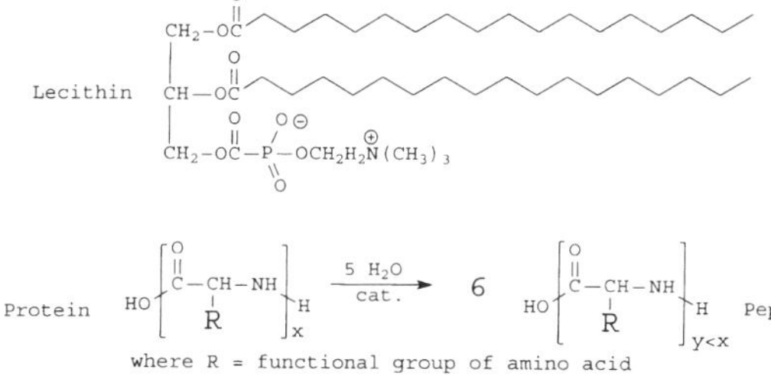

Fig. 3. Materials obtained from soybean.

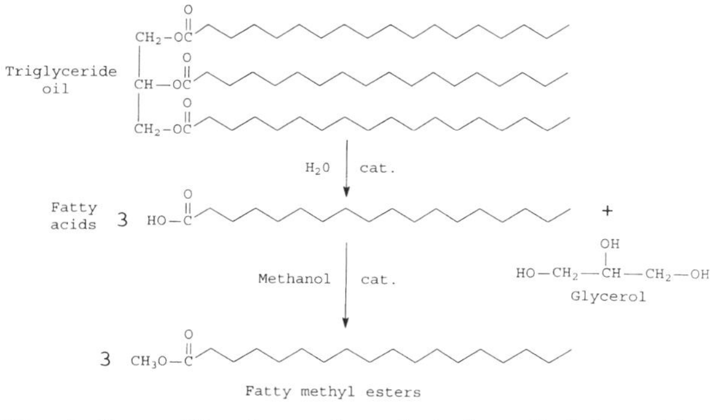

Fig. 4. Saponification and methylation of triglycerides.

polymerization like lactic acid or the complete polymer, such as poly(hydroxyalkyrate) from a carbon source.

Genetic engineering can also modify plant genomes, analogous to microbial genomes, to improve disease or drought resistance, or alter the product's chemical structure. In 1970, maize crops were devastated by a blight. Although the susceptibility was added through genetic engineering during an attempt to increase maize yields, genetic recombination with other cultivars yielded a disease resistant, high yield strain. Insect resistance, disease resistance, and plant yields, including the chemical structure of the product, are all affected by genome (Solbrig and Solbrig 1994). Transgenic crops are now a current trend in agricultural business (Rotman and Fairley 1995).

Lesquerella

Lesquerella is a winter annual crop native to the North American southwest and recently domesticated as an alternative crop. The plant seeds contain over 25% oil by weight where the primary fatty acid (~55–60%) content is the C20 lesquerolic hydroxy fatty acid. A second benefit of the pressed seed is its inherent nutrition, with an amino acid distribution high in lysine, methionine, and threonine similar to crambe meal. Commercialization problems are low seed yield, oil content, and hydroxy fatty acid content within the oil. Benefits of the hydroxy fatty acid are exemplified by castor oil products, which contain the C18 structural homologue ricinoleic acid. However, although ricinoleic acid comprises ~95% of castor oil fatty acids, the castor meal contains ricin, one of the most toxic substances known to man, and CB-1a, a potent allergen, both hazardous byproducts.

Products derived from lesquerella oil can be analogous to those obtained from castor. Dehydration of lesquerella oil produces a drying oil that is superior to dehydrated castor oil (Thames 1995). Hydrogenated waxes may be useful in lubricating greases, cosmetics such as lipstick, polishes, and inks (U.S.D.A. 1991). Pyrolysis and alkali fusion of lesquerolic acid yield tridecenoic acid and heptanal, and dodecanedioic acid and 2-octanol, respectively. The engineering nylons 13 and 6,12 can then be synthesized from the tridecenoic acid and dodecanedioic acid, respectively, similar to the nylons 11 and 6,10 obtained from ricinoleic acid. Stable, low isocyanate index polyurethane foams are prepared from alkoxylated lesquerella oils, and lesquerella oil has performed notably in cationically cured coatings (Thames, S.F. and H. Yu unpubl. results). The industrial future of lesquerella looks bright indeed.

Canola

Canola is a popular oilseed crop because of its high oleic acid content oil. A new, genetically engineered canola was recently planted near Dawson, Georgia that produces very high yields (40%) of lauric acid (Mounts 1995). Laurate canola is yet another example of agricultural diversification. Sodium dodecylsulfate, produced from lauric acid, is a major ingredient for toothpastes, shampoos, and laundry detergents. The C12 carbon, or dodecyl, structure of lauric acid is thus important for modern surfactant syntheses.

AGRICULTURAL NEW USES: POLYMER STRUCTURE/PROPERTY RELATIONSHIPS

The polymer industry is an underutilized industrial partner for agriculture where value-added monomer and additive markets present a viable means to commercializing new agricultural raw materials. Some acid

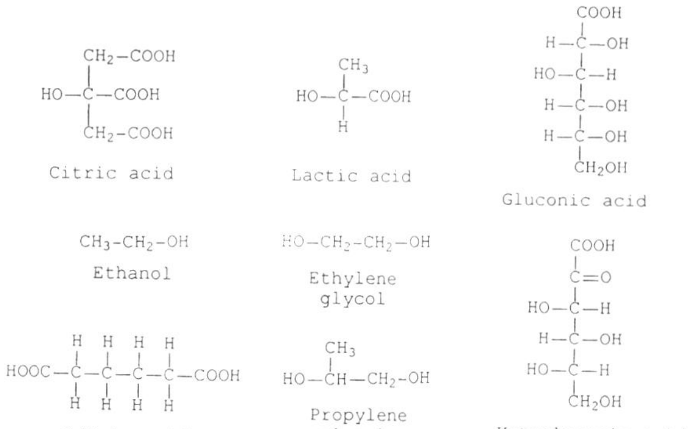

Fig. 5. Molecules obtained through fermentation.

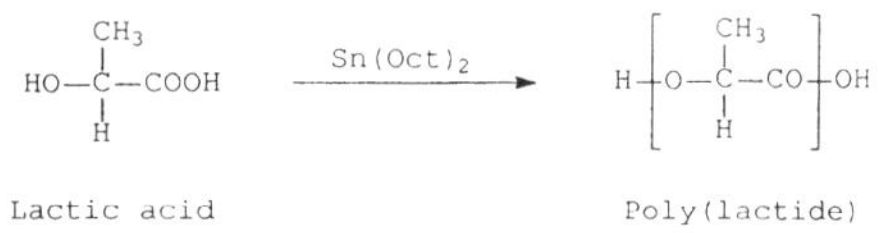

Fig. 6. Polymerization of lactic acid.

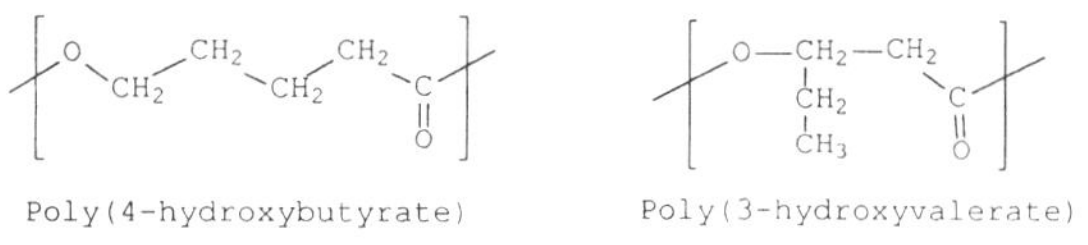

Fig. 7. Structures of poly(4-hydroxybutyrate) and poly (3-hydroxyvalerate).

structures important or potentially important to industry are erucic (C22-monounsaturated), lesquerolic (C20-OH), ricinoleic (C18-OH), oleic (C18-monounsaturated), lauric (C12), and caprylic (C8). The relevance of each to industry is entirely dependent on chemical structure. For instance, pyrolysis of erucic, lesquerolic, ricinoleic, and oleic acids and the products obtained are shown in Fig. 8; whereas alkali fusion of the same materials yields different products as demonstrated in Fig. 9. The products vary with carbon chain length, location of the double bond, and presence of other chemical functional groups, e.g., hydroxyl. The products can be further altered to produce amines or alcohols (Fig. 10), and other industrial raw materials. Fabricated small molecules, such as 2-heptanol, are useful as intermediates in various syntheses, as in flavors, perfumes, and pharmaceuticals. Difunctional products can be used for polymer synthesis (Fig. 11).

Nylons are a particularly useful class of high value polymers due to their moldability, toughness, and durability (Modern Plastics 1994). Nylons are a good example of the extent that physical properties are affected by structural variations. Nylons typically vary in structure in two ways (Fig. 12): by the carbon length of each unit segment, and by the segment form, -AB- or -AA-BB-, of the repeat unit where the "A" would stand for an amine and "B" for a carboxyl. Changes in physical behavior occur due to the ordering of the polymer chains as a crystal matrix, affecting the extent that nitrogen protons can align with carbonyl oxygens (Miyake 1960). Two forms of crystal dominate, the α and γ structures. For -AA-BB- polymers, even-even carbon unit lengths have the α form whereas odd-odd have the γ form. Even-odd and odd-even both have a γ form, except that solution processing may produce a third arrangement, the δ form. The -AB- polymers are more complex where even units tend to the α form and odd to the γ form, except nylons 8 and 10 which take on the γ pattern.

Nitrogen proton to oxygen interactions, indicated by crystal structure, influence the extent of hydrogen bonding which is a primary factor in nylon toughness and other physical properties. In general, hydrogen bonding increases the melting temperature of the polymer as the concentration of amide linkages increases. Nylon 12, an -AB- polymer formed from an ω-aminododecanoic acid, has a melt temperature of about 178°C compared to nylon 6, poly-ω-aminohexanoic acid, of 229°C. However, although nylon 11 has a more concentrated amide linkage than nylon 12 and therefore should have an increased melt temperature, the melt temperature actually decreases to 165°C. The odd carbon spaced nylon 11 segments, γ in crystal structure, are less able to align, decreasing interactions between chains and causing a corresponding decrease in melt temperature. This particular feature makes nylon 11, with nylon 12, useful to the powder coatings industry because powder coatings rely on good melting at a minimum temperature for film formation (Misev 1991).

Lengths of the carbon chain between the amide links also affect the physical properties of the bulk nylon. As observed above, more carbon atoms dilute the amide linkages, and decrease melt temperature and tensile modulus. Other related effects are increased polymer flexibility and elongation, while a constant tensile strength, nearly 83 MPa at break, is common for all carbon spacings. The exceptional performance of nylons commands premium prices (Modern Plastics 1994).

Another example illustrating the importance of structure is given by associative thickeners. Associative thickeners are used in coatings to control viscosity, or more appropriately, thicken on standing and thin on

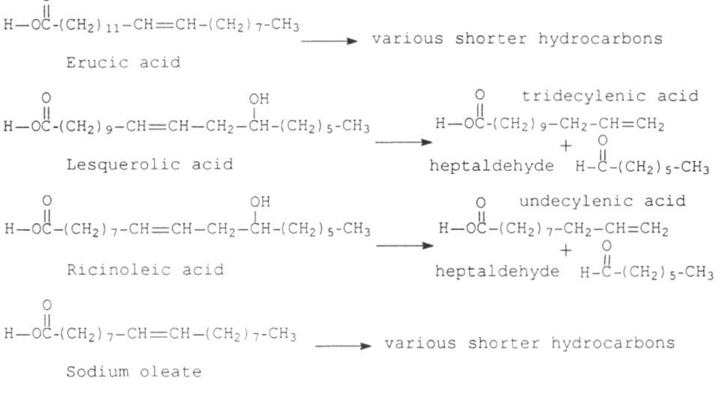

Fig. 8. Products from heat pyrolysis of erucic, lesquerolic, ricinoleic, and oleic acids.

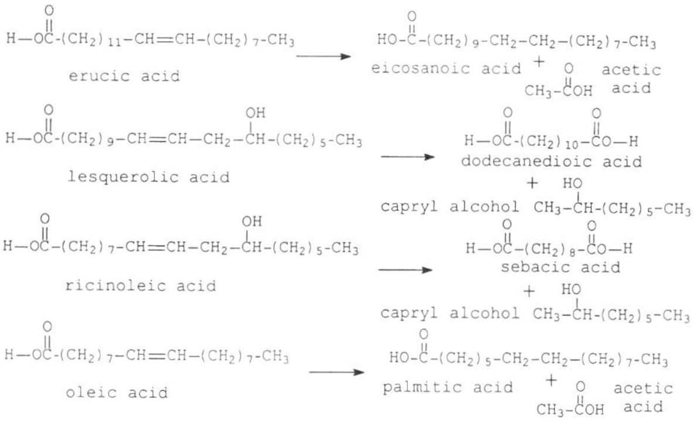

Fig. 9. Products from caustic fusion of erucic, lesquerolic, ricinoleic, and oleic acids.

shear. Shear thinning allows the coating to be easily applied by brush or spray while thickening allows good film build and one-coat hiding of the substrate.

Associative polymers have long, soluble segments with short, insoluble end groups (Fig. 13). Hydrophobic end groups are typical of modified cellulosics and hydrophobically-modified ethoxylated urethanes (HEURs). Performance of the thickener is dependent on the length of the hydrophobic end group as a ratio to the overall polymer length. As the terminal hydrophobe of a HEUR polymer varies from 0 to 16 carbons in length, the low shear viscosity increases from 100 to 100,000 times, corresponding to a terminal hydrophobe increase from 0 to 2.6 weight percent, over the molecular weight range of 100,000 to 16,000 (5% W/W solutions, Fig. 14). For example, extending the terminal hydrophobe length to increase the solution concentration of hydrophobe from 0 to 0.1% W/W increases the solution viscosity 100,000 times, from 3×10^{-2} poise to 2×10^{3} poise (Jenkins 1990). Thus, a very small change in polymer structure can create a very large change in physical properties.

SUMMARY

Society is currently dependent on highly productive agriculture for food/feed and to a very limited extent, industrial materials. However, farm productivity is held hostage by petroleum, for fertilizers, pesticides, herbicides, and fuel to power equipment. Other than petroleum, the only current alternatives are pre-Industrial Revolution fertilizer (manure) and horse drawn equipment—not likely prospects considering the additional quantity of grain that would be required to fuel the required livestock—and coal. Coal is a reportedly unlimited resource that can compete for water and fertile farmland.

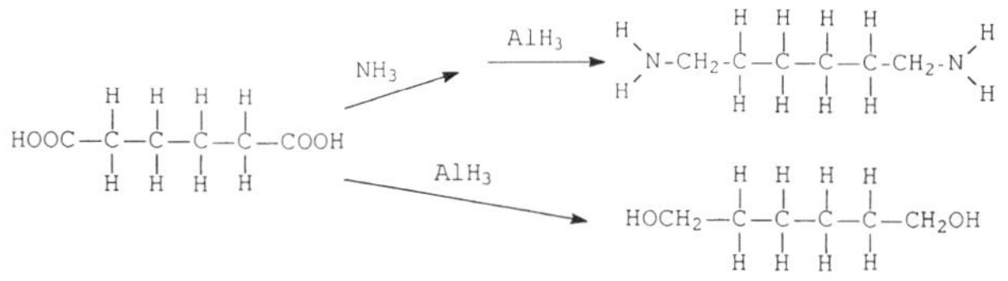

Fig. 10. Formation of 1,6-hexamethylenediamine and 1,6-hexanediol from adipic acid.

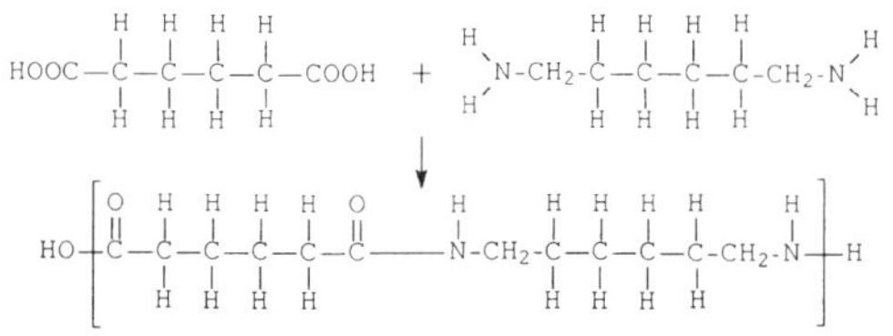

Fig. 11. Polymerization of adipic acid and 1,6-hexanediamine to form nylon 6,6.

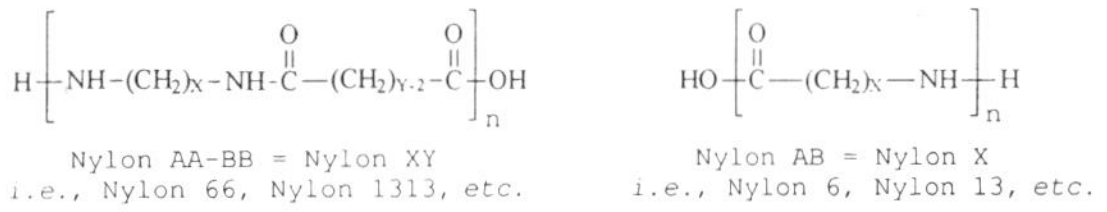

Fig. 12. Structure-derived nomencalture of the nylon polymer families.

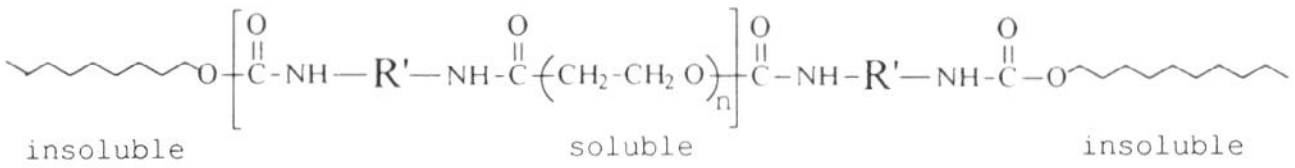

Fig. 13. Structure of the HEUR associative thickener.

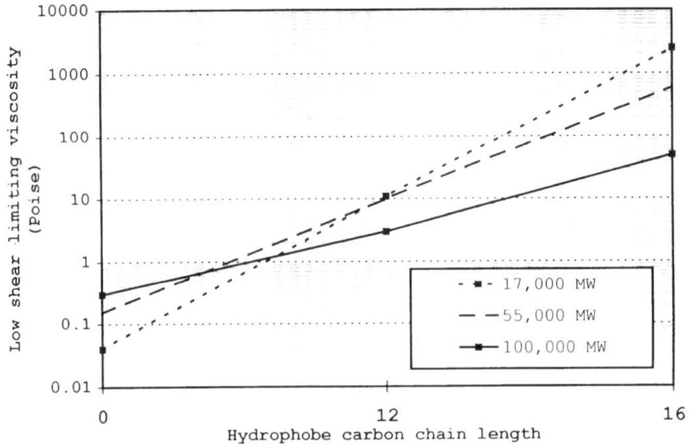

Fig. 14. Change in low shear viscosity as a function of the molecular weight and hydrophobe chain length of HUER associative thickeners.

New products from agriculture will be needed, not only from transgenic crops that treat resistance and production problems, but also industrial materials. Petroleum replaced the original ag materials on a cost and supply basis. Agricultural materials need to reciprocate, alleviating the demand on petroleum including fertilizer, pesticide, and fuel in support of a sustainable agriculture of maximum yield.

Pollution has historically been associated with population, threatening the very resources upon which life itself is contingent. However, pollution can be reduced through recycling and biodegradation while novel "green" technologies improve the global environment through reduction in VOCs, hazardous pollutants, and toxins (Kirschner 1994). These are cost driven markets that are ideal for new, alternative crops.

Today we consider maize, soybean, wheat, and sugarcane "old" crops, yet soybean, sugar, and wheat are actually new to North America having recently been introduced from southern Asia. It was through agriculture and domestication that plant diversity was given up by early man. Of more than 3,000 plant species used as food by different peoples throughout history, only about 200 are now domesticated while fewer than 20 are heavily cultivated within any region. Promotion of all crops, new or old, are probably best served through the new uses of their produce and service to society.

The question of new crops versus old crops is mute. The decision of which to use is entirely dependent upon raw material composition, processing, future risk (i.e., pollution cleanup), and product performance. An "old," overproduced crop can be instantly revitalized by the advent of innovative technologies. New raw materials can fill future needs; demands will expand for renewable raw materials, for disposal and pollution control. Clearly, the science of biodegradation will receive more consideration in the years ahead. The polymer industry plays a vital role in the development of cost effective, high performance polymers from agricultural materials, where small changes in structural design can greatly affect properties, and will further their industrial acceptance.

REFERENCES

Alexander, D.E. 1989. Maize. In: G. Röbbelen, R.K. Downey, and A. Ashri (eds.), Oil crops of the world. McGraw–Hill, New York.

Allen, H. 1949. The house of Goodyear; fifty years of men and industry. Corday & Gross, Cleveland, OH.

Babb, E.M. 1990. Marketing new crops. In: J. Janick and J.E. Simon (eds.), Advances in new crops. Timber Press, Portland, OR.

Brother, G.H. and L.L. McKinney. 1940. Protein plastics from soybean products. Ind. Eng. Chem. 32:1002–1006.

Brown, L.R. 1981. Building a sustainable society. W.W. Norton and Co., New York.

Brownback, S. 1992. Transforming the vision in reality: how to make it happen. In: New industrial uses, new markets for U.S. crops: status of technology and commercial adoption. Advance Issue Prepared Especially for Biobased Products Expo, St. Louis, MO., October 6–9. Hudson and Harsch, Maumee, OH.

Carrier, L. 1923. The beginnings of agriculture in America. p. 1–12. McGraw–Hill, New York.

Conrad, J.W. III. 1993. National new crops policy. In: J. Janick and J.E. Simon (eds.), New crops. Wiley, New York.

Davis, W.H. 1972. Overpopulated America. In: M.E. Adelstein and J.G. Pival (eds.), Ecocide and population. St. Martin's Press, New York.

Eggink, G., H. van der Wal, G.N.M. Huijberts, and P. de Waard. 1993. Oleic acid as a substrate for poly-3-hydroxyalkanoate formation in *Alcalingenes eutrophus* and *Psuedomonas putida*. Ind. Crops Prod. 1:157–163.

Gregor, H.F. 1970. Geography of agriculture. Prentice Hall, Englewood Cliffs, NJ.

Illman, D.L. 1994. Environmentally benign chemistry aims for processes that don't pollute. Chem. Eng. News 72(36):22–27.

Jackson, W. 1985. New roots for agriculture. Univ. Nebraska Press, Lincoln.

Jenkins, R.D. 1990. The fundamental thickening mechanism of associative polymers in latex systems: a rheological study. PhD Diss., Lehigh Univ., Lehigh, PA.

Johnson, L.A. and D.J. Meyers. 1994. Industrial uses for soybeans. In: D.R. Erikson (ed.), Practical handbook of soybean processing and utilization. Am. Oil Chem. Soc., Champaign, IL. and United Soybean Board, St. Louis, MO.

Kirschner, E.M. 1994. Environment, health concerns force shift in use of organic solvents. Chem. Eng. News 72(25):13–20.

Lambuth, A.L. 1989. Protein adhesives for wood. In: A. Pizzi (ed.), Wood adhesives. Marcel Dekker, New York.

Lusas, E.W. and K.C. Rhee. 1994. Soy protein processing and utilization. In: D.R. Erikson (ed.), Practical handbook of soybean processing and utilization. Am. Oil Chem. Soc., Champaign, IL. and United Soybean Board, St. Louis, MO.

Misev, T.A. 1991. Powder coatings: chemistry and technology. Wiley, New York.

Miyake, A. 1960. Infrared spectra and crystal structures of polyamides. J. Polymer Sci. 24:223–232.

Modern Plastics. 1994. Plastics handbook. McGraw–Hill, New York.

Mounts, T.L. 1995. Calgene marks planting of transgenic canola. INFORM 6:82–83.

Nakamura, S., Y. Doi, and M. Scandola. 1992. Microbial synthesis and characterization of poly(3-hydroxybutyrate-co-4-hydroxybutyrate). Macromolecules 25:4237–4241.

Narayan, R. 1994. Polymeric materials from agricultural feedstocks. In: M.L. Fishman, R.B. Friedman, and S.J. Huang (eds.), Polymers from agricultural coproducts. ACS Symposium Series. Am. Chem. Soc., Washington, DC.

Office of Technology Assessment. 1993. Biopolymers. U.S. Congress, United States Government Printing Office, Washington, DC.

Rotman, D. and P. Fairley. 1995. Plant biotech. Chem. Week 157(11):25–27.

Shimamura, E., M. Scandola, and Y. Doi. 1994. Microbial synthesis and characterization of poly(3-hydroxybutyrate-co-hydroxypropionate). Macromolecules 27:4429–4435.

Solbrig, O.T. and D.J. Solbrig. 1994. So shall you reap. Island Press, Washington, DC.

Thames, S.F., H. Yu, M.D. Wang, and T.P. Schuman. 1995. Dehydration of lesquerella oil. J. Appl. Polymer Sci. 58:943–950.

United Soybean Board. 1995. a) Status of current research projects. b) Soy industry news. United Soybean Board, Am. Soybean Assoc., St. Louis, MO.

United States Department of Agriculture. 1991. Lesquerella as a source of hydroxy fatty acids for industrial products. U.S. Govt. Printing Office, Washington DC.

United States Department of Agriculture. 1992. New industrial uses, new markets for U.S. crops: status of technology and commercial adoption. Advance Issue Prepared Especially for Biobased Products Expo, St. Louis, Missouri, October 6–9. Hudson and Harsch, Maumee, Ohio.

Vavilov, N.I. 1926. Studies on the origin of cultivated plants. Bul. Appl. Bot. Genet. Plant Breed. 16:1–248.

Wheaton, E.R. 1990. Industrial crops commercialization. In: J. Janick and J.E. Simon (eds.), Advances in new crops. Timber Press, Portland, OR.

Wilkinson, P.F. 1972. Current experimental domestication and its relevance to prehistory. In: E.S. Higgs (ed.), Papers in economic prehistory. Cambridge University Press, London.

Williams, J.T and N. Haq. 1993. International new crops policy. In: J. Janick and J.E. Simon (eds.), New crops. Wiley, New York.

Wilson, L.A. 1994. Soy foods. In: D.R. Erikson (ed.), Practical handbook of soybean processing and utilization. Am. Oil Chem. Soc., Champaign, IL. and United Soybean Board, St. Louis, MO.

Wit, D. de, L. Maat, and A.P.G. Kieboom. 1993. Carbohydrates as industrial raw materials. Ind. Crops Prod. 2:1–12.

Wolfe, J. 1991. Taking a fresh look at waste reduction. In: D.T. Smith (ed.), Agriculture and the environment. United States Government Printing Office, Washington, DC.

Wurtzburg, O.B. 1986. Converted starches. In: O.B. Wurtzburg (ed.), Modified starches: properties and uses. CRC Press, Boca Raton, FL.

New Industrial Crops for Europe

Anthony Capelle

When discussing new crops activities in Europe, it is appropriate to speak about industrial crops for nonfood use or renewable resources. In general, the activities can be divided into three categories: (1) existing crops for new markets; (2) new crops for existing markets; and (3) new crops for new markets. New crops for new markets is fundamental, strategic, and highly risky. Therefore, local and national governments see a function for themselves in stimulating these activities within the framework of their agricultural policy. In defining those respective policies, the following aspects play a role: contribution to solving agricultural problems; contribution to decreasing environmental impact; and contribution to the upkeep of rural structure.

Depending on the situation in each country, variations in policy can be observed. However, some general comments can be made that are valid for each European country:

1. The change in the common agricultural policy, coupled with the GATT agreement, has led to a substantial decrease in the income for arable farmers.
2. In the past, a rather inefficient structure was kept intact that protected small farmers in the Netherlands, parts of southern Germany and small parts of France. Support has shifted from product support to income support. Product support in the past led to a strong focus on increasing yields and, hence, to intensivation and overproduction. During the last 30 years, wheat production in the European Community (EC) more than doubled.
3. By taking land out of production (the crop area for each farm is linked to the volume of wheat produced), this overproduction is diminished towards a more self-supporting level. However, instead of letting this land lie fallow, it may and can be used for growing crops for nonfood use while keeping the income support intact.

It is this last factor, linked with the average area of farms in each country, that determines industrial crops activities. The relatively small scale of farms in the Netherlands and southern Germany means that the majority of the farms do not have to take land out of production, while in France and the U.K. we see large "non food producing" areas. Thus, the maximum theoretical area in the Netherlands that could be taken out of production is about 25,000 ha, but in reality only 10,000 ha is removed from production. In 1995, 250,000 ha of the 1.5 million "non food productive" ha in France were used for growing rapeseed for biodiesel.

UNITED KINGDOM

In the U.K., three agencies are actively engaged with renewable resources. Firstly, the Ministry of Agriculture, Food and Fisheries (MAFF); secondly, the Department of Trade and Industries (DTI); and thirdly, the Office of Science and Technology (OST). Within the framework of the LINK program, a mechanism for grants sponsoring precompetitive research cooperation between industry and research infrastructure (Department of Trade and Industry) there is a budget of £3 million (US$5 million).

Projects are judged by three criteria: (1) industrial involvement; (2) external review by independent experts; and (3) interest of the government itself. A majority of the activities deal with alternative energy sources, especially biomass and short-rotation tree crops for electricity generation. Biodiesel and bioethanol are rejected as there is no economical justification. A new policy and strategy is being worked out and expected to be in place in 1996.

GERMANY

The responsability for renewable resources rested with the Agricultural Ministry (Bundesministerium für Ernährung, Landwirtschaft und Forsten) and the Ministry for Research and Technology (Bundesministerium für Forschung und Technologie). However, in 1992, a new body with central coordination responsability and a budget of 55 million DM (US$35 million) was created, the Office for Renewable Resources (Fachagentur für nachwachsende Rohstoffe).

In 1995, 1.5 million ha were taken out of production, besides the assignment of 120,000 ha for nonfood oilseeds and 120,000 ha for nonfood starch and sugar. It is expected that oilseed production will double and reach about 1 million ha. This fact explains the high interest in Germany for biodiesel; rapeseed fits excellently in the crop rotation scheme. There is also a tendency to focus more on starch products and lubricant additives. Within the framework of EC research programs, some attention is given to "new" crops, such as marigold, crambe, and high-erucic rape.

FRANCE

In 1994, the responsability for renewable resources was shifted to a newly created organization, AGRICE (agricultural and chemical industry). This organization is a cooperation between the Ministry of Agriculture, Ministry of Industry, Ministry of Technology and Research, and the Ministry for the Environment, together with the French petroleum institute and the production organizations for wheat, sugar beet, and oilseeds. The budget is 65 million FF (US$12 million), of which one third is contributed by industry and agricultural production organizations.

In 1995, as in Germany, 1.5 million ha were taken out of food and feed production, with about 250,000 ha for rapeseed production (350,000 expected in 1996). As in Germany, attention is mainly focussed on biofuels, almost exclusively biodiesel, but about 8,000 ha of wheat are grown for bio-ethanol.

THE NETHERLANDS

The responsability for renewable resources lies with the Ministry of Agriculture and is coordinated by the Office for Agricultural Research (DLO). Over the last four years, the budget has been Hfl.90 million (US$50 million), which was spent over three large programs and several small ones. The budget for 1996 is still being discussed.

There is almost no surface taken out of production due to the relatively small scale of arable farms. The approach to new crops/industrial crops in the Netherlands is markedly different from the other European countries. At a rather early stage, the Council for Agricultural Research (NRLO) advised a strong market-driven approach with a focus on vegetable oils, essential oils, fibers, and starch. This resulted in three major programs: (a) hemp with a budget of Hfl.17 million (US$9 million); (b) oils with a budget of Hfl.9 million (US$5 million); and (c) caraway with a budget of Hfl.9 million (US$ 5 million). These program's terminated end of 1994.

One of the main conclusions drawn from these programs is that, despite the professed market pull/orientation, involvement of industry was too low. Commercialization was only initiated in those occasions, where industry saw a strategic advantage. This is happening with marigold as a source for resins for the paint and varnish industry, crambe as a source of erucic acid for the oleochemical industry, and caraway as a source for carvone for anti-sprouting agents.

It was also concluded that the presence of a processing industry even on a small scale, is an absolute necessity for providing industrial partners with enough material for industrial trial runs. Within the EC-sponsored VOSFA research program (Vegetable, Oils for Specific Fatty ocids), the following oil seed crops are studied in the Netherlands: marigold, dimorphotheca, lesquerella, euphorbia, and lunaria. Agronomics, processing of the seeds and application of the various oils are studied by a consortium of institutes and industries.

EUROPEAN COMMUNITY

Framework programs are the vehicles by which renewable resources are studied. Each framework program has a duration of four to five years. Between the subsequent programs there is an overlag of one year to safeguard continuity. The third framework program AIR II (Agro Industrial Research) consisted of shared-cost projects, demonstration projects and concerted actions. The fourth framework program FAIR (Fisheries Agro Industrial Research) is being defined now. The budget for the period 1995 to 1998 totals ECU620 million (US$750 million) for agricultural research in general. Of this amount, about ECU150 million (US$200 million) is earmarked for industrial crops/renewable resources. The Netherlands will, based on previous experience, receive about 10% of this sum. Most projects deal with fiber and wood.

For oilseeds, the VOSFA program started as a follow-up to the VOICI program this year. The research deals with breeding, growing, and processing of oilseeds aimed at lubricant additives, the paint and varnish industry, and cosmetics. The budget for this program is US$2.0 million. The earlier VOICI program (Vegetable Oils for Innovation in Chemical Industry) (budget US$2.5 million) focussed on crambe, limnanthes, and dimorphotheca.

SUMMARY

European new crop activities will be a booming business in the coming years, especially for the production of biodiesel. It is foreseen that in 1997 production of 1 million tonnes will be reached. In addition to biofuels, new oilseeds are being considered for use in the paint and varnish industry. Calendula looks very promising with a potential of several ten thousands of tons. Furthermore secondary metabolites have demonstrated their potential. Carvone from caraway for use as an anti-sprouting agent used during storage of potatoes. The market potential for use on table potatoes is estimated at several thousands of tons, equivalent to then thousand hectares. In the case of seed potatoes, where no chemical additive can be used the potential is even larger due to the reversible character. The anti-sprouting activity of carvone is now gradually replacing synthesized chemicals.

New Crops In The UK: From Concept to Bottom Line Profits

Francis H. Nicholls

Development of new crops for the large part appears to be a predominantly academic enterprise. Commercial entrepreneurs on the other hand can often seem to be a minority species. Where are those people who take ideas from off the drawing board and into the ground? John K King & Sons Ltd. have played a considerable role in the commercial introduction of new crops into the UK. This paper presents some of the company's philosophy, together with some of its failures and successes.

Kings is an agricultural supply company which, until eight years ago, had kept to a traditional rôle of selling basic farm commodities, such as seeds, fertilizers, and pesticides and trading in harvested commodities, particularly pulse crops and cereals. Since 1970, alongside our traditional trade, the company also worked on what were then referred to as "odd crops," and are now obligingly called "new alternative arable resources." This new work area fits our situation well because the company has its own farmland on which to trial new material.

By 1985, it was becoming obvious that reliance on trading basic farms goods would not, for ever, keep the wolf (or the bank manager) from the door, particularly with lower trading margins. It was at this time that a closer look was taken at what was already 15 years worth of experience with new crops and new uses for old crops. Over the last ten years the company's contracts for various new arable crops have grown to over 16000 ha (40000 acres) for the 1995 harvest.

New crop introductions into the UK are not always a success. For example, based on a theory that the UK could grow its own navy beans, Kings decided to take on Michigan and Ontario. In 1974, after three reasonably successful years of trials, the company contracted 280 ha of navy beans with local growers. The result was a total yield of less than 50 t. Yet there are four crops which merit highlighting: quinoa, borage, lunaria, and high erucic acid rapeseed (HEAR).

Before starting any project there are key questions that must be asked. (1) Is there a viable end use for the crop? (2) Will it perform satisfactorily in the field on a long term basis? (3) Will trials convert to economic field scale production? (4) Can the crop be mechanically harvested? (5) Will the end product have a quality that is stable, is able to be replicated, and give 100% confidence to the end buyer? (6) Is the new crop going to be as good as, or preferably better and/or cheaper than, what is already available?

This all needs time and investment for initial research and market development. There is no standard blueprint for success. Whilst a new arable crop is being tried and tested in the field, a working relationship must be developed with a potential end-user. The knot must be tied between the worlds of research and commerce. This is a skilled job, because the company is acting as mediator to the most exacting person in the world—the raw materials buyer.

Then at a critical stage, the new crop must be moved on from research trials and put under a contractual system, often throughout the UK, to produce continuity of supply and quality. The timing of this move is crucial. Over-production at a premature stage clearly does not make good commercial sense; equally, neither is the inability to supply an end-user's demand. Constant liaison is the only remedy.

Even when all has worked to plan, things can go awry because of elements out of our control, such as international market prices or new legislation. The whole field of new crop development requires initiative and fast reaction, good management–and a lot of luck.

QUINOA

The history of quinoa (*Quinoa chenopodium*) as a human consumption protein crop, originating from South America, is well documented (Galwey 1993). Washed or decorticated quinoa is now a widely sold product in health food outlets in Western Europe and North America. This crop has probably become our company's best example of making quick initial commercial use of a product and then utilizing the income from this to put towards R&D for wider industrial uses.

Our involvement with the species began in 1983 with a trial of a quinoa selection from Chile. A 10 m × 25 m plot was grown at our company trial ground in the Spring of 1983. It had a number of faults, not the least of which, was its uneven ripening and its lateness. What quinoa did do however was to: (1) appear to grow well under UK conditions; (2) produce good quality seed (albeit rather late); (3) grow to a height suitable for combine harvesting; and (4) attract a large number of wild pheasants and partridges. Quinoa also appeared to have potential as a gluten-free food with an excellent amino acid profile and the possibility of an unusually small starch granule size.

Quinoa for industrial use requires longterm research, so Kings looked at quinoa as a potential game cover crop for pheasant and partridges, (based on the initial chance information gained from the 1983 field trial). A blend of early, medium, and late maturing types were selected and the natural seed drop occurred throughout the European shooting season from Oct. to Jan. Thus our Kingmix® product evolved, containing a mixture of kale for cover and quinoa for feed. Kingmix® is now selling in 14 European countries and contributes considerably towards in-company R&D costs and bottom line profits.

Further investigations into quinoa led down two distinct paths. The first was quinoa for human consumption. Decorticated UK grown quinoa was sold by Kings in half kilo packs and initially retailed well. However, in 1992 an amendment to the UK Food Labelling Regulations was introduced and moved this retail human consumption operation out of King's sphere of normal trading activities. The regulations were onerous as to the correct level of saponin removal. This outlet for quinoa was therefore shelved until another day. Other industrial uses, such as the following are now being reviewed by ourselves and others, these include: a flow improver to incorporate into starch flour products, fillers in the plastics industry; anti-offset and dusting powders and complimentary protein for improving the amino acid balance of human and animal foods.

In 1993, Nutrasweet filed a European patent for a cream substitute with quinoa starch which highlighted the seriousness with which industry was taking this species. Saponins in themselves are also interesting, as potential insecticides, antibiotics and fungicides and also in the pharmaceutical industry as a mediator of intestinal permeability which could aid the absorbtion of specific drugs. Much of this work is outside the sphere of Kings' R&D activities. It is useful to be aware of the full scope of quinoa's potential even if it is likely to be a very long haul. Our expertise is in the supply of new selections of quinoa cultivars, which are adapted to UK cultivation and have higher yields. This R&D is financed by the income drawn from quinoa game cover products and is a good example of using an initial idea as a stepping stone to find other, more commercially sized opportunities.

BORAGE

Borage (*Borago officinalis* L.) is a crop which presents the farmer and his contracting merchant with the problem of producing what is essentially a garden flower on a field scale. It is widely recognized that borage, together with evening primrose, are the main arable sources of gamma-linoleic acid (GLA). (Galwey 1987). Whatever the claims and counter claims of the efficacy of GLA, the retail demand, certainly in Europe, for both evening primrose and borage oil is unquestionable.

In the mid-1980s the company helped to pioneer the commercial field production of borage. Early trials proved that borage likes heavier, well drained soil and dislikes most pesticides, which is a good thing for an oil crop mainly destined for the dietary health food trade. The crop appears to grow best in areas with moderate summer temperatures and reasonable moisture. Countries bordering on the North Sea offer the best European conditions for borage and the east coast of England has become the prominent European production area for the crop. The world production of borage is estimated by some to be between 1600 and 2000 ha. Thus in every sense this is a niche market. If however the use of GLA is widened, perhaps into food or pharmaceutical products, then the scope of borage would become even more attractive. The main world production areas are currently the UK, Canada and New Zealand.

Early trials proved that the borage crop grew well, but its complete aversion to chemical desiccants, resulted in swathing with a rapeseed swather becoming the standard practice. After 7 to 10 days, combine harvesting takes place, in most cases using a special pick-up reel. The growers have become experienced with the correct combine settings, aspiration and forward drive speed when combining into the swathed crop. The aim is to end up with a recleaned seed crop with about 30% oil and 23% to 24% GLA. In hot years, the GLA can drop below 20%. The price per tonne of seed to the grower in the mid-1980s was about US $4.25/kg. However, with improved growing techniques and higher yields (and also resultant early over production) this reduced to US $3.20/kg, which has been about the going rate for the past five years.

The first objective is to obtain the tonnage requirements of the end buyers for the next season. Orders are then covered with back-to-back contracts with experienced growers. An updated annual average yield is kept in order to evaluate the required acreage to fulfil orders. The second objective is to keep all growers alert to new techniques, ideal planting, swathing and harvesting times. Each grower can expect 3 to 4 visits from company representatives during the growing season, backed with information by telephone or company Newsletters.

Until now the gross margins have been, on average, higher than other spring-sown crops like cereals, pulses, or linseed. The recent EU Area Aid support and set-aside regime for main cereal and oilseed crops is likely to increase the contract borage seed price (and the resultant oil).

Borage has now settled into a regular UK farm rotation. It is a specialist crop requiring very special attention. Borage may well continue to be a Cinderella niche crop, but it has increasing commercial interest from international companies such as Hoffmann la Roche. Promotional material has also subtly changed the name borage to 'starflower.' Whatever the eventual nomenclature, borage has now been commercialized for growers in eastern England and has created a worldwide portfolio of buyers.

LUNARIA

Lunaria annua L. is a biennial garden flower which produces characteristic silver seed capsules often used in dry flower arrangements. The seed contains about 36% oil with an unusual distribution of fatty acids. The main feature is the simple lipid profile, the main constituents of which comprise 24% oleic acid (C18:1), 47% erucic acid (C22:1) and 21% nervonic acid (C.24.1). Nervonic acid is unusual in plant seed oils, but is present throughout mammalian nerve tissue—hence its name. Nervonic acid and its derivatives can be produced from lunaria oil by conventional techniques. A UK company, Croda Universal, is investigating its use in a variety of applications including its use in pharmaceuticals.

Since nervonic acid is an important constituent of brain and membrane phospholipids, it is vital that the body has an adequate provision of this fatty acid. If, for any reason, the body cannot make this lipid, then a dietary source is indicated. Two such situations are under active investigation; multiple sclerosis (MS), where there are indications that there is inadequate provision of nervonic acid; and premature birth, where the infant no longer benefits from maternal nutrition. It is of interest to note that human breast milk contains nervonic acid and that MS has been linked to the use of formula feeds. It is suggested that, breast milk appears to confer a resistance to developing the disease. Croda Universal have filed patents on the treatment of MS with nervonic acid and the use of nervonic acid as a lipid supplement for babies. Products containing lunaria oil are in clinical development and are expected to complete the initial phases of development within 2 to 3 years.

In 1991 Croda Universal approached Kings to grow this biennial garden flower on a field scale basis. The first problem was to recognize that growers would only be prepared to take on a biennial crop if they could derive two years' worth of income, if the land was to be occupied for two growing seasons. With potential recleaned seed yields of 60–100 kg/ha, this indicated a prohibitive end price. Lunaria had been tried as an open seeded crop over an 18 month period and it certainly looked good, with easier inter-row weed control. The end product price remained the major problem.

It was decided therefore that lunaria seed could be sown into a standing crop of winter wheat, with a special seed-drill, in mid-Mar. Lunaria seems to establish and thrive underneath the growing cereal. The wheat is harvested in Aug., leaving the young lunaria plants to grow away from within the stubble. It is by this method, that the grower is given the chance to produce a biennial crop, but with two distinct monetary returns over the two year period, the first from the winter wheat, the other from the undersown lunaria the following year.

Weed and disease problems have by no means been conquered but nevertheless *Lunaria* has been a reasonable success story in the UK. It really is a case of not being daunted by whatever is asked by an end-user. If an initial idea is taken steadily at the start and thought through, with a bit of luck, it is usually possible to respond with positive results.

HIGH ERUCIC ACID RAPESEED (HEAR)

In 1984 Kings specialist work with new arable crops came to the attention of the UK's main user of high erucic rapeseed oil (HERO) derived from *Brassica napus* L. A small UK production of HEAR was contracted with them in an initial effort to make them less dependent on Polish and East German supply. As it turned out, their strategy for a UK production was absolutely right but the actual reasons behind the strategy turned out to be wrong. Eleven years ago the conceived Western view of Russian policy was still one of suspicion and mistrust. The possibility of Russian tanks rolling westward over the plains of Germany towards the North Sea was a notion with which a UK user of HERO could not afford to feel entirely at ease with. As we know now, it was democracy which in fact flowed eastwards. As a result both Poland and the former East Germany quickly changed from HEAR into "double low" (low erucic acid, low glucosinolate) types. Thus our buyer's longterm defensive strategy, albeit based on political uncertainty, created a key position for HEAR production in the UK. From a 16 ha trial in 1985, the current production area now extends from Scotland down the East coast of England and into the south west. For the 1995 harvest it is estimated that 36,423 ha of HEAR were grown in the EU of which 13,234 ha were produced in the UK.

The principal end use of the HERO is erucimide, which is used as a slip agent on polythene film, for which there is an increasing world demand. The UK has an intensive oilseed rape production area, with average yields of just over 3 t/ha and average oil contents of 42% to 45% at 9% moisture content. HEAR is therefore an example of a new variation on an already accepted agricultural species. Due to the lower yield of HEAR (by about 250 kg/ha) compared to the "double low" types, Kings offer a premium to compensate growers for the lower yield. The premium also has to compensate for all the "hassle" factors such as field isolation and separate storage.

HEAR deliveries to the buyer are monitored so that product specification is realised in the resultant oil, i.e minimum 45% erucic acid. We have access to both winter-sown and spring-sown varieties. Winter varieties are currently "double high" (high erucic acid, high glucosinolate) types and spring are currently "high/low" (high erucic acid, low glucosinolate) types. It was found that the variation in the fatty acid profile of winter sown crops proved to be minimal. Individual spring sown crops have, however, shown much more variation but when bulked with other individual lots, the minimum specification has been achieved. The first winter "high/low" has just appeared on the EU market and we expect to progress down this road and eventually to HEAR hybrids. This improvement in yield must go hand in hand with the correct end product specifications and therefore is not necessarily a guaranteed smooth progression.

To counteract some of the negative comments about introducing new types of rape into UK farming rotations, Kings have also instigated a number of research projects, including joint work with Newcastle University to test the isolation requirements of oilseed rape. The results from the University and Kings showed that heavy, sticky rape pollen travels less than a few meters. Most cross fertilization occurs due to physical contact between plants and very limited cross pollination occurs by bees (Bilsborrow 1994). Through this work the Ministry of Agriculture (MAFF) was persuaded to reduce the isolation requirement to 50 meters.

There is also the question of volunteers. A field survey was instigated on actual numbers of volunteers of rape from 1 to 5 years after a HEAR crop had been grown. Due to problems with distinguishing between HEAR and "double low" cultivars in the field this study was made in the intervening cereal crops before herbicides were applied. It was found that the maximum volunteers occurred in the second year out of HEAR production and that they were under 1 plant/m^2 (Ramans 1995). From this work, advice could be passed to growers about the correct strategies to minimise volunteer contamination by good husbandry and rotational practices.

These investigations were necessary to facilitate a continuing confidence in HEAR in the UK. Sensitivity to the major political changes occurring within the EU also had to be shown. The introduction by the EU of the set-aside scheme provided a new challenge for the placement of the crop. Reductions in set-aside area over the last two seasons add to the challenge. GATT/Blair House considerations also effect and influence attitudes to non-food oilseed production.

Kings current area of winter and spring HEAR is contracted throughout the UK via Agents and more locally by the Head Office. The company and its buyer work together on the basis of mutual trust and confidence. The buyer brings the chance for Kings to participate in an expanding and secure end market. Kings provide the fine-tuning of many of the sectors of rapeseed agronomy and crop production. In addition, the correct buyer/seller contractual paperwork must be provided, under the obligations as participants within the EU Area Payment Scheme. Following the success of HEAR, Kings are now as well placed as any to participate in the expansion of other rapeseed types with their widely differing fatty-acid profiles.

SUMMARY

Having detailed some of the background agronomy and development of just four new crops, here is a summary of a few of the requirements for anyone trying to evaluate and commercialize new crops. By necessity, they are subjective, but are nevertheless a guide, based on experience.

- Commitment, time, and access to land for trials and access to entrepreneurial growers with a will to try something new.
- Full backing from the company you work for with the shortest possible line of communication within the company.
- An ability to think laterally, logically, and sometimes totally illogically. After all, logic is the art of being wrong and still going forward with confidence.
- A willingness to travel and to try to speak a few words of someone else's language.
- A sympathetic understanding of academics who must, by necessity, publish, which is often contrary to the traditional policy of commercial confidentiality.
- A willingness to delve into scientific topics and make some effort to understand them.
- When putting up a case for a new arable crop, don't just win the hearts and minds of the R & D section of a potential end-user. Meet the people who actually run the business, because without pressure and enthusiasm from the top, the commodity dealers will usually do their best to steer a wide path around you. After all, why should they change from the usual routine of picking up the telephone to purchase their recognized products from known dealers within a known market.

- An understanding of the rules, regulations, and laws, nationally and also if possible internationally.
- Harmonious working with Government officials.
- Be prepared to expect failures, which are numerous.
- Anticipate success by protecting oneself with Plant Breeders Rights, Patents, or whatever means you have to hand to hold onto that success for as long as possible.
- And finally if ones ideas appear to fail, take heart in the words of the founder of the Honda company in the early part of this century: "Many people dream of success. In fact, success can only be achieved by repeated failure. It is normally a fact that success will represent just 1% of a persons work. This therefore leaves 99% which people will always call failure."

If indeed a 1% success rate is anywhere near correct, one should be happy to accept this. The 99% balance is just part of the price of being fortunate enough to be living in these exciting and challenging times. These exiting times, however, can be best served by the greatest possible co-operation between academics and commerce.

REFERENCES

Bilsborrow, P.E., E.J Evans, and B.F. Bland. 1994. Pollen transfer between high and low erucic acid oilseed rape crops. Aspects Appl. Biol. 35:163–166.

Galwey, N.W. 1987. Breeding borage (*Borago officianalis*) as a seed crop for pharmaceutical uses. Report on Breeding programme for Kings, Dept. of Genetics, Cambridge University, England.

Galwey, N.W. 1993. The potential of quinoa as a multi-purpose crop of agricultural diversification: a review. Ind. Crops Prod. 1:101–106.

Ramans, M. 1995. Winter HEAR volunteer study in East Anglia 1993/94. 9th Int. Rapeseed Congress. vol. 4 p. 1369–1371.

Emerging Indigenous Crops of Australia

John A. Considine

The Australian bush has been regarded as a collection of weeds and something that interferes with the exploitation of the soil and mineral resources. Many urban people regard the bush as "foreign" and as a wasteland. A few see it as a unique living museum, to be preserved at all costs. For entrepreneurs and some government authorities, it is a resource for harvesting (mining) bioresources, as it was for aborigines.

Australia's genetic resources have been neglected. As in the case of our minerals, there is a need for a concerted program of exploration and research to identify and develop the resource. Conservation is also vital to protect our genetic heritage and maximise our contribution to the diversity of world agriculture.

THE NEED FOR ADDED DIVERSITY

World agriculture and horticulture is based on a tiny proportion of world plant genetic resources. For example Purseglove (1968) lists about 80 species of dicotyledonous food plants while Simmonds (1976) lists 94 plant species, both monocots and dicots (Table 1). Bailey (1951) lists 5,347 species of plants cultivated for food, decoration and amenity use. These figures compare with an estimated world diversity of about 250,000 species (Takhtajan 1969). Only 32 of the 94 species listed by Simmonds are significant crops and of these, less than 20 form the staple foods for the developed and much of the developing world and again only seven form the starch base. It is truly astonishing that in a world as diverse in peoples and cultures and with a population exceeding five billion that such a small selection of the total available resource is adequate.

In Western countries the "health food" industry is now a force driving the search for variety in foods. The vegetarian sector has been especially important in establishing a viable market for diverse plant foods in a trend that reflects the diversity of foods used by hunter-gatherer societies such as the Australian aborigine (Table 2, 3).

The need for diversity is readily apparent from a study of the countries of origin of the staple foods. Most are Mediterranean or tropical in origin (Tables 1, 2, 4). Thus the majority of our domesticated world crops are plants that are poorly adapted to the arid zones of Australia and other developed and developing nations with large tracts of arid, irrigable land (Cherry 1985).

The present emphasis on marginal lands for agriculture is being exacerbated by the loss of fertile lands to urban, non farming use. In the United States of America 3 million ha of prime land have been lost to agriculture (Nabhan and Felgar 1985). To address these problems substantial international effort is now being devoted to developing new cropping systems for arid regions of the world (Gonzalez 1978; Hall et al. 1979; Manassah and Briskey 1981).

Other points concern the water use efficiency of temperate and tropical plants grown in arid environments, the so called "transplanted" crops (Evans 1980). Certainly there can frequently be a substantial gain in quality due to minimal infestation with pests and disease in arid environments but there is a cost. This cost is the mining of water, the disruption of hydrogeologic cycles and salinization of soils and aquifers; a problem with which Australia is already severely affected (Peck et al. 1983). It has been demonstrated that "transplanted" crops grown under these circumstances use 20% to 40% more water to produce yields equivalent to those produced in more humid climates (Pimental et al. 1982). To complete the argument we have the viewpoint that "transplanted" crops have reached their physiological limit of yield (Austin et al. 1980; Evans 1980). There is the prospect also of global warming and a vastly altered climate that may need to be dealt with in the foreseeable future (WMO 1986).

Arguments in favour of a concerted effort to develop new crops are: (1) a depauperate range of foods in a diverse and populous world; (2) changing markets and demand for new products to meet special interest groups; (3) loss of prime farming land and drift to marginal lands unsuited to existing crops; (4) deple-

Table 1. Major food plants by continent of origin (Simmonds 1976).

Food type	Europe	Asia	Africa	Americas	Australasia	Total
Fruit	2	15	4	14	0	35
Vegetables						
Leafy	8	2	0	0	0	10
Fruit	1	2	0	0	0	3
Seeds & pods	4	4	0	3	1	12
Stems	0	0	0	0	1	1
Roots	6	1	0	2	0	9
Bulbs	0	2	0	0	0	2
Rhizomes	0	1	0	0	0	1
Tubers	0	0	0	1	0	1
Total	19	12	0	6	2	39
Seed meals	0	8	2	4	0	14
Non-seed meals[z]	0	2	0	4	0	6
Total	21	37	6	28	2	94

[z]Gruels or pastes of starchy, paranchymatous stem or root piths.

Table 2. Edible plant products documented as being used by Australian aboriginals (Issacs 1987).

Tree fruits

Acmena smithii, Adansonia gregorii, Ampelocissus acetosa, Amyema spp., *Antidesma bunis, A. dallachynum, A. ghaesembilla, Astroloma humifusum, Billardiera scandens, Buchanania arborescens, B. muelleri, B. obovata, Canthium latifolium, Capparis lasiantha, C. mitchellii, C. spinosa var nummularia, C. umbonata, Carisa lanceolata, Cassytha melantha, Cissus hypoglauca, Cucumis melo* spp *agrestis, Davidsonia pruriens, Dianella* spp., *Diospyros australis, Diploglottis campbellii, Eugenia grandis, Eugenia reinwardtiana, Exocarpus cupressiformis, E. latifolius, Ficus* spp., *F. superba* var. *henneana, Flueggea microcarpa, F. virosa* var. *melanthesioides, Ganophyllum falcatum, Grewia* spp., *Leptomeria aphylla, L. acida, Lysiana exocarpi, L. murrayi, Malaisia scandens, Morinda citrifolia, Musa acuminata, Pandanus spiralis, Parinari nonda, Persoonia falcata, Planchonia careya, Planchonella pohlmaniana, P. arnhemica, Pleiogynium timorense, Podocarpus lawrencei, Rubus parviflorus, Santalum acuminatum, S. lanceolatum, Solanum centrale, S. chippendalia, S. cleistogamum, S. ellipticum, S. esuriale, S. melanospermum, S. orbiculatum, S. phlomoides, S. laciniatum, S. liniarifolium, S. petrophilum, Syzygium forte, S. suborbiculare, Terminalia carpentariae, T. ferdinandiana, Uvaria* spp., *Vitex glabrata.*

Vegetables

Seeds & pods

Acacia coriacea, A. maritima, A. sophorae, A. stenophylla, A. tunida, Araucaria bidwilli, Avicennia marina, Beilschmiedia bancroftii, Canarium australianum, Elaeocarpus bancroftii, Endiandra palmerstonii, Etada phaseoloides, Eucalyptus gamophylla, E. leptopoda, Hicksbeachia pinnatifolia, Hovea spp., *Linum marginale, Lotus australis, L. corniculatus, Macadamia integrifolia, M. tetraphylla, Mallotus nesophyllus, Nelumbo nucifera, Nymphaea* spp., *Pandanus spiralis, Santalum acuminatum, Stercularia quadrifida, Terminalia grandiflora.*

Leaves & buds

Allocasuarina verticillata, Alpina spp., *Calamus caryotoides, Calandrinia balonensis, Disphyma clavellatum, Erythrina vespertilio, Hibiscus heterophyllus, Lepidium muellerii, Lepidium* spp., *Livingstona australis, L. eastonii, Lomandra longifolia, Portulaca oleracea, Typha spp.*

Flowers

Eremophila latrobei, Grevillea eriostachya.

Fruit

Cymbidium caniculatum, Hornstedtia scottiana, Leichardtia australis, Leichardtia spp., *Marsdenia viridiflora, Tamarindus indica.*

Stems & pith

Cymbidium canaliculatum, Hydriastele wendlandiana, Livistona humilis, Nymphaea spp., *Ptychosperma macarthurii*

Bulbs

Cyperus bulbosus, Hypoxis marginata, Microstema tuberosum, Nymphaea spp.

Rhizomes

Blechnum indicum, Nelumbo nucifera, Typha spp.

Tubers

Anguillaria dioica, Aponogeton elongatus, Arthropodium milleflorum, Bulbine bulbosa, Cartonema parviflorum, Dioscorea bulbifera, D. hastifolia, D. transversa, Eleocharis dulcis, E. sphacelata, Gastrodia sesamoides, Geodorum neoncaledonicum, Hypoxis hydrometrica, Operculina brownii, Orchidaceae, Portulaca pilosa, Pterostylis barbata, Triglochin procera

Roots

Alpina spp., *Ampelocissus acetosa, Arthropodium strictus, Boerhavia diffusa, Brachychiton populneum, Buchanania* spp., *Curcuma australasica, Dianella* spp., *Eriosema chinense, Erythrina vespertilio, Flemingia involucrata, Geranium solanderi, Ipomoea costata, Ipomoea* spp., *Lotus australis, L. corniculatus, Microseris scapigera, Murdannia graminea, Thysanotus tuberosus, Vigna lanceolata,*

Table 2. continued.

Meals (paste or gruel)
Seed
Acacia aneura, A. coriacea, A. cowleana, A. estrophiolata, A. kempeana, A. ligulata, A. murrayana, A. notabilis, A. pyrifolia, A. tetragonophylla, A. victoriae, Chenopodium rhadinostachum, Dactylotenium radulans, Eragrostis dielsii, E. leptocarpa, E. eriopoda, Oryza spp., *Panicum australiense, P. decompositum, P. effusum*
Non-seed
Alocasia macrorrhiza, Convolvulus erubescens, Curculigo ensifolia, Cyathea spp., *Dicksonia antarctica, Eucalyptus microtheca, Marsilea drummondii*

Table 3. Comparison of the spectrum of plants used for food by Australian aboriginals (excluding nectar and plant gums; Issacs 1987) and world-wide domesticated food crops (Simmonds 1976).

Food type	Number of plant species: Australian bush foods	World-wide domesticated foods
Vegetables		
Bulbs	4	2
Flowers	2	0
Fruit	6	3
Leaves & leafy buds	14	10
Rhizomes	4	1
Roots	20	9
Seeds & seed pods	28	12
Tubers	18	1
Total	96	38
Meals		
Seed	20	14
Non-seed	7	6
Total	27	20
Fleshy fruits	74	35
Totals	197	94

tion and degradation of water resources; (5) poor water use and nutrient use efficiency of present day crops; and (6) diminishing returns to breeding with existing crops.

The debate is developing now in order to define appropriate strategies to meet items 3 to 6 listed above; marginal lands, water resources and plant improvement. Alternative plant breeding strategies include introgression from wild relatives (Borlaug 1983) and the application of biotechnology to the insertion of foreign genetic material from whatever source seems appropriate (Qualset 1982; Barton and Brill 1983; Austin, 1988). However attractive these strategies are, another option deserving of equal attention is the domestication of minor crops, their wild relatives, and traditional wild foods (National Academy of Sciences 1975; Nabhan and Felgar 1985).

THE AUSTRALIAN RESOURCE

Australian plant resources have a unique heritage. Of particular significance is their evolutionary link to the early flora of Gondwana Land and the climatic pressures extant over the 100 to 150 million years of their evolution (Barlow 1981; Nix 1982). This, according to Nix, has led to development of plants adapted to climatic change with an emphasis in recent geological time on enhanced seasonality and aridity. Adaptability is a vital factor that argues strongly in favour of the domestication of selected species of Australian plants for commerce. Indeed their value has already been demonstrated by the many species in cultivation both here and on other continents.

Table 4. Some major food plants and their centres of diversity (Simmonds 1976).

Family	Crop	Center of diversity
Apiaceae	Carrot	Europe/Southwest Asia
	Celery	Eurasia
Arecaceae	Coconut	Asia/South America/Polynesia
Asteraceae	Lettuce	Mediterranean
Brassicaceae	Cabbage	Mediterranean
Chenopodiaceae	Beet	Europe
Convolvulaceae	Sweet potato	America
Cucurbitaceae	Cucumber	India
	Pumpkin/Squash	North America
Fabaceae	Beans	America
	Pea	Mediterranean/Asia
Lauraceae	Avocado	Central America
Liliaceae	Asparagus	Europe/Asia
	Onions/Garlic	Central Asia
Musaceae	Banana	Indonesia/Africa
	Banana	Indonesia/Africa
Oleaceae	Olive	Mediterranean
Poaceae	Barley	Middle East/Asia Minor
	Maize	America
	Rice	Asia (tropics)
	Rye	Middle East
	Sugar cane	New Guinea
	Wheat	Middle East
Rosaceae	Blackberry	Europe/Eastern U.S.
	Pome fruit	Asia Minor/China
	Stone fruit	Central Asia/China
	Strawberry	America
Rutaceae	Citrus	Southeast Asia
Solanaceae	Pepper	Central & South America
	Potato	South America
	Tomato	South America
Vitaceae	Grape	Central Asia/Europe/America

HISTORY OF TRANSFER OF GENETIC RESOURCES

The early history of the export and cultivation of Australian plants revolves around ornamental plants and forestry, rather than food plants. It is a good though narrow guide to those plants that combine desirable economic and agronomic features. That is, their products are desirable and the plants propagate easily and adapt to a wide range of geographic and edaphic conditions. This history is expressed in three areas: the identity of the collectors and the destination of the plants, the range and type of product, and modern history of collecting and harvesting from the bush.

Some of the key figures in early plant exploration and distribution of genetic resources were Sir Joseph Banks, Ludwig Leichhardt, and Baron Sir Ferdinand von Mueller (Table 5). Baron von Mueller's investigations and writings on the economic flora (Mueller 1879–84) were largely responsible for the establishment of eucalypt oil producing forests and factories in France, Portugal, Brazil, South Africa, and USSR (Elliot and Jones 1983).

The very intensity of the early harvest indicates the level of financial reward available to the avid collector and nurseryman. Elliot and Jones (1983) recorded 286 species exported to England both to botanic gardens and to private nurserymen (Fig. 1). Much of the botanic garden material was extended to private nurseries for subsequent sale to the public. These figures clearly understate the level of collecting because Leichhardt (1847) alone collected some 200 species. Nevertheless the figure indicates those species which met the combined criteria of desirability and suitable agronomic characteristics.

In recent times Elliot and Jones noted that one Midland nursery advertised some 300 species of Australian plants for sale while a second nursery had 125 species for sale. One Irish gardener, presently cultivates more than 50 species of eucalypt in a cool temperate environment indicating the adaptability of the flora (D. Robinson, pers. commun.), diversity rivaling that available in any retail or wholesale nursery or garden in Australia.

Table 5. Early collectors of Australian plants for horticultural purposes (Elliot and Jones 1983).

Collector/Explorer	Year	Capacity	Sponsoring country
Sir Josephs Banks and Daniel Solander	1770	Official	England
David Burton	1791	Private	England
Labillardiere	1792	Official	France
George Cayley	1800	Official	England
Leschenault de la Tour	1801	Official	France
Peter Good	1803	Official	England
Allan Cunningham	1817	Official	England
William Baxter	1823	Private	England
James Anderson	1826	Private	England
William Morrison	1828	Private	England
James Drummond	1829	Official	England
James Drummond	1837	Private	England
Mrs Molloy	1830	Private	England
Ludwig Leichhardt	1846	Private	England
James Kidd	1845	Official	England
Frederick Strange	1845	Private	England
Anon.	ca. 1850	Private	United States
Eugene Fitzalan	1859	Private	England
U.S. Consulate General	1860	Official	United States

The history of export to America is even more dramatic (Fig. 2) with nearly 800 Australian species now in cultivation (Elliot and Jones 1983). Many of these came via England as exchange material from Botanic Gardens although Baron von Mueller was very active in responding to a request for aid in providing advice for plants to rehabilitate the gold fields and provide on-site fuel and building materials. This he did, apparently in return for the U.S. Senate undertaking to publish his writing on the subject and to provide him with 50 copies (Elliot and Jones 1983).

The trade in "bush-picked" wildflowers and plant products is further evidence of the value of Australian genetic resources (Burgman and Hopper 1982). Data collected by the Department of Conservation and Land Management shows that about 21 million stems and about 4 tonnes of seed was harvested in 1980–81. At an

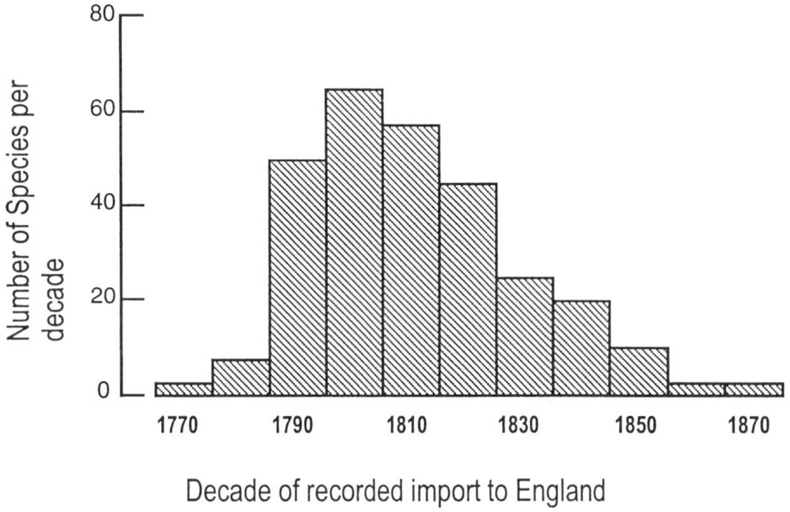

Fig. 1. Number of species of plant recorded in England from 1770 to 1880 (Elliot and Jones 1983). This data does not include the major exports for timber and essential oils which occurrred in the second half of the 19th century.

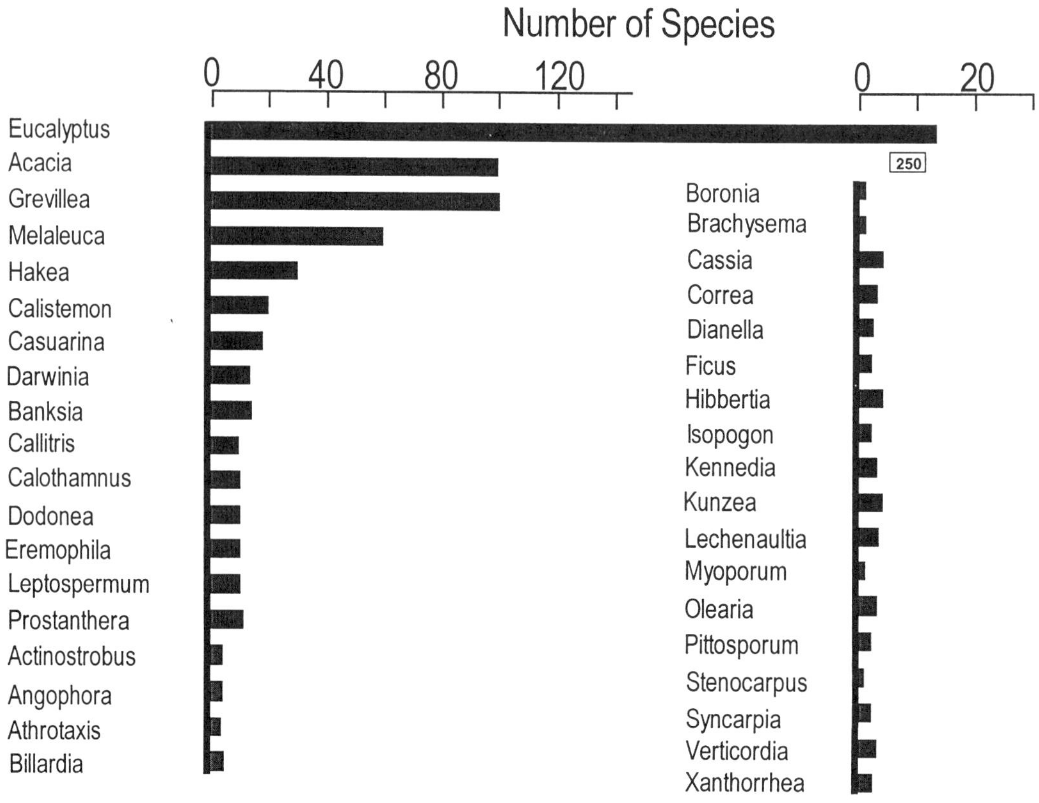

Fig. 2. Number of Australian plant species in cultivation in the U.S. (Elliott and Jones 1983).

average of 20 seeds per gram (guesstimate), this represents about 80 million seeds per annum at a cost of less than 1 cent each, a vastly greater scale of export than that undertaken by Botanic Gardens in their exchange programs. The range of species harvested is another important indicator of resource value and this has been as high as 1,119 (Rye et al. 1980).

Thus the number of species which are exploited presently is in the order of one to two thousand, about 8% of the total resource of 25,000 species. Probably no other continent has such a high proportion of its floral genetic resources being exploited commercially. It is important to note that no food plants are specifically included in the list of presently exploited plants.

AUSTRALIAN PLANT RESOURCES

Foods

Australian natural plant resources are distinguished from those of other lands not by their inate worth, but by the lack of effort devoted to domesticating and formalizing the harvesting of that resource as a renewable resource. Despite some evidence of rudimentary gardening (Kimber 1976; Latz and Griffin 1976; Hynes and Chase 1982), Australia, unlike other continents, does not have a record of active prehistoric agricultural development. Most food crops have a history of selection and improvement over a period of at least 5 to 10,000 years (Heiser 1979, 1981; Simmonds 1976; Roach 1988). The Australian aborigines were hunter-gatherers in a land that supplied all their needs. That aboriginal cultures have histories of 40,000 years or more (Issacs 1980) attests to the value of the native harvestable resources. This is reinforced by analyses of the nutritional value of their plant foods (Peterson 1978, 1979; Brand and Cherikoff 1985).

The diversity of foods used regularly by Australian aborigines is apparently far greater than that of Western societies (Table 2). Many fall into minor use categories (Peterson 1978). An estimated 207 species have been inventoried as aboriginal food plants (Table 3). This compares to 94 world food crops, many of which are also of minor use (Simmonds 1976).

The list of aboriginal food plants in Table 2 still understates the variety of foods in use. For example, Pate and Dixon (1982) list 36 species of Western Australian geophytes as being used for food (out of a total of 204 species) and Crawford (1982), in his investigation of the Kalumburu area of Western Australia, lists 39 species of food plant not included in Issac's (1987) account. The complete list is undoubtedly very extensive (Latz and Griffin 1978; Henshall et al. 1982).

The variety of edible foods contrasts markedly with the opinion of early explorers such as Burke and Wills (Anon. 1961) and early settlers, who considered the bush as a desert land lacking food resources. Leichhardt (1847) and Grey (1841) showed much more interest in "bush foods" but the early attitudes persist unto the present day. Maiden (1889), supported Hooker's quote that *"the products of many plants although 'eatable' are not 'fit to eat' and would never be employed as food except in the direst necessity,"* a view supported also by Pate and Dixon (1982).

This view is now changing and available evidence suggests strong selection in many existing crops for palatability and wholesomeness against an unpalatable and toxic background (Heiser 1981). We should take up the challenge of evaluating those judged most worthy by the aboriginal people themselves in collaboration with agronomists. Participation of aboriginal peoples is important from pragmatic, practical, and social viewpoints.

Industrial Crops

There are many other potential products for the other sectors of the plant market including timbers (Boland and Turnbull 1981), essential oils (Tjandra 1986), pharmaceuticals (Lassak and McCarthy 1983), and forage plants. Forage plants have received the greatest attention through efforts to improve productivity in the northern, tropical zones of Australia (Mott and Reid, 1985) and studies on the value of saltbush as a replacement for alfalfa seem to offer an exciting prospect (Petrikas 1987).

Ornamentals

With such an apparent diversity of material in cultivation overseas and being harvested from the bush locally we might easily conclude that there is no place for further expansion. That this is not the case is demonstrated by the size of the world market (ca. US$ 24.3 billion in 1987) and by the fact that no Australian plant has achieved a significant place in world floricultural trade. Market share is presently about 0.06%.

FUTURE PROSPECTS

It may be most appropriate that we direct our attention to the crops that form the bulk of world trade, the seed crops. Here the challenge is to develop harvestable, non-shattering forms. For while the shattering of seeds may assist ants to gather seed for bushland aborigines, it is the very character selected against by the early agronomists who sought to farm wild wheats (Heiser 1981).

Figure 3 outlines the interaction between research programs that should be carried out to successfully domesticate any unimproved wild species. The industry is presently founded on the "Eureka" syndrome which represents the search for immediately useful plants from the "wild." This is appropriate at the outset but as Fig. 3 shows, we are likely to ignore a substantial proportion of the useful genetic resource by such a superficial approach. Views on appropriate species and technologies have been presented by authors such as Latz and Griffin (1978), MacConochie (1985), and Maggs (1978).

In floriculture, the job is easier and hence the rapid and broadly based industry development (in terms of number of species). This is because the criteria for floricultural uses at the "novelty" level are not rigorous. If we are to produce a product of real world significance then it must be competitive in all agronomic, floricultural, and postharvest traits. Of these, development of technologies to control flowering is perhaps the most urgent item for the existing cultivars. This has been the key to the success of the rose, the carnation, and the chrysanthemum. For the desirable but as yet uncultivated species, developing selections that possess good agronomic characteristics may be the high priority issue. For any, development of a clear market strategy is vital.

The present industry is based on novelty and diversity. A concerted effort will be needed to develop a flagship product, one that will broaden the market base for floriculture as a whole and establish itself as a market leader along with roses, carnations and chrysanthemums. This is unlikely to be a banksia or a waratah but it may be a bush rose.

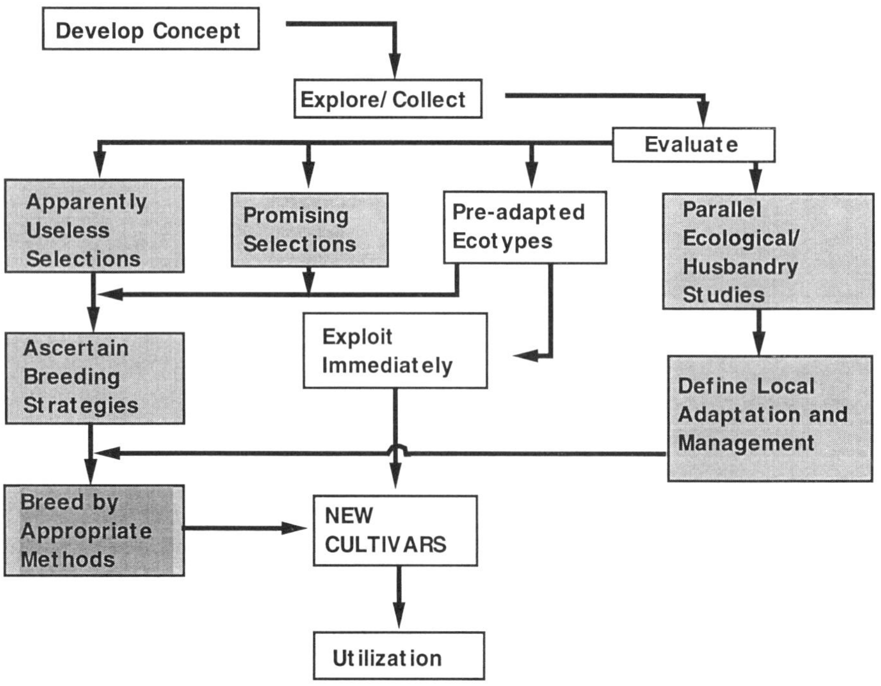

Fig. 3. Diagram illustrating the steps in adaptating a new species of plant. The shaded boxes indicate the steps that are often overlooked by collectors (adapted from Simmonds 1979, with permission).

REFERENCES

Anon. 1861. The Burke and Wills exploring expedition: An account of the crossing of the continent of Australia, from Cooper's Creek to Carpentaria. Reprinted from The Argus. Melbourne, Australia.

Austin, R.B. 1988. New opportunities in breeding. HortScience 23:41–45.

Austin, R.B., R.D. Bingham, R.D. Blackwell, L.T. Evans, M.A. Ford, C.L. Morgan and M. Taylor. 1980. Genetic improvements in winter wheat yields since 1900 and associated physiological changes. J. Agr. Sci. (Cambridge) 94:675–689.

Bailey, L.H. 1951. Manual of cultivated plants. Macmillan, New York.

Barton, K.A. and W.J. Brill. 1983. Prospects for plant genetic engineering. Science 219:671–676.

Barlow, B.A. 1981. The Australian flora: Its origin and evolution, p. 25–75. In: A.S. George (ed). Flora of Australia vol. 1. Australian Government Publishing Service, Canberra, Australia.

Boland, D.J. and J.W. Turnbull. 1981. Selection of Australian trees other than eucalypts for trials as fuelwood species in developing countries. Australian Forestry 44:235–246.

Borlaug, N.E. 1983. Contributions of conventional plant breeding to food production. Science 219:689–694.

Brand, J.C. and V. Cherikoff. 1985. The nutritional composition of Australian aboriginal foods: Plants of the desert regions. In: G.E. Wickers, J.R. Goodwin, and D.V. Fields (eds.), Plants for arid lands. Allen and Unwin, London.

Burgman, M.A. and S.D. Hopper. 1982. The Western Australian wildflower industry. Report No. 53. Perth, Western Australian Department of Conservation and Land Management.

Cherry, M. 1985. The needs of the people. In: G.E. Wickens, J.R. Goodwin, and D.V. Fields (eds.), Plants for arid lands. Allen and Unwin, London.

Crawford, I.M. 1982. Traditional aboriginal plant resoures in the Kalumburu Area. Records of the Western Australian Museum No. 15.

Elliot, W.R. and D.L. Jones. 1983. Encylopaedia of Australian plants suitable for cultivation. vol. 1. Lothian, Melbourne, Australia.

Evans L.T. 1980. The natural history of crop yields. Am. Sci. 68:388–397.

Gonzalez, N. (ed). 1978. Social and technological management on dry lands. Restview Press, Boulder, CO.

Grey, G. 1841. Journals of two expeditions of discovery in North-West and Western Australia, during the years 1837, 38, and 39, under the authority of Her Majesty's Government describing many newly discovered, important, and fertile districts, with observations on the moral and physical conditions of the aboriginal inhabitants, &c. &c. London.

Hall, A.E., G.H. Cannell, and H.W. Lawton. 1979. Agriculture in semi-arid environments. Springer Verlag, New York.

Heiser, C.B. 1979. Origins of some cultivated New World plants. Annu. Rev. Ecol. Syst. 10:309–326.

Heiser, C.B. Jr. 1981. Seed to civilization: The story of food. 2nd ed. W.H. Freeman & Co., San Franscisco.

Henshall, T.S., D. Jampijinpa, J.N. Spencer, and F.J. Kelly. 1982. Ngurrju maninja kurlangu yapa nyurnukurlangu: Bush medicine. Warlpiro Literature Production Centre, Yuendumu, Australia.

Hynes, R.A. and A.K.Chase. 1982. Plants, sites and domiculture: Aboriginal influence upon plant communities in Cape York Peninsula. Archeol. Oceania 17:38–51.

Isaacs, J. 1980. Australian dreaming: 40,000 years of aboriginal history. Lansdowne, Sydney, Australia.

Isaacs, J. 1987. Bush food: Aboriginal food and herbal medicine. Weldon, Sydney, Australia.

Kimber, R.G. 1976. Beginnings of farming? Some man-plant-animal relationships in central Australia. Mankind 10:142–150.

Lassak, E V. and T. McCarthy. 1983. Australian medicinal plants. Methuen Australia Pty Ltd, Sydney, Australia.

Latz, P.K. and G.F. Griffin. 1978. Changes in aboriginal land management in relation to fire and to foods in central Australia. In: B.S. Hetzel and H.J. Frith (eds), The nutrition of aborigines in relation to the ecosystem of central Australia. Commonwealth Scientific and Industrial Research Organisation, Canberra, Australia.

Leichhardt, L. 1847. Journal of an overland expedition in Australia from Morton Bay to Port Essington, a distance of 3000 miles, during the years 1844–1845. Boone, London.
MacConochie, J.R. 1985. Plants of the Australian arid zone—an undeveloped potential. In: G.E. Wickers, J.R. Goodwin and D.V. Fields (eds.), Plants for arid lands. Allen and Unwin, London.
Maggs, D.H. 1978. The potential for horticulture in central Australia. In: B.S. Hetzel and H.J. Frith (eds.), The nutrition of aborigines in relation to the ecosystem of central Australia. Commonwealth Scientific and Industrial Research Organisation, Canberra, Australia.
Maiden, J.H. 1889. The useful plants of Australia. Facsimile ed., 1975. Compendium, Melbourne, Australia.
Manassah, J. and E. Briskey (eds.). 1981. Advances in food producing systems for arid and semi-arid Lands. Vols A & B, Academic Press, London, New York.
Mott, J.J. and R. Reid. 1985. Forage and browse: The northern Australian experience. In: G.E. Wickers, J.R. Goodwin and D.V. Fields (eds.), Plants for arid lands. Allen & Unwin, London.
Mueller, Baron Sir F. von. (1879–1884). Eucalyptographia vols 1–10. Victorian Government Printers, Melbourne, Australia.
Nabhan, G.P. and R.S. Felgar. 1985. Wild desert relatives of crops: their direct uses as food. In: G.E. Wickers, J.R. Goodwin, and D.V. Fields (eds.), Plants for arid lands. Allen & Unwin, London.
National Academy of Sciences. 1975. Underexploited tropical plants with promising economic value. Washington, DC.
Nix, H. 1982. Environmental determinants of biogeography and evolution in Terra Australis. p. 47–66. In: W.R. Barker and P.J.M. Greenslade (eds.), Evolution of the flora and fauna of arid Australia. Peacock, Adelaide, Australia.
Pate, J. and K. Dixon. 1982. Tuberous, cormous and bulbous plants: Biology of an adaptive strategy in western Australia. Univ. of Western Australia Press, Perth, Australia.
Peck, A.J., J.F. Thomas and D.R. Williamson. 1983. Salinity issues: effects of man on salinity in Australia. Australian Government Publishing Service, Canberra, Australia.
Peterson, N. 1978. The traditional pattern of subsistence to 1975. In: The nutrition of aborigines in relation to the ecosystem of central Australia. B.S. Hetzel and H.J. Frith (eds.), Commonwealth Scientific and Industrial Research Organisation, Melbourne, Australia.
Peterson, N. 1979. Aboriginal uses of Australian Solanaceae. p. 171–189. In: J.G. Hawkes, R.N. Lester, and A.D. Skelding (eds.), The biology and taxonomy of the Solanaceae. Academic Press, New York.
Petrikas, D. 1987. Saltbush secret to drought survival. The Land, May 28 (3951).
Pimental, D., S. East, W.L. Choa, E. Stuart, O. Dintzis, G. Einbender, W. Schlappi, D. Andrew, and K. Broderick. 1982. Water resources in food and energy production. Science 32:861–866.
Purseglove, J.H. 1968. Tropical crops: Dicotyledons. Longman, Harlow.
Qualset, C.O. 1982. Integration of conventional and molecular genetics. Calif. Agr. 36:29–35.
Roach, F.A. 1988. History and evolution of fruit crops. HortScience 23:51–55.
Rye, B.L., S.D. Hopper, and L.E. Watson. 1980. Commercially exploited vascular plants native in western Australia: Census, atlas, and preliminary assessment of conservation status. Western Australian Department of Conservation and Land Management, Perth, Australia.
Simmonds, N.W. 1976. Evolution of crop plants. Longman, London.
Simmonds, N.W. 1979. Principles of crop improvement. Longman, London.
Takhtajan, A. 1969. Flowering plants: Origin and dispersal. Oliver and Boyd, Edinburgh.
Tjandra, J. 1986. Eucalyptus oil: Sources, analysis, and chemical properties. PhD Thesis, Murdoch Univ., Perth, Australia
W.M.O. 1986. Land use and agrosystem management under severe climatic conditions. World Meteorological Organization, Geneva, Switzerland.

New Crops as a Possible Solution for the Troubled Israeli Export Market

Y. Mizrahi and A. Nerd

In stimulating articles, Noel Vietmeyer (1986, 1990) pointed out that relatively few plant species, most of which were domesticated thousands of years ago, serve as food for humans and animals, as medicinal plants, and as industrial crops. As a result of atmospheric CO_2 increase and global warming, dryer conditions are expected in the future and many existing crop species will unable to survive. Many undeveloped and neglected species could be the new crops of the future, which will tolerate these changing climatic conditions. Research and Development on the development of tolerant crops should be initiated world wide to meet these challenges. New crops should have the potential to thrive in marginal, infertile, dry lands where common crops fail to provide the diversification required to enable sustainable agricultural systems in the future and offer viable commercial opportunities.

Israel is small country (~5 million people and 22,000 km^2) that is self sufficient in agricultural production. The local market is tiny and is subject to dramatic fluctuations in supply and demand. When extra few tonnes of fresh fruits and/or vegetables are dumped onto the local markets, prices plummet and farmers cannot make a living. In contrast the export market, mainly Europe, with hundreds of millions of consumers is unlimited from an Israeli point of view. Thus, Israel is basically an export oriented producer. At its peak, the export volume of fresh agriculture produce (mainly fruits, vegetables, and flowers) from Israel was valued around US$1 billion/year. In the 1960s the fruit export industry (mainly citrus, with the famous 'Jaffa' orange being the main product) constituted one of the main sources of foreign currency (out of a total export market of US$211 million agricultural produce comprised about US$64 million, i.e., 30.3% of the total!). Today, Israel is an industrial exporting country with total annual export value of US$14 billion while exports of agricultural fresh produce amount to US$547 million, a mere 3.9% of the total (Statistical Abstracts 1994). It is obvious that a small country such as Israel has to compete in world markets in terms of quality rather than quantity. As a result, millions of R&D dollars have been invested in the past two decades to enable Israel to compete in the world markets with the best possible products giving the highest possible return to the growers. Despite the huge input of R&D in the agricultural sector, the production of the citrus is diminishing (Fig. 1) due to lack of profitability (Fig. 2) and farmers are complaining. Citrus sales have diminished by 30.6% from 1990 to 1993 and profitability in terms of revenue's buying power has declined to 57.3% when inflation is taken into account (Fig. 2; Statistical Abstracts 1994; BDO 1995). The Israeli tomato export industry has stagnated at 8,000 t annually for the past few years, despite the enormous R&D efforts that have produced the

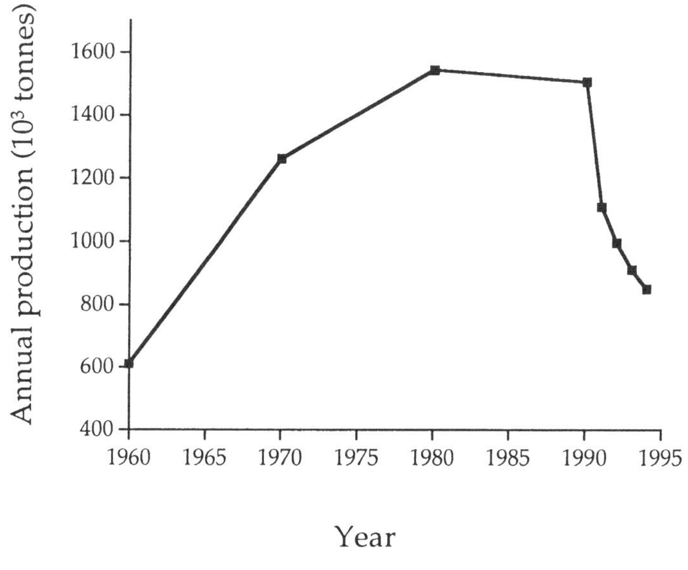

Fig. 1. Changes in annual citrus production in Israel 1960–1994. (Statistical Abstracts of Israel 1994).

world's finest quality (Statistical Abstracts 1994). Israel has additional cost limitations on its competitiveness. Water is a major limiting factor and its actual price is very high (US$0.22 per m^3) and the alternative price is even higher. Labor is very expensive, since farmers are in the upper middle class strata. More than 90% of the agricultural community in Israel is either a kibbutz (community farm) or a moshav (family farm) with minimum hired labor, where most of the work is performed by the farm owners and their families. Gross income below US$100 and even 150 US$/day is considered very low. Israel's competitors in the world markets pay to agricultural workers 10% and less of this sum! Finally, energy is expensive, since almost all the country's energy is imported.

These problems have led many experts in Israel to raise the possibility of giving up agricultural production, even for the local markets, and purchasing all the country's agricultural needs abroad, as do Singapore and Hong Kong. We subscribe to a totally different approach. We believe that supplying the new crops niche in the world markets will serve as a remedy for the troubled Israeli agricultural export industry.

PRODUCT PROFIT CYCLE

In a free market every product is following the profit cycle, as described in Fig. 3. When a totally new crop is first marketed, no profit is expected, since the market has yet to accustom itself to the product. At this stage, low prices are set to attract consumers. In many cases, the market does not respond to the new product, and it simply vanishes without being noticed by most consumers. But if the market likes the new product

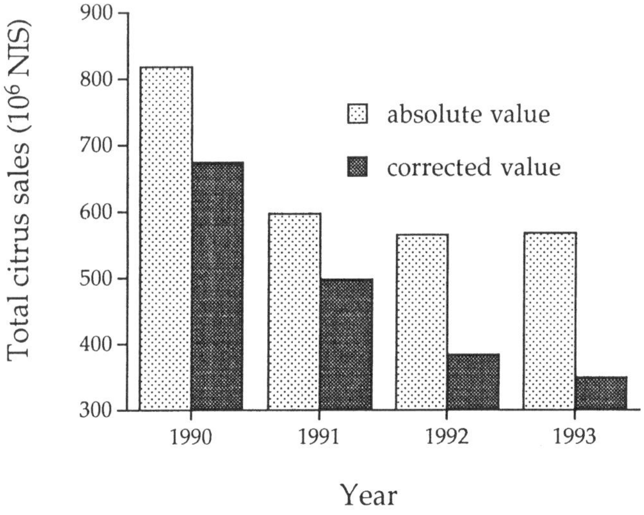

Fig. 2. Changes in sales volume and in buying power of citrus exporters in Israel 1990–1993. Data are taken from the Statistical Abstracts of Israel (1994). Production during this period decreased and prices increased but well below the consumer price index. The inflation rates were 17.6%, 18.6%, 9.4%, and 11.2% for 1990, 1991, 1992, and 1993, respectively (BDO Bavly Milner & Co. 1995). By the time farmers received their money, its value had been eroded due to inflation. The corrected value of their income is calculated from the actual value and the inflation rate.

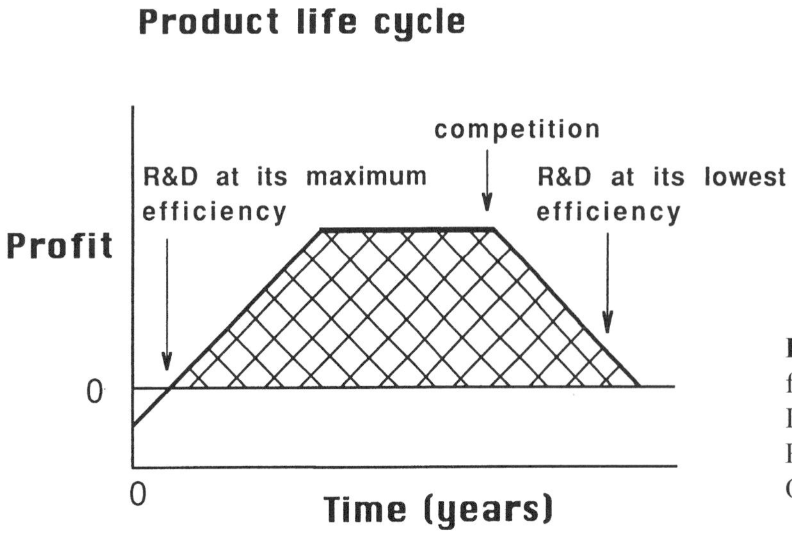

Fig. 3. Hypothetical product life cycle in a free market. For cut flowers exported from Israel to Europe the span is 17 years (D. Rymon, Agricultural Research Organisation, pers. commun.).

(perhaps as a result of good marketing strategy), profits will rise as a result of increasing demand and improved production efficiency (Fletcher et al. 1995). Maximum profitability will be achieved by the first producer as long as he is the only suplier in the market. Competition will then start, first with the most efficient and aware producers and from many others. As a result, profitability will decline, and as with all common crops, it will fall to a maginal level. At this stage, only the big and efficient producers can survive. It is thus evident that Israel cannot compete in the market for common crops. A good example of the scenario described above is the "iceberg lettuce episode." In the late 1970s, Marks and Spencer (from the UK) approached the Negev R&D Organization to test the feasibility of growing iceberg lettuce in the Israeli Negev Desert for their chain. Farmers in the Arava valley (Moshav Ein Yahav and Hatzeva) produced the best iceberg lettuce that Marks and Spencer had ever received. The company, which was ready to pay these farmes higher prices than those given their American counterparts in California, was mystified by the refusal of the Israeli farmers to accept the contract. The buyers from Marks & Spencer simply did not understand that these farmers consisted of very small farms (4 ha!), expensive water, and very high energy, and labor costs. In no way, could they compete with the large American producers.

On the other hand, when we start with totally new product, the high profits obtained for small quantities of exotic fruits and vegetables, creates a natural niche for Israeli farmers. These farmers, being highly educated, can easily adopt new crops and new technologies. Israel is also blessed with research institutions and extension services capable of carrying an efficient R&D programs, which are essential for the introduction of new crops. In addition, Israel has a single efficient marketing arm of agricultural products (AGREXCO), which can easily adapt to such programs.

Unfortunately, such a strategy has not yet been adopted by the Israeli R&D authorities. The main objection to such R&D programs is the fact that the market size and the prices of the new products are unknown. Our feeling is that such a niche does exist, and the only way to confirm its existence is to test it. Even the return from R&D is expected to be higher when dealing with new crops. We may ask the question: how much can new R&D increase the profitability of well-established crop at the edge of their profit cycle? We claim that such an effort can yield much more when new crops are investigated.

Unfortunately, in Israel as in most other countries, most agricultural R&D is conducted with money raised as levies from sales. Thus apple growers for example, would like to see their money invested back into R&D on apples. The bigger the crop and the greater the body of knowledge about its biological and agricultural aspects, more R&D is applied. As a result, money is allocated mainly to the "good old crops," while the new crops of the future are neglected. A change in policy is needed to promote more research in the direction of new crops.

NEW CROPS INITIATIVE

In the light of the dilemma discussed above, we initiated in 1984 an R&D project for the "Introduction and domestication of rare and wild fruit and nut trees as new crops to the Israeli Negev Desert" (Nerd et al 1990; 1993). This project includes about 40 different fruit tree species (Table 1) from all over the world that are considered to be potential new export crops. For the project, four sites were selected in the Negev Desert and one location in the Judean Desert, each site differing from the others in terms of climate, soil, and water (Nerd et al. 1990, 1993). The first stage, which lasted about 10 years, was devoted to assessing the survival, growth, phenology, yields, and quality of seedlings of the investigated species. For all the species we preferred to start with seedlings, which provide wide base of genetic backgrounds, rather than to concentrate on a very narrow base of vegetatively propagated preselected genotypes. Many species did not survive, and others are still at various stages of R&D, far away from any onsiderations of economic potential. Four types of fruit have already been moved to the second stage of this program, which will enable us to provide economic evaluation, such as the cost of various inputs per unit area and the output during the years up to the time that the orchards will reach the full production stage. In this second stage of the project, vegetatively propagated specimens are also being tested in a cultivar trial. The expected small quantities that will be produced at this stage (20–30 t/year/crop) will also enable us to evaluate the marketing and find solutions to post-harvest prob-

Table 1. List of candidate species.

Family Botanical name	Common name	Distribution
Apocynaceae		
Carrisa grandiflora A. DC.	Carrisa	Southern America
Anacardiaceae		
Sclerocarya birrea subsp. *caffra* Sounder	Marula (Morula)	Southern Africa
Spondias cytherea (*Spondias dulcis*) Forst	Ambarella	Polynesia
Bombacaeae		
Bombax glabra	Malabar nut	Central America
Cactaceae		
Acanthocereus tetragonus (L.) Humlk.	Acanthocereus	Mexico
Cereus peruvianus (L.) Miller	Apple cactus (Pitaya)	North South America
Escontria chiotilla (Weber) Britt & Rose	Pitaya (Jiotilla)	Mexico
Hylocereus costaricensis (Weber) Br. & R.	Pitahaya	Central America
Hylocereus paolyrhi (Weber) Br. & R.	Pitahaya	Central America
Hylocereus polyrhizus (Weber) Br. & R.	Pitahaya	Central America
Hylocereus purpusii (Weber) Br. & R.	Pitahaya	Central America
Hylocereus undatus (Weber) Br. & R.	Pitahaya	Central America
Myrtilloactus geometrizans (Mart.) Cons.	Pitaya	Mexico
Nopalea cochenillifera (L.) Salm-Dyck	Nopalito, Nopalea	Mexico
Opuntia ficus-indica (L.) Miller	Prickly pear	Tropical America
Opuntia streptacantha Lem.	Prickly pear	Tropical America
Pachycereus pringlei (Berger) Britt & Rose	Cardon pelon	Sonoran Desert
Selenicereus megalanthus (Schum.) Br. & R.	Pitaya	Columbia
Stenocereus griseus (Haw.) Buxb.	Pitaya	Oaxaca Mexico
Stenocereus gummosus (Engelm.) Gilbs.	Pitaya agria	Sonoran Desert
Stenocereus stellatus (Pfeiff.) Riccob.	Pitaya	Mexico
Stenocereus thurberi (Engelm.) Buxb.	Pitaya dulce	Sonoran Desert
Stenocereus thurberi var. *litoralis* (E.) B.	Pitaya dulce	Sonoran Desert
Caesalpiniaceae		
Cordeauxia edulis Hemsl.	Yehib	Northeast Africa
Ebenaceae		
Diospyros digyna Jacq.	Black sapote	South America
Diospyros discolor Willd.	Mabolo (Velvet apple)	Philline Islands
Diospyros mespiliformis Hocht.	Mmilo namibia	South Africa
Euphorbiaceae		
Ricinodendron rautanenii Schinz	Mongongo	Southern Africa
Guttiferae		
Rheedia madruno Triana & Planch.	Madrono	Central America
Flaucourtiaceae		
Dovialis caffra Warb.	Kei apple	Southern Africa
Leguminosae		
Tamarindus indica L.	Tamarind	Tropical Africa
Loganiaceae		
Strychnos cocculoides Backer	Monkey orange	Southern Africa
Strychnos spinosa Lam.	Monkey orange	Southern Africa
Strychnos pungens Solereder	Monkey orange	Southern Africa

Table 1. Continued.

Family Botanical name	Common name	Distribution
Mimosaceae		
Inga spp.	Ice cream bean	South America
Moraceae		
Artocarpus heterophyllus Lam.	Jackfruit	Asia
Rhamnaceae		
Ziziphus mauritiana Lank.	Ber	Old World Tropics
Rosaceae		
Prunus salicifolia H BK.	Capulin cherry	Mexico
Rubiaceae		
Vangueria infausta Burch.	Mmilo	Southern Africa
Rutaceae		
Casimiroa edulis Llave & Lex.	White sapote	Mexico, Central America
Santalaceae		
Santalum acuminatum (R. Br.) A. DC.	Quandong	Australia
Sapotaceae		
Argania spinosa L.	Argan	Morroco
Chrysophyllum cainito L.	Star apple	Central America
Manilkara zapota van Royen	Sapodilla	India, Africa, Central America
Mimusops angel Engler	Angel	Somalia
Mimusops zeyheri Sond.	Mmupudu	Southern Africa
Pouteria sapota (Jacq.) Merr.	Mammey sapote	Central America

lems. The market figures will enable farmers to take decisions whether or not to enter into the arena of these new crops.

The species that are currently in the second stage of the project include: (1) climbing trellised cacti growing in net houses—*Selenicereus megalanthus* and three species of *Hylocereus*; (2) the outdoor-grown cactus *Cereus peruvianus*, also known as apple cactus; (3) white sapote (*Casimiroa edulis*); and (4) Ber (*Ziziphus mauritiana*) also known as "desert apple," a species introduced from India. Two more species are under consideration for moving to the second R&D stage: (5) marula (*Sclerocarya birrea* subsp. *caffra*), for which 10 clones are currently being propagated and will be ready for planting next year, and (6) argan (*Argania spinosa*), a wild oil tree from Morocco, which will be promoted to the second stage after the selection of the current fruiting year.

Crawling Cacti

These species which are native to Central and north South America, climb on tree trunks in the tropics and may be epiphytic (Gibson and Nobel 1986). Their fruits have various sizes, tastes, shapes, and colors. Some have spines that abscise upon ripening and others have scales of various shapes and colors. The pulp also varies in color from white to various hues of red and purple, while the abundant seeds may be soft and edible (Mizrahi et al. 1996). The reproductive biology of these species is described in a review by Nerd and Mizrahi (1996). Five genotypes are already growing in an area of 2 ha, mainly in net houses since they require shade (Nerd et al. 1990; Raveh et al. 1993; Mizrahi et al. 1996). One clone of *Selenicereus megalanthus*, also known as yellow pitaya, is being cultivated (Weiss et al. 1995). Yellow pitaya is already an established crop that is being exported worldwide from Colombia (Arcadio 1986; Cacioppo 1990; Mizrahi et al. 1996). Other clones include one of *Hylocereus polyrhizus*, one of *H. undatus* and two of

Hylocereus sp. (Barbeau 1990), all our selections. These clones have been planted in two plantations, each of 0.5 ha. Each clone was planted in a different row to allow cross pollination from the neighboring rows (Weiss et al. 1994b). All were planted in the late summer of 1993 as rooted cuttings removed from the same mother plants, and all started to fruit in 1994. In these plantations two net houses were planted, one with 50% shade in the Arava valley, having a hot climate and saline water (EC 4 dS/m), and the other with both 30% and 60% shade sections in the Besor area, which is characterized by good quality water (EC 1 dS/m) and moderate temperatures with only rare frosts (Nerd et al. 1993). The second hectare was planted in the Yad Mordekhay area, with sub-freezing temperatures as low as –4°C; here, plastic houses were planted to accommodate selected and non-selected plants. All started to fruit one year after planting.

Cactus Apple

Of many columnar cacti tested by us as potential new crops, one species—*Cereus peruvianus*—grew the fastest. It started to flower and fruit four years after seeding (Nerd et al. 1993; Weiss et al. 1993). Rooted cuttings of seven clones of this cactus, selected from over 300 seedlings, were planted in the Arava valley and the Besor, with a total area of 2 ha. All cuttings were planted as a mixture of clones, since this species demonstrates self-incompatibility (Weiss et al 1994a). The reproductive biology of this cactus is also described in the review of Nerd and Mizrahi (1996). All clones started to flower and fruit two years after planting. Over 1,000 seedlings have been planted for further selection.

White Sapote

White sapote (*Casimiroa edulis*, Rutaceae) is an evergreen medium-size tree native to the highlands of Mexico and Central America. The fruits are green-yellow, with a thin skin and a creamy white-yellow sweet flesh (Morton 1987). Selected clones are available, mainly in Southern California (Chambers 1984; Morton 1987), and some effort has been made to introduce the species into New Zealand and Australia (Dawes and Martin 1988; George et al. 1988). A small commercial plantation (16 hectares) with selected cultivars is being grown in Carpenteria near Santa Barbara, California and the fruits can be found as an exotic item in the United States and Australia. Early tests in the Israeli Negev Desert demonstrated partial tolerance to salinity (Nerd et al. 1992). In autumn 1992 and spring of 1993, 21 grafted clones were planted in Qetura and Besor; 16 were introduced as bud-wood from Fallbrook, Southern California (from R.R. Chambers orchard), while the remaining five were propagated as grafted bud-wood from our own selections. Nine replications from each clone were planted in three blocks at each location. In 1995 some clones started to flower and set fruits in these two locations.

Desert Apple

Desert apple (*Ziziphus mauritiana*, Rahmnaceae), also known as ber or Indian jujube, is an evergreen, medium-size, thorny tree believed to be of African origin (Alexander, 1979). The fruits can reach plum size, turning yellow from green as ripening starts, and becomes sweet and sour in taste, both the flesh texture and taste being reminiscent of apples. The fruit has a unique aroma, similar to that of carob, which becomes too strong for "Western" tasters when fully ripe, at which stage the color turns brown. The fruit can be consumed dry, similar to its relative the "Chinese date" *Z. jujube*. Ber is grown commercially as a desert crop (hence the name desert apple) in India. Seedlings and introduced cultivars from India developed and yielded very well (over 100 kg/tree annually) in all our introduction orchards, including areas with frequent sub-freezing temperatures and highly saline water (Nerd et al. 1990). Three Indian cultivars were planted at Neot Hakikar, the lowest point on earth –400 m below sea level with 3,960 mm evaporation/year and saline water (EC 4 dS/m) with Na and Cl as the major ions (Nerd et al. 1993). Most of our introduced fruit tree species did not survive under these conditions, but ber has fruited heavily from very early ages. A semi-commercial plantation was planted by a farmer in 1993, and the first yield was sold in 1995 in the local market, mainly to immigrants from India who are familiar with the fruit.

Marula

Marula (*Sclerocarya birrea* subsp. *caffra*, Anacardiaceae) is a large, dioecious, deciduous tree, which grows wild in southern Africa. Female trees bear plum-sized fruits with a thick yellow peel and a translucent, white, highly aromatic sweet-sour fruit, which is eaten fresh, like a small mango, or used to prepare juices, jams, conserves, dry fruit rolls, and alcoholic beverages. The seeds, which are eaten as a delicate nut, are highly appreciated by the locals and hence the name "the kings nut." The nut has high nutritive value and a high oil content (56%) with very good dietetic ratio of saturated to unsaturated fatty acids (Weinert et al. 1990). Trees were established very well at introduction sites in the Negev Desert and produced abundant fruits from early ages, mainly when grown in a hot area with saline water (Qetura) (Nerd and Mizrahi 1993). Trees were badly damaged after a spell of sub-freezing temperatures of –6° and –7°C; all recovered but never set fruits, and thus this species is not recommended for areas with such low temperatures. At Qetura, some pistillate trees are bearing well, over 400 kg/tree annually, and we have moved the species to the second stage of our R&D program to test selected clones on a semi-commercial basis. Ten selected clones are being propagated and will be ready for planting in 1996.

Argan

Argan (*Argania spinosa*, also known as *A. sideroxylon*, Sapotaceae) is a medium, thorny, evergreen tree native to south western Morocco. The tree bears plum-sized fruits, which are eaten by goats which often climb the trees. The fruits have a bitter pericarp around a stone-like structure, containing one to three kernels with a high oil content (over 50%). The oil has high dietetic value, total unsaturated fatty acids/total saturated fatty acids being around 4.5, a ratio similar to that of olive oil (Morton and Voss 1987; Prendergast and Walker 1992). The oil has a unique aroma and is considered as the best culinary oil by Moroccans, who are the only people familiar with the oil. In Israel, where 600,000 immigrants from Morocco reside, imported argan oil is sold for US$43/liter in comparison with $4/liter for olive oil. Attempts to domesticate this wild tree in Israel started about 10 years ago. The species demonstrated adaptability to the hot hostile environment of the Arava valley when irrigated with brackish water; yields of oil per tree at Qetura were double those at Ramat Negev, which has much milder environmental conditions (Nerd et al. 1994). The oil yield of best specimens was around 1 kg/tree annually. Some seedlings died as a result of infection with *Fusarium oxysporum*. Until tolerant rootstocks can be found, we decided to plant grafted trees from the best yielding ones and to plant additional seedlings from various habitats in Morocco. Even though this species is not in as advanced stage of introduction as the marula, we consider it to be a high-priority species because of its rarity and the high demand in Israel for its oil.

CONCLUSIONS

It is interesting to note that in our earliest publication (Nerd et al. 1990), we mentioned six species as promising, an evaluation based on their early development and growth. These species included white sapote and ber, which are still considered promising, and marula, which has been moved to the second stage of the R&D, but they also include yehib (*Cordeauxia edulis*), mongongo (*Ricinodendron rautanenii*), and pitaya agria (*Stenocereus gummosus*), which did not meet our early optimistic expectations. The latter three species proved to be late yielders (pitaya agria); or exhibited sensitivity to the desert conditions of our introduction sites, such as salinity or sub-freezing temperatures; or were not as abundant in fruiting as the promising species described in this presentation. Other promising species emerged, such as *C. peruvianus*, with its adaptability to a spectrum of conditions, including slight salinity and sub-freezing temperatures. The species produces early and heavily, giving good-quality, tasty fruits. The most surprising successes were the various crawling cacti (*Hylocereus* species and *S. megalanthus*), which did not survive the outdoor Negev Desert conditions (Nerd et al. 1990), but once trellised and protected from high radiation by net houses, started yielding precociously and early to give some of the most beautiful fruits on earth.

Of the six promising species from the first R&D stage, three were moved into the second stage (white sapote, ber, and marula), while others, the apple cactus and the crawling cacti, emerged as promising and already planted as vegetative clones (moved to second stage). This evaluation was based on fruiting (both as yields and fruit quality) and early selection of good-performing specimens.

We anticipate that at least some of these newly introduced species will become export items with profit levels that will be sufficiently high to revive the fruit export industry and replace the old "dying-out" export crops. The high profitability of new fruit crops was demonstrated for kiwi fruit by New Zealand in the world market and for avocado by Israel in the European market. There is no reason why such new exotic fruits will not be the commodities of the future. We should not forget that no crop can stay at its peak forever, and low profits always loom in the future. Mr. Dan Rymon (pers. commun.), found that it took 17 years from the first sales of flower crops in the European markets until Israel was chased out by its competitors. With fruit trees, it may be much much longer, as was the case with the kiwi fruit from New Zealand (47 years) and the 'Shamuti' ('Jaffa') orange from Israel (80 years).

We should conclude with an evaluation of the benefits of our program to the world as a whole. Any species that can produce good yields in the Arava valley—the location of both Qetura and Neot Hakikar—can serve as a future crop species that can tolerate extremely high temperatures and salinity. Because of the unique situation in the Negev Desert of short distances between agriculturally different ecozones, we recommend that this area be used as a global laboratory for the introduction and acclimation of new desert crops. In 1994 the International Program for Arid Land Agricultural Crops (IPALAC) was initiated under the auspices of UNESCO. This program is aimed at R&D similar to that described in this paper, to be executed with all kinds of agricultural crops in representative desert areas around the globe.

REFERENCES

Alexander, D.McE. 1979. Rhamnaceae. p. 161–163. In: P.E. Page (compiler), Tropical trees for Australia. Queensland Department of Primary Industries Series Q183018.

Arcadio, L.B. 1986. Cultivo de la pitaya. Federacion de Cafeteros, Bogota, Colombia.

Barbeau, G. 1990. La pitaya rouge, un nouveau fruit exotique. Fruits 45:141–147.

BDO Bavly Millner & Co. 1995. Doing business in Israel 1995. Tel Aviv, Israel.

Cacioppo, O.G. 1990. Pitaya: una de las mejores frutas productivas por Colombia. Informative Agro Economico de Colombia. Feb., p. 15–19.

Chambers, R.R. 1984. White sapote varieties: progress report. California, Rare Fruit Growers Yearb. 16:56–64.

Dawes, S.N. and P.J. Martin. 1988. The casimiroa is close to commercial fruit status. Orchardist of New Zealand 61:72–74.

Fletcher, R.J., G.E.A. Ferguson, G.M. Kregor, and C.H. McConnell. 1995. Choosing new crops. Sixth Conference of the Australian Council on Tree and Nut Crops (ACONTAC - 95), Sept. 11–15, 1995. Lismore, NSW, Australia.

George, A.P., R.J. Nissen, and D.J. Wallace. 1988. The casimiroa. Queensland Agr. J. 114:57–62.

Gibson, A.C. and P.S. Nobel. 1986. The cactus primer. Harvard Univ. Press, Cambridge, MA.

Mizrahi, Y., A. Nerd, and P.S. Nobel. 1996. Cacti as crops. Hort. Rev. 18:(In press).

Morton, J.F. 1987. Fruits of warm climates. J.F. Morton. Miami, Fl. p 191–196.

Morton, J.F. and G.L. Voss. 1987. The argan tree (*Argania sideroxylon*, Sapotaceae), a desert source of edible oil. Econ. Bot. 41:221–233.

Nerd, A. and Y. Mizrahi. 1993. Domestication of marula (*Sclerocarya birrea* subsp. *caffra*) as a new crop for the Negev Desert of Israel. p. 496–499. In: J. Janick and J.E. Simon (eds.), New crops. Wiley, New York.

Nerd, A. and Y. Mizrahi. 1996. Reproductive biology of cactus fruit crops. Hort. Rev. 18:(in press).

Nerd, A., J.A. Aronson, and Y. Mizrahi. 1990. Introduction and domestication of rare fruits and nuts trees for desert areas. p. 355–363. In: J. Janick and J.E. Simon (eds.), Advances in new crops. Timber Press, Portland, OR.

Nerd, A., M. Lapidot, and Y. Mizrahi., 1992. White sapote (*Casimiroa edulis*); Performance under culture and environmental stress conditions in field studies. Scientia Hort. 51:213–222.

Nerd, A., E. Raveh, and Y. Mizrahi. 1993. Adaptation of five columnar cactus species to various conditions in the Negev Desert of Israel. Econ. Bot. 47:304–311.
Nerd, A., E. Eteshola, N. Borowy, and Y. Mizrahi. 1994. Growth and oil production of argan in the Negev Desert of Israel. Ind. Crops Prod. 2:89–95.
Prendergast, H.D.V. and C.C. Walker. 1992. The argan, multipurpose tree of Morocco. Kew Mag. 9:76–85.
Raveh, E., J. Weiss, A. Nerd, and Y. Mizrahi. 1993. Pitayas (genus *Hylocereus*): a new fruit crop for the Negev Desert of Israel. p. 491–495. In: J. Janick and J.E. Simon (eds.), New crops. Wiley, New York.
Statistical Abstracts of Israel. 1994. No 45, Central Bureau of Statistics, State of Israel.
Vietmeyer, N. 1986. Lesser known plants of potential use in agriculture and forestry. Science 232:1379–1381.
Vietmeyer, N.D. 1990. The new crops era. p. xviii–xxii. In: J. Janick, and J.E. Simon (eds.), Advances in new crops. Timber Press, Portland, OR.
Weinert, I.A.G., P.J. van Wyk, and L.C. Holtzhausen. 1990. Marula. p. 88–115. In: S. Nagy, P.E. Show, and W.F. Nardowsky (eds.), Fruits of tropical and subtropical origin. Florida Science Source, Lake Alfred.
Weiss, J., A. Nerd, and Y. Mizrahi. 1993. Development of the cactus apple (*Cereus peruvianus*) as a new crop to the Negev Desert of Israel. p. 486–491. In: J. Janick and J.E. Simon (eds.), New crops. Wiley, New York.
Weiss, J., A. Nerd, and Y. Mizrahi. 1994a. Flowering and pollination requirements in *Cereus peruvianus* cultivated in Israel. Israel J. Plant Sci. 42:149–158.
Weiss, J., A. Nerd, and Y. Mizrahi. 1994b. Flowering and pollination requirements in climbing cacti with fruit crop potential. HortScience 29:1487–1492.
Weiss, J., L. Scheinvar, and Y. Mizrahi. 1995. *Selenicereus megalanthus* (the yellow pitaya). A climbing cactus from Colombia. Cactus Succulent J. 67:280–283.

New Industrial Crops: Northwestern Argentina Regional Project

Ricardo Ayerza (h) and Wayne Coates

The goal of the Northwestern Argentina Regional Project is to identify and bring into commercial production new industrial crops which can help diversify agricultural production and increase profits for farmers in northwestern Argentina. Both private and government organizations in the United States and Argentina have been working cooperatively on this project since its inception.

The project started in 1991 with an agreement between Partners of the Americas Inc., a non-profit organization headquartered in Washington DC, and Agropecuaria El Valle S.A., an agricultural enterprise with offices in Buenos Aires and Catamarca, Argentina. This technical cooperation was made possible through the Farmer to Farmer program, financed by the Congress of the United States as a part of the 1990–95 Farm Bill (Public Law 480), and the Agency for International Development (USAID).

Subsequently, The University of Arizona, the University of California, the National University of Catamarca, the Rural Group Pulares, and the government of the province of Salta joined the project. In 1995 four grower organizations from the province of Jujuy also joined. These are CREA Los Lapachos, Union Caneros Independientes de Jujuy and Salta, Cooperativa de Tabacaleros de Jujuy, and La Camara del Tabaco de Jujuy.

NEW CROP COMMERCIALIZATION

The development of any new crop goes through four distinct phases: (1) market assessment; (2) identification of the most appropriate species; (3) domestication of the species; (4) commercial production.

Growers, processors and marketing agencies must be involved and share the risks in all four phases of this development process for it to be the most effective. Under these conditions all of the involved parties are unified through contracts, a common information base, and personal involvement. Because this concept of widespread involvement and responsibility is an integral part of the Northwestern Argentina Regional project, the growers, processors and marketing agencies have all assumed a degree of the cost, and hence risk, with it.

Ecosystems

Northwestern Argentina consists of five provinces: Salta, Jujuy, Tucuman, Santiago del Estero, and Catamarca. The region can be divided into three general ecosystems which run parallel to each other in a north and south direction. The eastern most ecosystem is called Chaco, and is a wide plateau approximately 300 to 700 m in elevation. The climate is warm and dry, and naturally supports an open woodland ecosystem. Directly to the west is a long sloping region which rises from 200 to 4000 m in elevation, over a distance of 50 to 100 km. This center ecosystem is called the Yungas, is frost free in most years, and receives more than 1200 mm of rainfall annually. This area is suitable for tropical fruit production. The western ecosystem is a high plateau called Puna, and is at an elevation of 4000 m or more. It is a cold and dry grassland, having no naturally occurring tree species.

The work reported in this manuscript took place in the Yungas and Chaco ecosystems, within the provinces of Catamarca, Jujuy, Salta, and Tucuman. The project began in 1991 and is planned for continuation at least through 1996, with the objective being to evaluate potential industrial crops and then bring them into production. Ancillary objectives are to bring about a better public understanding of how new crops can improve the economic situation in northwestern Argentina, and to promote closer cooperation between northwestern Argentina and the southwestern United States.

Education and Information Exchange

Establishing research plots, providing on-farm technical advice, and conducting workshops, conferences, and seminars for growers, university faculty, and technical personnel have all been part of the program. During 1995 alone, 12 different topics were presented in a series of workshops, conferences, and seminars held at various locations in northwestern Argentina.

Seventeen scientists, growers and company managers have been brought from the U.S. to Argentina to provide technical assistance. Areas of expertise have included agronomy, genetics, plant physiology, mechanization, processing, economics, and marketing.

Nineteen professors from the National University of Catamarca and three from the University of La Rioja travelled to the U.S. for specialized training programs lasting from 2 to 3 months. The time was spent at either the University of Arizona, or the University of California. In 1994, two members of a women's agricultural cooperative from Catamarca participated in the exchange program and spent time in California, Arizona and Northern Mexico, where they were able to exchange ideas on new crop production with growers and researchers.

IDENTIFICATION OF SUITABLE NEW CROPS

The research plots, as well as the demonstration plantings, have been located on private farms, thus directly involving the farming public in the project. The tests have been conducted in the four provinces of Catamarca, Jujuy, Salta, and Tucuman, (Table 1) providing wide public exposure to the project.

Through the trials, six species have been identified which appear to hold significant potential for the region. These are: chia, lesquerella, vernonia, and chan which are sources of industrial oil; guayule, a source of rubber, resins, and latex; and kenaf, a raw material for paper and newsprint. Of these crops, only lesquerella and chia have been grown commercially. For the others, trials are continuing to evaluate cultivars and/or sites in order to identify those crops and locations which hold the greatest commercial potential for northwestern Argentina. Some of the results are presented here.

Chan (*Hyptis suaveolens*, Labiatae)

This plant is native to southern Mexico and Central America, and was used as a food by pre-columbian people in the region. The seed contain 77%–80% linoleic acid, and little or no linolenic acid. Yields in Argentina have reached 1770 kg/ha (Coates and Ayerza 1995). The data in Table 2 summarize the results obtained from four-row plots which were harvested in June 1995. No statistically significant difference between sites, either in terms of total biomass or amount of seed harvested, was detected. This tends to indicate that either site would be equally suited for growing chan. Table 2 also shows that total oil content as well as percent α-linoleic fatty acid in the seed were similar at both locations. This also indicates that either site would prove equally suitable for growing chan.

The total oil yield is very low compared to other commercially grown oil seed crops. Because this is a new crop, it is necessary to conduct further tests and examine the analysis techniques which were used in the laboratory to determine oil quantity and distribution. Environmental factors which might have affected the results also need to be investigated before a sound conclusion can be drawn regarding the quantity and quality of oil contained in the seed.

Kenaf (*Hibiscus cannabinus*, Malvaceae)

Kenaf, a native of tropical Africa, is a fast-growing annual which is grown for its fiber. It can reach heights of 4 to 6 m in one growing season, and has yielded up to 17.5 t/ha of dry matter in Argentina (Ayerza and Cook 1996). Table 3 presents results from a series of kenaf trials that were conducted over a

Table 1. Locations where the new industrial crops have been evaluated or commercially grown.

Location	Province	South latitude	Elevation (m)	Rainfall (mm)
C. de Valle	Catamarca	28° 36'	454	437
Sumalao	Catamarca	28° 28'	546	394
Perico	Jujuy	24° 23'	936	600
Yuto	Jujuy	23° 35'	349	802
Chiocoana	Salta	25° 06'	1270	715
El Carril	Salta	25° 03'	1069	624
Metan	Salta	25° 30'	858	841
Pichanal	Salta	23° 17'	300	618
Pulares	Salta	25° 04'	1240	825
Guemes	Salta	24° 40'	734	507
Alberdi	Tucuman	27° 36'	369	1092

Table 2. Biomass, seed yield, oil content and composition of chan seed as affected by location.

Location	Biomass (kg/ha)	Seed (kg/ha)	Total oil (%)	Palmitic (%)	Palmitoleic (%)	Stearic (%)	Oleic (%)	Linoleic (%)	Linolenic (%)
Pichanal	2188a[z]	638a	11	8.4	0.0	2.5	7.8	81.9	0.4
Yuto	2125a	725a	12	8.9	0.0	1.8	8.6	80.4	0.3

[z]Mean separation in columns by Ryan-Einot-Gabriel-Welsch multiple range test, 5% level.

number of years, at five sites. Sumalao and Pichanal were both irrigated, while the other sites were not. Dry matter yields ranged from 4.3 to 13.3 t/ha, with the yields being quite variable. Yields have been influenced by both location and climate. One cultivar has not proven superior to the others at all locations, indicating that site influences are significant and will need to be taken into account when selecting cultivars for commercial production. These tests clearly indicate the need for additional cultivar and location trials, and consequently these are underway.

Kenaf was first planted in Argentina on a commercial scale in late 1995. Plans called for planting 25 ha using three cultivars, with the idea being to use the plantings not only to determine the production potential of the three cultivars which had demonstrated the best performance in the plots, but also to evaluate alternative harvesting technologies and determine which might be the most appropriate for the region. After harvest, the material was to be sent to a commercial operation to be pulped, and was then to be blended with bagasse pulp and made into paper. Unfortunately early 1996 was one of the driest summers on record in northwestern Argentina, and the extended trial had to be abandoned. As a consequence only two hectares were sown. Plans still call for this field to be used to test alternative harvesting methodologies, although in a more limited scale. Pulping will take place provided that a commercial entity can be located that is willing to work with a limited amount of material.

Vernonia (*Vernonia galamensis*, Asteraceae)

Vernonia, a member of the sunflower family, is native to tropical Africa. Its seed contain an unusual oil which, because of its low viscosity, can be used as a solvent in paint. Over 70 other potential uses for the oil have been identified. Table 4 presents the results from the vernonia trials conducted in 1995 at Pichanal and Yuto. At Pichanal significant differences in seed yield were detected among the cultivars, with more than a ten fold difference being found between the most and least productive. No statistically significant differences in 1000 seed weight were detected, and some cultivars had significantly lower oil content than others. As insufficient seed was available to plant a replicated trial at Yuto, only one block was sown. As a conse-

Table 3. Production of kenaf at five locations in northwestern Argentina.

	Location (no. of years)				
Cultivar	Sumalao[z] (t/ha)	Alberdi (t/ha)	Metan (t/ha)	Yuto[y] (t/ha)	Pichanal[z] (t/ha)
Tainung 2	13.3 (3 yr)	9.4 (2 yr)	7.7 (2 yr)	11.3 (1 yr)	6.0 (2 yr)
Cubano	10.9 (3 yr)	7.1 (2 yr)	7.2 (2 yr)	7.7 (1 yr)	5.9 (2 yr)
Cuba 108	10.0 (3 yr)	11.4 (1 yr)	6.9 (1 yr)	-.-	4.3 (1 yr)
Tainung 1	9.9 (3 yr)	6.8 (2 yr)	7.4 (2 yr)	8.4 (1 yr)	4.9 (2 yr)
15-2	9.5 (3 yr)	7.2 (1 yr)	9.8 (1 yr)	-.-	8.0 (1 yr)
Everglades 41	8.9 (3 yr)	8.28 (2 yr)	7.3 (2 yr)	9.8 (1 yr)	5.6 (2 yr)
SF 45-9	8.9 (3 yr)	6.4 (2 yr)	7.5 (2 yr)	10.6 (1 yr)	8.4 (2 yr)
Everglades 71	8.3 (3 yr)	7.5 (2 yr)	8.1 (2 yr)	10.4 (1 yr)	7.3 (2 yr)
SF-192	-.-	8.2 (2 yr)	8.2 (2 yr)	7.5 (1 yr)	7.7 (2 yr)
N-7	-.-	7.0 (2 yr)	5.6 (2 yr)	8.4 (1 yr)	5.2 (2 yr)
19-117-2	10.9 (2 yr)	-.-	-.-	-.-	-.-
Guatemala 51	11.6 (2 yr)	-.-	-.-	-.-	-.-
78-18 RS 10	7.8 (2 yr)	-.-	-.-	-.-	-.-

[z]Irrigated
[y]Two sites

quence only mean seed yields are presented in Table 4, and no statistical analysis of the data was conducted. The mean values, however, indicate that this site may be the best of the two for growing vernonia.

In general the vernonia yields are low, on a per hectare basis, as compared to other oilseed crops. The values were determined for rows planted 1 m on center. This spacing conforms to USDA test plots, and allows comparison with their data. In a commercial situation, row spacing would be on the order of 150–200 mm, significantly increasing yields. Pest and disease problems at both sites also decreased yield.

Guayule (*Parthenium argentatum* Gray, Asteraceae)

Guayule, a plant native to the Chihuahuan desert of Mexico, produces rubber that is almost identical to that from the rubber tree. It also produces several co-products, including resins, low-molecular weight rubber, and bagasse, each of which have potential uses. The rubber (or latex) from guayule can be used where synthetic materials are not satisfactory. Table 5 presents the analysis which was conducted on two year old guayule harvested in Catamarca in 1995. Significant differences in biomass yields were detected among varieties, however no differences in rubber yield were found on a dry matter basis. When the biomass yields are combined with the rubber content, rubber production ranged from 17 to 60 g per plant. Thus although percentage of rubber in a plant can influence processing costs, total rubber production on an area basis would probably be the governing factor used to select the best varieties.

Results of this study show that additional trials are required to determine not only the highest producing varieties, but also to assess yields as influenced by time of harvest. Plans call for these tests to be undertaken starting in June 1996.

Jojoba [*Simmondsia chinensis* (Link) Schneid., Simmondsiaceae]

Jojoba, which is native to the southwestern United States and northern Mexico, produces seeds which contain a liquid wax that has uses in the cosmetic industry, and it also serves as high quality lubricant. Jojoba has been grown commercially for a number of years in La Rioja and Catamarca provinces. The development of equipment to mechanize its production has been an integral part of the Northwestern Argentina Regional Project. Equipment has already been developed for transplanting cuttings, pruning the plants, and preparing

Table 4. Yield of vernonia grown at Pichanal and Yuto.

	Pichanal			Yuto
Cultivar	Seed yield (kg/ha)	Oil (%)	1000 seed wt (g)	Seed yield (kg/ha)
29E-OR2-14	229a[z]	43.8a	3.23a	457
35A-2-9	188ab	42.0a	3.19a	186
72A-1-2	158abc	42.6a	3.17a	257
48A-10	154abc	41.9a	3.17a	143
66C-1-9	139abcd	41.3ab	3.62a	193
15D-10-12	128abcd	41.2ab	3.68a	150
A0399	100bcd	42.7a	3.22a	186
14D-2-5	82bcd	43.2a	3.18a	157
AO382	81bcd	41.0ab	3.32a	NA[y]
35A-2-10	37d	36.6bc	3.12a	150

[z]Mean separation in columns by Ryan-Einot-Gabriel-Welsch multiple range test, 5% level.
[y]Not available; did not mature sufficiently prior to frost to permit harvesting.

Table 5. Guayule biomass and rubber content found in two year old plants grown in Catamarca.

Line	Biomass (g/plant)	Rubber content (%)
AZR1	1103a[z]	6.1a
O16-1	401b	8.8a
AZ-R2	368bc	8.0a
N9-5	337bcd	5.8a
N6-5	301cd	10.6a
P3-1	273d	14.1a
O16-3	268d	12.8a
AZ-R3	171e	11.0a

[z]Mean separation in columns by Ryan-Einot-Gabriel-Welsch multiple range test, 5% level.

the soil surface for harvest. A new type of jojoba harvester is under development, with a prototype having been built and tested.

Lesquerella (*Lesquerella fendleri*, Brassicaceae)

Lesquerella is an annual plant native to the southwestern United States. The seed of lesquerella contain an unusual fatty acid which is similar to that of castor bean. This fatty acid can be converted to many products which have both industrial and cosmetic potential. The meal remaining after the oil is removed contains a good amino acid balance, and can be used in livestock feed.

Chia (*Salvia hispanica*, Labiatae)

Like chan, chia is native to southern Mexico and has been used by pre-columbian inhabitants of Central America as a food source and a medicine. Oil from chia was also used in paints. The seed contain the highest known natural source of linolenic acid (60%), with linolenic acid having many uses in industry and cosmetics. The meal remaining after the oil is expressed is high in protein and fiber, and can be used for human and animal food. Seed yields in plots of up to 1602 kg/ha, and oil contents as high as 38.6% have been reported in Argentine trials (Ayerza, 1995).

Chia and lesquerella have been commercially produced in the provinces of Catamarca, Salta, and Tucuman. The seed produced has been exported to the United States, with contracts for its production being signed between the growers and commercial enterprises from Argentina and the U.S. Table 6 lists the sites and number of growers that have been involved in the commercial production of these crops.

Table 6. Number of growers and area commercially sown to chia and lesquerella in Northwestern Argentina.

Year	Location	Province	Growers (no.)	Area planted (ha)	Average size (ha)	Yearly total (ha)
Chia						
1992	C. del Valle	Catamarca	1	14	14	14
1993	C. del Valle	Catamarca	1	70	70	
	Alberdi	Tucuman	1	4	4	74
1994	C. del Valle	Catamarca	2	3	1.5	
	Sumalao	Catamarca	1	20	20	
	Alberdi	Tucuman	1	5	5	
	V. de Lerma[z]	Salta	1	3	3	31
1995	C. del Valle	Catamarca	2	3	1.5	
	Alberdi	Tucuman	1	5	5	
	V. de Lerma	Salta	5	40	8	
	Metan	Salta	1	10	10	58
1996	V. de Lerma	Salta	6	120	20	
	Guemes	Salta	1	5	5	
	Perico	Jujuy	2	20	10	145
Lesquerella						
1992	C. del Valle	Catamarca	1	12	12	12
1993	C. del Valle	Catamarca	1	2	2	2
1994	Sumalao	Catamarca	1	1	1	1
1995	Pulares	Salta	4	20	5	20

[z]Valle de Lerma: Includes Pulares, Chiocoana and El Carril

From Table 6 it can be seen that the number of farms and growers involved in the production of chia and lesquerella on a commercial basis has varied. The reasons for this have been the degree of success realized by each grower, and the level of satisfaction arising from being involved with the introduction of a new crop.

The commercial yields of both chia and lesquerella have varied from year to year, and from location to location. This is due to a number of factors including cultural practices, climate, weed infestations and harvesting techniques used. Cultural practices purposely have not been standardized across farms, as it is the intent of the project to allow farmers to use those cultural practices with which they are familiar to grow the new crops, while providing them with overall guidance as to how best to produce each crop.

Analyses of the chia seed which was commercially harvested in 1995 showed viability to range from 78% to 87%. Purity ranged from 84% to 97.5%. This is considered excellent, especially given the small size of the seed and the difficulties which were encountered with the harvesting and cleaning processes. These data demonstrate that chia can be commercially produced in northwestern Argentina.

Table 7 provides a comparison of the returns that can be realized from chia production, as compared to two traditional crops found in the region. The higher returns from chia have in large part prompted a significant increase in acreage planted in 1996.

LONG RANGE OUTLOOK

Over-production of traditional crops continues worldwide. It is therefore likely that prices and profit margins for these will remain very low, except when shortages arise because of war or climatic disasters. Furthermore, the consumption of many non-renewable resources continues, and environmental policies are already beginning to favor markets for many new industrial crops.

The success of the program is clearly demonstrated by the increasing numbers of hectares of chia being grown in northwestern Argentina. It is anticipated that this trend will continue and that the same thing will occur for lesquerella and kenaf. Such success is possible because the program is dynamic. Many organizations and growers are already a part of this project and they expect to take advantage of the conditions which exist to benefit their own enterprises. Others in the community benefit as a result of improved economics. The dynamic, open nature of the Northwest Argentina Regional Project means that other organizations and growers interested in the benefits of new industrial crops, including diversification of their cropping practices, are welcome to join the program at any time.

Table 7. Economic comparison of chia production to black and white bean production.

Crop	Yield (kg/ha)	Price ($/t)	Gross income ($/ha)	Production cost ($/ha)	Gross margin ($/ha)
Chia	1,000	800	800	220	580
Black beans	1,500	400	600	320	280
White beans	1,100	700	700	380	390

REFERENCES

Ayerza, R. (h). 1995. Oil content and fatty acid composition of chia (*Salvia hispanica*) from five northwestern locations of Argentina. J. Am. Oil Chem. Soc. 72:1079–1081.

Ayerza (h), R. and C.G. Cook. 1996. Potential of kenaf in Argentina: III—variety trial in Catamarca, 1991-1994. Proc. III Int. Conference on New Industrial Crops and Products, and the IX International Conference on Jojoba and Its Uses. Association for the Advancement of Industrial Crops, Catamarca, Argentina. Sept. 25–30, 1994. p 276–279.

Coates, W.E. and R. Ayerza (h). 1995. New crops for the Rio Bermejo Rivers Basin - Phase II. Final Report submitted to COREBE, Buenos Aires, Argentina.

Agricultural Opportunities and Challenges in China

Thomas A. Lumpkin*

China has only 7% of the world's cultivated land to feed 22% of the world's population. During the 45 years since the foundation of modern China, its population has more than doubled to 1.2 billion. While China has the world's third largest and most rapidly growing economy, per capita income is low; various estimates range between $1000 and $2600 for 1990. Food is thought to consume 50% of urban wages. This rate is high, not only because of low income but also because of China's food-centered culture, which places a high value on diversity and balance in eating. Thus, increasing affluence and the cultural importance of food in China are raising enormous opportunities and challenges for global production.

CONSUMPTION TRENDS

As a result of economic reforms starting in 1978 (Anon. 1989, 1994), agricultural production and consumption have changed considerably (Coyle 1995). Average daily food consumption has increased over 1000 calories since the famine years of 1961–62 (Fig. 1). Rising affluence has pushed the use of grain as feed for animals from only 8% in 1978 to 20% in 1990. The demand for feed grains is increasing at twice the rate of population growth. This trend will probably continue, using an increasing share of China's grains. Meat consumption, primarily pork, averages over 35 kg (Fig. 2, Table 1) and is increasing by about 15% per year,

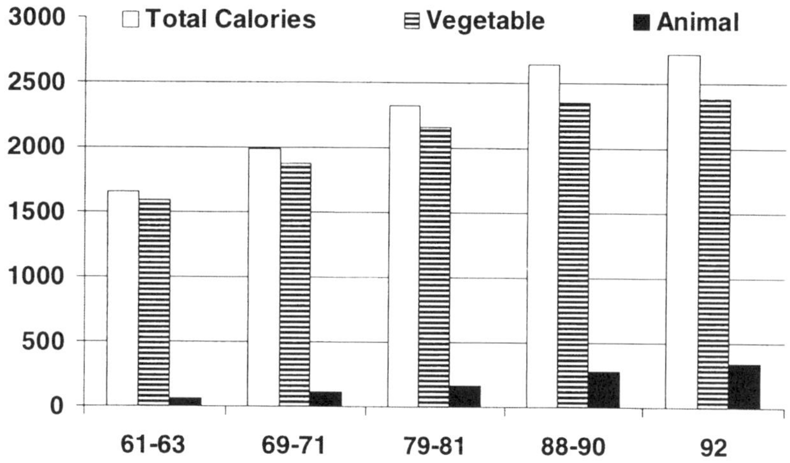

Fig. 1. Average total intake of food calories from animal and vegetable sources in China.

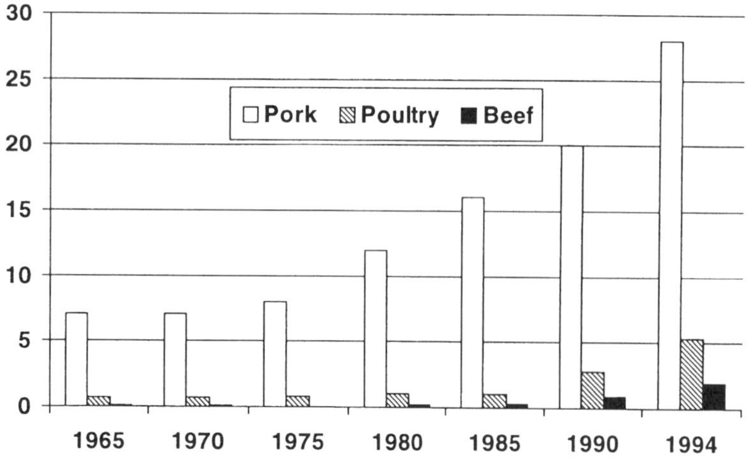

Fig. 2. Chinese consumption of meat from various sources, in kilograms.

*Research supported by WSU IMPACT Program.

about 40% of Taiwanese consumption. Consumption in Taiwan and Hong Kong, is now 60 and 90 kg of meat per capita per year, respectively. China has closer culinary affinity to Taiwan and Hong Kong than Korea or Japan. Similarly, edible oil consumption is about 6 kg per capita, only 22% of Taiwanese consumption (Fig.3), and is expected to continue increasing rapidly, partially through imports (Fig. 4). Milk consumption is low at about 4 kg per capita, but will never approach the 271 kg consumed per capita in the U.S.

Table 1. Per capita food consumption from various sources in four countries.

	Quantity (kg/person)								
Nations	Cereals	Soybean	Beef	Pork	Poultry	Milk	Eggs	Vegetables	Fruit
U.S.	1375	267.8	43.1	30.9	51.4	268	16.8	140	111
China	328	13.5	1.9	30.0	5.4	5	8.3	106	31
Japan	126	0.8	4.8	11.2	5.0	32	9.9	111	37
S. Korea	168	3.5	4.5	18.9	8.4	44	10.2	233	45

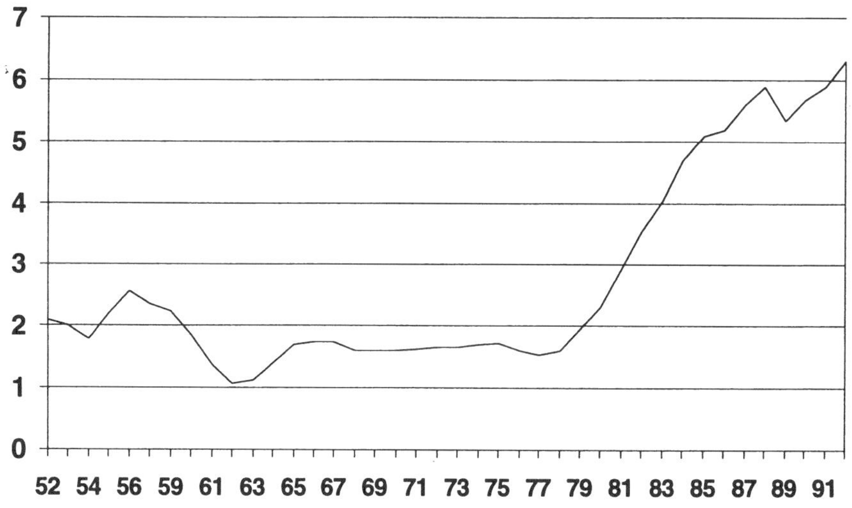

Fig. 3. Change in per capita Chinese consumption of edible vegetable oil.

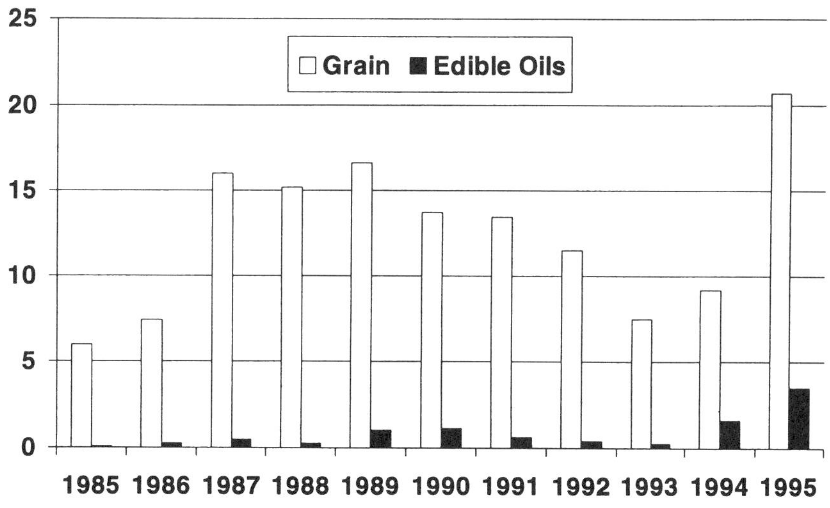

Fig. 4. Chinese imports of agricultural commodities, in millions of tons. Statistical Yearbook of China, 1995; Press release of State Statistical Bureau, February 1996.

FOOD PRODUCTION

China will follow, but to a lesser degree, the agricultural trends of Japan, Korea, Taiwan, and other more developed East Asian neighbors. The most notable of these trends is reduced self-sufficiency in food production. For example, according to the Food Balance Sheet released by the Ministry of Agriculture, Forestry, and Fisheries, Japan's self-sufficiency rate for food on a caloric basis was 46% in 1992 vs. 79% in 1960. This rate compares to 251% in Australia, 113% in the U.S., and 65% in Switzerland. More worrisome in regard to the future is the possibility of China following the trend of Japan's, S. Korea's, and Taiwan's low self-sufficiency in the production of grains; these countries are only 28%, 34% and 24% self-sufficient, respectively. Changing tastes for imported food forced reductions in rice production since consumption dropped from 118 kg per capita in 1962 to 69 kg in 1993 (Matsuoka 1995). If China follows a similar East Asian development path, including a westernization of its tastes, it will become a major importer of food on a scale that will dramatically affect world food trade patterns. The size of China's expected needs could push food prices back toward the record prices of 1973–74 (Fig. 5) and have devastating effects of the world's poor.

Japan's 1992 self-sufficiency rate for fruit was 59% and 90% for vegetables. Increased production of these high value crops is beginning to occur in China, where the area devoted to vegetable production increased from 2.7 million ha in 1970 to 8.7 million by 1994. China has the world's largest production of vegetables and is emerging as a major vegetable exporter. For example in 1991, China was the fifth largest vegetable exporting country ($1.9 billion), after the U.S. ($5.2), Spain ($4.8), Italy ($4.3), and France ($1.9).

China produces and consumes a different mix of the world's major food grains than western nations such as the U.S., the countries of the Former Soviet Union (Fig. 6), and developed Far Eastern nations such as Japan and Korea (Table 1). If the countries of the Former Soviet Union can successfully reconfigure their inefficient agriculture, they may be able to meet a large portion of China's expanding need for wheat, barley, and feed grains.

Lester Brown's 1995 book, *Who Will Feed China*, has raised considerable controversy in China among the highest levels of government officials and agricultural scientists. According to Brown, China will have a population of 1.6 billion by 2030 and will probably need to import 200 to 365 million t of grain. Most estimates for 2030 are higher than the total current world trade in grains at about 225 million t. Chinese scientists debated with Lester Brown about this issue while he was a guest of the World Development Institute of China's State Council's Development and Research Centre (Gao 1995). China already imports more than 20 million t of grain (Fig. 4) and will likely be importing 40 million t by the year 2000 (S. Rozelle unpubl.). The government of China has failed to keep investments in agriculture in line with increasing demand. As a result, grain production has plateaued in recent years (Fig. 7), while cultivated area per capita continues to fall alarmingly (Fig. 8).

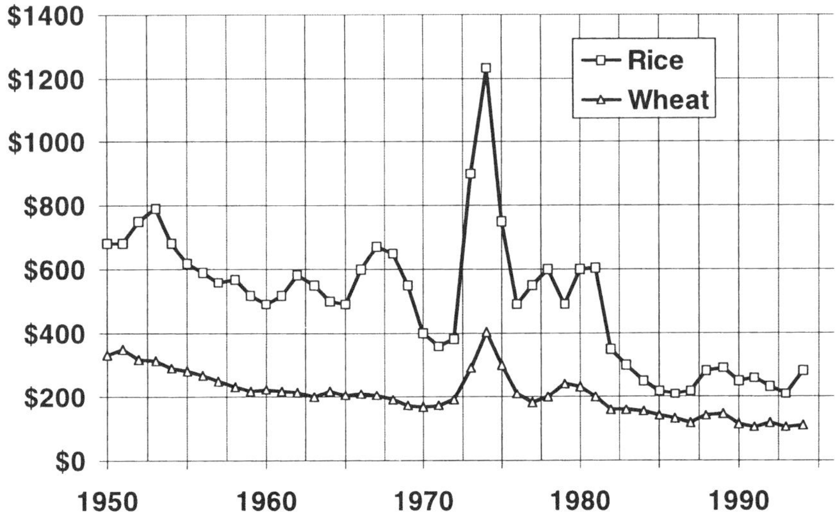

Fig. 5. Changes in the world price of a tonne of rice or wheat, in U.S. dollars.

However there are opportunities for increasing production. For example, Mei Fangquan, Chinese Academy of Agricultural Sciences, found from a satellite survey that China actually has 132 million ha of cropland not the 95.3 million ha officially reported. Thus China has an extra 36.7 million ha in production therefore the reported average yields of 4125 kg/ha is actually about 3000. Politics created an artificial situation where cultivated area was under-reported. This situation continued with central government demands that yields be increased. Production increases were easier to attain across more cultivated land than was actually reported.

In a 3-year analysis of world fertilizer consumption, average fertilizer use was thought to be at about 100% in China and 120%–150% in Japan, Korea, and Taiwan (Saleen Ahmed of the East West Center). However this discrepancy in reporting land leaves room for increased profitable application of fertilizer.

TRADE

The U.S. absorbs about 18.5% of China's exports and supplies about 10% of its imports. The U.S. currently runs a trade deficit of over $30 billion with China, second only to Japan, and the deficit is increasing at about 25% annually (Fig. 9). China exports to the world about twice the value of agricultural products ($15.9 billion, 1993) as it imports ($7.7 billion) and even has an agricultural trade surplus with the U.S. (Tuan 1994). Thus China can expect considerable political pressure from the U.S. to open agricultural markets and to purchase U.S. farm products in order to improve the trade imbalance.

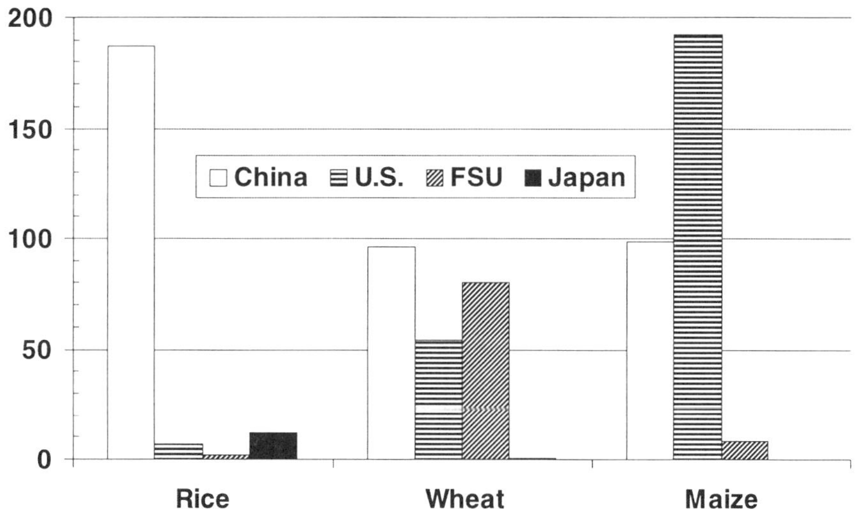

Fig. 6. Production of the three major cereals in China, the U.S., countries of the Former Soviet Union, and Japan, in million of tonnes.

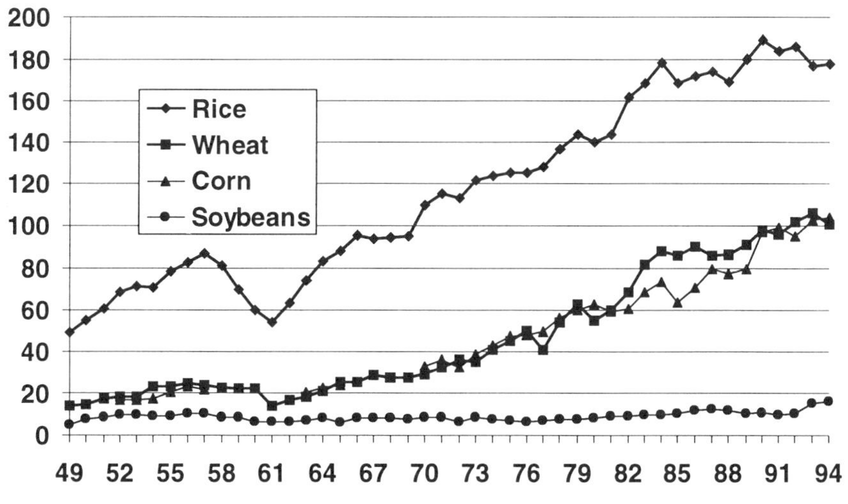

Fig. 7. Chinese production of major cereals and soybeans, in millions of tonnes (Anon. 1987 and FAO yearbooks).

China's leading agricultural exports to the world, in declining value are corn, sugar, pork, cotton yarn, tea, and silk. China exported $451 million worth of agricultural products to the U.S. in 1993 and imported about $376 million. Leading Chinese agricultural exports to the U.S., in declining value, are vegetables (Table 2), feathers and down, cocoa products, tea, and sugar products.

In 1993, China exported about $103 million in vegetables to the U.S. (Table 3) while importing only about $1.5 million worth (Table 4), though U.S. export of high value vegetable products to China is increasing rapidly (Crook 1994). China will be a major competitor with the U.S. for foreign vegetable markets in the coming years. Increased food exports from China to other countries have often come at the expense of U.S. market share—especially in Japan (Matsuoka 1995). About 65% of China's vegetable exports go to developed Asian countries. In 1992, $500 million worth of vegetable products were sold to Japan, $163 million to Hong Kong, $40 million to Singapore, and $26 million to Korea. Among the exported products were mushrooms, bean products, canned bamboo shoots, ferns, fungi, and asparagus.

U.S. garlic producers have taken a beating from China, even domestically. Garlic exports to the U.S. from China increased from 3273 tons in 1992 to 24,461 tons in 1993. The largest U.S. vegetable export item to China is frozen potatoes; 343,000 t in 1993 (Crook 1994).

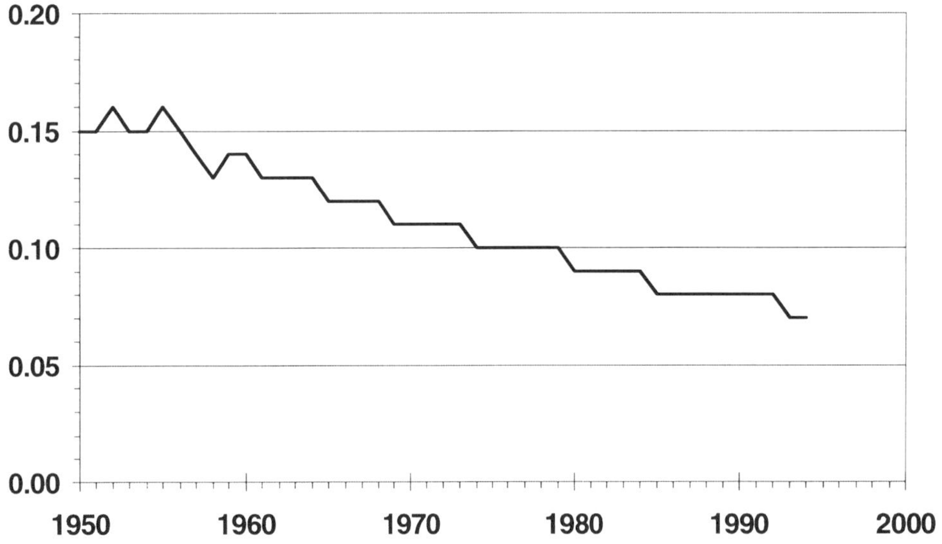

Fig. 8. Hectares per person of cultivated land in China.

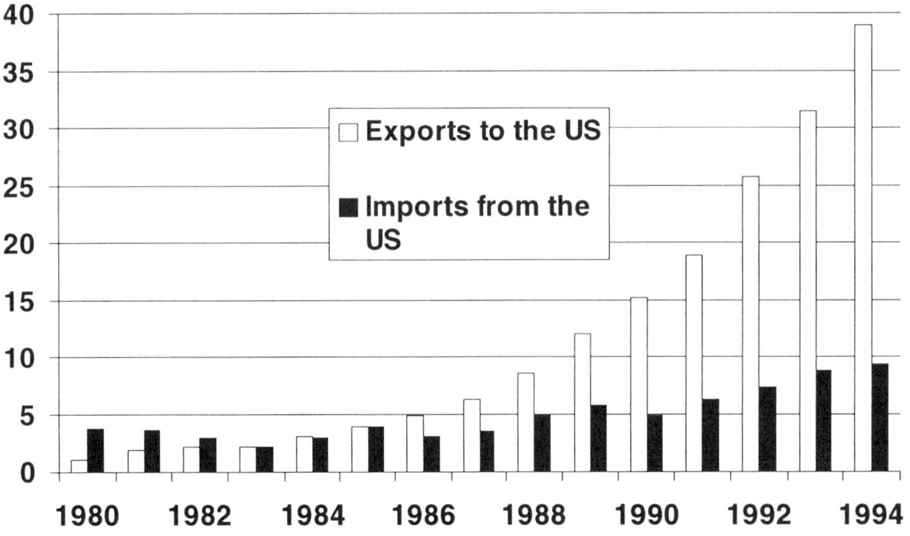

Fig. 9. China's trade surplus with the United States, in billions of dollars.

SUMMARY

China is developing a taste for western-style foods, especially those associated with the fast-food industry, e.g. potatoes, beef, chicken, wheat products, fruit juices, dairy products. As supermarkets develop and spread, China will also become a larger market for high value exotic foods and for high quality traditional Chinese foods. China may export lower quality types of foods while importing higher quality versions of the same foods. China will also expand production and consumption of medicinal plants, especially those enhancing vigor or stamina (*chi*).

In the short run major agricultural opportunities in China will develop as expressed by members of the Illinois Farm Bureau who made an autumn 1994 visit to China. They concluded that China will be a strong market for U.S. corn, soybeans, wheat, and meat. On the other hand, the U.S. will continue to develop as a major market for the export of Chinese vegetable and fruit products as health issues and the impact of China's growing economy and cultural influence affect American tastes. The developing U.S. market for Chinese foods will offer opportunities for U.S. farmers who understand how to produce Chinese crops (Larkcom 1991; Lumpkin et al. 1993a, b; Konovsky et al. 1994; Lumpkin and McClary 1994).

Table 2. China's major vegetable exports in 1992 (Crook 1994).

Item	Quantity (t)	Value (million $)
Asparagus, canned	39,938	60.4
Adzuki beans	51,919	26.2
Bamboo shoots, canned	65,336	68.2
Bamboo shoots, preserved in salt water	23,166	26.9
Beans, frozen	21,978	19.7
Broad beans	367,594	74.2
Cassava, dried	316,425	40.4
Ferns, dried	4,290	15.0
Gan bian beans	46,033	13.2
Garlic	128,200	67.6
Ketchup, canned	37,433	21.7
Kidney beans	335,262	118.4
Lichens, dried	2,842	23.1
Mung beans	53,432	30.6
Mushrooms	9,168	30.8
Mushrooms, canned	120,092	147.9
Mushrooms, hei mu er (black wood ear)	2,538	13.8
Mushrooms, other dried	10,320	45.9
Mushrooms, preserved in salt water	54,194	61.7
Peas, dried	64,295	11.2
Roots, other starch	30,147	21.3
Sweet potato[z], dried	546,929	72.8
Vegetables, other dried	35,775	59.4
Vegetables, other fresh and chilled	178,576	49.7
Vegetables, other preserved	93,034	48.8
Vegetables, other frozen	37,752	35.9
Vegetables, other simply preserved	67,134	46.2
Water chestnuts, canned fresh	33,838	17.2
Weicai gan	3,069	21.3

[z]Dried sweet potatoes and cassava used as animal feed.

Table 3. U.S. vegetable exports to China, 1992–93 (Crook 1994).

	Quantity (t)		Value ($1,000)	
Item	1992	1993	1992	1993
Cabbage, fresh	134	0	32	0
Hops	22	11	139	73
Peppers, fresh	18	0	20	0
Potatoes, frozen	242	343	167	246
Sweet corn, frozen	35	0	32	0
Vegetables, other frozen	18	0	32	0
Vegetables, other prepared and preserved	--	--	1,193	1,104
Total	--	--	1,616	1,514

Table 4. U.S. vegetable imports from China, 1992–93 (Crook 1994).

	Quantity (t)		Value ($1,000)	
Item	1992	1993	1992	1993
Asparagus, prepared	290	818	492	1,489
Bamboo shoots, preserved	2,395	2,429	2,513	2,513
Beans, fresh/frozen	36	232	54	307
Bean cake, miso	351	970	670	1,371
Beans and peas, dried	1,821	1,999	1,205	1,708
Cucumbers, preserved	2	12	2	10
Garlic	3,273	24,461	2,148	11,886
Garlic, dried	2,644	4,478	2,772	4,209
Hops	1,412	0	2,428	0
Mushrooms, canned	11,716	11,944	21,130	20,734
Mushrooms, dried	275	715	3,271	7,146
Mustard	12	6	26	7
Olives, prepared or preserved	8	13	25	46
Onions	37	25	36	15
Onions, preserved	33	620	84	880
Peas, including chick peas	3,724	5,169	5,871	7,434
Peppers	12	0	28	0
Peppers and pimentos, prepared	12	7	15	9
Potatoes, fresh or frozen	18	0	10	0
Water chestnuts	24,152	27,129	15,799	17,949
Total	--	--	78,592	102,726

REFERENCES

Anonymous. 1989. Changes and development in China 1949–1989, Beijing Rev., Beijing Review Press, Beijing. p. 15–18.

Anonymous. 1994. Farm reforms ushers in bright future. Beijing Rev. Oct. 24–30, Beijing Review Press, Beijing. p. 6.

Brown, L.R. 1995. Who will feed China? Wake-up call for a small planet. W.W. Norton & Company, New York.

China Agricultural Yearbook (*Zhonguo Nongye Nianjian*). 1983, 1985, 1986,1992, 1993. Agr. Publ. House, Beijing.

China's Customs General Administration. Summary Surveys of China's Customs Statistics. Knowledge Publishing House, Beijing, 1986–1992 issues.

Coyle, W.T. 1995. The future of U.S. agricultural trade with Asia. Proc. 1995 Grain, Fertilizer, and Transportation Outlook Seminar, Chicago, March 7, 1995.

Crook, F.W. 1994. Vegetable production and trade in China. p. 38–44. China. Situation and Outlook Series. Int. Agr. Trade Rep. USDA, ERS, WRS-94-4.

ERS-NASS. 1995. Products and Services from USDA's economics agencies(catalogue). USDA, Washington DC.

Gao Anming. 1995. China can boost grain production. China Daily. Aug. 4.

Konovsky, J., T.A. Lumpkin, and D.C. McClary. 1994. Chapter 15: Edamame: the vegetable soybean. p. 173–182. In: A.D. O'Rourke (ed.), Japanese food and agrimarket. Food Products Press of the Haworth Press, New York.

Larkcom, J. 1991. Oriental vegetables—the complete guide for the gardening cook. Kodansha Int., London.

Lumpkin, T.A. and D.C. McClary. 1994. Azuki bean: botany, production and uses. Commonwealth Agriculture Bureau, International. Oxon, UK.

Lumpkin, T.A., B.B. Dean, and D.A. O'Rourke. 1993a. IMPACT and the East Asian crops development program. p. 122–126. In: J. Janick and J. Simon (eds.), New crops. Wiley, New York.

Lumpkin, T.A., J.C. Konovsky, K.J. Larsen, and D. McClary. 1993b. Potential new specialty crops from Asia: Azuki bean, edamame soybean and *Astragalus*. p. 45–51. In: J. Janick and J. Simon (eds.), New crops. Wiley, New York.

Matsuoka, S. 1995. Food self-sufficiency rate falls to all-time low. Japan Agrinfo Newsl. 12(8):2.

Rozelle, S., A. Park, J. Huang, and H. Jin. 1996. Dilemmas in reforming state-market relations in China's grain economy. Series in the Chinese Rice Economy Project, Stanford Univ.

Tuan, F.C. 1994. China runs agricultural trade surplus with the United States. China. Situation and Outlook Series. Int. Agr. Trade Rep. USDA, ERS, WRS-94-4. p. 7–9.

USDA. 1993. China. USDA-ERS Situation and Outlook Series. Int. Agr. Trade Rep. RS-93-4, July 1993.

USDA. 1994a. Asia and Pacific rim. USDA-ERS Situation and Outlook Series. Int. Agr. Trade Rep. WRS-94-6, Oct. 1994.

USDA. 1994b. Wild card: China. U.S. Department of Agriculture. AgExporter, Jan. 1994. p. 9.

Wahl, T.I. (ed.). 1995. The China market: Data and information systems. Proc. WRCC-101, Washington DC, April 1995. IMPACT Center, Washington State University, Pullman, WA.

Wahl, T.I. and A.D. O'Rourke (eds.). 1994. China as a market and competitor. Proc. WRCC-101, Reno, NV, Nov. 1994. IMPACT Center, Washington State University, Pullman, WA.

Webb, Shwu-Eng, H. 1993. China: Agricultural trade prospects 2000. Asia, Aug. p. 27–37.

Underexploited Temperate Industrial and Fiber Crops

Richard J. Roseberg

Successful temperate new crops must either fit well into the rotation of established food, feed, or fiber crops, provide a product that is relatively more valuable than the current crop, or be better suited to growing on a given area. Proximity of growers to potential processors (who manufacture the refined product) is key to economic advantage for the new crop over prior raw material sources. In the case of arid industrial crops, the issues of survival in a harsh climate and the cost of water are paramount, while in the higher latitude temperate zones the constraints are frost tolerance, length of growing season, crop response to daylength, water requirement (rain-fed or irrigated), and weather patterns during harvest.

This paper examines the current status of several potential crops with an eye toward their requirements for success in temperate zones. Crops that are reaching or have reached active crop status include meadowfoam, fiber flax, and kenaf. Crops that are very promising, but require some further breeding, agronomy, or processing research include lesquerella, cuphea, euphorbia, and vernonia. Crops that are intriguing, but need more study to define their potential and value include grindelia and hesperaloe. Other potential crops, including those with application only in specific situations or those having received limited study and commercialization efforts are also briefly described.

The activity level for each crop is described using four new crop research and development categories put forth by L.J.M. van Soest (1993): I Plant exploration and evaluation; II Crop improvement (including plant breeding and agronomy; III Processing and application research; and IV Marketing, commercialization, and utilization. Estimating the crop's value or return to the farmer was difficult, and this difficulty increased the further away the crop was from commercialization. However, crop values were calculated using comparisons with currently available raw materials and estimates of possible differences in value.

MEADOWFOAM

The key early development for meadowfoam (*Limnanthes* species, Limnanthaceae) came out of the extensive USDA efforts of the late 1950s and early 1960s when many plant species were analyzed in a search for novel compounds. Out of these efforts at the National Center for Agricultural Utilization Research (NCAUR) laboratory, it was first recognized that 94% of the fatty acids in meadowfoam (*Limnanthes douglasii* R. Br.) seed oils had chain lengths of 20 carbon atoms or longer (Earle et al. 1959). Identification of these previously unknown long-chain fatty acids was made soon after (Smith et al. 1960; Bagby et al. 1961). Later interest shifted to *Limnanthes alba* Benth. due to its improved agronomic characteristics such as increased seed retention, upright growth habit, and plant height. Seed retention is no longer a problem. In fact, the cultivar Floral is considered difficult to thresh by growers. *L. alba* contains high levels of the same fatty acids as *L. douglasii* (Miller et al. 1964). Results of early agronomic and breeding work have been summarized by Jolliff et al. (1981). Results from additional crop and oil analysis were summarized by Purdy and Craig (1987).

Within the past two years the consistent involvement of industry, specifically Fanning Corp. (Chicago), has greatly stabilized the supply-demand situation, creating a steadily increasing demand. This is unlike the whipsawed supply, demand, and price history of the 1980s. Cooperation between the Oregon Meadowfoam Growers Association (OMGA), Oregon State Univ., Fanning Corp., and USDA–NCAUR has improved the coordination between crop research, product research, grower contracts, crop price, and crop area expansion.

Raw Material and Products

Meadowfoam seeds contain about 25% oil, 95% of which is made up of C:20 or C:22 monoene or diene fatty acids (Kleiman 1990). Such specificity in long chain fatty acids is rare in nature. Such fatty acids could be used in cosmetics (moisturizers, soaps, hair care), specialty lubricants and polymers (Purdy and Craig 1987; Bosisio 1989; Carlson et al. 1992).

Meadowfoam oil naturally occurs in the form of triglycerides. However, reports have described how C:40–C:44 wax esters similar to those of jojoba [*Simmondsia chinensis* [Link] Schneid.] and sperm whale oils could be produced from meadowfoam oil by reducing fatty acids to alcohols and then forming esters using unreacted fatty acids (Miwa and Wolff 1962; Miwa 1972; Nieschlag et al. 1977). Despite statements to the contrary (Cook 1971), meadowfoam oil is not a substitute for sperm whale oil. Unmodified meadowfoam oil is structurally quite different from sperm whale oil, and jojoba would appear to be a better source of these compounds (Kleiman 1990).

Competing Sources

Long chain fatty acids are currently produced from high erucic acid rape seed, crambe seed, and for some applications, fish oils (Jolliff et al. 1981; Purdy and Craig 1987). However, the erucic acid ($C22{:}1^{13}$) from crambe and rape seed is chemically different than the three other fatty acids ($C20{:}1^{5}$, $C22{:}1^{5}$, and $C22{:}2^{5,13}$) that make up about 85% of meadowfoam oil (Purdy and Craig 1987; Kleiman 1990).

Crop Status

Meadowfoam research and development efforts in the U.S. currently include categories I–IV. The only active crop research program is at Oregon State Univ. in cooperation with the OMGA and Fanning Corp. Product development is ongoing at the USDA–NCAUR in Peoria, Illinois, at Fanning Corp., and elsewhere. Fanning Corp. is leading market development efforts. With the current industrial demand remaining steady, the crop area is increasing from about 900 ha in 1995 to an estimated 2400 ha in 1996 in Oregon's Willamette Valley.

In Oregon, seed yields typically have ranged from 600–1200 kg/ha with current cultivars. Given the oil content and fatty acid composition, 1200 kg/ha would be worth $315/ha at rapeseed prices. However, the unique chemistry of meadowfoam oil has prompted recent contracts for $1.10/kg seed, resulting in a crop value of about $1320/ha at the high end of the normal yield range.

Limitations

The requirement for insect pollination has probably limited yields in commercial fields. Research plots can be saturated with honeybees, but in typical commercial fields only two of the five potential seeds in each flower matures. Thus, development of auto-fertile cultivars should improve yield. The range of adaptation should be further explored as new cultivars are developed. Little is currently known about range limitations.

Likely Commercial Production Areas

Meadowfoam growth on clay soils is usually acceptable, and in fact this characteristic has been one of the main reasons for its development in western Oregon. While native to western Oregon and northern California, meadowfoam should be well adapted to areas that have cool soils at planting, cool and moist weather during vegetative growth, and warm, dry harvest weather. In addition to the Willamette Valley in Oregon, such areas could include western Washington, northern Europe, New Zealand, and parts of southern Australia and southern Argentina.

FIBER FLAX

Flax [*Linum usitatissimum* L., Linaceae] is not a new crop. There are six references to women and flax in the Bible, indicating that flax spinning and weaving were household industries in antiquity. The center of origin of flax is thought to be in the Near East, but the exact area is a topic of debate. A highly selected cultivar dating from 5500–5000 BC was found in Iran (Dempsey 1975). Earlier flax types have been found both in Egypt and Switzerland. By 4000 BC the Egyptians had a highly developed flax industry. Other nations of the area also developed flax, and its domestication was well established in western Europe by the time of Charlemagne (742–814).

The first mechanized method for spinning flax yarn was developed in France. Between 1810 and 1820 Philippe de Girard was issued six patents relating to this new technology (Dempsey 1975). To maintain maximum fiber length and quality, flax is pulled from the ground at harvest, rather than cut. Until World War II flax was mainly pulled by hand. Since then this extremely labor intensive process has been mechanized, first by tractor-drawn machines, and then by self-propelled pullers. These machines, along with specialized turners and deseeders, were mainly developed in western Europe in the 1950s and 1960s, improving upon earlier U.S. developments during WWII and soon after. These technologies have greatly increased the speed and ease of harvest. They have also improved the utility of the inexpensive dew-retting process, whereby the pulled flax fiber is separated from cellulose and other stem parts by bacterial action in the presence of water (the dew) in the field. Field equipment and modern processing plant equipment have been further developed in western Europe during the 1980s and 1990s, allowing greater capture and separation of all classes of flax fibers while improving worker safety conditions (D. Ehrensing 1995, pers. commun.). Thus fiber recovery is currently about 25% of total biomass yield, up from about 20% forty years ago. Flax breeding was very active in the United States during the 1930s and until the 1950s, but ceased by the early 1960s (Calvert and Marks 1995). However, cultivar development has continued in Europe, especially in France, The Netherlands, and Belgium.

In the 1940s flax was grown on up to 7300 ha in Oregon (Hurst et al. 1953). The reintroduction of European flax after the end of World War II, the increase in cotton use in textiles, and the development of petroleum based fibers such as nylon combined to essentially eliminate the Oregon flax industry by the mid 1950s. Interest has recently been revived, mainly due to restrictions on stubble burning from grass seed production that have created problems for farmers in terms of weed control, insect, and disease cycles. Grass seed production occupies over 175,000 ha annually in Oregon, mainly in the Willamette Valley, (U.S. Dept. of Commerce 1993), and Oregon routinely produces over 90% of the world's perennial ryegrass (*Lolium perenne* L.), orchardgrass (*Dactylis glomerata* L.), and bentgrass (*Agrostis palustris* Huds.) seed. Because flax is a dicot it would provide a break in disease cycles and allow use of alternate herbicides while providing a cash crop for grass seed growers.

Raw Material and Products

Although all flax cultivars produce fiber in the stems and oil (linseed) in the seeds, fiber flax cultivars have been bred and selected specifically to produce large quantities of very long, high quality fiber, with oil production only a secondary consideration. Flax produces fibers of varying length. The longest fibers can be used in making fine linens for clothing, draperies, and furniture, medium fibers have been used for canvas and geotextiles, while short fibers have been used for paper and sacking.

Competing Sources

Fibers from oilseed flax, jute (*Corchorus capsularis* L.), sisal (*Agave sisalana* Perrine), hemp (*Cannabis sativa* L.), cotton (*Gossypium hirsutum* L.), and wood species are currently used for some of the applications suited to fiber flax. However, the high strength and quality of flax fiber makes it superior than other sources for some applications, such as linens.

Crop Status

Fiber flax research and development efforts in the United States currently include categories I–IV. In 1993 flax was grown for fiber in Russia (335,000 ha), Ukraine (127,000 ha), Belarus (120,000 ha), China (93,000 ha), and France (50,000 ha) (FAO 1994). Some fiber from oil seed flax production in Canada and northern United States (especially North Dakota) has been used recently in cigarette paper production (P.M. Carr 1995, pers. commun.).

Fiber yields in western Europe have recently been in the range of 1500–2000 kg/ha, while yields in Russia and eastern Europe have usually been less than half of those amounts (FAO 1994). This may have been due to use of oil varieties as well as poorer crop technology or management. Oregon farm fiber yields in 1995 were about 1200 kg/ha. Oregon statewide average yields from 1925 through 1951 ranged from 1.3 to 5.1 t/ha dry matter (or about 260 to 1020 kg/ha fiber) (Hurst et al. 1953).

Flax fibers that have been separated from the straw (skutched), but not combed, typically have ranged in value from $ 0.20–3.00/kg, with the longest fibers commanding the higher price (van Gelder et al. 1993; FAO

1994). Due to recent high demand, prices for short fibers have only been slightly less than those for long fibers (D. Ehrensing 1995, pers. commun.).

Limitations

Due to its long history of development, flax has few remaining problems, both in terms of agronomy and processing. The main hindrance to recommercialization in Oregon is lack of a processing plant. Estimated capital costs are $1.0 million (including the specialized field equipment) to process only short fibers, or $1.5 million to process the more valuable long fiber separately (D. Ehrensing 1995, pers. commun.).

Markets would likely expand if improved technology to use the strong flax fibers in combination with cheap, weaker fibers were developed. Flax could conceivably be mixed with excess grass seed straw or softwood fiber in composite boards or high quality papers, with cotton or polyester in clothing, or to reinforce plastics and composite materials.

Likely Commercial Production Areas

Fiber quality is enhanced by cool, moist spring weather followed by warm summers, with sufficient dew or light rain for field retting. Current production areas in northwest and eastern Europe are well-suited to fiber flax production, as are western Oregon and part of Michigan. Flax will grow in other climates, but fiber yield and quality are usually much poorer. For example, the oilseed flax production areas in north-central United States and south central Canada change from cool to hot weather rapidly, resulting in poorer fiber quality. Fiber flax production in the United States seems most likely to succeed in the Pacific Northwest due to its high fiber quality and the concentration of industry already producing fiber-based products there.

KENAF

Kenaf (*Hibiscus cannabinus* L., Malvaceae) has long been cultivated, probably as early as 4000 BC in Africa. Early research in the U.S. on using kenaf as a substitute for jute was begun in the 1940s due to the supply disruption from the Far East during World War II. This work continued into the 1950s, when more applied efforts, as part of the USDA Search for New Pulp Fibers program, began and continued into the 1970s (Taylor 1993). Details of this early work were well summarized by Dempsey (1975), and White et al. (1970). In 1977, the *Peoria Journal Star* was printed on kenaf newsprint, thus demonstrating the development of agronomic and processing technology for kenaf newsprint production. Official USDA and land grant university involvement in kenaf research was on hiatus from 1977 until 1986, when the Kenaf Demonstration Project was begun (Kugler 1988). In the meantime, several private sector efforts continued to demonstrate kenaf newsprint technology development, including the printing of several newspapers on kenaf newsprint at various times from 1977–1987 and the construction of a 200 t/day commercial pulp mill in Thailand that utilized kenaf fiber (Taylor 1993).

With the activation of the Kenaf Demonstration Project, many improvements in agronomic, processing, and newsprint production and use were made, building on the successes of the earlier efforts. An important commitment was the addition of Dr. Charles Cook to the USDA–ARS Weslaco, Texas station. Dr. Cook's primary research emphasis has been kenaf breeding and agronomy, and he has developed selections of kenaf having improved tolerance of root knot nematodes and associated pathogenic soil fungi (Cook and Mullin 1994). These pests have been significant problems in areas where kenaf was part of a cotton rotation. Development of effective harvesting, material handling, and fiber separation equipment was a breakthrough made primarily by the efforts of Harold Willett (Taylor 1993), although other separation designs also have been developed (Chen 1994). Summaries of developments since about 1987 have been published (USDA 1990; Taylor 1993: and D. Kugler elsewhere in this volume).

Raw Material and Products

The kenaf plant contains moderately long fibers in its outer stem and short fibers in its core. The outer stem (bark) makes up about 35%–40% of the stem weight, with the inner stem (core) containing the remaining 60%–65%. The fiber content of kenaf bark is about 50%–55%, increasing with plant population density, while the less valuable short fibers make up about 45%–50% of the inner core (Clark and Wolff 1969; Wood et al. 1983). Traditionally, the fiber has been used on several continents for rope, sacking, twine, and matting. However, its value will be greater if used for newsprint, carpet backing, and mixed into composite materials for boards or other structural materials (Taylor and Kugler 1992; Taylor 1993). The inner fiber has absorbent qualities that potentially could be used in products such as oil absorbents or poultry litter. Attributes of the kenaf plant were described in great detail by Dempsey (1975).

Competing Sources

For coarse fiber (lower value) applications, kenaf must compete with imported tropical monocots, chiefly jute. The higher value newsprint market in the United States is huge, and imports accounted for as much as 7.5 million tonnes, worth $4.5 billion in recent years (USDA 1993). In this market kenaf must compete with wood pulp. Increasing amounts of newsprint have been imported into the United States (up to 60% of consumption recently), mostly from Canada.

Crop Status

Kenaf research and development efforts in the United States currently include categories I–IV. Kenaf was grown on about 1660 ha in the U.S. during the early 1990s (USDA 1993). However, plans have been in place for several years to build a pulp mill requiring up to 2000 ha of kenaf mixed with recycled newspapers to produce 85 t newsprint per day (Taylor 1993). Construction has been delayed due to insufficient financing, but recent increased newsprint costs should tend to make such an enterprise more viable. The intent is still to complete the project, but details on timing and size are not yet public (C. Taylor 1996, pers. commun.). The potential area of U.S. kenaf cultivation could be as great as 0.4 to 2.0 million ha if the current upward trends in paper demand and price continue (USDA 1993; Stone 1995). Worldwide, kenaf was produced on about 200,000 ha, with major producers including China, Commonwealth of Independent States, Thailand, Cuba, India, and Mexico (FAO 1994).

Kenaf yields vary widely, not surprising given the range of areas where it has been grown and level of crop inputs. As plant density increases, stem diameters tend to decrease, but the proportion of bark tends to increase. However, the interactions between local climate, crop management, cultivar, stand density, and plant mortality make it difficult to predict stem and fiber yield without field testing (Clark and Wolff 1969; Higgins and White 1970; White et al. 1970; White et al. 1971; Dempsey 1975; Campbell and White 1982; Bhangoo et al. 1986; Scott et al. 1989). Commercial yields in the range of 9 to 22 t/ha biomass dry weight have often been reported. The higher yields were generally realized when growing conditions improved, typically as one moves from dry, high latitude locations to humid, lower latitude sites. In well adapted areas, such as the southeastern U.S., kenaf has typically yielded three to five times more fiber per year than southern pine, the typical pulping raw material source in that area (Wolff 1964; USDA 1993). Testing at several higher latitude temperate sites suggested that the adaptation of kenaf can change quite rapidly with a fairly small climatic change (White et al. 1970; Lauer 1990; Evans and Hang 1993). Results from a 1994 kenaf trial in southern Oregon showed this effect. The cultivar G4, although planted four weeks later than ideal, yielded an average 12.0 t/ha stem dry weight in the Rogue Valley, compared to 6.3 t/ha in the Willamette Valley (320 km to the north). Four cultivars that were planted four weeks before G4 yielded from 14.6 to 18.1 t/ha in the Rogue Valley.

Kenaf fiber in Thailand (where commercial kenaf pulping operations have been ongoing since 1981) were recently valued at $ 364/t (FAO 1994). To compete with wood as a pulp source in the northwestern U.S., where wood fiber is still relatively plentiful, a kenaf crop would generally sell in the range of $55–88/t (stem dry weight). For a yield of 13.5 t/ha and a price of $65/t, a grower would gross $878/ha. Part of the value of kenaf is due to the fact that pulping requires 15%–25% less energy and less chemical inputs than wood pulping (USDA 1993; Taylor and Kugler 1992).

An economic analysis for the Rio Grande Valley of Texas indicated that kenaf at $44/t would compare favorably to other crops commonly grown in that area (Scott and Taylor 1990). The overall demand for paper has continued to increase since that time. Pulp prices per tonne have rebounded from a low value of $453 in 1993 to an estimated $890 for 1995 (Brown 1995). Most analysts suggest that the dramatic price increases have been due to a combination of increased paper demand, unchanged production capacity, and increased cost and/or reduced availability of wood fiber, especially from federal timber lands in the U.S. (Hagler 1995; Stone 1995; Anon. 1995). These trends in paper demand and wood fiber availability are not expected to change. Thus, it appears that there could well be an increasing demand for quality paper fiber from non-wood sources, such as from kenaf.

Limitations

Storage and handling of annual crops like kenaf that are used in year-round manufacturing has always been somewhat of a problem. The problem is exacerbated if weather conditions or rotation requirement prevent the grower from leaving the crop standing in the field during the fall and winter after harvest. Also, while extended "vertical storage" in the field improves handling logistics, it also prevents use of the leaves as a forage (unless the plant tops were somehow harvested earlier in a separate operation). While ratios and procedures for using kenaf fiber in mixtures with other materials has been worked out in some cases (i.e. recycled newspapers for newsprint production), mixing and processing kenaf with materials such as straw, flax fiber, wood, and synthetics for composite board or other structural materials still need to be determined and optimized to broaden the applications for kenaf. Germplasm evaluation and cultivar development need to be continued, especially in terms of improved pest resistance and adaptation to higher latitudes (where summer days are longer, but seasons are shorter). Until recently, unfavorable economics had hindered the development of the kenaf industry in the United States, but research on the forage value of leaves, concurrent uses of both types of fiber produced, and increasing demand for fiber for all applications should improve the economic conditions affecting kenaf development.

Likely Commercial Production Areas

Kenaf has been grown in many areas, but highest yields have generally been observed under the following conditions: Warm soil and air (mean daily air temperatures between 22° and 30°C), sufficient moisture (monthly precipitation of 90–275 mm), fairly high relative humidity (65%–85%), a long frost free season, and fairly well drained soil which may otherwise vary greatly in texture and chemistry (Dempsey 1975). Good fertility contributes to higher yields (White et al. 1970; Dempsey 1975; Bhangoo et al. 1986). Kenaf was found to be moderately tolerant to saline irrigation water (Francois et al. 1990). Growing kenaf under irrigation modifies some of the general requirements listed above. Thus traditional growing areas in Asia, India, and the Caribbean/Central America should continue and new or increased production should be achievable in the southeast, southwest, central, and parts of the western U.S., as well as northern Australia, Africa, parts of South America, and southern Europe.

LESQUERELLA

The key early development for lesquerella (*Lesquerella* species, Brassicaceae) came out of the extensive USDA efforts of the late 1950s and early 1960s when many plant species were analyzed in a search for novel compounds. Out of this effort it was first recognized that *Lesquerella lasiocarpa* seed oils contained a high concentration of a hydroxyeicosenoic acid (Miwa et al. 1960). The structure of this C20:1-OH hydroxy fatty acid (HFA) was soon identified and given the trivial name of lesquerolic acid (Smith et al. 1961). Examination of 14 *Lesquerella* species showed that most had high concentrations of lesquerolic acid, while two species contained high levels of a hydroxyoctadecenoic acid (Mikolajczak et al. 1962). The structure of this C18:2-OH hydroxy fatty acid was soon identified and given the trivial name of densipolic acid (Smith et al. 1962). The relationship between native habitat and predominant HFA was recognized by Barclay et al. (1962), who noted that species native to the western U.S. usually had high concentrations of lesquerolic acid, while those native to the eastern U.S. had high concentrations of densipolic acid. However, *Lesquerella auriculata*, a native of Texas and Oklahoma, has seed oil with high concentrations of C20:2-OH (auricolic acid), similar in structure to densipolic acid, but having two additional carbons on the carboxyl end of the molecule (Kleiman et al. 1972). Based on morphological and agronomic characteristics, *Lesquerella fendleri* (Gray) Wats., a native of the southwestern U.S., was considered the most likely candidate for domestication (Barclay et al. 1962; Gentry and Barclay 1962). Early agronomic, germplasm collection, and selection studies were summarized by A.E. Thompson (1988).

Raw Material and Products

The HFAs found in high concentrations in the seed oil of several *Lesquerella* species could be used in producing specialty lubricants, heavy duty detergents, inks, and coatings due to the special properties of HFAs compared to other fatty acids, including higher viscosity and reactivity, caused by the presence of the hydroxyl group (Roetheli et al. 1991). Recent research has identified their utility in manufacture of plastics, additional coatings, extruded closed cell foam, facial soap, and cosmetics (Roetheli et al. 1991; Arquette and Brown 1993; Carlson et al. 1993; Thames et al. 1993a, b, 1996).

Competing Sources

Castor (*Ricinus communis* L.) oil contains the HFA ricinoleic acid (C18:1-OH). In recent years about 45,000 t (worth about $45 million) of castor oil are imported annually to the U.S. as its primary source of HFA. Because of this dependence on imports and the importance in certain applications, HFAs have been classified as a strategic material by the U.S. government. Because the chain length of lesquerolic acid is two carbons longer than ricinoleic acid, chemical reactions may provide more useful compounds than those from castor oil, such as the basic compound needed for nylon 1212 synthesis, whose current raw material source is petroleum (Kleiman 1990). The presence of two double bonds in densipolic and auricolic acids suggest their greater ability to undergo chemical reactions resulting in useful industrial materials. The low toxicity and low potential for irritation by lesquerella oil (unlike castor oil) suggests initial development in the high value cosmetics industry in products such as lotions, hair conditioners, and lipsticks (Arquette and Brown 1993).

Crop Status

Lesquerella research and development efforts in the U.S. currently include categories I–III. The crop has not been grown commercially, but for several years research-funded seed increases of *L. fendleri* have been contracted with growers in Arizona, New Mexico, and Texas, totalling 16–48 ha annually.

Based on plant morphology and limited experience with harvesting of wild stands in west Texas, Gentry and Barclay (1962) estimated seed yields of unimproved germplasm could exceed 1100 kg/ha. In more recent large field trials, seed yields in Arizona have often been between 500–1100 kg/ha, but between 1200–1800 kg/ha in small plots (Thompson et al. 1989; Thompson and Dierig 1988). In southwest Oregon, plot yields have generally been between 450–1000 kg/ha. *L. fendleri* seed typically contains about 25% oil of which 55%–60% is lesquerolic acid (Barclay et al. 1962; Mikolaczak et al. 1962). Castor seeds typically contain about 50% oil, of which about 85% is ricinoleic acid (Atsmon 1989). Between 1972 and 1990 castor oil prices ranged between a low of $0.51/kg in 1972 to a high of $1.60/kg in 1984, but castor oil prices have fluctuated dramatically, even within a single year (Roetheli et al. 1991). Based on these prices, calculated lesquerella seed value would be between $0.11 and $0.15/kg. If the seed yield were 1000 kg/ha, lesquerella crop value would range from $110 to 150/ha. However, lesquerella oil may be more valuable than castor oil due its applications in high value products such as cosmetics. Recent projections of seed value by potential industrial users have been in the range of $ 0.33 to 0.40/kg for improved seed containing 35% oil (Brown, 1994). Initial commercialization of lesquerella is dependent on developing such high value uses (Glaser et al. 1992; Arquette and Brown 1993).

Lesquerella fendleri is the species receiving the most research effort. Other species have been examined only at the category I and II level (exploration and crop improvement), but could also be valuable crops for several reasons. Some species produce seeds several times larger than *L. fendleri* (Barclay et al. 1962; Mikolajczak et al. 1962; Thompson 1988). Presumably, larger seeds would be easier to handle and possibly result in greater oil yields. Some species that produce densipolic or auricolic acid, unlike *L. fendleri* which produces lesquerolic acid, are native to areas outside the southwestern U.S., and thus may be adapted to production in these other locales. Some are perennials, and could be suitable for topography and cropping systems different than those that appear optimal for *L. fendleri*.

In two years of small plot studies in southwestern Oregon, *L. grandiflora* (lesquerolic acid type) plants grew well, were slightly larger than *L. fendleri*, and produced impressive amounts of seed. Small plots of *L. perforata* (densipolic acid) and *L. auriculata* (auricolic acid) have grown and yielded fairly well. *L. purpurea* and *L. pinetorum* (both lesquerolic acid) germinated but produced little or no seed. Plantings of *L. angustifolia*, *L. engelmannii*, *L. argyraea*, *L. gordonii*, *L. gracilis*, *L. lasiocarpa*, *L. ludoviciana*, (all lesquerolic acid), *L. densipila*, *L. lyrata*, and *L. stonensis* (all densipolic acid) failed to germinate, but the age of the seed (1972) probably had more effect than any differences in adaptation. Plantings in 1995 using several accessions (recently collected by D. Dierig) of *L. douglasii* (a perennial, lesquerolic acid type native to the Pacific Northwest) may further elucidate the cropping feasibility of species other than *L. fendleri* if grown in adapted regions.

Limitations

Because honey bees tend to very active in lesquerella plantings, it has been assumed that bee pollination enhances yield, although the degree to which insect pollination is required is not well understood. The generally lower yields seen in larger fields may be partially due to the difficulty and/or expense of achieving high bee densities and maximum seed set (Dierig and Thompson 1993). Thus, the successful domestication of lesquerella may, like meadowfoam, require the elimination of its insect-pollination requirement. Breeding work to increase oil and HFA concentrations should increase seed value.

Plant establishment has been a problem with lesquerella, due to its tendency toward seed dormancy and weak seedling vigor. Cultural practices that improve germination and stand establishment have been developed at New Mexico State Univ. (J. Fowler 1995, pers. commun.). Due to lesquerella's slow initial growth habit, competition with weeds has been a serious concern. However, several potential herbicides are good candidates for registration (Roseberg 1993; Foster et al. 1996; Roseberg 1996a), and experimental registration has been pursued in Arizona, New Mexico, and Texas (M. Foster 1995, pers. commun.).

Additional agronomic and breeding research on *L. fendleri* is ongoing at this time, mainly by scientists in Arizona (D. Ray, D. Dierig, J. Nelson, W. Coates, J. Brown, D. Hunsaker), New Mexico (J. Fowler), and Texas (M. Foster). Oil and gum analysis, processing, and product development research is ongoing at the USDA–NCAUR laboratory in Peoria, Illinois (T. Abbott, K. Carlson, and others), University of S. Mississippi (S. Thames), and scientists at Mycogen Corp., Lubrizol Inc., and International Flora Technologies (J. Brown and R. Kleiman).

Likely Commercial Production Areas

Lesquerella, especially *L. fendleri* and the other lesquerolic acid types, seem well adapted to semiarid or arid locations. Native stands of *L. fendleri* are typically found in well drained, calcareous soils (Gentry and Barclay 1962). Likely production areas may include the southwest and south-central U.S., western Australia, northern Argentina and Chile, and northern Africa. *L. grandiflora* and *L. douglasii* may be adapted to parts of the western U.S. and southern Europe.

CUPHEA

The key early development of cuphea came out of the extensive USDA efforts of the late 1950s and early 1960s when many plant species were analyzed in a search for novel compounds. Out of this effort it was first recognized that seed oils from certain *Cuphea* species (Lythraceae) contained high levels of several medium chain triglycerides (MCTs), but that the specific dominant MCT varied with species (Earle et al. 1960; Wilson et al. 1960; Miller et al. 1964; Litchfield et al. 1967). Later studies showed that many *Cuphea* species contain high levels of various MCTs (Wolf et al. 1983; Graham and Kleiman 1985). An excellent summary of the genus *Cuphea*, including botany and early work in seed oil analyses, was written by Graham (1989).

It was reported at the First National Symposium on New Crops in 1988 that two major barriers to the domestication of cuphea were seed dormancy and seed shattering (Knapp 1990). Because seed shattering and indeterminate flowering both had been universally present in cuphea, it was impossible to harvest a significant percentage of the seed with practical field equipment. Seed dormancy has been reduced by use of *C. lanceolata*, whose selections contain non-dormant traits, in interspecific hybridization with *C. viscosisima*. The *C. viscosisima* phenotype exhibits desirable agronomic characteristics due to its moderate plant size, good seed size, and self-fertile behavior (eliminating the need for insect pollination) (Hirsinger 1985; Knapp 1993). As the only species native to the U.S., *C. viscosisima* and its phenotypes should be well adapted (Graham 1989).

The biggest barrier (seed shattering) has been partially eliminated due to the discovery, under experimental conditions, of a natural mutation within *C. lanceolata* x *C. viscosisima* hybrids that has exhibited significantly reduced shattering (Knapp 1993). The non-shattering trait has been inherited and reproduced over several generations. Additional recent key developments include the development of lines that are both non-sticky and non-shattering, as well as lines that are both auto-fertile and non-shattering (Knapp et al. 1995).

Raw Material and Products

The genus *Cuphea* contains about 250 wild species native to Mexico, Central, and South America (Graham 1988, 1989). Although *C. viscosisima* is the only species native to the U.S., four introduced species exist in the wild. A fairly large number of cuphea species have seed oils that are rich in capric, lauric, caprylic, myristic, or other medium chain fatty acids (MCFAs) (Earle et al. 1960; Miller et al. 1964; Litchfield et al. 1967; Graham and Kleiman 1985). MCFAs are used in several industrial, medical, and food applications (Bach and Babayan 1983; Arkcoll 1988). Lauric acid is used in large amounts mainly in detergent and soap products. Caprylic, capric, and myristic acid are potentially very useful MCFAs in industrial or nutritional applications (Knapp 1992).

Competing Sources

Currently the U.S. MCT demand is entirely met by imports of coconut (*Cocos nucifera* L.) and oil palm (*Elaeis guineensis* Jacq.) (Arkcoll 1988), both of which contain high levels of lauric acid. From 1986 to 1988, annual imports of palm and coconut oil were about 723,000 t to the U.S. and 995,000 t to Europe (Mackie and Calhoun 1991; Knapp 1993). The supply (and price) of tropical oils has been unstable due to variations in weather, agronomic practices, and political climate in the main producing countries while demand continues to rise (Mackie and Calhoun 1991; Carlson et al. 1992). Demand for capric acid (present in some cuphea species and suitable for plasticizer and synthetic lubricant formulations) is currently met from modified petroleum sources (Thompson 1984; Hirsinger 1985; Carlson et al. 1992).

Crop Status

Cuphea research and development efforts in the U.S. currently include categories I–III. No commercial hectarage is cultivated at this time. Breeding and agronomic work is ongoing at Oregon State Univ., with product development research at Procter & Gamble Co., Cincinnati, Ohio and the USDA–NCAUR.

Until recently the seed shattering trait of cuphea made yield estimates difficult. Using either a modified cotton picker (vacuum) for shattering types or combine for the best available non-shattering types, seed yields in Oregon often ranged from 500–1500 kg/ha. Even so, the amount of seed loss suggests that potential seed yield is well over 2000 kg/ha. The slightly warmer climate in Medford, as compared to Corvallis, Oregon, has routinely resulted in more vigorous vegetative growth for a number of *Cuphea* species. However, recent seed yields of the *C. viscosisima* x *C. lanceolata* hybrids have been much lower in Medford than Corvallis. The effects of moisture stress during hot weather, including reduced growth and increased seed shattering, have been observed, but yield reductions have persisted even under well-watered conditions. Perhaps the improved hybrids are better suited to a slightly cooler climate than were the wild species grown previously.

Using a coconut oil price of $453/t (FAO, 1994) and adjusting for the differences in lauric acid content

(75% lauric acid in cuphea oil vs. 50% lauric acid in coconut oil) the value of cuphea oil would be $0.68/kg, or $0.24/kg seed containing 35% oil. At that price a 1500 kg/ha seed yield would be worth only $360/ha. However, as in the case of meadowfoam, the presence of unusual fatty acids (caprylic, capric, and myristic) in addition to the common (lauric), could add more value when used in new applications. This suggests that the crop should also be developed as a source of these unusual MCFAs rather than simply as a replacement for tropical oils (Knapp 1992, 1993).

Limitations

It would be desirable to combine the auto-fertile, non-sticky, and non-shattering traits into a single cultivar. Such a plant would allow a single harvest operation, possibly with a swathing operation followed several days later by threshing using a pickup-reel combine as is routinely used for grass seed harvests. The plant's stickiness can cause major equipment problems, especially if the crop is drought-stressed. However, the sticky resin may provide the plant with a protection against insects, and its removal may not be a completely positive development (Knapp 1993). As improved cultivars are developed, cultural practices must be continually evaluated. Research on product development must continue, especially in regards to the array of MCFAs available from different cuphea species.

Likely Commercial Production Areas

For the cuphea phenotypes closest to domestication, vegetative growth is favored by warm to hot weather with sufficient moisture, but seed yield is decreased under hot and dry conditions. The *C. viscosisima* phenotypes would likely grow well in the midwest and northwest U.S., southern Africa, southeastern South America, eastern and northern Australia, and much of Europe.

EUPHORBIA

The key early development came out of the extensive USDA efforts of the late 1950s and early 1960s when many plant species were analyzed in a search for novel compounds. Out of this effort it was first recognized that *Euphorbia lagascae* (Spreng.), Euphorbiaceae, was unique among the 58 euphorbs tested in that the seed oil contained high levels of vernolic (12,13 epoxy-cis-9-octadecenoic) acid. *E. lagascae* is a drought-tolerant native of Spain whose seed contains about 45%–50% oil, of which 60%–65% is vernolic acid (Kleiman et al. 1965; van Soest 1993; Vogel et al. 1993). Earlier, two other euphorbs (*Cephalocroton cordofanus* and *Cephalocroton peuschelii*) were shown to contain high levels of vernolic acid (Bharucha and Gunstone 1956; Gunstone and Sykes 1961). However, because they are both perennial shrubs their potential for cultivation was deemed less than that of the annual *E. lagascae* (Earle 1970).

The major problem with euphorbia that has hindered both breeding and agronomic research has been its violent seed shattering habit, making it difficult both to harvest and to measure seed yield. No wild accessions of euphorbia have contained a non-shattering trait (Vogel et al. 1993; Pascual-Villalobos et al. 1994). Recently, chemically induced, non-shattering mutants were developed in Spain (Pascual and Correal 1992; Pascual-Villalobos et al. 1994; Pascual-Villalobos 1996).

Raw Material and Products

Vernolic acid, the C:18 epoxy fatty acid (EFA) found in the seed oil of *E. lagascae*, would be useful in the paint and coating industry as a drying solvent in alkyd resin paints, a plasticizer or additive in polyvinyl chloride (PVC) resins (Riser et al. 1962; Carlson et al. 1981; Carlson and Chang 1985; Perdue 1986), and possibly in pharmaceutical applications (Ferrigni and McLaughlin 1984). Paints formulated with vernolic acid would greatly reduce volatile organic compound (VOC) air pollution that now occurs with volatilization of alkyd resins in conventional paints (Reisch 1989; Brownback and Glaser 1992; Anon. 1993). The Clean Air Act amendments of 1990 require the reduction of VOC pollutants, and regulations in California have been implemented earlier with greater effect upon the paint industry (Reisch 1989; Anon. 1993).

Competing Sources

Very few plants naturally produce high levels of vernolic acid in their seed oils (Kleiman 1990) and most of those that do have significant barriers to domestication and agronomic production (Earle 1970). The three species that appear to have the best chance for domestication include *Euphorbia lagascae* (Kleiman et al. 1965), *Vernonia galamensis* (Carlson et al. 1981; Perdue et al. 1986), and Stokes aster [*Stokesia laevis* Hill (Greene)] (Earle 1970; Campbell 1981).

Current sources of EFAs include epoxidized soybean oil, linseed oil (from oilseed flax), and processed petrochemicals (Carlson and Chang 1985; Perdue et al. 1986; Dierig and Thompson 1993). However, epoxidation of simple vegetable oils is an expensive process, and petrochemicals are a non-renewable and increasingly imported raw material. The U.S. currently uses about 18,000 t of EFAs in 1.2 billion liters/yr of paints and coatings alone (Reisch 1989; Anon. 1989).

Crop Status

Euphorbia research and development efforts in the U.S. currently include categories I–III, while research in Europe (especially Spain) has mainly been in categories I–II. There is no commercial hectarage at present. EFA product development is ongoing at the USDA–NCAUR and at private companies.

Four selection lines originating from the Spanish non-shattering mutants were grown for one year (1993) in Corvallis, Oregon. Seed yields in Corvallis were low, probably due to plant immaturity at the onset of cool, wet weather in the fall, but the non-shattering trait had persisted for the first generation. In 1994, four small, unreplicated, two-row plots were grown in Medford, Oregon using seed harvested in Corvallis. Plant growth was vigorous, and most plants were mature and beginning to senesce by early Oct. when they were carefully harvested by hand onto plastic tarps and dried indoors. Seed shattering was minimal, and calculated seed yields ranged from 1060 to 2800 kg/ha.

In 1995, using seed harvested in 1994, about 0.4 ha of euphorbia was grown in Medford in replicated plots including row spacing and harvest method variables. Unlike previous years, however, many plants were experiencing some degree of shattering by early Sept. Plots were either cut with an International 201 swather (draper type) or sprayed with diquat to dry the stems to allow direct combine threshing. Once dry, the euphorbia was combined in the field with a Hege 125C plot combine. Plot yields in 1995 ranged from 115–600 kg/ha. Differences were undoubtedly due to a combination of effects, including increased amount of seed shatter, use of field machinery, field drying, larger plots with minimal border effects, reduced irrigation in 1995, or other unknown factors. Based on the 1994 yields and observations of seed set in 1995, it is not unreasonable to believe that euphorbia could yield over 2000 kg/ha seed if breeding and agronomic problems were solved.

It has been observed in Europe that the percentage of vernolic acid in euphorbia seed oils varies little for different planting and harvest dates, or by differences in accession or environment (van Gelder et al. 1993; Vogel et al. 1993). This suggests that seed yield is a good predictor of fatty acid yield, and cultural practices developed to achieve maximum seed yield should also result in maximum fatty acid yield.

Average monthly prices for U.S. soybean oil ranged from $0.31 to 0.86/kg between 1982 and 1994 (Knight-Ridder Financial 1995). Based on a soybean oil price of $0.50/kg, and the value doubling or tripling after coversion to an EFA (Perdue et al. 1986), euphorbia oil would be worth $1.00 to 1.50/kg oil, or $0.50 to 0.75/kg seed. This value range was confirmed in a discussion with industry personnel. At $0.60/kg seed and 1500 kg/ha seed yield, the gross return to the farmer would be $900/ha. Of course, if euphorbia could be grown with reduced input costs, especially irrigation, its net return to the farmer would be greater than other crops having similar gross returns.

Limitations

Although the seed shattering trait has made progress on euphorbia difficult, availability of the non-shattering trait should prove to be very useful in breeding. Other characteristics present in some Spanish mutants would also increase seed yield, and could prove useful in breeding programs. Some mutants had four or five seeds per pod, instead of the three per pod observed with wild types (Pascual and Correal 1992; Vogel et al. 1993). While euphorbia's mainstem crown normally has three main terminal branches, some mutants in Spain had up to nine main terminal branches per mainstem crown, resulting in a greater number of inflorescences. Plants with such "mobheads" had shorter mainstems, possibly resulting in greater lodging resistance, and also may be more determinate in seed maturation (Pascual-Villalobos, 1996).

Breeding work has barely begun with euphorbia, but the availability of the non-shattering trait appears to be a breakthrough that can be exploited. *E. lagascae* is highly self-fertile, with pollen transfer occurring before insects can access the floral organs (Vogel et al. 1993). Therefore, outcrossing should be limited. Agronomic requirements of this crop have not been quantified, but now can be evaluated using non-shattering lines. Areas needing study include water and fertility requirement, planting and harvesting techniques, and seed cleaning and processing. Weed control research is also necessary, although some preliminary evaluations of potential herbicides have been made (Vogel et al. 1993). Due to the presence of latex and other potentially irritating compounds in the stems and petioles, it will be important to understand which safety precautions are necessary during harvest and processing. Processing chemistry and product development should continue on a larger scale as more seed becomes available.

Likely Commercial Production Areas

Euphorbia requires a long season for maximum seed production. It has appeared to be very tolerant to drought and heat (although the effect on yield is not known). Likely areas for production include much of the western U.S., northern Africa, southern Europe, northwest Argentina, and southern and western Australia. It should be able to compete well in situations where irrigation water is expensive or unavailable.

VERNONIA

The investigation of *Vernonia anthelmintica* (L.) Willd., Asteraceae, by Gunstone (1954) was the first report of a natural epoxy fatty acid in a seed oil (Earle 1970). As part of the extensive USDA efforts of the late 1950s and early 1960s to analyze many plant in search of novel compounds, it was confirmed that *V. anthelmintica* contained high levels of vernolic acid (Smith et al. 1959). Following that, a number of other *Vernonia* species were examined (Earle 1970). For a time, *V. anthelmintica* seemed to be the species with the greatest potential for domestication, and quite a bit of agronomic and utilization research was performed (summarized by Perdue et al. 1986). However, poor seed retention remained a major barrier to domestication. During late 1966 and early 1967, C.E. Smith Jr. collected some vernonia seeds in Africa from plants later identified as several sub-species of *Vernonia galamensis* [Cass.] Less. (Smith 1971). These contained about 30% oil, of which up to 78% was vernolic acid. A plant exploration trip to Africa by R.E. Perdue, Jr. in 1964 resulted in collection of seed from impressive mature plants that was later identified as a sub-species of *V. galamensis* different than that collected by Smith (Perdue et al. 1986). This seed contained about 42% oil, of which about 73% was vernolic acid (Carlson et al. 1981). Thus, this wild germplasm contained roughly 30% more vernolic acid than the best improved *V. anthelmintica* cultivars (Princen 1979). Based on the higher oil content, its observed seed retention, seed density, and seed size in the wild, it appeared *V. galamensis* had the greater potential for domestication.

One initial problem that discouraged domestication at temperate latitudes was short-day flowering requirement and frost intolerance. Thus, at temperate latitudes, the short days required for flowering were rapidly followed by freezing temperatures, preventing seed production (Perdue 1988; Phatak et al. 1989). A major breakthrough was the discovery that one accession of *V. galamensis* subsp. *galamensis* var. *petitiana* (Gilbert 1986) was day-neutral in flowering habit (Dierig and Thompson 1993) and auto-fertile. The day-neutral flowering habit has been retained by several generations of hybrids (Dierig et al. 1995). Although var. *petitiana* has several agronomic shortcomings, a breeding and selection program using hybrid crosses with var. *ethiopica* (which exhibited good plant vigor, large seed heads, large seeds, and good seed retention) has been ongoing with cooperators (including the author) in several states. Results suggest that further selection could result in vernonia hybrids that produce economically viable yields in several locations in the continental U.S. (D. Dierig 1995, pers. commun.).

Raw Material and Products

Seed oil of *V. galamensis* contains large amounts of vernolic acid (Perdue et al. 1986), an epoxy fatty acid identical to that found in euphorbia. However, about 45% of the vernolic acid in vernonia is in the trivernolin form, more than double that of euphorbia or stokes aster (Tallent et al. 1966; Earle 1970; Plattner et al. 1978; Carlson et al. 1981). For some applications this richer trivernolate content in vernonia may impute greater value than the vernolic acid from euphorbia or Stokes aster. In general, the product applications would be identical as those discussed for euphorbia.

Competing Sources

Competing sources for EFAs found in vernonia would be essentially the same as those listed for euphorbia.

Crop Status

Vernonia research and development efforts in the U.S. currently include categories I–III. No commercial hectarage is currently grown in the U.S. Small hectarage crops have been grown in Africa and South America using wild types to produce oil for analysis and processing research as well as improving agronomic knowledge (Anon. 1993; Carlson et al. 1992). Approximately 30% of vernonia's seed weight is vernolic acid, as in euphorbia. As such, the crop would have equivalent value, assuming extraction and processing requirements were similar. Reported seed yields in Africa using wild types have been as high as 2000 kg/ha, although details of experimental conditions have been limited. Day-neutral hybrids tested at several temperate North American sites in 1994 yielded between 166 and 1200 kg/ha (D. Dierig 1995, pers. commun.). If a seed yield of 1500 kg/ha were realized, the gross return to the farmer would be $900/ha at a value of $0.60/kg seed (the same as euphorbia). Involvement by USDA and university scientists, private industry, and the coordination of crop, processing, and product research has increased the chances for successful domestication of vernonia.

Limitations

While vernonia exhibits a dense canopy and thus competes well with weeds once it is established, early season weed control techniques will be essential during the seedling and early vegetative stage. Field tests have identified several herbicide compounds that should prove useful in vernonia cultivation (Roseberg 1996b). Seed retention and yield both need to be improved for vernonia to be commercialized. Ongoing breeding and selection programs increase the chances of improving these traits. Many questions regarding vernonia's agronomic requirements remain unanswered, including planting, water, fertilizer, and harvesting. Further evaluation and registration of herbicides or alternate weed control strategies is required. Some vernonia plants also suffer from a "sudden death syndrome," whereby an entire individual plant, at any growth stage, will wilt, senesce, and die, sometimes within a matter of hours. This has been observed at several locations in the U.S., and does not seem to be correlated to hybrid or management practice. It is assumed these are bacterial, fungal, or viral attacks, but a pathogen has not been positively identified. Worker safety during harvest and processing needs to be better understood due to the types of oils present in vernonia. Seed processing and product development research should continue in order to utilize vernonia's seed oils effectively.

Likely Commercial Production Areas

Vernonia seems to prefer well drained soil. It is also fairly drought tolerant once established, and requires a long season for maximum seed set. Production areas would likely include much of the temperate U.S., south and east Africa, parts of Argentina and southern Australia. Improved day-neutral cultivars would be required in all but the upland tropical areas of Africa and possibly parts of South America.

GRINDELIA

Several species of the genus *Grindelia* produce a diterpene resin (grindelic acid) and other resinous compounds on the surface of their flowers, leaves, and stems (Guerreiro et al. 1981; Bohlmann et al. 1982;

Timmermann et al. 1983). *Grindelia camporum* (Greene), Asteraceae, is a native of the western U.S., especially the central valley of California (Bailey 1976; Hoffmann et al. 1984; Hoffmann and McLaughlin 1986). The identification of grindelic acid as an important constituent of this resinous plant (Bohlmann et al. 1982), came out of a previous examination of 195 plant species as to their suitability for agriculturally derived crude oils in the arid southwestern U.S. (McLaughlin and Hoffmann 1982), and was a key factor in the initial development of this species. Diterpene resin acids constituted between 65% and 75% of the total crude resin in the plant (Hoffmann and McLaughlin 1986; McLaughlin 1986a, b). These diterpene acids have properties similar to wood rosin and its derivatives, and form the basis for grindelia's economic crop potential (Hoffmann 1985; Hoffmann and McLaughlin 1986). Studies on the heritability of resin production indicate that genetic improvement is feasible (Dunford 1964; McLaughlin 1986a, b; McLaughlin and Linker 1987). Glands on the surface of flowers are responsible for most of the resin production, with little resin produced on the stems or leaves (Hoffmann et al. 1984).

Raw Material and Products

The diterpene resins from grindelia are very similar to those obtained by grinding up stumps or tapping live old growth pine trees (Hoffmann and McLaughlin 1986; Thompson 1990) and are known as "naval stores." They consist of a large group of compounds, including turpentine, fatty acids, rosins, and their derivatives (Thompson 1990) once used to calk ships, but now used in paper sizing processes as well as producing rubber, chemicals, ester gums, and resins for many applications (Hoffmann and McLaughlin 1986). U.S. consumption of naval stores from old growth pine trees has been fairly static at about 500,000 t/yr (Hoffmann and McLaughlin 1986), but domestic production has almost disappeared, as supplies of old growth stumpage has decreased and costs of tapping the decreasing number of live trees has increased.

Crop Status

Grindelia research and development efforts in the U.S. currently include categories I–II, with no commercial production. *Grindelia camporum* has received the most interest in the U.S. due to its greater biomass production compared to other resin-producing U.S. natives, such as *G. squarrosa* (Pursh) Dunal., *G. latifolia* Kellogg, and *G. stricta* DC. (Bailey 1976; Ravetta et al. 1995). Recently, research has begun in Argentina using *G. chiloensis* (Cornelissen) Cabr., a South American native that exhibited a much higher resin content on a whole-plant basis (although lower total biomass) than *G. camporum* (Bailey 1976; Guerreiro et al. 1981; Ravetta et al. 1995). Large numbers of resin glands are found on the stems and leaves of *G. chiloensis* (D. Ravetta, 1995, pers. commun.), whereas resin glands occur mainly just on inflorescences of *G. camporum* (Hoffmann et al. 1984; Hoffmann and McLaughlin 1986).

In Arizona, experimental tetraploid lines of *G. camporum* have yielded up to 12.5 t/ha biomass per year, with a crude resin content of up to 11%, resulting in annual crude resin yields of up to 1.2 t/ha, using between 57 and 75 cm of irrigation water (Hoffmann and McLaughlin 1986; McLaughlin and Linker 1987). In more recent Arizona tests, *G. camporum* single plant biomass yields were about twice that of *G. chiloensis*, although resin content was about two-thirds higher for *G. chiloensis* (Ravetta et al. 1995).

In southern Oregon, plantings were made in 1992 and 1993 at two locations, one a deep sandy loam and the other a moderately deep cracking clay. In 1993 plants grown in the sandy loam yielded an oven-dry biomass average of 35.2 t/ha, vs. 6.8 t/ha in the clay. Analysis of the separated flower, leaf, and stem fractions resulted in an overall average resin yield of 3.70 t/ha in the sandy loam and 0.85 t/ha in the clay. In 1994 the same plantings yielded an oven-dry biomass average of 30.4 t/ha in the sandy loam vs. 28.3 t/ha for the clay, and an overall average resin yield of 4.42 t/ha in the sandy loam and 1.66 t/ha in the clay. In these experiments, the resin yield for flowers alone ranged from 11.6% to 26.6% on an individual plant basis, with immature flowers generally having higher resin content than mature flowers. Resin content of leaves ranged from 9.2% to 21.0%, stems ranged from 1.4% to 3.6%, and combined stems with leaves ranged from 3.3% to 7.3%. Resin content was much higher in all plant parts for plants grown in the clay soil in 1993, but the opposite was true in 1994. Due to different irrigation rates, water stress was greater in the clay in 1993, but greater in the sandy loam in 1994, which may have contributed to the observed differences in yield and resin content between the two sites and years. In natural sites, grindelia's preference for soils higher in clay content (S. McLaughlin 1992, pers. commun.) may be due to the greater water holding capacity of the soil in xeric climates rather than a preference for the drainage or texture of a clay soil *per se*.

Grindelia seed germination is enhanced under light and is optimum at 10° to 20°C (McLaughlin and Linker 1987; R. Roseberg unpub.). Dates of transplanting and planting density effects on biomass and resin yield, crop growth rates, and water use efficiency have been examined by McLaughlin and Linker (1987). Growth and yield data from Oregon and Argentina (D. Ravetta 1995, pers. commun.) indicate that grindelia may be better adapted to dry areas that are somewhat cooler than Arizona, and that plants with high biomass yields can also have high resin content.

One factor that favors growing domesticated industrial crops in place of some traditional food crops is that their raw materials (such as resins, gums, and waxes) have typically been more valuable than food on a unit weight basis (McLaughlin 1985). Resins similar to those found in grindelia rose in price from $0.59/kg in the late 1970s to $1.10 by the mid 1980s (Hoffmann and McLaughlin 1986). While rosin prices hit a low in 1993 (Gallagher 1994a), prices have increased since then due to reduced supply and increase demand (Santos 1994; Gallagher 1994b). The increased demand for fine paper for ink-jet, laser-jet, and copier applications as well as increased use of recycled paper have increased demand for paper sizing chemicals, including rosin (Killy 1994).

If grindelia resin were valued at $0.60/kg, a biomass yield of 30 t/ha with a crude resin yield of 3.0 t/ha would be worth $ 1800/ha. Due to resin extraction and processing costs, farm gate values for raw material would be estimated at 50%, or $900/ha. Reduced input costs would contribute to a farmer's net return from a grindelia crop, mainly due its presumed reduced irrigation requirement compared to other crops. Because grindelia is a perennial, establishment costs would be reduced, further adding to a farmer's net return. Thus, the return could prove attractive in areas where other crops were limited without large inputs of expensive irrigation water.

Limitations

Resin yields need to be increased to improve grindelia's economic feasibility (Hoffmann 1985; Hoffmann and McLaughlin 1986). Genetic analysis and preliminary selection trials indicate an increase to 15%–20% resin content is possible (McLaughlin 1986a, b). Areas of highest adaptation for *G. camporum* need to be identified. The high concentration of resin ducts on all parts of the *G. chiloensis* plant suggest that this species could be domesticated in adapted areas. Also, research in Argentina with interspecific crosses of grindelia species is underway to combine the biomass production of *G. camporum* with the ubiquitous resin gland morphology of *G. chiloensis* (D. Ravetta, 1995, pers. commun.).

Many agronomic requirements of grindelia and their effects on yield and resin production are still unknown. Results in Oregon suggest the possibility of achieving high biomass production without dramatically decreasing resin content. However, the plant's resinous nature could make harvest with standard cutting equipment difficult. Processing procedures also need to be developed. Product development using grindelia biomass as a raw material also is necessary.

Likely Commercial Production Areas

Grindelia is likely suited to areas that are dry and warm. Deep soils would decrease the need for irrigation, as would presence of some clay. It is frost tolerant and a perennial, making it useful in temperate climates. Likely areas would include western U.S., western Australia, northern Africa, southern Europe, central China, and southern areas in South America.

HESPERALOE

Interest in hesperaloe (*Hesperaloe funifera* [Koch] Trel., Agavaceae) arose from a recent study of the fiber properties of several agaves, including hesperaloe, that are native to the southwestern U.S. and northern

Mexico (McLaughlin and Schuck 1991). Hesperaloe leaves contained fiber cells that were both longer and narrower than those of sisal, and compared favorably with wood and non-wood fiber species for specialty paper applications. It had been noted that native peoples had cultivated *H. funifera* as a source of hard fibers (Trelease 1902; Dewey 1943) and several agavaceous species were already traded internationally due to their fiber quality, including sisal, henequen (*Agave fourcroydes* Lem.), and New Zealand flax (*Phormium tenax* J.R. & G. Forst.). It was assumed that species native to the arid southwest would be very drought tolerant. Finally, the value of pulp from hard fiber plants has typically been much higher than softwood pulps (Baker 1985; FAO 1994).

The first agronomic studies of *Hesperaloe funifera* were performed in Arizona. In addition to growth and yield data, preliminary information on water use, soil fertility requirements, regrowth yields and time requirements, and economic analysis were developed (McLaughlin 1993a, b, 1995). Determination that *H. funifera* was a crassulacean acid metabolism (CAM) plant further bolstered the observations of drought tolerance (Ravetta and McLaughlin 1993). Preliminary studies on floral biology have given insight into improved strategies for breeding and seed production (Ravetta and McLaughlin 1996).

Raw Material and Products

Hesperaloe produces hard fibers in the leaves that are very long, thin, and strong when compared with wood fibers, bast fibers, and even other hard fibers (McLaughlin and Schuck 1991; McLaughlin 1993a). Hard fibers have been successfully used to make thin, high value, specialty papers requiring strong tear resistance, such as are used in making currency, tea bags, filter papers, and specialty books (Clark 1965; Corradini 1979).

Competing Sources

Some of the world's important agricultural fibers are classified as bast fibers, obtained from phloem cells of dicot stems such as flax, kenaf, jute, ramie (*Boehmeria nivea* L.), and hemp. However, many important fibers are classified as hard fibers, obtained from the leaves of tropical monocots, such as abaca (*Musa textilis* Née), sisal, henequen, and New Zealand flax. Unlike other fibers, hesperaloe fiber is naturally white, thus eliminating the need for bleaching (Steve McLaughlin 1993, pers. commun.).

Crop Status

Hesperaloe research and development efforts in the U.S. currently include categories I–II. There is no commercial hectarage of hesperaloe. Until very recently all research plots have been small and located near Tucson, Arizona, but there are research plantings of several hectares at Maricopa, Arizona (McLaughlin 1995), and a small observation trial was planted near Medford, Oregon.

In Arizona, hesperaloe biomass accumulated to between 20 and 77 t/ha fresh weight after three years, depending on stand density (McLaughlin 1993a) and after five years of growth, yields increased to between 87 and 192 t/ha, again depending on stand density (McLaughlin 1993b). Under very good conditions, fresh weight yields in Arizona could reach about 200 t/ha (20 t/ha dry fiber) after five years (McLaughlin 1995). The growth rate accelerated after development of lateral rosettes (hesperaloe's growing point), which occurred mostly during the fourth year. Thus biomass accumulation after five years was greater than the sum of two harvests at three and five years (McLaughlin 1993b).

A small planting in southwest Oregon was started in 1992 and expanded in 1993 and 1994, to observe whether hesperaloe would survive cool, low light conditions during winter. Survival has been good for the 1993 and 1994 plants, with rodent damage during the dry summer causing more serious problems than winter weather. Growth, however, has been slow in Oregon. Yields after three years, estimated using the non-destructive technique developed by McLaughlin (1993a), were 9 t/ha (fresh weight) at densities of about 27,000 plant/ha, compared to 77 t/ha observed in Arizona. About 50% of the total biomass was added during the third year in the Arizona trial, while 90% was added during the third year in Oregon. Thus during the first and second years, survival is the main issue for hesperaloe in cooler, non-native areas. Whether or not the increased growth in subsequent years is enough to justify growing the crop will depend on relative value and return from other uses of the land.

Crop value for hesperaloe is very difficult to estimate until processing procedures are developed. However, preliminary market analysis suggested a potential processed value of $400–600/t dry fiber, corresponding to $40–60/t fresh weight (McLaughlin, 1993a). At $50/t fresh weight, a third year harvest in Arizona of 77 t/ha would be worth $3850/ha, and a fifth year harvest in Arizona of 200 t/ha would be worth $10,000 once processed. Processing costs could decrease the farm gate price to 50% of the processed fiber value, and until a market is developed, only crude estimates of crop price and value are possible.

Limitations

Although hesperaloe produces a high quality fiber, decortication of the leaves to produce dry, clean fiber has been difficult (McLaughlin 1995). There have been preliminary studies of hesperaloe genetics (Pinkava and Baker 1985; McLaughlin 1996), but the perennial habit of hesperaloe necessitates long-term efforts for breeding as well as agronomic studies. Its slow growth rate, especially in the first two years, makes weed control difficult on a large scale (S. McLaughlin 1995, pers. commun.). The range of adaptation is still unknown. The prospects and progress of hesperaloe domestication are discussed in more detail by Steven McLaughlin elsewhere in this volume.

Likely Commercial Production Areas

Hesperaloe funifera is found in the wild at low elevations in the east-central part of the Chihuahuan Desert in Coahuilla and Nuevo Leon, Mexico. *H. nocturna* Gentry, a species that produces similar fibers, is a native of the Sierra Madre Occidental area of northeastern Sonora, Mexico (McLaughlin 1993a; Ravetta and McLaughlin 1993). Based on research to date, hesperaloe production would seem limited to hot and dry locations near existing paper facilities, where its drought tolerance and high value fiber would make it competitive. The range of adaptation is still unknown, but its survival in southwest Oregon was somewhat surprising. Growth and yield is probably inadequate at higher latitudes, but there may be situations where it could be grown outside of Arizona and Mexico in locations less suited for conventional crops. Likely production areas include parts of the western U.S., Mexico, northern Africa, and western Australia. In any location, first year survival is the key to success, after which its perennial nature and regrowth from rosettes after cutting are advantageous.

HEMP

Hemp (*Cannabis sativa* L., Cannabaceae) has been grown in some parts of the world since ancient times, and it produces a high quality fiber (Dempsey 1975). Research programs have been most active recently in Europe, and production is a reality in eastern Europe and France (van Soest 1993). Restrictions on cultivation and research in the U.S. have been due to the potential for illegal drug production (marijuana). However, selections with low cannabinol levels have been developed (van Soest 1993), which may help decrease this problem. Progress in commercial production of hemp pulp for specialty papers in Europe is discussed by Anthony Capelle in more detail elsewhere in this volume.

SUNN HEMP

Sunn hemp (*Crotalaria juncea* L., Leguminosae) produces a bast fiber, similar to kenaf, that could be readily used in pulp and paper applications (Dempsey 1975; Cook and White 1995). Unlike kenaf, it is highly resistant to root-knot nematodes and thus can be grown in some areas where kenaf cannot (Cook and White 1995). Sunn hemp is fairly drought tolerant, can grow in marginal soils, and, being leguminous, has low nitrogen requirements. Sunn hemp is discussed by Charles Cook in more detail elsewhere in this volume.

HEAVY METAL HYPERACCUMULATORS

Various heavy metals may be present in soil naturally as well as due to mining or industrial operations, and areas of high metal concentrations occur throughout the U.S. (Holmgren et al. 1993). There are certain plants that "hyperaccumulate" various heavy metals such as Cd, Co, Cr, Cu, Mn, Ni, Pb, or Zn from the soil, taking them up in very large amounts that are unrelated to any known physiological need. Excellent summaries describing many hyperaccumulating species, representing several families, were written by Baker and Brooks (1989) and Baker et al. (1994a). A large number are in the Brassicaceae family, including *Thlaspi* and *Alyssum* species. Dr. Rufus Chaney of the USDA–ARS Environmental Chemistry Laboratory in Beltsville, MD, has examined numerous species regarding their heavy metal hyperaccumulation potential (Comis 1995). Some of these plants are very small and do not have any agronomic potential, except that this trait might be transferred to more agronomically suitable species. However, some of these plants are large enough that they aleady have potential as "biomining" crops, such as *Alyssum tenium* Halacsy and *Dichapetalum gelonioides* (Bedd.) Engl. subsp. *tuberculatum* Leenh. (Dichapetalaceae) for nickel, and possibly alpine pennycress (*Thlaspi caerulescens*, J. & C. Presl.) for Cd, Zn, and Pb (Homer et al. 1991; Baker et al. 1994a, b). Such crops would be valuable in reclaiming tailings from traditional mining operations, but also in mining the metal directly from naturally enriched soils, allowing the metal to be extracted from the crop biomass after cutting and ashing. This approach has the potential of allowing mining of heavy metals to occur in a more environmentally friendly way (Chaney 1983; Baker et al. 1994a; Comis 1995).

As an example, there are serpentine soils containing high concentrations of nickel in parts of southwestern Oregon and northwestern California, a region containing the only commercial Ni smelting operation in the U.S. (Helgerson and McNabb 1985). If one could grow a perennial bushy plant producing 40 t/ha biomass annually that contained 2% Ni, the metal yield would be 0.8 t/ha. At $7.00/kg, the refined value of the Ni would be $5600/ha (R. Chaney 1994, pers. commun.). Although processing and refining costs would be considerable (as they are in conventional mining operations), the potential to produce a crop in unproductive heavy metal-affected soil or even decontaminating mine spoils would provide a double benefit both to farmers and the environment.

OTHER POTENTIAL CROPS

There are many plants that naturally produce valuable and interesting compounds with potential industrial applications. An excellent overview of several of them was given by Robert Kleiman at the Association for the Advancement of Industrial Crops annual meeting in Catamarca, Argentina, during Sept., 1994. Those plants included: Stokes aster, *Dimorphotheca pluvialis* (L.) Moench; *Crepis biennis* and *C. alpina*; coriander (*Coriandrum sativum* L.); *Calendula officinalis* L.; and money plant (*Lunaria annua* L). Their development awaits the coordinated efforts of crop production, processing, product development, and marketing before domestication will become a reality.

REFERENCES

Anon. 1989. Vernonia—bursting with potential. Agr. Eng. 70(4):11–13.

Anon. 1993. Development of non-VOC diluent develops slowly. Chem. Eng. News 71(42):58. Oct. 18.

Anon. 1995. Month in statistics. U.S. paper and board production. Pulp & Paper, Oct. p. 11.

Arkcoll, D. 1988. Lauric oil resources. Econ. Bot. 42:195–205.

Arquette, J.G. and J.H. Brown. 1993. Development of a cosmetic grade oil from *Lesquerella fendleri* seed. p. 367–371. In: J. Janick and J.E. Simon (eds.), New crops. Wiley, New York.

Atsmon, D. 1989. Castor. p. 438–447. In: G. Robbelen, R.K. Downey, and A. Ashri (eds.), Oil crops of the world. McGraw–Hill, New York.

Bagby, M.O., C.R. Smith, Jr., T.K. Miwa, R.L. Lohmar, and I.A. Wolff. 1961. A unique fatty acid from *Limnanthes douglasii* seed oil: The C22 diene. J. Organic Chem. 26:1261–1265.

Bach, A.C. and V.K. Babayan. 1983. Medium-chain triglycerides: an update. Am. J. Clinical Nutr. 36:950–962.

Bailey, L.H. 1976. Hortus. (3rd ed.). Macmillan, New York.

Baker, A.J.M. and R.R. Brooks. 1989. Terrestrial higher plants which hyperaccumulate metallic elements—A review of their distribution, ecology, and phytochemistry. Biorecovery 1:81–126.

Baker, A.J.M., S.P. McGrath, C.M.D. Sidoli, and R.D. Reeves. 1994a. The possibility of *in situ* heavy metal decontamination of polluted soils using crops of metal-accumulating plants. Resources Conserv. Recycl. 11:41–49.

Baker, A.J.M., R.D. Reeves, and A.S.M. Hajar. 1994b. Heavy metal accumulation and tolerance in British populations of the metallophyte *Thlaspi caerulescens* J. & C. Presl. (Brassicaceae). New Phytol. 127:61–68.
Baker, D.M. 1985. Alternative uses for sisal fiber. p. 177–186. In: C. Cruz, L. del Castillo, M. Robert, and R.N. Ondara, (eds.), Biologia y aprovechamiento integral del henequen y otros Agaves. Centro de Investigacion Cientifica de Yucatan, A.C.
Barclay, A.S., H.S. Gentry, and Q. Jones. 1962. The search for new industrial crops II. *Lesquerella* (Cruciferae) as a source of new oilseeds. Econ. Bot. 16:95–100.
Bhangoo, M.S., H.S. Tehrani, and J. Henderson. 1986. Effect of planting date, nitrogen levels, row spacing, and plant population on kenaf performance in the San Joaquin Valley, California. Agron. J. 78:600–604.
Bharucha, K.E. and F.D. Gunstone. 1956. Vegetable oils. V. The component acids of *Cephalocroton cordofanus* (Muell.-Arg.) seed oil. J. Sci. Food Agr. 7:606–609.
Bohlmann, F., M. Ahmed, N. Borthakur, M. Wallmeyer, J. Jakupovic, R.M. King, and H. Robinson. 1982. Diterpenes related to grindelic acid and further constituents from *Grindelia* species. Phytochemistry 21:167–172.
Bosisio, M. 1989. Meadowfoam: Pretty flowers, pretty possibilities. Agr. Res. 37(2):10–11.
Brown, J. 1994. *Lesquerella fendleri*: Guidelines for cropping. Int. Flora Technologies, Ltd., Apache Junction, AZ. Oct.
Brown, M. 1995, (Aug. 22). Paper market monitor: Global industry report. S.G. Warburg Securities. Infotrac 3 General BusinessFile.
Brownback, S. and L. Glaser. 1992. Agriculture provides U.S. industry with diverse raw materials. p. 2–9. In: New crops, new uses, new markets. 1992 Yearb. Agr. U.S. Govt. Printing Office, Washington, DC.
Calvert, L.J. and J. Marks. 1995. Is flax back? Oregon's Agr. Progr. 41(4)–42(2), Fall, 1995. p. 16–21.
Campbell, A. 1981. Agronomic potential of stokes aster. p. 287–295. In: Pryde, E.H. et al. (ed.), New sources of fats and oils. Am. Oil Chem. Soc., Champaign, IL.
Campbell, T.A. and G.A. White. 1982. Population density and planting date effects on kenaf performance. Agron. J. 74:74–77.
Carlson, K.D. and S.P. Chang. 1985. Chemical epoxidation of a natural unsaturated epoxy seed oil from *Vernonia galamensis* and a look at epoxy oil markets. J. Am. Oil Chem. Soc. 62:934–939.
Carlson, K.D., S.A. Knapp, A.E. Thompson, J.H. Brown, and G.D. Jolliff. 1992. Nature's abundant variety: new oilseed crops on the horizon. p. 124–133. In: New crops, new uses, new markets. 1992 Yearb. Agr. U.S. Govt. Printing Office, Washington, DC.
Carlson, K.D., W.J. Schneider, S.P. Chang, and L.H. Princen. 1981. *Vernonia galamensis* seed oil: A new source for epoxy coatings. Am. Oil Chem. Soc. Monogr. 9:297–318.
Carlson, K.D., R.L. Cunningham, A. Shanahan, R. Kleiman, and M.O. Bagby. 1993. Products from new oilseed crops. Assoc. Adv. Ind. Crops, Annual Meeting Abstracts, p. 7. New Orleans, LA, Sept. 25–30.
Chaney, R.L. 1983. Plant uptake of inorganic waste constituents. p. 50–76. In: J.F. Parr, P.B. Marsh, and J.M. Kla (eds.), Land treatment of hazardous wastes. Noyes Data Corp., Park Ridge, NJ.
Chen, Lung-Hua and J.W. Pote. 1994. In-field separation of kenaf. p. 19–20. In: A summary of kenaf production and product development research 1989–1993. Mississippi Agr. Fors. Expt. Sta. Bul. 1011.
Clark, T.F. 1965. Plant fibers in the paper industry. Econ. Bot. 19:394–405.
Clark, T.F. and I.A. Wolff. 1969. A search for new fiber crops. XI. Compositional characteristics of Illinois kenaf at several population densities and maturities. Tappi 52:2111–2116.
Comis, D. 1995. Metal-scavenging plants to cleanse the soil. Agr. Res. 43(11):4–9.
Cook, A.A. 1971. Thar she grows!! In: The farm index, Oct. p. 4–6.
Cook, C.G. and B.A. Mullin. 1994. Growth response of kenaf cultivars in root-knot nematode/soil-borne fungi infested soil. Crop Sci. 34:1455–1457.

Cook, C.G. and G.A. White. 1995. *Crotalaria juncea*: A potential multi-purpose fiber crop. p. 43. Third National Symposium for New Crops Abstracts, Indianapolis, IN. Oct. 22–25.

Corradini, F.T. 1979. Companhia de Celulose da Bahia. p. 70–74. In: L.E. Hass (ed.), New pulps for the paper industry. Proc. Symp. New Pulps for the Paper Industry. Brussels, Belgium. May 1979. Miller Freeman Publ., San Francisco.

Dempsey, J.M. 1975. Fiber crops. Univ. Florida Press, Gainesville.

Dewey, L.H. 1943. Fiber production in the Western Hemisphere. USDA Misc. Publ. 518. U.S. Govt. Printing Office, Washington, DC.

Dierig, D.A. and A.E. Thompson. 1993. *Vernonia* and *Lesquerella* potential for commercialization. p. 362–371. In: J. Janick and J.E. Simon (eds.), New crops. Wiley, New York.

Dierig, D., T. Coffelt, D. Ray, H. Bhardwaj, M. Foster, R. Myers, R. Roseberg, R. Ayerza, and W. Coates. 1995. Performance of ten vernonia (*Vernonia galamensis*) lines in six environments. Assoc. Adv. Ind. Crops, Annual Meeting Abstracts, p. 21. Indianapolis, IN, Oct. 21–22.

Dunford, M.P. 1964. A cytogenic analysis of certain polyploids in *Grindelia* (Compositae). Am. J. Bot. 51:41–61.

Earle, F.R. 1970. Epoxy oils from plant seeds. J. Am. Oil Chem. Soc. 47:510–513.

Earle, F.R., E.H. Melvin, L.H. Mason, C.H. Van Etten, and I.A. Wolff. 1959. Search for new industrial oils. I. Selected oils from 24 plant families. J. Am. Oil Chem. Soc. 36:304–307.

Earle, F.R., C.A. Glass, C. Geisinger, I.A. Wolff, and Q. Jones. 1960. Search for new industrial oils. IV. J. Am. Oil Chem. Soc. 37:440–447.

Evans, D.W. and A.H. Hang. 1993. Kenaf in irrigated central Washington. p. 409–410. In: J. Janick and J.E. Simon (eds.), New crops. Wiley, New York.

FAO. 1994. 1993 production yearbook. United Nations, Rome.

Ferrigni, N.R. and J.L. McLaughlin. 1984. Use of potato disc and brine shrimp bioassays to detect activity and isolate piceatannol as the antileukemic principle from the seeds of *Euphorbia lagascae*. J. Nat. Prod. 47:347–352.

Foster, M. A., R. J. Roseberg, L. G. Kleine, and J. Moore. 1996. Herbicide tolerance and weed control strategies for lesquerella production. p. 328–330. In: L.H. Princen and C. Rossi (eds.), Proc. IX Int. Conf. Jojoba and its Uses and the III Int. Conf. New Crops and Products, Catamarca, Argentina, Sept. 25–30, 1994. Am. Oil Chem. Soc., Champaign, IL.

Francois, L.E., T.J. Donovan, and E.V. Maas. 1990. Salt tolerance of kenaf. p. 300–301. In: J. Janick and J.E. Simon (eds.), Advances in new crops. Timber Press, Portland, OR.

Gallagher, M. 1994a. TOFA market tightening raises end product prices. Chem. Mktg. Rptr. 245(21):10.

Gallagher, M. 1994b. CTO price increases in response to demand. Chem. Mktg. Rptr. 245(17):10.

Gentry, H.S. and A.S. Barclay. 1962. The search for new industrial crops III. Prospectus of *Lesquerella fendleri*. Econ. Bot. 16:206–211.

Gilbert, M.G. 1986. Notes on East African Vernonieae (Compositae). A revision of the *Vernonia galamensis* complex. Kew Bul. 41:19–35.

Glaser, L.K., J.C. Roetheli, A.E. Thompson, R.D. Brigham, and K.D. Carlson. 1992. Castor and lesquerella: Sources of hydroxy fatty acids. p. 111–117. In: New crops, new uses, new markets. 1992 Yearb. Agr. U.S. Govt. Printing Office, Washington, DC.

Graham, S.A. and R. Kleiman. 1985. Fatty acid composition in *Cuphea* seed oils from Brazil and Nicaragua. J. Am. Oil Chem. Soc. 62:81–82.

Graham, S.A. 1988. Revision of *Cuphea* section Heterodon (Lythraceae). Sys. Bot. Mono. 20:1–168.

Graham, S.A. 1989. Cuphea: A new plant source of medium-chain fatty acids. CRC Critical Rev. Food Sci. Nutr. 28:139–173.

Guerreiro, E., J. Kavka, J.R. Saad, M.A. Oriental, and O.S. Giordano. 1981. Acidos diterpenicos en *Grindelia pulchella* y *G. chiloensis* Cabr. Rev. Latinoam. Quim. 12:77–81.

Gunstone, F.D. 1954. Fatty acids. Part II. The nature of the oxygenated acid present in *Vernonia anthelmintica* (Willd.) seed oil. J. Chem. Soc. (London). May: 1611–1616.

Gunstone, F.D. and P.J. Sykes. 1961. Vegetable oils. IX. Application of reversed-phase chromatography to the analysis of seed oils. J. Sci. Food Agr. 12:115–123.

Hagler, R.W. 1995. Relative fiber scarcity may up prices, redefine markets. Pulp Paper Sept., p. 157–161.

Helgerson, O.T. and D.H. McNabb. 1985. Nickel mining and reclamation in the Klamath-Siskiyou Mountains. p. 263–267. In: Proc. Soc. Am. Foresters, National Convention, Fort Collins, CO, July 28–31, 1985. Soc. Am. For., Bethesda, MD.

Higgins, J.J. and G.A. White. 1970. Effects of plant population and harvest date on stem yield and growth components of kenaf in Maryland. Agron. J. 62:667–668.

Hirsinger, F. 1985. Agronomic potential and seed composition of cuphea, an annual crop for lauric and capric seed oils. J. Am. Oil Chem. Soc. 62:76–80.

Hoffmann, J.J. 1985. Resinous plants as an economic alternative to bioenergy plantations. Energy 10:1139–1143.

Hoffmann, J.J., B.E. Kingsolver, S.P. McLaughlin, and B.N. Timmermann. 1984. Production of resins by arid-adapted astereae, p. 251–271. In: B.N. Timmermann, C. Stellink, and F.A. Loewus (eds.), Phytochemical adaptations to stress. Plenum, New York.

Hoffmann, J.J. and S.P. McLaughlin. 1986. *Grindelia camporum*: Potential cash crop for the arid southwest. Econ. Bot. 40:162–169.

Holmgren, G.G.S., M.W. Meyer, R.L. Chaney, and R.B. Daniels. 1993. Cadmium, lead, zinc, copper, and nickel in agricultural soils of the United States of America. J. Environ. Qual. 22:335–348.

Homer, F.A., R.D. Reeves, R.P. Brooks, and A.J.M. Baker. 1991. Characterization of the nickel-rich extract from the nickel hyperaccumulator *Dichapetalum gelonioides*. Phytochem. 30:2141–2145.

Hurst, W.M., E.G. Nelson, J.E. Harmond, L.M. Klein, and D.W. Fishler. 1953. The fiber flax industry in Oregon. Oregon Agr. Expt. Sta. Bul. 531.

Jolliff, G.D., I.J. Tinsley, W. Calhoun, and J.M. Crane. 1981. Meadowfoam (*Limnanthes alba*): Its research and development as a potential new oilseed crop for the Willamette Valley of Oregon. Oregon Agr. Expt. Sta. Bul. 648.

Killy, R. 1994. Better days. Chem. Mktg. Reptr. 246(13):SR8–SR9. Sept. 26.

Kleiman, R. 1990. Chemistry of new industrial oilseed crops. p. 196–203. In: J. Janick and J.E. Simon (eds.), Advances in new crops. Timber Press, Portland, OR.

Kleiman, R., G.F. Spencer, F.R. Earle, and H.J. Nieschlag. 1972. Tetra-acid triglycerides containing a new hydroxy eicosadienoyl moiety in *Lesquerella auriculata* seed oil. Lipids 7:660–665.

Kleiman, R., C.R. Smith, and S.G. Yates. 1965. Search for new industrial oils. XII. Fifty-eight Euphorbiaceae oils, including one rich in vernolic acid. J. Am. Oil Chem. Soc. 42:169–172.

Knapp, S.J. 1990. New temperate oilseed crops. p. 203–210. In: J. Janick and J.E. Simon (eds.), Advances in new crops. Timber Press, Portland, OR.

Knapp, S.J. 1992. Modifying the seed oils of *Cuphea*—a new commercial source of caprylic, capric, lauric, and myristic acid. Am. Oil Chem, Soc. Monogr. Am. Oil Chem. Soc., Champaign, IL.

Knapp, S.J. 1993. Breakthroughs towards the domestication of *Cuphea*, p. 372–379. In: J. Janick and J.E. Simon (eds.), New crops. Wiley, New York.

Knapp, S.J., J.M. Crane, R.J. Roseberg, M.B. Slabaugh, D.T. Ehrensing, and J.M. Leonard. 1995. The domestication and commercialization of new industrial oilseed crops: A critical crossroads. p. 50–51. Third National Symposium for New Crops Abstracts, Indianapolis, IN. Oct. 22–25.

Knight-Ridder Financial, Commodity Research Bureau. 1995. The CRB commodity yearbook. Wiley, New York.

Kugler, D.E. 1988. Kenaf newsprint: Realizing commercialization of a new crop after four decades of research and development. USDA-Cooperative State Research Service, Washington, D.C.

Lauer, J.G. 1990. Kenaf production potential in northwest Wyoming. Wyoming Agr. Expt. Sta. Bul. B-932. Feb.

Litchfield, C., E. Miller, R.D. Harlow, and R. Reiser. 1967. The triglyceride composition of 17 seed fats rich in octanoic, decanoic, or lauric acid. Lipids 2:345–350.

Mackie, A.B. and S.D. Calhoun. 1991. World oilseed and products trade, 1962–1988. Agr. and Trade Div., Econ. Res. Serv., USDA. Statistical Bul. 819.

McLaughlin, S.P. 1985. Economic prospects for new crops in the southwestern United States. Econ. Bot. 39:473–481.

McLaughlin, S.P. 1986a. Heritabilities of traits determining resin yield in gumweed. J. Hered. 77:368–370.

McLaughlin, S.P. 1986b. Mass selection for increased resin yield in *Grindelia camporum* (Compositae). Econ. Bot. 40:155–161.

McLaughlin, S.P. 1993a. Development of *Hesperaloe* species (Agavaceae) as new fiber crops. p. 435–442. In: J. Janick and J.E. Simon (eds.), New crops. Wiley, New York.

McLaughlin, S.P. 1993b. Biomass production in *Hesperaloe funifera*: A five-year field plot study. Assoc. Adv. Ind. Crops, Annual Meeting Abstracts, p. 30. New Orleans, LA. Sept 25–30.

McLaughlin, S.P. 1995. Domestication of *Hesperaloe*: Prospects, problems, and progress. p. 43. Third National Symposium for New Crops Abstracts, Indianapolis, IN. Oct. 22–25.

McLaughlin, S.P. 1996. Intraspecific variability in agronomic traits and fiber properties of *Hesperaloe funifera*, p. 318–323. In: L.H. Princen and C. Rossi (eds.), Proc. IX Int. Conf. Jojoba and its Uses and the III Int. Conf. New Crops and Products, Catamarca, Argentina, Sept. 25–30, 1994. Am. Oil Chem. Soc., Champaign, IL.

McLaughlin, S.A. and J.J. Hoffmann. 1982. Survey of biocrude-producing plants from the Southwest. Econ. Bot. 36:323–339.

McLaughlin, S.P. and J.D. Linker. 1987. Agronomic studies on gumweed: Seed germination, planting density, planting dates, and biomass and resin production. Field Crops Res. 15:357–367.

McLaughlin, S.P. and S.M. Schuck. 1991. Fiber properties of several species of agavaceae from the southwestern United States and northern Mexico. Econ. Bot. 45:480–486.

Mikolajczak, K.L., F.R. Earle, and I.A. Wolff. 1962. Search for new industrial oils. VI. Seed oils of the genus *Lesquerella*. J. Am. Oil Chem. Soc. 39:78–80.

Miller, R.W., M.E. Daxenbichler, F.R. Earle, and H.S. Gentry. 1964. Search for new industrial oils. VIII. The genus *Limnanthes*. J. Am. Oil Chem. Soc. 41:167–169.

Miller, R.W., F.R. Earle, and I.A. Wolff. 1964. Search for new industrial oils. IX. *Cuphea*, a versatile source of fatty acids. J. Am. Oil Chem. Soc. 41:279–280.

Miwa, T.K. 1972. Gas chromatograms of synthetic liquid waxes prepared from seed triglycerides of *Limnanthes*, *Crambe*, and *Lunaria*. J. Am. Oil Chem. Soc. 49:673–674.

Miwa, T.K., K.L. Mikolajczak, F.R. Earle, and I.A. Wolff. 1960. Gas chromatographic characterization of fatty acids. Identification constants for mono- and dicarboxylic methyl esters. Analyt. Chem. 32:1739–1742.

Miwa, T.K. and I.A. Wolff. 1962. Fatty acids, fatty alcohols, and wax esters from *Limnanthes douglasii* (meadowfoam) seed oil. J. Am. Oil Chem. Soc. 39:320–322.

Nieschlag, H.J., G.F. Spencer, R.V. Madrigal, and J.A. Rothfus. 1977. Synthetic wax esters and diesters from crambe and *Limnanthes* seed oils. Ind. Eng. Chem., Prod. Res. Dev. 16:202–207.

Pascual, M.J. and E. Correal. 1992. Mutation studies of an oilseed spurge rich in vernolic acid. Crop Sci. 32:95–98.

Pascual-Villalobos, M.J., G.Robbelen, and E. Correal. 1994. Production and evaluation of indehiscent mutant genotypes in *Euphorbia lagascae*. Ind. Crops Prod. 3:129–143.

Pascual-Villalobos, M.J. 1996. Mutagenesis in *Euphorbia lagascae*: Induction of novel variability for breeding purposes. p. 393–397. In: L.H. Princen and C. Rossi (eds.), Proc. IX Int. Conf. Jojoba and its Uses and the III Int. Conf. New Crops and Products, Catamarca, Argentina, Sept. 25–30, 1994. Am. Oil Chem. Soc., Champaign, IL.

Perdue, R.E., K.D. Carlson, and M.G. Gilbert. 1986. *Vernonia galamensis*, potential new crop source of epoxy acid. Econ. Bot. 40:54–68.

Perdue, R.E., Jr. 1988. Systematic botany in the development of *Vernonia galamensis* as a new industrial oilseed crop for the semi-arid tropics. Acta Univ. Upsula, Symb. Bot. Ups. 28:125–135.

Phatak, S.C., C.A. Jaworski, and R.E. Perdue. 1989. Response of *Vernonia galamensis* to photoperiod. HortScience 24:746.

Pinkava, D.J. and M.A. Baker. 1985. Chromosome and hybridization studies of agaves. Desert Plants 7(2):93–100

Plattner, R.D., K. Wade, and R. Kleiman. 1978. Direct analysis of trivernolin by high-performance liquid chromatography. J. Am. Oil Chem. Soc. 55:381–382.

Princen, L.H. 1979. New crop developments for industrial oils. J. Am. Oil Chem. Soc. 56:845–848.

Purdy, R.H. and C.D. Craig. 1987. Meadowfoam: New source of long-chain fatty acids. J. Am. Oil Chem. Soc. 64:1493–1498.

Ravetta, D.A. and S.P. McLaughlin. 1993. Photosynthetic pathways of *Hesperaloe funifera* and *H. nocturna* (Agavaceae): Novel sources of specialty fibers. Am. J. Bot. 80:524–532.

Ravetta, D.A., A. Anouti, and S.P. McLaughlin. 1995. Evaluations of eight *Grindelia* accessions under field conditions in Arizona. Assoc. Adv. Ind. Crops, Annual Meeting Abstracts, p. 27. Indianapolis, IN. Oct. 21–22.

Ravetta, D.A. and S.P. McLaughlin. 1996. Floral biology and seed production in *Hesperaloe*: Implications for crop improvement and commercialization. p. 311–317. In: L.H. Princen and C. Rossi (eds.), Proc. IX Int. Conf. Jojoba and its Uses and the III Int. Conf. New Crops and Products, Catamarca, Argentina, Sept. 25–30, 1994. Am. Oil Chem. Soc., Champaign, IL.

Reisch, M.S. 1989. (Oct. 30). Demand puts paint sales at record levels (product report). Chemical and engineering news. 67(44):29–60.

Riser, G.R., J.J. Hunter, J.S. Ard, and L.P. Witnauer. 1962. *Vernonia anthelmintica* Willd. seed oil and salts of vernolic acid as stabilizers for plasticized poly (vinyl chloride). J. Am. Oil Chem. Soc. 39:266–268.

Roetheli, J.C., L.K. Glaser, and R.D. Brigham. 1991. Castor: Assessing the feasibility of U.S. production. USDA, Coop. State Res. Service, April.

Roseberg, R.J. 1993. Cultural practices for lesquerella production. J. Am. Oil Chem. Soc. 70:1241–1244.

Roseberg, R.J. 1996a. Herbicide tolerance and weed control strategies for lesquerella production. Ind. Crops Prod. (in press).

Roseberg, R.J. 1996b. Herbicide tolerance by vernonia grown in the temperate zone. Ind. Crops Prod. (in press).

Santos W. 1994. Tall oil rosin hikes spur price movement for resins. Chem. Mktg Rptr. 246(26):10. Dec. 26.

Scott, A.W., J.C. Garza, and T.G. Teague. 1989. Plant population density and dates of harvest on kenaf fiber production in the lower Rio Grande Valley of Texas. Proc. Assoc. Adv. Indust. Crops Annual Meeting, Peoria, IL, Oct. 5, 1989. p. 27–33.

Scott, A.W. and C. S. Taylor. 1990. Economics of kenaf production in the lower Rio Grande Valley of Texas. p. 292–297. In: J. Janick and J.E. Simon (eds.), Advances in new crops. Timber Press, Portland, OR.

Smith, C.E., Jr. 1971. Observations on Stengeloid species of *Vernonia*. Agr. Handb. 396. Agr. Res. Serv., USDA, Washington, DC.

Smith, C.R., Jr., M.O. Bagby, T.K. Miwa, R.L. Lohmar, and I.A. Wolff. 1960. Unique fatty acids from *Limnanthes douglasii* seed oil: The C20- and C22- monoenes. J. Org. Chem. 25:1770–1774.

Smith, C.R., K.F. Koch, and I.A. Wolff. 1959. Isolation of vernolic acid from *Vernonia anthelmintica* oil. J. Am. Oil Chem. Soc. 36:219–220.

Smith, C.R., Jr., T.L. Wilson, T.K. Miwa, H. Zobel, R.L. Lohmar, and I.A. Wolff. 1961. Lesquerolic acid. A new hydroxy acid from *Lesquerella* seed oil. J. Org. Chem. 26:2903–2905.

Smith, C.R., Jr., T.L. Wilson, R.B. Bates, and C.R. Scholfield. 1962. Densipolic acid: A unique hydroxydienoid acid from *Lesquerella densipila* seed oil. J. Org. Chem. 27:3112–3117.

Stone, R. 1995. Market pulp: What's different this time? Pulp Paper, Oct. p. 168.

Tallent, W.H., D.G. Cope, J.W. Hagemann, F.R. Earle, and I.A. Wolff. 1966. Identification and distribution of epoxyacyl groups in new, natural epoxy oils. Lipids 1:335–340.

Taylor, C.S. 1993. Kenaf: An emerging new crop industry. p. 402–407. In: J. Janick and J.E. Simon (eds.), New crops. Wiley, New York.

Taylor, C.S. and D.E. Kugler. 1992. Kenaf: Annual fiber crop products generate a growing response from industry. p. 92–98. In: New crops, new uses, new markets. 1992 Yearb. Agr. U.S. Govt. Printing Office, Washington, DC.

Thames S.F., M.O. Bautista, L.H. Edwards, and A.M. King. 1993a. Lesquerella oil and its coproducts in commercial applications. Assoc. Adv. Ind. Crops, Annual Meeting Abstracts, p. 13. New Orleans, LA, Sept. 25–30.

Thames S.F., M.D. Watson, and M.O. Bautista. 1993b. Products from new oilseed crops. Assoc. Adv. Ind. Crops, Annual Meeting Abstracts, p. 7. New Orleans, LA, Sept. 25–30.

Thames, S.F., M.D. Wang, and H. Yu. 1996. Dehydrated lesquerella oil in melamine-alkyd coatings. p. 346–349. In: L.H. Princen and C. Rossi (eds.), Proc. IX Int. Conf. Jojoba and its Uses and the III Int. Conf. New Crops and Products, Catamarca, Argentina, Sept. 25–30, 1994. Am. Oil Chem. Soc., Champaign, IL.

Thompson, A.E. 1984. Cuphea—a potential new crop. HortScience 19:352–354.

Thompson, A.E. 1988. Lesquerella—a potential new crop for arid lands. p. 1311–1320. In: E. Whitehead, C.F. Hutchinson, B.N. Timmermann, and R.G. Varady (eds.), Arid lands today and tomorrow. Proc. Int. Res. Dev. Conf., Tucson, AZ, Oct. 20–25, 1985. Westview Press, Boulder, CO.

Thompson, A.E. 1990. Arid-land industrial crops. p. 232–241. In: J. Janick and J.E. Simon (eds.), Advances in new crops. Timber Press, Portland, OR.

Thompson, A.E. and D.A. Dierig. 1988. Lesquerella—a new arid land industrial oil seed crop. El Guayulero 10(1&2):16–18.

Thompson, A.E., D.A. Dierig, and E.R. Johnson. 1989. Yield potential of *Lesquerella fendleri* (Gray) Wats., a new desert plant resource for hydroxy fatty acids. J. Arid Environ. 16:331–336.

Timmermann, B.N., D.J.Luzbetak, J.J. Hoffmann, S.D. Jolad, K.H. Schram, R.B. Bates, and R.E. Klenck. 1983. Grindelane diterpenoids from *Grindelia camporum* and *Chrysothamnus paniculatus*. Phytochemistry 22:523–525.

Trelease, W. 1902. The Yucceae. Missouri Bot. Gard. Annu. Rep. 13:27–133.

U.S. Department of Agriculture. 1990. Kenaf research, development and commercialization. Proc. Oct. 5 Session, Assoc. Adv. Ind. Crops Annu. Conf., Peoria, IL. Oct. 2–6, 1989. USDA–CSRS, Office of Agr. Industrial Materials, Washington, DC.

U.S. Department of Agriculture. 1993. Kenaf. p. 46–49. In: J. Harsch, (ed.), New industrial uses, new markets for U.S. Crops. USDA–CSRS, Office of Agr. Materials, Washington, DC.

U.S. Department of Commerce. 1993. Census of agriculture. Oregon state and county data. Vol.1, Part 37. U.S. Govt. Printing Office, Washington, DC.

van Gelder, W.M.J., F.P. Cuperus, J.T.P. Derksen, B.G. Muuse, and J.E.G. van Dam. 1993. Characterization and processing research for increased industrial applicability of new and traditional crops: A European perspective. p. 38–45. In: J. Janick and J.E. Simon (eds.), New crops. Wiley, New York.

van Soest, L.J.M. 1993. New crop development in Europe. p. 30–38. In: J. Janick and J.E. Simon (eds.), New crops. Wiley, New York.

Vogel, R., M.J. Pascual-Villalobos, and G. Robbelen. 1993. Seed oils for new chemical applications. Agnew. Bot. 67:31–41.

White, G.A., D.G. Cummins, E.L. Whiteley, W.T. Fike, J.K. Greig, J.A. Martin, G.B. Killinger, J.J. Higgins, and T.F. Clark. 1970. Cultural and harvesting methods for kenaf. USDA Prod. Res. Rep. 113. Washington, DC.

White, G.A., W.C. Adamson, and J.J. Higgins. 1971. Effect of population levels on growth factors in kenaf varieties. Agron. J. 63:233–235.

Wilson, T.L., T.K. Miwa, and C.R. Smith, Jr. 1960. *Cuphea llavea* seed oil, a good source of capric acid. J. Am. Oil Chem. Soc. 37:675–676.

Wood, I.M., R.C. Muchow, and D. Ratcliff. 1983. Effect of sowing date on the growth and yield of kenaf (*Hibiscus cannabinus*) grown under irrigation in tropical Australia. II. Stem production. Field Crops Res. 7:91–102.

Wolf, R.B., S.A. Graham, and R. Kleiman. 1983. Fatty acid composition of Cuphea seed oils. J. Am. Oil Chem. Soc. 60:103–104.

Wolff, I.A. 1964. New crop prospects: Paper from annual plant sources. Chemurg. Dig. 22(3):3–4.

Non-woody Fibers and the Future of Rural Economies

David Morris

Recent economic, technological and political changes are opening up vast new markets for agricultural waste fibers. These markets could provide the foundation for a re-industrialization of many rural economies.

Huge amounts of waste fibers exist (Fig. 1). The Institute for Local Self-Reliance (ILSR) estimates that over 150 million tonnes of economically and environmentally recoverable waste fibers are available each year. This is as much fiber as is contained in wood used for all purposes.

Until recently, these fibers have been thrown away or burned. But disposal costs are rising. State bans on burning rice or wheat straw after harvest, for example, have spurred farmers to seek alternative uses of straw. Wood fiber prices have increased since 1991, a result of increased worldwide demand and a reduced cutting on some national forests resulting from environmental restrictions (Fig. 2).

The result of these combined market and regulatory forces has been to direct entrepreneurial energies into converting waste fibers into pulp, paper, and construction materials. An ILSR study, *A New Industry Emerges: Making Construction Products from Cellulosic Wastes*, estimates that in the construction materials market alone in 1994 about 28,000 t of agricultural waste fibers and urban waste wood were used to make construction materials. By the end of 1996 this could rise to over 450,000 t, and by 2000 to 1.4 million t. By the year 2000 as much as 630,000 t of agricultural waste fibers could be used for making construction materials. By that year the agriboard industry could encompass 15 plants with over 3,000 employees representing a cumulative investment of some $750 million. Sufficient raw materials exist in the United States to build more than 1000 plants (Table 1).

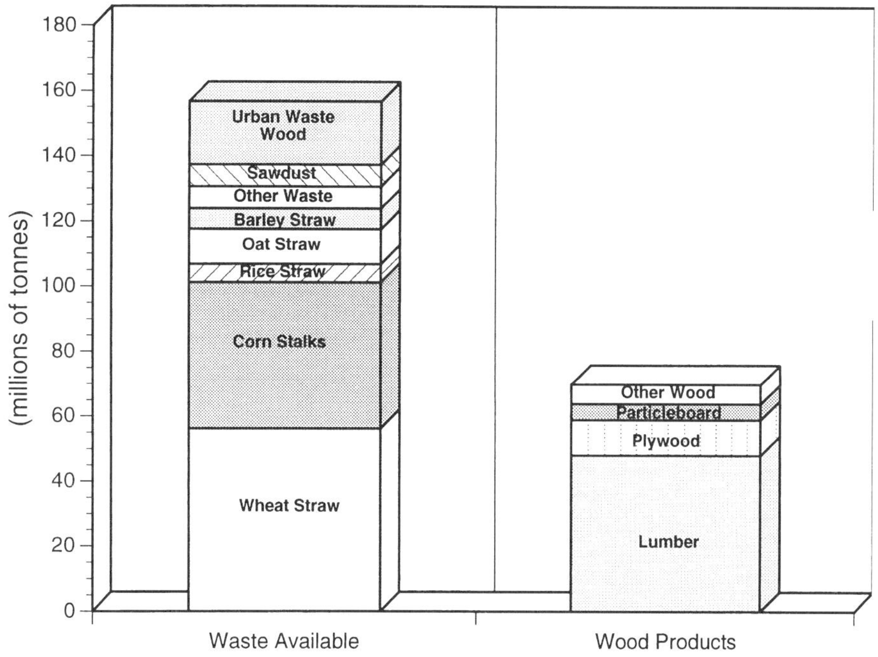

Fig. 1. Comparison of available wastes and production of wood products (1993). Estimates of waste availability indicate amounts that can be feasibly recovered without decreasing soil fertility (for straws) or are currently not utilized (for municipal wastes). "Other Waste" includes bagasse and nut shells, "Other Wood" includes oriented strand board, medium density fiberboard, hardboard, and insulation board. Sources: Morris and Ahmed 1992; Franklin Assoc. 1992; U.S. Dept. Commerce 1995; Chris Edwardson pers. commun.; Pettijohn and Lorenz 1995.

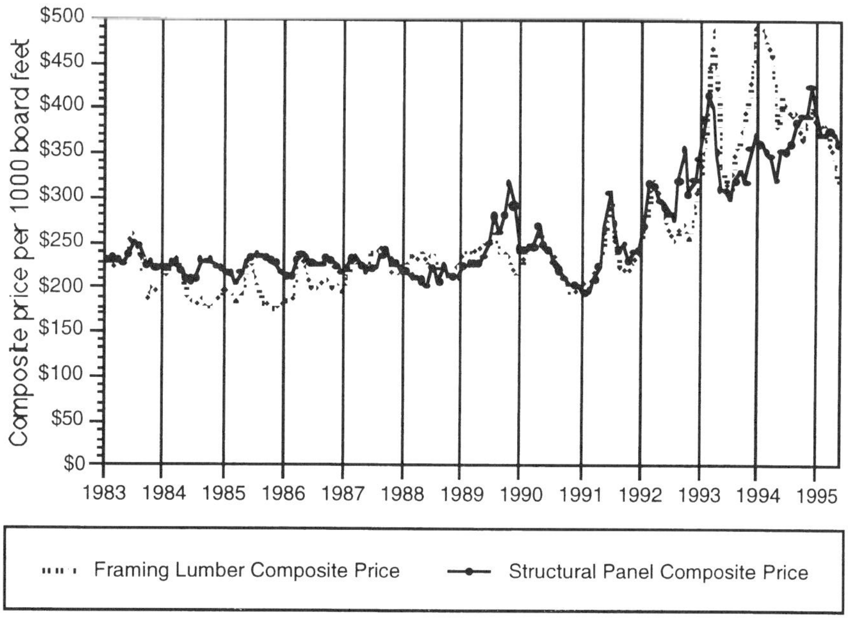

Fig. 2. Average wood prices, 1983–1995. One board foot = 1" × 1' × 1', standard unit of measurement for lumber. Source: Anon 1993, 1995.

Table 1. Quantity of cellulosic wastes used for construction materials (panel and composite products). Source: Morris and Ahmed 1992.

	Cellulosic wastes (t)		
Material	1994	1996 est.	2000 est.
Wheat straw	<9,000	229,050	630,000
Soy flour	<270	810	16,200
Recycled paper	70,200	76,500	108,900
Urban waste wood	18,900	237,150	777,150
Total	<98,370	543,510	1,532,250

Vegetable fiber based construction board plants are already in operation. Vegetable fiber based paper mills are in the planning stage, aside from a few specialized mills that have been operating for many years. About half a dozen such mills are on the drawing board, including a rice straw based pulp mill in Northern California, a wheat straw paper mill in Eastern Washington and a corn stalk based paper mill in Indiana.

Agricultural materials are bulky and expensive to transport. Therefore they often are processed near where the raw materials are located. That translates into well-paying jobs in rural areas. Moreover, agrifiber enterprises lend themselves to cooperative ownership. More than 4500 agricultural cooperatives currently exist with sales of over $80 billion. But until recently these cooperatives were used to reduce a farmer's input prices or increase his or her sale price. They rarely were involved in value-added processing.

In the last 15 years a new wave of cooperatively owned manufacturing facilities has emerged. Some two dozen already exist, from pasta and ethanol plants in the north to kenaf and pants factories in the south. The coming wave of vegetable fiber based construction materials and paper could lead to hundreds of locally owned and rooted community businesses.

The local economic impact of these plants could be profound. Consider the impact of one proposed particleboard facility on Pickert, North Dakota. The plant will create about 140 new jobs. Finley, the largest city in the area, has a population of about 800. The facility will generate almost $3 million in additional wages for the local economy. It can save the farmers an additional $1 to $2 million in disposal costs for their wheat straw. It will generate $1 million in additional income for area farmers through the sale of wheat straw. And finally, if the facility were to be cooperatively owned, the area's farmers could earn another $1 million in dividends per year.

Agricultural waste fibers are forming the basis for an emerging industry. This form of plant matter should be viewed as an important new crop in American agriculture.

REFERENCES

Franklin Associates, Ltd. 1992. Characterization of municipal solid waste in the United States, Prairie Villages, KS. Prepared for the U.S. Environmental Protection Agency.

Morris, D. and I. Ahmed. 1992. The carbohydrate economy: Making chemicals and industrial materials from plant matter. Institute for Local Self-Reliance,Washington, DC.

Pettijohn, D. and D. Lorenz. 1995. A new industry emerges: Making construction materials from cellulosic wastes. Institute for Local Self-Reliance, Washington, DC.

Anon. 1995. Random lengths yardstick. Random Lengths Publishing, Eugene, OR.

Anon. 1993. Random lengths 1993 factbook. Random Lengths Publishing, Eugene, OR.

U.S. Department of Commerce. 1995. 1994 Industrial outlook. Washington, DC.

Alternative Crops Research in Virginia

Harbans L. Bhardwaj, Andy Hankins, Tadesse Mebrahtu, Jimmy Mullins, Muddappa Rangappa,
Ozzie Abaye, and Gregory E. Welbaum

Virginia's land grant universities, Virginia State University (VSU), Petersburg, and Virginia Tech, Blacksburg, are both deeply involved in alternative crop research and development (Table 1). VSU conducts new crops research and extension activities to benefit small farmers in the state. Virginia Tech's program is geared toward large scale agriculture and more basic areas of research. In this presentation, we discuss new crops research activities that have occurred since the 1991 New Crop Symposium (Welbaum 1993).

ALTERNATIVE CROPS

Adzuki Bean

Adzuki bean (*Vigna angularis*) was grown in Blacksburg in 1992 and 1993. Seeds (Redwood Seed Co., Redwood City, California) were planted on June 20. The plants grew well vegetatively during the summer months but did not flower and set fruit until early Sept. Frost in mid-Oct. killed the plants before most seeds were fully mature. Adzuki is a cool season bean and summer temperatures were excessively high for reproductive growth. Fall production in parts of the southeastern U.S. may be possible.

Basil

An experiment with sweet basil (*Ocimum basilicum*) during 1994, indicated that yields of 11 lines were not different from each other 45 days after transplanting, but a significant difference existed for essential oil content. 'Genova', 'Cinnamon Basil', and 'Greek Miniature Scented Basil' had higher essential oil content than 'Lettuce Leaf', 'Purple', 'Opal', and 'Napolentano'.

Bitter Melon

Bitter melon or *foo gwa* (*Momordica charantia*) looks like a tapered, warted cucumber with waxy green skin, but cultivars with white skin are available. The requirements for bitter melon production are similar to pickling cucumbers. Bitter melon fruit production takes about 75 days from seed, requires well-drained soils, a constant supply of moisture, and bee pollination. The fruits turn brilliant orange when physiologically mature, but for commercial use they must be harvested when fully green or white as orange coloration reduces market value. The fruits grow rapidly during hot weather and, like pickling cucumbers, must be harvested daily to keep the fruits from becoming too large. The orange color can develop after harvest if fruits are stored for long periods, particularly at high temperatures. Fruits are sensitive to chilling injury and should not be stored below 13°C. Fruits will shrivel if stored under warm, dry conditions. The skin of the fruit is tender and can be easily damaged by abrasion. Trials of 'Spindle' (Takii Seed Co., Salinas, California) in Blacksburg in 1994 produced yields equivalent to 1840 kg/ha using black plastic mulch, drip irrigation, and production recommendations developed for pickling cucumber. In the Blacksburg trials, bitter melon appeared to be resistant to bacterial wilt and foliar diseases that affect other cucurbits. Bitter melon may also have potential as an ornamental plant when grown on a trellis because of its large yellow flowers, attractive green foliage, and brightly colored mature fruits and seeds.

Borage

Borage (*Borago officinalis*) is an oilseed crop that is also widely used as an herb. Borage oil is high in γ-linolenic acid which is used as a nutritional supplement in Europe and Japan. In 1992, seeds (John K. Kings Sons Limited, Coggeshall Colchester Essex, Great Britain) were sown on June 13 in four rows 6 m long and 45 cm apart with an in-row spacing of 2.5 cm. Fertilizer was broadcast prior to planting at a rate of 560 kg of 10N–4P–8.3K per ha. The plot was weeded by hand throughout the season, and no insecticides were applied.

Table 1. Summary of alternative crops research at Virginia State University and Virginia Tech.

Common name	Scientific name	Comments
Adzuki bean	*Vigna angularis* Willd.	Potential new cash-crop
Apricot	*Prunus armeniaca* L.	Late frosts are a problem, produces a full crop about one year in seven in Virginia
Asian pear	*Pyrus pyrifolia* Burm. f.	Some cultivars successful
Basil	*Ocimum basilicum* L.	Profitable herb
Bitter melon	*Momordica charantia* L.	Potential specialty crop
Borage	*Borago officinalis* L.	Shatters at maturity
Bottle gourd	*Lagenaria siceraria* Mol.	Potential specialty crop
Brambles	*Rubus* spp.	Trellis blackberry production improves yield
Canola	*Brassica* L. spp.	Potential new cashcrop
Castor	*Ricinus communis* L.	Potential in Virginia and other southern states
Chickpea	*Cicer arietinum* L.	Viable crop when planted in March/April
Chinese cabbage	*Brassica campestris* L.	Potential hydroponically grown greenhouse crop
Cilantro	*Coriandrum sativum* L.	Profitable herb
Cotton	*Gossypium hirsutum* L.	Production increasing in southeast Virginia
Cut Flowers	Various	Fresh and dried, must be handled properly
Dill	*Anethum graveolens* L.	Profitable herb
Durum wheat	*Triticum turgidum* L.	Yields lower than for winter wheat
Elephant garlic	*Allium ampeloprasum* L.	A successful cash crop
Fuzzy gourd	*Benincasa hispida* Thunb.	Potential specialty crop
Ginseng	*Panax quinquefolius* L.	High value crop that is difficult to grow
Globe artichoke	*Cynara scolymus* L.	'Imperial Star' can be successfully grown as an annual from seed
Goldenseal	*Hydrastis canadensis* L.	Potential high value crop
Grapes	*Vitis* L. spp.	Successful table and wine grapes identified
Grain sorghum	*Sorghum bicolor* L.	Promising grain for drought prone areas
Holly	*Ilex* L. spp.	Cut foliage and berries are valuable as ornamentals
Kenaf	*Hibiscus cannabinus* L.	Potential summer forage in Virginia and adjoining states
Kiwifruit	*Actinidia arguta* Siebold Zucc	Successful when grown near the coast in extreme southeastern Virginia
Lesquerella	*Lesquerella fendleri* Gray	May not be a successful replacement for castor
Mungbean	*Vigna radiata* L.	Potential new legume in Virginia and other southern states
Matua	*Bromus willdenowii* Kunth.	Potential spring and fall forage
Meadowfoam	*Limnanthes* R. Br. spp.	Not well adapted
Onion	*Allium cepa* L.	Long and short day types are under evaluation
Pepino	*Solanum muricatum* Ait.	Unsuited, breeding needed
Pigeonpea	*Cajanus cajan* L.	Promising new legume
Persimmon	*Diospyros kaki* L.	Promising specialty tree fruit crop
Purslane	*Portulaca* L. spp.	High yielding vegetable providing essential nutrients for humans and animals
Quinoa	*Chenopodium quinoa* Willd.	Too hot for seed set in most years
Sweet sorghum	*Sorghum bicolor* L.	Processed into syrup
Soybean	*Glycine max* L.	Developing vegetable types for human consumption and agronomic cultivars for export
Switchgrass	*Panicum virgatum* L.	Promising forage and biomass crop but seed dormancy is a problem
Tobacco	*Nicotiana tabacum* L.	Genetically altering tobacco to produce beneficial compounds
Winter melon	*Benincasa hispida* Thunb.	Potential specialty crop
Vernonia	*Vernonia galamensis* Cass.	Developing nonshattering lines

The actual stand obtained was 12 plants/30 cm. The plot was irrigated as needed using an overhead sprinkler system. Whole plants were hand-harvested on Sept. 22, placed in plastic bags, and dried for one week at 35°C. The seed was mechanically cleaned before the final grain weight was determined. The yield obtained was equivalent to 540 kg/ha, but at least half of the seeds were lost due to shattering. Overall, borage was easy to grow, and no major disease or pest problems were observed. It is unlikely that borage will become an important oil seed crop until non-shattering cultivars are developed.

Bottle Gourd

Mature fruit of *Lagenaria siceraria* are used as containers and made into utensils in many regions of the world. The immature fruits (*mo kwa*) can be used as a vegetable. Plants are often grown on a support as an ornamental, and dried small fruits are used as decorations in eastern Asia. The plants are very productive, and cultural requirements are similar to winter squash and pumpkin. The fruits last indefinitely if dried properly. In trials in Blacksburg, bottle gourds were easy to grow using the same production guidlines developed for other types of gourds.

Canola

Research conducted during 1992/93, 1993/94, and 1994/95 at Orange, Petersburg, and Suffolk has established that canola (*Brassica* spp.) has tremendous potential as a new cash-crop for Virginia farmers. The seed yields per hectare varied from 1696 to 2895 kg (1992/93), 1629 to 3147 kg (1993/94), and 1054 to 2975 (1994/95). The ideal planting time has been identified to be late-Sept. or early-Oct. Approximately 112 to 168 kg/ha N is needed for canola production, but application time (fall, spring, or split-application) did not affect yields.

Castor

During 1994, a collection of 74 accessions of castor (*Ricinus communis*) was planted in observation rows to evaluate agronomic performance and oil characteristics. All accessions were vigorous and produced abundant seed. Castor appears to have potential for production in Virginia and other southeastern states.

Chickpea

Research conducted during 1993 indicated that chickpea (*Cicer arietinum*) planted in Mar.–Apr. can be a viable crop. The yield of spring-planted 'desi' cultivars varied from 876 to 1400 kg/ha as compared to 307 to 1082 kg/ha for 'kabuli' type cultivars during 1993. Desi types (Aztec, ICC4948, ICC10136, C235, ICCC4, NEC1163, Garnet, and PI12074) as a group out yielded kabuli types (UC8532, UC85150, UC27, UC15, UC8624, UC85183, UC5, SR20I, UC8554, Surutato 77, Surutato, and UC8536), with mean seed yields of 1153 and 719 kg/ha, respectively. The new crops program of Virginia State University has been cooperating with ICARDA (Alleppo, Syria) to evaluate chickpea germplasm for cold-tolerance and agronomic performance.

Cilantro

In an effort to develop sustainable production technology, the nitrogen requirements of cilantro (green stage of coriander, *Coriandrum sativum*) are being studied. Results from two experiments during 1992 and one experiment during 1993 with three cultivars indicated that increased N fertilization, generally, did not increase yield. Soil nitrogen (approximately 14 kg/ha) plus 100 kg/ha of applied N was adequate for optimum growth of cilantro. During early growth (45 days after planting) 'C1410' had the highest fresh yield of 2.8, 2.4, and 1.3 kg/m, respectively, for experiments planted June 28, 1992, Aug. 26, 1992, and April 20, 1993, but 'C18135' yielded more in later plantings. The average protein content in cilantro foliage was 23%. Neither N fertilizer rates nor cultivars differed for protein and essential oil content, but a significantly higher protein content was produced during earlier growth.

Chinese Winter Melon

Benincasa hispida (also called wax melon or *ton kwa*) matures in about 120 days from seed and has cultural requirements similar to winter squash. The fruit is very distinctive because it is hairy and covered with a thick layer of white wax at maturity. The flowers are bee-pollinated, and the vines may grow to be 6 m long, so wide spacing is recommended. Both round- and oblong-fruited cultivars are available. In Blacksburg, both round and oblong cultivars have been successfully grown using production recommendations developed for winter squash. Winter melons have been stored as long as 6 months at 16°C and 75% relative humidity. Cultivars bred to be harvested when immature (called fuzzy gourds, Chinese squash, or *mo kwa*), have also been successfully grown. Fuzzy gourds are harvested when 15 cm long and are used like summer squash. Since the fruits are harvested when immature, they have a limited shelf life.

Cotton

Cotton (*Gossypium hirsutum*) has returned to prominence in Virginia. From a low of 40 ha in 1978, cotton crop area increased to over 5263 ha in 1993. 'Stoneville 453' is currently one of the leading cultivars. Current research is examining effects of cover crops and the use of conservation tillage techniques to improve the sustainablity of production.

Cut Flowers

Production and marketing of cut flowers has been evaluated in Virginia since 1988. The primary goal of this research is to discover which species of cut flowers might be successfully grown in various soil types and in various micro-climates in Virginia. A secondary goal, of equal importance, is to discover market demand and market specifications for any species that can be grown. Studies have included cultivar trials, fertilization, weed control through the use of woven plastic "weed barriers," disease control, use of plant growth regulators, and effective methods of drying everlasting flowers. Cut flowers have been test marketed to wholesale brokers, independent florist shops, gift shops, craft shops, and directly to the public at fairs and festivals. Approximately 90 Virginia farmers grow cut flowers as a supplemental source of income. Approximately 10 ha of cut flowers were harvested in 1995 generating gross income of approximately $300,000. The most successful species are: statice (*Limonium* spp.), gypsophila, gomphrena, and strawflowers (*Helichrysum* spp.).

Dill

An experiment to determine nitrogen fertilizer requirements of dill (*Anethum graveolens*) with two cultivars (Bouquet and Ducat) indicated that, increased nitrogen fertilization generally did not increase yield. The optimum rate of N fertilizer for maximum foliage yield was 25–50 kg/ha. The differences between dill cultivars for fresh yield were not significant. The average protein content in dill foliage was 30%. The contents of protein and essential oils in dill foliage were unaffected by P and K applications.

Durum Wheat

Durum wheat (*Triticum turgidum*) comprises approximately 8% of the worldwide wheat production. Durum wheat is produced in the same geographical area as hard red spring wheat and has a higher price. Most of the durum wheat produced in the United States is grown in North Dakota (76%), Montana, South Dakota, and Minnesota. Fifty durum wheat cultivars were grown at four Virginia locations to determine the feasibility of producing durum wheat in Virginia where soft red winter wheat is grown, but spring types did not reliably survive winters. Further research will focus on winter types of durum wheat. The yields of soft red winter wheat were at least 1.1 t/ha greater at all locations in Virginia.

Elephant Garlic

Applied research in production and marketing of elephant garlic (*Allium ampeloprasum*) has been conducted in Virginia since 1987. Control of one especially prevalent disease, yellow dwarf virus, has been accomplished through careful removal of infected plants during the growing season. Other research has included cultivar trials, fertilization studies, trials using black plastic mulch for weed control, and studies on curing methods. We have investigated marketing cases of single, large bulbs packaged in plastic-mesh bags to supermarkets. Approximately 130 Virginia farmers grow elephant garlic as a supplemental source of income. Approximately 20 ha of elephant garlic were harvested on small farms in Virginia in 1995 generating gross income of approximately $800,000.

Ginseng

Production research plots of American ginseng (*Panax quinquefolius*) have been established in woodland locations in ten counties in Virginia. The primary goal of this research is to determine plant survivability and yield of dried roots after seven years of growth in naturalized plant populations grown from seed. Other goals of this research include growth evaluation of American ginseng in various soil types and in various micro-climates of the piedmont region of Virginia, and to determine whether ginseng roots that have grown in a naturalized environment, without benefit of fertilization, weed control, disease control or irrigation, might be sold as wild ginseng. In 40 plant beds containing 500 ginseng seeds each planted in 10 piedmont counties in Nov. 1993, the average plant population per bed in 1995 was 156 plants. These plant populations are naturalized which means that no fertilization, irrigation, disease control, weed control, or other interventions have been made.

Globe Artichoke

Cynara scolymus is usually propagated vegetatively because plants grown from seed lack uniformity. Furthermore, in much of the United States only a small percentage of plants grown from seed flower during the first season due to insufficient chilling for vernalization. Artichokes cannot be reliably grown as perennials without winter protection where temperatures are consistently below –10°C. 'Imperial Star' and 'Talpiot', which reportedly produce uniform plants from seed and a high percentage of flower heads (capitulum) the first year with minimal chilling, were compared with the standard seed-propagated 'Green Globe Improved' and 'Grande Buerre'. Plants of each cultivar were tested over a three year period in Blacksburg or for one year in three other locations. Essentially all 'Imperial Star' and 'Green Globe Improved' plants flowered after receiving 1356 h of chilling at less than 10°C. With 205 h of chilling, 83% of 'Imperial Star' plants flowered compared to 25% for 'Green Globe Improved'. No 'Talpiot' or 'Green Globe Improved' plants flowered after receiving as much as 528 h of chilling. In the mountains of western Virginia, only 'Imperial Star' plants established in the field in early May received sufficient chilling to produce flower heads during the late summer and early fall. June transplants did not flower because sufficient chilling was not obtained for vernalization. In warmer areas of central and eastern Virginia, fall establishment for spring harvest may yield a higher percentage of flowering plants compared to spring planting and summer harvest.

Goldenseal

Goldenseal (*Hydrastis canadensis*) is being evaluated as an alternative medicinal crop. The focus of this research is development of production technology and education of growers regarding the tremendous income potential of this crop.

Kenaf

In order to sustain kenaf (*Hibiscus cannabinus*) until its use as a domestic source of pulp for newsprint becomes established and to educate Virginia farmers, kenaf is under evaluation as a summer forage. During 1993, the dry matter yield of 'Everglades 71' ranged from 7 t/ha at 70 days after planting to 16 t/ha 133 days after planting. The crude protein content declined with age but compared favorably with that of alfalfa during earlier stages of growth. During earlier stages, crude protein content ranged from 17% to 21% whereas at 140 days after planting the crude protein content ranged from 10% to 15%.

During 1994, investigations were undertaken to study the effects of multiple harvests on kenaf yield and quality. It was hypothesized that regrowth, following a first harvest around 80 days after planting (DAP), will be more nutritious as compared to unharvested plants of a similar age. The yields of seven kenaf cultivars

(Cubano, Everglades 41, Everglades 71, Guatemala 48, Indian, Tainung#1, and Tainung#2) were evaluated following four treatments (T1 = harvest at 85 DAP + final harvest of regrowth, T2 = harvest at 92 DAP + final harvest of regrowth, T3 = harvest at 99 DAP + final harvest of regrowth, and T4 = final harvest after the plants had died following a hard-freeze). The regrowth was sampled for crude protein and fiber content approximately 92, 100, and 107 DAP following first harvest, respectively. The dry matter yields per hectare, averaged over cultivars, ranged from 12,506 kg for T1 to 15,688 kg for T2. T1 had significantly lower yield than the other three treatments. The dry matter yield per hectare from T4 (14905 kg) was not different from that of T2 or T3 (14433 kg). The content of acid detergent fiber (ADF) was significantly less in regrowth as compared to first harvest with T1, T2, and T3. The ADF was reduced an average of 30.6%. The regrowth with T3 had significantly higher crude protein content as compared to the first harvest at 99 DAP.

A pilot study was conducted to assess the palatability of kenaf for meat-type goats and to evaluate possible differences in intake due to leaf shape. The experiment utilized eight yearling Spanish does, 30 to 40 kg body weight, individually housed in polydomes. A 90-day stand of kenaf was harvested fresh at two day intervals and the stem chopped by hand to 3 to 5 cm length. During a 10 day adaptation period animals were fed a pooled composite of whole plants with narrow and broad leaves. Intake was measured during a 6-day period with the two leaf shapes fed separately to four animals each. Feed was provided twice daily at 140% of the estimated intake determined during the adaptation period. Goats consumed 2.97 kg of kenaf daily, reflecting a dry matter intake of 0.87 kg. There was considerable daily variation in intake between individual animals 0.9 to 4.6 kg, while daily intake was unaffected by leaf shape; 2.93 kg/day of narrow and 3.00 kg/day of broad leaf types in this trial. These results indicate that kenaf has potential as a summer forage in Virginia and adjoining states.

Lesquerella

The U.S. is totally dependent upon imports to meet the needs of hydroxy fatty acids which have many defense and industrial applications. Lesquerella seeds contain an oil with a high content of a hydroxy fatty acid that is similar to castor oil. A bulk population of lesquerella (*Lesquerella fendleri*) was evaluated in Petersburg, during 1992, 1993, and 1994 for production potential. The results of numerous experiments (planting time, germplasm evaluations, fertilizer experiments) have been disappointing. The major hindrance has been germination and stand establishment. Since lesquerella was being studied as a replacement of castor and efforts to introduce lesquerella in Virginia may not succeed, it was decided to initiate research to evaluate a diverse collection of castor. The results of 1994 experiment with 74 accessions were encouraging.

Meadowfoam

Meadowfoam (*Limnanthes alba*) is an alternative oil seed crop that is an excellent source of C_{20} and C_{22} fatty acids. Small areas of meadowfoam are produced commercially in Oregon where it is grown as a winter annual. However, in trials in Idaho it lacked the hardiness to overwinter. Reportedly, meadowfoam is very tolerant of wet soil conditions. It also requires pollination to produce seeds.

Meadowfoam seed (John K. King & Sons Ltd., Coggeshall Colchester Essex, Great Britain) was planted at both the Northern Piedmont Agricultural Experiment Station in Orange County, Virginia and Blacksburg in early April 1992. However, there was no emergence at either location. Secondary seed dormancy has been reported at temperatures above 15°C which may explain the lack of field emergence. It may be possible to grow meadowfoam as a winter annual in eastern Virginia.

Mungbean

Approximately 7 to 9 million kg of mungbean are consumed annually in the United States and nearly 75% of this amount is imported. The research objective with mungbean (*Vigna radiata*) is to develop a new summer crop in Virginia. Five replicated experiments with seven entries ('Berken', 'Johnston's California', 'LSB-8205', 'Lincoln', 'M-12', 'OK-12', and 'Texas Sprout') were conducted during 1993 and 1994 to determine yields and suitable planting time. The seed yields during 1993 were 1567 and 1475 kg/ha, respectively, for experiments planted on June 9 and July 7. The mean seed yields during 1994 were 2706, 1975, and 902 kg/ha from experiments planted on May 17, June 16, and July 21, respectively. The highest yielding cultivars were LSB-8205, 'Johnston's California', and 'Texas Sprout'. Delayed planting resulted in reduced yield, plant height, and biomass. Based on these studies, mungbean seems to hold considerable potential as a new legume crop for Virginia and other areas in the southern United States.

Onion

In order to meet the local demand for pungent, high dry matter onions for processing and export, agronomic potential of both short- and long-day onion (*Allium cepa*) cultivars is being evaluated in a sustainable production system using green manure legumes and inorganic fertilizer. Preliminary results have been inconclusive but indicate potential for hairy vetch (*Vicia villosa*) as a good green manure source of nutrients for onion production.

Pepino

Solanum muricatum is perennial woody shrub native to South America. This crop is frost sensitive, so it must be moved to a greenhouse before winter. The pepino does not produce plants that are true-to-type from seed, so it must be propagated vegetatively. We have tested 'Golden Splendor', 'Temptation' and five numbered breeding lines from Dr. Carlos Quiros of the University of California, Davis for the past three years. The most promising accession for greenhouse production was line 88588 which produced egg-shaped fruit that turn yellow with purple stripes at harvest. For outdoor production, none of the plants set fruit during the summer months because the temperature was too high. Fruit set occurred in Sept., but only a few fruits matured before frost. The flavor of a mature pepino fruit is reminiscent of a casaba melon. The maximum sugar content was 9° Brix. Pepino fruit are very juicy but easily bruised during harvest and transport. In a survey at Virginia Tech, only 22% of the respondents who tried free samples of pepino said they would consider buying the fruit in a grocery store. Most people complained that the fruit tasted like a poor quality muskmelon and that fruit quality was extremely variable. We conclude that the pepino is not well adapted to Virginia conditions, and all the cultivars we tested lacked the quality necessary for commercial sale in the United States.

Pigeonpea

Since local and regional demand for green pods as well as green and mature seeds exists, research has focused on evaluation of pigeonpea (*Cajanus cajan*) as a grain and a vegetable crop in Virginia. In the experiments conducted during 1993 with three determinate pigeonpea lines, the green bean yield varied from 11888 to 15696 kg/ha with a moisture content varying from 79% to 84%, 130 days after planting. Shelling in these experiments ranged from 52% to 55%. The protein content of green seeds varied from 18% to 21%. Since most of the agronomic practices for production of pigeonpea are similar to soybean, pigeonpea seems to be a promising new legume crop for Virginia.

Purslane

Purslane *Portulaca* spp., is one of the vegetable crops eaten extensively in soups and salads in Mediterranean countries, where the incidence of both heart disease and cancer is low (Simopoulos and Salem 1986). Purslane is distributed widely in the tropical and subtropical areas of the world including many parts of the United States, where it is considered as a weed. Purslane is a rich source of omega-three fatty acids, that may have beneficial effects on coronary heart disease in humans. Topical application of an aqueous extract of the stems and leaves of purslane reduced muscle tone in individuals suffering from spasticity (Parry et al. 1988).

Eight purslane accessions [*Portulaca oleracea, P. sativa*, 'Golden Gelber' (Netherlands), 'Garden', (Netherlands), 'Golden' (England) and wild accessions from Greece, Beltsville, and Egypt] were planted in a splitplot design replicated four times on two planting dates (April and May) in 1992, 1993, and 1994 at the VSU Randolph Research Farm, Petersburg. The planting date was considered as the main plot and accessions as subplots. Each four row plot was 3 m long, with a spacing of 75 cm between rows and a seeding rate of 33 seeds per one meter of row.

There were significant difference for all growth traits studied among years and accessions. *Portulaca sativa* and 'Garden' showed the highest fresh yields, while 'Golden', 'Golden Gelber', and 'Egyptian' had the lowest yields. Plant height was significantly correlated to fresh yield ($r^2 = 0.52^{**}$). *Portulaca sativa* and 'Garden' appeared better adapted to Virginia soils and weather conditions. On a dry wt. basis, average leaf total protein was 22% and lipid content was 6%. Linolenic acid was the most abundant fatty acid in both leaves and seeds of purslane. The nutritional effects of purslane dietary supplements made from purslane leaves on blood metabolites and body composition were studied on growing rats. Diets containing 10% and 20% of freeze-dried purslane leaf powder supplements produced 26% and 17% reduction in total plasma cholesterol and 33% and 20% reduction in plasma triglycerides levels, respectively. These results suggest that purslane may provide a new dietary means for controlling high blood cholesterol levels and coronary heart disease in humans.

Quinoa

Quinoa (*Chenopodium quinoa*) is a pseudograin native to the Andes mountain region of South America. Although cultivated by the Incas for centuries, until recently quinoa has been virtually unknown elsewhere. Quinoa is an annual, broadleaved, dicotyledonous herb that reaches a height of about 1 to 2 m at maturity in about 100 days from seeding. Quinoa has been successfully grown at high elevations in Colorado and also in Oregon, Washington, western Canada, Sweden, and Great Britain. However, at other locations, particularly at lower elevations, the pollen has been sterile and plants do not develop seeds.

In 1992 and 1993, quinoa was grown near Blacksburg and at the Northern Piedmont Agricultural Experiment Station in Orange County. Seed was provided by John K. King & Sons Ltd., Coggeshall Colchester Essex, Great Britain. Fertilizer 10N–4.3P–8.3K was incorporated prior to planting at a rate of 560 kg/ha. No harvest data were available from the Orange County location, although plant growth and development appeared to be similar at both locations. At the Blacksburg location, a small plot of four 7 m long rows spaced 45 cm apart were seeded by hand in early June of both years with an inrow spacing of 10 cm. The plot was weeded by hand, no insecticides were applied, and irrigation was applied as needed using an overhead sprinkler system. The actual stand obtained was 1 plant/20 cm. Seed heads were handharvested in late Sept. in both years and dried for one week at 35°C. Seed was mechanically cleaned before the final grain weight was determined. In 1992, the average yield was 31 g per plant (2,804 kg/ha). In 1993, heads in both Blacksburg and Orange did not set seed. In 1992, summer temperatures were unseasonably cool which probably increased the percentage of seed set. We believe the 1993 results are more representative of the expected performance of quinoa in Virginia. Besides poor seed set, other production problems included poor emergence in crusted soil, inconsistent maturity, and lodging.

Sweet Sorghum

Applied research in production and marketing of syrup made from sweet sorghum cane (*Sorghum bicolor*) has been conducted in Virginia since 1992. The primary goal of this program is to introduce modern production and processing methods to experienced producers who are using inefficient, traditional practices. The introduction of 'Top 76-6', a sorghum cane cultivar developed by the University of Georgia, has increased yields and syrup quality.

Vegetable Soybean

Vegetable soybeans (*Glycine max*), long revered in East Asia cuisine, has had increased public acceptance in the U.S. Soybean breeders are developing improved vegetable-type soybeans for specific niches in the market. Genetic modification of the components of soybean seeds is the focus of VSU breeding programs. Soybean lines with increased protein content, low trypsin inhibitors, and lipoxygenase could be of particular interest for developing cultivars with improved nutritional quality, flavor, and increased yield. A total of 17 vegetable-type soybean genotypes representing a wide range of values for the characteristics of interest were planted in four-row plots, in a randomized complete block design, at Petersburg. Each genotype was evaluated at the R5, R6, and R7 stages of development for pod yield and size (hundred pod weight), protein, phytate, lipoxygenase, and trypsin inhibitor activities. Protein content was positively correlated with phytate and stage of harvest but negatively correlated with trypsin inhibitor. Thus, selection for high protein content will reduce trypsin inhibitor and increase phytate content, and selection of genotypes for high pod yield and size will significantly increase phytate content. Phytate binds with nutritionally important metals and contributes to nutritional deficiency in nonruminant animals and humans, while trypsin inhibitor reduces protein digestibility. Lipoxygenase, an antinutritional factor associated with undesirable flavors in soybean products, was negatively correlated with pod yield and size, and stage of harvest. These associations suggested that large-seeded and high yielding genotypes tend to have low lipoxygenase content. Several potential vegetable-type soybean breeding lines developed through hybridization at VSU are available for further yield and nutritional determination. An improved vegetable-type soybean could be an early cash crop similar to lima beans or peas. It would also be a valuable substitute for green beans in localities where the Mexican bean beetle prevents the growing of garden beans. In addition, vegetable-type soybean seeds are suitable for various soyfood products and may have a major potential as an export commodity to Japan and other Asian countries. In view of its great nutritive value, the vegetable-type soybean could have considerable potential for improving the diet of American families.

Vernonia

The seeds of vernonia, *Vernonia galamensis* contain high quantities of naturally epoxidized oil that has many industrial uses besides being a potential additive to oilbased paints to reduce the volatile organic compounds that contribute to air pollution. The vernonia research at VSU is focusing on identification of optimum cultural practices to facilitate vernonia production under Virginia conditions in cooperation with the USDA (U.S. Water Laboratory, Phoenix, Arizona). Evaluation and enhancement of vernonia germplasm for seed yield and oil quality are also a part of these endeavors. Considerable progress has been made towards development of adapted nonshattering lines. During 1994, significant variation existed among 10 entries (9 selections and a check) for plant height, ratio of mature to immature seedheads, and seed yield. Differences among entries for seed weight/seedhead were not significant. The plant height varied from 83 to 126 cm. The seed yield varied from 490 to 1288 kg/ha. Preliminary observations indicated that trifluralin (herbicide) might be suitable for weed control.

Other Crops

Various potential alternative crops are also being evaluated at Virginia Tech. Tobacco is being genetically transformed to produce high value compounds. Conventional breeding work with soybean has led to the release of the 'Vanatto' which is gown in Virginia and exported to Asia for processing into tofu and other soybased products. Switchgrass (*Panicum virgatum*) and matua (*Bromus willdenowii*) are promising forages for Virginia, and there is interest in growing switchgrass as biomass crop for the production of ethanol. Grain sorghum (*Sorghum bicolor*) is being evaluated as an alternative grain crop for nonirrigated, drought prone regions of Southern Virginia where soybeans and other traditional grain crops have been unsuccessful. There are over 30 commercial wineries in Virginia and extensive trials are underway to identify adapted wine and table grapes (*Vitis* spp.). The production of holly (*Ilex* spp.) foliage for ornamental use can be extremely profitable and production recommendations have been developed for foliage and berry production. There is interest in the commercial production of persimmons (*Diospyros kaki*), Asian pears (*Pyrus pyrifolia*), and apricots (*Prunus armeniaca*) and cultivar evaluations are ongoing. Kiwifruit (*Actinidia arguta*) has been successfully grown, but only in the extreme southeastern corner of the state. A trellis system has been developed that increases yield and simplifies the harvest of blackberries (*Rubus* spp.).

CONCLUSIONS

There are five investigators at Virginia State University and ten at Virginia Tech involved in alternative crop development activities. It is difficult to predict which crops have the potential to be more widely grown in the future. Some alternative crops have already increased in importance. In 1995, over 5,263 ha of cotton were grown in the state, up dramatically from 15 years ago. Canola can be grown successfully in Virginia, and hopefully a consistent market will cause further increases in crop area. Soybean cultivars developed at Virginia Tech are being exported to Japan for human consumption. Elephant garlic has become an important specialty crop. Kenaf is a versatile crop that can be grown as a source of fiber, and is also a high quality forage. Virginia's tourism industry attracts visitors willing to buy fruits, wine, and specialty items such as sorghum molasses directly from small farmers. Virginia's proximity to large population centers makes the production of specialty items such as exotic fruits, Asian vegetables, and herbs profitable alternatives to conventional crops.

REFERENCES

Parry, O., Okwuasaba, and C. Ejike. 1988. Effect of an aqueous extract of *Portulaca oleracea* leaves on smooth muscle and rat blood pressure. J. Ethnophormacol. 22:333–344.

Simopoulos, A.P. and N. Salem, Jr. 1986. Purslane: a terrestrial source of omega-three fatty acids. N. Engl. J. Med. 315:833.

Welbaum, G.E. 1993. Public sponsored new crops research and development projects in Virginia. p. 109–111. In: J. Janick and J.E. Simon (eds.), New crops. Wiley, New York.

PART II
POLICY & PROGRAMS

POLICY

Diversifying U.S. Crop Production*

Jules Janick, Melvin G. Blase, Duane L. Johnson, Gary D. Jolliff, and Robert L. Myers

The greatest service which can be rendered any country is to add a useful plant to its culture.

Thomas Jefferson, In Services to My Country (ca. 1800)

Neither my overseers nor manager will attend properly to anything but the crops they have usually cultivated; and, in spite of all I can say, if there is the smallest discretionary power allowed them they will fill the land with corn, although even to themselves there are the most obvious traces of its baneful effects. I am resolved, however, as soon as it shall be in my power to attend a little more closely to my own concerns, to make this crop yield [i.e., give way] in a degree to other grains, to pulses, and to grasses.

Letter from George Washington to Thomas Jefferson (October 4, 1795)

New Crops Development a Primary Goal when USDA was Established

Be it enacted by the Senate and the House of Representatives of the United States of America in Congress assembled. That there is hereby established at the seat of the Government of the United States a Department of Agriculture, the general designs and duties of which shall be to acquire and to diffuse among the people of the United States useful information on subjects connected with and distribute among the people new and valuable seeds and plants.

The opening section of the legislation establishing the new U.S. Department of Agriculture (May 15, 1862)

We are probably far from possessing, as yet, all the [crops] for which nature has fitted our country. To find out these, will require an abundance of unsuccessful experiments. But if, in a multitude of these, we make one or two useful acquisitions, it repays our trouble.

Letter from Thomas Jefferson to William Drayton (1786)

United States agriculture is based on crops once considered new**, and new crops research was championed by the founding fathers. The United States, despite its large landbase, is the source of only a few economically important crops. Even corn (maize), the major grain of Native Americans, resulted from introductions from Central and South America, as did beans, pumpkins, squash, tomatoes, and potatoes. Colonists introduced wheat, rice, barley, oats, forage, fiber, fruit, and vegetable crops from Europe and Asia. Some native crops such as sunflower and strawberry were developed abroad and reintroduced to American agriculture as profitable new crops. During the 200 years since Jefferson argued the critical importance of new crops to a strong American agriculture, many potentially profitable new crops have been introduced, only to languish, potential benefits unrealized, because public funds necessary for development were lacking. The twentieth century has continued to witness the introduction and/or the development of successful new crops such as

*Also published as Council for Agricultural Science and Technology (CAST) Issue Paper 6, February 1996.

**New crops include established crops not grown commercially in the United States, crops lost to U.S. agriculture, crops of ancient or primitive cultures, underexploited and specialty crops, and as yet undiscovered crops. In this report, new crops will refer to established crops not grown in the United States, underexploited crops, and species that remain to be domesticated.

soybean—now a major crop of the United States and of the world, avocado and pistachio in California, and pearl millet in the southeastern states. Regrettably, the United States has not taken a long-term, comprehensive, strategic approach to crop diversification.

The agrarian history of the United States is in many ways a chronicle of the rise and fall of new crops species. But through a process of introduction, trial, and error, U.S. agriculture has become based on a rather narrow group of crops. Almost 80% of annual row-crop acreage in the United States is planted to wheat, corn, and soybeans. As a result of this concentration, many growers have few alternatives, and low prices on major commodities have a disastrous economic effect.

To stabilize farm income, a system of federal programs including crop subsidies and cropland reduction developed. These programs have been enormously expensive. An estimate of costs from 1978 to 1994 is $291 billion (in 1987 dollars). Adding interest, lost crop wealth opportunities, and multiplier effects will more than double these astronomical costs to U.S. taxpayers (Jolliff 1996). Payments in turn are distributed unevenly among the various agricultural sectors (Carr 1992). Yet despite these programs, farm numbers, farm populations, and rural prosperity continue to decline ominously.

As these events have unfolded, a traditional American solution—that of searching for new opportunities through crop diversification—has ceased to be a major part of U.S. agricultural policy. Rather, the focus has been on increasing the yields and decreasing the production costs of traditional crops. The policy of concentrating research and farm support programs on a few major commodities runs counter to repeated recommendations from task forces, individual researchers, and government personnel to create policies encouraging the development of new crops and new products (Table 1). The key constraint seems to be that no single voice speaks for new crops whereas major commodity crops have established constituencies and therefore dedicated support.

Table 1. Previous statements on the need for diversification through new-crops development.

Year	Statement and reference
1786	Writings by Thomas Jefferson[z]
1862	Congressional Legislation Establishing U.S. Department of Agriculture (Rasmussen 1975)[y]
1894	Pepper Report, U.S. Senate Committee on Agriculture (Taylor and Taylor 1952)
1895	George Report, U.S. Senate Committee on Agriculture (Taylor and Taylor 1952)
1920s	Need for relief from surpluses (Holmes 1924; McMillen 1929)
1935	Chemurgic Movement founded (Barnard 1937; Benson 1937)
1957	New and Special Crops Report (Task Group on New and Special Crops 1957)
1983	Plants: The Potential for Extracts, Protein, Medicines and Other Useful Chemicals (U.S. Congress 1983)
1984	Development of New Crops: Needs, Procedures, Strategies, and Options (Council for Agricultural Science and Technology 1984)
1987	New Farm and Forest Products: Responses to Challenges and Opportunities Facing American Agriculture (Sampson et al. 1987)
1988	New Crops and New Farm Products (Congressional Research Service Report)
1989	Growing Industrial Materials: Renewable Resources from Agriculture and Forestry (U.S. Department of Agriculture Task Force Report)
1990	Alternative Opportunities in Agriculture: Expanding Output through Diversification (U.S. Department of Agriculture Economic Research Service Report #633)
1991	Agricultural Commodities as Industrial Raw Materials (Office of Technology Assessment Report)
1992	New Crops, New Uses, New Markets (U.S. Department of Agriculture 1992)

[z]Letter by Jefferson to William Drayton.
[y]New crops a primary purpose of the new USDA in legislation signed by Abraham Lincoln on May 15, 1862.

Recently, interest in new crops has intensified as the result of a number of interacting forces. The southern corn leaf blight epidemic of 1970 and the worldwide loss of biological diversity have been responsible for an upsurge of interest in germplasm diversity. Over much of the last four decades, low prices for major commodities—punctuated by brief periods of prosperity—have rekindled among growers centuries-old concerns about profitable crop alternatives. The continued strength of the environmental movement has spurred interest in a more sustainable and diversified agriculture while consumer demand for new foods and products has increased as a result of changing demography and health concerns. Finally, economic forces continue to attract innovators and entrepreneurs who see potential in new crops and products.

Federal funding for new crops research has been quite minimal in recent years (Jolliff 1989) and has occurred mainly as an offshoot of political interest in new uses of current commodity crops such as the use of corn for ethanol or for biodegradable plastics. The 1990 Food, Agriculture, Conservation and Trade Act created the Alternative Agricultural Research and Commercialization (AARC) Center to provide funds for the development of new crops, new uses, and new products; since the program's inception in 1992, however, only 15% of available funds have been awarded to new crops development. Other federal programs involved somewhat in new crops investment are found primarily in the Agricultural Research Service (ARS) or are supported by small grants administered by the Cooperative State Research, Education, and Extension Service (CSREES). The ARS has funded research on alternative oilseeds (National Center for Agricultural Utilization Research, Peoria, Illinois), on developing new industrial oilseeds and natural rubber and latex (Phoenix, Arizona), and on developing kenaf and other fiber crops (Weslaco, Texas).

Current state research efforts in the area of new crops have been scattered and fragmentary and have tended towards the short term as funding priorities have shifted. Support from agricultural experiment stations in all states has declined greatly as these institutions have focused on basic research and biotechnology funded by national grants programs and private industries. A number of states (Colorado, Indiana, Kansas, Minnesota, Mississippi, North Dakota, and Oregon) have small programs specializing in new crops, but funding is marginal or has been decreased. State departments of agriculture have focused primarily on commodity crops. A national organization, the Association for the Advancement of Industrial Crops (AAIC), was formed in 1988 to promote new nonfood crops. Small programs in private foundations are involved in alternative crop agriculture.

Occasionally, private companies are interested in a new crop as an opportunity; typically, however, they take the cautious approach of waiting until someone else develops the acreage before they commit company resources. The seed industry is reluctant to put resources into new crops development and prefers to devote breeding efforts to major crops.

Nonetheless, interest in new crops remains high in the research community and among many farmers who actively seek new crops despite the fact that initial risks can be high—especially in the absence of research and development support. Industry, especially the processing sector, often has an interest in new crops, but only if public funding can raise production above a certain threshold. Other governments have recognized this barrier and have provided sufficient funding to help boost new crops production to levels required to obtain industry participation. Canada successfully undertook this strategy with canola, and the European Community now is providing substantially increased funding for the development of several alternative crops, with an emphasis on industrial crops and biodiesel fuels. For the period 1995 to 1998, the European Community has budgeted $200 million for research on industrial crops and the Dutch government alone plans to spend $50 million over the same time period on alternative oil seeds (Capelle 1996).

Despite the successes of a number of new crops in the United States and abroad and despite increased grower and researcher interest, there is no concerted U.S. national policy focused on the introduction, development, and commercialization of new crops. Because of the complexity of new crops development, rapid progress is difficult, and the path to success for each new crop is neither simple nor easy. The development of crops was a process perfected by primitive peoples over untold generations of prehistory. New crops development is a process that cannot be achieved in a few years, or during the attention span of most funding agencies.

There are other roadblocks to new crops development that are created by federal policies and politics. Because federal programs such as crop support payments, production loans and crop insurance, and technical advice are limited to traditional crops, they represent a strong disincentive for farmers considering the risks and benefits of the production of new crops. Support for new crops is fragmented and is not a priority for the traditional commodity groups commonly influencing federal policy. Clearly, initial stages in the development of new crops require various forms of federal and state support and intervention (Jolliff and Snapp 1988). The Canadian government's methodical research and development program for low-erucic-acid rapeseed (canola) has provided a dramatic model of what is needed and what can be accomplished (Busch et al. 1994).

To capture the benefits of new crops in American agriculture, a coordinated national policy that encourages and supports new crops research and development programs should be created. This policy should foster regional and national cooperative efforts between farmers, industrialists, and researchers in fields as diverse as agronomy, botany, economics, food, nutrition and health, industrial engineering, and natural products chemistry. The consensus among new crops researchers is that a number of alternative crops will be successful, and some will be extremely successful; but the winners will depend on the mix of market acceptability, research information, income incentive, and enthusiastic and persistent crop champions. Federal government support of new crops programs would be a wise and long overdue economic investment in the future.

THE VALUE OF NEW CROPS

New crops offer U.S. agriculture many potential benefits for producers, rural communities, and industry. The soybean, for instance, contributed more than $500 billion to the U.S. economy from 1925 through 1985. New crops offer alternative means of increasing farm income by diversifying products, hedging risks, expanding markets, increasing exports, decreasing imports, improving human and livestock diets, and creating new industries based on renewable agricultural resources. New crops also can spur economic development in rural areas by creating local, rural based industries such as processing and packaging and by providing general economic stability. New crops currently serve the strategic interests of the nation by providing domestic sources for imported materials and by providing substitutes for petroleum based products. Promotion of new grain crops in the United States also would serve as a form of world food security because many of the preferred cereal grains, e.g., millet and teff, in food deficit areas such as Africa are not grown widely in the United States. The exploitation of crops that can be grown for fuel, fiber, and a wide array of industrial products would help decrease U.S. reliance on imports, decrease acreage devoted to feed grains in surplus, strengthen the economic base for American farmers, and promote new sustainable industries based on renewable resources.

New Crops for Diversification

Major benefits stem from the diversification of cropping systems. Those systems involving continuous crops or even two crops, e.g., the corn soybean rotation dominating the Corn Belt, have serious limitations. Such rotations suffer from pest buildups, which necessitate increased reliance on pesticides, which can have a detrimental environmental impact. Diversified crop rotations, which occur typically with new crops introductions, can limit the impact of pests and can provide other benefits to improve sustainability of farming systems.

Some new crops—winter canola for example—can offer excellent winter-time erosion control. When several crops are in production, each with different planting and harvest dates, the economic impact of weather extremes and disease epidemics is diminished and the demand for labor and equipment spread out. Both factors can improve farm efficiency. Finally, one of the major benefits of diversification is income stability in the face of low prices due to over-supply of a single commodity.

New Crops for New Industries

Although some new crops will substitute for current crops in the marketplace, many new crops will have little or no displacement effect on current commodities. Nonfood industrial crops can serve as a renewable resource base for domestic and export market needs by replacing petroleum based products or other imports (U.S. Department of Agriculture 1992). Potential new industrial uses include biofuels; lubricants; industrial chemicals; raw materials such as medium-chain, hydroxy, epoxy, and long-chain fatty acids; waxes; and rubber. Promising candidates include crambe, cuphea, guayule, jojoba, lesquerella, vernonia, meadowfoam, and

NEW-CROPS CASE STUDIES

A study of successful new crops indicates a variety of paths by which introduction and establishment occur. A number of features, however, are common to these paths. New crops usually require various forms of government support, especially funding for research and development in plant breeding, production, product development, and marketing. A number of players and institutions must be involved to solve problems transcending disciplinary boundaries and benefiting from timely efforts. New crops introduction demands dedicated leadership and coordination among various economic sectors including research, production, processing, and marketing. Crop champions are an essential component of all successful new-crops efforts.

Soybean

The soybean (*Glycine max*, Fabaceae) was introduced into North America in 1765 but until 1920 was grown primarily for forage. The U.S. Department of Agriculture (USDA) conducted limited testing for decades and consistently failed to recognize the potential of the crop as anything other than a forage. Research from publicly supported agricultural scientists working in collaboration with farmers and private industry was essential—especially the work of public breeders who developed new cultivars from germplasm introduced from China, Japan, and Korea. Major explorations were undertaken by the USDA from 1924 through 1931, after earlier seed collections were lost. Industrial support was essential, and in 1922 the Staley Company built the first major soybean processing facility, in Decatur, Illinois (Hymowitz 1990). In 1941, soybean acreage harvested for seed exceeded that harvested for forage and greatly increased during World War II and the next two decades.

In 1924, about 5 million bushels (bu) of soybeans were produced in the United States, with yields of 11 bu/acre; in 1994, about 2.6 billion bu were produced on 62 million acres, with yields averaging 42 bu/acre. To bring soybeans from a forage to a crop grown for oil and high-protein meal cost U.S. taxpayers an estimated $5 million from 1912 to 1941; the current annual value of the crop is estimated at more than $13 billion. Had there been no public-private partnerships or integrated efforts at production and marketing, soybeans still might be a minor crop, and many farmers in the United States would forfeit the income opportunity that soybeans now provide.

Canola

The development of canola (*Brassica napus*, Brassicaceae) is one of the great achievements in planned new crops development and represents a model for the modern introduction and development of new crops (Busch et al. 1994). Because it was considered an essential lubricant, rapeseed was grown first on a few acres on the northern Canadian prairie in 1942, as an emergency war measure. One million pounds (lb) were produced with a yield of 1,000 lb/acre. Although the initial lubricant markets decreased, other uses pushed rapeseed production to more than 4 million acres by 1970, with average yields of 1,123 lb/acre (Downey 1990).

Canadian and European nutritionists became interested in rapeseed oil because it has a significant fraction of long-chain monoenoic fatty acids. The identification of selections low in erucic acid resulted in the first release of low-erucic rapeseed in 1968. By 1970, nutritionists had demonstrated that the low-erucic rapeseed oil was nutritionally superior, and Canada converted its more than 4 million acres to low-erucic varieties (renamed canola) in 2 years. In 1994, there were 14.3 million acres of canola in Canada that were producing an annual value of approximately U.S. $1 billion.

Canola production and consumption is the fastest growing segment of the oilseed industry and is increasing at an annual rate of 8%. Canola production began in the United States in the mid-1980s, and by 1994, U.S. production was 447 million lb on 354,000 acres, with an annual value of $50 million. Expansion of the U.S. canola industry, however, is checked by restrictions on market access, unavailability of appropriate varieties, challenges regarding production, and lack of infrastructure.

Limited crushing capacity makes it necessary to transport most U.S. grown canola to Canada for processing. The current annual U.S. consumer demand for more than 400,000 tons of canola oil imports represents a major opportunity for U.S. producers, a demand that could support more than 2 million additional U.S. acres in canola.

Pearl Millet

This crop, the world's most important cereal in very hot and very dry climates, is grown as a food grain on almost 100 million acres in Africa and the Indian subcontinent (Andrews et al. 1993). As a result of breeding efforts by Dr. Glenn Burton, a USDA–ARS scientist in Tifton, Georgia, pearl millet (*Pennisetum glaucum*, Poaceae) was developed as a summer annual forage crop and now is grown on 1.5 million acres in the United States.

Agronomic studies and breeding efforts supported by the U.S. Agency for International Development through the Collaborative Research Support Program and conducted in Nebraska and Kansas have demonstrated that it is a promising grain crop for areas of the United States in which drought, soil type, short season, or excessive heat diminishes the yield potential of sorghum. Also, short-season pearl millet hybrids are a promising double-crop after wheat in the Midwest. Pearl millet is a preferred food grain in Africa and has been proved a superior feed for poultry, swine, and fish. In 1994, a newly developed grain-type pearl millet was planted on more than 15,000 acres in Georgia and Florida for poultry feed.

Kenaf

In the 1940s in the southern United States, kenaf (*Hibiscus cannabinus*, Malvaceae) was introduced to replace jute in cordage products. In the 1960s, a screening process led by USDA scientists identified kenaf as the most promising fiber crop from among 500 plant species considered for pulping and papermaking (Kugler 1990; Taylor 1993). Although this annual crop was tested successfully in 1977 and demonstrated commercially in 1987 for newsprint manufacture, paper market conditions drove development in other directions. Kenaf is coming into commercial use, with mechanical separation mills operating in Louisiana, Mississippi, and Texas. Stem fibers are being used to manufacture dry formed "instant lawn" mats and, through a New Mexico company, to manufacture printing- and writing-grade papers. Core fibers are being used in animal bedding, poultry litter, adsorbents for oil and chemical spills, and horticultural mixes replacing peatmoss. Current farm gate value is roughly $1 million and is expected to increase substantially in the next few years.

Pistachio

This delectable nut (*Pistacia vera*, Pistaciaceae) was introduced at the USDA Plant Introduction Station at Chico, California in the 1920s, but little interest in pistachio's potential occurred until the 1970s (Ferguson and Arpaia 1993). Key developments leading to successful crop commercialization were the selection of a suitable genotype ('Kerman') in 1957, the generation of research information by University of California researchers that was able to establish the suitability of California's Central Valley for production, the development of the Central Valley water project, and a tax structure that made it possible for growers to deduct orchard establishment costs.

The formation of the California Pistachio Association in 1976 levied growers and processors for funds to support production research and to educate growers and processors. The advent of the Iranian Hostage Crisis in 1979 and subsequent embargoes of Iranian pistachio imports helped stabilize U.S. pistachio prices. The pistachio industry now is a major California nut industry, with 74,000 acres planted, an acreage exceeded only by that of almonds or walnuts. In 1995, farm gate value of the crop (147 million lb) was $170 million, with a retail value of approximately $440 million.

NEW-CROPS CASE STUDIES

Quinoa

This little known Andean pseudograin initially was evaluated in Pennsylvania, where it was eliminated from consideration because of poor adaptability. In 1983, it was reintroduced in Colorado that has an environment similar to that of the Bolivian Altiplano, where the plant originated. Between 1983 and 1987, only $76,000 of taxpayer funds was invested in developing quinoa (*Chenopodium quinoa*, Chenopodiaceae) as a new U.S. crop; yet this investment has yielded a new crop with an estimated value exceeding $5 million in 1995. Interest in quinoa has increased worldwide because of unique starch properties suggesting a number of industrial uses.

Meadowfoam

Meadowfoam (*Limnanthes alba*, Limnanthaceae) is a winter-spring annual wildflower with unique seed oil that is native to northern California and southern Oregon. In the 1950s, USDA scientists discovered that its seed oil had unique properties making it a suitable source of a number of potentially useful derivatives. Crop domestication efforts (breeding, agronomy, seed processing, product development, and marketing) have been carried out at Oregon State University in cooperation with the USDA, industry, and private growers, in an effort to provide a potential new crop for the Willamette Valley. In 1995, about 3,000 acres of meadowfoam were in production.

Milkweed

Pods of this native weed were gathered and used as fill for life jackets during World War II, but such efforts were abandoned after the war. Research interest increased in the 1970s, when the crop was suggested as a source of biofuels, but the program was curtailed when the process proved uneconomical. An oil company executive became interested in the potential of milkweed (*Asclepias* sp., Asclepiadaceae) as a fiber crop and founded the Natural Fibers Corporation in 1987 to create a new agricultural industry based on milkweed (renamed syriaca) (Knudsen and Zeller 1993).

With the aid of federal support from a Small Business Innovation Grant (SBIG), a cooperative USDA grant to the University of Nebraska, and an AARC Center investment, the corporation created a means of using syriaca seed floss as a nonallergenic fill to replace imported duck and goose downs in comforters. Two hundred acres of milkweed now are being grown in Nebraska, and company sales have surpassed $1 million; expansion to 900 acres is expected for 1996. The strategic plan of the company calls for expansion into nonwoven uses such as batting made with other natural and synthetic fibers, as well as yarns, pulps, and papers.

Taxus for Taxol ® Production

Between 1960 and 1982, the National Cancer Institute screened 35,000 plant samples for anticancer activity. An anticancer drug registered as Taxol® (paclitaxel) and isolated from the bark of the Pacific yew (*Taxus brevifolia* and other *Taxus* spp., Taxaceae) proved effective for arresting ovarian and metastatic breast cancers and created a tremendous demand for the drug (Piesch et al. 1994).

Taxus brevifolia was found in old-growth forests in the Pacific Northwest, the habitat for many endangered species, but the tree had to be destroyed if the bark was to be extracted. A renewable source thus was needed to meet demand. Subsequently, precursors of paclitaxel were extracted from the needles of many *Taxus* species to produce semisynthetic Taxol®. The drug from this source was approved for use by the U.S. Food and Drug Administration in late 1994. Bristol-Myers Squibb, which was responsible for producing the drug, contracted with a number of companies including the Weyerhaeuser Corporation (Washington) and Zalenka Nursery (Michigan) to provide a renewable source of needles. Twenty to 25 million shrubs now are being cultivated to meet the demand for Taxol®.

Stokes aster. New woody and nonwoody fiber crops can be used to manufacture many of the diverse products now being imported.

Converting crop acreage into new industrial crops by decreasing production of major commodities often produced in surplus will decrease the amount of federal commodity support needed and will encourage the expansion of industry. New crops may decrease the importation of numerous crops and materials, many of which are of strategic importance. Examples include natural rubber and latex such as guayule, a number of waxes, resins, vegetable oils, gums, and sources of medium- and long-chain fatty acids. Most grains and oilseeds can be used for a variety of products over time, a fact illustrating the value of a diverse crop resource base.

New Crops for Improvements in Diet And Health

New crops represent potential sources of new human foods and livestock feeds with increased digestibility and decreased antidietary factors. New grains and oil-seeds—the foundation of swine and poultry feeds—and new forages for ruminants may improve the efficiency of animal agriculture. New medicinal crops could assist in the battle with diseases such as cancer and acquired immunodeficiency syndrome (AIDS). Examples of anti-carcinogens include vinblastine and vincristine from *Catharanthus (Vinca) rosea* and Taxol® (paclitaxel) from *Taxus brevifolia*. A number of new plant sources, hopefully effective against human-immunodeficiency virus (HIV) properties, are under investigation.

New Crops for Strengthened Rural Communities

An additional benefit of new crops development is its effect on rural communities. The most observable benefit may be the development of processing, packaging, and other value-added activities that, to decrease transportation costs, must be based in local agricultural communities. Because new crops by necessity are raised first on relatively small acreages, initial processing can be small-scale, typically with local entrepreneurial input. The increased farm income resulting from the planting of new crops will have other multiplier effects on local communities and will provide opportunities for input suppliers.

THE CASE FOR FEDERAL SUPPORT

America's cropland is a natural resource with the potential to produce renewable wealth, and the entire nation benefits directly or indirectly when a new crop is developed. Because of the complexity of development and the uncertainty regarding what sectors of agriculture and regions of the United States might gain most from investment in development, federal support and leadership are essential to a successful program of new crops introduction. Many obstacles make it unlikely that individuals can introduce, develop, and commercialize new crops successfully. Three obstacles will be discussed in depth.

1. The long-term nature of crop introduction. The aforementioned new crops case studies indicate that successful new crops development often is a lengthy process and that 10 to 40 years or longer often are necessary to discover, to domesticate, and to commercialize major new crops successfully. It is the long-term nature of new crops development that deters the private sector from including it in research and development programs. Nothing can be done without suitable germplasm and plant breeding research that is a long-term activity and must receive government and institutional support in the early stages. Thus, institutions such as the USDA–ARS and state agricultural universities and experiment stations must participate in new crops development.

2. High risk. New crops introduction, with its high failure rate, is inherently risky, at least over the short term. Bottlenecks are many and include problems with production, crop adaptability, and, most important, market forces. Because crop adaptation is involved, research must be multiregional. Often the tremendous benefits of success may not accrue to the originator, a fact that deters private sector investment. The geographic regions of environmental and economic adaptation for potential new crops often are unknown at the outset of work on domestication and improvement. For example, wild rice was improved substantially as a new crop in Minnesota, but California subsequently adopted the crop and now dominates production.

3. Coordination problems. Growers ordinarily are not interested in new crops without an assured market, and marketers will not handle new crops without an assured supply. Because of this dilemma, an independent and neutral party is needed to bring researchers, growers, processors, and marketers together within and between regions. Furthermore, the successful introduction of new crops involves solving problems that transcend disciplinary boundaries, and coordinating marketing problems that involve both public and private sectors of the economy. Transdisciplinary effort and cooperation are needed that extend well beyond the attempts most current institutions make with multidisciplinary research. For example, the Canadian government changed rail freight rates, created grading standards, invested heavily in production research, and examined the health and safety risks and benefits of canola.

PATHWAYS TO NEW CROPS DEVELOPMENT

The authors of this report believe that a successful new crops policy is essential to the productive capacity and sustainability of U.S. agriculture. New crops development is in the public interest because of the multiple benefits that may result. Obviously, the competition for limited public resources apportioned to agricultural research is intense and an arena in which established agricultural commodities are well represented and apply strong pressure. The poor record of past attempts to advance new crops development in the United States suggests that institutional innovation is needed. A consortium of policymakers, researchers, industry representatives, and farmers should be organized to create a permanent voice for new crops development.

Recommendations for new crops development have been made regularly for more than a century, but, in spite of these appeals, underinvestment continues at great economic, social, and environmental costs. An effective policy must be based on a reallocation of resources equal to the task. The authors believe that three steps must be taken if this objective is to be accomplished.

Renewing the Original USDA Commitment to New crops Development

The authors recommend a renewed focus on one of the primary congressional mandates to the USDA at the time of its establishment in 1861: "to acquire and to diffuse among the people of the United States useful information on subjects connected with and distribute among the people new and valuable seeds and plants." The authors also encourage the USDA (1) to examine how better to support new crops development through policies and programs and (2) to request additional funding from Congress for new crops development and/or to redirect current funding into this area.

To more strongly encourage USDA's efforts in new crops development, the authors suggest including under research priorities legislative language specifically supporting new crops:

> *It shall be national policy to diversify agriculture by the development of new crops, thereby providing farmers with more crop options, strengthening the renewable resource base, and stimulating additional rural economic development.*

Developing Incentives for Adoption of New Crops

Incentives would include but not be restricted to developing crop insurance for new crops, providing guaranteed loans for designated new crops, allowing experimental new crops to be grown on idled land, exclusively releasing new publicly supported cultivars to individuals and to organizations devoted to new crops, and creating a system of grants for farmers and small businesses interested in testing specific new crops.

Creating Institutional Innovation Through a Jefferson Initiative

To create a coordinated development effort for new crops, the authors propose the creation of a new crops initiative to be called the Jefferson Initiative in honor of the strong personal commitment of Thomas Jefferson to improving American agriculture through the development of new crops. The Jefferson Initiative would become the focal program for new crops development and would serve as an umbrella for many types of public-private partnerships aimed at commercialization.

An innovative new research and development entity is needed to provide the critical mass of talent and resources needed to boost new crops to a self-sustaining level and to catalyze the Jefferson Initiative. It is proposed that this entity be called The Thomas Jefferson Institute for Crop Diversification and that it be established with a progressive vision of how science can be employed on behalf of agriculture. The Jefferson Institute, comprised of a national research and development center and 8 to 10 cooperating regional centers (Fig.1), would employ interdisciplinary teams to address the interlocking problems of breeding, production, utilization, and marketing.

The benefit of a national center is that it would have the capacity to achieve major new crops introductions, to develop public awareness and interest, and to engage the participation of many partners, including major agribusiness firms. Regional centers would complement the national center by allowing optimal testing and development of varieties adapted to each region; by obtaining more complete involvement of researchers, extension specialists, and other agricultural experts across the country; and by developing the localized link to farmers, processors, and marketers that is needed to establish new crops in each region.

There are various models for funding the proposed Jefferson Institute. The authors of the document estimate that a minimum of $20 million/year is required to initiate the program. Funding must come primarily from the USDA, especially in the initiative's early years, but the Institute would be expected to use grants and contracts to leverage other funding from public and private sectors. As specific new crops moved into commercialization, packages of funding and resources could be organized from state and private sources to build on federal investments.

Organizationally, the Jefferson Institute could be set up as a nonprofit research institute with USDA and other funding, or by establishing institute components through cooperative agreements between the USDA and universities. Regardless of the exact model, the authors expect the Jefferson Institute to be affiliated with Land Grant universities and to have close ties to the ARS. If given sufficient flexibility, the institute would be able to cooperate with universities, government agencies, and the private sector in breaking down barriers to new crops development. The advisory panel with representatives from industry, farm groups, government, nonprofits, consumer groups, and research and extension should be able to guide institute centers in strategic planning for the new crops development effort.

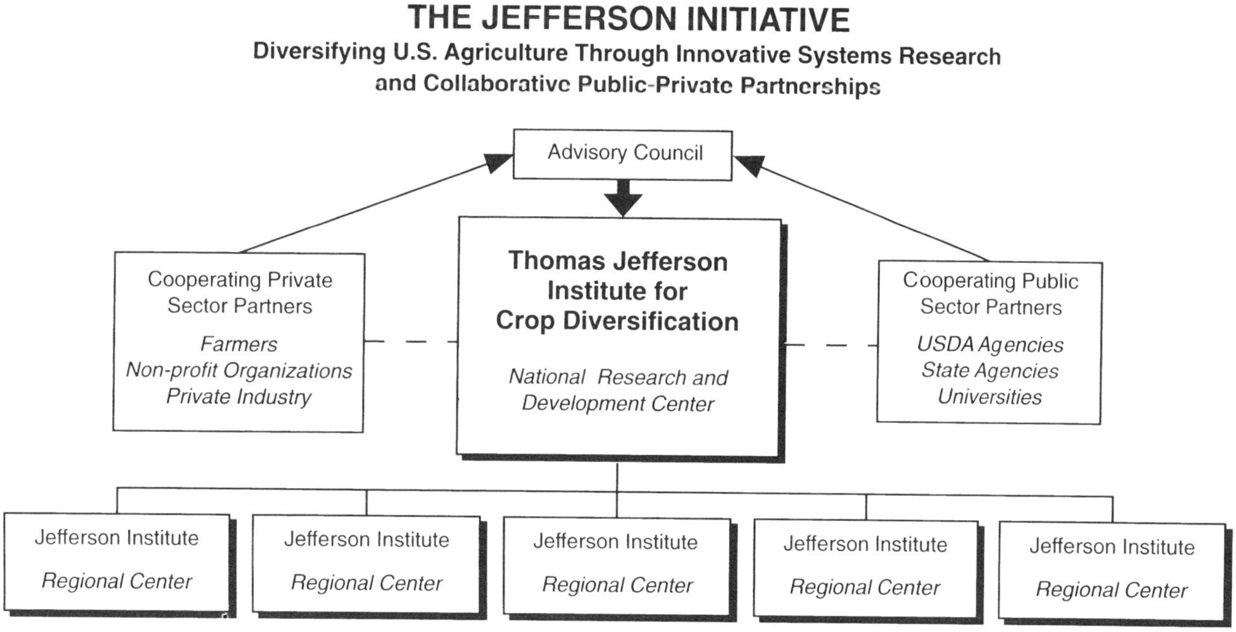

Fig. 1. The Jefferson Initiative to diversify crop production in the United States would be catalyzed by the Thomas Jefferson Institute, comprised of a national research and development center and cooperating regional centers. The Institute would stimulate partnerships with both public and private sector collaborators through grants and contracts derived from federal funding.

The authors of this report are convinced that a focused effort led by the Jefferson Institute, working with a multitude of cooperating partners under the banner of the Jefferson Initiative, is the most effective way of capturing the economic and environmental benefits of new crops in an increasingly diversified U.S. agriculture. The Jefferson Initiative proposes the kind of substantial, long-term, and coordinated frame-work necessary for the creation of a successful national strategic program in new crops development.

SUMMARY

For more than a century, U.S. agriculture has been beset by the pressures of crop price instability. These pressures have caused severe economic problems in rural America. In the past 60 years, federal programs, based on subsidies and acreage reduction, have attempted to stabilize returns to farmers but have failed to check agriculture's eroding economic viability. Moreover, federal crop support programs, at great cost to taxpayers, have served in an unexpected and unplanned way to limit the crop choices available to American agricultural producers.

Public concerns over food safety, commodity program costs, and agricultural sustainability have become important policy issues in the last two decades. Restriction of research funding and crop support payments to major commodity crops has undermined the potential of new crops to alleviate related concerns and pressures. Although support for the development of new and alternative crops has been proposed consistently, publicly funded research has been scarce and fragmented.

The authors of this document propose the *Jefferson Initiative* as a way to focus on new crops policy reflecting Thomas Jefferson's belief in the critical importance of new crops to American agriculture. The Jefferson Initiative would provide leadership in the search for and the development of new crops to improve the sustainability of U.S. agriculture through diversification. Coordinated support for new crops exploration, domestication, production, processing, utilization, and marketing offers tremendous potential benefits to all involved in the agricultural sector—producers, industries, rural communities, and consumers.

REFERENCES

Andrews, D.J., J.R. Rajewske, and K.A. Kumar. 1993. Pearl millet: New feed grain crop. p. 198–208. In: J. Janick and J.E. Simon (eds.), New crops. Wiley Press, New York.

Barnard, H.E. 1937. The background of the Farm Chemurgic Council and The Chemical Foundation. p. 54–55. Proc. Pacific Northwest Chemurgic Conference, Spokane, Washington, March 22–23, 1937. SDPW Program 101A–WPA A5101. Ernest N. Hutchinson, Secretary of State, State of Washington.

Benson, H.K. 1937. Closing remarks and adjournment. p. 120–121. Proc. Pacific Northwest Chemurgic Conference, Spokane, Washington, March 22–23, 1937. SDPW Program 101A–WPA A5101. Ernest N. Hutchinson, Secretary of State, State of Washington.

Busch, L., V. Gunter, T. Mentele, M. Tachikawa, and K. Tanaka. 1994. Socializing nature: Technoscience and the transformation of rapeseed into Canola. Crop Sci. 34:607–614.

Capelle, A. 1996. New industrial crops for Europe. p. 19–21. In: J. Janick (ed.), Progress in new crops. Am. Soc. Hort. Sci., Alexandria, VA.

Carr, E. 1992. Agriculture: The new corn laws. Economist Dec. 12, 1992. p. 3–18.

Council for Agricultural Science and Technology. 1984. Development of new crops: Needs, procedures, strategies, and options. Report 102. Council for Agricultural Science and Technology, Ames, IA.

Downey, R.K. 1990. Canola: A quality *Brassica* oilseed. p. 211–215. In: J. Janick and J.E. Simon (eds.), Advances in new crops. Timber Press, Portland, OR.

Ferguson, L. and M. Arpaia. 1993. New subtropical tree crops in California. p. 332–334. In: J. Janick and J.E. Simon (eds.), Advances in new crops. Timber Press, Portland, OR.

Holmes, C.L. 1924. The economic future of our agriculture. J. Polit. Economy 32:505–525.

Hymowitz, T. 1990. Soybean: The success story. p. 159–163. In J. Janick and J.E. Simon (eds.), Advances in new crops. Timber Press, Portland, OR.

Jefferson, T. ca. 1800. A memorandum: Services to my country. p. 702–704. In: M.D. Peterson (ed.), 1984. Thomas Jefferson Writings. Literary Classics of the United States, Inc., New York.

Jolliff, G.D. and S.S. Snapp. 1988. New crop development: Opportunity and challenges. J. Prod. Agr. 1:83–89.
Jolliff, G.D. 1989. Strategic planning for new crop development. J. Prod. Agr. 2:6–13.
Jolliff, G.D. 1996. New crops R&D: Necessity for increased public investment. p. 115–118. In: J. Janick (ed.), Progress in new crops. Am. Soc. Hort. Sci., Alexandria, VA.
Knudson, H.D. and R. D. Zeller. 1993. The milkweed business. p. 422–431. In J. Janick and J.E. Simon (eds.), New crops. Wiley Press, New York.
Kugler, D.E. 1990. Non-wood fiber crops: Commercialization of kenaf for newsprint. p. 289–292. In J. Janick and J.E. Simon (eds.), Advances in new crops. Timber Press, Portland, OR.
McMillen, W. 1929. Too many farmers: The story of what is here and ahead in agriculture. William Morrow and Company, New York.
Piesch, R., N.C. Wheeler, and P.P. Wyant-Wheeler. 1994. Large scale cultivation of *Taxus* species for Taxol ® production. Proc. 4th Int. Conf. Aromatic and Medicinal Plants, Lyons, France, Dec. 5–7, 1994.
Rasmussen, W.D. 1975. Agriculture in the United States: A documentary history. vol. 1. Random House, New York.
Sampson et al. 1987. New farm and forest products: Responses to the challenges and opportunities facing American agriculture. A report of the New Farm and Forest Products Task Force, U.S. Department of Agriculture, Washington, DC.
Task Group on New and Special Crops. 1957. Report to president's appointed bipartisan commission on increased industrial use of agricultural products. R.D. Lewis (chm.). Texas Agricultural Experiment Station, College Station.
Taylor, C.S. 1993. Kenaf: An emerging new crop industry. p. 402–407. In J. Janick and J.E. Simon (eds.), New crops. Wiley Press, New York.
Taylor, H.C. and A.D. Taylor. 1952. The story of agricultural economics in the United States, 1840–1932. p. 19–29. Iowa State College Press, Ames.
U.S. Congress. 1983. Plants: The potentials for extracting protein, medicines, and other useful chemicals. OTA–BP–F– 23. Office of Technology Assessment, U.S. Congress. U.S. Government Printing Office, Washington, DC.
U.S. Department of Agriculture. 1992. New crops, new uses, new markets. Yearbook of Agriculture. U.S. Department of Agriculture, Washington, DC.

Justification for a National New Crops Initiative

Peter Felker

We are in a world agricultural market in which too many farmers in too many countries are competing for the same market share of the same crop. The recent difficulties in negotiating the GATT treaty are indicative of this problem. For example, farmers in France, Canada, Argentina, Germany, India, and the United States are all competing for wheat markets. In the sub-tropics the fierce price competition among farmers from various countries for rice and sugar cane is an illustration of the need to diversify agricultural systems.

With U.S. price supports and subsidies for existing crops approaching $10 billion/year (Agricultural Statistics 1991) there is unquestionably a need to diversify agriculture to find new uses and products for these land resource bases. There is also an urgent need to find higher value crops for agricultural areas that have low economic returns, such as many areas of Texas.

Therefore it is in the national interest to find a long term permanent solution to the price support issue by finding alternative crops for those crops that require subsidies and price supports. Thus it seems reasonable to devote a small percentage of existing funds for price supports to develop agricultural products that will ultimately avoid the need for price supports. While this search for alternatives is underway, it seems reasonable and just that the United States should not abandon farmers currently receiving price supports overnight.

These new agricultural products should include anything that is legal that can be produced from the land resource base including pine straw bales for landscape, epiphytes from trees for interior design, alligator skins, cactus fruits, mesquite hardwood floors, wildlife, ecotourism, zapote fruits in the Arizona deserts, or native nuts and berries from the Appalachian and Rocky Mountains.

Perhaps the best example of the economic benefits of agricultural diversification is California agriculture. California is the leading agricultural state in the United States with gross farm income of $17.9 billion (Carter and Goldman 1992). Surprisingly California's agriculture industry is greater than that states defense, semi-conductor, or entertainment industries. No doubt part of this success is due to the more than 250 crops that are grown. California produces half of the nations nuts, fruits, and vegetables on less than 3% of U.S. farmland (Carter and Goldman 1992). Farmland in pasture and cash grains dropped by 2 million hectares from 1964 to 1991 while vegetable, melon, fruit, and nut acreage has increased by a third. This illustrates a trend away from extensive agricultural crops with more emphasis on intensive, high-value fruit, vegetable, and nut crops.

There is no compelling need to farm just because the land base exists. It makes little sense for a family to farm 400 ha of sorghum, netting only $62/acre to make a $25,000 income when high value crops on 1% of that area might yield an equivalent income. In the California example we see that when many small diversified, high value cropping systems are aggregated, the resultant agricultural industry produces more revenue than extensive, low-value commodity cropping systems.

NEED FOR NEW CROP CONSTITUENCY

In 1979, then California Governor Jerry Brown told a gathering of new crop specialists, "The problem with new crop development is that you don't have a political constituency." The cattle industry, the cotton industry, and all the other major farm crops have a strong presence in Washington to make their concerns known. However the new crops industry has no such constituency.

Thus it is imperative for New Crops people form a strong professional society to make their views known to Congress and the Executive Branch. It is critically important that this group have strong representation from the commercial sector including brokers, retailers, and land owners/farmers. It is unlikely that a group composed solely of academics from the universities and government research laboratories will be viewed seriously by policy makers. Thus it is critical that this group have strong representation from the commercial sector as well as from the scientific community.

The recent formation of the Professional Association for Cactus Development (PACD) is an example of such a well integrated constituency group. The PACD membership includes nationally recognized chefs and food editors, national brokers of exotic fruits, commercial growers, and scientists.

There is little doubt that many farmers are not pleased with low prices they are paid for major commodities such as sorghum, wheat, apples, or milk. It is to the farmers benefit to have an opportunity to diversify into higher value, specialty niche crops where there is less international competition. While not everyone can grow a niche crop, good farmers should constantly search for higher value alternatives and the government and the universities should fully assist them in this effort.

Any organization that represents a broad constituency of farmers such as the Farm Bureau or the Farmers Cooperatives should be asked to participate in a national program to identify more profitable products from the land that require less price support. It is indeed most fortunate and welcome that the Council of Agriculture Science and Technology (CAST) has initiated a policy paper on New Crop Development (Janick et al. 1996).

It is useful to contrast the grower base and political constituency of the mesquite industry with that of the leucaena industry. The mesquite producers and users organization Los Amigos del Mesquite consists of several hundred woodworkers, barbecue manufacturers, lumbermen, sawyers, and hobbyists. After 13 annual conventions this diverse group has significantly increased the visibility and demand for mesquite products. In

1993 they were successful in getting the National Hardwood Lumber Association to approve new grading rules for mesquite lumber that would be internationally recognized. They also developed a joint program between the Texas Beef Industry Council and the mesquite barbecue producers to co-market beef and mesquite. In contrast, a single man with a background in both the cattle and oil business financed the commercialization of the *Leucaena* industry with development of commercial planting, cultivating, harvesting, drying, and marketing of high-protein leucaena range cubes in less than 4 years with a $3 million investment. This was a directed, organized development plan using teams of engineers, farmers and agricultural consultants.

BARRIERS TO NEW CROP DEVELOPMENT

Attracting Human and Financial Capital

In rural areas it is often more difficult to obtain human capital to manage a budding new agricultural industry than to obtain funds for the development. The good business leaders already have their plate full. One potential investor's reason for withdrawing from a new crop development project was "because I can make or lose more money in the stock market in one afternoon than you will do in a year." Another major investor told a new crop developer that a $50,000 investment just isn't worth my management time. Very seldom are new crops sufficiently well developed that infusion of large amounts of capital will bring them to fruition, it takes time to develop equipment and production processes, supply streams, and markets.

Universities

There has been a lack of university participation in new crop development for a variety of reasons. Perhaps the two strongest reasons are (1) the strong agribusiness lobby for existing commodities i.e. cotton, sorghum, beef that demand faculty positions in those areas and (2) a national disciplinary scientific lobby to promote basic research in biotechnology and plant growth and development.

U.S. agricultural research mandates seem oblivious to the problems of excessive production of too few crops and continue to stress the need for new techniques (i.e. biotechnology) rather than for solutions to the major U.S. agricultural problems. For example a proposal to use genetic engineering technology to produce more of a crop that is already in surplus on a larger land resource base might be funded while a proposal to use non-genetic engineering techniques to replace the cropping system would not be funded. It is incredible that funding agencies continue to seek projects to increase the capability to produce more of the crops that are in surplus through biotechnology, while they have little interest in finding replacement agricultural systems to generate increased profitability for existing land resource base.

U.S. agriculture research is increasingly becoming involved in funding solutions to the significant scientific problems of the various disciplines i.e. plant physiology, soil science, stress resistance, genetic analysis of model systems, models for stress resistance—but not the problems of American Agriculture.

If scientists work on generic national problems they will be understood at national meetings and they are mobile for job opportunities between climatic regions. New crop development tends to be region specific and not broad enough to be in the national interest for all regions.

This region specificity makes it difficult to get grants funded that are not in the entire national interest. Without extramural grant support it is not possible to hire the graduate students and supplies necessary to generate the publications required for tenure.

Additional reasons why universities have not been involved in new crop development is that it is risky for new faculty members to base getting tenure on field projects. Lab work produces many more publications for the same effort than field work. Field trials must be repeated for several seasons (years) while lab work can be repeated in hours, days, weeks, or maybe months.

Universities also do not understand the need for taking their technologies to the market place. University researchers and administrators continue to believe that if research output has significant commercial potential, it will be adopted by the private sector. There are very few examples in the private sector of good products that have been developed without dedicated marketing efforts. Many excellent products developed in university research labs have not been adopted solely due to lack of a proactive approach. The "comfort zone" of university researchers is in the lab describing analytical results. It is much more difficult to achieve market penetration with new products. Furthermore achieving market penetration for new products is not the kind of activity that is considered for tenure or promotion.

It is absolutely necessary to fund resolution of the critical problems of the disciplines of agriculture science but not to the exclusion of direct approaches to resolution of critical problems of American agriculture especially the large expenditures to existing cropping systems.

Pesticide Registration

Perhaps the single most significant institutional barrier to new crop development is the EPA's position on pesticides for new crops. For example, EPA denied the special use permit request by the Texas Department of Agriculture for use of the preemergence herbicide norflurazon on a 150 ha *Leucaena* plantation that was to be used to produce high protein cattle feed. The request was denied in spite of the fact that (1) a refereed journal article identified norflurazon as the most effective herbicide, (2) norflurazon had been approved by the EPA for use in forage soybeans and alfalfa, and (3) USDA/EPA trials were being carried out to verify effectiveness and residue levels. EPA's position is that emergency use permits cannot be allowed for standard pests on new crops. They take the position that the investor should have understood the risks and if necessary initiated field trials with pesticides four to five years in advance of any commercial development. This EPA position is probably the single most significant institutional barrier to new crop diversification in the United States.

New Crop Equipment

There is a major problem getting machinery companies to produce equipment for new crops. Texas A&M University–Kingsville has contacted the engineering staff of several major equipment companies for assistance in adapting forage/hay equipment for harvesting the shrubby leguminous plant *Leucaena* for cattle feed. Most engineers at the national research centers express a genuine concern for the need to develop new equipment. However they are quick to point out that management has them fully assigned to fewer product lines. Thus there is very little help possible from equipment manufacturers even when using their equipment. It must be said that some of the engineers from smaller equipment companies, New Holland Equipment in Pennsylvania and Brown Bear Equipment of Corning, Iowa have been very helpful.

No doubt the difficulty in obtaining assistance in adapting existing agricultural equipment product lines to new crops is a result of the major mergers in world equipment dealers in the last decade. For instance the acquisition of Steiger by CASE, of Versatile by Ford, and of Farmall by International Harvester and then by CASE reduced the number of agriculture equipment dealers from nearly a dozen 15 years ago to four major companies today i.e. CASE, John Deere, Ford, and Agco. The resulting equipment companies want to only produce highly profitable equipment lines that have a proven high demand. The costs for expensive engineering staff required for new product development must show a high rate of return. Thus speculation on new equipment development for unproven markets is something that equipment manufacturers are very reluctant to do. The paucity of small equipment manufacturers is a major bottleneck for new equipment development. It is critically important that national programs for new crop development include representatives from large and small agricultural/forestry manufacturers. Legislation should be considered that would provide tax incentives for equipment manufacturers to compensate for staff time in diversifying our agricultural base.

PROPOSAL FOR NATIONAL NEW CROPS GRANTS PROGRAM

Due to the enormous financial burden of subsidies for existing crops, a long-term, permanent solution to these expenditures must be part of financial responsibility. Development of alternative crops for the land resource base that consume these subsidies is a rationale approach.

Funding for new crop development has come from some state governments, from AARC programs where the crops were non-food, non-feed and from earmarked funds in federal legislation. New crop development has not been funded by the USDA or NSF competitive grants as new crop development has been too applied.

Perhaps the greatest efforts in new crop development have arisen from small unsanctioned expenditures (i.e. bootlegged) of other programs. Many soft money research programs have struggled for years and even decades to keep field equipment repaired and graduate students funded. In summary funding for new crop development has proceeded without a national initiative and has only been possible due to the collective energies of soft-money researchers throughout the country.

Our experience with the parallel commercialization of mesquite, leucaena, and cactus over the last 16 years indicates that the total annual cost for the three crops is about $150,000. This is for a team with a PhD leader, graduate students, and field assistants, using laboratories, tractors, greenhouses and field supplies. Thus the average net cost/crop in a shared development program is about $50,000/yr, but overhead requirements can increase this to $75,000. This is expensive when viewed in perspective of university budgets, but small when compared to the cost of $30–100 million dollar capital improvements in industrial plant improvement, to the costs of a small overpass on the highway or to the $10 billion government subsidies to existing crops.

New crop development specialists have been asking for—and receiving far too little funding. By the same token, they can not be permitted to only describe potential new crops but must be required to take some of these new crops to commercial fruition.

Some staff aides of the Agriculture Committee of the U.S. House of Representatives are strongly opposed to using funds earmarked for farmer assistance to support "lab coat research in Universities." This is evidently due to the perception that this research is primarily for the professional advancement of the researchers and that the research is too long term to be useful in their life time. Clearly, new crop research must have immediate tangible benefits to be politically viable.

A national grants program for diversification is required for an applied program aimed at replacing crops that currently require subsidies. This diversification program should involve; plant exploration, genetic improvement trials, development of cultural practices (e.g. planting, weeding, harvesting), disease and pest management, getting approval for pesticides, market development, attracting investors, enhancement of grower base, and bringing product to commercial fruition.

It is suggested that $5 million/year ($100,000/yr for each state) be made available over a 5 year period. The Secretary of Agriculture has authority to create such a program under Section 5 of the Commodity Credit Corporation Charter. It is suggested that a national advisory board provide guidance and oversight to the program and that the advisory board be comprised of; successful food brokers, nationally known chefs and food editors involved in new crops, agronomists/breeders, EPA pesticide staff, chemical industry personnel, private sector investors, CEO's of new crop companies, USDA managers, and economic botanists/botanical garden directors.

The state programs should hold annual meetings to serve as the state focal point for new crop development. These annual meetings should feature new cultivars, cultural practices, exhibits of uses, and prizes for growers with the best new use.

It is appropriate for this program to originate from the states rather than from the central government to facilitate a more relevant bottom-up program for new crop development. Regionally based programs assumes that the management staff would be sufficiently familiar with the ecology, botany, and business climate to not overlook promising new crops. Clearly there must be coordination of state programs to avoid duplication of effort. However, to assure the broadest possible diversification of both crops and political constituency, state based programs would be preferred over regional programs. It would be inappropriate for the board to specify that the new crop must fit any category other than it be legal and that it replace existing crops that require subsidies or that the new crop provide greater economic returns/financial stability to existing crops. It is hoped that bipartisan support can be developed to create this program to create a permanent solution towards reduction in farm subsidies.

CONCLUSIONS

Two fundamental changes are required to take advantage of economic development of new crops. First, public policy must change to foster new crop development in federal agencies, state agencies, and in universities. Secondly, multi-disciplinary teams of active business people, botanists, generalists, and marketing specialists must be formed, given funding and held accountable for milestones and time lines for new crop development.

We cannot sit idly by and wait/hope for such a grants program. We must all communicate to our university administrators, state legislators, and Congressmen the advantages of a federal new crop program to provide a permanent solution to the budget crisis. The benefits to society in terms of reduced federal deficits, increased variety in our lives from more interesting food and fibers, and economic development of depressed regions will be the result of our efforts.

Key points of A National New Crops Grant Program are as follows.

- A National New Crops Grants Program is required to find long term, permanent solutions to the nearly 10 billion dollar payments for existing crops.
- The Secretary of Agriculture is requested to authorize such a program under Section 5 of the Commodity Credit Corporation Charter.
- An amount of $5 million/yr ($100,000/yr for each of the 50 states) be awarded over a 5 year period in a competitive grants program.
- No agriculture product be excluded, other than it be a legal commodity.
- A national review board be established that is comprised of food brokers, chefs and food editors involved in new crops, agronomists/breeders, EPA pesticide staff, private sector investors, agriculture equipment manufacturers, officers of new crop companies, USDA managers, and economic botanists/botanical garden directors to oversee and direct this program.
- As a guiding philosophy, the grants program should involve; plant exploration, genetic improvement trials, development of cultural practices, disease and pest management, facilitating approval for pesticides, market development, attracting investors, enhancement of grower base, and bringing product to commercial fruition.
- State departments of agriculture should assist this program by providing an annual fair/exhibition of new crop products and investment opportunities.

REFERENCES

Agricultural Statistics. 1991. United States Government Printing Office, Washington, DC.

Carter, H.O. and G. Goldman. 1992. The measure of California agriculture. Its impact on the state economy. Univ. California Publ. 21517, Oakland.

FAO. 1996. Cactus pear cultivation and uses. (Inglese et al. eds.) FAO Publishers, Rome.

Felker, P., R. Chamala, E. Glumac, C. Wiesman, and M. Greenstein. 1992. Forage production of *Leucaena leucocephala* and *L. pulverulenta* in mechanized farming systems in a semiarid region of Texas. Trop. Grasslands 25:342–348.

Felker, P, P. Anderson, D. Perino, and D. Miller. 1994. Grading mesquite lumber. Special Publ. Center Semi-Arid Forest Resources, Texas A&M Kingsville.

Janick, J., M.G. Blase, D.L. Jolliff, and R.L. Myers. 1996. Diversifying U.S. crop production. Issue Paper 6. Council for Agricultural Science and Technology. Ames, IA.

Russell, C.E. and P. Felker. 1987. The prickly pears (*Opuntia* spp): A source of human and animal food in semiarid regions. Econ. Bot. 41:433–445.

New-Crops R&D: Necessity for Increased Public Investment*

Gary D. Jolliff

Surplus production, and consequent low prices, of agricultural commodities has been a regular peacetime phenomenon in the United States for more than a century. It was a problem in the 1870s when the USDA was formed, in the 1890s when cotton producers were encouraged to diversify their cropping systems and introduce new crops (Taylor and Taylor 1952), and in 1933–1942, 1956–1973, and 1982–1995 (Fig. 1). National farm policy aimed at agricultural income stabilization was initiated in 1935 but the problem remains 60 years later despite large government outlays (Tweeten 1995) and several million farmers have been forced off the land because of insufficient economic opportunities. Peacetime cropland reduction, i.e. the cropland taken out of production by federal government programs (Crosswhite and Sandretto 1991) has trended upwards since 1933 (Fig. 1) and the absence of sufficient profitable new-crop alternatives has resulted in lost revenue opportunities from idled cropland, in addition to the costly government outlays to idle the land.

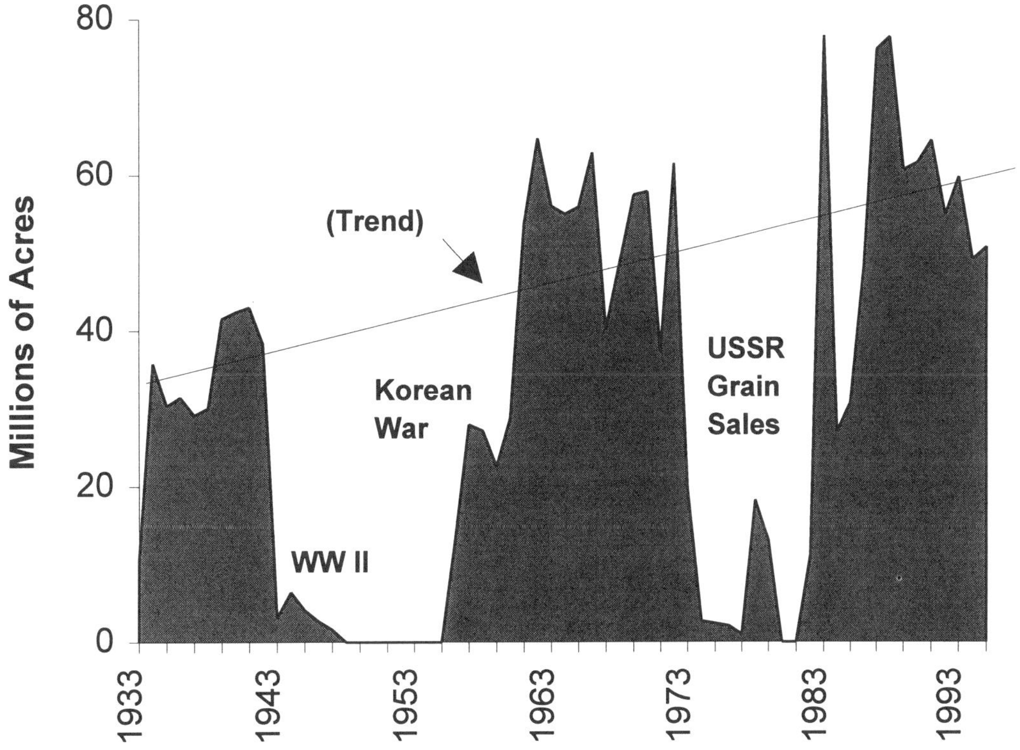

Fig. 1. United States cropland acreage reduction 1933–1994. Millions of acres of cropland taken out of production (idled) by federal government programs (trend line: approximate fit to three means: 1934–1942 = 35; 1957–1972 = 47; 1983–1993 = 58). Adapted from Crosswhite and Sandretto (1991) and updated from C. Sandretto (pers. commun. 1995).

*Oregon Agr. Expt. Station Technical Paper No. 10,963. Sincere appreciation is extended to Steve Dewhurst and Dennis Kaplan, USDA, Office of Budget and Program Analysis for providing historical budget data; to Carmen Sandretto, USDA, ERS for providing historic cropland reduction data used in Figure 1; to Dr. Abner Womack, and Gary Adams, Food and Agriculture Policy Research Institute, University of Missouri-Columbus, Missouri for providing the historic annual data on the seven crops for acreage reduction and participant returns per acre, and for reviewing the calculations; to Dr. James Cornelius, Department of Agricultural and Resource Economics, Oregon State University, and Dr. Rueben Buse, Department of Agricultural Economics, University of Wisconsin for guidance in preparation of the research, and reviewing the calculations.

Surplus production in the twentieth century is due to a combination of rapidly increasing technology and agricultural policy. From 1918 to 1940 surpluses were a consequence of the replacement of voracious draft animals by gas guzzling farm tractors and thereafter it was exacerbated by increases in yields caused by advances in genetic improvement, pest control, plant nutrition, mechanization, and agronomic management practices developed by state and federally supported research efforts combined with price support programs that encouraged planting of traditional crops. With the biological revolution currently underway, these yield improving technologies can be expected to continue and to spread at increasing rates throughout the world. In much of Asia increases in productivity have been sopped up by rapidly increasing populations, but crop increases have not kept up with demand in parts of Africa. The expanded productive capacity in North America and Europe however has not been compensated for by equivalent population increases and, until recently, lack of financial resources in the developing world have dampened exports and in many cases exported feed grains were either donated or subsidized. New crops development, particularly industrial crops (previously called chemurgic crops) have long been proposed as a solution to this problem and tremendous advances have been made in expanding new uses of traditional crops such as soybeans and maize. Clearly, the problem has not been solved suggesting that the neglected search for alternate and new crops be reconsidered.

There are obvious reasons why the private sector does not invest in new-crops research and development. New-crops development is: (1) long term, (2) high risk, and (3) may not yield benefits to the originators. Therefore, adapters and users intentionally wait until the profit potential is more obvious and nearer term before risking research and development investments for new crops. Although these factors justify government-funded intervention in agricultural research (Alston et al. 1994; Jolliff and Snapp 1988; Jolliff 1989), there has been extreme underinvestment in new-crops research and development compared to the established agronomic crops which are frequently in surplus (Jolliff 1989). As there is no voice for new crops, federal support has languished.

As a result of underinvestment in research, potentially viable new crops have not been developed to the point of profitability to the private sector. Determining the potential payoff to justify increased public-sector investments in new-crops development is needed. The objective of this paper is to estimate federal financial outlays and lost national revenue opportunities for the 17 year period, 1978 through 1994, associated with the absence of sufficient new-crop alternatives. This study was based on the assumption that science, technology, expertise, and plant materials have existed which could have been mobilized during the past century to provide profitable new-crops for U.S. farmers as alternatives to idling cropland and overproduction of surplus crops.

DETERMINING COSTS OF SURPLUSES

The total cost of crop overproduction for the 17 year period 1978 through 1994 was the summation of the total cost of federal outlays, plus an estimate of total lost opportunity costs.

Federal Outlays

Federal outlays included (1) cash outlays from Budget Function (BF) 351, (2) value of certificates, and (3) interest. Budget Function 351 of the federal budget is a budget category which includes many aspects of federal outlays for farm income stabilization. Much of the expenditure is for surplus commodity-related programs, such as loans, subsidies, and costs of acquiring and storing commodity stockpiles. Note that some surplus-commodity-related costs, such as foreign agricultural export losses, crop improvement research funded through the Agency for International Development, and domestic donations of food are not included in Budget Function 351 and are not estimated in this analysis. Also not included are some costs to the consumer and taxpayer such as inflated domestic prices (e.g. peanuts and sugar) as a consequence of government program quotas. The data source was the USDA Office of Budget and Program Analysis.

Certificates, which are non-cash outlays, include the $8.4 billion cost of the 1984 Payment-In-Kind (PIK) Program. The data source was the USDA Office of Budget and Program Analysis.

Interest on outlays (BF 351 and Certificates) are based on 3-year U.S. Treasury Security Interest Rates (Table B-72, Economic Report of the President, February, 1995). Interest was compounded annually. Sums for each year were adjusted to 1987 dollars using the appropriate deflator (Table B-3 of the Economic Report of the President, February, 1995). Conversion of 1987 dollars to 1995 dollars was done by multiplying the 1987 dollars by the 1995 GNP Implicit Price Deflator of 1.281.

Lost Wealth Estimates

Lost wealth opportunity estimates are based on estimated revenue losses from idled croplands plus the consequent lost multiplier effect based on the economic benefit of new wealth which was not produced. Calculation of lost wealth is based on idled acreage and returns/acre data from the Food and Agricultural Policy Research Institute, Univ. of Missouri–Columbia. Idled acres for each of seven crops (wheat, corn, sorghum, barley, oats, cotton, and rice) were multiplied by the return per acre of farmers who participated in the government program for the respective crop. This is an estimate of financial opportunity for alternative crops on idled cropland.

A multiplier effect of 2.0 was used to estimate the economic benefit to the economy from new wealth which could have been generated on idled land if profitable new crops had been developed for U.S. farmers. The values for federal outlays and lost wealth for the period 1978 to 1994 are shown in Table 1.

NEW CROP SOLUTIONS

The lack of profitable alternative crops for U.S. farmers during the period 1977 through 1994 has been extremely costly to taxpayers with severe social consequences to farmers and rural areas. Total costs for the 17-year period are estimated as $728 billion in 1987 dollars, equivalent to $932 billion in 1995 dollars. Clearly, an economic evaluation of the cost of crop surpluses requires careful examination in the context of new-crops development policy. The annual federal budget for agricultural research is approximately $2 billion. Plant productivity research, an important component, is almost completely devoted to established crops, with very little funding or institutional structure committed to new-crops development.

Agricultural research is rightly touted as yielding high returns on investment. There is question however, if the calculated rates of return in the literature took into account the high costs of government programs associated with the generation of crop surpluses. Also, arguments are being made with increasing frequency that environmental and social costs of activities, such as farm programs, should be considered (Van Dieren 1995). These types of accounting could strengthen the case for federal investments in new-crops development (Jolliff 1989). Federal appropriations would need to be new-crops-specific to assure resource use in the intended area. This implies that additional institutional innovation in national farm policy—beyond the USDA Alternative Agricultural Research and Commercialization Center—may be necessary to bring about needed change (Jolliff 1989). A suggested New Crops Initiative has been suggested in CAST Issue Paper 6 (Janick et

Table 1. Outlays and lost opportunities for crop agriculture, 1978–1994.

Category	Billions of 1987 Dollars
Federal outlay	
Budget function 351	254
Certificates	37
Interest (BF 351 + Certificates)	289
Sub total	580
Lost wealth opportunities	
Lost crop opportunities	74
Multiplier effect loss	74
Sub total	148
Grant total	728

al. 1996). Advances in new-crops development may be an important way to raise the standard of living in many countries of the world.

CONCLUSION

Historical fiscal evidence of the costs of surpluses justifies increased investment in the development of profitable and sustainable new-crop alternatives for U.S. farmers. A long-term strategic national effort is needed to make maximum use of genetic diversity in crop agriculture.

REFERENCES

Alston, J.M., P.G. Pardey, and H.O. Carter (eds.). 1994. Valuing UC agricultural research and extension. Pub. No. VR-1. Agricultural Issues Center, Div. Agricultural and Natural Resources. Univ. California, Davis.

Council for Agricultural Science and Technology. 1995. Sustainable agriculture and the 1995 Farm Bill. Council for Agricultural Science and Technology. Ames, IA.

Crosswhite, W.M. and C.L. Sandretto. 1991. Trends in resource protection policies in agriculture. p. 42–49. In: Agricultural resources: Cropland, water, and conservation. Situation and Outlook Report AR-23. Economic Research Service, U.S. Department of Agriculture. Washington, DC.

Janick, J., M.G. Blase, D.L. Johnson, G.D. Jolliff, and R.L. Myers. 1996. Diversifying U.S. crop production. Issue Paper 6, Council for Agricultural Science and Technology. Ames, IA.

Jolliff, G.D. 1989. Strategic planning for new-crop development. J. Prod. Agr. 2:6–13.

Jolliff, G.D. and S.S. Snapp. 1988. New crop development: Opportunities and challenges. J. Prod. Agr. 1:83–89.

Taylor, H.C. and A.D. Taylor. 1952. The story of agricultural economics in the United States, 1840–1932. Iowa State College Press, Ames.

Tweeten, L. 1995. The twelve best reasons for commodity programs: Why none stands scrutiny. Choices Second Quarter:4–7, 43–44.

Van Dieren, W. (ed.). 1995. Taking nature into account: A report to the Club of Rome. Copernicus. An Imprint of Springer-Verlag.

The Alternative Agricultural Research and Commercialization Center (AARC), U.S. Department of Agriculture

Robert Armstrong

On the seal of the U.S. Department of Agriculture, the Department's motto proclaims that, "Agriculture is the Foundation of Manufacture and Commerce." Aside from a few crops for fiber production, when was the last time you thought of agriculture as the basis of manufacturing?

The fact is that U.S. agriculture has become an enormously successful producer of food and feed. So successful, in fact, that it has become a victim of its own success. During the last three decades in this country, we have had acreage reductions/supply controls in place three out of four years, with an average of 40 million acres idled out of a 400 million acre cropland base (16 million of 160 million ha). The annual bill to taxpayers for idling 10% of our land ranges between $10–15 billion.

In addition, we are not the only ones capable of growing grain. In the 1980s, the rest of the world expanded its grain production, resulting in a decline of U.S. market share in the last ten years. (Maize down from 77% to 66%; wheat down from 44% to 32%; soybeans down from 79% to 66%.) Arguments about unfair trade practices aside, the bottom line is that the balance of global grain market power is shifting away from the U.S.

There is an alternative market for the tremendous productive capacity of U.S. agriculture, however. By recognizing that renewable materials produced on our nation's farms can satisfy more than just food and fiber needs, U.S. agriculture can supply the feedstock for a number of industries ranging from plastic substitutes to newsprint.

The result of developing such materials is nothing short of a complete shift in our nation's economy—away from its current orientation towards petroleum-derived products based on hydrocarbons to plant-derived products based on carbohydrates. It has been said that the United States is the Persian Gulf of carbohydrates.

Such an orientation is a return to the economy of 150 years ago—alluded to in the Department's motto. In recent times, the Chemurgic movement of the 1930s and 1940s championed such a return. That early work led to the formation of the four USDA regional laboratories.

With the peaking of the petroleum age in the 1970s, there is now a renewed interest in developing carbohydrate products. A number of factors have come together to make the use of farm and forest materials in industrial uses more viable in the 90s. They include: more efficient technologies for processing and manufacturing; environmental and regulatory requirements for clean air and water, landfill alternatives, disposability and renewability; availability of skilled labor and resources in rural America; and increased pressure to reduce agricultural subsidies. In its 1987 report to then-Secretary of Agriculture Block, the New Farm and Forest Products Task Force set as this country's goal the development and commercialization of an array of new farm and forest products. The report cited the goal as a 25 year effort that would use 150 million acres (60 million ha) of farmland, generate 750,000 jobs, increase farm income by $30 billion and add $100 billion in national economic activity.

In response to the 1987 report, and with bipartisan congressional support, the 1990 Farm Bill authorized the establishment of the Alternative Agricultural Research and Commercialization—AARC—Center. The Center's mission is to accelerate the commercialization of industrial products (non-food, non-feed) derived from farm and forestry materials—creating jobs and economic activity in rural America. Part of that mandate includes the development of new crops for new industrial materials.

In 1992, Congress authorized the first appropriations for the AARC Center and in March of that year then-Secretary Madigan appointed the Center's nine-member Board of Directors. In the spring of 1993, the Board conducted a series of eight hearings around the country to get advice from the public as to how they felt the Center could best accomplish its mission. In the fall of 1993, drawing on the recommendations from the public hearings, the AARC Center began to enter into its first agreements with private firms.

The AARC Center facilitates commercialization by forming private/public pre-commercial partnerships. This helps to initiate market penetration strategies early in the product/process development. It also places the private sector in the lead; the guiding philosophy being that the private sector has much more knowledge of the marketplace than a government-run organization. Moreover, the Center concentrates on the critical pre-commercialization stage because that is the most difficult stage for a company to secure funding. Most estimates cite the pre-commercialization expenses as being 100 times that of the expenses of basic laboratory research.

Monies provided by the AARC Center should not be viewed as a duplication of other programs by the Department. The unique nature of the AARC Center is that its activities are market driven. Other USDA organizations are technology pushed. Thus, the Center's activities are complementary to other work supported by the Department.

An additional aspect of the AARC Center is its reinvestment strategy. The Center is funded via a revolving fund and not on an annual "use it or lose it" appropriation. Money provided to a firm for commercializing a product is repaid, to include an appreciation for the investment risk taken by the Center. Drawing on these revenues, it is the goal of the Board to build up a fund of several hundred million dollars that can be used for financing future projects.

As of early 1996, the AARC Center is helping to finance nearly 60 projects—mostly small- and medium-sized firms—with some \$21 million of taxpayers' money. About \$73 million of private sector money is also invested in the projects. The projects not only generate demand for agricultural materials, but they also generate jobs in rural America; approximately 3,000 new jobs are expected to be created in the near-term.

For further information on the AARC Center program, contact the AARC Center via fax at (202)690-1655.

MARKETING AND COMMERCIALIZATION

Marketing New Crops to the American Consumer

Frieda Caplan

The well-constructed food trend superhighway that we are so used to these days started out in 1962 as a small pebbly road that few people in the food business cared to travel especially as it related to fresh fruits and vegetables. That's the time when produce like avocados, pineapples, papayas, and mangoes were considered "exotic" and were rarely used by chefs in chic restaurants.

It was tough selling such "exotic produce." The only outlets at that time were a few adventurous retailers who wanted to distance themselves from their competition, and to better meet their shoppers' needs, or those who served communities heavily populated with ethnic groups—mainly Hispanic, Indian and Filipino—who recognized these foods as common to their culture. It was also in 1962 that the company, presently known as Frieda's Inc., opened its doors.

THE FRIEDA STORY

The media, in recent years has given Frieda's the mantle of the "company that changed the way America eats" and it is an enjoyable crown to wear. However, it may be overstating the contribution we have made to what is now a mad rush to be the first to find, to predict, and to help introduce the latest food or cuisine to entice the American consumer. Ever since we first introduced—and named—kiwifruit in the early 1960s—which didn't become a trendy food until 18 years later—we have found ourselves as the marketers of all sorts of underdeveloped or under distributed produce items. Foods like alfalfa sprouts, blood oranges, cherimoyas, passion fruit, pearl onions, and the now very popular red seedless watermelon were amongst the produce specialties Frieda's introduced to supermarkets in our first couple of years in business in the early 1960s.

We were very lucky. Other produce operators, both locally and nationally, were only interested in high volume items such as oranges, apples, tomatoes, potatoes, lettuce, and strawberries. Fortunately for us, they did not hear opportunity knocking on their door. During the last few years, they have heard that knocking loud and clear and competition is now intense to discover, promote, and sell specialty fruits and vegetables and complementary items. This is a real plus for retailers and shoppers.

But in the late 1960s, when some of these specialty items did take off, as a result of our unique merchandising and marketing efforts, we became known, as trend setters—much to our surprise. We didn't realize that what we were doing, marketing wise, had never been done before in the produce industry. Frankly, we were just trying to find ways to move the products that we, as a new produce company, had an opportunity to market.

Getting Started

Everyone has a reason for choosing a certain career path. Mine, I admit, was a little different than most. Back in 1956, I chose to enter the produce industry primarily because of its most unusual hours. I wanted to breast feed Karen, my first born daughter but by necessity, I needed to continue working. Six years later, I was presented with an opportunity to start a produce company of my own. When I discussed this opportunity with friends and family, the general response was, "that sounds like a good business idea because everyone has to eat."

They were right, "everyone has to eat" but no one ever said that shoppers would consider eating jicama, spaghetti squash, elephant garlic, Sugar Snap Peas®, dried tomatoes, or even kiwifruit! You see, I started a produce business featuring specialty fruits and vegetables.

Early in 1962, the management of the Southern Pacific Railway, which at that time owned the Los Angeles wholesale produce market, suddenly had space available to lease! They were the ones that requested that I go into business for myself, even though they knew I had no funds and no experience running a business. These Southern Pacific Railway people told me they had watched me build fresh mushrooms into a high volume item over the previous five years for my several employers! Looking at it as an opportunity that would benefit their ownership of the Los Angeles produce market, they literally, in a ten day period, pushed me into my own business.

I thought they were real risk takers, but they insisted that I was destined to be a success. As I look back, these men were my first real mentors. They took a lot of extra steps to make my entry into this male-dominated and male-controlled industry relatively painless.

In addition to their support, I also started with a solid partnership already in place with the mushrooms growers of Southern California. This small network of mushroom growers minimized my risk of failure in a most unusual way. Knowing I had no funds to start a business, they agreed to consign their mushrooms to me. I didn't have to pay the growers until the mushrooms were sold and, the money collected. So, between the managers of the produce market and the mushroom growers and most important an incredibly supportive husband—you see by that time we had two small daughters (Karen was 6½ and Jackie would soon be 4)—I was launched on a career path in a business totally dominated by men!

So, on a Monday morning, the first week of April in 1962, we opened our doors with purple signs announcing the birth of a new produce enterprise, I literally started with nothing financially except grower support, plus encouragement from my father. He took me to his local bank and arranged for me to get a loan of $10,000 which in 1962 was a lot of money. When we left the bank he counseled me, "pay it back as quickly as possible, Frieda, so that you will have good credit standing should the need ever arise." The loan was paid back in four months!

The California mushroom growers didn't back me because they thought I was a nice young lady. They urged me to go into business because they needed help! The Taiwanese mushroom growers were invading the United States canned mushroom market. These California growers saw our new firm as an important vehicle for saving their industry. To survive, these growers had to expand the market for fresh mushrooms! They could no longer look to canning as an important outlet. They literally forced our new firm to do innovative thinking, to take a little known fresh specialty item—which mushrooms were at that time—and make them a must in supermarkets. Today, fresh mushrooms are amongst the ten top sellers in produce departments across the United States and Canada, a category that is very profitable for everyone concerned: retailers, growers, and wholesalers. We're proud we had a hand in this marketing achievement. The current media has overlooked the fact that before they dubbed me as the "queen of kiwifruit," early press coverage always referred to me as "the mushroom lady."

Because of the mushroom growers' need, we became first and foremost a merchandising/marketing organization. This was a totally new concept within the framework of the wholesale produce industry which up to that time just bought and sold produce as a commodity on a day to day basis. For Frieda's, Inc., mushrooms were just the beginning.

We first learned of the "Chinese gooseberry" shortly after we opened our business. The local Safeway buyer asked if we had ever heard of this fruit. A shopper in one of their stores who had visited New Zealand was looking for it. Well, we had never seen or heard of them. Two months later, when a produce broker mentioned he was offering a new fruit from New Zealand called "Chinese gooseberries," we immediately agreed to take all he had! After all, we knew we had that one customer. The broker was overwhelmed because everyone else he had approached on the wholesale market turned him down. They were only interested in high volume established items. So we had no qualms about bringing in the first 2400 lb of Chinese gooseberries which actually took us four months to sell. We quickly learned we had too much fruit for that one customer.

Soon after receiving our load of chinese gooseberries, we suggested to the New Zealand growers that they change the name of this egg-shaped/brown delicacy to kiwifruit, as it resembled so closely their national bird. They did, and the rest, of course, is marketing history. Not only does New Zealand credit Frieda's with introducing kiwifruit worldwide (we made the first sale of their fruit to Japan), but we are also given recognition for helping to launch the California kiwifruit industry in 1967.

Our willingness to accept the broker's offering of kiwifruit confirmed the reputation we had already earned at the time of being willing to try anything new. The men on the Los Angeles produce market had identified us as the only company around who would listen to small farmers with lesser known produce items such as limes, shallots, Jerusalem artichokes (which we eventually renamed and marketed as "sunchokes"), horseradish, papayas, pineapples, avocados—all considered specialties in 1962. Actually, our greatest source of new produce items at that time were the men on the produce market, too busy selling high volume produce to see the potential of these low volume items.

Growing

I'm often asked what I think my crowning achievement in the industry has been. It was not the introduction and marketing of a new produce specialty, but rather, the recognition that the time had come, in 1986, 24 years after the founding of Frieda's, for me to hand over the leadership of our company to my daughter Karen, the baby I nursed so many years earlier and who thus, was responsible for launching my produce career.

Karen is now president of Frieda's, Inc. With her exceptional administrative talents, which I knew was something I didn't possess, Karen took an already successful marketing organization doing between $10 and $11 million dollars a year in gross revenue and built it in a few short years into the 7th largest women-run business in Southern California, currently grossing $25 million annually.

Karen became my mentor during this leadership transfer. She was astute enough to recognize that I needed to expand my vision if I was to continue to be effective in the growth of Frieda's. One of her first steps was to enroll both of us in a course at UCLA taking companies from the entrepreneurial mode into the corporate environment. As I look back, I realize Karen didn't need to take this course but it was her way to make sure that I continued to grow and to become an effective support system.

Karen has been an equally effective mentor to her younger sister Jackie now vice-president of Frieda's. Recognizing her exceptional sales skills, Jackie has been given the responsibility of developing and expanding our sales program and personally servicing our major national accounts.

In July of last year, we left our 44,000 square foot distribution center in downtown Los Angeles and moved to Los Alamitos in Orange county to a facility nearly twice the size. We currently operate 24 hours a day, 5½ days a week, with a staff of just over 100 men and women. We service supermarkets through out the United States and Canada and in recent years, started selling internationally. Because of Karen's foresight, we have also developed a mail order business which finally this year may turn a profit!

Strategy

Why is our company considered successful? How did we make it grow? How did we gain retail acceptance which we had to have before consumers would buy our products? We think it is because of our communication strategy.

In 1965, when we renamed and introduced Sunchokes® as the first consumer package with recipes and descriptive information many of our retail clients laughed at us and said such a package wouldn't be accepted by their shoppers. They told us their shoppers wanted everything in bulk. When they learned that sales increased 600% for this item in the first two months after it's introduction, many of these retailers stopped laughing and started buying. This gave us a great deal of credability in the industry as an innovator although we developed this package trying to solve a bottom line problem for a customer: how to reduce his shrink at store level, enhance the shelf life of the product, and differentiate it from ginger, which visually it so closely resembled.

Suddenly, the food trend highway was becoming paved and widened and a lot smoother. It allowed us—through packaging or labeling—in the late 1960s to introduce and get acceptance for cocktail avocados, elephant garlic, feijoas, sugar cane batons, and the stunning, but slightly tangy, tamarillo which now, nearly 30 years later is just becoming an "in" fruit/vegetable (depending how you use it). Another burst of laughter greeted our introduction of hot house cucumbers. "Who in their right mind would pay the difference?" said growers and sellers of field grown cukes. They quickly learned that a whole new set of consumers developed, people who would eat cucumbers for the first time because they were burpless.

By the early 1970s, we hadn't as yet earned the support of chefs, who in years to come were to be stars of the food trend scene. However, some of the more advanced retailers and forward looking food media began to give us ideas and acquaint us with sources of new items. They supported our development efforts, the retailers with ads and the media with newspaper or magazine stories. Items like our field-ripened blackeyed peas, tofu, jicama, and fresh herbs came to us through these avenues.

Soon growers, nurserymen, seed companies, foreign embassies, trade commissions, and university farm advisors started to offer us unusual items: spaghetti squash, Asian pears, shiitake mushrooms, red bananas. One of our biggest triumphs was in 1979, when we worked hand-in-glove with Calvin Lamborn of the Gallatin Seed Company who had, after 25 years, developed the Sugar Snap Pea®. That marketing introduction took close to 10 years before it began to fly off produce stands.

And it was just about this time that we lucked out and met Noel Vietmeyer who had a dramatic impact on our business and set us on a new, more focused path. He made us aware of the abundance of underexploited produce just waiting for a market, items like the winged bean and quinoa, which are both still getting established, and taro root, which is much further along in market acceptance.

Back in 1972, when someone from our printing company mentioned that there was space available on the back side of one of our packages, it gave us an idea! That's when we added a paragraph inviting shoppers to write to us for more recipes. And write they did and write they still do. Since 1972, we've received well over a quarter of a million such letters. We hear from 300–500 consumers each week who tell us what Frieda items they bought and where they bought it. In response, we send attractive recipe brochures which go out within two to three working days. Requests for nutritional information or dietary needs are answered within 7 working days. If we receive a complaint, legitimate or not, we issue a full retail refund, plus postage, and often send a package of our dried starfruit or another one of our popular packaged items! Of course, we've also heard from every Frieda in the United States, wanting to know if the Frieda on the label is a real person.

In the early 1980s, I was invited to addresss a large gathering of the nation's top food media at the Food Marketing Institute's annual convention in Chicago, after which a real bonding took place with Frieda's. At the end of each year, thereafter, we would regularly get phone calls—and still do—asking us what's hot or new for the coming year. We take great pains to be very credible when we reply because we know that publicity can be very self fulfilling. If the chefs believe the press, they begin to feature the items on their menu and then retailers stock the products on their stands. This is why we feel we have had such success with the yellow finnish potato, dried red and yellow 'Roma' tomatoes, Kiwanos® (the African horned melon), kabocha squash, our incredibly eye-catching and super-delicious donut peaches (the peento peach), purple potatoes, and our fresh and dried habanero chiles, which when introduced in May of 1990, is given credit by many for launching the current hot chile craze.

Does making our produce and complementary items known to millions of consumers take a budget of a Sunkist or a Procter & Gamble? For Frieda's this is not really the case. For more than 33 years, our company has been courted by the media, and we find it has been a very successful and rewarding courtship. It blossomed full blown in the mid 1970s when the news gatherers began to recognize that fresh fruits and vegetables were newsworthy. Because the men on the produce market were scared to death of talking to the media, they would say, "go see Frieda, she will be glad to talk to you." I was never considered shy! That led to a two-year stint on the local ABC-TV outlet, where I appeared once a week on the eyewitness news, filmed on the market, promoting fresh produce of all types, not necessarily specialty produce. I was called the green grocer, until Joe Carcione threatened to sue ABC for using an identification he felt he owned.

Because of the nature of our particular produce business, we always have new items to talk about. Therefore, the media considers us a constant source of news. But what also continues to make us of interest to the press is that we are a women-owned and women-run company. The press seems to pay us special attention because they perceive Karen and Jackie as successful daughters of a successful mother, and believe me, it's not hard to take advantage of such a perception. When it comes to building a business you learn to use every advantage you can.

We were telemarketing long before the term was ever coined. For the first 12 years, I was the national sales staff, but I never had to make a sales call. Retailers called us as there was no other produce operation in this country recognized as a source of specialty produce. Then, in the early 1980s, competition began to surface. So we developed a team of sales people that telephoned clients and potential clients across the United States and Canada, selling and promoting our marketing program.

New Products

New product development is where I concentrate. We are constantly exploring what's hot and what's new in fruits and vegetables through the world. As some of the items we have introduced go mainstream—like conventional fresh mushrooms, kiwifruit, alfalfa sprouts, hot house cucumbers—they are no longer major profit centers for Frieda's. Although we continue to handle such items as a service to our clients, to gain continuous growth, we must concentrate our marketing efforts on new items that have high volume potential. In November of 1993, for instance, we introduced the first dried tabasco peppers and bird chiles to United States. In June of 1994, we developed the Camouflage Melon® and the Carnival Squash®, already one of our top selling squashes because of it's taste, colorful appearance, and excellent shelf life.

We realize that we can't just keep pace with consumer attitudes, but have to continue setting the trends. That is an important part of what makes Frieda's of interest to the media, but more importantly, to retailing decision makers. The fresh produce department is now recognized as the most profitable in the supermarket, so much so that many retailers are currently remodeling stores, giving produce a very prominent position, to enhance their image with their shoppers.

We seek the new, the untried and we are willing to risk failure. That's the attitude at Frieda's and that is our strategy for today and tomorrow. We operate with an open door and with open minds! And along the food trend highway, we have an awful lot of fun!

THE FUTURE

In a rather fragmented way, I've tried to paint a picture of where trends start and how we launch new products that propel some of these trends. It is not an overnight process, it takes a lot of time and effort, with heavy emphasis on research and market development. Keep in mind that kiwifruit didn't become trendy in the early 1960s when it was first introduced; it didn't make it until 1980 when innovative young chefs from New York to California began featuring it on their menus. This is why we call kiwifruit our 18-year overnight success story.

The food trend superhighway is now fully developed and flashy bill boards are on both sides of the road. If you look closely you will see the once-lowly cactus pear taking center stage around the world as 14 different countries join together in a professional organization to promote and develop both the fruit and it's pad, the nopalitos.

And as futurists, who can possibly overcome your food prejudices, you may have seen the article published last March in an edition of the Wall Street Journal which featured a story on the front page entitled "if you are what you eat, this story is food for thought," subtitled, "health restaurants in Asia prescribe dishes for ills." It was a story worthy of our attention. Although eating scorpians and snakes may not be our way of overcoming stress, we will watch what is going on in Hong Kong, because it certainly appears to be their cup of tea—a cup of tea we just might be sharing along the food trend superhighway in the not too distant future.

Frieda's™

The
Specialty Produce
People

Marketing Lost Crops™ of the Americas

Frieda Caplan

In 1994 at the annual Produce Marketing Association convention in San Antonio, Texas, Frieda's, Inc. introduced the "Lost Crops™ of the Americas," a line of plant based foods that are indigenous to South, Central, and North America. The concept of "Lost Crops™" came from a book published in 1989, entitled, *Lost Crops of the Incas*. Noel Vietmeyer, then senior program officer of the National Research Council was, at that time, the director and scientific editor of the Inca crops study that resulted in the publication of this book. With the permission, encouragement, and guidance of Noel Vietmeyer, we committed a great deal of our company's resources—financial, research, merchandising, and marketing—to introduce this program to the American consumer.

Since we opened our doors over 33 years ago, an important mission of Frieda's, Inc.—a merchandising and marketing company operating at the wholesale produce level—has been to discover, introduce, develop, and create markets for specialty fruits, vegetables, and complementary items. In the produce industry, we are known as creators of marketing excitement! We search out unique foods that have good taste, good shelf life, are good for you, and make the eating experience memorable. Through our packaging and the information we provide, we try to capture the imagination of the American consumer, but the first step is to gain the commitment of the retailer, the decision maker responsible for presenting our "discoveries" to their shoppers. Without the support of top management as well as their produce merchandisers and their buyers, we knew that our "Lost Crops™ of the Americas" concept would face many marketing hurdles.

It took nearly 18 years for the kiwifruit to start earning its stardom in the produce department. It took nearly ten years for both spaghetti squash and the Sugar Snap Pea® to reach this same level of acceptance. Can a program embracing a multitude of little known fruits, vegetables, grains, and nuts be a financial success—one that will favorably impact our bottom line as did kiwifruit, spaghetti squash, and the Sugar Snap Pea®? Only time will tell.

DEVELOPING THE LOST CROPS™ CONCEPT

How did we develop our intense interest in these "Lost Crops?" I actually tingle with excitement when I recall the seminar in early March of 1984 held at the offices of the National Research Council in Washington, D.C. that alerted us to the possibilites of such a produce line. There were around 20 people present including Hugh Popenoe, Donald Plucknett, and the late Calvin Sperling, people who were legends to me. Also present was George White, former plant introduction officer of the USDA/ARS and his staff botanist, Sharon Drexler who both turned out to be invaluable contacts. I was the only marketing person.

The primary purpose for the gathering was to draw attention to overlooked food crops of the Andes. The crops were not truly lost and most were well known in many areas of the Andes, especially among the indigenous populations. These crops, however, were lost to the mainstream of international science and to people outside of the Andes.

For Frieda's, that meeting planted a marketing seed that took nearly 10 years for fruition. We already had experienced some of the products of the Andes. These included oca (*Oxalis tuberosa* Mol.). Which we tasted in New Zealand some seven years earlier. California-grown cherimoyas (*Annona cherimola* Mill.),

LOST CROPS™ OF THE AMERICAS

which we began marketing soon after we started in business—passion fruit (*Passiflora edulis*) and tamarillos [*Cyphomandra betacea* (Cav.) Sendtn.] which we were already sourcing from New Zealand.

When the report of our ad hoc panel was published in 1989 replete with additional information and access to resources, that seed planted in our marketing minds five years earlier, began to sprout. In 1992, during the 500th anniversary celebration of the discovery of America by Christopher Columbus my daughter Karen, who is president of Frieda's, Inc. and Bess Petlak, our special projects coordinator, attended a unique series of events sponsored by the Smithsonian that brought a concrete marketing idea into focus, and out of it, two years later was born our "Lost Crops™ of the Americas" concept.

Soon after their return, Karen assembled an all-day meeting involving our key management personnel, our design firm, and our marketing consultants who specialized in new product marketing. The meeting was conducted by an outside facilitator who lead us through a brain-storming exercise after Karen presented her concept for this new program. We batted ideas around for hours. We quickly recognized that we had a tremendous amount of research, product sourcing, and packaging development to do before any such program could be implemented. Tasks were assigned. In June of 1994, nearly two years later, a formal marketing proposal was finalized.

> **Our objective:** *to increase and reinforce the leadership and innovative image of Frieda's, Inc. through the introduction of a new line of specialty fruit, vegetables, and grains. These products would be unique and their origins indigenous to South America, Latin America, and North America. Our intentions were to position these products as a department within a department to highlight their uniqueness and individuality.*

We were already aware that since the publication of *The Lost Crops of the Incas*, many national associations and academies of science in many countries began investigation, research, and development on the crops covered by this publication. Peru began investing and sponsoring a return to Incan agriculture and crop production. We also saw interest from American tobacco farmers as they began looking for alternative crops.

Seeing all this, we at Frieda's, Inc., wanted to capitalize on what we felt was the next frontier of innovative cuisine and specialty foods. We began the trademark process on the name "Lost Crops™ of the Americas" and the unique products it would represent.

Positioning the Product Line

Then came the positioning of the product line. These specialty fruits, vegetables, and grains needed to have a distinct and different appearance in comparison to common, domestically grown produce. The positioning of the products would have to reflect the elements of the cultures of the Incas, Mayans, Aztec, and southwestern Native Americans. Recipes using the "Lost Crops" products would have to be developed to replicate those recipes in the culture from which they originated. The line of items would need to appeal to those individuals who like to prepare unique menus and who would not be intimidated by extensive preparation and hard-to-find items.

Marketing

To begin the marketing of "Lost Crops," we made the decision to move some existing Frieda's items out of our regular product line and augment them with our new line of products as they became available. Our primary strategy was to develop very unique custom packaging which would set "Lost Crops" apart from other variety items and which would communicate the concept of "Lost Crops" visually. We decided that all packaging would be produced on recycled paper with a grain-like appearance and that we would use colors that reflect the Native American culture.

During the introductory period, we would try to position "Lost Crops" with key retailers: industry leaders who recognized the value and the message of these unique items and retailers who would want to capitalize on their marketability and "star quality" and who wanted to be viewed by their customers as innovators. As we got closer to finalizing all of the items to be in the line, we flew Noel Vietmeyer out to our facility in Los Alamitos, California to review our product mix, our copy, and our packaging and to confirm the authenticity of this new program.

After this peer review, confident that we had all our ducks in a row, we then developed a consumer brochure, a catalog sheet in full color displaying the line, fact sheets for our sales staff and for the media and continued our extensive recipe testing program that always goes along with anything introduced by Frieda's.

THE BOTTOM LINE

Has this venture been profitable for us? Not as yet. Tremendous expense was involved in launching such a line even before we got to packaging, labeling, and recipe development. There were years of development that involved our most talented people. So even with broad retail acceptance, a high volume of sales is necessary just to cover initial costs.

Has it reached our objective of increasing and reinforcing the leadership and innovative image of Frieda's, Inc.? The answer is yes! What we didn't fully anticipate was the incredible response from the media, both locally and nationally. The resulting consumer interest actually created early sales for us as selected retailers called us inquiring about product availability as a result of consumer requests.

The "Lost Crops" received a tremendous send-off shortly after three major retailers in different parts of the country agreed to introduce our line to their shoppers. Kings supermarkets in Irvington, New Jersey, Dominicks in Chicago, and H.E. Butt in San Antonio were featured in an article in the New York Times on November 1st 1994, headlined "America Embraces Its Native Foods." Since that article first appeared, we've been able to track no less than 122 news stories which appeared in papers across the nation with an estimated readership of over ten million people. There is a great deal effort and expense that goes into introducing a new crop or in this case a new line of products.

Frieda's has built its reputation on operating with an open mind and an open door. We welcome concepts and ideas that fit our parameters and marketing style. Please contact us so we can give close consideration to what you have to offer.

Kenaf Commercialization: 1986–1995

Daniel E. Kugler

The most recent ten years (1986–1995) have been marked in the Federal government by increasing acknowledgement of the benefits of collaborative work with the private sector for technology research and development, and, in some cases, commercialization assistance. This has certainly been the case for kenaf (*Hibiscus cannabinus* L., Malvaceae), a crop that the U.S. Department of Agriculture (USDA) has been working with for fifty years. Currently, in the United States, there are three mills that mechanically separate the core and bast fibers, then sell or use these fibers for manufacture of a variety of market products. Products range from high quality printing and writing paper from bast fiber to adsorbents and horticultural mixes from core fiber.

The Federal government, almost exclusively USDA, has played several roles in the decade of commercialization of kenaf. The decade was marked early on by changes in Federal law and policy encouraging closer cooperative R&D projects with private industry. This in turn encouraged collaboration on shorter term, targeted projects. Then came a return to more traditional, longer term agricultural research programs in breeding, genetics, and agronomy. Passage of the 1990 Farm Bill opened the potential for direct commercialization assistance from USDA just as the first kenaf businesses were beginning operations. It was also during this mid-decade period that kenaf became sufficiently well known and gained enough R&D background for other USDA funding programs to acknowledge and favorably rate proposals for support.

At the end of the decade of commercialization, there are four businesses, hopefully profitable businesses, leading the way to what could eventually become a new fiber industry in the United States. Natural Fibers of Louisiana (Jeanerette, Louisiana), Mississippi Delta Fiber Cooperative (Charleston, Mississippi), and Kenaf International (McAllen, Texas) run mechanical fiber separation mills and maintain a variety of business relations with other fiber product manufacturing firms. KP Products (Albuquerque, New Mexico) manages the manufacture and marketing of printing and writing grade paper made from kenaf bast fiber.

THE DECADE IN REVIEW

The Kenaf Demonstration Project began in the first months of 1986 as a cooperative effort between USDA and Kenaf International to commercialize kenaf for the manufacture of newsprint. The Project was planned and organized in three phases, with success in one phase a contingency for moving to the successive phase.

Phases I and II took place from 1986 to 1988. In Phase I, USDA and Kenaf International executed a full commercial scale newsprint demonstration with several partners. Kenaf was grown in Texas by Kenaf International and Rio Farms, and pulped in Ohio by C-E Sprout Bauer. Newsprint was manufactured by Canadian Pacific Forest Products in Quebec, Canada, and pressroom runs done in July 1987 by The Bakersfield Californian in California. Each component of Phase I was declared successful by industry.

Phase II aimed to remove another barrier to commercialization of kenaf. Harvest, handling and storage of kenaf had been accommodated by borrowing equipment from forage, cotton and sugar cane operations. A harvest system dedicated to kenaf was needed and the Kenaf Demonstration Project engaged the services of Willett and Associates, the eventual founder of Natural Fibers of Louisiana. Starting from seed and row spacing in the farm field and ending with 11–14 inch kenaf billets delivered to the fiber yard of a paper mill, a custom system was designed, built and successfully tested by the fall of 1988.

With newsprint shown to be a proven commodity and a custom harvest system in place, the Kenaf Demonstration Project entered Phase III, covering the years from 1988 to 1990. This period was characterized by a Kenaf Demonstration Project team decision to dissolve itself and move into more traditional operations. For USDA this meant opening breeding, genetics and agronomy programs at Agricultural Research Service stations in Mississippi, Oklahoma, and Texas. For the private partners, it meant pursuing plans to finance and build a newsprint mill in the Lower Rio Grande Valley of Texas.

As the planning for a 90 ton-per-day newsprint mill progressed in the years from 1990 to 1993, the paper industry entered a cyclic down turn characterized by over capacity to produce many paper grades and resulting lowering prices. Financing $40 million for a small newsprint mill to use a new nonwood fiber feedstock became untenable. This precipitated a shift by private entrepreneurs and USDA to search for less capital intensive, smaller scale markets for kenaf's bast and core fibers. Like newsprint, other kenaf fiber products had to be developed and readied for demand-driven market needs. Unlike newsprint, efficient and clean mechanical separation of the two fiber fractions was needed.

The aforementioned three separation mills began operations in 1990–1991. The Louisiana mill opened first and specialized in oil/chemical adsorbents and animal bedding made from core. Mississippi opened with a fiber supply push and eventually gained markets in adsorbents and horticultural mixes from core. The Texas operation was/is foremost in newsprint but specialized in composites from bast and was first to bring horticultural mixes to the market. A fourth mill operated for a two years in California, but closed down and moved its instant-lawn mat product made from bast to the Mississippi mill.

In 1993, KP Products began to have kenaf printing and writing grade paper manufactured from bast fiber by Ecusta Paper Company in North Carolina. Bast fiber is purchased from the mechanical separation mills. The paper, marketed as tree-free, chlorine-free, and acid-free, appeals to environmental organizations. Due to small volume and batch pulping, the kenaf paper is more expensive than bright-white, traditional bond paper made from wood.

Kenaf made its way into USDA competitive research and commercialization assistance programs in 1993 to 1995, the latter part of the decade of commercialization. The National Research Initiative, Small Business Innovation Research, and Alternative Agricultural Research and Commercialization programs added to the continuing research program conducted by the Agricultural Research Service.

USDA SUPPORT IN 1993/1994/1995

Agricultural Research Service

The Agricultural Research Service is USDA's in-house research agency for physical and biological sciences. The agency provides access to agricultural information and develops new knowledge and technology needed to solve technical agricultural problems of broad scope and high national priority.

Total support for kenaf research by the Agricultural Research Service in 1995 was $1.6 million for programs at four locations and headquarters. The Southern Regional Research Center in New Orleans, Louisiana, conducts research into non-woven technologies to minimize damage to fibers during mat formation. The Lane, Oklahoma, field station concentrates on agronomy in the northern reaches of the Cotton Belt.

The Weslaco, Texas, field station leads the breeding/genetics program and the uniform variety trials, as well as agronomy studies. Support is passed through the Agricultural Research Service to Mississippi State University for a program which currently concentrates on kenaf fiber processing and product applications. The headquarters office in Beltsville, Maryland, allocates some funding to a variety of projects, mainly in kenaf fiber quality assurance and product development.

Small Business Innovation Research

The Small Business Innovation Research (SBIR) is housed in USDA's Cooperative State Research, Education and Extension Service. SBIR is designed to (1) stimulate technological innovation in the private sector, (2) strengthen the role of small business to meet Federal R&D needs, and (3) increase private sector commercialization of innovations derived from USDA-supported R&D. Special consideration is given to foster and encourage participation by women-owned and socially/economically disadvantaged small business firms.

In 1994 and 1995, under the topic area Industrial Applications, SBIR sponsored two Phase I, proof-of-concept, studies for kenaf fiber products: a structural composite and an erosion control mulch mat. Over $100,000 of the $626,000 total for the Industrial Applications topic was provided to two small firms in Colorado and Wisconsin. Note that Phase II SBIR grants allow the small business to take the proof-of-concept to the point of commercial start-up.

Phase III calls for the small business to commercialize the product with other sources of funding and support.

National Research Initiative

The National Research Initiative (NRI) is a USDA-Cooperative State Research, Education and Extension Service program which competitively awards grants to research key problems of national and regional importance in biological, environmental, physical, and social sciences relevant to agriculture, food and the environment. Both fundamental and mission-linked research are sponsored in this $100 million program. The NRI division Enhancing Value and Use of Agricultural and Forest Products makes $6 million in awards each year. In 1993 a $150,000 award was made to the University of Wisconsin to use kenaf bast fibers to replace mineral and glass fibers in thermoplastic composites. Mississippi State University received $110,000 in 1994 to use kenaf bast fibers in woven and nonwoven structures for industrial (composites) and apparel.

Alternative Agricultural Research and Commercialization

The Alternative Agricultural Research and Commercialization (AARC) Center is located with the Under Secretary for Rural Development in USDA. AARC's mission is to expedite development and market penetration of industrial (nonfood, nonfeed) products from agricultural and forestry materials and animal co-products. Private individuals or companies may apply to AARC for repayable funds to commercialize research which may have been conducted with public or private funds. Preference is given to projects that benefit rural communities and are environmentally friendly. A total of $15.5 million was made available in 1993 and 1994 by AARC to join $43.0 million from the private partners. Repayments to AARC are put into a revolving fund and reinvested in new projects. AARC is governed by a nine person Board of Directors where eight members are from the private sector and one from USDA.

In 1993–1994, AARC funded four projects. Kenaf International received $100,000 for newsprint and for pulp for spaceboard. Gridcore Systems International's grant of $50,000 was for kenaf bast mixed with waste paper to manufacture spaceboard. Dry-formed mats impregnated with fertilizer and seed for instant lawns won $800,000 for Agro-Fibers. KP Products was awarded $100,000 to expand the market for printing and writing grade paper made form kenaf bast fiber.

SUMMARY

Federal support for kenaf research and development over the years 1986 to 1995 has expanded from targeted technology demonstrations to base research programs and competitive awards for fundamental research and commercialization assistance. In 1995, Federal support was approximately $2 million for work ranging from breeding and genetics research to end product commercialization. Commercial kenaf area in the United States in 1995 totaled about 2500 acres (1000 ha). Despite the small commercial acreage in 1995, the high level of Federal investment anticipates that new companies will come into operation and that the existing companies will expand.

REFERENCES

Kugler, D.E. Kenaf acreage and bast fiber production in the United States. Fact sheet prepared Sept. 12, 1995, for the United Nations Food and Agriculture Organization—Jute, Kenaf and Allied Fibres Statistics. Washington, DC.

International Kenaf Association. 1994/1995. Proc. Sixth and Seventh Annual Conferences. Ladonia, Texas

U.S. Department of Agriculture. 1993. New industrial uses, new markets for U.S. crops: Status of technology and commercial adoption. Cooperative State Research Service, Aug. 1993, Washington, DC.

U.S. Department of Agriculture. 1993/1994. Fact Sheets for Sponsored Commercialization Projects. Alternative Agricultural Research and Commercialization Center, Washington, DC.

U.S. Department of Agriculture. 1994/1995. National Research Initiative Competitive Grants Program: Abstracts of Funded Research, Fiscal Year 1994 and Fiscal Year 1995. Cooperative State Research, Education and Extension Service. Washington, DC.

U.S. Department of Agriculture. 1994/1995. Small Business Innovation Research Program: Technical Abstracts, Fiscal Year 1994 and Fiscal Year 1995. Cooperative State Research, Education and Extension Service, Washington, DC.

U.S. Department of Agriculture. 1995. Agency estimates of R&D support for kenaf for fiscal years 1993 to 1995. Agricultural Research Service, Sept. 21, 1995. Beltsville, Maryland.

U.S. Department of Agriculture. 1995. Kenaf newsprint: Realizing commercialization of a new crop after four decades of research and development. Cooperative State Research, Education and Extension Service, Aug. 1995 reprint of a June 1988 publ., Washington, DC.

Commercializing Mesquite, Leucaena, and Cactus in Texas

Peter Felker

Since 1977, Texas A&M University–Kingsville personnel have been involved in commercialization of the nitrogen fixing trees leucaena (*Leucaena leucocephala*), mesquite (*Prosopis* spp.), and cactus pears (*Opuntia* spp.). Fast-growing *Leucaena leucocephala* has been primarily developed as a high protein forage. Mesquite has been primarily promoted for "barbecue" cooking, as a luxury quality furniture wood, as a drought hardy ornamental tree, as a nitrogen fixing soil improver, a source of high sugar pods, and as a biomass fuel source. Cactus has been promoted as a fruit, forage, and vegetable.

MESQUITE

Mesquite is a nitrogen-fixing tree or shrub of the genus *Prosopis* (Fabaceae) that occurs on 30 million hectares in U.S. from Houston, Texas to Bakersfield, California. In southern New Mexico mesquite is typically a 2 m tall multistemmed shrub. However, substantial sized trees i.e. 50 cm in trunk diameter and 8 m tall frequently occur in California, Arizona, and New Mexico where there is permanent water within 5 m of the soil surface or where the rainfall is greater than 500 mm annually (Felker 1979). *Prosopis* has yellow or deep red pods that are typically 13% protein and 35% sucrose (Oduol et al. 1986). *Prosopis* has a soil improving effect with soils under the canopy typically having 0.05% N and 0.5% C versus concentrations of 0.02% N and 0.2% C outside the canopy (East and Felker 1993). *Prosopis* is also salt tolerant with some selections capable of growth in seawater salinity levels (Rhodes and Felker 1987).

In the United States there are four species: *P. glandulosa* var. *torreyana* (in California), *P. velutina* (in Arizona), *P. glandulosa* var. *glandulosa* (in Texas), and *P. pubescens* i.e. screwbean (minor distribution in all regions). There are 44 *Prosopis* species worldwide. *Prosopis* is prominent in Mexico, Peru, Chile, Argentina, Brazil, Sahelian Africa, Haiti, Pakistan, and the arid parts of India. In many these locations it is useful for firewood, livestock feed (the pods), and soil improvement. In Peru its pods are used in human food preparations and in Argentina and Texas fine furniture and flooring are made from its wood.

Prosopis wood is perhaps the most dimensionally stable of all woods in having equal radial and tangential shrinkage values of only 2%–3% and in having total volumetric shrinkage values of 4%–5%. Other fine woods i.e. oak, cherry, walnut, teak, mahogany and Indian rosewood have volumetric shrinkage values in the range of 8%–15% (Felker et al. 1994). The result of the low shrinkage value is that furniture made from mesquite will have less expansion and contraction when ambient humidity levels change, such as from summer to winter in northern regions. Thus furniture will not crack and drawers will not stick with changing humidity levels.

In many regions of southwestern United States, dense nearly impenetrable stands of mesquite have occurred on rangelands much to the displeasure of ranchers (Cornejo-Oveido et al. 1991; Felker et al. 1988). The worse stands have occurred on overgrazed rangeland sites or sites that have been subjected to mechanical manipulation. It has been hypothesized that a major reason for the spread of mesquite has been the competitive advantage of nitrogen fixing woody legumes on sites that have become impoverished in nitrogen. Measurements of tree spacing versus tree diameter over a range of 10,000 to 20 stems/ha revealed a highly significant logarithmic relationship (Felker et al. 1988). This suggested the presence of substantial intraspecific genetic variation and the possibility of preventing the spread of dense stands of young mesquites by competition from large mature mesquites.

Agroforestry experiments indicate that the growth of the trees was maximal in intercropped situations. Diameter growth rates of 1.25 cm/year were recorded over a three year period that are equal or greater to hardwood growth in more temperate regions of the United States (Cornejo-Oveido et al. 1991).

Given the fact that there are over 20 million ha of *Prosopis* just in Texas, it became important to convert what was viewed as a liability into an asset. Extensive anthropological data on human food uses of

Prosopis pods and food technological experiments have demonstrated conclusively the potential of mesquite pods in human food preparation. The limitation to food use is the cost of hand labor to gather the pods. Presumably equipment could be devised for use in managed orchards to economically harvest the pods, but without this equipment and orchards of mesquite in rows, this aspect of mesquite pod utilization is not economically possible.

The three main uses of *Prosopis* wood are (1) as a source of energy and/or chemical feedstock, (2) as barbecue chips and chunks, and (3) as lumber or flooring. Assuming a value of \$2 per million BTU delivered to a power plant for fuel, mesquite wood has a delivered value of \$34/t for fuel. Current wholesale prices for mesquite packaged in 2 kg plastic bags for the grocery trade are about \$400/t. If mesquite lumber is priced equivalent to cherry, walnut, and premium oak at \$3,000/thousand board ft, then mesquite is worth about \$1700/t. It is then clear that the highest and best use of mesquite is in solid wood applications.

A 300 hp mesquite combine has been under development for nearly 10 years to harvest small diameter mesquite and related brush for industrial fuel and/or chemical feedstock (McLauchlan et al. 1994). While great improvements have been made, this equipment is not yet commercially viable. The goal of this machine will be to harvest woody brush for about \$9/t (green) in the field.

Fortunately there is a good demand for smaller pieces (15 to 25 cm in diameter and 1 m long) by the barbecue industry that could result from thinning overstocked stands. Prices of about \$25/m^3 (\$90/cord) are being paid by mesquite barbecue processors for these small diameter logs. The hauling costs dictate that at \$25/cubic meter these logs can be hauled a maximum of about 150 km (100 miles).

The mesquite barbecue industry is experiencing rapid growth with the two top processors having product in over 8,000 retail outlets in all 50 states (J. Lawson and G. Wartsbaugh pers. commun.). In 1995, the volume of mesquite processed by the three largest mesquite barbecue manufacturers was about 14,000,000 kg which at \$0.44/kg is a \$6 million industry. Since there are about 600 kg/m^3 (5,000 lb/cord) and since the volume of mesquite ranges from 9 to 90 m^3/ha (1 to 10 cords/acre), the area harvested can be calculated to range from 250 to 2500 ha. As there are 20 million hectares in Texas and as raw material for the barbecue plants can only be economically hauled 150 km, there is obviously a need to greatly expand the demand for mesquite wood for barbecue uses. Fortunately the charcoal briquette industry is a \$400 million/year industry so there is great opportunity to increase the market share of this product.

As it is to the benefit of Texas cattlemen to create a market for beef and mesquite, mesquite wood barbecue producers of the trade organization, Los Amigos del Mesquite, conducted a beef/mesquite co-marketing program with the Texas Beef Council. Two mesquite processors contributed \$5,000 that was matched with \$20,000 by the Texas Beef Council to produce color displays with beef and mesquite in the meat section of 1700 grocery stores in Texas.

Los Amigos del Mesquite also worked with the National Hardwood Lumber Association to establish a new grading system applicable to mesquites short and narrow lengths. In 1996, Los Amigos del Mesquite began working with the Lumbermans Association of Texas which have many retail chains that sell wood products for joint marketing efforts. These latter 3 joint activities clearly point to the advantages of developing professional societies to work together in win-win situations.

In 1981, when Los Amigos del Mesquite was formed, there were only 3 mesquite sawmills in Texas and these sawmills only operated part time to produce end-grain flooring and rough, non-kiln dried lumber. In 1996, there were at least three full time companies producing mesquite flooring and lumber, one of which has developed into a million dollar per year business. Detailed information on the mesquite program can be obtained from Los Amigos del Mesquite through Texas A&M University-Kingsville. World-wide information of mesquite is also available in the *Prosopis* email network. To subscribe to the network, mail to Listserv@taiu.edu and type subscribe *Prosopis* and mail the message.

LEUCAENA

The fast growing nitrogen fixing tree *Leucaena* (Fabaceae) has been widely grown in the tropics for forage, fuelwood, and soil restoration. Of the dozen species of *Leucaena*, three are naturalized to Texas. *Leucaena retusa* is the slowest growing of the species, but is the most resistant to freezing weather in being able to tolerate –20°C temperatures for short durations with only minor tip damage (Glumac et al. 1987). *L.r etusa* rarely achieves heights greater than 2.5 m. *L. pulverulenta* is not damaged by short durations of –8°C, but is killed to ground level by freezes of –12°C. These trees often obtain heights of 10–12 m between 10 year record freezes in the most southern regions of Texas. Unfortunately the dry matter digestibility of the foliage of *L. pulverulenta* is only about 35%. *L. leucocephala* is the fastest growing species. With partial drip irrigation, *L. leucocephala* K636 grew from direct seed to 4 m tall in 5 months. Freezes of –5°C kill the top growth of *L. leucocephala* to ground level. However even in severe freezes such as Christmas 1989, when temperatures were below freezing for 60 h with absolute minimum of –15°C, all rootstocks of *L. leucocephala* resprouted the following year reaching heights of 4–5 m in 6 months (Glumac et al. 1987). *Leucaena leucocephala* has an excellent dry matter digestibility of about 70%.

It is significant that alfalfa cannot be grown in much of Texas due to susceptibility to cotton root and that *Leucaena* is 100% resistant to this pathogen (S. Lyda pers. commun.). In a 4 year trial, on two sites with four large (12 m by 36 m) replicates per site, we demonstrated that one-row commercial equipment could be used to plant, cultivate and harvest leucaena (Felker et al. 1991). The annual forage production ranged from 1.5 t/ha to 9 t/ha and the crude protein content ranged from 11.8 to 23.9% (Felker et al. 1991).

As a follow-on to this research, Mr. Edwin Singer and associates of Corpus Christi, Texas have established a 200 ha pivot-irrigated leucaena plantation and developed mechanized planting, harvesting, drying, and pelletizing operations. The largest demand for the pelletized product is in feed for wildlife operations. Mr. Singer has also held organizational meetings of scientists, growers, and ranchers. For further information contact Mr. Singer at Suite 2020, Mercantile Bank Tower, Corpus Christi, Texas 78477.

CACTUS

It is fortunate that the high water use efficiency of Crassulacean Acid Metabolism (CAM) is present in rapidly growing cactus species such as *Opuntia ficus-indica*, *O. megacantha*, and *O. amychlea* (Cactaceae) that produce forage for animals, vegetables, and 14% glucose fruits (Russell and Felker 1987; FAO 996). World wide interest in cactus is expanding rapidly as evidenced by the formation of the Cactus Pear network by the UN Food and Agricultural Organization, and the Professional Association for Cactus Development comprised of food brokers, chefs, ranchers, growers, and scientists. In addition a Cactus email network has been developed to enhance communication across several continents. To subscribe to the cactus network, contact Listserv@taiu.edu and type subscribe Cactus-L

In Texas the primary use of cactus has been in times of drought when the spines have been burned off the cactus to feed cattle. In thick stands of native *Opuntia lindheimerii*, it has been reported that one man can burn enough cactus for 100 cattle. It has been estimated that 1 liter of propane is sufficient for one mature cow and when propane is purchased in bulk for $0.10/liter, the daily feeding propane cost is about $0.10/animal. A daily ration of 40 kg of cactus, 0.5 kg of mineral salts, and 0.5 kg of protein supplement is sufficient to permit excellent liveweight gain, reproduction, and lactation from nursing cattle (Maltsberger 1991).

While the young tender spring vegetative growth of wild cactus (nopalitos) has been extensively consumed by Hispanics during Lent (Russell and Felker 1987), only recently have plantations of spineless cacti been established in Texas for nopalito production. An improved nopalito cultivar with greater year round production, lack of spines, lack of glochids (nearly microscopic spines), and low mucilage is available as 'Texas A&M 1308'. In contrast, spineless *Opuntia* plantations for nopalito production have been common in central Mexico (Milpa Alta) for many years (Russell and Felker 1987).

The fruit of native *Opuntias* weigh about 30 g and have a sugar content of about 9%. In contrast commercial cactus fruit varieties from Mexico weigh about 110 g with a sugar content from 13%–15% (Pimienta-

Barrios 1991). There is great diversity in colors, seediness, and date of ripening of the Mexican cactus fruits. The fruits may be lime green, orange, pink, red, or purple. Typically the lime-green fruits are the sweetest and most desirable in Mexico, but the red fruits are in most demand for export to northeastern United States and Canada.

A commercial cactus fruit orchard, operated by the D'Arrigo Bros, has been in existence for nearly 50 years near Salinas, California. This orchard produces red fruits that are sold to people primarily of Italian descent in northeastern United States (R. Bunch pers. commun.).

An *Opuntia* genetic improvement program is underway at Texas A&M University–Kingsville to find fruit producing cactus clones that are adaptable to Texas. The primary limitation to cactus fruit production in Texas is the presence of Arctic cold fronts about every 10 years in which temperatures may remain below freezing for several days with minimum temperatures of –12°C. *Opuntia* fruit cultivars have been collected in high elevation sites in Mexico's cold deserts with the goal of finding cultivars with good fruit quality and cold hardiness. In 1996 over 130 clones were being evaluated in field trials at the Texas A&M University–Kingsville field station.

The cactus new crop effort at Texas A&M University–Kingsville is about equally divided between development of new clones/cultural practices and working with all aspects of the commercial cactus industry to stimulate awareness and market demand.

CONCLUSIONS

New crop development in semi-arid Texas is undergoing substantial development with mesquite, leucaena, and cactus. The private sector has invested substantial sums in all three industries. The mesquite lumber and flooring industry is currently the largest of these three new crops with gross sales estimated at between $5 and 10 million annually. Fifteen years ago this industry did not exist. Over $2 million in private investments have occurred in the *Leucaena* industry. Excellent markets are now available for pellets, made from *Leucaena* leaves and stems, for the wildlife industry. Cactus use for cattle feed is the next largest "new crop industry" with at least two ranchers purchasing 50,000 liters of propane/year for use in burning spines from cactus for cattle feed. The nopalito industry is making substantial progress with pickled and fresh products.

The private/federal investiment of about $150,000/year into new crops R&D at Texas A&M University–Kingville from 1981 through 1995, has been closely associated with the development of the mesquite industry which grew from 0 to $5 million/year over the same time period. We believe in the medium term both cactus and leucaena will also develop into multimillion dollar per year industries. While these industries are very small by industrial standards, these revenues are being generated in poor rural areas in desperate need of economic development. In the case of mesquite, this industry has converted what was previously a weedy-liability into an asset. The university R&D effort facilitated creation of organizations of diversified users (i.e. chefs, brokers, barbecue manufacturers, farmers, physicians) that were critical to the development of the entire industry. Both from a financial and sociological perspective these R&D funds into new crop development in Texas were a sound investment. We believe that each state in the United States has similar valuable indigenous resources that could be developed into commercially viable new crops. It is our hope that proactive New Crop policies will change to obtain financial resources to create similar programs throughout the United States and the world.

REFERENCES

Cornejo-Oviedo, E., J.E. Meyer, and P. Felker. 1991. Thinning sapling high density mesquite (*Prosopis glandulosa* var. *glandulosa*) stands to maximize lumber production and pasture improvement. For. Ecol. Manag. 46:189–200.

East, R. and P. Felker. 1993. Forage production and quality of 4 perennial grasses grown under and outside the canopies of mature *Prosopis glandulosa* Torr. var *glandulosa* (mesquite). Agroforestry Systems 22:91–111.

FAO. 1996. Cactus pear cultivation and uses. (Inglese et al. eds.). FAO Publ., Rome.

Felker, P. 1979. Mesquite—An all purpose leguminous arid land tree. p. 89–132. In: G.A. Ritchie (ed.), New agricultural crops. American Association for the Advancement of Science Symposium.

Felker, P., J.M. Meyer, and S.J. Gronski. 1988. Application of self-thinning in mesquite *(Prosopis glandulosa* var. *glandulosa*) to range management and lumber production. For. Ecol. Manag. 31:225–232.

Felker, P., R. Chamala, E. Glumac, C. Wiesman, and M. Greenstein. 1991. Forage production of *Leucaena leucocephala* and *L. pulverulenta* in mechanized farming systems in a semiarid region of Texas. Trop. Grasslands 25:342–348.

Felker, P, P. Anderson, D. Perino, and D. Miller. 1994. Grading mesquite lumber. Special Publ. of Center Semi-Arid Forest Resources, Texas A&M Kingsville.

Glumac, E.L., P. Felker and I. Reyes. 1987. A comparison of cold tolerance and biomass production in *Leucaena leucocephala*, *Leucaena pulverulenta*, and *Leucaena retusa*. For. Ecol. Manag. 18:251–271.

Maltsberger, W.A. 1991. Feeding and supplementing prickly pear cactus to beef cattle. p. 104–118. Proc. 2nd Annual Texas Prickly Pear Council, Kingsville.

McLauchlan, R.A., A. Conkey, G. Scherer, P. Felker, and S. Brown. 1994. Development of a flail harvester for small diameter brush and coppiced trees to produce energy/chemical feedstock. p 31–42. In: B.J. Stokes and T.P. McDonald (eds.), Mechanization in short rotation intensive culture (SRIC) forestry. USDA/ U.S. Forest Service, Southern Forest Experiment Station.

Oduol, P.A., P. Felker, C.R. McKinley, and C.E. Meier. 1986. Variation among selected *Prosopis* families for pod sugar and pod protein contents. For. Ecol. Manag. 16:423–431.

Parish, J. and P. Felker. 1995. Evaluation of fruit quality and production of cold hardy *Opuntia* fruit clones. Proc. First Annual Professional Association for Cactus Development, San Antonio, TX.

Pimienta-Barrios, E. 1991. An overview of prickly pear cultivation in the central part of Mexico. p. 1–18. Proc. 2nd Annual Texas Prickly Pear Council, Kingsville, TX.

Rhodes, D. and P. Felker. 1987. Mass screening *Prosopis* (mesquite) seedlings for growth at seawater salinity. For. Ecol. Manag. 24:169–176.

Russell, C.E. and P. Felker. 1987. The prickly pears (*Opuntia* spp): a source of human and animal food in semiarid regions. Econ. Botany 41:433–445.

Establishing New Crops Industries: The Indiana Grape and Wine Industry Model

Bruce Bordelon, Theresa Browning, and Cheri Wagner

Establishing a new crops industry can be challenging for several reasons. Consumer awareness and acceptance of a new product must be developed before markets can be established. Producers must determine consumer demands and develop methods of production that meet those demands. Producers and marketers must work together to develop a product that will bring a reasonable return to the producers.

The Indiana grape and wine industry is a good model for other new crops industries. Grape and wine production is "new" to Indiana only in the sense that the industry is at a stage of rebirth. Indiana has a rich history of grape and wine production. Swiss settlers, led by John James Dufour, planted vineyards in 1802 in Vevay, Indiana and established one of the nation's first commercial wineries. Dufour subsequently wrote the first book on American viticulture, *The American Vine Dressers Guide*, published in 1826. French and German settlers established similar enterprises soon after in other parts of the state. Nicholas Longworth brought the Catawba grape to the Cincinnati region in 1825 and established a significant sparkling wine industry. These events lead to the development of a substantial grape and wine industry in the state. The 1880 Bureau of the Census shows that Indiana ranked 5th in grape area with 3,851 acres (1,559 ha) and 16th in wine production at 99,566 gallons (3,769 hL). Though the industry survived disease and severe weather, prohibition (1919–1934) finally ended its progress.

The new era of the Indiana wine industry began following legislation. In 1971, the wine trade lobbied for passage of the Small Winery Law. Under this law Indiana wineries are allowed to produce up to 50,000 gallons (1,894 hL) of table wine per year from grapes, other fruits, and honey produced in Indiana. The law allows wineries to serve complimentary samples and sell wine by the glass, bottle or case on the premises of the winery. They may also sell wine by the bottle or case directly to the public, or to holders of retail or wholesale licenses, bypassing the three-tier distribution system. Today, 90% of the wine is sold at retail from the wineries—about 70,000 gallons (2,650 hL) in 1994—ensuring the winemaker's best profit margin.

Following passage of the Small Winery law, grape area and the number of wineries increased. As many as 18 wineries were operating during the 1970s. Since then several have closed, but the good producers remain active today. Grape area dropped sharply during the early 1980s following several harsh winters and poor prices for grapes. Only recently has the area been on the increase. Renewed interest in alternative crops, increased demand for Indiana-grown grapes and a strong regional industry have rejuvenated the Indiana grape and wine industry.

In 1989, the wine trade and a significant group of amateur producers, represented by the Indiana Winegrowers Guild, lobbied for legislation to establish the Indiana Wine Grape Market Development Council, a program designed to enhance Indiana's economic development by establishing a successful wine grape industry through research and market development. The Director of Purdue's Agricultural Research Programs administers the program funded by $0.05 per gallon of the existing excise tax on all wine sold in Indiana. The bill allows for grape and wine research and extension programs and a market development program at Purdue University. These programs provide critical early support for a unique agricultural industry without the use of general revenue funds.

CHALLENGES

Indiana is mostly rural and agricultural society in which wine consumption with meals is not a familiar practice. Wine jargon and snobbery add to the problem by keeping consumers uncomfortable with their knowledge of wine. Per capita wine consumption in Indiana remains low: about one gallon per person annually as of 1994, ranking Indiana 31st in per capita consumption out of 51 markets surveyed. However, consumption for all alcoholic beverages is flat, and although wine has been relatively exempt from neo-prohibitionism, it is not immune to the effects or reduced alcohol consumption.

Grape producers face challenges of adverse weather—hot, humid summers, and cold harsh winters—which increase the cost of production. Though grape production is well suited to lands unsuitable for row-crop production, compatibility with current farming practices can be a problem because grapes can be sensitive to herbicide drift from corn and soybean production. Competition with other grape producing areas keeps prices low, and discourages expansion. The Small Wineries Act specifically dictates that grapes, other fruit, and honey can be imported for wine making purposes only if those products are not available from within the state. This law reads: "For the purpose of encouraging the development of domestic vineyards, a small winery permit may be issued.... ." Though this gives Indiana growers a marketing advantage, the law has never been tested. Instead of creating legal arguments and antagonism, new growers are working with wineries to fill niches by producing cultivars that are not readily available on the open market. Many of these cultivars are well adapted to the state and can be produced very economically. The Indiana wine industry is building a reputation for high quality wines from French hybrid and American grapes and other fruits such as blackberry, blueberry, and strawberry. Planting of premium European cultivars has increased in recent years to meet the demand for premium wines. Though not well adapted to the state's climate, special production techniques allow consistent production on the best sites. The superior wine quality produced from these cultivars increases the value of European grapes, making the extra effort and risk worth taking.

The wine industry's willingness to work with growers of grapes and other fruits has solidified support among the state's diverse fruit growers. There is a strong movement for home-grown produce in the state and it is catching on in the wine industry as well.

The Indiana Winegrowers Guild was established in 1975 to further the causes of grape and wine producers in the state. The existence of this agricultural association has been instrumental in promoting legislation that benefits the industry. Without this organization's commitment and support, the recent legislation establishing the Indiana Wine Grape Council may not have passed. By working with the producers of other fruits, the wine industry is winning allies throughout the state's horticultural sector, and vice versa.

EDUCATIONAL PROGRAMS

Efforts to develop the Indiana wine industry have focused on the following areas: consumer education, education for commercial growers and vintners, and legislation to increase the industry's market. Purdue faculty and staff have emphasized trade education through workshops, private consultation, and printed resources. Education is crucial because few enter this industry with any academic or professional training in wine making, grape growing, or marketing. Many have no farming or business experience. It is the "romance" of wine that motivates them. To help these novices avoid mistakes, five professionals assist the trade with marketing, public relations, enology, and viticulture advice. Beginners and established producers can receive help through start up packets with pertinent market and trade resources, and private consultation with the experts to evaluate planned or existing business practices. An equipment loan program for those starting new vineyards and wineries provides physical resources. Workshops given three to four times a year by the team of experts keep the industry abreast of the latest technologies. Quarterly marketing newsletters provide education on local matters not available in the commercial wine trade press. Growers and vintners newsletters provide timely information throughout the year.

CONSUMER EDUCATION

Indiana wineries are marketed as tourist destinations for tours, tasting, and leisure. The message is sent through media relations, event planning, brochure development/distribution, and public speaking. Tasting is a key component in promotions, as tasting ensures public acceptance as a quality product. Tasting occurs at large fairs and festivals in urban markets, largely Indianapolis, within a short driving distance of the majority of wineries. In addition, educational speeches before civic, social, and church groups provide some depth on wine appreciation techniques. Consumer newsletters—containing local winery news and general wine education—are sent three times annually to the public who have attended events and responded to advertising and public relations activities. Media relations is focused on the key urban markets of Louisville, Cincinnati, and Indianapolis because of their proximity to the majority of wineries, as well as secondary markets in Indiana, such as Lafayette, Anderson, Evansville, Bloomington, and other winery locales.

A wine appreciation class is taught at Purdue University to familiarize students (future consumers) with wines and their role as a food beverage. The course is very popular across campus and is especially useful for students in the Restaurant, Hotel, Institutional, and Tourism Management curriculum.

CONCLUSIONS

The efforts of the Indiana Winegrowers Guild and the Indiana Wine Grape Council have paid off. The Indiana Wine Grape Council was authorized in 1989 and began its programs in 1990. Prior to this time, Purdue University had paid little attention to the grape and wine industry in the state. Since the beginning of the Council programs, four new wineries have opened, three additional wineries are opening before the end of 1995, and five more are in the planning stages. Vineyard area has almost doubled and continues to expand as growers adopt modern production practices and plant French hybrid, American, and European cultivars that are in peak demand. Sales of Indiana wines have increased 74% since 1990.

Cooperation is the key to the development of a non-traditional crop in unfamiliar territory. Indiana wineries work together to promote their industry, and neighboring wine states' marketing programs share successes and failures. The Indiana grape and wine industry did not try to reinvent the wheel. We have worked with our neighbors and adopted their ideas. Regional collaboration in research and extension activities with adjoining states have strengthened the industry throughout the region. A strong trade association has promoted legislation that supports the industry. The expansion of the Indiana grape and wine industry is proof that a cooperative approach to market development will work.

REFERENCES

Baxevanis, J.J. 1992. The wine regions of America; geographic reflections and appraisals. Vinifera Wine Growers J., Stroudsburg, PA.

Bordelon, B.P. 1995. Growing grapes in Indiana. Purdue Univ. Coop. Ext. Publ. HO-45.

Bordelon, B.P. 1995. Grape varieties for Indiana. Purdue Univ. Coop. Ext. Publ. HO-221.

INFORMATION EXPLOSION

The NewCROP Electronic Network

James E. Simon, Jules Janick, and Amots Hetzroni

The New Crops Resource On-Line Program (=NewCROP) is an electronic information system that provides a window to new and specialty crops, crop experts, farm markets and highlights the activities of the Indiana Center for New Crops and Plant Products. NewCROP is composed of a large collection of computer files and databases that reside on a Macintosh computer server in our Center. This computer is connected to the internet and acts as a network server by allowing users from all over the world to connect simultaneously and read the information. The files on the server are formatted to conform with the format adapted by World Wide Web (WWW) browsers such as www, lynx, mosaic, Internet Explorer, and Netscape. Netscape is the preferred browser. One of the key features of this server is that users can easily link to other documents anywhere in NewCROP or externally to other sites on the internet which focus on new crops and plant products. The connection is established when the user selects, typically by clicking the "mouse," on highlighted words or pictures in the document.

SYSTEMS ENVIRONMENT

The NewCROP Server runs on a MacIntosh OS—Power PC 7100 with 40 megabytes RAM. The server is housed on a 1 gigabyte external hard-drive, currently occupying about 300 mb of hard-drive space and permits access to six outside users simultaneously. HyperText Mark-up Language (HTML) is written using editing programs such as Alpha, HTML Web Weaver, HTML editor and HTML SuperText. Construction of NewCROP WWW server began Oct. 1994 and the system went on-line March 15, 1995. The Universal Resource Locator (URL) or address of NewCROP is: http://www.hort.purdue.edu/newcrop/home

FORMAT

A successful connection into the NewCROP WWW server leads to the Home or Welcome Page which provides the user with links to other places of interest (Fig. 1). The topic choices are underlined bold text. By clicking the mouse over any of the bolded text (e.g. CropSEARCH), a new screen will appear with information pertaining to the selected topic. The available topics include CropSEARCH, indexed database of crop plants; CropREFERENCE, reference books and manuals; CropEXPERT, directory of new crop researchers; NewCropNEWS, newsletters of the New Crop Center; NewCropEVENTS, announcements and upcoming events; FarmMARKET, farmer's markets of the U.S. The server has links on every page providing easy access back to the Welcome Page. An extensive list of pointers provides access to external new-crops and plant products-related web sites, external Data-Bases, and Libraries.

CropSEARCH

The major feature of the server is CropSEARCH, an indexed database of crop plants that contains records with basic information on crop plants. This information can be accessed by the user by clicking the mouse on a common name or scientific binomial (Fig 2). The basic skeleton was based on about 580 descriptions of food and feed crops found in Magness et al. (1971) but we continue to supplement this list with additional information in the form of photographs, new plant listings, and new databases. In addition, crop descriptions are being enriched by the incorporation of FactSHEETS (Fig. 3), consisting of in-depth information including botany, crop culture, natural products, production guide, toxicity, listing of experts, sources of germplasm, and key literature. We are now in the process of adding a list of about 400 energy crops from an unpublished manuscript of James Duke and a list of 500 famine plants from Robert Freedman.

*This work was supported by grants from the Indiana Business and Modernization Technology Corporation and the Center for New Crops & Plant Products, Purdue University. We acknowledge the efforts of Ben Alkire, Paul Angers, Denys Charles, Terry McCain, Nissim Ozer, and Anna Whipkey.

Access to information is aided by a powerful search mechanism built into the server (AppleSearch) so that crops an be potentially accessed by a single word contained in crop descriptions in the data base (e.g. plant name, bulb, tuber, fruit). CropSEARCH is a dynamic resource which is continually being updated and expanded. We foresee scanning seminal papers for each crop into the database to provide even further information.

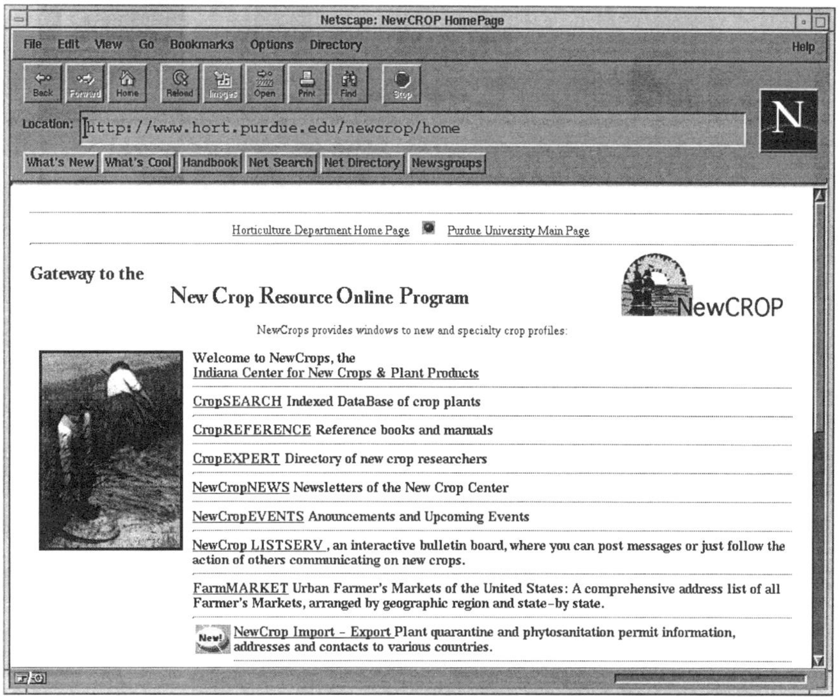

Fig. 1. NewCROP Home or Welcome Page.

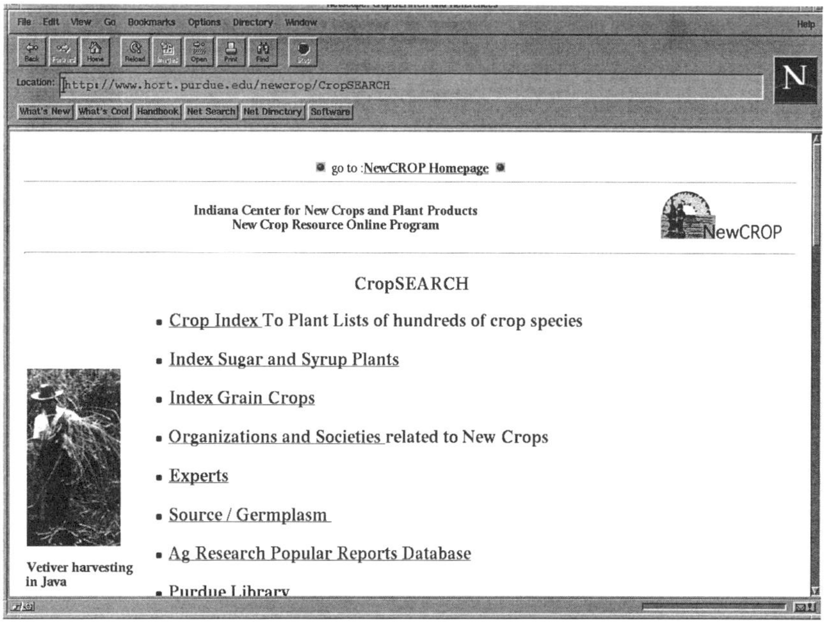

Fig. 2. CropSEARCH indexed crop database.

CropREFERENCE

This database presently consists of a list of monographs and general reference books on new crops. This list presently is subdivided (e.g. General, Legumes, Nuts, Medicinal Aromatic Spices & Herbs).

CropEXPERTS

This resource lists names of experts plus addresses, phone, and fax subdivided by basic crop groupings. These include: General; Cereals & Pseudocereals; Fiber Crops; Forages; Fruits (Subtropical, Tropical, Temperate); Industrial Crops; Legumes; Medicinals, Aromatic Spices & Herbs, Mushrooms; Nuts; Oilseeds; Ornamentals (Floral, Landscape); and Vegetables (Fruit, Leafy, Root).

NewcropNEWS

All the issues of the newsletter of the Indiana's Center for New Crops and Plant Products are available on the server and there is a link to the Australian New Crop Association Newsletter. We foresee that this grouping could be expanded to include other relevant newsletters including that from the Center for Alternative Crops and Produces, University of Minnesota.

NewcropEVENTS

This calendar of upcoming events includes national and international conferences and workshops, and is kept current. Many of the listings include detailed information about the program, some have direct WWW links to conference secretariats, and symposium organizers for more complete information.

(A) Crop Description

Pawpaw

Annonaceae *Asimina triloba* Dunal.

A. obovata, syn. *A. grandiflora* Dunal.

Source: Magness et al. 1971

These two American species are not grown commercially for their fruits, but some quantities of the fruit may be harvested from native trees. *A. triloba* occurs from the Mississippi Valley east to the Atlantic and from the Great Lakes southward. It is a small tree, occasionally to 40 feet. *A. obovata* is a shrub, up to 6 feet, native in South Georgia and Florida. Both of these pawpaws have fruits 2 to 6 inches long, half as wide, generally oval in shape, dark brown when mature. Up to several large seeds are surrounded by a soft pulp that is strongly aromatic, esteemed by some, but objectionable to others.

(B) Crop Factsheet

Pawpaw

Contributor: Desmond R. Layne

1. Common Names
2. Scientific Names
3. Uses
4. Origin and Botany
    1. Description of the Plant
5. Crop Status
6. Crop Culture
    1. Seedling Production
    2. Field Planting
    3. Vegetative Propagation
7. Germplasm
    1. Collections
    2. Commercial Seed Sources
    3. Commercial Seedling Sources
    4. Commercial Named Variety Sources
8. Key References
9. Selected Experts

Fig. 3. Pawpaw (A) article from Magness et al. (B) crop FactSHEET.

FarmMARKETS

This database is a comprehensive listing of farmers' markets operating in the United States at the end of 1993 (Johnson and Bragg 1994). The directory list all markets within each state by name, address, telephone numbers, and contact person when available, and whether the market operates seasonal or year-round. It also contains a "Gateway Link" to other market-related sites, e.g. Open-Air Market Network.

ELECTRONIC BULLETIN BOARD

The NewCROP Server is a subscribing location to the NewCROP LISTSERV which functions as an interactive electronic "bulletin board" for the world community of new crop workers. The NewCROP LISTSERV discussion group operates from an IBM mainframe computer at the Purdue University Computer Center. At present, there are about 300 subscribers, most from the U.S. and Canada but also from over 30 other countries. While many subscribers tend to be university-based researchers or educators and extension specialists seeking specialized information, most subscribers come from the commercial sector. The number of subscribers is dynamic but there has been a general trend upward. All LISTSERV communications are downloaded so that there is an archived listing of its use. A series of four tutorials provide directions and help for those seeking to both subscribe and use the LISTSERV. To subscribe to NewCROPS LISTSERV, send an e-mail message to: LISTSERV@vm.cc.purdue.edu

and then in the message area type: SUBSCRIBE NEWCROPS <Your name, Institution>

PERFORMANCE AND USE

There has been a strong demand for new crops information by the commercial and academic communities. In part, this may be due to the increased interest in new crops for crop diversification and to find new extractable chemicals and plant products from traditional and new crops. Since bringing the NewCROP server on-line, the number of files requested by users continues to increase. By the end of the first week of going on-line, where our only promotion was to announce the NewCROP in the new crops LISTSERV, more than 1,000 files were searched and sent to users. During the period from March 12, 1995 through July 15, 1995, file usage increased from 1,000 to almost 5,000 (Fig. 4).

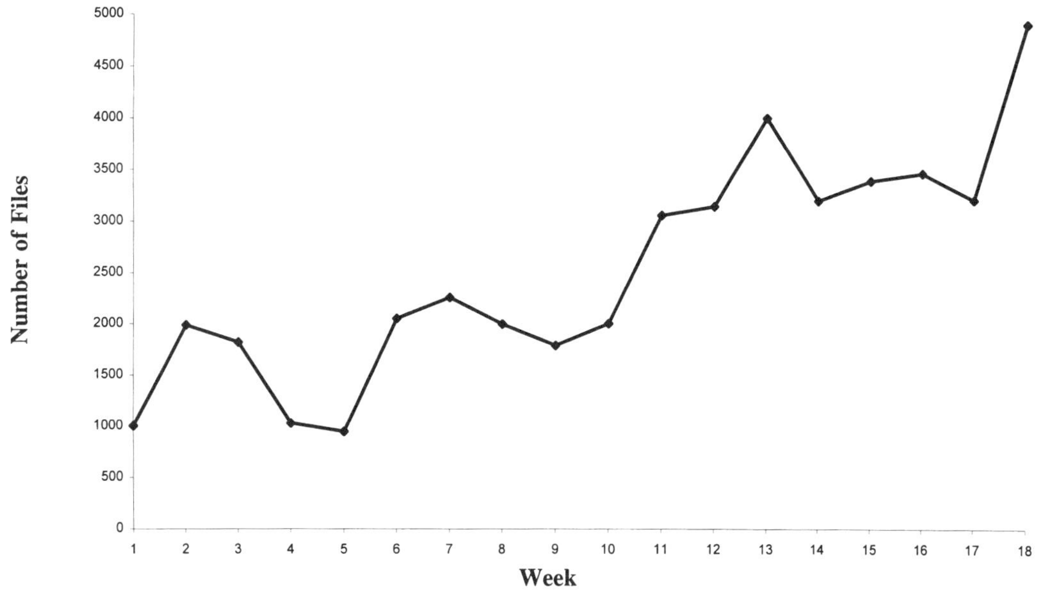

Fig. 4. Number of files requested and sent from Purdue NewCROP server each week from Mar. 12–July 15, 1995.

International demand for new crop information is very high. For example, during the week of July 16-24, 1995, users from 18 different countries accessed the NewCROP server and made 774 contacts (Table 1). This does not account for the number of files requested and transferred out of the system, only the number of times a user successfully connected into the server.

There also has been considerable domestic interest in new crops. There were 2,876 domestic contacts into the NewCROP server the week of July 16–24, 1995 (Table 2). While 41% of contacts originated from U.S. educational sources (presumably universities, colleges and extension), more than 52% originated from commercial sources, an observation which was unexpected. The 3% of contacts which came from U.S. government (e.g. USDA, state agencies), probably reflects new crop researchers. These data show that there is a considerable demand for such a new crop information system, and when coupled with the positive direct feedback received from users, we conclude that NewCROP has been an effective vehicle to reach thousands of professionals in the private and public sector, and can function as an outreach for extension and training. This electronic information network clearly can deliver production, processing and marketing information in an efficient and timely manner to both commercial and academic sectors. The challenge before us is to continue to build and maintain NewCROP to meet user needs.

Table 1. A one-week sample survey of international use and demand for NewCROP (A New Crops Resource On-Line Program). The total number of files requested and sent to on-line users from countries outside of the United States, listed by country domains. Week of July 16–24, 1995.

	Contacts	
Country	Number	% of total
Australia	86	11.1
Belgium	6	0.8
Canada	140	18.1
Finland	3	0.4
France	3	0.4
Germany	2	0.2
Italy	5	0.6
Japan	1	0.1
Mexico	3	0.2
Netherlands	59	7.6
New Zealand	54	6.9
Norway	13	1.7
Portugal	12	1.6
Singapore	5	0.6
Spain	11	1.4
Sweden	2	0.2
Switzerland	13	1.6
United Kingdom	66	8.5
Unresolved	290	37.5
Total	774	100.0

Table 2. A one-week sample survey of domestic use and demand for NewCROP. The total number of files requested and sent to on-line domestic users, listed by user's professional/industrial affiliation. Week of July 16–24, 1995.

	Contacts	
U.S. user	Number	% of total
Commercial	1506	52.4
Educational	1189	41.4
Government	75	2.6
Military	74	1.5
Organization	62	2.2
Total	2876	100.0

CONCLUSION

NewCROP became operational on March 15, 1995 and now can be viewed over the WWW by mosaic, netscape, lynx, and other software browsers that provide access into the internet. The electronic network involves interactive communication via a LISTSERV bulletin board and non-interactive communication via NewCROP. The non-interactive branch is composed of a virtual library based upon CropSEARCH, CropREFERENCE, CropEXPERT, NewCropNEWS, NewCropEVENTS, and FarmMARKETS. The server also has links to a multitude of other WWW servers related to new crops and plant products. NewCROP is being used domestically and internationally by both the private and public sectors and offers an effective vehicle as a new crop information center and distribution resource.

REFERENCES

Johnson, D.N. and E.R. Bragg. 1994. National Farmers' Market Directory. USDA Agr. Marketing Serv. Transp. and Marketing Division. USDA, Washington, DC.

Magness, J.R., G.M. Markle, and C.C. Compton. 1971. Food and Feed Crops of the United States. Interregional Research Project IR-4. Bul. 1. New Jersey Agr. Ext. Sta. Bul. 828.

The Morton Collectanea: An Information Center in Economic Botany

Julia Morton

The Morton Collectanea of the University of Miami, Coral Gables, Florida, is a specialized reference and research department devoted to the acquisition, classification, physical collation, and maintenance of subject-filed data in the field of economic botany, with special emphasis on the tropical and subtropical. This information is contained in 500 file drawers and includes approximately 15,000 species (Fig. 1).

Fig. 1. The Morton Collectanea.

HISTORY

The Morton Collectanea is an outgrowth of an effort to enhance the American food supply in the hungry days of the "Great Depression." When Kendal Morton, a Canadian, who preferred the climate of Florida, met Julia (who was raised on a big and beautiful farm), they began a search for literature on edible plants. Their collection grew from 1932 onward and was presented to the University of Miami in 1949 on the recommendation of the famed Plant Explorer Dr. David Fairchild.

The December issue of the Chemurgic Digest carried a plea by Wheeler McMillan and Harry J. Prebluda, eager growers, for exploration for new crops to aid agriculture and industry. The dissection of books and periodicals, and bulletins, yielded 90,000 items pertaining to food and food sources from 1933–1942. Work was suspended during the War Years—1944 and 1945—but continues, and Julia has been the Director since the loss of her husband in 1952.

ACQUISITION OF INFORMATION

In selecting literature, book, or periodical, for processing, primary attention is given materials having the broadest topical coverage. Publications devoted to highly developed or industrialized single commodities, which have substantial and readily accessible literature in their own right, are maintained for reference but are represented in the files mainly by bibliographic listings. The Collectanea is chiefly concerned with bringing together the more scattered data on a greater number of subjects and from many sources and areas. The mass dismemberment of appropriate literature and subject-sorting of its contents automatically reveals an array of subjects and data from unsuspected regional and other sources to which the researcher would ordinarily have no clue and would not ordinarily be directed—thus recovering much material not found by the usual searching processes.

Works such as Sturtevant's *Notes on edible plants*, Dalziel's *Useful Plants of West Tropical Africa*, Cook and Collins' *Economic Plants of Puerto Rico*, Wester's *Food Plants of the Philippines*, Maiden's *Useful Native Plants of Australia*, Safford's *Useful Plants of the Island of Guam*, Fernald and Kinsey's *Edible Wild Plants of Eastern North America*, Quisumbing's *Medicinal Plants of the Philippines*, Webb's *Guide to the Medicinal and Poisonous Plants of Queensland*, Hooper and Field's *Useful Plants and Drugs of Iraq and Iran*, Rose's *Notes on the Useful Plants of Mexico*, Dastur's *Useful Plants of India and Pakistan* (to name only a few) have provided excellent geographical coverage and a great diversity of species. As work has progressed, layer after layer of excerpted material has enriched these files, and folders for new species are constantly added. Publications of lesser topical range, but dealing with fruits, vegetables, nuts, fibers, fungi, palms, spices, and so forth, have swelled the folders of crops already under cultivation, as does the constant influx of periodical excerpts and bulletins and reprints covering individual subjects. Lengthy publications devoted to highly industrialized single commodities, which have substantial literature in their own right, are represented in the files mainly by bibliographic listings.

In addition to enriching the files, the Collectanea has inspired the publication of papers. Oral presentations with exhibits or color slides have been made to various audiences including the staff and health officials assembled at St. Elizabeth's Hospital, Curacao (in 1970 and 1971), and at the Pasteur Institute, Utrecht; specialists at the Institute for Tropical Hygiene, Amsterdam; International Cancer Congress, Cancer Demography (1970); Round Table on Cancer of the Esophagus; Pan American Medical Association's International Congress on Medicine and Surgery, Guayaquil, Ecuador; University of Florida and University of Miami groups, and others.

Utility of the Collectanea

Inclusion of the Collectanea in the National Science Foundation's *Directory of Specialized Science Information Services* and in Gale's *Directory of Special Libraries and Information Centers*, has brought requests for help from all over the world. For example: a query was received from North Rhodesia as to the edibility of dahlia tubers; an S.O.S. from a clinic in Iran for methods of making soybean milk for infants; and a missionary in the Soloman Islands needed to know how to create smudge with local materials to combat insects in coconut palms. There is a very active demand for information on cashew culture and processing. Considerable material on passion fruit raising has been supplied to pioneer growers in Venezuela, and on tea culture to experimenters in Costa Rica.

The Collectanea is primarily for the use of the University faculty in the course of teaching or research, and for students in the preparation of term papers or theses. However, materials are made available for perusal on the premises by researchers from other institutions, and photocopies of desired excerpts are furnished at cost plus a nominal handling fee. Reports or tabulations of medicinal plants (for instance, those used as diuretics) may be prepared for pharmaceutical companies, or data on the growing and utilization of bean sprouts may be assembled for a food concern. Special studies have been made of toxic plants which present an active problem in Florida, and frequent calls are received concerning external or internal reactions of people, cattle, ducks, and dogs to plant parts ingested, contacted, or inhaled. Some of these cases require plant identification and personal investigation.

During World War II, I as Director of the Collectanea, had conducted field work in tropical survival (at Maxwell Air Force Base in the Philippines, and in Thailand and Vietnam). Later, utilizing the resource facilities of the Collectanea, I prepared survival manuals for the Military. I have also engaged in an ethnobotanical and epidemiological survey for the National Institutes of Health in an effort to discover possible plant causes of esophageal cancer. The plant information resources of the Collectanea were an indispensable part of the study. Having this facility for instant characterization of numerous plant materials found to be ingested by populations under field study (including the people of Curacao, Costa Rica, Guatemala, Yucatan, Northwestern Venezuela, and coastal South Carolina), it has been possible to establish a correlation between the disease and the intake of plant products high in condensed tannin and associated phenols. These findings revealed a strong correlation between certain food habits and susceptibility to esophageal cancer in high-risk areas of Japan.

Field work in the Netherlands Antilles for several weeks each year (off and on from July 1965 to September 6, 1971) revealed esophageal cancer to be the leading tumor in females in Curacao; and it is second in importance in men. Aruba has scarcely more than 2 new cases per year. It is reasonable to expect that the results of these studies will provide clues to the causes of esophageal cancer in other parts of the world and of other tumors at other sites in the human body.

A number of national and international organizations have relied on the resources of the Morton Collectanea. This includes the National Institutes of Health for extensive plant information; the United States Department of Agriculture for urgently-needed information on agricultural crops; the Department of the Army and the Department of the Navy to produce plant information which could not be obtained as readily from any other source; and the Food and Agriculture Organization of the United Nations has appealed to the Morton Collectanea for help in filling "formidable gaps in knowledge." As a result, it has been possible to improve the quality and quantity of information disseminated by such agencies and also to forestall the issuance of misinformation.

A few examples illustrate the utility of the Collectanea files. A representative of a cargo airline sought the help of the Collectanea in ascertaining the consumer acceptability and import possibilities of a native South American fruit which he knew only by the local name "curuba." He had been greatly impressed with the flavor of the ice cream and sherbets made from it in Colombia and felt it might find a market among American ice cream manufacturers. The Collectanea's index led to the file on *Passiflora mollissima* that revealed an introduction of seeds of the species into the United States from Ecuador by the Bureau of Plant Industry in 1920 and another record of similar introduction from Colombia in 1922. Then appeared an excerpt with illustration, from Popenoe's *Economic Fruit-bearing Plants of Ecuador* (1924), a brief mention from Pope's *Propagation of Plants by Cuttings in Hawaii* (1934), and an excerpt from Arbelaez's *Plantes Utiles de Colombia* (2nd ed, 1947), and 2 full pages from the New Zealand *Journal of Agriculture*, March 1953, featuring the "banana passion fruit," the name under which the species is cultivated in Auckland, with nearly a dozen recipes for dessert use, and photographs of the vine, the fruit, and several of the cooked products. There existed no clue that would lead the seeker from Colombia to New Zealand.

A more dramatic instance of the utility of our files occurred when a child was hospitalized with serious symptoms of poisoning as the result of eating the seeds of a wild vine. Specimens of the seeds were brought to the Collectanea, botanically identified as *Abrus precatorius*, and a file of excerpts on that species from 51 sources (representing 33 different countries or territories) was placed before the pediatrician in charge of the case. Totally unfamiliar with the species at the outset, he became quickly conversant with its toxic properties, symptoms, recommended treatment, and a wealth of background material—all of which aided in saving the life of the patient. The Collectanea is often called upon to render similar service to physicians and hospitals. However, one might shudder at the literary fiascos that might ensue if taxonomic study was not given the excerpts prior to subject-filing.

Articles on the eggfruit of Queensland (the eggplant, *Solanum melongena*), might easily be placed with the "eggfruit" of the West Indies (the canistel, *Pouteria campechiana*). There are many such pitfalls in nomenclature, both common and scientific. A number of years ago, the Federal Standard Stock Catalog appeared with a series of newly drafted specifications on food commodities. One was an impressive 5-page document headed "CURRANTS." It was found to have been compiled by blending material on the so-called "currant" of the bakery trade (actually the dried Corinthian grape, a variety of *Vitis vinifera*) with material from agricultural bulletins on the red and black currants (*Ribes vulgare* and *R. nigrum*) respectively. The result was ludicrous; the document was found to be extant and uncorrected 7 years later!

SEARCHING THE COLLECTANEA

While the facilities of the Morton Collectanea are unique and provide a great advantage in the finding and assembling of requested plant data, the services rendered—that is, the outputs of information—are still dependent on human effort and ability. The Morton Collectanea, though efficiently organized, is operated totally on a manual basis. While there is instant access to literature by botanical species, there is no internal index to subject content. Manual search of the large and increasing number of excerpts for broad categories of data (such as toxic elements in plants) would be a lengthy task, even though much more efficient than from the unprocessed literature.

The information contained within each excerpt may be varied to a great degree. It may include common and scientific names of plants, geographical locations, chemical constituents, food, medicinal, or other utility for man or animal, and beneficial and/or toxic effects. Because of the necessity of filing under a single subject heading (or multiple, synonymous file headings) in the manual method, much of the knowledge contained on any one plant, or any geographical region, or any plant property or product, remains irretrievable *except* by laborious manual search, or foreknowledge of the species.

To optimize the use of the Morton Collectanea and to extract a broad range of data from its subject files, it is considered essential that a computerized information storage and retrieval system be installed. There are doubtless many approaches to such a task. The Morton Collectanea is an outgrowth of one approach that started in New York in 1933 with the original goal being the assembling of data on all known edible substances. This objective has since been liberalized to include all economic plants, edible and otherwise.

One basic factor in the Collectanea concept was the ideal of keeping to a minimum the skilled research that would be involved in such a large task by so "mechanizing" the effort that the maximum could be accomplished by unskilled clerical labor. After some years of experimentation, the project took form and has been carried on under the auspices of the University of Miami, Florida since early 1949.

In conclusion, the efforts of the Collectanea are directed toward the goal of rendering the subject content of economic-plant literature readily accessible and accurate. The Collectanea contains a tremendous wealth of difficult-to-find information. The development of an index is necessary to increase the usefulnes of this unique collection.

New Literature for New Crops

Jane Potter Gates

The United States government has long recognized the importance of agriculture in keeping people well-fed, well-clothed, and well-housed, as well as acknowledging the role of science in helping to achieve and sustain those conditions through exploration and research. In the pursuit of new crops, the importance of bibliographic information in the success of such exploration and research cannot be over-emphasized. Since 1862, the U.S. Department of Agriculture (USDA), through the National Agricultural Library (NAL), has collected, compiled, and disseminated information relevant to that mission.

LIBRARIES, CENTERS, AND SERVICES

The National Agricultural Library (NAL), part of the USDA's Agricultural Research Service (ARS), is the largest agricultural library in the world. NAL has 10 subject specific information centers that provide customized services to the agricultural community and others. All of the Centers can be accessed electronically through the NAL gopher and the NAL homepage (see addresses below). The Centers include: Agricultural Trade and Marketing, Alternative Farming Systems, Animal Welfare, Aquaculture, Biotechnology, Food and Nutrition, Plant Genome Database. Rural Technology Transfer, and Water Quality.

Established even earlier than the NAL, the U.K.'s Centre for Economic Botany was originally conceived by Sir William J. Hooker, in 1847, to "render great service, not only to the scientific botanist, but to the merchant, the manufacturer, the physician, the chemist, the druggist, the dyer, the carpenter, and the cabinet maker and artisans of every description, who might here find the raw materials employed in their several professions correctly named." Dr. James Morley, Director of the Centre, states that "almost 150 years later the significance of this statement is more apparent than ever, given increasing awareness of the importance of plants as sources of useful raw materials." The Centre houses more than 75,000 botanical artifacts.

Appropriate Technology Transfer for Rural Areas (ATTRA) is a national sustainable agriculture information service designed to be used by commercial farmers. ATTRA also offers technical assistance and sustainable agriculture information free of charge to Extension agents, agricultural support groups, researchers, educators, and agribusinesses. In addition to providing customized research, ATTRA offers three types of standard materials: "Information Packages, Summaries, and Resource Lists." Located in Fayetteville, Arkansas, ATTRA can be reached by phone: 1-800-346-9140.

DATABASES

AGRICOLA (AGRICultural OnLine Access)

This resource of the National Agricultural Library is a bibliographic database consisting of literature citations for journal articles, monographs, serials, theses, patents, proceedings, audiovisual materials, software, computer files, and technical reports relating to all aspects of agriculture since 1965. Librarians and experienced information specialists at the NAL will perform searches of AGRICOLA upon request. There is no charge for this service unless the search requires more than one hour, in which case there is a cost-recovery fee. Many people prefer to do their own searching, either online or through use of AGRICOLA on diskette. AGRICOLA is available for search and retrieval through commercial vendors. (Some commercial vendors: Knight-Ridder, 2440-El Camino Real, Mountain View, CA 94040; OCLC/EPIC, 6565 Frantz Road, Dublin, OH 43017; Silver Platter, 100 River Ridge Drive, Norwood, MA 02062).

Citations in AGRICOLA include selected journal articles. NAL receives over 25,000 journals, and 1,410 titles are indexed into the database. Some titles are indexed comprehensively, some selectively, and both categories include some foreign language publications. The *List of Journals Indexed in AGRICOLA 1995* is available from NAL in hardcopy, and electronically on both NAL's gopher (gopher.nal.usda.gov) and its worldwide web homepage (http://www.nal.usda.gov).

AGRIS International

In addition to AGRICOLA, other databases available through commercial vendors are important sources of information on new crops. Countries other than the U.S. are covered more comprehensively in the database AGRIS International, which corresponds in part to AgrIndex, published monthly by the Food and Agriculture Organization (FAO) of the United Nations. The Agris International database serves as a comprehensive inventory of worldwide agricultural literature that reflects research results, food production, and rural development. Subject coverage focuses on many topics.

CAB Abstracts

CAB Abstracts is a comprehensive file of agricultural and biological information and contains all records in the 26 main abstract journals published by CAB International. Over 8,500 journals in 37 different languages are scanned for inclusion, as well as books, reports, theses, conference proceedings, patents, annual reports, and guides. Significant papers are abstracted, while less important works are reported with bibliographic details only. The journals included in CAB cover all aspects of agriculture and animal science.

CRIS

The best source for information about current USDA research is the Current Research Information Service (CRIS) database. It is available electronically: http://cristel.nal.usda.gov:8080

TropAg

TropAg database covers literature related to practical aspects of agriculture in tropical and subtropical regions from 1975.

U.K. Centre for Economic Botany Bibliographic Database

The Centre Economic Botany Bibliographic Database contains over 150,000 literature references, covering the uses of plants from around the world (excluding major crop species), with references dating back to the mid-19th Century. It provides access to detailed information—by species, vernacular names, geographical area, uses and/or properties, materials (timber, fibre, textiles, paper, oils, waxes, cosmetics, etc.), medicines, and fuels. The Database, only one activity of the Centre, is unfortunately not available on disk. However, Dr. J. Morley will discuss the Centre's fee policy and following mutual agreement as to terms will run a search on a particular species, use, country, or any combination thereof. Contact: J.Morley@rbgkew.org.uk

NEW CROP MONOGRAPHS 1980–1995

All of the resources listed above are sources of information about agriculture, including new crops. Recent information on new crops also include a number of monographs in the NAL collection that have been published since 1980. These are listed below (NAL call numbers listed in brackets).

Anthony, K.R.M., J. Meadley, and G.Robbelen. 1993. New crops for temperate regions. 1st ed. Chapman & Hall, London, New York. [SB160.N49-1993]

Arkel, H. 1982? The identification and introduction of a new crop in the highlands of Kenya. v. 1. Wageningen, The Netherlands. [SB187.K4A6]

Beech, D. and F. Brisbane. 1992. Crop economics: a program for assessing the economic viability of new crops.: CSIRO Division of Tropical Crops and Pastures. [SB193.3.A8T7-no.77]

Council for Agricultural Science and Technology. 1984. Development of new crops: needs, procedures, strategies, and options. Ames, IA. [S22.C6-no.102. S22.C6-no.102]

CSIRO Division of Tropical Crops and Pastures. 1992. Cropsieve: A computer program to select out potential new crops. St. Lucia, Brisbane. [SB193.3.A8T7-no.71]

Dailey, C. L., K.M. Abotteen, and J.D. Nichols. 1980. Corn/soybean decision logic: Improvements and new crops. USDA. Lyndon B. Johnson Space Center, NASA, Tech. Rep. Houston, TX. [S494.5.R4D353]

Dary, C. 1991. Traditional women and new crops. (Mujeres tradicionales y nuevos cultivos.) FLACSO, Guatemala. [HD6073.A292G93-1991]

Hoyle, F.C. 1992. A review of four potential new crops for Australian agriculture: *Amaranthus caudata*, *Chenopodium quinoa*, *Grindelia camporum* and *Stevia rebaudiana*. Division of Plant Industries, Dept. of Agriculture, South Perth, W.A., Australia. [SB87.A82W4-1992]

Janick, J. and J.E. Simon (eds.). 1990. Advances in new crops. Timber Press, Portland, OR. [SB160.N38-1988]

Janick, J. and J.E. Simon (eds.). 1993. New crops. Wiley, New York. [SB160.N38-1991]

Jessop, R.S., and R.L. Wright. 1991. New crops: agronomy and potential of alternative crop species. Inkata Press, Melbourne, Australia. [SB160.N48-1991]

Karow, R.S. 1990. Cropping alternatives: a questionnaire for evaluating a new cropping enterprise. Oregon State Univ. Extension Service, Corvallis. [100-Or3M-no.838-1990]

Karow, R.S., W.R. Rogers, and R. Penhallegon. 1989. Cropping alternatives: A questionnaire for evaluating a new cropping enterprise. Oregon State Univ. Extension Service. Corvallis. [100-Or3M-no.838]

Kontaxis, D.G. 1989. Capers: A new crop for California? Small Farm Center, Univ. California, Coop. Ext., Davis. [SB306.U6K65-1989]

Knox, E.G. and A.A. Theisen. 1981. Feasibility of introducing new crops: Production—marketing—consumption (PMC). Soil and Land Use Tech. Rodale, Emmans, PA. [SB176.U5F43]

Kugler, D.E. 1988. Kenaf newsprint: realizing commercialization of a new crop after four decades of research and development: A report on the Kenaf Demonstration Project. United States. Cooperative State Research Service. Special Projects and Program Systems. Washington, DC. [aHD9156.K662U55]

Naqvi, H.H., A.I. Estilai, and I.P. Ting. 1992. Proceedings New industrial crops and products. Int. Conference on New Industrial Crops and Products. Association for the Advancement of Industrial Crops. Riverside, CA. [SB160.I57-1990]

National Research Council (U.S.). 1985. Jojoba: new crop for arid lands, new raw material for industry. Advisory Committee on Technology Innovation. Ad Hoc Panel. Washington, D.C. National Academy Press. [SB299.J6J62]

North Carolina State Univ. Dept. of Crop Science. 1987. New crops memo. North Carolina State Univ., Raleigh. [SB187.U6N48]

Okojie, J.A., D.U.U. Okali. 1993. Lost crops of Nigeria: Implications for food security: Proc. Seminar Univ. Agriculture, Abeokuta, Nigeria, 21st–22nd May 1992. [SB16.N6L67-1993]

Oregon Agricultural Experiment Station. 1987. Alfalfa, corn, melons, mint, small grains and new crop research. Oregon State Univ. Corvallis, OR. [100-OR3M-no.814]

Russell, J.S. 1992. A global market analysis of FAO data as a guide to selection of potential new crops for Australia. CSIRO Division of Tropical Crops and Pastures. Brisbane. [SB193.3.A8T7-no.75]

Thalacker, B. 1992. Cultivation methods for newly introduced vegetable species. Anbauverfahren fur Neu Eingefuhrte Gemusearten. Institut fur Gemuse und Zierpflanzenbau Grossbeeren Erfurt. Braunschweig. [SB323.A52-1992]

Tisdell, C.A. 1988. Diversification and stability implications of new crop varieties: Theoretical and empirical evidence. Dept. of Economics, Univ. Newcastle, Resource and Development Group. N.S.W., Australia. [HC611.A5N4-no.158]

United States. Dept. of Agriculture. 1992? New crops, new uses, new markets: industrial and commercial products from U.S. agriculture. Industrial and commercial products from U.S. agriculture. Office of Publishing and Visual Communication. Washington, DC. [1-Ag84Y-1992]

Univ. Missouri. Agricultural Experiment Station. 1990. Economic development via new crops/products from agriculture. Report of seminar on agricultural marketing and policy, College of Agriculture and Extension Division Univ. Missouri, Columbia. [100-M693Sp-no.422]

Vienna, A. Nebraska Cooperative Extension Service. 1989. Alternative crops for Nebraska. (1st ed.). Plant domestication by induced mutation. Proc. Advisory Group Meeting on the Possible Use of Mutation Breeding for Rapid Domestication of New Crop Plants.Cooperative Extension Service, Institute of Agriculture and Natural Resources, Univ. Nebraska-Lincoln, Int. Atomic Energy Agency. [S494.5.A65A47-1989. SB123.A3-1986]

Wickens, G.E., N. Haq, and P. Day. 1989. New crops for food and industry. Chapman and Hall, London. [SB91.N48]

Zuang, H. 1991. Agenda for new vegetable species. (Memento nouvelles especes legumieres.) Centre Technique Interprofessionnel des Fruits et Legumes, Paris. [SB320.9.Z82-1991]

PART III
NEW CROPS, NEW TECHNOLOGIES

CEREALS AND PSEUDOCEREALS

Alternative Wheat Cereals as Food Grains: Einkorn, Emmer, Spelt, Kamut, and Triticale

G.F. Stallknecht, K.M. Gilbertson, and J.E. Ranney

The U.S. Dept. of Agriculture food guide pyramid recommends 6–11 daily servings of the "grain" group. The growing interest in home bread making along with the development and increased availability of alternative cereal grain flours has encouraged the development of various tasty bread products. Many bread products varying in taste and texture can be developed utilizing alternative grain flour alone or in various proportions with common bread wheat flour. The tremendous potential of grains to produce a variety of bread products has been summarized by Pomeranz (1983).

The objective of this paper is to generate the interest of the home baker in the use of cereal grain flours other than common wheat. Presently Kamut and spelt flours are readily available. Triticale flour availability is limited, and the authors are still pursuing agronomic and quality evaluations of einkorn and emmer PI accessions from the USDA-ARS National Small Grain Germplasm Research Facility (NSGGRF), Aberdeen, Idaho.

EINKORN

Origin and Taxonomy

Einkorn is thought to have originated in the upper area of the fertile crescent of the Near East (Tigris–Euphrates regions). Wild einkorn *Triticum boeoticum* includes both the single grain *T. aegilopoides* and the two grain *T. thaoudar* and *T. urartu*. Cultivated einkorn is *T. monococcum*, and like wild einkorn has the genome constitution AA (Table 1). In cereal crops the head (inflorescence) if unbranched is called a spike. The spike consists of flowers (spikelets) arranged on the rachis (which is an extension of the stem). The flowers (spikelets) arise from nodes along the rachis which are called rachilla. The spikelet is enclosed by bracts, the glumes, or chaff. The kernels within the spikelet as enclosed in bracts, the lemma, and palea. As an example, kernels of free threshing wheats thresh free of the bracts; barley threshes free of the glumes, while lemma and palea make up the hull of the kernel; einkorn, emmer, and spelt thresh with the complete spikelet intact. A classification and description of *Triticum* sp. is outlined by Briggle and Reitz (1963). The wild and cultivated einkorn are differentiated by the brittleness of the rachis. The rachis of wild einkorn is brittle and the spikelets readily disarticulate when mature, whereas the rachis of cultivated einkorn is less fragile and remains intact until thrashed.

Einkorn along with emmer and spelt are often referred to as "the covered wheats," since the kernels do not thresh free of the glumes or the lemma and palea when harvested (Fig. 1). In contrast to the free threshing wheats, the spikes of einkorn disarticulate at threshing (the seed head breaks apart into intact spikelets). The spikes disarticulate with the rachilla apex attached to the base of the spikelet. Einkorn has long narrow glumes which are awned. Cultivated einkorn generally has one kernel per spikelet.

Einkorn became widely distributed throughout the Near East, Transcaucasia, the Mediterranean region, southwestern Europe, and the Balkans, and was one of the first cereals cultivated for food.

Harlan (1981), cites information suggesting that wild einkorn grain was harvested in the late Paleolithic and early Mesolithic Ages, 16,000–15,000 BC. Confirmed finds of wild grain remains have been dated to the early Neolithic (Stone Age) 10,000 BC. (Helmqvist 1955; Zohary and Hopf 1993). Cultivated einkorn continued to be a popular cultivated crop during the Neolithic and early Bronze Age 10,000–4,000 BC giving way to emmer by the mid-Bronze Age. Einkorn cultivation continued to be popular in isolated regions from the Bronze Age into the early 20th century. Today, einkorn production is limited to small isolated regions within France, India, Italy, Turkey, and Yugoslavia (Harlan 1981; Perrino and Hammer 1982).

Agronomy and Production

Historically, einkorn was cultivated in cool environments on marginal agricultural land through the Mid-east and southwestern Europe. Einkorn is still cultivated in harsh environments and poor soil in Italy (Perrino and Hammer 1984). Einkorn selections produced protein and yield equal to or higher than barley and durum wheat when grown under adverse growing conditions (Vallega 1979). Evaluations of 15 einkorn accessions grown in Italy indicate that the yields were significantly lower than that of modern wheats when grown under intensive cropping management (Vallega 1992). However, in this study several progeny of selected einkorn crossings (while lacking in several desirable agronomic traits) produced yields comparable to the modern wheats. Eighty einkorn PI accessions from (NSGGRF) have been evaluated for yield, straw characteristics, and date of heading at the Southern Agricultural Research Center, Huntley, Montana (SARC) from 1992 to the present.

Table 1. Genomic constitution of wheats and their relatives.

Species	Genome constitution	Common name
Diploid ($2n = 14$)		
Triticum boeoticum, Boiss. emend. Schiem.	AA	Wild einkorn
T. aegilopoides Link Bal.		Single grain einkorn
T. thaoudar Reuter		Two grain einkorn
T. urartu Tuman		Two grain einkorn
Triticum monococcum L.	AA	Einkorn (cultivated)
Triticum speltoides (Tausch) Gren. Ex Richter	BB	Wild grass
Triticum tripsacoides (Jaub. & Spach) Bowden	BB	Wild grass
Triticum searsii (Feldman & Kislev) Feldman, comb. nov.	BB	Wild grass
Triticum tauschii (Coss) Schmal. (*Ag. squarerosa*)	DD	Wild grass
Secale cereale L.	RR	Rye (cultivated)
Tetraploid ($2n = 28$)		
Triticum turgidum L. group *dicoccoides* (Korn, in litt. in Schweinf.) Bowden	AABB	Emmer (wild)
Triticum turgidum L. group *dicoccum*, (Schrank) Thell.	AABB	Emmer (cultivated)
Triticum turgidum L. group Durum Desf.	AABB	Durum
Triticum turgidum, ssp. *turanicum*[z]	AABB	Kamut
Triticum turgidum, ssp. *polonicum*	AABB	Polish wheat
Hexaploid ($2n = 42$)		
Triticum spelta, L.	AABBDD	Spelt (cultivated)
Triticum aestivum L. (em. Thell.)	AABBDD	Common wheat (cultivated)
×*Triticosecale* (Wittmack)[y]	AABBRR	Triticale

[z]Suggested species, sub-species, Kamut® registered trademark of Kamut International.
[y]Most triticale cultivars are hexaploid but can be octaploids or tetraploids

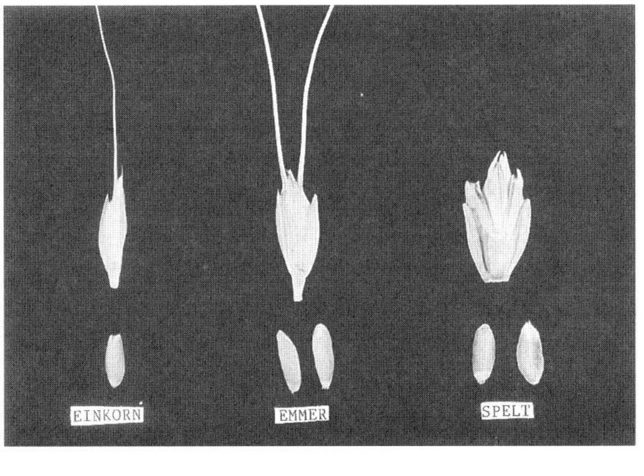

Fig. 1. Spikelet and kernels of einkorn, emmer, and spelt.

The yields of einkorn ranged from 4160 to 120 kg/ha, 1992; 1290 to 130 kg/ha, 1993; 2160 to 220 kg/ha, 1994; and 2400 to 720 kg/ha in 1995. The 1995 yield range represents 25 final PI accession selections (based on yield record and straw strength) of which five produced total yields higher than oats and three higher total yields than the barley and wheat included in the trial. Einkorn grain yields in comparison to spring wheat under dryland cropping were dependent upon growing season environment (Table 2). The protein content of einkorn when threshed in the hull varied from 10% to 26% higher, and the grain from 50% to 75% higher than the protein content (12.5% to 13.5%) of the hard red wheats. Agronomic production practices for spring grains would be applicable to einkorn, which has a tendency to mature later than spring wheat. Einkorn may be most suitable for cropping in lower moisture environments similar to the northern Great Plains area of Montana. The einkorn accessions tested had only moderate straw strength, averaged 109 cm in height, and would be susceptible to lodging in high moisture environments. The susceptibility to diseases is unknown and may be expressed in high moisture environments.

Marketing and Utilization

In the U.S., einkorn production is presently limited to evaluations of PI accessions for agronomic yield and quality traits, and or germplasm sources for plant breeders to improve protein and disease resistance in the development of modern wheats. However recent studies in Europe and Canada emphasized the nutritional quality of einkorn. Grain protein of einkorn accessions and progeny of einkorn crossings were consistently significantly higher than modern wheats (Vallega 1992). The data also indicate that given the significant increase in yields of the progeny and the higher grain protein, progeny lines produced significantly more protein/ha than the modern wheats. The amino acid composition of einkorn was found to be similar to wheat, irrespective of very large variations in total grain protein among the einkorn accessions tested (Acquistucci et al. 1995). The composition and nutritional characteristics of a selected spring einkorn were compared to spelt and hard red spring wheat grown in Canada (Abdel-Aal et al. 1995). The einkorn accession was considered more nutritious than the hard red wheat, based on the higher level of protein, crude fat, phosphorous, potassium, pyridoxine, and beta-carotene. The gluten of the einkorn accession had a gliadin to glutenin ratio of 2:1 compared to 0.8:1 for durum and hard red wheat. Flour and dough characteristics of gluten from 12 einkorn accessions were compared to durum and common wheats (D'Egidio et al. 1993). The einkorn flours were characterized by high protein, high ash, a very high carotene content, and small flour particle size when compared to the modern bread wheats. Dough characteristics of the einkorn accessions were significantly inferior to the modern wheats. The gluten strength was similar to that of soft wheats, but remained sticky, with a low water retention capacity. While breads made from einkorn were considered to be inferior to emmer or spelt breads (LeClerc et al. 1918), Bond (1989) states that breads made from einkorn in France had a light rich taste which left common bread wheat products tasteless and insipid by comparison. Bond also indicated that similar to ancient civilizations the einkorn grains were used in various food dishes such as soups, salads, casseroles, and sauces. The consideration that flour from *T. monococcum* may be non toxic to individuals with celiac disease (Favret et al. 1984, 1987) as cited by D'Egidio et al. (1993), and Abdel-Aal et al. (1995) suggest that given the nutritional advantage of einkorn and possible consumption by individuals allergic to common wheats, an increased interest will be given to the diploid wheats.

EMMER

Origin and Taxonomy

The sites of origin of emmer are considered to be similar to einkorn, within the regions of the Near East (Nevo 1988). Wild emmer *T. dicoccoides*, like wild einkorn is distinguished by the brittleness of the rachis, which disarticulate when mature. The rachis of cultivated emmer *T. dicoccum* is less fragile and tends to remain intact until threshed. The genomic constituents of emmer are described in Table 1. The genomic constitution AA of emmer is thought to be derived from *T. monococcum*. Various sources of the BB genome have been suggested, *T. speltoides, T. searsii*, and *T. tripsacoides* (Morris and Sears 1967; Kimber and Sears

1987). Emmers are predominantly awned with spikelets consisting of two well developed kernels. Emmer glumes are long and narrow with sharp beaks.

The use of emmer as a cereal food is considered to be contemporary with that of einkorn. Similar to einkorn, the earliest civilizations initially consumed emmer as a porridge prior to developing the process of bread making.

Remnants of wild emmer in early civilization sites date to the late Paleolithic Age 17,000 BC (Zohary and Hopf 1993). Cultivated emmer emerged as the predominant wheat along with barley as the principal cereals utilized by civilizations in the late Mesolithic, and early Neolithic Ages 10,000 BC (Helmqvist 1955; Harlan 1981; Zohary and Hopf 1993). Cultivated emmer dispersion and use by early civilizations greatly exceeded that of einkorn. Due to the addition of the BB genome cultivated emmer could be grown in a wider range of environments including regions having high growing season temperatures. Cultivated emmer became the dominant wheat throughout the Near and Far East, Europe, and Northern Africa from the Neolithic (Stone Age) through the Bronze Age 10,000–4,000 BC. Emmer utilization continued through the Bronze Age 4,000–1,000 BC, during which the naked wheats, primarily the tetraploid species slowly displaced emmer. However, emmer continued to be popular in isolated regions such as south central Russia into the early 1900s. Presently emmer remains an important crop in Ethiopia and a minor crop in India and Italy (Harlan 1981; Perrino and Hammer, 1982).

Agronomy and Production

Several thousand hectares of emmer were grown throughout the midwest and western states in the early 1900s (Martin and Leighty 1924). During this period 5 cultivars and 3 selections were identified and evaluated for yield and quality traits. Emmer yields exceeded yields of barley, oats, and wheat cultivars in years which were characterized by less than favorable growing seasons, and produced equal or lower yields when growing conditions were more suitable for cereal production. Three hundred and 50 PI accessions from the NSGGRF were evaluated for yield and straw strength beginning in 1987 at the SARC. During the past nine years emmer yields (as threshed) averaged from 3,700 to 225 kg/ha. The yields of emmers in comparison to barley, oats, and spring wheat were variable, dependent upon growing season environments and locations. Higher yielding emmer PI accessions would equal barley yields and out yield oats in growing seasons which were less ideal for grain production. Advanced selections of emmer PI accessions produced grain kernel yields ranging from 48% to 74% of spring wheat under dryland cropping (Table 2). Emmer PI 254148, a black-chaffed accession with moderate height and straw strength outyielded barley in low moisture years and consistently outyielded oats in a five year study conducted from 1988–1992 at SARC (Schulz-Schaeffer et al. 1995).

Agronomic practices for emmer production are similar to oats. Emmer test weights are similar to oat test weights (360–440 kg/m^3) when grown under dryland cropping at the SARC. Emmer seeding rates are similar to oats, (76 kg/ha) in low rainfall and 100 kg/ha in high rainfall regions. Emmer should be swathed prior to threshing to prevent head shatter loses.

Marketing and Utilization

Presently limited amounts of spring emmer are grown in scattered areas throughout Montana and North Dakota. Two unidentified emmer selections, Cenex emmer and common emmer are grown and available in limited quantities from a small number of seed grain dealers. Emmers marketed and grown in Montana and North Dakota are often mistakenly referred to as spring spelt. The Cenex emmer selection significantly out yields the common emmer which is usually sold by individual farmers. The emmers are grown for grain and used as cattle feed, replacing either oats or barley in feedlot rations. Protein levels of emmer as threshed ranged from 5%–35% higher than oats or barley, while the protein of the grain kernels ranged from 18.5%–21.5% in trials grown in Montana.

Milling and baking studies conducted by LeClerc et al. (1918) indicated that while emmer flour was satisfactory to produce a good loaf of bread, the quality was not equal to breads made from common wheat. Breads produced from whole grain flour of emmers grown at SARC were heavy textured with a pleasing taste

Table 2. Yield comparisons of spring wheat, emmer, and einkorn under dryland cropping in south central Montana, 1992 to 1994.

Crop Year	Dehulled grain yield[z] (kg/ha)	% of wheat[y]
Emmer		
1992	2550	48
1993	1880	84
1994	1540	59
Einkorn		
1992	4220	79
1993	1420	64
1994	2160	82
Wheat		
1992	5370	
1993	2250	
1994	2630	

[z]Emmer and einkorn grain yields were estimated at 60 percent of hulled grain.
[y]'Newana' hard red spring wheat was used as the check wheat against the 5 highest yielding emmer and einkorn selections.

that was milder than breads made from rye flour. A recent study at the Centro Internacional de Mejoramiento de Maiz y Trigo (CIMMYT) in Mexico evaluated 137 emmer accessions using gel electrophoresis to identify selections which contain glutenin fractions known to contribute to bread quality (Pena et al. 1993). Evaluation of over 800 lines of wild emmers resulted in the identification of lines which had high protein and high molecular weight sub-units characteristic of gluten having good bread baking quality (Blum et al. 1984). Presently breeding and selection for superior emmer cultivars is being conducted at the Northwest Plant Breeding Co., Pullman, WA (C. Konzak, pers. commun. 1995).

SPELT

Origin and Taxonomy

The origin of spelt is controversial. While general agreement exists on the origin, extent, and utilization of wild and cultivated einkorn and emmer, archaeobotanists and cereal geneticists have proposed two primary hypothesis for the origin of spelt. One hypothesis suggests a single site of origin in the geographic region of present day Iran. The second suggests two independent sites of origin, the Iranian region and a southeastern European region. Suggested dates for the Iranian origin range from the mid-late Neolithic (Stone Age) 6,000–5,000 BC (Zohary and Hopf 1993). Some authors (Dorofejev 1971; Belea 1971) discuss the possibility of a much later independent European origin, others using genetic markers (McFadden and Sears 1946; Kuckuck 1964); seed protein profiles (Johnson 1972); and genes for resistance to rust (Kema 1992); cite data for support of a common origin such as the Iranian region. While the majority of evidence indicates the single site of origin, possible evidence for both sites are reviewed by Harlan (1981), Kema (1992), and Zohary and Hopf (1994) who reviewed 19 and 21 references by Zohary and Hopf respectively, specific to the origin of cultivated crops. The majority of evidence indicates that the origin of spelt must have occurred when either wild or cultivated emmer (AABB) dispersed to regions where *T. tauschii* (*Ae. squarrosa*) (DD) was an indigenous wild grass species.

The genomic constitution of *T. spelta* is described in Table 1. Similar to the advantage of emmer AABB over einkorn AA, the addition of the genome DD contributed by the wild grass *T. tauschii* resulted in increasing the adaptation of spelt to an even wider range of environments. Spelt represents the hexaploid series of the *Triticum* genome constitution (AABBDD) which like the diploid and tetraploid einkorn and emmer is characterized as a "covered wheat," the kernels do not thresh free of the glumes, lemma, and palea. The spelt spikelets contain two well developed kernels, and are characterized by glumes which have wide square shoulders and short obtuse beaks. The spelt spikelets are most often awnless, however many awned selections also exist. While the rachis of the spelt seed head like cultivated einkorn and emmer is fragile, the point at which the spikelet disarticulate is distinctly different. In contrast to einkorn or emmer, which break apart with the rachilla attached to the base of the spikelet, spelt spikelets break apart with the rachilla remaining attached to the face of the spikelet rather than at the spikelet base (Fig. 1).

Spelt was widely distributed from the Near East origin during the Bronze Age (4,000–1,000 BC), throughout the Balkans, Europe, and transcaucasia. Some of the earliest recordings of spelt appear in the Bible (Exodus 9:30, Isaiah 28:25, and Ezekiel 4:9). The first reference to spelt is found in the "Edict of the Roman Empire Dioletian," in 301 (Flaksberger 1930). Along with the free threshing wheats, spelt may have played a role in the first politically established welfare system in Rome, beginning in 59 BC when after food riots, grain was distributed free to the Roman citizens (Harlan 1981). The wide distribution of spelt was facilitated by the northern and southern route migrations of early civilizations westward. Spelt production continues to be a major cereal crop in isolated regions throughout southeastern Europe, primarily in Germany and Switzerland.

Agronomy and Production

U.S. production of emmer and spelt peaked in the early 1900s and declined steadily thereafter. The first recorded U.S. production information on emmer and spelt, records 233,000 ha, primarily in the states of North and South Dakota, Kansas, Nebraska, and Minnesota. Limited production of spelt also occurred in Wisconsin, Michigan, Iowa, Illinois, Indiana, Montana, Wyoming, and Texas. Production decreased rapidly during the next ten years to 68,000 ha in 1919. This included predominantly winter spelt and spring emmers. Martin and Leighty (1924) documented the production and utilization of the "covered wheats" einkorn, emmer, and spelt in the U.S. and cited over 70 publications dating from 1899 to 1924. Five spelt cultivars were known to be grown in the U.S. during the 1900s. Martin and Leighty (1924) review 52 studies conducted from Texas to Canada on the yield of emmer and spelt in comparison to barley, oats, and wheat. The inconsistent yield potential and higher protein advantage of spelt could not compete with the progress of breeding programs which improved the yields and quality of barley, oats, and the free threshing wheats. Factors such as limited availability of adapted cultivars, low test weight 465–310 kg/m^3 in addition to time and expense of dehulling (for grain use) also contributed to the loss of interest in the covered wheats.

Environmental conditions, particularly growing season precipitation, significantly affected the yield competitiveness of spelt. Winter spelt often outyielded spring oats and barley when early growing season temperatures are cold and moisture is limited. Studies conducted in Germany indicate that the hull of spelt provided an advantage to the seed germination (Ruegger et al. 1990a) and provided protection against soil borne pathogens (Riesen et al. 1986), in conditions unfavorable to germination. Rates of C^{14} assimilation into developing spelt and wheat kernels were evaluated by Ruegger et al. (1990b). Results indicated that low temperatures had less effect on C^{14} assimilation into the spelt kernels as compared to wheat.

Yields and agronomic traits varied significantly among the spelt PI accessions grown under dryland at the SARC. Studies (1990–1995) of 1000 PI accessions indicated wide variations in yield (7000–1000 kg/ha), test weight (462–315 kg/m^3), days to heading (154–166 days), plant height (97–140 cm), and kernel protein content (15.8%–19.2%). Winter spelt data (average of the five highest yielding selections) is compared to winter wheat (Table 3). Spelt yields are given as harvested with the kernel in the hull, and kernel yield only as estimated for a 60% kernel weight thresh out. Percent kernel weight thresh out during the 4 year study

Table 3. Winter spelt and hard red winter wheat multi-year yields under dryland cropping in south central Montana, 1991 to 1994.

Year	Spelt[z] as harvested w/hulls (kg/ha)	Spelt yield[y] (kg/ha)	Wheat yield[x] (kg/ha)	Spelt yield as % of wheat
1991	6070	3640	3760	97
1992	3650	2190	3430	64
1993	7070	4240	5910	72
1994	3480	2090	3830	55

[z]Yield of the 5 highest yielding spelt selections.
[y]Spelt grain yield was estimated at 60% of hulled grain when dehulled.
[x]'Tiber' winter wheat.

ranged from 55%–75%, thus 60% is a conservative estimate. Yield percentage of spelt grain in comparison to the hard red winter wheat check varied from 55%–97% during the 4 year study. Total harvested yields (hull and grain) of spelt grown in Montana were often higher than the total weight of wheat grain harvested. The protein content of the covered wheats when threshed in the hull varied from 10%–26% higher than the protein content (12.5%–13.5%) of hard red wheats, thus offering a potential feed advantage when used for livestock growing rations. However, if used for high concentrate fattening rations, the feed-to-gain ratio is less than barley or maize due to the high percentage of fiber (hull portion) of the covered wheats. Feeding studies with dairy cattle and poultry indicated that the feed value of spelt was similar to oats (Arscott and Harper 1962; Ingalls et al. 1963). Yields and protein content of winter spelt harvested for forage were significantly higher than traditional hay barley or spring oat cultivars (Stallknecht and Gilbertson 1995).

In Montana, some cattle producers in regions of low growing season precipitation, plant spelt in preference to spring oats due to the yield advantage of the winter spelt. Spelt production has however varied significantly. In 1987, 200 Montana farms grew 7300 ha, compared to 1992 when spelt production was recorded as 25 farms and 700 ha (Census of Agr. 1992). Limited spelt production occurs in Pennsylvania, Michigan, Indiana, Kansas, and North Dakota. At present, major spelt production in the U.S. centers in the Midwest, specifically Ohio, which has over 12,000 ha.

'Champ' (a spelt/wheat cross) developed by Ohio State University (Lafever 1988) was the first spelt cultivar released in the U.S. Lafever also released 'GR900', a head selection from a mixed spelt population. Lafever has continued to develop spelt cultivars, for Sunbeam Extract Co. (H. Lafever pers. commun. 1995). Six advanced spelt lines, developed by Sunbeam Extract Co., were included in a yield trial conducted at the SARC in 1995. These lines produced 25%–40% greater yields than the highest yielding spelt PI accessions or spelt cultivars included in the 1995 trials. Northwest Plant Breeders Co., Pullman, Washington is also involved in the development of potential spelt cultivars (C. Konzak pers. commun. 1995). Proprietary European cultivars of spelt, utilized specifically for human food are presently grown in the U.S. for Arrowhead Mills, Texas, and Purity Foods, Michigan (B. Cater pers. commun. 1995; D. Stinchcomb pers. commun. 1995). Foundation and certified seed of Champ, GR900, and an old standard German cultivar, 'Oberkulmer' are available from Frenchs Inc., Waheman, Ohio (L. French pers. commun. 1995).

General information and practical production guides are available for producers interested in spelt production (Lafever and Campbell 1976; Oplinger et al. 1990). The suggested seeding rates for spelt in the Midwest are 90–112 kg/ha. Information generated at SARC, indicated no differences in spelt yields when planted at 67 or 100 kg/ha on dryland or at 100 or 134 kg/ha under irrigation. Seeding of the large hulled spelt seed can be accomplished by use of grain drills which have adjustable openings of sufficient size to accommodate the large pointed seed, and allow for the planting of adequate seeding rates. Smooth drop tubes are desired to prevent seed from lodging and plugging the tube. Midwest studies suggest lower nitrogen fertility rates for spelt in comparison to wheat to compensate for susceptibility to lodging. However, SARC studies have identified selections with excellent straw strength. The advanced semi-dwarf types from Sunbeam Extract Co. have excellent resistance to lodging under higher nitrogen levels which increase the yield potential of spelt. Spelt harvest is generally accomplished by swathing the grain when the stem has not completely turned color. Delayed harvest can result in significant head shatter at maturity.

Studies on the nutritional aspects of spelt report wide variability in the chemical constituents of the grain. Ranhotra, et al. (1995) present data which show few differences between a hard red wheat cultivar and a Canadian spelt selection. The grains were evaluated for gluten traits, chemical composition, amino acid composition, and protein efficiency. The data suggests possible validity to the claim that spelt may be easier for humans to digest than wheat. Recent studies have reported variations in protein, lysine, vitamins, crude fat, minerals, and gliadin/glutenin ratios among spelt selections (Abdel-Aal et al. 1995; Ranhotra et al. 1995, 1996a). A study was initiated by SARC in 1994 to evaluate the performance of three spelt selections, and two hard red wheat cultivars for yield, protein, lysine, fiber, and carbohydrate content over five environments in Montana and North Dakota (Ranhotra et al. 1996b). Results indicate that while variable among locations, the protein content of all spelt selections grown at all locations was consistently higher (18%–40%) than that of the hard red wheats. Lysine content was lower in spelt compared to the wheat, and was inversely related to percent protein. The inverse relationship between percent protein and lysine content of spelt has been reported previously. Variations in protein and lysine content and the inverse protein/lysine relationship were recorded for 164 spelt selections grown over a three year period in Belgium (Clamot 1984). The results on nutritional constituents of the preceding study indicate that variations in the protein content of the grain for a given species is highly dependent upon cropping practices and environmental conditions.

Marketing and Utilization

Spelt is the only "covered wheat" species grown and marketed in the U.S. for human food. Stimulated by market promotions, spelt planted for human consumption increased from less than 40 ha to over 3200 ha between 1987 and the present. Organic and commercial spelt are grown under contract and graded for test weight and percent protein (B. Carter pers. commun. 1995; Stinchcomb pers. commum. 1995). Spelt products are available through organic health food outlets as grain, whole grain and white flours, and processed products. Processed products include assorted pasta, cold and hot cereals, and pre-packaged bread, muffin, and pancake mixes. Baking qualities of spelt cultivars available in the early 1900s were evaluated by LeClerc et al. (1918). The authors reported that good loaves of bread could be produced from spelt flours. Evaluations of spring spelt accessions for bread and pasta products have been conducted in Canada (Hucl et al. 1995). Results indicated that spelt flours treated with an oxidant produced loaf volumes similar to bread wheats. The Canadian researchers anticipate releasing a spring spelt cultivar in 1996.

The suggested attributes of spelt relative to wheat are ease of digestion, taste, and that individuals with certain allergies to common bread wheats can consume spelt. The success of The Berlin Natural Bakery, Berlin, OH, a major commercial bakery of spelt products is based on the attributes given to spelt (H. Graves pers. commun. 1995). In Europe spelt harvested in the hard dough stage and roasted is called "Grunkern," and is considered a "gourmet" food to be used in breads, cereal, soups and casseroles.

KAMUT®

Origin and Taxonomy

The origin of Kamut is thought to have occurred contemporary with the free threshing tetraploid wheats, and is considered to be an ancient relative of modern durum wheats (R. Quinn pers. commun. 1995). The identification of Kamut has been confusing, at least two scientists identify it as *Triticum turgidum,* ssp. *polonicum* another as *Triticum turgidum,* ssp. *turanicum*. More recently, taxonomists specializing in wheat from the U.S.

and Russia have identified Kamut as *T. turgidum*, ssp. *durum*, genomic constitution AABB (Table 1), or similar to an Egyptian cultivar 'Egiptianka'. The inflorescence is somewhat less dense than wheat. The spikelet lemmas have strong long black awns and the glumes have a distinct black acuminate beak. The stem immediately below the inflorescence is characterized by a distinctive wavy morphological trait (Fig. 2). Kamut kernels are twice the size of wheat kernels and are characterized by a distinctive hump shape.

Agronomy and Production

Kamut arrived in the U.S. approximately 40 years ago, when a U.S. airman mailed 36 kernels from Egypt to his father in Montana. The seed was increased and produced commercially for a few years, but was discontinued due to lack of markets and yield averages which were lower than wheat. In 1977, the Quinn family secured a quart jar of remnant seed from which they selected and propagated a specific seed type that was registered as QK 77, and named Kamut, a word thought to mean wheat in ancient Egypt. Kamut production in the U.S. is determined through exclusive contracts with Montana Flour and Grains, Big Sandy (R. Quinn, pers. commun, 1995). All contracts require organic certification of the crop and agronomic practices for production are outlined by Montana Flour and Grains. Results of yield comparisons between Kamut and hard red spring wheats are similar to the results observed in the comparisons between "the covered wheats" and other free threshing wheats. Kamut will out yield spring wheats when environmental stresses are experienced during the growing season, and yield equal to or lower in more ideal growing seasons. Kamut plant height reaches 127 cm with good to excellent straw strength.

Marketing and Utilization

Kamut grains represent one of the most extensive specialty cereal product lines available, and is marketed primarily through health food outlets. Presently 70 plus processors list over 100 Kamut products in the U.S. and Canada under regulation of the Kamut Association of North America (KANA), and the Kamut Association of Europe (KAME). Kamut products include whole grain flour, breads, hot and cold cereals, pastas, and chips, in addition to a green plant dehydrated product. Kamut grains are generally higher in protein when compared to wheat when grown under similar environments. Kamut products made from whole grain flours have a mild, nutty flavor. Individuals who experienced certain types of allergic reactions to products made from common wheat are able to eat Kamut products.

TRITICALE

Origin and Taxonomy

The first wheat/rye cross is considered to have occurred in Scotland in 1875. Several publications outline the historical progress of triticale in the 20th century (Stoskopf 1985; National Research Council 1989; Villareal et al. 1990). The initial crosses between wheat and rye were sterile, with the first fertile crosses made in Germany in 1888. The name triticale first appeared in literature published in Germany in 1935. The first

Fig. 2. Seed heads of einkorn, emmer, spelt, and Kamut.

release of a commercial triticale cultivar occurred in Europe, while 'Rosner' a Canadian release was the first triticale cultivar developed in North America.

Triticale (×*Triticosecale*) genomic constitution AABBRR or AABBDDRR (Table 1) is an artificial cereal crop genus created from crosses between wheat *Triticum* sp. and rye (*Secale cereale* L.) (Table 1). Triticale can be octoploids or tetraploids, however most triticale cultivars are hexaploids. Hexaploid triticale synthesized from wheats (AABB) and rye (RR) are called primary hexaploids, while hexaploid triticale synthesized from crosses of hexaploid triticale and/or hexaploid wheats or octoploid triticale are called secondary hexaploid triticale (Lukaszewski and Gustafson 1987). One advantage of the secondary hexaploid triticale is the increased genomic diversity, including the insertion of portions of the D genome from the hexaploid wheats. Triticale have either winter or spring growth habit, vary significantly in plant height, tend to tiller less, and generally have larger inflorescence in comparison to wheat. The majority of triticale cultivars have prominent awns, however recently, a limited number of both spring and winter types exhibiting awnless traits (less than 5 mm) have become available which have increased potential for use as a hay forage for livestock.

Agronomy and Production

The history of triticale breeding for cultivar development has been an agronomic success story. The first triticale cultivars were characterized by low yields, tall weak straw, shrunken and shriveled kernels, high susceptibility to ergot [*Claviceps purpurea* (Fr.) Tul.], high protein, and high levels of the amino acid lysine. The advantage of high protein and high lysine in swine and poultry rations were nullified by the poor yield performance and the high incidence of ergot. However, triticale cultivars released in recent years have improved agronomic traits including high yields, resistance to lodging and ergot, plump kernels, and lysine levels higher than other cereal grains (Skovmand et al. 1984). The higher yield potential and plumper kernels of modern triticale cultivars have resulted in lower kernel protein levels which are similar to common bread wheats. Triticale cultivars at SARC, Montana, have consistently produced yields higher than spring and winter wheats grown under irrigated and dryland cropping (Table 4). Test weight of triticale averaged 90 kg/m^3 (less than wheat), while heading date and percent protein were similar to the wheats. Research evaluations in Germany have also shown that modern triticale selections produce higher yields and equal or lower protein levels as compared to winter wheat cultivars (Karpstein-Machan and Heyn 1992). The authors concluded that triticale utilized soil and applied nitrogen more efficiently than winter wheat.

Table 4. Comparisons of yield and quality of spring and winter triticale and hard red spring and winter wheat standard varieties grown in south central Montana, 1990–1993.

Cultivar	Yield (kg/ha)	Test wt (kg/m^3)	Head date (Julian days)	Protein content (%)
		Dryland		
'Tiber' hard red winter wheat	4232	766	155	13.4
'Newana' hard red spring wheat	3360	755	166	14.6
Winter triticale^z	4620	659	153	12.8
Spring triticale^z	3758	646	162	13.7
		Irrigated		
'Newana' hard red spring wheat	6047	775	165	11.4
Spring triticale^x	7400	669	163	10.4

^{z}Triticale data is the average of the three top yielding varieties in each respective year.

Emphasis on triticale development has steadily declined in the U.S. in recent years due in part to limited marketing opportunities, and to the U.S. Government wheat and barley support programs. Presently the majority of triticale grown in the U.S. is harvested as forage for livestock feed. Triticale cultivar development, production and utilization has been extensively reviewed during the past 15 years (Lorenz 1974a; Forsberg 1985; Nat. Res. Council 1989; Villareal et al. 1990). More recently gene transformation techniques are being used in an attempt to develop triticale cultivars which have hard white kernels, similar to the increased interest in the development of the hard white common bread wheats (R. Metzger pers. commun. 1995).

Several production guides outline cultural practices and utilization of triticale (Oelke et al. 1989; Salmon and Jedel 1993). However, many of the triticale cultivars discussed in these guides have been replaced by improved cultivars in recent years. Cultural practices at SARC indicate that seeding rates and fertility requirements are similar to wheat when grown for grain, however triticale seeding rates for forage use must be increased to 80 kg/ha low moisture dryland and 107 kg/ha under irrigated cropping, to compensate for reduced tillering as compared to other cereals.

Marketing and Utilization

Triticale grains, flours, and prepared products are available through both health food and commercial outlets on a limited basis. Triticale is often included in prepared mixed-grain hot and cold cereals, and muffin flours. Triticale bread and cracker products were available in the early 1980s to western Canada consumers when triticale was grown to the advantage of farmers under wheat grain marketing programs. While consumers demand was high, the triticale products became unavailable due to the lack of farm production as a result of changes in the wheat marketing program (D. Salmon, pers. commun. 1995). Detailed studies on the nutritional composition and baking quality of triticale have been conducted during the past 20 years (Lorenz 1974b, 1982; Nat. Res. Council 1989). The data indicate that while the nutritional quality of triticale is considered superior to wheat, the higher ash content, lower milling yields of flour, and inferior loaf volume and texture distract from commercial baking use of triticale. In 1991, nine of the twenty two papers presented at the 2nd International Triticale Symposium related specifically to the advances in the nutritional, and baking quality of triticale, an indication of significant interest in the promotion of triticale (CIMMYT 1991). Recent studies by Pena and Amaya (1992) indicate that triticale flour blends of up to 50% with wheat flours produce breads with quality similar to breads made from wheat flours only.

FUTURE POTENTIAL

Consumers are becoming increasingly aware of the benefits of including a variety of cereal grains as a major portion of their diets. Increased consumption of cereals should spawn consumer interest to seek out breads and products made from cereal grains other than from common bread wheat cultivars. The key factor in producing light textured breads is gluten quality of the flour. While the desired gluten traits have been successfully obtained in common bread wheats through many years of intensive cultivar development, little or no effort was applied to the alternative cereal crops described in this publication. Recently, however, studies have been directed to the gluten quality of einkorn, emmer, spelt, and triticale. Studies on gluten quality of common wheat cultivars suggest that characteristics of high molecular weight glutenin subunits, which are controlled by specific genes, are responsible for baking quality (MacRitchie et al. 1990). High molecular weight glutenin subunits considered to have good bread making qualities have been identified in emmer (Pena et al. 1993), spelt (Rodriquez-Quijano et al. 1990), and triticale (Bittle and Gustafson 1991; Smith et al. 1994). Cultivar development for the specific improvement of spelt protein quality for baking quality is also in progress (H. Lafever, pers. commun. 1995). Genetic studies for development of improved einkorn, emmer, and spelt are in progress, Crop Development Center, Univ. of Saskatoon, SK (P. Hucl pers. commun. 1996).

Presently, the dough characteristics of flours from the alternative cereals lack some of the traits required by commercial bakeries to produce light texture and adequate loaf volumes. Bread loaves produced entirely from whole grain flours of emmer, spelt, Kamut, and triticale, while having a range of pleasing flavors, tend to have a heavy dense texture. Bread having lighter textures can be made from whole grain flour of the alternative cereals utilizing dough additives to increase loaf volume. Surfactant (lecithin), oxidants (ascorbic acid), and lipids (shortening), enhance the texture and freshness of baked products. Home bakers may also consider the sponge-and-dough procedure as described by Hoseny (1992) which produces a softer, well flavored bread loaf.

Readers interested in the processes involved in bread making, may refer to a condensed publication (Hoseney 1992) or a comprehensive book on cereal chemistry and products (Hoseney 1994) which details the chemistry of making bread. An excellent book on baking describes the use of cereal crop species (other than common bread wheat), pseudo cereals, and tuber flours for use in the baking of both yeast and non-yeast breads (Dumke 1992). The combination of both gluten and non-gluten ingredients for use in non-traditional breads is near endless.

One of the most alluring aspects of the alternate cereals is the consideration that food products from einkorn, emmer, spelt, and Kamut are relatively hypoallergenic in comparison to food products made form the common bread wheats. Individuals who suffer certain allergic reactions to common bread wheat products state that the reactions are absent when consuming either kamut or spelt products. Research results suggest that a gliadin fraction of the wheat gluten may be responsible for the allergic reactions (Auricchio et al. 1982, 1985). Differences in gliadin proteins between spelt and common bread wheats have been reported (Federmann et al. 1992; Hucl et al. 1995). However, the factors which are responsible for the variations in allergic responses remains unknown. Recent studies also suggest that durum flours may offer an alternative to individuals with allergies to common wheat flour (Boyacioglu and D'Appolonia 1994a,b).

REFERENCES

Abdel-Aal, E.S., P. Hucl, and F.W. Sosulski. 1995. Compositional and nutritional characteristics of a spring einkorn and spelt wheats. Cereal Chem. 72:621–624.

Acquistucci, R., M.G. D'Egidio, and V. Vallega. 1995. Amino acid composition of selected strains of diploid wheat, *Triticum monococcum* L. Cereal Chem. 72:213–216.

Arscott, G.H., and J.A. Harper. 1962. Use of spelt in chick and poult rations. World's Poultry Sci. J. 18:278–284.

Auricchio, S., G. DeRitis, M. DeVincenzi, P. Occorsio, and V. Silano. 1982. Effects of gliadin-derived peptides from bread and durum wheats on small intestine cultures from rat fetus and coeliac children. Pediatr. Res. 16:1004–1010.

Auricchio, S., G. DeRitis, M. DeVincenzi, and V. Silano. 1985. Toxicity mechanisms of wheat and other cereals in coeliac disease and related enteropathies. J. Pediar. Gastr. Nutr. 4:923–930.

Belea, A. 1971. Are there other possibilities for the origin of *T. spelta* or *T. aestivum*. Acta Agronomica Academiae Sci. Hungaricae 20:227–229.

Bittle, D.C., and J.P. Gustafson. 1991. High molecular weight glutenin from wheat for triticale flour improvement. Proc. 2nd. Int. Triticale Symp. Mexico D.F. CIMMYT. p. 550–553.

Blum, A., Z.K. Gerechter-Amitai, A. Grama, J.G. Moseman, E.L. Sharp, and C.F. Konzak. 1984. The collection, evaluation, and utilization of wild emmer (*Triticum dicoccoides*) for wheat breeding. BARD, Final Report. Project No. 1-63-79, The Volcani Center, ARO, POB 6, Bet Dagan, Israel.

Bond, A. 1989. Discovering einkorn in Haute Provence, France. Cerealist 1:1, 6–7. (Publ. KUSA Soc.)

Boyacioglu, M.H. and B.L. D'Appolonia. 1994a. Characterization and utilization or durum wheat for breadmaking. I. Comparison of chemical, rheological, and baking properties between bread wheat flours and durum wheat flours. Cereal Chem. 71(1):21–28.

Boyacioglu, M.H., and B.L. D'Appolonia. 1994b. Durum wheat and bread products. Cereal Foods World 39:168–174.

Briggle, L.W. and L.P. Reitz. 1963. Classification of *Triticum* species and of the wheat varieties grown in the United States. U.S. Dept. Agr. Tech. Bul. 1278.

Census of Agriculture. 1992. United States, Geographic area series, Vols. 1, Part 51, No. AC92-A-51. U.S. Dept. Commerce, Econ. Statistics Adm. Bur. Census.

CIMMYT. 1991. Proc. 2nd. Int. Triticale Symp. Mexico, D.F. CIMMYT.

Clamot, G. 1984. Genetic variability of the protein and lysine content or spelt (*Triticum spelta* L.) Z. Pflanzenzucht. 93:106–114.
D'Egidio, M.G., S. Nardi, and V. Vallega. 1991. Quality of diploid wheat, *Triticum monococcum* L. In: D.J. Martin and C.W. Wrigley (eds.), Proc. Cereals Int. Conf. 205–208. Brisbane, Australia.
D'Egidio, M.G., S. Nardi, and V. Vallega. 1993. Grain, flour, and dough characteristics of selected strains of diploid wheat *Triticum monococcum* L. Cereal Chem. 70:298–303.
Dorofejev, W.F. 1971. Die weizen transkaukasiens und ihre bedeutung in der evolution der gattung *Triticum* L. Z. Pflanzenzuchtg. 66:353–360.
Dumke, N.M. 1992. Allergy cooking with ease. Starburst Publ. Lancaster, PA.
Federmann, G.R., E.U. Goecke, and A.M. Steiner. 1992. Detection of adulteration of flour of spelt (*Triticum spelta* L.) with flour of wheat (*Triticum aestivum* L. emend. Fiori et Paol.) by electrophoresis. Plant Var. Seeds 5:123–125.
Flaksberger, C. 1930. Ursprungszentrum und geographische verbreitung des spelzes (*Triticum spelta* L.). Angew. Bot. 12:86–99.
Forsberg, R.A. (ed.). 1985. Triticale. CSSA Special Pub. No. 9. Crop Sci. Symp. Proc. Am. Soc. Agronomy, Madison, WI.
Harlan, J.R. 1981. The early history of wheat: Earliest traces to the sack of Rome. In: L.T. Evans and W.J. Peacock (eds.), Wheat science today and tomorrow. Cambridge Univ. Press, Cambridge UK.
Helmqvist, H. 1955. The oldest history of cultivated plants in Sweden. Opera Bot. 1:1–186.
Hoseney, R.C. 1992. Principles of cereal science and technology. Am. Assoc. Cereal Chemists, St. Paul, MN
Hoseney, R.C. 1994. Bread baking. Cereal Foods World 39:180–183.
Hucl, P., E.S.M. Abdel-Aal, and F.W. Sosulski. 1995. Specialty wheats. Proc. Cropportunities III:40–47. Univ. Saskatchewan, Saskatoon, Canada.
Ingalls, J.R., D.E. Morgan, J.W. Thomas, and C.F. Huffman. 1963 Nutritive value of spelt (*Triticum sativum spelta*) for dairy cattle. J. Dairy Sci. 46:1085–1088.
Johnson, B. 1972. Seed protein profiles and the origin of the Hexaploid wheats. Am. J. Bot. 59:952–960.
Karpenstein-Machan, M., and J. Heyn. 1992. Yield and yield structure of the winter cereals triticale and wheat in the middle mountain areas of northern Hessen. Agribiol. Res. 45:88–96.
Kema, Gert H.J. 1992. Resistance in spelt to wheat yellow rust: III. Phylogenetical considerations. Euphytica 63:225–231.
Kimber, G. and E.R. Sears. 1987. Evolution in the genus *Triticum* and the origin of cultivated wheat. In E.G. Heyne (ed.), Wheat and wheat improvement. Agron. Monog. 13 (2nd ed.) Am. Soc. Agron., Madison, WI.
Kuckuck, H. 1964. Experimentelle untersuchungen zur entstehung der kulturweizen. Z. Pflanzenzucht. 51:97–138.
Lafever, H.N., and L.G. Campbell. 1976. Spelt: An obscure crop finds a home. Report 61:41–43 Ohio Agricultural Research and Development Center, Wooster.
Lafever, H.N. 1988. Registration of 'Champ' spelt. Crop Sci. 28:377–378.
LeClerc, J.A., L.H. Bailey, and H.L. Wessling. 1918. Milling and baking tests of einkorn, emmer, spelt, and polish wheat. J. Am. Soc. Agron. 10:215–217.
Lorenz, K. 1974a. The history, development, and utilization of Triticale. Critical Rev. Food Tech. 5:175–280.
Lorenz, K. 1974b. Triticale, a promising new cereal grain for the baking industry. Baker's Dig. 48:24–60.
Lorenz, K. 1982. Triticale processing and utilization: Comparison with other cereals. In: CRC Handbook of Processing and Utilization in Agriculture Plant Products 2:277–327. CRC Press, Boca Raton, FL.
Lukaszewski, A.J. and J.P. Gustafson. 1987. Cytogenetics of Triticale. Plant Breed. Rev. 5:41–93.
MacRitchie, F., D.L. duCros, and C.W. Wrigley. 1990. Flour polypeptide related to wheat quality. Adv. Cereal Sci. Technol. 10:79–146.
Martin, J.H., and C.E. Leighty. 1924. Experiments with emmer, spelt, and einkorn. USDA. Dept. Bul. 1197.
McFadden, E.S. and E.R. Sears. 1946. The origin of *Triticum spelta* and its free-threshing hexaploid relatives. J. Hered. 37:107–116.

Morris, R. and E.R. Sears. 1967. The cytogenetics of wheat and its relatives. In: K.S. Quesenberry and L.P. Reitz (eds.), Wheat and wheat improvement. Agron. Monog. 13. Am. Soc. Agron., Madison, WI.

National Research Council. 1989. Triticale: A promising addition to the worlds cereal grains. Nat. Acad. Press. Washington, DC.

Nevo, E. 1988. Genetic resources of wild emmer wheat revisited: genetic evolution, conservation and utilization. Proc. Int. Wheat Genetics Symp.(7th). Cambridge. p. 121–126.

Oelke, E.A., E.S. Oplinger, and M.A. Brinkman. 1989. Triticale. In: Alternative field crops manual. Univ. Wisconsin, CES, Madison, WI, Univ. Minnesota CES, St. Paul.

Oplinger, E.S., E.A. Oelke, A.R. Kaminski, J.D. Doll, B.R. Durgan, and R.T. Schuler. 1990. Spelt. In: Alternative field crops manual. Univ. Wisconsin, CES, Madison, Univ. of Minnesota CES, St. Paul.

Pena, R.J., and A. Amaya. 1992. Milling and breadmaking properties of wheat-triticale blends. J. Sci. Food Agr. 60:483–487.

Pena, R.J., B. Skovmand, and A. Amaya. 1993. Glutenin (HMW- and LMW) subunit composition, quality characteristics and their relationship in *Triticum dicoccon* populations. In: Z.S. Li and Z.Y. Xin eds., Proc. 8th. Int. Wheat Genetics Symp. p. 781–786, July 20–25, 1993. Bejing China, China Agri. Scientech Press

Perrino, P. and K. Hammer. 1982. *Triticum monococcum* L. and *T. dicoccum* Schubler (Syn of *T. Dicoccon* Schrank) are still cultivated in Italy. Genet. Agr. 36:343–352.

Perrino, P. and K. Hammer. 1984. The Farro: further information on its cultivation in Italy, utilization, and conservation (1). Genet. Agr. 38:303–311.

Pomeranz, Y. 1983. From the land of 200 breads and 1200 small baked goods. Cereal Foods World 28:387–390.

Ranhotra, G.S., J.A. Gelroth, B.K. Glaser, and K.J. Lorenz. 1995. Baking and nutritional qualities of a spelt wheat sample. Lebnsm. Wiss. Technol. 28:118–122.

Ranhotra, G.S., J.A. Gelroth, B.K. Glaser, and K.J. Lorenz. 1996a. Nutrient composition of spelt wheat. J. Food Comp. Anal. 9:81–84.

Ranhotra, G.S., J.A. Gelroth, B.K. Glaser, and G.F. Stallknecht. 1996b. Nutritional profile of three spelt wheat cultivars grown at five different locations. Cereal Chem. (submitted).

Riesen, Th., H. Winzeler, A. Ruegger, and P.M. Fried. 1986. The effect of glumes on fungal infection of germinating seed of spelt (*Triticum spelta* L.) in comparison to wheat (*Triticum aestivum* L.) J. Phytopath. 115:318–324.

Rodriquez-Quijano, M., J.F. Vazquez, and J.M. Carillo. 1990. Variation of high molecular weight glutenin subunits in Spanish landraces or *Triticum aestivum* ssp. *vulgare* and ssp. *spelta*. J. Genet. Breed. 44:121–126.

Ruegger, A., H. Winzeler, and J. Nosberger. 1990a. Die ertagsbildung von dinkel (*Triticum spelta* L.) und weizen (*Triticum aestivum* L.) unter verschiedenen umweltbedingungen im Freiland. J. Agron. Crop Sci.164:145–152.

Ruegger, A., H. Winzeler, and J. Nosberger. 1990b. Dry matter production of ^{14}C-assimilates of spelt (*Triticum spelta* L.) and wheat (*Triticum aestivum* L.) as influenced by different temperatures before and during grain filling. J. Agron. Crop Sci. 165:110–120.

Salmon, D.F., and P.E. Jedel. 1993. Triticale, Agri-Fax, Alberta Agr. Food & Rural Devel. Edmonton, Alberta, Canada.

Schulz-Schaeffer, G.F. Stallknecht, D.M. Wesenberg, and K.M. Gilbertson. 1995. Karyological characterization of a high yielding emmer wheat (*Triticum dicoccum* L.) line with two reciprocal homozygous T4BS-5AL and T5AS-4BL translocation chromosome pairs. Biol. Zent. 114:266–272.

Skovmand, B., P.N. Fox, and R.L. Villareal. 1984. Triticale in commercial agriculture: progress and promise. Adv. Agron. 37:1–45.

Smith, R.L., M.E. Schweder, and R.D. Barnett. 1994. Identification of glutenin alleles in wheat and triticale using PCR generated DNA markers. Crop Sci. 34:1373–1378.

Stallknecht, G.F. and K.M. Gilbertson. 1995. Spelt is alternate cereal food crop. Montana Farmer 83(2):12–13.
Stoskopf, N.C. 19885. Chapter 20: In: Cereal grain crops. Reston, Reston, VA.
Vallega, V. 1979. Field performance of varieties of *Triticum monococcum*, *T. Durum*, and *Hordeum vulgare* grown at two locations. Genet. Agr. 33:363–370.
Vallega, V. 1992. Agronomic performance and breeding value of selected strains of diploid wheat, *Triticum monococcum*. Euphytica 61:13–23.
Villareal, R.L., G. Varughese, and O.S. Abdalla. 1990. Advances in spring triticale breeding. Plant Breed. Rev. 8:43–90.
Zohary, D. and M. Hopf. 1993. Domestication of plants in the Old World, the origin and spread of cultivated plants in West Asia, Europe, and the Nile valley. 2nd ed. Oxford Univ. Press, New York.

Advances in Grain Pearl Millet: Utilization and Production Research*

David J. Andrews, Wayne W. Hanna, John F. Rajewski, and Victoria P. Collins

Pearl millet [*Pennisetum glaucum* (L) R. Br.] ranks as the world's fourth most important tropical food cereal with 26 million ha (64 million acres) being grown mostly in semi-arid west Africa and India (Andrews and Bramel-Cox 1994). While early maturing and well adapted to droughty, sandy acid soils of low fertility, pearl millet is highly responsive to fertilizer and moisture on well drained soils.

Until recently, pearl millet has only been grown in the United States as a hybrid forage crop, on about 0.6 million ha (1.5 million acres). Grain breeding programs have been established at Kansas State University (Hays), USDA/ARS (Tifton, Georgia) and the University of Nebraska–Lincoln as previously described by Andrews et al. (1993). Grain hybrid HGM 100™, an exclusive release from Tifton, Georgia, has been marketed in the Southeast by Agra Tech Ltd., of Atlanta since 1991. The grain has been purchased at the current maize price and is used in poultry and animal rations.

UTILIZATION RESEARCH

From a nutritional standpoint, pearl millet is an attractive feed grain. Several studies indicate that metabolizable energy of pearl millet for non-ruminant animals is approximately equal to that of maize (Abate and Gomez 1984; Fancher et al. 1987; Amato and Forrester 1995). When compared to maize on a weight basis, pearl millet is 8%–60% higher in crude protein, 40% richer in lysine and methionine, and 30% richer in threonine (Burton et al. 1972). Therefore, substitution of pearl millet for maize reduces the need for high protein feed ingredients and supplemental amino acids.

Pearl millet has fewer anti-nutritional factors than most grain crops. In contrast to rye and sorghum, pearl millet grain is low in tannins, which limit palatability and inhibit protein digestion. There appears to be no need for heat treatment of pearl millet to destroy protease inhibitors or other harmful factors. However, pearl millet can contain saponin anti-metabolites at levels up to 200 ppm (Sodipo and Arinze 1985). Saponins are known to damage membranes in the digestive tract, so Burtle and Newton (1995) have suggested caution when feeding high levels of pearl millet to fish species, which may be especially sensitive to saponin toxicity.

In the United States, mycotoxins are not likely to cause major problems in pearl millet. Wilson and coworkers (1993) found that this crop appears to be resistant to *Aspergillus flavus* infestation. Pearl millet is susceptible to infection by *Fusarium* fungi. When harvest was delayed 50 days post anthesis, Wilson et al. (1995) isolated *Fusarium* species in 71.5% of the grain samples. However, levels of *Fusarium* toxins remained very low.

*Pearl millet research at the University of Nebraska is funded in part by USAID Grant No. DAN1254-G-00-0021-00 through INTSORMIL, the International Sorghum and Pearl Millet Collaborative Research Support Project.

Poultry are likely to be the major consumers of pearl millet in the United States and a Georgia poultry cooperative has already provided a market for this grain. In 1993, 6300 ha (15,500 acres) of the first commercially available strain of pearl millet bred for grain, HGM 100™, were grown in Georgia and Florida for broiler feed. In 1994, 4000 ha (9,900 acres) were planted (Grabow and Wilson 1995). Several studies have compared isocaloric, isonitrogeneous broiler diets based on maize and pearl millet (Fancher et al. 1987; Smith et al. 1989; Amato and Forrester 1995). All have shown that when pearl millet diets are formulated with correct estimates of protein and energy, broiler performance is equivalent to that obtained with maize (Table1).

Supplementation of pearl millet–soy diets with lysine or sulfur amino acids appears to be unnecessary (Amato and Forrester 1995). Because the need for protein supplementation is reduced, the feed cost of per unit gain is about 3% lower for millet than maize (Bramel-Cox et al. 1995).

Pearl millet is also a satisfactory feed for laying hens. Kumar et al. (1991) found increased egg size and better feed conversion when pearl millet was substituted for maize at 60% by weight. The higher methionine and energy content of pearl millet might explain these results. Using more recent estimates of nutrient content, Collins et al. (1995) found millet and maize gave equivalent egg production and feed efficiency. However, the fatty acid profile of the eggs produced by millet-fed hens differed significantly from those from maize-fed hens. Feeding millet produced eggs which were higher in mono-unsaturated and omega-3 polyunsaturated fatty acids (PUFA), and lower in omega-6 fatty acids than when feeding any other common cereal (Table 2). Since mono-unsaturated and especially omega-3 PUFA may offer health benefits to consumers, pearl millet may offer one route to a more healthy "designer" egg.

Swine may represent a second major market for pearl millet, although few studies have been reported. Calder (1955) found pigs fed 50% or 75% pearl millet reached slaughter weight 10 days earlier than maize-fed pigs. Murray and Lewis (1995) found "optimal growth performance" in pearl millet-fed pigs. Lawrence and co-workers (1995) and Adeola and Orbain (1995) found that replacement of maize with pearl millet on a weight basis gave equivalent gain and feed efficiency, but the pearl millet diets used in these studies were higher in protein than the maize diets. In contrast, Dove and Myer (1995) found that replacement of more than 67% of maize with pearl millet resulted in an increased feed/gain ratio. They suggested that methionine may be less available in pearl millet than in other grains, but this is not supported by the above studies with laying hens, which are very responsive to methionine levels. Undigested whole millet seeds may have been responsible for the poorer performance in this trial.

Ducks have even faster growth rates and higher protein requirements than broiler chickens, so this species is very sensitive to nutrition. Adeola et al. (1994) estimated conservatively that pearl millet is equivalent to maize in feeding value for ducks. They calculated that pearl millet is 6% higher than maize in metabolizable energy.

Wild birds can cause significant crop losses in pearl millet, and this suggests its use in wildlife plantings for game birds. In these situations, a tendency to lodge might even be useful because it gives the birds easy

Table 1. Final body weights at 41 days and feed conversion data for broilers where pearl millet was substituted for maize and soybean in isonitrogenous diets (Amato and Forrester 1995).

Treatment (% millet)	Final body wt. (kg)	Feed conversion
0	1.91	1.68
15	1.92	1.66
30	1.96	1.65
45	1.98	1.66

Table 2. Effect of substituting pearl millet for maize on egg fatty acids (Collins et al. 1995).

Substitution of maize with pearl millet (%)	Content (% of total fatty acids)			
	Mono-unsaturated	Poly-unsaturated	DHA (C22:6n3)	n-6/n-3 ratio
0	44.9a[z]	20.3a	0.52a	13.1a
50	47.6b	17.9b	0.64b	10.1b
100	48.7c	15.7c	0.78c	8.6c

[z]Mean separation in columns by Duncans multiple range test, 1% level.

access to feed. Iler and Hanna (1995) found pearl millet gave good cover to birds in the summer and easy access to hunters in the fall. Savage (1995) fed pearl millet to newly hatched bobwhite quail, and found pearl millet fed chicks consumed more and wasted far less feed than chicks fed maize. Fourteen-day weight gains were improved 8%–22% and mortality was halved when chicks were fed a pearl millet based diet instead of maize (Table 3). Pearl millet may thus be ideal for the specialty game bird market.

At University of Georgia, channel catfish showed equivalent weight gain and feed efficiency when either maize or pearl millet was fed at 30% of total diet. However, diets containing pearl millet: maize ratios of 1:2 or 2:3 gave significantly better gain and efficiency than either grain alone (Table 4), (Burtle and Newton 1995).

For beef cattle, pearl millet grain appears to be superior to sorghum but inferior to maize in feeding value. Christiansen and co-workers (1984) found pearl millet was 4% higher in net energy than sorghum and pearl millet gave better calf performance. Hill et al. (1990) evaluated pearl millet grain for finishing beef cattle in both metabolism and feedlot trials. Both experiments showed that the usable energy value of pearl millet is 85%–90% of that of maize, when equivalent protein/energy ratios are fed. These authors also reported that careful grinding of pearl millet seems to improve nutrient utilization.

Processing and handling pearl millet feeds requires some practice, but no special equipment. Harvesting equipment should be carefully adjusted to thresh the small seeds completely, because poorly threshed millet has lower feeding value (Collins et al. 1995). Grinding or rolling is probably necessary, because experiments with cattle (Hill and Hanna 1990) and swine (Dove and Myer 1995) indicate that small unground seeds may not be fully digested. Amato and Forrester (1995) found that pearl millet shows satisfactory behavior in pelleting processes. However, ground feeds made with pearl millet tend to flow much more easily than traditional maize-based feeds. This trait may be an advantage with some feeder designs, but it could cause wastage with other feeding systems.

In summary, pearl millet appears to have significant potential as a feed grain, especially for non-ruminant species. Best results will be obtained with careful attention to harvesting and processing the grain and to correct formulation of feeds.

PRODUCTION RESEARCH

Research has been conducted on planting date, row spacing, weed control and fertilizer response. Experiments in Florida (Woodruff 1995) with HGM 100™ showed the highest yields, 5430 kg/ha (6080 lb/acre), were obtained with the closest row spacing, 13 cm (5 inch) compared to 4380 kg/ha (4900 lb/acre) at 38 cm (15 inch) and 3730 kg/ha (4180 lb/acre) from 76 cm (30 inch) rows. Wiatrak et al. (1995) also obtained the highest yields at close row spacings, recording grain yields of 2840, 1760, and 1870 kg/ha (3180, 1970 and 2090 lb/acre) from 13, 38, and 76 cm rows, respectively. A contributory factor may be better weed competition from the narrow rows. Experiments in Nebraska (Boggs et al. 1992) on planting date × row spacing × Atrazine applications indicated that early planting (before June 7), 61 cm (24 inch) rather than 91 cm (36 inch) rows, and Atrazine 1.12 kg/ha (1.0 lb/acre) gave the highest yields with early hybrid 79-2068A x 57028R (Table 5). Yields from Atrazine treated plots were as good as those from hand cultivated, and both were significantly better than no weed control. There was no advantage of 30 cm over 61 cm rows.

Table 3. Body weight and mortality of bob-white Quail chicks fed pearl millet and maize-based diets (Savage 1995).

	Weight (g)			Mortality (%)		
Crop	16 days	34 days	48 days	16 days	34 days	48 days
Pearl millet	28.7	88.7	117.5	4.6	5.7	14.3
Maize	23.2	80.1	110.6	22.3	27.7	31.1

The lack of herbicide registrations[z] and the physiological similarities between pearl millet and annual grass weeds need to be considered for chemical weed control in pearl millet production. Although Banvel and 2,4-D are the only herbicides registered for use on millet, there is no federally approved registration of these two herbicides or any other herbicide on pearl millet as a grain crop. Most grass herbicides damage pearl millet but some cause less injury at reduced application rates from which pearl millet recovers depending on the soil and field conditions. Tests conducted by Dowler and Wright (1995) in Georgia and Florida on sandy loam soils using hybrid HGM 100™ gave variable results but some general trends appeared (Table 6). Atrazine and Prowl caused less damage than Dual. Ramrod caused moderate damage. Postemergence Atrazine at 2.24 kg/ha (2 lb/acre) gave moderate grass control and good broadleaf control with light damage to pearl millet and no significant yield reduction. Preemergence 0.56 or 0.84 kg/ha Prowl (0.5 or 0.75 lb/acre) + 0.56 kg/ha 2,4-D (0.5 lb/acre) gave good grass and broadleaf control with little crop damage and no yield reduc-

Table 4. Performance and survival of channel catfish fed diets where pearl millet was substituted for maize (Burtle and Newton 1995).

Diet: Millet (%)	Diet: Maize (%)	Gain/fish (g/day)	Feed/ gain	Survival (%)
0	30	2.27a[z]	1.28a	91.5
10	20	2.44b	1.21b	91.0
20	10	2.47b	1.18b	93.4
30	0	2.15a	1.32a	89.2

[z]Mean separation in columns by Duncan's multiple range test, 5% level.

Table 5. Row spacing and planting date effects on grain pearl millet yields (kg/ha) in Nebraska, 1991 (with Atrazine 1.12 kg/ha) (Boggs et al. 1992).

	Row Spacing		
Planting time	30 cm	61 cm	91 cm
Early (May 20)	4310	4380	3880
Mid (June 7)	3530	3430	3040
Late (June 24)	2140	2510	2480

Planting date LSD, (5%) = 890
Row spacing LSD, (5%) = 220

Table 6. Herbicide test in Quincy, Florida, 1993, on grain pearl millet (Dowler and Wright 1995).

Treatment	Rate (kg/ha)	Control: Grass	Control: Broadleaf	Injury (%)	Grain yield (kg/ha)
Atrazine[z]	2.2	90	100	11	3160
Dual + 2, 4-D	1.7 + 0.6	90	96	96	1340
Ramrod + 2, 4-D	3.4 + 0.6	98	85	45	2430
Ramrod + Atrazine	3.4 + 1.1	98	97	31	2360
Prowl + 2, 4-D	0.6 + 0.6	95	97	16	3430
Prowl + Atrazine	0.6 + 1.1	99	97	10	2630
Hand Weeded	--	100	100	0	3140
Non-Treated	--	0	0	0	3120
LSD 5%		7	5	7	

[z]Postemergence

[z]Mention of herbicides and other products is for identification only and does not imply endorsement by USDA or Universities.

tion. Herbicide rates used would depend on soil type. Herbicide studies on silt clay loam at Nebraska (Rajewski et al. 1987; Rajewski et al. 1995) indicate that surface and incorporated preemergence applications of Ramrod, Lasso, Lorox, Mowdown, Eradicane-X, Prowl, Treflan, or Sonalan can reduce pearl millet plant populations by 40% or more. Incorporated applications of Evik, Alanap, Atrazine, and Propazine did not reduce pearl millet emergence. Postemergence applications of Ramrod, Propazine, Atrazine, Basagran, Buctril, or Banvel at the 4-leaf stage of pearl millet development did not reduce plant numbers or cause significant leaf injury.

Research programs at Kansas and Nebraska have routinely used 1.12 kg/ha Atrazine (1.0 lb/acre) or 2.8 kg/ha Propazine (2.5 lb/acre) as preemergence grass and broadleaf weed control in breeding nurseries and hybrid yield tests on silt clay loam soils. After pearl millet emerges and reaches 2–3 leaf stage, the Nebraska program surface applies 3.36 kg/ha (3 lb/acre) Ramrod to prevent further emergence of grassy weeds. Basagran or Buctril, at normal recommended rates, is used as needed to provide postemergence control of broadleaf weeds.

Early planting (May, June in the Southeast; early June in the Midwest) are advantageous for growing pearl millet. Until new rust resistant cultivars are available in the Southeast, early planting is necessary to avoid losses to rust. However, warm soil (18° to 24°C, 65° to 75°F) is needed for good emergence which determines the earliest planting date in the Midwest. The need for warmer temperatures also allows the early germinated annual grass weeds to be removed prior to planting. Lower temperatures in conjunction with heavy rains risk soil crusting and fungal damping off which severely affects emergence. Experimental plantings made in early July in Indiana have shown that pearl millet has potential as an early maturing double crop after wheat harvest (J.D. Axtell 1994, pers. commun.). Optimum plant populations are about 150,000 plants/ha (60,000 plants/acre) but pearl millet can compensate well for much lower or irregular populations because of high tillering capacity. Among insects, chinch bug and European corn borer can cause damage, and varietal differences in susceptibility have been noted. In some locations, birds can be a severe problem on small isolated fields requiring timely harvest at grain maturity.

Pearl millet's feed advantages partly depend on consistently high grain protein levels. On soils of low inherent fertility, as in the Southeast, depending on soil tests, 90–134 kg/ha (80–120 lb N/acre) are needed (Woodruff 1995). In the Midwest, a side dressing of 45–90 kg/ha (40–80 lb N/acre) is worthwhile on the heavier fertile soils, depending on rotation with soybean and soil moisture availability. Production of grain with consistent protein levels is needed by the feed industry (especially for poultry).

BREEDING RESEARCH

Inbred parents for producing new hybrids are being developed by all three programs. Releases from the Kansas and Nebraska programs due this year are likely to be unrestricted public releases. The next releases from Tifton will carry rust and blight resistance. A grain pearl millet regional test with about 20 entries including sorghum checks has been conducted at 5 to 9 locations each year since 1988, (Rajewski and Andrews 1995a). This has allowed breeders to determine regional adaptabilities and cooperators to assess hybrids from different breeding programs. Pearl millet yields have exceeded sorghum in some locations, principally where the season is short, rainfall low, or soils sandy. The location average of all millet entries in 1994 trials ranged from 620 to 4370 kg/ha (550 to 3900 lb/acre). The best individual pearl millet entry yields were 5340, 5590, and 5430 kg/ha (4770, 4990, and 4850 lb/acre) at Tifton, Georgia; Lafayette, Indiana; and Mead, Nebraska, respectively (Rajewski and Andrews 1995b).

Population improvement has continued at Nebraska for herbicide tolerance to a higher preemergence application rate of Ramrod (5.6 kg/ha; 5 lb/acre) with Atrazine (1.12 kg/ha; 1 lb/acre). Population releases NPM-1, NPM-2, and NPM-3 all show potential for tolerance selection to the herbicides at rates recommended for grain sorghum. Four additional populations of different genetic backgrounds are being developed for future release with Ramrod tolerance.

Until the recent discovery of useful male fertility restorers for the A_4 system of cytoplasmic male sterility (CMS) found by Hanna (1989), most hybrid grain breeding research had been based on developing maintainers (for seed parents) and restorers (pollen parents) for the A_1 CMS system. Although the A_1 system has been widely used in India, and is currently used in the U.S. to produce forage millet hybrids and HGM 100™, it does produce some pollen shedding revertants and causes problems in breeding because of the occurrence of high levels of partial sterility (Andrews and Rajewski 1994). The expression of male sterility and fertility in the A_4 system is complete and no instances have yet been noted where minor genes or environment have modified this expression. Because almost all existing germplasm lines can, with a high probability of success, be converted to A_4 seed parents, the A_4 CMS system greatly increases the opportunities for breeding superior hybrids. Restorer (R_4) genes, though generally infrequent, are readily identified and much more widely distributed than initially thought. The first grain type source population for A_4 cytoplasm restorers, NPM-3, was released in 1994, (Andrews and Rajewski 1995). Rapid progress has been made in the University of Nebraska program, both in converting good A_1 seed parents to A_4 seed parents and in developing completely new A_4 hybrid parents.

FUTURE NEEDS

Apart from higher yield potential, which is always needed, several obvious research and development needs still exist. These include: (1) better resistance to rust and blight for the South; (2) better germination, cool tolerance and seedling vigor for the Midwest/High Plains to permit earlier planting; (3) better herbicide tolerance, new herbicide evaluation, registration and label approval of available herbicides for grain production; (4) better resistance to lodging due to shallow rooting, weak stalk, or corn borer susceptibility; (5) good threshability; (6) release of new hybrid parents; and (7) continued end-use research and market development.

CONCLUSION

In the last four years interest in pearl millet has grown and substantially more is now known about how to produce pearl millet and its comparative feed values for a number of end-uses. In 1993, following commercial tests that showed pearl millet was equivalent to maize in feed value for broilers, pearl millet grain production commenced in Georgia and Florida. Utilization research has shown that pearl millet is better than maize in broiler and pig rations and equivalent to maize for ducks. While pearl millet is at least as good as maize for egg production, it produces a decrease in the egg n6:n3 fatty acid ratio. In human diets, lower n6:n3 ratios are associated with reduced incidence of cardiovascular disease. Pearl millet was found to be superior to maize in bob white chick and channel catfish diets.

Production research has determined that higher grain yields can be obtained from 61 cm (24 inch) or closer rows. Pearl millet is generally more sensitive to grass herbicides than soybeans, but adequate weed control has been obtained with some herbicides including Atrazine, Propazine, and Prowl. Except for double cropping situations, planting should be as early as consistent with soil temperature reaching 18° to 24°C (65° to 75°F).

Parental lines for early maturing hybrids for the midwest are due for release from the Kansas and Nebraska breeding programs early in 1996. Breeding research is well advanced to producing hybrids in the A_4 cytoplasmic male sterile system, which offers advantages in both breeding efficiency and in the commercial production of hybrid seed.

In summary, recent research work has shown that pearl millet is a high quality feed grain, especially for poultry, and is well adapted to situations that may involve sandy soil, short season, heat, and low moisture. Markets need to be developed and more hybrid options made available.

REFERENCES

Abate, A.N. and M. Gomez. 1984. Substitution of finger millet (*Eleusine coracana*) and bulrush millet (*Pennisetum typhoides*) for maize in broiler feeds. Anim. Feed Sci. Tech. 10:291–299.

Adeola, O. and J.I. Orban. 1995. Chemical composition and nutrient digestibility of pearl millet (*Pennisetum glaucum*) fed to growing pigs. J. Cereal Sci. 22:177–184.

Adeola, O., J.C. Rogler, and T.W. Sullivan. 1994. Metabolism and nutrition: pearl millet in diets of White Pekin ducks. Poultry Sci. 73:425–435.

Amato, S.V. and R.R. Forrester. 1995. Evaluation of pearl millet as a feed ingredient for broiler rations. p. 125–128. In: I.D. Teare (ed.), Proc. 1st Nat. Grain Pearl Millet Symp., Univ. Georgia, Tifton.

Andrews, D.J. and P. Bramel-Cox. 1994. Breeding cultivars for sustainable crop production in low-input dryland agriculture in the tropics. Chapter 28. p. 211–222. In: D.A. Buxton (ed.), Int. Crop Sci. I. CSSA, Madison, WI.

Andrews, D.J. and J.F. Rajewski. 1994. Male fertility restoration and attributes of the A_4 cytoplasmic-nuclear male sterile system for grain production in pearl millet. Int. Sorghum Millets Newsl. 35:64.

Andrews, D.J. and J.F. Rajewski. 1995. Registration of NPM-3 grain pearl millet. Crop Sci. 35:1229.

Andrews, D.J., J.F. Rajewski, and K.A. Kumar. 1993. Pearl millet: A new feed grain crop. p. 198–208. In: J. Janick and J. Simon (eds.), New crops. Wiley, New York.

Bramel-Cox, P.J., K. Anand Kumar, J.H. Hancock, and D.J. Andrews. 1995. Sorghum and millets for forage and feed. Chapter 11. p. 325–364. In: D.A.V. Dendy (ed.), Sorghum and the millets: Chemistry and technology. Am. Assoc. Cereal Chem. St Paul, MN.

Boggs, L.L., J.F. Rajewski, and D.J. Andrews. 1992. Planting date, row spacing, and weed control effects on grain yield of a dwarf pearl millet hybrid. Agronomy. Abstr. 137–138. (Abstr.)

Burtle, G.J. and G.L. Newton. 1995. Catfish performance on pearl millet grain. p. 116–118. In: I.D. Teare (ed.), Proc. 1st Nat. Grain Pearl Millet Symp., Univ. Georgia, Tifton.

Burton, G.W., A.T. Wallace, and K.O. Rachie. 1972. Chemical composition and nutritive value of pearl millet. Crop Sci. 12:187–188.

Calder, A. 1955. Value of munga (pearl millet) for pig feeding. Rhod. Agr. J. 52:161–170.

Christensen, N.B., J.C. Palmer, H.A. Praeger, W.D. Stegmeier, and R.L. Vanderlip. 1984. Pearl millet: A potential crop for Kansas. Keeping up with Research #77. Kansas Agr. Expt. Sta., Manhattan.

Collins, V.P., A.H. Cantor, A.J. Pescatore, M.I. Straw, and M.J. Ford. 1995. Effect of pearl millet in laying hen diets on egg fatty acids. Poultry Sci. 74(supp. 1):4.

Dove, C.R. and R.O. Myer. 1995. Swine performance on HGM™ 100 pearl millet grain. p. 110–113. In: I.D. Teare (ed.), Proc. 1st Nat. Grain Pearl Millet Symp., Univ. Georgia, Tifton.

Dowler, C.C. and D.L. Wright. 1995. Weed management systems for pearl millet in the Southeastern United States. p. 64–71. In: I.D. Teare (ed.), Proc. 1st Nat. Grain Pearl Millet Symp., Univ. of Georgia, Tifton.

Fancher, B.I., L.S. Jensen, R.L. Smith, and W.W. Hanna. 1987. Metabolizable energy content of pearl millet. Poultry Sci. 66:1693–1696.

Grabow, D. and L. Wilson. 1995. Commercial perspective on marketing pearl millet for grain. p. 149–151. In: I.D. Teare (ed.), Proc. 1st Nat. Grain Pearl Millet Symp., Univ. Georgia, Tifton.

Hanna, W.W. 1989. Characteristics and stability of a new cytoplasmic-nuclear male sterile source in pearl millet. Crop Sci. 29:1457–1459.

Hill, G.M. and W.W. Hanna. 1990. Nutritive characteristics of pearl millet grain in beef cattle diets. J. Anim. Sci. 68:2061–2066.

Iler, A. and W.W. Hanna. 1995. Pearl millet in wildlife plantings. p. 124. In: I.D. Teare (ed.), Proc. 1st Nat. Grain Pearl Millet Symp., Univ. Georgia, Tifton.

Kumar, A.M., V.R. Reddy, P.V. Reddy, and R.S. Reddy. 1991. Utilization of pearl millet (*Pennisetum typhoides*) for egg production. Poultry Sci. 32:463–469.

Lawrence, B.V., O. Adeola, and J.C. Rogler. 1995. Nutrient digestibility and growth performance of pigs fed pearl millet as a replacement corn. J. Anim. Sci. 73:2026–2032.

Murry, A.C. and R.D. Lewis. 1995. Growth performance: Mineral balance, serum mineral concentration and bone status of growing pigs fed microbial phytase in a pearl millet-soybean meal diet. J. Anim. Sci. 73:(supp. 1):173.

Rajewski, J.F. and D.J. Andrews. 1995a. Grain pearl millet performance and adaptability in six years of regional tests. p. 14–17. In: I.D. Teare (ed.), Proc. 1st Nat. Grain Pearl Millet Symp., Univ. Georgia, Tifton.

Rajewski, J.F. and D.J. Andrews. 1995b. Pearl millet regional grain yield trials. Dept. Agronomy, Univ. Nebraska, Manhattan. mimeo.
Rajewski, J.F., A.R. Martin, and D.J. Andrews. 1987. Effects of grass herbicides on grain sorghum and pearl millet. Sorghum Newsl. 30:52.
Rajewski, J.F., A.R. Martin, and D.J. Andrews. 1995. Effects of pre-emergence and post-emergence herbicides on pearl millet. p. 74. In: I.D. Teare (ed.), Proc. 1st Nat. Grain Pearl Millet Symp., Univ. Georgia, Tifton.
Savage, S. 1995. Quail performance on pearl millet grain. p. 121–123. In: I.D. Teare (ed.), Proc. 1st Nat. Grain Pearl Millet Symp., Univ. Georgia, Tifton.
Smith, R.L., L.S. Jensen, C.S. Hoveland, and W.W. Hanna. 1989. Use of pearl millet, sorghum and triticale grain in broiler diets. J. Prod. Agr. 2:78–82.
Sodipo, A.O. and H.U. Arinze. 1985. Saponin content of some Nigerian foods. J. Sci. Food Agr. 36:407–408.
Wiatrak, P.J., D.L. Wright, W.W. Hanna, J.A. Pudelko, J. Spitalniak, and I.D. Teare. 1995. Plant populations and seeding rates. p. 32–37. In: I.D. Teare (ed.), Proc. 1st Nat. Grain Pearl Millet Symp., Univ. Georgia, Tifton.
Wilson, J.P., W.W. Hanna, D.M. Wilson, R.W. Beaver and H.H. Casper. 1993. Fungal and mycotoxin contamination of pearl millet grain in response to environmental conditions in Georgia. Plant Dis. 77:121–124.
Wilson, J.P., H.H. Casper and D.M. Wilson. 1995. Effects of delayed harvest on contamination of pearl millet grain with mycotoxin-producing fungi and mycotoxins. p. 61–63. In: I.D. Teare (ed.), Proc. 1st Nat. Grain Pearl Millet Symp., Univ. Georgia, Tifton.
Woodruff, J.M. 1995. Pearl millet for grain production guidelines. Coop. Ext. Serv. Univ. Georgia, Tifton.

Evaluation of Pearl Millet for Swine and Ducks*

Olayiwola Adeola, Dale King, and Bradley V. Lawrence

Pearl millet (*Pennisetum glaucum*) is a cereal grain with good drought tolerance and hardiness commonly grown in the semiarid regions of Africa and Asia on an estimated 26 million ha primarily for human consumption (Andrews and Kumar 1992). In addition to drought tolerance and hardiness, pearl millet has a relatively short growing season, which makes it a potential crop that can be double-cropped after wheat has been harvested so that valuable soil is not allowed to lay idle prior to the following cropping season. Pearl millet could fill the niche during periods when maize production is affected by unseasonable weather. The energy density of pearl millet is relatively high, arising from the higher oil content of this grain relative to maize, wheat, or sorghum (Hill and Hanna 1990; Sullivan 1990; Haydon and Hobbs 1991). Sullivan (1990) also reported that pearl millet usually has a superior amino acid profile and a higher protein content than maize. Research investigating the utility of pearl millet in livestock diets is limited. In broiler diets, pearl millet could replace maize without adversely affecting weight gains or feed efficiency (Singh and Barsaul 1976; Sharma 1979; Smith 1989; Sullivan 1990). Pearl millet also has the potential for use in diets of swine (Haydon and Hobbs 1991) and beef cattle (Hill and Hanna 1990). This communication summarizes the results of a series of studies conducted to evaluate the nutritional value of pearl millet in diets of pigs (Lawrence 1995) and ducks (Adeola 1994).

---

*This research was supported by the Indiana Business Modernization and Techology Corporation.

NUTRIENT PROFILE OF MAIZE VERSUS PEARL MILLET

Nutrient profiles of maize and pearl millet grain used in studies conducted at Purdue is presented in Table 1. Pearl millet contains 27% to 32% more protein than maize, higher concentrations of essential amino acids, two times as much ether extract, and a higher gross energy value than maize. Pearl millet has been shown to have a higher protein content and higher essential amino acid, and ether extract compared with maize (Burton 1972; Ejeta 1987; Smith 1989) sorghum, wheat, and triticale (Ejeta 1987; Smith 1989; Haydon and Hobbs 1991). The higher oil content would account for the higher gross energy value of pearl millet (Table 1).

SWINE

Digestibility of Nutrients in Pearl Millet for Pigs

In nutrient digestibility studies with 20-kg pigs, dry matter digestibility was higher for maize (87%) compared with pearl millet (77%). Protein digestibility (70%) and retention (57%) were similar between the maize and pearl millet. The ileal digestibility of most amino acids from maize seems to be similar to the values obtained for pearl millet (Haydon and Hobbs 1991), thus supporting the similar protein retention values for maize and pearl millet. Energy digestibility was lower for pearl millet (75%) than for maize (86%); and in conjunction with a lower dry matter digestibility, resulted in lower digestible and metabolizable energy values for pearl millet (2.92 and 2.84 kcal/g, respectively) than for maize (3.17 and 3.08 kcal/g, respectively). Haydon and Hobbs (1991) reported that the energy value of pearl millet was lower than that of soft red wheat, while Lin (1987) reported that the energy digestibility of hard red wheat was lower than maize, suggesting, that when fed to pigs, maize has the highest digestible and metabolizable energy values, with wheat having a lower value than maize, but a higher value than pearl millet. With increasing pearl millet inclusion, the gross energy

Table 1. Nutrient composition of maize and two samples of pearl millet on a dry matter basis

		Pearl millet	
Variable	Maize	Nebraska[z]	Indiana[y]
Dry matter (%)	88.9	89.2	89.1
Ether extract (%)	3.0	6.9	6.2
Neutral detergent fiber (%)	14.5	18.2	15.3
Acid detergent fiber (%)	2.9	4.1	4.1
Ash (%)	5.7	6.7	6.1
Gross energy (kcal/g)	4.33	4.53	4.52
Crude protein (%)	8.4	11.5	12.5
Essential amino acid (%)			
Histidine	.21	.27	.29
Isoleucine	.39	.59	.68
Leucine	.95	1.03	1.16
Lysine	.23	.35	.36
Methionine	.17	.23	.23
Phenylalanine	.40	.54	.59
Threonine	.27	.41	.44
Tryptophan	.07	.19	.21
Valine	.39	.58	.64

[z]Pearl millet grown at the University of Nebraska, Lincoln, Nebraska.
[y]Pearl millet grown at Purdue University, West Lafayette, Indiana.

value of the diet would increase while the supply of digestible energy and metabolizable energy would decline. The higher protein/amino acid content of pearl millet would indicate that equal-weight substitutions of pearl millet for maize could result in an increase in the supply of total amino acids for growth.

Growth Performance Response of Nursery and Growing Pigs

Growth performance responses were investigated for 28 days in nursery (10-kg) and growing (20-kg) pigs fed diets in which pearl millet was substituted for maize on a weight-for-weight basis. Growth rate of during the first 14 days and over the entire 28-day period was unaffected by dietary weight-for-weight replacement of maize with pearl millet in nursery pigs. From day 14 to 28, however, growth rate and feed intake increased quadratically to increases in substitution of pearl millet for maize with the maximum rate of weight gain achieved when 25% of the maize was replaced with pearl millet. The substitution of pearl millet for maize on a weight-for-weight basis in growing pigs did not affect growth rate, feed intake, or feed efficiency. Body weights were similar across dietary treatments at the end of the 28-day experiment.

The value of a higher amino acid composition of pearl millet relative to maize would be better tested when pearl millet- and maize-based diets are compared on an equal digestible amino acid basis. Because amino acids and energy are closely related, lysine, the first-limiting amino acid in cereal-based diets for pigs is more appropriately expressed in terms of lysine:digestible energy ratio. Lawrence (1994) determined the optimum ratio to be 3 g of lysine/Mcal of digestible energy. Based on 80% ileal lysine digestibility, this translates to 2.4 g of digestible lysine/Mcal of digestible energy. For optimum pig performance, dietary essential amino acids should be at the proper ratio to lysine. Based on this premise, another study was conducted to evaluate the performance of growing pigs fed two different ratios of threonine (65% or 70% of lysine) and total sulfur amino acids (60% or 65% of lysine) to lysine in maize-based compared with pearl millet-based diets.

Four diets were formulated to contain 2.4 g of digestible lysine/Mcal of digestible energy. Diet 1 was maize-based and formulated to contain digestible valine, isoleucine, tryptophan, total sulfur amino acids, and threonine at the optimum ratio (68%, 60%, 18%, 60%, and 65%) to digestible lysine developed by Chung and Baker (1992). Diet 2 was pearl millet-based and formulated to contain digestible valine, isoleucine, tryptophan, total sulfur amino acids, and threonine at the optimum ratio (68%, 60%, 18%, 60%, and 65%) to digestible lysine developed by Chung and Baker (1992). In Diets 1 and 2, the ratios of the digestible levels of other essential amino acids to digestible lysine exceeded the optimum ratio in the pattern for pigs (Chung and Baker 1992). Diets 3 and 4 were maize- and pearl millet-based diets, respectively; and formulated to contain digestible SAA and threonine at ratios of 65% and 70% of lysine, respectively.

The average daily gain, feed intake and feed efficiency are given in Table 2. Increasing the threonine and sulfur amino acid relative to lysine had no effect on the average daily gain, feed intake, feed efficiency, or days on feed. However, average daily gains were higher ($P < 0.05$) in the maize-based diets (1.07 and 1.08 kg/d vs. 1.01 and 1.02 kg/d) than pearl millet-based diets and pigs took between 1.5 and 2 days longer to reach the 50 kg live weight with the pearl millet-based diets. This apparent superiority of maize-based diets over pearl millet-based diets in supporting weight gains may be due to palatability of the feeds as pigs on the maize-based diets consume in general more feed per day. Furthermore, feed efficiency was found to be similar ($P>0.05$) between maize-based and pearl millet based diets.

In this study, regardless of whether diets were maize- or pearl millet-based, pigs within the 20% to 50 kg growth phase performed better on the diets formulated to contain digestible threonine and sulfur amino acids at 65% and 60% of digestible lysine than on the diets contain digestible threonine and sulfur amino acids at 70 and 65% of digestible lysine. In a previous experiment Wang and Fuller (1990) investigated the effect of the plane of nutrition on the optimum dietary amino acid pattern for growing pigs. They used six diets based on maize and soy bean meal. They reported the optimum ratio of four amino acids for 25–50 kg pigs as lysine 100, threonine 64, methionine + cystine 61 and tryptophan 20. The present study is in agreement with observations reported by Wang and Fuller (1990). It is our opinion that the lower feed intake of pearl millet-based diets when compared to maize-based diets accounted for the lower gains since feed efficiency was similar. Higher intakes of pearl millet-based diets will have the added advantage of requiring less amino acid supplementation over maize-based diets.

DUCKS

Utilization of Nutrients in Pearl Millet for Ducks

Nutrient utilization of diets containing pearl millet was investigated with 22-day-old ducks that weighed 1.2 kg. Dry matter in diets containing maize was less well utilized than that in the diets containing pearl millet. The diets containing maize had lower energy retention values than those containing pearl millet. Thus, higher apparent metabolizable energy values were observed for diets containing pearl millet than those containing maize (3.3 vs. 3.1 kcal/g). The higher metabolizable energy value of the diets containing pearl millet is in conformity with its higher ether extract content compared to maize. Feed consumption was similar across diet comparisons, but the caloric value of diets containing pearl millet was higher than those containing maize. This observation is an indication that a compensatory increase in feed intake to overcome a decrease in energy content of the diet does not always occur.

The apparent metabolizable energy content of the diets containing maize determined in the nutrient utilization study are similar to values calculated using published (NRC 1984) values for composition of feedstuffs. Using NRC (1984) values for pearl millet underestimated the apparent metabolizable energy contents of diets containing pearl millet by ca 0.77 kcal/g. In an experiment on feed grain utilization in broiler diets, Sharma (1979) reported that pearl millet had lower metabolizable energy than maize, an observation that is contrary to the results of the present nutrient utilization studies. The fact that broilers were used in their experiments, that the ether extract content of maize (3.7%) and pearl millet (4.3%) were similar, and that the level of protein meal used in their experiments were higher than that used in the present experiments may be responsible for the disparity.

Growth Performance and Carcass Composition of Ducks Fed Diets Containing Pearl Millet

In 3-week growth performance studies with day-old ducks, maize and pearl millet were compared on an equal-protein basis or on an equal-weight basis. At the end of the second week of the study, the maize diets compared on an equal-protein basis were significantly superior to the millet diets in promoting weight gain. When compared on an equal weight, the pearl millet diets were equal to the maize diets in promoting growth

Table 2. Performance of growing pigs fed two different amino acid patterns in maize-based and pearl millet-based diets.

Diet[z]	Initial weight (kg)	Final weight (kg)	Daily gain (kg/d)	Daily feed (kg/d)	Gain/feed (kg/kg)	Days on feed	Number of pigs per diet
1	19.47	52.33	1.07a[y]	2.45ab	.44	30.75bc	12
2	19.55	52.00	1.01b	2.22b	.46	32.25ab	12
3	19.25	51.98	1.08a	2.54a	.46	30.05c	12
4	19.23	52.27	1.02b	2.21b	.46	32.50a	12
SD	.6	1.29	.07	.30	.05	2.04	

[z] Diet 1 was maize-based and formulated to contain digestible valine, isoleucine, tryptophan, total sulfur amino acids, and threonine at 68%, 60%, 18%, 60%, and 65%, respectively of digestible lysine. Diet 2 was pearl millet-based and formulated to contain digestible valine, isoleucine, tryptophan, total sulfur amino acids, and threonine at 68%, 60%, 18%, 60%, and 65%, respectively of digestible lysine. Diets 3 and 4 were maize- and pearl millet-based diets, respectively; and formulated to contain digestible total sulfur amino acids and threonine at ratios of 65 and 70% of lysine, respectively.

[y] Mean separation in column by Neuman-Keuls' test, 5% level.

during the first 2 week of the experiment. Furthermore, there were no significant differences in weight gain over the 3-week period between the maize and the pearl millet diets. Feed intake during the 3-week period of the experiment was higher in ducks that received maize, than those that received pearl millet diets. In a second 3-week growth performance experiment, ducks that received pearl millet were more efficient in converting feed to body weight gains than those that received the maize diets. This is similar to the observations reported by Smith (1989) in broiler experiments where replacement of maize with pearl millet improved the efficiency of feed conversion, an improvement that was attributed to an increase in the metabolizable energy contents of diets that contained pearl millet. Carcass protein of ducks that received the diets containing maize was greater than in ducks that received the diets containing pearl millet. The maize-pearl millet comparison for other carcass components (dry matter, ether extract, and ash) was not statistically significant.

SUMMARY

Grain pearl millet is higher in protein, amino acids, ether extract, and gross energy than maize. However the digestible and metabolizable energy values are lower in pearl millet than maize when fed to pigs. Replacement of maize with pearl millet on an equal-weight basis in the diets of young pigs has no adverse effect on growth performance, other than slight increases in gain and feed intake. The use of pearl millet in diets formulated on an equal digestible amino acid basis results in slower growth rate than maize-based diets. In duck diets, pearl millet appears to have a similar feeding value to maize when substituted for maize on an equal-weight basis. In contrast to pigs, pearl millet diets have higher metabolizable energy than those containing maize. Equal-weight replacement of maize with pearl millet in the diets of ducks has no adverse effects on weight gains.

REFERENCES

Adeola, O., J.C. Rogler, and T.W. Sullivan. 1994. Pearl millet in diets of White Pekin ducks. Poult. Sci. 73:425–435.

Andrews, D.J., and K.A. Kumar. 1992. Pearl millet for food, feed, and forage. Adv. Agron. 48:89–139.

Burton , G.W., A.T. Wallace, and K.O. Rachie. 1972. Chemical composition and nutritive value of pearl millet [*Pennisetum typhoide* (Burm.) Staph and E. C. Hubbard] grain. Crop Sci. 12:187–188.

Chung, T.K. and D.H. Baker. 1992. Ideal amino acid pattern for 10 kilogram pigs. J. Anim. Sci. 70:3102–3111.

Ejeta, G., M.M. Hassen, and E.T. Mertz. 1987. In vitro digestibility and amino acid composition of pearl millet (*Pennisetum typhoides*) and other cereals. Proc. Nat. Acad. Sci. (USA) 84:6016–6019.

Haydon, K.D. and S.E. Hobbs. 1991. Nutrient digestibilities of soft winter wheat, improved triticale cultivars, and pearl millet for finishing pigs. J. Anim. Sci. 69:719–725.

Hill, G.M. and W.W. Hanna, 1990. Nutritive characteristics of pearl millet grain in beef cattle diets. J. Anim. Sci. 68:2061–2066.

Lawrence, B.V., O. Adeola, and T.R. Cline. 1994. Nitrogen utilization and lean growth performance of 20- to 50-kilogram pigs fed diets balanced for lysine:energy ratio. J. Anim. Sci. 72:2887–2895.

Lawrence, B.V., O. Adeola, and J.C. Rogler. 1995. Nutrient digestibility and growth performance of pigs fed pearl millet as a replacement for corn. J. Anim. Sci. 73:2026–2032.

Lin, F.D., D.A. Knabe, and T.D. Tanksley, Jr. 1987. Apparent digestibility of amino acids, gross energy and starch in maize, sorghum, wheat, barley, oat groats and wheat middlings for growing pigs. J. Anim. Sci. 64:1655–1663.

National Research Council, 1984. Nutrients requirements of poultry. 8th ed. National Academy Press, Washington, DC.

Sharma, B.D., V.R. Sadagopan, and V.R. Reddy. 1979. Utilization of different cereals in broiler diets. Brit. Poult. Sci. 20:371–378.

Singh, S.D., and C.S. Barsaul. 1976. Replacement of maize by coarse grains for growth and production in White Leghorn and Rhode Island Red birds. Ind. J. Anim. Sci. 46:96–99.

Smith, R.L., L.S. Jensen, C.S. Hoveland, and W.W. Hanna. 1989. Use of pearl millet, sorghum, and triticale grain in broiler diets. J. Prod. Agr. 2:78–82.

Sullivan, T.W., J.H. Douglas, D.J. Andrews, P.L. Bowland, J.D. Hancock, P.J. Bramel-Cox, W.D. Stegmeier, and J.R. Brethour. 1990. Nutritional value of pearl millet for food and feed. p. 83–94. In: Proc. International Conference on Sorghum Nutritional Quality, Purdue Univ., West Lafayette, IN.

Wang, T.C. and M.F. Fuller. 1990. The effect of the plane of nutrition on the optimum dietary amino acid pattern for growing pigs. Anim. Prod. 50:155–164.

Foxtail and Proso Millet

David D. Baltensperger

Millet is a general term for a wide range of small seeded cereals (Marcon 1994). The millets of the Poaceae (Gramineae) are small seeded annual grasses, including a wide range of cereals, that may be used both as grain and for forage (Schery 1972). Millets can be successfully grown in a wide range of environmental conditions, being better adapted than most crops to hot, dry regions. They are of potential value particularly in semiarid regions because of their short growing seasons. They can either tolerate drought and intense heat or avoid these conditions by growing to maturity very quickly. The millets can play an important role in the economy of many less-developed countries of the Old World because they are consumed directly for human food. However, the millets in the western world are grown primarily for birdseed, livestock feed, hay, or as an emergency catch crop (Casey and Lorenz 1977).

CLASSIFICATION AND DOMESTICATION

Millets are thought to have been among the first cultivated crops, being one of the stable foods in central and eastern Asia (mainly in China, India, and Russia), Europe and some parts of Africa during the very early ages (Wietgrefe 1990). Millets are diverse enough in type that they occur in two tribes of the grass family (Rachie 1975) as shown in Table 1.

Eleusine coracana, known as finger or African millet, is the only species of economic importance in the tribe Chlorideae. Finger millet has been grown predominantly in India and Africa, both for grain and forage. In contrast to most millet, it requires mild weather and a considerable amount of rainfall to mature (Haq 1989). The tribe Paniceae is the largest tribe of the Poaceae, containing 71 genera with approximately 1400 species distributed mainly in tropical and also temperate regions (Roshevits 1980). Economically important culti-

Table 1. Classification of millets of the Poaceae. Adapted from Rachie (1975).

Tribe	Species	Common names
Paniceae	*Setaria italica*	Foxtail millet, Italian millet, German millet, Hay millet
	Panicum miliaceum	Proso millet, Common millet, Hershey millet, Hog millet, White millet, Broomcorn millet
	Brachiaria ramosum	Brown top millet
	Digitaria exilis	Fonio millet
	Echinocloa colonum	Shama millet
	Decompositum	Australian millet
	Frumentaceae	Japanese millet, Barnyard millet
	Paspalum scrobiculatum	Kodo millet
	Pennisetum americanum	Pearl millet
Chlorideae	*Eleusine coracana*	Finger millet

vated millets in the tribe Paniceae are: pearl millet (*Pennisetum americanum* L.) (Andrews et al. 1993) foxtail millet (*Setaria italica* L. Beauv.), proso millet (*Panicum miliaceum* L.), and Japanese millet (*Echinocloa frumentaceae* L.).

The genus *Panicum* is one of the largest genera of the grasses, including more than 400 species distributed primarily in the tropics and subtropics and also warmer parts of temperate zones (Roshevits 1980). Within this large genus, two species are of economic importance; proso millet and to a lesser extent little millet (*P. miliare* L.). The genus *Panicum* is poorly represented in drier tropics, apart from a few more or less xerophytic species (Cobley 1976).

Proso, an ancient Slav name used in Russia and Poland (Candolle 1964), has also been called common millet, hog millet, broom corn, yellow hog, hershey, and white millet. Although the origin of proso millet has not been ascertained, it is one of the first cultivated cereal grasses, most likely prior to wheat. Proso millet has been known for many thousands of years in Eastern Asia including China, India, and Russia. The cultivation of proso is more generally thought to come from central and eastern Asia and spread to India, Russia, the Middle East, and Europe (Roshevits 1980). Proso is still extensively cultivated in India, China, Russia, in the Middle East including Iran, Iraq, Syria, and Turkey, and also in Afghanistan and Romania (Martin et al. 1976). According to Matz (1986), however, proso is probably a native of Egypt and Arabia and has been cultivated in Asia Minor and southern Europe since prehistoric times. Candolle (1964) also states that Egypto-Arabian origin of proso millet is very probable.

The genus *Setaria* is widely distributed in warm and temperate areas. Foxtail millet is the most economically valuable of the genus. Foxtail millet is one of the world's oldest cultivated crops. Its planting has been mentioned in Chinese records as early as 2700 BC (Vinall 1924). *S. italica* was the most important plant food in the neolithic culture in China, and its domestication and cultivation was the earliest identifiable manifestation of this culture, the beginning of which has been estimated at over 4,000 years ago (Chang 1968). In Russia, foxtail millet has been cultivated since ancient times and there is evidence that it was grown as long as 1,500 years ago (Movsisyants, 1950). Malm and Rachi (1968) thoroughly reviewed the domestication of foxtail millets and the taxonomy. They report the use of subspecies by Scholz (1960) based on papillae on the epidermal cells of the glumes (*moharia*) and no papillae (*maximas*). Perhaps of more practical value was the breakout by Coleman (1939) based on head type and seed color.

BIOLOGY AND ADAPTATION

Proso millet is well adapted to many soil and climatic conditions. Being a short-season crop with a low water requirement, it grows further north (up to 54°N latitude) than the other millets and also adapts well to plateau conditions and high elevations (Theisen el al. 1978; Matz 1986). Proso is found high in the mountains; in the former USSR up to 1200 m and in India up to 3500 m (Roshevits 1980).

The proso plant is considered a short-day plant and usually an erect annual, 30 to 100 cm tall with few tillers and an adventitious root system. However, Anderson (1994) showed that 'Cope' proso millet development can be related to temperature by using growing degree days (GDD). For proso, he used a base temperature of 10°C. He showed that 'Cope' began tillering at 430 GDD and initiated flowering at 750 GDD. This relationship needs to be confirmed for other cultivars.

Proso stems and leaves are covered with slight hairs. The leaves may be up to 30 cm long with a short ligule but no auricles (Cobley 1976). The stem is terminated by a drooping panicle 10 to 45 cm long that may be open or compact (Hulse et al. 1980). Proso ($2n$ = 36) is considered a self-pollinated crop, but natural cross-pollination may exceed 10% (Popov 1946). Proso seeds are smaller than pearl millet and grain sorghum, generally oval in shape and about 3 mm long and 2 mm wide (Cardenas et al. 1983). The seeds vary in color as white cream, yellow, orange, red, black to brown. Growth and development characteristics of proso millet were explained in detail by Cardenas et al. (1983). They distinguished three phases; vegetative, reproductive, and ripening, which may be further sub-divided into physiologically distinct stages. The vegetative phase covers the period from germination to panicle initiation, depending on the cultivar used and climate in the area, may be completed 16 to 20 days after planting. An increase in number of leaves, tiller buds, and plant height are characteristics of this phase. The period, about 20 to 25 days, from panicle differentiation to flowering of the main culm is the reproduction phase. This phase initiates when the panicle primordium is greater than 0.5 mm. Rapid elongation of stem internodes and an increase in leaf area accompanied by more tillers are noticed in this phase. The ripening phase starts at flowering or blooming and continues to the end of physiological maturity, which covers a period of 20 to 30 days. Throughout this period, the plant actively accumulates dry matter, particularly in the grains. Rodriques et al. (1990) studied the dry matter accumulation pattern of two cultivars of proso millet with optimum nutrient and water availability. From anthesis to maturity, dry matter accumulated at a rate of 0.5 mg/ha-day. Panicles at maturity accounted for 55% of the total plant dry matter, which is higher than for wheat and grain sorghum. The authors concluded that since proso accumulates higher dry matter in the reproductive parts compared to wheat, maize, and sorghum, it may be advantageous to grow proso in dry and short growing seasons. Physiological maturity proceeds from top to bottom of the panicle. The ripening of the seed is not uniform throughout the panicle and delay in harvesting may cause losses due to shattering (Theisen et al. 1978; Baltensperger et al. 1995a). The grains from the main panicle reach maximum dry weight and a small dark layer at the hilar region of the seed is formed. At maturity, the grain generally includes about 20% or less moisture.

Proso plants generally mature between 60–90 days after planting and can be grown successfully in poor soil and hot dry weather. Proso is an easy crop to grow and it seems to be better adapted than most crops to primitive agricultural practices. Under high-input conditions, however, it may be a poor choice because of low response to high fertility and abundant water (Baltensperger et al. 1995a).

Proso millet requires very little water, possibly the lowest water requirement of any cereal, and converts water most efficiently to dry matter/grain (Theisen et al. 1978; Hulse et al. 1980). According to Arnon (1972), this is not because of its drought-resistance but because of its short growing season. Proso has a very low transpiration ratio, which may be attributed in part to the C4 photosynthetic mechanism (Martin et al. 1976). The low straw grain ratio of proso also contributes to its water use efficiency and adaptation to moisture-limited areas. Proso is very shallow rooted and does not grow well under water stress, nor under excess moisture. Plants seem to tolerate more cold than most other C4 crops, especially sorghum, but will not tolerate frosts. Moderately warm weather is necessary for germination of the seed and growth of the plant. Proso germinates well at temperatures of 10° to 45°C, but does not germinate at 5° or 50°C. The highest rate of germinations is between 35° and 40°C (Theisen et al. 1978).

The feed value of proso has been studied with several classes of livestock. Luis et al. (1982a, b), investigated both broiler and turkey diets with proso. The turkey poults were significantly heavier at the end of the 28 day feeding period with proso than those on either corn or sorghum diets. Broilers fed balanced diets of the three cereals were not significantly different in gains. When proso was evaluated in layer diets it was concluded that "proso millet, either ground or whole, is an excellent ingredient for layer diets" (Luis et al. 1982c). Kies et al. (1975) investigated various millet mixes in human diets and discussed their potential. The feed value of proso millet for cattle and swine is generally considered to equal that of grain sorghum or milo and maize when less than 50% of the maize in the ration is replaced (Table 2).

Table 2. Feed value of maize, proso millet, and grain sorghum in beef cattle (Baltensperger et al. 1995).

Feed	Crude protein (%)	Net energy for maintenance (Mcal)	Net energy for gain (Mcal)	Total digestible nutrients (%)
Maize, dry rolled	10.0	102	70	90
Proso	12.9	93	64	84
Sorghum, dry rolled	10.0	93	64	84
Wheat, hard dry rolled	12.5	99	68	88

Growth and development of foxtail millet is well described by Malm and Rachie (1971). A progressive shortening of the vegetative period occurs with later plantings, and stem elongation is determined by the length of the vegetative period. Stem elongation was hastened by late sowing and was retarded by early sowing. The transition from vegetative growth into inflorescence takes place very rapidly. In the lower third of the cone of growth, initial humps of secondary axes appear following enlarged growth of the growing point, with the middle and upper parts still undifferentiated. Later, secondary axes differentiate up to the very top into points of growth. The inflorescence of foxtail millet has a main stalk with shortened side branches bearings spikes and bristles. The first flowers of foxtail millet may open when three fourths of the panicle emerges from the sheath, or as many as five days after full emergence. Flowering proceeds from the top of the head downward in each of the panicle branches. A large head may take 8 to 16 days to compete flowering. A single floret may remain open about 30 min, and about 80 min are required for the complete blooming process, which is hastened by high temperatures and low humidity (Malm and Rachie 1971). The world's foxtail millet crop is grown largely in China, India, Russia, and the United States. Foxtail millet is by far the most important millet species in China and is grown in Hopei, Horank, and Shansi provinces to a large extent and in most northern areas of China (Anderson and Martin 1949). Foxtail millet grows in mountainous as well as plains regions. It provided approximately 17% of the total food consumed in China (Malm and Rachie 1971). Werth (1937) considered China or central Asia to be the center of origin of the cultivated varieties. Many related species, with $2n$ levels as multiples of the 18 found in foxtail, occur in China. In Russia foxtail cultivars are grown up to 50°N. Foxtail millet is grown primarily for hay in the United States. The primary area is the Northern and Western Great Plains or High Plains, including Colorado, Kansas, Nebraska, Wyoming, South Dakota, and North Dakota. It has been grown at elevations over 1500 m in New Mexico. Seed supplies for hay production are grown in the more southern portion of this region including Texas. The seed can yield as much as proso but foxtail millet seeds are generally much smaller than proso millet seeds. Foxtail millet has about 500,000 seeds/kg, or more than twice that of proso (Wietgrefe 1990). Foxtail millet and proso millet both grow from a height of 0.5 to 2 m, but make little or no regrowth once they are cut. They are generally cut for hay during the early heading stage. Research on forage quality shows foxtail is similar in quality to other C4 annual grasses. It is finer-stemmed and easier to cure than pearl millet or forage sorghum (Baltensperger 1992). Malm and Rachie (1971) give tables of the relative feed value of foxtail millet and discuss its use as a human food and on animal performance. A small amount of foxtail millet is grown and hand picked for millet sprays for the caged bird seed industry (Wietgrefe 1989).

BREEDING

The floral morphology and breeding behavior of foxtail and proso millet make them two of the more difficult species to work with. Nearly all foxtail millet cultivars grown in the United States are the result of selections from land races rather than designed crosses and selections. Nelson (1981) selected and released the last registered cultivar, 'Sno-Fox' for grain, but it has not been grown extensively. A recent breeding program was initiated at the Univ. of Nebraska to incorporate wheat streak mosaic virus resistance into adapted material. Melicio Siles, a graduate student working on the project has made many foxtail millet crosses by hand (Siles, et al. 1995). However, in China, India, and the former USSR a great deal of foxtail millet breeding work has been done. Genetic studies havebeen conducted on several morphological characters and disease response (Malm and Rachie 1971). More recent work in the region has not been well reviewed in English literature, but numerous Chinese, Indian, and Russian scientists are working on foxtail breeding programs at the current time.

Proso breeding in the United States has been conducted at primarily two locations: the USDA program at Akron, Colorado, which was discontinued when Dr. Greg Heinz left in the late 1980s, and the Univ. of Nebraska program at the Panhandle Research and Extension Center. Dr. Lenis Nelson developed crossing techniques for proso at this location and produced the first cultivar from a hand-made cross ('Rise'), in 1984 (Nelson 1984a, b). Since that time, several cultivars have been released from this program that are currently grown on the majority of U.S. millet hectorage (Nelson 1990a; Baltensperger et al. 1995a,b).

Recent work has emphasized the development of increased seed size (Kara 1992). A wider choice of proso cultivars is available today than ever before in the U.S., but most of the high yield, white seed, strong straw types have 'Dawn' as one parent in their pedigree, so germplasm base may be narrower than expected. Nearly all proso grown in the major production areas is white-seeded. Red-seeded proso has some demand, but probably is best grown away from the usual production areas and with a contract or specific market identified. A yearly update of cultivars can be found in the Nebraska Cooperative Extension Publication, EC 107, (Nebraska Proso and Sunflower Variety Tests). The average proso yield in Nebraska is approximately 2 t/ha (Baltensperger et al. 1994).

Three cultivars (Sunrise, Huntsman, and Earlybird) were released in 1994 and 1995. 'Sunrise' is a joint release of the Univ. of Nebraska and USDA (Baltensperger et al. 1994). It is large, white-seeded, with excellent yield potential, intermediate in maturity, excellent lodging tolerance, and is expected to replace 'Sunup' in most growing areas. 'Huntsman' is also a joint release of the University of Nebraska and USDA. It is a large, white-seeded cultivar with excellent yield potential, late in maturity, excellent lodging tolerance, and is expected to replace 'Cope' in most growing areas. 'Huntsman' is shorter than 'Cope' (Baltensperger et al. 1995c). 'Earlybird' was released by the University of Nebraska. It is a large, white-seed cultivar with excellent yield potential. It is early in maturity, has excellent lodging tolerance, and is expected to replace 'Dawn' and 'Rise' in most growing areas (Baltensperger et al. 1995b). 'Sunup' is a 1989 release from Nebraska (Nelson 1990b). It is a white-seeded cultivar similar to 'Dawn' and 'Rise'. 'Sunup' is as lodging resistant as 'Dawn' and 'Rise' and is larger seeded than 'Rise'. It is less susceptible to seed shattering than the common proso cultivars. 'Sunup' has replaced 'Rise' as the most commonly grown cultivar in the United States. The newer releases are all larger seeded, but have a similar panicle type. Several other distinct types of proso are available. 'Panhandle' from Nebraska, 'Minco' from Minnesota, and 'Abarr' from Colorado are all similar to the original common white (Baltensperger et al. 1995a). Although these cultivars differ slightly in height, yielding ability, and maturity, they have white seed and are readily marketed though normal channels in the proso growing areas. They all have fairly open panicles and suffer from lodging and seed shattering.

A second type with acceptable white seed is 'Minsum', which was developed in Minnesota as an early maturing cultivar with a very loose panicle type. Colorado State Univ. developed a tall, late maturing cultivar, 'Cope', that has acceptable white seed. Due to its maturity, 'Cope' is likely best adapted to relatively longer growing seasons. University of Nebraska released 'Dawn', a very short, very early cultivar with a tight panicle. It has a superior white grain, but is inferior agronomically because of its height (Nelson 1976). Because of its earliness, shatter resistance, seed size, and lodging tolerance it has been used extensively as a parent in breeding programs.

If red seed is desired, the two choices are 'Cerise', an early maturing cultivar from Nebraska, or 'Red Leonard', a tall, late maturing cultivar from Colorado. 'Red Leonard' has superior yield if planted early. Both have smaller seed size and are less acceptable as a feed due to a higher tannin content. Bird seed manufacturers use small amounts of red proso to improve eye appeal of the final product.

CULTIVATION

One of the most critical periods in growing either proso or foxtail millet is the two weeks after planting. Both millets grow very slowly during the first few weeks and they are relatively poor weed competitors resulting in unsatisfactory yield on weedy land. Control of weeds prior to planting by herbicides or tillage is critical to good yields. A good seedbed preparation is important in growing these millets and the seedbed should be moist, firm, weed free, and have adequate nutrients to permit a good stand and seedling development. The advantage to the no-till system is greatest at this time as reseeding because crusting is seldom required.

Proso and foxtail millet usually follow in the spring after wheat has been harvested. Both may be planted late as catch crops to replace winter wheat that has been destroyed due to winterkilling, blowing, or hail. Hail storms usually occur after May 15. An attempt is made to plant as soon as the soil is dry enough. Three problems prevail under these circumstances. First, wheat has taken a considerable amount of moisture from the soil profile; second, decaying wheat residue interferes with proso growth; and third, residual herbicide from the wheat crop may persist and injure the millet. Generally, in a wheat-fallow rotation it is more desirable to plant millet into fallow intended for wheat planting in the fall and to utilize the hailed wheat land for fall winter wheat planting. In yield studies conducted in the Nebraska Panhandle, proso yields were more than doubled by using available fallow ground for seeding compared with planting into hailed-out wheat land. The increase of more than 1,000 kg/ha would increase gross returns more than $120/ha using a price of $0.12/kg. The advantage of using the fallow ground is the soil moisture at planting. The danger with this strategy is that hailed wheat land may not store sufficient moisture to provide successful seeding of winter wheat in the fall (Baltensperger et al. 1995a).

The millets can also replace summer crops late in the season when they have poor germination or are destroyed due to early drought, late spring frost, hail, or insects, providing a good chance of obtaining a crop. In moisture-limited areas of the central Great Plains, proso is often well adapted to winter wheat-proso-fallow rotation and gives an extra cash crop every three years. Shanahan et al. (1988) compared three different crop rotations; millet-millet, winter wheat-millet, and fallow-millet for the productivity and water use of proso in the central Great Plains. Over the 5-year period of study, the wheat-millet and fallow-millet out yielded the millet-millet 50% and 116% in grain yield, respectively. Lyon and Baltensperger (1995) and Higgins et al. (1994) showed the value of including millet in a rotation with winter wheat for the control of winter-annual grass weeds.

Several methods of seedbed preparation, as outlined by Baltensperger et al. (1995a) are available to farmers growing proso millet. Work by Nelson and Fenster (1983) and Anderson (1990) show that management within a technique is probably more important than the technique. No-till systems were becoming dominant in the High Plains prior to proso being dropped from the atrazine label. The dropping of the highly tolerant crop from the label is representative of the problems in working with minor crops. Atrazine at the use rates found in proso (less than 0.5 kg/ha) is seldom an environmental problem in dryland areas with more than 20 m to the water table. Alternative strategies to grow no-till proso require several applications of alternative herbicides. The loss of atrazine labeling in proso has increased management demands and the cost of proso production (Baltensperger et al. 1995a).

In general, proso and foxtail millet do not respond consistently to the addition of nitrogen fertilizer. If a full year of fallow precedes the millet, fertilizer is not generally recommended except on the poorest soils (soil test with less than 25 ppm in upper 0.5 m). But, when proso follows wheat, it is generally recommended that 45 kg/ha of nitrogen be applied (Baltensperger et al. 1995a). Chadhari and Rai (1982) suggest that 40 kg N, 20 kg P and 20 kg K/ha used for obtaining a good crop of proso under irrigated conditions in India. Higher rates typically lead to lodging. Proso currently has an upper yield plateau around 5 t/ha because of a lack of straw strength. Yield limits may be higher in more tropically adapted foxtail millets.

Proso millet has a very long planting period and it is often used for replacing dead winter crops or unsuccessful early spring crops. If good weed control is practiced, taking better advantage of spring rains, the earlier dates will produce better yields. In western Nebraska, proso can produce a seed crop when planted as early as May 15 or as late as July 5. Nelson (1990a), using three different cultivars of proso, evaluated influence of planting date on grain yield and other agronomic traits of proso millet at six-year location sites in western Nebraska. The results indicated that planting date of May 15 and June 1 were optimal for grain yield. Foxtail millet cultivars are more variable in maturity than available proso cultivars and some will not mature if planted after June 15 ('White Wonder', 'Golden German' and others), but since most are grown for hay, the practical planting window may be extended to July 15 and still produce a good hay crop.

It is generally accepted that narrow rows (<30cm) are beneficial in reducing weed competition and increasing yield (Nelson 1977; Agdag 1995). The equipment used to plant millet is usually dictated by the equipment available for other crops. The most common drill is one used for winter wheat, usually a deep furrow or hoe-opener type. This is not the most desirable type since it is difficult to maintain a shallow depth with it. When a hoe drill is used, the rate is set high (17 kg/ha) and the depth is quite shallow. The packerwheels can be very beneficial if planting into a loose seedbed, but that combination is most susceptible to crusting from a heavy rainstorm (Baltensperger et al. 1995). The most desirable drill for planting proso where stubble is not a problem is double-disk drill. It can place the seed into a shallow slot and firm the soil around it. It leaves very little furrow to wash in and crust in the event of a heavy rainstorm. The final drill type is one designed for no-till wheat. Regardless of the type of equipment used, it is necessary to have the seed firmly packed and covered with 1 to 2 cm of soil. Nebraska studies showed that choice of drills did not greatly affect yield if a good job of seeding was accomplished (Nelson 1990a).

Ripening is not uniform throughout the panicle in proso. Frequently the seeds in the upper portion of the panicle are ripe while the lower seeds are still green (Matz 1986). A delay in harvesting until the grain has dried sufficiently for threshing will cause considerable losses due to shattering. It is generally recommended that swathing be done when most of the panicle has lost its green color. Threshing can the be delayed until the grain is below 13% moisture (Baltensperger et al. 1995a). Foxtail heads ripen more uniformly and are less subject to shatter, but frequently have tillers that are significantly different in maturity. Foliage is still green at maturity so it is generally swathed prior to combining

The primary limitation on foxtail millet production in the High Plains is not actually a serious problem for foxtail millet itself. It is that it serves as a carrier for both the wheat curl mite, *Eriophyes tullipae* Keifer, the carrier for wheat streak mosaic virus and the virus itself. While adapted cultivars of foxtail are not seriously impacted by the disease, the crop serves as an over-summering host and adjacent wheat fields are frequently severely impacted. Recent research by Marcon (1994) has identified germplasm with much higher levels of resistance that may help to expand the utilization of foxtail millet in the region. Proso has been immune to this disease and is the primary reason it is utilized on a larger scale. However, one of the weeds in many proso fields is green foxtail *Setaria viridis* which can cause the same problems in wheat fields adjacent to proso fields. Management can be utilized to reduce the severity of the problem by harvesting the proso early enough that no living material is present for at least 2 weeks prior to wheat planting.

MARKETS

Proso millet is the primary millet in the world import and export market (Wietgrefe 1989). The bulk of proso sold in the cash trade is marketed through elevators in the counties where it is grown most extensively. The grain is cleaned further, processed, and used for bird seed. Both domestic and wild bird seed is packaged by adding other grains for color and nutrition. Some proso goes through a dehulling process and supplies both human and animal needs. Some is exported and some used for specialty purposes, such as mushroom production. Proso is the only millet of quantity involved in world trade. Proso prices have historically been higher

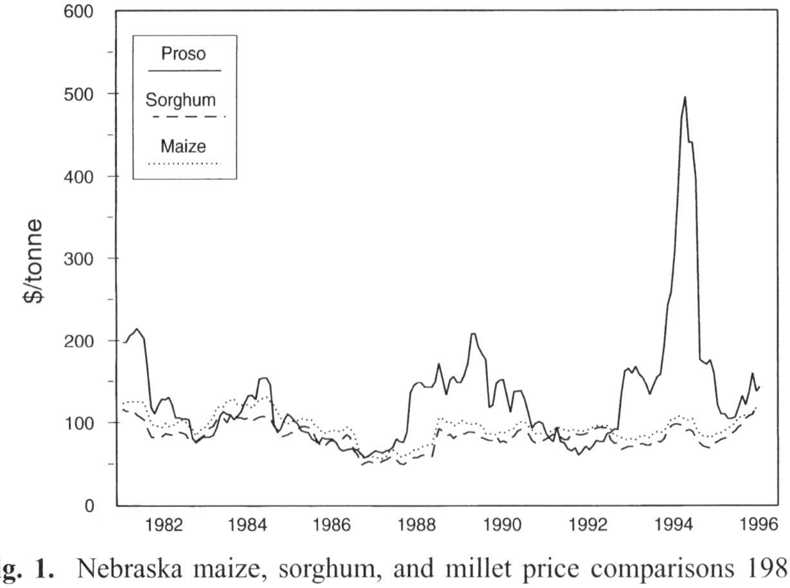

Fig. 1. Nebraska maize, sorghum, and millet price comparisons 1981–1995. Prices shown are $/tonne.

than maize or sorghum (Fig. 1) but this varies dramatically from season to season. When the premium human food and bird seed markets are saturated, the price quickly drops to feed grain values. Proso prices have ranged from $0.05 to $0.55 per kg over 5-year period. In a period of increased crop area, more grain is produced than the markets can absorb. This excess also includes grain produced outside the traditional marketing area. The hay market is more local because of transportation costs for both foxtail and proso millet. The potential for foxtail as a dryland feed grain is severely underdeveloped.

The International Sorghum and Millet Project has specifically excluded research on these two species. As a result, even though foxtail and proso millet have been more significant in U.S. agriculture and are a significant food source for millions of people, much less research has occurred on them during the past ten years than on pearl millet in the United States. The politics of minor crops continues to be a significant factor to overcome in their development, but the potential remains high for both foxtail and proso millet.

REFERENCES

Agdag, M.I. 1995. Row spacing in proso millet. Univ. Nebraska, Lincoln. M.S. Thesis.

Anderson, R.L. 1990. No-till proso millet production. Agron. J. 82:577–580.

Anderson, R.L. 1994. Planting date effect on no-till proso millet. J. Prod. Agr. 7:454–458.

Anderson, E. and J.H. Martin. 1949. World production and consumption of millet and sorghum. Econ. Bot. 3:265–288.

Andrews, D.J., J.F. Rajewski, and K.A. Kumar. 1993. Pearl millet: New feed grain crop. p. 198–208. In: J. Janick and J.E. Simon (eds.), New crops. Wiley, New York.

Arnon, I. 1972. Crop production in dry regions. Vol. II. p. 136–138. Leonard Hill, London.

Baltensperger, D.D. 1992. Summer annual forage grasses in western Nebraska. Univ. Nebraska Panhandle Research and Extension Center Guide. 92-1. p. 14.

Baltensperger, D.D., G. Frickel, M. Swanson, T. Homan, R. Klein, J. Krall, and R. Anderson. 1994. Nebraska proso, safflower, sunflower and amaranth variety tests. Univ. Nebraska E.C. 94-107-A.

Baltensperger, D., D. Lyon, R. Anderson, T. Homan, C. Stymiest, J. Shanahan, L. Nelson, K. DeBoer, G. Hein, and J. Krall. 1995a. Producing and marketing proso millet in the high plains. EC 95-137-C. Univ. Nebraska, Lincoln.

Baltensperger, D.D., L.A. Nelson and G.E. Frickel. 1995b. Registration of earlybird proso millet. Crop Sci. 35:1204–1205.

Baltensperger, D.D., L.A. Nelson, G.E. Frickel, and R.L. Anderson. 1995c. Registration of huntsman proso millet. Crop Sci. 35:941.

Candolle, A. 1964. Origin of cultivated plants. p. 376–377. Hafner Pub, New York.

Cardenas, A., L. Nelson, and R. Neild. 1983. Phenological stages of proso millet. MP45. Univ. Nebraska, Lincoln.

Casey, P. and K. Lorenz. 1977. Millet: Functional and nutritional properties. The Bakers Dig. p. 45–51.

Chang K. 1968. Archeology of ancient China. Science 162:519–526.

Chaudhari, L.B. and B. Rai. 1982. Production technology for proso millet. Indian Farming 82:13–16.

Cobley, L.S. 1976. An introduction to the botany of tropical crops. 2nd ed. Longman, London.

Coleman, F.B. 1939. A classification of millets. Queensland Agr. J. 52:655–657.

Haq, N. 1989. Crop plants: Potential for food and industry. In: G.E. Wickens, N. Haq, and P. Day (eds.), New crops for food and industry. Chapman and Hall Ltd., London.

Higgins, R.K., D.J. Lyon, and D.D. Baltensperger. 1994. Crop rotations for winter annual grass control in winter wheat. Agron. Abstr. 94:81. (Abstr.).

Hulse, J.H., E.M. Laing, and O.E. Pearson. 1980. Sorghum and the millets. Their composition and nutritional value. Academic Press. New York.

Kara, M. 1992. Effectiveness of mass selection for seed and plant attributes in proso millet. M.S. Thesis, Univ. Nebraska, Lincoln.

Keis, C., H.M. Fox, and L.A. Nelson. 1975. Triticale, soy-tvp and millet based diets as protein suppliers for human adults. J. Food Sci. 40:90–93.

Luis, E.S., T.W. Sullivan, and L.A. Nelson. 1982a. Nutrient composition and feeding value of proso millets, sorghum grains, and corn in broiler diets. Poultry Sci. 61:311–320.
Luis, E.S., T.W. Sullivan, and L.A. Nelson. 1982b. Nutritional value of proso millets, sorghum grains, and corn in turkey starter diets. Poultry Sci. 61:321–326.
Luis, E.S., T.W. Sullivan, and L.A. Nelson. 1982c. Nutritional value of proso millet in layer diets. Poultry Sci. 61:1176–1182.
Lyon, D.J. and D.D. Baltensperger. 1993. Proso millet *Panicum miliaceum* L. tolerance to several post emergence herbicides. Weed Tech. 7:230–233.
Lyon, D. and D.D. Baltensperger. 1995. Cropping systems control winter annual grass weeds in winter wheat. J. Prod. Agr. 8:535–539.
Malm, R.N. and K.O. Rachie. 1971. Setaria millets: A review of the world literature. S.B. 513. p. 133. Univ. Nebraska, Lincoln.
Marcon, A.E. 1994. Wheat streak mosaic virus resistance in foxtail millet, *Setaria italica*, L. Beauv., and factors related to resistance. p. 78. M.S. Thesis. Univ. Nebraska, Lincoln
Martin, J.H., W.H. Leonard, and D.L. Stamp. 1976. Principles of field crop production. Macmillan, New York.
Matz, S.A. 1986. Millet, wild rice, adlay, and rice grass. p. 225–229. In: Cereal science. AVI, Westport, CT.
Movsisyants, A.P. 1950. Foxtail millet—A valuable forage crop. Sotsialist. Zhivotn. 12:60–63.
Nelson, L.A. 1976. Registration of dawn proso millet. Crop Sci. 16:379.
Nelson, L.A. 1977. Influence of various row widths on yields and agronomic characteristics of proso millet. Agron. J. 69:351–353.
Nelson, L.A. 1981. Registration of sno-fox Italian millet. Crop Sci. 21:71.
Nelson, L.A. and C.R. Fenster. 1983. Seedbed preparation and planter comparisons for proso millet following wheat. Agron. J. 75:9–13.
Nelson, L.A. 1984a. Technique for crossing proso millet. Crop Sci. 24:205–206.
Nelson, L.A. 1984b. Registration of rise proso millet. Crop Sci. 24:383.
Nelson, L.A. 1990a. Influence of planting dates, seeding rates, and cultivars on grain yield and other agronomic traits of proso millet. J. Prod. Agr. 3:184–189.
Nelson, L.A. 1990b. Registration of sunup proso millet. Crop Sci. 30:746–747.
Popov, G.I. 1946. The Importance of diversity in millet. Agrobiologiya 2:28–43.
Rachie, K.O. 1975. The millets. Importance, utilization and outlook. Int Crops Res Inst. Semi-Arid Tropics. Hyderabad. India.
Rodriguez, J.B., G.A. Peterson, and D.G. Westfall. 1989. Calibration of nitrogen and phosphorous soil tests with yield of proso millet. Soil Sci. Soc. Am. 53:1737–1741.
Roshevits, R. Yu. 1980. Grasses: An introduction to the study of fodder and cereal grasses. Indian National Scientific Doc. Center. New Delhi.
Schery, R.W. 1972. Plants for man. 2nd ed. p. 440–442. Prentice-Hall, Englewood Cliffs, NJ.
Scholz, H. 1960. A contribution to knowledge of cultivated plants. Ber. Dtsch. Bot. Ges. 73:471–476.
Shanahan J.F., R.L. Anderson, and B.W. Greb. 1988. Productivity and water use of proso millet grown under three crop rotations in the central Great Plains. Agron. J. 80:487–492.
Siles, M.M., L.A. Nelson and D.D. Baltensperger. 1995. Technique for artificial hybridization of foxtail millet. Agron. Abstr. 95:84 (Abstr.).
Theisen, A.A., E.G. Knox, and F.L. Mann. 1978. Feasibility of introducing food crops better adapted to environmental stress. Vol. II. p. 168–172. Individual Crop Reports. National Sci. Found.
Vinall, H.N. 1924. Foxtail millet: Its culture and utilization in the United States. USDA Farmers Bul. 793.
Werth, E. 1937. Geography and history of millets. Angewandte Botanik. 19:41.
Wietgrefe, G.W. 1989. Proso millet. A trade summary. Wietgrefe, Pierre, SD.
Wietgrefe, G.W. 1990. How to produce proso millet. A farmer's guide. Wietgrefe, Pierre, SD.

Variation in Electrophoregrams of Total Seed Protein and *Wx* Protein in Foxtail Millet

Muhammad Afzal, Makoto Kawase, Hiroki Nakayama, and Kazutoshi Okuno

Foxtail millet (*Setaria italica* Beauv.) was one of the crops which was domesticated in the early stages of agriculture and has been cultivated across Eurasia continent from tropical to temperate regions. Today it is a staple food in India and northeastern China.

Intraspecific variation of foxtail millet and its geographical distribution has been discussed on the basis of protein polymorphism by de Wet et al. (1979). They suggested a mechanism for domestication due to germplasm exchange based on a comparison of the variation in seed storage proteins of wild and cultivated species of *Setaria* from western Europe, Afghanistan, and East Central China. Jusuf and Pernes (1985) studied the genetic diversity of landraces collected from East Asia, India, the former U.S.S.R and Europe using discriminant analysis of data scored for 10 isozyme loci and found clear regional differences among samples tested.

Starch granule bound polypeptide, *Wx* protein, is responsible for amylose production and has been detected as a single band of approximately 60 kDa in maize (Echt and Schwartz 1981), rice (Sano 1984), grain amaranths (Konishi et al. 1985) and potato (Hovenkamp-Hermelink et al. 1987). In foxtail millet, accessions showing low amylose content in endosperm starch were found in Japan, Taiwan, Philippines, and Indonesia (Takei et al. 1989). Nakamura et al. (1990) detected *Wx* protein in a few Japanese landraces of foxtail millet, which were glutinous, non-glutinous and low amylose strains. A close relation between amylose content and level of *Wx* protein was recognized.

This report deals with intraspecific variation and the geographical distribution of total seed protein and *Wx* protein in foxtail millet landraces.

METHODOLOGY

A total of 217 landraces of foxtail millet were used in this study (Table 1). SDS-polyacrylamide gel electrophoresis (SDS-PAGE) of crude proteins extracted from seed samples followed the method of Laemmli (1970). Starch granule bound protein (*Wx* protein) was isolated and analyzed using SDS-PAGE (Echt and Schwartz 1981). Amylose content in endosperm starch was colorimetrically determined on a dry weight basis using Autoanalyzer II (BRAN+LUEBBE).

RESULTS

Intraspecific Variation and Geographical Distribution of Total Seed Protein

Variation in electrophoregrams of total seed protein was observed in the range of 20 to 30 kDa and about 60 kDa. Based on the combination of 4 protein bands (Fig. 1 a,b,c and d) detected between 20 and 30 kDa, landraces were classified into six different types based on electrophoregrams (Fig. 2). They were designated protein type A (with bands a,b,c), type B (b,c), type C (a,b), type D (c), type E (b), and type F (b,d). A geographical cline of SDA-PAGE electrophoregram was observed, despite the limited number of landraces from Thailand, Myanmar, and Indonesia (Table 1). Distribution of type A is confined to India and countries to the west. Type B, the second most frequent type, is distributed in East Asia, except for Japan and Korea, and west of Pakistan. Type C is found in Japan and Europe. Type D is distributed in China, Taiwan, and Indonesia. Type E is the most common type and widespread throughout Eurasia. Type F occurs in Japan, Korea, and China. Landraces from India and Pakistan were relatively uniform for protein types.

Intraspecific Variation and Geographical Distribution of *Wx* Protein

All landraces analyzed could be classified into three groups based on the level of *Wx* protein. One group had a thick band, a second group a thin band (Fig. 3) and a third group no band. Amylose content of each group was 20 to 35% (non-waxy), 8.5 to 15.6% (intermediate), and close to 0% (waxy) respectively, and corresponded to the visual estimate of level of *Wx* protein. The intermediate and waxy groups were found only in East Asia and Southeast Asia (Fig. 4), which is in agreement with the previous report by Sakamoto (1979) and Takei et al. (1989).

Table 1. Variation in SDS-PAGE electrophoregram of total seed protein of foxtail millet.

Origin	No. of strains examined	Electrophoregrams[z] A	B	C	D	E	F
Japan	12	--	--	2	--	7	3
Okinawa (Japan)	4	--	--	--	--	4	--
Korea	9	--	--	--	--	8	1
China	14	--	5	--	1	7	1
Taiwan	14	--	2	--	4	8	--
Philippines	5	--	3	--	--	2	--
Indonesia	1	--	--	--	1	--	--
Thailand	2	--	--	--	--	2	--
Nepal	4	--	--	--	--	4	--
India	10	1	--	--	--	9	--
Pakistan	120	1	108	--	--	11	--
Afghanistan	5	--	1	--	--	4	--
Central Asia	6	1	2	--	--	3	--
Europe	11	2	5	2	--	2	--
Total	217	5	126	4	6	71	5

[z]A (bands a,b,c), B (b,c), C (a,b), D (c), E (b), F (b,d) respectively.

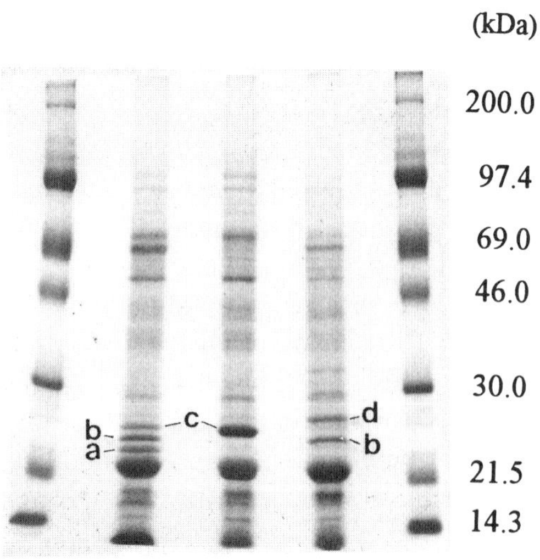

Fig. 1. Four polymorphic band analyzed in electrophoregram of total seed protein.

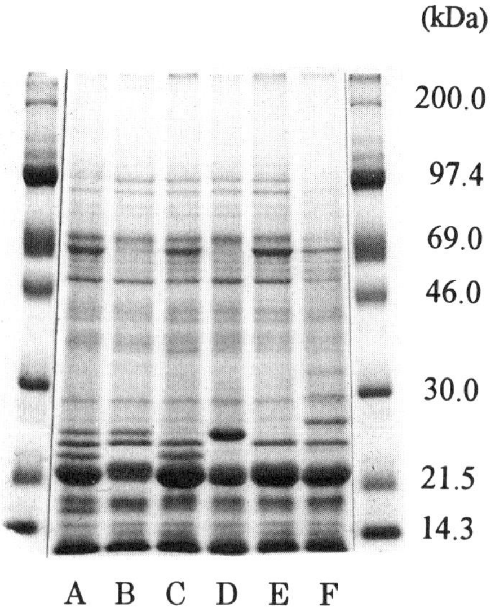

Fig. 2. Intraspecific variation of total seed protein in foxtail millet.

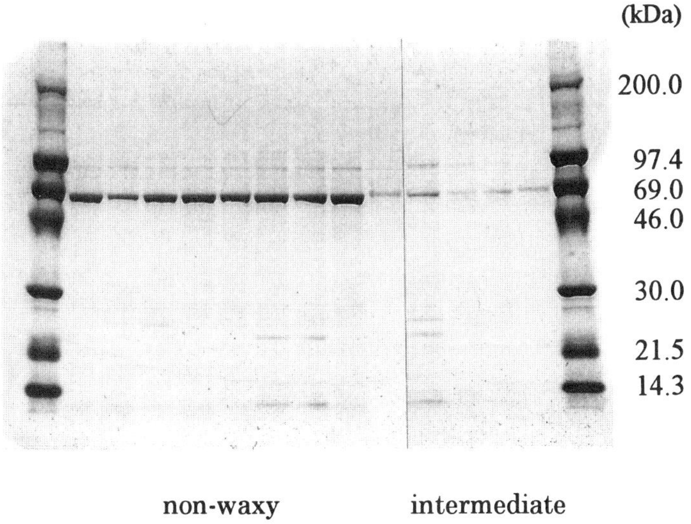

Fig. 3. Electrophoregram of *Wx* protein from non-waxy and intermediate type of endosperm.

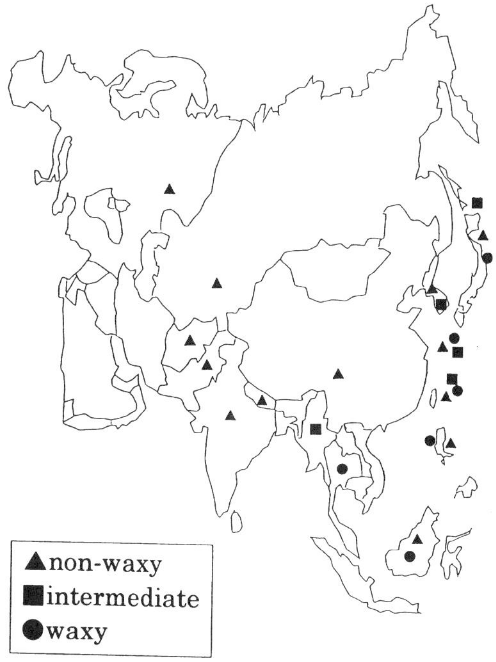

Fig. 4. Geographical distribution of three types of *Wx* protein in foxtail millet landraces.

Table 2. Observed protein phenotypes based on the variation in total seed protein and level of *Wx* protein.

	Electrophoregrams						
Wx protein level	A	B	C	D	E	F	Total
Waxy	0	2	2	3	17	3	27
Intermediate	0	3	0	1	3	0	7
Non-waxy	4	114	1	0	38	1	158[z]
Not determined	1	7	1	2	13	1	25
Total	5	126	4	6	71	5	217

[z]120 accessions were from Pakistan.

In addition, the distribution of these groups is similar to the geographical distribution of waxy endosperm found in six other crop species described by Sakamoto (1982). The regulatory mechanism of the intermediate type is not clear. A regulatory gene (*du*) lowering *Wx* protein level and amylose content has been found in rice (Okuno et al. 1983). Lowering amylose content is considered to be controlled by alleles at *wx* locus or any independent loci of *wx* locus. Gene(s) controlling intermediate level of *Wx* protein and low amylose content (whether *du*-like *trans*-element or *Wx* allele) should be genetically examined.

Intraspecific Variation and Geographical Distribution of *Wx* Protein

All landraces analyzed could be classified into three groups based on the level of *Wx* protein. One group had a thick band, a second group a thin band (Fig. 3) and a third group no band. Amylose content of each group was 20 to 35% (non-waxy), 8.5% to 15.6% (intermediate), and close to 0% (waxy) respectively, and corresponded to the visual estimate of level of *Wx* protein. The intermediate and waxy groups were found only in East Asia and Southeast Asia (Fig. 4), which is in agreement with the previous report by Sakamoto (1979) and Taker et al. (1989). In addition, the distribution of these groups is similar to the geographical distribution of waxy endosperm found in six other crop species described by Sakamoto (1982). The regulatory mechanism of the intermediate type is not clear. A regulatory gene (*du*) lowering *Wx* protein level and amylose content has been found in rice (Okuno et al. 1983). Lowering amylose content is considered to be controlled by alleles at *wx* locus or independent loci of *wx* locus. Gene(s) controlling intermediate level of *Wx* protein and low amylose content (whether *du*-like *trans*-element or *Wx* allele) should be genetically examined.

Variation and Geographical Distribution in Foxtail Millet in East Asia

Based on the polymorphism of total seed protein, the center of diversity could not be clearly identified. As mentioned above, types A,D and F distributed particularly in the western half of Eurasia, Taiwan and Philippines, and East Asia, respectively. Out of 18 possible phenotypes based on the variation in total seed protein and level of Wxprotein, thirteen types were observed (Table 2) and five or six of all the phenotypes existed in Taiwan, Japan, and China. Positive type for phenol coloration was primarily found in Taiwan, Philippines, Nepal, and India (Kawase and Sakamoto 1982 Table 2). Jusuf and Pernes (1985) suggested six regional groups including China-Korea-Japan, Taiwan-Okinawa islands of Japan, India-Kenya, former U.S.S.R., Central Europe, and France, according to discriminant analysis for 10 isozyme loci. Furthermore, most landraces from China, Korea, and Japan (except for the Okinawa islands) belong to the same group for hybrid sterility, while major part of those from Taiwan and Okinawa were included in different group (Kawase and Sakamoto 1987).

SUMMARY

Variation in electrophoregrams of total seed protein was observed in the range of 20 to 30 kDa and about 60 kDa in 217 landraces of foxtail millet collected across the Eurasia. Based on the combination of 4 protein bands (Fig. 1 a,b,c, and d), landraces were classified into six different types (Fig. 2). A geographical cline of electrophoregrams was observed (Table 1). Based on the level of starch granule bound polypeptide (*Wx* protein), these 217 landraces weredivided into, non-waxy, intermediate, and waxy starch types (Fig. 3). The intermediate and waxy types were found only in East Asia and Southeast Asia (Fig. 4).

REFERENCES

De Wet, J.M.J., L.L. Oestry-Stidd, and J.I. Cubero. 1979. Origins and evolution of foxtail millets. J. d'Agr. Trad. Bota. Appl. 26:53–64.

Echt, C.G. and D. Schwartz. 1981. Evidence for the inclusion of controlling elements within the structural gene at waxy locus in maize. Genetics 99:275–284.

Hovenkamp-Hermelink, J.H.M., E. Jacobsen, A.S. Ponstein, R.G.F. Visser, C.H. Vos-Scheperkeuter, E.W. Bijmolt, J.N. de Vries, B. Witholt, and W.J. Feenstra. 1987. Isolation of an amylose free starch mutant of the potato (*Solanum tuberosum* L.). Theor. Appl. Genet. 75:217–221.

Jusuf, M. and J. Pernes. 1985. Genetic variability of foxtail millet (*Setaria italica* Beauv.): Electrophoretic study of five isoenzyme systems. Theor. Appl. Genet. 71:385–391.

Kawase, M. and S. Sakamoto. 1982. Geographical distribution and genetic analysis of phenol color reaction in foxtail millet, *Setaria italica* (L.) P. Beauv. Theor. Appl. Genet. 63:117–119.

Kawase, M. and S. Sakamoto. 1987. Geographical distribution of landrace groups classified by hybrid pollen sterility in foxtail millet, *Setaria italica* (L.) P. Beauv. Japan. J. Breed. 37:1–9.

Konishi, Y., H. Nojima, K. Okuno, M. Asaoka, and H. Fuwa. 1985. Characterization of starch granules from waxy, nonwaxy, and hybrid seeds of *Amaranthus hypochondriacus* L. Agr. Biol. Chem. 49:1965–1971.

Laemmli, U.K. 1970. Cleavage of structural proteins during the assembly of the head of bacteriophage T4. Nature 227:680–685.

Nakamura, T., M. Yamamori, and I. Yamaguchi. 1990. Variation of starch-granule bound protein (*Wx* protein) in foxtail millet and Japanese barnyard millet (in Japanese). Japan. J. Breed. 40(Suppl.1):418–419.

Okuno, K., H. Fuwa, and M. Yano. 1983. A new mutant gene lowering amylose content in endosperm starch of rice, *Oryza sativa* L. Japan. J. Breed. 33:387–394.
Sakamoto, S. 1979. Variation and geographical distribution in foxtail millet. 2. On non-glutinous and glutinous endosperm characters (in Japanese). Japan. J. Breed. 29(Suppl.2):192–193.
Sakamoto, S. 1982. Waxy endosperm and perisperm of cereals and grain amaranth and their geographical distributions (in Japanese with English summary). J. Jap. Soc. Starch Sci. 29:41–55.
Sano, Y. 1984. Differential regulation of waxy gene expression in rice endosperm. Theor. Appl. Genet. 68:467–473.
Takei, E., H. Taira, and S. Sakamoto. 1989. Geographical variation of amylose content in foxtail millet (*Setaria italica* P. Beauv.) (in Japanese). Japan. J. Breed. 39(Suppl.2):260–261.

Buckwheat: Pseudocereal and Nutraceutical

Steven Edwardson

Buckwheat (*Fagopyrum esculentum* Moench, Polygonaceae) is a crop which holds tremendous agronomic and nutritional benefits. Buckwheat is native to temperate east Asia, and has proven itself to be widely adapted around the world. Buckwheat must be further evaluated for its contribution to diversifying cropping systems, enhancing human nutrition, and contributing to regional (rural) economies. This provides a base of information for enhancing buckwheat management in the short term, while simultaneously determining avenues for long term research enhancement, especially in the area of cultivar improvement.

CLASSIFICATION AND ADAPTATION

Buckwheat, native to temperate east Asia, where it was grown in China before 1000 AD (Robinson 1980) is adapted in many areas of the world. Although buckwheat production is concentrated in China, Japan, and North America, it is also produced in Europe, India, Tibet, Tasmania, Australia, Argentina, Bhutan, and numerous other countries (Joshi and Paroda 1991). The name "buckwheat" comes from the Anglo-Saxon words *boc* (beech) and *whoet* (wheat), because the seed resembles a small beech nut (Robinson 1980).

Buckwheat seeds are the fruits (matured ovaries) of the plant. The indehiscent fruit (seed) is an achene because the three sided pericarp (hull) encloses only one true seed (Robinson 1980). The seed is similar in size and weight to barley.

Until recently, only two cultivated and seven wild species of buckwheat were believed to exist (Onishi 1995). However, seed collection and classification by Onishi (1995) has resulted in a total of 14 species of buckwheat, with new discoveries occuring every year. The genetic origins of buckwheat are believed to occur in the Yunnan and Sichuan provinces of China (Onishi 1995). The recent discoveries of new wild species have provided valuable genetic material for advancing cultivar development. The most notable of these discoveries is *Fagopyrum homotropicum*. This species is self pollinated and has many characteristics similar to *F. esculentum*, thus allowing plant breeders to improve and develop new varieties with desirable production and consumption characteristics (Campbell 1995). Table 1 lists the various species of buckwheat described by Onishi (1995).

PRODUCTION

Production practices for buckwheat are well documented in North America (Marshall 1969; Robinson 1980; Stiles 1982; Auld et al. 1986; Campbell and Gubbells 1986; Oplinger et al. 1989; Smith et al. 1989; Berglund and Schneiter 1992). Most of the information available addresses buckwheat production on a regional basis. In general, buckwheat production practices (i.e. planting, rotation, weed control, fertility, in-

Table 1. Identification of species of *Fagopyrum* (Onishi 1995).

Species	Comments
F. esculentum Moench	Most widely utilized form of buckwheat
F. homotropicum Onishi	Recently discovered, can be crossed with *F. esculentum*
F. cymosum	Perennial wild species
F. tartaricum	Also known as "bitter buckwheat"
F. statice	Perennial specie with well established root system
F. urophyllum	Ochrea is green, not transparent
F. ue	Recent discovery, classification in process
F. leptopodum	Ovate leaf blades
F. lineare	Linear leaf blades
F. callianthum	Erect plant, erect growth habit
F. uf	Recent discovery, classification in process
F. pleioramosum	Has many creeping branches
F. gracilipes	Known for its drooping inflorescence
F. capillatum	Erect branches, cordate or ovate blade

sects, disease, harvest, storage, and yield expectations) deviate only slightly from one region to another in North America. Table 2 provides a summary of buckwheat production practices for North Dakota. Similar practices are utilized in other states as well.

Planting and Rotation

As farmers continue to seek rotation crops under reduced tillage and no till farming systems, crops such as buckwheat will need to be evaluated for their contribution to the cropping system. Research is needed to determine how well buckwheat performs in no till cropping systems, with specific emphasis placed upon (1) planting depth and stand establishment; (2) rooting depth and water use; and (3) influence on yield of the following crop. This will assist farmers in determining the contribution of buckwheat in a cropping system.

Weed Control

Minor crops in general have limited options with respect to weed control. Buckwheat, however, is fortunate in that it is very competitive with other plants (Robinson 1980). Numerous organic farmers include buckwheat in their rotation for weed suppression. Nonetheless, post emergence weed control options in buckwheat are lacking, and research is needed in this area. Currently, there are no herbicides registered for use on buckwheat in the United States. Research is underway to obtain registration for sethoxydim (Poast) for grass control in buckwheat under the IR-4 program. Weed control research in buckwheat is needed so that herbicide options are available when crop rescue is necessary due to excessive weed pressure.

Fertility

Most buckwheat production guides indicate that high levels of nitrogen result in excessive vegetative development, lodging, and reduced seed set (Marshall 1969; Robinson 1980; Berglund and Schneiter 1992). It is also recognized that buckwheat requires fairly high levels of phosphorous in order to set seed and mature (Stiles 1982; Berglund and Schneiter 1992). However, little is understood about the interaction of nitrogen and phosphorous on buckwheat production. Micronutrient nutrition research is also lacking in buckwheat. Considerable effort is needed in this area to improve crop management.

Disease

Downy mildew, powdery mildew, and rhizoctonia root rot are the major diseases reported in buckwheat (Marshall 1969; Joshi and Paroda 1991). Disease problems in buckwheat are reported infrequently, and are typically of minimal concern in most years. Varietal screening to maintain and improve disease tolerance in breeding programs is the most logical area for disease research.

Insects

Wireworms and aphids may attack buckwheat, but usually do not cause serious economic losses (Marshall and Pomeranz 1982). Grasshoppers can cause defoliation on field borders when their population is high and other food sources are scarce. Japanese beetles can cause serious damage to buckwheat flowers, but are usually of minor importance (Marshall and Pomeranz 1982). In general, little research has been conducted on insect pests of buckwheat in North America, because insect related problems have been minimal.

Crop Development and Physiology

Buckwheat exhibits a rapid growth habit. Emergence can occur as early as 4 days after planting. The plant typically matures in 75 to 90 days, depending upon environmental conditions. Buckwheat is quite sensitive to periods of drought stress. Temperatures in excess of 25°C can significantly reduce seed set and yield (Obendorf et al. 1991).

Table 2. Summary of typical buckwheat production practices used in North Dakota.

Crop development:	Emerges 5 to 7 days after planting. Begins flowering about 5 weeks after planting. Optimum seed set period is during weeks 5 to 7 after planting. Makes many flowers, but not all will set a seed. Wilts in hot weather, not real drought tolerant. Matures in 75 to 90 days.
Rooting depth:	45 to 60 cm.
Soils:	Adapts to a wide range of soil types. Prefers medium texture soils. Peforms well on poor soils. Soils that crust easily can cause emergence problems.
Planting rate:	50 to 70 kg ha^{-1}; 15 to 17 cm row spacing.
Planting date:	May 25 to June 15. Planting up to July 1 has occurred in North Dakota, but yields are reduced. Buckwheat is not frost tolerant at any stage.
Planting depth:	2 to 5 cm.
Fertility:	Nitrogen: 55 kg N per 1000 kg ha^{-1} yield goal. Phosphorous use is similar to wheat (use of a drill row starter is a good practice).
Weed control:	No chemicals are registered in the United States. Buckwheat is usually quite competitive.
Chemical carryover:	Sensitive to carryover of trifluralin, atrazine, and sulfonylurea products (e.g. Glean, Ally).
Rotation:	Usually placed between small grains (e.g. wheat and barley).
Insects:	Aphids and grasshoppers. Insects are usually not a major problem in buckwheat production.
Diseases:	Subject to Rhizoctonia root rot, alternaria, and downy mildew. These diseases are not reported often.
Harvest:	Swath when 75% of seeds in upper 1/3 of plant are ripe. Swath when field is damp to reduce shatter loss. Swath immediately after a frost to reduce shatter loss. Stores safe at 14% to 16% moisture content when seed is clean of foreign material.
Cultivars:	Mancan, Manor, and Giant American (all are large seeded).
Market:	Used in Japan for milling into flour. Flour is used in buckwheat (soba) noodles. Also used in pancake mixes.
Typical yields:	1100 to 1700 kg ha^{-1}. Yields over 2200 kg ha^{-1} have been reported. Yields are very dependent upon environment in a given region.

Buckwheat is self incompatible, so cross pollination is essential for production. The plants flower profusely, but only 10% to 20% of the flowers set seed (Obendorf et al. 1991). A given population is divided equally between pin (long pistils, short stamens) and thrum (short pistils, long stamens) flowers. Legitimate cross pollination occurs only between pin and thrum flowers. Buckwheat displays considerable plant to plant variability, which can cause considerable yield variability. Some plants set well over 200 seeds, while some plants set only 10 to 20 seeds.

The development rate of buckwheat is heat unit driven and closely related to air temperature. Gorski (1986) found that buckwheat development was highly correlated with average daily temperature from planting to initiation of flowering ($r^2 = 0.84$), but correlation between average temperature and maturity was much weaker ($r^2 = 0.53$). This was due to the fact that buckwheat can have ripe seeds, developing seeds, and flowers on the plant all at the same time. Edwardson (1995d) correlated buckwheat development with growing degree days to assist in estimating optimum windrowing time (i.e. when 75% of the seeds are ripe as per Gubbels and Campbell 1984). Buckwheat ripening is highly correlated with growing degree days ($r^2 = 0.93$), and growing degree days offer a fairly precise method for estimating optimum windrowing time (Edwardson 1995d). Fig. 1 shows the relationship between accumulated growing degree days (AGDD) and the percentage of ripe seeds (Edwardson 1995d). Unpublished data from Edwardson indicates that a "typical" buckwheat plant is 100 cm tall, has 15 leaves on the main stem, has 3 to 5 axillary branches, and matures at an accumulated level of 1200 growing degree days (with base temperature at 5°C). In general, buckwheat follows a fairly disciplined pattern of development that is more reflective of a determinate crop, as opposed to the indeterminate classification of buckwheat reported in most production guides (Edwardson 1995; Obendorf et al. 1991). Reducing the variability of seed set and yield is a major issue for cultivar improvement.

Harvest and Storage

Buckwheat should be windrowed when 75% of the seeds are ripe (Gubbels and Campbell 1984). Due to the plant to plant variability of buckwheat, this can be difficult for the producer to estimate. Gubbels and

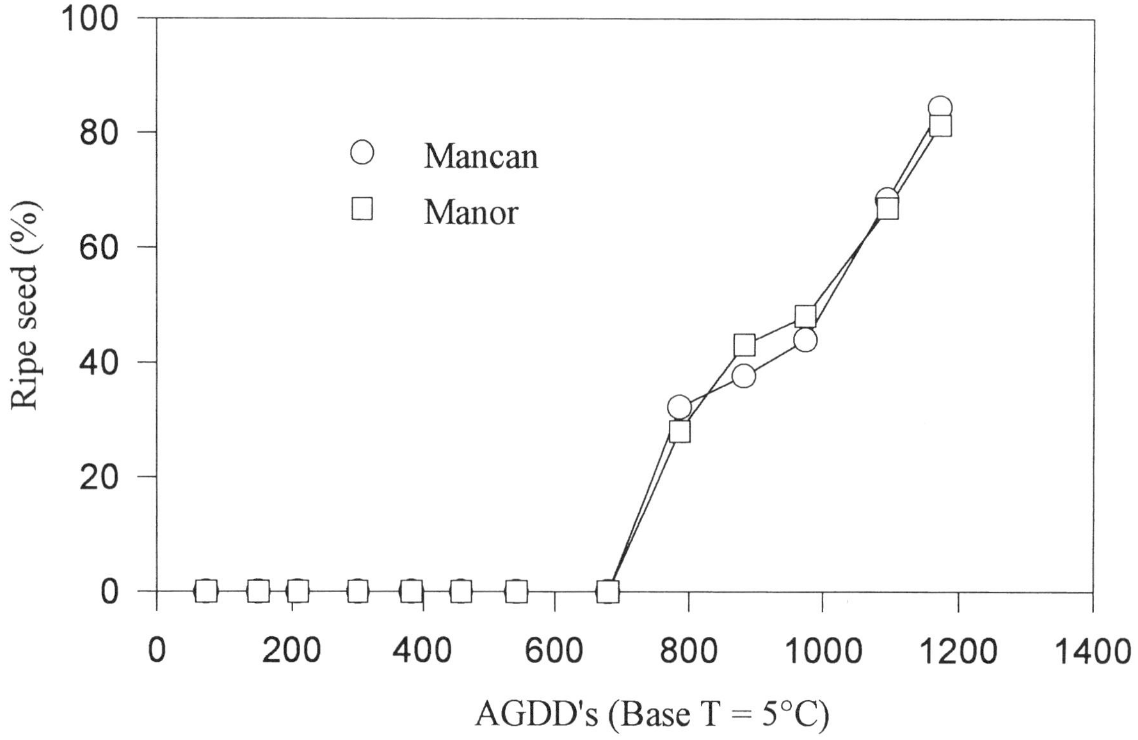

Fig. 1. Relation between accumulated growing degree days (AGDD) and percent ripe seed in buckwheat.

Campbell (1984) and Edwardson (1995d) describe visual methods and climatic models, respectively, to assist in estimating when the crop should be windrowed. This is important in order to minimize shatter loss, and maximize crop quality and yield. Improving the uniformity of seed set and yield through breeding efforts will greatly assist in improving buckwheat harvest.

Buckwheat will store safely at 14% to 16% moisture content. In order for the crop to be marketed, it is important for the groat to maintain a green color. Over time, oxidation will cause the groat to turn red, which reduces milling quality. Research targeted at maintaining the green color of the groat is currently in progress. Results from this research will be very beneficial in maintaining crop quality during storage.

Yields, Cultivar Improvement, and Production Areas

Little improvement has been made in buckwheat yields in the past century. Ujihara (1995) reports that average buckwheat yields in Japan have varied from 500 to 1,100 kg/ha since 1880. The extreme plant to plant variability of buckwheat, coupled with increasing and more stable yields in other crops, have left buckwheat at a disadvantage with respect to cultivar improvement. The discovery of new plant genetic resources (Onishi 1995) should assist breeders in developing new cultivars that exhibit greater uniformity in yield. Interspecific hybridization between *F. homotropicum* and *F. esculentum* is underway (Campbell 1995).

Typical buckwheat yields in North America range from 1,000 to 3,000 kg/ha. Unfortunately, the variability can be quite extreme, with yields as low as 200 kg/ha, to over 3,000 kg/ha. Table 3 presents buckwheat yields at various agricultural experiment stations in North Dakota. Yields in other states are typical of yields presented in Table 3.

The few buckwheat cultivars available in North America are of two basic types: small seeded and large seeded. 'Tokyo', 'Tempest', and 'Common' are small seeded, whereas 'Manor', 'Mancan', 'Manisoba', and 'Giant American' are large seeded. Processors prefer the large seeded cultivars, with 'Mancan' and 'Manor' being the most widely utilized, and breeding research is currently focused on these cultivars (Campbell 1995).

Buckwheat is produced extensively in North Dakota. Approximately 13,355 ha are produced in the state each year (North Dakota State Consolidated Farm Service Agency records, 1991 to 1994). Buckwheat pro-

Table 3. Buckwheat yields in North Dakota, 1985 to 1994.

Cultivar	Yield (kg/ha)										
Location	1985	1986	1987	1988	1989	1990	1991	1992	1993	1994	Avg.
Manor											
Carrington				215	1,990	1,434		1,430	1,955	608	1,272
Dickinson				436		570				1,569	858
Fargo	4,417	2,148									3,283
Hettinger	582		1,243		1,243	211	218				699
Langdon			1,108	1,760	394	2,294	2,248		339	2,044	1,455
Minot	1,800	2,296	2,676		547	662	901	657	1,057	792	1,265
Williston		1,203	1,997		538	638				1,431	1,162
Mancan											
Carrington				169	1,835	1,584		1,433	1,869	688	1,263
Dickinson				380		451				1,819	883
Fargo	4,217	1,892									3,054
Hettinger	533		885		472	472	283				529
Langdon			1,643	1,991	463	2,102	2,255		318	1,902	1,525
Minot	1,800	1,831	2,379		547	557	781	739	1,499	847	1,220
Williston		1,088	134		548	608	253	799	1,697	1,355	810

duction is widespread, with concentrated production regions in the south central and southwest regions of the state. It is estimated that total U.S. buckwheat output is 17,000 to 22,000 t annually (Peterson et al. 1992). Buckwheat is also produced in South Dakota, Minnesota, Montana, Washington, Pennsylvania, and New York. Production estimates were not available for these states, since USDA production records for buckwheat ceased in 1964 (Robinson 1980).

Buckwheat production has been decreasing in North Dakota in recent years (Table 4) because more farmers are shifting their acreage to canola, which has a more thoroughly developed market, and wet conditions often interfere with planting. Nonetheless, buckwheat is a viable crop for numerous producers. Crop area will become more stable when cultivar development and crop management result in improved yield uniformity. At present, buckwheat requires more management expertise than other crops.

Buckwheat production and consumption is relatively widespread in Europe (Matano and Ujihara 1995; Borghi 1995). Although no formal statistics on area and quantity of production are currently available (Matano and Ujihara 1995), buckwheat is produced in Poland, Germany, Italy, Slovenia, and the Ukraine. Each of these countries have established buckwheat research programs (Matano and Ujihara 1995). Most of the production in these countries is consumed internally, and consequently exports from each country are relatively limited (Matano and Ujihara 1995). Quantification of crop area, production, and consumption patterns in each of these countries will assist in determining potential barriers to increasing North American buckwheat exports in Europe.

ECONOMICS AND MARKETS

Crop budget data presented in Table 5 indicate that buckwheat is a profitable alternative crop in comparison with unsubsidized spring wheat (Swenson and Aakre 1992, 1993, 1994, 1995). Buckwheat is a relatively low input crop that is complementary in small grain rotations to break disease cycles. It should also be noted that buckwheat prices are typically higher than spring wheat prices. Similar cost of production analyses by Edwardson (1989, 1990, 1991a, 1992a, 1993a, 1995b) indicate that buckwheat is a profitable competitor with numerous alternative crops.

About 95% of the buckwheat produced in North Dakota and surrounding states is exported to Japan (Harris Peterson, Minn-Dak Growers Ltd., pers. commun.). The Japanese mill buckwheat into flour for use in soba noodles, which are a staple part of their diet (Udesky 1992). Japan consumes over 100,000 t of buckwheat every year, but only produces 20% to 30% of their needs domestically (Ujihara 1995). Although North American buckwheat exporters enjoy solid relationships with the Japanese market, it is desireable to develop additional markets, especially with respect to value-added buckwheat.

Table 4. Buckwheat production by county, North Dakota, 1991 to 1994.

Year	Total area[z] (ha)	Yield[y] (kg/ha)	Total output (t)	Farm price[y] ($/t)	Farm gate value ($)
1991	15,729	725	11,228	198.45	2,228,196
1992	15,390	1,000	15,389	231.52	3,562,938
1993	11,219	756	8,477	275.63	2,336,473
1994	13,428	791	10,627	242.55	2,577,579

[z]Data was obtained from North Dakota Consolidated Farm Service Agency records, 1991 to 1994.
[y]Yield and price data was obtained from Minn-Dak Growers Ltd. company records.

PROCESSING, UTILIZATION, AND NUTRITIONAL CHARACTERISTICS

Buckwheat is primarily used for human consumption. The dehulled seed, or groat, is used in breakfast cereals and milled into grits (Robinson 1980). Roasted groats, which are called kasha, are sold in whole and granulated forms. Both kasha and groats can be baked, steamed, or boiled for nutritious alternatives to potatoes and rice.

Buckwheat flour has numerous uses. It is used in pancake mixes as well as in various breads. It is often blended with wheat flour for use in bread, pasta products, and some breakfast cereals (Robinson 1980). The Japanese mill buckwheat groats into flour for use in the production of soba noodles, which are a main part of the Japanese diet (Udesky 1992). In Eastern Europe, buckwheat flour is used in cooking similar to wheat flour. Bread, cakes, and dumplings are made with the addition of wheat flour. Buckwheat flour is used alone in some dishes, such as polenta and zganci.

Buckwheat fields in bloom can serve as a valuable source of nectar for bees. Honey produced from buckwheat is typically dark and has a stronger flavor than honey produced from clover, and is preferred by some consumers.

Studies have shown that up to 60% buckwheat flour mixed with wheat flour produced an acceptable bread (Pomeranz 1983). In general, buckwheat flour in bread mixes comprises only 30% to 40% of the total. Buckwheat flour may also be used in deserts, ice cream cones, dietetic foods, pancake mixes, canned meat products, canned vegetable products, and dried breakfast cereals.

Pasta produced from a mixture of wheat and buckwheat flour, with the addition of egg yolk, vital gluten, and skimmed powdered milk has been characterized to possess shorter cooking time. This provides buckwheat with numerous potential applications in nutritious foods that could be packaged for rapid preparation.

Buckwheat can be used to produce extruded cereal and snack products. Extruded buckwheat products are of very high nutritional quality when compared with products extruded from maize, wheat, or barley alone. The specific character of the proteins allows buckwheat to potentially be used in extruded products targeted to special nutrition needs.

Buckwheat is one of the best sources of high quality, easily digestible protein in the plant kingdom. It has over 90% of the value of non-fat milk solids and over 80% of whole egg solids (Udesky 1992). The balanced amino acid profile and a high level of essential amino acids allow buckwheat to be used in human diets, especially where shortages of lysine and sulfur containing amino acids appear.

Buckwheat is quite complementary to cereal flours, and can be used to improve their nutritional quality, since it is high in essential amino acids as evidenced in Table 6. The main protein solubility fraction in buckwheat is globulin. In addition, the content of prolamines is low. It is relatively high in potassium and phos-

Table 5. Costs and returns for buckwheat and spring wheat in southwest North Dakota.

Crop	Year	Yield (kg/ha)	Price ($/t)	Gross revenue ($/ha)	Variable costs ($/ha)	Return over variable costs ($/ha)
Buckwheat	1992	1,121	185.22	94.16	38.67	55.49
	1993	1,121	240.35	122.19	34.28	87.91
	1994	1,121	240.35	122.19	34.28	87.91
	1995	1,121	240.35	122.19	39.29	82.90
Spring wheat	1992	1,211	106.50	58.48	42.43	16.05
	1993	1,278	106.50	61.72	36.74	24.99
	1994	1,345	110.25	67.26	41.77	25.49
	1995	1,412	119.51	76.56	43.27	33.28

phorus, and contains 150% more vitamin B than wheat. Buckwheat has no more calories than wheat products or most other grains. It is gluten free, thus making it a valuable nutrient in the diets of people who are sensitive to gluten.

Starch is the major carbohydrate in buckwheat, comprising 51% to 67% of the seed (Pomeranz 1983). Robinson (1980) reports that buckwheat flour is very high in starch (Table 7). Buckwheat starch has a water binding capacity of 103.7%, blue value of 0.35 and amylose content of 25% (Pomeranz 1983). Amylograph data showed that the starch had an initial pastin temperature of 64°C. This starch forms satisfactory fillings, but not acceptable quality cakes.

Total lipids in whole buckwheat grain are 1.5% to 3.7%. Buckwheat oil contains 16% to 20% saturated fatty acids, 30% to 45% oleic acid, and 31% to 41% linoleic acid. Palmitic (19.3% to 22.9%), oleic (29.1% to 31.6%), linoleic (19.1% to 34.8%) and linolenic acids (4.7% to 6.8%) account for about 95% of buckwheat fatty acids. Phospholipids comprise ~3.6% of the buckwheat flour and are ~67% of the conjugated lipids (Pomeranz 1983).

The mineral composition of buckwheat is also quite complementary to cereal crops (Table 8). The inclusion of buckwheat in food products that contain other cereals would assist in improving nutritional quality.

Table 6. Essential amino acid composition of buckwheat and commonly utilized cereal grains (Thacker et al. 1984).

	Content as % of whole grain						
Amino acid	Wheat	Maize	Barley	Oats	Rye	Triticale	Buckwheat
Arginine	0.52	0.40	0.40	0.65	0.56	0.56	0.90
Histidine	0.32	0.25	0.24	0.22	0.26	0.33	0.33
Isoleucine	0.46	0.33	0.40	0.37	0.40	0.50	0.46
Leucine	0.91	1.20	0.79	0.73	0.74	0.95	0.84
Lysine	0.39	0.26	0.42	0.40	0.45	0.47	0.77
Methionine	0.18	0.18	0.15	0.15	0.15	0.16	0.19
Phenylalanine	0.64	0.47	0.59	0.49	0.54	0.65	0.56
Threonine	0.40	0.33	0.39	0.35	0.40	0.43	0.49
Valine	0.56	0.44	0.55	0.50	0.54	0.59	0.60

Table 7. Average nutritional composition of buckwheat seed and groat (Robinson 1980)[z].

	Content (%)				
Product	Protein	Carbohydrate	Fat	Fiber	Ash
Buckwheat seed	12.3	73.3	2.3	10.9	2.1
Buckwheat groats	16.8	67.8	3.2	0.6	2.2
Buckwheat dark flour	14.1	68.6	3.5	8.3	1.8
Buckwheat light flour	11.7	72.0	2.5	1.6	1.8
Buckwheat white flour	6.4	79.5	1.2	0.5	0.9
Bread wheat flour	11.8	74.7	1.1	0.3	0.4

[z]Based on a 12% moisture basis. Divide by 0.88 to convert to a dry matter basis. Flour data was obtained from commercial sources, and from Watt and Merrill 1963.

Table 8. Comparison of mineral composition of buckwheat flour with corn meal, semolina, wheat flour, and soybean flour.

	Content (mg/100 g)								
Product	Ca	Fe	Mg	P	K	Na	Zn	Cu	Mn
Buckwheat flour	41	4	251	337	577	0	3.12	0.09	0.099
Corn meal (whole grain)	6	3.45	127	241	287	35	1.82	0.193	0.498
Semolina	17	1.23	47	136	186	1	1.05	0.189	0.619
Wheat flour	15	1.17	22	108	107	2	0.70	0.144	0.682
Soybean flour	206	6.37	429	494	2515	13	3.92	2.92	2.275

The nutritional characteristics of buckwheat place this crop in an excellent position to be included in a variety of food products. New developments in medical research are providing new avenues for consideration in advancing the utilization of buckwheat.

BUCKWHEAT AS A NUTRACEUTICAL

Americans are increasingly concerned with eating healthy foods. Numerous food labels indicate lower cholesterol levels and reduced fat. Because of this trend, the nutritional attributes of buckwheat place this crop in an excellent position for value added processing and increased utilization.

A recently introduced term in the food ingredients industry is nutraceutical. A nutraceutical is defined as any substance that is a food or part of a food and provides medical or health benefits, including the prevention and treatment of disease (DeFelice 1994). Since nutraceuticals will no doubt transcend a broad range of food products, their value will need to be supported by good clinical research data.

Numerous research articles are published on buckwheat as a food ingredient. Both Food Science and Technology Abstracts (FSAT) and the MEDLINE medical database contain a relatively large amount of research based information on the food ingredient applications and medicinal benefits of buckwheat. Buckwheat enjoys the following nutritional and health benefits (Udesky 1992):

1. Buckwheat contains vitamin P, which contains the flavonoid rutin. Rutin is known for its effectiveness in reducing the cholesterol count in the blood. In addition, buckwheat is an effective preventative measure against high blood pressure. Rutin is known to keep capillaries and arteries strong and flexible. The effectiveness of rutin in buckwheat is strengthened with the addition of vitamin C.
2. Japanese publications suggest that buckwheat "is a very nourishing food which works to relax the body, helps relieve and prevent inflammation, excessive perspiration, nosebleed, or high blood pressure; and helps maintain a good functioning of the intestines. . . ."
3. Buckwheat flour requires little cooking and has good shelf life.
4. The human body can utilize 74% of the available protein in buckwheat. Although other food products are also high in available protein, many contain high levels of fat. Buckwheat is virtually fat free. It is also gluten free.
5. Buckwheat has an excellent amino acid composition that is very complementary to cereal grains. It is very high in lysine, nearly twice the amount found in wheat and white rice.
6. Buckwheat contains considerable amounts of vitamins B1 and B2. Vitamin B1 is important in rekindling energy by facilitating the working of the nerves. Vitamin B2 assists the lipids in their work. Combining these vitamins with vitamin E results in an effective precaution against hardening of the arteries (arteriosclerosis). Consuming 100 g of buckwheat can supply 40% of the daily adult requirement of vitamin B1.

7. Buckwheat is very effective in helping the human body rid itself of unwanted cholesterol. Sterols in buckwheat prevent cholesterol from increasing in the blood serum.
8. Potassium, magnesium, phosphate, and iron are abundant in buckwheat flour. Buckwheat is higher in iron than cereal grains. These minerals play an essential role in the prevention of hypertension and anemia.
9. Buckwheat contains choline, which facilitates the working of the liver.
10. Combining small amounts of egg, chicken, or vegetable with buckwheat noodles provides a nutritional meal that approaches the ideal balance of nutrients.

Regular consumption of 30 g of buckwheat has been shown to lower blood pressure regardless of other factors such as age and weight. In a study conducted in cooperation with the Johns Hopkins Medical Institute, Jiang et al. (1995) reported that subjects who consumed the greatest amount of buckwheat had the lowest blood pressures.

The addition of buckwheat to human diets could also help cut calories and keep blood sugars at optimal levels. Buckwheat grain is digested more slowly than other carbohydrates (Buckwheat diet 1989). Consequently, the inclusion of buckwheat in a meal leaves people feeling full longer, thus reducing the urge to snack. The slow uptake of buckwheat also has potential to prevent adult-onset diabetes, as well as improve glucose tolerance in those who have developed the disease (Buckwheat diet 1989).

Incorporating present knowledge of medicinal and health benefits with targeted market research will provide the framework necessary to advance the utilization of buckwheat.

ADVANCING THE UTILIZATION OF BUCKWHEAT

Advancing the utilization of buckwheat requires an integrated research effort across production, processing, and marketing levels. Although buckwheat is still considered a minor crop, there is a growing amount of published research information across all 3 levels. Numerous research based articles are found in major computer databases (e.g. FSAT, CAB, AGRICOLA, MEDLINE).

Every three years, the International Buckwheat Research Association (IBRA) assists in organizing an international buckwheat research symposium. Symposiums have been held in Slovenia (1980), Miyazki, Japan (1983), Pulawy, Poland (1986), Russia (1989), Taiyuan, China (1992), Ina City, Japan (1995), and is slated for Winnipeg, Manitoba, Canada in 1998. Proceedings have been published from each international buckwheat symposium. The 1995 symposium had a large portion of the proceedings devoted to food science and human health (Matano and Ujihara 1995).

Previous information suggests that buckwheat has nutritional characteristics which place this crop in an excellent position for expanding utilization. Increasing the utilization of buckwheat as a food ingredient will require targeted market research. Since the cost of introducing new food products into the market is substantial, buckwheat processors will need to focus on the health benefits of buckwheat, then target those products in which buckwheat would be complementary. Present market expansion research indicates that buckwheat is well positioned to be included in specialty breads, pasta, snack foods, and ready to eat cereals (Edwardson 1991b, 1992b, 1993b, 1994, 1995a).

At present, there are very few buckwheat researchers in North America. Research efforts have typically focused on buckwheat production practices within a given state, while research on processing and utilization has been somewhat fragmented. In July 1994, buckwheat researchers and processors from North America gathered in Carrington, North Dakota to discuss current research efforts in buckwheat, and to develop a plan for advancing the utilization of this crop. Japanese buckwheat millers and traders also participated in this conference. This conference resulted in the formation of the North American Buckwheat Consortium (NABC). The industry representatives of this group agreed to jointly fund specific research efforts on buckwheat with the public sector. Proceedings from this conference outlined needed research efforts at the production, processing, and marketing levels (Edwardson 1995c).

Until 1995, Agriculture Canada had the only active buckwheat breeding program in North America. This program was led by Dr. Clayton Campbell at the Agri-Food Diversification Research Center at Morden,

Manitoba. Reductions in research funding caused Agriculture Canada to eliminate the buckwheat breeding program. Fortunately, private sector traders and processors decided to privatize this breeding program so that cultivar development would continue. Efforts to formalize a new buckwheat breeding company in Canada are in progress.

CONCLUSIONS

Information presented indicates that buckwheat is not only a viable alternative crop, but is also a commodity which holds excellent opportunity for expanded utilization. Current knowledge, coupled with an integrated research effort, will provide the framework necessary for increasing the utilization of this crop.

Buckwheat is produced extensively in many regions of the world. Production practices for buckwheat are well documented, and the crop is relatively easy to produce across a wide range of environmental conditions. Buckwheat is complementary to crop rotations, and is also nutritionally complementary to cereal grains.

Although current buckwheat use is limited to soba noodles, pancake mixes, and kasha, the potential for including buckwheat in other products (e.g. pasta, ready to eat cereals, extruded snacks, and specialty breads) is tremendous. The nutraceutical characteristics of buckwheat offer substantial dietary benefits to people with high blood pressure and other ailments.

There is a wealth of published information on producution, processing, and utilization of buckwheat. The international buckwheat research symposiums have been very beneficial in presenting current advances in buckwheat research. The formation of the North American Buckwheat Consortium further demonstrates a committment from both the public and private sector to advance the utilization of this crop.

In summary, buckwheat is: (1) relatively easy to produce; (2) readily available; (3) very nutritious; (4) complementary to cereal grains; (5) functional in many food products; and (6) underutilized. Continued integration of research across production, processing, and marketing levels will greatly assist in developing the future for buckwheat.

REFERENCES

Auld, D.L., R.L. Mahler, and K.D. Kephart. 1986. Production of buckwheat in northern Idaho. Current Inf. Series 780, Univ. Idaho Coop. Ext. Serv., Moscow.

Berglund, D.R. and A.A. Schneiter. 1992. Buckwheat production. A-687 (revised). North Dakota State Univ. Ext. Serv., Nov. 1992, Fargo.

Borghi, B. 1995. Buckwheat in Italy, state of the art and preliminary results. In: Current advances in buckwheat research, Vol. I. Proc. 6th Int. Symp. Buckwheat, Aug. 26–29, 1995, Shinshu Univ., Nagano, Japan.

Buckwheat diet. AGWEEK News, Aug. 14, 1989.

Campbell, C.G. 1995. Inter-specific hybridization in the genus *Fagopyrum*. In: Current advances in buckwheat research, Vol. I. Proc. 6th Int. Symp. Buckwheat, Aug. 26–29, 1995, Shinshu Univ., Nagano, Japan.

Campbell, C.G. and G.H. Gubbels. 1986. Growing buckwheat. Tech. Bul. 1986-7E, Ag. Canada Res. Sta., Morden, Manitoba.

DeFelice, S.L. 1994. Food companies must pursue nutraceutical R & D—now! Food Eng. Dec. 1994.

Edwardson, S.E. 1989. Estimated cost of production budgets for selected small grain and specialty crops in southwest North Dakota. CMTR No. E-1, March 1989, MINN-DAK Growers Ltd., Grand Forks, ND.

Edwardson, S.E. 1990. Estimated production costs for selected North Dakota specialty crops: production system reference guide. CMTR E-5, Feb. 1990, MINN-DAK Growers Ltd., Grand Forks, ND.

Edwardson, S.E. 1991a. Estimated crop budgets for selected North Dakota specialty crops for 1991: production system reference guide. CMTR E-10, March 1991, MINN-DAK Growers Ltd., Grand Forks, ND.

Edwardson, S.E. 1991b. Potential opportunities for enhanced commercialization of MINN-DAK buckwheat. CMTR RP-6, May 1991, MINN-DAK Growers Ltd., Grand Forks, ND.

Edwardson, S.E. 1992a. Crop enterprise budgets for 1992. Summarized production costs for selected small grain and specialty crops in North Dakota: Production system reference guide. CMTR E-11, Feb. 1992, MINN-DAK Growers Ltd., Grand Forks, ND.

Edwardson, S.E. 1992b. Domestic and international market assessment of raw and processed buckwheat. Phase I progress report. CMTR MR-1, March 1992, MINN-DAK Growers Ltd., Grand Forks, ND.

Edwardson, S.E. 1993a. Crop production costs—1993. Crop enterprise budgets for selected small grain and specialty crops in North Dakota. CMTR E-12, Feb. 1993, MINN-DAK Growers Ltd., Grand Forks, ND.

Edwardson, S.E. 1993b. Revitalization of buckwheat market research for value added processing. CMTR MR-4, Dec. 1993, MINN-DAK Growers Ltd., Grand Forks, ND.

Edwardson, S.E. 1994. A strategy for marketing value added buckwheat. MR-6, MINN-DAK Growers Ltd., Grand Forks, ND.

Edwardson, S.E. 1995a. Advancing the utilization of buckwheat. A report on market research for processed buckwheat. CMTR MR-8, March 1995, MINN-DAK Growers Ltd., Grand Forks, ND.

Edwardson, S.E. 1995b. Crop production costs - 1995. Crop enterprise budgets for selected small grain and specialty crops in North Dakota. CMTR E-14, March 1995. MINN-DAK Growers Ltd., Grand Forks, ND.

Edwardson, S.E. 1995c. A strategic alliance for integrated buckwheat research. Alternative crop production research, a progress report, April 1995, North Dakota State Univ., Fargo.

Edwardson, S.E. 1995d. Using growing degree days to estimate optimum windrowing time in buckwheat. Vol. II. Proc. 6th Int. Symp. Buckwheat, Aug. 26–29, 1995, Shinshu Univ., Nagano, Japan.

Gorski, T. 1986. Buckwheat yield dependency on climatic conditions. In: Buckwheat research 1986. Proc. 3rd Int. Symp. on Buckwheat, Pulawy, Poland, July 7–12 1986, p. 169–179.

Gubbels, G.H. and C.G. Campbell. 1984. Visual indicators for determining the optimum windrowing time for buckwheat. Can. J. Plant Sci. 65:207–209.

Jiang, H., M.J. Klag, P.K. Whelton, J.P. Mo, J.Y. Chen, M.C. Qian, P.S. Mo, and G.Q. He. 1995. Oats and buckwheat intakes and cardiovascular disease risk factors in an ethnic minority in China. Am. J. Clin. Nutr. 61:366–372.

Joshi, B.D. and R.S. Paroda. 1991. Buckwheat in India. National Bureau of Plant Genetic Resources, Regional Station, Phagli, Shimla, India.

Marshall, H.G. 1969. Description and culture of buckwheat. Bul. 754, February 1969, Pennsylvania State Univ., College of Agriculture, Ag. Exp. Sta., University Park.

Marshall, H.G. and Y. Pomeranz. 1982. Buckwheat: description, breeding, production, and utilization. Adv. Cereal Sci. Tech. 5:157–210.

Matano, T. and A. Ujihara. 1995. Current advances in buckwheat research. Proc. 6th Int. Symp. on Buckwheat, Aug. 26–29, 1995, Shinshu Univ., Nagano, Japan.

Obendorf, R.L., D.P. Taylor, and J. Slawinska. 1991. Seed set and cessation of seed growth in buckwheat. Annual report, Dept. of Soil, Crop, and Atmospheric Science, Cornell Univ., Ithaca, NY. Research sponsored in part by MINN-DAK Growers Ltd., Grand Forks, ND.

Onishi, O. 1995. Discovery of new *Fagopyrum* species and its implication for the studies of evolution of *Fagopyrum* and of the origin of cultivated buckwheat. Vol. I. Proc. 6th Int. Symp. Buckwheat, Aug. 26–29, 1995, Shinshu Univ., Nagano, Japan.

Oplinger, E.S., E.A. Oelke, M.A. Brinkman, and K.A. Kelling. 1989. Buckwheat. In: Alternative field crops manual, Univ. Wisconsin Ext. Serv., Madison.

Peterson, H.A., S.E. Edwardson, and M.D. Bohn. 1992. International and domestic assessment of buckwheat processing and market expansion through cooperative market research. CMTR MR-3, June 1992, MINN-DAK Growers Ltd., Grand Forks, ND.

Pomeranz, Y. 1983. Buckwheat: Structure, composition, and utilization. Crit. Rev. Food Sci. Nutr. 19(3):213–250.

Robinson, R.G. 1980. The buckwheat crop in Minnesota. Agr. Exp. Sta. Bul. 539, Univ. Minnesota, St. Paul.

Smith, H.A., C.H. Rust, and D. Baldridge. 1989. Buckwheat: a potential Montana specialty crop. MONTGUIDE MT 8909, Montana State Univ. Ext. Serv., Bozeman.
Stiles, D.G. 1982. Commercial buckwheat production in Maine. Coop. Ext. Serv., Univ. Maine at Orono.
Sure, B. 1955. Nutritive value of proteins in buckwheat and their role as supplements to proteins in cereal grains. J. Ag. Food Chem. 3:793–795.
Swenson, A. and D. Aakre. 1992. Estimated 1992 crop budgets, southwest North Dakota. Farm Management Planning Guide, Section VI, region 4, Dec. 1992, North Dakota State Univ. Ext. Serv., Fargo.
Swenson, A. and D. Aakre. 1993. Estimated 1993 crop budgets, southwest North Dakota. Farm Management Planning Guide, Section VI, region 4, December 1993, North Dakota State Univ. Ext. Serv., Fargo.
Swenson, A. and D. Aakre. 1994. Estimated 1994 crop budgets, southwest North Dakota. Farm Management lanning Guide, Section VI, region 4, December 1994, North Dakota State Univ. Ext. Serv., Fargo.
Swenson, A. and D. Aakre. 1995. Estimated 1995 crop budgets, southwest North Dakota. Farm Management Planning Guide, Section VI, region 4, December 1995, North Dakota State Univ. Ext. Serv., Fargo.
Thacker, P.A., D.M. Anderson, and J.P. Bowland. 1984. Buckwheat as a potential feed ingredient for use in pig diets. Pig News Inform. 5(2):77–81.
Udesky, J. 1992. The book of soba. Harper & Row, New York.
Ujihara, A. 1995. Historical changes of cropping area and yield of common buckwheat during over a century in Japan. Vol. III, Proc. 6th Int. Symp. on Buckwheat, Aug. 26–29, 1995, Shinshu Univ., Nagano, Japan.
Watt, B.K. and A.L. Merrill. 1963. Composition of foods. USDA Agr. Handb. 8. Washington, DC.

Amaranth: New Crop Opportunity

Robert L. Myers

Grain amaranth (*Amaranthus* spp.) is a broadleaf pseudocereal (Fig. 1), used by the Aztecs and earlier societies, which began the slow process of coming into modern use and practice in the 1970s. Researcher and consumer interest grew in the 1980s, and formed the basis of continued development of the crop in the 1990s. Consumer interest in amaranth as a food ingredient has been primarily due to the positive nutritional charac-

Fig. 1. Grain amaranth research plots at Colombia, Missouri. Grain heads, or inflorescences, are fully developed at this stage of the season, in late Aug.

teristics of the grain, although some of the interest is undoubtedly due to the colorful history of the crop as a "lost" grain. Market demand for amaranth has fluctuated over the last decade, but there has been steady use of the crop for breakfast cereals, snack foods, and more recently, in mass produced multi-grain bread products. Most typically, amaranth products are sold in the health food sections of grocery stores, in specialty food stores, or through direct marketing. Amaranth to-date has only appeared in "mainstream" products when used as a minor component of multigrain foods.

The U.S. has been the leading commercial producer of grain amaranth in recent years, although production has been less than 2000 ha annually since U.S. farmers started growing the crop in the early 1980s. Most U.S. production has been in the upper Midwest and Great Plains, particularly western Nebraska, with widely scattered fields in other parts of the U.S. Amaranth grain species are widely adapted, and have good potential for moisture limited areas. Other species of amaranth have been domesticated primarily for leafy vegetable or ornamental use.

Amaranth's development in the U.S. has benefitted by the formation of private companies that have focused exclusively or in part on amaranth products, and have done much of the work to build public awareness of the crop. The Amaranth Institute, a small non-profit organization of scientists, growers, and agribusiness, has also supported development of the crop through information exchange, promotional activities, and annual meetings.

This paper provides an overview of some of the history, characteristics, uses, and market dynamics for amaranth, and includes a summary of 12 field studies and 2 controlled environment studies with amaranth carried out in Missouri by the author during the period 1990–1994. Information concerning amaranth germplasm and amaranth grain characteristics can be found in previous review papers on amaranth (Kauffman 1992a, b; Stallknecht and Schulz-Schaeffer 1993).

HISTORY

Sauer (1993) evaluated the evidence on amaranth's domestication, and concluded that amaranth came into use as a grain at least 6000 years ago in Central America. The oldest seeds found are those of *Amaranthus cruentus*. Another species, *A. hypochondriacus*, was in use by at least 1500 years ago, and later became the species used by the Aztecs. Although the earliest seeds of *A. cruentus* and *A. hypochondriacus* were both found in caves at Tehuacan, Mexico, Sauer observes they could easily have been domesticated far away, and possibly much earlier than those finds indicate. The origin of the third grain-type species, *A. caudatus,* is not well understood, but Sauer believes the species was first grown as a grain in the Andean mountain valleys of South America. Amaranth was used as a grain over a wide region, from at least the southwestern U.S., down through Central and South America to Argentina. Amaranth may have had some use as a grain by Native Americans in other parts of the U.S. Although there is no recorded history of its use, *A. hypochondriacus* seeds found in the Ozarks of the south central U.S. have been dated to 1100 AD.

The most significant historical use of amaranth as a grain crop was in central Mexico during the Aztec civilization (the Aztec name for amaranth was *huahtli*). Amaranth was both an important food grain for the Aztecs, and a part of their religious practices. Annual grain tributes of amaranth to the Aztec emperor were roughly equal to corn tributes. The reasons for decline in amaranth production following Spanish conquest of the Aztecs in the 1500s are not well understood. Amaranth was significantly used in Aztec religious ceremonies, and some have speculated that the Spanish conquistadors discouraged production of amaranth as part of their overall efforts to suppress the Aztec culture and religion.

Although the production of amaranth in Latin America diminished dramatically subsequent to Spanish rule, the positive attributes of the crop led to its adoption in other areas of the world. By the 1700s, amaranth had spread throughout Europe for use as a herb and ornamental. In the late 1800s, amaranth was reportedly being grown in mountain valleys of Nepal and parts of east Africa. During the 20th century it has been grown in China, India, Africa, and Europe, as well as North and South America.

Although the U.S. has been the leading producer of grain amaranth used in retail food products, the largest production area in the last decade is believed to have been in China (the Chinese amaranth production is reportedly based on cultivars developed by the Rodale Research Center in Pennsylvania). The main Chinese use of amaranth is reportedly to feed the forage to hogs, rather than harvesting the grain.

USES

Although grain amaranth has a variety of potential uses, its only commercial market in the U.S. to date has been for food products. A few small companies have been the primary buyers and processors of amaranth, including Arrowhead Mills (Texas), Health Valley (California), and American Amaranth, now reorganized as Amaranth Resources (Minnesota). These companies sell amaranth both in processed foods, such as cereals, breakfast bars, crackers, and cookies, and also as amaranth flour and whole grain. Beginning in the early 1990s, amaranth also began to be purchased by larger food processors interested in developing multigrain products. Although this trend has led to a broader use of amaranth, market demand has not risen dramatically, due to the small fraction of amaranth used in many commercial multi-grain products, primarily sandwich breads and cereals.

The small but gradually growing market for amaranth as a food grain is based on its nutritional characteristics, and to some extent, its historical interest as the "lost crop of the 'mystic' Aztecs." The nutritional characteristics are indeed positive, with protein content ranging from 12%–17%, and a well balanced amino acid profile, including a relatively high amount of lysine. Amaranth is also high in fiber, and low in saturated fat. The amaranth grain can be popped or flaked, and works well in mixes with flours of other grains, including for extrusion processing.

Although amaranth leaves have been used in both human and livestock diets, little is know about the potential of grain amaranth species for livestock use. When the leaves, stem, and head are used for forage, the product will range from 15% to 24% protein. Palatability is not well known. Within the available amaranth germplasm are genotypes that show potential for significant forage production, but more research in this area is needed.

Amaranth, like most grains, has potential for use in industrial products. The oil fraction of the grain is unusually high in squalene, a chemical that sells for thousands of dollars per ton. However, the percent of squalene in the grain is still small, and may not be economical to extract. Greater promise lies with the starch fraction of the grain. Amaranth, like quinoa (*Chenopodium quinoa*), has very small, micro-crystalline starch granules, about one-tenth the diameter of maize (*Zea maize*) starch. The physical characteristics of the starch grains have been cited as being of potential value for both industrial and food product uses, though none has been commercialized to date.

YIELD AND ECONOMIC RETURN

Yields of grain amaranth, like any new crop, are highly variable. Good yields, in relative terms, would be considered to be 1000 kg/ha or better, and these have been achieved in research plots in a number of states throughout the United States (Myers 1994). With a yield of 1000 kg/ha, a farmer with market contracts can easily make a profit, given the typical market price of $0.90 to $1.00/kg for amaranth. However, it is not uncommon for the crop to be a complete failure (little or no stand establishment), or a near complete loss at harvest due to shattering and/or lodging. On the positive side, replicated yields of 3000 kg/ha have been achieved in small research test plots in a few locations. Genetic potential for even higher yields seems to exist, given the productivity of some individual plants. A yield of 3000 kg/ha would be dismal compared to maize, or even sorghum, but for a food crop with a comparable protein level such as wheat, yields of 3000 kg/ha would be respectable, although not outstanding.

Amaranth costs of production are modest, being similar to sorghum. Costs are kept low particularly because fertilizer needs are minimal, and registered pesticides are not available. Although seeding rates are low (1 to 2 kg/ha), seeding costs can be above average (up to $20/acre, or $50/ha) if certified seed is used. Farmers use either certified seed, or, for lower cost, bin run seed. Planting and cultivation costs are equivalent to other grain crops since the same equipment can be used, with some modifications to a planter to handle the small seed size. The other cost that can be high is in the harvest and post-harvest phase. While properly equipped and adjusted grain combines can effectively harvest amaranth, such modifications do add to production costs, from a few hundred to a few thousand dollars. More notably, distance to market is a significant factor for most amaranth producers, since there are only a few amaranth processors and delivery points in the United States (outside of direct marketing). Most growers are faced with shipping the crop hundreds of kilometers by truck, which can lead to a sizable shipping cost for the grain, as opposed to hauling the grain a few kilometers to the local grain elevator. Growers may also experience higher cleaning and handling costs with amaranth than with other grain crops. Altogether, cost of amaranth production per hectare can be expected to be in a range of ±20% that of maize, with the variability mainly coming from postharvest handling and shipping costs.

The factor that has buoyed the success of amaranth to-date is the relatively high price of the grain ($0.90 to $1.00/kg), which has been up to 10× higher than maize and up to 5× higher than wheat on a per weight basis. Unfortunately, this high price received by farmers has also been a major factor in limiting use of the grain to "health" food products that sell for a premium, or to use as a tiny fraction of a multi-grain mainstream product. Discussions by the author with one major food processor and retailer indicated that for amaranth to gain broader use in the food marketplace, it would have to be sold for about 1/3 to 1/2 of its current price (still a premium compared to other grains), be available in adequate supply (i.e., more production area), and have broader consumer recognition and demand. Certainly amaranth is subject to the same price fluctuations as most grains, when supply overwhelms demand. If amaranth is to ultimately have a much larger and steadier market demand, it will have to follow the path of higher yields and lower prices.

AGRONOMIC RESEARCH

While the United States has led research efforts during the past two decades, grain amaranth has also been receiving research attention in China, India, and Mexico, and by a few scientists in other countries. Most United States scientific work with amaranth during the last two decades has been on production methods, grain characterization and use, and breeding improved cultivars. Although breeding efforts with amaranth in the United States have diminished in recent years, production and utilization research have continued. Cultivar testing has been conducted on a regional basis since the mid-1980s, at over a dozen locations in the United States and Canada. Production research has been conducted in several states, especially Minnesota, North Dakota, and Missouri. Multiple field studies have also been conducted in Colorado, Iowa, Montana, Nebraska, and Pennsylvania. Most of the production research has focused on practical questions such as seeding rates, planting dates, row widths, and fertilizer response. Some evaluation of water use by amaranth has also been conducted in the field. Both insect and disease pests of amaranth have been evaluated.

The predominant source of amaranth cultivars used in the past decade in the United States was the Rodale Research Center in Kutztown, Pennsylvania. Initial breeding was done by Charles Kauffman, with followup work by Leon Weber (Kauffman and Weber 1988). Weber served as the initial distributer of lines for regional testing. Coordination of the regional test was later carried out by Dan Putnam (1990–91), then at the Univ. of Minnesota, followed by R.L. Myers (1992–94) at the Univ. of Missouri. The current regional test coordinator is David Baltensperger at the Univ. of Nebraska. A summary of the regional variety trial results for the period 1985 to 1993 was published in the 1994 annual issue of *Legacy,* newsletter of the Amaranth Institute (Myers 1994).

Early utilization research with amaranth focused on the grain's nutritional characteristics (Teutonico and Knorr 1985; Breene 1991; Bressani et al. 1992). More recently, work has been conducted on the potential industrial uses of amaranth, and on potential health benefits from amaranth in the diet. The industrial use research has focused on the starch or squalene content of the seeds. The health implications of consuming naturally occurring tocotrienols in amaranth has received preliminary evaluation. Recent food utilization research has included extrusion processing of the grain. Only a few studies have evaluated amaranth cultivars for forage use, and no substantive research has been done on the potential of amaranth grain as a livestock feed.

Production Management

Optimum production practices with amaranth are beginning to be better understood, having been researched in a few different states, particularly North Dakota, Minnesota, and Missouri. What follows are summaries of research studies conducted at Univ. of Missouri during the period 1991–1995, by the author and project staff. Unless otherwise specified, the following management practices were used. Amaranth plots were typically planted in late May or early June, in 76 cm (30 inch) wide rows, at a seeding rate of 2.2 kg/ha. Planting depth was usually 1–2 cm, and planting was done with a 4-row Winter Steiger Precision Plot Planter. Plot size was normally 3 m × 9 m, with each treatment combination replicated four times. Only the middle two rows of the 4-row wide plots were harvested, and plot ends were also trimmed before sampling, all to eliminate border effects. Plots were mechanically harvested with a Winter Steiger Elite Nurserymaster combine. Grain was cleaned, weighed and moisture tested, with recorded yields adjusted to a standard of 10% moisture. Typical data collected besides yield were mature plant height, lodging score, vigor ratings during the season, stand establishment ratings, and as needed, plant population and seed weight at maturity. Occurrence of disease, insects, or other management problems were noted. Amaranth test plots normally were fertilized with 90–100 kg N/ha, with P and K applied according to soil test recommendations for sorghum. Weed control was through pre-plant tillage, row cultivation, hand weeding, and on occasion, Poast herbicide for grass weed control (this is a post-emerge broadcast herbicide legal for research plots, but not labeled for use in commercial fields).

Nitrogen Rate

The effect of nitrogen fertilizer rate on amaranth development and yield was evaluated at two central Missouri sites (Columbia and New Franklin) in 1991 and 1992. The lines used in this study were D136-1, K266, and 'Plainsman'. Rates of 0, 45, 90, 130, and 180 kg N/ha, broadcast preplant as ammonium nitrate, were used. Results indicated that only 45 to 90 kg N/ha were required to reach maximum yield across cultivars, but varieties differed in responsiveness. For the soils tested, it was clear that amaranth does not need as much N fertilizer as maize, or even sorghum. The responsiveness of amaranth to N in Missouri was very comparable to similar research by Elberhri et al. (1993) in Minnesota.

Plant height and lodging were increased by addition of N fertilizer, factors which negatively affected harvesting. Mature plant population and seed weight were unaffected by nitrogen fertilizer rate. Yield compensation was due to higher seed numbers per plant, rather than changes in seed weight, which is a trait that usually held true in other studies on amaranth management (there are differences in seed size between varieties, however).

Seeding Rate

Amaranth response to seeding rate was tested at one central Missouri site (Columbia) in 1991, 1992, and 1993. The lines used in this study were D136-1, K266, and 'Plainsman'. Seeding rates of 0.28, 0.55, 1.1, 2.2, and 4.4 kg/ha were planted (2.2 kg/ha is the most common seeding rate used), all in 76 cm rows. Grain yield was not different for any of the seeding rates. This was due to a combination of two factors, which were the ability of the plants to compensate in seed production per plant at reduced population levels, and also self-thinning of the amaranth stands when planted densely. The lack of response to seeding rates that varied by 16-fold from low to high rate is rather remarkable, given that most crops show a yield response when seeding rate is varied by a factor of 2, and some respond when rates vary by as little as 10% (i.e., maize).

Planting Date

The effect of planting date was studied at one central Missouri site (Columbia) in 1991, 1992, and 1994. The lines used in this study were D136-1, K266, and 'Plainsman'. Three to four planting dates were used each year with about 10–14 days separating each planting (rainfall and soil moisture conditions dictated the dates possible to plant on). Mid-May to mid-June plantings were not different in yield, but planting in early July reduced yield 10%–60%, depending on the cultivar and year of the test. Other studies evaluating amaranth as a double crop (see below) confirmed that planting in early July (after either fallow, or a winter crop like wheat or canola is harvested in late June) reduced amaranth yield. However, such yield reductions also occur with soybeans, used as a double crop on a half-million acres in Missouri. In two studies that did not directly evaluate planting date, amaranth was seeded in the first week of May in central Missouri, with successful stand establishment and good yields. Overall, amaranth has a fairly wide planting window in Missouri, especially if time of maturity and harvest is not a concern. Many northern growers and researchers working with amaranth have tried to time planting so that the crop matures shortly before frost, with a frost serving to accelerate plant dry down and facilitate mechanical harvesting. Missouri tests over 5 planting seasons (1990–94) demonstrated for several lines that amaranth can be successfully harvested without frost, and the plants can dry down at maturity without excessive seed loss, provided no severe storms occur during dry down which exacerbate seed shatter or lodging.

Row Width

Most growers have used wide rows when planting amaranth to allow for mechanical weed control cultivation. However, there has certainly been success with crops such as soybeans in narrow row drilling (albeit in part due to the availability of postemerge herbicides for soybeans). In the early 1990s, no data was available on the response of amaranth to narrow row widths. The effect of row width was evaluated at the Columbia location in central Missouri in 1992 and 1993. Row spacings of 19, 38, and 76 cm were used (76 cm is the standard row width). Narrow row spacing (19 cm) provided good early season weed control, but excessive self competition led to reduced plant height, earlier flowering and maturity, and reduced yield. These responses also occurred, but in a reduced fashion, at the intermediate row width of 38 cm, as compared to the standard with of 76 cm. The excessive plant competition in the narrow row widths was not the result of a higher seeding rate, but reflected the fact that more plants seemed to fully develop in the narrow rows. The reduced self-thinning in narrow rows contrasts with the typical dynamic of wide rows where a smaller percentage of seedlings fully develop, due to self-thinning and in-row plant suppression. This observation is consistent with data reported by Henderson et al. (1993). It might be possible to find a reduced seeding rate that would work combine effectively with a narrow row spacing, but based on the results of these two years of data, narrow rows are not recommended when using a standard seeding rate of 2 kg/ha.

Crop Rotation

A study to evaluate amaranth and canola (*Brassica napus*) as alternative crops in various rotation combinations with maize (*Zea mays*), soybeans (*Glycine max*), and wheat (*Triticum aestivum*) was begun at Columbia, Missouri, in the fall of 1990. 'Plainsman' amaranth was placed in rotations both as a full season substitute for maize and soybeans, and a double crop substitute for soybeans following wheat. Amaranth was grown in continuous plantings for four years (the studied ended in 1994), in two year rotations with either maize or soybeans, in three year rotations with maize and soybeans, and in a five crop, four year rotation that involved all of the crops above. Amaranth was tested as a double crop after both wheat and canola, in comparison with soybeans (see also double crop study below).

The study had been intended to go for 8 years, but it was terminated after 4 years due to funding cuts, which limited the conclusions that can be drawn. Still, there were some clear findings. For all of the rotations, amaranth had no noticeable allelopathic effects on the following crop, and amaranth crop residue did not present a physical problem for planting and stand establishment of the following crop. Residue levels from amaranth after a winter of decomposition in Missouri are more than soybeans, but less than maize. There did not appear to be any disease or insect accumulation in amaranth grown continuously in this study, but plants were not systematically sampled for disease, nor were insect populations closely tracked. The most obvious problem occurring with continuously planted amaranth was volunteer plants from the previous season(s). Amaranth volunteers are readily controlled in maize and soybeans with tillage and herbicides, but the degree of volunteers in a new amaranth plot led to ragged stands, and also allowed pigweed (a weedy relative of grain amaranth) to build up. Mechanical cultivation to control volunteer grain amaranth and pigweed in a new stand helped, but did not deal with in-row problems. In this rotation study, amaranth planted as a double crop did not perform as well as in a later study that focused on double crop systems.

Double Cropping

From planting date studies, and limited double crop evaluation in the rotation study, it appeared that amaranth had potential as a double crop in Missouri. (In more northern states, amaranth has been strictly limited to use as a full season crop.) To evaluate more fully the potential of amaranth and other alternative double crops, a study was initiated at Columbia (central Missouri) in 1992–93, and expanded in 1993–94 to two additional sites: New Franklin (central Missouri), and Portageville (southeast Missouri) (Pullins 1995). In the fall, large blocks were seeded to canola, wheat, or tilled and left fallow (control). After harvest of the winter crops in mid to late June, the double crops planted were planted no-till, usually by July 1. 'Plainsman' amaranth was planted, along with buckwheat (*Fagopyrum esculentum*), pearl millet (*Pennisetum glaucum*), sunflower (*Helianthus annuus*), and soybean, the standard double crop in Missouri. A partial factorial design was used to also provide comparisons of till and no-till establishment for double crop amaranth, and to evaluate residue effects. Canola and wheat residue were removed from some plots, and applied to selected fallow plots prior to planting the amaranth.

At all three locations, amaranth was successfully established and harvested as a double crop. Amaranth plant height was reduced following winter crops versus fallow, and establishment no-till into wheat residue was difficult, but possible, with the planting equipment available. In two out of the four site-years, amaranth yielded better after fallow than after wheat or canola. Amaranth yields after wheat versus canola were not different in any of the site-years. Double crop amaranth yields ranged from a low of 500–700 kg/ha at Portageville, where lodging was a problem, to a high of 1849 kg/ha following fallow at Columbia in 1993. In two of the four site-years, amaranth yielded 1200 kg/ha or better following wheat and canola, which would provide a substantial profit at current amaranth prices. The top yielding double crop was sunflower, with yields of 1500 to 3000 kg/ha, followed by soybeans, with 600 to 2100 kg/ha. However, amaranth would have provided the greatest profit in this study.

Intercropping

In 1991 and 1992, an intercrop study with amaranth was conducted at Columbia and New Franklin (Clark and Myers, 1994). Amaranth was planted in various row arrangements as an intercrop with cowpea (*Vigna unguiculata*). The amaranth was compared to pearl millet for intercrop success with cowpea. Row arrangements were alternating rows 38 cm apart, two row strips (on 76 cm spacing), and six row strips (also 76 cm). The cowpeas were planted approximately 10 days before amaranth and millet at each site to allow them to get the cowpea partially established before the faster growing amaranth and millet emerged. In a split application, plots were also treated with broadcast rates of 0, 56, and 112 kg N/ha to gain some information on N dynamics in the intercrop systems.

Amaranth intercropped with cowpea in alternate row fashion had a higher land equivalent ratio (better yield response) than cowpea-millet, but for other row arrangements the intercrop systems were similar. Millet generally outyielded amaranth: amaranth in 2-row strips yielded 973 to 1791 kg/ha, compared to 1750 to 3964 kg/ha for millet; when monocropped, amaranth yields were 1089 to 1591 kg/ha, versus 1849 to 3718 kg/ha for millet. When no nitrogen was applied, amaranth yields were higher in alternate row intercrop with cowpea than when grown as a monocrop, so N and/or other resource complementarity was occurring. An overall conclusion from the study was that amaranth and cowpea can be effectively intercropped, as a variation on the pearl millet and cowpea intercrop system common in parts of Africa.

Cover Crop Effects

Leguminous cover crops have been gaining attention for their ability to help control erosion, reduce weed pressure, and serve as a biological source of N. All these potential benefits are needed in amaranth production, especially since most of the market for amaranth is based on organic production methods. To evaluate the response of amaranth to cover crops, four common cover crops were planted: Austrian winter pea (*Pisum sativum*), crimson clover (*Trifolium incarnatum*), hairy vetch (*Vicia villosa*), and rye (*Secale cereale*), the last as a non-legume comparison (Fisk 1993). Control plots were kept fallow. Plots received either 0, 45, or 90 kg N/ha, to partially evaluate the N effect of the cover crop. Soil inorganic N was also sampled at three points during the growing season. Austrian winter pea and control (fallow) plots were further split into till and no-till treatments. Cover crops were controlled with paraquat, then residue was mowed down before planting amaranth in early June. Three locations were planted in 1992, the only year of the test, but only two of those were ultimately used for data collection. Those sites, Columbia, and Novelty (northeast Missouri), had much different cover crop growth, since Novelty had a shortage of rain during late spring. Cover crop growth in Columbia was good, but cover crops at Novelty had only modest biomass under the dry conditions.

Overall, Austrian winter pea outperformed the other cover crops, boosting amaranth yields an average of two-fold versus the control plots (from about 600 to 1200 kg/ha). Pea produced the most biomass, and had the highest level of inorganic N during July sampling of the upper soil profile. Amaranth grown after pea and hairy vetch were taller and had higher vigor ratings compared to rye and control plots. On the negative side, lodging was increased in plots following pea. Rye had the expected effect of reducing amaranth vigor compared to control plots. This reduction in vigor was partly, but not completely, offset by the addition of N fertilizer. Surprisingly, amaranth following rye had much higher stand populations in late season than those in control plots or following legumes. It is possible that the reduction in vigor caused by rye also reduced self-competition among amaranth plans, in turn reducing the amount of self-thinning of plants that normally occurs in an amaranth stand.

This study was also the first attempt at planting amaranth no-till in Missouri, an approach later used in double crop studies as well. In this instance, where amaranth was planted following control of growing cover crops by chemical burn down and mowing, no-till establishment worked reasonably well. Seed zone moisture for the amaranth can be helped or hindered by the use of cover crops, depending on rainfall timing preceding amaranth planting.

Control of *Lygus* with Alfalfa Strips

The most common, and usually most severe, insect pest of grain amaranth is *Lygus lineolaris*, or tarnished plant bug. This small, brown, sucking insect, slightly smaller than a lady bug, is a general feeder that has many hosts and feeds on a variety of plant parts, but is particularly attracted to the amaranth inflorescence (Wilson 1989). The insect does little damage to amaranth leaves or stems, but can reduce seed weight by up to 70%, and damaged as much as 20%–30% of seeds in Missouri test plots. Worse, the insects damage flowers, and often the vascular tissues feeding a cluster of flowers, preventing the development of sometimes large numbers of flowers into seeds. Pyrethrin based products can provide at least some control of *Lygus*, but are not normally labeled in a way that allows for use on amaranth.

One non-chemical method of controlling *Lygus* in cotton has been to plant strips of alfalfa through the cotton fields, using selective mowing to keep parts of the alfalfa in bloom, and thus attracting *Lygus*. This system of alfalfa strips was duplicated in 1991 and 1992 in amaranth test plots to evaluate its effectiveness at controlling *Lygus* in amaranth (Clark and Myers 1995). 'Plainsman' amaranth was planted in plots 18 m wide, alternating with 6 m wide strips of alfalfa in a replicated design. Regular sweep net insect surveys were made in the center and edge of alfalfa strips, and at varying distances from the alfalfa into the amaranth strips. Insect counts were made at several times during the season to determine the buildup of *Lygus* in the two crop strips.

Alfalfa did attract a high percentage of the *Lygus* relative to amaranth early in the season. However, once amaranth began flowering, the *Lygus* populations built-up in the amaranth instead, and it was clear the alfalfa did not provide effective control.

PLANT GROWTH AND DEVELOPMENT

Although careful observation has revealed many traits of amaranth's growth and development, there is still much to be learned. In general, a field of amaranth develops the way one would expect for a semi-domesticated crop with little modern plant breeding. Currently available cultivars are highly variable, leading to differences in rate of emergence and growth among neighboring plants, differences in inflorescence morphology and development, and differences in maturity. Amaranth seeds usually germinate quite readily, and when planted shallow (e.g. 1 cm) will emerge within 3 to 4 days in warm soil (≥20°C). However, the seedlings are weak, and a crusting soil, heavy rains, or blowing soil particles can easily reduce or prevent stand establishment. During initial development, some plants develop much more rapidly, and vigor differences are accentuated by plant-to-plant competition within a row. Although a mature amaranth canopy is 1 to 2 m in height, individual plants within a row may be much smaller, down to 5 to 10 cm, with tiny grain heads.

When planted in early June in Missouri, inflorescences will begin to emerge by late July, in a pattern that is daylength sensitive for the available cultivars. Anthesis begins to occur after an inflorescence is only partially formed, and continues for parts of the inflorescence until the grain "head" reaches full size (typically 20 to 30 cm top to bottom). Seed development is thus not simultaneous within a head. Seeds that form early begin shattering well before all the seeds on a head are mature, but the seed heads are compact enough to hold most shattered seed in the head, provided strong wind or rain does not occur. (Note: nonshattering germplasm has been identified by D. Brenner, Plant Introduction Station, Iowa State Univ., Ames.) Seeds change in appearance from a glossy, or translucent appearance, to a dull, or opaque appearance as they mature (see research discussion on seed development and maturation).

In states such as Nebraska, North Dakota, and Minnesota, where most amaranth has been commercially grown, amaranth is normally killed by frost before it has a chance to naturally senesce. Plants killed by frost can be harvested a week or so later, depending on drying conditions, which can improve the opportunity to combine the crop before too many seed fall to the ground. However, in more southern areas, such as Missouri, the crop will normally senesce and dry down in Sept., before the first hard frost (in central Missouri, average frost date is end of Oct.). With current cultivars, the plants will turn a medium brown in the grain head and leaves during senescence, then drop almost all leaves. Once senescence begins, stalk strength declines, and plants are more susceptible to stem breakage from high winds.

Because relatively little has been known about many aspects of amaranth growth and development, a series of studies at Univ. of Missouri over a five year period sought to evaluate amaranth's germination requirements, seedling vigor, response to moisture stress, and seed development and maturation. These studies are briefly described below.

Germination

Seed germination tests were conducted in a controlled environment germinator for a variety of light and temperature conditions. Seeds exposed to a 16-h light, 8-h dark cycle, generally had better germination than seeds kept continuously in the dark. To evaluate temperature response, a series of trials was conducted and duplicated at the following day/night temperature regimes: 27°/24°, 24°/21°, 21°/18°, 18°/15°, 15°/12°, and 12°/9°C. Germination percentage and speed of germination declined when temperature was reduced to the 21°/18°C regime, and germination percentage was greatly reduced at 18°/15°C or below. Amaranth seeds normally sprout within 24 hours at warmer temperatures, and the hypocotyl elongates enough within 72 hours for emergence from a 1 cm depth, which corresponds to typical field emergence of 3–4 days at shallow (1–1.5 cm) planting depths.

Seedling Vigor

It has been repeatedly noted that amaranth seedlings grow at different rates, and that in mature stands, plants vary from 5 to 200 cm in height. It has also seemed that some seedlings are eliminated through self-thinning. A simple study was designed In an attempt to determine if these vigor differences are genetic, competition induced, or both. Seeds were planted in bedding plant flats, with each unit approximately 3 × 5 cm across and 5 cm deep, with about 1 cm spacing between adjacent units in a flat. Either 1 or 3 seeds was planted in each unit, with some flats having all units seeded, and others having only alternate units seeded. In this way a simple comparison of seed competition could be done, both with seedling roots interlacing within a unit, and with physical barriers between individual seedlings. Also, above ground competition could be compared based on whether adjacent units were planted or empty. Results indicated both genetic variability for seed vigor, and competition effects when plants were as little as 3 cm tall. The taller the seedlings grew (they were observed until the tallest were about 10 cm in height), the more pronounced the competition effects became relative to the genetic differences. Number of leaves was slightly reduced in plants that were being outcompeted. What is unusual about amaranth, compared to more established grain crops, is the inconsistent pattern of seedling vigor. Most grain crops, when planted densely, would have all plants more or less equally reduced in vigor, but with amaranth, some plants just "outrun" the others, suggesting amaranth cultivars are not homogeneous.

Indicator of Seed Maturity

In most research plots and commercial fields of amaranth grown in the 1980s, harvest timing was simply tied to the first hard frost that killed the plants, allowing 5–7 days for dry down, and then combining the crop. In a more southern region such as Missouri, however, amaranth cultivars such as 'Plainsman' often dry down before frost, making timing of harvest more important. Although mechanical harvesting is easiest when plants are almost completely dry, it is likely seed shattering will be significant by that time point. Also, the dry down period for grain amaranth in Missouri was sometimes up to 3–4 weeks. It was observed in 1990 field plots that amaranth seeds seemed to turn from translucent to opaque as the plant matured. To evaluate if the change in seed appearance corresponded to physiological maturity, seed samples were collected in a fashion intended to evaluate the relationship of seed coat appearance with their seed maturity.

In 1991, heads were collected and threshed from replicated plots near Columbia, Missouri. Comparisons were made of seed appearance (translucent vs. opaque), seed weight (fresh and dry), and seed germination, for three cultivars ('Plainsman', D136, and K256), two planting dates (June 3 and 21), and three sampling dates. The samples were taken at approximately two week intervals. In 1992, a more limited sampling was done of 'Plainsman', D136, and 'Amont'.

Results indicated the percent of opaque seeds increases as the plant ages. The dry weight of the opaque seeds was higher than translucent seeds, and opaque seeds were generally lower in percent seed moisture at time of sampling. Based on this data, it was clear that opaqueness in amaranth seed is a suitable indicator of maturity. However, opaqueness was not an exact measure of physiological maturity (physiological maturity is defined as the point when seeds reach maximum dry weight, and generally coincides with the seeds being developed enough to germinate). Opaque seeds did have a higher percentage of germination than translucent seeds, but some seeds rated as translucent were still able to germinate.

Other characteristics of opaque seeds were noted which distinguished them from translucent seeds. Opaque seeds are harder and more brittle. It was also noted, under microscopic examination of seed cross sections, that opaque seeds have a perisperm with a white, fine granular appearance inside, as opposed to the glossy, more globular perisperm appearance of translucent seeds.

Response to Moisture Stress

Amaranth has often been described in the literature as a drought tolerant crop, but the veracity of this claim had not been validated by any scientific tests. Two field studies involving comparative analysis of multiple grain crops, both alternative and traditional, provided some insights into the relative drought tolerance of amaranth, and the crop's physiological response to moisture limited conditions. One study evaluated several alternative and traditional grain crops in soils with varying depths of flood-deposited sand, while the other study evaluated the same set of crops in constructed field plots of varying soil depth.

Response to sandy soils. Eight different grain plots were planted into replicated plots in the Missouri River bottoms near Jefferson City in 1994. The intent of this field test was to evaluate the relative performance of traditional and alternative grain crops on newly created sandy soils versus 'regular' alluvial soil in the river valley. The sand deposits were the result of record flooding along the Missouri river in 1993, which covered approximately 150,000 ha of bottomland fields with 15 cm or more of sand. A site was selected that had sand ranging in depth from 30 cm to 120+ cm. Two adjacent field areas were used, one with "shallow" sand (30–45 cm) and one with "deep" sand (>75 cm). A nearby plot of ground with no sand deposits was chosen as a check plot area, and designated as "regular" soil (silty loam topsoil, formed as floodplain alluvium).

The eight crops chosen were soybeans and sorghum as traditional crops, and sunflowers, pearl millet, cowpeas, mung beans (*Vigna radiata*), amaranth, and sesame (*Sesamum indicum*) as alternative crops. Crops were selected on the basis of their suitability for planting in early summer, and expected tolerance for moisture limited conditions created by sandy soil conditions.

Stand establishment of all eight crops was good in the regular soil plot area. In the sandy plots, establishment was delayed by dry conditions until the first rainfall, which came a week after planting. Only amaranth germinated prior to rainfall, which was rather remarkable considering the very low moisture content in the seed zone of the sand. After rainfall, emergence was slow for some crops due to inconsistent planting depth in the sand. Ultimately, good stands were obtained of all crops except sunflowers and amaranth, which had partial stands. The partial stands of amaranth were probably due to the very dry seed zone conditions when the seed was planted, and began to germinate. Although some amaranth seedlings became established, other seeds probably only imbibed water, or barely sprouted, before running out of moisture. The other crop seeds which sat idle until the first rainfall a week after planting, had much more moisture available during establishment. The tendency of amaranth seeds to imbibe moisture and start germinating even under rather limited moisture conditions has been noted in regular field plots, and is somewhat of a problem, occasionally leading to ragged stands, where some plants emerge quickly, and others emerge later after rainfall, or not at all.

It was expected that tap-rooted crops like sunflower and amaranth would do well under the soil conditions of this study, where flood-deposited sand was layered over silt loam soil. However, both sunflower and amaranth tap-roots were found to go only as deep as the wetting front in the sand. Relatively modest rains of 1 cm or less would wet the sand down to about 10–15 cm. The tap-roots would go that deep, then turn sharply at a right angle (running parallel to the surface) as if they had hit an impermeable barrier. The relative advantage of the pearl millet and sorghum (grass) crops was their ability to send out a mass of fibrous roots laterally that could capture a larger portion of the rainfall that had occurred. The tap rooted crops (sunflowers and amaranth) would likely have been much more successful if early rainfall had wetted the sand all the way through to the underlying silt loam, allowing the tap-roots to reach that depth. It was noted that weedy amaranths (pigweed), established themselves quite successfully from rainfalls that occurred during the month preceding the test planting on June 14 of amaranth and the other grains. Pigweed was the dominant weed by far in the newly deposited sand areas, which overall were surprising free of weeds in the first year after flooding (weed seed probably floated off while the sand dropped out of flood waters as a heavy material; the sand was deepest near levy breaks).

Response to soil rooting depth. In 1994, a study evaluating response to rooting depth was planted at the Univ. of Missouri Agronomy Research Center, Columbia. The same 8 crops as in the sand test above, plus foxtail millet (*Setaria italica*), were planted into a unique constructed plot area that had been created in a previous field test. The site consisted of 6 × 60 m soil pits running parallel to each other, and separated by about 4 m. The pits had been dug out to varying depths and bottom-lined with a heavy grade, impermeable black plastic. On top of the plastic was laid a small drain tile. The pit was then filled with top soil to the surface level. Thus, the deepest pit was filled with about 94 cm of topsoil, which is certainly a better root zone through that depth than the regular soil profile of that site, with topsoil down to only about 20 cm, and somewhat of a claypan underneath. Not surprisingly, some of the crops planted on the 94 cm deep plot (thus with 94 cm of topsoil) actually had better vigor and more height than those planted on the regular soil, at least until moisture stress became limiting in early Aug.

The crops were planted in replicated 3 m wide strips across the field perpendicular to the direction of the soil pits, creating plot units 3 × 6 m in size. The soil depths varied from 33 to 94 cm (Fig. 2).

The rainfall pattern in May through August at the test location was conducive to gaining some valuable insights into the response of these crops to moisture deficiency. During May and early June, there was adequate rainfall to establish the crops to a height of 30–60 cm. From late June through early Aug., there was a dry spell broken only by a couple of minor rainfalls of about 2 mm each. During this dry spell, crops began to wilt, first in the shallow plots, then successively in the deeper and deeper plots marching across the field. By early Aug., after almost no rain for almost 6 weeks, even the crops on 94 cm of topsoil were affected, some severely. By contrast the crops on the undisturbed soil showed no wilting or other obvious adverse affects from the dry conditions.

The progressive wilting across the field in successively deeper soils was no surprise, but provided a chance to observe wilting patterns in some crops, including amaranth, for which little information is available. What was surprising was the way amaranth dealt with moisture stress compared to eight other grains. Before the test, it was presumed that amaranth would not fair particularly well in shallow soils, based on the idea that amaranth is tap-rooted like sunflower (sunflower gets its drought tolerance by sending its tap-root down deeply, as opposed to being efficient in water use). It was a complete surprise that amaranth was the first of the nine crops to show wilting symptoms. The wilting pattern of amaranth was quite dramatic, with plants going from full turgor one day to completely limp (leaves hanging straight down) the next. However, after almost 10 days of the plants in the shallow plots being completely wilted, the amaranth plants made a stunning recovery to full turgor and apparent health (no leaf lesions or brown leaf edges) after a small rainfall of 2 mm. This pattern repeated itself in deeper plots, after further dry conditions, and another 2 mm rainfall several days later. In contrast to amaranth's ability to rebound so completely from wilting, sunflower had substantial leaf tissue death, as did other crops, even though they were wilted for fewer days than amaranth, and perked up to some extent after the small rains. These observations indicated amaranth may owe part of its reputed drought tolerance to an ability to shut down transpiration through wilting, then recovering easily when moisture is available.

Amaranth was not noticeably affected by moisture limitation in the deepest plot (94 cm), unlike soybean, which used up the moisture available in the profile and completely died by mid-Aug. (soybeans on adjacent regular soil were seemingly unaffected by the dry conditions). Sunflower wilted severely in the 94 cm plots before rain finally came in mid-Aug. In other words, amaranth ran out of moisture first in 33 cm of soil, but for some reason was better able to get by on the moisture in 94 cm of soil than crops such as sunflowers and soybeans. This would seem to indicate amaranth has some water use efficiency compared to those other crops, at least during later stages of growth.

There is still much to be learned about amaranth's water use patterns, especially if the crop is to be promoted for moisture-limited agricultural areas around the world. The only other significant evaluation

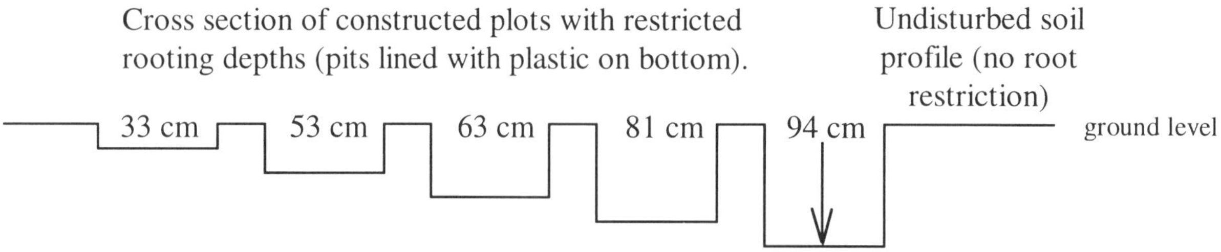

Fig. 2. Cross section diagram showing how restricted root zones of varying depths were developed by digging pits or trenches across a field, with the bottoms lined with impermeable plastic, and then backfilled with topsoil. Pits ran 60 m east-west, and were 6 m wide. Crops were planted north-south perpendicular to the length of the restricted root zone areas. In the undisturbed soil, a claypan is prevalent in the B horizon.

of amaranth water use in the United States was conducted by Henderson et al. (1992) in North Dakota. Neutron probes were used to evaluate seasonal water use in four grain amaranth cultivars. Data indicated amaranth water use extended down to about 1.2 m under their soil conditions (Prosper, North Dakota), with no cultivar differences in water use. Total water use was typically about 27–32 cm in 1990–91, but was half of that during 1992, which was an unusually cool season with limited plant growth.

CONCLUSIONS

The current status of amaranth is as a crop which has great potential, a variety of possible uses, and a decade-plus of research behind it. However, as with most alternative crops, cultivar improvements are needed, production and utilization research challenges remain, and major barriers exist in market development. More specifically, further breeding should improve seedling vigor, reduce lodging and seed shatter, improve ease of harvesting, and improve yield. Insect pests need to be better understood, and control methods developed. Likewise, disease symptoms and prevention are poorly understood with amaranth. Although information on amaranth response to basic management practices had been developed, including in the studies reported on in this paper, additional work could be done, on factors such as pH, NPK interaction, effects of soil type and seedbed preparation methods. Amaranth response to moisture stress deserves more study. From a utilization standpoint, a fair amount has been learned about general amaranth grain characteristics. However, the response of factors such as grain protein to management and environment is only partially understood. Also, specific high value uses of amaranth, such as the starch fraction, merit additional investigation.

Markets remain relatively small and undeveloped, in part because there is a general lack of familiarity with amaranth in the public and private sector. To achieve a higher level of market penetration, amaranth will have to become more publicized, prices will have to fall (although a premium could still be commanded), and availability will have to be increased. Distance to buyers is a problem for many current amaranth growers. Special markets such as the starches or other seed components could lead to increased marketing opportunities.

On the positive side, amaranth is widely adapted, tolerant of dry conditions, and diverse germplasm is available for breeding to improve the crop. Amaranth has relatively good yield potential for a high protein grain crop, especially considering the lack of breeding with the crop. It can be grown successfully with conventional grain crop equipment, usually with only minor modifications, and has a production cost comparable to other grain crops. The colorful appearance of the crop and its colorful history continue to generate interest in the crop, and its good nutritional characteristics combined with its variety of potential uses illustrate the importance of continued work with this "rediscovered" crop.

REFERENCES

Breene, W.M. 1991. Food uses of grain amaranth. Cereal Foods World 36:426–430.

Bressani, R., A. Sanchez-Marroquin, and E. Morales. 1992. Chemical composition of grain amaranth. Food Rev. Int. 8:5–21.

Clark, K.M., W.C. Bailey, and R.L. Myers. 1995. Alfalfa as a companion crop for control of *Lygus lineolaris* in amaranth. J. Kansas Entomol. Soc. 68:143–148.

Clark, K.M. and R.L. Myers. 1994. Yield response in intercropping of pearl millet, amaranth, cowpea, soybean, and guar. Agron. J. 86:1097–1102.

Elbehri, A., D.H. Putnam, and M. Schmitt. 1993. Nitrogen fertilizer and cultivar effects on yield and nitrogen-use efficiency of grain amaranth. Agron. J. 85:120–128.

Fisk, J.W. 1993. Cover crop effects on establishment, growth, and N needs of amaranth. M.S. Thesis, Univ. Missouri, Columbia.

Henderson, T.L., A.A. Schneiter, and B.L. Johnson. 1993. Production of amaranth in the northern Great Plains. p. 22–30. In Alternative crop research: a progress report. North Dakota State Univ., Fargo.

Henderson, T.L., A.A. Schneiter, and A. Riveland. 1993. Row spacing and population effects on yield of grain amaranth in North Dakota. p. 219–221. In: J. Janick and J.E. Simon (eds.), New crops. Wiley, New York.

Kauffman, C.S. 1992a. Realizing the potential of grain amaranth. Food Rev. Int. 8:5–21.
Kauffman, C.S. 1992b. The status of grain amaranth for the 1990's. Food Rev. Int. 8:165–186.
Kauffman, C.S. and L.E. Weber. 1990. Grain amaranth. p. 127–139. In: J. Janick and J.E. Simon (eds.), Advances in new crops. Timber Press, Portland, OR.
Myers, R.L. 1994. Regional amaranth variety test. Legacy 7:5–8.
Pullins, E.E. 1995. Agronomic and economic performance of wheat and canola-based double-crop systems. M.S. Thesis, University of Missouri, Columbia.
Sauer, J.D. 1993. Historical geography of crop plants: a select roster. CRC Press, Boca Raton, FL.
Stallnecht, G.F. and J.R. Schulz-Schaeffer. 1993. Amaranth rediscovered. p. 211–218. In: J. Janick and J.E. Simon (eds.), New crops. Wiley, New York.
Teutonico, R.A. and D. Knorr. 1985. Amaranth: composition, properties, and applications of a rediscovered food crop. Food Technol. 39:49–60.
Wilson, R.L. 1989. Studies of insects feeding on grain amaranth in the Midwest. J. Kansas Entomol. Soc. 62:440–448.

Grain Amaranth Harvest Timeliness in Eastern North Dakota

S.A. Fitterer, B.L. Johnson, and A.A. Schneiter

The majority of the grain amaranth (*Amaranthus* spp., Amaranthaceae) grown in the world is hand-harvested. Harvesting by combine is necessary if amaranth grain production is to compete economically with other crops (Krishnan et al. 1987).

Crop harvest management is critical, as amaranth shatters readily (Myers and Putnam 1988). Preliminary studies indicate that losses caused by seed shattering can decrease yields of some cultivars over 1100 kg/ha (Shroyer et al. 1990). Tucker (1986) reported that mechanical harvesting recovered only 50% of the potential yield.

Amaranth is a plant with high plasticity, and adjusts to a range of environmental conditions (Putnam 1990). Plants grown in thin stands branch more, have thick stalks, and mature late while plants grown at high densities are smaller with thin stalks and are better suited for mechanical harvest. Increased lodging often occurs at high plant populations.

Harvesting should begin as soon as practical to minimize grain shattering and losses due to wind. Direct combine harvest of grain amaranth prior to frost can be complicated by variation of grain maturation in the inflorescence and by high stem and leaf moisture at the time of grain maturity (Krishnan et al. 1987). Shattering losses can be severe, particularly when a light frost kills portions of the flower, but does not kill the entire plant (Weber and Kauffman 1990). If plants are harvested while wet, seeds often adhere to wet plant tissue and may be non-harvestable (Tucker 1986).

The seed of most grain amaranth commercially produced are white or light tan. Some immature seeds turn dark brown when harvested or after frost (Baltensperger et al. 1992). The seed has a lenticular shape, with the embryo located in a ring around the perisperm on the longitudinal axis (Bressani 1990). The placement of the embryo around the endosperm makes it vulnerable to mechanical damage during combine harvest. Combine cylinder speed should be reduced to minimize grain damage and loss in seed germination (Weber and Kauffman 1990). Krishnan et al. (1987) reported mechanically harvested seed at a cylinder speed of 437 rpm had 62% germination, while hand-harvested seed had 97% germination.

METHODOLOGY

The harvest study used 'Plainsman', a cross between *A. hypochondriacus* L. and *A. hybridus* L. (Baltensperger et al. 1992) first released by the Rodale Research Center in 1985 as 'K-343', and renamed

'Plainsman' in 1991, by the Univ. of Nebraska (Brenner 1992). During the juvenile stages, 'Plainsman' has a red marking in the center of its leaves which fades as the reproductive stages approach; flowers are red and upright, white seeds are early maturing; plant is short (average 1.6 m) and has good stem strength after frost.

Before planting, soil fertility of all plot land, based on fall soil tests, was brought to a level suitable for a 5700 kg/ha wheat crop (Danke et al. 1992). Weeds were controlled with sethoxydim and hand weeding when needed.

Amaranth was sown at Prosper, North Dakota, the first week of June, in 1993 and 1994. The soil series at Prosper are Perella (fine-silty, frigid, mixed Typic Haplaquoll) and Bearden (fine-silty, frigid, mixed Aeric Calciaquoll). Each plot consisted of six rows, 6.1-m long at a 30-cm row spacing. Seeds were planted with an Almaco planter at the rate of 3.4 kg/ha. Established plots were thinned, at the three to four leaf stage, to a population of 173 000 plants/ha. The experiment was a randomized complete block design with four replicates. Treatments consisted of five harvest dates at six-day intervals after first seed shatter.

A Hege 125B plot combine was used to harvest the plots. Harvested samples were placed in cloth sacks, weighed before drying, and force air-dried at a temperature <38°C. A 90-g moisture sample was removed before samples were placed in the dryer. The moisture samples were placed in an oven at 110°C for approximately 48 h.

Data collected included (1) seed and foreign material moisture content at each harvest, (2) seed yield, (3) seed germination percentage after harvest, and (4) percent dockage at harvest. Moisture was calculated by dividing the moisture loss (g) by the fresh sample weight (g). Percent dockage was calculated by dividing the weight (g) of the dried harvest sample (before cleaning) by the weight (g) of the foreign material removed.

RESULTS

Precipitation during both the 1993 and 1994 growing seasons at Fargo and Prosper was above normal. Soil moisture at seeding was adequate and timely rains occurred after seeding resulting in good stands. Temperatures were below normal during the 1993 growing season at Fargo and Prosper. Temperatures at Fargo and Prosper were slightly above normal during the 1994 growing season.

Date 1 was harvested at first seed shatter which occurred on Sept. 15 and Oct. 14, in 1993 and 1994, respectively. Date 2, in 1994, was not harvested due to a heavy (76 mm) rain that did not allow for combine access to the plots. Therefore, date 2 was not included in the analysis. The harvest date-by-environment interaction was significant for grain yield, dockage, seed germination, and moisture.

Grain yields in 1993 were 20% to 40% of those at the same harvest date in 1994. Grain yield decreased significantly after dates 3 and 1, in 1993 and 1994, respectively (Table 1), probably due to seed shatter. Seed loss at date 4 occurred because whole branches of the floral cyme fell off due to plant dry down. Significant

Table 1. Mean grain yield, percent dockage, percent seed germination at harvest, and percent moisture of the uncleaned seed plus trash for four harvest dates at the Prosper, North Dakota, in 1993 and 1994, environments.

Harvest			Grain yield (kg/ha)		Dockage (%)		Germination (%)		Moisture (%)	
No.[z]	1993	1994	1993	1994	1993	1994	1993	1994	1993	1994
1	Sept. 15	Oct. 14	480a[y]	1346a	64ab	35b	86a	58c	32a	60a
3	Sept. 27	Oct. 20	425ab	1131c	66a	42a	85a	74a	16b	60a
4	Sept. 31	Nov. 1	376a	1248b	63b	42a	74b	68b	10c	21b
5	Oct. 6	Nov. 7	253c	1191bc	62b	37b	69c	65b	11c	18c

[z]Harvest 2 abandoned due to heavy rain in 1994.
[y]Mean separation in rows by Duncan's multiple range test, 5% level.

differences in grain yield occurred after date 1 in 1994. Grain yields decreased as harvest was delayed beyond first seed shatter.

The significant harvest date-by-environment interaction for percent dockage at harvest was due to the magnitude of differences between 1993 and 1994 (Table 1). In 1993, dockage, across all harvest dates averaged 25% greater than in 1994. This change may have been due to loss of seed to shattering and the ability of the plot combine to remove the empty floral parts from the cyme. The unusually long growing season in 1994 may have increased the seed:foreign material ratio by allowing more seed to mature before harvesting.

Germination percentages differed among environments for seed harvested at different dates. Germination decreased as harvest date was delayed, in 1993. Germination of seed harvested at dates 1 and 3 were significantly higher than seed from dates 4 and 5. Germination of seed from harvest date 1 in 1994 was significantly lower than that of seed from dates 3, 4, and 5. This may have been due to higher seed injury by the plot combine, as seed had not been frosted and may have been at a more immature state. Germination of seed from harvest date 3 was significantly higher than from dates 4 and 5. Seed weathering and seed injury during threshing may have reduced seed germination at the later harvest dates.

Percent moisture of the harvested samples, including seed and plant material, from Prosper in the 1993 and 1994 growing seasons. In 1993, a killing frost occurred before seed shatter was observed. A significant decrease in percent seed moisture for dates 1, 3, and 4 in 1993 was observed. This was probably due to the periodic occurrence of frost and a dry fall, both of which encourage plant dry down.

Percent moisture of the harvested sample did not differ significantly for dates 1 and 3 during the 1994 harvest. Frost had not occurred before date 3, and plant material had not dried down. The lush vegetative plant material in the harvest sample provided an explanation for the high moisture values for dates 1 and 3. A killing frost occurred after date 3, and the moisture values for dates 4 and 5 decreased significantly.

In 1994, comparisons were made between hand-harvest and plot combine harvest. Grain yield and percent seed germination were compared at each date. Mean values for harvest date effect are presented in Table 2. Plot combine harvested plots yielded 70% to 75% of the hand-harvested plots. Although considerable (25% to 30%) seed loss occurred due to mechanical harvesting, this loss was less than the 50% reported by Tucker (1986). This difference may be due to the use of a plot harvester rather than a commercial harvester.

Percent seed germination after harvest was significantly higher in hand-harvest plots than in combine harvested plots. Germination of hand-harvested grain ranged from 82% to 93% across harvest dates while combine harvested seed ranged from 58 to 74% across harvest dates. The decrease in percent germination was probably due to seed damage caused by the plot combine.

Table 2. Mean grain yield and percent seed germination of plot combine harvested and hand-harvested seed at several dates at Prosper, North Dakota, in 1994.

Harvest			Grain yield (kg/ha)		Germination (%)	
No.	1993	1994	Combine	Hand	Combine	Hand
1	Sept. 15	Oct. 14	1346a[y]	1890a	58c	93a
3	Sept. 27	Oct. 20	1131c	1616b	74a	85b
4	Sept. 31	Nov. 1	1248b	1634b	68b	91a
5	Oct. 6	Nov. 7	1191c	1640b	65b	82b

[y]Mean separation in rows by Duncan's multiple range test, 5% level.

SUMMARY

Seed shatter was observed before the first killing frost. Grain yields in 1993, which was cooler than 1994, were 20% to 40% of those at the same harvest dates. Grain yield decreased as harvest was delayed beyond first seed shatter, especially after a killing frost.

Percent dockage was high at all harvest dates. Seed germination was found to decrease with the later harvest dates, possibly due to weathering and mechanical damage. Percent moisture of the harvested sample was very high before a killing frost and decreased rapidly after a killing frost.

Amaranth should be harvested as soon as possible after a killing frost to reduce yield losses and maximize seed germination. Harvest before a killing frost was not feasible due to a high moisture content of the harvest sample that makes storage difficult and increases drying costs. Arrangements must be made to remove plant material to avoid grain spoilage. The harvest sample may need to be dried so seed can be separated from wet plant tissues.

Hand-harvested plots yielded 25% to 30% higher than plot combine harvest yields, due to seed losses caused by mechanical harvesting. Germination was approximately 20% higher for hand-harvest seed than for plot combine harvest seed.

REFERENCES

Baltensperger, D.D., L.E. Weber, and L.A. Nelson. 1992. Registration of 'Plainsman' grain amaranth. Crop Sci. 32:1510–1511.

Brenner, D. 1992. The Plainsman story. Legacy 5:1.

Bressani, R. 1990. Grain amaranth: Its chemical composition and nutritive value. p. 19-34. In: Proc. Fourth National Amaranth Symp., Minneapolis, MN. 23–25 Aug. 1990. Minnesota Ext. Serv., Univ. of Minnesota, St. Paul.

Danke, W.C., C. Fanning, and L.J. Swenson. 1992. Fertilizing wheat and rye. North Dakota State Univ. Ext. Ser. Bul. SF-712. Fargo.

Krishnan, P., P. Kasturi, B. Beeson, A.G. Berlage, L.E. Weber, and C.S. Kauffman. 1987. Effect of threshing cylinder speeds on amaranth seed germination. Paper 87-6509. Am. Soc. of Agr. Eng., St. Joseph, MI.

Myers, R.L., and D.H. Putnam. 1988. Growing grain amaranth as a specialty crop. Station Bul. AG-FS-3458. Minnesota Ext. Serv., Univ. Minnesota, St. Paul.

Putnam, D.H. 1990. Agronomic practices for amaranth. p. 151–162. In: Proc. Fourth National Amaranth Symposium. Minneapolis, MN. 23–25 Aug. 1990. Minnesota Ext. Serv., Univ. Minnesota, St. Paul.

Shroyer, J.P., W.D. Stegmeier, and M. Mikesell. 1990. Grain amaranth. MF-953. Coop. Ext. Service, Kansas State Univ., Manhattan.

Tucker, J.B. 1986. Amaranth: the once and future crop. BioScience 36(1):9–13.

Weber, L.E., and C.S. Kauffman. 1990. Amaranth grain production guide. Rodale Press, Emmaus, PA.

Field Evaluation of Grain Amaranth in Chile*

Marisol Berti, Humberto Serri, Rosemarie Wilckens, and Inés Figueroa

Grain amaranth (*Amaranthus* spp.) is a high protein and high lysine pseudocereal which originated in Central and South America (Bressani 1989; Lehman 1989; Henderson et al. 1993). Amaranth has been rediscovered in several countries, among them the United States and Mexico (Stallknecht and Schulz-Schaeffer 1993; Early 1990; Kauffman and Weber 1990) but is not grown in Chile. Chilean agriculture is going through

*Contribution from the Plant Production Department, Universidad de Concepción, Chillán, Chile. This project has been financed by the Research Division of the University of Concepción.

an economic crisis, especially, for traditional crops such as wheat, oats, barley, and sugarbeets. Consequently the crop alternatives for a rotation are limited. In Chile, previous research on amaranth has been quite limited. Amaranth has been selected as a possible alternative for South-Central Chile because it grows in a wide range of environments, latitudes 0° to 47°S, at a range of elevations (NRC 1984). Amaranth also has a high yielding potential, and is an interesting crop nutritionally.

METHODOLOGY

During 1994 and 1995, a replicated evaluation study was conducted in Chillán, Chile (36°36'S; 72°10'W and 144 m altitude) which included 6 genotypes of *A. cruentus*, 4 of *A. hybridus* x *A. hypochondriacus*, 1 of *A. hypochondriacus*, and 7 of *A. caudatus* plus 30 accesions of *A. cruentus*, *A. hypochondriacus*, *A. caudatus*, and *A. tricolor*. Trials were sown on Oct. 26 1995. The experimental design was a randomized complete block with four replications. Experimental plots had three 5 m long rows, spaced 40 cm apart. Seeding rates and depth were 2 g per row and 1.5 cm, respectively. Plots were hand weeded. The whole experiment was irrigated approximately every 10 days depending on daily evaporation. The central row of each plot was harvested by hand on March 20 for the earlier maturing genotypes and on March 31 for the later ones. Data was collected on number of plants emerged, days and growing degree days to panicle emergence, days and growing degree days to initiation of anthesis, number of plants at harvest time, height at harvest, seed yield, and panicle color at anthesis. Panicle emergence and initiation of anthesis were considered when 50% of the plants in the central row were at this stage. Growing degree days to panicle emergence and initiation of

Table 1. Days from planting to panicle emergence and anthesis initiation of eighteen amaranth genotypes grown during 1994–1995 season in Chillán, Chile.

Species[z]	Genotype or cultivar	Panicle emergence (days from planting)	Initiation of anthesis (days from planting)	Height at harvest (cm)	Seed yield (kg/ha)
AhxAh	Plainsman	50	59	133	4084
A.c.	Concepcion	58	69	165	3650
AhxAh	K433	53	60	113	3481
AhxAh	K432	53	61	111	3394
A.c.	R158	57	70	156	3284
AhxAh	K593	50	60	121	3134
A.c.	K266	58	66	144	3055
A.c.	Canada	58	71	165	2928
A.c.	K283	57	67	144	2883
A.c.	A200D	58	68	144	2408
A.h.	Seleccion DED-H1	86	99	209	1047
A.ca.	ICTA01-0012	77	107	241	617
A.ca.	Oscar Blanco	84	107	231	183
A.ca.	Ataco	87	107	247	114
A.ca.	Linea-10 C	71	92	220	94
A.ca.	Iniap-Alegria	72	100	240	74
A.ca.	Linea-41F	74	91	223	63
A.ca.	Noel Vietmeyer	69	89	230	50
LSD (0.05)		4	4	20	808
C.V. (%)		5	3	7	29

[z]AhxAh = *A. hybridus* x *A. hypochondriacus*, A.c. = *A. cruentus*, A.h. = *A. hypochondriacus*, A.ca. = *A. caudatus*

anthesis were calculated by adding the mean between maximum and minimum daily temperatures minus the base temperature of 8°C (NRC 1984).

RESULTS

Amaranthus cruentus and *A. hybridus* x *A. hypochondriacus* were significantly different from all *A. hypochondriacus* and *A. caudatus* for days from planting and thermal time to panicle emergence and initiation of anthesis (Table 1). The accession Seleccion DED-H1 needed at least 300 growing degree days more than *A. cruentus* and *A. hybridus* x *A. hypochondriacus* to reach the stages already mentioned. The cultivar 'Plainsman' required the least growing degree days for both stages. Seed yields were significantly different among

Table 2. Days from planting to panicle emergence and anthesis initiation of 30 amaranth lines grown during 1994–1995 season in Chillán, Chile.

Species	Accession	Country of origin	Days from planting to panicle initiation	Days from planting to anthesis	Height at harvest (cm)	Seed yield (kg/ha)
A. cruentus	Ames 8272	Central Afr. Rep.	62	79	168	7578
A. cruentus	Ames 2243	Mexico	61	74	159	6488
A. cruentus	MT-5	USA	55	71	163	5428
A. hypochondriacus	Ames 1972	Nigeria	70	85	178	5132
A. cruentus	Ames 2216	Mexico	58	70	182	4866
A. cruentus	Ames 5356	Chile	57	72	139	4047
A. cruentus	MT-3	USA	57	70	156	3694
A. cruentus	Ames 3216	USA	59	69	140	3203
A. cruentus	K283	USA	56	64	146	2803
A. cruentus	K266	USA	53	65	157	2669
A. cruentus	Ames 5385	Mexico	68	88	180	2575
A. hypochondriacus	Ames 2061	Nepal	68	83	172	2506
A. cruentus	Ames 5386	India	67	85	152	2234
A. cruentus	PI 511719	Guatemala	67	79	167	2191
A. cruentus	Ames 2262	USA	62	76	147	2173
A. cruentus	R-158	USA	58	65	112	1709
A. cruentus	Ames 1981	China	67	82	152	1528
A. cruentus	Ames 5648	Mexico	58	74	161	1462
A. cruentus	Ames 2151	USA	75	82	175	1401
A. caudatus	PI 511679	Argentina	107	117	249	1284
A. cruentus	Ames 2266	USA	59	63	116	953
A. hypochondriacus	Ames 2260	Mexico	94	107	222	928
A. cruentus	Ames 5606	USA	58	82	65	469
A. cruentus	Ames 2263	USA	59	59	98	338
A. *cruentus*	Fotete	Benin	61	79	112	269
A. cruentus	Lenga-lenga	Zaire	60	75	121	200
A. cruentus	Yendi15	Ghana	63	79	116	100
A. cruentus	Ames 1973	Nigeria	59	78	111	81
A. cruentus	Ames 5362	Unknown	58	75	90	44
A. tricolor	Red amaranth	China	63	75	82	19

accesions from *A. cruentus*, *A. caudatus,* and *A. hypochondriacus*. A. *hypochondriacus* x *A. hybridus* were similar to *A. cruentus* (Table 1).

It is very clear that *A. hypochondriacus* and *A. caudatus* do not adapt well to Chillán environment. Plants flower too late in the season to set enough seed. The other two species have potential. All produced seed yields over 2400 kg/ha. 'Plainsman' had the highest seed yield, over 4000 kg/ha. Interestingly, this cultivar required the fewest growing degree days for panicle initiation and anthesis.

Panicle color at anthesis included red-maroon (K593, A200D, R-158, K283, Plainsman), golden green (K266, K433, K432, Noel Vietmeyer, Linea 10-C, LINEA 41-F), golden orange (Canada), yellow-pink (Concepcion), and red-pink (Ataco, Oscar Blanco, Seleccion DED, ICTA 01-0012).

Results from the 30 accessions were similar. *A. cruentus* genotypes reached panicle emergence and anthesis at least 100 growing degree days before *A. tricolor*, *A. caudatus,* and *A. hypochondriacus* (Table 2). *A. cruentus* genotypes had the highest seed yields with the exception of eight genotypes that had low yield, but mainly because of low plant stands (Table 2). The line originated in the Central African Republic had the highest seed yield (7578 kg/ha), but this genotype was not the earliest maturing one like 'Plainsman' in the first study.

Only one *A. hypochondriacus* genotype had good seed yields, but according to plant characteristics, this genotype may have been wrongly classified. It resembles much more *A. cruentus* than the earlier one.

FUTURE PROSPECTS

Grain amaranth has shown a yield potential of up to 7500 kg/ha under irrigated experiments in Chillán, Chile, a yield unreported in any other environment in the world. These results suggest that further study of this crop to South-Central Chile is warranted.

REFERENCES

Bressani, R. 1989. The proteins of grain amaranth. Food Rev. Int. 5:13–38.

Early, D.K. 1990. Amaranth production in Mexico and Perú. p. 140–142. In: J. Janick and J.E. Simon (eds.), Advances in new crops. Timber Press, Portland, OR.

Henderson, T., A. Schneiter, B. Johnson, N. Riveland, and B.G. Schatz. 1993. Production of amaranth in the northern Great Plains. p. 22–30. In: Alternative crop and alternative crop production research. A progress report. May 1993. North Dakota State Univ., Fargo.

Kauffman, C.S. and L.E. Weber. 1990. Grain amaranth. p. 127–139. In: J. Janick and J.E. Simon (eds.), Advances in new crops. Timber Press, Portland, OR.

Lehman, J. 1989. Proteins of grain amaranth. Legacy 2:3–6. Am. Amaranth Inst. Bricelyn, MN.

National Research Council. 1984. Amaranth: Modern prospects for an ancient crop. National Academy Press, Washington, DC.

Stallknecht, G.F. and J.R. Schulz-Schaeffer. 1993. Amaranth rediscovered. p. 211–218. In: J. Janick and J.E. Simon (eds.), New crops. Wiley, NY.

GRAIN LEGUMES

Acacia in Australia: Ethnobotany and Potential Food Crop*

Peter R Lister, Paul Holford, Tony Haigh, and David A. Morrison

Acacia is the second largest genus in Australia comprising more than 700 species (Harden 1991; Morrison and Davies 1991) and occurs in almost all habitat types. Species range in size from small shrubs to large trees and are ecologically important as 'pioneer' species where they rapidly establish cover following major natural disturbances such as fire (Christensen et al. 1981). *Acacia* species are commonly known simply as acacias or as wattles and *Acacia pycnantha* has been adopted as the Australian national floral emblem. Wattles are frequently grown as ornamentals, some are harvested for timber, while others are a source of gums or bark used in various tanning processes. Like the majority of legumes, acacias utilize rhizobia to fix atmospheric nitrogen in the soil enabling them to grow in relatively poor soils (Harden 1991).

BOTANY

In Australia there are two commonly accepted schools of thought regarding the taxonomy of legumes (order Fabales) and their treatment varies with institution and state. For example, The Flora of Australia (George, 1981) recognizes three distinct families, Mimosaceae, Caesalpiniaceae, and Fabaceae while the Flora of New South Wales (Harden 1991) recognizes a single family, Fabaceae (sometimes still referred to as Leguminosae by some workers) with three subfamilies, Faboideae (or Papilionoideae), Caesalpinioideae, and Mimosoideae. There has also been some debate about the placement of Australian wattles in the genus *Acacia*. Between 1977 and 1989, Pedley of the Queensland herbarium proposed splitting *Acacia* into several segregate genera supported by findings of workers in France during the late 1960s and 1970s (Pedley 1987). Pedley also proposed the reinstatement of the genus *Racosperma* for the majority of Australian wattles. To date, this has not been accepted, based on the argument that the name Racosperma was not validly published in 1835 (Hall and Johnson 1993) and the need for further research into the systematics of this large taxon (Maslin 1989).

THE ETHNOBOTANY OF ACACIA IN AUSTRALIA

Wattleseed, an Historical Perspective

Archaeological evidence demonstrates the presence of Aboriginal people in Australia for at least 50000 years (Flood 1990) and during this time there has been considerable change in the spatial distribution of vegetation. This has resulted, not only from a changing climate, but also as a consequence of megaherbivore extinction and Aboriginal burning practices (Flannery 1994). This unnatural, increased fire frequency has favored those species able to cope with such a regime. The proportions of grasses, "pioneer" species, annuals and pyrophilic species have increased compared with fire sensitive taxa (Flannery 1994). Such fire adaptive plants usually produce large quantities of seed and increasing the population size of those plants utilized for their seed will naturally increase the food supply. Aborigines deliberately burnt areas to achieve this aim (Flannery 1994; Latz 1995).

Seeds form a staple food among many indigenous peoples and plants native to Australia are no exception. Of all the plant foods in central Australia, seeds are by far the most important. Seeds are usually high in proteins, carbohydrates, and fats and are easily collected, providing a high energy food for the expenditure of relatively small amounts of energy (Latz 1995). Although Australian plants generally produce small seeds they are produced in large quantities. In arid Australia, seed supply is widely available, somewhat predictable and dependable (Flood 1990). These plant products form the dietary staple in that they represent greater than 50% of the total diet and often would constitute 70% to 80%. Hiatt's data compiled from several sources, and describing the proportions of hunting, gathering and fishing performed by various indigenous peoples,

---

*Funding to establish the trial at Umuwa has been made available from a grant by the University of Technology Sydney–Jumbunna Centre for Australian Indigenous Studies, Education and Research for which we are most grateful.

lists three central Australian linguistic groups, the Dieri, Arrernte, and Walpiri (Hiatt 1978). In all three cases, 70% of the diet consists of gatherable foods and 30% from hunting. Women are the sole providers of gatherable foods and men the sole hunters and as such, women provide 70% of the total diet of these people in arid Australia. The northern half of the Northern Territory possesses some 40 species of *Acacia* and although 19 species are useful to Aboriginal people, only one species, *A. difficilis* has seed that is eaten (Brock 1988). There are other more readily available carbohydrate sources such as yams that require less preparation.

Many grasses provide large amounts of soft seed and have been heavily utilized as a staple food by Aboriginal people, especially throughout arid and semi-arid Australia. The grain was collected, ground to a flour using millstones and water was added to form a paste which was eaten raw or cooked as a damper (unleavened bread) in the ashes. Particular wattle seed was similarly collected, prepared and eaten throughout central Australia. Although these skills still survive (Nganyintja 1985), the use of processed wheat flour has largely replaced these traditional practices (Bryce 1983). These seed grinding practices appear to be a relatively recent technological development. Archaeological excavations in central Australia at Puntutjarpa date the oldest millstones to 3500 years; at Puritjarra they are present for the last 2000 years and at Intirtekwerle, they constitute 10 percent of the artifacts in the last 700 years of deposits. The stratigraphy at two of these sites suggests a massive build-up in the level of the sandplain, the sediments having originated in the Simpson Desert dunefield. This suggests that sites in central Australia older than 5000 years may be deeply buried (Flood 1990). Furthermore, this indicates that Aboriginal people in central Australia have been grinding grass or wattle seed for no more than 4000 years. There are older sites closer to the coast in semi-arid country where the development of such practices became a possibility as a result of the drier climate and in creasing fire frequency. Archaeological evidence from the earliest of these sites, Lake Mungo, in the Willandra Lakes system in western New South Wales, demonstrates the presence of a seed grinding economy over the last 16000 years (Flood 1990).

Of the sixty or so species of *Acacia* in central Australia, Latz (1995) states that some 50% were, or still are, eaten by Aboriginal people and it is not only the seed which is consumed. Several species exude an edible sugary gum from wounds in the stem or branches which supplies a source of energy. Others are fed upon by insects which themselves secrete an edible substance while species such as *A. kempeana* are the host for various edible grubs (Kalotas and Goddard 1985) often referred to by non-Aboriginal people as witchetty grubs.

Toxicity

Not all wattleseed was used for food. Many coastal and some arid species contain toxic compounds. *A. longifolia* is one of the few species recorded as having been eaten in coastal eastern Australia (Kohen 1992), similarly, *Acacia georginae* seed reportedly contains sodium fluoroacetate the major constituent of 1080, a widely used rodenticide (P. Latz pers. commun.).

A. ligulata, umbrella bush, is a widespread and common semi-arid species. A. Kalotas (pers. comm. 1994) noted that there are mixed reports of the consumption of this seed. During his research near Warburton (eastern Western Australia—approx. 750 km WSW of Alice Springs) with Ngaanyatjara people in 1981–82, this species was recorded as one, the seed of which was commonly consumed. Anecdotal evidence from Yankunytjatjara speakers (approx. 600 km ESE of Warburton), suggests it was a species only eaten when no other seed was available as it caused hair loss, the hair regrowing sometime later (Kalotas 1985). Pintupi people (approx. 400 km north of Warburton) also say it was regularly consumed but said nothing of hair loss (A. Kalotas pers. commun. 1994). It may be that the alopecia (hair loss) resulted from a combination of factors rather than the action of *A. ligulata* seed alone. If it was consumed amongst the Yankunytjatjara only when other foods were scarce, then malnourishment may have played a role in the loss of hair. The tropical American legume genera, *Leucaena* and *Mimosa*, both closely related to *Acacia*, cause hair loss, when consumed, as a result of the presence of the amino acid, mimosine (Mabberley 1987; Windholz et al. 1983). When *Leucaena* was first used as a stock feed in northern Australia it caused problems with cattle. This was remedied when a bacterium was isolated from the gut of cattle in Java and introduced into drinking troughs in Australia (A. Kalotas pers. commun. 1994). It is possible that similar toxic compounds are present in Australian acacias and care needs to be taken in the choice of species and their subsequent screening as a part of any development of a new crop. Brand and Maggiore (1991) state that testing for the presence of toxic compounds is mandatory if these plants are to be developed as new food products. Many legume seeds contain a variety of toxic compounds that are usually denatured by the application of heat. These compounds, if untreated, can disrupt intestinal absorption of nutrients and produce growth retardation (Brand and Maggiore 1991).

Non-Food Uses

Although the main use of wattles is as a food source (Goddard 1992; Kalotas and Goddard 1985; Isaacs 1987; Kean 1991; Kohen 1992; Latz 1982,1995; Meggitt 1962; Tindale 1972) they have a number of other uses. These include provision of shelter (Nganyintj a 1985), preparations used as medicine (Anon. 1993), and the manufacture of various tools, musical instruments and weapons (Meggitt 1962; Moyle 1979; Brokensha 1978). Today they are largely used for artifact manufacture (Brokensha 1978) and more recently in various revegetation practices (Last 1990).

Linguistics

Until the mid 1900s, an apparent lack of collaborative research between anthropologists, botanists and zoologists led to difficulties in understanding Aboriginal natural resource use and management. Little traditional ecological knowledge was recorded, but much attention was paid to recording traditional botanical and zoological nomenclature. Anthropologists such as Meggitt (1962) recorded the use of particular plants or animals and the indigenous names for each, but it appears specimens were not collected for accurate identification by botanists or zoologists at a later date. As Aboriginal languages were oral, there was no standard spelling, and so anthropologists and linguists made attempts at recording such names phonetically. Older texts will spell the languages discussed here as "Alyawara" for Alyawerre; "Anmatjarra" and "Anmadjarra" for Anmatyerre; "Aranda" and "Arunta" for Arrernte; "Bindubi" and "Pintubi" for Pintupi; "Bidjandjara", "Pidjandjara", "Pitjandjara" and "Pitjantjara" for Pitjantjatjara; and "Walbiri" for Walpiri, with many other minor variants. There are several sounds for which there are no English equivalents and although spelling standards were established, they were, at times, inadequate and have been revised several times. Pitjantjatjara, for example, has been written since the early 1940s and despite this being a comparatively recent development, there are still inconsistencies in the way some of the sounds are written (Eckert and Hudson 1988). There are various lists of Aboriginal names for plants and animals which do not necessarily match with currently recorded names and so identifying species from these early records is problematic. We have managed to identify wattles that Meggitt (1962) recorded as being used by Walpiri people by cross referencing modern names and taking into account changes in the spelling conventions of linguists during this time (Table 1).

Table 2 indicates the number of species of acacia used as a seed food by particular language groups and, conversely, the number of language groups utilizing each species, in central Australia. There are many names for particular species in common between languages, the main reason being the linguistic affinities represented in the table. The Alyawarre, Anmatyerre and Arrernte languages all belong to the Arandic group while Pintupi and Pitjantjatjara are of the Western Desert group. The Walpiri is a member of the Ngarrkic Group but even so, Walpiri names for *Acacia aneura*, *A. kempeana*, *A. murrayana*, *A. stipuligera* and *A. tetragonophylla* are similar to Pintupi and Pitjantjatjara names. The name for *A. coriacea*, pangkuna or pungkuna, is common amongst all the languages. *A. adsurgens* and *A. stipuligera* are very similar in appearance, as are *A. pachyacra* and *A. murrayana*, and bear the same Alyawerre and Pitjantjatjara names, respectively. Although *A. macdonelliensis* is common throughout central Australia, only the Alyawerre used it, while *A. maitlandii* is widespread but rare and was never used by the Walpiri (Latz 1995). Of the 30 species and languages listed, both the Arandic and Western Desert groups ate 20 species, while the Ngarrkic group ate 21 species. *A. aneura*, *A. coriacea*, *A. cuthbertsonii*, *A. estrophiolata*, *A. kempeana*, *A. murrayana*,

A. tetragonophylla and *A. victoriae* are the most common and more widespread species and are, therefore, the most widely consumed species. *A. victoriae* is present throughout the entire area, is common and used by all language groups.

THE BUSHFOOD INDUSTRY & CROP POTENTIAL

In recent years, there has been an upsurge of interest in novel food products and in particular those products branded as "bushfoods." Bushfoods can be defined as native plant or animal products used by indigenous Australians as either a traditional or contemporary foodstuff. This interest has led to the widespread use of bushfoods in the restaurant industry both in Australia and overseas. In some instances, these products have become souvenir items, marketed as "uniquely Australian cuisine" or as having been "wild collected by Aboriginal people on their traditional homelands, just as they have done so for thousands of years." Some of these products such as emu are now farmed commercially to meet growing demands.

Wattleseed is in high demand for use as a ground product in pastries and breads and also as a flavoring in desserts, especially ice-cream. It is also used to produce a high quality coffee-like beverage. Wattleseed is one bushfood product collected almost exclusively by Aboriginal people from wild populations throughout its

Table 1. Aboriginal names given to *Acacia* species used by the Walpiri as listed by Meggitt (1962) with nomenclatural additions and corrections (Latz 1995). Species names in bold are those identified and/or corrected by the authors. One species remains unidentified and the identity of Meggitt's "waralga" as *A. ligulata* is uncertain.

	Aboriginal names		
Acacia species	Former spelling[z]	Current spelling[y]	Uses[z]
ancistrocarpa	birauru	pirraru	No use recorded
aneura	mandja	manja	Edible seeds; wood for implements
coriacea	bangguna	pangkuna	Edible seeds; wood for implements
dictyophleba	bilbirinba	pilpirrinpa	Leaves used medicinally
estrophiolata	jadanbi	yajarnpi	Wood for implements and sacred objects
farnesiana	budunari	putunarri	No use recorded
kempeana	ngalgiri	ngalkirdi	Edible seeds; trunk harbours witchetty grubs
notabilis syn. *pruinocarpa*[x]	mandala	marntarla	Edible seeds, gum; wood for implements
dictyophleba	badudu	patutu	Wood for spear shafts
unidentified	bilingarba		No use recorded
adsurgens	budjubanda	puju-parnta	Edible seeds
cowleana	ganalarambi	kanarlarrampi	Wood for spear shafts
aff. aneura[x]	jabiljaru	yapilyardu	Trunk harbours witchetty grubs
adsurgens or *tenuissima*	minjana	minyana	Edible seeds; wood for implements
ligulata?	waralga	wardarrka?	No use recorded
spondylophylla	bundalji	puntaltji	Trunk harbours witchetty grubs
tetragonophylla	gurara	kurara	Edible seeds
victoriae	ganabargu	kanaparlku	No use recorded

[z]Meggitt 1962.
[y]Latz 1982, 1985.
[x]Nomenclature based on Latz 1995.

Table 2. *Acacia* species traditionally used as edible seed by various central desert linguistic groups and their traditional names. The names are given only where the species was used for food as determined by Latz (1982, 1995); Goddard (1992); Kalotas and Goddard (1985); Meggitt (1962) and Tindale (1972). The lack of use of an otherwise edible species can be due to the absence of that species within that particular linguistic area and does not necessarily suggest the species was regarded as inedible, toxic or of inferior quality. NIA indicates the species is not in the area. Alternative spellings for Alyawarre (Alyawerr), Anmatyerre (Anmatyerr) and Arrernte words are included for the benefit of readers with limited access to recent Australian linguistic works.

	Aboriginal linguistic group						
Acacia species	Alyawarre	Anmatyerre	Arrernte	Arrernte (southern)	Pintupi	Pitjantjatjara	Walpiri
acradenia	ampwey mpwiya		NIA	NIA	NIA	NIA	ngardurrkura ngarulkurra
adsurgens	ilkirta ilkert ilwerreny alirrinya	atiyipinha ateyepenh ilwerreny lirrinytja	NIA	NIA	NIA	NIA	minyana puju-parnta mintirlpiri kulaki
ammobia (syn. *doratoxylon*)	NIA	NIA	NIA	NIA	NIA	utjalpara	NIA
ancistrocarpa							wartarurru pirraru
aff. aneura (?syn. *aneura* var. *latifolia*)			ititja- ilpatjata irtetye- irlpelharte		lililiyi		yarlpiyardu
aneura (syn. *brachystachya*)	artitja artety	artitja artety	ititja irtetye manytja		wanari kurrku manytja mantja	wanari kurku kalpilya puyukara wintalyka	yulnantji? wartiji manja wanajiti
colei (syn. *holosericea*)	alerrey aliriya alyari	alkart alkarta	NIA	NIA	kuna-kuna?	kilkiti kuna-kuna?	kalkardi
coriacea	awenth ntjirrima	akiyrlpirra awenth ntjirrima	pungkuna		irrkili yirrkili pangkuna	kunapuka mulupuka	pangkuna kunarnturu wakirlpirri
cowleana	aliriya alerrey	alkarta alkart	NIA	NIA	kilkiti	NIA	kanarlarrampi kalkardi parrapi
cuthbertsonii	alhanker irley pirley	pilhi perley ulyuya lywey			yalpirri piliyi	alpiri kalirma	pirliyi

Table 2. Continued.

Acacia species	Aboriginal linguistic group Alyawarre	Anmatyerre	Arrernte	Arrernte (southern)	Pintupi	Pitjantjatjara	Walpiri
dictyophleba	ulupula ulunkurra alhanker alhepalh	ulkurnarra lkwernarr paturta partwert		ilpakilparra ilpakilparre	minytju mulyati yurrtjanpa utjanypa	mintju ngarkalya	wurpardi yinjirtingu yurrpardi pilpirrinpa patutu matutu marlarntarrpa
estrophiolata	athiyimpa athimp athinga atheng	tjarnpa tywarnpe atjarnpa atyarnp athenga tunga	tjwarnpa tywarnpe		walakarri	utjanypa tjau	walirri yajarnpi wajarnpi ajarnpa
hemignosta	NIA	NIA	NIA	NIA	NIA	NIA	mirrirn- mirrirnpa? luwiluwilta
inaequalatera (syn. *pyrifolia*)	NIA	NIA	NIA	NIA	NIA	NIA	janjirnngi janjinki
jennerae	NIA	lalkirrika lalkerrek	NIA	NIA	NIA	NIA	walalyirrki
kempeana	atnyima atnyem	utnyima atnyem	tnyima tnyeme		yilykuwarra ilykuwarra iripili piyanpa	ilykuwara	ngarlkirdi yiripili
ligulata					nyukurrka? watarrka	watarka	
lysiphloia	awurrnga awernng	NIA	NIA	NIA	NIA	NIA	murlurrpa
macdonelliensis	irrara irrar						
maitlandii (syn. *patens*)			ilupa-lupa lwepe-lwepe				
murrayana (syn. *frumentacea*)	arrilya arrely	arrilya arrely	irrilya irrelye		nyurrinpa	tjuntala tjuntjula	juntala
olgana						kaliwara	NIA
oswaldii						wilpiya waltari	
pachyacra						tjuntala	
pruinocarpa (syn. *notabilis*)	NIA	NIA				itawara	
ramulosa	NIA	NIA	NIA	NIA	NIA	pakuta palpa	NIA

Table 2. Continued.

	Aboriginal linguistic group						
Acacia species	Alyawarre	Anmatyerre	Arrernte	Arrernte (southern)	Pintupi	Pitjantjatjara	Walpiri
stipuligera	mpwiya ampwey	NIA	NIA	NIA	tjilpirinpa tjirrpirinypa wilpurra	NIA	jirrpirinypa kurapuka wirlpurpa ngirnti-yirrpi
tenuissima	antjulinya antywerleny artepwel	antjulinya antywerleny	NIA	NIA		NIA	minyana kuwiyangayi watiyawarnu kulaki nyintirriyilpi watiyawarnu
tetragonophylla	alkitjirra arlketyerr	alkitjirra arlketyerr	ilkitjirra arlketyerre		wakalpuka	wakalpuka kurara kurungantiri	kurarra
victoriae	arlupa arlep	arlupa arlep	tupurla urlupa urlepe	tuperle	pulkuru	aliti ngatunpa	kanaparlku yalupu yarlirti

natural range. The species most commonly collected is *Acacia victoriae* Benth. as it is generally regarded as having a superior flavor. *A. victoriae* is widespread over much of central Australia and fruits during December and January. Yield is unpredictable and is influenced by climatic conditions and, as such, is extremely variable. Wattleseed is not yet grown on a commercial scale and the demand far exceeds the supply. Despite this, small quantities of wattleseed are exported to the U.S., Canada, UK, France, Japan and SE Asia.

Several species of *Acacia* indigenous to central Australia are planted to revegetate or rehabilitate degraded land predominantly on Aboriginal communities throughout central Australia. Species commonly used throughout Pitjantjatjara Lands include *A. victoriae*, *A. murrayana* and *A. kempeana* (pers. obs. 1995). These are relatively fast growing species adapted to low rainfall and extreme temperatures and are planted to provide windbreaks, reduce erosion and to revegetate damaged sites. As the plants reach maturity they are often used for other purposes such as firewood or artifact manufacture (Last 1990) but less commonly for food. Edible grubs (maku) are extracted from the roots of *A. kempeana* at any opportunity but seed is not usually collected for food (M. Last pers. commun. 1995). These plants have potential as an informal crop, in that they possess a variety of uses which could form an additional source of seed for the bushfoods industry.

Nutritional Analysis

Acacia seeds are highly nutritious and contain 26% protein, 26% available carbohydrate, 32% fiber and 9% fat (Brand and Maggiore 1992). The fat content is higher than most legumes with the aril providing the bulk of fatty acids present. These fatty acids are largely unsaturated which is a distinct health advantage although it presents storage problems as such fats readily oxidize (Brand and Maggiore 1992). The mean total carbohydrate content of 55.8±13.7% is lower than that of lentils, but higher than that of soybeans while the mean fiber content of 32.3±14.3% is higher than that of other legumes such as lentils with a level of 11.7% (Brand and Maggiore 1992). The energy content is high in all species tested, averaging 1480±270 kJ per 100g. Wattle seeds are low glycaemic index foods. The starch is digested and absorbed very slowly, producing a small, but sustained rise in blood glucose and so delaying the onset of exhaustion in prolonged exercise (Brand and Maggiore 1992).

Research at UWSH

The research at UWS–Hawkesbury is centered upon two species, *Acacia victoriae* and *A. murrayana* F. Muell ex Benth. The main aim is to develop quality plants with higher and more consistent yields and with an ease of cultivation that will allow Aboriginal people in arid areas to cultivate, harvest, process and market wattleseed to the world. In addition to this, we are seeking to gain an understanding of the agronomy of these plants through greenhouse and field trials.

A. murrayana is being studied as it has a very different growth habit to *A. victoriae*. Unlike *A. victoriae*, it is a spineless species which is a distinct advantage when harvesting seed by hand. In addition, it has potential for soil stabilization and land rehabilitation projects as it is a species capable of regeneration from its roots. This means it can regenerate vegetatively following fire or clearing. The possibility also exists that if crop yields fall due to senescence, the plants could be cut back to ground level without disturbing the soil and the subsequent regrowth should retain the growth and yield characteristics for which it was originally selected.

Work thus far has been aimed primarily at establishing field trials to examine the variation within these plants and the plants' responses to irrigation and fertilizers. Experiments to determine how the plants respond to nitrogen and potassium fertilization and rhizobial innoculation are also in progress.

There are two field trials planted on campus with another to be located at Umuwa in the Musgrave Ranges of northern South Australia. A visit was made to Umuwa in April 1995 to select a site for planting in collaboration with the Pitjantjatjara community.

Studies concerning floral and fruit development and also pollination are planned. Genetic analysis will be performed as part of the examination of variation within these species.

CONCLUSION

Acacia seed in Australia was, and in some areas still is, used as a food source by Aboriginal people. It is now popular with the emergence of the bush foods industry as a new product with a variety of culinary applications. Wild populations are harvested for their seed, but the plants have potential as a commercial crop. It is hoped that the production of seed for food use is adopted by Aboriginal communities. Roasted Acacia seeds offer an exciting new flavor for pastries and ice-creams and a caffeine free beverage.

REFERENCES

Aboriginal Communities of the Northern Territory. 1993. Traditional Aboriginal medicines in the Northern Territory of Australia. Conservation Commission of the Northern Territory of Australia, Darwin.

Brand, J and P. Maggiore. 1992. The nutritional composition of Australian *Acacia* seeds. In: A.P.N. House and C.E. Harwood (eds.), Australian Dry-zone Acacias for Human Food Proceedings of a workshop held at Glen Helen, Northern Territory, 7–10 Aug. 1991, CSIRO Division of Forestry & Australian Tree Seed Centre, Canberra.

Brock, J. 1988. Top End native plants. A comprehensive guide to the trees and shrubs of the Top End of the Northern Territory, John Brock Publ., Darwin.

Brokensha, P. 1978. The Pitjantjatjara and their crafts. Aboriginal Arts Board Australia Council, Sydney.

Bryce, S. 1983. The role of bush tucker in nutrition education. In: O'Dea, K. (ed.), Proc. Aboriginal Bushfoods Workshop. p. 20–23.

Christensen, P., H. Recher, and J. Hoare. 1991. Responses of open forests (dry sclerophyll forests) to fire regimes. p. 367–393. In: A.M. Gill, R.H. Groves, and I.R. Noble. (eds.), Fire and the Australian Biota. Australian Academy of Science, Canberra.

Eckert, P. and J. Hudson. 1988. Wangka Wiru: A handbook for the Pitjantjatjara language learner. Univ. of South Australia, Adelaide.

Flannery, T.F. 1994. The future eaters: an ecological history of the Australasian lands and people. Reed Books, NSW, Australia.

Flood, J. 1990. The riches of ancient Australia. University of Queensland Press, Queensland.

Goddard, C. 1992. Pitjantjatjara/Yankunytjatjara to English Dictionary. Institute for Aboriginal Development, Alice Springs, NT, Australia.

Hall, N. and L.A.S. Johnson. 1993. The names of acacias of New South Wales with a guide to pronunciation of botanical names. Royal Botanic Gardens, Sydney, Australia.

Harden, G.J. (ed.). 1991. Flora of New South Wales, Vol. 2. New South Wales Univ. Press, NSW, Australia.

Hiatt, B. 1978. Woman the gatherer. In: F. Gale, Woman's role in Aboriginal society (ed.), Australian Institute of Aboriginal Studies, Canberra.

Isaacs, J. 1987. Bush food. Aboriginal food and herbal medicine. Weldons, Sydney, Australia.

Kalotas, A. and C. Goddard. 1985. Punu, Yankunytjatjara plant use. Institute for Aboriginal Development, Alice Springs, NT, Australia.

Kean, J. 1991. Aboriginal-acacia relationships in central Australia. Records of the South Australian Museum, 24(2):111–124.

Kohen, J.L. and A.J. Downing. 1992. Aboriginal use of plants on the western Cumberland Plain. Sydney Basin Naturalist, No. 1, p. 1–8. Australasian Naturalist Pub., Sydney, Australia.

Latz, P.K. 1982. Bushfires and Bushtucker: Aborigines and Plants in Central Australia. MA (Hons) thesis, Univ. of New England, NSW.

Latz, P.K. 1995. Bushfires and bushtucker: Aboriginal plant use in Central Australia. IAD Press, Alice Springs, NT, Australia.

Mabberley, D.J. 1987. The plant-book. A portable dictionary of the higher plants. Cambridge Univ., Cambridge.

Meggitt, M.J. 1962. Desert people: A study of the Walbiri Aborigines of Central Australia. Angus and Robertson Publishers, Australia.

Morrison, D.A. and S.J. Davies. 1991. Acacia. p. 327–328. In: G.J. Harden, (ed.), Flora of New South Wales, vol. 2. New South Wales Univ. Press, Australia.

Moyle, R.M. 1979. Songs of the Pintupi: Musical life in a central Australian society. Australian Institute of Aboriginal Studies, Canberra.

Nganyintja. 1985. Mayi Wiru, Part 1, Winter foods. Angatja Video in association with Riverbed Productions, South Australia.

Pedley, L. 1987. Generic status of *Acacia sensu lato*, Australian Systematic Botany Society Newsletter, 53(Dec. 1987):87–91.

George, A.S. (ed.), 1981. Flora of Australia Volume 1, Introduction. Australian Government Publishing Service, Canberra.

Tindale, N.B. 1972. The Pitjandjara. p. 217–268. In: M.G. Bicchieri (ed.), Hunters and gatherers today. Holt, Rinehart & Winston, London.

Windholz, M., S. Budavari, R.F. Blumetti, and E.S. Otterbein (eds.). 1983. The Merck index. An encyclopedia of chemicals, drugs and biologicals. 10th ed. Merck & Co., Rahway, NJ.

Winter Survival of Austrian Winter Pea and Annual Medic on the Western High Plains

James Krall, Robin W. Groose, and Jonathan Sobels

Hard red winter wheat (HRWW) is the most commonly grown non-irrigated crop in eastern Wyoming. Approximately 81,000 ha of HRWW are grown annually in a wheat fallow system (Wyo. Agr. Stat. Serv. 1995). The predominance of HRWW is due largely to climatic patterns of this region. The latitude (41 to 43 N) and altitude (1220 to 1830 m) result in a mean frost-free period of 128 days. Mean annual precipitation is 390 mm with 66% occurring between the months of April and August. This can result in mid-winter conditions of infrequent snow cover and below freezing temperatures.

Current HRWW production utilizes primarily native organic N which is depleted as N is mineralized and soil organic matter is lost (Schuman et al. 1994). N applied as commercial fertilizer is also used which adds to production costs and public concern about environmental pollution.

In some regions of the world annual legumes have replaced fallow in cereal agroecosystems. Self-regenerating pastures of annual medics (*Medicago* spp.) grown in rotation with wheat contribute to sustainable agriculture in the cereal/pasture zone of southern Australia by providing organic N, conserving and building soil, and providing a break from cereal pests and diseases.

This wheat/sheep system is a short-term annual system of pasturing in a 330 to 450 mm annual precipitation zone which occupies as much as 30 million hectares (Crawford et al. 1989). The legumes naturally regenerate annually from seed. They are prostrate in growth habit, making them relatively easy to manage in the wheat crop. Pasture phases are short term (1 to 3 years) followed by wheat. Soil nitrogen and water-stable soil aggregates increased with each pasture phase and decreased with each wheat phase (Reeves 1987); however, in the integrated system both nitrogen and water-stable aggregates are maintained above the previous continuous cropping levels.

Pest control benefits that result from crop rotation can also have a significant economic and environmental impact. For example, in Australia, cereal cyst nematode (*Heterodera avenae* Woll.) can be a serious problem. Rotation to annual grass-free legume pasture can effectively control this pest as well as the disease, take-all (*Gaeumannomyces graminis* var. *tritici* Walker) (Tow and Schultz 1991). On the Great Plains 2- or 3-years rotations with legumes will result in control of Cephalosporium stripe [*Cephalosporium gramineum* Nis. & Ika. (syn. *Hymenula cerealis* Ell. & Ev.)] (Wiese 1977). Additionally, winter annual grass weeds such as downy brome (*Bromus tectorum* L.) can be effectively managed in legume rotations.

In northern Idaho, fall-planted Austrian winter pea [AWP, *Pisum sativum* spp. *arvense* (L.) Poir] has also been shown to have the potential to improve soil quality, producing in excess of 10 t ha^{-1} of organic matter (Auld et al. 1982). On the Western High Plains, winter annual legumes could play a similar role in rotation with wheat by partially replacing the typical 14 month fallow period if materials can be identified with sufficient cold-tolerance to survive the harsh winters of this region. The objective was to determine winter survival potential of cultivars and/or experimental lines of annual medic and Austrian winter pea on the Western High Plains.

METHODOLOGY

Replicated field trials at various locations in southeastern Wyoming were conducted starting in 1992 beginning with AWP. Field trials contained 3 to 4 replications. Plots were 1.5 by 7.6 m, except for the 1994/95 medic trial which were 15 by 60 cm microplots. Trials were established during Aug.–Sept. To assess winter survival, stand counts were made in Nov. and again in Apr. Means were calculated and are presented in plants/m^2.

RESULTS AND DISCUSSION

Annual Medics

Commercial Australian annual medic cultivars failed the winters of 1993/94 and 1994/95 (Table 1 and 2). Experimental lines with potential cold tolerance were established in fall 1994 along with commercial cultivars. On Apr. 1, 1995 survival was as follows: *M. polymorpha* cv. Santiago (0%), *M. polymorpha* line SC03077 (3%), *M. truncatula* cv. Paraggio (3%), *M. truncatula* line SCO7078 (5%), *M. rigidula* line SCO3075 (55%), and (80%) for the control *M. sativa* cv. Ladak, a very winter hardy alfalfa (Abernethy and Thiel 1985). Plants of *M. rigidula* flowered beginning Apr. 28 and produced mature seed pods by July 6.

Our results indicate that *M. rigidula* offers the greatest potential for winter survival of all the annual medics evaluated in this region. Our study confirms results of Abd El Moneim and Cocks (1986) and Cocks and Ehrman (1987) at the International Center for Agricultural Research in the Dry Areas, Allepo, Syria. These researchers found that in two severe winters (1983 and 1984), *Medicago rigidula* exhibited more cold tolerance than other annual *Medicago* spp.

Recently *M. rigidula* has been differentiated into the two species *M. rigidula* (L.) All. and a new species, *M. rigiduloides* E. Small. (Small et al. 1990; Small 1990). Subsequently, research at the South Australian Research and Development Institute (R.W. Groose, R. Ballard, N. Charman, A.W.H. Lake, unpublished re-

Table 1. Fall and spring plant populations of commercial Australian annual medic cultivars along with alfalfa in Wyoming 1993/94, values are means based on four locations.

	Population (plants/m^2)	
Genotype	Fall	Spring
M. littoralis cv. Harbinger AR (strand medic)	15	0
M. truncatula cv. Caliph (barrel medic)	44	0
M. truncatula cv. Paraggio (barrel medic)	9	0
M. sativa cv. Ladak[z] (alfalfa)	28	11
M. scutellata cv. Sava (snail medic)	16	0

[z]Very winter hardy.

Table 2. Fall and spring plant populations of experimental lines and commercial Australian annual medic cultivars along with alfalfa in Wyoming 1994/95.

	Population (plants/m^2)	
Genotype	Fall	Spring
M. polymorpha cv. Santiago (burr medic)	260	0
M. polymorpha SC03077 (burr medic)	260	7
M. rigidula SCO3075	260	143
M. sativa cv. Ladak[z] (alfalfa)	260	208
M. truncatula cv. Paraggio (barrel medic)	260	7
M. truncatula SCO7078 (barrel medic)	260	14

[z]Very winter hardy.

Table 3. Location, year, fall and spring populations of Melrose Austrian winter peas planted in Wyoming.

Location	Year	Population (plants/m²)	
		Fall	Spring
Archer	1990/91	28	4
Torrington	1990/91	13	7
Archer	1992/93	108	68
Chugwater	1992/93	104	96
Hawk Springs	1992/93	116	56
Pine Bluffs	1992/93	144	40
Torrington	1992/93	60	52
Chugwater	1993/94	69	52
Hawk Springs	1993/94	62	45
Pine Bluffs	1993/94	71	19
Chugwater	1994/95	62 (56)[z]	51(48)
Hawk Springs	1994/95	67 (61)	54(58)
Pine Bluffs	1994/95	65 (57)	49(46)
Torrington	1994/95	60 (57)	55(49)
Archer	1994/95	22 (20)	17(16)
Mean		70 (50)	44(43)

[z]Parenthetical values are for 'Common' Austrian winter pea.

sults) has demonstrated that the two species have very different *Rhizobium* requirements. Commercial strains of *Rhizobium* recommended for various annual medic species in South Australia effectively nodulated *M. rigidula* lines but failed utterly with lines of *M. rigiduloides*. Conversely, some experimental *Rhizobium* strains effectively nodulated *M. rigiduloides* but were only weakly effective with *M. rigidula*.

Lines of *M. rigidula* and *M. rigiduloides* from high elevations and the northern limits of the ranges of these species in Europe and Asia have been selected and used as parents in a breeding program to produce winter-hardy medics adapted to the Western High Plains that will combine enhanced cold-tolerance, optimum *Rhizobium* specificity, reduced pod spininess and high productivity.

Austrian Winter Pea

Mean winter survival of AWP cv. Melrose across 15 location-years was 63% (Table 3). Survival of AWP cv. Common across 5 location-years was 86%. These findings compare favorably to those reported for Moscow, Idaho and Bozeman, Montana (Auld et al. 1983). Our findings indicate that commercial sources of AWP should contain good winterhardiness potential for the Western High Plains. A screening program is under way with hopes of identifying lines with even greater winter survival potential.

REFERENCES

Abd El Moneim, A.M. and P.S. Cocks. 1986. Adaptation of *Medicago rigidula* to a cereal-pasture rotation in north-west Syria. J. Agr. Sci. Camb. 107:179–186.

Abernethy, R.H. and D.S. Thiel. 1985. Selection of alfalfa varieties. p. 6–11. In: D.W. Koch (ed.), Profitable alfalfa production. Wyoming Coop. Ext. Serv. B 867.

Auld, D.L., B.L. Bettis, M.J. Dial, and G.A. Murray. 1982. Austrian winter and spring peas as green manure crops in northern Idaho. Agron. J. 74:1047–1050.

Auld, D.L., R.L. Ditterline, G.A. Murray, and J.B. Swensen. 1983. Screening peas for winterhardiness under field and laboratory conditions. Crop Sci. 23:85–88.
Cocks, P.S. and T.A. Ehrman. 1987. The geographic origin of frost tolerance in Syrian pasture legumes. J. Appl. Ecol. 24:673–683.
Crawford, E.J., A.W.H. Lake, and K.G. Boyce. 1989. Breeding annual *Medicago* species for semiarid conditions in southern Australia. Adv. Agron. 42:399–437.
Reeves, T.G. 1987. Pastures in cropping systems. p. 501–515. In: J.L. Wheeler, et al. (eds.), Temperate pastures: their production, use and management. Australian Wool Cooperation/CSRIO, Sydney, Australia.
Schuman, G.E., J.D. Reeder, and R.A. Bowman. 1994. Short-term changes in soil quality in response to a grass community established on marginal croplands. p. 265–266. In: Trans. 15th World Cong. Soil Sci., Vol. 66, Comm. V, Acapulco, Mexico, July 1994.
Small, E., B. Brookes, and E.J. Crawford. 1990. Intercontinental differentiation in *Medicago rigidula*. Can. J. Bot. 68:2607–2613.
Small, E. 1990. *Medicago rigiduloides*, a new species segregated from *M. rigidula*. Can. J. Bot. 68:2614–2617.
Tow, P.G. and J.E. Schultz. 1991. Crop and crop-pasture sequences. p. 55–75. In: V. Squires and P. Tow (eds.), Dryland farming—a systems approach. Sydney Univ. Press, Sydney, Australia.
Wiese, M.V. 1977. Compendium of wheat diseases. Am. Phytopath. Soc., St. Paul, MN.
Wyoming Agricultural Statistics Service. 1995. Wyoming Agricultural Statistics. Wyo Dept. Agr., Cheyenne, WY.

Grain Yield of Oat-Pea Intercrop

Patrick M. Carr, Eric D. Eriksmoen, Glenn B. Martin, and N. Rick Olson

Spring oat (*Avena sativa* L., Poaceae) is grown for both forage and grain in temperate regions at upper latitudes. Oat is sometimes intercropped with spring pea (*Pisum sativum* L.) to enhance forage quality. Oat-pea mixtures have been considered for grain and seed production, though not to the extent as for forage. Generally, total grain yield has not been increased by intercropping (Carter and Larson 1964; Cowell et al. 1988).

Oat-pea intercrops have been established by sowing both crops at rates less than those used when sole-cropping either crop. This investigation was conducted to determine how grain yield of oat-pea mixtures was affected by sowing oat and pea components at or above the sole-cropped rate in North Dakota, U.S.

METHODOLOGY

A field experiment was conducted at Dickinson (46°53' N, 102°49' W) in southwestern North Dakota under dryland management during the 1993 and 1994 growing seasons, and at Hettinger (46°00' N, 102°30'W) under dryland management during the 1993 growing season. Average daily temperature ranges from –10°C during the winter (Dec. 1 to Feb. 19) to 19°C during the summer (June 1 to Aug. 31) at Dickinson, and –9° to 20°C at Hettinger. Precipitation averages 422 mm annually at Dickinson and 407 mm at Hettinger, with slightly less than two-thirds occurring between Apr. 1 and July 31.

The experiment was established in a seedbed that had been fallowed the previous year at Dickinson in 1993 and 1994, and at Hettinger in 1993. The seedbed was prepared by disking and then firming with a field cultivator and harrow. The experiment was also established in an annually-cropped, no-tillage seedbed at Dickinson in both years.

'Dumont' oat was sown alone and mixed with 'Trapper' field pea on Apr. 29 at Dickinson and Apr. 22 at Hettinger in 1993, and May 2 at Dickinson in 1994. 'Dumont' is grown for both grain and forage and is a popular oat cultivar in the northern Great Plains region of North America. 'Trapper' is a small sized pea that is suited to intercropping with oat (Chapko et al. 1991).

Oat was sown at 185 kernels/m² (pure live seed) in sole-crops at Dickinson in 1993, and at 93, 185, and 278 kernels/m² in 1994. Oat was sown at 99, 198, and 297 kernels/m² at Hettinger in 1993. Pea was sown at 80 seeds/m² (pure live seed) in sole-crops at Dickinson and Hettinger in both years, and also at 40 and 120 seeds/m² at Hettinger in 1993 and at Dickinson in 1994. Oat was intercropped with pea at all sole-crop rates used at each location in 1993 and 1994.

Table 1. Total yield of sole-cropped and intercropped 'Dumont' oat and 'Trapper' pea during 1993 in a fallowed environment located at Hettinger (1993H) and in fallowed and annually-cropped environments located at Dickinson in southwestern North Dakota in 1993 and 1994, and the Land Equivalent Ratio (LER) of intercropped treatments.

	Yield					LER				
	Fallow			Annually-cropped		Fallow			Annually-cropped	
Seeding rate (seeds/m²)	1993	1994	1993H[z]	1993	1994	1993	1994	1993H	1993	1994
					Sole-crop					
93 (oat low)	--	1712	3871	--	1796	-.-	-.-	-.-	-.-	-.-
185 (oat med.)	2777	2272	4229	2825	2386	-.-	-.-	-.-	-.-	-.-
375 (oat high)	--	2742	3871	--	2280	-.-	-.-	-.-	-.-	-.-
40 (pea low)	--	1816	1839	--	1814	-.-	-.-	-.-	-.-	-.-
80 (pea med.)	1768	1888	1860	1636	2229	-.-	-.-	-.-	-.-	-.-
160 (pea high)	--	1897	2489	--	2115	-.-	-.-	-.-	-.-	-.-
					Intercrops[y]					
Oat low/pea low	3399	1733	3959	2982	2046	1.37	0.82	0.96	1.29	0.89
Oat low/pea med.	2601	1942	3745	2842	2164	1.08	0.95	0.94	1.31	0.95
Oat low/pea high	2513	2017	3913	2814	2217	1.08	1.02	1.01	1.33	0.98
Oat med./pea low	2932	2419	4041	3138	2298	1.11	1.09	0.96	1.24	0.98
Oat med./pea med.	3084	1877	4264	2936	2360	1.19	0.90	1.02	1.26	1.03
Oat med./pea high	2926	2416	4051	2924	1961	1.14	1.15	0.99	1.32	0.86
Oat high/pea low	2991	2653	4377	2975	2304	1.11	1.19	1.05	1.18	0.99
Oat high/pea med.	2868	2533	4103	3043	2191	1.08	1.16	0.99	1.27	0.95
Oat high/pea high	3010	2359	3829	3372	2216	1.14	1.10	0.91	1.42	0.96
Mean	2806	2152	4021	2863	2159	-.-	-.-	-.-	-.-	-.-
CV(%)	19.8	16.0	9.0	17.5	16.1	-.-	-.-	-.-	-.-	-.-
LSD (0.05)	803	492	NS[x]	722	496	-.-	-.-	-.-	-.-	-.-

[z]Sowing rates at Hettinger for oat were: oat low = 99, oat med. = 198, and oat high = 297 kernels/m².
[y]Yield is combined yield of oat and pea.
[x]NS = not significant.

A randomized complete-block design was used at all locations. Blocks were replicated four times. Plot dimensions were 1.8 by 8 m at Dickinson and 1.6 by 8.5 m at Hettinger.

Yield was determined by collecting mixed grain and seed with a research combine. Grain and seed samples were cleaned and dried at 60°C for 48 to 72 h and weighed. Split pea seed could not be separated from oat kernels in oat-pea mixtures, so a 200 g subsample was collected from each intercropped sample and separated by hand so crop composition of the grain mixture could be estimated.

The land equivalent ratio (LER) index was used to evaluate intercrop efficiencies with respect to sole-crops. The LER defines yield as a function of area (LER = $Io/So + Ip/Sp$) where I and S refer to intercrop and sole-crop yields, respectively, and the subscripts o and p indicate the oat and pea component crops in the mixture.

Data were analyzed for each location but not across years or environments because some sole-cropped treatments were included in 1994 but not in 1993 at Dickinson. In addition, the recommended sole-crop rate for oat differed between Dickinson and Hettinger.

Table 2. Yield of sole-cropped and intercropped 'Dumont' oat and 'Trapper' pea during 1993 in a fallowed environment located at Hettinger (1993H) and in fallowed and annually-cropped environments located at Dickinson in southwestern North Dakota in 1993 and 1994.

	Oat grain yield (kg/ha)					Pea seed yield (kg/ha)				
	Fallow			Annually-cropped		Fallow			Annually-cropped	
Seeding rate (seeds/m^2)	1993	1994	1993H[z]	1993	1994	1993	1994	1993H	1993	1994
					Sole crop					
93 (oat low)	--	17121	3871	--	1796	--	--	--	--	--
185 (oat med.)	2777	2272	4229	2825	2386	--	--	--	--	--
375 (oat high)	--	2742	3871	--	2280	--	--	--	--	--
40 (pea low)	--	--	--	--	--	--	1816	1839	--	1814
80 (pea med.)	--	--	--	--	--	1768	1888	1860	1636	2229
160 (pea high)	--	--	--	--	--	--	1897	2489	--	2115
					Intercrops					
Oat low/pea low	2695	1136	3871	2084	1002	704	596	88	898	1044
Oat low/pea med.	1882	833	3548	1674	563	719	1109	197	1168	1602
Oat low/pea high	1673	552	3620	1524	356	839	1464	293	1290	1861
Oat med./pea low	2642	2108	4014	2624	1540	289	310	27	514	757
Oat med./pea med.	2703	1061	4229	2070	1092	380	815	35	866	1268
Oat med./pea high	2498	1400	3942	1826	541	428	1016	109	1098	1420
Oat high/pea low	2823	2437	4337	2464	1609	169	216	40	511	695
Oat high/pea med.	2646	2057	4050	2285	1106	223	475	53	759	1085
Oat high/pea high	2723	1619	3799	2506	1192	287	741	30	866	1024
Mean	2506	1661	3948	2188	1289	581	1029	588	961	1410
CV(%)	18.0	21.9	9.0	21.5	28.1	36.4	11.5	54.0	14.8	19.4
LSD (0.05)	653	523	502	684	520	307	170	464	207	393

[z]Sowing rates at Hettinger (1993H) for oat were 99, 198, and 297 kernels/m^2.

RESULTS

Yield was similar between intercrop treatments and sole-crop oat treatments where oat was sown at the recommended rate at each location (Table 1). More grain and seed generally was produced by some intercrop treatments than seed by a sole-crop of pea. These data indicate that intercropping may not reduce total yield compared with a monoculture of oat if oat is sown at the sole-crop rate or a heavier rate in the mixture. Total yield can be increased by intercropping compared with a sole-crop of pea.

More of the mixed grain and seed was oat as the oat seeding rate was increased and the pea seeding rate was decreased (Table 2). Pea composed more of the mixed grain and seed as the pea seeding rate was increased and the seeding rate for oat was reduced. No intercrop treatment in this investigation produced mixed grain that was comprised of equal proportions of oat and pea at all locations. However, roughly equal proportions of oat groats and pea was produced at Dickinson across environments and years when oat was sown at half the sole-crop rate and pea at the sole-crop rate.

The LER suggested a potential advantage for intercropping in both environments at Dickinson in 1993, and for some intercrop treatments in the fallowed environment in 1994 (Table 1). An advantage for intercropping was not suggested by LERs of intercropped treatments at Hettinger in 1993 and in the annually-cropped environment at Dickinson in 1994. The largest LER value for an intercrop treatment across locations and years occurred when both oat and pea were sown at rates heavier than those recommended for a sole-crop.

CONCLUSION

Total yield generally was maintained or enhanced when oat was intercropped at a sole-crop rate with pea compared to growing oat alone in this investigation. Previous work indicates that crude protein content is increased by intercropping oat with pea compared with growing oat alone (Robinson 1960; Johnston et al. 1978). Since total yield can be maintained when intercropping oat and pea, we suggest that oat-pea mixtures are a higher N-grain alternative to a monoculture of oat, and should be considered if the mixed grain is grown for direct consumption by livestock.

REFERENCES

Carter, J.F. and K.L. Larson. 1964. Oats, peas, and vetch for hay or silage in North Dakota. North Dakota Ag. Expt. Sta. Fargo. Bul. 14 Agr. 9–5.

Chapko, L.B., M.A. Brinkman, and K.A. Albrecht. 1991. Oat, oat-pea, barley, and barley-pea for forage yield, forage quality, and alfalfa establishment. J. Prod. Agr. 4:486–491.

Cowell, L.E., E. Bremer, and C. Van Kessel. 1988. Yield and N_2 fixation of pea and lentil as affected by intercropping and N application. Can. J. Soil Sci. 69:243–251.

Johnston, H.W., J.B. Sanderson, and J.A. MaCleod. 1978. Cropping mixtures of field peas and cereals in Prince Edward Island. Can. J. Plant Sci. 58:421–426.

Robinson, R.G. 1960. Oat-pea or oat-vetch mixtures for forage or seed. Agron. J. 52:546–549.

FORAGES

Annual Forages: New Approaches for C-4 Forages

Jeffrey F. Pedersen

The current agricultural paradigm in the U.S. is heavily biased towards the production and marketing of crops as commodities. This paradigm is kept in place by grain handling and marketing infrastructure, as well as government farm programs, designed for crops as commodities. Sorghum (*Sorghum bicolor* L. Moench) and maize (*Zea mays* L.) grown for grain certainly fit into this current paradigm.

A new approach to agricultural production and marketing, identity preserved products, is gaining in importance. This is made possible through new technologies and markets demanding products designed specifically for their needs. Examples of investment in identity preserved products include Pioneer Hi-Bred's new research facility for grain quality (Johnson 1995), DuPont's current emphasis on value-added maize (Freiberg 1994), and Iowa State University's value-added grain marketing program (Wrage 1995). However, marketing strategies for such new grain products must be outside the current norm of sale to a local elevator and resale of a bulk commodity to grain consumers.

Sorghum and maize grown for forage are already outside of the current marketing structure for grain crops. They are typically grown, stored, and fed on the same farm. Management decisions including choice of hybrids can be based on end use. As such, it could be argued that the current paradigm for production and utilization of forages in the United States, especially annual forages, is very favorable for the development and marketing of identity preserved hybrids and varieties designed for specific end users.

ATTRIBUTES OF SORGHUM AND MAIZE FORAGE

Both sorghum and maize produce extremely high yields of high quality forage. Typical forage sorghum yields in the central U.S. (Nebraska, 1991–1993) are 27 Mg/ha, and forage maize yields are 31 Mg/ha (USDA Agricultural Statistics 1994). When grown for silage, grain content often approaches 50% ensuring high energy content (Irlbeck et al. 1993). Both sorghum and maize offer year-to-year management flexibility since they are annuals. Both also require considerably more management inputs than other forages. However, forage sorghum being very heat and drought tolerant is usually grown without irrigation. Nearly all forage maize is harvested, stored, and utilized as silage. Forage sorghum is grazed, baled for hay, green chopped, or ensiled.

Quality of the non-grain components of the forage has been historically viewed as relatively poor. To date, breeding emphasis has been for high grain yields in both sorghum and maize. High grain content has been relied upon to provide high energy content in ensiled crops. For grazing and hay type forage sorghum, harvestable dry matter yield has been emphasized. However, if 50% or more of the dry matter of ensiled forages, and up to 100% of dry matter production of forage sorghum is non-grain, quality enhancement of the forage component should contribute significantly to improved animal performance.

Users of maize silage as a forage source have always been relatively quality conscious. Producers of sorghum for forage often utilize a very different decision tree. The decision to plant forage sorghum is often based on many factors which include government programs, anticipated rainfall, pasture conditions, availability of harvesting equipment, etc. However, the decision to select a particular hybrid or variety of forage sorghum is increasingly made on perceived quality. Often these perceptions are based on animal acceptance or palatability and can outweigh yield in the process of selecting a hybrid or variety.

NEW APPROACHES TO FORAGE SORGHUM IMPROVEMENT

New genetic discoveries can typically take 10–15 years to move through testing and integration into useful lines, to commercially available products. *Brown midrib* mutants were originally induced and described in sorghum 17 years ago (Porter et al. 1978). This trait is associated with reduced lignin content and up to a 55% concomitant increase in forage digestibility expressed in both sorghum and sudangrass (*Sorghum bicolor* L. Moench var. *sudanense*) (Fritz et al. 1981) and their hybrids (Cherney et al. 1986; Fritz et al. 1990).

Such increases in digestibility are associated with increased animal performance. Lusk et al. (1984) fed a *brown midrib* sorghum silage and a corn silage to dairy cows and showed milk and fat corrected milk production to be equal (normal sorghum is usually associated with lower animal performance). More recently in our lab, milk production from cows fed silage of a pre-commercial *brown midrib* sorghum, normal sorghum, maize, and alfalfa was measured (Grant et al. 1995). Milk production was similar for cows fed the *brown midrib* sorghum, alfalfa, and corn silage. Cows fed the normal sorghum silage produced approximately 23% less milk than those fed *brown midrib* silage (Table 1). Milk fat production was similar for cows fed the *brown midrib* sorghum, alfalfa, and corn silage. Cows fed the normal sorghum silage produced approximately 35% less milk fat than cows fed *brown midrib* sorghum, alfalfa, and corn silage. Incorporation of the *brown midrib* trait into sorghum clearly improves animal performance.

Although not widely documented in the literature, it is common knowledge among sorghum breeders that the *brown midrib* trait is also associated with decreased agronomic performance. This is being overcome with applied breeding effort. Although replicated comparisons were not made in our above study, the pre-commercial *brown midrib* sorghum hybrid and the normal sorghum hybrid both yielded nearly 9000 kg/ha dry matter with no apparent differences in lodging or disease noted at harvest. Regretfully, the fate of this *brown midrib* sorghum hybrid is currently unknown due to changes in ownership and management of the company that provided the hybrid. However, commercial availability of another *brown midrib* sorghum hybrid is anticipated for the crop year 1997 (Jerry O'Rear, Garrison & Townsend, Inc. pers. commun.). Considerable effort is currently being invested in development of *brown midrib* seed parent (female) inbred lines for use in hybrid production by this and other public sector research programs.

Another new approach to forage sorghum improvement involves the use of alternate sources of cytoplasmic male sterility. Stephens and Holland (1954) originally described a system, now known as the A1 system, for hybrid sorghum seed production based on the interaction of the nuclear and cytoplasmic genes of two races of sorghum, milo, and kafir. In this system, sorghum lines are classified based on their ability to restore fertility or maintain male sterility in A1 cytoplasm. This limits potential heterotic combinations, but because of A1 cytoplasm's stability and ease of use, it has been adopted as the commercial standard and is used for essentially all commercial hybrid sorghum seed production. Alternate sources of cytoplasmic sterility do, however, exist and offer unique opportunities (Schertz and Ritchey 1978).

As alluded above, many new heterotic combinations are possible using alternate sources of cytoplasmic male-sterility. One alternate male-sterile cytoplasm, A3, has very few known restorers of male fertility. This allows the production of male-sterile forage hybrids with a much reduced risk of volunteer outcrosses with weedy sorghums the subsequent year. Although male-sterile hybrids would obviously not be acceptable for grain production, such hybrids may offer improvements in quality for grazing applications by removing seed as a strong carbohydrate sink, and increasing the non-structural carbohydrates in the forage.

A common forage sorghum hybrid is A1 cytoplasmic male-sterile 'Redlan' (a grain sorghum seed parent line) pollinated with 'Greenleaf' sudangrass (Harvey 1977). We recently obtained A1 and A3 male sterile

Table 1. Performance of midlactation cows as influenced by forage source (Grant et al. 1995).

	Yield (kg/d)	
Diet	Milk	Fat
Normal sorghum	20 b[z]	0.7 b
bmr sorghum	26 a	1.1 a
Alfalfa	30 a	1.1 a
Maize	26 a	1.1 a

[z]Mean separation in columns by Duncan's Multiple Range Test, 10% level.

versions of 'Redlan' and three other seed parent lines, produced hybrid seed, and conducted an experiment to compare hybrid forages in these two cytoplasms. Preliminary results showed equivalent dry matter yields, days to 50% anthesis (a measure of maturity), and height between A1 and A3 versions of the same hybrids (Table 2).

NEW APPROACHES TO FORAGE MAIZE IMPROVEMENT

Recent changes in approach to forage maize improvement in the U.S. have focused on variation in the forage quality of the stover portion of the plant. Most previous breeding efforts have emphasized increasing grain yield as a means of increasing the energy content of the overall forage since it is very digestible and makes up a large proportion of the whole-plant dry matter. Surprisingly, correlations between maize grain yield and whole plant yield, stover yield, and whole-plant digestibility are low (Table 3). Increases in the forage quality of maize will, therefore, probably focus on improvement of stover quality.

Brown mid-rib mutants have been known in maize since at least 1926 (Eyster). However, widespread

Table 2. Mean values for total seasonal dry matter production, days to anthesis, and plant height of A1 vs. A3 sorghum x sudangrass hybrids (Toy et al. 1993).

Seed parent	Dry matter yield (kg/ha)	Days to anthesis (days)	Height (cm)
A1 Martin	10600[z]	77	162
A3 Martin	12300	77	169
A1 Redbine58	12500	76	168
A3 Redbine58	12200	76	172
A1 Wheatland	10400	82	152
A3 Wheatland	10700	82	151
A1 Ks24	11000	79	159
A3 Ks24	11100	79	165

[z]No significant differences were shown for the single degree of freedom contrasts shown in this table at 5% level.

Table 3. Correlation of maize grain yield with whole-plant dry matter yield, stover yield, and whole-plant digestibility in two testcrosses (Wolf et al. 1993)

Variable	Maize grain yield correlation (r) Testcross A	Testcross B
Dry matter yield		
whole plant	0.65	0.54
stover	0.53	0.46
Whole plant digestibility	–0.03	–0.02

Table 4. Extreme variation for in vitro digestible dry matter among 45 maize inbred lines (Lundvall et al. 1994)

Inbred	In vitro dry matter digestibility (g/kg) Early-harvest stems	Leaves	Late-harvest stems
B57	727	676	576
bm1bm1	654	665	497
LAN232	666	589	502
B52	513	632	461
L289	557	583	262
LSD (5%)	31	20	49

use of these mutants in commercial hybrids has not occurred because they are also associated with poor grain and stover yield, and susceptibility to lodging (Miller et al. 1983; Lee and Brewbaker 1984).

Several recent studies have investigated the relationships among various forage quality traits in maize (Allen et al. 1991; Wolf et al. 1993). Jung and Buxton (1994) utilized a multiple regression approach to identify the principle components of maize polysaccharide degradability (quality), and concluded that because of the complex interactions in cell-wall organization, no single cell-wall component, or simple combination, can accurately predict degradability of maize cell walls. Significant genetic variation for maize stover digestibility has been shown by several researchers (Hunt et al. 1992; Lundvall et al. 1994) (Table 4). Although as pointed out by Deinum and Bakker (1981), "breeding for stover quality will not be simple as a good correlation between stover quality and stover morphology has not yet been found," it appears that as concluded by Wolf et al. (1993) "future efforts to improve whole-plant digestibility should concentrate on stover digestibility."

New Markets for C-4 Forages

The above discussions centered on approaches to design sorghum and maize forages for the major existing market, i.e. livestock. If we move past this "old" marketing paradigm, some exciting new markets for these forage materials open up. The first and possibly the potentially largest is energy production. Lueschen et al. (1991) reported ethanol yields of 3048 liters/ha from fermentable carbohydrates in 'Keller' sweet sorghum sap with no nitrogen fertilization (residual soil N was approximately 90 kg/ha prior to the experiment). Although the authors indicate that the study was inconclusive, reduced N inputs compared to maize grain could make energy balances favorable. New technologies to convert the fiber to fermentable carbohydrates will make yields and energy balances even more favorable, and are discussed in another paper in this volume (Vogel and Shearman 1996).

Other potential new markets for sorghum and maize forages includes uses in the paper and construction industries. Sorghum stalk residue following grain harvest have been combined with high strength adhesives to produce a product very similar to wafer board (Barbara Kliment, Nebraska Sorghum, Board pers. commun.). This product demonstrates a viable alternate use of sorghum or maize stalks. However, it has not reached the commercial marketplace.

Another promising alternate use of sorghum stalk residues, the production of paper products, is receiving current market interest due to high pulp and paper prices (Tim Lust, National Sorghum Board, pers. commun.). The short cellulose fibers of sorghum stalks are similar to hardwood fibers. Production options include fine white paper, magazine print, copy paper, newsprint, and heavy unbleached paper for use in corrugated boxes.

Table 5. Dry matter yield, nitrogen accumulation, and in vitro dry matter digestibility (IVDMD) of six types of sorghum grown on a site containing high soil nitrogen (Pedersen et al. 1995).

Type	Dry matter yield (Mg/ha)	Nitrogen accumulated (kg/ha)	IVDMD (%)
Grain sorghum	13.5	204	63
Sudangrass	15.6	208	50
Sorghum × sudangrass	19.5	251	52
Forage sorghum	17.1	240	61
Sweet sorghum	15.8	203	60
Tropical sorghum	19.4	266	54
SE	0.7	11	1

We have also investigated the potential of using forage sorghum for remediation of chemically (N) contaminated soils, as well as genetic variation for this character (Pedersen et al. 1995). In general, sorghum has a very high N-uptake potential and an extensive root system which is very effective in scavenging N from the soil. Three hybrids each of six types of sorghum (tropical, forage, sudangrass, sorghum x sudangrass, grain, and sweet) were evaluated for N accumulation on a municipal biosolids disposal site. Tropical sorghum and sorghum x sudangrass had the highest dry matter production and accumulated the most nitrogen. Biomass produced was adequate for beef maintenance diets, and may be a useful biomass source for ethanol production from cellulose (Table 5).

CONCLUSIONS

New approaches for genetic improvement of forage sorghum and forage maize are being used to develop superior products for livestock feeding, the major existing market. These products are good current examples of a defined market need supporting the development of identity preserved, value added hybrids. These forage products are already marketed outside of the normal commodity grain/grain elevator paradigm common to much of U.S. agriculture. As the marketplace demands new raw materials from agriculture, crops such as sorghum and maize can be genetically modified to meet those specific needs. Successful integration of new approaches and new raw products into the marketplace will depend in part on new marketing structures which will include identity preserved value-added, agricultural products which must be developed specifically for targeted markets.

REFERENCES

Allen, M.S, K.A. O'Neil, D.G. Main, and J. Beck. 1991. Relationship among yield and quality traits of corn hybrids for silage. J. Dairy Sci. 74:(suppl. 1):221.

Cherney, J.H., K.J. Moore, J.J. Volenec, and J.D. Axtell. 1986. Rate and extent of digestion of cell wall components of brown-midrib sorghum species. Crop Sci. 26:1055–1059.

Deinum, B. and J.J. Bakker. 1981. Genetic differences in digestibility of forage maize hybrids. Neth. J. Agr. Sci. 29:92–98.

Eyster, W.H. 1926. Chromosomes VIII in maize. Science 64:22.

Freiberg, B. 1994. Will DuPont's muscle speed up value-added ag production? Seed Crops Ind. December:4–36.

Fritz, J.O., R.P. Cantrell, V.L. Lechtenberg, J.D. Axtell, and J.M. Hertel. 1981. Brown midrib mutants in sudangrass and grain sorghum. Crop Sci. 21:706–709.

Fritz, J.O., K.J. Moore, and E.H. Jaster. 1990. Digestion kinetics and cell wall composition of brown midrib sorghum x sudangrass morphological components. Crop Sci. 30:213–219.

Grant, R.J., K.J. Moore, and J.P. Pedersen. 1995. Brown midrib sorghum silage for midlactation dairy cows. J. Dairy Sci. 78:1970–1980.

Harvey, P.H. 1977. Sorghum germplasm base in the U.S. Proc. 32nd Annual Corn and Sorghum Research Conference, Am. Seed Trade Assn., Chicago. p. 168–198.

Hunt, C.W., W. Kezar, and R. Vinande. 1992. Yield, chemical composition, and ruminal fermentability of corn whole plant, ear, and stover as affected by hybrid. J. Prod. Agr. 5:286–290.

Irlbeck, N.A., J.R. Russell, A.R. Hallauer, and D.R. Buxton. 1993. Nutritive value and ensiling characteristics of maize stover as influenced by harvest maturity and generation, plant density and harvest date. Anim. Feed Sci. Tech. 41:51–64.

Johnson, K.L. 1995. From seed to feed testing. Farm Ind. News 28:30.

Jung, H.-J.G. and D.R. Buxton. 1994. Forage quality variation among maize inbreds: relationships of cell-wall composition and in-vitro degradability for stem internodes. J. Sci. Food Agr. 66:313–322.

Lee, W.C. and J.L. Brewbaker. 1984. Effects of brown midrib-3 on yields and yield components of maize. Crop Sci. 24:105–108.

Lueschen, W.E., D.H. Putnam, B.K. Kanne, and T.R. Hoverstad. 1991. Agronomic practices for production of ethanol from sweet sorghum. J. Prod. Agr. 4:619–625.

Lusk, J.W., P.K. Karau, D.O. Balogu, and L.M. Gourley. 1984. Brown midrib or corn silage for milk production. J. Dairy Sci. 67:1739–1744.
Lundvall, J.P., D.R. Buxton, A.R. Hallauer, and J.R. George. 1994. Forage quality variation among maize inbreds: in vitro digestibility and cell-wall components. Crop Sci. 34:1672–1678.
Miller, J.E., J.L. Geadelmann, and G.C. Martin. 1983. Effect of the brown midrib-allele on maize silage quality and yield. Crop Sci. 23:493–496.
Pedersen, J.F., K.J. Moore, S. Schroth, and D.T. Walters. 1995. Nitrogen accumulation of six groups of sorhgum grown on a municipal biosolids use site. Water Environ. Res. 67:1076–1080.
Porter, K.S., J.D. Axtell, V.L. Lechtenberg, and V.F. Colenbrander. 1978. Phenotype, fiber composition, and in vitro dry matter disappearance of chemically induced brown midrib (*bmr*) mutants of sorghum. Crop Sci. 18:205–208.
Schertz, K.F. and J.J. Ritchey. 1978. Cytoplasmic-genic male-sterility systems in sorghum. Crop Sci. 18:890–893.
Stephens J.C. and R.F. Holland. 1954. Cytoplasmic male-sterility for hybrid sorghum seed production. Agron. J. 46:320–323.
Toy, J.J., J.F. Pedersen, K.J. Moore, and R.D. Lee. 1993. Fertility and forage yield of sorghum hybrids in A1 and A3 cytoplasm. Proc. 1993 American Forage and Grasslands Conference, Mar. 29–31, Des Moines, IA. p. 192–195.
USDA National Agricultural Statistics Service. 1994. USDA Agricultural Statistics. U.S. Government Printing Office, Washington, DC.
Vogel, K.P. and R. Shearman. 1996. Perennial grasses: New applications and uses. p. 263–270. In: J. Janick (ed)., Progress in new crops. Am. Soc. Hort. Sci., Alexandria, VA.
Wolf, D.P., J.G. Coors, K.A. Albrecht, D.J. Undersander, and P.R. Carter. 1993. Forage quality of maize genotypes selected for extreme fiber concentrations. Crop Sci. 33:1353–1359.
Wrage, K. 1995. NIR scanners may get value-added grain market rolling. Seed Crops Ind. February:6–26.

New Uses for Alfalfa and Other "Old" Forage Legumes

Joe H. Bouton

Legumes are a large, diverse, and agriculturally important family of plants. Most legume species are trees and shrubs as were their ancestral forms which probably arose about 135 million years ago during the Cretaceous period (Heywood 1971). To humans, however, the most important legume species belong to a small group of herbaceous crop and forage species. The main trait common to these legumes, and the trait of most importance to us, is the ability to fix atmospheric nitrogen and convert it to a useable form for plant growth (Allen and Allen 1981). This fixed nitrogen leads to a higher protein concentration in its various plant parts which in turn enhances our diet and can also be recycled into the environment as a form of fertilizer (e.g. green manure).

The role of grasses as the agricultural centerpiece in the ascent of human civilization is well documented. However, Isely (1982) makes a compelling argument that the archeological records show the domestication of grasses and legumes took place together. His reasons for this conclusion are that species of grasses and legumes as important and currently used food crops are specific to the early agricultural centers: e.g. *Glycine* with *Oryza* in China; *Pisum* and *Lens* with *Triticum* and *Hordeum* in the Middle East; *Phaseolus* with *Zea* in the New World; and *Sorghum* with *Vigna* in Africa. The exchange of legumes between the Old and New Worlds is also important in the development of our current civilization. In this exchange, *Phaseolus*, *Arachis*, and others went from the New World to the Old while *Pisum*, *Lens*, *Vigna*, *Glycine*, and others went from the Old World to the New.

Although the first written records of legume rotations trace back to the Roman Empire, it was not until the European feudal times that the practice of legume rotation was practiced to a great extent. This practice of using grasses and legumes as both rotation crop and livestock fodder reached its zenith in the ley farming system of Europe (Heath and Kaiser 1985). This practice is still the most economical and environmentally friendly method of farming.

Ley farming, therefore, is simply grassland farming. In today's grassland farming operations, grasses are the dominant plant component with legumes usually comprising a small percentage (<25%) of the sward. There are many methods of grassland farming now practiced throughout the world, but each has common objectives of using grasses and legumes to stabilize land from erosion, improve soil tilth, and allow livestock production either as a rotation with grain crops or as permanent pasture. Whatever method is practiced, the main purposes of legumes are to provide an input of nitrogen through its N_2-fixation ability and to upgrade the nutritive value of the grazed pasture.

TRADITIONAL ROLE OF LEGUMES IN GRASSLAND FARMING

The legume family, Fabaceae (Leguminosae), has three subfamilies, Caesalpinioideae, Mimosoideae, and Faboideae (or Papilionoideae), which may be considered confluent (Allen and Allen 1981; Heywood 1971). The Caesalpinioideae are primarily tropical trees and shrubs with some rare herbaceous forms and are probably the most primitive and diversified subfamily. They possess a well developed corolla which may be regular or irregular but not papilionoid. *Cassia* and *Swartzia* are representative of the Caesalpinioideae subfamily. The Mimosoideae are mainly trees, shrubs, and woody vines which probably evolved from the Caesalpinioideae. It is characterized by a reduction of the individual flowers and their congregation in compact heads or spikes. These infloresences are their most noticeable feature with the reduced corolla and long, filiform stamens providing it with a showy presence. Mimosa tree, *Acacia*, and *Leucaena* are representative types. The Papilionoideae possess the most familiar and important forage and crop species with trees, shrubs, and herbaceous types being commonly found. They are intergradient with the Caesalpinioideae and possess a distinct, papilionoid flowers. Representative genera are *Trifolium* (clovers), *Medicago* (medics), *Arachis* (peanut), *Glycine* (soybeans), *Phaseolus* (beans), *Lupinus* (lupine), and *Pisum* (pea).

Today, forage legume species are assigned different roles in grassland farming depending on their plant structures and abilities (Ball et al. 1993; Heath et al. 1985). The traditional roles of three major forage legume species along with the characteristics which allow them to assume a particular role are shown in Table 1. From this, one sees that low growing, stoloniferous, and competitive species such as white clover (*Trifolium*

Table 1. Widely used forage legume species, their major role in forage production, and the characteristics which allow this role.

Species	Role	Characteristics
Alfalfa, *Medicago sativa* L.	Hay and silage	Crown former, high yield, upright growth for easy harvesting, difficult to establish in grasses and non-tolerant of intensive grazing.
Red clover, *Trifolium pratense* L.	Hay and pasture	Crown former, easy to establish in grasses, high yield, some tolerance to intensive grazing.
White clover, *Trifolium repens* L.	Pasture	Stoloniferous, easy to establish in grasses, competitive, tolerant of intensive grazing.

repens L.) are used as a component of grazed pasture while high yielding, upright species such as alfalfa (*Medicago sativa* L.) are best suited to produce hay and silage in a monocropping system (Blaser et al. 1973).

Another important role for forages is in soil conservation (Shiflet and Darby 1985). In these systems, forage grasses and legumes are used in watershed management and to reduce erosion runoff. This includes their use as cover crops and their role in conservation tillage systems. A legume, when used as a cover crop, can contribute a large amount of nitrogen to the rotation. This is possible because legumes fix nitrogen in symbiosis with *Rhizobium* bacteria. The different species of *Rhizobium* are very specific to the legume species they infect (Table 2). This specificity may also be used to classify the different *Rhizobium* species (Heichel 1985).

The amount of nitrogen produced by cover crop legumes depends on the species used, the success of seed inoculation at planting time, and the total biomass produced. In a study in Georgia, adapted winter cover crops such as hairy vetch (*Vicia villosa* Roth) and crimson clover (*Trifolium incarnatum* L.) were found to replace all of the fertilizer nitrogen needed to produce good non-irrigated yields of grain sorghum [*Sorghum bicolor* (L.) Moench] and almost two-thirds of the nitrogen required for good yields of maize, *Zea mays* L. (McVay et al. 1989). There was also an improvement in soil physical properties such as water soluble aggregates and water infiltration rate when plots with forage legumes were compared with fallow areas.

What has emerged, therefore, is a traditional or "old" role for temperate forage legumes in grassland and conservation farming. Pertinent questions now being asked are: (1) can some of these species be modified to assume new roles within grassland agriculture, and (2) can forage legumes have major roles outside of traditional grassland agriculture?

The reasons for using old forage legumes in new roles are: (1) farmers and the lay public are already familiar with them, (2) each species is generally served by a successful seed industry, and (3) good cultivars and clear methods of successfully planting and managing them as crops are already available and practiced. These characteristics allow traditional forage legumes to assume any new role quicker and with less problems than a newly introduced species.

NONTRADITIONAL ROLES FOR ALFALFA

The forage legume with the greatest ability to assume many roles both within and outside of traditional grassland agriculture is alfalfa. Its main use is as a hay and silage crop, but when dried, ground, and pelleted, alfalfa is marketed as a dehydrated feed for many animals (Conrad and Klopfenstein 1988). Dehydrated or dehy alfalfa is used as either a protein supplement or the sole dietary ration for cattle, horses, and other small animals such as rabbits, mice, gerbils, etc. which are often used as pets. This dehy industry is now an important economic consideration in many areas of the United States.

Direct Grazing

Since tall, erect species like alfalfa require strict and intense grazing management to insure plant survival (Blaser et al. 1973), its widespread use for direct grazing was not practiced because of fear of losing stands.

Table 2. *Rhizobium* species and their legume host species (from Heichel 1985).

Species	Legume host
Rhizobium leguminosarum biovar *trifolii*	Clovers
Rhizobium leguminosarum biovar *phaseoli*	Bean
Rhizobium leguminosarum biovar *viceae*	Vetch
Rhizobium meliloti	Alfalfa, sweetclover
Rhizobium loti	Trefoil, lupine, chickpea
Bradyrhizobium japonicum	Soybean, cowpea, peanut

What was needed to broaden the pasture use of alfalfa was development of grazing-tolerant cultivars. These cultivars would give the livestock producers more flexibility for using alfalfa in many ways.

After World War II, development of grazing-tolerant alfalfa cultivars was a high priority for several alfalfa breeding programs (Heinrichs 1963). The approach for developing grazing cultivars was to identify a trait, which if enhanced in the germplasm, should lead to grazing persistence. In other words, find morphological features which should enhance grazing survival and incorporate these traits into alfalfa cultivars useful for grazing. The main trait identified and researched during that time was creeping rootedness (Heinrichs 1963). This trait is not really a true rhizome, but since a plant possessing it can originate a shoot from the root system, it was felt to provide the plant with a rhizome-like ability to regenerate in stands. However, creeping rootedness was found to be associated with low forage regrowth (Busbice and Hanson 1969), was erratic in its ability to truly creep and regenerate in pastures (Leach 1978), and most important of all, germplasms selected for the trait were not found to be any more grazing persistent than adaptive, hay-type checks (Gramshaw et al. 1982; Smith et al. 1989). Others traits thought to be useful in providing grazing persistence in alfalfa are deep-set crowns (Piskovatski and Stepanova 1980), subsurface budding (Lorenz et al. 1982), broad crowns (Kehr et al. 1963), prolific and nonsynchronous budding (Leach 1977), maintenance of root total nonstructural carbohydrates (Counce et al. 1984), and pest resistance (Leach and Clements 1984; Lodge 1991). However, no data was available which demonstrated that enhancement of any of these traits alone would increase grazing persistence.

Our approach at the Univ. of Georgia to develop and test grazing persistent alfalfa cultivars has been not to concentrate on individual trait enhancement, but to use the grazing animal during development and testing. Early screening of elite, multiple-pest resistant parental germplasms in grazed paddocks (sometimes with grass competition) was found to be important. During each cycle of selection, all germplasm is placed in replicated small plots within grazing paddocks and subjected to intensive, continuous stocking with beef cattle (Bouton et al. 1992; Smith and Bouton 1993). After this initial selection has eliminated material not adapted to the pasture environment, then reselection based on other specific limiting traits can be practiced. The importance of this approach of initially bringing all stresses to bear at the same time was seen with the success of 'Alfagraze' alfalfa (Bouton et al. 1991). 'Alfagraze' was found to possess excellent grazing tolerance (Fig. 1) and yield (Smith et al. 1989) and to be very competitive with grasses when grazed (Smith et al. 1992). Further work showed this same selection approach of achieving grazing tolerance was successful across a range of cultivars and also indicated that resistance to major alfalfa diseases played more of a secondary role in enhancing grazing persistence (Smith and Bouton 1993).

The importance and farmer interest of using alfalfa for grazing was recently seen at a research and industry conference held in Nashville, Tennessee and dedicated solely to grazing alfalfa (Certified Alfalfa Seed Council 1994). That an entire conference would be conducted and so well attended on this specific subject was noteworthy. The common message from both the scientists and producers speaking at the conference was using alfalfa as a grazing crop represented a very economical and efficient system for production of meat,

Fig. 1. Survival of 'Alfagraze' alfalfa, a grazing tolerant cultivar, when compared to an intolerant cultivar after intensive grazing.

milk, and wool. The current and future development of grazing tolerant cultivars was also highlighted as being instrumental in driving this interest and removing the central management problem of stand loss.

Alfalfa Sprouts

Beginning in the 1970s, alfalfa sprouts found a major use in human diets. When placed on salads or sandwiches, it supplies significant quantities of minerals, protein, and vitamins. In 1988, sprout production represented about 7% of the total alfalfa seed produced in the U.S. (Bass et al. 1988).

Production of Industrial Enzymes

A very important consideration for alfalfa is its adaptability in many areas of biotechnology research. It is easily grown and regenerated in cell culture and is receptive to transformation vectors such as the Ti plasmid (Austin et al. 1993). These traits allow it to be considered as a good plant for engineering with unique genes (e.g. trangenes). In this regard, using transgenic alfalfa containing genes which code for production of industrial enzymes has been proposed by a research group at the Univ. of Wisconsin (Austin et al. 1993, 1995). Most industrial enzymes are produced by microbes in large fermentation vessels, but an alternative system would be to produce enzymes such as alpha-amylase and manganese dependent lignin peroxidase (Mn-P) in alfalfa plants where they could be recovered by extraction. Since alpha-amylase is used in industrial starch processing and Mn-P has potential for lignin degradation and as a biopulping bleaching agent, this approach has great industrial potential. These investigators found that production of these enzymes in transgenic alfalfa was possible and strategies for extraction and processing of "juice" containing these enzymes from the alfalfa plant were also reported. There was a range of expression among transgenic genotypes for Mn-P (Fig. 2A), but in most cases, production of Mn-P suppressed plant yield (Fig. 2B) (Austin et al. 1995). Mn-P was also shown to segregate in sexual progeny from the transgenic genotypes. However, transgenic plants expressing alpha-amylase showed no effect on plant performance. These researchers concluded this technology could be used for inserting genes and extracting products for other value added traits in alfalfa (Austin et al. 1993).

Pulp Production

Another potential use for alfalfa is as a source of pulp for paper manufacturing. In Sweden, a basic feasibility study for an "Agro-Fibre" project designed to investigate the use of high yielding grasses and alfalfa instead of trees for production of pulp for paper was recently reported (Berggren 1993). Since the fiber yield per unit of land of alfalfa and grasses such as elephant grass (*Miscanthus*) was so much higher than the traditional birch trees, this report was fairly optimistic about the future of Agro-Fibre production. However, it was noted that two major limitations would need to be overcome before their use in production of pulp fiber becomes a reality: (1) the lack of a chemical process capable of economically handling these forages on a large scale and (2) risk factors. Since the investment in a new chemical pulp mill is always high, it is economi-

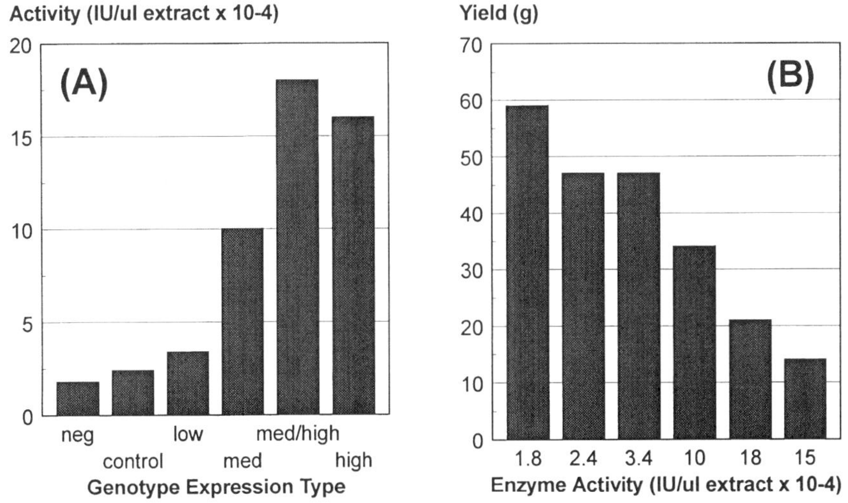

Fig. 2. (A) Manganese-dependent lignin peroxidase (Mn-P) activity of different alfalfa transgenic genotypic classes and (B) the relationship of Mn-P with plant yield (Austin et al. 1995).

cally safer for an individual company to use well-know processes and raw materials. Therefore, more time and research will be needed.

Fuel and Bioremediation

Finally, new projects are being investigated by USDA and Univ. of Minnesota scientists to use alfalfa as a fuel for generating electricity and as a bioremediation system to remove leached nitrates and prevent nitrate contamination (J. Lamb 1995, pers. commun.). The Minnesota Agri-Power project estimates a potential for converting 700,000 t of alfalfa per year into enough electricity to power a city the size of St. Cloud, for a year. In the proposed process, the alfalfa stems would be separated and used to create gas for fueling a 75-megawatt power plant while the leaves would be processed into dehydrated animal feed.

For the nitrate bioremediation to be successful, it was important to develop an alfalfa which possessed a more extensive rooting pattern yet was ineffective in establishing the nodulation and nitrogen fixation process with *Rhizobium* bacteria. This extensive rooting, ineffective alfalfa was developed, and when tested, was found to remove 140 and 700 kg ha^{-1} more nitrogen than reed canarygrass (*Phalaris arundinacea* L.) and switchgrass (*Panicum virgatum* L.), respectively (Fig. 3).

NONTRADITIONAL ROLES FOR OTHER FORAGE LEGUMES

Grain and Oil Seed Uses

When first introduced into the United States, soybean (*Glycine max* Jacq.) was used as a forage crop mainly being grown for hay, silage, fattening pasture for hogs and sheep, and green manure (Probst and Judd 1973). In fact, as late as 1939, approximately 60% of the acreage planted to soybeans in the United States was for forage and 43 of the 108 cultivars available were forage types (Hartwig 1973). However, due to intensive research efforts in breeding and management, soybean is now one of the most important oil seed crops in the United States and the world.

In the United States, lupines (*Lupinus* spp.) were traditionally grown alone or in mixtures with cereals for grazing and crop rotation until cheap nitrogen and diseases reduced their importance (Hoveland and Townsend 1985). With the advent of "sweet" lupines which show reduced alkaloids in the seed, these species have assumed a new role as a grain legume in Australia, New Zealand, and Europe. Much like the above story for soybeans, white lupine (*Lupinus albus* L.) is currently being researched as a grain crop for the southern United States (Noffsinger and van Santen 1995). White lupine is reported to possess grain characteristics which compare very favorably with other grain legumes as shown in Table 3 (E.S. van Santen 1995, pers. commun.). Its yield and production also take place in the winter months allowing it to be produced during the "off-season" of a crop like soybean.

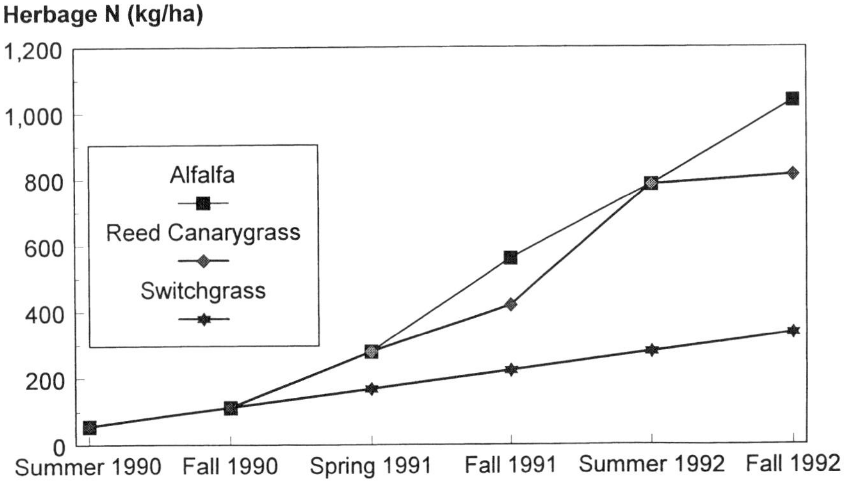

Fig. 3. Cumulative nitrogen removal of an ineffectively nodulating alfalfa compared to reed canarygrass and switchgrass (J. Lamb 1995, pers. commun.).

Table 3. Grain characteristics of white lupine compared with lima bean and soybean (E.S. van Santen 1995, pers. commun.).

Crop	1000 Seed wt (g)	Protein (%)	Oil (%)
White lupine	245–609	34–45	10–15
Lima bean	450–2000	19–30	1–3
Soybean	120–180	39–42	20–22

Ornamental Use

Due to their showy and colorful flowers, some forage legumes have potential for use as ornamentals. Recently, two white clover germplasms were described which possess genes having potential in this area. One germplasm showed corolla color change from white to red (Pederson and McLaughlin 1995) while another demonstrated expression of a large red leaf mark (Pederson 1995). Combining these two traits should give a red flowered, red leaved but still stoloniferous plant which has great potential for ornamental use as both hanging baskets and perennial ground cover.

The inherent colorful flowers of crimson clover, birdsfoot trefoil (*Lotus corniculatus* L.), and crown vetch (*Coronilla varia* L.) along with their ability to reseed and grow under stressful conditions has made them a natural for low-maintenance highway beautification (Grant and Marten 1985; Hoveland and Knight 1985; Hoveland and Townsend 1985). Other legume species with these same characteristics also have great potential as low maintenance ornamentals.

CONCLUSION

The forage legume with the greatest ability to assume many roles both within and outside of traditional grassland agriculture is alfalfa. It is the most widely planted and used forage legume species in the world, yet is an excellent model system for biotechnology especially in the area of genetic transformation. Its traditional use as hay, silage, and dehy crop has recently been expanded with the release of grazing tolerant cultivars to include a larger role as pasture for direct livestock consumption. Its use as a grazing crop has been one of the driving forces in moving some sectors of the United States livestock industry into more efficient pasture systems with a legume base. At the same time, nontraditional roles such as sprouts for salads and nutritional supplements for human diets have increased for alfalfa. It is also being investigated as a fuel for use in generating electricity, a bioremediaton system for removal of harmful nitrates, a source of pulp for paper manufacturing, and a "factory" for production of industrial enzymes.

Other forage legumes are finding a use as ornamentals. Recent successes with developing white clover to be more colorful and showy for perennial ground cover and captilizing on the inherent colorful flowers of several species for low-maintenance highway beautification demonstrate what can be accomplished.

Soybean was an important forage crop in the United States. However, due to intensive research efforts, soybean is now mainly used as an oil seed crop. Similarly, the recent development of lupines into a grain crop further demonstrates that traditional forage legumes can become important grain and oil seed crops.

The reasons for using old forage legumes in new roles are: (1) farmers and the general public are already familiar with them, (2) each species is generally served by a successful seed industry, and (3) clear methods of successfully planting and managing them as crops are already available and practiced. These characteristics allow old legumes to assume any new role quicker and with less problems than a newly introduced species.

REFERENCES

Allen, O.N. and E.K. Allen. 1981. The leguminosae: A source book of characteristics, uses, and nodulation. Univ. of Wisconsin Press, Madison.

Austin, S., E.T. Bingham, R.G. Koegel, D.E. Mathews, M.N. Shahan, R.J. Straub, and R.R. Burgess. 1993. An overview of a feasibility study for the production of industrial enzymes in transgenic alfalfa. Ann. New York Acad. Sci. 721:234–244.

Austin, S., E.T. Bingham, D.E. Mathews, M.N. Shahan, J. Will, and R.R. Burgess. 1995. Production and field performance of transgenic alfalfa (*Medicago sativa* L.) expressing alpha-amylase and manganese-dependent lignin peroxidase. Euphytica. 85:381–393.

Ball, D.M., C.S. Hoveland, and G.D. Lacefield. 1991. Southern forages. Potash & Phosphate Inst. Atlanta.

Bass, L.N., C.R. Gunn, O.B. Hesterman, and E.E. Roos. 1988. Seed physiology, seedling performance, and seed sprouting. p. 961–983. In: A.A. Hanson, D.K. Barnes, and R.R. Hill, Jr. (eds.), Alfalfa and alfalfa improvement. Am. Soc. Agron., Madison, WI.

Berggren, H. 1993. Project Agro-Fibre: an investigation of the possibility for using agricultural crops for pulp production in Sweden. p. 143–153. In: K.R.M. Anthony, J. Meadley, and G. Robbelen (eds.), New crops for temperate regions. Chapman & Hall, London.

Blaser, R.E., D.D. Wolf, and H.T. Bryant. 1973. Systems of grazing management. p. 581–595. In: M.E. Heath, R.F. Barnes, and D.S. Metcalfe (eds.), Forages, the science of grassland agriculture. 3rd. ed. Iowa State Univ. Press, Ames.

Bouton, J.H., S.R. Smith, Jr., and C.S. Hoveland. 1992. Selection and testing of alfalfa cultivars for grazing tolerance. p. 166–170. In: W. Faw (ed.), Proc. Am. Forage and Grassland Council, Grand Rapids, MI. 5–9 April 1992. AFGC, Georgetown, TX.

Bouton, J.H., S.R. Smith, Jr., D.T. Wood, C.S. Hoveland, and E.C. Brummer. 1991. Registration of 'Alfagraze' alfalfa. Crop Sci. 31:479.

Busbice, T.H. and C.H. Hanson. 1969. Selection for improving creeping-rooted characteristics in alfalfa. Crop Sci. 14:783–787.

Certified Alfalfa Seed Council. 1994. Proc. National Alfalfa Grazing Conference. Certified Alfalfa Seed Council, Davis, CA.

Conrad, H.R. and T.J. Klopfenstein. 1988. Role in livestock feeding—greenchop, silage, hay, and dehy. p. 539–551. In: A.A. Hanson, D.K. Barnes, and R.R. Hill, Jr. (eds.), Alfalfa and alfalfa improvement. Am. Soc. Agron., Madison, WI.

Counce, P.A., J.H. Bouton, and R.H. Brown. 1984. Screening and characterizing alfalfa for persistence under mowing and continuous grazing. Crop Sci. 24:282–285.

Gramshaw, D., H.G. Bishop, and D.H. Ludke. 1982. Performance of grazed creeping-rooted lucernes on two soils in central Queensland. Austral. J. Expt. Agr. Anim. Husb. 22:177–181.

Grant, W.F. and G.C. Marten. 1985. Birdsfoot trefoil. p. 98–108. In: M.E. Heath, R.F. Barnes, and D.S. Metcalfe (eds.), Forages, the science of grassland agriculture. 4th ed. Iowa State Univ. Press, Ames.

Hartwig, E.E. 1973. Varietal development. p. 187–210. In: B.E. Caldwell (ed.), Soybeans: Improvement, production, and uses. Am. Soc. Agron., Madison, WI.

Heath, M.E., R.F. Barnes, and D.S. Metcalfe. 1985. Forages, the science of grassland agriculture. 4th ed. Iowa State Univ. Press, Ames.

Heath, M.E. and C.J. Kaiser. 1985. Forages in a changing world. p. 3–11. In: M.E. Heath, R.F. Barnes, and D.S. Metcalfe (eds.), Forages, the science of grassland agriculture. 4th ed. Iowa State Univ. Press, Ames.

Heichel, G.H. 1985. Symbiosis: nodule bacteria and leguminous plants. p. 64–71. In: M.E. Heath, R.F. Barnes, and D.S. Metcalfe (eds.), Forages, the science of grassland agriculture. 4th ed. Iowa State Univ. Press, Ames.

Heinrichs, D.H. 1963. Creeping alfalfas. Adv. Agron. 15:317–337.

Heywood, V.H. 1971. The leguminosae: A systematic purview. p. 1–29. In: J.B. Harborne, D. Boulter, and B.L. Turner (eds.), Chemotaxonomy of the Leguminosae. Academic Press, London.

Hoveland, C.S. and C.E. Townsend. 1985. Other legumes. p. 146–153. In: M.E. Heath, R.F. Barnes, and D.S. Metcalfe (eds.), Forages, the science of grassland agriculture. 4th ed. Iowa State Univ. Press, Ames.

Hoveland, C.S. and W.E. Knight. 1985. Arrowleaf, crimson, and other annual clovers. p.136–145. In: M.E. Heath, R.F. Barnes, and D.S. Metcalfe (eds.), Forages, the science of grassland agriculture. 4th ed. Iowa State Univ. Press, Ames.

Isely, D. 1982. Leguminoseae and *Homo sapiens*. Econ. Bot. 36:46–70.

Kehr, W.R., M.A. Conrad, M.A. Alexander, and F.G. Owen. 1963. Performance of alfalfas under five management systems. Nebraska Agr. Expt. Sta. Lincoln. Res. Bul. 211.

Leach, G.J. 1978. The ecology of lucerne pasture. p. 290–308. In: J.R. Wilson (ed.), Plant relations in pastures. CSIRO, Melbourne, Australia.

Leach, G.J. and R.J. Clements. 1984. Ecology and grazing management of alfalfa pastures in the subtropics. Adv. Agron. 37:127–149.

Lodge, G.M. 1991. Management practices and other factors contributing to the decline in persistence of grazed lucerne in temperate Australia: a review. Austral. J. Expt. Agr. 31:713–724.

Lorenz, R.J., R.E. Ries, C.S. Cooper, C.E. Townsend, and M.D. Rumbaugh. 1982. Alfalfa for dryland grazing. USDA, U.S. Gov. Print. Office, Washington, DC. Agr. Inf. Bul. 444 .

McVay, K.A., D.E. Radcliffe, and Hagan. 1989. Winter legume effects on soil properties and nitrogen fertilizer requirement. Soil Sci. J. 53:1856–1862.

Noffsinger, S.L. and E. van Santen. 1995. Yield and yield components of spring-sown white lupin in the southeastern USA. Agron. J. 87:493–497.

Pederson, G.A. and M.R. McLaughlin. 1995. Registration of MSRedF1 red-flowered white clover germplasm. Crop Sci. 35:596.

Pederson, G.A. 1995. Registration of MSRLM red leaf mark white clover germplasm. Crop Sci. 35:1235–1236.

Piskovatski, J.M. and G.V. Stepanova. 1980. Evaluation of pasture alfalfa hybrids in grazing and grazing simulation. p. 70–73. In: D.K. Barnes. (ed.), Report 27th Alfalfa Improvement Conf., Madison, WI. 8–10 July 1980. USDA-ARS, Peoria, IL.

Probst, A.H., and R.W. Judd. 1973. Origin, U.S. history and development, and world distribution. p. 1–15. In: B.E. Caldwell (ed.), Soybeans: Improvement, production, and uses. Am. Soc. Agron., Madison, WI.

Shiflet, T.N. and G.M. Darby. 1985. Forages and soil conservation. p. 21–32. In: M.E. Heath, R.F. Barnes, and D.S. Metcalfe (eds.), Forages, the science of grassland agriculture. 4th ed. Iowa State Univ. Press, Ames.

Smith, S.R., Jr. and J.H. Bouton. 1993. Selection within alfalfa cultivars for persistence under continuous stocking. Crop Sci. 33:1321–1328.

Smith, S.R., Jr., J.H. Bouton, and C.S. Hoveland. 1989. Alfalfa persistence and regrowth potential under continuous grazing. Agron. J. 81:960–965.

Alfalfa Stems: Potential Biofuel for Woodstoves

A. Gray, C. Anderson, E. Koppelman, B. Bjornsen, K. Frank, and M. Siedell

Irrigated alfalfa (*Medicago sativa* L., Fabaceae) is a high-yielding, nitrogen-fixing perennial when produced on high pH soils in the arid Western United States. For most soil types in this region, phosphorus is the only soil amendment required for alfalfa production. Irrigated alfalfa is well adapted to the high elevation (1070–2200 m) growing conditions of Wyoming. At least three harvests are possible (Gray et al. 1992) during the growing season for areas located at elevations up to 1680 m. Livestock is the most important commodity in Wyoming's forage-based agriculture. Hay continues as Wyoming's most economically important agronomic crop, but unstable demand and price fluctuations have piqued interest in the use of hay as a value-added, renewable biofuel. This prompted an evaluation of alfalfa stems as an environmentally friendly, biofuel alternative for woodstoves.

METHODOLOGY

Samples from eleven randomly-selected harvest lots of alfalfa hay from each of first, second, and third cuttings were subjected to a leaf-stem separation process. Hay bales from each lot were sampled with a coring tool (50.8 mm diam) designed to sample baled wool. For each harvest lot, a hay sample was divided into two equal fractions with a riffle-splitter. One fraction was leaf-screened once with a Number 32 sieve (12.7 mm diam) and then twice again with a Number 12 (3.2 mm diam) to separate leaves from stems.

Leaf and stem separations, including whole hay (unseparated) samples, were analyzed for energy, fixed carbon, ash, volatile gases, and sulfur (American Society for Testing and Materials 1993). Hay samples were also evaluated for crude protein (CP), acid and neutral detergent fiber (ADF and NDF) , and relative feed value (RFV) to determine analytical characteristics (AOAC 1990; U.S. Alfalfa Hay Quality Committee 1986) important in ruminant nutrition. Differences among cuttings, leaves, stems, and whole samples were determined with a 3 by 3 factorial experimental design. An economic evaluation of leaf-stem separation processes or the market potentials of separation co-products is not addressed in this study.

The combustion characteristics of alfalfa co-products were compared with those for bituminous coal. Combustion reference values for coal from the Powder River Basin (PRB) of northeastern Wyoming and for coal from eastern U.S. appear in Table 1. The PRB coal is considerably lower in sulfur when compared to coal from eastern U.S. On the other hand, a major consideration in the cost of shipping is the moisture content of PBC at 30% compared to 3%–13% for coal from the eastern U.S. The moisture content of pelletized alfalfa might vary but would be less than 10%. All combustion and nutritional quality characteristics in this study are reported on a dry matter basis.

RESULTS

Combustion Characteristics of Alfalfa Plant Components

Ash is an approximate measure of the mineral content and other inorganic matter which remains as residue after the combustion of organic material. Ash presents a disposal problem, particularly in biofuels with a higher ash content. Stems had a lower ($P = 0.01$) ash content than leaves (Table 1). In general, the ash content of alfalfa components was quite similar to the ash content of PRB coal (Table 1).

The test for volatiles determines the percentage of gaseous products (exclusive of moisture) as a result of combustion. High values resulting from a volatile test indicate a higher potential to produce smoke. Volatile gases produced from alfalfa components were considerably greater than reference values for volatiles produced from coal (Table 1). Volatiles are sometimes used to rank or determine the market value of coal. However, in the case of an alfalfa biofuel product, volatiles might not be important because wood pelletstoves are practically smokeless. Significant differences in volatile gas contents were not detected among cuttings or plant parts.

The Environmental Protection Agency standard for sulfur allows 0.523 kg of sulfur dioxide emission per million British Thermal Units (BTU). Thus, the lower sulfur dioxide emission potential of alfalfa is attractive even with the lower BTU content (Table 1). Compared to stems, the sulfur dioxide and elemental sulfur content of leaves were considerably greater (P = 0.01) and probably relate to higher levels of crude protein (CP) in leaves.

Coal is marketed on the basis of moisture ash free (MAF) BTU content. On a dry matter basis, coal has a considerably higher MAF BTU content than alfalfa hay, but the PRB coal also has a moisture content of about 30%. The cost of removing moisture or transporting a high moisture product can be considerable, but the cost of milling and pelleting hay with an air dry moisture content of 7 to 13% can also be significant. In this study, alfalfa leaves had a slightly greater (P = 0.01) MAF BTU content than stems.

Fixed carbon, an indication of the ratio of combustible to incombustible constituents, was greater (P = 0.01) for stems than leaves. Levels of fixed carbon in alfalfa are considerably lower than levels in coal, as reflected by the higher BTU content of coal.

Nutritional Quality of Alfalfa Separation Co-Products

Compared to first and second cutting alfalfa hay, the nutritional quality of third cutting was considerably better (Table 1). For all cuttings, stems were lower (P = 0.01) in crude protein and relative feed value and higher (P = 0.01) in both acid and neutral detergent fiber than were leaves. Both high and low quality hay are feasible biofuels, but the value-added potentials of high quality hay, third cutting in particular, are more likely to be realized by addressing dairy, horse, or other speciality markets that pay cash bonuses for prime quality hay (Gray et al. 1992; Gray and Hill 1994).

Table 1. Combustion and forage quality variables of alfalfa components[z]

Components	Harvest	Ash (%)	Volatiles (%)	Fixed C (%)	MAF[x] BTU/kg	Sulfur (%)	SO_2 kg/MBTU[x]	CP[x] (%)	ADF[x] (%)	NDF[x] (%)	IVDMD[x] (%)	RFV[x] index
Whole plant	1	7.5	77.8	14.5	3848	0.24	0.28	18.6	30.1	39.3	63.0	158
	2	7.5	77.5	15.0	3882	0.26	0.29	18.3	31.9	41.7	61.0	145
	3	8.4	77.7	13.8	3972	0.29	0.32	20.4	23.7	31.5	68.3	212
Stems	1	6.4	77.5	16.5	3831	0.16	0.19	15.2	36.9	47.6	57.1	121
	2	6.2	77.4	16.4	3798	0.16	0.19	14.5	39.5	50.2	57.5	110
	3	7.1	77.7	15.2	3904	0.19	0.21	17.2	29.1	37.3	65.3	171
Leaves	1	9.1	77.9	13.0	3997	0.34	0.38	23.6	20.4	28.1	69.1	245
	2	9.4	78.1	12.5	3969	0.32	0.37	24.4	19.9	27.0	68.8	257
	3	10.1	77.4	12.5	4031	0.36	0.41	25.2	15.8	21.8	72.3	328
LSD[y] 5%		1.0	0.8	1.2	83	0.06	0.07	1.9	3.6	3.9	4.0	27
LSD[y] 1%		1.3	1.1	1.6	110	0.08	0.09	2.5	4.7	5.2	5.3	36
Reference Coal Values												
Powder River Basin		7.4	47.7	45.1	5763	0.45	0.35					
Eastern U.S.		9.8	37.8	52.4	6574	2.16	1.54					

[z]All variables reported on a dry matter basis.
[y]Variables within columns may be significantly different at 1 or 5% probability levels: least significant difference (LSD).
[x]Abbreviations: MAF BTU, moisture ash free British Thermal Units; SO_2 lbs/million British Thermal Units; CP, crude protein; ADF, acid detergent fiber; NDF, neutral detergent fiber; IVDMD, invitro digestible dry matter; RFV, relative feed value.

Depending on the cost of milling and pelleting hay, we conclude that weathered or other low quality hay might be an appropriate biofuel in an oversupplied, price-depressed hay market. Alfalfa leaves are high in nutritional quality, regardless of hay quality, and would be less likely than stems to be utilized as a biofuel feedstock. When compared to stems, alfalfa leaves have more value-added potential in high quality specialty rations for livestock. And finally, a future study should evaluate the economics of milling, separating, and densifying alfalfa co-products as well as the competitiveness of these potential biofuels with conventional heat-producing alternatives such as coal, wood, natural gas or electricity.

SUMMARY

Alfalfa plant components were evaluated as an environmentally friendly, biofuel alternative for woodstoves. The ash content of alfalfa plant components and PRB coal was similar, but stems had a slightly lower ash content than leaves. Volatile gases produced from alfalfa components were considerably greater than reference values from coal. However, in the case of an alfalfa biofuel product, volatiles might not be too important because wood pelletstoves are practically smokeless. The sulfur dioxide emission potential and the BTU content of alfalfa were lower than coal. Hay is a feasible biofuel, but the value-added potential of high quality hay is more likely to be realized by addressing dairy, horse, or other speciality markets that pay cash bonuses for prime quality hay. Depending on the cost of milling and pelleting hay, weathered or other low quality hay might be an appropriate biofuel in an oversupplied, price-depressed hay market.

REFERENCES

Am. Soc. for Testing and Materials. 1993. 1916 Race Street, Philadelphia, PA.

Gray, A., J. Adams, D. Cooperrider, J. Buk, S. Hinninger, R. Hybner, J. Langbehn, J. Nix, M. Schwope, and K. Drake. 1992. Hay evaluation improves quality and marketing. J. Nat. Resour. Life Sci. Educ. 21:37–40.

Gray, A. and M. Hill. 1994. Returns for hay marketed on relative feed value depended on analytical procedure. J. Nat. Resour. Life Sci. Educ. 23:109–112.

U.S. Alfalfa Hay Quality Committee. 1986. Hay testing laboratory certification manual. Oregon State Univ., Corvallis.

Perennial Grasses: New Applications and Uses

Kenneth P. Vogel and Robert C. Shearman

Perennial grasses provide humans with food and fiber via ruminant animals and areas for recreation. Recently, new applications and uses are being found for grasses and some species are being genetically altered and enhanced for these purposes. These include using specific grasses to meet specific or unique livestock production goals, developing grasses to meet specific turfgrass and ornamental requirements, utilization and development of grasses for specific conservation requirements, and developing grasses into industrial crops.

PASTURE AND RANGELAND GRASSES

Grasses are a principal source of food energy for ruminant animals including beef cattle and sheep in the United States. In most beef cattle and sheep production systems, there are critical times of the year when there are shortages of high quality forage in pastures. For example in the midwestern states, the beef cattle industry largely uses cool season grasses such as smooth bromegrass, *Bromus inermis* Leyss. (Casler and Carlson 1995) or tall fescue, *Festuca arundinaceae* Schreb. (Sleper and Buchner 1995) grown in pastures during the spring, summer, and fall months followed by grazing of crop residues during the winter months (Fig. 1). There is a production slump in mid-summer because the cool-season grasses are not very productive during this period. Native warm-season grasses such as switchgrass (*Panicum virgatum* L.) and big bluestem (*Andropogon gerardii* Vitman.) are beginning to fill this production niche in many areas in the Midwest (Moser and Vogel 1995). The use of these grasses results in a production system (Fig. 2) that provides high quality forage over a larger part of the grazing season than the conventional system (Sharp and Gates 1986; Jung 1986). These are not new grasses in this region since they were components of the original tallgrass prairie, but their use as cultivated grasses in this region is new. It is likely that in 20 years, one-third of the pastureland in this region will be seeded to native warm-season grasses.

New applications for perennial forage species as pasture or rangeland species include introduced cool grasses such as the wheatgrasses (*Agropyron* spp.) to provide early spring and late fall grazing in traditional warm-season native rangeland areas of the western United States such as the Nebraska Sandhills. In this area,

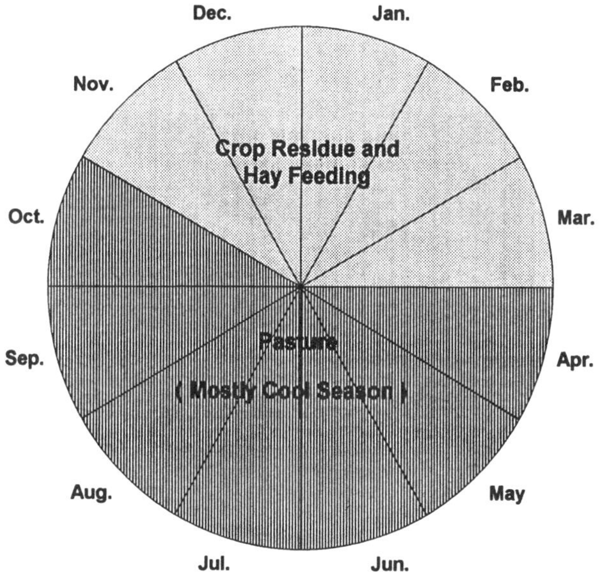

Fig. 1. Typical cattle production system in Corn Belt States.

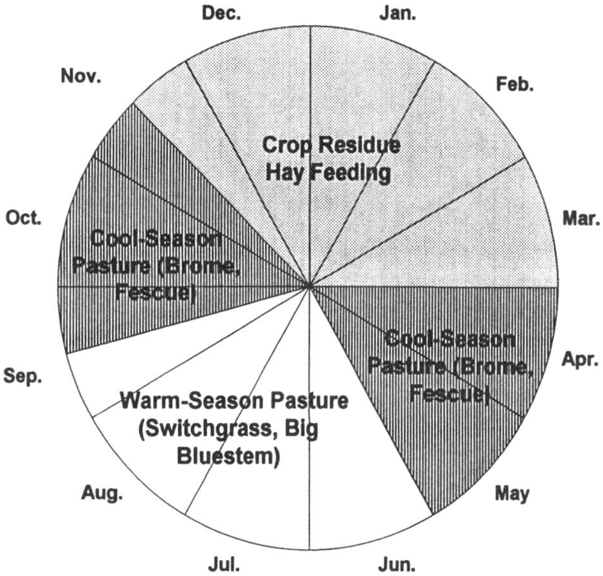

Fig. 2. Integrated cattle production system in Corn Belt States.

the native warm-season rangeland is not ready to graze until mid-May to early June and goes dormant at the first killing frost which is about the first of Sept. This results in a seven month period where livestock producers maintain beef cattle herds on winter range with supplemental hay feeding (Fig. 3). The use of cool-season grasses six weeks in the spring and six weeks in the fall (Fig. 4) can greatly improve the profitability of ranching operations by reducing the amount of hay feeding and by increasing livestock performance due to the high quality forage produced by the cool-season grasses during the spring and fall (Asay 1995; Bauer et al. 1995). Research to match specific grasses to specific livestock production needs is in progress at several locations in the United States.

The use of pastures by dairy cows was once a common practice in the United States. This practice was largely discontinued by the end of the 1950s. Harvested feeds were economical because of mechanization and cheap energy costs and because the highest milk yields per cow was obtained with specific rations in feeding stations (Elbehri and Ford 1995). Improved fencing and watering systems have reduced pasturage costs and the use of intensive rotational grazing systems have improved milk production and reduced production costs to the extent that the use of pastures in dairy production systems can often improve the profitability of dairy farming (Elbehri and Ford 1995). The trend for increased use of pastures in dairy production systems will likely continue.

The use of pastures to produce broiler chickens is another new use of perennial grasses. Joel Salatin (1993), an innovative farmer in the Shenandoah Valley of Virginia produces broiler chickens by raising newly hatched chicks in brooder houses until they are two to four weeks old and then transfers then to portable mini-pasture pens in a pasture. He has about 100 broilers per pen that are provided water and are fed a growing ration that comprises about 70% of their diet. The remaining 30% of the ration comes from grazing. The broilers are provided a fresh, clean grazing area daily by moving the pens. The broilers that he markets are produced under a type of "free ranging" system that appeals to consumers who want and are willing to pay a premium for a more "natural" product. His system requires intensive management but net return is over $3000/ha. He also markets his technology via his book and videos. Farmers in other areas are using his system to produce chickens on pasture.

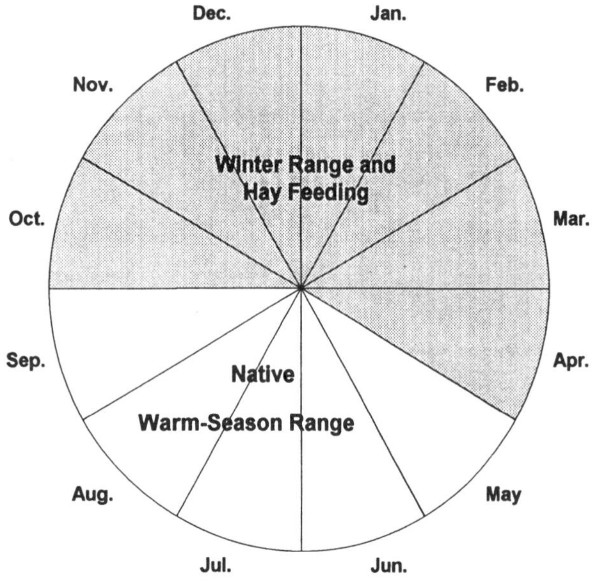

Fig. 3. Typical rangeland cattle production system in the Central Great Plains.

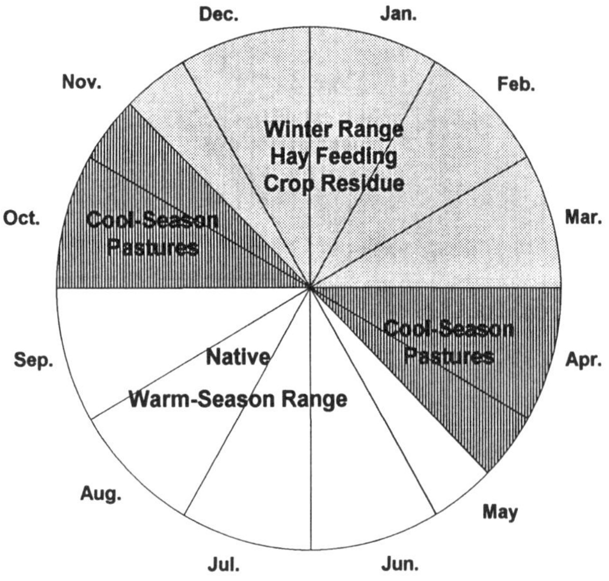

Fig. 4. Integrated cattle production system in the Central Great Plains.

TURFGRASSES

Native perennial grasses, like blue grama [*Bouteloua gracilis* (H.B.K.) Lag. ex Steud.], buffalograss [*Buchloe dactyloides* (Nutt.) Englem], and saltgrass [*Distichlis spicata* (L.) Greene] have considerable potential to be developed as reduced maintenance turfgrasses and they have been periodically evaluated for use as turfgrasses since the 1930s. Buffalograss is a low-growing, drought resistant, warm-season perennial grass species that is native to the Great Plains of North America (Stubbendieck et al. 1991). It can be found growing in an area extending from Central Mexico to Southern Canada. Buffalograss is a dioecious species that spreads by branching stolons; tolerates low mowing and intense grazing; and withstands cold, heat and drought stress.

Buffalograss has been used on a limited basis as a low maintenance, utility turfgrass for lawns, grounds, and roadsides (Beard 1973). Until recently, common and 'Texoka' were the most frequent types. They produced a light, gray-green turf of low density and variable texture and were not considered highly desirable for most turfgrass uses. Recently, there has been an increasing interest among turfgrass managers for turfgrasses that require less inputs, such as fertilizer, pesticides and labor, and conserve water (Shearman 1994). Buffalograss is a species with a potential to meet these interests. 'Prairie' is a turf-type cultivar developed by Milton C. Engelke and released by Texas A&M University that offers improved turfgrass characteristics over 'Texoka' (Riordan et al. 1993).

The United States Golf Association (USGA) is supporting breeding projects that are intended to develop turfgrass cultivars, requiring less inputs and conserving water for golf course roughs, tees, fairways, and greens (Kenna 1994). One project that the USGA has supported since 1984 is a buffalograss breeding project at the University of Nebraska, under the leadership of Terrance P. Riordan with four primary objectives: (1) selecting clones with turf-type characteristics and breeding of cultivars to be planted vegetatively or by seed; (2) determining the range of adaptation of turf-type buffalograsses; (3) developing propagation techniques that improve vegetative establishment and seedling vigor; and (4) identifying management practices that allow buffalograss cultivars to perform in turfgrass situations. The Nebraska program has developed and released three vegetative cultivars ('609', '315' and '378') that are commercially available and are performing well in home lawns, grounds, and golf course roughs. 'Cody' and 'Tatanka' are seeded cultivars released from this project in 1994. Seed of these cultivars was commercially available in 1995.

The vegetative cultivar, '609', was released in cooperation with Crenshaw and Doguet and has consistently out performed 'Prairie' in the buffalograss trials. Sales of '609' sod exceeded $1.5 million in 1994 and its performance has been excellent in the National Turfgrass Evaluation Program Buffalograss Cultivar trials (Table 1). Its principal market has been in Texas where its turf quality in comparison to other adapted grasses such as St. Augustine grass (Fig. 5) has created a high demand for adapted buffalograss cultivars. '609' is considered to be a southern-type that is less cold tolerant than '315', '378', 'Cody' or 'Tatanka', but it forms a dense, low-growing, medium-green turf in its areas of adaptation. '315' and '378' are northern-types that green-up earlier in the spring and go dormant earlier in the fall than '609' or 'Prairie' (Riordan et al. 1993).

It is apparent that buffalograss is a promising species for low input, water conserving turfs. As a species, buffalograss is quite variable and offers considerable opportunity for improvement by breeding and selection. Research results indicate that cultivars can be developed with more desirable turf color, quality, and overall performance, when compared to common-type selections and 'Texoka'. With increased concerns over environmental quality and water conservation, buffalograss may offer a means to maintain turfgrass function and quality, while reducing inputs and minimizing negative impacts on the environment (Shearman 1994). Other species of perennial grasses will undoubtably be developed to meet specific turfgrass requirements.

Perennial grasses including native prairie grasses are being included in landscaping plans by innovative gardeners and landscape architects. The native prairie grasses being used in landscaping include big bluestem, little bluestem [*Schizchyrium scoparium* (Michaux) Nash], switchgrass, and others. In many cases, cultivars developed as pasture or reclamation plantings are being utilized, but at a few locations including the Univ. of Nebraska, selection work is being conducted to developed strains and cultivars specifically for use as ornamentals.

CONSERVATION GRASSES

Perennial grasses have long been used to stabilize landscapes including roadsides, construction areas, and agricultural lands. There are still new conservation uses being developed for perennial grasses. The uses of perennial grasses to form living terraces and windbreaks has been successfully tested and evaluated in the Northern Great Plains (Aase et al. 1985; Aase and Pikul 1995) and are currently being evaluated with switchgrass in the midwestern states. The grasses are seeded in narrow rows either on the contour or perpendicular to prevailing winds at intervals 15 m or more in width so that conventional agriculture equipment can be used to farm between the grass strips. The grass strips provide many of the same benefits as terraces built with heavy construction equipment and are much cheaper to implement (Aase and Pikul 1995). The vegetative strips stop or reduce water flow and trap sediment (Fig. 6) and can also reduce wind erosion if the grass used in the strips is tall and rigid enough to remain erect during the winter months. The U.S. Government's requirement that farmers have conservation compliance plans in operation before farmers can participate in farm programs will likely lead to the increased use of perennial grasses by farmers to inexpensively achieve specific conservation requirements such as the establishment of terraces.

Table 1. Mean turfgrass quality ratings of buffalograss cultivars for each month grown at nineteen locations in the United States, 1994 data. Source: National Turfgrass Evaluation Program. National Buffalograss Test—1993 (Morris 1994).

	Turfgrass quality ratings, 1 to 9 ideal basis												
Name	Jan	Feb	Mar	Apr	May	Jun	Jul	Aug	Sept	Oct	Nov	Dec	Mean
NE 85-378	5.2	4.5	4.3	5.9	5.8	6.6	6.2	5.8	5.7	4.8	4.3	4.3	5.8
609 (NE 84-609)	5.0	4.7	4.6	4.6	4.9	6.0	6.1	5.6	6.1	5.4	5.8	5.1	5.6
NTG-4	4.8	4.5	4.7	6.1	5.2	6.1	5.9	5.7	5.6	4.9	4.8	4.1	5.6
NTG-5	5.3	5.0	4.2	6.3	5.2	6.1	5.9	5.6	5.3	4.9	4.9	3.6	5.5
315 (NE 84-315)	5.2	4.7	4.7	4.8	5.9	6.4	5.7	5.5	5.5	4.7	4.2	3.4	5.5
NTG-2	5.5	4.2	4.6	6.0	5.1	5.9	5.7	5.6	5.5	4.9	4.8	3.6	5.5
NE 84-436	5.0	5.2	4.7	4.8	4.8	6.1	5.8	5.7	5.6	4.7	5.1	3.4	5.4
NTG-3	5.0	5.0	4.2	4.3	5.1	6.0	5.9	5.6	5.5	4.9	5.2	3.6	5.4
AZ 143	5.5	4.7	4.4	4.9	4.9	6.1	5.6	5.6	5.4	4.7	4.4	3.9	5.4
Tatanka (NTG-1)	4.8	4.8	4.2	4.8	4.9	6.0	5.8	5.1	5.3	4.7	4.9	3.9	5.3
Texoka	5.2	4.3	4.3	4.8	4.6	5.7	5.6	5.2	5.2	4.8	4.8	3.8	5.2
Bison	5.2	4.3	4.8	5.4	4.8	5.1	5.4	5.0	5.5	4.8	5.3	3.6	5.1
Sharps Improved	4.8	4.7	4.8	4.2	4.6	5.7	5.4	5.0	5.3	4.8	5.0	3.7	5.1
Top Gun (BAM 101)	5.0	4.7	4.3	5.4	4.6	5.6	5.5	5.0	5.1	4.6	4.9	3.3	5.0
Plains (BAM 202)	4.8	4.3	4.7	4.8	4.4	5.5	5.4	4.8	5.1	4.8	5.2	3.8	5.0
Prairie	5.2	5.0	4.3	4.3	4.1	5.6	5.3	4.9	5.5	5.1	5.2	4.4	5.0
Buffalawn	5.0	5.0	4.1	4.1	3.7	5.5	5.4	5.4	5.7	5.0	5.3	4.2	4.9
NE 84-45-3	5.0	4.2	4.1	4.1	4.3	5.3	5.2	4.6	4.4	4.0	4.2	3.1	4.6
Highlight 25	5.2	4.8	3.9	3.9	3.6	5.3	4.7	4.9	5.1	5.7	5.2	4.1	4.5
Highlight 4	5.2	4.7	3.8	4.0	3.5	5.0	4.9	4.8	5.1	5.0	5.4	4.1	4.5
Highlight 15	5.0	4.2	4.0	4.1	3.4	4.8	4.6	4.4	4.9	5.1	5.3	4.3	4.4
Rutgers	5.3	4.2	3.4	3.8	3.3	4.8	4.7	4.5	5.0	5.1	5.0	4.0	4.3
LSD value	1.6	1.1	1.9	1.7	0.9	0.8	0.7	0.8	0.7	0.9	1.3	1.9	0.6

GRASSES FOR INDUSTRIAL USES

The U.S. Department of Energy (DOE) is interested in developing renewable energy sources for use as fuels. At the present time, ethanol is produced by fermenting the starch in grains using classical procedures. Ethanol could be made from other plant products. Forage crops excel in the production of plant cell walls, the most abundant plant material. Plant cell walls are primarily the macromolecules, cellulose and hemicellulose, which are comprised of simple sugars, glucose and xylose. However, plant cells walls are not fermentable using classical fermentation procedures.

Molecular genetics research has made significant advances in making ethanol production from biomass feasible. Recently, Zhang et al. (1995) reported producing a recombinant bacteria that can anaerobically ferment both xylose and glucose sugars to ethanol at yields exceeding 86% of theoretical yield. Research also is being conducted on procedures to break cellulose and hemicellulose down into simple sugars. Research has reduced the cost of producing ethanol from biomass from \$0.95 per liter in 1982 to \$0.36/liter in 1992 (Wyman

Fig. 5. A buffalograss turf (right side) growing next to an St. Augustine turf (left) in Dallas, Texas. Note the finer leaf texture of the buffalograss turf.

Fig. 6. Sediment trapped at the base of switchgrass plants growing in a living terrace. Photograph courtesy of Dr. John Gilley, USDA-ARS, Lincoln, Nebraska.

1992). With improvements in both conversion technology and biomass plant productivity, it is estimated that it may be feasible to produce ethanol at $0.16 per liter by 2010 from herbaceous biomass which would make it equivalent in cost to petroleum fuels at $25/barrel for crude oil (Wyman 1992). In a series of evaluation trials with cooperating state universities and the USDA, the Department of Energy (DOE) has identified switchgrass as the most promising species for development into a herbaceous biomass fuel crop. Its desirable attributes include broad adaptation, high yields, stress tolerance and it is harvestable with conventional hay-making equipment. It can produce high yields on marginal lands that are unsuitable for row crop production due to high erosion potential. DOE, USDA, and cooperating state experiment stations are currently conducting breeding and management research to improve switchgrass as a potential biomass crop.

As part of this research, currently available switchgrass cultivars and experimental strains were evaluated in trials in Nebraska, Iowa, and Indiana for their potential as biomass fuel crops (Hopkins et al. 1995). The research plots were seeded in 1991 and harvested in 1991 and 1992. They were fertilized with 112 kg/ha N per acre. The only other cultural practices were the harvesting operations. The highest yielding strains produced over 14 Mg/ha dry matter per acre (Table 2). Turhollow (1994) estimated that switchgrass would have to sell for $43/Mg to $60/Mg to be competitive with corn in the midwest. At a price of $55/Mg ($50/ton), which is a typical price for grass hay in many years, the gross return per ha would be over $770 (Table 2). Assuming 75% conversion of the constituent cellulose and hemicellulose to ethanol, these yields would result in ethanol production of over 4600 l/ha (Table 2).

Table 2. Forage yields of the five highest yielding switchgrass strains at three midwestern locations in 1991 and 1992.[z]

Location	Cut 1 (mg/ha)	Cut 2 (mg/ha)	Total (mg/ha)	Gross return[y] ($/ha)	Ethanol[x] (liters/ha)
Mead, Nebraska	13.4	0.9	14.3	786	4720
Ames, Iowa	12.1	2.5	14.6	803	4790
West Lafayette, Indiana	13.7	2.0	15.7	864	5170

[z]Yield data from Hopkins et al. 1995.
[y]Biomass priced at $55.00/mg ($50/ton).
[x]Assumes conversion rate of 330 l/mg (79 gallon/ton) of biomass (Turhollow et al. 1988).

Table 3. Non-irrigated maize yields and gross return for three counties in which switchgrass trials were conducted in the Midwest, U.S. in 1991 and 1992.

Location	Yield kg/ha[z] 1991	Yield kg/ha[z] 1992	Corn price[y] ($/kg) 1991	Corn price[y] ($/kg) 1992	Gross return ($/ha)	Ethanol yield[x] (liters/ha)
Saunders County, (Mead), NE	6459	8340	92.03	80.62	635	3217
Story County, (Ames), IA	7713	10159	90.46	76.69	741	3885
Tippecanoe County, (W. Lafayette), IN	4766	9532	96.36	78.66	605	3108

[z]National Agricultural Statistics Service databases, National Agricultural Statistics Service, U.S. Dept. of Agriculture.
[y]Agricultural Statistics 1993.
[x]Assumes conversion of 1 liter/2.3 kg (2.9 gallons of ethanol/bushel) of corn (Turhollow et al. 1988).

Average maize yields for the counties in which the switchgrass trials were located ranged from 4760 to 10100 kg/ha (Table 3). In the Midwest, 1991 was a drier year than average while in 1992, the growing season was cooler than average. The average gross return per ha for maize for the three counties averaged $605 to $741 (Table 3). Assuming a conversion rate of 1 liter of ethanol per 2.3 kg of maize, the average ethanol yield from maize from these three Corn Belt counties would have been 3100 to 3800 liters per ha (Table 3). Ethanol feedstocks cost expressed in terms of $/liter would likely be cheaper for ethanol produced from switchgrass (Tables 2 and 3).

Herbaceous biomass fuel plants such as switchgrass would have to be grown east of 100° W longitude because of precipitation requirements for economic yields. In this region, which is east of a north to south line 320 km west of Omaha, NE, there are 8 to 16 million ha of land that could be converted to biomass crop production without significant displacement of crops (Graham 1994).

Use of marginal lands for biomass fuel production would produce onsite and offsite benefits similar to the Conservation Reserve Program. If the biomass can be converted to liquid fuels as economically as predicted and adequate yields can be obtained, this conversion of land to energy production could be achieved without the need for massive federal subsidies. Land use for biomass fuel production would be reverting land to its prior use at the early part of this century when the nation was powered by herbaceous biomass via over 25 million head of horses and mules on farms and over 2 head million in cities (Census of Agriculture 1900, 1920). It would also revert to energy production some of the 80 million acres of land that was switched from feed production for horses. The decrease in the numbers of draft animals released approximately 80 million acres of land for other purposes (Census of Agriculture 1954).

Perennial grasses can be high producers of lignocellulosic compounds. These compounds are utilized in industry for paper productions and the production of products such as fiberboard. Currently wood products provide the raw materials for these products. It is feasible that mature perennial grasses could be used in the future as a commercial source of lignocellulosic compounds.

REFERENCES

Aase, J.K., F.H. Siddoway, and A. Black. 1985. Effectiveness of grass barriers for reducing wind erosion. J. Soil Water Conserv. 40:354–357.

Aase, J.K. and J.L. Pikul, Jr. 1995. Terrace formation in cropping strips protected by tall wheatgrass barriers. J. Soil Water Conserv. 50:110–112.

Asay, Kay H. 1995. Wheatgrasses and wildryes: The perennial triticeae. p. 373–394. In: R.F. Barnes, D.A. Miller, and C.J. Nelson (eds.), Forages, 5th ed. Vol. I: An introduction to grassland agriculture. Iowa State Univ. Press, Ames.

Agricultural Statistics 1993. National Agricultural Statistics Service. U.S. Government Printing Office, Washington, DC.

Bauer, D., L.E. Moser, and K.P. Vogel. 1995. Grazing evaluation of cool-season grasses in the Nebraska Sandhills in replicated ranch trials. Agron. Abst. p. 165.

Beard, J.B. 1973. Turfgrass science and culture. Prentice–Hall, Englewood Cliffs, NJ.

Casler, M.D. and I.T. Carlson. 1995. Smooth Bromegrass. p. 313–324. In: R.F. Barnes, D.A. Miller, and C.J. Nelson (eds.), Forages, 5th ed. Vol. I: An introduction to grassland agriculture. Iowa State Univ. Press, Ames.

Census of Agriculture. 1900, 1920, 1954. U.S. Bureau of the Census. U.S. Government Printing Office, Washington, DC.

Elbehri, A. and S.A. Ford. 1995. Economic analysis of major dairy forage systems in Pennsylvania: The role of intensive grazing. J. Prod. Agr. 8:501–507.

Graham, R.L. 1994. An analyses of the potential land lase for energy crops in the conterminous United States. Biomass Bioenergy 6:175–189.

Hopkins, A.A., K.P. Vogel, K.J. Moore, K.D. Johnson, and I.T. Carlson. 1995. Genotype effects and genotype × environment interactions for traits of elite switchgrass populations. Crop Sci. 35:125–132.

Jung, G.A. 1986. How warm-season grasses perform in the Northeast. p.19–22. In: W.C. Sharp and T.N. Shiflet (eds.), Warm-season grasses. Balancing forage programs in the Northeast and Southern Corn Belt. Soil Conserv. Soc. of America, Ankeny, IA.

Kenna, M.P. USGA: 1994 Turfgrass research summary. United States Golf Association, Golf House, Far Hills, NJ.

Morris, K.N. NTEP: 1994 National Buffalograss Progress Report. National Turfgrass Evaluation Program, Beltsville Agr. Res. Ctr., Beltsville, MD.

Moser, L.E. and K.P. Vogel. 1995. Switchgrass, big bluestem, and indiangrass. Chapter 32. p. 409–420. In: R.F. Barnes, D.A. Miller, and C.J. Nelson (eds.), Forages, 5th ed. Vol. I: An introduction to grassland agriculture. Iowa State Univ. Press, Ames.

Riordan, T.P., S.A. de Shazer, J.M. Johnson-Cicalease, and R.C. Shearman. 1993. An overview of breeding and development of buffalograss for golf course turf. p. 816–823. In R.N. Carrow, N.E. Christians, and R.C. Shearman (eds.), Int. Turfgrass Soc. Research J. (Vol. 7). Intertec Publishing Corp., Overland Park, KS.

Salatin, J. 1993. Pastured poultry profits. Polyface, Inc., Swoope, VA.

Sharp, W.C. and G.A. Gates. 1986. Why use warm-season grasses in the Corn Belt and Northeast? p. 7–9. W.C. Sharp and T.N. Shiflet (eds.), Warm-season grasses. Balancing forage programs in the Northeast and Southern Corn Belt Soil Conserv. Soc. of America, Ankeny, IA.

Shearman, R.C. 1994. Turfgrass. Encyclopedia of Agricultural Science. Academic Press, Vol. 4, p. 413–420.

Sleper, D.A. and R.C. Buchner. 1995. The fescues. p. 345–356. In: R.F. Barnes, D.A. Miller, and C.J. Nelson (eds.), Forages, 5th ed. Vol. I: An introduction to grassland agriculture. Iowa State Univ. Press, Ames.

Stubbendieck, J., S.L. Hatch, and C.H. Butterfield. 1991. North American range plants. Univ. of Nebraska Press, Lincoln.

Turhollow, A.F., J.W. Johnson, and J.H. Cushman. 1988. Linking energy crop production to conversion: The case of herbaceous lignocellulosic crops to ethanol. RERIC Int. Energy J. 10:41–49.

Turhollow, A. 1994. The economics of energy crop production. Biomass Bioenergy 6:229–241.

Wyman, C.E. 1990. The DOE/SERI ethanol for biomass program. Ethanol from Biomass Annu. Rev. Meeting—Draft Proc. Sept. 12, 1990, Lincoln, NE. National Renewable Energy Lab, Golden, CO.

Zhang, M., C. Eddy, K. Deanda, M. Finkstein, S. Picataggio. 1995. Metabolic engineering of a pentose metabolism pathway in ethanologenic *Zymomonas mobilis*. Science 267:240–243.

OILSEEDS

Selecting Winter Hardy Oilseed Rape for the Great Plains

C.L. Rife and J.P. Salgado

Canola is an important oil crop worldwide and has significant potential for the central Great Plains. Both spring and winter types have been developed, but only those with the winter growth habit have suitable yield potential for the region. All commercial cultivars have inadequate winter hardiness and dormancy to survive consistently in the plains and production of the crop will not be economically feasible until these deficiencies are remedied. Sometimes the plants are killed by cold temperatures during the late fall-early winter (lack of winter hardiness), and other times the plants break dormancy during warm spells in the late winter-early spring and are killed by cold temperatures later (lack of spring survival). A crop of a hardy cultivar will survive some winters in Kansas, but attempts to grow canola consistently have failed. Seed yields of winter canola in Kansas test plots have been variable, ranging from near zero to over 3,000 kg/ha (Canola Production Handbook 1990). Low yields generally have been attributed to winter-kill. The wide diversity of genetic materials available should allow improved varieties with winter hardiness to be developed.

Breeding for better winter rapeseed began with population improvement and mass selection. A large number of plants were selected from a population based on plant phenotype and seeded together. This method is effective for highly heritable characters that can be determined before flowering, including winter hardiness in areas with more severe winters (Thompson and Hughes 1986). Field survival is the most commonly used method of evaluating winter plants for winter hardiness. This method is simple, requires minimum labor and is the ultimate test for winter hardiness (Fowler et al. 1981). Unfortunately, field trials are often inconclusive due to either complete death or complete survival of all or most plants within a trial (Fowler and Gusta 1979). Even when differential kill does occur, it is often irregular with large variations occurring within very short distances (Fowler 1979).

Generally, insufficient cold hardiness is also the main factor limiting the survival of cereals in regions with harsh winter environments (Fowler et al. 1981). Winter survival of crop plants is also influenced by many other factors (Levitt 1972). A University of Wisconsin study indicated winter survival in rapeseed was dependent on many morphological and physiological characteristics of the plant, soil conditions and weather fluctuations and that freezing tolerances differed between cultivars (Teutonico et al. 1993). In a winter canola study in Canada, sites with low winter survival suffered from late emergence, autumn drainage problems, freezing winter rains, or spring flooding conditions (Beaulieu and Hume 1987). As a result, screening methods that allow for accurate and precise assessment of cold hardiness potential are critical to winter crop research programs.

The inherent difficulties with field trials have stimulated interest in the development of cold hardiness prediction tests to complement field screening. Controlled freeze tests that employ a single medium temperature (Fowler et al. 1973), prolonged freezing (Thomas et al. 1988), or a range of temperatures to measure lethal dose temperature (Fowler et al. 1981) have been used to identify differences in plant cold hardiness. Controlled freeze tests have the advantage over field testing in that they are faster, give greater control of environmental conditions, and provide the opportunity for replication over time (Pomeroy and Fowler 1973). The most important advantage is that controlled freezing tests accelerate selection and provide positive results. They can be used year around to cause the differential cold injury that is needed for progress in breeding, where field trials can be used only once per year and frequently do not cause differential injury. A rapid screening test must be highly correlated with field survival, simple, repeatable, rapid, nondestructive, and require only a single plant for analysis (Fowler et al. 1981).

MATERIALS AND METHODS

Field Studies

Advanced experimental lines were derived from a number of field selections and advanced from segregating populations supplied from the University of Idaho in the late 1980s. These selections were originally made at Hays and Colby, Kansas, and Columbia, Missouri. An extremely diverse testing area was needed and established to increase the potential of having multiple locations were differential winter kill would occur (Fig. 1). Cooperators in nine states in the Great Plains region are growing this uniform yield test. The 1994–95 Advanced Canola Nursery (ACN) included 21 experimental lines and seven check cultivars. The 1995–96 ACN includes 12 experimental lines and 11 check cultivars. Tests are replicated three times and planted in a RCBD. Planting recommendations include seeding the tests six weeks before the first killing frost but individual cooperators use past experience and local conditions for making most management decisions. Data is collected for fall stand, winter survival, 50% bloom date, maturity date, plant height, lodging, shattering, seed yield, moisture content, and test weight. Total oil concentrations are also obtained.

Laboratory Freeze Tests

Seeds form nine lines were planted in the greenhouse to conduct one test. Seeds were placed in moist soil in each 2.5 by 5 cm cell of plastic six pack pots. After all plants were well established, each cell was thinned to one plant. Forty-eight plants of each line were established and maintained in eight six-pack pots in order to give eight experimental units (two replications of each line by four freeze combinations). After four weeks in the greenhouse, they were cold hardened in a growth chamber with an 11 h light period and the following temperature settings: Week 1, 7°C day and 4°C night; Week 2, 4°C and 1°C; Week 3, 1°C and –1°C; Week 4, 7°C and 4°C. At the end of the hardening period, the plants were placed in incubators for the freeze treatment. The temperature was set at –1°C for 10 h and then reduced at the rate of 1°C/h. When the temperature reached –9°C, that temperature was maintained. Plants were removed at 0, 12, 24, and 36 h. After one day recovery in mild temperatures, the plants were transferred to the greenhouse. After three week, survival counts were taken. In order for a plant to be considered alive, both active leaf and root growth was present. Percent survival for each experimental unit was calculated. LT_{50} (the time needed at –9°C to observe 50% death loss) were calculated using simple linear regression models. The same lines tested in this procedure were also grown in the field at three locations (Hays, Hutchinson, and Manhattan, Kansas) over the 1993–94 season and significant differences in winter survival were observed at all locations. Results obtained from this procedure (LT_{50} and mean percent survival) were compared to field data (Field Survival Indexes (Fowler and Gusta 1979) and percent winter survival). This procedure was repeated in order to test all lines grown in the field.

In addition to testing the cold hardiness of different lines, this procedure has been modified to select individual plants with increased cold tolerance. Segregating populations were exposed to –9°C for 30 h. Surviving plants were grown to seed and the procedure repeated. Seed obtained in this manner was planted along with seed obtained from plants of the same populations that were not exposed to lethal temperatures and the parents of the populations. This experiment was planted at two locations in Kansas in the fall of 1995 and survival data should be available in the summer of 1996.

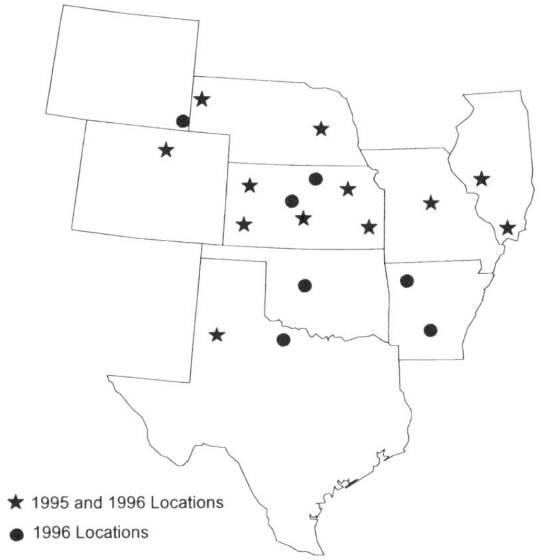

Fig. 1. Test sites of the 1995 and 1996 ACN.

The Effect of Light Duration During Hardening

Six rapeseed lines ('Cascade', 'Aspen', 'Cathy', PI458919, PI469737, and PI469892) were planted in six pack pots as described above. Three flats of 48 seedlings were established for each line. The lines were divided into three different treatments and planted at different times so plants would be at the same developmental stage for the freezing treatment. Four weeks after establishment, the cold hardened treatment (Chill) was placed in a vernalization chamber for four weeks. Three weeks after establishing the low light treatment (LLight), it was placed in the growth chamber. The light was reduced to 8 h with a day/night temperature regime of 18°/9°C. The unacclimated treatment (GrnHs) remained in the greenhouse (greenhouse lights were on 18 hours during the entire time, including the setup of the previous treatments). All treatments were thinned to 1 plants per pot two weeks after planting.

At the end of the acclimation periods, all treatments were placed in an incubator set at 0°C. After 5 h, the temperature was lowered to –1°C for 12 h. Chambers were then set to decrease 1°C/h until the temperature reached –5°C. They were then held at that temperature for 30 min and the –5°C treatment was removed. The temperature was reduced at the same rate and held at –8°C for 30 min and the –8°C treatment was removed. The temperature was decreased again and held at –11°C before the –11°C treatment was removed. The plants were initially placed in the growth chamber with 12 h light and 10°C day and 5°C night to recover. Plants were transferred to the greenhouse after two days. Plant counts were completed after three weeks.

RESULTS AND DISCUSSION

Field Studies

Although the winter of 1994–95 was milder than normal, seven of 12 locations still observed differential winter kill (Table 1). Over these seven locations, mean winter survival ranged from 90.1% for KS-M3579 to 62.2% for Cathy with an over all mean of 83.5%. Four additional locations had 100% survival and one location had near total deathloss. Very few lines had stands that were not harvestable at any location. Fourteen of the 15 lines in the first LSD group were experimental selections advanced in the Great Plains. Winter survival from two location in Kansas for 1993–94 of the lines grown in those tests is also included in Table 1. Conditions were much more severe at these tests and mean survival for all entries was 36.7%.

Although no experimental lines have proven themselves in field conditions, progress has been made in increasing winter survival by selecting in our environment. All released cultivars previously grown in the Great Plains have been developed in Europe, Canada, or the Pacific Northwest. These are areas with very different environments from that of the Great Plains. One experimental line KS3579, was released in the fall of 1995 as a germplasm. Canola quality breeder seed will be increased for four other lines with a decision on increasing foundation seed made in 1996.

Laboratory Freeze Tests

Significant differences in both freeze time and genotype were detected in both tests (Table 2). In all lines tested, as time at –9°C was increased, survival was decreased. Over all, survival averaged 76.9% at 0 h, 32.1% at 24 h, and 18.4% at 36 h. Survival ranged from highs of 100% at 0 h and 50% at 36 h to lows of 66.6% at 0 h and 0% at 36 h. Significant correlations were detected for both tests when freeze test data was compared to data from the field (Table 3).

The laboratory freeze test has the potential to be an important tool in screening germplasm for cold tolerance. The 1993–94 growing season was one where the primary death loss came over the winter and was due to cold temperatures. This may play a part in the strong correlations between the laboratory and the field data. This test was replanted in 1994–95 but due to mild conditions, significant death loss was not observed at any location. Many factors, both physiological and environmental, are involved in winter survival and cold tolerance is only one important factor. Advancing seed from surviving plants in segregating populations using this procedure should select for genes responsible for cold tolerance. Two generations per year are possible using this technique. In the field, at most, one generation is possible and factoring in the winters where survival is near 100% or near 0%, improvement will be made much less frequently.

Table 1. Mean results of the 1994–95 Advanced Canola Nursery planted at 12 locations and winter survival (WS) results from 1993–94 planted at 2 locations.

Genotype	Yield (kg/ha) 10 Loc	Yield (%Avg) 10 Loc	WS (%) 7 Loc	Bloom (April) 7 Loc	Height (cm) 7 Loc	Shatter (%) 5 Loc	Lodge (%) 5 Loc	Test Wt (kg/m3) 6 Loc	Moist. (%) 9 Loc	Oil (%) 3 Loc	WS (%) 1993–94
KS-M3579	1417	108	90*	15E	110	5.8	10*	533	9.6*	33.0	88*
KS-C3505	1388	106	89*	22L	129t	6.2	8*	565*	11.2	34.3	--
KS-C3504	1439	110	89*	21L	131t	7.5	15*	546	11.4	34.3	--
KS-M3311	1200	92	88*	19	117	4.5	9*	515	10.8	35.0	75*
KS-M3580	1593*	122*	88*	19	120	5.7	12*	556*	9.9*	34.5	80*
KS-C3104	1204	92	88*	20	119	6.7	12*	569*	10.0*	34.6	--
KS-C806	1369	105	87*	18	109s	5.0	14*	519	9.9*	33.2	--
MO-503-1	1509*	115*	87*	19	123	9.1	11*	556*	10.5	33.6	--
KS-C3208	1192	91	87*	20	120	4.1	10*	539	10.4	33.7	--
MO-503-24	1419	108	86*	20	124	5.7	9*	546	10.0*	35.1	--
KS-M3635	1477	113	86*	20	124	6.7	10*	550	10.8	34.2	68*
MO-513-9	1370	105	85*	19	122	7.7	12*	550	10.6	35.2	--
KS-C1701	1340	102	85*	20	125	4.6	5*	534	11.0	35.0	--
Ceres	1672*	128*	85*	21	123	4.5	8*	555*	11.7	34.2	30
MO-503-9	1343	103	85*	20	124	8.0	16*	529	10.5	34.6	--
MO-503-2	1327	101	84	19	123	6.4	15*	552	9.9*	36.0*	--
KS-C2504	1308	100	83	16	110	5.9	23	528	10.4	33.6	--
KS-M3723	920	70	83	20	109s	8.8	25	529	9.8*	34.7	65*
Glacier	1312	100	83	21	128t	5.7	18	562*	10.7	34.8	58
KS-M3403	1151	88	83	20	117	6.8	15*	543	10.3*	33.1	70*
KS-C2403	1139	87	82	18	113	5.1	18	529	10.4	36.1*	--
Winfield	1161	89	82	17	114	5.7	18	528	9.6*	37.0*	--
KS-M3346	1282	98	81	19	118	3.9	10*	535	9.8*	35.0	78*
Bridger	1062	81	81	17	109s	8.3	37	538	10.1*	35.4*	58
KS-M3314	1250	96	79	19	116	6.1	15*	541	11.4	34.4	53
Liborius	1445	110	78	21L	125	6.0	9*	573*	11.0	36.2*	--
KayStar 11	1575*	120*	73	21L	126t	5.7	10*	579*	11.4	36.2*	--
Cathy	811	62	62	20	105s	5.8	37	537	9.4*	34.0	--
Mean	1309	100	84	19	119	6.2	15	544	10.4	34.7	37
LSD (.05)	193	15	6	1	5	NS	11	26	1.0	1.6	23
CV	29	29	12	8	7	87	102	7	17	5	31

Table 2. Survival from two freeze and three field tests.

Line	Freeze test		Field tests	
	LT_{50}[z]	Survival (%)	FSI[y]	Survival (%)
		Test 1		
Bridger	24	42	66	39
PI531284	23	33	67	40
PI305282	22	33	59	33
PI535849	19	30	63	36
PI469793	16	29	64	34
PI535868	14	27	53	26
PI458956	11	23	52	20
Ceres	9	24	59	31
		Test 2		
A112	–4	17	25	7
PI458959	42	65	59	31
Ceres	37	71	59	31
PI458925	28	68	50	23
PI535871	23	60	58	32
PI305282	23	57	59	33
PI470007	23	55	59	33
PI535849	20	55	63	36

[z]LT_{50}=Time in hours at –9°C where 50% death is observed.
[y]FSI=Field Survival Index from Hays, Hutchinson, and Manhattan Kansas, 1993–94.

Table 3. Correlation between field and laboratory data for two freeze tests.

Freeze Test	Field	Test 1		Test 2	
		r	p	r	p
% Survival	% Survival	0.86	0.002	0.77	0.022
% Survival	FSI	0.79	0.010	0.83	0.008
LT50	% Survival	0.92	0.000	0.82	0.010
LT50	FSI	0.90	0.001	0.88	0.003

The Effect of Light Duration During Hardening

Significant differences were detected for temperature, hardening method, genotype, and a temperature by hardening method interaction. All hardening methods survived well at –5°C (96%) but the Chill and LLight had a 48% and 37% increase in survival over no hardening at –8°C. At –11°C, only the Chill treatment had any survival (12%) (Table 4). Although a significant interaction involving lines was not present, PI469737 and PI469892 did tend to have increased survival in the LLight treatment as compared to the other lines tested.

Table 4. Survival from the artificial freeze test. Different pre-freeze hardening methods include chilling treatment; low light at growing temperatures, and greenhouse conditions.

		Survival (%)			
Hardening		Freeze temperature			
method	Line	–5°C	–8°C	–11°C	Mean
Chill	Casca	100	75	8	61
Chill	Aspen	100	55	8	54
Chill	Cathy	100	33	5	46
Chill	Spr19	100	50	0	50
Chill	Wnr37	100	75	25	67
Chill	Wnr92	100	30	25	52
Chill	Mean	100	53	12	55
LLight	Casca	78	25	0	34
LLight	Aspen	75	40	0	38
LLight	Cathy	100	42	0	47
LLight	Spr19	92	18	0	36
LLight	Wnr37	92	58	0	50
LLight	Wnr92	100	75	0	58
LLight	Mean	89	42	0	44
GrnHs	Casca	100	14	0	38
GrnHs	Aspen	100	0	0	33
GrnHs	Cathy	100	0	0	33
GrnHs	Spr19	83	0	0	28
GrnHs	Wnr37	100	17	0	39
GrnHs	Wnr92	100	0	0	33
GrnHs	Mean	97	5	0	34
LSD(.05)	Hard & Temp	51	51	51	40
LSD(.05)	Lines	28	28	28	16

These two lines also showed increased survival in the field in Texas when a mild fall was followed by an abrupt cold front (Auld et al. 1992). This evidence supports the role that photoperiod may play in hardening oilseed rape lines and that there are genetic differences in a rapeseed line's ability to harden with changes in the photo period.

SUMMARY

Progress in increasing winter hardiness has been made for the Great Plains region. Areas of the world currently growing winter rapeseed are environmentally different and no cultivars adapted to the Great Plains have yet to be identified. Populations that have sufficient cold tolerance and dormancy that are adapted to the winter photoperiod south of the 40th parallel need to be developed and identified. This will only be accomplished by incorporating traits form different rapeseed lines that has been developed in different parts of the world. Using the tools outlined should help speed the development of populations that are stable and consistent in the environmental conditions of the Great Plains.

REFERENCES

Auld, D.L., N.W. Hopper, R.E. Zartman, M.G. Hickey, and F. Gaitan-Gaitan. 1992. Adaptation and use of *Brassica* species in West Texas. Agron. Abs. 440.

Beaulieu, G.C. and D.J. Hume. 1987. Adaptation of winter rapeseed in Ontario. Can. J. Plant Sci. 67:675–684.

Canola Production Handbook. 1990. Coop. Ext. Serv., Kansas State Univ., Manhattan. Publ. C-706 revised.

Fowler, D.B. 1979. Selection for winterhardiness in wheat. II. Variation within field trials. Crop Sci. 19:773–775.

Fowler, D.B. and L.V. Gusta. 1979. Selection for winterhardiness in wheat. I. Identification of genotypic variability. Crop Sci. 19:769–772.

Fowler, D.B., L.V. Gusta, and N.J. Tyler. 1981. Selection for winterhardiness in wheat. III. Screening methods. Crop Sci. 21:896–901.

Fowler, D.B., D. Siminovitch, and M.K. Pomeroy. 1973. Evaluation of an artificial test for frost hardiness in wheat. Can. J. Plant Sci. 53:53–59.

Levitt, J. 1972. Responses of plants to environmental stresses. Academic Press, New York.

Pomeroy, M.K. and D.B. Fowler. 1973. Use of lethal dose temperature estimates as indices of frost tolerance for wheat cold acclimated under natural and controlled environments. Can. J. Plant Sci. 53:489–494.

Teutonico, R.A., J.P. Palta, and T.C. Osborn. 1993. In vitro freezing tolerance in relation to winter survival of rapeseed cultivars. Crop Sci. 33:103–107.

Thomas, J.B., G.B. Schaalje, and D.W.A. Roberts. 1988. Prolonged freezing of dark-hardened seedlings for rating and selection of winter wheats for winter survival ability. Can. J. Plant Sci. 68:47–55.

Thompson, K.F. and W.G. Hughes. 1986. Breeding and varieties. In: D.H. Scarisbrick and R.W. Daniels (eds.), Oilseed rape. Collins Pub., London.

Evaluation of Planting Date for Winter Canola Production in Indiana

Ellsworth P. Christmas

Winter canola (*Brassica napus* L.) is a winter annual and as such is seeded in late summer or fall. When commercial production was first considered in Indiana, site location and planting date were thought to be the two most important cultural decisions. Since winter wheat is commonly grown in Indiana, winter wheat areas were considered most likely to be well suited for winter canola. This has proven to be the case. Therefore, good sites for winter canola are easy for producers to identify through their experience with winter wheat production. The second most important production decision relates to planting date. Fribourg et al. (1989) reported that Sept. plantings of winter canola in Tennessee yielded significantly higher than Oct. plantings. Planting date studies were also conducted in western Kentucky by Herbek and Murdock (1989) with plantings on Sept. 1, 15, and Oct. 1, 1987 and on Sept. 2, 15, 30, and Oct. 14, 1988. The Sept. 15 planting date in 1987 produced a significantly higher yield than either the Sept. 1 or Oct. 1 plantings. In 1988, the Sept. 2 and 15 planting dates yielded significantly more that the Sept. 30 and Oct. 14 planting dates. Teo et al. (1988) reported that early seeding of spring canola resulted in significantly greater seedling infection from *Rhizoctonia solani*. The objective of the study was to evaluate the effect of planting date on winter survival and yield of selected cultivars of winter canola when grown under Indiana climatic conditions.

METHODOLOGY

Three sites were selected for the experiment on planting date, each representing a distinctly different climate in terms of severity of winter temperatures. The Northeast Purdue Agricultural Center (NEPAC) near Fort Wayne, was selected as the site representing the most severe winter weather. The Southwest Purdue Agricultural Center (SWPAC), near Vincennes, represented the least severe winter weather and the Throckmorton Purdue Agricultural Center (THPAC), near Lafayette, being intermediate between the other two sites in terms of winter weather severity.

Four planting dates were selected for each site ranging from mid Aug. for the initial date at NEPAC to mid Oct. for the last date at SWPAC. The goal was to plant the first date at NEPAC about Aug. 15 with THPAC one week later and SWPAC two weeks later. The succeeding three dates were to be planted at 10 to 14 day intervals following the initial date. The actual date of the initial planting as well as the number of plantings was dependent upon the soil conditions at each of the sites.

The experiment utilized a split plot design having four replications with planting date as the whole plot and cultivar as the split plot. Four cultivars of winter canola were selected for planting in 1990 and 1992 and five cultivars in 1994.

The previous crop on the plot area selected was either wheat or watermelons and was prepared conventionally with a plow or chisel plow and a disc. Prior to planting 67 kg/ha of P was incorporated and 28 to 34 kg/ha of N was applied in the fall with an additional 135 kg applied in the spring as a topdress. The plots were seeded with a drill either 2.1 or 1.5 m wide to a length of 15.3 m.

Stand counts were taken in late fall or early winter and again in Feb. or March as the plants broke dormancy and began to regrow. The plots were end trimmed to a length of 12.3 m and harvested when the seed moisture level reached 10% or less with a Wintersteiger Harvest Master Elite plot combine equipped with a 1.5 m header. The weight and moisture of the production of each plot was determined and converted to a kg/ha yield corrected to 9% moisture.

RESULTS

1990 Planting

Analysis of the data indicated a significant difference at the 5% level for planting date at all three sites. The Oct. 15 planting date at SWPAC and the Oct. 5 planting date at THPAC resulted in total death loss of all four cultivars. However, the Oct. 2 planting date at NEPAC resulted in a significant yield reduction even though the spring stand count showed a near perfect stand. As a result of limited plant growth in the fall, the Oct. 2 planting date (NEPAC) resulted in serious plant heaving which may explain the yield reduction.

The Sept. 6 planting date (NEPAC) and Sept. 4 (THPAC), resulted in a significantly higher yields at 1957 and 2501 kg/ha, receptively (Table 1). At SWPAC, the Sept. 13 and 25 planting dates were not significantly different.

There was a significant difference for cultivar at all three sites. 'Winfield', was the overall poorest performer while 'Liborius' gave the best performance at NEPAC and THPAC, 'Ceres' performed best at SWPAC.

1992 Planting

The weather conditions for the fall and winter of 1992–93 were near normal. The last planting dates of Sept. 29 (NEPAC), Sept. 30 (THPAC), and Oct. 1 (SWPAC), each gave totally different results with a 100% death at NEPAC, a 100% survival at SWPAC, and combination of death and full survival at THPAC depending on cultivar. The first planting date at each site either gave the highest yield or was not significantly different from the second planting date. The yields produced by the three planting dates at SWPAC were not significantly different, while the NEPAC and THPAC planting dates produced yields that were significantly different at the 1% level.

'Liborius' was the overall poorest yielder at all three sites in 1992–93. 'Ceres' was the higher yielder at NEPAC and 'Touchdown' at THPAC, with no significant difference between cultivars at SWPAC.

Table 1. Yield of winter canola by cultivar, year, and date of planting at three climatically different sites in Indiana

Location Cultivar	Yield (kg/ha) 1990–91			1992–93			1994–95			
Northeast (NEPAC)	Sept 6	Sept 20	Oct 2	Sept 4	Sept 16	Sept 29	Aug 18	Aug 29	Sept 9	Sept 20
Accord							2711	2789	2800	2873
Ceres	1733	1339	703	3805	3498	0				
Doublol	2073	1850	945	3432	2777	0	2543	2623	2902	2690
Falcon							2601	2511	2736	2752
Liborius	2270	1806	1114	3482	2452	0	2345	2474	2732	2571
Touchdown				3232	3518	0	2560	2620	2669	2644
Winfield	1752	1655	1023							
Mean	1957	1663	946	3488	3061	0	2552	2603	2768	2705
Central (THPAC)	Sept 4	Sept 26	Oct 5	Sept 3	Sept 15	Sept 30	Aug 24	Sept 13	Sept 22	Oct 3
Accord							2598	2910	2862	1964
Ceres	2559	2119	0	1912	2160	962				
Doublol	2651	2385	0	2601	2724	0	2668	2648	3252	2102
Falcon							2856	3264	3266	2788
Liborius	2693	2536	0	2195	2184	238	2549	2984	3003	2113
Touchdown				2840	2664	2169	2896	3270	3106	2423
Winfield	2104	2095	0							
Mean	2501	2284	0	2388	2433	842	2714	3128	3098	2278
Southwest (SWPAC)	Sept 13	Sept 25	Oct 15	Sept 14	Sept 24	Oct 1	Aug 26	Sept 12	Sept 21	
Accord							202	599	1110	
Ceres	2042	2069	0	2509	2665	2668				
Doublol	1795	1865	0	2627	2651	2850	243	819	1206	
Falcon							258	635	1124	
Liborius	1901	1781	0	2214	3165	2435	150	927	1193	
Touchdown				2810	2423	2903	623	1184	2512	
Winfield	1443	1752	0							
Mean	1800	1867	0	2540	2726	2714	295	833	1204	

1994 Planting

The fall of 1994 was very warm and considered to be a very late resulting in temperatures warmer than normal well into November. The first planting dates in the fall of 1994 of Aug. 18 (NEPAC), Aug. 24 (THPAC) and Aug. 26 (SWPAC) resulted in significantly lower seed yields than the Sept. planting dates. The planting dates of Sept. 9 (NEPAC), Sept. 13 (THPAC), and Sept. 21 (SWPAC), produced the highest yields in 1995.

The weather conditions at the SWPAC site were so mild that all of the planting dates were negatively impacted. The plant stems of the first planting date elongated to the point where the terminal bud was about 20 cm above ground level when the plants became dormant. In 1994–95, there was no consistent pattern to cultivar performance. 'Touchdown' at SWPAC, remained a poor performer.

SUMMARY

Each of the three years included in this study exhibited distinctly different weather conditions. The fall of 1994 was very warm and late resulting in excessive growth and significant winter injury particularly with the early planting dates. The late Sept. or Oct. planting dates in 1990 and 1992 resulted in significant death loss and no or greatly reduced yields. From this study, it can be concluded that winter canola can be successfully grown in northern Indiana if planted between mid Aug. 25 and Sept. 20. Plantings which are made after Sept. 20 will result in significant yield reductions or total loss of the crop. Data from southern Indiana indicate that plantings made during the month of Sept. are most likely to be successful with both Aug. and Oct. plantings resulting in death loss or yield reductions.

Cultivar performance varied by site indicating that some cultivars may be more tolerant to adverse weather than others. 'Touchdown' appears to be the best performer under adverse conditions while 'Liborius' was the poorest performer.

REFERENCES

Fribourg, H.A., C.R. Graves, G.N. Rhodes Jr., J.F. Bradley, E.C. Bernard, G.M. Lessman, M.A. Mueller, R.B. Graves, M.L. Thornton, B.A. Latka, and A.M. Plouy. 1989. Rapeseed: a potential new crop for Tennessee. Univ. Tennessee Agr. Expt. Sta. Bul. 669.

Herbek, J. and L. Murdock. 1989. Canola production guide and research in Kentucky. Univ. Kentucky College of Agriculture.

Teo, B.K., S.M. Yitbarek, P.R. Verna, and R.A.A. Morrall. 1988. Influence of soil moisture, seeding date, and *Rhizoctonia solani* isolates (AG 2-1 and AG 4) on disease incidence and yield of canola. Canadian Plant Path. 10:151–158.

Potential of Canola as a Dryland Crop in Northeastern Colorado

David C. Nielsen

Reduced tillage systems have increased precipitation storage efficiencies and increased the amount of available water for crop production in the central Great Plains (Greb et al. 1970; Smika and Unger 1986; Nielsen and Anderson 1993). Increased available water affords producers the opportunity to diversify and intensify their production systems from the traditional wheat-fallow system (Halvorson and Reule 1994; Peterson et al. 1994; Halvorson et al. 1994). Precipitation timing and amounts exhibit wide year-to-year variation, producing variations in timing and severity of water stress. The production potential for any alternative crop grown under dryland agricultural production systems needs to be evaluated with regard to this variable water availability.

Canola (*Brassica napus* L.) is an oil seed crop that may have production potential in the central Great Plains. A market is readily available due to the existence of processing facilities that currently handle sunflower oil production and consumer demand for low saturated fat oil. Producers would be able to use their existing wheat production equipment for tillage, spraying, planting, and harvesting of canola. Sims et al. (1993) reported that canola yields in Montana increased greatly with increased availability of water, but that increased water lowered mean oil content. Canola production in Alberta is reported to be about 1008 kg/ha for 203 mm of water use, and to increase by 59.5 kg/ha for each additional 10 mm of water used (Anonymous

1985). Shafii et al. (1992) reported that four winter canola cultivars grown in 1988 in Kansas yielded from 1170 to 1550 kg/ha with oil contents ranging from 37.7% to 40.0%. They provided no precipitation or water use data. Francois (1994) reported that the oil content of irrigated canola (cv. Westar) grown in Brawley, California averaged 40% in a 2-year study. He also reported that the long-term average oil content for 'Westar' grown in Canada was 43%. Wright et al. (1988) reported that severe evironmental stresses during the rapeseed growing season caused intense competition for assimilates, pod abortion, and seed loss.

Evaluations of the response of crops to varying water availability and water stress can easily be accomplished by calculating the Crop Water Stress Index from crop temperatures obtained with an infrared thermometer (Gardner et al. 1992a, b). This calculation requires knowledge of the relationship between crop temperature, air temperature, and vapor pressure deficit for a non-water-stressed crop (the non-water-stressed baseline). This relationship has not been determined for canola.

The objectives of this study were to determine: (1) a water use/seed yield production function for spring canola; (2) the sensitivity of yield components, oil content, and leaf area development to water deficits at various growth stages; (3) canola rooting depth; (4) canola production potential from the long-term precipitation record at Akron, Colorado; and (5) a non-water-stressed baseline for future water stress evaluations of canola.

METHODOLOGY

Two studies using 'Westar' canola were conducted during both the 1993 and 1994 growing seasons at the USDA Central Great Plains Research Station, 4 miles east of Akron, Colorado (45° 09' N, 103° 09' W, 1384 m). The soil type is a Rago silt loam (fine, montmorillonitic, mesic Pachic Argiustoll). In both studies evapotranspiration was calculated by the water balance method using measurements of soil water content and assuming runoff and deep percolation were negligible. The measurements of soil water content in the 0–30 cm layer were made by time-domain reflectometry. Measurements of soil water content at 45, 75, 105, 135, and 165 cm were made with a neutron probe.

Experiment 1

This experiment was used to determine a water use/seed production function for canola. Canola was planted on May 3, 1993 and Apr. 22, 1994 using a grain drill with double disk openers. Seeding rate was approximately 2,223,900 seeds/ha in rows spaced 20 cm apart. Prior to planting, the plot area was fertilized with 69 kg/ha N and 14 kg/ha P in 1993 and 94 kg/ha N and 17 kg/ha P in 1994. Treflan (trifluralin) was applied at a rate of 1.7 kg ai/ha and disk-incorporated prior to planting.

Irrigations were applied to the plot area with a gradient line-source solid-set irrigation system, with full irrigation next to the irrigation line, and linearly declining water application as distance increased from the line. Four replications of four irrigation levels existed along the line-source system, with a soil water measurement site and irrigation catch gage at each of the 16 locations. Irrigations were applied weekly to replace evapotranspiration losses from the measurment sites closest to the irrigation line. These were considered the fully irrigated, non-water-stressed plots.

Canopy temperatures were measured on six dates from June 21 to July 27, 1993 and five dates from June 9 to July 5, 1994. Measurements were taken every 45 min from 1000 to 1700 MDT on the fully irrigated plots from the southeast and southwest corners of the plots following the methods described by Gardner et al. (1992a, b). These data provided a range of temperature and vapor pressure deficit conditions from which to construct the non-water-stressed baseline for canola.

Plots were harvested for seed yield on Aug. 6, 1993, and July 18 and 27, 1994. Two harvest dates were used in 1994 due to differences in development rate associated with the gradient application of water.

Experiment 2

This experiment was used to determine the effect of water stress timing on canola yield components. Canola was hand-planted in rows 30 cm apart on Apr. 20, 1993 and Apr. 7, 1994 into 12 small plots (2.74 m by 2.67 m) which could be covered by an automated rainout shelter during precipitation events. The twelve plots were arranged in a randomized complete block design of three replications of four water treatments (Table 1). All plots received the same amount of water over the growing season, but at different times. The 15-week growing season was divided into a 5-week vegetative period (V), a 5-week reproductive period (R), and a 5-week grain-filling period (GF). Long-term average precipitation during the 15 week growing season is 234 mm. This amount of water was applied in equal weekly amounts as shown in Table 1.

Following emergence, plots were thinned to a stand of about 1,092,000 plants/ha. Leaf area was measured periodically during the growing season with the LAI-2000 Plant Canopy Analyzer (Li-Cor, Inc., Lincoln, Nebraska). Prior to planting, plots were fertilized with 67 kg/ha N. Plots were hand-weeded as needed throughout the experiment. Final seed yields were taken on July 29 and Aug. 4, 1993 and July 11, 1994.

RESULTS

The results of the gradient irrigation treatments are shown in Fig. 1. The linear regression fit to the combined data for the two years indicates that 7.73 kg/ha of seed are produced for every mm of water used after the first 158 mm of water use. The yields ranged from 538 kg/ha with 249 mm of water use to 3416 kg/ha with 521 mm of water use. A similar yield function for winter wheat grown in northeastern Colorado shows a much higher water use efficiency for wheat, with 17.21 kg/ha produced for every mm of water use after the first 172 mm of water use (Nielsen 1995).

The change in soil water content between the beginning and ending soil water readings is shown in Fig. 2a (rainout shelter plots, trt. 2) and Fig. 2b (solid set irrigation plots, low end of the irrigation gradient). The data show that water extraction by canola occurred from depths down to 180 cm, but 92% to 95% of growing season water use comes from growing season precipitation and water extracted from the 0–120 cm soil layer. Under the extreme water deficit condition of trt. 2 in the rainout shelter (no water applied during the last 5 weeks of development), canola was able to extract water out of the soil down to a volumetric water content of 0.08 m^3/m^3.

Water stress during the vegetative growth stage (trt. 4) limited early leaf area development, but plants recovered and produced more leaf area as water became available later in the growing season (Fig. 3). Water

Table 1. Irrigation treatments to determine effect of timing of water stress on canola production.

Treatment	Water withheld[z]	Water applied[z]	No. irrigations	Weekly irrigation amount (mm)	Total water applied (mm)
1	--	V, R, GF	15	15.7	234
2	GF	V, R	10	23.4	234
3	R	V, GF	10	23.4	234
4	V	R, GF	10	23.4	234

[z]V = vegetative stage, R = reproductive stage, GF = grain-filling stage

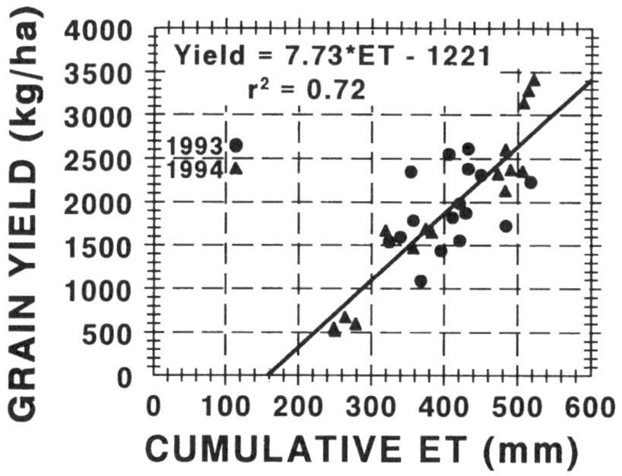

Fig. 1. Water use/seed yield production function for canola grown at Akron, Colorado, during 1993 and 1994 growing seasons.

stress during the grain-filling stage (trt. 2) resulted in a more rapid loss of leaf area than water stress occurring during other growth stages. Water stress during the reproductive growth stage (trt. 3) was the most restrictive to leaf area development, with maximum leaf area development 64 to 68% of that observed when water stress did not occur until the grain-filling period (trt. 2) (Fig. 4).

In neither 1993 nor 1994 was there a statistically significant effect of water stress timing on seed yield, although the trend in 1993 was for the lowest yield to occur when water stress occurred during the grain-filling period (trt. 2) (Table 2). This was a result of fewer branches/plant and pods/branch, and smaller seeds. The seed yields ranged from 629 kg/ha when water stress occurred during grain-filling to 1018 kg/ha when water stress occurred during the vegetative period. Yields were much lower for all four treatments in 1994, for which we have no explanation. Plants showed no visual signs of insect or disease problems. There was no trend for any particular treatment to result in higher or lower yields than the other treatments. Water stress during grain-filling (trt. 4) did result in fewer branches/plant than the other treatments, as in 1993.

The highest water use in both years occurred with water stress during grain-filling (trt. 2). The larger leaf area that developed early in the growing season and maintained itself during the reproductive stage was the probable cause of this higher water use. This higher water use resulted in a statistically nonsignificant trend for lowest water use efficiency in trt. 2. Water use efficiencies from Experiment 1 (line source) ranged from 2.20 to 4.41 kg/ha/mm between the evapotranspiration range of 254 to 381 mm, similar to the values obtained from Experiment 2 (rainout shelter) in 1993 (1.58 to 3.08 kg/ha/mm). The low yields in Experiment 2 in 1994 resulted in extremely low water use efficiencies (0.81 to 1.04 kg/ha/mm).

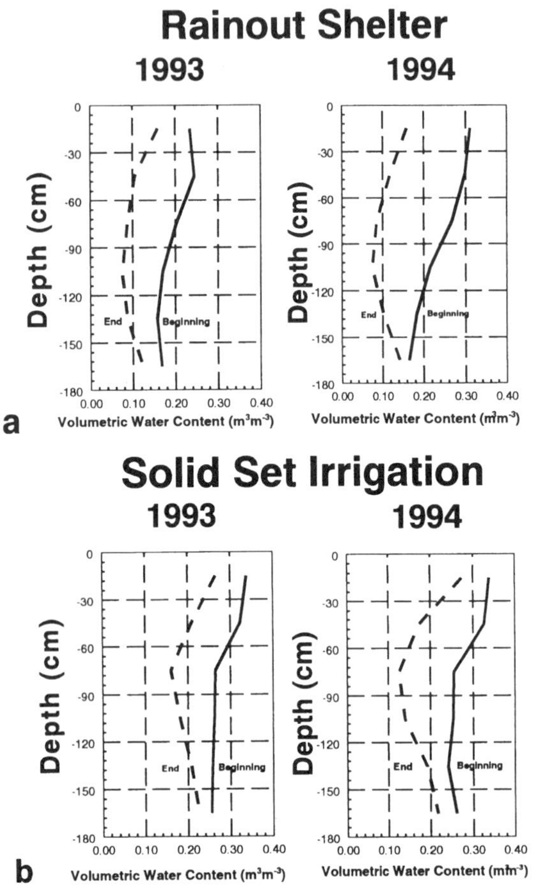

Fig. 2. Soil profile volumetric water content at the beginning and end of the canola growing season in (a) the rainout shelter and (b) the solid set irrigation area.

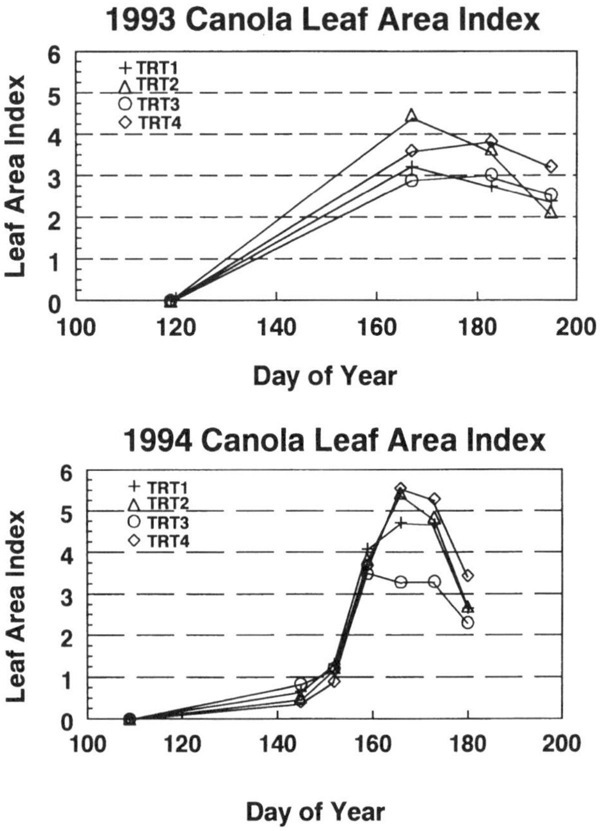

Fig. 3. Seasonal development of leaf area index.

There was a small reduction in oil content with water stress during grain-filling (trt. 2) (Fig. 5). The oil contents in Experiment 2 in the rainout shelter ranged from 34% to 39%, with higher contents in 1994. Oil contents in Experiment 2 under the solid set gradient irrigation were also higher in 1994 than in 1993. These data showed a strong trend for increasing oil content with increasing level of irrigation, with values ranging from 37% for the low irrigation level in 1993 to 44% for the high irrigation level in 1994.

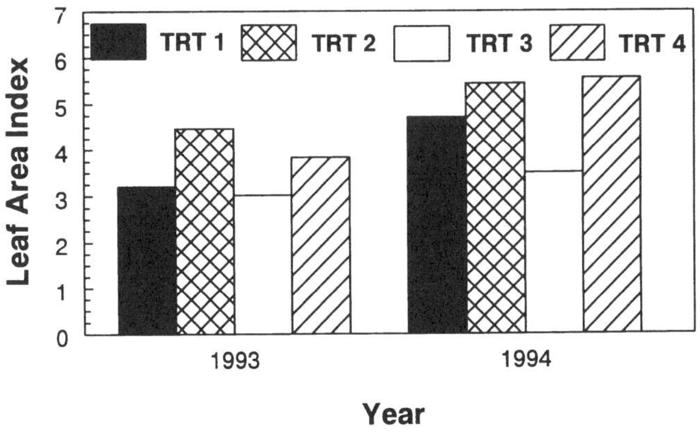

Fig. 4. Maximum leaf area index as affected by timing of water stress.

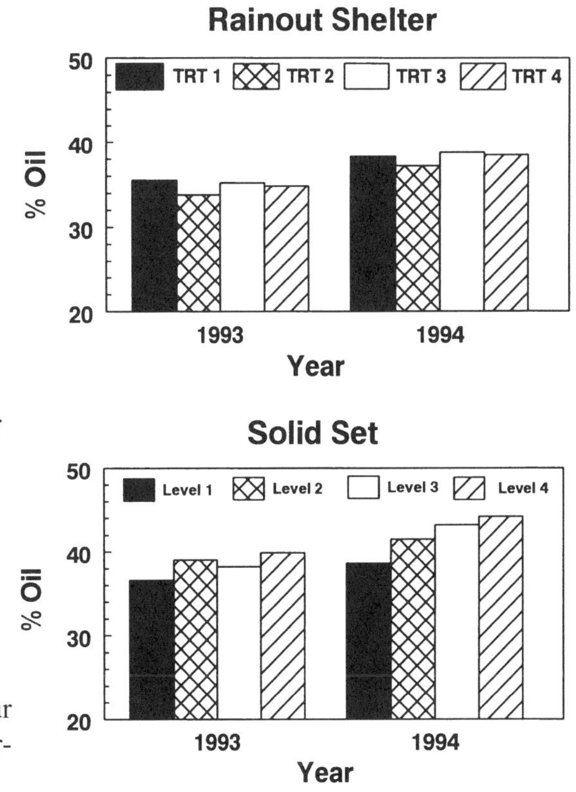

Fig. 5. Percent oil content for canola grown under four water stress timing treatments (rainout shelter) and four irrigation application levels (solid set).

Table 2. Yield component analysis for water stress timing treatments imposed in Experiment 2 (rainout shelter).

Irrig. trt.	No. branches/ plant	No. pods/ branch	No. seeds/ pod	1000 seed wt (g)	Seed yield (kg/ha)	Evapotranspiration (mm)	Water use efficiency (kg/ha/mm)
				1993			
1	4.55	6.65	10.0	3.19	942	358	2.63
2	3.51	5.61	10.6	2.70	629	399	1.58
3	4.61	6.01	8.9	3.44	930	302	3.08
4	4.69	8.68	7.7	2.90	1018	333	3.06
p	0.058	0.009	0.374	0.145	0.343	0.001	0.179
				1994			
1	2.95	8.34	3.9	2.93	412	396	1.04
2	2.78	8.34	5.1	2.67	371	460	0.81
3	2.20	7.44	3.8	3.00	310	358	0.87
4	3.45	7.68	4.2	3.22	392	419	0.94
p	0.093	0.597	0.391	0.134	0.490	0.010	0.400

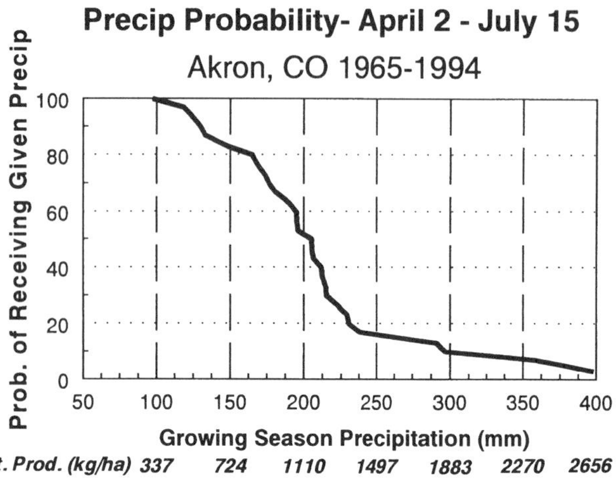

Fig. 6. Probability of receiving at least a given amount of precipitation during the period of Apr. 2 through July 15 at Akron, Colorado.

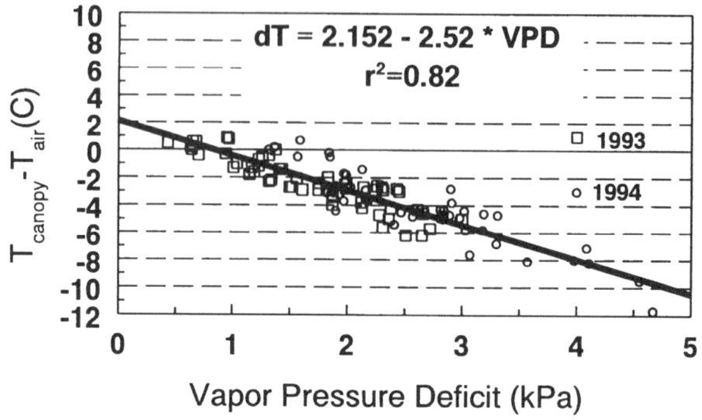

Fig. 7. Non-water-stressed baseline for canola.

In order to assess the long-term yield potential for canola in the central Great Plains, the precipitation records were examined for the 15-week growing season of Apr. 2 to July 15 over the 30-year period of 1965 to 1994 (Fig. 6). These data show that 50% of the years have growing season precipitation of less than 203 mm. Assuming, conservatively, that canola could extract 102 mm of soil water from the profile during the growing season, and applying the water use/seed yield production function given in Fig.1, 50% of the years would have seed production less than 1133 kg/ha. The predicted range of seed production over the past 30 years was 314 to 2643 kg/ha, averaging 1142 kg/ha.

Fig. 7 shows the relationship between vapor pressure deficit and canopy temperature minus air temperature (the non-water-stressed baseline). The data over the two growing seasons shows a linear response over the vapor pressure deficit range of 0.5 to 4.6 kPa. Infrared thermometry can be used with the non-water-stressed baseline to reliably quantify water stress in canola in future studies of water stress effects on canola production.

SUMMARY

Canola exhibits a linear response of seed yield to water use with approximately 7.73 kg/ha of seed produced for every mm of water used after the first 158 mm of water use. Soil water extraction comes primarily from the top 120 cm of the soil profile. Canola is most sensitive to water deficits during grain-filling, and least sensitive during vegetative development. Oil contents ranged from 34% to 44% for the various water treatments in the two years of this study. Average canola production under the dryland conditions of the central Great Plains would likely average about 1142 kg/ha with a range of 314 to 2643 kg/ha. This average level of seed production would not make canola a profitable dryland crop at current market prices (about $0.22/kg). Water stress effects on canola development can be quantified with infrared thermometry measurements of canopy temperature.

REFERENCES

Anonymous. 1985. Canola production in Alberta. Alberta Agr. Field Crops Branch AGDEX 149/20–1.

Francois, L.E. 1994. Growth, seed yield, and oil content of canola grown under saline conditions. Agron. J. 86:233–237.

Gardner, B.R., D.C. Nielsen, and C.C. Shock. 1992a. Infrared thermometry and the Crop Water Stress Index. I. History, theory, and baselines. J. Prod. Agr. 5:462–466.

Gardner, B.R., D.C. Nielsen, and C.C. Shock. 1992b. Infrared thermometry and the Crop Water Stress Index. II. Sampling procedures and interpretation. J. Prod. Agr. 5:466–475.

Greb, B.W., D.E. Smika, and A.L. Black. 1970. Water conservation with stubble mulch fallow. J. Soil Water Conserv. 25:58–62.
Halvorson, A.D., R.L. Anderson, S.E. Hinkle, D.C. Nielsen, R.A. Bowman, and M.F. Vigil. 1994. Alternative crop rotations to winter wheat-fallow. p. 6–11. In: J. Havlin (ed.), Great Plains Fertility Conference, Vol. 5, March 7–8, 1994. Kansas State Univ., Manhattan.
Halvorson, A.D. and C.A. Reule. 1994. Nitrogen fertilizer requirements in an annual dryland cropping system. Agron. J. 86:315–318.
Nielsen, D.C. 1995. Water use/yield relationships for central Great Plains Crops. Central Great Plains Res. Sta., Akron, Colorado, Conservation Tillage Fact Sheet #2-95.
Nielsen, D.C. and R.L. Anderson. 1993. Managing residue and storing precipitation. Central Great Plains Res. Sta., Akron, Colorado, Conservation Tillage Fact Sheet #2-93.
Peterson, G.A., D.G. Westfall, L. Sherrod, R. Kolberg, and B. Rouppet. 1994. Sustainable dryland agroecosystem management. Colorado State Univ. Ag. Expt. Sta. Tech. Bul. TB94-1.
Shafii, B., K.A. Mahler, W.J. Price, and D.L. Auld. 1992. Genotype × environment interaction effects on winter rapeseed yield and oil content. Crop Sci. 32:922–927.
Sims, J.R., D.J. Solum, D.M. Wichman, G.D. Kushnak, L.E. Welty, G.D. Jackson, G.F. Stallknecht, M.P. Westcott, and G.R. Carlson. 1993. Canola variety yield trials. Montana State University Ag. Expt. Sta., Bozeman, Montana AgResearch 10:15–20.
Smika, D.E. and P.W. Unger. 1986. Effect of surface residues on soil water storage. p. 111–138. In: B.A. Stewart (ed.), Advances in soil science, Vol. 5. Springer-Verlag, NewYork.
Wright, G.C., C.J. Smith, and M.R. Woodroofe. 1988. The effect of irrigation and nitrogen fertilizer on rapeseed (*Brassica napus*) production in south-eastern Australia: I. Growth and seed yield. Irrig. Sci. 9:1–13.

Canola Production in Virginia*

David E. Starner, Harbans L. Bhardwaj, Anwar A. Hamama, and Muddappa Rangappa

Use of canola (*Brassica* spp.) oil is increasing steadily among health-conscious consumers due to its lowest content of saturated fatty acids (5% to 8%) among edible oils (Downey 1990; Sovero 1993). The United States is dependent upon imported canola oil to meet the consumer demands. During 1993–94 and 1994–95, approximately 409,508 and 425,852 t, respectively, of canola oil were imported. The imports of canola oil during 1995–96 are expected to be approximately 500,000 t (Foreign Agricultural Service, 1995). Domestic production of canola would offset costly imports, enhance the productivity of American farms, and diversify agriculture.

Evaluation of rapeseed as a new crop has been continuing since 1982 at the Northern Piedmont Agricultural Research and Extension Center in Orange, Virginia. Spring and winter types were evaluated for their yielding potential by planting during both fall and spring seasons from 1983 to 1986 at Orange, Virginia. The spring types planted during fall did not survive due to lack of winter-hardiness whereas the yield of spring-types planted in the spring was very low as compared to winter types.

Commercial canola production in Virginia has been declining since 1990–91 when canola was planted on about 600 ha. During 1991–92 and 1992–93, approximately 300 and 200 ha were planted with canola. Pro-

*This research was supported by a project funded by U.S. Department of Agriculture/CSREES/National Canola Research Program.

duction during 1994–95 was practically non-existent. This situation, however, is changing due to considerable positive interest on the part of local marketing concerns and some producers who are planning to install small crushers.

In order to develop an agronomic system for canola production in Virginia, research was initiated during 1992–93 at three locations: Orange (Northern Piedmont region), Petersburg (Southern Piedmont region), and Suffolk (Coastal Plain region), to identify suitable cultivars and ideal agronomic practices (planting dates, seeding rates, and fertility requirements). The main objective of this research, is to establish canola as an alternate cash crop for Virginia.

METHODOLOGY

Cultivar Evaluations

Many cultivars have been evaluated under the National Winter Rapeseed and Virginia State Canola Variety Evaluation programs. These evaluations were conducted at Orange, Petersburg, and Suffolk as replicated experiments with three-row plots at Orange and four-row plots at Petersburg and Suffolk with a row spacing of approximately 30cm. These plots received 100 kg/ha of nitrogen, 44 kg/ha phosphorus, and 80 kg/ha potassium.

Date of Planting/Seeding Rate

During 1993–94, replicated experiments with four canola cultivars ('Cascade', 'Ceres', 'Cobra', and 'Doublol') were conducted at Orange, Petersburg, and Suffolk to determine ideal planting time. These experiments were planted on four dates at Petersburg and Suffolk, and on three dates at Orange. The experimental plots consisted of three rows at Orange and four rows at Petersburg and Suffolk with a row spacing of about 30 cm. During 1994–95, three planting times were evaluated at each location but the first and second planting experiments at Petersburg and Suffolk were unsuccessful. However, the third planting (Nov 8, 1994 at Suffolk and Dec 9, 1994 at Petersburg) was successful. In this planting, four seeding rates (approximately 4.5, 9.0, 13.5, and 18.0 kg/ha) were evaluated for each of the four cultivars. Three seeding rates (approximately 4.5, 9.0, and 13.5 kg/ha) were evaluated at Orange in a separate experiment. These plots received 100 kg/ha of nitrogen, 44 kg/ha phosphorus, and 80 kg/ha potassium.

Fertility Evaluations

Replicated experiments, using 'Ceres', were conducted at Orange, Petersburg, and Suffolk during 1993–94 and 1994–95 to determine the effects of 50, 100, 150, and 200 kg/ha of N on canola yield. In experiments at Petersburg and Suffolk, three N application times (all applied in fall as top-dress, all applied in spring as top-dress, and half applied in fall and the other half applied in spring) were also evaluated with 'Ceres'. Calcium nitrate (15.5%) was used as the nitrogen source at Petersburg and Suffolk. At Orange, ammonium nitrate (34% N) was used as nitrogen source. All plots received 44 kg/ha phosphorus and 80 kg/ha potassium.

Plot yields from all experiments were converted to t/ha with a moisture of 8.5%. One t/ha is approximately equal to 18 bushels/acre.

RESULTS

Mean canola yields (t/ha) during the 1992–93 season varied from 1.7 to 2.2 at Orange, 1.7 to 2.5 at Petersburg, and 2.4 to 4.0 at Suffolk (Table 1). During 1993–94, the mean canola yields (t/ha) varied from 1.5 to 2.3 at Orange, 1.6 to 2.5 at Petersburg, and 1.7 to 3.1 at Suffolk (Table 1). During 1994–95, the mean canola yields (t/ha) varied from 1.4 to 2.8 at Orange, 1.0 to 1.9 at Petersburg, and 1.1 to 3.0 at Suffolk (Table 1). These results indicate that canola yields in Virginia can be competitive with yields from other states where canola yields generally average 2.0 to 2.2 t/ha. The average state yields in Virginia during 1992–93, 1993–94, and 1994–95, were 2.1, 2.1, and 1.8 t/ha, respectively, however, yields of two highest yielding cultivars in each year averaged to be 2.7, 2.4, and 2.2 t/ha for 1992–93, 1993–94, and 1994–95, respectively. Based on these yield levels, canola can successfully compete with winter wheat production which has an aver-

Table 1. Performance of canola cultivars during 1992–93, 1993–94, and 1994–95 in Virginia.

Cultivar	Yield(t/ha) Orange	Petersburg	Suffolk	Mean
		1992–93		
NPZ-01	-.-	2.5	3.3	2.9
HN120-91	2.2	1.7	4.0	2.6
HN057-92	2.2	1.8	3.7	2.6
HN031-91	2.1	2.0	3.0	2.4
Ceres	2.2	1.9	2.8	2.3
Falcon	2.2	1.9	2.6	2.3
Accord	-.-	1.8	2.7	2.2
HN029-91	2.1	1.8	2.6	2.2
HN020-91	1.8	1.9	2.4	2.0
NPZ-101	1.7	-.-	-.-	1.7
Mean	2.1	1.9	3.0	2.1
LSD(.05)	0.5	0.8	1.0	0.5
		1993–94		
Jetton	2.3	2.5	2.5	2.4
HN057-92	2.0	2.1	3.1	2.4
HN9324	2.3	2.0	2.3	2.2
HN090-91	2.0	2.0	2.3	2.1
A731	2.1	1.7	2.2	2.0
Accord	1.5	1.6	1.7	1.6
Mean	2.1	2.0	2.4	2.1
LSD(.05)	0.4	0.3	0.6	0.4
		1994–95		
Jetton	2.4	1.3	3.0	2.2
HN031-91	2.7	1.9	1.8	2.1
HN074-91	2.5	1.3	2.4	2.1
HN020-91	2.4	1.9	1.6	2.0
HN120-91	2.8	1.6	1.4	1.9
HN042-92	2.4	1.2	2.1	1.9
Falcon	2.4	1.2	1.9	1.9
Casino	2.2	1.3	2.1	1.8
HN033-91	2.0	1.4	2.0	1.8
Cobra	2.6	1.4	1.1	1.7
Cascade	2.0	1.6	1.1	1.6
Pendleton	1.8	1.1	1.7	1.6
Ceres	2.0	1.0	1.7	1.6
Onyx	1.4	1.2	1.3	1.3
Mean	2.3	1.4	1.8	1.8
LSD(.05)	0.5	0.6	0.9	0.6

Table 2. Effects of planting time on canola yield in Virginia.

Location	Planting date	Yield (t/ha)
	1993	
Petersburg	Sept. 29	1.6a[z]
	Oct. 6	1.6a
	Oct. 13	1.2b
	Oct. 20	0.4c
Suffolk	Oct. 1	2.4a
	Oct. 8	2.0b
	Oct. 15	1.8b
	Oct. 22	1.2c
Orange	Oct. 4	0.8a
	Oct. 15	0.2b
	Oct. 26	0.0b
	1994	
Orange	Sept. 13	2.5a
	Sept. 28	2.4a
	Oct. 7	1.6b

[z]Means separation within locations by Duncan's Multiple Range Test (5% level).

Table 3. Effects of nitrogen rates on canola yield in Virginia.

Treatment rate (kg/ha)	Yield(t/ha) Orange	Petersburg	Suffolk	Mean
		1993–94		
50	1.2b[z]	1.6c	1.7b	1.5b
100	1.6a	1.8b	2.1a	1.8a
150	1.6a	2.0a	2.1a	1.9a
200	1.5a	2.0a	2.4a	2.0a
		1994–95		
50	2.7a	2.5b	1.7b	2.3b
100	2.5a	3.0a	1.9ab	2.5a
150	2.2a	2.4b	2.3a	2.3b
200	2.3a	2.4b	2.2ab	2.3b

[z]Means separated within locations and years by Duncan's Multiple Range Test (5% level).

age yield of 3.7 t/ha ($367/ha) (Virginia Agricultural Statistics, 1993). Based on canola yields of 2.7, 2.4, and 2.2 t/ha, the returns will be $605, $531, and $483/ha, respectively. The oil content of canola seeds varied from 32% to 49% with a mean of 40% during 1993–94 season. The bird damage to maturing canola has been observed at research farms at Orange and Petersburg due to proximity to trees and small plots but has not been an economic problem in commercial canola fields or at the research farm located in Suffolk.

Based on the results in Table 2, the ideal planting time for canola is late Sept. or early Oct. Since a delay in canola planting beyond the first week of Oct. resulted in significant yield loss, it may be suitable for planting only in rotation with early maturing crops other than soybean. The results presented in Table 3 indicate that under Virginia conditions canola needs approximately 100 to 150 kg of N/ha. The residual N approximates 14 kg/ha at Petersburg and Suffolk indicating that in most situations application of 100 kg/ha of nitrogen should be optimum. The results from application time experiments indicated lack of differences when nitrogen was applied in fall, spring, or split into fall and spring. The experiments at Petersburg and Suffolk with four seeding rates (approximately 4.5, 9.0, 13.5, and 18.0 kg/ha) each with 'Ceres', 'Cascade', and 'Cobra' and three seeding rates with 'Ceres' (4.5, 9.0, and 13.5 kg/ha) at Orange (results not presented) indicated that seeding rates did not affect canola yield.

Even though canola has been shown to have potential as an alternative cash crop in Virginia, lack of crushing facilities and/or marketing outlets have been a limitation in its development as a viable crop. This situation, however, is changing for the better. A farmer in Eastern Virginia installed an extruder type crusher that is capable of crushing canola produced on approximately 150 to 200 ha during 1995–96 season. Some other individuals in Virginia are also planning to establish such small crushers with an approximate cost of less than $100,000 each and market the oil rather than the whole seed. Such endeavors might prove to be a boost in establishing canola in Virginia.

These studies have identified suitable cultivars and optimum cultural practices. The yield potential of Virginia-grown canola compares favorably with that of other states. The only hurdle in the way of canola becoming a successful cash crop in Virginia is the availability of market access.

REFERENCES

Downey, R.K. 1990. Canola: A quality brassica oilseed. p. 211–215. In: J. Janick and J.E. Simon (eds.), Advances in new crops. Timber Press, Portland, OR.

Sovero, M. 1993. Rapeseed, a new oilseed crop for the United States. p. 302–307. In: J. Janick and J.E. Simon (eds.), New crops. Wiley, New York.

Foreign Agricultural Service. 1995. Oilseeds world market and trade bulletin. U.S. Department of Agriculture, Washington, DC.

Virginia Agricultural Statistics. 1993. Virginia Department of Agriculture and Consumer Services. Richmond.

INDUSTRIAL CROPS

Germplasm Use in Arid Lands Industrial Crops

Dennis T. Ray and David A. Dierig

Domestication is the unique process by which a plant species is adapted to a life of intimate association with and for the advantage of humans. Charles Darwin (1868) stated this quite well in *The Variation of Plants and Animals Under Domestication*.

> *"...although man did not cause variability and cannot prevent it, he can select, preserve, and accumulate the variations given him by the hand of nature almost any way he chooses; and certainly produce a great result...."*
>
> *"...domestic races of plants often exhibit an abnormal character, as compared with natural species; for they have been modified not for their own benefit, but for that of man."*

Although he did not understand how diversity arises, Darwin understood that we can alter germplasm as needed by preserving and accumulating the variations found in nature. These are the major goals of any germplasm program.

A diverse germplasm is necessary for breeding both established and new crops. Thus, the improvement of new industrial crops is not appreciably different from the enhancement and breeding of more established food, feed, and fiber crops (Thompson 1990a; White et al. 1994). In both, the plant breeder takes the extant germplasm and searches for genetic variability in desired traits. The major differences are that new crops plant breeders often are working with a species that is not as yet domesticated, they are unfamiliar with the species, and are starting with a limited and frequently un-evaluated or exotic germplasm base (Thompson 1990a).

In conventional crops the barriers to domestication were overcome thousands of years ago. Yet access to a diverse germplasm is still essential for the enhancement of specific traits (e.g. a specific disease resistance) in already well adapted and high yielding lines. In new crops a diverse germplasm may be even more important, since genetic diversity must be available in order to overcome the barriers to domestication, as well as, increase yield and produce lines adaptable to cultivation in diverse environments (Thompson et al. 1992). Yet, since little work has been done on these species, often relatively few accessions have been collected and are available to plant breeders.

We identify five major elements in new crops research and development programs, that encompass all aspects from domestication to commercialization. These are (1) germplasm collection, (2) germplasm evaluation, (3) germplasm enhancement and development, (4) cultivar development and evaluation, and (5) commercialization (Thompson 1990a). The first three are concerned with germplasm use and are the emphasis of this paper, in which we examine three potential new crops (vernonia, lesquerella, and guayule) for use in arid regions. The fourth step involves the utilization of improved germplasm in cultivar development for specific environments and cultural systems, and the fifth is the acceptance by industry and consumers and scale-up of production. The last two steps will not be discussed in this paper.

Germplasm collection should involve an understanding of the taxonomy and ecology of the prospective crop plant, so the collector can identify potential sources of genetic variability. After seed and plants are collected and documented, the germplasm is increased and must be maintained and stored under proper conditions before distribution. Established crops have the advantage of a long history of collecting, and their germplasm base has a higher probability of reflecting the natural genetic diversity. New crop germplasm is often unmethodical in its collection either because it was available while a collector was in the region or collected deterministically, seeking a specific trait. The problem with deterministic collecting is that breakthroughs often occur in hybrids of very divergent genetic stocks, in which neither line contained the desired characteristic (Knapp 1993; Ray 1994; Thompson et al. 1994b).

Germplasm evaluation is critical in new crop development. Rarely can a plant species be taken from the wild and converted to a commercial crop without considerable manipulation of their genetic resources. In germplasm evaluation the barriers to domestication must be identified early and the germplasm searched for genetic diversity by which to overcome these obstacles. These barriers might include adaptation, growth habit, pollination and reproduction, mode of propagation and distribution (which include seed shattering and dormancy), and pest and stress tolerance. After the barriers to domestication are identified, initial productivity and yield estimates can be made, potential chemical and nutritional utilization evaluated, and perhaps even an initial economic assessment.

In germplasm enhancement and development the barriers to domestication are addressed through plant breeding. This begins with the identification, isolation, and selection of desired traits from the available germplasm. These genes, assuming the differences are genetic, are transferred by whatever means available. These techniques might be genetic or cytogenetic, including intraspecific and interspecific hybridization, tissue culture, embryo culture, protoplast fusion, and possibly even genetic engineering technology, although this is doubtful at such an early stage in crop development. Once improved lines are developed, they should be released so that they are available to other breeders, and can become the basis upon which cultivars are developed.

VERNONIA

In the United States over 63 million kg of epoxy compounds are used in coatings and adhesives annually. Essentially all of the industrially used epoxy compounds are either produced by chemical modification of petrochemicals or by epoxidation of soybean and linseed oils (Ray 1994). Vernonia produces a natural epoxy fatty acid and has been the focus of a major commercialization effort in this country since 1990. The main use of vernonia oil would be in the reformulation of oil-based or alkyd-resin paints, of which, over 1 billion liters are produced annually. Volatile organic solvents evaporate in production and use, reacting with other chemicals in the atmosphere, contributing to smog. Vernonia oil can replace substantial quantities of these solvents (Thompson et al. 1994a).

In 1954, Gunstone discovered that seed oil of *Vernonia anthelmintica* contained high amounts of epoxy fatty acids. However, breeding and cultural work stopped when essentially no genetic variability was identified for seed retention. Perdue at al. (1986) reported the finding of *V. galamensis* subsp. *galamensis* var. *ethiopica*, which had high oil content and good seed retention. The problem was that var. *ethiopica* is a short-day plant and does not flower in the continental United States (Dierig and Thompson 1993).

In order to achieve full domestication of vernonia for cultivation and production in the United States, several barriers must be overcome. The desired traits are day-neutrality, autofertility, non-dormant seed germination, good seed retention, increased uniformity of seed maturity, and high oil and vernolic acid contents (Dierig and Thompson 1993). The first major breakthrough came when it was found that some *V. galamensis* subsp. *galamensis* var. *petitiana* germplasm was day-neutral (Thompson et al. 1992). *V. galamensis* subsp. *galamensis* var. *petitiana* was grown at six locations across the United States and highly significant differences were found in oil content (34.5% to 44.3%), vernolic acid (61.0% to 80.0%), and seed weight (1.87 to 2.92 g/1000) (Thompson et al. 1994a).

Dierig and Thompson (1993) found while evaluating germplasm, that var. *petitiana* was self-sterile, allowing for ease of crossing with var. *ethiopica*. Crosses were made forming intraspecies hybrids, and selections were performed for self-fertility, day-neutral photoperiodic response, good seed retention, non-dormant seed, and high yield (both for seed and oil) (Thompson et al. 1994b). The top 20% of the selections were grown in a winter nursery in 1993, and this seed was grown at 10 locations across the United States in 1994.

The intraspecific hybrids have been very successful, but only a limited number of accessions have been used in the crosses to date (Thompson et al. 1994b). Only 30 plants from one accession of var. *petitiana* were used as seed parents in the formation of hybrids. Pollen parents for all hybrids consisted of one accession (18 plants) of var. *ethiopica*, two accessions of var. *galamensis* (3 and 4 plants, respectively), one accession (3 plants) of a variety that has yet to be determined, and finally one plant from one accession of subsp. *mutomoensis*. This narrow genetic base is a product of the limited germplasm, since in the National Plant Germplasm System (NPGS) there are only five accessions of var. *petitiana*, one accession of var. *ethiopica*, and four accessions of var. *galamensis* (White et al. 1994). The danger here is that the genetic base will be narrowed too quickly while making selections in only a few environments.

LESQUERELLA

Lesquerella spp. are a good source of lesquerolic, densipolic, and auricolic fatty acids (Thompson 1990b). The seed oil of *Lesquerella* spp. may serve as a replacement for castor oil in the United States, with annual imports of some 40 to 45 thousand tonnes at a cost of $40 to 45 million. In addition, lesquerella seed oil is believed to have a broader utility than castor oil, and may serve as an essential chemical feed-stock for production of a wide array of high performance lubricants, plastics, protective coatings, surfactants, cosmetics, and pharmaceuticals (Thompson 1990b). There are currently commercial fields of lesquerella under production, and the oil is being refined by a single processor.

Lesquerella fendleri has been targeted as the primary species for domestication, with variation in lesquerolic acid from 55% to 60%. Seed yields of 1,400 kg/ha have been obtained from an essentially unselected population (Thompson et al. 1992), and through selections seed yields have increased to 1,700 kg/ha, with 30% oil and 59% lesquerolic acid. To date, all of the selected families used in cultural and processing research have a very narrow genetic base (Thompson and Dierig 1994).

Early selections were primarily to improve plant height, vigor, and seed yield. Now the breeding program is looking for autofertility, although there appears to be limited variation in the available germplasm, determinate growth habit, and early flowering. Up to 15% of *L. fendleri* plants in seed production populations have been found to be male sterile (Thompson and Dierig 1994). The inheritance of male-sterility is being investigated and is thought to have real potential for use in crosses to increase the genetic diversity in breeding populations.

The need for acquisition and utilization of new sources of lesquerella germplasm was recognized and addressed by workers at the U.S. Water Conservation Laboratory in Phoenix, Arizona. Collections were made in 1993 and 1994. In 1993, 44 *L. fendleri* accessions, 38 accessions from nine other *Lesquerella* species, and two different *Physaria* species were added to the working germplasm collection at Phoenix, Arizona. In 1994, another 41 *L. fendleri* accessions and 51 other accessions from 10 other *Lesquerella* species were added to the collection (Dierig et al. 1996). Evaluation of this germplasm is presently taking place, and new variation for seed weight and growth habits have been found. Selecting for autofertility is still a high priority.

GUAYULE

The United States is totally dependent upon importation of natural rubber for industrial purposes, at an annual cost of about $1 billion (Thompson and Ray 1989). Although guayule has been tested extensively for use in the tire industry, its market niche today appears to be in hypoallergenic latex for the medical industry (Siler and Cornish 1994).

Unlike in *Hevea*, where latex flows in continuous ducts, in guayule the whole plant is harvested, ground, and the latex extracted chemically (Thompson and Ray 1989). Harvest of native stands and initial use of guayule as a source of natural rubber began in the late 1800s and became a major source for the United States and Mexico in the early 1900s. Unfortunately, native stands were rapidly depleted and much genetic diversity was undoubtedly lost.

Most guayule germplasm consists of apomictically reproducing triploids ($3n = 54$) and tetraploids ($4n = 72$), which have received the most attention in the breeding programs. Sexually reproducing, largely self-incompatible diploids ($2n = 36$) are found in a very restricted area in Mexico. Most related species are also diploid, although some have a polyploid series (Thompson and Ray 1989). A recent germplasm collection of new diploids, and their use in a recurrent selection program, along with use of interspecific hybridization is adding dimension to the total breeding effort (Estilai and Ray 1991).

Most of the current germplasm being utilized in breeding programs originated from material developed from two major collections during the Emergency Rubber Project. Most of this material traces back to a small number of accessions collected in a very limited area in the Mexican state of Durango. Twenty-two of the original 26 lines in the National Plant Germplasm System (NPGS) are from Durango. In fact, 15 of the accessions are descended from a bulk seed collection of only five plants at one location, and the sexual diploid accession was also from only five plants at one location.

Surprisingly a large amount of variability for rubber and resin quantity and quality and plant growth characteristics have been shown to exist within the apomictically reproducing polyploid germplasm (Ray et al. 1990). The facultative nature of apomixis in the polyploid material apparently serves as a mechanism for both conservation and propagation of a wide array of genetic and chromosomal variation. Many new apomictic single plant and line selections have been made with significantly increased yields (Thompson and Ray 1989; Estilai and Ray 1991).

Today there are 29 guayule accessions that have PI numbers in the NPGS. Two of these are interspecific hybrids and all are polyploid except two diploid accessions. Another 158 selections have been entered in the Germplasm Resources Information Network (GRIN) system, 111 of which are guayule. Much additional germplasm resides in the working collections in Arizona, California, and Texas. There may also be as many as 3,000 accessions collected from 310 locations in six Mexican states at the Universidad Autonoma Agraia in Saltillo, Coahuila (Thompson et al. 1992).

Selection has been primarily for rubber yield, with some work in disease resistance, stress tolerance, and multiple uses. Significant gains have been made in the sexual diploid populations, however, diploid lines generally yield lower than polyploid lines. The facultative nature of apomixis in polyploid guayule has been essentially ignored in the breeding programs. Breeding within polyploids has generally been by either mass selection or half-sib families of single-plant selections. Since apomixis is facultative, it is difficult to identify plants that are genetically different, and once improved lines are obtained, keeping them pure breeding.

Yields have been significantly increased (Thompson and Ray 1989; Estilai and Ray 1991), now the emphasis should turn to the barriers keeping guayule from becoming fully domesticated. The primary barrier is apomictic reproduction, and emphasis should be upon aggressively selecting separately for both obligate sexual and apomictic reproduction. Several barriers are concerned with seed production, and include dormancy, viability, genetic uniformity, and uniform maturity. Additionally, shortening the length of time until harvest should be seriously considered.

CONCLUSIONS

Potential new or alternative plant sources of industrial feedstocks, by their nature, have had little or no breeding work toward improvement of desired traits. Thus, unlike conventional crops, there are very minor, if any differences between "exotic" and "elite" germplasm. In addition, there are often few accessions from which to begin a breeding program. Through germplasm evaluation, needed characteristics are identified and new collections are often deterministic, where specific traits are sought rather than extensive collections to sample all of the natural genetic diversity.

Progress in breeding programs is scale dependent, and new crops programs have few individuals working on domestication and improvement. Unfortunately, there are often very short time frames during which new crops have a real chance of industrial acceptance. Because of these short time frames, and the number of plant breeders involved, new crops germplasm is often used directly without major improvements. These germplasm lines are really little more than landraces, thus it is very hard to anticipate yields and needed cultural practices in different environments.

REFERENCES

Darwin, C. 1868. The variation of plants and animals under domestication. 2 vols. London.

Dierig, D.A. and A.E. Thompson. 1993. *Vernonia* and *Lesquerella* potential for commercialization. p. 362–367. In: J. Janick and J.E. Simon (eds.), New crops. Wiley, New York.

Dierig, D.A., A.E. Thompson, J.P. Rebman, R. Kleiman, and B.S. Phillips. 1996. Collection and evaluation of new *Lesquerella* and *Physaria* germplasm. Ind. Crops Prod. 5:53–63.

Estilai, A. and D.T. Ray. 1991. Genetics, cytogenetics, and breeding of guayule. p. 47–92. In: J.W. Whitworth and E.E. Whitehead (eds.), Guayule natural rubber. Office of Arid Lands, Univ. Arizona, Tucson.

Gunstone, F.D. 1954. Fatty acids, part II. The nature of the oxygenated acid present in *Vernonia anthelmintica* L. (Willd) seed oil. J. Chem. Soc., p. 1611–1616.
Knapp, S.J. 1993. Breakthroughs towards the domestication of *Cuphea*. p. 372–379. In: J. Janick and J.E. Simon (eds.), New crops. Wiley, New York.
Perdue, R.E., Jr., K.D. Carlson, and M.G. Gilbert. 1986. *Vernonia galamensis*, a potential new crop source of epoxy acid. Econ. Bot. 40:54–68.
Ray, D.T., D.A. Dierig, and A.E. Thompson. 1990. Facultative apomixis in guayule as a source of genetic diversity. p. 245–247. In: J. Janick and J.E. Simon (eds.), Advances in new crops. Timber Press, Portland, OR.
Ray, D.T. 1994. Development of new oilseed crops in the USA. p. 26–33. In: S. Hennink, L.J.M. van Soest, K. Pithan, and L. Hof (eds.), Alternative oilseed and fibre crops for cool and wet regions of Europe. European Cooperation in the Field of Scientific and Technical Research COST 18.
Siler, D.J. and K. Cornish. 1994. Hypoallergenecity of guayule rubber proteins compared to *Hevea* latex proteins. Ind. Crops Prod. 2:307–313.
Thompson, A.E. 1990a. Breeding new industrial crops. p. 100–103. In: J. Janick and J.E. Simon (eds.), Advances in new crops. Timber Press, Portland, OR.
Thompson, A.E. 1990b. Arid-land industrial crops. p. 232–241. In: J. Janick and J.E. Simon (eds.), Advances in new crops. Timber Press, Portland, OR.
Thompson, A.E. and D.A. Dierig. 1994. Initial selection and breeding of *Lesquerella fendleri*, a new industrial oilseed. Ind. Crops Prod. 2:97–106.
Thompson, A.E. and D.T. Ray. 1989. Breeding guayule. Plant Breed. Rev. 6:93–165.
Thompson, A.E., D.A. Dierig, and R. Kleiman. 1994a. Variation in *Vernonia galamensis* flowering characteristics, seed oil and vernolic acid contents. Ind. Crops Prod. 3:175–183.
Thompson, A.E., D.A. Dierig, and G.A. White. 1992. Use of plant introductions to develop new industrial crop cultivars. p. 9–48. In: H.L. Shands and L.E. Wiesner (eds.), Uses of plant introductions in cultivar development, Part 2. Crop Sci. Soc. Am. Special Pub. 20, Madison, WI.
Thompson, A.E., D.A. Dierig, E.R. Johnson, G.H. Dahlquist, and R. Kleiman. 1994b. Germplasm development of *Vernonia galamensis* as a new industrial oilseed crop. Ind. Crops Prod. 3:185–200.
White, G.A., J.C. Gardner, and C.G. Cook. 1994. Biodiversity for industrial crop development in the United States. Ind. Crops Prod. 2:259–272.

Economic Evaluation of New Oilseed Crops for The Netherlands

Gerard Boswinkel, Johannes T.P. Derksen, and F. Petrus Cuperus

Because of the limited variation in oils and fats feed-stocks available for the production of fatty acids and derivatives for non-food products, there is an increasing industrial interest in the Netherlands in the development of new oilseed crops that are optimized for specific applications. The new crops of interest include those that contain a higher percentage of a desirable fatty acid or a lower percentage of undesirable fatty acids and those that contain unique fatty acids. The latter unusual fatty acids could on the one hand replace raw materials from petrochemical origins with renewable resources, and on the other hand, expand the existing range of raw materials available and potentially lead to novel end products. Moreover, consumer products made from renewable resources may also carry an appealling environment-friendly or "green" label.

The development of new crops to provide renewable resources to industry is not only welcomed by oleochemical industries but also very much so by agriculture. A new or increased demand for agricultural products can be beneficial to farmers incomes, but could also diversify the existing range of cash crops, resulting in environmental benefits. In particular the introduction in agricultural practice of plant species that were not previously exploited as crops can lead to a broadening of existing narrow crop rotation schedules, thus reducing the need for pesticides to keep production levels up.

It has been recognized that a major bottleneck in introducing a new plant species as an agricultural crop is the limited knowledge on the proper handling and processing of the agricultural products to industrial raw materials and in the characterization of these raw materials. Although through plant breeding a large number of plant cultivars can be generated, the industry is not so much interested in a field of attractive plants *per se*. What needs to be added to succesfully introduce a new plant species or cultivar as a novel renewable resource is the know-how to process the plant products to useful and economically attractive specialties or commodities.

As part of the mission of the Agrotechnological Research Institute (ATO-DLO) of Wageningen, The Netherlands to carry out research on increasing the added value of agricultural products in the postharvest phase we have studied the processing of novel oilseed crops. This has resulted in knowledge on the storage and extraction of these new oilseeds, as well as on optimal refining and oleochemical processing of their oils, including potential pitfalls (see references).

However, information on the economic aspects of processing these new seed oils is virtually nonexistent. Therefore we assembled cost estimates for the production of certain oils and compounds from new vegetable oilseed crops. In particular, we have studied the production costs, starting from the oilseeds, of crambe oil, methylvernolate, vernolic acid, dimorphecolic acid, and supercritical CO_2-extracted dimorphotheca oil (SCE-DMO). This selection was based on information received from oleochemical and (fine) chemical industries on which feedstocks and intermediates would be of value to the industry and, if produced economically competitive, would find a ready market. A summary of this desk study is presented below.

ASSUMPTIONS

It was assumed that the feedstock for the production of oleochemical intermediates was the cleaned and, if applicable, dehulled oilseed. Therefore it is important to note that costs associated with growing the oilseed crops, harvesting the seeds and cleaning the seeds are not included in the cost estimates given below. For estimating the processing costs of the different products we made the following assumptions:

1. Three capacities of seed: 10, 60, and 400 t a day. These capacities are at the low end of the production scale, a scale that can be anticipated to be the most relevant during the introduction stage of new agrochemical raw materials, but which provide an indication of the economy of scale.
2. One year contains effectively 300 operating days, therefore 65 days per year of production capacity "down time" have been taken into account.
3. Prices were set at US$ 0.07/kWh for electricity, $15.00/t for steam, $0.74/m^3 for tap water, 0.15/m^3 for natural gas, and 0.35/kg for hexane.
4. Where solvent extraction of oilseeds was relevant, the hexane loss during the extraction process was estimated at 3 kg/t cake.
5. The selection of the optimal unit operations for processing of the new crops and seed oils as well as the efficiency of these operations are based on experimental studies performed by the ATO-DLO on new oilseed processing over the past several years.
6. Costs associated with the purchase of new land and housing to establish a processing plant was not taken into account, since these costs can vary tremendously with location. Moreover, for small scale processing it was assumed that such a plant would find a place in an existing facility, in conjunction with the processing of conventional oilseeds rather than as a stand alone operation. Therefore these cost would be associated only in part with the new oilseed processing facility.

RESULTS

A possible flow-sheet for the production of selected oleochemical intermediates is based on applicable processing unit operation steps as resulting from prior research at ATO-DLO, is depicted in Fig 1. On the basis of the possible processing routes suggested by the flow-sheet, a choice was made on the preferred processing route to arrive at each oleochemical product under consideration. In Table 1 this preferred flow-sheet is given as a succession of unit operation steps as taken from Fig 1. This preferred flow sheet, tailored to each product, was used as a basis to estimate the production costs of the different oils and derivatives in relation to the processed amount of seeds per day. Higher capacities were not considered to be relevant for SCE DMO. Note that our calculation is limited to processing costs. Costs for facilities have not been incorporated. Our calculations show clearly for all products that increasing the production capacity leads to a decrease in costs per kg of product.

These data should be put in perspective. In Germany the production costs for rapeseed methylester ("biodiesel") are calculated to be US$ 0.25/kg, which is about half that of the US$ 0.47/kg calculated for methylvernolate. However, it should be taken into account that distillation is necessry for the production of pure methylvernolate and this additional process step is largely responsible to the higher costs.

The results also show clearly that the production costs for SCE DMO are almost 6 times higher than the production costs for oil produced by expelling. However the quality of SCE DMO is much higher (Muuse et al. 1994). The production costs for SCE DMO can be decreased by increasing the maximum solubility of the oil in the supercritical CO_2.

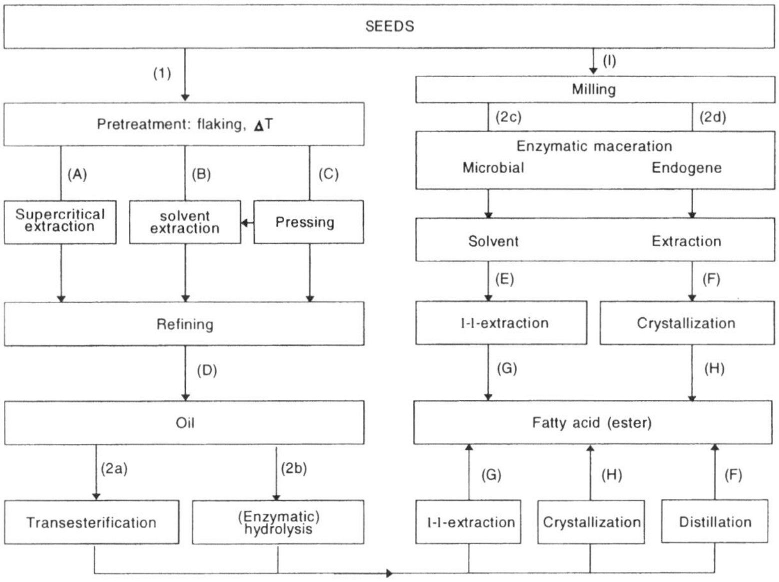

Fig 1. Flow-sheet showing possible production routes for making vegetable oils or fatty acids derivatives.

Table 1. Flow-sheet and production costs for products derived from new seed oils.

Crop	Preferred flow-sheet (Fig. 1)	Product	Production costs (US$/kg) per day 10 t seed	60 t seed	400 t seed
Crambe	1→B+C→D	Refined oil	0.18	0.12	0.07
Euphorbia	1→B+C→2a→F	Methylvernolate	0.96	0.67	0.47
	1→B+C→2b→G	Vernolic acid	0.33	0.24	0.16
Dimorphotheca	1→A	Refined oil	3.75	-.-	-.-
	1→B→D→2b→H	Dimorphecolic acid	0.63	0.44	0.24

It is expected that new oilseed crops can be processed at cost levels that are not much different from those for conventional oilseed crops. However, agronomic factors still have a large impact on the raw material, (oilseed costs), and therefore on final product costs.

REFERENCES

Derksen, J.T.P., G. Boswinkel, and W.M.J. Van Gelder. 1991. Lipase-catalyzed hydrolysis of triglycerides from new oil crops for oleochemical industries. p. 377–380. In: L. Alberghina, R.D. Schmid, and R. Verger (eds.), GBF Monograph 16: Lipases, structure, mechanism and genetic engineering. VCH Publishers, New York.

Derksen, J.T.P., A.-M. Krosse, and F.P. Cuperus. 1993. Lipase-catalyzed production of functionalized fatty acids from *Dimorphotheca pluvialis* seed oil. Med. Fac. Landbouww. Univ. Gent, 57/4a, p. 1741–1747.

Derksen, J.T.P., B.G. Muuse, F.P. Cuperus, and W.M.J. Van Gelder. 1993. New seed oils for oleochemical industry: Evaluation and enzyme-bioreactor mediated processing. Ind. Crops Prod. 1:133–139

Derksen, J.T.P., B.G. Muuse, and F.P. Cuperus. 1993. Processing of novel oilseeds and seed oils. p. 253–281. In: D.L. Murphy (ed.), Designer oilseed crops. VCH Publ., Weinheim, Germany.

Hayes, D.G., R. Kleiman, D. Weisleder, R.O. Adlof, F.P. Cuperus, and J.T.P. Derksen. 1995. Occurence of estolides in processed *Dimorphotheca pluvialis* seed oil. Ind. Crops Prod. 4:295–301.

Muuse, B.G., F.P. Cuperus, and J.T.P. Derksen. 1992. Composition and properties of oils from new oilseed crops. Ind. Crops Prod. 1:57–65.

Muuse, B.G., F.P. Cuperus, and J.T.P. Derksen. 1994. Extraction and characterization of *Dimorphotheca pluvialis* seed oil. J. Am. Oil Chem. Soc. 71:313–317.

Van Gelder, W.M.J., F.P. Cuperus, J.T.P. Derksen, B.G. Muuse, and J.E.G. Van Dam. 1993. Characterization and processing research for increased industrial applicability of new and traditional crops: A European perspective. p. 38–45. In: J. Janick and J. Simon (eds.), New crops. Wiley, New York.

Engineering New Oilseed Crops from Rapeseed

H. Maelor Davies

The engineering of new rapeseed oils has made tremendous progress in the last four years. In the II New Crops Conference I reported on the isolation of an unique enzyme involved in the biosynthesis of a valuable 12-carbon fatty acid, laurate (12:0), from the seeds of an undomesticated tree species. We had cloned the corresponding gene and were able to show that its expression in *E. coli* resulted in the production of large amounts of laurate by the bacterial culture (Davies et al. 1993). At that time there was eager anticipation of the forthcoming transgenic plant data. I can now report that laurate was produced in the transgenic seeds, and also that the high-lauric rapeseed line developed from them is now a new oilseed crop in commercial agriculture. The first commercial production of the resulting specialty oil, Laurical™, which contains 40 mol% laurate, took place last year (1995).

In parallel with this rapid progress in the development of a high-lauric oil, there have been significant advances with other kinds of oil modification. Many important genes in fatty acid and oil biosynthesis have been cloned, and several of them have been introduced into rapeseed. As a result there now exist some very interesting transgenic rapeseed plants which make new and different kinds of seed oils, and which will serve as prototypes for the development of commercial cultivars. Within a few years we can expect a whole series of new rapeseed crops to be developed, each producing a different oil for specific applications, and each representing a new opportunity for rapeseed growers which diversifies and augments, rather than competes with, their existing production. We feel that this will be a very positive development for the grower, oilseed processor, oil user, and for the consumers of products that contain plant oils.

Several recent reviews have presented comprehensive coverage of the biochemistry and molecular biology of plant lipid metabolism (Ohlrogge 1994; Ohlrogge and Browse 1995), as well as recent oilseed engineering work world-wide (Töpfer et al. 1995). In this paper I will briefly review the major progress in the engineering of plant oils at the Oils Division of Calgene Inc. Our vehicle for the production of novel oils is rapeseed (*Brassica napus*), and I will be referring to the low-erucic type called "canola" unless I indicate otherwise. The chapter organization is based on the new crop applications under development.

LAURICAL™

The 12:0-ACP thioesterase from California bay (*Umbellularia californica*) interacted successfully with fatty acid biosynthesis in the plastids of rapeseed embryos, with the result that considerable laurate accumulated (Voelker et al. 1992; Voelker et al. 1996). Furthermore this laurate was found almost entirely in the reserve oil, indicating that the "foreign" fatty acid was exported from the plastids and utilized by the enzymes of the Kennedy pathway for the assembly of triglycerides (TAGs). The initial transformants exhibited wide variation in the laurate content of the oil, resulting from the customary variation in expression of the transgene and consequential wide-ranging thioesterase activity. Non-destructive analysis of the oil composition of single seeds enabled high-lauric lines to be selected from the best events, and through subsequent breeding and performance trials a cultivar has been developed. This cultivar is known as 'Laurate Canola'; a considerable area was planted in Georgia in the late-1994 season, ultimately producing a very acceptable yield of an oil in which 40% (by weight) of the fatty acyl groups are laurate (Table 1). Known as Laurical™, this unique oil finds applications in the manufacture of soaps and detergents. The possibility of its use as a food ingredient is also under investigation.

HIGH-LAURIC OIL

The laurate enrichment of the oil in these transgenic plants is correlated with the amount of 12:0-ACP thioesterase activity expressed in the seeds, up to about 60 mol% laurate. Although further increases in thioesterase activity are possible through the introduction of additional gene copies or new genomic locations of those genes, very high activities do not have much additional effect on the laurate content of the oil (Voelker et al. 1996). We are investigating several factors which may contribute to the apparent "ceiling" of 60 mol%. One which has already received considerable attention is the inherent limitation in the use of saturated substrates by the second acyltransferase reaction of the Kennedy pathway. In rapeseed, lysophosphatidic acid acyltransferase (LPAAT) which catalyzes this *sn*-2 acylation step exhibits a strong preference for 18:1-containing substrates (for example, Oo and Huang 1989). As a result, rapeseed TAGs have an almost exclusively unsaturated acyl composition at the *sn*-2 position. We found that this was still true of the TAGs in the transgenic plants accumulating laurate. For example, the oil from a transgenic line containing 52 mol% laurate has only 5 mol% laurate at the *sn*-2 position of the TAGs (Voelker et al. 1996). One can calculate that the average laurate content at the *sn*-1 and *sn*-3 positions is therefore 76 mol%. It seems likely that the development of an oil enriched very highly in laurate will require the presence of an additional LPAAT activity able to incorporate saturated, and particularly medium-chain, acyl groups at the *sn*-2 position.

To locate an LPAAT having such a preference for medium-chain substrates we again turned our attention to seed tissues which accumulate oil rich in laurate. We identified an appropriate activity in the immature endosperm of coconut (Fig. 1), and we have since solubilized this membrane-associated enzyme (Davies et al. 1995). Sufficient purification was undertaken to enable a candidate protein to be identified, and the corresponding gene was then cloned (Knutzon et al. 1995). Expression of this gene in *E. coli* produced abundant medium-chain LPAAT activity, confirming its identity. Rapeseed plants expressing this LPAAT under the control of the napin promoter have been crossed with a laurate-producing transgenic line. Preliminary analysis of the oil accumulated by the seeds of the F_1 progeny plants shows that the introduced coconut LPAAT is incorporating laurate into the *sn*-2 position (Fig. 2). Laurate enrichments at *sn*-2 of up to 30 mol% have been seen in individual seed oils averaging 40 mol% laurate content. These results represent the first demonstration of the manipulation of TAG structure, sometimes referred to as the production of "structured triglycerides," by the genetic engineering of an oilseed crop. The F_1 seeds from the above crosses were segregating for both the thioesterase and LPAAT transgenes, so we must now examine successive generations of selfed plants to determine whether the introduction of laurate at *sn*-2 results only in a re-arrangement of the same amount of laurate or whether the total laurate enrichment of the oil is increased.

Table 1. Seed fatty acyl composition of 'Laurate Canola' and canola seed.

Fatty acyl group	Composition (wt %)	
	Laurate canola	Control canola
12:0	40.0	0
14:0	3.9	0
16:0	2.9	6.2
18:0	1.1	1.6
18:1	29.8	60.8
18:2	13.3	9.2
18:3	6.8	9.2
20:0	0.3	0.3
20:1	0.6	0.9

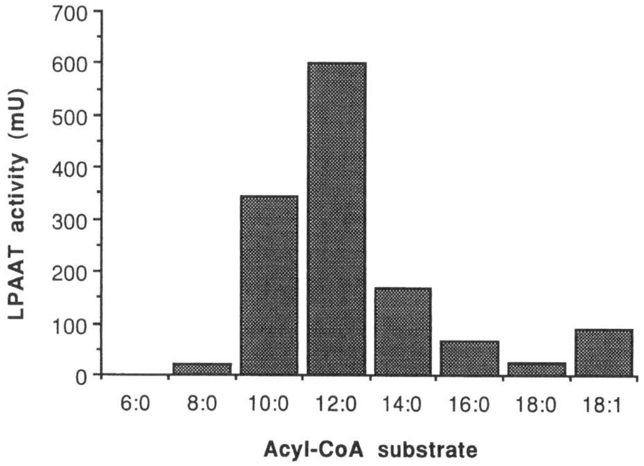

Fig. 1. Substrate specificity of lysophosphatidic acid acyltransferase (LPAAT) from immature coconut endosperm. A crude membrane fraction was prepared from a homogenate of the tissue, the activity was solubilized with detergent and then recombined with phospholipids for assay. The acyl donor substrates were those indicated, and the acyl acceptor substrate was 12:0-lysophosphatidic acid throughout. Very similar results were obtained using 18:1-lysophosphatidic acid as acceptor substrate. For assay (mU = milliunits activity) and other details see Davies et al. (1995).

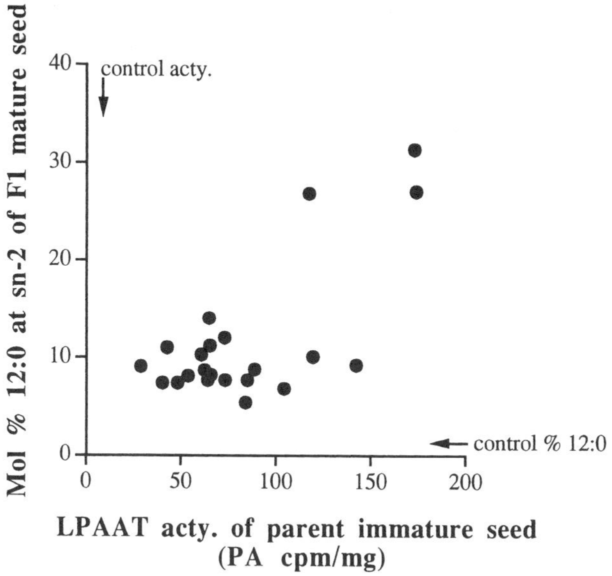

Fig. 2. Combination of the engineered traits for laurate production (via bay 12:0-ACP thioesterase) and medium-chain LPAAT activity, in rapeseed. Plants expressing the medium-chain-preferring coconut LPAAT activity were crossed with a laurate-producing line in order to examine the effect of the specialized LPAAT on the laurate content of the oil. Each point indicates the result from crossing an independent LPAAT-expressing "event" with the laurate-producing line. From each cross a pool of 20 mature F_1 seeds was obtained and analyzed for *sn*-2 laurate content. Each of these results is plotted against the LPAAT enzyme activity in pooled immature seeds from the corresponding LPAAT parent, measured using 12:0-containing substrates. The F_1 progeny are segregating for both the LPAAT activity and the overall laurate content, hence the considerable scatter in the data. Control data, shown by the arrows, were generated by crossing the same laurate-producing line with untransformed canola. Note that all the LPAAT "events" produced F_1 seed in which the laurate content at *sn*-2 was elevated relative that in the control cross, and that three events produced progeny with markedly increased *sn*-2 laurate contents.

OTHER MEDIUM-CHAIN FATTY ACIDS

By taking advantage of sequence relatedness amongst acyl-ACP thioesterases, T.A. Voelker, K. Dehesh, J. Kridl, L. Yuan and colleagues at Calgene have cloned many of these genes from a diverse range of species. Some examples are listed in Fig. 3, which shows the sequence relatedness between the respective open reading frames, and summarizes the principal substrate specificities of the corresponding enzyme activities (determined by expression in *E. coli*). The sequences fall into two distinct classes, "A" and "B," and we believe that representatives of both classes exist in all higher plants (Jones et al. 1995). All examples of class A thioesterases cloned to date appear to be the ubiquitous oleoyl-ACP thioesterase of plant fatty acid biosynthesis. The members of class B all exhibit a preference for saturated acyl-ACP substrates, and it is within this class that the specialized medium-chain-preferring types which are expressed specifically in certain oilseeds are found. Class B also includes thioesterases that exhibit preference for 14:0- and 16:0-substrates, cloned from plants which do not accumulate these fatty acids in their seed oils (e.g. *At* FatB1 in Fig. 3). The function of these thioesterases, which are quite possibly ubiquitous in higher plants and expressed in many plant tissues, is not yet known. (For a review of these thioesterases and a discussion of their evolutionary origins see Jones et al., 1995).

Several of the class B thioesterases acting on medium-chain substrates have now been expressed in rapeseed. Oils have been obtained containing 8:0, 10:0, and 14:0, and various combinations of these fatty acids in

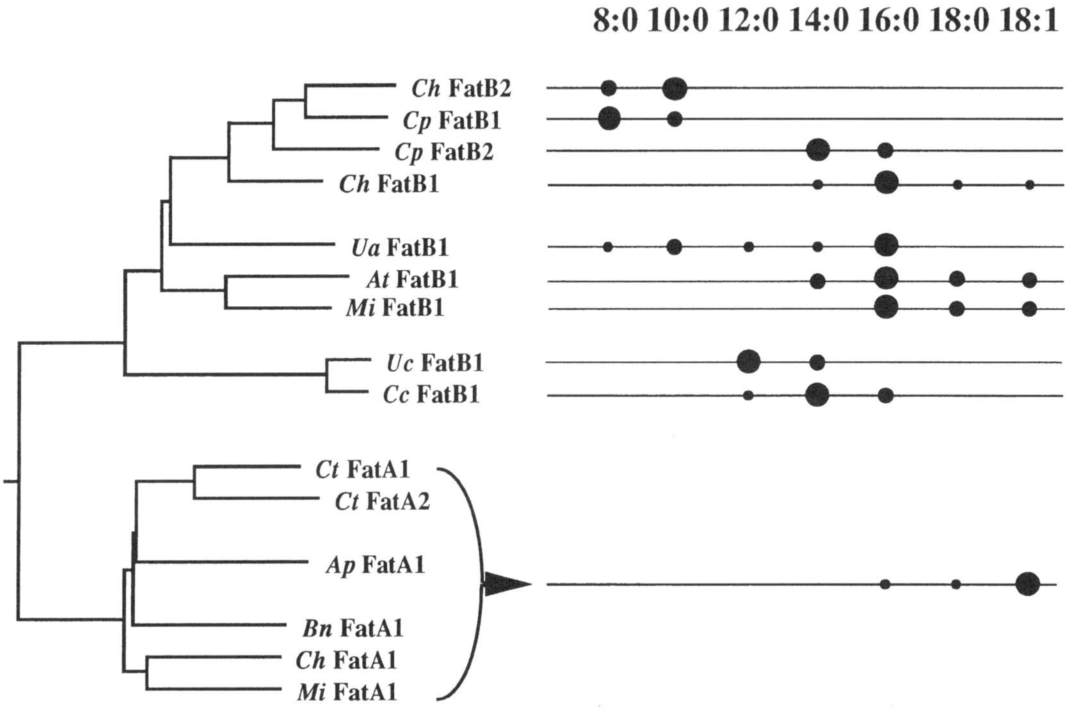

Fig. 3. Examples of homologous, higher plant, acyl-ACP thioesterases. The left-hand side of the figure shows a phylogenetic tree of some of the acyl-ACP thioesterases cloned to date, constructed from an alignment of amino acid sequences. [For a more complete listing, and for additional details and references, see Jones et al. (1995). For details of the thioesterase *At* FatB1 see Dörman et al. (1995)] The right-hand diagram summarizes the substrate specificities of the corresponding enzyme activities, semi-quantitatively. The activity on each acyl-ACP substrate is represented by a dot below the corresponding substrate acyl group (8:0 through to 18:1), and the diameter of the dot is an approximate indication of relative activity. The sources of these thioesterases are as follows [for terminology see Jones et al. (1995)]: Ap, *Allium porrum* (leek); At, *Arabidopsis thaliana*; Bn, *Brassica napus* (rapeseed); Cc, *Cinnamomum camphora* (camphor); Ch, *Cuphea hookeriana*; Cp, *Cuphea palustris*; Ct, *Carthamus tinctorius* (safflower); Mi, *Mangifera indica* (Mango); Ua, *Ulmus americana* (elm); Uc, *Umbellularia californica* (California bay).

significant amounts. For example, the expression of thioesterase *Ch* FatB2 from *Cuphea hookeriana*, a plant which accumulates an oil highly enriched in 8:0 and 10:0, has produced a rapeseed oil containing up to 10 mol% 8:0 and 25 mol% 10:0. Expression of other class B thioesterases has resulted in rapeseed oil containing up to 40 mol% 14:0. Further improvements in the amounts of the medium chains can be anticipated as these prototype lines are developed through breeding and selection. These fatty acids and the respective TAGs have diverse applications in the food ingredients, pharmaceutical, and lubricant industries.

ALTERED SATURATION

There are many food quality and health considerations that encourage the development of oils containing altered ratios of saturated/unsaturated fatty acids. For example, solid glycerides from highly saturated (high-stearic) oils may have novel functionalities, will contain no trans fatty acids, and may be less cholesterolemic than hydrogenated fats. Significant alterations to increase, and decrease, the degree of saturation have been achieved by D.S. Knutzon, G.A. Thompson, J. Kridl, and co-workers via manipulation of the level of stearoyl-ACP desaturase activity (Knutzon et al. 1992, and unpublished work). This enzyme is responsible for the first of the desaturation reactions, by forming oleoyl-ACP from stearoyl-ACP during fatty acid biosynthesis. The introduction of additional activity in canola by transformation with the safflower stearoyl-ACP desaturase gene reduced the saturate content of the oil from c. 2% (by weight) to around 1%. Expressing the rapeseed desaturase gene in the inverse, or "antisense" orientation decreased the native activity and as a result the stearate (18:0) content of the oil increased up to 40%.

Another strategy for elevating the stearate, and hence the saturate, content of the oil is to express an acyl-ACP thioesterase having a higher activity on 18:0-ACP than the incumbent rapeseed enzyme (which shows strong preference for 18:1-ACP). J. Kridl and co-workers recently cloned the gene for such a thioesterase from mangosteen (*Garcinia mangostana*), whose seeds contain 50% stearate. The expression of this gene in canola raised the stearic content of the oil to 20%. These transgenic lines will now be crossed with those expressing the antisense desaturase construct to combine the traits.

HIGH-ERUCIC OIL

Erucic acid (22:1) for the manufacture of industrial lubricants is currently obtained from high-erucic (HEAR) varieties of rapeseed. These TAGs lack erucoyl residues in the *sn*-2 "position," and there has been considerable interest in raising the erucate levels further by overcoming this compositional limitation. Such an oil would not only serve as a higher-yielding source of erucoyl residues, but would also provide trierucin for certain clinical applications. Lassner, Metz and colleagues at Calgene have approached this objective in two parallel ways. First they cloned a gene responsible for the initial reaction of the cytoplasmic fatty acyl elongation system, i.e. ketoacyl-CoA synthase (KCS), from the jojoba plant, *Simmondsia chinensis*. In its native species this enzyme is part of the "elongase" system that produces the C20, C22, and C24 acyl groups that predominate in the stored wax esters. The introduction of this gene into canola resulted in the production of TAGs containing up to 58% of their acyl groups as these very long-chain fatty acids (VLCFAs). Previous work had suggested that a deficiency in the cytoplasmic fatty acyl elongation system was responsible for the near-absence of VLCFAs in canola relative to HEAR (Stumpf and Pollard 1983). This transformation experiment conclusively showed that the metabolic deficiency is a lack of functional KCS. The KCS gene will now be used to isolate the homologous gene from HEAR, in order to over-express it and so to obtain an erucate content higher than the typical HEAR value of 40%–50% .

The second part of the strategy is to correct the limitation at the *sn*-2 position by the introduction of another specialized LPAAT. Making use of sequence homologies with the coconut LPAAT discussed above, Lassner et al. (1995) have cloned the gene for an appropriately erucate-preferring LPAAT from the meadowfoam plant (*Limnanthes alba*). The introduction of this gene into a HEAR line resulted in the incorporation of erucoyl residues into the *sn*-2 position of the TAGs, the oil containing a significant amount of trierucin as a result (Lassner et al. 1995; Fig. 4).

These genes for specialized KCS and LPAAT enzymes are now being combined in a HEAR background with the intention of producing a new, super-erucic rapeseed line.

ALCOHOLS AND WAX ESTERS

All of the above projects are concerned with altering the fatty acyl composition of rapeseed TAGs. A more ambitious plan is the replacement of the TAGs with another kind of reserve lipid, namely a wax ester having industrial applications such as that found in jojoba seeds. This will require the production of significant amounts of very-long-chain fatty acids (VLCFAs), the reduction of part of the VLCFA-CoA pool to form long-chain alcohols, and the condensation of those alcohols with more of the VLCFA-CoAs to form the wax esters. Lassner, Metz, and co-workers have cloned the jojoba acyl-CoA reductase responsible for formation of the alcohol moiety and have expressed that gene in HEAR (Metz et al. 1994). Approximately 4% of the acyl groups in the seed oil were reduced to alcohol groups. Most interestingly, an as-yet uncharacterized rapeseed enzyme activity esterified at least some of this alcohol so that about 8% of the weight of the final seed oil comprised wax esters (Fig. 5). Work is progressing on the cloning of the jojoba "wax synthase" gene to increase the rate of ester formation.

LOOKING AHEAD

Within a few years we can expect the rapeseed grower to be planting a whole series of new rapeseed cultivars that will have sufficiently different applications to represent entirely new crops. These lines will be engineered for the production of diverse oils having well-defined uses in the food, detergent, and lubricants industries. Specialty oils may also be developed with pharmaceutical and chemical feedstock applications in mind. The technology and the prototypes for many of these new crops are here today, and the engineering strategies are already moving from the relatively simple, single-gene transformations of a few years ago to two-gene projects and the introduction of longer biosynthetic pathways. The specialized genes which are required have come from the germplasm of diverse, uncultivated plant species, as well as from crop plants. In the future we can expect non-plant genes to feature as well. Along the way, much has been learnt about the basic mechanisms of the biosynthesis of fatty acids, TAGs etc. and we can continue to look forward to a productive synergy between the applied projects and the underlying basic science. The genetic engineering of oilseed plants promises to be one of the most impactive and impressive sources of new crops for both new and existing applications.

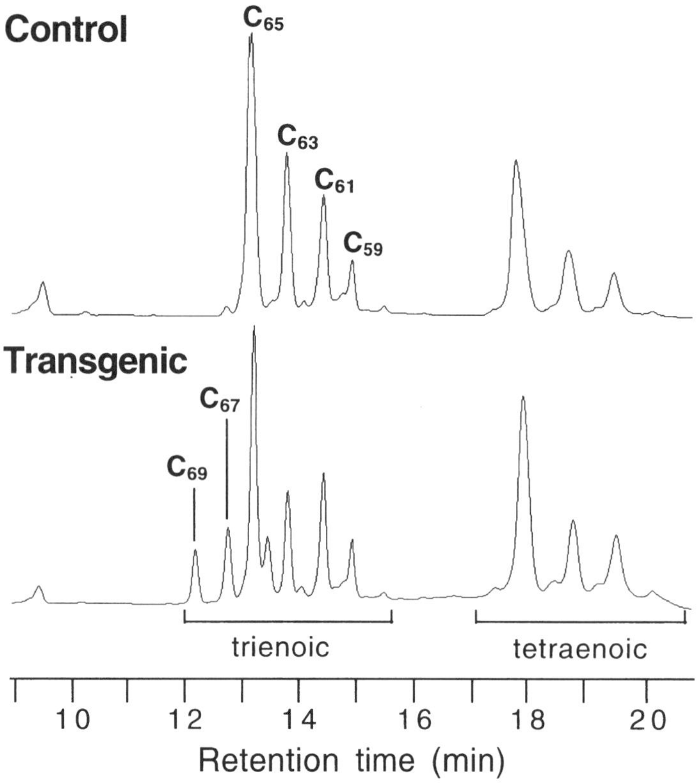

Fig. 4. Resolution of triglycerides from rapeseed (HEAR) oil of a control plant and a transgenic plant containing the meadowfoam LPAAT gene. Triglycerides were separated by silver-phase HPLC [for details see Lassner et al. (1995)] according to their degrees of unsaturation. Portions of the chromatograms occupied by trienoic and tetraenoic TAGs are shown, and peaks of interest are labeled with the total carbon content of the corresponding TAGs. Trierucin has a total carbon content of 69. The introduced LPAAT also enabled the formation of 22:1/20:1/22:1 TAG (C67).

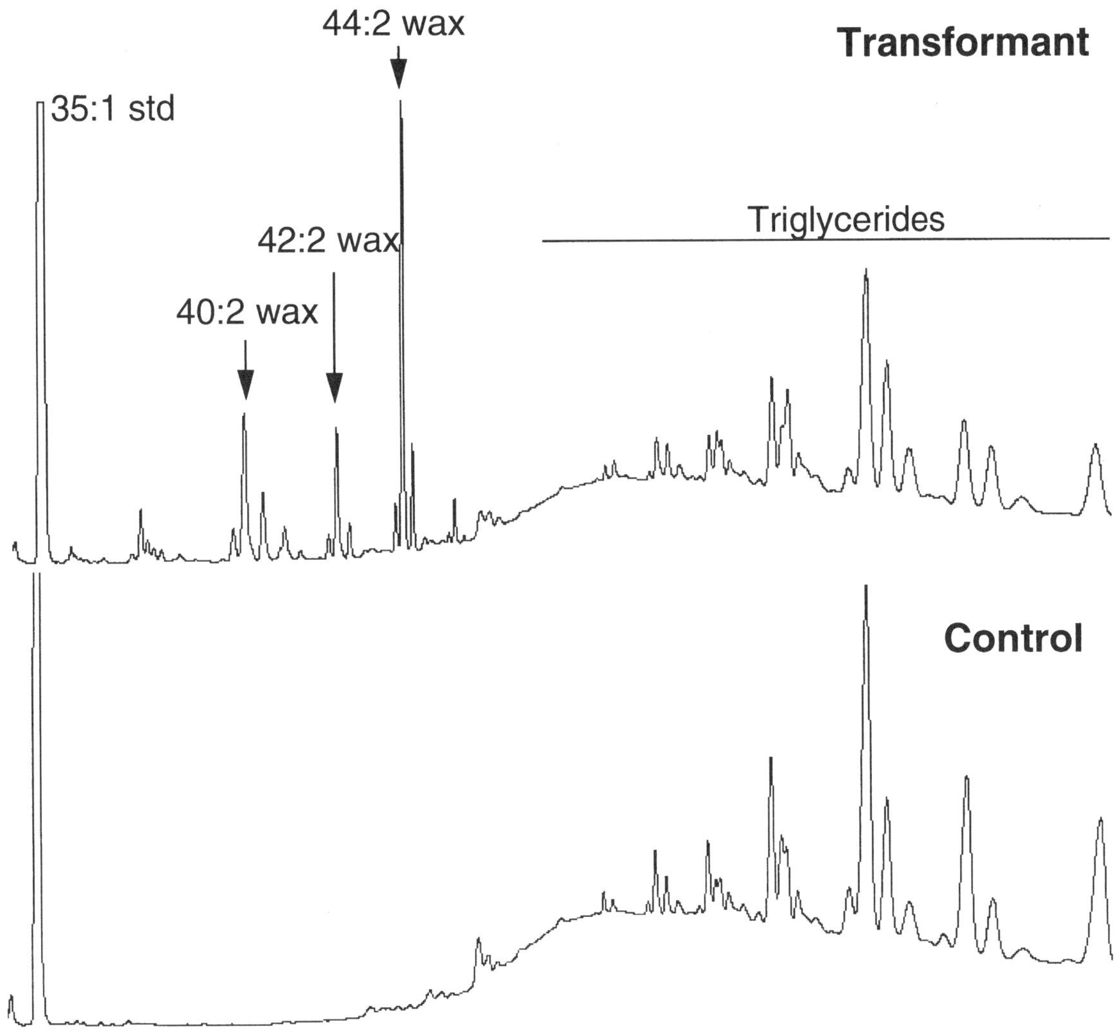

Fig. 5. Resolution of triglycerides and wax esters from rapeseed (HEAR) oil of a control plant and a transgenic plant containing the jojoba acyl-CoA reductase gene. Triglycerides and wax esters were separated by high-temperature GLC [for details see Metz et al. (1994)]. The wax esters contain alcohol and acyl moieties derived from 20:1 and 22:1 fatty acyl groups as indicated (e.g. 40:2 contains 20:1 alcohol and 20:1 acyl moieties). The 35:1 peak is from an internal standard wax ester.

REFERENCES

Davies, H.M., L. Anderson, J. Bleibaum, D.J. Hawkins, C. Fan, A.C. Worrell, and T.A. Voelker. 1993. Fatty acid synthesis genes: Engineering the production of medium-chain fatty acids. p. 176–181. In: J. Janick and J.E. Simon (eds.), New crops. Wiley, New York.

Davies, H.M., D.J. Hawkins and J.S. Nelsen. 1995. Lysophosphatidic acid acyltransferase from immature coconut endosperm having medium chain-length substrate specificity. Phytochemistry 39:989–996.

Dörman, P., T.A. Voelker, and J.B. Ohlrogge. 1995. Cloning and expression in *E. coli* of a novel thioesterase from *Arabidopsis thaliana* specific for long chain acyl-acyl carrier proteins. Arch. Biochem. Biophys. 316:612–618.

Jones, A., H.M. Davies and T.A. Voelker. 1995. Palmitoyl-acyl carrier protein (ACP) thioesterase and the evolutionary origin of plant acy-ACP thioesterases. The Plant Cell 7:359–371.

Knutzon, D.S., G.A. Thompson, S.E. Radke, W.B. Johnson, V.C. Knauf, and J.C. Kridl. 1992. Modification of *Brassica* seed oil by antisense expression of a stearoyl-acyl carrier protein desaturase gene. Proc. Nat. Acad. Sci. USA 89:2624–2628.

Knutzon, D.S., K.D. Lardizabal, J.S. Nelsen, J.L. Bleibaum, H.M. Davies and J.G. Metz. 1995. Cloning of a coconut endosperm cDNA encoding a 1-acyl sn-glycerol-3-phosphate acyltransferase which accepts medium-chain substrates. Plant Physiol. *in press*.

Lassner, M.W., C.K. Levering, H.M. Davies, and D.S. Knutzon. 1995. Lysophosphatidic acid acyltransferase from *Limnanthes alba* mediates insertion of erucic acid at the *sn*-2 position of triacylglycerol in transgenic rapeseed oil. Plant Physiol. 109:1389–1394.

Metz, J.G., M.R. Pollard, and M.W. Lassner. 1994. Fatty acyl reductases. U.S. Patent 5,370,996.

Ohlrogge, J.B. 1994. Design of new plant products: Engineering of fatty acid metabolism. Plant Physiol. 104:821–826.

Ohlrogge, J. and J. Browse. 1995. Lipid biosynthesis. Plant Cell 7:957–970.

Oo, K.-C. and A.H.C. Huang. 1989. Lysophosphatidate acyltransferase activities in the microsomes from palm endosperm, maize scutellum, and rapeseed cotyledon of maturing seeds. Plant. Physiol. 91:1288–1295.

Stumpf, P.K., and M.R. Pollard. 1983. Pathways of fatty acid biosynthesis in higher plants with particular reference to developing rapeseed. p. 131–141. In: J.K. Kramer, F. Sauer, and W.J. Pigden (eds.), High and low erucic acid rapeseed oils. Academic Press.

Töpfer, R., N. Martini, and J. Schell. 1995. Modification of plant lipid synthesis. Science 268:681–686.

Voelker, T.A., A.C. Worrell, L. Anderson, J. Bleibaum, C. Fan, D.J. Hawkins, S.E. Radke and H.M. Davies. 1992. Fatty acid biosynthesis redirected to medium chains in transgenic oilseed plants. Science 257:72–74.

Voelker, T.A., T.R. Hayes, A.M. Cranmer, J.C. Turner and H.M. Davies. 1996. Genetic engineering of a quantitative trait: metabolic and genetic parameters influencing the accumulation of laurate in rapeseed. Plant Journal 9:229–241.

Crambe: New Crop Success*

Kenneth D. Carlson, John C. Gardner, Vernon L. Anderson, and James J. Hanzel

Crambe was commercialized in the 1990s through the efforts of a dedicated team of farmers, agribusiness people, and scientists to develop a reliable domestic supply of erucic acid. Crambe (*Crambe abyssinica*), a plant of Mediterranean origin, produces an abundance of erucic acid in its seed oil. Erucic acid has an established market in erucamide, a preferred slip and antiblock agent for polyolefin films. In 1990, North Dakota farmers teamed up with National Sun Industries (NSI) and North Dakota State University (NDSU) to produce 900 ha of crambe, and within four years that team had nearly 24,000 ha of crambe under cultivation. NSI crushed the seed in their Enderlin, ND, mill and marketed the oil and the coproduct defatted seed meal. Concurrently, the High Erucic Acid Development Effort (HEADE) team sponsored and conducted production,

*The use of brand or trade names may be necessary to report factually on available data. The USDA neither guarantees nor warrants the standard of the product, and the use of the name by USDA implies no approval of the product to the exclusion of others that may also be suitable. All programs and services of the USDA are offered on a nondiscriminatory basis without regard to race, color, national origin, religion, sex, age, mariatal status, or handicap.

breeding, processing, product development, and marketing research, including critical feeding experiments at NDSU that affirmed the efficacy of using crambe meal in beef rations. Crambe was fed at up to 16.9% of diet dry matter to growing steers and at 4.2% to finishing steers. In creep feeds, crambe was included at up to 10% of the diet, and was supplemented at up to 9% of dry matter intake in cow rations. No pattern of significant effects was observed on intake, digestibility, blood metabolites, thyroid hormones, or performance of animals due to increasing levels of crambe meal. Acceptance of crambe meal by commercial feed companies and feeders was based both on attractive price and satisfactory animal performance. Objectives of a crambe breeding program established at NDSU include higher yielding cultivars, with more oil and erucic acid, with lower seed glucosinolate content, and with improved plant resistance to diseases, insects and seed shatter. Genotypes that perform better than currently grown varieties in all desirable traits have been identified, as have several lines with acceptable levels of glucosinolates in the meal. The HEADE approach to commercializing crambe probably can be applied to the development of other new crops, especially when modeled with a regional perspective. The importance of a strong market pull and dedicated private sector partners cannot be over emphasized. A crambe growers group currently has commitments that will enable sustainable production while new markets for the seed oil and meal are developed.

BACKGROUND TO COMMERCIALIZATION

Since the end of WWII, U.S. agriculture's splendid capacity to overproduce has been apparent to even the most casual observer. Multidisciplinary research and America's unmatched capacity for technology development and efficient transfer of that technology to all sectors of the agricultural community contributed to making the U.S. the food basket of the world. Even with huge outlays of money, various combinations of support programs have been only partly successful in matching production capabilities with market demand and in maintaining profitability for the producers of the major commodities.

Efforts to replace some commodity acreage with specialty food crops have succeeded because of the pull of the market place and consumer acceptance of new foods. However, attempts to introduce new industrial crops, i.e., those with non-food uses, have been very slow, often haphazard, and inefficient. Intentional efforts in the last ten years to create new industrial crops and to discover new industrial uses for existing commodity crops have met with a modicum of success, acceptance, and support. More movers and shakers are seeing the value of this effort in reducing commodity surpluses while maintaining a strong agricultural economy and undergirding an industrial infrastructure for the future.

One small piece of this effort was advanced at a conference in Kansas City, MO, in December 1986, when a small group of university scientists, private sector representatives, and government scientists and managers focused on rapeseed and crambe (*Crambe abyssinica)* as candidate industrial crops for the U.S. (Carlson and Van Dyne 1992). Both crops produce seed oils rich in erucic acid (EA), which is used to manufacture a key additive for the plastics industry, erucamide, a slip and anti-block agent critical to manufacture and use of polyolefin films. Plastic films such as polyethylene are produced in a plethora of forms for such familiar items as bread wraps, shopping and garbage bags, and shrink wraps and plastic sheeting for an array of commercial applications. At the time of the Kansas City meeting, domestic needs for erucic acid were entirely satisfied by imports of high erucic acid rapeseed (HEAR) oil, originating largely in northern and eastern Europe and in Canada. This niche industrial market for high erucic oils had been known for over 20 years. As a source of edible oil, HEA (industrial) rapeseed had been introduced on a large scale to the North American continent in the 1960s and 1970s through the splendid research of Canadian agricultural scientists. Their equally elegant research converted the crop (1970s and 1980s) to low erucic acid cultivars (LEAR) and ultimately to canola cultivars without significant erucic acid content. With the widespread acceptance of LEAR and canola worldwide, HEAR oil appeared to be moving from a widely available commodity oil to a contracted specialty oil.

By December, 1986, the University of Idaho and the U.S. Department of Agriculture had conducted substantial research on rapeseed and crambe, respectively, and came to Kansas City with significant data bases relating to the production of the two crops in the U.S. However, annual production of several thousand ha of HEAR in Idaho, which was marketed to Pacific Rim countries, and very sporadic production of crambe, though funneled to domestic erucamide producers, had failed to instill confidence in either crop's reliability as a domestic source of erucic acid. Faced with a rather dismal marketing and production track record, participants at the meeting still recognized the potential of, and critical need for, erucic acid's industrial niche, and the significance of developing a domestic source of erucic acid. Thus was laid the groundwork for a high erucic acid development effort (HEADE), which would evolve into an informal consortium of university scientists, industry participants and government scientists and managers focused as a team on the commercialization of crambe and industrial rapeseed as domestic sources of erucic acid.

HIGH ERUCIC ACID DEVELOPMENT EFFORT (HEADE)

HEADE Origination

The U.S. Department of Agriculture, through its Cooperative State Research, Education, and Extension Service's (CSREES) Office of Agricultural Materials and its Agricultural Research Service's (ARS) National Center for Agricultural Utilization Research (NCAUR), and the University of Missouri's Department of Agricultural Economics were instrumental in organizing the Kansas City meeting. They, along with Iowa State and Kansas State Universities and the Kansas Board of Agriculture, provided leadership for establishing initial management structure, setting goals and developing action plans, primarily for commercializing crambe. In 1989, the "crambe project" sponsored a crush of about 150 t of crambe seed produced in Iowa and successfully marketed the oil, which immediately heightened both public and private sector interest in high erucic oils. In 1990, the project was broadened to include industrial rapeseed when several additional universities joined the project, HEADE was born, and its management structure was functionally established. A key event that year occurred when National Sun Industries of Minneapolis, MN, committed to commercializing crambe and worked closely with the North Dakota State University (NDSU) Research Extension Center and HEADE from 1990 to 1995.

Table 1 lists the institutional partners and some areas of expertise that they brought to the HEADE project during the period 1986–95. Contributions made by each institution depended on the disciplines of numerous individual scientists, and ranged from chemistry, biochemistry and engineering to multiple aspects of plant, insect, and animal sciences to crop production and oilseed processing to technology transfer, marketing, and

Table 1. High Erucic Acid Development Effort (HEADE) participants and contributions by discipline or expertise. USDA = U.S. Department of Agriculture. CSREES = Cooperative State Research, Education, and Extension Service. OIM = Office of Industrial Materials. ARS =Agricultural Research Service. NCAUR = National Center for Agricultural Utilization Research.

Institutional partner	Contributions by discipline and acquired knowledge
USDA, CSREES, OIM	Management, oversight, budgeting
USDA, ARS, NCAUR	Management, chemistry, oil & meal processing, markets
University of Georgia	Rapeseed: breeding, production sciences
University of Idaho	Rapeseed: breeding, plant & animal sciences, crop production, economics
University Illinois	Animal science, marketing
Iowa State University	Crambe: plant science, processing, economics
Kansas Board of Agriculture	Marketing, economics, management skills
Kansas State University	Plant & animal science, processing
University of Missouri	Management, plant & animal science, economics
University of Nebraska	Plant & animal science, process engineering
New Mexico State University	Crambe: breeding, plant science, crop production
North Dakota State University	Crambe: breeding, plant and animal science, chemistry, marketing, extension service to growers

economics. Each contribution was critical to the evolutionary path that HEADE took in successfully bringing crambe to significant commercial production.

Management and Operation

HEADE functioned under a Management Committee consisting of the state experiment station directors (or their representatives) and two USDA designates. The Management Committee elected a chairperson and appointed a project coordinator, who together established agendas, conducted regular meetings of the Committee, and coordinated and oversaw HEADE activities between meetings. Subcommittees, consisting of multidisciplinary teams of scientists, directed HEADE-approved activities involving production, processing and co-products, and marketing and economics. Committee functions and responsibilities are summarized in Table 2.

An annual line item appropriation from the Congress provided partial funding for HEADE activities (<$500,000). The funds were administered through the CSREES Office of Agricultural Materials (USDA) to the participating universities under cooperative agreements.

Each member university also provided funds and in-kind support for the research and commercialization projects. The HEADE management committee established RFP (request for proposals) and peer review procedures (filtered through the subcommittees) to select 15 to 20 projects for funding each year.

When line item funding for HEADE ended in FY 95, the HEADE management committee completed remaining funded research projects, and encouraged project members to continue their institutional commitments to crambe and rapeseed commercialization. Internal funding and development of cooperative interactions with seed, biotech and chemical companies and with growers groups have enabled several universities to continue breeding, agronomic, animal feeding, and chemical research on crambe and rapeseed. For example, North Dakota State University and the Universities of Georgia, Idaho, and Nebraska are funding projects through their plant and animal sciences, chemistry and engineering, and extension departments. Numerous companies and private sector individuals formally and informally partnered with HEADE, and their contributions were especially important to HEADE's decade of work to commercialize high erucic acid crops. The discussion that follows covers HEADE efforts on crambe seed production, meal utilization and renewed breeding studies.

CRAMBE PRODUCTION

Though natives to the Middle-east, rapeseed (both *Brassica napus* and *B. campestris*) and crambe (*Crambe abyssinica*) have emerged as commercially viable crops adapted to a broad range of growing regions in the U.S. and world. Between the two crops there exists winter and spring growth habits, a variety of pest resis-

Table 2. High Erucic Acid Development Effort (HEADE) management structure.

Organizational unit	Unit functions and responsibilities
Management Committee[z]	Provide leadership, establish budget, set priorities, allocate resources, coordinate research, promote communication, review progress, report results
Subcommittees:	
Production	Research and advice for: seed increase/certification, agronomy, plant diseases, insect & weed control, and breeding; commercial seed production assistance
Processing & Co-products	Research and advice for: seed preparation and oil extraction, oil and meal quality, meal feeding studies
Marketing & Economics	Market development, product development, economics, private sector interactions

[z]Chairperson and Coordinator elected for 'day-to-day' management, conducting meetings, coordinating activities, appointing subcommittees, and handling communications and paper work.

tance mechanisms, the ability to enhance rotations with traditional crops, and potential to better conserve soil resources, all ecological characteristics important to their success. Older EA-containing rapeseed varieties are being replaced world wide by canola types (no EA), because of the latter's popularity in the edible oil markets. EA-containing rapeseed varieties now are specialty crops in Canada and Europe. In the U.S., industrial rapeseed was the dominant HEA non-food oilseed before 1990, when its acreage was largely replaced by canola, and crambe became the dominant HEA crop.

In comparison to the cultivation of rapeseed, which is usually dated to the 20th century BC, crambe was introduced a little more than 50 years ago through a coordinated network of federal, state, and private interests. While still far from being considered a common crop, its path from potential crop designee to commercial production in such a short period of time has made crambe an interesting case in new crop development. During the rapid increase in crambe production in the 1990s, scientists at NDSU strove to reach beyond traditional small plot research to better identify why and how farmers respond to new crops, which ultimately determines whether they will be successfully incorporated into new farming systems.

Early Attempts at Commercialization

Crambe has been evaluated as a potential oilseed crop since about 1932 (Hirsinger 1989). Interest first surfaced in the U.S. in 1957 when the USDA began a program of systematically screening a large number of plant species as potential new crops (Wolff and Jones 1958; Princen 1983). Among some 8,000 species evaluated by NCAUR between 1957 and 1983, crambe emerged as a promising oilseed crop due to its high content of erucic acid. This long-chain fatty acid has many potential industrial uses including use as lubricants, coatings, slip-agents, plasticizers, polymers, and nylon precursors (Carlson and Van Dyne 1992).

Since this early promise was identified, various state and federal agencies have evaluated crambe for brief periods. North Dakota investigated crambe at several research centers in the late 1960s and early 1970s. Hoag and Geiszler (1970) reported a five year mean yield at Minot, ND, on fallow of 1698 kg ha^{-1}. At Williston, ND, on dryland, a mean yield of 842 kg ha^{-1} was obtained during the same 1965–1969 period. In reporting results with crambe, North Dakota was no different than several other state and national efforts at the time, all exploring production feasibility (White and Higgins 1966). The majority of production research, however, came from the Midwest cornbelt and was championed by breeder Koert Lessman at Purdue University (Lessman and Meier 1972). Crambe appeared drought tolerant, widely adapted, and among the most promising of several potential new crops.

As with most crops, coordinating and balancing supply with demand has historically been a problem with crambe. The USDA initiated field production of crambe for several years in the late 1970s to build seed supply. This effort ultimately led to the first private sector promoted attempt to commercially produce crambe (Princen 1983). In the spring of 1981, 405 ha of crambe were contracted with Humko (now Witco Chemical Co.) to be grown in western Kentucky (Durwood Beatty pers. commun.). A price of 25 cents per kg^{-1} was agreed upon and the crop was seeded in March with the hopes of a late June or early July harvest in time to double crop soybeans. The warm, dry spring gave crambe a good start, but also resulted in a wide-spread aphid epidemic, which destroyed ca. 80% of the crambe plantings. Harvested seed was destroyed in a warehouse fire later that year. Such an egregious result left little private sector interest in crambe, and some observers at the time of the Kansas City meeting in Dec., 1986, questioned the wisdom of advancing the cause of high erucic acid crops for the U.S. However, others believed that a systems approach to breeding, production, processing and marketing of both industrial rapeseed and crambe was not only appropriate but a right avenue to successful domestication (Carlson and Van Dyne 1992).

North Dakota Commercialization

By 1990, HEADE promoted research showed that crambe seemed particularly viable in North Dakota. Commercialization was again attempted in the spring of 1990 by spreading responsibilities and risks via a three way partnership of HEADE, National Sun Industries (NSI, an oilseed processor at Enderlin, ND), and thirty-eight North Dakota farmers. HEADE, through NDSU's Research and Extension Center in Carrington, solicited farmers in central North Dakota to grow crambe, provided 'Meyer' crambe seed and production information to the growers, and served as a communications link between the farmers and NSI. Farmers delivered the crop to NSI's Enderlin plant within a narrow window timed to coordinate with NSI's other oilseed deliveries. NSI paid farmers for the seed upon delivery and inspection (16.5 to 23.1 cents per kg). Following recommendations based on previous USDA Research (Carlson et al. 1985), NSI processed the seed via prepress/solvent extraction technology, and sold the oil and marketed the meal to feed formulators (cattle feed lots).

Successful production of crambe in North Dakota has continued (Table 3), but not without difficulties. NSI's attempts to produce crambe in the Central Great Plains, near Goodland, KS and also in the San Luis Valley of Colorado in 1993 were disappointing, and commercial production of crambe remains concentrated in North Dakota. In North Dakota, production began in the central and eastern portions of the state, but wet conditions during the past few years have moved production west. Today, the three principal growing regions surround the counties of Barnes, Foster, and Stutsman in the central region, Ward in the northwest, and Hettinger in the southwest portions of the state. Despite attempts to balance supply and demand, over-production limited the 1995 crop to seed increase until oil stocks could be consumed. Also, NSI decided to leave the oilseed crushing business, and leased their processing facilities in late 1994. Perhaps as evidence of its success, HEADE also lost its federal funding in 1994. Therefore, North Dakota crambe growers have organized into the American Renewable Oil Association (AROA), which has taken on the responsibilities of crop production, seeking processing facilities, and marketing oil and meal. At the time of this writing, AROA anticipates 16,000 ha of crambe production in 1996 (7–9 million kg of oil).

Production Practices

Given the manner in which crambe has been introduced in North Dakota, unique opportunities have existed for cooperative research among farmers, university researchers, and the private sector processors/marketers. Beyond its HEADE role, NDSU has assisted the farmers with an introduction to crambe as a crop,

Table 3. Commercial crambe production statistics from North Dakota, 1990–1995.

Statistic	1990	1991	1992	1993	1994	1995
Hectares	955	1,812	8,532	22,469	17,790	538
Net yield (kg ha^{-1})	1,456	1,499	1,275	1,132	1,456[z]	1,568[z]
Net crush (tonnes)	1,057	2,340	11,128	21,948	na[y]	0
Foreign matter[x] (%)	14.8	10.3	9.9	13.8	na	na
Moisture[x] (%)	8.5	9.0	9.6	10.4	na	na
Test weight[x] (kg m^3)	327	341	332	324	na	na
Oil yield (%)	28.5	32.2	31.1	32.3	na	--[w]
Meal yield[v] (%)	67.6	66.4	61.5	56.8	na	--
Seed oil (%)	29.1	32.7	33.1	31.9	na	--
Meal moisture (%)	8.8	9.5	8.3	9.6	na	--
Meal protein (%)	32.1	30.4	30.5	30.5	na	--
Meal fiber (%)	20.3	19.4	21.4	19.9	na	--
Erucic acid (%)	57.2	57.6	57.1	55.3	na	--

[z]Estimate.
[y]Not available.
[x]North Dakota Grain Inspection Service (NDGIS). Test wt. mean=331 kg m^{-3} (25.7 lb bu^{-1}).
[w]Not determined.
[v]Crop was partially dehulled before crushing beginning in 1992.

suggested production alternatives, supplied seed, and been available to help solve specific problems. The university has also helped the processor identify potential growing regions, supplied information on processing the oil, and provided quality information on both the oil and the meal in order for these new products to best seek their true market value.

Through a combination of formal surveys and informal observations, North Dakota crambe production practices have been documented in a series of reports published by NDSU (Gardner 1991–1995). Crambe, usually seeded in May and harvested in Aug., has typically followed a small grain crop in rotation (>85%), but also followed flax, fallow, and various other crops. Most growers rate the crop that followed crambe to have performed equivalent to, or better, than when that crop followed a small grain. Volunteer crambe is manageable in succeeding crops through either use of tillage, herbicides, or a combination of both; however, seed dormancy studies suggest that germination may be delayed for extended periods under field conditions. Crambe has been seeded successfully with a wide variety of planting equipment into tilled and untilled seedbeds. It is nearly always seeded into narrow rows (<20 cm) to increase it's ability to compete with weeds early in the season. Herbicides labeled and used are much the same as those used with canola, with soil-applied dinitroanalines applied pre-plant the most popular. The most troublesome weeds are often those that mature late, once the crambe crop is drying down. Though competition with these weeds may not directly reduce yield, it makes harvest more difficult. Crambe can be swathed to dry in the field, but most growers prefer to harvest crambe direct with combine headers commonly used with wheat. Unlike canola, crambe cannot be swathed before the majority of the seeds are physiologically mature, since erucic acid content of the oil is lower in immature seeds. Crambe stubble is easily managed, and requires little fall tillage. The residue is valuable to reduce wind erosion and is an efficient snow trap during the winter.

In comparison to other Northern Plains crops, crambe is relatively inexpensive to produce. Certified seed costs are generally less than \$12 ha^{-1} for the typical 22 kg ha^{-1} seeding rate. The fertilizer requirements have been found to be no greater than wheat. Dinitroanaline and post-emergent grass herbicides currently labeled are also among the most inexpensive available. Insecticides have not been needed, since, unlike rapeseed or canola, crambe is resistant to flea beetles, the most troublesome insect pest among crucifers in the Great Plains (Anderson et al. 1992). During a typical harvest, crambe seed does not require drying, and due to it's low test weight, is easily aerated should moisture content need to be reduced for storage. Over all, crambe is easily grown with traditional wheat equipment.

Production Problems and Opportunities

As with most new crops, unanticipated production problems have been the most challenging. Crambe possesses several unique characteristics, which have been the source of the majority of special problems and opportunities.

Low test weight. Crambe, unlike rapeseed and most other crucifers, has each seed encapsulated in its own hull. This reduces test weight by more than 50%, to about 322 kg m^{-3} (25 lb bu^{-1}). Such a bulky commodity increases the cost of transportation from the farmgate to the processing facility, and misleads the grower in visually estimating dockage, or foreign material, in the harvested seed. During the first year of commercial production, the overall mean dockage content of crambe delivered to the processing plant was 14.8%, significantly adding to already substantial transportation costs of the grower. This initially kept commercial production no further than 240 km from the processing plant. Since 1990, above average rainfall across North Dakota has shifted the production region west to avoid diseases such as *Alternaria*, which limited production in the eastern portion of the state for the first time in 1992. Greater distance from the processing plant has encouraged the grower organization to arrange for group rail shipments. Limited investigations into farmgate dehulling have been discouraging since it can lead to a reduction in oil yield and quality.

Seed dormancy and quality. Barely domesticated, it is not surprising that crambe exhibits a post-harvest dormancy. Such mechanisms are typical in undomesticated species to improve longevity and success. In an annual crop, however, such a trait makes it difficult to accurately estimate germination percentages and can lead to volunteer plants emerging in the field several years after the crop is harvested. Crambe has suffered from both problems. In addition, with abnormally greater rainfall the last few seasons, *Alternaria* has been observed among the eastern-most growers. *Alternaria* can infect the seed, causing a subsequent reduction in germination and vigor when later planted (Kilpatrick 1976). The growers group has responded to these problems through establishment of a network of certified seed growers, regular testing, and encouragement of further research at NDSU and elsewhere (Gutormson et al. 1993).

Pesticide availability. Like all new and minor crops, not only must new growers accept the risk of dealing with an unknown crop, but they must also do it with few, or no, labeled pesticides to help aid in weed, disease, and insect management. Current federal registration policies, and an industry that increasingly limits pesticide development even for the few major crops, combine to put new crops at a significant disadvantage in terms of production risk. Through the coordinated work of federal and state researchers, crambe has been coupled with rapeseed/canola in terms of federal pesticide registration, but this still requires the manufacturers to list the specific oilseed (canola, rapeseed, crambe) on the label (U.S. Govt. 1992). Perhaps foreshadowing a requirement for future new crops, crambe was successfully commercialized, because of its inherent ability to compete with weeds, ward off insects, and escape diseases without help from pesticides.

Market Size and Familiarity

Capturing market share with a new product is always challenging. Crambe has been successful to date because of its premium high erucic acid oil (>55% erucic) in the steady niche market for surfactants and slip agents. Current world consumption of high erucic acid oil is growing, but is still only estimated to be about 57 million kg annually, of which only about 16 million kg is used in the U.S. (Charles Leonard pers. commun.). Even if all domestic consumption were crambe, this would still represent less than 40,000 ha on an annual basis. Producing a unique oilseed on this scale presents difficulty in finding a willing processing partner, especially in North Dakota where the oilseed processing industry operates only large facilities, each capable of processing 40,000 ha of crambe in less than three weeks. In addition to the small crush volume, crambe meal has received slow acceptance by the feeding industry, despite encouraging research (Carlson 1992; Anderson 1993; Anderson et al. 1993a, b). Given these problems, crambe was successfully commercialized in North Dakota because it inherited the right production climate, and had willing partners in the processing industry. Jeff Berkow, President, National Sun Industries, and his able and enthusiastic staff, were champions essential to establishing the crop.

CRAMBE MEAL

A Question of Markets

National Sun Industries produced tens of thousands of tonnes of crambe meal between 1990 and 1995. Initially, there was general unfamiliarity with crambe meal among animal nutritionists and in the feed industry, in spite of crambe's listing as a feed ingredient by the Association of American Feed Control Officials (AAFCO 1989) and its clearance by FDA for use as a protein supplement in beef cattle rations (Federal Register 1981; Carlson and Tookey 1983). The latter, however, restricted crambe meal to feedlot cattle at only 4.2% of their total ration. This restricted use seriously impacted NSI's ability to market the meal with a reasonable economic return, and the company asked HEADE for help in marketing the meal. HEADE produced a brochure (Carlson 1992) detailing the composition, merits, and uses of crambe meal, which was widely distributed to animal nutritionists and feed manufacturers and formulators to help with the awareness issue. Further, HEADE sponsored and promoted a series of carefully controlled cattle feeding studies with the commercial meal to gather new and specific information on the safe use and economic value of crambe meal (Table 4). As discussed below, these studies all pointed to possibly higher levels of use and wider application of crambe meal in beef cattle rations. The following discussions provide additional research perspective on the use of crambe meal in animal feeds.

Defining the Product

Crambe fruits are borne in single-seeded pods (pericarp), and the pods normally remain intact with the seed when harvested. For various reasons, including partial harvest dehulling and intentional process dehulling, the proportion of pericarp to seed meats can vary significantly as processed in the oilseed mill, and therefore the composition of the defatted meal can also vary. Table 5 shows that the protein content of the *whole* seed meal contains 22% to 28% protein, whereas *partially dehulled meal* as produced by NSI in 1991 contained 34 to 35% protein. If *completely dehulled*, crambe meal contains 48% to 50% protein (Kirk et al. 1966; Mustakas et al. 1968; Baker et al. 1977; Carlson et al. 1985; Anderson et al. 1993b). Crude fiber levels also vary widely with the amount of pod present in the processed seed, from >20% in the *whole seed* to <10% in *fully dehulled* seed (Table 5). Mineral composition of 1991 commercial crambe meal (Table 6) shows higher levels of calcium, sulfur, chlorine, cobalt and selenium with lower levels of potassium, iron and copper in comparison to soybean meal (Anderson et al. 1993b). Amino acid composition of crambe meal is compared to soybean meal in Table 7, and has higher levels of sulfur-containing amino acids, cystine and methionine.

As a member of the mustard family of plants, crambe seeds and meal contain several glucosinolates recognized as sources of potentially antinutritional compounds. Similar compounds are responsible for the sharp flavors of certain condiments and food vegetables, such as radish, horseradish, cabbage, brussels sprouts, broccoli, and mustard. Of the glucosinolates in crambe, the major one accounts for ca. 95% of the total and is commonly called *epi*-progoitrin [(S)-2-hydroxy-3-butenyl glucosinolate]. These compounds and their metabolites may interfere with iodine metabolism by reducing the ability of the thyroid to take up iodine. Most re-

Table 4. Crambe meal beef feeding studies promoted by the High Erucic Acid Development Effort (HEADE). UNE = University of Nebraska-Lincoln. NDSU = North Dakota State University. WS = whole seed meal. DHS = partially dehulled seed meal. Source of crambe meal: National Sun Industries, Inc., Enderlin, North Dakota.

Year	Study	Location
1990–91:	Commercial feedlot study	Foxley Cattle Co. with UNE
1991–92:	Feedlot study with WS	NDSU, Carrington, ND
1992:	Pilot study, lactating cows	NDSU, Carrington, ND
1992–93	Feedlot study with DHS	NDSU, Carrington, ND
1993:	Digestion study (high forage)	NDSU, Fargo, ND
1993–94:	Creep feed, beef calves	NDSU, Carrington, ND

Table 5. Major nutrients in defatted crambe meals.[z]

	Content (% DM)						
Defatted seed meal	Protein	Crude fiber	Crude fat	Acid deterg. fiber	Ash	N-free extract	Glucosinolates (µmol g^{-1})
Whole seed	27.7	22.0	na	na	7.7	40.0	45–70
Partially dehulled	34.6	na[y]	0.8	34.7	8.4	na	56.0
Totally dehulled	49.5	6.5	na	7.5	9.9	35.5	80–100

[z]Source: Anderson et al. 1993a; Kirk et al. 1966; Liu et al. 1994a, b; Mustakas et al. 1968; Baker et al. 1977; Carlson and Tookey 1983.

[y]Not available.

ports conclude that oilseed meals containing glucosinolates reduce feed intake and animal performance when fed to non-ruminant animals, but ruminants tolerate glucosinolates at much higher levels without negative effects. Typical levels of individual and total glucosinolates found in commercially processed crambe meal, and in high and low glucosinolate rapeseed meals, are given in Table 8. Glucosinolate contents of defatted meals produced in the 1990s (50–60 µmol/g) generally were higher than in meals produced in the 1970s (25–45 µmol/g), although harvested 1990s seed had lower glucosinolate contents (90–100 µmol/g) than seed processed in the 1970s (125–155 µmol/g). Also, the one glucosinolate breakdown product consistently found in processed crambe meals, 1-cyano-(S)-2-hydroxy-3-butene, relates to glucosinolate destruction during processing and was present in 1990s meals at levels about one half that found in 1970s meals. These results suggest better process control and milder operating conditions in NSI's Enderlin mill during the 1990s, leading to nutritionally better products (Carlson and Tookey 1983; Carlson et al. 1985).

Meal Efficacy

Digestibility. In vitro digestibility of the organic matter in crambe hulls and meal using rumen fluid varied considerably (Steg et al. 1994). Hulls and whole seed meal were poorly digested at 44.5 and 57.3%, respectively, whereas dehulled seed meal was much more digestible at 85.0%. In situ ruminal organic matter digestibilities for dehulled crambe meal, crambe hulls, rapeseed meal, and soybean meal were 96%, 43%, 85%, and 95%, respectively (Liu 1994; Liu et al. 1994a). Total tract organic matter digestibility for crambe meal and rapeseed meal was reported at 98% and 92%, respectively. Crambe meal organic matter, crude protein and neutral detergent fiber degraded more rapidly in the rumen than soybean meal (Liu 1994; Liu et al. 1994a). However, heat treatment or toasting of crambe meal could substantially decrease rumen degradability.

Table 6. Minerals in crambe meal and soybean meal (DM Basis).

Mineral	Crambe meal[z]	Soybean meal[y]
Calcium (%)	1.26	0.33
Phosphorus (%)	0.88	0.71
Potassium (%)	-.-	2.14
Sulfur (%)	1.26	0.47
Chloride (%)	0.70	-.-
Magnesium (%)	0.51	0.30
Sodium (%)	0.04	0.03
Iron (ppm)	110	142
Boron (ppm)	67	-.-
Zinc (ppm)	44	61
Manganese (ppm)	43	32
Copper (ppm)	15	30
Cobalt (ppm)	1.35	0.10
Selenium (ppm)	1.07	0.14

[z]National Sun Industries, Inc.
[y]National Research Council, Nutrient Requirements of Beef Cattle (NRC 1984).

Table 7. Amino acids in crambe meal and soybean meal (DM basis).

	Content (g/16g N) DM basis	
Amino acid	Crambe meal[z]	Soybean meal[y]
Alanine	3.8–4.2	4.29
Arginine	5.7–7.3	7.27
Aspartic acid	6.0–7.6	11.78
Cystine	2.6–2.8	0.93
Glutamic acid	14.2–17.0	18.63
Glycine	4.7–5.3	4.30
Histidine	2.2–2.7	2.55
Isoleucine	3.7–4.1	4.58
Leucine	5.9–6.8	7.75
Lysine	4.9–5.7	6.43
Methionine	1.6–1.9	1.13
Phenylalanine	3.4–4.0	5.01
Proline	5.5–6.2	-.-
Serine	3.5–4.1	5.45
Threonine	3.1–4.6	3.93
Tyrosine	2.5–3.0	3.75
Valine	4.5–5.6	4.58

[z]Source: National Sun Industries, Inc., Enderlin, North Dakota.
[y]Cavins et al. 1972.

Table 8. Glucosinolates in crambe meal and in high (HGRS) and low (LGRS) glucosinolate rapeseed meals.

		Glucosinolate content (μmol g^{-1})		
Common name	Chemical name	Crambe meal[z]	HGRS meal[y]	LGRS meal[x]
Progoitrin[w]	2-Hydroxy-3-butenyl	56.3	22.5	5.2
Gluconapin	3-Butenyl	0.3	31.2	4.5
Sinigrin	Allyl or 2-Propenyl	0.4	--	--
Glucobrassicanapin	4-Pentenyl	0.8	22.9	3.9
Gluconapoleiferin	2-Hydroxy-4-pentenyl	0.4	3.8	1.3
Neoglucobrassicin	1-Methoxy-3-indolylmethyl	--	12.3	12.5
Total glucosinolates		59.4	93.1	27.4

[z]Partially dehulled 1992 crambe meal. Source: National Sun Industries, Enderlin, ND. Analysis by POS Pilot Plant, 118 Veterinary Road, Saskatoon, SK, Canada.
[y]*B. campestris,* cultivar 'Torch' (Bell 1984).
[x]*B. campestris*, cultivar 'Candle' (Bell 1984).
[w]Sterioisomers: *epi*-progoitrin in crambe and progoitrin in rapeseed (Daxenbichler et al. 1968.)

Finally, rate of in situ protein disappearance in the rumen was not different for reciprocal amounts of crambe meal and soybean meal fed as protein supplements with a grass hay diet (Caton et al. 1994).

Crambe meal in ruminant diets. Since growth and profitability of crambe as a crop is limited by commercial acceptance and use of the meal as a feed component, positive results in recent HEADE promoted feedlot trials with commercial meals are highly encouraging (Table 4).

Feeding nearly 900 feedlot cattle in 1990 on the first NSI produced crambe meal, Stock et al. (1993) reported an initial decrease in feed intake, but within 3 days intake was equal to control animals. Slightly more crambe ration was consumed over the first 80 days on feed compared to controls, but slightly less over the last 40 days on feed. Duncan and Milne (1991) reported a 6 day adaptation period by ruminant microflora, which may or may not affect initial feed intake when glucosinolate containing meals are introduced to ruminant diets (Carlson and Breeze 1984; Darroch et al. 1991). Anderson et al. (1993b) found that beef cows offered only crambe meal did not consume it until mixed with as little as equal parts of any other feed, such as corn silage. Enhancing palatability of high glucosinolate meals with flavor enhancers has not been extensively studied, but Ingalls and Sharma (1975) reported a non-significant trend to increased intake with flavor or molasses added to high glucosinolate rapeseed diets. Also, no differences in intake of high glucosinolate rapeseed meal in diets based on grain sorghum were observed in steers (Heidker and Klopfenstein 1990). In addition to blending crambe meal with other feed components, pelleting the rations is important to prevent cattle from sorting diets (Lambert et al. 1970; Perry et al. 1979).

In a 63-day creep feeding trial (Table 4), crambe meal fed as a protein source in creep feed at 0%, 5%, and 10% did not affect feed intake, gain, feed efficiency, or thyroid hormones (T3 and T4) measured in blood serum of nursing beef calves (Anderson and Trautman 1995). Perry et al. (1979) previously observed that younger animals may have a higher tolerance for crambe meal. Comparison with more extensive feeding studies with high glucosinolate rapeseed are useful. Ingalls and Sharma (1975) reported that lactating cows were not affected by feeding high glucosinolate rapeseed meal at up to 10% of the diet. Papas et al. (1979) found that feed intake, milk yield, protein, fat, and solids were not affected by replacing soybean meal with rapeseed meal containing either 94 or 50 μmol/g of glucosinolates. However, the higher glucosinolate meal reduced iodine content of the milk and tended to reduce plasma thyroxine in cows. When calves were fed control, low, or high glucosinolate diets, feed intake, weight gain, hemoglobin blood cell volume, and erythrocyte count were unaffected except at the higher glucosinolate levels, which also resulted in more pronounced changes in the thyroids of the calves (Papas et al. 1979).

In two feedlot studies, Anderson et al. (1993a, b) compared four pelleted protein supplements using crambe meal as the protein source to replace 0%, 33%, 67%, or 100% of soybean meal. No differences were observed in feed intake, feed conversion, or rate of gain during an 84-day growing period with crambe meal fed at a maximum of 16.9% of dry matter intake and a 96-day finishing period with crambe meal fed at a maximum of 4.1% of dry matter intake (Table 9, combined feeding results). The growing period study utilized the highest published use level for crambe meal in preparing beef cattle for finishing at the prescribed FDA level of 4.2%. Beef cattle showed no adverse response to crambe meal at levels up to 8.5% in a previous study by Perry et al. (1979), using what is believed to have been a nutritionally inferior meal to the commercial meal used by Anderson et al. (1993a, b).

Earlier research, to support the petition to FDA to clear crambe meal for use in beef cattle rations (Federal Register 1981), determined that no glucosinolate products accumulated in beef fat or tissues (to 1 ppm) when dehulled, defatted crambe meal was fed to steers at 10% of their diets (Van Etten et al. 1977).

Effect of crambe meal on reproduction. Higher levels (25% to 32%) of crambe meal were used in yearling heifer rations than in most feedlot trials without detriment to ovarian cyclic activity or behavioral estrus. Long term studies with postpartum bovines, however, suggest a significant increase in days from calving to conception and increased services per conception with low levels of glucosinolate (31 g/day) in the diet (Cheeke 1987; Cheeke and Shul 1987). Lactating beef cows on 10% crambe meal produced equal calf gains and similar post partum interval to other protein supplement treatments (Anderson 1993). Pregnant ewes have shown no effect from high glucosinolate rapeseed meal on reproduction (Cheeke 1987; Cheeke and Shul 1987).

Implications. Crambe meal as currently produced will continue to be used in ruminant feeds (beef cattle), and be excluded from nonruminant feeds because of reported problems of growth depression, weight loss, toxicity, and organ pathology in monogastric animals (Hesketh et al. 1963; Van Etten et al. 1965, 1969; Fenwick et al. 1983; Wallig et al. 1988). An economical process for removal of glucosinolates and aglucons from crambe meal or low or no glucosinolate crambe germplasm would enhance the potential for use of crambe meal in nonruminant animal feeds. There is no assurance that breeding programs will develop low glucosinolate crambe cultivars. In fact, there are arguments for retaining glucosinolate levels in crambe, since they appear

Table 9. Performance of steers fed partially dehulled crambe meal (CM).[z,y]

Statistic	Control 100 SB[w]	Crambe meal diets[x] 33 CM	67 CM	100 CM	Urea[v]	SE
Initial wt. (kg)	294	294	294	293	294	2.44
Final wt. (kg)	520	540	538	520	529	4.15
Avg. days on feed	162	161	160	160	159	1.32
Avg. daily gain (kg)	1.40	1.53	1.52	1.42	1.48	0.09
Feed/gain (kg/kg)	2.97	2.84	2.91	3.06	2.95	0.57
DM intake (kg/d)	9.19	9.55	9.71	9.54	9.61	0.19

[z]Growing (84-d) and finishing (96-d) trials (Anderson et al. 1993a, b).
[y]1992 Crambe meal. Source: National Sun Industries, Inc., Enderlin, ND.
[x]Percent of supplemental protein from crambe meal replacing supplemental protein from soybean meal.
[w]100% of supplemental protein from soybean meal.
[v]Urea diet supplemental protein provided by crambe meal (50%), soybean meal (23%), and urea (27%).

to play a role in crambe's insect resistance (Bell and Charlwood 1980). Research on crambe meal needs to continue with simultaneous basic and applied studies. For example, a series of trials conducted according to FDA guidelines to address the current restrictions for use of crambe meal are needed, wherein the first objective would be to increase the allowable use level or to remove the current limitation on use of crambe meal in feedlot beef cattle. A second objective would be to follow with clearances for crambe meal use in other phases of beef production. Research sponsored and encouraged by HEADE, and recent results obtained at NDSU, provides strong support for these objectives.

CRAMBE BREEDING

Although crambe has been grown successfully on a larger scale in the 1990s than in the past, further genetic improvement is required to insure future advancement and competitiveness with crops that produce similar end products. With this in mind, a vigorous crambe breeding effort is being continued (Table 10).

Past Breeding Efforts

Lessman (1990) reviewed the early breeding work with the crop at Purdue University. This research culminated with the release of the three cultivars, 'Prophet', 'Indy', and 'Meyer', and the accumulation of valuable information concerning the inheritance of agronomic and quality traits. Lessman suggested that improvement in crambe may be obtained by crossing among diverse germplasm, followed by selection, in addition to selection within and among existing germplasm.

Campbell et al. (1986a, b) developed and released the cultivars, 'BelAnn' and 'BelEnzian', and the lines C-22, C-29, and C-37, by introgressing genetic material from wild populations into Indy. The five releases performed as good as or better than 'Prophet', 'Indy', and 'Meyer' in several locations.

Lessman continued his breeding effort at New Mexico State University during the 1980s and developed a group of elite lines that approached his earlier releases in performance. With Lessman's retirement in 1991, HEADE arranged to establish a crambe breeding program at North Dakota State University. Crambe germplasm gathered for the NDSU program included previously released cultivars, 103 accessions from the world collection maintained at the North Central Plant Introduction Station at Ames, Iowa, and breeding material from Lessman's Purdue and New Mexico State University programs.

Current Objectives

The objectives of the NDSU crambe breeding program include: (1) increase seed yield; (2) increase oil content of the seed to a level near that of rapeseed; (3) lower glucosinolate content to acceptable levels or eliminate them altogether; (4) increase erucic acid content of the oil to its maximum level; (5) and develop cultivars that are resistant/tolerant to potentially harmful diseases, *Alternaria* and *Sclerotinia,* and insects, such as the diamond back moth. Other characteristics being considered include resistance to lodging and seed shatter, and a short seed dormancy period.

Table 10. Crambe breeding programs.

Breeder/ Institution	Era	Results
K.J. Lessman		
Purdue Univ.	1960s & 1970s	Prophet, Indy, & Meyer cultivars
K.J. Lessman		
New Mexico State Univ.	1980s	Elite lines (possible male sterility)
T.A. Campbell		
USDA, Beltsville, MD	1980s	BelAnn & BelEnzian cultivars
J.J. Hanzel		
North Dakota State Univ.	1990s	Elite lines (possibly lower glucosinolates)

Breeding Approaches

These objectives are being approached in two ways: (a) evaluation of existing germplasm; and (b) selection among and within breeding populations developed through hybridization among existing germplasm and subsequent selfing. A bulk-pedigree method is being used as inbred lines are originally derived from F_2 plants and seed is bulked in the succeeding generations. By using a two-generation greenhouse season and a winter nursery, $F_{2:5}$ lines are available for evaluation two years after the crosses are made. Preliminary yield evaluations are being initiated in the $F_{2:6}$ generation.

Recent Progress

Performance evaluations conducted at numerous locations in North Dakota in the years 1992–94 have indicated that many breeding lines have the potential to perform better than the currently grown cultivars, 'Meyer', 'BelAnn', and 'BelEnzian', in all important characteristics (Hanzel et al. 1993; Hanzel and Montgomery 1994, 1995). However, no one breeding line contains all traits expressed at the desired levels. Breeding efforts are continuing to combine all desirable traits. Oil content has still averaged well below the desired level, and, of thousands of lines evaluated in the three-year period, only several have possessed glucosinolate levels that approach acceptable levels. Non-conventional breeding techniques are being considered to modify the expression of these traits.

SUMMARY

Crambe has found a home in the Northern Plains where it has been successfully grown, processed, and marketed on a commercial scale since 1990. It's success was fostered by the sharing of risk among producers and processors, each with visions of the potential benefits of commercializing this alternative crop. Ecologically, crambe offers a unique opportunity for farmers to diversify their crop rotations. Crambe shares few pests with small grains and exhibits considerably more field tolerance to diseases common among other broadleaf crops, such as sunflower and dry beans. Crambe also exhibits tolerance to a fairly wide range of insects and is not generally eaten by birds, which are often a problem near water in the Northern Plains, a major North American flyway.

Economically, crambe has also been advantageous. Investment in more equipment is not required, since crambe can be grown with equipment traditionally used to produce small grains. It is produced with a minimum of expense, and requires few pesticides. Being an industrial oilseed, crambe's price is somewhat independent of the traditional edible oilseed complex, thus offering farmers additional marketing alternatives. And, because of it's relatively low test weight, crambe will tend to be processed locally. In short, crambe has been a welcome addition to the repertoire of crops grown in the Northern Plains.

Crambe oil's major use in the well-defined erucamide market is welcome, but innovative research to develop additional markets is needed to sustain and expand domestic production of crambe and crambe oil. Perhaps the HEADE team's greatest disappointment was its inability to garner sufficient funding to pursue additional research and development in potential new product areas. The proprietary nature of some of the product development research sponsored by HEADE makes it inappropriate to discuss in this open forum. Also, though not discussed in this paper, HEADE sponsored some significant oilseed processing research, which aided the commercialization of crambe and has been or will be published elsewhere.

REFERENCES

AAFCO. 1989. Official Publication, Association of American Feed Control Officials, Inc., Atlanta GA.

Anderson, M.D., C. Peng, and M.J. Weiss. 1992. Crambe, *Crambe abyssinica* Hochst., as a flea beetle resistant crop (*Coleoptera: Chrysomelidae*). J. Econ. Entomol. 85:594–600.

Anderson, V.L. 1993. Low input drylot beef cow/calf production with alternative crop products. NDSU Carrington Research/Extension Center Beef Cattle Proc. 16:1.

Anderson, V.L., P. Berg, and S.L. Boyles. 1993a. Dehulled crambe meal as a protein source for feedlot steers. NDSU Carrington Research/Extension Center. 16:22.

Anderson, V.L., S.L. Boyles, P. Berg, K.B. Koch, and J.S. Caton. 1992. Crambe meal as a protein supplement

for feedlot steers. NDSU Carrington Research/Extension Center Beef Cattle Proc. 15:15
Anderson, V.L., W.D. Slanger, S.L. Boyles, and P.T. Berg. 1993b. Crambe meal is equivalent to soybean meal for backgrounding and finishing beef steers. J. Anim. Sci. 71:2608.
Anderson, V.L. and B. Trautman, 1995. Crambe meal in creep feed for beef calves. NDSU Carrington Research/Extension Center Beef Cattle Proc. 18:6.
Baker, E.C., G.C. Mustakas, M.R. Gumbmann, and D.H. Gould. 1977. Biological evaluation of crambe meals detoxified by water extraction on a continuous filter. J. Am. Oil Chem. Soc. 54:392.
Bell, E.A. and B.V. Charlwood (eds.). 1980. Secondary plant products: glucosinolates. In: Encyclopedia of plant physiology. Vol 8. Springer-Verlag, Berlin.
Bell, J.M. 1984. Nutrients and toxicants in rapeseed meal, a review. J. Anim. Sci. 50:996.
Campbell, T.A., J. Crock, J.H. Williams, A.N Hang, R.E. Sigafus, A.A. Schneiter, E.F. McClain, C.R. Graves, D.F. Woodley, R. Kleiman, and W.C. Adamson. 1986a. Registration of 'BelAnn' and 'BelEnzian' crambe. Crop Sci. 26:1082–1083.
Campbell, T.A., J. Crock, J.H. Williams, A.N. Hang, R.E. Sigafus, A.A. Schneiter, E.F. McClain, C.R. Graves, D.F. Woodley, R. Kleiman, and W.C. Adamson. 1986b. Registration of C-22, C-29, and C-37 crambe germplasm. Crop Sci. 26:1088–1089.
Carlson, J. and R.G. Breeze. 1984. Ruminal metabolism of plant toxins with emphasis on indolic compounds. J. Anim. Sci. 58:1040.
Carlson, K.D. (ed.). 1992. Crambe meal: protein supplement for beef cattle. Univ. Nebraska, Lincoln.
Carlson, K.D., E.C. Baker, and G.C. Mustakas. 1985. Processing of *Crambe abyssinica* seed in commercial extraction facilities. J. Am. Oil Chem. Soc. 62:897.
Carlson, K.D. and H.L. Tookey. 1983. Crambe meal as a protein source for feeds. J. Am. Oil Chem. Soc. 60:1979.
Carlson, K.D. and D.L. Van Dyne. 1992. Industrial uses for high erucic acid oils from crambe and rapeseed. Univ. Missouri, Columbia.
Caton, J.S., V.I. Burke, V.L. Anderson, L.A. Burgwald, P.L. Norton, and K.C. Olson. 1994. Influence of crambe meal as a protein source on intake, site of digestion, ruminal fermentation, and microbial efficiency in beef steers fed grass hay. J. Anim. Sci. 72:3238.
Cavins, J.C., W.F. Kwolek, G.E. Inglett, and J.C. Cowan. 1972. Amino acid analysis of soybean meal: interlaboratory study. J. Assoc. Off. Anal. Chem. 55:685.
Cheeke, P.R. (ed.). 1987. Toxicants of plant origin: Glycosides. Vol. 11:21 CRC Press, Inc., Boca Raton, FL.
Cheeke, P.R. and L.R. Shull. (eds.). 1987. Natural toxicants in feeds and poisonous plants. AVI, Westport, CT.
Darroch, C.S., J.M. Bell, D.I. McGregor, and J.H.L. Mills. 1991. The effects of a linear increase in dietary indole glucosinolates on growth and physiology of mice. Can. J. Anim. Sci. 71:887.
Daxenbichler, M.E., C.H. Van Etten, and I.A. Wolff. 1968. Diastereomeric epi-sulfides from *epi*-progoitrin upon autolysis of crambe seed meal. Phytochemistry 7:989.
Duncan, A.J. and J.A. Milne. 1991. Rumen microbial degradation of allyl cyanide as a possible explanation for the tolerance of sheep to *Brassica*-derived glucosinolates. J. Sci. Food Agr. 58:15.
Federal Register. 21 CFR Part 573. 1981. Food additives permitted in feed and drinking water of animals: Crambe meal, heat toasted. 46:108.
Fenwick, G.R., R.K. Heaney, and W.J. Mullin. 1983. Glucosinolates and their breakdown products in food and food plants. CRC Critical Rev. Food Sci. Nutr. 18:187.
Gardner; J.C. (ed.). 1991–1995. Alternative crop production research. North Dakota State Univ., Fargo.
Gutormson, T.J., K.L. Larson, and T. Waqar. 1993. Germination of freshly harvested and after-ripened crambe (*Crambe abyssinica* Hochst. ex. R.E. Fries) seed. J. Seed Tech. 16:1–23.
Hanzel, J.J., J.W. Mitchell Fetch, M.L. Montgomery, B.G. Schatz, and B.K. Hanson. 1993. Breeding of crambe and *Brassica* oilseeds. p. 38–44. In: J.C. Gardner (ed.), Alternative crop production research. North Dakota State Univ., Fargo.

Hanzel, J.J. and M.L. Montgomery. 1994. Breeding of crambe and *Brassica* oilseeds. p. 52–58. In: J.C. Gardner (ed.), Alternative crop production research. North Dakota State Univ., Fargo.
Hanzel, J.J. and M.L. Montgomery. 1995. Breeding of crambe and *Brassica* oilseeds. p. 33–45. In: J.C. Gardner (ed.), Alternative crop production research. North Dakota State Univ., Fargo.
Heidker, J.I. and C.F. Klopfenstein. 1990. The use of high glucosinolate rapeseed meal as a supplemental protein source in finishing cattle diets. Feed Manag. 41(3):8.
Hesketh, H.R., C.R. Creger, and J.R. Couch. 1963. *Crambe abyssinica* meal as a protein source for broilers. Poult. Sci. 42:1276.
Hirsinger, F. 1989. New annual oil crops. p. 518–532. In: G. Robbelen, R.K. Downey, and A. Ashri (eds.), Oil crops of the world. McGraw-Hill, New York.
Hoag, B.K. and G.N. Geiszler. 1970. Crambe as a potential crop for western North Dakota. ND Farm Res. 27(6):16–18.
Ingalls, J.R. and H.R. Sharma. 1975. Feeding of Bronowski, Span and commercial rapeseed meals with or without addition of molasses or flavor in rations of lactating cows. Can. J. Anim. Sci. 55:721.
Kilpatrick, R.A. 1976. Fungal flora of crambe seeds and virulence of *Alternaria brassicola.* Phytopathology 66:945–948.
Kirk, L.D., G.C. Mustakas, and E.L. Griffin, Jr. 1966. Crambe seed processing: improved feed meal by ammoniation. J. Am. Oil Chem. Soc. 43:550.
Lambert, J.L., D.C. Clanton, I.A. Wolff, and G.C. Mustakas. 1970. Crambe meal protein and hulls in beef cattle rations. J. Anim. Sci. 31:601.
Lessman, K.J. 1990. Crambe: A new industrial crop in limbo. p. 217–222. In: J. Janick and J.E. Simon (eds.), Advances in new crops. Timber Press, Portland, OR.
Lessman, K.J. and V.D. Meier. 1972. Agronomic evaluation of crambe as a source of oil. Crop Sci. 12:224–227.
Liu, Yong-Gang. 1994. Crambe meal: evaluation, improvement and comparison with rapeseed meal. Ph.D. Thesis. Department of Animal Nutrition, Wageningen Agricultural Univ., Haagsteeg 4, 6708 PM Wageningen, the Netherlands.
Liu, Yong-Gang, A. Steg, and V.A. Hindle. 1994a. Rumen degradation and intestinal digestion of crambe and other oilseeds by-products in dairy cows. Anim. Feed Sci. Technol. 45:409.
Liu, Yong-Gang, A. Steg, B. Smits, and S. Tamminga. 1994b. Crambe meal: removal of glucosinolates by heating, with chemicals and water extraction. Anim. Feed Sci. Technol. 48:273.
Mustakas, G.D., L.D. Kirk, E.L. Griffin, Jr., and A.N. Booth. 1976. Crambe seed processing: removal of glucosinolates by water extraction. J. Am. Oil Chem. Soc. 53:12.
Mustakas, G.C., L.D. Kirk, E.L. Griffin, Jr., and D.C. Clanton. 1968. J. Am. Oil Chem. Soc. 45:53.
NRC. 1984. Nutrient requirements of beef cattle, National Academy Press. Washington, DC.
Papas, A., J.R. Ingalls, and L.D. Campbell. 1979. Studies on the effects of rapeseed meal on thyroid status of cattle, glucosinolate and iodine content of milk and other parameters. J. Nutr. 109:1129.
Perry, T.W., W.F. Kwolek, H.L. Tookey, LH. Princen, W.M. Beeson, and M.T. Mohler. 1979. Crambe meal as a source of supplemental protein for growing-finishing beef cattle. J. Anim. Sci. 48:758.
Princen, L.H. 1983. New oilseed crops on the horizon. J. Econ. Bot. 37:478–491.
Steg, A., V.A. Hindle, and Yong-Gang Liu. 1994. By-products of some novel oilseeds for feeding: laboratory evaluation. Anim. Feed Sci. Technol. 50:87.
Stock, R., R. Britton, T. Klopfenstein, K. Karges, C. Krehbiel, and R. Huffman. 1993. Feeding value of crambe meal. Nebraska Beef Cattle Rep. p. 51–53.
US Government. 1992. Definitions and interpretations: Rapeseed. Rules and regulations, Federal Register, vol. 57, no. 219, p. 53569, Nov. 12, 1992.
Van Etten, C.H., M.E. Daxenbichler, J.E. Peters, I.A. Wolff, and A.N. Booth. 1965. Seed meal from *Crambe abyssinica*. J. Agr. Food Chem. 13:24.

Van Etten, C.H., W.E. Gagne, D.J. Robbins, A.N. Booth, M.E. Daxenbichler, and I.A. Wolff. 1969. Biological evaluation of crambe seed meals and derived products by rat feeding. Cereal Chem. 46:145.
Van Etten, C.H., M.E. Daxenbichler, W. Schroeder, L.H. Princen, and T.W. Perry. 1977. Tests for *epi*-progoitrin, derived nitriles, and goitrin in body tissues from cattle fed crambe meal. Can. J. Anim. Sci. 57:75.
Wallig, M.A., D.H. Gould, and M.J. Fettman. 1988. Nutritive value of rapeseed meal:effects of individual glucosinolates. J. Sci. Food Agr. 37:1197.
White, G.A. and J.J. Higgins. 1966. Culture of crambe: a new industrial oilseed crop. USDA/ARS Res. Rep. 95.
Wolff, I.A. and Q. Jones. 1958. Cooperative new crops research: what the program has to involve. Chemurgic Dig. 17:4.

Isolation and Functional Properties of Proteins from *Crambe abyssinica* Oil Seeds*

E. Massoura, J.M. Vereijken, P. Kolster, and J.T.P. Derksen

In recent years a number of new oilseed crops have been considered for introduction in agriculture as a result of the need to broaden the narrow crop rotation schedule and to contribute to a reduction in agricultural surpluses of traditional arable crops. *Crambe abyssinica* is considered to be a promising new oil crop because of both its agronomic performance (Erickson and Bassin 1990; Lazzeri et al. 1993) and the specific properties of its oil. Crambe oil is one of the richest known sources of erucic acid which makes up 55% to 60% of the oil glycerides (Princen and Rothfus 1984; Lazzeri et al. 1994). The high erucic acid oils have a wide diversity of applications (Nieschlag and Wolff 1971) for products such as rubber and plastic additives, lubricants, coatings, and as raw material for the synthesis of nylon. However, for profitable cultivation, exploration of other major seed components, such as the protein fraction, would be of major importance.

In contrast to the oil, very little is known about the proteins of Crambe. Crude protein levels range from 20% in the whole seed to nearly 50% in the dehulled, defatted meal (Carlson and Tookey 1983). Since crambe protein has a well balanced amino acid pattern (Baker et al. 1977; Steg et al. 1994), most research has been concerned with the use of the meal as feed (Kirk et al. 1971; Mustakas et al. 1976; Baker et al. 1977; Carlson and Tookey 1983; Steg et al. 1994). The suitability of the proteins for applications with a higher added value (food and non-food applications) is not yet known. The use of proteins in food and non-food industries is determined by their functional properties. Therefore, development of a protein isolation procedure and assessment of the functional properties of the isolated proteins is of prime importance to strengthen the economics of crambe cultivation.

METHODOLOGY

Crambe seeds were provided by Cebeco-Handelsraad B.V. (The Netherlands). Dehulling was performed by cracking the seeds in roller mills. Removal of the hulls was obtained by air classification. The ratio of hulls to meal was 35:65. For milling, whole or dehulled seeds were mixed with liquid nitrogen in an

*Financial support by the Commission of the European Communities (DG XII) is gratefully acknowledged.

analytical grinder and ground for 30 sec. Defatting was performed by stirring a suspension of whole or dehulled meal in petroleum ether (ratio meal to solvent 1:10 w/v) for 45 min at room temperature. After phase separation the supernatant was decanted and the pellet was resuspended in the solvent and the same procedure was repeated. The remaining pellet was left to dry overnight at room temperature.

Fat content was determined by the Soxtec system HT 1043 extraction unit. Protein content of seeds, meal, and of freeze dried materials was determined by Kjeldahl analysis and in solutions by using the Bio-Rad *DC* protein assay.

Protein extractability was studied as a function of pH. A suspension of defatted meal in water (1:10 w:v) was stirred for 5 min. The pH was then adjusted to the desired value by adding drops of dilute sodium hydroxide or hydrochloric acid solution. Stirring continued for 1 h, while the pH was checked every 10 min and readjusted, if necessary. After centrifugation in a Beckman J-21 C centrifuge at 10.960 *g* for 45 min, the supernatant was filtered to remove floating particles. Aliquots of the supernatant were freeze dried and their protein content was determined by Kjeldahl analysis.

Protein precipitation with and without the addition of carboxylmethylcellulose (CMC) was studied as a function of pH. First, proteins were extracted at pH 11 from defatted-dehulled meal. Two aliquots of this extract were diluted with water to give a concentration of approximately 1.5 mg protein per ml. The pH was then readjusted to 11. In one of the two solutions, CMC was added to a ratio of 0.166 CMC/protein (w/w); this ratio was found to yield the highest precipitation of rapeseed proteins (Gillberg and Törnell 1976). Samples of varying pHs were prepared from both solutions and kept at ambient temperature for 1 h, after which they were centrifuged for 15 min at 16,000 *g*. The protein content of the supernatants was quantified using the Bio-Rad assay. Solubility at pH 11 was assumed to be 100% and the results were expressed relative to this value. Proteins that remained soluble after isoelectric precipitation were concentrated by ultrafiltration. The protein solution (pH 5.5) was circulated through a membrane (Nephross Andante HF/Organon Technica) of molecular weight cut-off (MWCO) 5000 at a pressure of 0.5 Bar. The solution was ultrafiltrated to a concentration factor (initial volume/final volume) of 20. The retentate was then freeze dried.

In order to determine protein solubility freeze-dried protein concentrates were dissolved in water at pH 11 (concentration of approximately 1.5 mg protein per ml). Samples of varying pHs were prepared from this solution by acidification. After standing 1 h at room temperature they were centrifuged for 15 min at 16,000 *g*. The protein content of the supernatants was determined by the Bio-Rad assay.

To assess emulsifying properties a protein sample was dispersed in water (0.1% w/v). After stirring for 1 h the protein dispersion was mixed with commercial soybean oil. The mixture was homogenised for 1 min by means of an ultra-turrax at 13,500 rpm. An aliquot of the emulsion (1 ml) was diluted in water (250 ml) and the absorbance was measured at 500 nm. This value was taken as a measure of emulsifying activity. For assessment of the emulsifying stability, the emulsion was left to rest for 30 min. An aliquot (1 ml) of the bottom half of the emulsion was diluted in water (250 ml) and the absorbance of this solution was measured at 500 nm. The emulsifying stability was expressed as a percentage of the emulsifying activity.

Foaming properties were determined according to Patel et al. (1988) with a slight modification. The protein sample was stirred for 30 min in 0.1 M phosphate buffer (pH 7) instead of distilled water.

RESULTS

Isolation of Proteins

Table 1 shows the crude fat and crude protein content of crambe meals. Crude protein levels ranged from 18.3% in the whole seed to nearly 40% in the dehulled, defatted meal. The efficiency of oil extraction was 85.7% and 87.8% for the whole and dehulled meal, respectively. This result indicates that the presence of hulls did not interfere with oil removal, at least at a laboratory scale. Defatting was not complete, probably due to the procedure being performed at room temperature to avoid protein denaturation.

Fig. 1 shows protein extractability as a function of pH for dehulled and non-dehulled meal. The shape of both curves was similar with three solubility minima namely at pH 4, 6.5, and around 8. This could be an indication of the presence of proteins with different isoelectric points. For both types of meal, the extraction efficiency was higher at alkaline than at acid pH values. In comparison with the whole meal the extraction efficiency for the dehulled meal was higher at almost all pH values tested.

Highest protein extractability was obtained from dehulled meal at pH 11. Double extraction at this pH increased protein yield to 78%. This extract was used in further experiments. Its protein content was 50% (dry wt basis). Therefore, further treatment was necessary in order to decrease the amount of non-protein components present in the extract. Precipitation of proteins, with or without addition of precipitating agent, from alkaline extracts by acidification is a procedure often applied for the preparation of protein concentrates/isolates from oilseeds (Gillberg and Törnell 1976; Lásztity et al. 1993; Mansour et al. 1993; Xu and Diosady 1994). To determine the optimal pH value for precipitation, the protein precipitation efficiency was studied as a function of pH with and without the presence of CMC (Fig. 2). Addition of CMC clearly affected protein precipitation. Without CMC, maximum precipitation (55%) occurred at pH 5–6. Addition of CMC resulted in a more narrow curve, with a maximum precipitation (75%) at pH 4.4. This displacement of the precipitation maximum towards lower pH values is in agreement with the results of Gillberg and Törnell (1976). The protein contents of the precipitates (dry wt basis) recovered at pH 4.4 (with CMC) and pH 5.5 (without CMC) were 64% and 60%, respectively. Since this difference was rather small and because CMC might affect the functional properties of the proteins, precipitation was performed in further experiments at pH 5.5 without addition of CMC.

A proportion of the oil originally present in the dehulled meal was not extracted. It was found that most, if not all, of the unextracted oil was present in the precipitate. Therefore, the precipitate was cold-defatted which increased its protein content to 70% (dry wt basis). This protein concentrate is denoted as fraction 1.

Protein content of the supernatant after precipitation at pH 5.5 was 42% (dry wt basis). To concentrate the proteins present in this solution, ultrafiltration was applied. In this way, a concentrate of 84% protein (dry wt basis) was produced (fraction 2). The functional properties of both fractions were determined.

Functional Properties

Fig. 3 shows that there were marked differences in the solubility profiles of fraction 1 and 2. Proteins present in fraction 1 showed minimal solubility at pH 5.5–6, where approximately 80% of the proteins

Table 1. Crude fat and protein content of crambe meals (calculated on a dry basis).

Constituent	Crude fat (%)	Protein (N×6.25) (%)
Whole meal	34.6	18.3
Dehulled meal	45.4	25.1
Defatted whole meal	7.0	26.2
Defatted dehulled meal	9.2	39.6

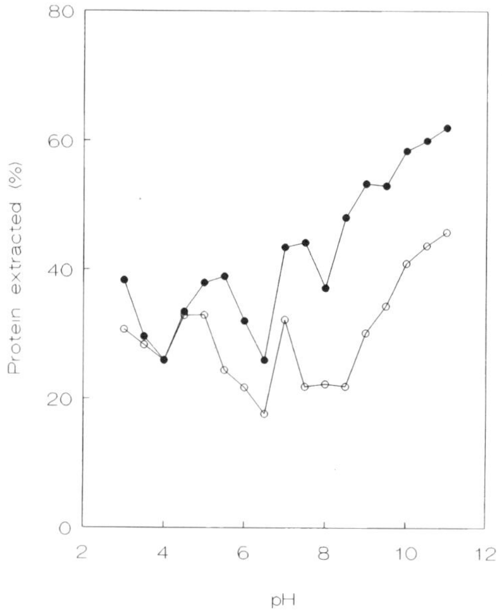

Fig. 1. Protein extractability as a function of pH: (○) whole meal; (●) dehulled meal.

were insoluble whereas high solubility occurred at pH values <3.5 and >8. The solubility of proteins from fraction 2 was high (>80%) over the whole range of pH tested. Similar results were obtained by Xu and Diosady (1994), who used comparable preparation techniques (precipitation-ultrafiltration) for rapeseed protein isolates.

Figs. 4 and 5 show the emulsifying activity and stability of fractions 1 and 2. Both properties were markedly affected by the pH. For both fractions, lowest emulsifying activity and stability occurred near pH 6. For fraction 1, this could be due to the low solubility at this pH; for fraction 2 the reason is unclear. Both fractions were more efficient in emulsifying the oil at alkaline than at acid pH values. Similar pH dependence of emulsification properties has been reported for other proteins (Crenwelge et al. 1974; Ramanatham et al. 1978).

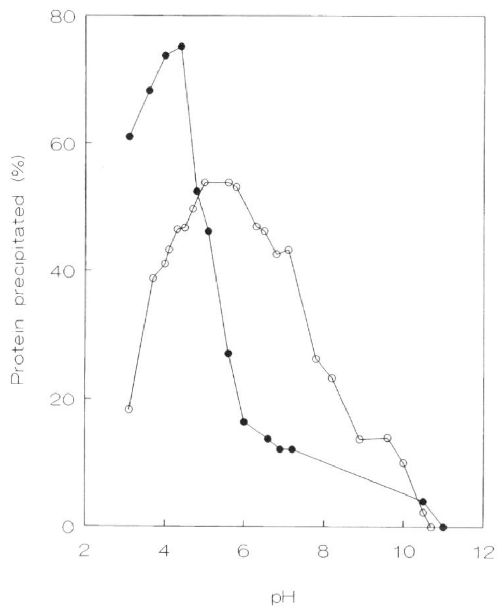

Fig. 2. Protein precipitation as a function of pH: (○) without CMC; (●) with CMC.

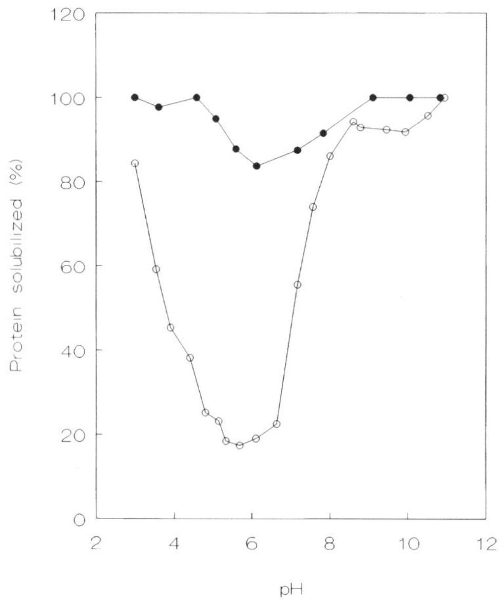

Fig. 3. Solubility of both fractions: (○) fraction 1; (●) fraction 2.

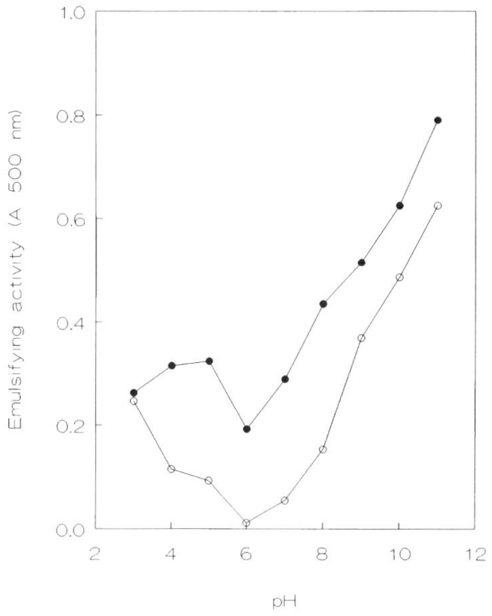

Fig. 4. Emulsifying activity as a function of pH: (○) fraction 1; (●) fraction 2.

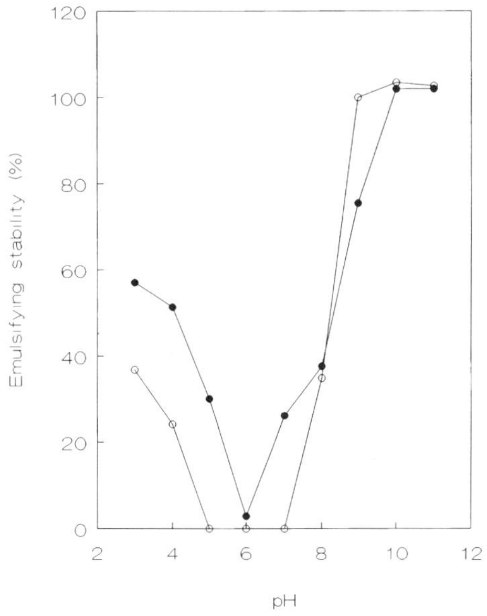

Fig. 5. Emulsifying stability as a function of pH: (○) fraction 1; (●) fraction 2.

Table 2. Foaming properties at pH 7 0.1 M phosphate buffer

Type of protein	Foam expansion (%)	Foam volume stability (%)	Liquid drainage (%)	Foam liquid stability (%)
Hen egg white	420	80	62	37
Fraction 1	500	90	59	40
Fraction 2	620	80	80	20

The foaming properties of fraction 1 and 2 are presented in Table 2. In comparison with hen's egg white, the foam expansion of both fractions was higher and foam volume stability was similar. Proteins from fraction 1 exhibited foam liquid stability comparable to that of hen's egg white while, proteins of fraction 2 exhibited lower values.

CONCLUSIONS

The proteins from crambe seeds were efficiently recovered by alkaline extraction followed by isoelectric precipitation/ultrafiltration. The fractions obtained have distinct solubility characteristics and very interesting foaming properties. At alkaline pH, crambe proteins show good emulsifying properties.

REFERENCES

Baker, E.C., G.C.Mustakas, M.R. Gumbmann, and D.H. Gould. 1977. Biological evaluation of crambe meals detoxified by water extraction on a continuous filter. J. Am. Oil Chem. Soc. 54:392–396.

Carlson, K.D. and H.L. Tookey. 1983. Crambe meal as a protein source for feeds. J. Am. Oil. Chem. Soc. 60:1979–1985.

Crenwelge, D.D., C.W. Dill, P.T. Tybor, and W.A. Landmann. 1974. A comparison of the emulsification capacities of some protein concentrates. J. Food Sci. 39:175–177.

Erickson, D.B. and P. Bassin. 1990. Rapeseed and crambe: alternative crops with potential industrial uses. Kansas State Univ., Manhattan, Agr. Expt. Sta. Bul. 656,

Gillberg, L. and B. Törnell. 1976. Preparation of rapeseed protein isolates: Precipitation of rapeseed proteins in the presence of polyacids. J. Food Sci. 41:1070–1075.

Kirk, L.D., G.C.Mustakas, E.L. Griffin, Jr., and A.N. Booth. 1971. Crambe seed processing: decomposition of glucosinolates (thioglucosides) with chemical additives. J. Am. Oil Chem. Soc. 48:845–850.

Lásztity, R., M. Goma, S. Tömösközi, and J. Nagy. 1993. Functional and nutritive properties of sunflower seed protein preparations. In: Proc. World Conference on Oilseed Technology and Utilization. AOCS Press, Champaign, IL.

Lazzeri, L., L.S. Conte, and S. Palmieri. 1993. Agronomical and technological characterization of Italian-grown *Crambe abyssinica*. Agr. Med. 123:55–64.

Lazzeri, L., L. Leoni, L.S. Conte, and S. Palmieri. 1994. Some technological characteristics and potential uses of *Crambe abyssinica* products. Ind. Crops Prod. 3:103–112.

Mansour, E., E. Dworschák, J. Perédi, and A. Lugasi. 1993. Preparation and functional properties of proteins from rapeseed and pumpkin seed. In: Proc. World Conference on Oilseed Technology and Utilization. AOCS Press, Champaign, IL.

Mustakas, G.C., L.D. Kirk, E.L. Griffin Jr., and A.N. Booth. 1976. Crambe seed processing: removal of glucosinolates by water extraction. J. Am. Oil Chem. Soc. 53:12–16.

Nieschlag, H.J. and L.A. Wolff. 1971. Industrial uses of high erucic oils. J. Am. Oil Chem. Soc. 48:723–727.

Patel, P.D., A.M. Stripp, and J.C. Fry. 1988. Whipping test for the determination of foaming capacity of protein: a collaborative study. Int. J. Food Sci. Technol. 23:57–63.

Princen, L.H. and J.A. Rothfus. 1984. Development of new crops for industrial raw materials. J. Am. Oil Chem. Soc. 61:281–289.
Ramanatham, G., L.H. Ran, and L.N. Urs. 1978. Emulsification properties of groundnut protein. J. Food Sci. 43:1270–1273.
Steg, A., V.A. Hindle, and Y.-G. Liu. 1994. By-products of some novel oil-seeds for feeding: laboratory evaluation. Anim. Feed Sci. Technol. 50:87–99.
Xu, L. and L.L. Diosady. 1994. Functional properties of Chinese rapeseed protein isolates. J. Food Sci. 59:1127–1130.

Hypoallergenic Guayule Latex: Research to Commercialization

Katrina Cornish and Deborah J. Siler

A fresh look at guayule has led to the discovery and development of hypoallergenic latex to supply a major, new market. Guayule (*Parthenium argentatum*, Gray, Asteraceae), a new industrial crop candidate, is a desert shrub native to the Chihuahuan desert of Texas and Mexico. It is the only species other than *Hevea* (the Brazilian rubber tree, *Hevea brasiliensis* Muell. Arg.) that has been used for rubber production on a commercial scale. Guayule has been known to produce rubber for at least a millennium, but modern commercialization efforts have been unsuccessful. The resurgence of guayule research since the early 1970s has resulted in higher yielding lines and improved agronomic practices, but the low price of imported *Hevea* natural rubber from tropical countries has continued to bar guayule rubber from the marketplace.

An extraordinary series of events has reversed guayule's dismal prospects, culminating with the emergence of an important natural rubber market in which *Hevea* cannot be used, *i.e.* hypoallergenic latex products. This superb opportunity for guayule has arisen because *Hevea* latex products, due to increased usage and manufacturing short-cuts, are causing serious and widespread health problems. Proteins present in *Hevea* latex products can trigger immediate (Type I) hypersensitivity, which in its most serious manifestation, leads to life-threatening anaphylaxis. Synthetic alternatives to many *Hevea* latex products do not have the required range of physical properties that encompass resilience, strength, elasticity, and viral impermeability. The unique properties of natural rubber, together with concerns for public safety, make the development of an alternative, hypoallergenic source of natural rubber imperative. Guayule makes high quality rubber which can be produced as a latex suitable for hypoallergenic product manufacture. Guayule latex products should be safe for use by even those with severe Type I latex allergy. Hypoallergenic latex presents a high margin, rapidly-expanding market for guayule rubber latex products. This market provides a sufficient premium to permit the immediate development of guayule as a major domestic crop. Large scale trials have produced low protein, hypoallergenic guayule latex, which we have shown suitable for product manufacture.

BACKGROUND

Natural rubber is a vital raw material used in enormous quantities by commercial, medical, transportation, and defense industries. The United States is wholly dependent upon imports from developing countries. Although over 2,500 plants are known to produce natural rubber, few synthesize commercial quality or quantities. At present, all commercial natural rubber comes from a single, tropical, plant species, *Hevea brasiliensis*, largely from plantation-grown, bud-grafted, clonal material. Continuity of the natural rubber supply is endangered by many factors, including diminishing acreage (as growers switch to more profitable crops, such as palm oil), increasing global demand, changing political climes, and crop disease. The recent widespread occurrence of life-threatening "latex allergy" to *Hevea* rubber products makes development of an alternative source of natural rubber imperative. A domestic natural rubber crop would guarantee sufficient natural rubber to supply existing United States strategic demands as well as the new, major, health-related, hypoallergenic rubber market. A sustainable, domestic rubber industrial crop also would enhance rural economies, utilize semi-arid lands, and bring retired farmlands back into profitable production.

In any development of a new crop, however worthy the underlying rationale, the new crop must have an existing market and be profitable for the grower and the buyer. Approaches to achieving commercial competitiveness may be grouped in two categories, (1) increasing the yield and (2) adding value to the agricultural commodity. Both of these approaches have been attempted with guayule.

HISTORY OF GUAYULE COMMERCIALIZATION

Guayule commercialization has been ventured at several different times during the twentieth century (as reviewed by Bonner 1991), but without lasting success. In the 1900s, the Continental Mexican Rubber Company harvested and processed the native guayule stands. At the height of this endeavor (1910), rubber production was 10,000 t/year and constituted 24% of U.S. rubber imports. From 1906 to 1912, the United States imported 42,000 t of guayule rubber from Mexico. Production ended due to depletion of the native stands and civil unrest. A fresh enterprise began in Texas with the appearance of the Intercontinental Rubber Company and the cultivation of guayule. Rubber production reached 1,400 t/year, with commercial plantings of over 7,500 acres, but ended in 1929 with the onset of the Great Depression. Guayule then had to await the Second World War and the United States inauguration of the Emergency Rubber Project to supply the strategic raw material in light of the war-induced dearth of *Hevea* natural rubber supplies. This enormous undertaking employed 1,000 scientists and technicians, 9,000 laborers and 13,000 ha of guayule under cultivation in three States. The project had produced its first 1,400 t of rubber when the war ended. The guayule plantings were destroyed due to the reavailability of *Hevea* rubber, a perhaps over-optimistic view of the utility of synthetic rubber, and international political pressure.

Guayule then waited until the oil shortage of the 1970s. High oil prices caused increased costs of synthetic rubber and led to renewed interest in guayule as an alternative source of natural rubber. Scientific conferences took place which led to extensive lobbying and, in 1978, the Native Latex Commercialization and Economic Development Act (PL 95-592) (Huang 1991). The Gila River Indian Community Guayule Project got underway in 1981, and 1984 saw the act reauthorized as the Critical Agricultural Materials Act (PL 98-284). Considerable progress was made in improving guayule's commercial viability through plant breeding and agronomy. As has been described in detail (Ray 1991), rubber yields were between 220 and 560 kg/ha during the 1950's, and by the Second Guayule Regional Variety Trials (1985–1988) annual yields had been increased to between 600 and 900 kg/ha (Thompson and Ray 1989). Yields of over 1,100 kg/ha have been achieved in breeding plots (Gathman et al. 1989). The advanced lines also have been selected for their ability to regrow after pollarding (trimming the shrubs to about 10 cm above soil level, leaving the trunk and branch stubs). First harvest by pollarding at three years, followed by regrowth and reharvest at two year intervals maximizes the profitability of the current best lines (Thompson and Ray 1989; Estilai and Ray 1991). The affect of cultural practices on rubber yield has been extensively reviewed (Thompson and Ray 1989; Foster and Moore 1991; Nakayama 1991) and suitable agricultural machinery for modern mechanized farming has been successfully developed (Coates 1991). Direct seeding methodologies also have been developed (Foster and Moore 1992; Foster et al. 1993) which greatly reduce the costs involved in the transplanting procedures previously required for good stand establishment. Of course, growing guayule in areas climatically unsuitable for its cold-induced rubber biosynthetic pathway (Appleton and van Staden 1989; Ji et al. 1993; Cornish and Siler 1996) will not give acceptable rubber yields.

During this research and development period, the Gila River Indian Community Guayule Project planted 117 ha of guayule (1981 to 1986) predominantly of line Gila-1 (or AZ101). Unfortunately, this line proved to be a hybrid between *P. argentatum* and the large, but virtually rubber-free, *P. tomentosum*. Unsurprisingly, the hybrid proved to be a rapidly-growing but low-yielding line. Nonetheless, a processing plant based on a continuous simultaneous solvent extraction and rubber fractionation process, was built on the Gila River Indian Reservation (Wagner and Schloman 1991). The guayule rubber produced was manufactured into airplane and truck tires. Performance testing has shown that guayule rubber truck tires are of comparable quality to *Hevea* rubber tires (pers. commun. Wayne Lucas, Automotive Systems Engineering Branch, Yuma Proving Grounds).

This commercialization effort targeted the production of natural rubber in competition with imported *Hevea* bulk rubber, essentially for the tire market. Thus, guayule was confined to competition with the cheapest end of the natural rubber market. The profitably increases achieved through plant breeding and improved cultural practices were not sufficient to support commercialization. Approximately a doubling of yield or price was still required (Wright et al. 1991).

THE CHANGING MARKETPLACE

The value of an agricultural commodity may be increased through the discovery of a specific new use or market, or through increased demand and/or decreased supply leading to lower competition. The global demand for natural rubber is on the increase (Reisch 1995) even though area has declined, resulting in substantial price rises. Also, latex rubber is more expensive than bulk (solid) rubber, which costs usually about 70% the latex price. Guayule has been extensively investigated (Whitworth and Whitehead 1991) with the hope of its commercialization as a true competitor in natural rubber markets currently occupied by *Hevea* bulk rubber (d'Auzac et al. 1989). As mentioned earlier, guayule rubber has been used for fabricating automobile tires (Hammond and Polhamus 1965; pers. commun. Wayne Lucas) but has not been used for dipped or extruded goods since their manufacture requires the raw rubber to be in latex form. Unlike laticiferous species, such as *Hevea*, guayule produces rubber in individual cells, located predominantly in the root and stem bark parenchyma. None of the established extraction methods used to produce bulk rubber from guayule, whether solvent or water-based (Whitworth and Whitehead 1991), have led to competitively-priced product, even from the best bred lines. Guayule, as a source of more valuable latex rubber, has much greater commercial potential than bulk rubber.

Latex and Latex Products

Natural rubber is used in over 40,000 different products, including more than 300 medical applications. High-value dipped and extruded products are manufactured directly from the *Hevea* latex. A life-threatening "latex allergy" has developed, in recent years, that is triggered by proteins present in latex. Synthetic alternatives to *Hevea* latex products do not have the required range of physical properties that encompass resilience, strength, elasticity, and viral impermeability. The unique properties of natural rubber, together with concerns for public safety, make the need for an alternative, hypoallergenic source of natural rubber imperative. This special need creates an excellent opportunity for the development of a new, hypoallergenic, natural rubber crop.

In this section, we review the feasibility of ameliorating and circumventing Type I latex allergy caused by the presence of *Hevea* proteins in latex products. Also, we report on the development of hypoallergenic guayule latex.

Latex Allergy (Type I, IgE-Mediated)

Latex products had been safely used for many decades with only infrequent contact reactions (Type IV) occurring generally in response to chemical additives. However, in the 1980s, life-threatening Type I allergy appeared, which is triggered by proteins present in *Hevea* latex. A hypersensitive individual must take care to avoid contact with all natural rubber products made from *Hevea* latex. Allergic reactions to *Hevea* rubber products include local urticaria, systemic urticaria, rhinitis, conjunctivitis, edema, bronchospasm, tachycardia, anaphylaxis, and death (Slater 1989; Slater et al. 1990; Morales et al. 1989; Tomazic et al. 1992). The Food and Drug Administration responded to this public health problem by issuing a medical alert (March 1991) concerning use of latex products. Most, but not all, severe reactions occur from direct patient contact with latex gloves and other devices during medical and dental procedures.

Causes of Latex Allergy

The surge of Type I latex allergy coincided with sudden world-wide increased demand for latex gloves in response to institution of universal precautions to prevent transmission of the human diseases caused by HIV and Hepatitis B. New and inexperienced glove manufacturers entered the market, and short-cuts in manufacturing became common, in order to supply the increased demand (Bodycoat 1993). Over-production of these gloves led to a drop in price which, combined with the increasing demand, induced some other manufacturers to reduce costs through curtailed processing (Russell-Fell 1993). Altered manufacturing processes included reduction and sometimes elimination of the leaching step normally used to wash the latex products (Bodycoat 1993; Russell-Fell 1993). The washing process removes soluble latex components, as well as chemical additives, and underwashed products contain high levels of *Hevea* latex proteins.

The use of high protein latex products, especially of single-use latex gloves, as well as their extensive deployment throughout society, led to widespread development of Type I latex allergy. The problem was compounded by requirements for health-care workers to frequently change gloves. Proteins in gloves can bind to glove powder. When gloves are removed, the powder becomes a source of air-borne, respirable, latex allergens (Tarlo et al. 1994; Tomazic et al. 1994).

The Latex-Hypersensitive Population

The first incidents of latex allergy, in the United States, were reported in 1988. Numbers had increased to at least 500,000 by 1992. Published estimates, based on medical testing, indicate that 17,000,000 adults in the general U.S. population are affected by Type I latex allergy, 10%–40% of health care workers, and up to 60% of multiple surgery cases including spina bifida children (Alenius et al. 1993; Hamann 1993; Ownby et al. 1994). Many other countries around the world also have serious problems with latex allergy. The number of latex-hypersensitive people continues to climb due both to increased usage of latex products and to the continued sale of high protein latex gloves and other products. Manufacturing methods known to reduce product protein levels are available, but it is important to realize that even the best processing practices currently available will not produce latex products that are protein-free.

Latex Allergy Amelioration

Since Type I latex allergy arose because of high protein latex products, the manufacture and use of low protein latex products should greatly diminish the incidence of new allergy cases. Various methods for removing protein from latex and latex products are available (Pailhories 1993). However, future latex products will have to be shown to be low protein.

Hevea latex and latex products are difficult materials to assay for protein since latex is a cytoplasm that also contains about 30% rubber in the form of microscopic particles. Latex proteins fall into several groups with some soluble (48%), some associated with organelles (26%), and some bound to the surface of rubber particles (26%) (Kekwick 1993). Allergic patients can react to a large number of different latex protein allergens, which can originate from all of these protein groups. At least 57 different latex proteins and peptides, out of approximately 240, have been shown to be human allergens (Alenius et al. 1994). These range in size from 2 to 100 kD (Hamann 1993; Yeang et al. 1996). Nevertheless, soluble latex proteins, and others not directly associated with rubber particles, should be easier to remove from latex and/or latex products than the rubber particle-bound proteins (Siler and Cornish 1992). Their removal would substantially lower the overall protein level, thus reducing the allergenicity of the final latex product for people who have not yet developed latex allergy.

Latex protein assays must effectively measure all classes of protein allergens in latex, or latex products, and eliminate interference from nonproteinaceous substances in the latex and additives such as ammonia. A suitable method for assaying protein levels in latex has been developed (Siler and Cornish 1995) and an ASTM panel has developed an assay for use with latex products (ASTM 1995).

Latex Allergy Circumvention

As discussed above, it is possible to produce low protein latex products that should moderate the incidence of new allergy cases. However, trace amounts of allergen can induce serious systemic reactions in a hypersensitive person. Thus, production of latex products safe for use by allergic individuals requires the elimination of *all* latex allergens.

Allergens that are not associated with the rubber particles can be removed from the latex, prior to product manufacture, by repeated purification cycles using centrifugation/flotation (Siler and Cornish 1994a). Certain types of dipped products, such as some surgical gloves, electricians' gloves, and condoms, are made from doubly-centrifuged latex. This is produced by centrifuging standard latex (30%–35% dry rubber), diluting to 20%–25% with water, and centrifuging again. This process can remove about 70% of the non-rubber substances (Pailhories 1993). However, 25% of the proteins are bound to the rubber particles. Even extensive rubber particle purification does not remove them and these proteins would persist in the final product.

We tested the immunogenicity of various latex materials using antibodies, raised in rabbits, against proteins extracted from *Hevea* latex films. These *Hevea* latex protein antibodies recognized many of the proteins in complete *Hevea* latex (Siler and Cornish 1994b). As expected, purified rubber particle preparations, from which the soluble latex proteins were removed, contained far fewer proteins than latex (Siler and Cornish 1992, 1994b). Nevertheless, these highly purified *Hevea* rubber particle preparations still contained numerous proteins many of which are immunogens (Siler and Cornish 1994b). Also, a report that IgE antibodies from serum of latex-sensitized individuals recognized the 14.6 kD protein "rubber elongation factor" implicates this abundant rubber particle-bound protein as a major allergen in *Hevea* latex (Czuppon et al. 1993).

Thus, complete removal of both rubber particle-bound and soluble latex protein allergens is required to produce a *Hevea* latex material safe for use by a latex-hypersensitive person. Removal of the rubber particle-bound proteins would necessitate drastic treatment e.g. with proteases and/or detergents, that would not only probably be prohibitively expensive, but also likely would adversely affect the performance characteristics and quality of the resulting latex products.

Some manufacturers have introduced so-called "hypoallergenic" *Hevea* latex gloves into the market. Although some of these gloves are low protein, tests demonstrated that many contain substantial amounts of proteins that bind IgE antibodies from sera of *Hevea*-hypersensitive patients (Yunginger et al. 1994). Low protein hypoallergenic gloves would be unlikely to induce allergies in individuals not already allergic to *Hevea* latex proteins. However, these gloves would *not* be safe for use by people who are already *Hevea*-hypersensitive.

Suggestions have been made to remove the allergens from *Hevea* latex by breeding or by using genetic engineering techniques (antisense technology) to eliminate the allergenic proteins. These approaches are biologically unsound. This is primarily because many different latex proteins have been shown to be allergens (see above). It is very unlikely that they could be removed from the *Hevea* tree without dire effects on latex production and plant health. Furthermore, the genetic base of cultivated *Hevea* is extremely narrow which would likely preclude a plant breeding approach. We confirmed that clonal selection is unlikely to eliminate latex antigens in experiments that demonstrated that *Hevea* latex protein antibodies recognized many proteins in samples from three different commercial lines of *Hevea* (Siler and Cornish 1994b). The proteins recognized by the antibodies showed similar patterns in all three clones.

HYPOALLERGENIC GUAYULE LATEX

Different rubber-producing species may contain proteins distinct from those of *Hevea* and perhaps provide a source of hypoallergenic rubber (Siler and Cornish 1992). Guayule rubber is of high molecular weight and of comparable quality to *Hevea* rubber. However, guayule had not been seriously considered as a source of latex rubber (this liquid rubber form is essential for the manufacture of dipped and extruded products) since it does not produce its rubber as a tapable latex. Unlike the laticiferous species *Hevea*, guayule rubber particles are compartmentalized in individual bark parenchyma cells. Nonetheless, methods have been developed to extract guayule rubber in a latex-like form (Jones 1948; Madhavan and Benedict 1984; Cornish and Backhaus 1989; Cornish 1993). It seemed likely that the *Hevea* protein allergens would be absent from guayule latex since the *Hevea* and guayule are evolutionarily-divergent species and their rubber particle protein complements are dissimilar. We have proven this to be true. Anti-*Hevea* latex protein antibodies did not recognize latex proteins from guayule using ELISA (enzyme-linked immunosorbent assay). Proteins extracted from purified rubber particle preparations of guayule showed no reaction, even at concentrations at least 2 or 3 orders of magnitude above the *Hevea* rubber particle protein concentration that elicited a positive reaction (Siler and Cornish 1994b). Western blot analyses of the rubber particle proteins from guayule with the *Hevea* antibodies confirmed the lack of cross-reactivity (Siler and Cornish 1994b). These results indicate that *Hevea* latex allergy can indeed be circumvented using rubber from guayule.

Our results were confirmed in preliminary clinical trials on humans. In one trial, individuals with proven hypersensitivity to *Hevea* latex were skin-prick tested with latex samples from different sources. All these subjects reacted strongly to *Hevea* latex (wheals and flare), whereas none responded to guayule latex (Carey et al. 1995). In a second trial, 56 "at risk" people (such as health care workers, spina bifida patients, and multiple surgery cases), ages 16 months to 72 years, and three controls, were tested both by skin prick and by in vitro RAST (radioallergosorbent assay) assays (Ber et al. 1993). The RAST assay detects the presence of human IgE antibodies specific to *Hevea* latex proteins. 25 reacted to *Hevea* materials, 25 reacted in the skin prick test, and 30 in the RAST. None reacted to guayule latex, confirming the earlier clinical test results, and our experiments using the rabbit anti-*Hevea* latex protein antibodies.

Guayule rubber latex should be suitable for the manufacture of high quality, hypoallergenic natural rubber products safe even for the *Hevea*-hypersensitive individual. Some other rubber-producing plant species also produce high molecular weight rubber but currently none are suitable for commercial cultivation. In contrast, guayule agronomics are well understood (see History of Guayule Commercialization above).

Thus, diverse immunological tests have demonstrated that guayule latex is truly hypoallergenic and that guayule latex products could be used safely by *Hevea* latex-sensitive people. Of course, it is essential that guayule latex products be maintained as low protein so that allergy problems don't redevelop later. The most difficult proteins to completely remove are those bound to the rubber particles themselves. Guayule rubber particles have less protein than *Hevea* (Cornish et al. 1993) and, therefore, guayule latex could be produced with lower protein levels than will ever be possible with *Hevea* latex.

Production of Hypoallergenic Guayule Latex

We have developed processing methods to extract intact guayule rubber particles and generate an artificially-produced purified latex (Cornish 1993). These methods also are suitable for the production of latex from other rubber producing species.

Guayule hypoallergenic latex presents a major new, high value, rapidly expanding market because hypersensitive people cannot safely use *Hevea* latex products. The U.S. market for latex gloves alone had a retail value of over $3,000,000,000 in 1993. Hypoallergenic natural rubber latex products would provide a sufficient premium to permit the immediate development of guayule as a new crop even at its current rubber yield (about 10% on a dry weight basis).

During the Spring of 1994, a production run was carried out in Arizona, where 2,300 kg of field-grown guayule shrub were processed. A processing system was constructed and used (Coates and Cornish, unpublished results) to generate ammoniated guayule latex for allergy testing, performance testing, and manufacturing trials. This batch of guayule latex was less highly purified than the latex used in the allergy trials described in the last section. Nevertheless, immunological experiments (RAST and Western blot analysis) to investigate the ability of IgE antibodies from *Hevea* latex allergic patients, and IgG antibodies from hyperimmunized mice, to detect proteins in the guayule latex displayed a total lack of cross-reactivity (Siler et al. 1996). Several additional trials have been completed on this material demonstrating that guayule latex (1) may be compounded using similar procedures to *Hevea* latex (Schloman et al. 1996), (2) can be used to produce quality latex medical products (H.F. Bader, and K. Cornish, unpublished data), and

(3) prototype dipped films provide an impermeable barrier to virus transmission (D. Lytle and K. Cornish, unpublished data).

Commercialization Position

Hypoallergenic latex reflects a technological breakthrough that provides a new impetus to the commercialization of this new crop. However, guayule latex would be years away from successful commercialization but for the diverse and numerous successes of the scientists involved in guayule research in recent years, as described earlier. Their work has led to the advanced guayule lines and refined, mechanized agronomic practices which provide an essential foundation for guayule latex commercialization.

CONCLUSIONS

Large-scale production of guayule latex is feasible and the high quality of guayule natural rubber makes it suitable for hypoallergenic product manufacture. These products should be safe for use by the *Hevea*-hypersensitive population. Clinical trials on humans showed that even severely *Hevea*-hypersensitive patients had no reaction to guayule latex. Guayule latex products must be produced with low protein levels so that allergy problems don't develop with the new material. Guayule rubber particles have much less protein than those of *Hevea*, and guayule latex can be produced with lower protein levels than even highly purified *Hevea* latex. Upon the generation of sufficient supplies, hypoallergenic guayule latex products first would become available to hypersensitive patients, and the at-risk health-care profession, and then to the general population. Guayule latex provides the first, natural rubber solution to the urgent global need for a reliable hypoallergenic source of natural rubber latex. Concerted and dedicated efforts should make safe guayule latex products a reality and the guayule crop a major feature of the farming landscape of the south-western United States. Continuing research also should lead to the introduction of genetically-engineered guayule lines with enhanced rubber yield and extended growing range.

REFERENCES

Alenius, H., T. Palosuo, Y. Kelly, V. Kurup, T. Reunala, S. Makinen-Kiljunen, K. Turjanmaa, and J. Fink. 1993. IgE reactivity to 14-kD and 27-kD natural rubber proteins in latex-allergic children with spina bifida and other congenital anomalies. Int. Arch. Allergy Immunol. 102:61–66.

Alenius, H., V. Kurup, Y. Kelly, T. Palosuo, K. Turijanmaa, and J. Fink. 1994. Frequent occurrence of IgE antibodies to a cluster of 11 latex proteins in patients with spina bifida and histories of anaphylaxis. J. Lab. Clin. Med. 123:712–720.

Appleton, M.R. and J. van Staden. 1989. The relationship between season, growth and isoprenoid biosynthesis in *Parthenium argentatum*. J. Plant Physiol. 134:524–523.

ASTM Designation: D 5712-95. 1995. Standard test method for analysis of protein in natural rubber and its products. Annual Book of ASTM Standards 1402:1–5.

Ber, D.J., K.H. Munemasa, R. Gilumilion, K. Cornish, D.E. Klein, and G.A. Settipane. 1993. Allergy to latex. J. Allergy Clin. Immunol. 91:271.

Berndt, J. 1963. The biosynthesis of rubber. U.S. Government Res. Rep. AD-601:729.

Bodycoat, I. 1993. Manufacture of hypoallergenic rubber products. p. 43–46. In: Latex protein allergy: The present position. Crain Communications Ltd., Rubber Consultants, Brickendonbury, Hertford, U.K.

Bonner, J. 1991. The history of rubber. p. 1–6. In: J.W. Whitworth and E.E. Whitehead (eds.), Guayule natural rubber. Office of Arid Lands Studies, Univ. Arizona, Tucson.

Carey, A.B., K. Cornish, P.J. Schrank, B. Ward, and R.A. Simon. 1995. Cross reactivity of alternate plant sources of latex in subjects with systemic IgE mediated sensitivity to *Hevea brasiliensis* latex. Ann. Allergy, Asthma Immunol. 74:317–320.

Coates, W. 1991. Guayule harvesting equipment. p. 241–260. In: J.W. Whitworth and E.E. Whitehead (eds.), Guayule natural rubber. Office of Arid Lands Studies, Univ. Arizona, Tucson.

Cornish, K. 1993. Hypoallergenic natural rubber products from *Parthenium argentatum* (Gray) and other non-*Hevea brasiliensis* species. U.S. Patent pending S.N. 08/145,456.
Cornish, K. and R.A. Backhaus. 1990. Rubber transferase activity in rubber particles of guayule. Phytochemistry 29:3809–3813.
Cornish, K. and D.J. Siler. 1996. Advances in alternative natural rubber production. In: G. Fuller, D.D. Bills, and T.A. McKeon (eds.), Agriculture as a renewable source of raw materials. ACS Books, Washington, DC. (in press).
Cornish, K., D.J. Siler, O.K. Grosjean, and N. Goodman. 1993. Fundamental similarities in rubber particle architecture and function in three evolutionarily divergent plant species. J. Nat. Rubber Res. 8:275–285
Czuppon, A.B., Z. Chen, S. Rennert, T. Engelke, H.E. Meyer, M. Heber, and X. Baur. 1993. The rubber elongation factor of rubber trees (*Hevea brasiliensis*) is the major latex allergen in latex. J. Allergy Clin. Immunol. 92:690–697.
d'Auzac J., J-L. Jacob, and H. Chrestin (eds.). 1989. Physiology of rubber tree latex. CRC Press, Boca Raton, FL.
Estilai, A. and D.T. Ray. 1991. Genetics, cytogenetics and breeding of guayule. p. 47–92. In: J.W. Whitworth and E.E. Whitehead (eds.), Guayule natural rubber. Office of Arid Lands Studies, Univ. Arizona, Tucson.
Foster, M.A. and J. Moore. 1991. Land preparation. p. 107–144. In: J.W. Whitworth and E.E. Whitehead (eds.), Guayule natural rubber. Office of Arid Lands Studies, Univ. Arizona, Tucson.
Foster, M.A. and J. Moore. 1992. Direct-seeding techniques for guayule stand establishment in west Texas. J. Prod. Agr. 5:163–167.
Foster, M.A., L.G. Kleine, and J. Moore. Response of direct-seeded guayule to preemergence herbicides. 1993. J. Am. Oil Chem. Soc. 70:1239–1240.
Gathman, A.C, D.T. Ray, and M. Livingston. 1992. Comparison of three stability measures in guayule. Industr. Crops Prod. 1:67–74.
Hamann, C.P. 1993. Natural rubber latex sensitivity in review. Am. J. Contact Dermatitis 4:4–21.
Hammond, B.L. and L.G. Polhamus. 1965. Research on guayule (*Parthenium argentatum*): 1942–1959. USDA Tech. Bul. 1327.
Huang, H.T. 1991. Introduction. p. xv–xix. In: J.W. Whitworth and E.E. Whitehead (eds.), Guayule natural rubber. Office of Arid Lands Studies, Univ. Arizona, Tucson.
Ji, W., C.R. Benedict, and M.A. Foster. 1993. Seasonal variations in rubber biosynthesis, 3-hydroxy-3-methylglutaryl-coenzyme A reductase, and rubber transferase activities in *Parthenium argentatum* in the Chihuahuan desert. Plant Physiol. 103:535–542.
Jones, E.D. 1948. Recovery of rubber latex from guayule shrub. Ind. Eng. Chem. 40:864–874.
Kekwick, R.G.O. 1993. Origin and source of latex protein allergy. p. 21–24. In: Latex protein allergy: the present position. Crain Commun. Ltd., Rubber Consultants, Brickendonbury, Hertford, U.K.
Madhavan, S. and C.R. Benedict. 1984. Isopentenyl pyrophosphate *cis*-1,4-polyisoprenyl transferase from guayule (*Parthenium argentatum* Gray). Plant Physiol. 75:908–913.
Morales, C., A. Bascomba, J. Carrieira, and A. Sastre. 1989. Anaphylaxis produced by rubber gloves contact: case reports and immunological identification of the antigens involved. Clin. Expt. Allergy 19:425–430.
Nakayama, F.S. 1991. Influences of environment and management practices on rubber quantity and quality. p. 217–240. In: J.W. Whitworth and E.E. Whitehead (eds.), Guayule natural rubber. Office of Arid Lands Studies, Univ. Arizona, Tucson.
Ownby, D.R., H.E. Ownby, J.A. McCullough, and A.W. Shafer. 1994. The prevalence of anti-latex IgE antibodies in 1000 volunteer blood donors. J. Allergy Clin. Immunol. 93:282.
Pailhories, G. 1993. Reducing proteins in latex gloves. Clin. Rev. Allergy 11:391–402.
Ray, D.T. 1991. Guayule: A source of natural rubber. p. 338–342. In: J. Janick and J.E. Simon (eds.), New crops. Wiley, New York.

Reisch, M.S. 1995. Rubber consumption is rising, but producers' profits are squeezed. Chem. Eng. News August 14, 11–14.

Russell-Fell, R. 1993. Introductory paper. p. 3–6. In: Latex protein allergy: the present position, Crain Commun. Ltd., Rubber Consultants, Brickendonbury, Hertford, U.K.

Schloman, W.W. Jr., F. Wyzgoski, D. McIntyre, K. Cornish, and D.J. Siler. 1996. Characterization and performance testing of guayule latex. Rubber Chem. Tech. (in press).

Siler, D.J. and K. Cornish. 1992. Potential for allergy amelioration by rubber particle purification from *Hevea brasiliensis*, and for allergy circumvention by use of alternate rubber-producing plants, # 34. p. 46. In: Proc. Int. Latex Conf., Sensitivity to latex in medical devices, FDA, Washington, DC.

Siler D.J. and K. Cornish. 1994a. Identification of *Parthenium argentatum* rubber particle proteins immunoprecipitated by an antibody that specifically inhibits rubber transferase activity. Phytochemistry 36:623–627.

Siler, D.J. and K. Cornish. 1994b. Hypoallergenicity of guayule rubber particle proteins compared to *Hevea* latex proteins. Ind. Crops Prod. 2:307–313.

Siler, D.J. and K. Cornish. 1995. Measurement of protein in natural rubber latex. Anal. Biochem. 229:278–281.

Siler, D.J., K. Cornish, and R.G. Hamilton. 1996. Absence of cross-reactivity of IgE antibodies from *Hevea brasiliensis* latex allergic subjects with a new source of natural rubber latex from guayule (*Parthenium argentatum*). J. Allergy Clin. Immunol. in press.

Slater, J.E. 1989. Rubber anaphylaxis. N. Engl. J. Med. 320:1126–1130.

Slater, J.E., L.A. Mostello, C. Shaer, and R.W. Honsinger. 1990. Type I hypersensitivity to rubber. Ann. Allergy 65:411–414.

Tarlo, S.M., G. Sussman, A. Contala, and M.C. Swanson. 1994. Control of airborne latex allergens by use of powder-free gloves. J. Allergy Clin. Immunol. 93:985–989.

Thompson, A.E. and D.T. Ray. 1989. Breeding guayule. Plant Breed. Rev. 6:93–165.

Tomazic, V.J., T.J. Withrow, B.R. Fisher, and S.F. Dillard. 1992. Latex-associated allergies and anaphylactic reactions. Clin. Immunol. Immunopath. 64:89–97.

Tomazic, V.J., E.L. Shampaine, A. Lamanna, T.J. Withrow, N.F. Adkinson, and R.G. Hamilton. 1994. Cornstarch powder on latex products is an allergen carrier. J. Allergy Clin. Immunol. 93:451–758.

Wagner, J.P. and W.W. Schloman, Jr. 1991. Processing. p. 261–286. In: J.W. Whitworth and E.E. Whitehead (eds.), Guayule natural rubber. Office of Arid Lands Studies, Univ. Arizona, Tucson.

Whitworth J.W. and E.E. Whitehead (eds.). 1991. Guayule natural rubber: a technical publication with emphasis on recent findings. Guayule Administrative Management Committee and USDA Cooperative State Research Service, Office of Arid Lands Studies, Univ. Arizona, Tucson.

Wright, N.G., S. Fansler, and R.D. Lacewell. 1991. Guayule economics. p. 351–365. In: J.W. Whitworth and E.E. Whitehead (eds.), Guayule natural rubber. Office of Arid Lands Studies, Univ. Arizona, Tucson.

Yeang, H.Y., K.F. Cheong, E. Sunderasan, S. Hamzah, N.P. Chew, S. Hamid, R.G. Hamilton, and M.J. Cardosa. 1996. The 14.6 kD (REF, Hev b I) and 24 kD (Hev b III) rubber particle proteins are recognized by IgE from spina bifida patients with latex allergy. J. Allergy Clin. Immunol. in press.

Yunginger, J.W., R.T. Jones, A.F. Fransway, J.M. Kelso, M.A. Warner, and L.W. Hunt. 1994. Extractable latex allergens and proteins in disposable medical gloves and other rubber products. J. Allergy Clin. Immunol. 93:836–842.

Identification of Guayule Regions in Northern Mexico, Based on Rubber Yield and Coproducts Quality

Diana Jasso Cantú, José Luis Angulo Sánchez, and Raúl Rodriguez García

The development of guayule as a commercial crop has encountered several problems in México, despite a great amount of research and development carried out in the 1970s and mid 1980s (Angulo and Lopez 1978; Lopez and Kuruvadi 1985; Kuruvadi 1985; Lopez and Kuruvadi 1987; Kuruvadi 1988). During this period, a pilot plant established technical feasibility evaluation for the rubber recovery process, and a commercial unit for rubber production was designed (Campos et al. 1978). However, the commercial production units have not been established up to this moment. Among the problems preventing guayule commercial development are the lack of raw product and the low international price of *Hevea* rubber.

The present guayule areas are based on wild stands, the plants are scarce, widely distributed, and rubber content is variable, ranging from 5% to over 17% (Jasso et al. 1993). This fact makes difficult to pinpoint the potential rubber yields per unit area, hence there is no solid basis to calculate profitability. Furthermore, fluctuations in the rubber price make cost estimations uncertain. Previously, we proposed to consider guayule as a source of more than one product, and to defining new applications in order to improve commercial exploitation (Angulo and Jasso 1995). In addition, we have considered that selection and reproduction of high yielding clones is critical to develop guayule as a commercial crop.

We established a productivity index (PI) in order to identify plants with outstanding features. This index is used in this work to designate high yielding areas and plants, within the Mexican guayule region. Iodine value was determined because the resin might be used for varnish and adhesives (Belmares et al. 1980; Schloman 1988; Schloman and Wagner 1991; Thames and Wagner 1991). This parameter reflects the presence of double bonds required for these applications. Because biomass would be used as a feedstock, bagasse quality was tested by determining protein content after resin and rubber extraction. Although some data on the protein content in the guayule bagasse have been reported (Schloman and Wagner 1991), only one previous work on the resin iodine value determination was found (Belmares et al. 1980). There are no reports on the possible correlation between rubber or resin production in guayule and the iodine value. We suggest that the chemicals possessing double bonds (including fatty acids, terpenoids, and other chemical species) may be related to rubber biosynthesis.

In this study, three zones, in two states, Coahuila and Zacatecas, formerly identified as high rubber yielding areas (Naqvi and Hanson 1983; Lopez and Kuruvadi 1985), were revisited, guayule shrubs samples collected, and analyzed. Biomass, rubber, resin, bagasse, iodine value, and protein were evaluated along with common morphologic characteristics (Kuruvadi 1985; Lopez and Kuruvadi 1987) such as shrubs height and top spread.

METHODOLOGY

Collecting Sites and Sampling

Three sites were selected for plant acquisition: Atenco (site 1), close to Saltillo, Rocamontes (site 2), Coahuila, and Noria de Guadalupe (site 3) in the state of Zacatecas (Fig 1). These sites have been subjected to rubber yielding studies and the molecular weight characteristics reported. Plant acquisition was carried out during the first week of Sept. 1991. Twenty-five plants, between 35 and 75 cm height, were randomly selected from each site. Measurements of the shrub height and spread were obtained in the field, the results are presented in Table 1. Afterwards, the plants were cut at ground level and transported to the laboratory, where the total weight was determined. The leaves were cut from the shrub and the defoliated plants immersed in liquid nitrogen and grounded in a Wiley laboratory mill with 2 mm screen.

Rubber, Resin, and Biomass Measurements

Part of the grounded material was used for biomass quantification. The tissue was oven dried at 60°C until constant weight was achieved and total dry weight calculated (Table 2). Resin and rubber content measurements were performed on 5 g samples, by soxhlet extraction, using acetone and toluene as the respective solvents. A resin quality determination was effectuated by double bonds titration using the Wijs method. This was carried out by dissolving the resin in chloroform and titrating with iodine-bromine. After resin and rubber extraction, the protein content was determined in the bagasse obtaining the nitrogen percent by the Kjeldahl method. The protein content (%Pr) was calculated using the nitrogen percent (%N_2) and the equation (Dintzis

Fig 1. Map of the guayule growing region in Mexico and localization of the three analysis sites.

Table 1. Guayule morphologic characteristics at the three sites.

		Height (cm)		Spread (cm)	
Site	Locality	Mean	Range	Mean	Range
1	Atenco, Coahuila	50.5	42–61	57.1	46–71
2	Rocamontes, Coahuila	52.3	48–58	55.4	48–63
3	Noria, Zacatecas	53.5	42–64	57.9	37–66
Significance		nonsignificant		nonsignificant	

Table 2. Mean values for guayule physiologic characteristics at different sites.

Site	Total dry weight (g)	Rubber (%)	Resin (%)	Iodine value (%)	Bagasse (%)	Protein (%)
1	492	10.6	12.9	29.0	76.4	6.5
2	723	9.6	13.7	35.7	76.8	6.2
3	761	9.7	11.9	42.8	78.4	6.9
Significance	**	NS	**	**	**	**

** NS = Significant at 1% level (**) or nonsignificant (NS).

et al. 1988): %Pr = 6.25 (%N_2). The results from the morpho-physiologic characteristics were analyzed using a statistical random design with three treatments (localities) and 25 replications (plants).

RESULTS

Morphologic Characteristics

Table 1 shows the results for shrubs height and spread, parameters commonly used for plant selection. The range for these parameters is different within sites, but the mean values are very similar. These results reflect the different plants composition of wild guayule stands and are in agreement with several other reports.

Physiologic Characteristics

The results for dry weight, percent rubber and percent resin, iodine value and protein content in the bagasse are shown in Table 2. The weight difference between plants growing in the three sites, contrasted strongly with the similarity observed in the height and spread data. Site 3 (Noria de Guadalupe) and site 2 (Rocamontes) have a difference of only 5%, but that between site 3 and site 1 (Atenco) is approximately 50%, site 3 having the highest plant weight. This behavior is due to shrub thickness, the plants from site 3 and site 2 have thicker stem and branches than those in site 1. Despite height, spread, and weight differences, rubber content are very similar between sites. Large differences in iodine value between the sites were found.

A relationship between the resin content and the iodine value was noted (Fig 2). As resin content increases, iodine value diminishes. The chemicals constituting the resin are a complex mix (Belmares et al. 1980; Schloman and Wagner 1991) including terpenoids, fatty acids triglycerides (i.e. linoleic, linolenic, stearic, and palmitic), very low molecular weight (VLMW) rubber, and a drying oil, "shellac" type. Several of these chemicals contain double bonds, for example linoleic and linolenic acids, the terpenoids and the VLMW rubber. Iodine value is an indicator of the concentration of these compounds, but it was not possible to differentiate between them with this technique. There was no correlation between rubber and resin content. However when a ternary diagram was constructed with the rubber, resin and iodine value (Fig 3) all the samples lie within a narrow region. This suggests a relationship between these three variables and rubber biosynthesis.

In order to consider differences in all the physiologic parameters a global productivity index (PI) was defined, as $PI = [\%Rb + (\%Rs \times I_2) + (\%Bg \times Pr)]W$, where Rb = rubber, Rs = resin, I_2 = iodine number, Bg = bagasse, and Pr = protein content. Resin and the bagasse are multiplied by a "quality factor" which is the double bonds or protein content, respectively. The results indicate a productivity index of 12.8 for site 1, 23.3 for site 2, and 29.0 for site 3.

Site 3 possessed the highest index based on rubber yielding and resin and bagasse quality. Differences in the parameters did not follow the same trend in the three sites. Data in Table 2 indicated that for rubber

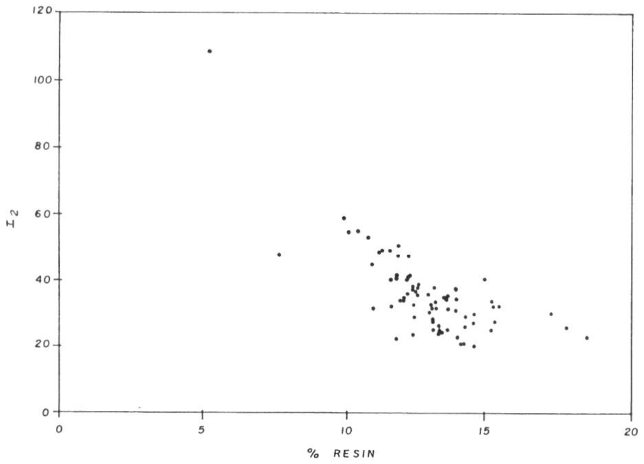

Fig 2. Relationship between resin content and iodine value for the plants from the three sampling sites.

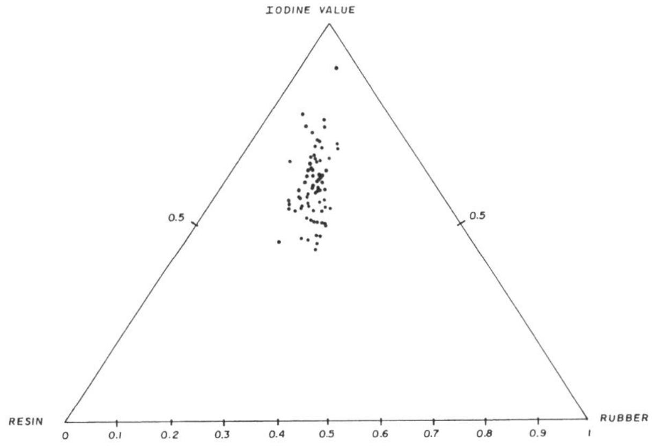

Fig 3. Ternary diagram correlating rubber, resin and iodine value.

content site 1 > site 3 > site 2, for iodine value site 3 > site 2 > site 1, and for resin content site site 2 > site 1 > site 3. However, the PI results show clearly that when all factors are considered the difference in value between the sites is wide. Site 3 has the highest value due to its higher mean total weight and iodine value; protein and resin content showed smaller differences but its contribution was limited. Rubber content was similar, and hence is not a decisive factor. The rubber molecular weight has been considered as a quality parameter for the rubber. However, previous studies show that differences in MW are not significant, for plants reaching this height. We consider that PI might be a valuable selection index to detect outstanding shrubs for genetic improvement. Although the numbers allow the identification of high yielding plants (rubber and coproducts) we do not know the highest or lowest values that will define, the maximum PI or the acceptability limit for guayule. The analysis of plants from regions with high and low rubber production will provide this information. According to these results, shrub height and spread seem not to be the best parameters for plant selection, intended for integral commercial use, as the product and coproducts yielding is very different for similar sized plants.

REFERENCES

Angulo Sánchez, J.L. and D. Jasso de Rodríguez. 1996. An overview of outstanding species from semi arid lands. p. 176–186. In: L.H. Princen and C. Rossi (eds.), Proc. Ninth Int. Conference on Jojoba and its uses and of the Third Int. Conference on New Industrial Crops and Products, Assoc. Adv. Ind. Crops.

Angulo Sánchez, J.L. and E. Campos Lopez. 1978. Geographic influence on guayule rubber. p. 177–190. In: Guayule: Reencuentro en el desierto. Centro de Investigación en Química Aplicada. Saltillo, México.

Belmares Sarabia, H., L.L. Jimenez Valdes, and M. Ortega. 1980. New rubber peptizers and coating derived from guayule resin (*Parthenium argentatum* Gray). Ind. Eng. Chem. Prod. Res. Dev. 19:107–111.

Campos Lopez, E., E. Neavez Camacho, and R. Maldonado García. 1978. Guayule: present state of knowledge. p. 375–410. In: Guayule: reencuentro en el desierto. Centro de Investigación en Química Aplicada. Saltillo, México.

Dintzis, F.R., J.F. Cavins, E. Graff, and T. Stahly. 1988. Nitrogen to protein conversion factors in animal feed and fecal samples. J. Anim. Sci. 66:5–11.

Jasso de Rodriguez, D., J.L. Angulo Sánchez, F. Ramirez Godina, R. Morones Reza, and R. Rodriguez García. 1993. J. Am. Oil Chem. Soc. 70:1229–1234.

Kuruvadi, S. 1985. Evaluation of genetic resources of guayule in Mexico. Guayulero 7(1&2):24–26.

Kuruvadi, S. 1988. Identification and evaluation of consistently high rubber yielding genotypes in native guayule populations. Guayulero 10(3&4):10–12.

Lopez Benitez, A. and S. Kuruvadi. 1985. Variation in yield components and correlations in guayule. Guayulero 7(1&2):19–23.

Lopez Benitez, A. and S. Kuruvadi. 1987. Variability in rubber content of three guayule populations in Durango, Mexico. Guayulero 9:(1&2):3–7.

Naqvi, H.H., and G.P. Hanson. 1983. Observation on the distribution and ecology of native guayule populations in Mexico. p. 145–154. In: E.C. Gregg, J.L. Tipton, and H.T. Huang (eds.), Proc. 3rd. Int. Guayule Conf., Pasadena, California, U.S.A. 27 April–1 May. 1980. Guayule Rubber Soc., Publ. Univ. California, Pasadena.

Schloman W.W., Jr. 1988. Rubber additives derived from guayule resin. Ind. Eng. Chem. Res. 27:712–716.

Schloman W.W., Jr. and J.P. Wagner. 1991. Rubber and coproducts utilization. p. 287–310. In: J.W. Withworth and E.E. Whitehead (eds.), Guayule natural rubber. Office of Arid Lands Studies. Univ. Arizona, Tucson.

Thames, S.F. and J.P. Wagner. 1991. Recent advances in guayule coproduct research and development. p. 311–350. In: J.W. Withworth and E.E. Whitehead (eds.), Guayule natural rubber. Office of Arid Lands Studies. Univ. Arizona, Tucson.

Development of Castor Bean Production in France

Françoise Labalette, André Estragnat, and Antoine Messéan

Castor bean (*Ricinus communis* L., Euphorbiaceae) is a plant of tropical origin which produces seeds containing up to 55% of a natural oil rich in ricinoleic acid. Among all the vegetable oils, castor oil is distinctive because of its high level of ricinoleic acid (over 85%), a fatty acid consisting of 18 carbons, a double bond between the ninth and tenth position, and a hydroxyl group attached to C12. Ricinoleic acid is responsible for castor oil interest, the highest and most stable viscosity index among all the vegetable oils combined with high lubricity, especially under low temperature conditions.

Castor oil has various applications in different industrial sectors: paintings and coatings, polyurethane coating, plastics, transport, cosmetics, textiles, and leathers. One of the major products derived from castor oil is the Rilsan B, developed by Atochem (France). This 100% castor-based product has numerous applications such as rotating glass car-wipers, ski boots fixatives, and for use in air-brake systems on trucks. Many new uses, based on biodegradability of castor oil derived products, are expected in the future.

World annual production of castor oil is about 460,000 t (1.1 million tonnes of seeds); the main producers are India, Brazil, and China. The European Union uses about 90,000 t of castor oil and imports oil as well as about 30,000 t of seeds which is mainly crushed in Germany. Due to new outlets for castor oil (such as anticorrosive products or odorant captivators), European demand should increase over the next ten years.

DEVELOPMENT OF DOMESTIC PRODUCTION

In order to meet this demand, the French oilseed organization launched domestic production in 1986. The project, extended to Europe, has the objective of producing 100,000 t, which corresponds to about 100,000 ha.

First Phase (1986–1990)

After preliminary trials during the 1950s, and as a result of high oil prices in the middle of the 1980s ONIDOL (Organisation Interprofessionnelle des Oléagineux) launched local production. In order to adapt this perennial plant to local conditions, three research thrusts were developed: breeding of early and productive varieties (Protosemences), crop management under French conditions (CETIOM), and design of a combine harvest (CEMAGREF).

A Cooperative Research Program at the European Level

As a result of encouraging results and evolution of European policy, a demonstration project (1991–1995), partially funded by the European Union, was initiated in 1991. Eight members from Germany, Italy, Portugal, Greece, and France, joined a European Group of Economic Interest (EURORICIN), the main objectives of this project included:

- Adaptation of a combine harvester including design of a specific header and modification of the drum in order to harvest seeds without husks.
- Breeding of adapted lines (cultivars Rica and Venda) and seed production.
- Crop management (weed control, water management, establishment methods).
- Development of field production with various cooperatives (about 400 ha were cultivated every year between 1992 and 1994).
- Development of new uses for castor oil (grease for mineral oil substitution, use of fractions derived from Rilsan production such as C7 (anti-corrosive agents) or C11 (anti-odor products) polyurethane.

A 3-year research project was also initiated in 1995 under the European Agro-industry Research framework. Specific objectives include: identification of main pests and pathogens, physiological mechanisms, crop management, genetics and breeding, and new chemical applications.

PRODUCTION

Castor bean is mainly produced on set-aside land and yields remain low. Under dry conditions, yields are about 1.0 to 1.2 t/ha but reach 1.5 to 1.8 t under irrigation. However, irrigation is not planned for set-aside land. The discrepancy between experimental yields and actual field yields indicates that progress can still be obtained through better crop management and optimal harvesting conditions (Table 1).

Crop Establishment

Soil tillage is important for a spring crop which requires high water resources from the soil. Ploughing is necessary. Fertilization requirements are similar to those of sunflower (40 to 60 kg of N and 50 kg of PK /ha).

Weed control is based on a pre-sowing application of trifluraline and a pre-emergence application of linuron. Due to lack of selectivity, both products are usually applied by farmers before sowing.

Sowing date must be as early as possible and generally is similar to the maize establishment date. Emergence requires about three weeks and, thus, sowing depth must be about 4 to 7 cm.

The seeding rate is about 50,000 seeds/ha. Row spacing depends on the combine harvester used (80 cm in our conditions). The cycle duration for current cultivars is about 130–150 days or 1200–1300 degree-days.

Fig. 1 compares Photosynthetic Area Index (PAI) and available light during the crop cycle. Note that maximum light interception occurs when light is decreasing. Earlier genetic material should thus be required in order to optimize photosynthesis.

Table 1. Yields of castors in France.

	Yields (t/ha)				
		Actual field			Oil content
Year	Trials	Mean	Minimum	Maximum	(%)
1991	2.4	1.0	0.5	2.2	51
1993	2.6	1.7	1.1	2.5	52
1994	2.4	1.2	0.6	2.2	51

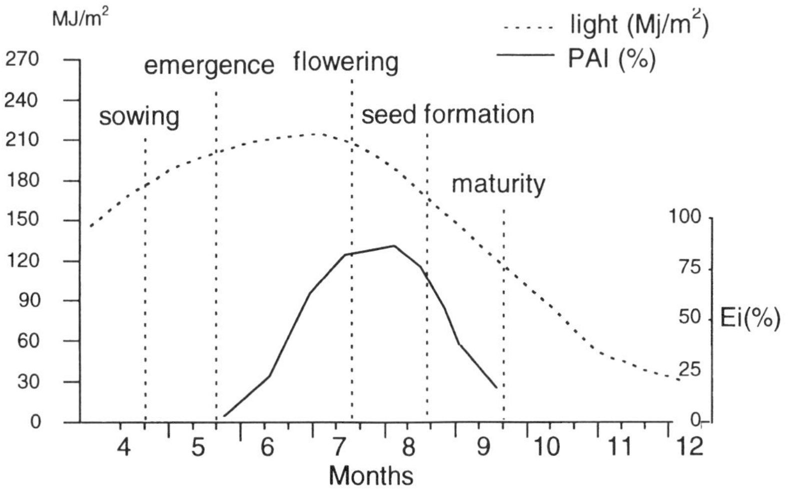

Fig. 1. Castor bean crop cycle

Crop Protection

Harrowing is generally carried out when plants have between 2 and 5 leaves in order to improve chemical weed control. Irrigation is appropriate if water is available. The optimal period for irrigation is after flowering in order to maintain leaf surface and improve seed production.

Pests and diseases remained minor problems except bugs *(Nezara viridula)* where high sensitivity occurs at the beginning of flowering. However, we suspect that some insects (Lysus) and pathogens (such as Alternaria) induce adverse effects on yield.

Defoliation speeds up maturity and makes the harvest easier. Diquat or glyphosate can be used. Breeding of earlier cultivars will avoid this application under some conditions.

FUTURE PROSPECTS

Methods to renew the castor industry in United States has been reviewed by Brigham (1993). In France yields need to be increased by improved harvesting practices and improvement of earliness. The first hybrids show a reduction of about 10–15 days of the crop cycle duration and no longer require defoliation.

A new system for dehusking the seeds during harvesting has been designed. About 90% of the husks are removed by the combine harvester. Although such a system avoids an additional cost of 400 FF/t ($ 80/t), the cost of the harvester modification remains too high and harvesting capacity too low. Improvement of the harvester is currently going on but an increase of the crop area should be necessary to reduce costs.

At the present time, no profitable return is possible without subsidies in the current European context. Furthermore, extra costs due to collecting, storage, and transport must be reduced in order to allow a farmer return price of about 2000 FF/t ($ 400/t).

REFERENCES

Brigham, R.D. 1993. Castor: Return of an old crop. p. 380–383. In: J. Janick and J.E. Simon (eds.), New crops. Wiley, New York.

Evaluation of Castor Germplasm for Agronomic and Oil Characteristics*

Harbans L. Bhardwaj, Ali I. Mohamed, Charles L. Webber, III, and Gilbert R. Lovell

The castor (*Ricinus communis* L., Euphorbiaceae) oil is used by American industries to manufacture many products such as nylon-11, hydrogenated oil, dehydrated oil and its fatty acids, sulfated and sulfonated oil, sebacic acid, ethoxylated oil, polyurethanes, and oxidized and polymerized oil. Castor oil is also used in a wide range of cosmetics, toiletries, and transparent soaps. Food-grade castor oil is a standard lubricant in food processing plants where its uses ranges from an anti-stick agent in candy molds to lubricant for trolleys and conveyors that carry food (CasChem 1982). Castor oil and its derivatives are also used in lubricating formulations (Roetheli et al. 1991a).

Commercial production of castor existed in the central United States as early as 1850. By 1959, Texas was the leading producer of castor with over 30,000 ha of castor being grown in the late 1960s. During this period, over 23 crushing mills reportedly were operating in Texas (Brigham, 1989). U.S. castor production ceased in the early 1970s due to: (1) contract negotiation problems between oil buyers and

*This research was supported by funds allocated to Virginia State University by USDA/CSREES/Evans-Allen program.

the farm cooperative involved in crushing castor, (2) low world prices for castor oil, (3) competition between castor and other crops for which high prices were being paid, and (4) discontinuation of government price supports in 1972.

The seed, leaves, and stem of castor are poisonous to humans and livestock if consumed (Zimmerman et al. 1958). Ingestion of even one seed can be fatal to humans (Naughton 1979). Ricin, a very lethal protein, is the major toxic component. Ricinine, a poisonous alkaloid, exists at very low levels and presents little problem (Horton and Williams 1989). In addition, castor contains an allergin (known as CB-1A) which can cause an allergic reaction in humans (Roetheli et al. 1991a).

The United States is totally dependent upon imports to meet the industry and defense needs for castor oil. A total dependence on imported castor oil places the United States in a vulnerable position should the supply be interrupted because of world events. A domestic crop would provide a stable source of supply. To replace imports, the United States would need about 45,000 ha of castor to supply domestic users (Roetheli et al. 1991b). Considerable importance must also be given to potential exports of U.S.-produced castor oil to France, nations of former USSR, Germany, and UK.

Since seeds of *Lesquerella fendleri* (L.) contain oil that is similar to castor oil, it is a potential replacement for castor (Roetheli et al. 1991a). Under the new crops program of Virginia State University, research was conducted during 1992, 1993, and 1994 to evaluate the potential of *Lesquerella fendleri* (L.) under Virginia conditions and to develop it as a domestic source of hydroxy fatty acids and as a replacement for castor. Field and greenhouse experiments to evaluate lesquerella germplasm, planting times, and fertilizer rates, indicated that lesquerella production under Virginia conditions may not be feasible, largely due to lack of adequate germination/stand establishment and biomass production.

Based on these negative observations, it was decided, in 1994, to initiate research with castor. The objective of this research was to determine feasibility of castor production under Virginia conditions and to evaluate castor germplasm for agronomic and oil characteristics and to lay the foundation for a castor breeding program.

METHODOLOGY

During 1994, a collection of 72 castor accessions was received from Southern Region Plant Introduction Station, Griffin, Georgia. This collection included 5 accessions from Argentina, 18 from India, 7 from Iran, 2 from Morocco, 4 from Pakistan, 3 from Russia, 20 from South Africa, 6 from Turkey, and one each from Afghanistan, Egypt, Indonesia, Israel, Kenya, Peru, and Zaire. The seed of two U.S. cultivars ('Hale' and 'Lynn') was provided by Dr. R.D. Brigham (Texas A&M Univ., Lubbock, Texas).

Five hills (3 to 5 seeds/hill) of each accession were planted on May 13, 1994 in rows spaced 1.5 m apart. Upon germination, each hill was thinned to one plant/hill. The distance between plants was 60 cm. The seedheads (racemes) were hand-harvested from each plant twice (Sept. 28 and Dec. 6, 1994). No yield data were available for accessions PI250026 and PI250881 due to extremely sever shattering. The capsules were shelled by hand.

Plant height (cm) was determined at final harvest. Seed yield per plant was converted to tonnes/ha. Oil content in the open-pollinated seeds and percentage of ricinoleic acid in the oil were determined by standard chemical techniques. The ricin was extracted from castor seeds using the method of Lis and Sharon (1972). The agglutinating ability of ricin, determined by using the quantitative method described by Turner and Liener (1975), was used to calculate the ricin content in mg/g of meal.

RESULTS

All accessions were vigorous (Table 1) and produced abundant seed. Plant height varied from 64 to 242 cm, and 100-seed weight varied from 10 to 44 g. The oil content (dry weight basis) varied from 22% to 41% whereas the content of ricinoleic fatty acid in the oil varied from 58% to 92%. Significant variation existed among 51 accessions for the poisonous compound, ricin. The ricin content varied from 2.9 to 5.5 mg/g of meal except for PI250226 which contained 10.8 mg/g of meal.

Table 1. Performance of 72 accessions and two castor cultivars at Petersburg, Virginia during 1994.

Entry	Country	Yield (t/ha)	Height (cm)	100-seed wt(g)	Oil (%)	RAcid[z] (%)	Ricin (mg/g)
PI257457	S. Africa	2.6	191	27.6	40.9	61.4	4.0
PI248950	India	2.2	240	22.4	26.0	88.1	-.-
PI257462	S. Africa	2.0	140	33.9	35.3	88.3	-.-
PI221049	Argentina	1.9	138	29.0	28.0	88.1	-.-
PI257461	S. Africa	1.9	181	38.5	35.0	88.4	3.3
PI250226	Pakistan	1.8	231	30.3	30.8	89.1	10.8
PI257455	S. Africa	1.8	161	33.2	39.2	74.5	3.6
PI257458	S. Africa	1.7	196	28.1	35.4	87.7	3.3
PI192949	Kenya	1.7	171	33.1	39.1	87.8	3.9
PI257459	S. Africa	1.6	142	31.9	35.4	86.3	3.5
PI274773	S. Africa	1.6	128	31.5	30.8	85.2	3.6
PI221471	Afghanistan	1.5	167	43.5	32.3	87.1	4.4
PI170682	Turkey	1.5	175	27.2	36.2	87.8	3.8
PI257653	Russia	1.5	149	34.3	24.2	86.2	3.9
PI167288	Turkey	1.4	224	31.9	41.3	77.8	4.1
PI257456	S. Africa	1.4	181	30.9	35.0	86.4	3.4
PI257451	S. Africa	1.4	173	30.4	27.6	87.1	3.8
PI252001	Turkey	1.4	157	36.8	33.8	88.3	3.5
PI167287	Turkey	1.4	178	26.8	35.4	88.2	4.2
PI221698	Indonesia	1.4	121	28.8	28.2	87.2	-.-
Hale[y]	USA	1.4	82	31.8	38.2	88.1	-.-
PI250623	Pakistan	1.4	199	29.5	34.8	87.6	3.8
PI257445	S. Africa	1.3	155	33.8	30.5	58.5	3.4
PI248945	India	1.3	223	29.0	26.8	90.0	-.-
PI221048	Argentina	1.2	117	29.0	33.0	88.1	3.8
PI257446	S. Africa	1.2	154	31.9	32.0	88.2	3.7
PI248947	India	1.2	220	23.8	27.7	87.5	-.-
PI179029	Turkey	1.2	153	37.8	31.0	87.7	4.2
PI257454	S. Africa	1.2	170	39.8	35.0	84.5	4.0
PI219771	Argentina	1.1	154	23.4	39.1	88.4	-.-
PI257453	S. Africa	1.1	180	38.8	30.5	77.3	4.1
Lynn[y]	USA	1.0	64	27.9	39.2	85.3	-.-
PI229786	Iran	1.0	177	29.1	31.6	86.6	3.9
PI170684	Turkey	1.0	186	22.1	34.1	89.1	4.3
PI215775	Peru	0.9	139	29.4	34.1	85.7	-.-
PI254406	India	0.9	165	38.8	26.5	92.3	2.9
PI257460	S. Africa	0.9	163	29.5	41.2	89.9	-.-
PI248501	S. Africa	0.9	199	29.4	28.7	88.6	-.-
PI257452	S. Africa	0.9	161	31.0	28.6	69.6	4.7
PI254405	India	0.9	136	40.6	30.0	88.6	3.8
PI248393	India	0.9	149	29.8	22.3	80.7	4.3
PI257447	S. Africa	0.9	175	28.1	31.1	88.6	-.-
PI253620	Morocco	0.9	192	30.0	31.5	87.9	-.-

Table 1. continued

Entry	Country	Yield (t/ha)	Height (cm)	100-seed wt(g)	Oil (%)	RAcid[z] (%)	Ricin (mg/g)
PI250398	Pakistan	0.9	170	23.2	32.1	88.3	5.5
PI250574	Egypt	0.9	171	22.2	36.2	87.4	-.-
PI257449	S. Africa	0.8	171	32.0	30.9	85.3	3.7
PI248967	India	0.8	202	26.7	30.8	89.5	4.6
PI221046	Argentina	0.8	128	29.0	35.4	85.8	3.5
PI250397	Pakistan	0.8	190	24.2	26.4	88.8	-.-
PI182987	India	0.8	175	22.7	36.6	87.7	-.-
PI248424	Zaire	0.7	131	27.0	28.4	85.8	4.6
PI253422	Israel	0.7	109	23.0	34.4	88.4	4.8
PI257654	Russia	0.7	141	26.2	33.4	88.8	3.7
PI202668	India	0.7	174	18.7	30.6	87.1	4.1
PI248939	India	0.7	153	34.8	25.7	88.4	3.5
PI202667	India	0.7	183	23.5	40.1	88.0	5.7
PI257652	Russia	0.7	135	25.2	32.5	86.4	4.3
PI248970	India	0.6	204	23.2	23.9	82.5	5.5
PI257444	S. Africa	0.6	181	27.4	28.3	88.4	3.8
PI248391	India	0.6	162	43.8	25.1	70.7	3.5
PI248941	India	0.5	202	26.1	25.8	87.7	4.1
PI248946	India	0.6	230	33.9	26.8	89.0	3.6
PI248390	India	0.5	153	39.3	28.1	89.2	-.-
PI253621	Morocco	0.5	103	24.8	30.6	89.0	4.4
PI248943	India	0.5	242	28.9	31.1	85.1	3.8
PI219776	Argentina	0.3	108	30.4	35.7	87.5	-.-
PI250027	Iran	0.3	195	21.4	31.1	89.1	-.-
PI248392	India	0.2	230	37.4	22.9	88.7	3.4
PI247095	Iran	0.2	142	20.6	29.1	88.1	4.6
PI250942	Iran	0.1	162	16.5	35.4	89.4	4.6
PI248500	S. Africa	0.1	158	22.1	28.8	85.3	5.2
PI245105	Iran	0.1	130	16.8	25.7	88.6	-.-
PI250026	Iran	-.-	176	9.6	24.4	88.6	-.-
PI250881	Iran	-.-	167	26.7	34.4	72.0	-.-
Mean		1.0	167	29.1	31.5	85.7	4.2
LSD (.05)		1.0	57	12.2	9.0	13.9	2.2

[z]RAcid = Ricinoleic fatty acid
[y]U.S. cultivar

The mean seed yield (calculated by using single plant yield) of 70 accessions was 1 t/ha with a range from 0.1 to 2.6 t/ha. 'Hale' yielded 1.4 and 'Lynn' yielded 1 t/ha. The highest yielding accession, PI257457, from South Africa, had a seed yield of 2.6 t/ha.

Significant positive correlation existed between yield and 100-seed weight and yield and oil content (Table 2). The content of ricinoleic acid had a negative correlation with yield. A significant negative correlation (–0.30) existed between seed weight and ricin content. The existence of variation for ricin and the lack of

Table 2. Correlation among various characteristics of castor grown at Petersburg, Virginia during 1994.

	Correlation coefficients (r)				
Character	100-seed weight	Plant height	Oil content	Ricinoleic acid content	Ricin content
Yield	0.35**	0.09	0.33**	–0.19	–0.02
100-seed weight		–0.03	–0.02	–0.20	–0.30*
Plant height			–0.10	0.00	0.22
Oil content				–0.11	–0.01
Ricinoleic acid content					0.10

*, ** Significant at 5% (*) and 1% (**) levels.

a significant relationship with yield, oil content and ricinoleic acid indicated that a breeding program could be successfully undertaken to reduce the ricin content, thus, reducing or eliminating the problems associated with this toxic compound.

Our results indicate that commercial castor production in Virginia is feasible. Furthermore, the results suggest that a high potential of success if breeding efforts are undertaken to develop improved cultivars of castor combing high yields of seed with high oil content and quality.

REFERENCES

Brigham, R.D. 1989. Castor. Genetics, breeding, and production of *Ricinus communis* L. Notes from presentation on castor at USDA Water Resources Laboratory, Phoenix, AZ, March 9, 1989. Texas A&M University, Lubbock, TX.

CasChem. 1982. Urethanes, castor oil, chemical derivatives. Tech. Bull. 100. CasChem, Inc., Bayonne, NJ.

Horton, J. and M.A. Williams. 1989. A cooker-extruder for deallergenation of castor bean meal. J. Am. Oil Chem. Soc. 66:227–231.

Lis H. and N. Sharon. 1972. Soybean (*Glycine max*) agglutin. Methods Enzymol 288:360–365

Naughton, F.C. 1979. Castor oil. In: Kirk-Othmer (ed.), Encyclopedia of chemical technology, Vol. 5, 3rd ed. Wiley, New York.

Roetheli, J.C., K.D. Carlson, A.E. Thompson, L.K. Glasser, M.G. Blase, and J. Goodell. 1991a. Lesquerella as a source of hydroxy fatty acids for industrial products. USDA\CSREES, Office of Agricultural Materials. Washington, DC.

Roetheli, J.C., L.K. Glasser, and R.D. Brigham. 1991b. Castor: Assessing the feasibility of U.S. Production. USDA\CSREES, Office of Agricultural Materials. Washington, DC.

Turner, R.H. and I.E. Liener. 1975. The use of glutaldehyde treated erythrocytes for assaying the agglutinating activity of lectins. Anal. Biochem. 68:651–653.

Zimmerman, L.H., M.D. Miller, and P.F. Knowles. 1958. Castorbeans in California. Calif. Agr. Expt. Sta., Davis, Cir. 468.

Lesquerella and Vernonia: Oilseeds for Arid Lands

David A. Dierig, Terry A. Coffelt, Francis S. Nakayama, and Anson E. Thompson

Two oilseed crops, lesquerella and vernonia, with excellent potential for commercialization are presently being developed in the arid Southwest. *Lesquerella fendleri* (Gray) S. Wats., Brassicaceae, is a source of hydroxy fatty acids, for use in lubricants, plastics, protective coatings, surfactants, and cosmetics. In addition to the seed-oil, the seed-coat gums and seed-meal would also add considerable value. Lesquerella has progressed from small plot research to successful cultivation on larger farm areas. *Vernonia galamensis* is a source of epoxy fatty acids, for use in formulations of protective coatings and paints with low volatile organic compounds (VOC). This is an important consideration in order for industry to comply with environmental laws that limit VOC emissions. Other uses for vernonia include thermoset resins and coatings, polymer blends, dibasic acids, adhesives, and epoxy composite materials. Breeding and cultural management improvements have been made on these crops and both are suited for production in arid climates. The successful adoption of these new oilseeds can provide high economic return to growers, processors, and industrial users compared to the expenditure for the necessary agricultural research.

LESQUERELLA

Classification and Domestication

All species of *Lesquerella* are native to North and South America with 83 of 95 known species indigenous to North America (Rollins 1993). Many are concentrated in the southwestern United States and Mexico. Interest in developing lesquerella began in the 1960s when three new hydroxy fatty acids (HFA), lesquerolic, densipolic, and auricolic, were discovered in *Lesquerella* species (Smith et al. 1961; Gentry and Barclay 1962; Mikolajczak et al. 1962; Kleiman et al. 1972). Most species occurring in the western U.S. contain lesquerolic acid (C20:1-OH) as the primary fatty acid in its seed-oil (Barclay et al. 1962; Dierig et al. 1996). This HFA is very similar to ricinoleic acid, the primary HFA of castor. Most species of the eastern U.S. contain densipolic acid (C18:2-OH) as their primary seed-oil component (Barclay et al. 1962). Only two species are known to contain auricolic acid (C20:2-OH) (Hayes et al. 1995). A closely related genus, *Physaria*, also contains significant quantities of lesquerolic acid. All three HFA have many potential markets (Roetheli et al. 1991). Commercialization interests are presently focused on *L. fendleri* because of its growth and yield characteristics (Thompson et al. 1989).

Biology and Adaptation

L. fendleri is a perennial with densely pubescent leaves. The leaf trichomes, fused half or more in length, and the lack of pubescence on the capsule are used as a distinguishing characteristic from other *Lesquerella* species (Rollins and Shaw 1973). It is found throughout the southwestern U.S. including Arizona, Colorado, New Mexico, Texas, Utah, and northern Mexico. Many of these native populations are found on limestone outcroppings. The chromosome number is $2n = 12$, although populations of $2n = 14$ and $2n = 24$ have been identified (Rollins and Shaw 1973). Seeds are contained in capsules or pods, referred to as siliques, that have a globose shape. Each capsule normally contains between 6 and 25 seeds. Seed weight varies between 0.5 and 1.2 g/1000.

In the field, bees and other insects are observed frequently visiting the flowers. We have estimated outcrossing rates in this species to be between 86% and 89%. Self-incompatibility occurs and is assumed to be a sporophytic multiple allele system as generally found in the Brassicaceae. Self-incompatibility limits production of selfed seed. Male sterility also occurs frequently and is characterized by varying degrees of reduced (vestigial) anthers without pollen production. The genetic inheritance of these traits is being investigated at our laboratory. Interspecific hybridization is not common among native populations found in the western U.S. and difficult to obtain from controlled greenhouse crossing. We have produced punitive hybrids between

L. fendleri x *L. lindheimeri* and *L. fendleri* x *L. gracilis*. These three species have the same chromosome number. However, based on general affinities, Rollins and Shaw (1973) place *L. fendleri* into a different phylogenetic group than *L. gracilis* and *L. lindheimeri*. We have not determined if the hybrids are fertile.

Germplasm Evaluation and Breeding

Lesquerella fendleri is a very productive species in regards to seed yield. Unselected accessions yielded between 1000 and 1350 kg/ha of seed in replicated field plots compared to yields in a synthetic population of 1880 kg/ha (Thompson and Dierig 1993). A 25% decrease in seed yield was observed and attributed to inbreeding depression in half-sib family progenies. There are no major barriers to domestication of this plant. The goals of our breeding program are relatively simple and deal with increasing various yield related components that remain stable over different environments.

This species possesses a great amount of variability. Payson (1922) referred to *L. fendleri* as a remarkably polymorphic species. Even within populations of native stands, there are large amounts of variability. Originally, germplasm available to our breeding program were from relatively few accessions, and did not represent the geographical range of the species (Thompson and Dierig 1993). In 1993, we initiated the collection and evaluation of as many accessions of this species as possible throughout the U.S., and incorporated useful traits into elite lines.

In the future, other species of *Lesquerella* and/or *Physaria* may also be domesticated as a source of one of the other HFA or another source of lesquerolic acid for other geographic locations where *L. fendleri* is not adapted. Over the past three years, we collected germplasm of these genera throughout the U.S. This includes 216 accessions of 31 different species of *Lesquerella* and three accessions of two species of *Physaria*. Individual numbers of accessions and species, along with the state of their origin, are listed in Table 1. Many of these have not been previously included in the National Plant Germplasm System. Most accessions have only small amounts of seed from the original collection. Over the 1994–96 growing seasons, we increased seed amounts and evaluated the *L. fendleri* accessions by planting them in caged field plots supplied with honey bees to prevent cross-pollination between accessions.

Increases in oil and fatty acid contents have been achieved through recurrent selection populations. In 1993, seed from 20 open-pollinated plants selected for high oil content, high lesquerolic acid content, and a third population selected for both high oil and lesquerolic acid yield were field grown. The oil content in 1993 ranged from 22% to 25% in the high oil population. Lesquerolic acid content in 1993 for the high lesquerolic population ranged from 58% to 64%. The high lesquerolic yield population ranged in an index for lesquerolic acid yield (oil content multiplied by lesquerolic acid content) between 15 to 18. These are the base populations which are recurrently selected each year for the three seed-oil traits. Means for subsequent years are expected to be lower since this is an open pollinated population. The ranges of values in following years should be wider because of new genetic recombinations being produced. Seeds with high values are then selected for the next generation. Means and ranges of values for oil characteristics in 1994 and 1995 populations are shown in Table 2. An unselected population was used both years for comparison.

The comparison of population means between years shows modest increases for oil and lesquerolic acid content. However, differences are greater when comparing check and recurrent populations in the same year. Environmental influences are more prevalent between years. An important result of these populations is the extremes of the ranges of values have increased each year.

Within these populations, we are concentrating on selecting for other desirable characteristics such as autofertility and upright growth habit. Autofertility is important since it would eliminate the expense of supplying bees in the field to improve seed-set. Variability for this trait is present within the available germplasm. Upright growth habits would improve combine harvesting efficiency. Yellow seed coat lines are also being developed with the prospect of use in the cosmetic industry. Oil from seed with normal orange-brown seedcoats has a pigmentation that must be removed for their applications. The yellow seed coats may have less or no pigmentation and could eliminate the extra processing step.

Cultivation

Although *Lesquerella fendleri* is a perennial plant, it is cultivated as a winter annual. Lesquerella does best in production on well drained soils. In the southwestern U.S., planting is done in Oct. by direct seeding with a broadcast planter such as a Brillion used to sow alfalfa. Sowing the seed by aircraft has also worked successfully. The most critical element for a good plant stand is keeping a moist soil surface to allow germination. Emergence occurs 8 to 14 days after planting. Seedlings remain very small until early Feb. when tem-

Table 1. Species of *Lesquerella* and *Physaria* collected from various states from 1993 to 1995.

Species	State	No. populations
L. arizonica	Arizona	3
L. angustifolia	Oklahoma	3
L. argyraea	Texas	16
L. auriculata	Oklahoma	1
L. cinerea	Arizona	5
L. densiflora	Texas	2
L. densipilia	Tennesse	2
L. douglasii	Washington	7
L. fendleri	Arizona	23
L. fendleri	New Mexico	22
L. fendleri	Texas	41
L. gracilis var. *nutt.*	Oklahoma	4
L. globosa	Tennesse	2
L. gordonii	Arizona	7
L. gordonii	New Mexico	2
L. gordonii	Oklahoma	9
L. grandiflora	Texas	6
L. intermedia	Arizona	4
L. kaibabensis	Arizona	1
L. lasiocarpa	Texas	7
L. lescurii	Tennesse	2
L. lindheimeri	Texas	4
L. lyrata	Alabama	2
L. mcvaughiana	Texas	1
L. perforata	Tennesse	2
L. pinetorum	Arizona	1
L. purpurea	Arizona	2
L. ovalifolia ssp. *alba*	Oklahoma	3
L. ovalifolia ssp. *oval*	Oklahoma	6
L. ovalifolia ssp. *oval*	Arizona	10
L. ovalifolia ssp. *oval*	Texas	7
L. recurvata	Texas	7
L. sessilis	Texas	2
L. stonensis	Tennesse	3
L. wardii	Arizona	2
P. floribunda	New Mexico	1
P. newberryii	Arizona	2

Table 2. Means and ranges from three recurrent populations of lesquerella harvested in 1994 and 1995. The respective populations include high oil content, high lesquerolic acid content, and high lesquerolic acid yield.

		Mean (and range)		
Population	Year	Oil (%)	Lesquerolic acid (%)	Lesquerolic yield (%)
High oil population				
	1994[z]	26.7 (18.0–32.0)		
	1995[y]	26.5 (14.0–36.7)		
High lesquerolic population				
	1994[x]		53.3 (42.1–59.8)	
	1995		54.5 (42.9–64.2)	
High lesquerolic yield population				
	1994[w]			14.4 (9.7–19.0)
	1995			15.3 (10.8–19.8)

[z]Control population was 22.1% oil with a range of 19.0 to 24.4%
[y]Control population was 23.4% oil with a range of 20.0 to 25.6%
[x]Control population was 53.3% lesquerolic acid with a range of 42.1 to 59%
[w]Control population was 11.8% lesquerolic yield with a range of 7.9 to 15.1%

peratures start to increase. Rapid vegetative development takes place and full soil cover is attained within two months. Seeding rates should be between 6 and 8 kg/ha.

Lesquerella has been successfully cultivated on both raised and flat fields. However, salts may be better managed on raised beds. Flowering begins in Feb. and seeds develop and mature between Mar. and late May. Nitrogen fertilizer applied at rates of 60 to 120 kg/ha resulted in increased dry matter and seed yields (Nelson et al. 1996). A good strategy for irrigation scheduling of lesquerella in southern Arizona is to irrigate about once every 15 to 20 days starting in late Feb. through mid-Apr., then once every 10 days between late Apr. through May.

Irrigation is stopped in mid-May and the plant is allowed to dry until seed moisture has reached about 12% before harvesting. Plants should be harvested from mid to late June to avoid rainstorms. Conventional combines with suitable sieves are used for seed harvesting. Seed losses can be as little as 5% with properly equipped and operating combines. Seed yields of 1800 kg/ha have been obtained in breeding test plots. Large scale field trials have yielded about 900 kg/ha.

Weed control has been a problem with lesquerella production fields. Special use permits have been applied for using Treflan and Goal in Arizona. Permits will be obtained for Fusilade, Balan, and paraquat.

VERNONIA

Classification and Domestication

Vernonia galamensis (Cass.) Less., Asteraceae is an annual native to Africa. Taxa include six subspecies. One subspecies, *galamensis*, is divided into four botanical varieties (Gilbert 1986). Literature sometimes still refers to this species as *V. pauciflora* (Willd.) Less. Gilbert (1986) revised the taxonomic treatment and included three taxa that had not previously been described. Eastern Africa appears to be the center of diversity.

Development of an epoxy fatty acid source from plants began in the late 1950s to mid-1960s when *Vernonia anthelmintica* (L.)Willd., native to India, was seen as a possible candidate. Vernolic acid in this species was discovered and isolated by Gunstone (1954). Excessive seed shattering prevented further development of this species. Collections of *V. galamensis* and subsequent evaluation (Carlson et al. 1981; Thompson et al. 1994b, c) indicated that this species is substantially better because of the quantity and quality of the seed-oil and better seed retention. The plant is native to equatorial Africa. To develop this crop for temperate zones, the short day-length flowering requirement needed to be altered (Dierig and Thompson 1993).

Biology and Adaptation

V. galamensis is an annual, ranging in plant height from 0.2 m to 5.0 m depending on the subspecies and the geographic location (Perdue et al. 1986; Thompson et al. 1994a). Plantings in cultivated field plots in the U.S. range in plant height between 0.2 and 2.0 m. The two centers of diversity are Kenya and northern Tanzania with only one botanical variety occurring outside eastern Africa. *V. galamensis* differs from other annual species of *Vernonia* in leaf form, and /or pappus, and involucre form and size.

The seed head (capitula) contains hermaphroditic, protandrous florets. The range in number of florets per capitula, also representing the number of potential seeds per capitula, is between 50 and 150. The pistil is completely surrounded by an anther sheath, which dehisces as the stigma is emerging through. As the stigma opens and becomes receptive, the under side is covered with pollen by rubbing past this sheath. However, self-incompatibility is prevalent. The chromosome number of all subspecies is $2n = 18$. Subspecies readily hybridize among themselves (Thompson et al. 1994a).

Although the arrangement of the flower structure makes controlled crossing and selfed seed production difficult, we have been able to accomplish crossing with a washing process. Central florets are chosen from the seed head. The anther sheath is gently removed from each floret. The remaining stigmata are washed with distilled water from a water bottle. Stigmata at this stage of development are still in an unforked position. We manually separate the lobes of the stigma approximately three-fourths to the base. Stigmata are then washed again and dried after one to three minutes. They can then be pollinated and bagged.

Since *V. galamensis* is native to equatorial areas, flowering in most of the germplasm is controlled by short day-lengths. As a result, most of the native germplasm cannot be grown in the U.S. for seed production. Plants will remain vegetative during the entire growing season. Plants begin to flower as days become shorter, which is also accompanied by colder temperatures, and senescence occurs. Day-neutral germplasm in one of the accessions, *V. galamensis* ssp. *galamensis* var. *petitiana*, flowers and produces seed during the normal growing season, but lacks other desirable characteristics. Transfer of the day-neutral flowering trait from this variety into var. *ethiopica* and other germplasm resulted in more desirable plant growth characteristics. These hybrids have been evaluated and selections made over the past four years at various locations across the U.S.

Germplasm Evaluation and Breeding

Short-day flowering response of *Vernonia galamensis* was the major barrier to overcome for successful seed production in the U.S. Original collections from Africa were not suited for the continental U.S. because of the flowering day length response. The growing season of vernonia is during the long-days of summer months in the U.S. when plants will not flower. An accession of var. *petitiana* from Kenya was found to be day-neutral. However, these accessions had fewer seeds per flower head, smaller seed weights, and more seed shattering than other germplasm. The day-neutral characteristic was introduced into the working germplasm at ARS-USWCL by hybridizing with short-day flowering plants. Fertile hybrids were retrieved with day-neutral flowering. Flowering response of plants from the two parents and three hybrid lines is listed in Table 3. Daylength was controlled by placing a dark cloth over container grown plants in the greenhouse at Phoenix, Arizona during Aug. and Sept. under natural long-day conditions. The plants under the short-day treatment were kept in the dark for 14 h and then exposed to daylight for the following 10 h. The long-day treatment received 14 h of daylight and 10 h of dark. Plants were six weeks-old when treatments began. The experiment continued until flower-buds were apparent. The parent var. *petitiana* flowered under both treatments, var. *ethiopica* flowered only under short-day treatments, and hybrids flowered under both long and short-days (Table 3).

Petitiana is not strictly day-neutral since flowering increases under short-days compared to long-day treatment. The hybrids followed the same trend since flowering significantly decreased under long-days, and when compared to *petitiana* under this treatment, flowering was less. Our breeding objective now is to increase the quantity of flower heads per plant and maintain other characteristics of the short-day germplasm. These include larger flower-head size, which corresponds to the number of seeds per flower-head, increased seed weight, seed retention, and removing seed dormancy.

In 1994, seeds were planted at five locations in the U.S., and one in Argentina to evaluate oil content, seed weight, and seed yields. Eight hybrid lines and one parent line *petitiana* were compared. The U.S. locations included Phoenix, Arizona; Fort Stockton, Texas; Medford, Oregon; Columbia, Missouri; Petersburg, Virginia; and Salta, Argentina. The results are summarized by line across the six locations in Table 4 and combined lines results from each location in Table 5. Severe white fly and disease pressure at the Phoenix

Table 3. Response of two parent lines and three hybrid lines of *Vernonia galamensis* subspecies *galamensis* to photoperiod.

		Mean number of flower heads per plant			
		14 h day		10 h day	
Line	No. plants	Mean	Range	Mean	Range
Var. *petitiana* (day-neutral parent)	7	28.1	13–64	60.5	38–101
Var. *ethiopica* (short-day parent)	9	0	0	4.9	2–5
Hybrid 21J-7-10	7	6.1	1–13	54.0	26–72
Hybrid 5E-10-19	7	15.2	4–41	51.0	15–93
Hybrid 66C-1-9	7	11.7	2–34	55.1	40–72

Table 4. Mean vernonia yield, seed weight, and seed oil content of eight hybrid populations and the AO399 parent from five locations in 1994.

Line	Yield (kg/ha)	Seed wt (g/1000)	Oil (%)
14D-2-5	397	2.76	37.1
15D-10-12	435	3.01	38.9
29E-OR2-14	575	2.96	40.2
35A-2-9	434	2.74	38.3
35A-2-10	395	2.69	36.3
48A-10	376	2.68	38.6
66C-1-9	504	3.05	37.7
72A-1-2	432	2.65	36.5
AO399	496	2.62	38.6
LSD	115	0.13	1.4

Table 5. Mean vernonia seed yield, seed weight, and seed oil content of five locations in 1994.

Location	Yield (kg/ha)	Seed wt (g/1000)	Oil (%)
Virginia	820	2.82	38.1
Missouri	325	2.76	37.2
Texas	303	2.55	37.8
Oregon	645	2.53	35.2
Argentina	155	3.31	41.7
LSD	86	0.10	1.0

location resulted in only flowering data being collected at this location. Results showed that location affected all traits with Virginia and Missouri higher in yield, and Argentina higher in 1000 seed-weight and total oil content. Line 29E-OR2-14 and 66C-1-9 had higher yields; 15D-10-12, 29E-OR2-14 and 66C-1-9 had larger seed-weights; and 29E-OR2-14 had higher oil content (Table 4). Significant location × entry interactions for number of plants flowering at 81 days after planting, yield and total oil content were present indicating variability in stability for these traits among the populations. Lower yields, higher oil contents, and 1000 seed weights in Argentina may be due to the harvest method of picking ripe seed instead of combining plants.

Cultivation

Little information is available on cultural management of vernonia since only small research plots have been grown in the U.S. Larger production fields have been grown in Zimbabwe, and other equatorial countries where plants are not limited by long day growing season. Seed yields from 1345 kg/ha in 1985 to 2494 kg/ha in 1987 has been reported by Perdue (pers. commun.).

A planting density study at Maricopa, Arizona was made in 1994 with hybrid lines by varying plant spacing within rows, which were 1 m apart. Plant populations were established at 15,000, 30,000, and 60,000 plants/ha using spacings of 15, 30, and 60 cm, respectively in a randomized complete block design with four replications. Although plants flowered earlier in the 0.60 m spacing, these plants appeared to be the poorest performers toward the end of the season. As plants mature, the stems become brittle. There is an advantage to space plants closer to provide support for other plants. The closer spacing may also force flowering on the top and outside canopy, achieving better plant architecture for harvesting.

CONCLUSION

Significant improvements in lesquerella have been made. Oil and lesquerolic acid contents have increased in recurrent selection populations. These lines should be available for public release in the near future. New accessions of *L. fendleri* are available for evaluation and breeding. Other species of *Lesquerella* occurring in the U.S. which have been collected also have commercial potential. These new accessions with the available passport data are being entered into the National Plant Germplasm System. Many of these are becoming increasingly rare and are candidates for endangered species lists.

Hybrid lines of *Vernonia galamensis* have been identified that will flower under both short and long-days. Further improvements are still necessary for commercialization. Backcross populations have successfully increased the number and size of flower-heads per plant. Comparison of F_1 plants to hybrids from later generations for flower head and seed size indicates that some inbreeding depression occurs. High yielding lines adapted to various locations should soon be available.

REFERENCES

Barclay, A.S., H.S. Gentry, and Q. Jones. 1962. The search for new industrial crops II. *Lesquerella* (Cruciferae) as a source of new oilseeds. Econ. Bot. 16:95–100.

Carlson, K.D., W.J. Schneider, S.P. Chang, and L.H. Princen. 1981. *Vernonia galamensis* seed oil: A new source for epoxy coatings. p 297–318. In: E.H. Pryde, L.H. Princen, and K.D. Mukherjee (eds.), New sources of fats and oils. Am. Oil Chem. Soc., Champaign, IL,

Dierig, D.A. and A.E. Thompson. 1993. *Vernonia* and *Lesquerella* potential for commercialization. p 362–367. In: J. Janick and J.E. Simon (eds.), New crops. Wiley, New York.

Dierig, D.A., A.E. Thompson, J.P. Rebman, R. Kleiman, and B.S. Phillips. 1996. Collections and evaluation of new *Lesquerella* and *Physaria* germplasm. Ind. Crops Prod. 5:53–63.

Gentry, H.S. and A.S. Barclay. 1962. The search for new industrial crops III: Prospectus of *Lesquerella fendleri*. Econ. Bot. 16:206–211.

Gilbert, M.G. 1986. Notes on East African *Vernonieae* (Compositae). A revision on the *Vernonia galamensis* complex. Kew Bul. 41:19–35.

Gunstone, F.D. 1954. Fatty acids, Part II. The nature of the oxygenated acids present in *Vernonia anthelmintica*

(Willd.) seed oil. J. Chem. Soc. (London) May:1611–1616.
Hayes, D.C., R. Kleiman, and B.S. Phillips. 1995. The triglyceride composition, structure, and presence of estolides in the oils of *Lesquerella* and related species. J. Am. Oil Chem. Soc. 72:559–569.
Kleiman, R., G.F. Spencer, F.R. Earle, H.J. Nieschlag, and A.S. Barclay. 1972. Tetra-acid triglycerides containing a new hydroxy eicosadienoyl moiety in *Lesquerella auriculata* seed oil. Lipids 7:660–665.
Mikolajczak, K.L., F.R. Earle, and I.A. Wolff. 1962. Search for new industrial oils. IV. Seed oils of the genus *Lesquerella*. J. Am. Oil Chem. Soc. 39:78–80.
Nelson, J.M., D.A. Dierig, and F.S. Nakayama. 1996. Planting date and nitrogen effects on lesquerella production. Ind. Crops Prod. (In press).
Payson, E.B. 1922. Monograph of the genus *Lesquerella*. Ann. Missouri Bot. Garden 8:103–236.
Perdue, R.E., K.D. Carlson, and M.G. Gilbert. 1986. *Vernonia galamensis* potential new crop source of epoxy acid. Econ. Bot. 40:54–68.
Roetheli, J.C., K.D. Carlson, R. Kleiman, A.E. Thompson, D.A. Dierig, L.K. Glaser, M.G. Blase, and J. Goodell. 1991. *Lesquerella* as a source of hydroxy fatty acids for industrial products. USDA-CSRS Office of Agricultural Materials Series (unnumbered). Washington DC.
Rollins, R.C. 1993. The Cruciferae of continental North America, Stanford Univ. Press, Stanford, CA.
Rollins, R.C. and E.A. Shaw. 1973. The genus *Lesquerella* (Cruciferae) in North America. Harvard Univ. Press. Cambridge, MA.
Smith, C.R, T.L Wilson, T.K. Miwa, H. Zobel, R.L. Lobmar, and I.A. Wolff. 1961. Lesquerolic acid. A new hydroxy acid from *Lesquerella* seed oil. J. Org. Chem. 26:2903–2905.
Thompson, A.E. and D.A., Dierig. 1993. Initial selection and breeding of *Lesquerella fendleri*, a new industrial oilseed. Indust. Crops Prod. 2:97–106.
Thompson, A.E., D.A. Dierig, and E.R. Johnson. 1989. Yield potential of *Lesquerella fendleri* (Gray) Wats., a new desert plant resource for hydroxy fatty acids. J. Arid Environ. 16:331–336.
Thompson, A.E., D.A. Dierig, E.R. Johnson, G.H. Dahlquist, and R. Kleiman. 1994a. Germplasm development of *Vernonia galamensis* as a new industrial oilseed crop. Ind. Crops Prod. 3:185–200.
Thompson, A.E., D.A. Dierig, and R. Kleiman. 1994b. Characterization of *Vernonia galamensis* germplasm for seed oil content, fatty acid composition, seed weight, and chromosome number. Ind. Crops Prod. 2:299–305.
Thompson, A.E., D.A. Dierig, and R. Kleiman. 1994c. Variation in *Vernonia galamensis* flowering characteristics, seed oil, and vernolic acid contents. Ind. Crops Prod. 3:175–183.

High Value-Added Applications from Vernolic Acid*

F. Petrus Cuperus and Johannes T.P. Derksen

In 1994 the total world production of vegetable oils amounted to approximately 68 million tonnes, of which 80% originated from five crops: soybean, oil palm, rapeseed, sunflower, and coconut. Although the exact composition of the oil from these crops can vary, the number of different fatty acid species available from these crops is rather limited. For food uses this is not a major problem, however, for non-food applications the limited range of oil and fat feed-stocks available for the production of fatty acids and derivatives can present a severe constraint. Therefore there is an increasing industrial interest in the development of new crops, which are optimized for specific applications. The new crops of interest include those that contain a higher percentage of a desirable fatty acid or a lower percentage of undesirable fatty acids and those that contain unique fatty acids. Such unusual fatty acids could on the one hand replace raw materials from petrochemical origins with renewable resources, and on the other hand expand the existing range of raw materials available and potentially lead to novel end products. Moreover, consumer products made from renewable

*The authors wish to thank Drs. G. Kramer, R. Stienstra, and A. Mekkering for their contribution to this study.

resources may also carry an appealling environment-friendly or "green" label.

Currently, many oils are converted through more or less extensive oleochemical processing to value-added products, such as lubricants, polymer additives or surfactants. In most cases these products are based on fatty acids, their methyl esters or fatty alcohols as intermediates, because the reactive headgroup of these compounds is a ready candidate for chemical derivatization.

The limited availability of fatty acids that are functionalized (e.g., oxygenated) in the fatty acyl chain is especially noteworthy in this respect. In fact, the to date only commercially available, naturally occurring functionalized fatty acid is ricinoleic acid, which is recovered from castor bean oil. However, because of the chemical versatility of such functionalized fatty acids, researchers around the world are studying new oilseed species that may add to the oleochemists' palette of raw materials. In particular, oilseed crops that can supply new, hydroxylated (*Lesquerella* spp., *Dimorphotheca* spp.) or epoxidated fatty acids (*Vernonia* spp, *Euphorbia lagascae*) are of interest. However, the transformation of a new oilseed crop from an experimental plant species to an industrial commodity would be greatly facilitated if during the introductory phase a commercially viable specialty or niche application could be found in which agronomical (e.g., yield, harvesting techniques) and economic issues (e.g., available volume, raw materials costs) are less paramount than in bulk applications.

Therefore, as part of the mission of the Agrotechnological Research Institute (ATO-DLO) in The Netherlands to increase the added value of agricultural products in the post-harvest phase, preferably in close cooperations with industries that are knowledgeable in the field of specific applications, industrial scale processing and marketing, we have started to explore the oleochemistry and potential use of two unusual, functionalized fatty acids, i.e. vernolic and dimorphecolic acid, in a variety of high value products. In this paper attention will be focussed mainly on vernolic acid (12,13-epoxy, 9c-octadecenoic acid), which contains an epoxy group in combination with a *cis* double bond and which makes this fatty acid an interesting building block for specialty products.

METHODOLOGY

Starting from locally grown *Euphorbia lagascae* seeds, oil was recovered by cold pressing with a Komet single screw oil expeller, yielding a 94% recovery of good quality oil, as described before (Derksen et al. 1993). From this oil vernolic acid by enzymatic hydrolysis of the oil, using *Candida rugosa* lipase, followed by purification of the resulting vernolic acid by wiped-film vacuum distillation of the oily phase. The purity of vernolic acid recovered was approximately 95%.

The resulting vernolic acid was subsequently used to explore several synthesis routes on a lab scale towards new fine chemical products. High value substances synthesized include bombykol, bombykal, traumatinic acid, and lactones. Synthesis routes employed are indicated in the results section.

RESULTS

Vernolic acid is a natural epoxy fatty acid that is present in new oil seed crops like *Euphorbia lagascae* and *Vernonia* spp. Their seed oils consist over 60% vernolic acid allowing a relatively easy recovery of this versatile starting material for the synthesis of a wide range of (fine) chemicals. We have explored several synthesis routes for the production of selected specialty chemicals in which the vernolic acid epoxy group serves as a key functionality. A few examples are described below.

Bombykol

Bombykol, 10(E),12(Z)-hexadienol, has been identified as a sex pheromone of the silkworm *Bombyx mori* and can be used in pest control. It has been synthesized on a gram scale according to the route indicated in Fig. 1. Starting from vernolic acid (**1**), first the epoxy group was ring-opened, resulting in a diol product (**2**). This diol could be selectively cleaved to the aldehyde (**3**) without affecting the carbon-carbon double bond and leaving hexenal (**4**) as a side-product, that in itself may find useful applications in the flavor industry. Protecting the carboxylic acid group and isomerization of the double bond yielded a reactive intermediary product [12-oxo-10(E)-dodecenoic acid (**5**)] which contained a carboxylic ester and also an alde-

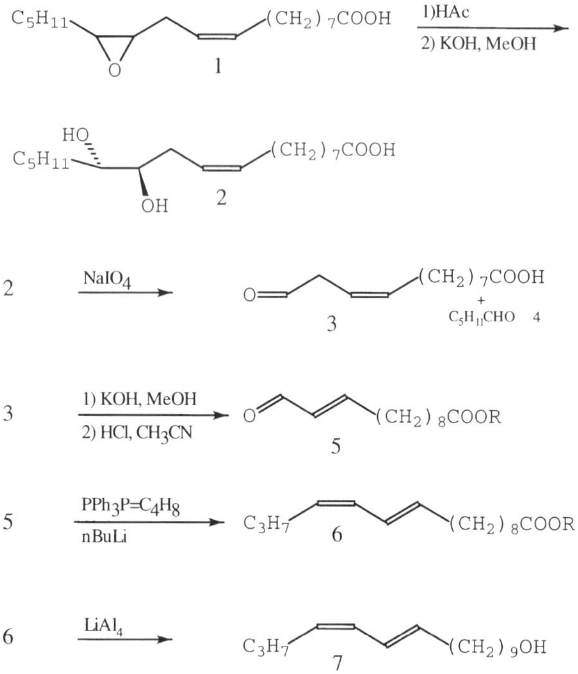

Fig. 1. Synthesis route from vernolic acid to bombykol.

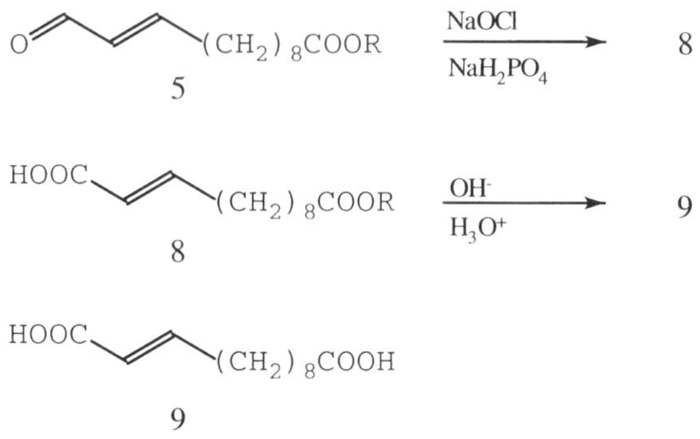

Fig. 2. Synthesis of traumatic acid from the vernolic acid-derived intermediate 12-oxo-10-E-dodecenoic acid.

hyde functionality in conjugation with the carbon-carbon double bond. Subsequent Wittig olefination of (**5**) leads to the 10(E), 12(Z)-product (**6**), which is selectively reduced to bombykol (**7**).

Traumatic Acid

Traumatic acid, (2(E)-dodecenoic acid, is active as wound hormone of plants and can be used as an intermediate in prostaglandine synthesis. The first part of the synthesis (Fig. 2) is identical to the synthesis of bombykol. Mild oxidation of the reactive aldehyde 12-oxo-10(E)-dodecenoic acid (**5**) leads to the formation of a carboxylic acid functionality while leaving the carbon-carbon double bond unaffected [product (**8**)]. Cleavage of the ester part of (**8**) yields traumatic acid (**9**) a base for certain pharmaceuticals.

CONCLUSIONS

The synthesis described above, using functionalized fatty acids as a renewable starting material, provide good alternatives for existing syntheses (Dasardhi et al. 1991) and can reduce the amount of hazardous chemicals required. On-going research will extend the range of products derived from vernolic acid far beyond the products mentioned here, leading already in the near future, to an increased interest in natural starting materials in general and specifically in unusual fatty acids for the chemical syntheses of, particularly, fine chemicals.

The successful introduction of a new agricultural raw material must start with the identification of initial "niche" markets of high added-value products. When a raw material–product combination is established, further development of the source material and processing technology is warranted for further market expansion, including towards large volume–low added-value markets. Ideally, this will then result in a firm foothold for the new crop in question.

REFERENCES

Dasardhi, P., P. Neelakantan, S.J. Rao, and U.T. Bhalerao. 1991. The oxidation of bombykol to bombykal. Synth. Comm. 21:183.

Derksen, J.T.P., B.G. Muuse, F.P. Cuperus, and W.M.J. van Gelder. 1993. New seed oils for oleochemical industry: evaluation and enzyme-bioreactor mediated processing. Ind. Crops Prod. 1:133–139.

Improvement of *Camelina sativa*, an Underexploited Oilseed

Johann Vollmann, Astrid Damboeck, Anna Eckl, Heinrich Schrems, and Peter Ruckenbauer

Camelina [*Camelina sativa* (L.) Crtz., *Brassicaceae*], known as false flax or gold-of-pleasure is a spring-planted crop species. Although camelina has been cultivated in Europe since the Bronze Age (Schultze-Motel 1979), it is an underexploited oilseed crop at present. Recent interest in the species is mainly due to the demand for alternative low-input oilseed crops with the potential for a non-food utilization of the seed oil (Seehuber 1984; Putnam et al. 1993). Camelina oil has a unique fatty acid pattern and is characterized by a linolenic acid (C18:3) content of 30% to 40% and an eicosenic acid (C20:1) content of around 15%, with less than 4% erucic acid (Seehuber 1984; Marquard and Kuhlmann 1986; Budin et al. 1995), which suggests a utilization of the seed oil as a drying oil with environmentally safe painting and coating applications similarly to linseed oil (Luehs and Friedt, 1993). Moreover, a rather low content of glucosinolates has been found in camelina as compared to other brassicaceous species (Lange et al. 1995), which makes the utilization of meals easier.

Agronomic investigations of camelina have been carried out both in Europe (Zimmermann and Kuechler 1961; Marquard and Kuhlmann 1986) and North America (Plessers et al. 1962; Robinson 1987). Unique agronomic features such as the compatibility with reduced tillage and cover crops, competitiveness with weeds or winter surface seeding have been emphasized by Putnam et al. (1993) with regard to the suitability of camelina for sustainable agriculture systems. Furthermore, the species has a potential as a low-cost crop for green-manuring. Breeding research and genetic improvement of camelina have been initiated in Germany during the 1980s; apart from the collection and evaluation of germplasm (Seehuber 1984), agronomically advanced lines have been developed (Seehuber et al. 1987), which are available as base population for further improvement (Seehuber and Dambroth 1987).

In the present investigation, results from an agronomic evaluation of new lines of camelina are reported. The genotypes tested were derived from a recombination program, which mainly focused on the improvement of agronomic performance of camelina as an alternative and low-input oilseed crop for Central Europe.

MATERIALS AND METHODS

Genetic Entries

Camelina populations CA13X and CA2X were derived from crosses BGRC 51558 x BGRC 51572 and Voronezskij 339 x BGRC 51558, respectively. BGRC genotypes were obtained from the genetic resources collection of the FAL Institute of Agronomy (Braunschweig, Germany), Voronezskij 339 is an accession obtained from the N.I. Vavilov Institute (St. Petersburg, Russia). Segregating populations were advanced to the F_4-generation by a bulk breeding method. In the F_4, single plants were harvested separately and F_4-derived lines were selected visually from F_5-single row plots for further testing.

Performance Trials

Homozygous breeding lines from the populations described above were evaluated in 1993 ($F_{4\text{-}6}$) and 1994 ($F_{4\text{-}7}$) at the two Austrian locations Gross Enzersdorf (10 km east of Vienna, 48°12' N, 16°32' E) and Reichersberg (60 km west of Linz, 48°20' N, 13°20' E) in replicated field trials. Cv. Iwan (Saatbau Linz, Reichersberg, Austria) was used as a check cultivar. Plot size was 5 × 1.25 m in Gross Enzersdorf and 8 × 1.25 m in Reichersberg. Agronomic parameters were set as described by Seehuber et al. (1987). All experiments were arranged as lattice designs in two replications. As particular genotypes were discarded after the 1993 experiments due to the necessity of selection, a complete set of data was available for 10 genotypes from both locations and for both years for a combined analysis of variance. Another set of 32 genotypes was analyzed across 3 environments (Gross Enzersdorf 1993, 1994; Reichersberg 1994). In general, adjusted entry means from the lattice analysis were used for the combined analyses of variance; in the experiment at Gross Enzersdorf 1993, seed yield data were adjusted by a neighbor analysis procedure (Vollmann et al. 1996), which was most efficient in controlling the spatial variation present in the particular trial field. In order to identify genotypes superior to the check cultivar, LSI (= least significant increase, one-tailed t-statistic) was used as a statistical testing procedure (Petersen 1994). Oil content of seed was determined non-destructively by near-infrared reflectance spectroscopy (NIRS) and is reported in g kg^{-1} on a dry weight basis.

Selection for Large Seed Size

For an improvement of seed size (mass), sub-populations of crosses CA2X and CA13X were established by sieving out small fractions (< 0.1 %) of largest seeds from the F_2- (CA2X-1S, CA13X-1S) or from both the F_2- and F_3-bulks (CA2X-2S, CA13X-2S). F_4-derived lines of these sub-populations were evaluated for 1000-seed weight in the F_5-generation using single row plots with two replications (Gross Enzersdorf 1992) and a selected number of large-seeded genotypes was tested for agronomic performance (Gross Enzersdorf 1993) as described above.

RESULTS

Environmental Effects

In the set of 10 genotypes of camelina evaluated in two locations for two years, both seed yield and oil content were highly influenced by year and location effects as well as by a highly significant year × location interaction. Genotype effects were also significant, but genotype × year and genotype × location interactions were of minor importance for seed yield and not significant for oil content indicating a similar response of the different genotypes to environmental changes. Seed yields were in the range of 1050 to 1700 kg ha^{-1} in 1993 and from 1450 up to 3250 kg ha^{-1} in 1994. In both locations, oil contents were high in 1993 (400 to 455 g kg^{-1}) and clearly lower in 1994 (385 to 425 g kg^{-1}), as shown in Fig. 1.

Genotype Performance

Genetic variation in different agronomic characters was highly significant (F-test) in the set of 32 genotypes tested across 3 environments (Gross Enzersdorf 1993, 1994; Reichersberg 1994). The check cultivar was significantly outyielded in both seed and oil yield by several of the new genotypes evaluated (Table 1), an improved oil content of seed was however found in only one genotype. In the populations investigated, seed yield and oil content were positively correlated (Fig. 2), whereas a tightly negative correlation was found be-

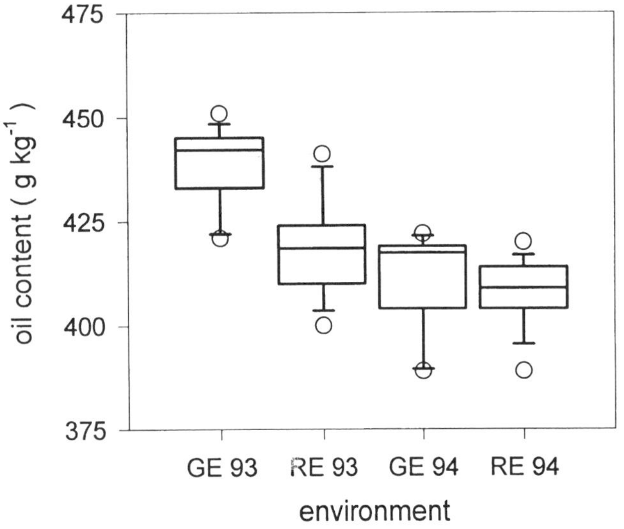

Fig. 1. Oil content of a set of 10 genotypes of camelina as affected by environmental conditions (locations: GE = Gross Enzersdorf, RE = Reichersberg; years: 1993, 1994).

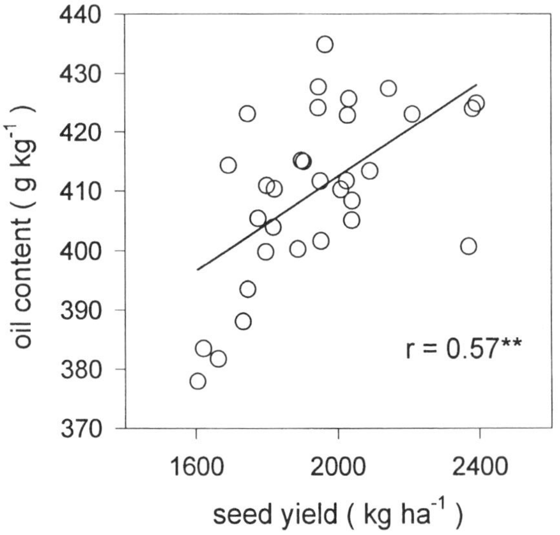

Fig. 2. Relationship between seed yield and oil content in 32 genotypes of camelina (genotype mean values across 3 environments).

tween 1000-seed weight and oil content (Fig. 3); a negative correlation was also detected between 1000-seed weight and yield. For the genotypes listed in Table 1, estimates of heritability were rather high in characters such as oil content or 1000-seed weight and lower for plant height (Table 2).

In particular performance trials (Gross Enzersdorf 1993), spatial variation in the trial field affected oil content (apart from the influence on seed yield) in the range of –8 to +7 g kg^{-1} seed oil. A clear trend in variation of oil content was found using a neighbor analysis covariate (Fig. 4), which was also helpful in order to adjust individual plot values for field heterogeneity during analysis of variance (results not shown).

Table 1. Agronomic performance of 32 genotypes of camelina (mean values across 3 environments).

No.	Genotype name	Seed yield (kg ha^{-1})	Oil content (g kg^{-1})	Oil yield (kg ha^{-1})	1000-seed weight (g)	Plant height (cm)
1	CA13X-11	2091	413.3	858	1.26	72.5
2	CA13X-17	2381	424.0	1002	1.21	70.0
3	CA13X-20	2032	425.7	861	1.16	64.2
4	CA13X-6	2143	427.4	911	1.16	71.7
5	CA13X-1S-9	2392	424.9	1011	1.13	79.2
6	CA13X-2S-23	1947	427.7	828	1.21	69.2
7	CA13X-2S-44	2210	423.1	927	1.19	75.0
8	CA13X-2S-96	2370	400.7	940	1.34	75.8
9	CA2X-2S-29	2040	405.1	820	1.28	70.8
10	Iwan (check)	2029	422.9	853	1.29	71.7
11	CA13X-13	1965	434.9	847	1.20	72.5
12	CA13X-1S-19	1951	411.6	797	1.28	78.3
13	CA13X-2S-17	1799	410.9	735	1.35	70.8
14	CA13X-2S-20	1818	404.0	731	1.43	76.7
15	CA13X-2S-22	1746	393.5	677	1.41	70.8
16	CA13X-2S-26	1904	415.0	783	1.24	74.2
17	CA13X-2S-29	2041	408.4	824	1.32	75.0
18	CA13X-2S-4	1692	414.3	698	1.31	70.0
19	CA13X-2S-53	1945	424.2	822	1.26	70.8
20	CA13X-2S-56	1746	423.1	733	1.22	67.5
21	CA13X-2S-6	1775	405.4	715	1.42	75.0
22	CA13X-2S-63	1898	415.2	786	1.40	71.7
23	CA13X-2S-69	1953	401.7	782	1.47	73.3
24	CA13X-2S-7	2025	411.7	828	1.34	73.3
25	CA13X-2S-75	2009	410.3	819	1.31	74.2
26	CA13X-2S-85	1821	410.4	747	1.33	70.8
27	CA13X-2S-95	1797	399.8	714	1.36	70.8
28	CA2X-1S-1	1622	383.5	617	1.68	72.5
29	CA2X-1S-14	1888	400.3	769	1.47	67.4
30	CA2X-2S-17	1663	381.8	633	1.62	72.5
31	CA2X-2S-20	1605	378.0	604	1.66	70.2
32	CA2X-2S-22	1733	388.1	661	1.57	72.5
	Total mean	1939	410.0	792	1.34	72.0
	$LSD_{0.05}$	215	8.8	88	0.06	5.0
	$LSI_{0.05}$	183	7.5	75	0.05	4.5

Selection for Seed Size

In population CA2X, seed size of F_{4-5}-lines was clearly improved by sieving of F_2- or F_3-bulks and lines with a 1000-seed weight of up to 2 g were identified within sub-population CA2X-2S (Fig. 5), whereas 1000-seed weights of the respective parent genotypes were only 0.9 and 1.2 g. In population CA13X, selection response was less pronounced with respect to seed weight. In a preliminary assessment of agronomic performance of large-seeded genotypes (96 F_{4-6}-lines, Gross Enzersdorf 1993), seed yield was positively affected by 1000-seed weights of up to 1.5 g, whereas seed yield was reduced in genotypes with larger seed size (Fig. 6). Moreover, seed size and oil content were negatively correlated, and large-seeded genotypes had a reduced number of seeds per pod, malformation of pods as well as higher lodging scores (data not presented).

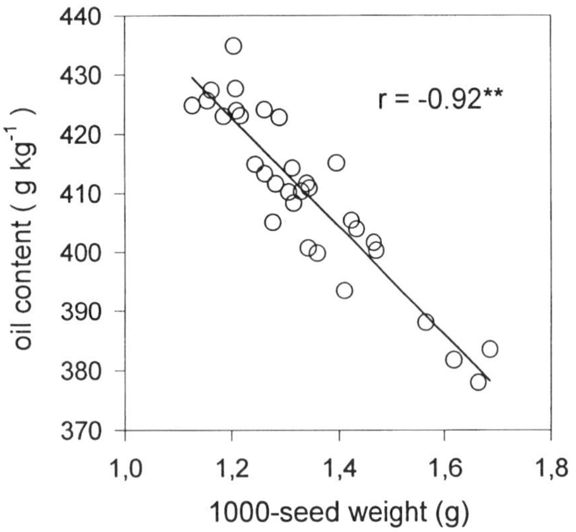

Fig. 3. Relationship between 1000-seed weight and oil content in 32 genotypes of camelina (genotype mean values across 3 environments).

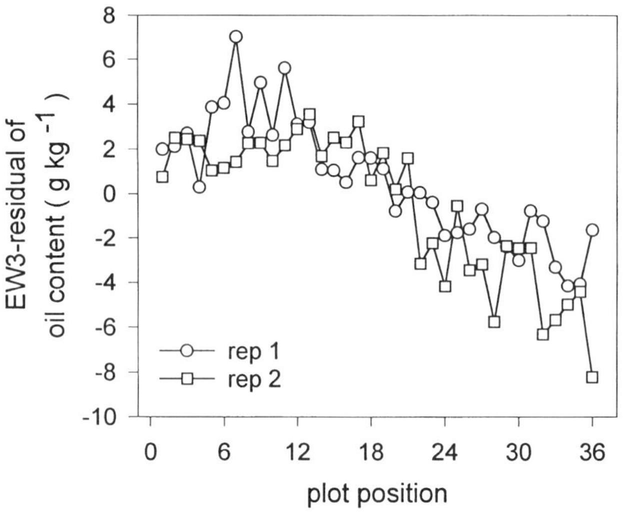

Fig. 4. Trends of oil content due to spatial variation in the trial field (Gross Enzersdorf 1993) as visualized by a neighbour analysis covariate (residual EW3).

Table 2. Estimates of heritability calculated from components of variance for different agronomic characters of camelina (32 genotypes tested in 3 environments).

Character	Heritability (%)
Seed yield	86.5
Oil content	95.6
Oil yield	90.5
1000-seed weight	97.6
Plant height	62.5

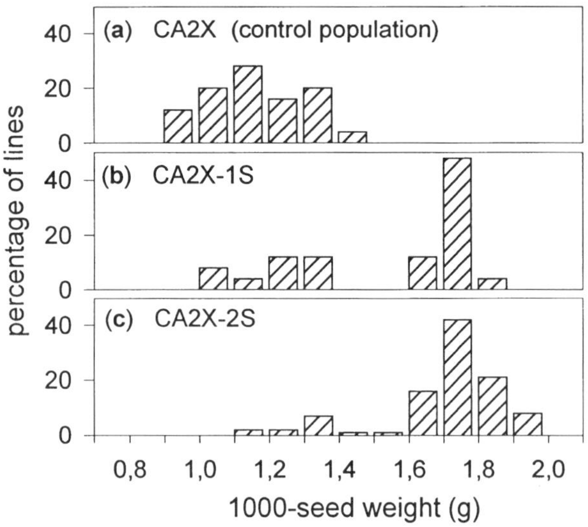

Fig. 5. Distributions of seed size (1000-seed weight) in F_{4-5} lines of population CA2X. a: unselected control population (n = 25 lines); b: F_2-bulk sieved for large-seededness (n = 25); c: both F_2- and F_3-bulks sieved (n = 100).

DISCUSSION

A considerable agronomic potential has been detected in newly developed lines of camelina during the present study. Furthermore, strong year × location interactions for seed yield and oil content indicated that specific weather conditions within a location could modify performance considerably. In the 1993 experiments, a severe drought during the flowering phase limited plant development and yield potential at both locations, whereas sufficient rainfall during the seed filling period resulted in the subsequent expression of high oil contents (Fig. 1). In the 1994 experiments, water conditions were not limiting plant development throughout the growing season, which allowed seed yields of up to 3250 kg ha^{-1}, whereas oil contents were lower as a response to the high yield potential. Similar ranges of seed yield and oil content have also been reported from experiments in Germany (Seehuber 1984; Seehuber et al. 1987), whereas both seed yield and oil content were lower in North American studies (Plessers et al. 1962; Robinson 1987; Putnam et al. 1993), which might partly be due to a lack of environmental adaptation. The strong influence of environmental conditions on seed yield has also been reported for other spring-sown oilseed crops (Diepenbrock et al. 1995).

Sufficient genetic variation was present within the germplasm available, which allowed the selection of lines with improved agronomic features (Tables 1 and 2). A broad genetic variation and moderately high heritabilities of important agronomic characters have also been identified by Seehuber et al. (1987) investigating similar genetic materials. The positive correlation between seed yield and oil content indicates that a simultaneous improvement of both characters is possible. An improvement of seed oil content would be of particular interest to make camelina more competitive to other oilseed crops exhibiting higher oil contents (Plessers et al. 1962; Putnam et al. 1993). Screening of large numbers of entries for oil content can be accomplished easily by non-destructive techniques such as near-infrared reflectance spectroscopy or nuclear magnetic resonance (Thies and McGregor 1989); the presence of spatial field variations affecting oil content (Fig. 4) can however reduce the efficiency of selection if not controlled properly (Ball et al. 1993; Vollmann et al. 1996).

In camelina, improvement of seed size would be of interest for a rapid field emergence and crop establishment under less favorable growing conditions; moreover, solvent extraction of oil is generally more efficient with large-seeded cultivars. A considerable increase of 1000-seed weight of up to 2 g has been achieved during the present selection experiment, as compared to the unselected control population (Fig. 5) as well as to variation in seed size reported previously (Seehuber 1984; Seehuber et al. 1987; Putnam et al. 1993). However, the obtained improvement of 1000-seed weight seems to be of low immediate value due to the drastic reduction of both oil content and seed yield, which makes further cycles of recombination necessary. Within smaller ranges of seed size, a significant correlation betweeen seed size and oil content was not found in

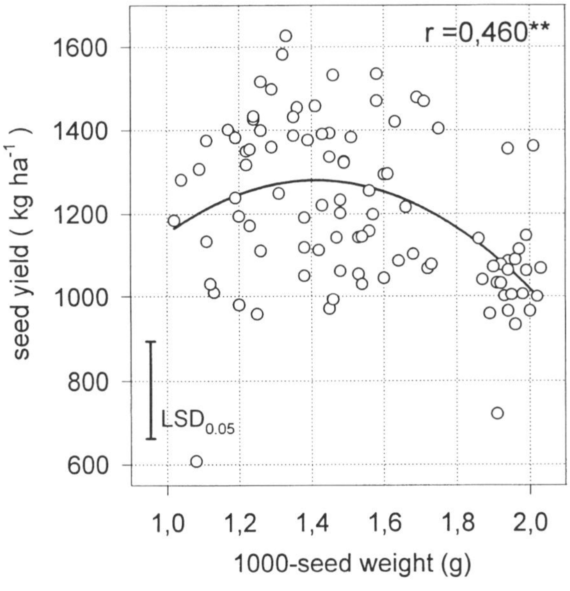

Fig. 6. Relationship between 1000-seed weight and seed yield in a set of 96 genotypes selected for large seed size (environment: Gross Enzersdorf 1993; regression line and coefficient of correlation according to 2nd order regressional function).

camelina (Seehuber 1984), whereas positive correlations between 1000-seed weight and oil content have been reported for particular crosses in oilseed rape (Engqvist and Becker 1993).

The results obtained from the present study suggest that a considerable agronomic potential is present in newly developed lines of camelina. However, high oil content and a specific fatty acid profile would be the key requirements to re-establish camelina as an industrial oilseed crop. These quality characters should therefore deserve most attention in future breeding programs.

REFERENCES

Ball, S.T., D.J. Mulla, and C.F. Konzak. 1993. Spatial heterogeneity affects variety trial interpretation. Crop Sci. 33:931–935.

Budin, J.T., W.M. Breene, and D.H. Putnam. 1995. Some compositional properties of camelina (*Camelina sativa* L. Crantz) seeds and oils. J. Am. Oil Chem. Soc. 72:309–315.

Diepenbrock, W.A., J. Leon, and K. Clasen. 1995. Yielding ability and yield stability of linseed in Central Europe. Agron. J. 87:84–88.

Engqvist, G.M. and H.C. Becker. 1993. Correlation studies for agronomic characters in segregating families of spring oilseed rape (*Brassica napus*). Hereditas 118:211–216.

Lange, R., W. Schumann, M. Petrzika, H. Busch, and R. Marquard. 1995. Glucosinolates in linseed dodder. Fat Sci. Technol. 97:146–152.

Luehs, W. and W. Friedt. 1993. Non-food uses of vegetable oils and fatty acids. p. 73–130. In: D.J. Murphy (ed.), Designer oil crops, breeding, processing and biotechnology. VCH Verlagsgesellschaft, Weinheim, Germany.

Marquard, R. and H. Kuhlmann. 1986. Investigations of productive capacity and seed quality of linseed dodder (*Camelina sativa* Crtz.). Fette-Seifen-Anstrichmittel 88:245–249.

Petersen, R.G. 1994. Agricultural field experiments, design and analysis. Marcel Dekker, New York.

Plessers, A.G., W.G. McGregor, R.B. Carson, and W. Nakoneshny. 1962. Species trials with oilseed plants, II. Camelina. Can. J. Plant Sci. 42:452–459.

Putnam, D.H., J.T. Budin, L.A. Field, and W.M. Breene. 1993. Camelina: a promising low-input oilseed. p. 314–322. In: J. Janick and J.E. Simon (eds.), New crops. Wiley, New York.

Robinson, R.G. 1987. Camelina: A useful research crop and a potential oilseed crop. Minnesota Agr. Expt. Sta., Univ. Minnesota. Bul. 579.

Schultze-Motel, J. 1979. Die Anbaugeschichte des Leindotters, *Camelina sativa* (L.) Crantz. Archaeo-Physika 8:267–281.

Seehuber, R. 1984. Genotypic variation for yield- and quality-traits in poppy and false flax. Fette-Seifen-Anstrichmittel 86:177–180.

Seehuber, R. and M. Dambroth. 1987. Development of basic populations of plant species suitable for the production of fatty acids, especially considering linseed, false flax and poppy. Landbauforsch. Voelkenrode (Germany) 37:219–223.

Seehuber, R., J. Vollmann, and M. Dambroth. 1987. Application of the single-seed-descent method in false flax to increase the yield level. Landbauforsch. Voelkenrode (Germany) 37:132–136.

Thies, W. and D.I. McGregor. 1989. Analytical methods for the selection of oil content and fatty acid composition. p. 132–164. In: G. Roebbelen, R.K. Downey, and A. Ashri (eds.), Oil crops of the world, their breeding and utilization. McGraw–Hill, New York.

Vollmann, J., H. Buerstmayr, and P. Ruckenbauer. 1996. Efficient control of spatial variation in yield trials using neighbour plot residuals. Exp. Agr. 32:185–197.

Zimmermann, H.-G. and M. Kuechler. 1961. Die Ertraege von Leindotter und Oellein und Untersuchungen ueber den Einfluss der Saatstaerke auf den Anbauerfolg bei einer Landsorte und Zuchtstaemmen des Leindotters [*Camelina sativa* (L.) Cr.]. Albrecht-Thaer-Archiv (Germany) 5:622–636.

Intercropping Stokes Aster: Seedling Growth under a Soybean Canopy*

E.J. Callan and C.W. Kennedy

The perennial Stokes aster (*Stokesia laevis*, Asterallae) has the potential to become a new industrial oilseed crop. The vernolic acid (12,13-epoxy-*cis*-9-octadecenoic acid) contained in its achenes is easily converted to epoxy oil, a material widely used in the chemical industry (Perdue et al. 1986). Stokes aster's favorable agronomic traits, such as an upright growth habit and achenes in bracts that do not shatter, could facilitate its use as a crop (Campbell 1981; Gunn and White 1974). The fact that it is a perennial could be a potential advantage, as the crop would not need to be replanted for up to 5 years (Campbell 1981, 1984). However, Stokes aster will not flower or produce an economic yield during the first year of growth (Campbell 1981).

One management technique to overcome first-year expenses would be to intercrop Stokes aster seedlings with an annual such as soybean. Success with such a system would depend on the ability of Stokes aster plants to tolerate shade. Controlled studies of shading help distinguish the different effects of an intercrop system on the component crops (Newman 1986). By separating out the effect of light on the crops, the performance of the intercrop in the field can be better understood.

Determining appropriate relative planting dates for the intercrop species is also important. Optimum planting dates for the components of an intercrop appear to be highly dependent on the specific intercrop. A roselle (*Hibiscus sabdariffa* L.) and oilseed intercrop performed better when the oilseed was planted later (Roy et al. 1990). However, a rice (*Oryza sativa* L.) and legume (various species) intercrop performed best when legumes were planted early, so the rice was not overly competitive with the legumes (Mandal et al. 1990). A cotton (*Gossypium hirsutum* L.) and mungbean [*Vigna radiata* (L.) R. Wilczek] intercrop was most successful using late maturing cotton cultivars, as early season bean growth caused less damage to these compared to short-season cotton varieties (Sulistyowati 1989).

Our objective was to evaluate the growth of Stokes aster seedlings of different ages during and after exposure to the shade of an intercrop by (1) measuring Stokes aster's growth under a soybean canopy and (2) evaluating Stokes aster's recovery from shade effects after the soybean canopy was removed.

METHODOLOGY

Culture

A greenhouse experiment was conducted at the Louisiana Agricultural Experiment Station in Baton Rouge from Feb. 11 to Aug. 21, 1992 and repeated from Sept. 1, 1992 to Mar. 12, 1993. Germination of achenes of USDA Stokes aster accession BSLE2 was initiated at two different times: 30 days before soybean seed germination was initiated [early planting (EP)] and the same day soybean seed germination was initiated [concurrent planting (CP)]. Seeds were started in germination paper, then transferred to styrofoam seedling trays. Forty-two days after planting seedlings were transferred to 3-liter buckets filled with a 0.2-strength Hoagland's solution (Hoagland and Arnon 1950) changed every 10 days. There were three seedlings per bucket, and aeration was achieved by pumping humidified air through tubing.

Maturity group V 'Delta Pine 105' soybean seeds were inoculated with *Bradyrhizobium japonicum* and germinated in germination paper, transferred to peat press pots then transplanted to 22-liter tubs filled with gravel and sand and covered with a thin layer of potting soil. Plants were spaced at 50-mm intervals in a single row with a total of 9 plants per tub. Two tubs were aligned to produce a soybean row 0.91 m long. Three such rows were spaced to produce 2 interrows with a width of 0.76 m each. Soybeans were supplied with a nutrient solution without NO_3 consisting of 1 mM $MgSO_4$, 1 mM KH_2PO_4, 0.5 mM NH_4SO_4, 1 mM $CaCl_2$, 1 ml/liter stock micronutrients, and 1 mg/liter Fe as iron chelate. This solution was later doubled in

*Approved for publication by the director of the Louisiana Agr. Expt. Sta. as manuscript no. 94-09-8412.

concentration. The nutrient solution was applied by trickle irrigation every 30 minutes and was recycled from a 19-liter bucket, the contents of which were replaced every 2 to 5 days. The 3-liter buckets containing Stokes aster were placed in the interrow spaces between the soybean rows at the time of plant transfer from seedling trays. The height of all containers was about the same.

The soybean canopy closed approximately 50 days after planting, and to simulate conditions within a field row of soybeans, 55% neutral density shade cloth was placed across the ends of the rows to prevent extra diffuse light from reaching the Stokes aster plants. Fifty-two days after canopy closure, soybean plants were removed to simulate the duration of a soybean field canopy (Hicks 1978) and recovery of the shade treated plants was measured. Stokes aster plants were harvested 60 days after the soybean canopy was removed.

Environmental Conditions

To prevent premature soybean flowering in the greenhouse, natural daylength was extended during short days with 2-1000W metal halide lamps. To determine photosynthetic photon flux density (PPFD) above and below the soybean canopy, light measurements were taken with a one meter-long quantum sensor at solar noon. Day and night temperatures varied widely. A maximum of 40°C day and a minimum of 18°C night were recorded.

Data Collection

Leaf number per plant and fresh weight (FW) of plants per pot were determined every 10 days. True leaves longer than 10 mm were counted. Fresh weight was measured rather than dry weight (DW) as an experiment large enough to allow destructive measurements was not within reason. The three plants per pot were blotted dry with paper towels and weighed with the plastic cover in which they were growing. Cover weights were then subtracted and the average weight taken to the nearest 0.01 g. Relative leaf production rate (RLPR) was calculated as the number of leaves produced per day per leaf (Garner 1992), leaf production rate, as leaves per day, and FW relative growth rate (RGR) as the increase in fresh weight per day per unit of fresh weight. Harvest measurements of individual plants included leaf area (measured on a Li Cor LI-3000 portable area meter to the nearest 10 mm^2, total leaf length (measured to the nearest 10 mm), leaf number, shoot dry weight, and root dry weight (weighed after drying at 70°C for 1 week, to the nearest 0.01 g).

Experimental Design and Statistical Analysis

The experiment was a complete block with a split split plot treatment arrangement and four replications. The repeated experiment was the main plot, light condition (shade or sun) was the sub plot and date of planting (EP or CP) was the sub-subplot. Both repetitions of the experiment were combined as there was no significant interaction between experiments. Leaf production rate, RGR, or RLPR were found using a piecewise multiple regression (Neter et al. 1989), the two parts being during canopy cover and after the canopy was removed. Days after canopy closure and treatment and their interaction were independent variables and means of the three plants making up each replication were dependent variables. Significant differences between growth rates were calculated using standard errors (SE). At each sampling time, SE was also computed for graphing purposes. The General Linear Model procedure was used to analyze harvest measurements and means were separated using the LSD procedure ($p < 0.05$).

RESULTS

Stokes aster seedling growth was greatly reduced by dense shade. Average light intensity under the soybean canopy and available sunlight approximately 30 days after canopy closure was 20 and 440 $\mu mol\ m^{-2}\ s^{-1}$ PPFD, respectively. It was apparent Stokes aster plants were only achieving metabolic stasis, at best, in this low light condition (Fig. 1 and 2).

After the canopy was removed, intercropped plants responded with almost a six-fold increase in FW RGR, an eight-fold increase in leaf production rate, and a seven-fold increase in RLPR. During this same time, plants continuously grown in available sunlight underwent a 50% decrease in FW RGR, a 40% decrease in RLPR, and only a 1.25-fold increase in leaf production rate (Table 1). After canopy removal, intercropped

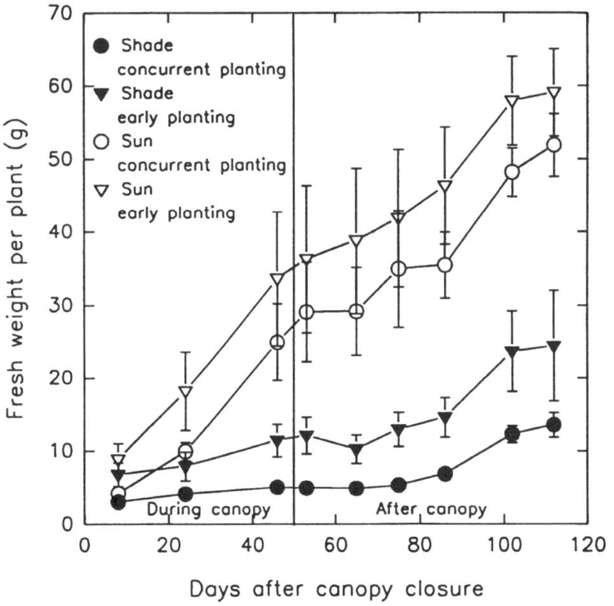

Fig. 1. Fresh weight of Stokes aster plants grown under a soybean canopy (shade) compared to plants grown in available sunlight (sun). Stokes aster seed were initially germinated either 30 days earlier than or concurrently with soybean seed. Measurements were continued after the soybean canopy was removed (after canopy). Vertical bars indicate SE. Average of two experiments.

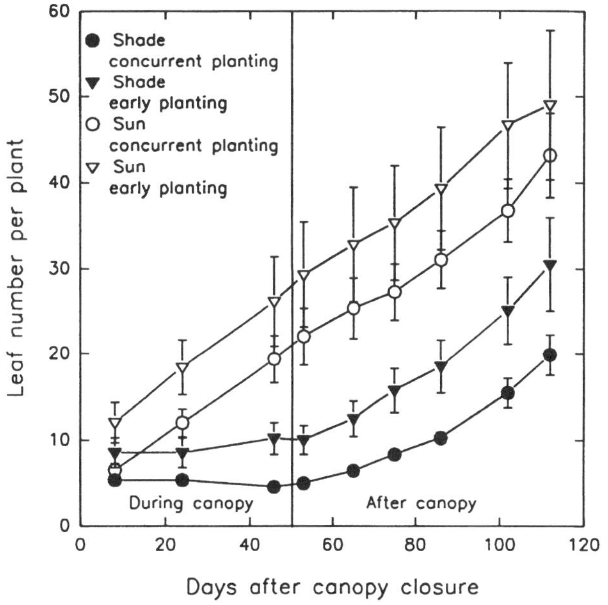

Fig. 2. Leaf number of Stokes aster plants grown under a soybean canopy (shade) compared to plants grown in available sunlight (sun). Stokes aster seed were initially germinated either 30 days earlier than or concurrently with soybean seed. Measurements were continued after the soybean canopy was removed (after canopy). Vertical bars indicate SE. Average of two experiments.

plants had non significant but higher average FW RGR and RLPR than sun plants, but they did not reach the size of sun plants during the time of the experiment. Moreover, the effect of shade on FW may be more long term than its effect on leaf number as FW measurements revealed less recovery than leaf number measurements (Fig. 1 and Table 1). Even after recovery from the canopy, absolute growth parameters for intercropped plants averaged 42% of EP sun plants and 32% of CP sun plants (Table 2).

Introducing intense shade to very young Stokes aster seedlings had only a slightly greater effect than introducing shade to slightly older seedlings. Average FW RGR, leaf production rate, and RLPR of CP and EP shade plants were similar, but average FW and leaf numbers were initially larger for EP plants which gave them a slight numerical advantage during most of the growing season. (Fig. 1 and 2 and Table 1). Altering the planting time in relay intercrops has previously resulted in light interception patterns that benefitted one of the components (Steiner and Snelling 1994; Cenpukdee and Fukai 1992). In this case, however, the growth of Stokes aster seedlings is so slow that a difference of 30 days in planting did very little towards increasing the potential to compete with a fast growing annual species for available sunlight.

In field conditions, it seems Stokes aster would recover at least partially from the shade of a soybean canopy. Plants have been shown to adapt to a change in light intensity over a period of weeks because new organs being initiated and differentiated are adjusted to the ambient light level during their formation (Larcher 1983). We have previously found Stokes aster grown in a light intensity of 120 μmol m^{-2} s^{-1} PPFD to have, when exposed to higher light intensities, photosynthetic rates similar to rates of plants grown at higher light intensities (1010 μmol m^{-2} s^{-1} PPFD) (Callan and Kennedy 1995). This indicated plants grown in low light had excellent capability to rapidly recover when exposed to full sunlight. In this study, however, recovery was not as complete as expected. Perhaps the extremely low level of light available to Stokes aster under the soybean canopy in this study slowed recovery subsequent to overstory removal. Additionally, ambient light intensity in the greenhouse was less than would be expected under field conditions, perhaps also slowing recovery.

Table 1. Fresh weight (FW) relative growth rate (RGR), relative leaf production rate (RLPR), and leaf production rate of plants grown in available sunlight (Sun) compared to plants grown under a soybean canopy (Shade). Average of two experiments.

Treatment	Time period[z]	FW RGR[y] ($g\ g^{-1}\ d^{-1}$)	RLPR[x] ($lf\ lf^{-1}\ d^{-1}$)	Leaf production rate ($lf\ d^{-1}$)
Early planting[w]				
Shade	During canopy	0.005b[v]	0.002bc	–0.029bc
	After canopy	0.013ab	0.019a	0.329ab
Sun	During canopy	0.021ab	0.015ab	0.286abc
	After canopy	0.013ab	0.010ab	0.357a
Concurrent planting[w]				
Shade	During canopy	0.002b	–0.005c	–0.056c
	After canopy	0.018ab	0.023a	0.237abc
Sun	During canopy	0.033a	0.022a	0.276abc
	After canopy	0.013ab	0.012ab	0.348a

[z]Parameters were calculated from replication means during canopy closure and after the soybean canopy was removed.
[y]Fresh weight relative growth rate; grams per gram per day.
[x]Relative leaf production rate; leaves per leaf per day.
[w]Early planting = Stokes aster seed germination began 30 days earlier than soybean seed; concurrent planting = Stokes aster seed and soybean seed germination began on same day.
[v]Mean separation in column by LSD test, 5% level. All values were determined by piecewise regression on data taken at 10 day intervals.

Table 2. Comparison of Stokes aster previously grown in the shade of a soybean canopy to Stokes aster continually grown in available sunlight. Results are from harvest data after the soybean canopy had been removed and plants recovered in available sunlight for 60 days. Average of two experiments.

Treatment	Leaf area (cm^2/plant)	Leaf length (cm/plant)	Leaf number (no./plant)	Leaf dry wt. (g/plant)	Root dry wt. (g/plant)
Previously Shaded					
Early planting[z]	333b[y]	296b	29bc	1.60b	0.83c
Concurrent planting[z]	216b	198b	20c	0.96b	0.44c
Sun					
Early planting	808a	654a	51a	4.29a	2.73a
Concurrent planting	704a	573a	44ab	3.87a	2.06b

[z]Stokes aster seed germination began either 30 days earlier than or concurrently with soybean seed germination.
[y]Mean separation by LSD test, 5% level.

Campbell (1981) suggested a size requirement for successful floral induction of Stokes aster. Though it appears Stokes aster can recover from shade, quantification of this size requirement is needed to determine if Stokes aster's growth after overstory removal would provide for successful floral induction.

CONCLUSIONS

From these experiments, we have found Stokes aster to be adversely affected by dense shade in the range of 20 μmol m^{-2} s^{-1} PPFD. However, when the overstory was removed, Stokes aster plants recovered from shade; FW RGR and RLPR of intercropped plants after overstory removal were slightly greater but statistically equivalent to rates of plants grown in sun.

Introducing intense shade to younger seedlings had a slightly more adverse effect than introducing shade to slightly older seedlings. Planting Stokes aster 30 days earlier than soybeans, if feasible, would marginally improve its growth potential after overstory removal. Under field conditions, however, optimum date of planting is dependant on many environmental conditions, of which light is only one.

Although we did not find Stokes aster to thrive under a very dense soybean canopy, we did find it could survive such a canopy for a period of at least 52 days. In addition, Stokes aster plants recovered well from dense shade. Overall, Stokes aster shows promise as the understory crop of an intercrop depending on the amount and duration of canopy closure of the overstory crop, the duration of growing season after overstory removal, and, marginally, the seeding time of each component.

REFERENCES

Callan, E.J. and Kennedy, C.W. 1995. Intercropping Stokes aster: Effect of shade on photosynthesis and plant morphology. Crop Sci. 35:1110–1115.

Campbell, T.A. 1984. Responses of Stokes aster achenes to chilling. J. Am. Soc. Hort. Sci. 109:736–741.

Campbell, T.A. 1981. Agronomic potential of Stokes aster. In: E.H. Pryde, L.H. Princen, and K.D. Mukherjee (eds.), Am. Oil Chem. Soc. 9:287–295.

Cenpukdee, U. And S. Fukai. 1992. Agronomic modification of competition between cassava and pigeon pea in intercropping. Field Crop Res. 30:131–146.

Garner, E. 1992. Growth analysis of congeneric annual and perennial grass species. J. Ecol. 80:665–675.

Gunn, C.R. and White, G.A. 1974. *Stokesia laevis*: taxonomy and economic value. Econ. Bot. 28:130–135.

Hicks, D.R. 1978. Growth and development. p. 17–41. In: A.G. Norman (ed.), Soybean physiology, agronomy and utilization. Academic Press, New York.

Hoagland, D.R. and Arnon, D. I. 1950. The water culture method for growing plants without soil. California Agr. Expt. Sta. Cir. 347, Berkely, CA.

Larcher, W. 1983. Physiological plant ecology. Springer Verlag, New York.

Mandal, B.K., M.C. Dhara, B.B. Mandal, S.K. Das, and R. Nandy. 1990. Rice, mungbean, soybean, peanut, ricebean and blackgram yields under different intercropping systems. Agron. J. 82:1063–1066.

Neter, J., W. Wasserman, and M.H. Kutner. 1989. Applied linear regression models, 2nd ed. Richard D. Irwin, Inc., Boston, MA.

Newman, S.M. 1986. A pear and vegetable interculture system: land equivalent ratio, light use efficiency and productivity. Expt. Agr. 22:385–392.

Perdue, R.E. Jr., K.D. Carlson, and M.G. Gilbert. 1986. *Vernonia galamensis*, potential new crop source of epoxy acid. Econ. Bot. 40:54–68.

Roy, A.R., B.C. Sasmal, and A.K. Bhattacharjee. 1990. Effects of intercropping oilseeds and pulses in roselle (*Hibiscus sabdariffa*). Expt. Agr. 26:407–411.

Steiner, J.J. and J.P. Snelling. 1994. Kura clover seed production when intercropped with wheat. Crop. Sci. 34:1330–1335.

Sulistyowati, H.E. 1989. Performance of cotton varieties under intercropping with mungbean. Indonesian J. Crop Sci. 4:15–25.

Matthiola incana: Source of Omega-3-Linolenic Acid*

Zohara Yaniv, D. Schafferman, M. Zur, and I. Shamir

Current research in nutritional medicine indicates that the omega-3 fatty acids are receiving more and more attention as essential components of the human diet (Adam et al. 1986; Simopoulos 1986). The observed low incidence of arteriosclerosis and chronic inflammatory heart disease in Greenland Eskimos has been attributed to their traditional ethnic diet, consisting in great part of marine foods rich in two omega-3 fatty acids: C22:6 and C20:5.

Dietary fish oils containing omega-3 fatty acids are increasingly recommended for their antithrombic and hypolipidemic effects to persons consuming typical western diets (Phillipson et al. 1985). Additional benefits include improving immunologic function and fighting allergies (Leaf and Weber 1988). Omega-3 fatty acids from vegetable oils could provide health benefits without any concomitant intake of cholesterol (Hunter 1990).

Based on preliminary evaluations (Yaniv et al. 1991) we have chosen to investigate germplasm of *Matthiola incana* (Brassicaceae) with an average oil content of 20%–24% in the seeds and maximum levels of 65% omega-3-linolenic acid (Ω3LA) of the total fatty acids in the oil (Ecker et al. 1992). This is one of the highest known content of Ω3LA in a plant species (Table 1). A second important aspect of this vegetable oil is its quality as drying oil, due to the high content of Ω3LA. Various *Matthiola* lines were tested and evaluated as a potential new-oil crop for dietary supplement and industrial uses.

METHODOLOGY

Cultivation

Five lines of *M. incana* selected from a collection of commercial lines for flower production, having a high content of Ω3LA were tested. Plants were grown in the experiment stations at Bet Dagan, Ramat haGolan, and Jerusalem, Israel, representing three different climatic regions, during the 1991/92 growing season. Each line was replicated four times in a random block design. Seeds of each line were sown in small plots of 1.2 m^2 consisting of four rows with 30 cm between rows. Basic fertilization was done at the time of soil preparation, at rates of 2N–2P–1K. 'Trifluralin' (2.5 kg/ha) was applied as a herbicide. The plots were irrigated until seedling establishment. Mean monthly temperatures at the three sites during the growing period (Nov. 1991 to June 1992) are presented in Fig. 1.

Table 1. Comparison of the fatty acid composition of five seed lipids.

	Fatty acid composition (%)				
	Palmitic 16:0	Stearic 18:0	Oleic 18:1	Linoleic 18:2	Linolenic Ω318:3
Canola oil	6.0	1.0	62.0	20.0	11.0
Cocoa butter	26.4	31.0	37.7	3.8	T
Safflower oil	7.0	2.3	12.8	78.0	T
Soybean oil	15.0	T	24.0	54.0	7.0
Matthiola oil	9.0	2.0	13.0	11.0	65.0

*Contribution No. 1521-E, 1995 series from the Agricultural Research Organization, The Volcani Center, Bet Dagan, Israel.

Observations were made on plant height, percent of fertile plants, and seed yield (number of pods/plant, number of seeds/pod, and weight of 1000 seeds). Plants were harvested at each location according to the time of full maturity. Fully mature seeds from each line were oven-dried overnight at 50°C and analyzed for oil content and fatty acid composition (Yaniv et al. 1991).

Lipid Extraction

Seeds were dried overnight at 50°C and ground into powder in a Moulinex coffee grinder; 5 g of powder was mixed with 100 cc petroleum ether (40° to 60°C), and the lipid fraction was extracted in a Soxhlet apparatus for 16 h at 60°C. The solvent was evaporated, and the lipid fraction residues were weighed (Yaniv et al. 1991).

Direct Transesterification from Seeds

Seeds (200 mg) were dried overnight at 50°C and ground into powder with a mortar and pestle, after which 0.3 cm^3 of dichloromethane and 2.0 cm^3 of 0.5 N sodium methoxide (MeONa) were added. The tube was heated to 50°C and shaken for 30 min. The reaction was stopped by adding 5.0 cm^3 of water containing 0.1 cm^3 of glacial acetic acid. The esterified fatty acids were extracted with 2.0 cm^3 of petroleum ether (40° to 60°C). The clear fraction was kept at –20°C until further analysis. Samples of 2.0 mm^3 were injected into the gas chromatograph for fatty acid analysis.

Gas Chromatography of Methylated Fatty Acids

A Megabore column (DB-23, 0.5 mm film thickness, 30 m × 0.54 mm, J&W Scientific) was used in a gas chromatograph equipped with a flame ionization detector (Varian model 3700 GC) and an automatic area integrator (3390A HP). The flow rate of N_2 (carrier gas) was 30 cc/min and the oven temperature was 135° to 200°C, programmed at a rate of 4°C min. The following fatty acids were identified using known standards (Supelco): C16:0, palmitic; C18:0, stearic; C18:1, oleic; C18:2, linoleic; C18:3, linolenic; C20:1, eicosenoic; and C22:1, erucic acid.

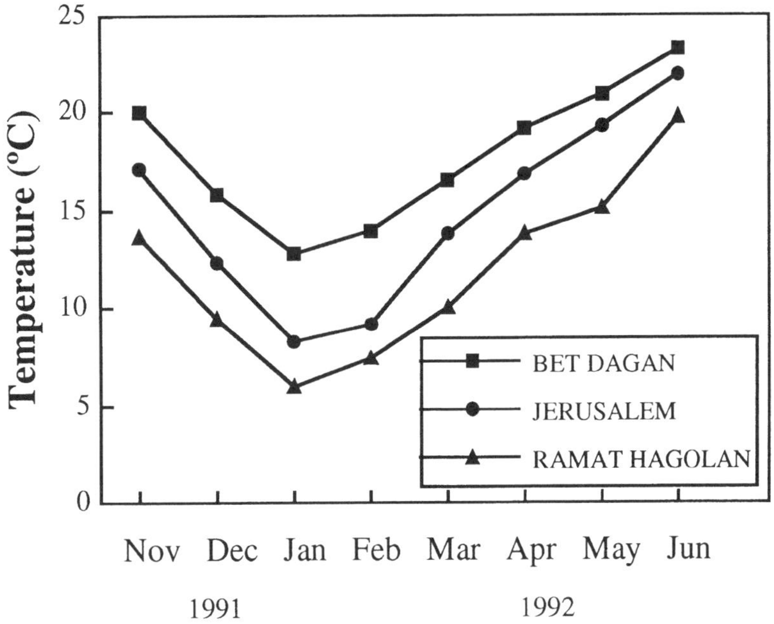

Fig. 1. Mean temperatures (°C) measured at the Bet Dagan, Jerusalem and haGolan Experiment Stations, during the 1991/1992 growing season.

RESULTS AND DISCUSSION

Agronomy

Table 2 summarizes the results of the evaluation of five lines of *M. incana* cultivated in the Bet Dagan (BD), Jerusalem (JM), and Ramat haGolan (RG) experiment stations during autumn 1991 to spring 1992. The best location was BD with all lines showing higher yield potential at that site. Yield per plant was by far the highest at BD (3.5 g vs. 0.9 g at JM and 0.7 g at RG). Number of pods/plant was 73.3 in BD as compared with 36.7 in JM and 26.7 in RG. Number of seeds/pod, plant height, pod length, and 1000 seed weight were also slightly higher at BD than at the two other sites. The best performing lines were ROZ 45 and ROZ 19. At BD, both lines showed the highest yield potential. Under the low temperature conditions prevailing in winter at JM and RG (Fig. 1), ROZ 45 and ROZ 19 maintained the highest yield. ROZ 45 and ROZ 19 had the highest fertility rate (close to 90%) at all sites tested.

Oil Content and Quality

Table 3 summarizes the oil content and fatty acid evaluation of the three test sites. Oil quantity ranged from 21% (V6) to 28%–29% (ROZ 19 and ROZ 46), with the best sites at Jerusalem and Ramat haGolan.

Seeds of ROZ 46 and ROZ 19 accumulated 29% oil in JM and RG, as compared to 25% at BD. Temperatures during seed maturation were lower at JM and RG than at BD (Fig. 1) and it is known that low temperatures during seed development have a positive effect on oil quantity (Canvin 1965; Yaniv et al. 1989). Difference in oil quantity of the three sites is probable due to temperature.

Table 2. Yield parameters (mean of 4 replications) of *Matthiola incana* cultivated at three locations in Israel during the 1991/1992 growing season.

Site[z]	Line	Plant ht. (cm)	Fertility (%)	Pods/plant	Seeds/pod	1000 seeds weight (g)	Yield/plant (g)
BD	V-6	123 d[y]	37 b	68 c	48 b	1.9 ab	1.8 d
	ROZ17	139 b	30 b	87 a	49 b	1.9 ab	3.4 bc
	ROZ19	164 a	90 a	79 b	58 ab	2.0 ab	5.2 a
	ROZ45	140 b	86 a	66 c	63 a	2.4 a	4.5 ab
	ROZ46	132 c	84 a	66 c	55 ab	1.7 b	2.6 c
	Avg	140	66	73	54	2.0	3.5
JM	V-6	82 c	44 c	26 c	43 c	1.6 b	0.5 b
	ROZ17	84 c	49 c	36 b	45 c	1.7 b	0.5 b
	ROZ19	98 a	87 a	38 b	62 b	2.0 ab	1.1 ab
	ROZ45	96 a	93 a	43 a	59 a	2.3 a	1.4 a
	ROZ46	93 ab	68 b	41 a	52 b	1.8 b	0.9 ab
	Avg	91	68	37	50	1.9	0.9
RG	V-6	56 c	40 b	18 c	40 b	1.2 c	0.3 c
	ROZ17	70 b	41 b	26 b	46 ab	1.5 b	0.4 c
	ROZ19	83 a	88 a	31 a	54 a	1.7 b	0.8 b
	ROZ45	82 a	92 a	27 b	52 a	2.2 a	1.4 a
	ROZ46	81 a	81 a	32 a	50 a	1.6 b	0.6 bc
	Avg	74 c	66 a	27 c	48 b	1.6 b	0.7 b

[z]BD = Bet Dagan; JM = Jerusalem; RG = Ramat haGolan

[y]Mean separation in columns by Duncan's multiple range test, 1% level.

Table 3. Fatty acid composition of *Matthiola incana* seeds cultivated at three locations in Israel (4 replications).

Line	Fatty acid composition (%)					
	C16:0	C18:0	C18:1	C18:2	18:3	% oil
			Bet Dagan			
V6	9.7 a[z]	3.0 ab	18.4 b	11.9 b	56.3 a	21.1 c
ROZ17	9.5 a	2.5 c	21.1 a	15.6 a	50.7 c	22.5 b
ROZ19	8.6 b	2.6 bc	18.9 b	13.2 b	56.1 ab	24.7 a
ROZ45	8.8 b	3.1 a	17.7 b	12.8 b	57.1 a	25.3 a
ROZ46	8.9 b	3.1 a	19.3 ab	13.0 b	55.1 b	25.6 a
			Jerusalem			
V6	9.9 a	3.5 a	18.6 a	13.1 b	54.4 c	24.9 c
ROZ17	9.4 ab	2.7 c	16.4 b	14.2 a	56.8 b	25.7 bc
ROZ19	9.1 b	3.0	17.3 ab	14.1 a	55.9 b	28.9 a
ROZ45	9.0 b	3.2 ab	16.0 b	11.5 c	59.8 a	27.5 ab
ROZ46	9.1 b	3.3 ab	17.9 a	13.4 b	55.8 b	29.1 a
			Ramat haGolan			
V6	11.1 a	4.1 a	21.8 a	20.3 a	41.0 c	21.6 b
ROZ17	10.4 b	3.2 b	18.7 b	20.6 a	46.3 b	26.2 a
ROZ19	9.4 c	3.2 b	18.6 b	17.7 ab	50.4 a	28.2 a
ROZ45	9.2 c	3.3 b	18.2 b	14.9 b	53.6 a	26.0 a
ROZ46	9.1 c	3.5 b	20.0 b	15.3 b	51.4 a	28.8 a

[z]Mean separation in columns by Duncan's multiple range test, 0.01% level.

Our main goal is to obtain a high concentration of Ω3LA (C18:3) in the seed oil. The levels obtained ranged from 50% to 60%, with ROZ 45, the highest at all three locations. The best locations were JM and BD. Temperature plays a major role in the relative concentration of unsaturated fatty acids of seed oils (Mazliak 1988), and cooler conditions usually favor the production of polyunsaturated fatty acids (Yaniv et al. 1995). When *Matthiola* were grown under controlled temperature in a Phytotron, the content of seed unsaturated Ω3LA was 69% at 12°C night 17°C day as compared to 58% at 22°/27°C (Yaniv et al. 1992). However, in spite of the fact that the coolest temperatures prevailing during seed maturation were in RG (Fig. 1), the highest content of C18:3, was obtained at JM and BD, and did not go above 60% (Table 3). It is important to note that temperatures never dropped below 15°C during the maturation period (May–June). It could be that in order to induce a significant increase in the level of C18:3, maturation should occur under a much cooler temperature regime than tested by us in the three sites.

Potential Yield

A yield of 750 kg seeds/ha, calculated on the basis of 17 plants/m^2, was obtained at BD. With 20% to 25% oil in the seed, this yield is equivalent to 150 liters oil/ha. A 50% content of Ω3LA in the oil will yield 75 liters/ha pure Ω3LA. Due to the fact that the research is in an early stage, we can't predict the profit per ha of the producte. This will be evaluated in during the second stage of the research.

REFERENCES

Adam, O., G. Wolfmann, and N. Zollner. 1986. Effect of α-linolenic acid in the human diet on linolenic acid metabolism and prostaglandin biosynthesis. J. Lipid Res. 27:421–426.

Canvin, D.T. 1965. The effect of temperature on the oil content and fatty acid composition of the oils from several oil seed crops. Can. J. Bot. 43:63–69.

Ecker, R., Z. Yaniv, M. Zur, and D. Schafferman. 1992. Embryonic heterosis in the linolenic acid content of *Matthiola incana* seed oil. Euphytica 59:93–96.

Hunter, J.E. 1990. N-3 Fatty acids from vegetable oils. Am. J. Clin. Nutr. 51:809–814.

Leaf, A. and P.C. Weber. 1988. Cardiovascular effects of n-3 fatty acids. New England J. Med. 318:549–557.

Mazliak, P. 1988. Environmental effects on fatty acid quality. p. 57–71. In: N.J. Pinfield and A.K. Stobart (eds.), Plant lipids: Targets for manipulation. Britist Plant Growth Regulator Group.

Phillipson, B.E., D.W. Rothrock, W.E. Connor, W.S. Harris, and D.R. Illingworth. 1985. Reduction of plasma lipids, lipoproteins and apoproteins by dietary fish oils in patients with hypertriglyceridemia. New England J. Med. 312:1210–1216,

Simopoulos, A.P. (ed.). 1986. Health aspects of polyunsaturated fatty acids in seafoods. p. 3–29. Academic Press, New York.

Yaniv, Z., R. Ecker, and D. Schafferman. 1992. Earliness and lateness of flowering in *Matthiola incana* and their relationship to oil quantity of mature seeds. Israel. J. Bot. 41:279–284.

Yaniv, Z., Y. Elber, M. Zur, and D. Schafferman. 1991. Differences in fatty acid composition of oils of wild cruciferae seed. Phytochemistry 30:841–843.

Yaniv, Z., C. Ranen, A. Levy, and D. Palevitch. 1989. Effect of temperature on the fatty acid composition and yield of evening primrose (*Oenothera lamarkiana*) seeds. J. Expt. Bot. 40:609–613.

Yaniv, Z., D. Schafferman, and M. Zur. 1995. The effect of temperature on oil quality and yield parameters of high-and low-erucic acid Cruciferae seeds (rape and mustard). J. Ind. Crops Prod. 3:247–251.

Dimorphotheca pluvialis: A New Source of Hydroxy Fatty Acid

Lysbeth Hof

In the search for alternative crops for Dutch agriculture, *Dimorphotheca pluvialis* L. (Mnch), Asteraceae, is being considered as a potential new crop with industrial applications. Its seeds contain oil with 60% to 65% dimorphecolic acid (δ9-hydroxy,t10,t12-octadecadienoic acid): a hydroxy fatty acid with two conjugated double bonds. This feature provides dimorphecolic acid with a unique functionality and properties that are totally different from other known hydroxy fatty acids as ricinoleic and lesquerolic acid. The chemical structure suggests that the molecule should be very reactive, and hence suitable for a wide range of industrial products such as surface coatings, surfactants, plastic foams, or as additive in plastics (Knowles et al. 1965; Muuse et al. 1992). New applications and markets need to be developed, possibly leading to new products with a high added value.

Dimorphotheca pluvialis is a herbaceous annual native to South West Africa (Norlindh 1977). As is common in the Asteraceae, the capitulum (flower head) bears two types of florets. The species *Dimorphotheca* is characterized by hermaphrodite disc florets and female-fertile (male-sterile) ray florets. Both types of florets produce distinctly different types of seeds (achenes). Seeds produced by ray florets are small, angular, while those of the disc florets are flattened and have winged margins (Barclay and Earle 1965). The ray florets have one large white petal, which is often colored purple at the base, giving the appearance of a "ring" in the inflorescence.

The species is well adapted to the maritime climate of Northern and Western Europe, and fits in a rotation system with annual crops. Although it is known as a garden ornamental, *Dimorphotheca* is considered an undomesticated species showing many primitive characteristics. Populations collected from the natural habitat in general have a long, unsynchronized flowering period and show poor seed retention. These factors together account for severe yield losses prior to and during harvest. Realized yields at trial fields range 500–1500 kg/ha, with potential yields of at least 2000–2500 kg/ha. As it is sensitive to frost, in the Netherlands, *Dimorphotheca* is grown as a summer annual. It is sown in April, flowers in July and can be harvested in August (van Dijk et al. 1993). The average oil content of collected populations is 21%, which is too low for mechanical expelling. At present, oil recovery should be done by solvent extraction or preferably with supercritical carbon dioxide extraction (Muuse et al. 1992).

Dimorphotheca was first introduced in the Netherlands by the Dutch Gene Bank (CGN) in 1986, and since 1990 has been studied extensively in the framework of three large multidisciplinary projects (the Dutch National Oilseeds Program, and the EC-projects VOICI and VOSFA). In these projects expertise from the whole production chain was brought together including germplasm collection, evaluation, cultivation, breeding, crop physiology, pathology, harvest techniques, oil recovery, processing of the oil, application, and market research (van Soest and Mulder 1993).

At the Centre for Plant Breeding and Reproduction Research (CPRO-DLO), research in *Dimorphotheca* has been focusing on improvement of synchronization of flowering and seed ripening, oil content, and plant architecture. Furthermore, research is carried out to study pollination and mating system.

EXPERIMENTAL

In order to determine optimal selection strategies for synchronization of flowering and oil content, variation and heritability of these characters were estimated. Flowering synchronization was considered particularly important. Large differences in time of flowering make it difficult to detect slight differences in seed retention, hence making selection for this character at present almost impossible. More synchronized populations are expected to have a shorter flowering and seed ripening period, facilitating the determination of the optimal harvest date. A too early harvest causes reduction of seed yield because of a large proportion of immature seeds. If harvest is carried out too late, yield is severely reduced because of seed losses due to shattering.

For synchronization of flowering two distinct characteristics were distinguished:

1. The synchronization between plants, which was defined as the period of time between sowing and the first flower to open. A population flowers synchronously if all plants start flowering at the same time.
2. The synchronization within plants, which was defined as the period of time in which individual plants produced 90% of their total number of flowers. A plant flowers synchronously if it produces its flowers in a relatively short period of time.

Both the synchronization between plants of a population, as well as the synchronization within plants of that population are considered to have a large effect on the synchronization of flowering of a population.

To establish the variation for synchronization of flowering (both between and within plants) and oil content, two experiments were carried out on loam soil at location Lelystad in the Netherlands. Heritabilities were estimated by means of parent-offspring regression in the following year.

Synchronization Between Plants

In 1992, 350 plants of population 883168 were sown at a wide density (50 × 50 cm), allowing scoring and harvesting of individual plants. Time of flowering was scored every Monday and Thursday, and for further analyses expressed as the number of days from sowing until first open flower. The mean time of flowering was 78 days after sowing (range: 68–90 days). Although most plants started to flower after 75 to 80 days, some started to flower one week earlier, others as much as three weeks later. This means there was a difference of four weeks between the first and last plant to open its first flower.

To estimate the heritability of this character in this population, 40 plants were selected and their seeds collected. *D. pluvialis* is considered a predominantly outcrossing species, and the progenies of the selected plants are considered to be half-sib families. In 1993, a trial field was sown with 24 (three rows of eight) plants of each of the 40 families, and time of flowering was scored.

The relationship between selected plants and the mean of their progenies for time of flowering is presented in Fig. 1. It is clear that selected plants and their corresponding families showed considerable resemblance, this despite difference in weather conditions in 1992 and 1993.

Narrow sense heritability (h^2_n) is described as the ratio between genotypic and phenotypic effects. High heritabilities indicate that the genetic component in the observed phenotype is relatively important to the environmental component. This means that high heritabilities for a character usually result in a quick response to selection. With low heritabilities the response to selection is not necessarily lower, but more time consuming. From the linear regression of offspring (Y) on female parents (X), expressed as Y = a + bX, the (narrow sense) heritability can be estimated by $h^2_n = 2b$ (Falconer 1989). The estimated heritability for time of flowering in this experiment was 0.94, which is very high.

Synchronization Within Plants

For estimation of duration of flowering of individual plants a similar experimental lay out was used. In 1992, a field was sown with 220 plants of an unselected population (879585). Plant spacing was approx. 100 × 100 cm. Twice a week the number of open flowers per plant (NOF) was scored. At the end of the growing season plants were removed from the field and the total number of heads per plant (TNH) was counted. Knowing the total number of heads per plant, a correction could be made for the error caused by counting open flowers twice on two consecutive counting dates. In general, flowers stay open about 4 to 6 days. For each plant a correction term (CT) was estimated, being the ratio between the actual total number of heads (TNH) and the mathematical sum of counted open flowers on the counting dates (MNF). This ratio is 1 if no flowers are counted twice. In this experiment, in most plants the ratio was estimated between 1.0 and 2.0, meaning that 0% to 100% of the flowers were indeed counted twice on two consecutive counting dates. The number of newly opened flowers per counting date (NNOF) was estimated by the product of the number of counted open flowers (NOF) and the estimated correction term (CT): NNOF = NOF × CT, where CT is TNH/MNF.

The cumulative numbers of open flowers per plant plotted against time fitted a logistic curve ($Y = c/[1 + e^{-b(X - m)}]$). In this curve, *c* represents the upper asymptote (being the total number of heads, TNH), *b* the "slope parameter," and *m* the inflexion point of the curve, which is also the date at which the maximum number of open flowers was counted: peak bloom. The flowering of each individual plant was characterized by these three parameters. In this experiment this model on average accounted for 99.6% of the observed variation, indicating that it described the flowering of individual plants well.

Using this model, the period in which the plants produced 90% of their total number of flowers (the duration of flowering) could be calculated. The 90% interval, and not 100%, was chosen because slight deviations from the model occurred at beginning and end of flowering, accounting for relatively large aberrations in estimates of duration of flowering when using the 100% interval.

Duration of flowering ranged from 11 to 63 days, with a mean of 27 days. Since plants did not start flowering at the same time, environmental factors may have had a considerable effect. Therefore, also for this character heritability was estimated by means of parent-offspring regression.

From the population grown in 1992, 20 plants were selected showing much variation for duration of flowering. In 1993, ten plants per progeny were sown in a complete randomized block design, and flower counts were made in the same way as before. The explained variation for the fit of the logistic model on the data was 99.8%.

The relationship between female parents and the mean of their corresponding families is shown in Fig. 2. The calculated regression line explained only 13% of the variation. This means that it leaves 87% of the variation still to be accounted for, and seems drawn rather arbitrarily through a cloud of data points. The estimated heritability (based on the slope of this regression line) of 0.27 can therefore be regarded as unreliable.

Oil Content

The same trial field as described in the first experiment was used to assess the variation and heritability for oil content. Of all plants seeds were harvested and separated in winged and unwinged seeds. Oil content of the winged seeds was measured with Near InfraRed Spectroscopy (NIRS) equipment (InfraAlyzer 500, Bran+Lübbe). Oil content of individual plants ranged from 15% to 29%, with a mean of 21.5%. From this experiment 40 plants were selected representing almost the whole range. Progenies were tested in 1993 (same experimental lay out as in time of flowering experiment). Seeds were collected and oil content was measured with NIRS. The relationship between the female parents and the mean of the corresponding offspring is presented in Fig. 3. The estimated heritability was 0.36 but the regression line explained only 17% of the observed variation.

Pollen Transfer

Dimorphotheca pluvialis is considered a predominantly outcrossing species, however little is known on the mode of pollen transfer. The influence of insects on seed set was investigated in a complete randomized block design with three replications, three populations, and three treatments. The treatments consisted of: (1) plots in open air (free insect visitation), (2) plots with cages open at the North side (free insect visitation + shading effect of cage), and (3) plots with cages (no insects, shading effect). The second treatment was included as a control, to determine possible effects of the shading caused by the cages on growth and development of the crop. Plot size was 3 × 3 m. The number of open flowers was counted weekly on two subplots of 0.25 m^2. Seed set was determined by picking 20 flowers randomly, and counting the number of winged and unwinged seeds.

Analyses of variance revealed no difference between treatments 1 and 2 for crop development, seed set and thousand seed weight. Apparently the light shading did not effect these characters (Table 1). Population 879585 flowered slightly earlier than the other two. No population × treatment interaction was found for any of these characters.

Seed yield of plots with open cages was lower than yield of open fields. This could not be explained by a lower seed set, lower number of flowers, or lower thousand seed weight. In the harvest bags of this treatment moths were found, which might have caused severe damage.

Exclusion of insects led to a prolonged flowering of the crop, and a severely reduced seed set and yield. Thousand seed weight was higher.

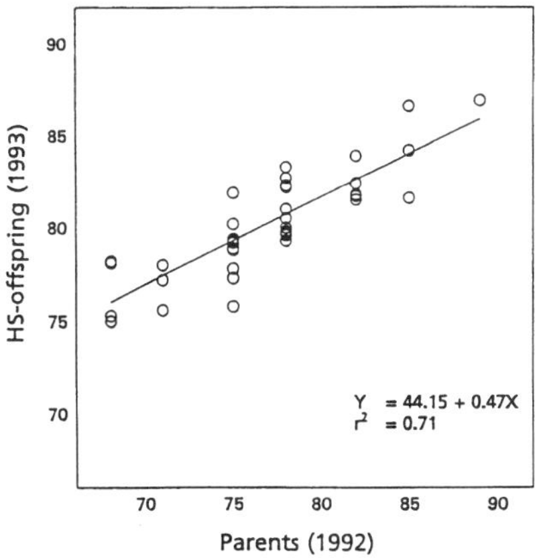

Fig. 1. Parent-offspring relationship for beginning of flowering in a population of *Dimorphotheca pluvialis* (days after sowing).

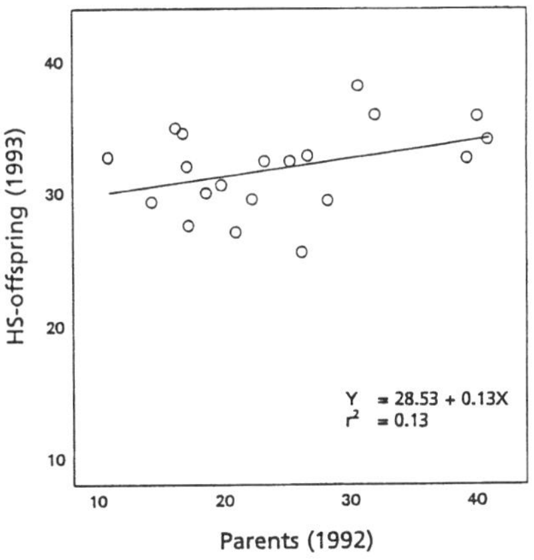

Fig. 2. Parent-offspring relationship for duration of flowering in a population of *Dimorphotheca pluvialis* (days).

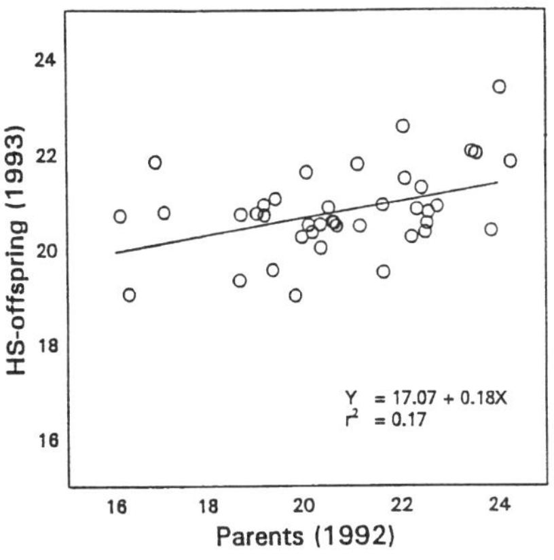

Fig. 3. Parent-offspring relationship for oil content in a population of *Dimorphotheca pluvialis* (% oil in the seed).

Table 1. Effect of exclusion of insects during flowering on crop development and yield characteristics of *Dimorphotheca pluvialis*.

Variable	No. of open flowers/0.25 m² June 30	July 7	July 14	July 21	July 28	Seed yield (g/m²)	No. unwinged seeds/ flower	No. winged seeds/ flower	Thousand seed weight unwinged seeds (g)	winged seeds (g)
Population										
879127	1.2	28.3	140.9	126.0	22.8	70.7	10.7	29.0	3.39	2.26
879731	0.7	29.6	134.6	107.0	26.1	85.7	10.3	30.2	3.50	2.28
879585	2.2	48.3	195.3	74.1	30.9	79.8	11.1	28.4	3.48	2.25
Sign.[z]	**	***	***	***	x	*	NS	NS	NS	NS
Cage treatment[y]										
1	1.4	40.6	148.9	82.9	7.5	125.9	14.4	41.3	2.90	2.14
2	0.9	33.8	142.1	71.6	4.0	83.0	15.4	41.2	3.03	2.20
3	1.7	31.9	179.7	152.6	68.4	27.4	2.4	5.1	4.46	2.46
Sign.[z]	NS	NS	*	***	x	***	***	***	***	***

[z]NS = effect not statistically significant, ***significant 0.1%, ** at 1%, * at 5%.
[y]Treatments: 1 = open plots (insects yes, shading no), 2 = partly opened cages (insects yes, shading yes), 3 = closed cages (insects no, shading yes).
[x]Distribution of residuals not Normal, ANOVA performed on √-transformed data.

CONCLUSIONS

For time of flowering, duration of flowering, and oil content of seeds, sufficient variation was found to enable improvement by means of selection. Heritability for time of flowering appears to be high, indicating that selection for this character will show quick response. For duration of flowering and oil content of the seeds, heritability estimates by means of parent-offspring regression were questionable, but most likely these heritabilities are not very high. In this case the result of selection does not necessarily have to be less, but is more time consuming. The environmental component of the observed phenotype is relatively large, and therefore may conceal the genotypic component.

For these experiments it was assumed that random mating and complete cross pollination has taken place. Furthermore, interaction effects (epistasis, genotype-year, year-location, and genotype-location) were considered negligible. It is likely that some of these assumptions were incorrect, and may have affected the outcome. Year and location effects can only be studied when experiments are carried out at several locations in several years. The presented results on heritability estimates are therefore preliminary, but nevertheless give an indication of what can be expected from selection.

Presence of insects during flowering is essential for a good seed set. Exclusion of insects may result in yield losses up to 75%.

Dimorphotheca pluvialis is as yet not ready for commercialization. Several agronomic constraints are recognized, but most can be overcome given time. Other problems still lay in the area of oil recovery and purification. However, the unique structure of dimorphecolic acid justifies further studies.

REFERENCES

Barclay, A.S. and F.R. Earle. 1965. The search for new industrial crops V. The South African *Calendulae* (Compositae) as a source of new oil seeds. Econ. Bot. 19:33–43.

Dijk, N van, H.P. de Bie, and H. Breemhaar. 1993. Voorlopige teelthandleiding nieuwe olieproducerende gewassen. (in Dutch), Int. Report PAGV-DAT Lelystad, CPRO-DLO Wageningen and IMAG-DLO Wageningen.
Falconer, D.S. 1989. Introduction to quantitative genetics. 3rd ed. Longman, London.
Knowles, R.E., L.A. Goldblatt, G.O. Kohler, S.J. Toy, and J.R. Haun. 1965. Oil seed composition of two species of *Dimorphotheca* grown at five locations in the United States. Econ. Bot. 19:262–266.
Muuse, B.G., F.P. Cuperus, and J.T.P. Derksen. 1992. Composition and physical properties of oils from new oilseed crops. Ind. Crops Prod. 1:57–65.
Norlindh, T. 1977. Calendulae—systematic review. p. 961–987. In: Heywood, V.H., J.B. Harborne, and B.L. Turner (eds.), Biology and chemistry of the compositae.
Soest, L.J.M. van and F. Mulder. 1993. Potential new oilseed crops for industrial use. Lipid Tech. 5:60–65.

Food Intake Inhibitory Activity of Simmondsin and Defatted Jojoba Meal: Dose-Response Curves in Rats

Marnix M. Cokelaere, Gerda Flo, Paul Daenens, Eddy Decuypere, Maurits Van Boven, and Sabien Vermaut

Jojoba (*Simmondsia chinensis*) is a shrub indigenous to the Sonoran desert of Arizona, California, and Mexico. The seeds contain liquid wax esters used as a high-temperature lubricant and in cosmetics. The meal by-product remaining after wax extraction cannot be used as animal feed because of the presence of several cyanide-containing glycosides, such as simmondsin, simmondsin 2'-ferulate, and several minor simmondsin derivatives (Booth et al. 1974; Elliger et al. 1973, 1974a, b; Van Boven et al. 1994a, b, c, 1995). Simmondsin and simmondsin-containing jojoba meal induce food intake inhibition, emaciation and, occasionally mortality, and because of this, simmondsins have been considered toxic (Booth et al. 1974; Ngoupayou et al. 1982; Verbiscar et al. 1980, 1981). However, long-term administration of lower doses of simmondsin or defatted jojoba meal to growing rats, which induced a sustained food intake inhibition of about 20%, showed no toxic effects (Cokelaere et al. 1993a, b), although, real toxicity, especially at higher doses cannot yet be ruled out. Furthermore, it has been demonstrated that food intake inhibition in rats can be reversed by the cholecystokinin receptor antagonist, devazepide (Cokelaere et al. 1995a, b), suggesting that the anorexia seen following simmondsin administration is due to stimulation of the cholecystokinin satiation system. Other studies have shown that the food intake reduction induced with lower doses of defatted jojoba meal is due to satiation (Cokelaere et al. 1995c). Some authors claim that the anorexia induced by defatted jojoba meal is caused by its bitter taste due to the presence of simmondsin 2'-ferulate and tannins, the latter being found in the skin of the jojoba seed (Medina et al. 1988, 1990; Ngoupayou et al. 1985; Verbiscar et al. 1981). Simmondsin itself is tasteless (Verbiscar et al. 1981).

In the present study, the anorexic effects of simmondsin and simmondsin 2'-ferulate are compared and dose-response curves constructed for food containing pure simmondsin or defatted jojoba meal (from commercially available presscake or deskinned jojoba seeds) in order to discriminate between the satiating effects of the simmondsins and other factors present in jojoba seeds.

The effects of jojoba meal from Israel and the United States will be compared.

METHODOLOGY

Adult male Wistar rats, weighing 350 to 425 g, were housed in iron wire cages under normal laboratory circumstances (22°C, 40% to 60% relative humidity, light from 08:00 h to 20:00 h, free access to water and food). Food was provided as meal in specially designed mangers to avoid spilling (Scholz, Overijse, Belgium).

Experiment 1

In the first experiment, 10 rats (5 groups of 2) were used to compare the food intake inhibitory effects of food supplemented with pure simmondsin (S) and simmondsin 2'-ferulate (SF), the two major simmondsins of jojoba meal. The control daily food intake was measured for 3 days (group C), then the rats were given food supplemented with 0.5% (w/w) of pure simmondsin for another 3 days (group SM) and the food intake measured daily. After a recovery period of 2 weeks, they were given normal food supplemented with 0.5% (w/w) of simmondsin 2'-ferulate for 3 days (group SMF) and the food intake again measured daily.

Experiment 2

In a second series of experiments, the food intake inhibitory effect of food supplemented with increasing doses of simmondsin (0.1%, 0.25%, 0.50%, 0.75%, and 1.00%) (group SIM) was compared with the effect of food supplemented with increasing doses of 4 different preparations of defatted jojoba meal (Table 2). These jojoba meals were obtained from commercially available presscake (1) from Israel (Jojoba Israel, Kibbutz Hatzarim) (group PCI) or (2) from the United States (International Flora Technologies, Apache Junction) (group PCU), from deskinned jojoba seeds from (3) Israel (group DI), and (4) from the United States (group DU). The Israeli seeds were deskinned by peeling after cooking for 30 min, while American seeds were deskinned using razor blades. The seeds were then pressed in a hand screw press to extrude most of the jojoba oil. The presscakes and pressed deskinned Israeli seeds were Soxhlet-extracted using n-hexane for 8 h, while the pressed deskinned American seeds were defatted by stirring in a container with n-hexane at room temperature. Pure simmondsin was prepared and the content of S and SF determined as described previously (Van Boven et al. 1993, 1994b).

Fifty rats, caged in pairs, were used for each experiment, 10 for each simmondsin (S) or jojoba meal concentration. Rats were provided normal food for 7 days and the food intake was measured daily to obtain the control food intake; they were then given the S- or jojoba meal-supplemented foods for a further 7 days, and the daily food intake was recorded. The amount of food eaten during treatment was then expressed as a percentage of their own control intake.

RESULTS AND DISCUSSION

Experiment 1

The daily food intake of control group C was 20.7±0.4 g. 0.5% pure S or SF mixed in the food reduced food intake to 7.6±1.4 g or 12.5±1 g, respectively; in terms of food intake reduction, on a weight basis, SF was therefore only 62.5% as effective as S. However, the molecular weight of S is only 68% that of SF (375 and 551, respectively) and this therefore suggests that a similar food intake reduction is obtained with equimolar amounts of SF and S. These results contradict the suggestion of Weber (1978, quoted by Verbiscar et al. 1981) that SF has no significant involvement in anti nutritional aspects of jojoba meal.

The sum of the contents of S and SF $\times$ 0.68 (in %), of defatted jojoba meal was used as indicator of simmondsin activity (SA). The concentrations of defatted jojoba meals to be mixed in the food for the second experiment were calculated taking their SA into account.

Experiment 2

The S and SF contents of the different meals are shown in Table 1 and the concentrations to be mixed with the food in Table 2. The control daily food intake was 21.05±0.45 g (pooled for the 5 subgroups).

Fig. 1 shows the food intake inhibitory effect of food supplemented with increasing doses of pure simmondsin (SIM) or defatted jojoba meal obtained after deskinning jojoba seeds from Israel (DI) and the United States (DU). There was a clear, and similar, dose-response effect for SIM, DI, and DU up to a SA of 0.75% and it was only at a SA of 1%, that a greater degree of anorexia was seen in DI and DU animals compared with the SIM group. Food intake reduction was identical in the DI and the DU groups, although the jojoba meal concentration was significantly higher for the DI group (Table 2). It can thus be concluded that, up to a SA of 0.75%, the anorexic effect of food containing defatted jojoba meal, prepared from deskinned jojoba seeds, can be predicted from its SA rather than from the concentration of the jojoba meal mixed in the diet, and that it is only at higher concentrations that other factors, such as other minor simmondsins or taste effects, may play an additional role. The SF content of food supplemented with high concentrations of jojoba meal may produce a bitter taste, as SF is the major bitter principle of jojoba meal (Medina et al. 1990). An anorexic effect of other simmondsin analogues, present at lower concentrations in defatted jojoba meal, has also been suggested (Verbiscar et al. 1981) but remains to be proven. On the basis of the present results, they may have only a small additional effect. This confirms the results of Abbott et al. (1990) who fed mice with

Table 1. Simmondsin (S) and simmondsin 2'-ferulate (SF) concentrations of different jojoba meals. (SA = calculated simmondsin activity; DI = defatted jojoba meal from deskinned seeds from Israel; DU = defatted jojoba meal from deskinned seeds from the United States; PCI = defatted complete jojoba meal from Israel, PCU = defatted complete jojoba meal from the United States.

Defatted jojoba meal	Concentration (%) S	SF	SA
DI	4.0	1.4	4.9
DU	7.2	1.8	8.5
PCI	4.8	1.4	5.6
PCU	4.6	1.2	5.4

Table 2. Concentrations of defatted jojoba meal mixed in the food of rats to obtain supplemented food with an SA equivalent to that of pure simmondsin. (DI = defatted jojoba meal from deskinned seeds from Israel; DU = defatted jojoba meal from deskinned seeds from the United States; PCI = defatted complete jojoba meal from Israel, PCU = defatted complete jojoba meal from the United States.

Defatted jojoba meal	Amount of defatted jojoba meal (%) equivalent to stated simmondsin amount				
	Simmondsin 0.10%	0.25%	0.50%	0.75%	1.00%
DI	2.0	5.0	10.0	15.0	20.0
DU	1.2	3.0	5.9	8.6	11.9
PCI	1.7	4.4	8.7	13.0	17.4
PCU	1.9	4.7	9.3	14.0	18.6

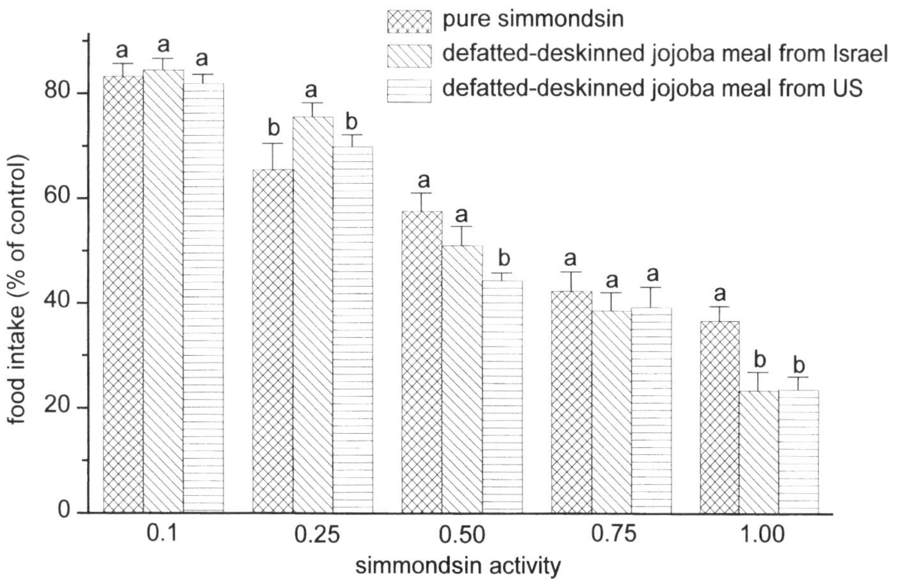

Fig. 1. Food intake as percent of control food intake (±SEM), at increasing concentrations of pure simmondsin and defatted jojoba meal from deskinned jojoba seeds (Israel and United States), n = 10; comparison of means within concentrations by Duncan multiple range test, 5% level.

jojoba meal, detoxified enzymatically or with microorganisms that degraded simmondsin and simmondsin 2'-ferulate but did not attack the other simmondsin derivatives. Mice did well on this treated jojoba meals.

Fig. 2 shows the food intake reducing effect of food containing defatted jojoba meal produced from commercially available presscake. Up to a SA of 0.25%, PCI and PCU had a similar anorexic effect as SIM. At higher concentrations, food intake reduction became more pronounced, especially for PCU, although the difference in concentration of defatted jojoba meal used to obtain similar SAs in the PCI and PCU groups, were very small. This difference in anorexic effect probably results from a more pronounced influence of other factors than SA in jojoba meal. Taste could be one such factor since the skin of jojoba seeds (about 20% of the presscake, data not shown) contains high concentrations of tannins and taste very bitter (Verbiscar and Banigan 1978; Medina et al. 1990). The American jojoba seeds used were much smaller than those from Israel, so the presscake made from complete American jojoba seeds contained a higher percentage of skin than that from Israel and probably caused a more pronounced aversion to the supplemented food. This effect, additional to the anorexic effect of the simmondsins, was already evident at a lower SA content than in the experiments using deskinned seeds, which again points to a supplementary taste effect of skin components.

The present results confirm the results of Medina et al. (1990) in rats using jojoba meal made from complete or deskinned seeds, showing that debittering and tannin-extraction reduces the anorexic effect of jojoba meal. The dose-response curves obtained in the present experiment are in good agreement with previous separate results in rodents (Cokelaere et al. 1993a, b; Verbiscar et al. 1981), rabbits (Ngoupayou et al. 1985), cattle (Swingle et al. 1985), or chicken (Arnouts et al. 1993; Ngoupayou et al. 1982). We therefore conclude that, in rats, the anorexic effect of food containing S is dose-related, sustained, and predictable. On a molar base, SF and S are equipotent. Using the combined % S + % (SF × 0.68) content, designated as SA, the anorexic effect of food containing defatted jojoba meal prepared from deskinned jojoba seeds can be predicted from its SA up to a SA of 0.75%. At higher concentration of jojoba meal, a more pronounced anorexic effect is observed. Food mixed with defatted jojoba meal containing the skin of jojoba seeds, at a SA higher than 0.25%, has a more pronounced anorexic effect than can be predicted from its SA. This additional anorexic effect is probably due to taste or to other factors present in the skin.

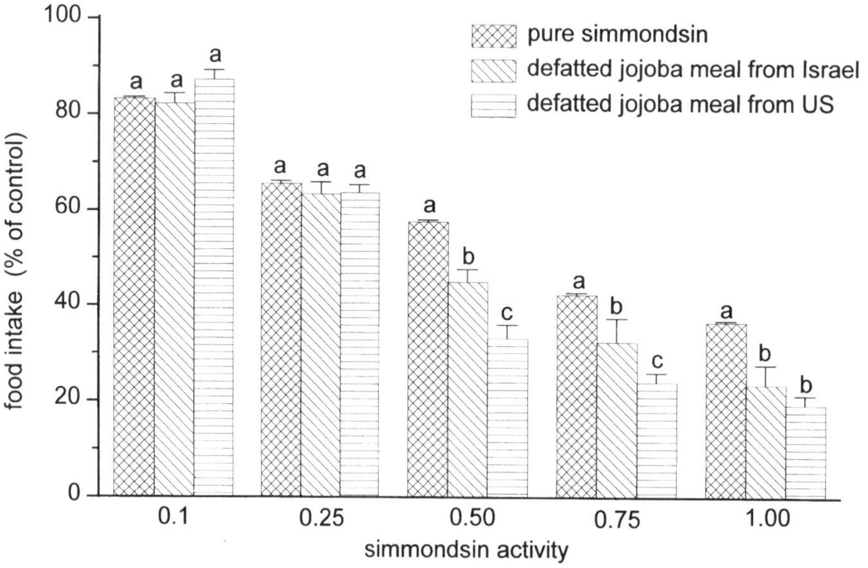

Fig. 2. Food intake as percent of control food intake (±SEM) at increasing concentrations of pure simmondsin and defatted jojoba meal from commercial presscake (Israel and United States), n = 10; comparison of means within concentrations by Duncan multiple range test, 5% level.

SUMMARY

Over a 3-day period, the anorexic effects of pure simmondsin (MW=375) and simmondsin-2'-ferulate (MW = 551) were compared in adult male rats. On a weight base, the anorexic effect of simmondsin-2'-ferulate was only ±68% of that of simmondsin, but, on molar base, they were equipotent. The sum of the concentrations, expressed as percent, of simmondsin plus (simmondsin 2'- ferulate × 0.68) in jojoba meal was taken as an indicator of simmondsin activity (SA).

Dose-response curves of food intake inhibition, produced by increasing concentrations of pure simmondsin mixed in normal food (0.1% to 1.0%), were obtained in adult rats over a 7-day period and compared with those for increasing concentrations of defatted jojoba meal. The anorexic effect of pure simmondsin was dose-related up to a concentration of 1% in the diet (15% to 70% food intake reduction). Up to concentrations giving a SA of 0.75% in the diet, hexane-defatted jojoba meal, obtained from deskinned seeds, gave a similar dose-response curve to pure simmondsin and the anorexic effect of these jojoba meal containing mixtures could be predicted from their SA.

Hexane-defatted jojoba meal, obtained from commercially available jojoba presscake, had a similar food intake reducing effect to simmondsin up to concentrations giving a SA of 0.25%, but, at higher concentrations food intake was reduced to a greater extent than expected from the SA. This supplementary effect is probably due to the presence of the skins in the press cake, causing a bitter taste.

REFERENCES

Abbott, T.P., L.K. Nakamura, T.C. Nelsen, H.J. Gasdorf, G.A. Bennett, and R. Kleiman. 1990. Microorganisms for degrading simmondsin and related cyanogenic toxins in jojoba. Appl. Microbiol. Biotechnol. 34:270–273.

Arnouts, S., J. Buyse, M.M. Cokelaere, and E. Decuypere. 1993. Jojoba meal (*Simmondsia chinensis*) in the diet of broiler breeder pullets: physiological and endocrinological effects. Poultry Sci. 72: 1714–1721.

Booth, A.N., C.A. Elliger, and A.C. Waiss Jr. 1974. Isolation of a toxic factor from jojoba meal. Life Sci. 15:1115–1120.

Cokelaere, M., J. Buyse, P. Daenens, E. Decuypere, E.R. Kühn, and M. Van Boven. 1993a. Influence of jojoba meal supplementation on growth and organ function in rats. J. Agr. Food Chem. 41:1444–1448.

Cokelaere, M., J. Buyse, P. Daenens, E. Decuypere, E.R. Kühn, and M. Van Boven. 1993b. Fertility in rats after long term jojoba meal supplementations. J. Agr. Food Chem. 41:1449–1451.

Cokelaere, M., P. Busselen, G. Flo, P. Daenens, E. Decuypere, E.R. Kühn, and M. Van Boven. 1995a. Devazepide reverses the anorexic effect of simmondsin in the rat. J. Endocrinol. 147:473–477.

Cokelaere, M., P. Daenens, G. Flo, E.R. Kühn, M. Van Boven, S. Vermaut, J. Buyse, and E. Decuypere. 1995b. Simmondsin: physiological effects and working mechanism in rats. In: L.H. Princen (ed.), Proc. 9th Int. Conf. on Jojoba and its uses. Sept. 25–30, 1994. Catamarca, in press.

Cokelaere, M., G. Flo, E. Decuypere, S. Vermaut, P. Daenens, and M. Van Boven. 1995c. Evidences for a satiating effect of defatted jojoba meal. Ind. Crops Prod. 4:91–96.

Elliger, C.A., A.C. Waiss Jr., and R.E. Lundin. 1973. Simmondsin, an unusual 2-cyanomethylenecyclohexyl glucoside from Simmondsia californica. J. Chem. Soc. Perkin Trans. 19:2209–2212.

Elliger, C.A., A.C. Waiss Jr., and R.E. Lundin. 1974a. Structure and stereochemistry of simmondsin. J. Org. Chem. 39:2930–2931.

Elliger, C.A., A.C. Waiss Jr., and R.E. Lundin. 1974b. Cyanomethylenecyclohexyl glucosides from *Simmondsia californica*. Phytochemistry 13:2319–2320.

Medina, L.A. and A. Trejo-Gonzales. 1990. Detoxified and debittered jojoba meal: biological evaluation and physical-chemical characterization. Cereal Chem. 67:476–479.

Medina, L.A., A. Trejo, and M. Sanchez-Lucero. 1988. Elimination of toxic compounds, nutritional evaluation and partial characterization of proteins from jojoba meal. p. 423–429. In: A.R. Baldwin (ed.), Proc. 7th Int. Conf. on Jojoba and its uses. Champaign, IL.

Ngoupayou, J.D.N., P.M. Maiorino, and B.L. Reid. 1982. Jojoba meal in poultry diets. Poultry Sci. 61:1692–1696.

Ngoupayou, J.D.N., P.M. Maiorino, W.A. Schurg, and B.L. Reid. 1985. Jojoba meal in rabbit diets. Nutr. Rep. Int. 31:11–19.

Swingle, R.S., M.R. Garcia, F.J. Delfino, and F.L. Prouty. 1985. An evaluation of *Lactobacilli acidophillus*-treated jojoba meal in beef cattle diets. J. Anim. Sci. 60:832–838.

Van Boven, M., N. Blaton, M. Cokelaere, and P. Daenens. 1993. Isolation, purification and stereochemistry of simmondsin. J. Agr. Food Chem. 41:1605–1607.

Van Boven, M., S. Toppet, M.M. Cokelaere, and P. Daenens. 1994a. Isolation and structural identification of a new simmondsin ferulate from jojoba meal. J. Agr. Food Chem. 42:1118–1121.

Van Boven, M., P. Daenens, M.M. Cokelaere, and G. Janssen. 1994b. Isolation and structure elucidation of the major simmondsin analogues in jojoba meal by two dimensional NMR spectroscopy. J. Agr. Food Chem. 42:2684–2687.

Van Boven, M., P. Daenens, M. Cokelaere, and E. Decuypere. 1994c. Extraction and liquid chromatographic method for the determination of simmondsin in plasma. J. Chromat. 655:281–285.

Van Boven, M., P. Daenens, and M. Cokelaere. 1995. New simmondsin 2'-ferulates from jojoba meal. J. Agr. Food Chem. 43:1193–1197.

Verbiscar, A.J. and T.F. Banigan. 1978. Composition of jojoba seed and foliage. J. Agr. Food Chem. 26:1456–1459.

Verbiscar, A.J., T.F. Banigan, C.W. Weber, B.L. Reid, R.S. Swingle, J.E. Trei, and E.A. Nelson. 1981. Detoxification of jojoba meal by *Lactobacilli*. J. Agr. Food Chem. 29:296–302.

Verbiscar, A.J., T.F. Banigan, C.W. Weber, B.L. Reid, J.E. Trei, E.A. Nelson, R.F. Raffauf, and D. Kosersky. 1980. Detoxification of jojoba meal. J. Agr. Food Chem. 28:571–578.

FIBER CROPS

Hemp: Specialty Crop for the Paper Industry*

Anthony Capelle

In 1994 a four year research program to evaluate the feasability of the cultivation of hemp (*Cannabis sativa* L., Cannabinaceae) as a raw material source for the paper industry was finalized. Three aspects, primary production, pulp processing, and paper production, based on an integrated chain model from farm to factory were studied (Fig. 1).

This research program was part of a broader business concept study as a basis for starting a commercial hemp pulp processing unit. In this paper a summary of the research results will be presented. For more detailed information see van Berlo (1993).

PRIMARY PRODUCTION

Genetic Collection/Optimization

Since 1986, a collection of commercial available germplasm including old cultivars and wild material was initiated. The research objectives which culivar was best suited for paperpulp production under Dutch condition. The stem composition and content of cannabinoids were used as the selection criteria.

Based on earlier studies it was concluded that a late flowering resulted in a higher yield. In field trials carried out in successive years, several cultivars (Kompolti Hybrid TC, Kompolti Sárgaszáru, Kompolti Hyper Elite, Kozuhara, and Fedrina 74) were evaluated. 'Kompolti Sárgaszáru' and 'Kompolti Hyper Elite' proved to be the best suited in regards to yield and fiber quality. Yield in dry matter varied between 11 t/ha for peat soil and 17 t/ha for clay soils. The bast fiber content varied between 12% and 35%, the woody core varied between 50% and 75%. The total of the two fractions remained more or less constant at 85% of the stem dry substance. Cannabinoids content of the different accessions varied between 0.06% and 0.22% for THC, in comparison with 10% THC for drug types.

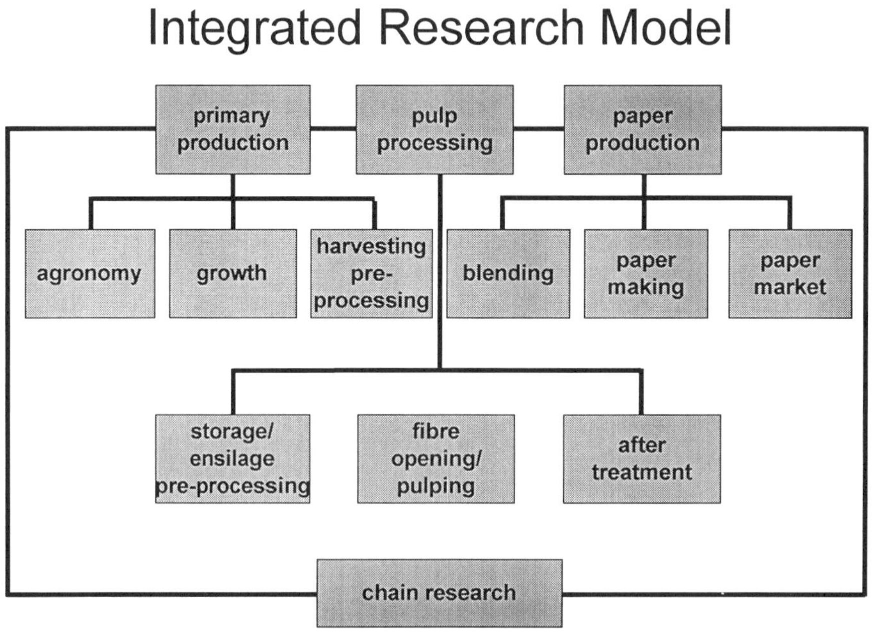

Fig. 1. Integrated research module for hemp.

*Special thanks to Mrs. M.J.J.M. van Kemenade for ATO-DLO at Wageningen, the Netherlands for making available all relevant information.

Cultivation

Growth trials were carried out at three different locations (peat soil, sandy clay, and heavy clay) in a three year period (Fig. 2). Hemp gives the best results when cultivated on light soil at a soil pH of at least 5. Fertilization of 12 kg N, 15 kg K_2O and 4 kg P_2O_5 is needed per ton of dry matter. Seeding should start from mid Apr., as soon as soil conditions make it possible. On heavy clay soils compaction must be prevented as this severely hinders the growth. Hemp can be grown as a row crop with optimum distance 12.5 cm between the rows and a plant density of 900,000 plants/ha. At a seed weight of 19 mg and an emergence of 85% a quantity of 20.1 kg seed/ha is required for planting.

No herbicides are necessary as the crop suppresses weeds. On weed infested soils, plant density should be relatively higher and the row distance less. No major diseases have been identified, however during wet seasons, Sclerotinia and Botrytis, result in lower yields. Spraying with fungicides had no effect.

Harvesting/Postharvest Treatment

The best harvest time is from the beginning to mid Sept. depending on the cultivar.

Harvesting and postharvest treatment with existing equipment is a decisive factor in the production of hemp pulp. If the farmer can carry out most of the needed activities with already available equipment, the direct costs will be low thereby promoting a succesful adoption of this crop. The following harvesting techniques were studied: topping, cutting, windrowing, chopping, press baling, and ensilage. In topping, most of the leaves and flowers from the top of the plant are removed prior to harvesting. The cutting has to be done in such a way that as little sand as possible is picked up, while maintaining optimum yield. It is best carried out with a so called row distance independant cutter bar in front of a forage harvester.

After chopping the hemp can be ensiled at the farm. To optimize the processing of hemp into pulp over the year drying and ensilage was studied as a conservation storage technique. Ensilage is appreciable cheaper than drying and more reliable. The paper quality of ensiled hemp is slightly less than of the dried hemp, but still of acceptable quality.

Fig. 2. Hemp field in the Netherlands.

PULP PROCESSING

Preprocessing

The hemp stem consist of about 65% corefibers and 35% bast. The corefiber consist of 40% cellulose, 24% hemicellulose and 22% lignin, the length is 5.5 mm. The bastfiber is 75% cellulose, 9% hemicellulose and 5% lignin with a length of average 20 mm. As the differences between these two fractions merit a separation into two separate raw materials, research was carried out into this aspect.

Separation can be carried out before of after storage at farm level or at the pulping facility. As harvest takes place in Sept., a rather wet period, dry separation was discarded. Two options were selected, sieving and flotation. Flotation proved to be a feasible technique with a selectivity of more than 90% for bastfiber.

Pulping

Four pulping processes were investigated for hemp (Table 1). Depending on the end application of the pulp, thermomechanical pulping (TMP) of the whole stem is possible. A more optimal result was reached when the core fraction from the earlier in core and bast fraction separated stem was thermomechanically pulped and milled in a refiner and the bast fraction, together with an alkaline (pre)treatment, was extruded in a double screw extruder (Fig. 3, 4). Extrusion has the advantage of 30% less energy consumption, while the length of the fiber was better controlled by fine tuning of the extrusion conditions. A patent was filed for this process by ATO-DLO.

Table 1. Comparison of four pulping processes investigated for hemp.

Pulping process	Temp. (°C)	Time (h)	Fiber/ liquid	Comments
Thermomechanical (TMP)	120	2	10:1	
Akaline TMP (CTMP)[z]	120			Extruder <4% NaOH
Soda	165	0.5	10:1	2% NaOH
Organosolv	195	0.5	9:1	Ethanol/water = 1.2

[z]CTMP = chemithermomechanical pulping.

Fig. 3. Pulping hemp fibers.

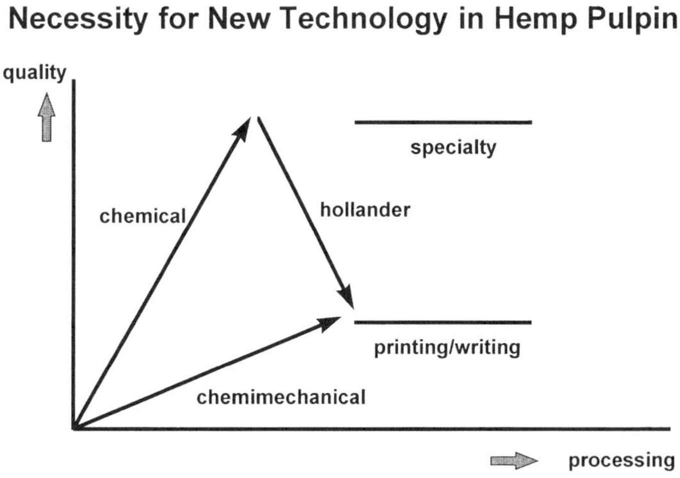

Fig. 4. Necessity for new technology in hemp pulping.

In most cases, especially for short lived applications, there is no lignin removal. The desired color of the paper can be adjusted by bleaching with peroxide (Fig. 5); when pure cellulose is the end product, an ozone treatment is possible.

After Treatment

An important factor to be considered in pulping (Table 1) is the cost of waste water treatment. In waste water treatment, biological oxygen demand, aquatic toxicity, and the color of the water are important factors. In principle the waste water contains easily degradable components such as sugars, fatty acids, and alcohols while lignin an almost non-degradable component is responsible for the color. A combined anaerobic-aerobic treatment with an upfront dilution process to remove the resins, effectively solves the problems.

PAPER PRODUCTION

Different types of paper were produced from the pulped hemp. The traditional applications for hemp papers vary from cigarette paper to art papers and represent a total volume of about 120,000 t/year worldwide, making up about 0.05% of the world pulp volume. However, for these applications a chemical pulping is necessary. For wider bulk applications, such as printing and writing papers and corrugated board, cost is the decisive factor (Table 2).

The properties of chemomechanical core and bast pulp compare very favorable with the traditional wood pulps (Table 3, 4).

CONCLUSION

We conclude that hemp pulp is very competetive with the usual pulps when printing and writing paper application is taken into consideration (Table 5).

The building of a pilot unit with the capacity of about 5,000 t/year for upscaling studies is now under consideration.

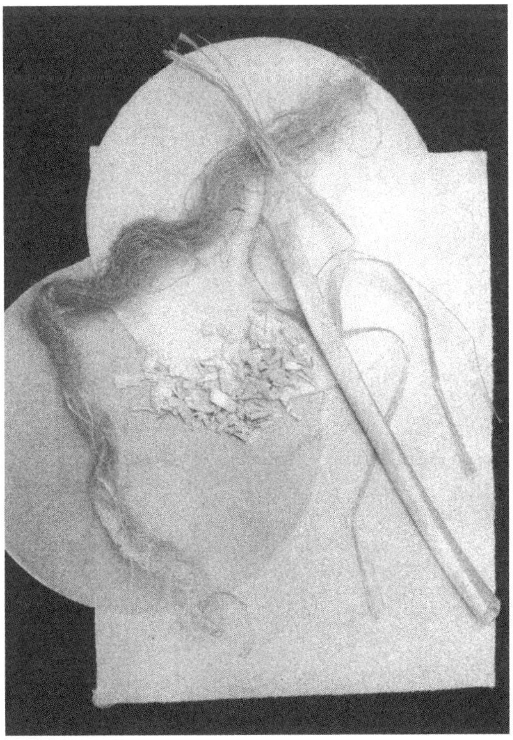

Fig. 5. Bleached hemp products.

Table 2. Cost of chemical and chemomechanical processes for hemp vs. chemical processing for softwoods.

	Cost (US$/t)		
	Hemp		Softwood
Process	Chemical	Chemomechanical	Chemical
Pulping	2,100	530	820
Papermaking	1,900	800	800
Total cost	4,000	1,330	1,620

Table 3. Chemomechanical characteristics of bast pulp vs wood pulps for printing and writing grade blends.

Fiber	Beating degree (SR)[z]	Density (kg/m^3)	Tensile strength (Nm/g)	Burst strength (kPa.m^2/g)	Tear strength (mN.m^2/g)	Brightness (ISO)[z]	Opacity (ISO)
Softwood sulphate	30	720	91	4.0	11	90	75
Hemp bast	50	550	48	4.2	12	82	76

[z]SR = standard reflection, ISO = international standardization organization.

Table 4. Chemomechanical characteristics of core pulp vs wood pulps for printing and writing grade blends.

Fiber	Beating degree (SR)	Density (kg/m^3)	Tensile strength (Nm/g)	Burst strength (kPa.m^2/g)	Tear strength (mN.m^2/g)	Brightness (ISO)	Opacity (ISO)
Aspen APMP	56	709	53	2.5	2.2	82	73
Hemp core refiner	59	714	59	2.7	2.8	78	75

Table 5. Economics of hemp assuming high yield, small module scale, and no recovery system. Investment is calculated for a 5,000 ton/year pilotplant.

Operation	Annual cost per tonne (US$)
Investment estimates	
refit	480
greenfield	600
Operational costs	
fiber crop	130
pulping operation	400
Total	530
NBSK[z]	950

[z]Non-bleached standard kraftpulp, price is used as a reference.

REFERENCES

van Berlo, J.M. Dec. 1993. Papier uit hennep van Nederlandse grond. ATO/DLO, Wageningen.

Crotalaria juncea: A Potential Multi-Purpose Fiber Crop

Charles G. Cook and George A. White

Sunn hemp (*Crotalaria juncea* L.) is one of the earliest and most distinctly named fibers of India. Past research efforts have shown that the soft, lignified fibers produced in the stem of sunn hemp could be utilized in the manufacturing of pulp and paper, and more recent efforts have indicated that other potential products can be developed from these fibers. Additional characteristics that enhance the value potential of sunn hemp as a nonwood, fiber crop are low nitrogen fertilization requirements, the ability to fix nitrogen and to grow in marginal soils, drought resistance, and resistance to root-knot nematodes. Unlike kenaf (*Hibiscus cannabinus* L.), another potential nonwood fiber crop, which produces consistently greater yields and is less susceptible to lodging, sunn hemp is highly resistant to root-knot nematodes and the stalks dry out more rapidly after a killing frost prior to harvest. Past and present research efforts have identified sunn hemp as a potential nonwood fiber source that could be used in the manufacturing of several products, including newsprint, specialty papers, and as a component of commercial nursery potting media. This paper reports on the potential use of *Crotalaria juncea* as a nonwood source of fiber.

ECONOMIC IMPORTANCE

Sunn hemp, a member of the legume family (Fabaceae), has great potential as an annually renewable, multi-purpose fiber crop. It is the most important species of the *Crotalaria* genus, which is comprised of over 350 species located in the tropics and subtropics of both hemispheres. As one of the most widely grown green manure crops throughout the tropics, sunn hemp is often grown in rotation with several different crop species (Kundu 1964; White and Haun 1965; Lai et al. 1967; Purseglove 1968; Srivastava and Pandit 1968; Barros Salgado et al. 1972; Mascarenhas et al. 1980; Rotar and Joy 1983). The stems of sunn hemp are composed of two fibers, the bast and woody core. The bast fibers, which are located in the outer bark, are much longer than the core fibers, but the two fiber widths are very similar (Cunningham et al. 1978). The proportion of bark in the total stalk, by dry weight, generally ranges from 15% to 20%. Kundu (1964) reports that the actual proportion of bast fiber in dry stalks ranges from 6.4% to 10.5%.

Research with sunn hemp has been conducted in the United States since the 1930s, where it was reported to be an excellent soil-improving crop. It produced high organic matter yields, was able to fix nitrogen, and could reduce the build-up of root-knot nematode populations (Breitenbach 1958; Dempsey 1975). However, the difficulty in producing seed caused many farmers to abandon the growing of this crop. Interest in growing sunn hemp was renewed during World War II, as *C. juncea* was added to the list of critical war materials in 1942 because of its potential use as cordage fiber, especially for Marine Oakum. Attention again focused on sunn hemp from the late 1950s through the 1960s. This was the result of efforts to identify annually renewable sources of nonwood fiber which could be used in the manufacture of paper and pulp. Although research showed that sunn hemp possessed good pulping characteristics and that high macerate yields could be obtained (Nieschlag et al. 1960; Nelson et al. 1961), most research eventually concentrated on kenaf. Kenaf received more attention because of its ability to produce consistently greater yields and have less lodging, which could reduce yields and harvest efficiency. A more recent report by Cunningham et al. (1978) indicates that sunn hemp may be superior to kenaf for some fiber properties, including bast fiber length (3.79 vs. 2.62 mm) and width (24.3 vs. 19.7 μm). They concluded that sunn hemp possessed the following properties that made it an excellent candidate for papermaking: (1) good yields of bleachable sulfate pulps, (2) pulp strength properties that are equal to or greater than those of mixed southern hardwood pulp, and (3) bast fiber length to width ratio that is greater than those of wood fibers. A study by Scott and Cook (1994) indicated the potential yield advantage of sunn hemp compared to kenaf when grown in soils infested with the southern root-knot nematode [*Meloidogyne incognita* (Kofoid and White) Chitwood]. In tests conducted under growing conditions in South Texas, Kansas, and South Carolina total stalk yields have been reported to range from 14,800 to 18,450 kg/ha (White and Haun 1965; Scott and Cook 1994; Scott et al. 1991).

Most of the present-day sunn hemp production is located in India, Bangladesh, and Brazil, where it is grown as a green manure crop, a fodder crop, or for the bast fibers. The bast fibers are utilized for the manufacture of cordage and high quality paper. In addition to the uses as a green manure crop and as a fiber source, research is being conducted in Texas to determine if the shorter core fibers of sunn hemp can be utilized in the manufacture of soil-less potting media for commercial nursery application.

ORIGIN

Crotalaria juncea is generally considered to have originated in India, where it has been cultivated since prehistoric times (Montgomery 1954). The genus name *Crotalaria* means rattle and is indicative of the noise made by the seeds shaken in the mature pods (White and Haun 1965). The species name *juncea* was given to this plant by Linnaeus because of its resemblance to *Spartium junceaum* L., the Spanish broom of the Mediterranean region with its green rushlike, scantily-leaved branches (Kundu 1964).

ADAPTATION

Crotalaria juncea is the most important and fastest growing species of the *Crotalaria* genus. Although generally considered to be a tropical or subtropical crop, it is drought resistant and has a wide range of adaptation to soil types. For fiber production, a light loam, moderately rich, well-drained soil is preferred (Montgomery 1954; Kundu 1964). Vigorous growth can be achieved on clay or low lying soils; however, the bast fiber may be coarser and yields lower. The major sunn hemp growing areas of India and Brazil are characterized by high humidity, an average temperature of 23.0° to 29.4°C, and a rainfall of 170–200 mm during the growing season (Dempsey 1975). Sunn hemp is typically photoperiod-sensitive and flowering occurs in response to short days (White and Haun 1965).

BOTANICAL DESCRIPTION

Sunn hemp is a short-day, erect shrubby annual, generally 1 to 4 m in height. The stems are cylindrical and ribbed. Branching in the upper portion is minimized with dense plantings. The simple, elliptic to oblong shaped leaves, are spirally arranged on the stem. The root system is characterized by a long, strong taproot, well developed lateral roots, and much branched and lobed nodules, up to 2.5 cm in length. The inflorescence is a terminal open raceme to 25 cm in length with deep yellow flowers. Flowering is indeterminate. Extensive cross-pollination occurs in sunn hemp and self-pollination takes place after the stigmatic surface has been insect or mechanically stimulated (Purseglove 1968). Seeds are small, flattened, kidney-shaped, and contain approximately 35% protein. Due to cultivar and environment, seed weight is highly variable, ranging from 18,000 to 30,000 seed per kg (Dempsey 1975). 'Tropic Sun', a Hawaiian cultivar, was reported to have 30,000 to 35,000 seed per kg (Rotar and Joy 1983).

CULTIVATION AND MANAGEMENT

Good land preparation should be made before planting sunn hemp for fiber production (White and Haun 1965; Ghumary and Bisen 1967; Barros Salgado et al. 1972). Although reports on fertilization requirements vary, additions of P are generally recommended for low phosphorous soils (Barros Salgado et al. 1972; Rotar and Joy 1983). Sunn hemp is fast growing and generally suppresses weed populations due to dense canopy shading (Burnside and Williams 1968). However, early season weed control has been shown to improve yields when sunn hemp is grown for fiber (White and Haun 1965). No herbicides are currently registered for use in sunn hemp production. However, several preemergence herbicides (most notably clomazone at 1.38 kg a.i./ha) recently have been identified as causing minimal phytotoxicity and providing efficient weed control (J.R. Smart and C.G. Cook, unpubl.).

Recommendations for seeding rates vary greatly among production areas. In the 1995 South Texas commercial planting, seeds were sown at 17 kg/ha in double-drilled rows spaced 1.02 m apart (G. Kinney pers. commun). White and Haun (1965) recommended seeding rates of 16.8 to 22.4 kg/ha in rows spaced 30.5 to 35.6 cm apart. For fiber production in Brazil, Lovadini et al. (1970) advocated 60 kg/ha of seed in rows spaced 50 cm apart, while Barros Salgado et al. (1972) suggested a seeding rate of 35 to 40 kg/ha.

Ghumary and Bisen (1967) report that the broadcast seeding rate in India is 66.5 to 88.5 kg/ha. However, when drill-planted on rows spaced 30 cm apart, they reported that a seeding rate of 6.6 kg/ha can yield 10% more fiber than the higher broadcast rate. Preliminary results from studies conducted at Weslaco, Texas indicate that plant height and total stalk dry yield was not changed for four different seeding rates of ca. 4.9, 6.8, 8.7, or 10.8 kg/ha on rows spaced 1.02 m apart.

Planting dates also differ among locations; however, adequate soil moisture and frost-free, warm weather conditions will provide rapid emergence and the highest yields (Kundu 1964; White and Haun 1965). In Kansas, White and Haun (1965) reported a 40% decrease in yield due to a two-week delay in a 1962 planting. However, a two-week delay in 1963 did not affect yields. Preliminary results in South Texas also indicate that significant decreases in yield, plant height, and stalk diameter may occur when planting is delayed by two, four, or six weeks. Later harvest dates do not appear to improve yields of late planted crops and delaying harvest may actually result in lower yields.

HARVESTING

To obtain the highest quality fiber, harvesting should be done at the optimum time. Although opinions differ, the general recommendation is that harvesting should take place during the seed pod stage (Medina et al. 1961; Kundu 1964; Purseglove 1968). Mechanized harvests will be necessary for large-scale commercial plantings. White and Haun (1965) reported that although forage harvesters had been successfully used, precision cutting was difficult to obtain and the chopped material was bulky to handle. A whole stalk kenaf harvester that was designed using sugarcane technology will be used to harvest the South Texas sunn hemp crop. This harvesting system is adjusted to cut the stalks slightly above ground level and to remove the very upper foliar portion of the stalk, which may be necessary if the fiber is to be utilized in the manufacture of paper. The experimental harvesting of small research plots in 1994 with this machinery indicated that this method of harvest should be successful. However, more efficient, rapid, and specifically designed machinery for harvesting and processing may be necessary for large-scale commercial operations, since it was reported by Scott et al. (1991) that increasing the growing season beyond six months may not be beneficial and may actually result in declining fiber yields.

SEED PRODUCTION

Sunn hemp is generally photoperiod-sensitive and flowering commences in response to short days (White and Haun 1965). Because of the lateness in flowering, with the exception of Texas, very little seed production has occurred in the continental United States. The primary sources of seed for the sunn hemp grown in the continental United States are from Hawaii and South America (most notably Brazil and Columbia). The short-day, frost-free environments of these locations allow for reliable and excellent yields of good quality seed to be produced. Small-scale seed production has occurred in South Texas, but the possibility of early frost always poses a threat of reducing yields of seed crops grown in the continental United States. Recommended seed rates vary from 3 to 5 kg/ha in Hawaii (Rotar and Joy 1983) to 17 to 22 kg/ha in South Africa (DeToit 1946).

Seeds are ready for harvest when pods begin to turn yellow in color and the seeds rattle in the pods (DeToit 1946; Rotar and Joy 1983). For mechanical harvest, a combine with the header raised as high as possible without leaving the seed pods on the stalk should be used. This technique will result in cleaner seed, less moisture, and maintain the speed of the combine harvesting operations. Seed yields are variable between locations. DeToit (1946) reported 450 to 900 kg/ha could be produced in South Africa, while Baird et al. (1957) found that yields in Columbia ranged from 555 to 1000 kg/ha. Rotar and Joy (1983) indicated that 1460 to 2240 kg/ha could be produced in Hawaii.

Many species of *Crotalaria* contain toxic pyrrolizidine alkaloids in the seeds. In *C. juncea*, trichodesmine was identified as the principal toxic alkaloid (Zhang 1985). Studies have indicated that poisoning can occur in both horses (Nobre et al. 1994) and pigs (Zhang 1985) when *C. juncea* seed are ingested.

MAJOR DISEASES AND OTHER PESTS

No serious diseases have developed in South Texas and few have been reported for the United States. However, the soils of South Texas are naturally infested with Phymatotrichum root rot, incited by *Phymatotrichum omnivorum* Duggar [= *Phymatotrichopsis omnivora* (Duggar) Hennebert], a fungal pathogen that attacks over 2,000 plant species (Streets and Bloss 1973). Sunn hemp is reported to be susceptible to attack from this fungal pathogen (Streets and Bloss 1973; Cook and Hickman 1990). However, laboratory studies have indicated that sunn hemp is more resistant to *P. omnivorum* than kenaf (Cook and Hickman 1990). Control of this pathogen can be achieved through cultural practices and crop rotations (Streets and Bloss 1973). Other pathogens of sunn hemp that have been reported from the United States are a powdery mildew, caused by *Microsphaera diffusa* Cook and Peck, and a root and stem rot, incited by *Sclerotium rolfsii* Sacc. (Farr et al. 1989).

Serious diseases of sunn hemp grown in India are anthracnose, caused by *Colletotrichum curvatum* Briant and Martyn (Mitra 1934; Whiteside 1955) and a wilt, caused by *Fusarium udam* E.J. Butler f. sp. *crotalariae* (G.S. Kulkarni) Subramanian (Mitra 1934; Kundu 1964; Purseglove 1968). Fungicide seed treatments and crop rotations are the most recommended and practiced disease control measures (Mitra 1934; Whiteside 1955). Sources of resistance to anthracnose have been reported by Dey et al. (1990), indicating the potential to reduce disease losses through the development of anthracnose resistant varieties. A fungus (*Ceratocystis fimbriata* Ellis & Halst) that causes a wilt disease of sunn hemp in South America has been reported by Barros Salgado et al. (1972) and Malaguti (1951). In addition to implementing crop rotations for disease control, sunn hemp germplasm which possesses resistance to *C. fimbriata* has been reported (Ribeiro et al. 1977).

The major insect pests of sunn hemp in the United States are reported to be the lima bean pod borer, *Etiella zinckenella* Treit. and bella moth, *Utetheisa bella* L. (Seale et al. 1957). In Florida, these insect pests were reported to attack the fruiting pods, with little to no seed being produced. In India, the two major insect pests of sunn hemp are the sunn hemp moth (*Utethesia pulchella* L.), which feed on the leaves and seed pods, and the top-shoot borer (*Laspeyresia pseudonectis* Meyr.), which bores into the apical tip of the plant, causing excessive branching and cessation of growth (Kundu 1964). Dempsey (1975) reports that significant shoot borer resistance has been identified.

Other insect pests that are reported to periodically attack sunn hemp are the leaf feeding catapillars, *Argina cribraria* Clerck and *A. syringa* Cramer. Additional stem or shoot borer species include *Laspeyresia tricenta* Meyr., *Cymotricha tetraschema* Meyr., and *Selinas monotropa* Gaert. Dempsey (1975) and Reddy (1956) also report that the sunn hemp mirid (*Ragmus importunitas* Distant), flea beetle (*Longitarsus belgaumensis* Jac.), and stink bug (*Nezara viridula* L.) will attack sunn hemp. In 1995, the silverleaf whitefly (*Bemisia argentifolii* Bellows & Perring) was observed to feed and reproduce on the lower surface of sunn hemp leaves. Although no actual damage was quantified, premature leaf defoliation appeared to occur.

GENETICS AND BREEDING

Sunn hemp, $2n = 16$, is generally reported to be a self-incompatible plant. Cross-pollination is extensive and self-pollination occurs only after the stigmatic surface has been stimulated by insects or some other means (Kundu 1964). More recently, successful efforts in breeding for self compatibility have been reported (Ribeiro et al. 1977). The development of self compatible germplasm should accelerate the development of true breeding, stable pure lines.

According to Purseglove (1968) and Dempsey (1975), most of the sunn hemp cultivars have originated from selection of improved types suited to specific localities, rather than by breeding procedures. These selections generally focused on early maturity, improved fiber yield, and resistance to pests. Dempsey (1975) provides lists of improved sunn hemp cultivars that have been selected for the growing conditions of India, Brazil, and Taiwan. Past attempts to develop strains from interspecific crosses among *Crotalaria* have not been successful (Kundu 1964).

In more recent years, genetic research and breeding procedures have been conducted on sunn hemp in Brazil (Ribeiro et al. 1977; Miranda et al. 1989; Miranda 1991). Miranda (1991) conducted a genetic study in 1990 to determine the inheritance pattern of seed yield and plant height for sunn hemp. Results indicated that the estimation of the dominant effect was nine times larger than the additive effect. Since pollination generally can be controlled in sunn hemp due to the self incompatibility factor, these findings indicate the possibility of exploiting heterosis. Ribeiro et al. (1977) described the breeding methodology used in the development of sunn hemp germplasm possessing both self compatibility and resistance to *C. fimbrata*. The presence of a self compatibility factor in sunn hemp will allow breeders to treat sunn hemp like a self pollinated crop, ie. single, individual plant selections can be made and the potential exists for a more rapid development of true breeding, pure line cultivars. Preliminary results indicate that final plant height and basal stem diameter are positively correlated with total stalk dry matter, indicating selections for these traits could result in higher yielding cultivars.

SUMMARY AND FUTURE

As the world faces an increased need for fiber, sunn hemp, a nonwood fiber crop, has the potential to be grown on a large commercial scale. Present known industrial uses for sunn hemp fiber include the manufacture of pulp, specialty papers, and as a component of a soil-less potting media. In the past, sunn hemp bast fiber was used for cordage and for making specialty paper. Today, in countries outside of the United States, sunn hemp is often grown in crop rotations as a green manure crop to improve soils and reduce root-knot nematode infestations. Future efforts should focus on the expansion of product research and development, the development of harvesting and processing equipment, and the identification of more efficient crop management strategies. The future of sunn hemp appears very promising given the fact that worldwide demands for fiber are increasing and current wood fiber supplies cannot continue to meet the industrial market demands of the future. Research has shown that sunn hemp can be successfully grown as a fiber crop and that these fibers can supplement many of the current fiber market needs.

REFERENCES

Baird, G.B., M. Rodriguez, B. Martinez, and P. Sanchez. 1957. *Crotalaria juncea*-a green manure. DIA Boletin de Divulgacion 2. Havana, Cuba.

Barros Salgado, A.L., L.A.C. Lovadini, M. Pimental, and W. Gimenez. 1972. Instrucoes para a cultura da *Crotalaria juncea*. Instituto Agronomico, Campinas, Brazil. Secao de Plantas Fibrosas Boletim 198.

Breitenbach, C.A. 1958. Land management practices for kenaf production. p. 97–101. In: Proc. World Conf. on Kenaf. ICA, Havana, Cuba.

Burnside, O.C. and J.H. Williams. 1968. Weed control methods for kinkaoil, kenaf and crotalaria. Agron. J. 60:162–164.

Cook, C.G. and M.V. Hickman. 1990. Response of kenaf and sunn crotalaria to *Phymatotrichopsis omnivora*. Guayulero 12:4–9.

Cunningham, R.L., T.F. Clark, and M.O. Bagby. 1978. *Crotalaria juncea*-annual source of papermaking fiber. TAPPI 61:37–39.

Dempsey, J.M. 1975. Fiber Crops. The University Presses of Florida, Gainesville, FL.

Detoit, J.J. 1946. Sunn hemp, a valuable fodder crop and soil renovator. Farmer's Weekly (S. Africa) 72:90–91.

Dey, D.K., K. Banerjee, R.D.N. Singh, and S.A.K.M. Kaiser. 1990. Sources of resistance to anthracnose disease of sunnhemp. Environ. Ecol. 8:1217–1219.

Farr, D.F., G.F. Bills, G.P. Chamuris, and A.Y. Rossman. 1989. Fungi on plants and plant products in the United States. The American Phytopathological Society, St. Paul, MN.

Ghumary, S.R. and M.S. Bisen. 1967. Sunn hemp (Sunn crotalaria) has satisfied M.P. farmers. Indian Farming 16:17–18.

Kundu, B.C. 1964. Sunn-hemp in India. Proc. Soil Crop Soc. Florida. 24:396–404.

Lai, M.L., C.C. Yang, and P.S. Ching. 1967. Introduction and selection of the dual-purpose varieties of sunn hemp and their fiber extraction and paper manufacturing experiments. Taiwan Agr. Quart. 3:34–46.

Lovadini, L.A.C., A.L.B. Salgado, and S. Miyasaka. 1970. Efeito de epoca de plantio e da poda na producao de massa verde e sementes de *Crotalaria juncea* L. Bragantina 6:25–29.

Malaguti, G. 1951. Una podredumbre del tallo de *Crotalaria juncea*, causada por *Ceratostomella fimbrata*. Agronomia Tropical 1:287–292.

Mascarenhas, H.A.A., N.R. Braga, M.A.C. de Miranda, C.T. Feitosa, and O.C. Bataglia. 1980. Efeito de adubos verdes e organicos na producao de soja. Instituto Agronomico, Campinas, Brazil. Boletim Tecnico 63.

Medina, J.C., D. Ciaramello, and A. Pettinelli. 1961. Resultados experimentals com a cultura de *Crotalaria juncea* L. Como planta produtora de celulose para papel. Bragantia 20:659–668.

Miranda, M.A.C. de. 1991. Adequacao modelo aditivo-dominante em dois caracteres de crotalaria. Bragantia 50:195–202.

Miranda, M.A.C. de, E.A. Bulisani, J.P.F. Teixeira, and H.A.A. Mascarenhas. 1989. Haranca da pigmentacao com antocianina em *Crotalaria juncea* L. Bragantia 48:87–94.

Mitra, M. 1934. Wilt disease of *Crotalaria juncea* Linn. (sunn-hemp). Indian J. Agr. Sci. 4:701–714.

Montgomery, B. 1954. Sunn fiber. p. 323–327. In: H.R. Mauersberger (ed.), Mathew's textile fibers. 6th ed. Wiley, New York.

Nelson, G.H., H.J. Nieschlag, M.E. Daxenbichler, I.A. Wolf, and R.E. Perdue. 1961. A search for new fiber crops. III. Laboratory scale pulping studies. TAPPI 44:319–325.

Nieschlag, H.J., G.H. Nelson, I.A. Wolf, and R.E. Perdue. 1960. A search for new fiber crops. TAPPI 43:193–201.

Nobre, D., M.L.Z. Dagli, and M. Haraguchi. 1994. *Crotalaria juncea* intoxication in horses. Vet. Human Toxicol. 36:445–448.

Purseglove, J.W. 1968. Tropical crops. Wiley, New York.

Reddy, D.B. 1956. Sunn hemp and its insect fauna. Proc. Tenth Int. Cong. Entomol. 3:439–440.

Ribeiro, I.J.A., M.C.A. de Miranda, E.A. Bulisani, L.D'A. Almeida, L.A.C. Lovadini, M.H. Sugimori, and O.P. Filho. 1977. Melhoramento da crotalaria I-autocompatibilidade e resistancia a murcha de *Ceratocystis fimbriata*. Bragantia 36:291–295.

Rotar, P.P. and R.J. Joy. 1983. 'Tropic Sun' sunn hemp, *Crotalaria juncea* L. Univ. of Hawaii, College of Tropical Agr. and Human Resources, Institute of Tropical Agr. and Human Resources Research Extension Series 36.

Scott, A.W., Jr. and C.G. Cook. 1994. Kenaf, crotalaria, roselle—two year comparison in South Texas. p. 97–101. In: Proc. Int. Kenaf Assoc. Conf.

Scott, A.W., Jr., L.N. Namken, and J. Garza. 1991. Kenaf/crotalaria variety study—1989. p. 44–46. In: Rio Farms, Inc. 50th Anniversary Research Report.

Seale, C.C., J.F. Joyner, and J.B. Pate. 1957. Agronomic studies of fiber plants. Fla. Agr. Expt. Sta. Bul. 590:16–17.

Srivastava, S.C. and S.N. Pandit. 1968. Relative role of sunn hemp for tops and roots in contributing to the green-manuring benefits of sugarcane. Indian J. Agr. Sci. 38:338–342.

Streets, R.B. and H.E. Bloss. 1973. Phymatotrichum root rot. Monogr. 8. The American Phytopathological Society, St. Paul, MN.

White, G.A. and J.R. Haun. 1965. Growing *Crotalaria juncea*, a multi-purpose fiber legume, for paper pulp. Econ. Bot. 19:175–183.

Whiteside, J.O. 1955. Stem break (*Colletotrichum curvatum*) of sunn hemp in southern Rhodesia. Rhodesian Agr. J. 52:417–425.

Zhang, X.L. 1985. Toxic components of the seeds of *Crotalaria juncea* and their toxicity to pigs. Chinese J. Vet. Sci. Tech. 7:13–17.

Domestication of *Hesperaloe*: Progress, Problems, and Prospects

Steven P. McLaughlin*

Representatives of a major United States pulp and paper company, the James River Corporation (JRC), approached the University of Arizona in late 1985 about establishing a domestic sisal (*Agave sisalana*) industry. At that time JRC was a large, diversified pulp and paper company; sisal was one of the raw materials used by their specialty paper business. Sisal is a tropical hard fiber crop originally domesticated as a source of cordage fibers. When pulped, sisal cordage yields long, thin fibers that command a high price in the paper industry (Table 1). JRC initially was interested in developing a domestic sisal industry in order to obtain a more reliable supply of this fiber at a low cost.

JRC representatives were advised that sisal could not be successfully grown in Arizona because of its lack of frost tolerance. Bulbils of sisal obtained from the Huntington Botanical Garden were planted in Tucson as a demonstration and they deliquesced and died immediately after temperatures first dropped to 0°C in Dec. 1985. Sisal was selected for domestication because it made good rope, not because it made good paper, but many Southwestern relatives of sisal have frost tolerance and were used by indigenous peoples for cordage products (McLaughlin and Schuck 1991). We jointly decided therefore to screen various members of the Agavaceae to determine if any (1) possessed high-quality fibers and (2) were suitable for production in the temperate climates of the United States.

During 1986 leaf samples were collected from the wild and from several botanical gardens, including the Desert Botanical Garden in Phoenix, the Huntington Botanical Garden in San Marino, California, and the Boyce-Thompson Southwestern Arboretum in Superior, Arizona. The screen included over 100 collections of 62 species from six genera in the Agavaceae: *Agave*, *Dasylirion*, *Furcraea*, *Hesperaloe*, *Nolina*, and *Yucca*. (Many systematists today place *Dasylirion* and *Nolina* in a separate family, the Nolinaceae). All leaf samples were sent to James River's Neenah Technical Center in Wisconsin, where they were pulped and made into paper samples that were tested for strength properties.

The best prospects to emerge from the screen were *Hesperaloe funifera* and *Hesperaloe nocturna*, both because they had superior fiber characteristics and because they had favorable agronomic traits (McLaughlin 1993). The results of this screening study were never published. In fact, JRC consistently discouraged any disclosure of their participation in the research and development of *Hesperaloe*. Populations of several species were subsequently resampled and fiber lengths, fiber widths, and cell-wall dimensions were determined at the University of Arizona (McLaughlin and Schuck 1991). Fibers of both *Hesperaloe* species were longer and thinner than those of any other species examined. Indeed, the length-width ratio of fibers from *Hesperaloe funifera* is greater than those of most other paper-making fibers and comparable to that of abaca (*Musa textilis*), the premium paper-making fiber (Table 2).

Based on the results of the 1986 screen, JRC and the University of Arizona selected *Hesperaloe funifera* and *H. nocturna* for further research and development. JRC worked on pulping, paper-making, and product development while the University of Arizona worked on various aspects of the agronomy and biology of the plants. Although both *Hesperaloe* species were not common in nature, we obtained sufficient seed of *H. funifera* to initiate long-term biomass production studies in 1988 but similar studies could not be started with *H. nocturna* until 1990.

---

*Many coworkers at the University of Arizona have contributed to the research and development efforts on *Hesperaloe*. I thank Damian Ravetta, Abed Anouti, Susan Schuck, Wayne Coates, John Nelson, Bill McCloskey, and Ramon Cinco-Castro. Financial support was received from the James River Corporation, the USDA AARC Center (93AARC20030), and the USDA CSREES (94-COOP-1-0036). Mr. Alfred Wong of Arbokem, Inc., has played a major role in defining the pulping and paper-making characteristics of *Hesperaloe* fibers, as well as in maintaining private-sector involvement in this project.

AGRONOMY

Growth and Biomass Production

Experimental details of the long-term growth and biomass studies have been presented previously (McLaughlin 1993, 1995). We estimated fresh weights on a sample of 20 plants of *Hesperaloe funifera* at 4-week intervals over a period of five years (Fig. 1). *Hesperaloe* species are acaulescent and their leaves are tightly packed into clusters of several rosettes. In its first year *H. funifera* grew very slowly. During years 2 and 3 growth was most rapid in the latter half of the year with maximum growth rates about 10 g FW (fresh weight) plant^{-1} day^{-1}. Approximately half of the plants flowered during their third year. This was a critical stage because lateral rosettes were also initiated at about the time of flowering. More rapid growth followed the development of these lateral rosettes during years 4 and 5, when maximum growth rates were about 40 g FW plant^{-1} day^{-1} (Fig. 1). Within-stand variation in plant size appeared to reach a maximum in year 4 and decline in year 5.

Biomass production for *Hesperaloe funifera* has been examined in six 300-m^2 plots for seven years. The plots were established at three densities and have been monitored by randomly sampling plants from within the plots at the end of each growing season (McLaughlin 1995). The plots at the highest density (27,000 ha^{-1}) produced a yield of 192 t FW ha^{-1} after five years (Fig. 2). Very little aboveground growth was observed during the first year. Plots harvested at the end of year 5 regrew slowly in year 6; growth rates in year 7 were comparable to those observed in year 5 prior to harvest.

The relationship between stand density and individual plant biomass is presented in Fig. 3. There was little evidence of competition among *Hesperaloe funifera* plants in these plots during their first three years, i.e., there was no discernable relationship between stand density and mean plant size. However, effects of

Table 1. Prices for pulps, 1991 (from Judt 1993).

Pulp type	US $/Mg
Well-cleaned, bleached fiber pulps (abaca, hemp, flax, sisal)	$1800–$2400
Not so-well cleaned, unbleached fiber pulps (hemp, flax, kenaf, cotton lint)	$1200–$1800
Special softwood pulps	$750–$850
Normal softwood pulps	$550–$750
Normal hardwood pulps	$450–$550

Table 2. Dimensions of paper-making fibers. Data for *Hesperaloe funifera* are from McLaughlin and Schuck (1992) for accession A-184; data for other species are from Hurter (1991).

Fiber	Mean length (mm)	Mean width (μm)	L/W Ratio
Abaca	6.00	20	300:1
Hesperaloe funifera	3.20	15	215:1
Sisal	3.03	17	180:1
Cotton linters	3.50	21	165:1
Kenaf bast fiber	2.74	20	135:1
Wheat straw	1.48	13	110:1
Softwoods	3.00	30	100:1
Jute	2.00	20	100:1
Hardwoods	1.25	25	50:1

crowding were evident in years 4 and 5 after plants flowered and produced their first group of lateral rosettes. The highest density used in this study was achieved using a 61 cm row spacing. Arizona cotton growers most often use 102-cm rows. *Hesperaloe* planted on 46 cm centers within 102 cm rows would correspond to a density of approximately 21,000 plants ha^{-1}. From Fig. 3 it can be estimated that a stand of 21,000 plants ha^{-1} should have an average plant size of 8.5 kg FW $plant^{-1}$ and a standing crop at first harvest of 180 t FW ha^{-1}.

The crop cycle for *Hesperaloe* is still unclear. Biomass yields can be optimized by delaying the initial harvest until the end of year 5. It appears now that the first reharvest would be at year 8, not year 7 as previously suggested (McLaughlin 1995). A second reharvest might be achieved at year 10 (Fig. 4). Plants should regrow more rapidly after the second cut at year 7 than after the first cut at year 5 because they will have more meristems (rosettes) from which leaves can be produced.

Physiology

Damian Ravetta studied gas exchange in *Hesperaloe* species (Ravetta 1994; Ravetta and McLaughlin 1993), and determined that these species have Crassulacean Acid Metabolism (CAM). CAM plants close their stomata during most of the daylight hours but open them at night, absorbing CO_2 and transpiring water

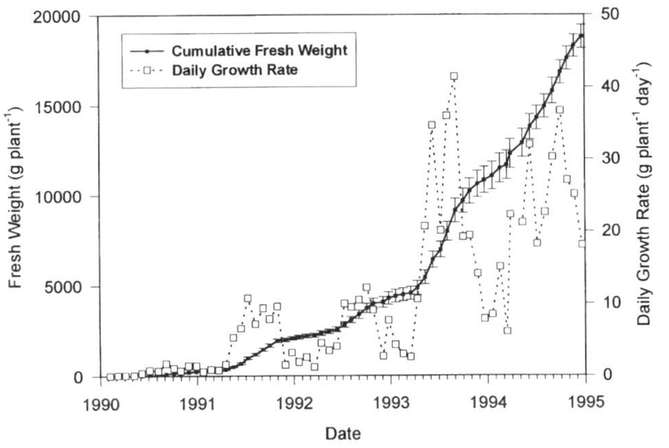

Fig. 1. Cumulative biomass production and daily growth rates in fresh weight measured over a 5-year period in a sample of 20 plants of *Hesperaloe funifera*.

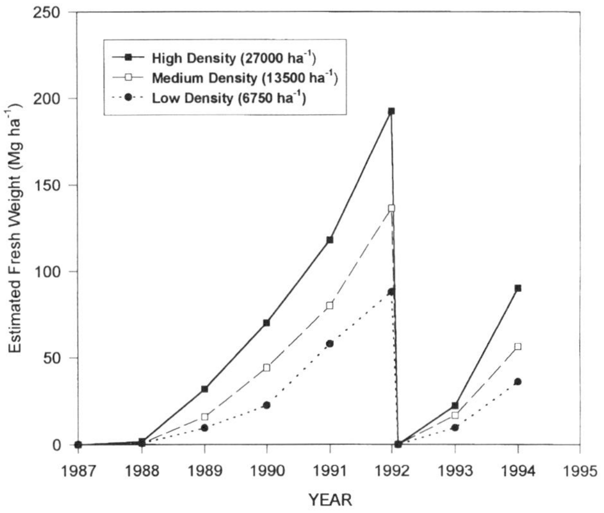

Fig. 2. Biomass production of *Hesperaloe funifera* at three stand densities measured over a 7-year period. Plots were completely harvested at the end of 1992 after 5 years of growth.

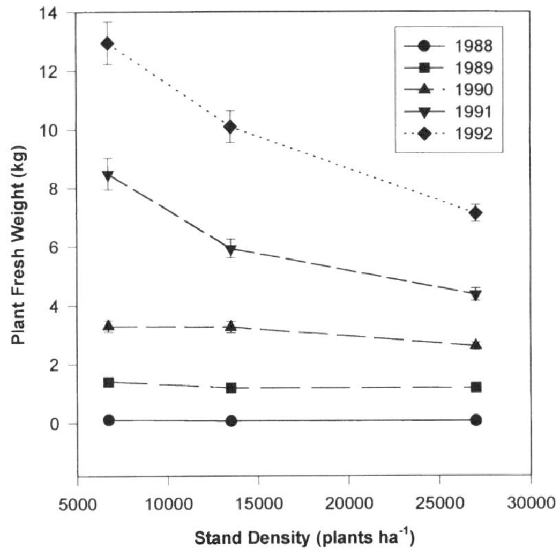

Fig. 3. Relationship between plant size and stand density in *Hesperaloe funifera*.

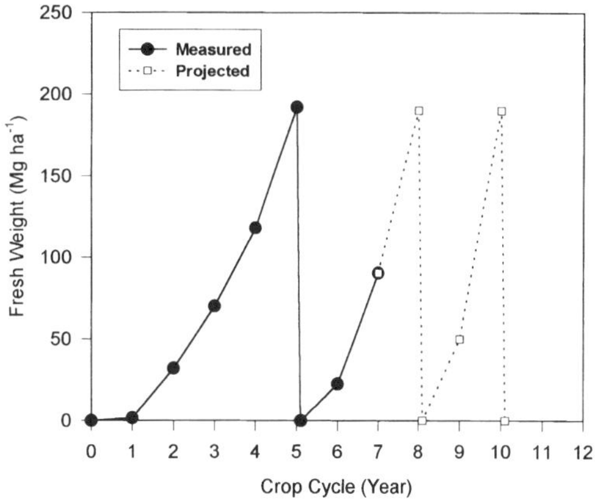

Fig. 4. Potential crop cycle for *Hesperaloe funifera*.

when temperatures are lower and relative humidity is higher. *Hesperaloe* species were found to have low maximum photosynthesis rates, low transpiration rates, and very high water-use efficiency (WUE), all consistent with CAM. The new leaves of *H. funifera* emerge from the central bud at a 90° angle to the ground surface and remain at high angles for 2–3 years. This vertical-leaf orientation is also consistent with CAM, because vertical leaves absorb little solar radiation at midday when their stomata are closed, thus avoiding heat stress.

Ravetta (1994) found that photosynthetic rates in *H. funifera* were highest during the fall months. High fall photosynthetic rates are consistent with the high daily growth rates that are observed in the fall (Fig. 1). Solar angles and night temperatures are lower in the fall than in the summer months and this may promote the higher photosynthetic rates. There may also be enhanced sink capacity in the fall when the lateral rosettes which emerged during the summer are growing at a rapid rate.

The agronomic significance of CAM is high WUE, and our initial biomass production trials confirmed that the high physiological WUE of *Hesperaloe funifera* did indeed translate into low water requirements (McLaughlin 1995; Table 3). Water-use efficiency was very low during the first year of stand establishment from transplants and very high during the fifth year (0.29 to 0.44 t DW cm^{-1}). For comparison with the WUE values shown in Table 3, alfalfa grown in Arizona has a WUE of about 0.09 t DW cm^{-1}. Averaged over 5 years, including year 1 when the crop uses all resources inefficiently, WUE in *H. funifera* is about twice that of C3 crops. Regrowing stands are expected to have a higher WUE than newly established stands.

Genetic Variation

A collection of 10 accessions of *Hesperaloe funifera* was assembled from populations from the wild and from cultivation (McLaughlin and Schuck 1992; McLaughlin 1994). These accessions were grown in small, replicated plots in a common-garden study in Tucson. There were significant differences among these populations in plant basal circumference, leaf lengths, leaf widths, plant fresh weights, number of rosettes at year 3, fiber lengths, fiber widths, and cell-wall thickness. No accession combined the best agronomic traits with the best fiber traits. The phenotypic variation observed in the common garden indicated a potential for improving the crop through classical breeding and selection. Because of the long life cycle of *Hesperaloe*, genetic improvement would be a long-term process that would likely follow, rather than precede, successful commercialization. Repeated measurements from individual plants showed good correlations between early plant fresh weights before first flowering (year 2), at an initial harvest (year 4), and at reharvest (year 6). Thus it may be possible to make early selections in a breeding program. For example, a 3-year cycle for crop improvement

Table 3. Water applications, biomass production, and water-use efficiency (WUE) in *Hesperaloe funifera*, from planting until first harvest at year 5. MD = medium density (13000 plants ha^{-1}); HD = high density (27000 plants ha^{-1}). Dry weights are estimated as 0.325 × fresh weight.

	Water applied (cm)		Biomass production (Mg DW ha^{-1} y^{-1})		WUE (Mg DW cm^{-1})	
Year	MD	HD	MD	HD	MD	HD
1	30	42	0.3	0.5	<.01	<.01
2	37	46	4.9	9.8	0.13	0.21
3	46	74	9.2	12.5	0.20	0.17
4	69	114	11.7	15.5	0.17	0.13
5	41	83	18.3	24.2	0.44	0.29
Mean	44.6	71.8	8.8	12.5	0.20	0.17

could involve making selections at year 2 followed by culling and open-pollination of selected plants in year 3.

COMMERCIALIZATION EFFORTS

JRC sold its specialty paper business in 1991; however, they continued to investigate *Hesperaloe* fibers for other applications. By the end of 1992 it was established that (1) fibers of *Hesperaloe funifera* have exceptional strength and texture; (2) high biomass yields can be obtained at year 5, and (3) the crop does have low water requirements consistent with CAM, which should make it appealing to potential growers. At this same time the newly established USDA Alternative Agriculture Research and Commercialization (AARC) Center issued its first call for proposals. The University of Arizona, JRC, and Sundance Farms of Coolidge Arizona submitted a proposal to the AARC Center in 1993 to establish a demonstration farm and conduct pulping and paper-making on a pilot scale. AARC Center funds would have gone into the demonstration farm and into development of appropriate agricultural machinery. JRC was proposing to manufacture a line of premium tissue and towel products which they had developed using *Hesperaloe* fibers (Reeves et al. 1994). Softness, absorbency, and thickness are usually inversely related to fiber strength. However, sanitary papers made with a blend of *Hesperaloe* fibers have both greater strength and softness (low coarseness), and also greater bulk, flexibility, and improved absorbency (Reeves et al. 1994).

This proposal was reviewed favorably by the AARC Center. However, during subsequent negotiations, JRC decided not to commit to a new major business venture at that time. The primary reason was the price of pulp, which hit an industry-wide low in the 4th quarter of 1993 (Fig. 5); major pulp manufacturers were not making a profit with softwood kraft pulp selling for less than $500/t.

Based on preliminary statements of interest from several paper makers specializing in higher-value products, the AARC Center agreed to fund the University of Arizona for a one-year period (Jan.–Dec. 1994) to develop a consortium of private-sector parties to take over the commercialization of *Hesperaloe*. The AARC Center stipulated that this consortium should consist of no fewer than five companies in the pulp and paper industry. A consortium was successfully organized with Arbokem, Inc., a specialty pulp manufacturer, as the lead organization. Six paper makers initially agreed to participate in the consortium.

The consortium submitted a second proposal to the AARC Center in August 1994, but this proposal was not accepted. The new AARC Center Board of Directors, under pressure for early commercial successes, felt that commercialization of *Hesperaloe* was too much of a long-term project (B. Crain pers. commun.). Following the AARC Center's rejection of this proposal, several of the paper makers involved decided to terminate their participation in the consortium.

We are currently working with a single pulp manufacturer (Arbokem, Inc.), and a single paper maker; the latter company prefers not to disclose its participation at this time. Under the direction of Wayne Coates of the University of Arizona, enough *Hesperaloe funifera* biomass will be harvested to ship approximately 2–3 t

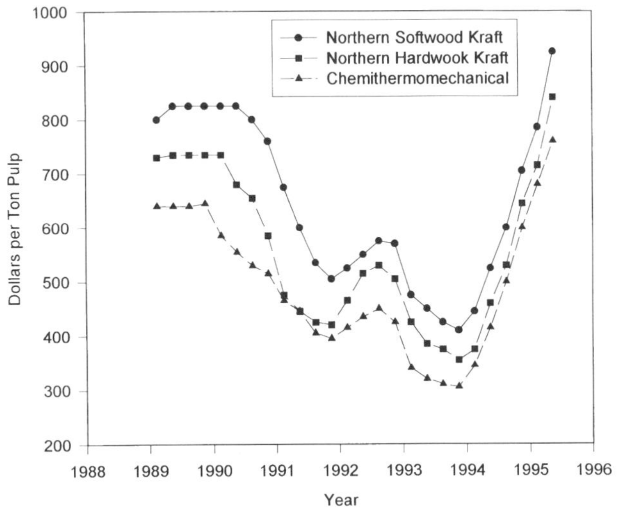

Fig. 5. Prices of market pulps, 1989–1995; adapted from Presley (1995).

of decorticated fibers to Arbokem for pilot-scale pulping trials. The pulp will then be used for pilot-scale paper-making trials. If the results of these pilot-scale trials are satisfactory, a decision to move forward with a commercialization program could be made.

The major limitation on commercialization at this time is seed supply. We have been harvesting and cleaning seed from our experimental plots and have accumulated enough seed to plant about 150 ha. The first plantings of *Hesperaloe funifera* on demonstration farms will have to accomplish several objectives, including further development and testing of best management practices, seed production, and biomass production. Commercial-scale fiber production, approximately 20,000 to 25,000 t bone-dry pulp per year, would require about 5000 ha of cropland. However, assuming that the production cycle for *Hesperaloe* entails harvests at ages 5, 8, and 10, it is difficult to devise a planting scenario that would produce the exact quantity of biomass required each year by the pulper. One potential scenario is shown in Fig. 6, which assumes a yields of 180 t FW ha^{-1}; an initial harvest at 5 years, and a productive lifetime for the stand of just 10 years. Under this plan, full production for a single processing plant would be reached at year 8. There would be excess production in years 13, 15, and 20, and sufficient biomass production to double the pulping capacity in year 26. Many different commercialization plans could be devised, but they would all have to share 2 basic features: an initial period of demonstration-scale production (ca. 50–200 ha) for increased seed production, followed by staggered plantings at full commercial scale. Sizes of plantings, planting dates, harvest dates, and the number of years that plantings remain in production could all be varied to adjust the annual supply of raw materials.

CONCLUSIONS

Progress

Much progress has been made in understanding how *Hesperaloe* species grow and develop, in determining likely biomass yields, and in defining the likely production cycle for this crop. We have also learned much about the physiology and reproductive biology of these species. Sufficient seed has been produced to move from the experimental scale to a demonstration-farm scale. Considerable research and development work also has been carried out by the private sector, although none of this research has been published in the scientific or technical literature. Efforts have concentrated on determining the critical strength, texture, and brightness properties of paper samples made from fibers of *Hesperaloe*; work has also been done on pulping and devel-

Planting:	No. 1	No. 2	No. 3	No. 4	No. 5	No. 6		
Hectares:	100	100	100	1200	1200	1200		
Year planted:	1	2	3	4	5	6	Total	
Year replanted:	11,21	12,22	13,23	19	20		area	Annual
Ha replanted:	1200	1200	1200	1200	1200		planted	production
YEAR	Production (Mg pulp/ha)						(ha)	(Mg pulp)
1	0						100	0
2	0	0					100	0
3	0	0	0				100	0
4	0	0	0	0			1500	0
5	1800	0	0	0	0		2700	1800
6	0	1800	0	0	0	0	3900	1800
7	0	0	1800	0	0	0	3900	1800
8	1800	0	0	21600	0	0	3900	23400
9	0	1800	0	0	21600	0	3900	23400
10	1800	0	1800	0	0	21600	3900	25200
11	0	1800	0	21600	0	0	5000	23400
12	0	0	1800	0	21600	0	6100	23400
13	0	0	0	21600	0	21600	7200	43200
14	0	0	0		21600	0	6000	21600
15	21600	0	0			21600	4800	43200
16	0	21600	0				3600	21600
17	0	0	21600				3600	21600
18	21600	0	0				3600	21600
19	0	21600	0	0			4800	21600
20	21600	0	21600	0	0		6000	43200
21	0	21600	0	0	0		6000	21600
22	0	0	21600	0	0		6000	21600
23	0	0	0	21600	0		6000	21600
24	0	0	0	0	21600		6000	21600
25	21600	0	0	0	0		6000	21600
26	0	21600	0	21600	0		6000	43200
27	0	0	21600	0	21600		6000	43200
28	21600	0	0	21600	0		6000	43200
29	0	21600	0		21600		4800	43200
30	21600	0	21600				3600	43200

Fig. 6. An agricultural production plan for commercialization of *Hesperaloe*. Initial demonstration farm plantings will provide the seed for full commercial scale-up. A sequence of plantings will be required to provide a uniform supply of materials to a pulping facility. Each box represented a planting lasting 10 years and producing 18 t pulp ha^{-1} (from 180 t FW ha^{-1}) at years 5, 8, and 10.

opment of products. There is one patent on the use of *Hesperaloe* fibers in tissue and towel products (Reeves et al. 1994).

Problems

The comparatively large size and perennial habit of *Hesperaloe funifera* have posed major problems for conducting agronomic research with this plant. The large plant size necessitates large plot sizes and thus substantial commitments of land and effort for agronomic experimentation. The slow development and long production cycle of the crop necessitate that experimental plantings be evaluated for several years. One cannot redesign and replant a study the next year based on the current year's results, as with annual crops. In fact, we are continuing to learn more about the growth, biomass production, and reproduction of *H. funifera* from the first studies we established. Another 3 years of observation and monitoring will be needed just to determine the optimum growth cycle (Fig. 4).

We established expanded plantings of *Hesperaloe funifera* at the University of Arizona Maricopa Agricultural Center this past year. We have encountered many problems, including production of vigorous transplants, adequate preparation of beds, effective transplanting with available equipment, and weed control. The slow growth rate of *Hesperaloe* during its first year following transplanting (Fig. 1, 2) make this crop a very poor competitor with weeds. All weed control during stand establishment will have to be accomplished using cultivation and chemical control. William McCloskey and Ramon Cinco-Castro have conducted preliminary work on herbicide evaluations. All pre-emergent herbicides tested proved to be phytotoxic to *Hesperaloe funifera*, but the plant did appear to tolerate several post-emergent herbicides.

Over the near term, lack of adequate supplies of seed and the long period of stand establishment (5 years to first harvest) will limit how rapidly commercialization of *Hesperaloe* can proceed. The private sector, like the AARC Center, may balk at investing in a project with a comparatively long payback period. Growers, processors, and paper makers will have to find a mutually acceptable way to share the risks and establishment costs.

Prospects

In general the prospects for increased utilization of nonwood fibers, both worldwide and in the United States, are favorable (Atchison 1995; Hurter 1991; Judt 1993). Annual crops such as kenaf, hemp, roselle, and others are likely to play an important role in meeting the growing worldwide demand for paper products. Novel perennial fiber crops will have to have special properties in order to succeed commercially. The unusual combination of improved strength and texture provided by *Hesperaloe* fibers (Reeves et al. 1994) should guarantee a market if an adequate supply can be developed at a reasonable cost.

Basic research on *Hesperaloe*, including continued studies on biomass production, agricultural engineering, water use, and weed control, is currently being supported by the USDA Cooperative State Research, Education, and Extension Services (CSREES). Lack of longer-term AARC Center support for commercialization is probably not a major obstacle to successful commercialization. Potential support from the AARC Center provided a good incentive for many private-sector parties at a time when market pulp prices were very low (Fig. 5). With an economically healthier pulp and paper industry, federal support for commercialization is less critical.

REFERENCES

Atchison, J.E. 1995. Nonwood fiber could play a major role in future U.S. papermaking furnishes. Pulp & Paper (July 1995) 125–131.

Hurter, A.M. 1991. Utilization of annual plants and agricultural residues for the production of pulp and paper. Nonwood Plant Fiber Pulping, Progress Rep. 19, TAPPI Press. p. 49–70.

Judt, M. 1993. Non-wood plant fibres, will there be a come-back in paper-making? Ind. Crops Prod. 2:51–57.

McLaughlin, S.P. 1993. Development of *Hesperaloe* species (Agavaceae) as new fiber crops. p. 435–442. In: J. Janick and J. E. Simon (eds.), New crops. Wiley, New York.

McLaughlin, S.P. 1994. Intraspecific variability in agronomic traits and fiber properties of *Hesperaloe funifera*, a new crop for specialty fibers. Paper presented at 3rd Int. Conference on New Industrial Crops and Products, Catamarca, Argentina, Sept. 29, 1994.
McLaughlin, S.P. 1995. Morphological development and yield potential in *Hesperaloe funifera* (Agavaceae), an experimental fibre crop for dry regions. Expt. Agr. 31:345–353.
McLaughlin, S.P. and S.M. Schuck. 1991. Fiber properties of several species of Agavaceae from the southwestern United States and northwestern Mexico. Econ. Bot. 45:480–486.
McLaughlin, S.P. and S.M. Schuck. 1992. Intraspecific variation in fiber properties in *Yucca elata* and *Hesperaloe funifera* (Agavaceae). Econ. Bot. 46:181–186.
Presley, R. 1995. Market pulp price "boom" driven by worldwide growth in demand. Pulp & Paper (August 1995) 97–100.
Ravetta, D.A. 1994. Plant morphology and gas exchange in *Hesperaloe*: influence on its adaptation for cultivation. Ph. D. Diss., Univ. Arizona, Tucson.
Ravetta, D.A. and S.P. McLaughlin. 1993. Photosynthetic pathways of *Hesperaloe funifera* and *H. nocturna* (Agavaceae): novel sources of specialty fibers. Am. J. Bot. 80:524–532.
Reeves, R.H., J.D. Plantikow, L.J. Smith, T.P. Oriaran, A.O. Awofeso, and G.L. Worry. 1994. Soft high strength tissue using long-low coarseness hesperaloe fibers. U.S. Patent No. 5320710.

Adaptation of Kenaf to Temperate Climatic Zones

Brian S. Baldwin

Kenaf (*Hibiscus cannabinus* L., Malvaceae) is related to okra (*Abelmoschus esculentus* L.), cotton (*Gossypium* spp.) and Rose-of-Sharon (*Hibiscus syriacus* L.). Kenaf plants tend to grow as a single unbranched stem when planted at production densities (170–220,000 plants/ha). Stem color varies from pure green to deep burgundy. Flowers are borne singly, with five yellow petals and a red "blood spot" in the center of the flower. Leaves generally take two forms: deeply lobed and entire more or less resembling cotton leaves. The fiber is derived from the stem. Bast, or bark fibers are composed of phloem bundles and used extensively in cordage. The core, or woody fibers can be utilized in the production of paper, absorbants, animal bedding, and soilless potting media (Goforth and Fuller 1994).

Between the 1940s and 1960s, the feasibility of utilizing kenaf as a source of cellulose fiber for use in pulp and cordage was investigated by the USDA-ARS (White et al. 1970). It has been reported that whole stalk kenaf could be utilized in the production of paper (Nieschlag et al. 1960). However, pulping processes in the 1950s and 1960s as well as the availability of cheap wood caused interest in kenaf to wane. White and his group (1970) made an extensive investigation of the potential of growing kenaf in the southeastern United States as well as three mid-western states as far north as Nebraska. Of the 12 locations tested, only southern Florida was free of frost late enough in the year to provide viable seed for almost all of the varieties tested. Cultivars from Cuban (Cubano, C108, C2032), Floridian (E41 and E71), and Guatemalan (G4) ancestry were utilized because each showed some degree of resistance to anthracnose (*Colletotrichum hibisci* Polacci) (Wilson et al. 1965). Most kenaf cultivars are short-day plants. Dempsey (1975) reports that most improved kenaf cultivars remain vegetative until the daylight period falls below 12.5 h. Plantings of these cultivars north of 20° latitude results in strong vegetative growth, but induction of flowering occurs too close to the frost date for the production of viable seed. As a result, much of the seed planted in the United States is produced in Florida, south Texas, the Caribbean, and Latin America. Like its relatives, okra and cotton, kenaf attracts a number of pollinators to its flowers. Because of the moderate level of cross-pollination, seed obtained from tropical sources is frequently not true-to-type. This means that seed labelled as a single cultivar may contain plants with differing characteristics. Selections from these mixed seed lots and other germplasm has allowed researchers to develop lines that flower and set seed in Mississippi.

MISSISSIPPI KENAF PROJECT

The kenaf project was initiated in 1989 with the purchase and contract growth of several cultivars of seed for planting in the delta region of Mississippi. Survey of the literature indicated that the Cuban, Floridian, Guatemalan, and Taiwanese cultivars were probably best suited to Mississippi. Initial selections were made from several of these cultivars in 1993. Initial selection criteria was based solely on flowering date. Seven single plant selections were made from E71. In 1994, 0.20 ha plots were planted of G4, G45, G48, E41, E71, Cubano, C108, and C2032. Flowering date as well as *Cristulariella pyramidalis* Waterman & Marshall (a defoliating leaf spot) resistance were the selection criteria. Initial selections were crossed and selfed in the greenhouse and subsequent seed planted to the field the following spring. Germplasm was obtained from Charles Cook (USDA-ARS, Weslaco, TX) that were believed to be determinant. Somaclonal variants from Nancy Reichert's research (MSU/MAFES, Starkville, MS) were incorporated into the field nursery and observed for deviation from their parental lines. Additional selections were made from Indian and Russian germplasm. Crosses were made by utilizing the rooted meristems of the improved short-day types (that initiated flowering in the field) and germinated plants of the determinant types, or crossing rooted cuttings of a shorter short-day plant with a longer short-day plant. Agronomic potential of each selection and cross was evaluated when sufficient seed was generated. Evaluations were made based on plant height, plot yield, date of flowering and *Cristulariella* resistance.

DEVELOPMENT OF ADAPTED CULTIVARS

Currently, kenaf lines developed in Mississippi initiate flowers from the last week in June (May planting date) until frost. A study by Baldwin (1994) indicated that kenaf seed developing in north Mississippi needs 43–45 day after pollination to mature. With an average frost date of Nov. 7 (Starkville, MS), that study determined a target of Sept. 21–25 as the date of first open flower.

Our goal is not to be a major seed production center, but to have the ability to manipulate the genetics of the crop should the need arise. *Cristulariella* is a disease that strikes kenaf during humid summers. It defoliates the entire plant from 0.3 m below the apical meristem and remaining leaves are damaged by necrotic spotting. *Cristulariella* has been reported in North Carolina, Maryland, Alabama, (Pollack and Waterworth 1969), and Mississippi (Lawrence, MSU Plant Pathologist pers. commun.). The disease can be severe in the southeastern U.S., but typically is not a problem in the less humid western states. Southeastern growers cannot expect western breeders to select resistance for this disease. We must work with our own germplasm in this case.

Special interest is being paid to lines descended from determinant germplasm. The weather patterns of the Southeast and much of the Midwest result in heavy rains during the winter months. Frost-killed kenaf takes roughly 3–4 weeks to dry to 30% moisture (safe for harvest and storage). Therefore, the dependence on a killing frost to desiccate kenaf has meant that commercial production has had to stand in the fields until Mar. or Apr. Several selections from the determinant stock senesce and die as early as the last week in Sept., however, yield is severely compromised. Evaluations are now being made to find lines that senesce closer to the frost date, so that dry-down time will be decreased and fall harvest may be possible. Indeed, cultivars show differential rate of dry-down after kill. Investigations are under way to determine which cultivars reliably desiccate rapidly. Mississippi State University experimental lines MOP6 and MOP6E show the greatest promise for this, drying to 30% moisture 2–3 weeks before most other cultivars.

PROSPECTS

Variation existing in seedlots from kenaf production fields has allowed plant breeders to make selections for lines that are best adapted to each production region. Efforts in Mississippi have focused on successful seed set, disease resistance, and dry-down time to avoid winter rains and late harvest. Currently there are nearly 300 lines in the Mississippi program that flower and set seed at 33°N latitude. The best of these lines are being evaluated for yield and disease resistance at three locations in Mississippi. Release of germplasm and cultivars are expected in Spring 1997.

REFERENCES

Baldwin, B.S. 1994. Physiological maturity determination and germinability of kenaf seed. Sixth Annu. Internat. Kenaf Conf. New Orleans, LA.

Dempsey, J.M. 1975. Kenaf. p. 203–304. In: Fiber crops. Rose Printing Co., Tallahassee, FL.

Goforth, C., and M.J. Fuller 1994. A summary of kenaf production and product development research, 1989–1993. Mississippi Agr. & For. Expt. Sta., Mississippi State. Bul. 1011.

Nieschag, H.J., G.H. Nelson, I.A. Wolf, and R.E. Perdue. 1960. A search for new fiber crops. TAPPI 43(3):193–201

Pollock, F.G. and H.E. Waterworth. 1969. A leafspot disease of kenaf in Maryland associated with *Cristulariella pyramidalis*. Plant Dis. Rep. 53:810–812.

White, G.A., D.G. Cummins, E.L. Whiteley, W.T. Fike, J.K. Greig, J.A. Martin, G.B. Killinger, J.J. Higgins, and T.F. Clark. 1970. Cultural and harvesting methods for kenaf: An annual crop source of pulp in the southeast. USDA-ARS Production Research Rep. 113. Washington, DC.

Wilson, R.D., T.E. Summers, J.G. Joyner, D.W. Fishler, and C.C. Sealer. 1965. 'Everglades 41' and 'Everglades 71' two new varieties of kenaf (*Hibiscus cannabinus* L.) for fiber and seed. Fla. Agr. Expt. Sta. Circ. S-168.

Response of Kenaf to Nitrogen Fertilization

Charles L. Webber III

As a result of inconsistent responses to supplemental nitrogen fertilization and the general absence of kenaf (*Hibiscus cannabinus* L.) fertility recommendations, research was conducted at Lane, OK, to determine the effect of N application rates of 0 (control) 56, 112, 168, and 224 kg/ha on kenaf yield components. Kenaf cultivars 'Tainung 1' and 'Everglades 41' were planted in May of 1989 and 1990 on a Bernow fine sandy loam soil with a 0%–3% slope. Nitrogen applications did not affect plant height, plant density, or stalk biomass compared to the control. Stalk diameters at 1 m increased with N application rates of 112, 168, and 224 kg/ha compared to the control. Though not significantly different, there was a trend of increased stalk yields up to 168 kg N/ha, while 224 kg N/ha significantly decreased stalk yields compared to 168 kg N/ha. This research demonstrated the importance of determining N requirements for specific production areas without assuming the need for high N applications to achieve maximum yields. It also demonstrated that excess N can have a detrimental effect on stalk yields. Future research should examine the effect of other essential nutrients on fiber quantities and qualities produced on specific soil types.

Kenaf (*Hibiscus cannabinus* L.), a warm-season annual, was first identified as a potential jute substitute in the manufacturing of rope, twine, carpet backing, and burlap (Wilson et al. 1965). Research demonstrated that kenaf is also a very promising fiber source for production of paper pulp (Nieschlag et al. 1960; White et al. 1970). Kenaf fibers were processed into both newsprint and bond paper (Clark et al. 1971; Bagby et al. 1979). Other potential commercial uses for kenaf are being investigated, such as animal feed (Webber 1993), poultry litter (Tilmon et al. 1988), bulking agent for sewage sludge (Webber 1992), and for oil absorption (Tiller et al. 1994).

As with other crops proper fertility maintenance, especially for supplemental nitrogen application, is needed to optimize kenaf yields, and minimize production cost. Reports so far are inconsistent relative to the effects of N on kenaf stalk yields (White and Higgins 1965); researchers in Georgia have reported both positive (Adamson et al. 1979) and no benefits (Massey 1974). Studies in Florida demonstrated that the positive response to N applications on stalk yields were dependent on soil type (Joyner et al. 1965), where kenaf grown on a sandy soil responded to N and did not respond to N on a peat soil. Bhangoo et al. (1986) in California, and Sij and Turner (1988) in Texas, increased stalk yields with the addition of N to soils with low available nitrogen. Stalk yields in Missouri (Ching and Webber 1993) on a silty clay soil and in Nebraska (Williams 1966) on a silty clay loam soil did not benefit from N applications. Stalk yields have also responded differently to N at the same location and soil between years (Hovermale 1993). In this study the effect of five nitrogen application rates (0, 56, 112, 168, and 224 kg/ha) on kenaf yield components was investigated in order to increase available information in kenaf products.

METHODOLOGY

A two-year field-plot study was conducted at Lane, Oklahoma, on a Bernow fine sandy loam, 0%–3% slope, (fine-loamy, siliceous, thermic Glossic Paleudalf). Each year the N fertilizer treatments (56, 112, 168, and 224 kg/ha) were applied in the form of ammonia nitrate. In addition, the entire research area also received a broadcast fertilizer application of 0N–72P–139K kg/ha, which was incorporated prior to planting. Plots were 3 m wide (four 76-cm rows) and 6 m long. Kenaf cultivars 'Tainung 1' and 'Everglades 41' were planted 2 May 1989, and 23 May 1990. All plots received a pre-emergence application of metolachlor [2-chloro-N-(2-ethyl-6-methlphenyl)-N-(2-methoxy-1-methylethyl) acetamide] at a rate of 1.12 kg ai/ha at planting. The herbicide was applied using a tractor-mounted sprayer delivering 187 liters/ha at 275 kPa pressure. Fan-tip sprayer nozzles were 0.5 m apart on a 3-m boom. All plots were kept weed free by handweeding.

Kenaf plots were hand-harvested on 21 Oct. 1989, 172 days after planting (DAP), and on 19 Oct. 1990, 149 DAP. A 2.25 m^2 (1.50 m by 1.50 m) quadrate was harvested from the center rows. Plant population and moisture content were also determined in the harvest area. The harvested plants were cut at ground level and fresh weights were determined. Three plants were randomly selected at 105 DAP and at harvest to determine plant heights. The harvested plants used to determine heights were also measured with calipers to determine stalk diameters at 1 m above ground level.

Leaves, flowers, and flower buds were removed from the stalks and weighed separately before and after the samples were oven dried at 66°C for 48 h. Fresh and oven-dry weights of the three plants were used to determine the percent moisture of the plants and the percent stalks by weight. The plant moisture content and percent stalks were used to convert kenaf fresh weight to dry weight. Stalk yields were based on oven-dry weights. Precipitation was measured at the Lane, OK, research location.

In both years, treatments were arranged in a randomized complete block design with four replications. When the F-test indicated statistical significance at the $P = 0.05$ level, the least significant difference (LSD) test was used to separate means. In addition, coefficient of linear correlation (r) analysis was used to establish the correlation between various yield components. Since no significant year by treatment by cultivar interactions were indicated results are reported averaged across years.

RESULTS

Total precipitation was 664 mm for the 1989 and 623 mm for the 1990 growing season. The 20-yr average rainfall for the location from 1 May to 31 Oct. is 611 mm. Precipitation during the first half of the 1989 growing season (455 mm) was 214 mm greater than that of the first half of the 1990 growing season (241 mm).

Plant heights, plant populations, and stalk biomass percentages for the N application levels were not different from the control (Table 1). Stalk yield and diameter were affected by N application rates. Stalk yield tended to increase as N applications rates increased up to 168 kg N/ha, and at 224 kg N/ha, a significant reduction in stalk yield occurred compared to the 168 kg N/ha level. Thus, excess N application can be detrimental to stalk yield. Stalk diameter at 1 m was increased at the 112, 168, and 224 kg/ha N rates compared to the control (Table 1). Massey (1974) and Hovermale (1993) also reported that N significantly increased stalk diameter without increasing stalk yield.

Regressional analysis showed that the only consistent difference between the r values between rates was for the stalk biomass percentage (Table 2). As N application rates increased the correlation between stalk yields and stalk biomass percentages decreased (Table 2). At the 0 N rate, yields were significantly correlated,

Table 1. Yield components of kenaf as influenced by nitrogen fertilization, cultivars, and years, means for 1989 and 1990.

Variable	Plant height(cm)		Plant population (1000 plants/ha)	Stalk diameters (mm)	Stalk biomass (%)	Stalk yield (t/ha)
	Early	Harvest				
Nitrogen[z] (kg/ha)						
0	185	265	148	12.9	80.2	16.3
56	183	261	161	13.2	81.0	16.3
112	187	264	165	13.9	79.8	17.2
168	183	267	156	13.9	78.8	17.7
224	186	262	159	13.9	80.2	15.3
LSD (0.05)[y]	NS	NS	NS	0.9	NS	2.0
Cultivar[x]						
Tainung 1	195	275	153	14.2	80.0	17.0
Ev. 41	175	252	163	13.0	80.0	16.2
LSD (0.05)	4	8	NS	0.6	NS	NS
Year[w]						
1989	211	289	152	15.2	83.3	22.5
1990	159	239	164	12.0	76.7	10.6
LSD (0.05)	4	8	NS	0.6	1.4	1.3

[z]Nitrogen (kg/ha) means averaged across cultivars and years.
[y]Least significant difference test level of $P < 0.05$.
[x]Cultivar means averaged across nitrogen rates and years.
[w]Year means averaged across nitrogen rates and cultivars.

($r = 0.63$) at $P > 0.01$ level, whereas at the 56 kg N/ha and 112 kg N/ha rate the positive correlation was at the $P > 0.05$ level. At the 168 kg N/ha and 224 kg N/ha rates there was no significant correlation between stalk yields and stalk biomass percentages.

Plant height, and stalk diameter were greater for 'Tainung 1' than 'Everglades 41', but no differences were detected between cultivars for plant population, stalk biomass percentage or stalk yield. 'Tainung 1' plants were 20 cm taller than 'Everglades 41' at 105 DAP and this trend continued until the final harvest when 'Tainung 1' was 23 cm taller than 'Everglades 41' (Table 1). The height advantage for 'Tainung 1' was achieve prior to 105 DAP and was maintained until the end of the growing season. These results were consistent with research in Missouri, Oklahoma, and Mississippi (Ching et al. 1993).

All yield components, except for plant population, were greater in 1989 than 1990 (Table 1). Plants in 1989 were 52 cm taller at 105 DAP, and 50 cm greater at harvest than in 1990. Stalk diameter was the only plant parameter significantly affected by N rates, cultivars and years, whereas stalk biomass percentage was only affected by years. Stalk yield was more than twice as great in 1989 (22.5 t/ha) than to 1990 (10.6 t/ha) (Table 1). Greater plant height, stalk diameter, stalk percentage and yield in 1989 may have resulted in part from an earlier planting date in 1989 (21 days) and greater early seasonal precipitation.

CONCLUSIONS

This research demonstrated that kenaf in loamy soils was relatively unresponsive to N application of 168 kg/ha and lower, and at high N-level there was a potential risk of decreasing stalk yield. Future research should examine the effect of other essential nutrients on fiber quantity and quality produced on specific soil types.

Table 2. Coefficients of correlation (r) for yield components as related to stalk yield combined across cultivars and years.

N (kg/ha)	Coefficients of correlation (r)			
	Plant ht.	Plant population	Stalk diam.	Stalk (%)
Combined rates	0.74**	–0.31**	0.75**	0.47**
0	0.81**	–0.36	0.90**	0.63**
56	0.67**	–0.24	0.59**	0.50*
112	0.71**	–0.07	0.66**	0.56*
168	0.72**	–0.63**	0.76**	0.43
224	0.81**	–0.37	0.89**	0.46

*, ** Significant at 5% (*) or 1% (**) level.

REFERENCES

Adamson, W.C., F.L. Long, and M.O. Bagby. 1979. Effect of nitrogen fertilization on yield, composition, and quality of kenaf. Agron. J. 71:11–14.

Bagby, M.O., R.L. Cunningham, G.F. Touzinsky, G.E. Hamerstrand, E.L. Curtis, and B.T. Hofreiter. 1979. Kenaf thermomechanical pulp in newsprint. p. 111. TAPPI/NPFP Committee Progress Report 10. Atlanta, GA.

Bhangoo, M.S., H.S. Tehrani, and J. Henderson. 1986. Effect of planting date, nitrogen levels, row spacing, and plant population on kenaf performance in the San Joaquin Valley, California. Agron. J. 78:600–604.

Ching, A., Jr. and C.L. Webber III. 1993. Effect of fertilizer applications on kenaf photosysthesis, growth and yield. p. 17–23. Proc. Fourth Int. Kenaf Conf., Int. Kenaf Assoc., Ladonia, TX.

Ching, A., C.L. Webber III, and S.W. Neill. 1993. Effect of location and cultivar on kenaf yield components. Ind. Crops Prod. 1:191–196.

Clark, T.F., R.L. Cunningham, and I.A. Wolff. 1971. A search for new fiber crops. TAPPI 54(1):63–65.

Hovermale, C.H. 1993. Effect of row width and nitrogen rate on biomass yield of kenaf. p. 35–40. Proc. Fourth Int. Kenaf Conf., Int. Kenaf Assoc. Ladonia, TX.

Joyner, J.F., D.F. Fishler, and F.D. Wilson. 1965. Fertility studies in kenaf. p. 105–118. Proc. Second Int. Kenaf Conf., Palm Beach, FL.

Massey, J.H. 1974. Effects of nitrogen levels and row widths on kenaf. Agron. J. 66:822–823.

Nieschlag, H.J., G.H. Nelson, I.A. Wolff, and R.E. Perdue Jr. 1960. A search for new fiber crops. TAPPI 43(3):193–201.

Sij, J.W. and F.T. Turner. 1988. Varietal evaluations and fertility requirements of kenaf in Southeast Texas. Texas Agr. Expt. Sta. Bul.

Tilmon, H.D., R. Taylor, and G. Malone. 1988. Kenaf: an alternative crop for Delaware. p. 301–302. In: J. Janick and J.E. Simon (eds.), Advances in new crops. Timber Press, Portland, OR.

Tiller, F.M., F. Zhou, and S.L. Tarng. 1994. Agricultural fibers as aids in separation: coalescence, combustible filter aids, deep-bed filtration, oil absorption. p. 161–185. Proc. Int. Kenaf Assoc. Conf., Int. Kenaf Assoc., Ladonia, TX.

Webber, C.L., III. 1992. Kenaf (*Hibiscus cannabinus* L.) yield components as affected by sewage sludge applications. p. 57–62. Proc. Pulping Conf., TAPPI, Atlanta, GA.

Webber, C.L., III. 1993. Crude protein and yield components of six kenaf cultivars as affected by crop maturity. Ind. Crops Prod. 2:27–31.

White, G.A., D.G. Cummins, E.L. Whiteley, W.T. Fike, J.K. Greig, J.A. Martin, G.B. Killinger, J.J. Higgins, and T.F. Clark. 1970. Cultural and harvesting methods for kenaf. Product Research Report 113. USDA, Washington, DC.
White, G.A. and J.J. Higgins. 1965. Growing kenaf for paper. p. 27–40. Proc. Second Int. Kenaf Conf., Palm Beach, FL.
Williams, J.H. 1966. Influence of row spacing and nitrogen levels on dry matter yields of kenaf (*Hibiscus cannabinus* L.). Agron. J. 58:166–168.
Wilson, F.D., T.E. Summers, J.F. Joyner, D.W. Fishler, and C.C. Seale. 1965. 'Everglades 41' and 'Everglades 71', two new varieties of kenaf (*Hibiscus cannabinus* L.) for the fiber and seed. Florida Agr. Exp. Sta., Belle Glade. Cir. S-168.

Potential for Kenaf Improvement via Somaclonal Variation

Nancy A. Reichert and Brian S. Baldwin

Kenaf (*Hibiscus cannabinus* L., Malvaceae) is a potential fiber crop adapted to the southern United States. As part of the kenaf improvement program at Mississippi State University, tissue culture is being employed as a means for introduction of new or altered traits into kenaf. Regenerants often display altered phenotypes, termed somaclonal variation (Larkin and Scowcroft, 1981). The exploitation of heritable somaclonal variants has been used in various plant improvement strategies (Larkin and Scowcroft 1981; Evans 1989; Larkin et al. 1989; Phillips et al. 1994) but success has been sporadic.

Our objectives were to: (1) develop an adventitious regeneration protocol for kenaf and, (2) analyze the regenerants (R_0) and subsequent progeny for trait differences. Various traits were assessed over the course of two to three generations (R_1 and R_2) and compared to control plants.

METHODOLOGY

Internodal stem sections were obtained from field-grown plants of 'Everglades 41' ('E41'), surface-disinfested, then placed on a medium containing Murashige and Skoog (1962) salts, supplemented with (per liter): 1.0 mg thiamine-HCl, 0.5 mg pyridoxine-HCl, 0.5 mg nicotinic acid, and 100 mg myo-inositol, 3% sucrose, 0.8% phytagar, plant growth regulators [0.1 mg 1-naphthaleneacetic acid (NAA) plus 3.0 mg thidiazuron (TDZ)] with pH 5.8 for shoot initiation (Reichert and Liu 1994). Shoots were rooted on MS medium containing no plant growth regulators and transplanted into peat cubes for acclimatization. 'E41' R_0 regenerants were transferred to the greenhouse for one week, then transplanted into the field at the Plant Science Research Center (PSRC), at Mississippi State Univ. in 1993. All plants were allowed to self-pollinate. R_1 seeds were harvested by plant, then germinated in peat plugs in the greenhouse and transplanted into the field at the PSRC in 1994. R_1 seedlings were compared to 'E41' seedlings produced in the same year.

In 1994, the regeneration protocol was found to be more efficient using leaf sections. Seeds of various kenaf cultivars were germinated aseptically and maintained in vitro via nodal sections on hormone free medium. These are referred to as seed clonal lines (SCL). Leaf sections were explanted on MS-based shoot induction medium containing (per liter) 0.1 mg NAA plus 3.5 mg TDZ (Reichert and Liu 1994). We obtained 150 R_0 regenerants from 'E41', 'Guatemala 45' ('G45') and 'G48' for field testing at the PSRC in June–July, 1994. In addition, 20 clonal control plants (nodal section-propagated) from each SCL were transplanted into the field concurrently. These clonal plants were used to determine: (1) the normal range of variation within a clonal line as a result of environmental influences and (2) potential heterozygosity and subsequent segregation in the original seed. (No segregation was observed.) Both checks were necessary to determine altered phenotypes were actually due to the leaf regeneration protocol. Additional regenerants produced after July 1st were maintained in the greenhouse for the production of R_1 seed.

In 1995, R_2 seeds harvested from individual 'E41' R_1 plants (generated in 1994; from R_0s planted in 1993) as well as R_1 seeds harvested from each R_0 regenerant (field planted in 1994) were germinated in peat plugs in the greenhouse. Eight days later, each line was evaluated for altered seedling phenotypes then, 15 seedlings from each line were space planted in sib rows at the PSRC. Selfed seedlings obtained from SCL lines were space planted adjacent as controls. Plant height, stem girth, stem color, leaf shape, days to first flower bud, days to first open flower, and flower shape/morphology were determined in the field in 1994 and 1995.

RESULTS

Adventitious Regeneration

The leaf regeneration protocol initially developed for 'E41' was successful on six other cultivars: E71, G4, G45, G48, G51, and Tainung 1. Adventitious shoots were not obtained from 'Cuba 2032', 'Cubano', or 'GR2563'. However, sufficient numbers of plantlets for field trials were generated only from E41, G45, and G48.

R_0 and R_1 Analyses for 'E41', 'G45' and 'G48'

R_0 kenaf regenerated adventitiously from leaves displayed somaclonal variation for stem color based on a five point color scale. Stems were classified in discrete classes based upon the relative amount of anthocyanins present. Individual plants fell into one of the following color classes: green, light pink, dark pink, red, or burgundy. 'G45' regenerants displayed greater variation in stem color than the SCL control plants. The SCL controls displayed stem colors varying from light pink to dark pink due to environmental effects. 'G45' R_0 plants displayed stem colors spanning from green all the way to red; displaying less or more anthocyanins in the stem than the SCL controls. 'E41' R_0 plants displayed greater amounts of anthocyanins in the stem (dark pink to red) than the corresponding SCL controls (light pink to dark pink). No differences were detected in 'G48' between the R_0 and SCL plants for stem color.

R_0 flowers were observed for differences in flower shape/morphology. Normal kenaf flowers have a crimson eye with five very wide light yellow flower petals which overlap slightly. Most R_0 plants displayed normal flowers, but there were some notable exceptions. In 'E41', flowers containing narrow, non-overlapping petals were observed. Most were composed of five petals, but one plant consistently displayed flowers which contained six and seven petals and one 'G48' regenerant displayed flowers with 10 petals. One 'E41' R_0 regenerant had flowers that were consistently male sterile (vestigial anthers) and one 'G48' R_0 regenerant produced flowers with indehiscent anthers. One R_1 plant from a 'G45' R_0 line was observed to produce flowers that were pink due to "bleeding" of the crimson eye into the petals.

R_1 and R_2 Analyses for 'E41'

R_1 seedlings of all R_0 plants of 'E41' displayed variation in seedling characteristics and when mature expressed variation in days to flower as compared with 'E41' controls. Germination of the R_1 seed revealed abnormal seedling traits. These characteristics were compared to 'E41' seedlings generated from control plants grown in the field, adjacent to the R_0 during the same season (Table 1). There was a higher percentage of abnormal seedlings from the R_1 plants than from the 'E41' control seedlings. An albino (chlorophyll deficient) mutation was displayed in 5% of the R_1 seedlings, but was not observed in the controls and was never detected in the millions of seeds planted in the kenaf cultivar trial conducted in 1994 at the PSRC.

At maturity 'E41' (R_1) plants were not compared to controls for stem color or leaf shape since the controls were not SCL. Therefore, environmental influences as well as potential outcrossing which could contribute to variation in stem color could not be ruled out. Flowering dates for R_1 plants were compared to normal field grown 'E41' plants (Table 2). The first flower bud appeared four weeks earlier and the first open flower appeared more than three weeks earlier in the R_1 population versus field-derived 'E41' plants. The dates when 50% of the plants had flower buds or open flowers occurred approximately two weeks earlier in the R_1 (Table 2). No differences were detected between R_1 and field-derived 'E41' plants for

Table 1. Seedling traits in control and R_1 kenaf ('E41').

Treatment	n	Distribution (%) Normal seedling	Empty seed	Crinkled cotyledon	Chlorophyll deficient	Single cotyledon
Control[z]	250	93	5	2	0	0
R_1[y]	176	57	27	10	5	1

[z]Control seeds were from a commerial seedlot grown under the same conditions.
[y]Seeds harvested from R_0 plants in 1993 and pooled.

Table 2. Days to flower for control vs R_1 kenaf ('E41').

Treatment	n	Days after planting[z] First flower bud	50% plants w/ flower buds	First open flower	50% plants open flower
Control[y]	135	122	130±6[x]	136	149±5
R_1[w]	139	94	106±5	112	125±7

[z]June 3 was designated as day 1.
[y]Control plants were initiated from seed in the field on May 19.
[x]Numbers ±1 SD, where applicable.
[w]Plants were initiated from seed on May 23 and transplanted to the field on June 3.

plant height, stem girth, and flower shape. R_2 seed harvested from one 'E41' R_1 plant displayed the albino (chlorophyll deficient) trait upon germination at a frequency of 3%.

R_0, R_1, and initial R_2 data confirm that the adventitious regeneration protocol developed for kenaf results in regenerants displaying new or altered phenotypes. Some visible traits that appeared in kenaf regenerants and subsequent progeny (albino, crinkled cotyledons, different colored reproductive structures, male sterility) were similar to those previously observed as single gene mutations in progeny of maize (*Zea mays* L.) somaclones (Phillips et al. 1994). Early flowering date is probably the most important variant phenotype to come from this work. However, altered flower color, polypetally, and male sterility are also desirable characteristics for the kenaf breeding program. All of these altered phenotypes are indicators that heritable changes can be made in the genome of kenaf via this leaf regeneration protocol. This protocol has proven to be successful on seven cultivars to date. Continued screenings will determine if any traits will be useful to the kenaf breeding program.

REFERENCES

Evans, D.A. 1989. Somaclonal variation: Genetic basis and breeding applications. Trends Genet. 5:46–50.

Larkin, P.J. and W.R. Scowcroft. 1981. Somaclonal variation: A novel source of variability from cell cultures for plant improvement. Theor. Appl. Genet. 60:197–214.

Larkin, P.J., P.M. Banks, R. Bhati, R.I.S. Brettell, P.A. Davies. S.A. Ryan, W.R. Scowcroft, L.H. Spindler, and G.J. Tanner. 1989. From somatic variation to variant plants: mechanisms and applications. Genome 31:705–711.

Phillips, R.L., S.M. Kaeppler, and P. Olhoft. 1994. Genetic instability of plant tissue cultures: breakdown of normal controls. Proc. Nat. Acad. Sci. (USA) 91:5222–5226.
Reichert, N.A. and D. Liu. 1994. Protoplast culture and *in vitro* regeneration of kenaf. Conference Proc., Sixth Int. Kenaf Assoc. p. 61–65.

Growth of Bedding Plants in a Kenaf-Based Potting Medium

Nancy A. Reichert and Brian S. Baldwin

Kenaf (*Hibiscus cannabinus* L., Malvaceae) is an annual, tropical fiber crop that is currently being grown in the southern U.S. The outer bast (bark) fiber has been used in newsprint production and is being explored for use in textiles and other products (Goforth and Fuller 1994). However, few markets have been developed for the inner core fiber, which comprise 65%–70% of the stalk by weight. Studies have been conducted on the use of ground kenaf core as an amendment for greenhouse potting media in concentrations up to 100% (v/v). In that study, Wang (1994) determined that three foliage plants [*Brassaia actinophylla* Endl. (schefflera), *Hibiscus rosa-sinensis* L. (tropical hibiscus) and *Pittosporum tobira* (Thunb.) Ait.] grown in a 70 kenaf: 30 peat (v/v) medium were superior with regard to height, width and dry weight compared to those grown in two commercial media (Metro Mix 250 or Sunshine #1). Since kenaf core is a renewable biomass and is less expensive than sphagnum peat moss we initiated a study to determine if annual bedding plants could be successfully grown in a potting medium containing ground kenaf core as the primary component.

METHODOLOGY

Finely ground fresh kenaf core was obtained from Core Products, Charleston, Mississippi. A kenaf-peat potting medium composed of 70% (v/v) kenaf core and 30% (v/v) sphagnum peat moss contained the following amendments: 5.22 kg gypsum, 5.22 kg superphosphate, 4.75 kg dolomitic limestone, 0.95 kg Micromax Plus micronutrients (Scott Company), and 0.59 kg AquaGro 2000•G/m^3 (Aquatrols of America). Slo-Cote slow release fertilizer (14% N–14% P_2O_5–14% K_2O; Bonus Crop Fertilizer, Inc.) was added at 6.41 kg/m^3 to the kenaf-peat medium and a commercial peat-based medium (Sunshine #1; Fisons Horticulture).

Twenty-three cultivars representing 17 ornamental and vegetable plant species (Tables 1 and 2) were grown in kenaf-peat medium and commercial peat-based medium. Plants were started from seed and transplanted or purchased as plugs. All were transplanted into 72-cell bedding flat inserts. Each cell measured 3.7 $cm^2 \times 5$ cm deep, and one pack of 6 cells represented one replication (12 packs per sheet, 72 cells per tray). For each cultivar, there were six to 12 replications per medium. Supplemental nitrogen at 200 mg/l was provided to all plants twice weekly. Harvest dates were at 18 days (*Capsicum*, *Limonium*, *Lycopersicon*, *Solanum*, *Tithonia*), 4 weeks (*Begonia*, *Celosia*, *Coleus*, *Impatiens*, *Ocimum*, *Petunia*), 6 weeks (*Salvia*), 8 weeks (*Tagetes*), and 10 weeks (*Ageratum*, *Catharanthus*, *Dianthus*, *Senecio*) after transplanting.

RESULTS

There were no differences for total numbers of flowers or numbers of open flowers between the kenaf-peat medium and the peat-based medium with one exception—a single cultivar of *Impatiens wallerana* displayed greater numbers of open flowers in kenaf-peat medium (Table 1). Therefore, within all plant species tested, plants grown in kenaf-peat medium produced as many (or more, in one case) flowers as the peat-based medium. No differences were noted for plant heights in 15 of the plant species, but two (*Celosia argentea* and *Lycopersicon esculentum*) displayed greater heights in peat-based medium (Table 2). Shoot dry weights were similar for ten of the 17 plant species, but when different, they were greater for plants grown in peat-based

Table 1. Growth of seven ornamental plant species in kenaf-peat and peat mix.

Cultivar	Potting[z] medium	Plant ht. (cm)	No. open flowers	Total flowers	Plant dry wt. (g)
	Begonia ×semperflorens cultorum Hort. (wax begonia)				
Ambassador	Peat	10	9.0[y]	--[x]	1.3
Scarlet	Kenaf–peat	8	8.5	--	1.2
Vodka	Peat	10	5.8	--	1.5
	Kenaf–peat	10	8.8	--	1.5
	Catharanthus roseus (L.) (vinca)				
Peppermint	Peat	14	8.0	15.0	3.2
Cooler	Kenaf–peat	12	6.8	14.8	2.9
	Dianthus chinensis (L.) (pinks)				
Princess Mix	Peat	23	8.5	14.2	7.5
	Kenaf–peat	22	8.8	13.6	6.4
Telstar Mix	Peat	18	2.5	6.0	--
	Kenaf–peat	15	4.5	6.5	--
	Impatiens wallerana Hook. f. (impatiens)				
Deco Orange	Peat	21	6.8	14.3	2.5*
	Kenaf–peat	14	9.2	13.2	1.3
Super Elfin White	Peat	19	3.8	9.5	2.6
	Kenaf–peat	17	3.5	12.7	2.1
Tempo Orange	Peat	20	17.0	34.7	3.2
	Kenaf–peat	18	21.8	42.0	2.9
	Petunia ×hybrida Hort. Vilm.-Andr. (petunia)				
Celebrity Ice Mix	Peat	24	5.2	6.4	4.8*
	Kenaf–peat	19	3.5	4.8	2.6
	Salvia splendens F. Sellow ex Roem. & Schult. (salvia)				
Flare	Peat	27	1.8	3.5	6.7*
	Kenaf–peat	23	1.7	2.2	4.7
	Tagetes erecta L. (marigold)				
Inca Yellow	Peat	15	1.6	5.5	4.1
	Kenaf–peat	16	1.9	6.0	4.5
Bonanza Orange	Peat	15	7.8	10.3	4.0
	Kenaf–peat	15	8.1	12.3	3.3
Lemondrop	Peat	11	3.3	5.2	2.3
	Kenaf–peat	11	3.1	5.9	2.1

[z]Peat based mix = Sunshine #1; kenaf–peat = 70% kenaf core plus 30% peat.
[y]In *Begonia*, numbers represent the number of flower stalks per 6-cell pack.
[x]Measurement not taken.
*Paired numbers within a column differ by greater than one standard deviation.

medium (Tables 1 and 2). At harvest increased root masses for all plant species grown in kenaf-peat medium were observed. Wang (1994) also noted more vigorous root systems on the plants grown in kenaf-based media compared to commercial peat-based media. The emphasis on shoot *and* root growth for plants grown in kenaf-peat medium may account for the slightly lower shoot dry weights displayed by some plants species. Had dry

Table 2. Growth of ten ornamental and vegetable plant species in kenaf-peat and peat mix.

Cultivar	Potting[z] medium	Height (cm)	Dry wt. (g)
	Ageratum houstonianum Mill. (ageratum)		
Danube	Peat	16	4.2
	Kenaf–peat	13	3.2
	Capsicum annuum L. (bell pepper)		
Blockbuster Hybrid	Peat	13	1.5*
	Kenaf–peat	11	1.0
	Celosia argentea L. (cockscomb)		
Treasure Chest	Peat	5*	2.9
	Kenaf–peat	4	1.4
	Coleus blumei Benth. (coleus)		
Wizard Golden	Peat	15	3.6*
	Kenaf–peat	14	2.1
	Limonium sinuatum (L.) Mill. (statice)		
Petite Bouquet	Peat	20[y]	2.3
	Kenaf–peat	19	1.9
	Lycopersicon esculentum Mill. (tomato)		
Heatwave Hybrid	Peat	23*	2.4
	Kenaf–peat	19	2.1
	Ocimum basilicum L. (sweet basil)		
Purple Ruffles	Peat	13	--
	Kenaf–peat	14	--
	Senecio cineraria DC. (dusty miller)		
Silver Dust	Peat	9	3.9*
	Kenaf–peat	7	1.8
	Solanum melongena L. (eggplant)		
Burpee Hybrid	Peat	9	1.0
	Kenaf–peat	8	1.0
	Tithonia rotundifolia (Mill.) S.F. Blake (Mexican sunflower)		
Goldfinger	Peat	10	-
	Kenaf–peat	9	-

[z]Peat based mix = Sunshine #1; kenaf–peat = 70% kenaf core plus 30% peat.
[y]Measurements represent the average of two plant widths.
*Paired numbers within a column differ by greater than one standard deviation.
All but one were initiated as seedling transplants (*Coleus* was initiated from plugs).

weights been determined on whole plants instead of just shoot (above-ground) growth, differences in weights may not have existed. Various plant species grown in the two media were graded by the Mississippi State University Horticulture Judging Team. All received similar grades (very good to superior) regardless of the growing medium.

Annuals grown in the kenaf-peat medium, in general, performed as well as those grown in peat-based medium. Cost comparisons indicated that the kenaf-peat medium was much less expensive than the peat-based medium. New formulations of kenaf-based media are currently being developed to give similar or superior results when compared to other peat-based potting media. The lower cost of kenaf coupled with the decreasing availability of peat should make kenaf-based media an attractive alternative to conventional greenhouse potting media.

REFERENCES

Goforth, C.E. and M.J. Fuller. 1994. A summary of kenaf production and product development research 1989–1993. Mississippi Agr. & Forestry Expt. Sta., Mississippi State. Bul. 1011

Wang, Y.-T. 1994. Using ground kenaf stem core as a major component of container media. J. Am. Soc. Hort. Sci. 119:931–935.

FRUITS AND NUTS

Mediterranean Fruits

Joan Tous* and Louise Ferguson

The Mediterranean basin has long been a site of temperate fruit and nut production. Grapes, olives, figs, almonds, dates, and carobs have been cultivated there since early times. This area has both active consumption and commerce in these crops. Mediterranean countries are also rich sources of plant germplasm with the potential for new crops, and the revival of old crops. Recently interest in traditional diets, particularly in the Mediterranean diet, has increased among the public and scientific health communities. At a recent international symposium (Tree Nuts, Health and the Mediterranean Diet), in San Francisco, scientists presented several lines of evidence indicating the Mediterranean diet has the potential to prevent heart disease and other chronic diseases (Drescher et al. 1995).

Grains, legumes, fruits, vegetables, olive oil, wine, seeds, and tree nuts are a part of the "Traditional Healthy Mediterranean Diet Pyramid" (Fig 1), a diet now associated with the good health and high adult life expectancy of the Mediterranean people (Sacks 1995). Traditional Mediterranean fruit and nut crops include grapes, olives, figs, almonds, hazelnuts, pistachios, pomegranates, apricots, and citrus. One important aspect is the fat in olive and nut oils is mostly unsaturated, good for the prevention of heart disease. These oils are very high in monounsaturated fat (oleic acid), and secondarily high in polyunsaturated fat (linoleic acid).

Olive oil, an important part of the Mediterranean diet, has been object of renewed interest in recent years. Other Mediterranean crops, such as mandarins, figs, loquats, persimmons, pomegranates, pistachios, carob

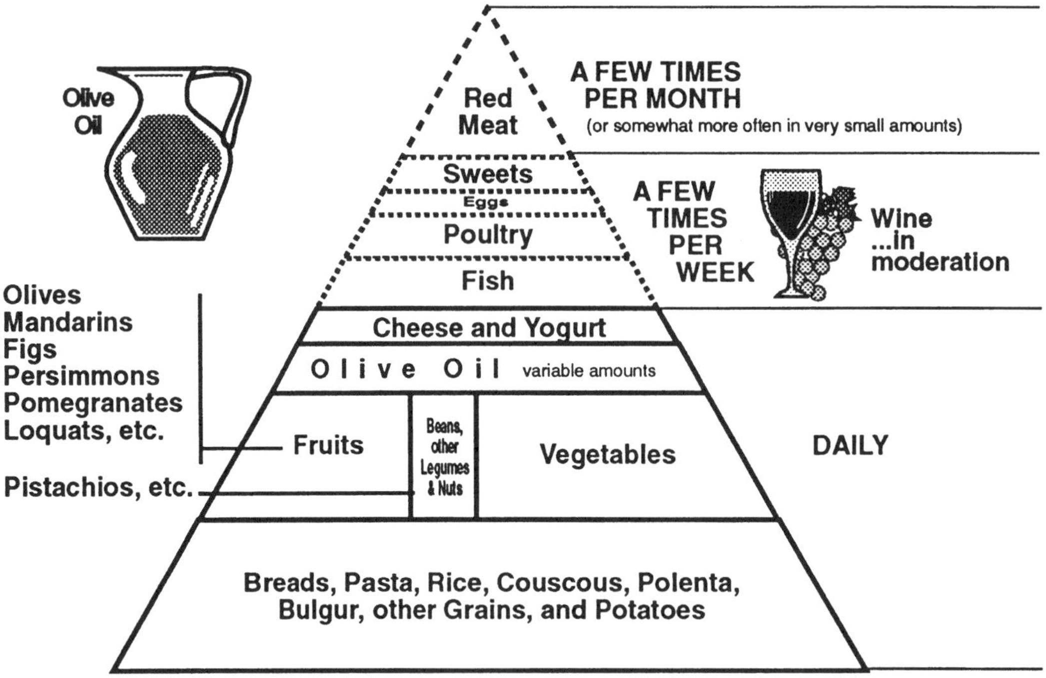

Fig. 1. This Mediterranean Diet Pyramid was co-developed and jointly released in 1994 by Oldways Preservation & Exchange Trust, the World Health Organization (WHO) European Regional Office, and the WHO/FAO Collaboration Center in Nutritional Epidemiology at Harvard School of Public Health.

*The first author greatly acknowledges the financial support of the Instituto Nacional de Investigacion y Tecnologia Agraria y Alimentaria (INIA) of Spain and the University of California, Davis, Department of Pomology, Kearney Agricultural Center.

pods, and cactus pear, have received little attention up until now but are now being re-emphasized in areas with Mediterranean climates for diversification and revitalization of local agriculture. These crops are important in many Mediterranean countries: Spain, Portugal through Southern France to Italy, Greece, Turkey, and the Middle East through Morocco and Tunisia to Egypt. They are also being introduced in other areas of the world such as California, Australia, and South America. The economic importance of these Mediterranean crops is shown in Table 1.

Mediterranean crops covered are a diverse group, ranging from those of major international importance supporting large industries to locally important species grown only in home-gardens. They vary in their contribution to local diets (Table 2). While the majority are high in carbohydrate and supply some vitamins and minerals, many are of low nutritional value. Some, such as carob and cactus pear, make a significant contribution to carbohydrate intake to local diets; these multi-purpose plants have a wide range of uses in medicine, industry or agriculture, agroforestry and soil conservation, and for production in marginal lands in semi-arid environments.

The current state of genetic erosion of these crops is quite high. This is a result of many factors, including the origin and distribution of the crop, its breeding system, the propagation methods used, commercial policies, and lack of research. In recent years several countries of Southern Europe have started surveys of native material of these crops in collections and gene pools (IBPGR 1986; Bettencourt and Konopka 1989; Lionakis 1994; Llacer et al. 1994; Monastra et al. 1994). Traditionally morphological and physiological characteristics have been used to identify these cultivars (Zielinski 1955). Some of them are strongly influenced by the environment, and their use as indicators of performance may lead to unreliable or erroneous results. For this reason isozyme gel electrophoresis is increasingly being used for cultivar identification in many tree crops (Arulsekar and Parfitt 1986; Ryugo et al. 1988; Tous et al. 1992).

MEDITERRANEAN FRUIT CROPS

The current situation of nine temperate Mediterranean fruits and nuts in the world, will be discussed in descending order of economic importance.

Olive

The cultivated olive (*Olea europaea*, Oleaceae) is a long-lived, evergreen tree native to the Mediterranean basin. It is valued for its fruit and oil. Mediterranean countries account for around 95% of the world's

Table 1. Production, trend and leading countries of some Mediterranean fruit crops in the world. Sources: Knight (1980), IBPGR (1986), Sadhu (1990), George and Nissen (1990), Tous and Batlle (1990), IOOC (1994), and INC (1995).

Fruit crops	Commercial production (thousand tonnes)	Trend	Leading countries
Olives	9,940[z]	Increasing	Spain, Italy, Greece, Tunisia
Mandarins	9,570	Increasing	Spain, Japan, China, Brazil
Figs	1,050	Stable	Turkey, Egypt, Morocco, Spain
Persimmons	1,200	Stable	China, Japan, Italy, Korea
Pomegranates	800	Stable	Tunisia, Turkey, Iran, Egypt
Pistachios in shell	258	Increasing	Iran, U.S., Turkey, Syria
Carobs	330	Stable	Spain, Italy, Portugal, Morocco
Cactus pear	300	Increasing	Mexico, Italy, Spain, Egypt
Loquats	150	Stable–increasing	Spain, Algeria, Japan, China

[z]90% of total olive production for olive oil.

Table 2. Nutritional composition of Mediterranean crops (per 100 g of edible portion). Source: Goulart (1980); Sawaya et al. (1983); Fernandez Diez (1983); IBPGR (1986); Morton (1987); Cantwell (1994).

					Carbohydrates			Minerals					Vitamins				
Crop/Product	Water (%)	Cal.	Protein (g)	Fat (g)	Total (g)	Fiber (g)	Ash (g)	Ca (mg)	P (mg)	Fe (mg)	Na (mg)	K (mg)	A (IU)	Thia-mine (mg)	Ribo-flavin (mg)	Niacin (mg)	Ascorbic acid (mg)
Olive (ripe pulp)	70.8	163	1.2	18.6	--	1.7	2.1	79	19	0.9	760	48	200	0.01	0.18	0.1	3
Mandarin	87.0	45	0.8	0.1	13.0	0.5	0.3	30	23	0.4	5	140	30	0.08	0.03	0.2	45
Fig, fresh	78.0	80	1.3	0.3	20.3	2.0	0.6	50	22	0.6	2	194	80	0.06	0.05	0.4	2
Fig, dried	23.0	274	4.3	1.3	69.0	5.6	2.3	126	77	3.0	34	640	100	0.10	0.10	0.7	0
Persimmon	79.0	77	0.7	0.4	19.6	1.2	--	6	26	0.3	6	174	2710	0.03	0.02	0.2	15
Pomegranate (pulp)	82.3	65	0.9	0.3	16.4	0.3	0.5	3	8	0.7	3	259	Tr	0.02	0.03	0.3	4
Pistachio nut	5.3	594	19.3	54.0	19.0	2.0	2.7	131	500	7.3	--	972	230	0.70	0.20	1.4	0
Carob flour	11.2	180	4.5	1.4	80.7	7.7	2.2	352	81	5.0	--	950	50	0.03	0.05	2.5	--
Cactus pear (fruit pulp)	85.0	38	0.5	0.1	11.0	1.8	1.6	60	34	0.8	0.8	161	40	0.01	0.02	0.3	30
Loquat	86.5	168	1.4	0.7	43.3	0.9	--	70	126	1.4	--	348	2340	--	--	--	3

olive cultivation (8,702,000 ha). In the Mediterranean 90% of the olive trees are grown for the oil. World olive oil production is currently increasing; the commercial crop during 1990–1994 was 1,796,900 t (IOOC 1994), with the European Community (EC) producing around 1,337,000 t (Spain 40%, Italy 33%, and Greece 22%), Tunisia 194,000 t and Turkey 61,500 t of oil. World table olive production in 1990–1994 was 954,5000 t (IOOC 1994). Spain is the world's largest producer (23%) and exporter of table olives, follow by Turkey (12%), U.S. (11.4%), Morocco (8.5%), Syria (7.5%), and Greece (7.3%). In Australia, Chile, China, Mexico, New Zealand, and South Africa, olives are considered a new crop.

The olive oil and table olive industries play an important role in the agricultural and processing sectors of the major olive producing countries. Most olive oil is consumed within the Mediterranean countries; only 18% of production enters world trade. On average from 1988 through 1991 the world olive oil marketed represented 6% of the quantity and 23% of the value of the world trade in fluid edible oils (UNCTD 1993). To a large extent olive oil does not compete with other vegetable oils but occupies a specialty niche market.

The olive tree has a wide range of adaptability. It requires a mild climate with warm summers and cold winters. The tree requires substantial chilling for good fruiting (Martin et al. 1994), but is injured when temperatures fall below –10°C. Olive is considered a drought-resistant species because it thrives in areas where water stress is frequent: Mediterranean climates. It has been postulated that the minimum water requirement for olive is 2,000 m^3/ha year, mainly during flowering and fruit setting in late spring, and again in the summer as the fruit increases in size (Tous 1990; Bongi and Palliotti 1994). Olive trees will grow on poor soils and rocky hillsides, but deep soils produce the best quality fruit. They tolerate saline or alkaline soils and those with a high lime content. Their root system is relatively shallow, and will not tolerate waterlogged soils.

Olives are wind pollinated, thus flowering during rain, high temperatures and dry wind conditions are deleterious to good fruit set (Griggs et al. 1975). Mature olive trees produce huge numbers of flowers, but the fruit set is normally below 5% (Lavee 1986; Martin et al. 1994). Most olive cultivars will set some fruit in a monocultivar culture. However, they benefit greatly from cross-pollination (Lavee 1986). Olives are picked late in autumn or winter, as the oil content and fruit characteristics change with ripening. Olive cultivars usually fall into one of two commercial uses (Jacoboni and Fontanazza 1981; Barranco and Rallo 1984; Tous and Romero 1993): "Oil" and "Table." "Oil" cultivars predominate. The most famous are: 'Picual', 'Arbequina', 'Cornicabra', 'Hojiblanca', and 'Empeltre' in Spain; 'Frantoio', 'Moraiolo', 'Leccino', and 'Pendolino' in Italy; 'Koroneiki' in Greece; 'Chemlali' in Tunisia; 'Ayvalik' in Turkey; 'Mission' in California and Australia, etc. The "table olive" cultivars include 'Manzanilla' and 'Gordal' (syn. 'Sevillano' in California and 'Queen' in Australia) from Spain; 'Kalamata' from Greece; 'Ascolano' from Italy; 'Barouni' from Tunisia, etc. Alternate bearing is a problem with some cultivars.

Tree management of this crop has undergone many changes in recent decades including increased planting densities (100 to 300 trees/ha), new methods of propagation, improved cultural practices and mechanical harvesting, are some examples. Growth regulators, such as naphthaleneacetic acid (NAA), are used for fruit thinning to increase the fruit size and control the alternate bearing in some table olive orchards (Lavee 1986; Martin et al. 1994). Olives are propagated primarily by cuttings, truncheons or by budding seedling rootstocks. The trees propagated by rooting semi-hardwood cuttings come into bearing within three to four years after planting. The type of harvest depends upon fruit use. Oil olives are harvested by hand or by mechanically shaking the tree. Those for pickling must still be harvested by hand as the fruit bruises easily. The major pest problems are the following: olive fly (*Dacus oleae,* mainly in the Mediterranean basin), black scale (*Saissetia oleae*), olive leaf spot (*Spilocaea oleaginea*), olive knot (*Pseudomonas savastanoi*), and *Verticillium dahliae*.

Olives are nutritious (Table 2), the oil component includes unsaturated fatty acid (70% to 80% oleic acid and 7% to 12% linoleic acid), and small amounts of polyphenols, tocopherols, sterols, and many aromatic compounds (Mataix and Martinez 1988; Tous and Romero 1993).

Olive oil and table olive industries play an important role in the agricultural and processing sectors of the major olive producing countries of the Mediterranean basin. World olive oil production and exports are projected to increase slightly in the next few years, while table olive production will remain stable. While Italian olive oil production is expected to stabilize or decrease, Spain, Tunisia, Morocco, and other countries (Argentina, Australia) are expected to achieve substantial increases in this crop (UNCTD 1993). In addition to culinary uses, olive oil, an unsaturated fat, has recently become more valued for its health benefits. Mainly for this reason, through the last decade the olive oil consumption has increased in several non-Mediterranean countries (USA, Australia, Japan, etc.).

Mandarin

The mandarin (*Citrus reticulata*, Rutaceae) is the largest and most varied group of edible citrus. It is a native of China and Southeast Asia, and was known and extensively planted in China and Japan in the 16th century, but was not introduced to Europe and North America until the 19th century (Webber 1967; Saunt 1990). World production of mandarins is currently increasing; the commercial crop in 1993 was 9,570,000 t, 12.5% of the total world citrus production (FAO 1993). Spain and Japan each account for 16% of the world production. Other major producers in descending order of production are China (8%), Brazil (7%), Korea (6.5%), Italy (5%), United States (4.7%), and Morocco (4%). While production by Japan has decreased around 40% over the last decade, that of Mediterranean countries has increased by 50% (Saunt 1990; Ferguson 1992).

The mandarin has a wide range of adaptability and is grown under desert, semi-tropical and sub-tropical Mediterranean climatic conditions. Mandarin trees and their hybrids are usually the most cold-resistant of all commercially grown citrus. However, mandarin fruits suffer more frost damage than most oranges and grapefruit. The climate in which mandarins are grown has a great influence on fruit quality. Most commercial mandarins are irrigated, but excessively wet conditions should be avoided as this may lead to increased disease. Windbreaks are commonly planted to protect crops from high winds. Light, fertile loams are ideal, although a wide range of soils will produce good crops.

Mandarins are called "slip-skin" oranges "zipper skin" or "easypeeler" because they have loose skins and are easy to peel and separate into segments. The fruit has long been appreciated for its distinctive and sweet flavor and aroma. They are used primarily for eating out-of-hand, as fruit sections, in fresh juice, and to a limited extend for processing (Young 1986). Mandarin cultivars usually fall into one of three commercial groups (Bono et al. 1985; Saunt 1990): "Satsumas" ('Owari', 'Okitsu', and 'Clausellina'); "Clementine" ('Arrufatina', 'Clemenules', 'Esbal', 'Fina', 'Hernandina', 'Marisol', and 'Oroval'); and "other mandarins and hybrids" ('Dancy', 'Fairchild', 'Fortune', 'Kara', 'Nova', and 'Wilking'). Careful selection of three or four mandarin cultivars per orchard can yield fresh fruit for an eighth-month ripening period, beginning in mid-Sept. with 'Okitsu' and 'Clausellina' and extending into Apr.–May with late cultivars such as 'Fortune'. Some cultivars such as 'Comun mandarin', 'Dancy', 'Fairchild', 'Fortune', require cross-pollination for best fruit production (Eti and Stosser 1992).

This evergreen tree requires close plantings (2 × 5 m) and minimum heights for easier harvesting; in this dense planting system it is customary to hedge prune mechanically. Growth regulators are used to control alternate bearing and substitute for pollination, which produces excessive seeds. Mandarins are propagated by budding selected scions on to nucellar seedling rootstocks; a commonly used rootstock is 'Cleopatra' mandarin (Samson 1980). However, any of the rootstocks noted for producing high quality fruit are suitable. The tree comes into bearing three to four years after planting. Most of the pests that attack citrus may be present in mandarin orchards. The major pests are aphids, scale insects, whiteflies, citrus thrips, fruit flies and mites.

Overall, world mandarin production and exports are projected to lightly increase in the next few years. While Japanese production is expected to stabilize or decrease, Spain, Morocco, and other Mediterranean countries are expected to achieve substantial increases in this crop. The excellent quality and characteristic flavor of the new mandarin cultivars are highly prized by some, and if seedless varieties of larger size can be obtained, their popularity will greatly increase.

Fig

The fig (*Ficus carica*, Moraceae) probably originated in Western Asia, and spread to the Mediterranean. Today, the fig is a moderately important world crop, with an estimated annual production of one million tons of fruit (Sadhu 1990). Approximately 40% of this crop is sold as dried fruit. About 30% of the crop is produced by Turkey (300,000 t). Other major producers in descending order are Egypt (160,000 t), Morocco (57,000 mt), Spain (50,000 t), Greece (50,000 mt), California (43,000 t), Italy (40,000 t), Algeria (38,400 t), Syria (37,000 t), Tunisia (35,000 t), and Libya, Iraq, and Portugal (Sadhu 1990; CIHEAM 1994). While production by Italy and Spain has decreased over the last decade, that of Turkey, Syria, and Brazil has increased (IBPGR 1986).

The fruit usually is consumed fresh locally or in dried, canned, and preserved form. Dried figs and those unfit for human consumption, can be used as animal fodder. Several countries import dried figs or the paste. The main exporters of dried figs and paste are Turkey and the United States. Of California's production, 85% is marketed as dried figs, 12% as canned figs and fig juice, and 3% as fresh fruit (Storey 1975). The nutritional value of fresh figs is comparable to that of many other fruits. They are high in calcium. Dried figs, with only 20% water are nutritious relative to other fresh fruits (Table 2).

The typical fig-producing regions have mild winters and hot dry summers. The fig tree is tolerant of a wide range of environmental conditions, has a low chilling requirement, will withstand some frost and is tolerant of drought, although it grows most vigorously with abundant water. A frost of –5° to –10°C may kill the plant down to ground level. Figs can be grown on a wide range of soils, including heavy clays, loams, and light sands, but ideally the soil should be well-drained. The plant is moderately tolerant of high salinity (IBPGR 1986).

There are two main commercial types of figs, the "common fig" that produces fruit without pollination, and the "Smyrna fig" that requires pollination by a fig wasp (*Blastophaga* spp.), that lives in the "caprifig" (male fig), to set fruit. The "common-type" (self-pollinated) fig is more commonly grown. These cultivars bear one or two crops per year. The most famous are: 'White Adriatic' (syn. 'Verdone'), 'Black Mission', 'Kadota' (syn. 'Dottato' in Italy), and 'Conadria' in California (Storey 1975; Ferguson et al. 1990); 'Cuello de Dama Blanco' and 'Napolitana Negra' in Spain (Llacer et al. 1994); 'Kalamon' in Greece (Lionakis 1994); 'Sultani' in Egypt and Tunisia (Mansour 1994; Mars 1994); 'Rhouddane' in Morocco (Loudyi 1994), etc. The "Smyrna-types" include the popular Turkish cultivar known as 'Sarilop' in Turkey and 'Calimyrna' in the United States and other cultivars of Algeria 'Taranimt' and 'Tameriout', that have high dried fruit quality (Rebour 1955; Aksoy 1994).

The tree is propagated by rooting 20 cm cuttings of one to three year old wood taken during the dormant season. Trees are normally planted 4 m apart with 5–6 m between rows. Fruiting begins after three years. Regular fertilization will increase yields without reducing fruit quality. Figs carry two crops in a year, the main crop normally being the second crop in late summer and autumn. The first, or "breba" crop, is produced from flowers initiated in the preceding late summer and maturing from May to June. The second crop is produced from flowers on the current season's growth. Pruning may be required to maintain a balance between new and old wood, as well as to remove suckers and to keep the tree's canopy to a reasonable size for easy harvesting.

Fresh figs are picked when they begin to soften and the color change indicates maturity. When picking, gloves should be worn to prevent damaging the fresh fruit and to prevent the skin irritation caused by the white sap that contains ficin exuding from the broken stem. Since fresh figs ripen irregularly, picking should be done daily or weekly during the long harvest period (4–6 weeks). In California most figs are grown for drying. They are mechanically harvested by sweepers from the ground during Sept. and Oct. (Obenauf et al. 1978). After harvest, the dried figs are washed and can be stored for a few days at 0° to 1°C. Fruit is dried in the sun or by using an electric dryer at a temperature of 60° to 70°C before processing as dried figs.

Figs are not usually seriously affected by pests except in high rainfall areas. In these areas and during the rainy season, fruit cracking usually occurs and fungicide sprays may be necessary to control surface rot (*Alternaria alternata*), smut (*Aspergillus niger*), and mold of fig (*Botrytis* spp., *Penicillium* spp.). Aphids, birds, fruit flies, and scale insects are occasionally a problem. Figs are highly susceptible to nematodes and should not be planted in infested soils.

The economic importance of fig production is likely to continue into the future. In the world market, there is an increasing demand for fresh figs and a stable demand for dried figs. The most important trade aspects of this species are the short commercial life of the fresh fruits, and for the dried fruits, the market competition of Turkish production, where production costs are lower than for other countries (Europe). Fig production in some European countries is slowly decreasing. Turkey and Mexico are expanding their fig production. At present, evaluation of fresh cultivars in Europe and 'Calimyrna' improvement in the United States, combined with improved cultural practices and better fresh fruit postharvest practices are opening new prospects for this crop.

Persimmon

The persimmon (*Diospyros khaki*, Ebenaceae) is native to East Asia, most likely China. It has been cultivated in the Mediterranean basin for more than a century (Morton 1987). Europeans refer to the persimmon by the Japanese name *kaki*. World production is about 1,200,000 t annually from an area of approximately 235,000 ha, with China producing around 57% and Japan 27% of the crop. Smaller, but expanding persimmon industries are being developed in Italy (70,000 t), Korea (50,000 t), Brazil (45,000 t), Israel (10,000 t), Spain (8,000 t), the U.S. (7,000 t), New Zealand, and Australia (George and Nissen 1990; CIHEAM 1994; Mowat and George 1994). Persimmons are a deciduous, monoecious plant, and set fruit parthenocarpically or with pollination. They grow well in warm-temperate and subtropical areas, and tolerate the hot and humid coastal areas. They can tolerate winter temperatures of –12°C but require no winter chilling. Persimmons are grown on a wide range of soils but produce best on deep, fertile, medium-textured soil with good drainage. They produce better than most fruit trees on heavy clay soils. Ferguson and Arpaia (1990) showed that, under California San Joaquin Valley conditions (145–367 mm), they require at least 9,000–12,000 m^3/ha per year of irrigation in addition to annual rainfall.

The fruit is oblong-conical and shows wide variation in size, shape and color. Fruit may be yellow, orange, or red when ripe. Fruit is normally harvested from late Sept. to early Dec. for the fresh market or for drying. The fruit is sweet but most cultivars have an astringent taste until fully ripe. The cultivars are broadly classified for horticultural purposes into two major groups (Itoo 1980; Kitagawa and Glucina 1984; Morton 1987): "astringent" and "non-astringent," based on their flavor at maturity. The most important cultivars are 'Hiratanenashi', 'Hachiya', 'Tanenashi', 'Triumph', 'Kaki Tipo', and 'Rojo Brillante' (astringent); 'Fuyu', 'Jiro', 'Hana-Fuyu', and 'Gosho' (non-astringent). Until recently, most cultivars were commercial the astringent type. These can be treated with ethylene, CO_2 and other chemicals to remove the astringency while the fruit is still firm (Itoo 1980; LaRue et al. 1982).

The trees are trained to a modified central leader and require 7–10 years to reach full bearing. Propagation is by budding or grafting onto seedling rootstocks. Persimmons are relatively free of pests and diseases. The production problems depend on the growing areas. In Mediterranean countries the most pressing pest is the fruit fly (*Ceratitis capitata*), while in others (Japan, California, etc.) fungi such as anthracnose (*Gloesporium kaki*) and grey mold (*Botrytis cinerea*) are common diseases. Physiological problems include cracks radiating from the calyx, sunburn, post-flowering fruit drop, and alternate bearing.

The kaki or Japanese persimmon is mainly eaten fresh, but can be frozen, canned or dried, and is sometimes used in oriental cooking. They can be stored for up to 6 months in modified or controlled atmospheres. Persimmons are high in vitamin A, and are a moderate source of ascorbic acid (Table 2). Some astringent cultivars contain an average of 1.4% soluble tannins in skin cells when not fully ripe.

The persimmon has long been appreciated in the Far East as China and Japan together produce 80% of the world crop. The overall production trend is stationary. While Japanese production is expected to decrease, Brazil, Israel, Spain, Australia, and New Zealand are expected to increase plantings of this crop. The future of this fruit depends on the selection of high quality cultivars and specific marketing techniques to ensure a firm non-astringent persimmon reaching the consumer (George and Nissen 1990).

Pomegranate

The pomegranate (*Punica granatum*, Punicaceae), native to Persia, is one of the oldest known edible fruits. The total world crop was estimated at 800,000 t of fruit (IBPGR 1986). The tree has been cultivated extensively in the Mediterranean countries (Tunisia, Turkey, Egypt, Spain, and Morocco), Iran, Afghanistan, India, and to some extent in the United States (California), China, Japan, and Russia. The fruits normally are sold in the local markets of these countries, except Spain and Turkey which also export. The crop requires a long hot summer to mature the fruit, yet can withstand low temperatures of –9° to –12°C in the winter and is extremely drought-tolerant. Pomegranates grow naturally as a bushy shrub or a small tree with spiny branches, in subtropical areas with arid summers. The pomegranate bears heavily and regularly every year, and requires very little care. The tree adapts to a wide variety of soils.

The fruit is a false berry, subglobose in shape, shiny red, yellowish green or whitish, crowned with a persistent calyx, and has a hard rind. Internally, the fruit has few to many locules containing many seeds. The testa of the seed is filled with a whitish pulp (aril) and an acid juice. The seeds may vary in size and hardness. Fruits with hard seeds possess poor eating quality and, therefore, cultivars producing soft seeded fruits are preferred. Hard seeded fruits, having higher juice content and an intense red color are utilized for processing (Kumar 1990). Several cultivars are grown commercially in the world: 'Wonderful' in California and Israel; 'Mollar' and 'Tendral' in Spain; 'Schahvar' and 'Robab' in Iran; 'Hicaz' in Turkey; 'Zehri' and 'Gabsi' in Tunisia; 'Alandi' and 'Ganesh' in India, etc. (LaRue 1980; Morton 1987; Patil and Karale 1990; Aksoy 1994; Llacer et al. 1994; Mars 1994). The pomegranate is both self-pollinated and cross-pollinated by insects. The fruit is consumed fresh or can be processed into juice, a syrup (grenadine), jams, or a type of wine. The juice content of pomegranate fruit accounts for about 45% to 65% of the whole fruit or 76% to 85% of the arils. Table 2 lists general compositional data of pomegranate pulp. Pomegranates are also valued for their therapeutic properties (Kumar 1990).

The tree requires minimal pruning, except for removing suckers at the base of the plant and giving some shape to the canopy. It is propagated mainly by hardwood or softwood cuttings. The tree comes into bearing three to four years after planting. The fruit is harvested by hand from mid-Sept. to early Nov. The most important problems cause cracking of fruits that affects the pomegranate quality in storage. In India and Egypt the fruit borer (*Virachola isocrates*) is a serious pest (IBPGR 1986; Patil and Karale 1990). Different postharvest diseases such as *Aspergillus niger*, *Botrytis cinerea,* and *Alternaria* sp. cause the fruit to split before ripening. The practice of picking early in the season to avoid fruit cracking is known to result in poor quality fruit (Kumar 1990). The pomegranate is resistant to the oak root fungus, *Armillaria mellea.*

Crop production is stable. Turkey and India may increase the area planted to pomegranate. The future of this fruit depends on the selection of high quality cultivars with soft seeds, higher total soluble solids and richer color (Kumar 1990) and fruits resistant to cracking and fruit borers.

Pistachio

The pistachio, (*Pistacia vera*, Anacardiaceae), probably originated in Central Asia, though it has often been referred in early histories of Western Asia, Asia Minor, and the Mediterranean countries. The fruits are classified as drupes with edible seeds. World production of pistachios is currently increasing, and the commercial crop in 1993/94 was about 258,000 t in-shell (INC 1995). The main producers are Iran (43%), the United States (26.5%, Sacramento and San Joaquin Valleys in California), Turkey (19.5%), Syria (8.5%), Greece (1.5%), and Italy (1%). Lesser quantities are produced in Lebanon, Tunisia, and Australia. Spain is a big importer of pistachio nuts, almost 10,000 t in-shell (Vargas 1990). There are good market possibilities for expansion of the pistachio crop in Turkey, Spain, and California (Chang 1990; Galan-Sauco and Rallo 1993).

The native range of pistachio is characterized by long, hot, dry summers and moderately cool or cold winters. In California the tree is grown in areas where winter temperatures accumulate 1,000 h under 7°C, the conditions required to break the rest period for normal growth and fruiting of 'Kerman' pistachio (Crane and Maranto 1988). Although pistachios will grow in rocky shallow soils, productivity is enhanced in a deep well-drained soils and by proper cultural practices. They tolerate saline or alkaline soils and those with a high lime content. Though originally from a xerophytic environment, they require more water than most fruit trees to produce well, approximately 10,000 m^3/ha per year under California Central Valley conditions (Ferguson and Arpaia 1990). In other areas of the Mediterranean basin, such as Southern Italy, the recommended minimal irrigation of 500–600 m^3/ha/year in the summer maintains good tree performance (Caruso 1990).

The pistachio is a moderately sized deciduous and dioecious tree (3 to 8 m). In commercial orchards the ratio of staminate trees to pistillate trees is 1 to 8. Pistachios have an extended juvenile period and require 5 years of training to establish a modified central leader canopy and 7 to 10 years to reach full bearing under good irrigated conditions. It is also a strongly biennial-bearing species, as fruit embryo growth induces the abscission of already induced floral buds (Crane and Iwakiri 1981). For example in Turkey the ratio of "on" to "off" trees is 1:1 (Kaska 1994). Cultivars differ among countries and are budded on seedlings of different *Pistacia* rootstocks. California's industry consists of one pistillate cultivar, 'Kerman', one staminate cultivar ('Peters') and four rootstocks (*P. terebinthus, P. atlantica, P. integerrima,* and the hybrid *P. atlantica* x *P. integerrima*). 'Kerman' was imported from Iran (then Persia) and named here for its province of origin and Peters for the Fresno grower who selected it. Other cultivars growing in others countries are 'Momtaz', 'Owhadi', and 'Kalehghouchi' in Iran; 'Uzun' and 'Kirmizi' and Turkey; 'Red Aleppo' in Syria; 'Aegina' in Greece; 'Bianca' in Italy; 'Mateur' in Tunisia; 'Larnaka' in Cyprus; and 'Sirora' in Australia; (Vargas 1985; Crane and Maranto 1988; Vargas et al. 1994).

The nuts are borne in clusters that can be harvested by hand, by knocking with a pole or by shaking. The crop matures in early autumn. Mechanical harvesting and processing procedures have been well developed in California. Aflatoxin contamination by the fungus *Aspergillus* is generally not a problem in California, as all pistachios are processed within 24 h of harvest (Ferguson and Arpaia 1990). No serious diseases of pistachio have been reported in Turkey or Iran. In California, however, many orchards have been seriously threatened by Verticillium wilt (*Verticillium dahliae*). Recent orchard surveys in California have revealed other diseases, such as Botrytis blossom and shoot blight (*Botrytis cinerea)*, and Botryosphaeria panicle and shoot blight caused by *Botryosphaeria dothidea* are increasing in California (Michailides et al. 1988).

Pistachio nuts may be bought roasted and salted, shelled or unshelled, and are used in ice creams, sweets, cakes, and meat dishes. The kernels are nutritious (Table 2), with a fat component including 69% oleic acid and 19.8% linoleic acid (Woodroof 1979). Pistachio nut oils are also used in the cosmetic and pharmaceuticals industries (Caruso 1990).

World production of pistachios is currently increasing. The United States (California), Iran and several countries of the Mediterranean basin (Turkey, Greece) are expected to increase their planted hectares. The most serious problems of this tree are alternate-bearing, the soil borne fungus *Verticillium dahliae*, and the long juvenility.

Carob

The carob (*Ceratonia siliqua*, Fagaceae) is a perennial leguminous tree native to the Mediterranean basin and southwestern Asia (Smith 1976). Its cultivation began before written history. This species is also known as "St. John's Bread," "algarrobo" in Spain, "carrubo" in Italy, and "caroubier" in France. The beans and kernels are useful for a variety of purposes including food, fodder, and as a source of industrial products such as gums, sugar, and alcohol. The tree is also used for shade and erosion control (NAS 1979). World production of carob pods was estimated about at 330,000 t (Tous and Batlle 1990), mainly in Spain (45%), Italy (16%), Portugal (9%), Morocco (7.5%), Cyprus (6%), Greece (5%), and Turkey (4.5%). Lesser quantities are produced in Algeria, Tunisia, and Israel. Spain is also the world's largest exporter of carob beans. The only significant recent plantings outside the Mediterranean region are in Australia (Esbenshade and Wilson 1986; Tous 1995).

The tree grows best in a Mediterranean-type climate with cool, not cold, winters, mild to warm springs, and warm to hot summers with little rain. A frost of –4°C or below may kill the young trees, or shoots and flowers in mature trees. Although drought-resistant, carob trees do not bear commercial crops unless they get enough rainfall, a minimum average rainfall of about 500–550 mm is needed to ensure a profitable crop (NAS

1979). The tree survives in a wide variety of soils, like pistachio, but must have good drainage. Although it is a legume, it is unable to fix nitrogen (Martins-Loucao 1985).

The carob is a large (7 to 10 m) trioecious tree with staminate, pistillate and hermaphrodite inflorescences in different trees. Pollination is by insects (bees, flies, etc.) and the wind (Tous and Batlle 1990). In new commercial orchards the ratio of staminate or hermaphrodite trees to pistillate trees should be 1:8. In traditional orchards male branches are grafted on female trees to provide pollen in situ. The carob, like the pistachio, matures slowly and most budded cultivars begin to bear in the 5–6th year after planting. It is also a strongly alternate-bearing species, particularly when growing conditions are poor. Today, commercial cultivars differ in pulp quantity or yield of seeds. The most important cultivars are: 'Negra', 'Rojal', 'Matalafera', and 'Duraio' in Spain; 'Gibiliana' and 'Amele' in Italy; 'Mulata' and 'Galhosa' in Portugal; 'Tylliria' in Cyprus; 'Hemere' in Greece; 'Sfax' in Tunisia; and 'Santa Fe' in California (Tous and Batlle 1990).

The tree requires little pruning, and minimal cultural care. Carobs respond to N fertilization by increasing pod yield (Lloveras and Tous 1992). They are propagated mainly by budding or grafting onto seedling rootstocks. The flowers, which are borne in small lateral racemes or clusters, bloom during Sept.–Nov. on older wood. The carob bean is a dark-brown, fleshy edible pod, 10–30 cm long, containing 5–15 hard seeds, and ripening in late summer, during bloom. Carob seeds comprise about 10% to 20% of the weight of the pod. Fruits can be harvested by hand, by knocking down with a long pole, or by shaking. No serious problems of carob have been reported in the Mediterranean area. The major pests are the stem borer (*Zeuzera pyrina*) and carob moth (*Myelois ceratoniae*), and only one fungus, *Oidium ceratoniae*, is occasionally a problem in the leaves and young pods.

The carob pod is very rich in sugar (up to 45%) and low in protein. Mixed with other feeds higher in protein, it is good source of stock feed (horses, calves, goats, milk cow). Finely-ground pods make a sweet nutritious human food (Table 2), and are used in a similar manner to cocoa in cakes, drinks, puddings, breakfast cereals, and candy. The carob powder has more vitamins, minerals, and fewer fats, carbohydrates, and calories than chocolate made with cocoa. Additionally carob powder contains no caffeine and theobromine (Whitheside 1981). The carob gum or locust bean gum (LBG) surrounding the endosperm is an important commercial stabilizer and thickener in many food industries (bakery goods, ice creams, baby foods, sauces, jams, cheese, jellies, and canned meats), as well as in pet foods, cosmetics, and pharmaceuticals (NAS 1979; Morton 1987; Di Lorenzo 1990; Tous and Batlle 1990).

The overall production trend is stationary. Several countries, such as Spain, Portugal, and Australia, are expected to have new plantings of this crop. In the world market, there is an increasing demand for carob seed and a stable demand for carob pulp. In the last 5 to 10 years, because of the rising prices of the carob seed derivatives (LBG), the carob has been the object of renewed interest (Tous and Batlle 1990).

Cactus Pear

The cactus pear (*Opuntia ficus-indica*; *Opuntia* spp., Cactaceae) is a fleshy bush or small tree, native to the desert zones of Northwestern Mexico and the Southwestern United States (Knight 1980). The plant was brought to Europe by the first Spanish colonists from Mexico and has been cultivated along the Mediterranean coast since the late 17th century (Retamal et al. 1987). This fruit is also known as "barbary fig," "tuna," and "Indian fig." The total world crop was estimated at 300,000 t of fruit. Mexico's estimated current production of 200,000 t of fruit on an area of 50,000 ha (Flores and Gallegos 1994). The other important production region is the Mediterranean basin (Italy, Spain, Egypt, Morocco, and Israel) and to a more limited extent South America, South Africa, Southwestern Asia, the U.S., and Australia. The fruits normally are sold in the local markets of these countries, except for Mexico, Italy, and Chile which are also exporters. The cactus pear is not cultivated as a regular commercial crop, but rather is planted as fences, windbreaks, and gardens. The only significant plantings are in Mexico, Italy, Spain, Chile, Israel, and South Africa. The plant may reach a height of 5 m and prefers calcareous soils and semi-arid climate with temperatures ranging from 18° to 26°C.

The *Opuntia* is a cactus in which the branches or stems ("cladodes" or "nopalitos") have the appearance and function of a leaf. The fruit is generally pear-shaped and has a number of small spines, with a soft juicy pulp that contains numerous small black seeds. Fruits ripen and are harvested by hand from July to Dec.

The plant requires minimal cultural care. It is propagated mainly by vegetative means (cladodes or shoots). Fruiting begins at 2 or 3 years and reaches full production around 7 or 8 years. A mature plant can yield 100 to 200 fruits (Knight 1980). Many local cultivars have been developed, and generally are named for their fruit color at maturity, particularly in Mexico ('Reyna', 'Rojo Pelona', 'Esmeralda', etc.), Italy ('Gialla', 'Rossa', and 'Bianca') and Spain ('Verdales', 'Morados', 'Sanguinos', and 'Blancos') (Barbera et al. 1992; Flores and Gallegos 1994; Llacer et al. 1994). Future demand for this fruit depends upon developing high quality cultivars without spines and few seeds. Due to its xerophitic habitat the cactus pear does not have serious problems with pests and diseases. However in some areas the fruit fly (*Ceratitis capitata*) causes damage to the fruits, particularly during years with mild weather during autumn (Barbera et al. 1992; Lionakis 1994).

The vegetable stems ('nopalitos') and fruits of cactus pear are useful for a variety of purposes including food (fresh fruit, paste, jam, salads, and refreshing drinks), fodder (auxiliary feed for cattle, sheep, and goats), medicinal (antidiabetic agent) and for industrial products such as alcohol, soap, pigments, pectins, and oils (Lakshminarayana 1980; Borrego et al. 1990; Hegwood 1990). Table 2 lists general compositional data of cactus pear fruit pulp.

The present trend of the production of this fruit is increasing. In the future, Mexico, Italy, and South Africa are expected to increase the production of this fruit. Millions of hectares of marginal land in arid parts of the world, where irrigation cannot be supplied, can come into their own as food producers when improved cultivars of cactus pear are available (Knight 1980).

Loquat

The loquat (*Eriobotrya japonica*, Rosaceae) is a small, evergreen tree native to central-eastern China, introduced into Japan in very early times. In Europe it was planted in the 18th century (Morton 1987). It is grown both as an ornamental and for its fruit. The total world crop was estimated as 150,000 t. It has been cultivated extensively in the Mediterranean basin (Spain, Algeria, Turkey, Israel, and Italy), Japan and China, to some extent in India and Brazil, and in a more limited fashion in Chile and the United States. Spain is the world's largest producer (40,000 t) and exporter of loquats, follow by Algeria (22,000 t), Japan (18,000 t), China, and Brazil (Lupescu et al. 1980; Morton 1987; Llacer et al. 1994). The tree grows best in a subtropical to warm-temperate climate. It does well on a variety of soils but does best on clay loams with good drainage.

The fruits, which are borne in large loose clusters, are commonly round, oval, or pyriform and in the best cultivars may reach a length of 7 cm. They vary in color from pale yellow to deep orange and have a tough plumlike skin. The flesh is white to orange, firm or soft, juicy, and flavorful. From 1 to 4 smooth, brown seeds are commonly found in each fruit. The seeds comprise about 20% to 30% of the weight of the whole fruit (Insero et al. 1990). Loquats are consumed largely as fresh fruit, although small amounts are used in jams, jellies, syrups, and pies (Shaw 1980). Loquats are high in vitamin A and minerals (Table 2). For the fresh market they should not be picked before full maturity; otherwise they are too acid. If properly handled they can be shipped to distant markets.

Generally, the loquat tree blooms in the autumn with fruits ripening in early spring (Apr.–May). They are normally pollinated by bees, but some cultivars such as 'Akko 13' from Israel and 'Golden Yellow' from India are not self-fertile, and others such as 'Advance' and 'Tanaka' are partially self-fertile. It has been observed that cross-pollination generally results in 10% to 17% increased production over self-pollination (Morton 1987). A list of important cultivars are: 'Advance', 'Algerie', 'Akko 13', 'Champagne', 'Magdall', 'Premier', 'Saint Michel', 'Tanaka', 'Thales' (syn. 'Gold Nugget'), etc. (Knight 1980; Morton 1987; Pathak and Gautam 1990).

The tree requires a minimal pruning, but some thinning of fruits may be required for optimum size. It is propagated mainly by budding or grafting onto seedling rootstocks; quince root can be used if a dwarf tree is desired. The tree comes into bearing in three to four years and the yield in the new orchards is very high (25 t/ha, Blumenfeld 1994). They are hand-harvested by clipping bunches of fruit and are sorted and graded by hand. The most important problems cause damage to the fruit rind, which downgrades fruit quality. In Europe two disorders damage the fruit rinds (Caballero 1993): sunburn ('purple spot'), a physiological disorder directly related to the calcium content in the fruit tissues, and the fungal black spot (*Fusicladium eriobotryae*). In California two diseases that sometimes create problems are fireblight (*Erwinia amylovora*) and loquat scab (*Spilocea pyracanthae*) (Ogawa and English 1991).

The overall production is steadily increasing. Several countries, such as Spain, Brazil, and India, are expected to increase commercial plantings of this crop. An interesting commercial aspect of loquat is that it ripens in early spring before other fruits (cherries, apricots, peaches, and plums) appear in the market. The most important problems (physiological and disease) result in a damaged fruit rind, which downgrades the fruit quality. Also, there is a need to reduce the labor costs of hand thinning and manual harvesting.

FUTURE PROSPECTS

In recent years, perhaps because of the crises that many agricultural sectors have experienced, and because of the need for new crop alternatives, and also for health reasons, interest has developed in traditional Mediterranean fruits. At present, olives, mandarins, figs, persimmons, and pistachios are receiving more research interest than other Mediterranean crops because these crops are sufficiently extensive and profitable enough to support university based research. As long as these crops continue to be profitable money for research support and market development will be available. However, the minor crops such as pomegranates, cactus pear, carob, and loquat do not have organized industries to support their research and market development. It now falls to nascent industry organizations, rare fruit grower associations, national agricultural services and individuals to develop these crops until they become more popular. Information about these less popular "new" crops is available online through such World Wide Web (WWW) home pages as the University of California Fruit and Nut Crop Research and Information Center and NewCROP (New Crop Resource Online Program) developed by Purdue University in Indiana.

REFERENCES

Aksoy, U. 1994. Present status and future prospects of underutilized fruit production in Turkey. First Meeting CIHEAM Cooperative Research Network on Underutilized Fruit Trees. Zaragoza, Spain. p. 84–94.

Arulsekar, S. and D.E. Parfitt. 1986. Isozyme analysis procedures for stone fruits, almond, grape, walnut, pistachio and fig. HortScience 21:928–933.

Barbera, G., F. Carimi, and P. Inglese. 1992. Past and present role of the indian-fig prickly pear (*Opuntia ficus-indica* L. Miller, Cactaceae) in the agriculture of Sicily. Econ. Bot. 46(1):10–20.

Barranco, D. and L. Rallo. 1984. Las variedades de olivo cultivadas en Andalucia. Ministerio de Agricultura-Junta de Andalucia, Madrid.

Bettencourt, E.J. and J. Konopka. 1989. Directory of germplasm collections-6.II. Temperate fruits and tree nuts. Int. Board for Plant Genetic Resources (IBPGR). Rome.

Blumenfeld, A. 1994. Underutilized fruit trees in Israel. First Meeting CIHEAM Cooperative Research Network on Underutilized Fruit Trees. Zaragoza, Spain. p. 22–27.

Bongi, G. and A. Palliotti. 1994. Olive. p. 165–188. In: B. Schaffer and P.C. Andersen (eds.), Handbook of environmental physiology of fruit crops. Volume I: Temperate crops. CRC Press, Boca Raton, FL.

Bono, R., J. Soler Aznar, and L. Fernandez de Cordoba. 1985. Variedades de agrios cultivadas en Espana. Generalitat Valenciana, Valencia.

Borrego, E.F., M. Murillo, V. Flores, V. Parga, and E. Guzman. 1990. Industrial potential of fifteen varieties of nopal (Prickly pear, *Opuntia* spp.). p. 215–218. In: H. Naqui, A. Estilai, and I.P. Ting (eds.), New industrial crops and products. Riverside, CA.

Caballero, P. 1993. El nispero y su expansion, posibilidades y limitaciones. Fruticultura Profesional 54:35–40.

Cantwell, M.I. 1994. Composition and nutritive value of cactus fruits and stems. p. 61–74. In: P. Felker and J.R. Moss (eds.), Proc. Fifth Annual Texas Prickly Pear Council. Kingsville, TX.

Caruso, T. 1990. Pistacchio. L'Italia Agricola 3(7–9):167–170.

CIHEAM. 1994. Proc. First Meeting of the CIHEAM Cooperative Research Network on Underutilized Fruit Trees. CIHEAM Instituto Agronomico Mediterraneo de Zaragoza. Zaragoza, Spain.
Chang, K. 1990. Market prospects for edible nuts. p. 47–86. In: Nut production and industry in Europe, Near East and North Africa. FAO, REVR, Technical Series 13. Yalova, Turkey.
Crane, J.C. and B.T. Iwakiri. 1981. Morphology and reproduction in pistachio. Hort. Rev. 3:376–393.
Crane, J.C. and J. Maranto. 1988. Pistachio production. Div. Agr. & Natural Resources. Public. 2279.
Di Lorenzo, R. 1990. Carrubo. L'Italia Agricola 3(7–9):123–134.
Drescher, G., K.D. Gifford, and S. Baer-Sinnot. 1995. Tree nuts, health and the Mediterranean diet. A summary report. Oldways Preservation & Exchange Trust. San Francisco, CA.
Esbenshade, H.W. and G. Wilson. 1986. Growing carobs in Australia. Goddar and Dobson, Victoria, Australia.
Eti, S. and R. Stosser. 1992. Pollen tube growth and development of ovules in relation to fruit set in mandarines cv. 'clementine' (*Citrus reticulata*). Acta Hort. 321:621–625.
FAO. 1993. Yearbook production. Vol 47. FAO, United Nations, Roma.
Ferguson, L. 1992. Mandarins. California's next citrus sensation. California Grower 16(12):37–38.
Ferguson, L. and M. Arpaia. 1990. New subtropical tree crops in California. p. 331–337. In: J. Janick and J. Simon (eds.), Advances in new crops. Timber Press, Portland, OR.
Ferguson, L., T.J. Michailides, and H. Shorey. 1990. The California fig industry. Hort. Rev. 12:409–490.
Fernandez Diez, M.J. 1983. Olives. Biotechnology 5:380–397.
Flores, C.A. and C. Gallegos. 1994. The production of prickly pear in the north-central region of Mexico. p. 13–28. In: P. Felker and J.R. Moss (eds.), Proc. Fifth Annual Texas Prickly Pear Council. Kingsville, TX.
Galan-Sauco, V., and Rallo, L. 1993. Opportunities in new fruit crops in Spain, p. 91–106. In K.R.M. Anthony, J. Meadley and G. Robbelen (eds.). New Crops for Temperate Regions. Chapman & Hall, London.
George, A.P. and R.J. Nissen. 1990. Persimmon. p. 469–487. In: T.K. Kose and S.K. Mitra (eds.), Fruits: Tropical and subtropical. Naya Prokash, Calcutta.
Goulart, F.S. 1980. Cooking with carob: the healthful alternative to chocolate. Garden Way Publ. Bul. A-48. Pownal, VT.
Griggs, W.H., H.T. Hartmann, M.V. Bradley, B.T. Iwakini, and J. Whisler. 1975. Olive pollination in California. Calif. Agr. Expt., Sta. Bul. 869.
Hegwood, D.A. 1990. Human health discoveries with *Opuntia* sp. (prickly pear). HortScience 25:1515–1516.
IBPGR. 1986. Genetic resources of tropical and sub-tropical fruits and nuts. Int. Board Plant Genetic Resources (IBPGR). Rome.
INC (International Nut Council). 1995. The cracker. No. 1 (April).
Insero, O., F. Monastra, and G. Paesano. 1990. Nespolo del Giappone. L'Italia Agricola 3(7–9):149–152.
IOOC (International Olive Oil Council). 1994. International market conditions and trends for olive oil and table olives. Olivae 54(12):14–21.
Itoo, S. 1980. Persimmon. p. 442–468. In: S. Nagy and P.E. Shaw (eds.), Tropical and subtropical fruits. Composition, properties and uses. AVI, Westport, CT.
Jacoboni, N. and G. Fontanazza. 1981. Cultivar. REDA, L'Olivo, Roma. p. 7–52.
Kaska, N. 1994. Pistachio nut growing in Turkey. First Int. Symposium on Pistachio Nut. ISHS-FAO. Adana, Turkey. Abstr. Book p. 1.
Kitagawa, H. and P.Glucina. 1984. Persimmon culture in New Zealand. Dept. Science and Industrial Research. DSRI Information Series No. 159. Wellington.
Knight, R. 1980. Origin and world importance of tropical and subtropical fruit crops. p. 1–120. In: S. Nagy and P.E. Shaw (eds.), Tropical and subtropical fruits. Composition, properties and uses. AVI, Westport, CT.

Kumar, G.N. Mohan. 1990. Pomegranate. p. 328–347. In S. Nagy, P.E. Shaw, and W.F. Wardowski (eds.), Fruits of tropical and subtropical origin. Florida Science Source.

Lakshminarayana, S. 1980. Sapodilla and prickly pear. p. 415–441. In: S. Nagy and P.E. Shaw (eds.), Tropical and subtropical fruits. Composition, properties and uses. AVI, Westport, CT.

LaRue, J.H. 1980. Growing pomegranates in California. Univ. Calif. Div. Agr. Sci. Leaflet 2459.

LaRue, J.H., K.W. Opitz, and J.A. Beutel. 1982. Growing persimmons. Univ. Calif. Div. Agr. Sci. Leaflet 21277.

Lavee, S. 1986. Olive. p. 261–276. In: P.S. Monselive (ed.), Handbook of fruit set and development. CRC Press, Boca Raton, FL.

Lionakis, S.M. 1994. Present status and future prospects of the cultivation in Greece of the plants: fig, loquat, persimmon, pomegranate and barbary fig. First Meeting of the CIHEAM Cooperative Research Network on Underutilized Fruit Trees. Zaragoza, Spain. p. 14–21.

Llacer, G., R. Martinez-Valero, P. Melgarejo, M. Romero, and F. Toribio. 1994. Present status and future prospects of underutilized fruit tree crops in Spain. First Meeting of the CIHEAM Cooperative Research Network on Underutilized Fruit Trees. Zaragoza, Spain. p. 63–75.

Lloveras, J., and J.Tous. 1992. Response of carob tree to nitrogen fertilization. HortScience 27:849.

Loudyi, W. 1994. Quelques especes fruitieres d'interet secondaire cultivees au Maroc. First Meeting of the CIHEAM Cooperative Research Network on Underutilized Fruit Trees. Zaragoza, Spain. p. 37–43.

Lupescu, F., T. Lupescu, A. Khelil, and G. Tanislaw. 1980. Performances agro-biologiques de quelques varietes d'*Eriobotrya japonica*, cultivees a la Station horticole de l'Institut national agronomique d'Alger. Fruits 35:251–261.

Mansour, K.M. 1994. Underutilized fruit crops in Egypt. First Meeting of the CIHEAM Cooperative Research Network on Underutilized Fruit Trees. Zaragoza, Spain. p. 8–13.

Mars, M. 1994. La culture du granadier (*Punica granatum* L.) et du Figuier (*Ficus carica* L.) en Tunisie. First Meeting of the CIHEAM Cooperative Research Network on Underutilized Fruit Trees. Zaragoza, Spain. p. 76–83.

Martin, G.C., L. Ferguson, and V.S. Polito. 1994. Flowering, pollination, fruiting, alternate bearing, and abscission. p. 51–56. In L. Ferguson, G. Steven Sibbett, and G.C. Martin (eds.), Olive production manual. Univ. California. Div. Agr. Natural Resources, Oakland, CA. Publ. 3353.

Martins Loucao, M.A. 1985. Estudos fisiologicos e microbiologicos da associacao da alfarrobeira (*Ceratonia siliqua*, L.) com bacterias de Rhizobiacease. Ph.D. Diss., Univ. Lisboa, Portugal.

Mataix, F.J. and E. Martinez. 1988. El aceite de oliva. Bases para el futuro. Consejeria de Agricultura y Pesca de la Junta de Andalucia. Sevilla.

Michailides, T.J., J.M. Ogawa, J.D. MacDonald, and L. Ferguson. 1988. Symptoms of fungal diseases of pistachios in California. California Pistachio Industry. Grower's Leaflet 3.

Monastra, F., O. Insero, and G. Tamponi. 1994. The present situation in Italy of some underutilized fruit crops. First Meeting of the CIHEAM Cooperative Research Network on Underutilized Fruit Trees. Zaragoza, Spain. p. 28–36.

Morton, J.F. 1987. Fruits of warm climates. Julia F. Morton, Miami, FL.

Mowat, AD. and P. George. 1994. Persimmon. p. 209–232. In: B. Shaffer and P.C. Andersen (eds.), Handbook of environmental physiology of fruits crops. I: Temperate crops. CRC Press, Boca Raton, FL.

NAS (National Academy of Sciences). 1979. Tropical legumes. Resources for the future. Washington, DC.

Obenauf, G., M. Gerdts, G.Leavitt, and J. Crane. 1978. Commercial dried fig production in California. Univ. Calif. Agr. Ext. Leaflet 21051.

Ogawa, J. and H. English. 1991. Diseases of temperate zone tree fruit and nut crops. Univ. California, Div. Agr. and Natural Resources, Oakland, CA. Publ. 3345.

Pathak, R.K. and H.O. Gautam. 1990. Loquat. p. 720–731. In: T.K. Kose and S.K. Mitra (eds.), Fruits: Tropical and subtropical. Naya Prokash, Calcutta.

Patil, A.V. and A.R. Karale. 1990. Pomegranate. p. 614–631. In: T.K. Kose and S.K. Mitra (eds.), Fruits: Tropical and subtropical. Naya Prokash, Calcutta.

Rebour, H. 1955. Le verger mediterraneen. Tome II, Arboriculture Speciale, Paris.

Retamal, N., J.M. Duran, and J. Fernandez. 1987. Seasonal variations of chemical composition in prickly pear (*Opuntia ficus-indica*). J. Sci. Food Agr. 38:303–311.

Ryugo, K., C.A. Shroeder, A. Sugiura, and K. Yonemori. 1988. Persimmons for California. California Agr. vol. 42(4):7–9.

Sacks, F. 1995. Forward. p. 4. Tree Nuts, Health and the Mediterranean Diet Symposium. Oldways Preservation & Exchange Trust, Boston, MA.

Sadhu, M.K. 1990. Fig. p. 650–663. In: T.K. Kose and S.K. Mitra (eds.), Fruits: Tropical and subtropical. Naya Prokash, Calcutta.

Samson, J.A. 1980. Tropical fruits. Longman, New York.

Saunt, J. 1990. Citrus varieties of the world. Sinclair Int., Norwich, England.

Sawaya, W.N., H.A. Khatchadourian, and W.M. Safi. 1983. Chemical characterization of prickly pear pulp, *Opuntia ficus-indica,* and the manufacturing of prickly pear jam. J. Food Technol. 18:183–193.

Shaw, P.E. 1980. Loquat. p. 479–491. In: S. Nagy and P.E. Shaw (eds.), Tropical and subtropical fruits. Composition, properties and uses. AVI, Westport, CT.

Smith, P.M. 1976. Minor crops. p. 311. In: N.W. Simmonds (ed.), Evolution of crop plants. Longman, New York.

Storey, W.B. 1975. Figs. p. 568–589. In: J. Janick and J.N. Moore (eds.), Advances in fruit breeding. Purdue Univ. Press. West Lafayette, IN.

Tous, J. 1990. El olivo. Situacion y perspectivas en Tarragona. Diputacio de Tarragona, Tarragona, Spain.

Tous, J. 1995. El algarrobo en Australia. Boletin Agropecuario 'La Caixa' 35(1–3):43–49.

Tous, J. and I. Batlle. 1990. El algarrobo. Ed. Mundi-Prensa. Madrid.

Tous, J., C. Olarte, M.J. Truco, and P. Arus. 1992. Isozyme polymorphisms in carob cultivars. HortScience 27:257–258.

Tous, J. and A. Romero. 1993. Variedades del olivo. Fundacion 'La Caixa'-AEDOS. Barcelona.

UNCTD (United Nations Conference Trade Development). 1993. International cooperation on olive and table olives. TD/OLIVE OIL. 9/3. United Nations.

Vargas, F.J. 1985. El pistachero. Algunos aspectos importantes del cultivo. Diputacio de Tarragona. Publ. CAMB, no 33.

Vargas, F.J. 1990. Nut trees in Spain: Almond, hazelnut, walnut and pistachio. p. 755–780. In: Nut production and industry in Europe, Near East and North Africa. FAO, REVR, Tech. Series, 13. Yalova, Turkey.

Vargas, F.J., M. Romero, J. Plana, M. Rovira, I. Batlle. 1994. Characterization and behavior of pistachio cultivars in Catalonia (Spain). First Int. Symposium on Pistachio Nut. ISHS-FAO. Adana, Turkey. Abstr. Book p. 16–17.

Webber, H.J. 1967. History and development of the citrus industry. p. 1–39. In: The citrus industry. vol. I Univ. California, Berkeley.

Whiteside, L. 1981. The carob cookbook. Thorsons Publ. Ltd., Wellingborough, Northamptonshire.

Woodroof, J.G. 1979. Tree nuts: production processing products. 2nd ed. AVI, Westport, CT.

Young, R.H. 1986. Fresh fruit cultivars. p. 101–126. In: W.F. Wardowski, S. Nagy, and W. Grierson (eds.), Fresh citrus fruits. AVI, Westport, CT.

Zielinski, Q.B. 1955. Modern systematic pomology. W.M.C. Brown Company, Dubuque, IA.

South American Fruits Deserving Further Attention

Richard J. Campbell

Considerable attention has been given in recent years to fruits with potential for further economic development (Clement 1983; Donadio 1983; Arkcoll 1990; Ferguson and Arpaia 1990; Lamberts and Crane 1990; Nerd et al. 1990; Campbell 1990; Silva 1991; Crane 1993). South America has been the focus of many of these previous studies due to the sheer number of edible fruit crops originating on the continent, which is considered an important center of diversity for fruit crops. Our study takes another look at South American fruit crops with potential for further economic development, emphasizing fruit crops not previously discussed in recent studies. Discussions for each individual fruit are preceded by a general overview of the family to which each fruit belongs, highlighting the characteristics of the family and some of its commercial members. The fruit crops discussed include crops presently grown on a limited commercial scale, as well as minor fruit which at present are of importance only on a subsistence level within their native range.

SPECIFIC FRUIT CROPS

Anacardiaceae

This family consists of 850 species, including fruit crops of great economic importance throughout the tropical and subtropical regions of the world. The mango, *Mangifera indica* L. (Asia) and the cashew, *Anacardium occidentale* L. (South America) are produced in most tropical and subtropical countries. Another genus, *Spondias*, consisting of 8 to 10 fruit species of American and Asian origin (Popenoe 1979) is also prevalent throughout the tropics. *Spondias purpurea* L. and *Spondias cytherea* Sonn. are both commercial fruit crops in the American and Asian tropics. Other *Spondias* species native to South America are of great local importance as subsistence crops.

Spondias tuberosa Arruda. Umbú. The umbú is native to the dry plains of northeastern Brazil. This fruit has been described as perhaps the best flavored among all of the *Spondias* species by Popenoe (1920). The tree can attain a height of 6 m, although it usually forms a low, spreading tree when left unmanaged (Martin et al. 1987). Fruit are oval, averaging 4 cm in length, with a 2 cm stone. When fully ripe the flesh is almost liquid, with a sweet, aromatic flavor. The fruit are sour if eaten before they are fully ripe. In its native region the umbú is consumed fresh, used in preserves, made into juices, or sweetened and mixed with milk to make "imbuzada," a typical drink of the region. Popenoe (1920) discussed the importance of this fruit to the people of northeastern Brazil, who consumed vast numbers of umbú during the fruiting season. The same can be said in the region today, as wild umbú trees are protected and visited by local residents while the trees are fruiting.

The umbú offers potential for arid tropical regions due to its production in the harshest of conditions. Annual production has been up to 300 kg/tree with good selections (Cavalcanti and Abilio de Queiroz 1992). There has been limited work on the identification and selection of superior clones of umbú; however, within Brazil clones have been identified which weigh nearly 90 g and have a pulp to seed ratio of 80% (Cavalcanti and Abilio de Queiroz 1992). Popenoe (1920) questioned why the umbú had not yet attained a greater status in the world due to its superior flavor, and today the same sentiment is commonly expressed. However, due to its short shelf life and delicate texture when mature, it is doubtful that the umbú could be a viable consideration as a fresh fruit beyond the local level. However, as a flavoring and/or juice crop for the arid, hot tropics, the potential is much greater.

Little is known about its adaptability to different climates. Young trees have been killed by freezes in the state of Sao Paulo, Brazil (Donadio 1983) and in Florida (Campbell et al. 1977). The growth of seedling umbú trees in the calcareous soils of South Florida has also been poor (Campbell and Sauls 1980). Umbú is graft compatible with other *Spondias* species (Popenoe 1920), possibly widening its adaptation to other climates and soil conditions.

Spondias mombin L. Yellow mombin. The yellow mombin is native to Central America and northern South America (Popenoe 1979) and can be found under semi-wild cultivation in most lowland areas of the American Tropics. The tree can attain a height of 10 m and is tolerant of most soil types and rainfall patterns. The fruit are ellipsoid about 2.5 to 4 cm in length and hang singly on the tree (Martin et al. 1977). There is great variation in quality among fruit from region to region, some being sweet and pleasant and others quite disagreeable in flavor (Martin et al. 1977; Popenoe 1920). The fruit can be eaten fresh, used in preserves or as a flavoring. Although common in most of the lowland tropics, the yellow mombin has not been highly commercialized. It is generally considered inferior in quality to *S. purpurea*, which is grown in the same environments. Yellow mombin has been introduced to most tropical locations and performs well under varied conditions. Trees are severely damaged by freezing temperatures (Campbell et al. 1977). Trees are generally grown from vegetative cuttings, but patch budding has been used as well (Coelho Pedrosa et al. 1991). The fruit are consumed fresh on a local level; however, its potential is more as a processed fruit. It offers an adaptive advantage over other species of *Spondias* in some climates and soil types.

Annonaceae

This family is comprised of 2,050 species, many of which are cultivated for their fruit, both on a commercial and subsistence level (Leon 1987). The cherimoya (*Annona cherimola* Miller), guanabana (*A. muricata* L.), sugar apple (*A. squamosa* L.) and atemoya (*A. cherimola* x *A. squamosa*) are the major commercial members of the family in tropical and subtropical regions. Due to their unique appearance and flavor, many other members of this family from Central and South America have potential for further commercialization.

Rollinia deliciosa A. DC. Biribá. The biribá is native to northern South America and the Caribbean (Leon 1987). The tree is small, usually reaching a height of 6 to 10 m, with a dense, multiple-branched growth habit. Biribá is one of the most common home garden trees in the state of Para, Brazil (Clement 1983), where there is also limited commercial production (Cavalcante 1974). Biribá fruit from this region can sometimes be found in markets as far away as Rio de Janeiro. The fruit are globose and can weigh up to 1350 g (Cavalcante 1974). The mature fruit is yellow with multiple protuberances, which turn black as the fruit are handled. The flesh is translucent, juicy, and sweet with a somewhat mucilaginous consistency, which is objectionable to many. It is usually consumed fresh, but it is also used as a juice, a preserve, or a flavoring. Within its native region it is considered to be of the highest quality among the Annonaceae. Outside of northern Brazil, however, the biribá is much less common and not as highly esteemed. It can be found in germplasm collections throughout the lowland tropics, but the fruit encountered are generally of inferior-quality seedlings.

Production of biribá is limited to the hot, humid lowland tropics. Temperatures of –1° to –2°C can kill young trees (Campbell et al. 1977). The trees grow quickly and bear fruit from seed in 4 to 6 years. There is great variation in fruit quality among seedling trees. Biribá trees produce well without hand pollination, which is an advantage over many other Annonaceae. In order to increase the commercialization of the biribá, superior cultivars would need to be identified and selected. Research would also be required on storage, shipping, and handling procedures.

Caricaceae

In terms of fruit production, the Caricaceae is a small (31 species), yet important family throughout both the American and Asian tropics (Badillo 1993). The most widely known member of the family is the papaya (*Carica papaya* L.), which is cultivated commercially throughout most of the lowland tropics. There are, however, other species currently produced on a limited commercial scale in the South American highlands which have potential for greater exploitation, both within their present production areas and beyond.

Carica pubescens (A. DC.) Solms-Laub. Chamburo. This fruit is native to northern South America and is cultivated from Panama to Bolivia at elevations above 1,000 m. The plants often attain a height of more than 10 m and have a similar appearance to papaya. They can be distinguished by their variable leaf shape and pubescence, which covers the leaves and flowers. The fruit are from 5 to 20 cm in length, turning yellow or orange at maturity. The flesh is yellow and tart, even when fully ripe. Trees are grown from seed, and there are no widely recognized superior selections. The fruit are usually not consumed fresh; instead, they are processed into juices or preserved. Thinly-sliced chamburo flesh is preserved in sugar solutions and eaten as an accent at meals (e.g. with cheese) or used in cooking. Lizana et al. (1978) conducted experiments with the production of dehydrated slices of chamburo as a potential new product in Chile.

Carica ×heilbornii var. *pentagona* Heilborn. Babaco. The babaco is a sterile hybrid between *C. pubescens* and *C. stipulata*. It is cultivated mostly in Ecuador at elevations above 1,000 m. The fruit range in diameter from 6 to 12 cm and can reach a length of 30 cm. The plants are precocious in cultivation and usually remain less than 4 m in height. The pulp is white, and like the chamburo is tart when ripe. Babaco is important locally as a juice or preserve and is extensively used in cooking. For commercial production, cuttings are used, as no seed are available due to sterility of the clone. There are only a few recognized cultivars of babaco and these selections are not widely distributed. Annual production at elevations of 1,500 to 2,000 m is up to 46 t/ha, with production beginning 10 months after planting (Camacho and Rodriquez 1982).

Throughout the highlands of South America, products made from chamburo and babaco are readily available in markets. Also, local residents use these fruit in all forms of cooking. Babaco has been tested as a commercial fresh fruit throughout the American tropics and in New Zealand, but consumer acceptance has been poor. The preserved products made from both fruit have a mild, widely acceptable flavor. If these two fruit are to significantly expand in production, these processed products will need to be successfully marketed; however, some competition can be expected from similar, readily available products such as green papaya preserves. However, given the precocity and heavy production of these fruit at elevations above 1,000 m, they remain candidates for further development in the highland tropics. Also, these two fruit are resistant to the papaya ringspot virus, a major limiting factor for papaya production in the tropics. The chamburo has been investigated as a potential source of resistant genes for imparting virus resistance to papaya (Moore and Litz 1984).

Chrysobalanaceae

This family has over 460 species, but few among them yield edible fruit. The icaco (*Chrysobalanus icaco* L.) is perhaps the most widely grown, with subsistence production throughout the lowland tropics, particularly in coastal regions.

Licania platypus Fritsch. Sunsapote. The sunsapote is native throughout the moist lowlands of Central America and northern South America. It forms a large tree up to 30 m and is usually found in secondary forests under semi-wild cultivation, or in small plantings of a few individual trees. The trees produce 1 to 5 large, oblong fruit per panicle that range from 10 to 15 cm in diameter and 15 to 20 cm in length, and weigh up to 900 g (Leon 1987; Martin et al. 1977). The fruit has a fibrous skin that is easily removed. The pulp is dry and sweet. The seed is large in most clones, often with fibers that protrude into the flesh. The fruit are generally consumed fresh, but can also be used as a flavoring. The fruit are highly esteemed in localized areas of Central and South America, but in northern Brazil, they are not preferred (Cavalcante 1974).

Trees are grown from seed and can require up to 10 years to come into production. There have been only limited trials of this fruit in orchard conditions within the lowland tropics. Trees grew well in Florida until damaged by freezing temperatures (Campbell et al. 1977). Fruit quality within the markets of Central and South America is highly variable because the fruit are often collected from seedling trees maintained in semi-wild cultivation. Martin et al. (1977) considered this fruit to have little potential for further commercialization. However, the fruit are large and can withstand handling, making them a candidate as a fresh fruit for the lowland humid and seasonally dry tropics. In order to improve the commercialization of this fruit, a concerted effort into the selection of superior cultivars would be needed, as well as research on production, and handling procedures.

Guttiferae

This is a large family (1350 species) that contains many fine-flavored fruit crops. Popenoe (1920) went so far as to proclaim one member of this family, the mangosteen (*Garcinia mangostana* L.), as perhaps the finest flavored of all of the fruit in the world. Other species in the family from South America also possess a superior flavor and may actually have greater potential for further commercialization than the mangosteen due to superiority in adaptation to diverse climates.

Platonia esculenta (Arruda) Rickett and Stafleu. Bacurí. The bacurí is native to northern South America and is now grown extensively throughout the Amazonian lowlands. The tree can attain a height of 25 m under optimal conditions. Fruit production is reported to be quite heavy in comparison to other Guttiferae (Cavalcante 1974), although specific yield records are not available. The fruit are yellow, with a leathery shell enclosing a creamy white flesh, which is usually divided into 6 sections (similar to mangosteen). The flavor is excellent, being sweet and aromatic and highly appreciated. Care must be taken when eating the fruit because the leathery shell contains a yellow latex that is quite bitter. The fruit range from 300 to 900 g and are 10 to 12 cm in diameter (Donadio 1983). There can be up to 6 seeds per fruit, weighing about 20 to 40 g each. Often the seeds abort, and edible flesh fills the space which would otherwise be occupied by the normal-sized seed. In contrast to mangosteen, the tree is tolerant of many different environmental conditions, including poor drainage (Martin et al. 1987). Trees are quite sensitive to temperatures below 0°C and to desiccating winds. Propagation is usually by seed, but bacurí is graft compatible with other *Garcinia* and *Rheedia* species.

Rheedia macrophylla Planch. et Triana. Bacuripari. The bacuripari is native to the Amazonian lowlands, where it grows as an understory tree. The tree can grow to 9 m, forming an attractive, pyramidal canopy (Campbell 1983). Trees are propagated by seed and may require 7 to 10 years to come into production. Fruit are variable in shape, averaging 4 to 5 cm in diameter and 5 to 6 cm in length. The fruit have a thick, hard outer wall containing a bitter latex, as in bacurí. Inside the hard shell is a white, creamy flesh surrounding 3 to 4 large seeds. The flesh is scanty in comparison to mangosteen or bacurí. The bacuripari is outstanding because it grows and produces a significant crop in shaded conditions (Campbell 1983). The trees are also tolerant of full sun and wind exposure, making them more adaptable to varied climates than the mangosteen. There is considerable variation in fruit quality among bacuripari from different regions of South America, and there may be different species involved.

Wherever bacurí, bacuripari or other *Rheedia* sp. are grown, the flavor is considered excellent. Although not superior to mangosteen in terms of flavor or edible flesh percentage, these other species have better adaptation to varied climatic and edaphic conditions, allowing for their production in many regions. The latex in both of these fruit can be a major obstacle to commercialization, because those unfamiliar with the consumption of these fruit are likely to ingest it, leading to an unpleasant taste experience. Silva (1991) reports that bacurí fruit can be stored a few days after harvest to reduce the amount of latex in the fruit. There has been little selection for superior clones among either bacurí or bacuripari, although there is considerable variation present among seedling trees.

Malpighiaceae

This family of 1100 species is best represented in fruit crops by the acerola (*Malpighia glabra* L.) which is widely cultivated on a commercial scale throughout the tropics as a fresh fruit, juice, and natural source of vitamin C. Other members of this family from South America are also widespread throughout the tropics as food sources on a local scale.

Byrsonima crassifolia (L.) H.B.K. Nance. This fruit is native from the Caribbean through Central America and throughout most of South America. It has one of the widest native ranges of all fruit crops in Tropical America. Trees are tolerant of a wide range of environments, from the coastal Caribbean, the semi-desert regions of northeastern Brazil, the humid tropical lowlands and the middle elevations (1,000 m) of Central and South America. Throughout their range, nance trees are left when forests are cleared, and maintained in a state of semi-wild cultivation. The fruit from these trees are harvested by local residents, consumed, preserved, or sold to local markets.

The tree can attain a height of 10 to 15 m. While in flower the tree is quite ornamental, with showy orange and yellow inflorescences. The fruit are variable, ranging in size from 2 to 5 cm throughout its range. The skin is usually yellow, with a yellowish, translucent flesh and a single seed. The flavor is sweet and aromatic, sometimes with an oily or musky flavor. The nance is consumed fresh, as a preserved product, a juice or a liquer. In the markets of the lowland tropics, it is commonplace to find nance packed in water in glass containers.

Propagation is generally by seed, but the trees are easily grafted and in some locations (Yucatan, MX) superior clones are commercially propagated by veneer or cleft grafting. Fruit of superior clones may be yellow or red, and up to 6 to 7 cm in diameter. These clones typically have a superior flavor to the wild types, and are commonly consumed as a fresh fruit. The trees are sensitive to cold, but have survived repeated freezes in South Florida (Campbell et al. 1977). Due to the wide-spread familiarity with this fruit there is the potential for the marketing of fresh fruit of superior cultivars. In addition, with its adaptation to varied climates, the nance could become an important processed fruit for the lowland tropics if it could be successfully marketed.

Bunchosia armeniaca Rich. The bunchosia is native to South America, and is uncommon in most other locations. The trees are found from low to middle elevations, producing a small, attractive tree up to 10 m. The trees are precocious, fruiting within 3 years from seed. The trees flower and fruit throughout most of the year. Fruit are ellipsoid and borne in clusters. The red or yellow fruit are from 3 to 4 cm in length with a cream-colored flesh. The flavor is sweet, but often astringent. Even in areas where the tree is common, the fruit are not highly esteemed for fresh consumption. They are more commonly used as a flavoring. Bunchosia is a common addition to the home garden, but only rarely used as a commercial crop. The trees are tolerant of freezes, being slightly damaged by temperatures of –2°C in Florida (Campbell et al. 1977).

Martin et al. (1977) finds the bunchosia to have little potential for further commercialization. However, it could have potential, given its precocity and adaptive nature if superior cultivars could be identified. New cultivars not withstanding, the tree has good potential as an ornamental in the low to middle elevations throughout the tropics (Donadio 1983), where it would be a pleasant addition to the home garden landscape. With the commercial importance of ornamental horticulture throughout the world, this aspect of the tree should not be ignored.

Myrtaceae

Of all of the tropical fruit families, the Myrtaceae has attracted perhaps the most attention for possible increased commercialization (Campbell 1977; Clement 1983; Donadio 1983; Arkcoll 1990). Among the 3,850 species within this family, many produce edible fruit of superior quality. These same species often possess unusual growth forms, making them good candidates for ornamentals as well. The guava (*Psidium guajava* L.) and the jaboticaba (*Myrciaria cauliflora* Berg) are two of the best known members of this family in terms of commercial use. There are, however, many others (particularly from South America) that deserve greater attention.

Eugenia luschnathiana Klotzsch ex. O. Berg. Pitomba. The pitomba is native to Brazil and is relatively uncommon outside of this region. As with many of the Myrtaceae, its growth rate is slow, particularly in calcareous soils, where micronutrient deficiencies are often problematic (Campbell 1977). In acid soils, the growth is much faster, and plants generally have a better nutritional status. The pitomba forms a bush or a small tree to 8 m. The fruit are yellow or orange, 2 to 3.5 cm in diameter, with orange flesh. The flavor is sweet and aromatic. The fruit can be eaten fresh, but more commonly are made into preserves or juices. Dorsett et al. (1917) commented that the tree had value as an ornamental in the correct environment, although Campbell (1977) noted that in Florida the tree is not as attractive as many other members of this family. Trees are grown from seed and can require 7 to 10 years to produce fruit. Grafting can be used to reduce the time to fruiting. The pitomba is not as productive as some Myrtaceae (Dorsett et al. 1917).

Marliera edulis Cambess. Cambucá. The cambucá is native to the coastal rain forests of Sao Paulo and Rio de Janeiro states, Brazil. It has long been known and consumed locally, but is uncommon today within its native region, and even lesser known outside of Brazil. The tree is attractive, but slow growing, eventually reaching heights of 5 to 12 m depending on the environmental conditions. The bark has the same attractive mottled appearance as a jaboticaba tree. The fruit, from 50 to 70 g in weight, are born on the trunk and larger limbs of the tree. They are yellow, with longitudinal ridges. The skin is leathery with a translucent and juicy flesh and 1 to 2 large seeds, leaving little flesh to eat. The flavor is excellent. In recent years this fruit has been described by many as superior to jaboticaba in flavor, but Dorsett (1917), describing the diversity of fruit available at that time in Brazil, considered the taste inferior to jaboticaba. There has been, however, little work done on selection due to the slow growth of the tree and the narrow genetic base.

The pitomba and the cambucá are both good examples of fruit which have truly excellent flavors, but are not likely to be developed in the future as fresh fruit. Instead, development will depend on some form of a niche market, taking advantage of their excellent flavor for ice creams, or juices. As with most of the Myrtaceae, the fruit of these two species are small and easily damaged by handling. Even if their excellent flavor can be exploited, their commercial future will depend on the selection of better clones and improvements in propagation and production techniques, allowing for feasible economic production.

Sapotaceae

The Sapotaceae, with 1,000 species, is a prevalent fruit crop family, particularly in the Caribbean. The sapodilla [*Manilkara zapota* (L.) P. Royen] and the mamey sapote [*Pouteria sapota* (Jacg.) H. Moore and Stearn] are commercial fruit crops throughout the Caribbean and Central America. In South America there are also representatives of this family which have potential for much wider cultivation than at present.

Pouteria caimito Radlk. Abiu. The abiu is native to the warm, moist lowlands of South America. Trees have been introduced to many other locations, but they are still relatively uncommon. The trees attain a height of 12 m under optimal conditions. The fruit are ellipsoid to spherical and can range from 4 to 10 cm in length. The skin color is yellow when ripe, with a translucent flesh surrounding 1 to 5 seeds. The skin is leathery, containing a sticky white latex. The abiu is common in local markets throughout South America, where fruit quality is highly variable, with round, oblong, pointed, and dorso-ventrally flattened fruit types.

The abiu has good potential for commercial development in warm, moist tropical climates due to its precocity and heavy production. Yet, propagation of this crop is still predominantly by seed. In order to further develop this crop there will need to be selection of superior clones. Clement (1983) discussed clones in the eastern Amazon of up to 1,000 g, but these are not widely available. In Australia several named varieties have recently been selected which are precocious and productive, and have good fruit quality. These new selections have been accepted in the Southeast Asian marketplace, and should also have potential in the American Tropics. As with the Guttiferae, the latex of the abiu could be a hindrance to the development of this fruit.

Pouteria obovata Baehni. Lucuma. The lucuma is native to the cool highlands of South American, above 1,000 m. Lucuma grows best in cool climates, and is difficult to cultivate in the lowlands. The trees attain a height of 12 m and yield an ovoid to ovate fruit 4 to 8 cm long. The fruit are yellow at maturity with a dry, yellow flesh. While the fruit are immature, it contains a bitter white latex. The fruit can be eaten fresh when ripe, but is generally consumed as a drink or a flavoring. In Chile and Peru the lucuma is a significant commercial crop (Lizana et al. 1986), and specialized grafting techniques to improve precocity are practiced by some nurseries. Within Chile and Peru, the lucuma fruit are usually dehydrated and ground into a fine powder and used as an additive to milk.

This fruit does have some potential for fresh consumption, and there has been research conducted on the harvest indices and storage characteristics (Lizana et al. 1986). The lucuma could fill the same niche in the tropical highlands as the canistel [*Pouteria campechiana* (Kunth) Baehni] does in the lowlands. However, canistel has not been successful as a fresh fruit in the U.S. market to date. There is little reason to assume that the lucuma would succeed as a fresh fruit when the canistel has failed. The greater potential for lucuma is probably as a processed product, such as the powder used as an additive to milk.

Sterculiaceae

This family with over 1500 species is best known for cacao (*Theobroma cacao* L.), used for chocolate and the cola nuts (*Cola* sp.) used for caffeine production. Within South America there are several other members of this family with potential. One of these species, the cupuaçu (*Theobroma grandiflorum* Schumann) was previously discussed in detail by Arkcoll (1990) and Cabral Velho et al. (1990).

Theobroma bicolor Bonpl. Mocambo. This fruit is commonly used as a beverage fruit within its native range of Central and South America, but is uncommon outside of this area. The trees are propagated by seed and can attain a height of 12 m, although they are usually found as smaller understory plants. The fruit are ellipsoid from 15 to 20 cm long and 10 to 15 cm wide, with a felty brown exterior. The pulp is sweet and pleasant and the seeds can be roasted for consumption. The odor of the fruit can be disagreeable to many. In Central America it is traditionally mixed with achiote (*Bixa orellana* L.) and sugar to make a sweet desert. As with cupuaçu, the mocambo will never have potential as a fresh fruit, but the flesh could be marketed as an additive for beverages as proposed for cupuaçu (Cabral Velho et al. 1990).

CONCLUSIONS

The commercial status of many of the fruit described in this paper has not changed dramatically from when they were described over 75 years ago by Dorsett et al. (1917) and Popenoe (1920). A narrow range of environmental adaptations, the lack of superior selections, difficulties in propagation, exotic looks and tastes, and ignorance about the proper use of the fruit may have doomed these fruit to obscurity to date. They were then, and remain today as "under-exploited" fruit crops.

Regardless of the reason for their continued obscurity, it is clear that a superior taste and an exotic appearance alone was not enough to insure greater economic development of these crops. There was not sufficient economic incentive for their development. It may require innovative methods of marketing and promotion to introduce the new tastes and sights, and gain greater acceptance of these products. Perhaps the time is right to market these fruit crops and their products as the "fruits of the rainforest," taking advantage of the present sentiment that they are the salvation of these endangered ecosystems (Schemo 1995).

The greatest challenge lies in the promotion of fruit that are not consumed "out-of-hand," but are better used as a juice or flavoring. A change is needed in the consumption habits of consumers to further develop these markets. Fresh or frozen juices will have to at least partially replace soft drinks and juices which include only small percentages of fruit juice. Prices of fresh fruit or frozen pulp will have to be affordable to allow the economical use of the fruit within the home. The promotion of these products as "all-natural" could also be a plus for their marketing.

The potential value of fruit crops in the home landscape should also not be ignored, providing both beauty and a nutritious product for the home. Campbell (1973) discussed the use of fruit trees in the home garden within metropolitan areas. South Florida is an excellent model of how fruit crops can be used profitably in this manner (Crane 1993; Lamberts and Crane 1990). With the continuing economic development of Latin America, the use of fruit crops as ornamentals and home landscape components will become more important.

Increased demand through the promotion of these fruits must be accompanied by the selection of clones that meet the needs of the market. Whether it is size, color, self-fruitfulness, or precocity, a concerted effort into the genetic improvement of these crops is vital. Finally, there must also be an effort put forth for trials in other locations and research into propagation and production for the fruit crop. Regardless of the fruit crop considered, further development will depend on sound economic principals (Campbell 1990). If the production of the fruit is not profitable within a particular region, its development will not proceed.

REFERENCES

Arkcoll, D. 1990. New crops from Brazil. p. 367–371. In: J. Janick and J.E. Simon (eds.), Advances in new crops. Timber Press, Portland, OR.

Badillo, V. 1993. Caricaceae: segundo esquema. Universidad Central de Venezuela. Maracay, Venezuela.

Cabral Velho, C., A. Whipkey, and J. Janick. 1990. Cupuassu: A new beverage crop for Brazil. p. 372–375. In: J. Janick and J.E. Simon (eds.), Advances in new crops. Timber Press, Portland, OR.

Camacho, S. and V. Rodriguez. 1982. El cultivo comercial del babaco (*Carica pentagona* Heilb.) en Ecuador. Proc. Trop. Region Am. Soc. Hort. Sci. 26:17–20.

Campbell, C.W. 1973. The potential for tropical fruits in metropolitan areas. Proc. Florida State Hort. Soc.

86:357.
Campbell, C.W. 1977. Cultivation of tropical fruit of the Myrtaceae in Southern Florida. Proc. Trop. Region Am. Soc. Hort. Sci. 21:3–7.
Campbell, C.W. 1983. The bacuripari: A shade-tolerant tropical fruit for Southern Florida. Proc. Florida State Hort. Soc. 96:219–220.
Campbell, C.W. 1990. Techniques for producing export quality tropical horticultural crops. HortScience 25:31–33.
Campbell, C.W. and J.W. Sauls. 1980. Spondias in Florida. Florida Coop. Ext. Serv., Fruit Crops Fact Sheet FC-63.
Campbell, C.W., R.J. Knight, and N.L. Zareski. 1977. Freeze damage to tropical fruits in Southern Florida in 1977. Proc. Florida State Hort. Soc. 90:254–257.
Cavalcante, P.B. 1974. Frutas comestiveis da Amazonia. Publicacoes Avulsas 17. Museu Paraense Emilio Goeldi, Belem, Para, Brasil.
Cavalcanti, J. and M. Abilio de Queiroz. 1992. Umbú gigante. Informativo, Sociedade Brasileira de Fruticultura.
Clement, C.R. 1983. Underexploited Amazonian fruits. Proc. Trop. Region Am. Soc. Hort. Sci. 27A:117–141.
Coelho Pedrosa, A., I.E. Lederman, J.E. Fernandes Bezerra, A. Pinheiro Dantas, and L.G. Neto. 1991. Metodos de enxertia do umbuzeiro (*Spondias tuberosa* Arr. Cam.) em viveiro. Rev. Brasileira Fruticultura 13:59–62.
Crane, J.C. 1993. Commercialization of carambola, atemoya and other tropical fruit in South Florida, p. 448–460. In: J. Janick and J.E. Simon (eds.), New crops. Wiley, New York.
Donadio, L.C. 1983. Review of South American Fruits. Proc. Trop. Region Am. Soc. Hort. Sci. 27A:21–42.
Dorsett, P.H., A.D. Shamel, and W. Popenoe. 1917. The navel orange of Bahia: with notes on some little known Brazilian fruits. U.S. Dept. Agr. Bul. 445.
Ferguson, L. and M. Arpaia. 1990. New subtropical tree crops in California. p. 331–337. In: J. Janick and J.E. Simon (eds.), Advances in new crops. Timber Press, Portland, OR.
Lamberts, M.L. and J.H. Crane. 1990. Tropical fruits. p. 337–355. In: J. Janick and J.E. Simon (eds.), Advances in new crops. Timber Press, Portland, OR.
Leon, J. 1987. Botanica de los cultivos tropicales. IICA, San Jose, Costa Rica.
Lizana, A., F. Kiger, and R. Cruz. 1978. Influencia del estado de madurez y temperaturas en la deshidatacion de papayas (*C. candamarcensis* Hook F.). Proc. Trop. Region Am. Soc. Hort. Sci 22:48–61.
Lizana, L.A., G. Reginato, and J.C. Lavanderos. 1986. Caracterizacion de la maduracion de Lucuma (*Lucuma bifera* Mol., L. obovata HBK). Proc. Int. Soc. Trop. Hort. 30:105–120.
Martin, F.W., C.W. Campbell, and R.M. Ruberte. 1987. Perennial edible fruits of the tropics. U.S. Dept. Agr. Handb. 642.
Moore, G.A. and R.E. Litz. 1984. Biochemical markers for *Carica papaya*, *C. cauliflora* and plants from somatic embryos of their hybrids. J. Am. Soc. Hort. Sci. 109:213–218.
Nerd, A., J.A. Aronson, and J. Mizrahi. 1990. Introduction and domestication of rare and wild fruit trees for desert areas. p. 355–363. In: J. Janick and J.E. Simon (eds.), Advances in new crops. Timber Press, Portland, OR.
Popenoe, J. 1979. The genus *Spondias* in Florida. Proc. Florida State Hort. Soc. 92:277–279.
Popenoe, W. 1920. Manual of tropical and subtropical fruits. MacMillan, New York.
Schemo, D.J. 1995. Hope for the Amazon rainforest: New fruit. New York Times, Sept. 24, 1995.
Silva, S.P. 1991. Frutas-Brasil. Empresa das Artes. Sao Paulo, Brazil.

Sapodilla: A Potential Crop For Subtropical Climates

Michael V. Mickelbart

Sapodilla (*Manilkara zapotilla*, Sapotaceae) is native to Central and South America, specifically from the Yucatan Peninsula of Mexico to Costa Rica, where the largest population of native trees still exists (Gilly 1943). It is now widespread throughout the tropical regions of the world, including Central and South America, the West Indies, India, and Florida in the United States.

Sapodilla is the source of chicle, the principle ingredient in chewing gum. The chicle is extracted from the trunk of the tree as a white latex exudate. Today, sapodilla is cultivated for its fruit in most areas. Although synthetic gums are primarily used, some countries such as Mexico, Venezuela, and Guatemala, still grow sapodilla for chicle.

Sapodilla is grown on a commercial basis in India, the Philippines, Sri Lanka, Malaysia, Mexico, Venezuela, Guatemala, and some other Central American countries. India is the largest producer of sapodilla fruit with current production around 24,000 ha (Chadha 1992). Sapodilla is widely planted in south Florida, where the fruit is marketed locally and shipped to northern and eastern U.S. markets. The fruit, however, is not commonly seen in the United States. In southern Mexico and Central America where sapodilla is native, it is considered to be one of the best of the tropical fruits.

The future of sapodilla appears to be promising, given the attention the crop is receiving from growers and consumers in many countries. Indian production of sapodilla continues to grow and there is an active research program in that country with specific goals toward improving storage, transport, and marketing strategies. Sapodilla has been identified by the Ministry of Agriculture in Malaysia to be promoted under the program for development of its fruit industry (Bakar and Abdul-Karim 1994). The fruit is also gaining popularity as a specialty fruit in restaurants in North America and production of sapodilla as a commercial crop seems to be a possibility in areas where environmental conditions are mild.

NOMENCLATURE

The name sapodilla is derived from the Spanish word *zapotilla*, meaning "small sapote." The fruit, commonly referred to as sapodilla in the U.S., is also known as chicku, chiku (India), chicopote, chicozapote (Mexico), dilly (Bahamas), kauki (Southeast Asia), mespel (Virgin Islands), mispu, mispel, mispelboon (Surinam), muyozapot (El Salvador), naseberry (British West Indies), nispero (Puerto Rico), sapatija, sapodilla plum, sapodille (Dutch West Indies), sapota, sapote (Cuba), sapoti or sapotilha (Brazil), sapotille, sapotillier (French West Indies), zapote, zapote chico, or zapotillo. The profusion of names for the fruit is probably due to the many small, isolated growing areas it occupies, where names are often derived from local words. The small-market characteristics of the fruit have helped to keep it in obscurity.

The botanical name of the fruit is no less confusing. Both *Manilkara* and *Achras* are commonly used as generic names and there appears to be no agreement among botanists or horticulturists as to the proper term. *Sapota* (*zapota*) or *sapote* (*zapote*) are commonly used as the species name, although this too is variable among regions and authors. Gilly (1943) addressed this problem of confused nomenclature. It seems the generic name *Achras*, given by Linnaeus, was based upon a plate and description by the botanist Plumier. Unfortunately, the plant described by Plumier is not sapodilla, leading to the misnaming. Gilly suggests that *Manilkara zapotilla* (Jacq.) Gilly is the only proper name since *Manilkara* is the earliest recorded name of the group to which sapodilla belongs and *zapotilla* was specifically applied to sapodilla at the time of its publication. Still, the nomenclature of this species remains confused.

TREE DESCRIPTION AND MANAGEMENT

The tree is a medium to large evergreen tree reaching 12 to 18 m in the tropics, although some trees may reach 40 m. Tree size would likely be much smaller in subtropical regions such as Southern California, where cool winters would slow growth for a portion of the year. The canopy is dense, often with a rounded crown, but sometimes with a more pyramidal shape. The trunks of the trees are reported to attain a diameter of 2 to

3.5 m in Central America. The glossy, dark green leathery leaves clustered toward the tips of the branches are ovate-elliptic to elliptic-lanceolate, 5 to 12.5 cm long, 6 cm wide.

Being a strictly tropical tree, the sapodilla is limited in the United States to the southern coastal region of Florida and possibly some southern coastal areas of California. Young trees are injured or sometimes killed at temperatures of –1° to 0°C, while mature trees can withstand temperatures as low as –2° to –3°C with only minor damage. Temperatures above 41°C during flowering or fruiting can cause flower abortion or fruit scalding.

Sapodilla has proven to be tolerant of dry conditions, and its ability to thrive on poor soils makes it an ideal fruit tree for less-than-optimum growing areas. The tree has shown ability to withstand extended periods of waterlogging, and trees are grown on most soil types, from clay soils to almost pure limestone. Sapodilla is remarkably tolerant of high levels of root zone salinity (Mickelbart and Marler 1996), a rare characteristic in tropical fruit species. It has also proven tolerant of salt spray off the coast of Florida, indicating it may thrive on the subtropical coasts of other regions as well. Sapodilla appears to perform better, in fact, in the coastal regions of the areas in which it is grown. Still, the fact that sapodilla has, at least initially, been successfully grown in the hot, arid desert regions of India suggests that the tree may be grown in some desert areas of California, although the effect of extremely high temperatures on such physiological events as flowering and fruit set may limit production.

Sapodilla is a shallow-rooted tree, with more than 80% of the roots located within the top 75 cm of soil, concentrated within an area half the width of the canopy (Avilan et al. 1981; Bhuva et al. 1991). About 66% of the moisture extracted from the soil is in the first 75 cm. This root morphology suggests that irrigation may be economically feasible in areas of low rainfall. Pruning does not appear to be necessary for at least the first ten years of growth, except possibly skirting of the lower branches.

Young trees require irrigation, especially during lengthy dry periods, and mature trees bear more consistent, higher quality crops with regular irrigation. Trees four years of age and older are generally able to cope with extended dry periods. Irrigation of mature trees increases fruit yield and canopy volume (Bhuva et al. 1990), although water use efficiency decreases with increased irrigation (Bhuva et al. 1991), suggesting that irrigation may not be economically feasible in areas which receive sufficient rainfall.

Fertilization appears to have a favorable effect on fruit yield (Bhuva et al. 1991) and on fruit quality characteristics such as total soluble solids and pulp:seed ratio (Durrani et al. 1982). Potassium is taken up in larger amounts than any other nutrient in sapodilla (Rao 1979; Avilan et al. 1980; Sulladmath 1983). Potassium fertilization seems to have a favorable effect on fruit set (Laborem et al. 1981), especially when done in conjunction with an application of phosphorus (Durrani et al. 1982). In fact, Avilan et al. (1980) found a linear relationship between fruit yield and potassium uptake.

Insects and disease are rarely a problem on sapodilla, although leaf miner and stem borer are sometimes minor problems. *Phytophthora palmivora* (Butler) may cause fruit rot in the lower fruits on the canopy if water directly contacts the fruit, especially during periods of elevated temperatures and high humidity. Leafspot (*Phaeophleospora indica* Chinnappa) is also reported to be a problem in India (Balasubramanian et al. 1988).

PROPAGATION

Sapodilla is vegetatively propagated, due to the great variability in progeny from seed, as well as the length of time from planting to fruit set, often taking as much as 6 to 10 years. Air layering is the most common means of propagation in India and there has been a fair amount of research regarding optimal treatment of air-layered plants. Workers in India have found that using recently-produced branches, along with etiolation and application of a growth regulator such as IBA or NAA results in optimal growth of air-layered plants (Chadha 1992).

Veneer grafting resulted in the highest percent survival and moderate root growth in a study comparing various propagation techniques including side grafting, side inarching, tongue grafting, whip grafting, and saddle grafting (Hussain and Bukhari 1977). Tongue grafting resulted in the lowest percent survival, although the plants produced had significantly larger root systems than those produced by the other methods. Malo (1967) describes a successful method for veneer grafting young sapodilla plants designed to efficiently propagate large numbers of sapodilla.

Softwood grafting was successful (ca. 80% survival) in a study conducted in India (Kulwal et al. 1985). The greatest success was obtained when scion wood was defoliated 8 days prior to removal and when grafting was done during the summer months.

While sapodilla seedlings are often used as rootstocks for grafted plants, some other species may be suitable and contribute to the management of sapodilla orchards. Wild dilly [*Manilkara emarginata* (Bakar) Lam. & Meeuse] (Ogden and Campbell 1980) and *Manilkara hexandra* (Roxb.) (Chandler 1958) have been proposed as possible dwarfing rootstocks for sapodilla.

FLOWERING AND POLLINATION

The small white flowers of sapodilla are borne on short pedicels in the leaf axils. Piatos and Knight (1975) found sapodilla to be self-incompatible, indicating that cross-pollination might be necessary. While production of some cultivars such as 'Prolific' is high in areas such as Florida, fruit yield among cultivars is variable, a characteristic which may be due to self-incompatibility. Gonzalez and Feliciano (1953) examined flowering and fruit set in 'Ponderosa' sapodilla and found that the trees could not be self-pollinated, either naturally or artificially. The best method of pollination was determined to be cross-pollination by hand, which resulted in a 39.6% fruit set. Open pollination resulted in only 5% fruit set in that study. Similar results were observed in 'Kalipatti' sapodilla by Relekar et al. (1991). Mulla and Desle (1990) found that the highest percent fruit set was observed in trees cross-pollinated with pollen of the same cultivar and ranged from 20% to 34%. There was no sign of parthenocarpy and self-pollination resulted in only negligible fruit set in some of the cultivars examined.

Reddi (1989) conducted a series of studies in which he showed that thrips (*Thrips hawaiiensis* Morgan and *Haplothrips tenuipennis* Bagnall) are the principle pollinators of sapodilla in India. The thrips apparently take shelter in the flowers and live on the pollen grains, nectar, and stigmatic exudations. They collect pollen grains while feeding on these components and transfer then to other flowers when the food reserves are exhausted. Bees have been observed in flowering sapodilla plantings (Sambamurty and Ramalingam 1954), although examination of the insects indicated that they were not carrying pollen.

Reddi's study showed several traits which indicated that wind is not an important pollinating agent, nor are large insects. Sambamurty and Ramalingham (1954), however, suggest that wind is an important factor in sapodilla pollination. Reddi also showed that sapodilla flowers are not self-pollinating, although pollen transfer is generally limited to a single tree unless trees are closely spaced. This information would provide evidence that sapodilla is not self-incompatible as previously suggested. Pollen size and viability are quite variable between cultivars, possibly causing the varied results in fruit set (Minhas and Sandhu 1985).

Much like avocado, sapodilla produces many more flowers than developed fruits. The great variability in fruit set may be due to differences in flower abortion or fruit drop, although differences in this phenomenon between cultivars has not been specifically examined in sapodilla. The major period of fruit drop occurs in the first five weeks following fruit set and as little as 1.6% of the flowers produced by a tree may develop into fruit (Relekar et al. 1991). Fruit set is highly variable, even within a cultivar. Gonzalez and Feliciano (1953) suggested that tree vigor may be related to flower production and fruit set.

FRUIT DESCRIPTION AND USES

The fruit of sapodilla is small, ranging from 5 to 9 cm in diameter with a round to egg-shaped appearance, from 75 to 200 g in weight. The rough brown skin, which gives the fruit a somewhat unattractive appearance, encloses a soft, sweet, light brown to reddish-brown flesh. The flesh is often gritty, much like a pear, and it can hold up to 12 flat, smooth black seeds, although some fruit are seedless. Superior strains have a fine smooth texture with a slightly fragrant and sweet flavor.

Fruit development follows a sigmoidal pattern (Sulladmath et al. 1979; Abdul-Karim et al. 1987). The initial growth phase is due to cell division and involves maturation of the embryo within the fruit. A phase of greatly reduced growth follows, until a second rapid growth phase occurs, during which time growth is due to cell enlargement. This second growth phase is the time when maximum growth occurs, between 5 and 7.5 months from fruit set (Lakshminarayana and Subramanyam 1966). The fruit are suitable for harvesting after the first growth phase, although higher quality fruit are obtained if they are harvested following the second growth phase, when there is a dramatic increase in sugar content of the fruit.

Fruit maturity occurs anywhere from 4 to 10 months following fruit set, depending on variety, climate, and soil conditions. In south Florida and the Virgin Islands, the fruit appears throughout the year, with a peak season from May to Sept. Fruit is also produced throughout the year in Malaysia. In the desert regions of India, the fruit ripens primarily in July and Aug. Seasonal variation in fruit shape and size is not uncommon and great variation exists among seedling fruits.

Because immature sapodilla fruit contain latex, harvesting fruit at full maturity is critical to quality. Judging maturity in sapodilla is extremely difficult. Even within a single cluster, fruit maturity may vary greatly, although fruits generally mature from flowers produced at the base of the cluster to the tip. Sundararajan and Rao (1967) suggest using total soluble solids as a measure of maturity in sapodilla, although the variation in fruit age within a tree may require that each fruit is judged individually for maturity. Abdul-Karim et al. (1987) found that fruit length and width were better indicators of maturity than weight and volume, or firmness. Some varieties of sapodilla maintain the remnants of the flower style until maturity (George 1982), which may also be a good indicator. Ripeness can occur anywhere from 9 to 13 days after harvest (Lakshminarayana and Subramanyam 1966).

Sapodilla fruit is most commonly consumed fresh, when fully ripe. It is primarily a dessert fruit in most areas. Flavor of the ripe fruit is improved by chilling just prior to eating. Sapodilla fruit is also used for making ice cream, but is not typically used for jam or canning. The fruit is also eaten as a dried fruit in India.

POSTHARVEST STORAGE

Sapodilla respiration pattern follows that of a climacteric fruit (Broughton and Wong 1979; Selvaraj and Pal 1984; Abdul-Karim et al. 1987; Brown and Wong 1987). Respiration may be slowed by growth retardants such as gibberellic acid (GA), kinetin and silver nitrate (Gautam and Chundawat 1990b).

Sapodilla may be stored under controlled conditions for a short period of time. Singh and Mathur (1954) found that optimum cold storage was obtained at 35° to 38°F with a relative humidity of 85%–90%. Under these conditions, fruit could be stored for up to eight weeks. Broughton and Wong (1979) found that holding the fruit at 4°C before storing at 20°C extended the storage life of the fruit although exposure to 4°C for longer than ten days resulted in chilling injury. Storage of sapodilla under high CO_2 concentrations, provided CO_2 was less than 20% (v/v), and low ethylene concentrations, also prolonged the storage life of the fruit. Upon reaching full maturity, sapodilla fruit deteriorates rapidly, lasting only 2 to 10 days (Brown and Wong 1987).

Gautam and Chundawat (1990a) examined the effects of various growth retardants on postharvest changes in sapodilla. Application of GA, kinetin, and silver nitrate resulted in up to a two-day increase in storage time due to reduction of catalase and pectin methyl esterase activity, and reductions in respiratory activity and ethylene production. The application of these compounds appears to reduce the rate at which fruit ripens as well as affecting fruit quality characteristics such as total sugars, acidity, ascorbic acid, and starch (Gautam and Chundawat 1990b).

Gibberellic acid prolongs the time which sapodilla fruit can be stored before rot occurs, as well as prolonging fruit softness and fruit skin shrinkage (Kumbhar and Desai 1986). Gautam and Chundawat (1990a, b) used a 300 ppm GA solution, although Kumbhar and Desai (1986) found a 75 ppm solution to be the most effective in a range of concentrations from 75 to 225 ppm.

Both wax coating and 2,4-Dichloro-phenoxy acetic acid (2,4-D) have been shown to retard the ripening process in sapodilla, while 2-Chloroethyl phosphonic acid (ethrel) (Ingle et al. 1981; Suryanarayana and Goud 1984) and ethylene (Sastry 1970) greatly accelerate ripening. Polyethylene bags can also reduce weight loss in sapodilla by about 50% (Kumbhar and Desai 1986).

Because of the high moisture and nutrient content of the fruit, sapodilla is especially prone to postharvest diseases. Common diseases include sour rot (*Geotrichum candidum*), Cladosporum rot (*Cladosporum oxysporium*), and blue mold rot (*Penicillium italicum*). Bakar and Abdul-Karim (1994) found benlate (methyl-N-1-butylcarbomoyl), a commonly used fungicide for postharvest treatment of sapodilla, to best control both fungal and bacterial pathogens of sapodilla. Although several non-chemical treatments have been tested, none have proven to be successful against postharvest pathogens of sapodilla.

Table 1. Description of some named cultivars of sapodilla.

Cultivar	Origin	Fruit size (g)	Yield/tree	Location	Reference
Addley	Bahamas		Very poor	Florida	Campbell and Malo 1973
Adelaide	Bahamas		150 kg	Venezuela	Avilan et al. 1980
Badam		45		India	Sundarajan and Rao 1967
Baramasi		118		India	Sundarajan and Rao 1967
Big Pine Key	Florida		Very poor	Florida	Campbell and Malo 1973
Black	Florida		Very poor	Florida	Campbell and Malo 1973
Brown Sugar	Florida	133–170	901 fruit	Venezuela	Avilan et al. 1980
Brown Sugar			125–200 kg	Florida	Campbell and Malo 1973
Calcutta Round		98		India	Sundarajan and Rao 1967
Cricket Ball		142	93 kg	India	Chundawat and Bhuva 1982
Dwarapudi		90		India	Sundarajan and Rao 1967
Gavarayya		112		India	Sundarajan and Rao 1967
Guthi		56		India	Sundarajan and Rao 1967
Jamaica No. 4	Jamaica		Very poor	Florida	Campbell and Malo 1973
Jamaica No. 5	Jamaica		Very poor	Florida	Campbell and Malo 1973
Jantung	Malaysia	100		Malaysia	Abdul-Karim et al. 1987
Kalipatti		98	160 kg	India	Chundawat and Bhuva 1982
Kirtabarti		84		India	Sundarajan and Rao 1967
Long Oval		140		India	Sundarajan and Rao 1967
Martin	Florida		Very poor	Florida	Campbell and Malo 1973
Modello	Florida	227–340	50–100 kg	Florida	Campbell and Malo 1973
Mohangottee		102	107 kg	India	Chundawat and Bhuva 1982
Oval		84		India	Sundarajan and Rao 1967
Pala		31		India	Sundarajan and Rao 1967
Pilipatti		82	115 kg	India	Chundawat and Bhuva 1982
Prolific	Florida	170–225	520 fruit	Venezuela	Avilan et al. 1980
Prolific			150–225 kg	Florida	Campbell and Malo 1973
Russell	Florida	284–454	544 fruit	Venezuela	Avilan et al. 1980
Russell			25–100 kg	Florida	Campbell and Malo 1973
Saunders	Florida		Very poor	Florida	Campbell and Malo 1973
Seedless	Florida	small	<12.5 kg	Florida	Campbell and Malo 1973
Tagarampudi		84		India	Sundarajan and Rao 1967
Tikal	Mexico	113–170	170–225 kg	Florida	Campbell and Malo 1973
Tikal		120	150–175 kg	Florida	Campbell et al. 1987
Vavivalasa		98		India	Sundarajan and Rao 1967
Zumakhia		57	55 kg	India	Chundawat and Bhuva 1982

CULTIVARS

A number of sapodilla cultivars are available at nurseries in the United States specializing in tropical fruit species (Table 1). Because of the lack of information in the literature on many of the cultivars, the table is incomplete. 'Prolific' appears to be the best producer in Florida. It is extremely precocious and produces fruit of good flavor. Grafted trees bear fruit within 3 to 5 years of planting in Florida and mature trees may bear 150 to 225 kg of fruit. The fruit is easily damaged and is therefore susceptible to some postharvest diseases. Another successful variety in Florida is 'Brown Sugar'. As its name implies, the fruit is very sweet, with a slightly granular flesh. Trees bear crops 3 to 5 years following planting, with mature trees bearing crops of 125 to 200 kg. In Florida, 'Tikal' fruits up to four months earlier than other sapodilla cultivars (Campbell et al. 1987). Many cloned Florida cultivars have proven to be unfruitful in successive years. Some examples include 'Martin', 'Adelaide', and 'Seedless'.

'Kalipatti' is the most planted cultivar in the Gujarat region of India, accounting for about 99% of the acreage there. It also appears to be the highest yielding cultivar of those tested in India (Chundawat and Bhuva 1982) and, therefore, will likely continue to be the most widely planted. Average yields are 160 kg per tree, with fruit of medium size and low seed count. 'Russel' was considered to be the best variety in a five-year Puerto Rico study.

The major obstacle to sapodilla producers appears to be opening foreign markets. At present, most sapodilla production is limited to local consumption regardless of where the fruit is produced. Past attempts at establishing a European market have failed (Chadha 1992), and the future of sapodilla is uncertain.

REFERENCES

Abdul-Karim, M.N.B., S.A. Tarmizi, and A.A. Bakar. 1987. The physio-chemical changes in ciku (*Achras sapota* L.) of Jantung variety. Pertanika 10(3):277–282.

Avilan, R.L., E.G. Laborem, M. Figueroa, and L. Rangel. 1980. Absorcion de nutrimentos por una cosecha de nispero (*Achras sapota* L.). Agronomia Tropical 30(1–6):7–16.

Avilan, R.L., L. Meneses, R. Sucre, and M. Figueroa. 1981. Distribucion del sistema radical del nispero [*Manilkara achras* (Mill.) Fosberg]. Agronomia Tropical 31(1–6):247–255.

Bakar, F.A. and M.N.B. Abdul-Karim. 1994. Chemical treatments for microbial control on sapota. ASEAN Food J. 9(1):42–43.

Balasubramanian, P., V. Ponnuswami, and I. Irulappan. 1988. A note on susceptibility of sapota varieties and hybrids to leaf spot disease (*Phaeophleospora indica* Chinnappa). South Indian Hort. 36(1–2):72–73.

Bhuva, H.P., J.S. Katrudia, and R.G. Patel. 1990. Effect of different levels of moisture regime and nutrient on growth, yield and economics of sapota (*Achras sapota* L.) cv. Kalipatti. Haryana J. Hort. Sci. 19(1–2):71–78.

Bhuva, H.P., J.S. Katrodia, and R.G. Patel. 1991. Consumptive use, water use efficiency and moisture extraction pattern of sapota (*Achras sapota* L.) as influenced by varying levels of irrigation and fertilizers. Indian J. Hort. 49(4):291–295.

Broughton, W.J. and H.C. Wong. 1979. Storage conditions and ripening of chiku fruits *Achras sapota* L. Scientia Hort. 10:377–385.

Brown, B.I. and L.S. Wong. 1987. Postharvest changes in respiration, ethylene production, firmness and ripe fruit quality of sapodilla fruit [*Manilkara zapota* (L.) Van Royen] of different maturities. Singapore J. Pri. Ind. 15(2):109–121.

Campbell, C.W. and S.E. Malo. 1973. Performance of sapodilla cultivars and seedling selections in Florida. Proc. Trop. Reg. Am. Soc. Hort. Sci. 17:220–226.

Campbell, C.W., S.E. Malo, and J. Popenoe. 1987. 'Tikal', an early-maturing sapodilla cultivar. Proc. Florida State Hort. Soc. 100:281–283.

Chadha, K.L. 1992. Strategy for optimisation of productivity and utilization of sapota [*Manilkara achras* (Mill.) Forberg.]. Indian J. Hort. 49(1):1–17.

Chandler, W.H. 1958. Evergreen orchards. Lea & Febiger, Philadelphia.

Chundawat, B.S. and H.P. Bhuva. 1982. Performance of some cultivars of sapota (*Achras sapota* L.) in Gujarat. Haryana J. Hort. Sci. 11(3–4):154–158.

Durrani, S.M., V.K. Patil, and B.A. Kadam. 1982. Effect of N, P and K on growth, yield, fruit quality and leaf composition of sapota. Indian J. Agr. Sci. 52(4):231–234.

Gautam, S.K. and B.S. Chundawat. 1990a. Postharvest changes in sapota cv. 'Kalipatti': I—Effect of various postharvest treatments on biochemical changes. Indian J. Hort. 46(3):310–315.

Gautam, S.K. and B.S. Chundawat. 1990b. Post-harvest changes in sapota cv. 'Kalipatti': II—Effect of various post-harvest treatments on physio-chemical attributes. Indian J. Hort. 47(3):264–269.

George, C. 1982. Growing mesples (sapodillas). College of the Virgin Islands Coop. Ext. Ser. Gardeners Factsheet 24.

Gilly, C.L. 1943. Studies in the Sapotaceae. II. The sapodilla-nispero complex. Trop. Woods 73:1–22.

Gonzalez, L.G. and P.A. Feliciano, Jr. 1953. The blooming and fruiting habits of the Ponderosa chico. Philippine Agr. 37(7):384–398.

Hussain, A. and M.A. Bukhari. 1977. Performance of different grafting methods in chiku (*Achras sapota*). Pakistan J. Bot. 9(1):47–57.

Ingle, G.S., D.M. Khedkar, and R.S. Dabhade. 1981. Ripening studies in sapota fruit (*Achras sapota* Linn.). Indian Food Packer 35(6):42–45.

Kulwal, L.V., G.S. Tayde, and P.P. Deshmukh. 1985. Studies on soft-wood grafting of sapota. P.K.V. Research J. 9(2):33–36.

Kumbhar, S.S. and U.T. Desai. 1986. Studies on the shelf-life of sapota fruits. J. Maharashtra Agr. Univ. 11(2):184–186.

Laborem, G., M. Figueroa, O. Verde, L. Rangel, and L. Bandres. 1981. Effecto de la fertilization con N, P y K sobre los rendimientos de nispero (*Manilkara achras*) en suelos del orden entisol. Agronomia Tropical 31(1–6):31–36.

Lakshminarayana, S. and H. Subramanyam. 1966. Physical, chemical and physiological changes in sapota fruit [*Achras sapota* Linn. (Sapotaceae)] during development and ripening. J. Food Sci. Tech. 3:151–154.

Malo, S.E. 1967. A successful method for propagating sapodilla trees. Proc. Florida State Hort. Soc. 80:373–376.

Mickelbart, M.V. and T.E. Marler. 1996. Photosynthesis, water relations, and mineral content of sapodilla foliage as influenced by root zone salinity. HortScience 31:230–233.

Minhas, P.P.S. and A.S. Sandhu. 1985. Investigations on floral biology of sapota (*Achras zapota* L.) under ludhiana conditions. II. Time of anthesis, dehisence, pollen studies, stigma receptivity and fruit set. J. Res., Punjab Agr. Univ. 22(2):270–276.

Mulla, A.L. and G.Y. Desle. 1990. Pollination studies in sapota cultivars. J. Maharashtra Agr. Univ. 15(2):266–268.

Ogden, M.H. and C.W. Campbell. 1980. Wild dilly as a potential rootstock for sapodilla. Proc. Trop. Reg., Am. Soc. Hort. Sci. 24:89–92.

Piatos, P. and R.J. Knight. 1975. Self-incompatability in the sapodilla. Proc. Florida State Hort. Soc. 88:464–465.

Rao, S.R. 1979. Elemental concentration in fruits and leaves of chicku and mango under natural environmental conditions. Proc. Indian Acad. Sci. 88(II):175–182.

Reddi, E.U.B. 1989. Thrips pollination in sapodilla (*Manilkara zapota*). Proc. Indian Natl. Sci. Acad. B55(5–6):407–410.

Relekar, P.P., A.G. Desai, J.C. Rajput, and M.J. Salvi. 1991. Fruit production in sapota cv. Kalipatti. Current Research, Univ. of Agr. Sci., Bangalore 20(6):104–106.

Sambamurty, K. and V. Ramalingham. 1954. A note on hydridization in the sapota. Indian J. Hort. 11:57–62.

Sastry, M.V. 1970. Biochemical studies in the physiology of sapota. Part IV—Ripening and storage studies. Indian Food Packer 24:24–26.
Selvaraj, Y. and D.K. Pal. 1984. Changes in the chemical composition and enzyme activity of two sapodilla (*Manilkara zapota*) cultivars during development and ripening. J. Hort. Sci. 59(2):275–281.
Singh, K.K. and P.B. Mathur. 1954. A note on the cold storage of sapotas (*Achras sapota*). Indian J. Agr. Sci. 24:149–150.
Sulladmath, V.V. 1983. Accumulation of mineral elements at different stages in developing sapodilla fruit. Scientia Hort. 19:79–83.
Sulladmath, V.V., M.M. Rao, and M.R. Advani. 1979. Studies on the pattern of growth of development of 'Kalipatti' sapota [*Manilkara achras* (Mill.) Forsberg] fruit. J. Maharashtra Agr. Univ. 4(1):55–60.
Sundararajan, S. and V.N. Madhava Rao. 1967. Studies on fruit development and fruit quality in some varieties of sapota (*Achras zapota* L.). South Indian Hort. 15:52–57.
Suryanarayana, V. and P. Veeranna Goud. 1984. Effect of post-harvest ethrel treatment on ripening of sapota fruits. Andhra Agr. J. 31(4):308–311.

Native Cultivars of Cactus Pear in México

C. Mondragon-Jacobo and S. Perez-Gonzalez

Cactus pears or prickly pears (*Opuntia* spp. Cactaceae) are omnipresent plants of the Mexican landscape and have great historical and ethnobotanical significance (Fig. 1). Ripe fruits and tender cladodes (pads), have been collected from the wild since preColumbian times, and the mature cladodes are used as a forage for livestock. *Opuntia* originated in Central Mexico and some parts of the Caribbean region (Benson 1982; Russel and Felker 1987). Today cactus pear is a well recognized fruit crop in Mexico cultivated throughout the central semiarid highlands.

A vast array of species are still found in the wild, while backyards contain mostly cultivars selected by farmers. Production at commercial level started in the 1960s and is based on outstanding plants taken from rural households where they provide food for the family (Bravo 1978; Pimienta 1990; Barbera et al. 1992). Orchards used foundation material obtained from plants bearing large fruit size, high color, and good flavor. The clones were named based on specific fruit and plant traits. It is difficult to identify plants based only on vegetative traits and most farmers and scientists still use fruit characteristics for descriptive purposes.

Official figures report that Mexico has 42,000 ha of cactus pear (SARH 1994) distributed in the central highland (2000 m) semiarid temperate part of the country, all of them under rainfed conditions. Mean annual precipitation ranges from 400 to 700 mm with a bimodal pattern. Soils are shallow to medium in depth, poor in organic matter, and acidic to lightly alkaline.

As a plant with crassulacean acid metabolism (CAM), cactus pear is able to withstand drought. It is cultivated in small tracts traditionally devoted to maize and beans production, with low revenues. Medical research has found value in cacti as a raw material for products to treat hypoglukemia, diabetes, high blood cholesterol levels, and obesity (Hegwood 1990). Recent research revealed that this plant is also able to thrive in high CO_2 environments (Nobel and Israel 1994). Cacti have received attention as a crop to fight global pollution and desertification.

Growers expertise generated by trial and error and recent agronomic advances account for high yields of cactus pear. Production depends on intensity of farming practices, orchard size, and environmental factors such as late frosts and rainfall. Small orchards northwest of Mexico City obtain 8–10 t/ha with minimum effect of alternate bearing. However, growers in semiarid regions obtain less than 5 t/ha, with low labor inputs and low crop management.

Due to dependance on a narrow base of cultivars, there is a temporary (three months) market saturation, which is associated with falling prices and low returns. It is desirable to extend the harvest season either with earlier or later ripening cultivars or crop management practices.

At least 95% of production in Mexico is for domestic comsumption. Cactus pears are eaten as a snack, and sold peeled and slightly chilled by street vendors. Harvest season starts late in June and extends through Sept. However it is possible to widen the market window if other cultivars are marketed (Fig. 2). High soluble solids, white flesh, and juiciness are praised by Mexican consumers, while foreign markets prefer yellow and red-fleshed cultivars.

In this paper we attempt to provide a reference for growers, brokers and researchers about cultivars currently grown and marketed in Mexico. For many areas of the world cactus pear still remains an unknown fruit crop.

METHODOLOGY

The first step of this study included field trips to the main cactus pear growing regions in the semiarid highlands of Central Mexico to gather information directly from growers and family orchards about the origin of their plant material, as well as uses and cultivation practices. The second stage was based on the collection of a wide range of germplasm which was planted in northern Guanajuato from 1987 to 1990. Data on several horticultural traits were collected from 1991 to 1994 and used to describe each accession.

Fig. 1. A) Mature cactus pears in full production on a family plot in central México. B) 'Naranjona', a yellow-orange fleshed cultivar of cactus pear. C) 'Reyna', a white-fleshed cultivar. D) 'Cristalina', a white-fleshed cultivar.

Table 1. Average quantitative traits of ripe fruits of 12 cultivars of Méxican cactus pears.

Cultivar	Color		Fruit			Edible portion (%)	Seed wt (g/fruit)	Aborted seed (%)	Peel thickness (cm)	Soluble solids (°Brix)	Flesh juciness (1 = dry 4 = juicy)
	Peel	Flesh	Weight (g)	Length (cm)	Width (cm)						
Amarilla Montesa	Yellow-orange	Yellow	112	8.5	4.9	64.0	3.4	21.7	0.40	15.0	3
Amarilla Redonda	Yellow-orange	Yellow	160	7.8	6.3	57.7	4.2	5.0	0.50	14.0	3
Burrona	Green	Light green	169	6.0	5.4	62.0	7.5	4.2	0.60	13.2	3
Cardona	Purple	Pink-purple	83	6.1	4.8	45.7	2.9	4.1	0.50	15.5	1
Chapeada	Green w/blush	Light green	133	7.6	5.4	51.1	5.0	4.6	0.50	14.0	4
Charola	Red-purple	Purple	67	5.5	4.4	56.4	2.8	5.3	0.30	15.0	2
Cristalina	Green	Light green	216	9.5	6.5	53.6	5.6	17.8	0.60	15.4	4
Esmeralda	Green	Light green	136	8.5	5.2	63.0	4.9	8.8	0.40	15.0	3
Fafayuco	Green-yellow	Light green	148	7.7	5.6	48.0	4.4	6.3	0.70	16.0	3
Naranjona	Yellow-orange	Orange	136	7.7	5.5	59.0	5.0	7.6	0.40	15.0	4
Reyna	Green	Light green	141	8.6	5.8	67.0	5.2	16.7	0.47	14.8	4
Roja Pelona	Green-reddish	Deep red	180	9.4	6.2	61.0	5.6	2.1	0.40	15.0	2

Twelve of the most important cultivars are described in Table 1, ten are cultivated and the rest are subjected to recollection from wild populations or small backyard orchardss. Characters measured included fruit color and shape, peel and flesh color, seed/pulp ratio, juiciness, and soluble solids.

RESULTS AND DISCUSSION

Central Mexico has the widest variability of *Opuntia* species and botanical varieties. Almost every family orchard in this region grows a mixture of cactus, with one or several uses, from fruit production to hedging. In contrast, cultivated orchards use only a small number of clones.

The most productive area near Mexico city is known as Las Piramides, because it is located around the monuments (pyramids) erected by the ancient inhabitants of the Teotihuacan Valley. This area relies mostly in a single cultivar 'Reyna' or 'Alfajayucan' grown under intense management in commercial orchards. A different situation is observed in the north central area (Guanajuato, Zacatecas, and San Luis Potosi states) where it is common to observe mixed orchards of several commercial cultivars. Cactus pear grow wild in feral land and in semidomesticated way in family plantings.

In any small grove it is possible to find a sample of cultivars as well as wild un-named materials, some of which could be readily included as commercial cultivars (Fig. 2). This allows for the availability of Opuntia products for an extended season. The reservoir for a formal breeding program is enormous and almost untapped.

The cultivars reported in Table 1 are the most well known and marketed. Six of them 'Reyna', 'Cristalina', 'Naranjona', 'Chapeada', 'Amarilla Montesa', and 'Roja Pelona' share about 90% of production.

In Mexico it is possible to provide freshly harvested fruits for almost six months under present culture practices (Table 1). Earliness and late maturing traits are important to extend the market. Cultivars such as 'Charola' and 'Fafayuco' are appropiate to extend end of season sales, but they are not commercially important yet. The possibility of widening the harvest season for the most preferred clones such as 'Reyna' (Queen) needs to be explored. This cultivar is the most important in Mexico, with an estimated production of 75,000 t annually. It ripens from early June to Aug. (Mondragon and Perez 1994).

Fruit quality is determined by several visual and organoleptic traits. Mexican consumers prefer white-fleshed types and 'Reyna' is the standard of qualiy. According to our analysis favored traits for primary selection were: large fruits (>120 g); juicy flesh; high soluble solids (sugar content >15%); edible portion >50%; and thin peel (<0.5 cm); despite the fact that thin peel is inconvenient for handling. Low seed content, as well as small, soft seeds are also considered high quality traits, because seeds are swal-

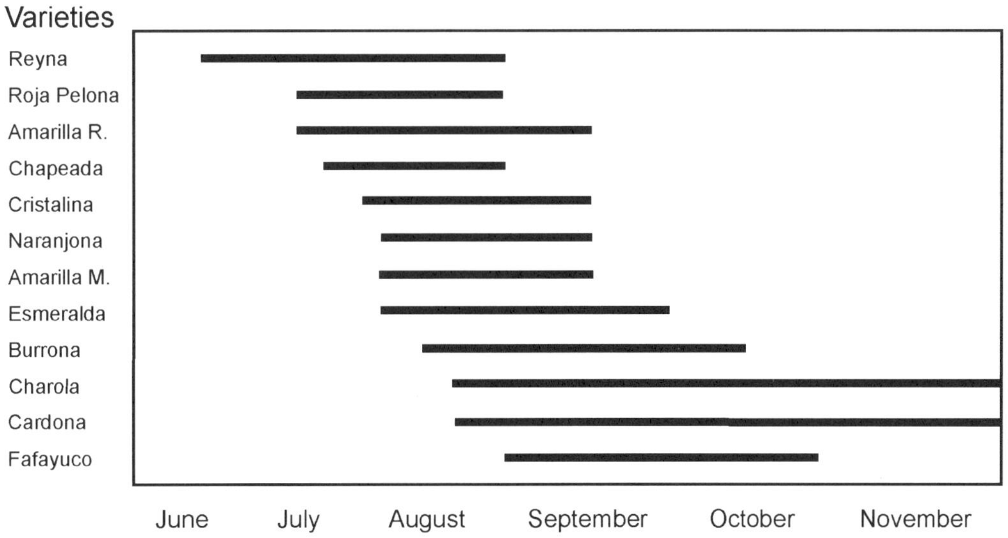

Fig. 2. Harvest season of cactus pears in central México.

lowed along with the pulp, total seed weight <6 g/fruit is acceptable. A high ratio of aborted seeds is desirable beacuse they are smaller and softer than the normal seeds. Parthenocarpy in cactus pear needs to be exploited.

The fruit size and productivity of 'Cristalina' is remarkable; fruits averaged 213 g, and yield up to 25 t in commercial groves. Another interesting trait is the presences of blush in 'Chapeada' which is atractive for some consumers. Even though it is not as juicy as 'Reyna' the 'Roja Pelona' and similar clones are gaining importance, due to its attractive flesh color, which ranges from light red to purple. There is some evidence that this cultivar could be marketed internationally.

Cultivars with yellow flesh are presently of secondary importance but this may be due to insufficient publicity. For quality traits such as seed and solid soluble content are well in the range of acceptability which is enhanced by striking peel and pulp color. 'Naranjona' is the most consumed yellow-fleshed variety but it is being replaced by 'Amarilla Montesa' because of its higher ratio of aborted to normal seeds.

'Charola' and 'Cardona' are still collected from wild stands and are highy appreciated in local markets. Both cultivars are able to remain in the plant for up to three months, and are also highly tolerant to handling and resistant to pests and diseases. Yellow and red fruits are under represented in commercial orchards but the trend is to diversify orchards in order to meet future demands of foreign markets.

REFERENCES

Benson, L. 1982. The cacti of the United States and Canada. Stanford Univ. Press, Stanford, CA.

Bravo-Hollis, H. 1978. Las cactaceas de Mexico. Univ. Nal. Autonoma de Mexico. Mexico.

Barbera, G., Carimi, and P. Inglese. 1992. Past and present role of the Indian-fig prickly pear [*Opuntia ficus indica* (L.) Miller, Cactaceae] in the agriculture of Scicily. Econ. Bot. 46:10–20.

Hegwood, A.D. 1990. Human health discoveries with *Opuntia* spp., Cactaceae (Prickly Pear) HortScience 25:1515–1516.

Mondragon, J.C. and G.S. Perez. 1994. Reyna (Syn. Alfajayucan) is the leading cactus pear cultivar in Central Mexico. Fruit. Var. J. 48:134–136.

Nobel, S.P. and A.A. Israel. 1994. Cladode development environmental responses of CO_2 uptake and productivity for *Opuntia ficus-indica* under elevated CO_2. J. Expt. Bot. 45:295–303.

Pimienta, B.E. 1990. El nopal tunero. Coleccion tiempos de ciencia. Univ. de Guadalajara. Guadalajara Mexico.

Pimienta, B.E. 1991. An overview of prickly pear cultivation in Central Mexico. In: Proc. 2nd. Annual Texas Prickly Pear Council. Aug. 16–17. McAllen TX.

Russell, Ch. and Felker, P. 1987. The prickly pears (*Opuntia* spp. *Cactaceae.*) A source of human and animal food in semiarid regions. Econ. Bot. 41:433–445.

Secretaria de Agricultura y Recursos Hidraulicos. 1994. Sistema producto nopal-tuna. Datos Basicos Mexico.

Guava Production in Georgia under Cold-protection Structure

Umedi L. Yadava

Guava the "poorman's fruit" or "apple of the tropics" is a popular tree fruit of the tropical and subtropical climates and is native to the tropical America stretching from Mexico to Peru. Guava is commercially cultivated in most of Latin America and also thrives well in the wild. In some countries and certain growing regions, guava is naturalized to the extent that this plant is increasingly being considered as a noxious weed. It is called: *amarood, sapari, jamphal, jamrukh* in India; *guyava* and *guayaba* in Spain; *guyave* (the plant *goyavier)* in France; *guyaaba* in Holland; *goiaba* in Portugal; *kuawa* in Hawaii; *abas* in Guam; *araca* in Brazil; *jambu batu* in Malaya; *bayabas* in the Philippines; whereas the Indians of Mexico and Central America designated it by such names as *pichi, posh, enandi,* etc. (Jaiswal and Amin 1992; Martin et al. 1987; Morton 1987). Despite its origin in tropical America, guava is presently cultivated in every tropical and subtropical country around the world (Samson 1986). Although quite inexpensive in countries of its production, guava is a delicious fruit which is very nutritious and exceptionally rich in ascorbic acid and several minerals useful for the human health (Wilson 1980). To those fruit lovers who become familiarized with its penetrating aroma, guava is considered as one of the most delectable and fascinating fruits (Menzel 1985). Besides its exceptionally high nutritive values, guava is also a prolific and regular bearer that could produce fruits year-round (Thonte and Chakrawar 1982). Guava plants exceed the majority of tropical and subtropical fruit trees in adaptability, productivity, and tolerance to mild cold and light frosts. The guava is successfully cultivated in a wide range of environmental and edaphic conditions owing to its tolerance to moisture stress and soil salinity as compared to most of the warm climate fruit plants (Samson 1986). It is cultivated in many countries including India, South Africa, Brazil, Cuba, Venezuela, New Zealand, the Phillippines, and Haiti. Only California, Florida, and Hawaii produce guava in the United States. For the past several years, guava has been successfully grown in Georgia at the Fort Valley State College (FVSC) Agricultural Research Station, under a protective structure in the field (Yadava 1992, 1996). Seedling trees of guava grown in the field, have been producing once a year a good crop of large fruits of pleasant eating quality. Harvesting of fully mature tree-ripe guavas in the Central Georgia area extends from late Aug. to middle of Feb. Studies on guava underway at FVSC emphasize production management under cold-protection structure.

Guava is important in international trade and domestic economy of several countries in warmer climates (Menzel 1985). Because of its easy cultivation under variable soils and climates, high nutritional value, and popular uses in such processed products like juice, preserves, and dairy or bakery items, guava is favorite of billions of people in the tropical and subtropical countries, but not so much in the temperate regions including the United States. Due to their astringent properties, mature guava fruits, leaves, roots, bark, and immature fruits, are used in local medicines to treat gastroenteritis, diarrhea, and dysentery (Morton 1987; Purseglove 1968). Guava consumption has been reported to significantly reduce serum total cholesterol, triglycerides, and blood pressure with the opposite effect (an explicit increase) in high-density lipoprotein (HDL) or good cholesterol (Singh et al. 1992). Furthermore, high concentrations of pectin in guava fruit may play a significant role in the reduction of cholesterol and thereby decrease the risk of cardiovascular diseases.

BOTANY

The guava (*Psidium guajava* L., Myrtaceae), is one of 150 species of *Psidium* most of which are fruit bearing trees native to tropical and subtropical America. The common guava is a diploid ($2n = 22$), but natural and artificial triploids ($2n = 33$) and aneuploids exist. Triploids generally produce seedless fruits (Jaiswal and Amin 1992), but most are shy bearers (Menzel 1985). Seedling trees of most guava cultivars vary in vigor and size, bearing habit, and fruit yield, shape, size, quality, maturity season, and storage ability. Natural cross pollination so common in guava cultivars, reaching up to 35% in some cases (Purseglove 1968; Menzel 1985), is responsible for the variability observed in seedling trees.

Morphology

The guava plant grows symmetrically dome-shaped with broad, spreading, low-branching canopy and a shallow-rooted small tree of 3 to 10 m in height, branching close to the ground and often heavily suckering from the base of the trunk. The green to reddish-brown and smooth bark on older branches and trunk peels off in thin flakes. The four-angled young twigs of guava are easily distinguished. The simple leaves of guava are opposite, 10 to 15 cm long, oval to oblong-elliptic, smooth, and light green in color. The perfect epigynous flowers 25 to 30 mm in diameter with four incurved white petals and a large tuft of white stamens with yellowish anthers, are borne solitary or in clusters of 2–3 in leaf axils on new growth from mature wood. Self-pollination is conspicuous (60% to 75%) since even isolated trees produce good crop; however, the distribution of cross-pollination by insects, is about 35% (Purseglove 1968; Menzel 1985). Based on the cultivar, guava fruit could be an ovoid, spherical or pyriform berry topped by calyx lobes. Generally, guava fruits measure 4 to 10 cm in diameter and weigh from 100 to 450 g; however, the largest guavas harvested from our plots at Fort Valley, Georgia, during 1992, 1993, and 1994 seasons, were 597, 756, and 862 g, respectively. Guava has a blend of sweet and acid flavors when fully ripe and characteristically a rather penetrating aroma (Rathore 1976; Wilson 1980; Thonte and Chakrawar 1982; Menzel 1985; Morton 1987; Yadava 1994, 1996). From fruit set to maturity guava takes 150 days. Most of the famous cultivars contain numerous small, hard, yellowish-cream colored seeds which are imbedded in the soft pulp. However, seedless cultivars available in many countries, have great potential to become very popular in the United States in the future.

Bearing Habit

Guava produces flowers on new vegetative shoots arising from mature wood. Thus, in equatorial climate, it can flower during every month of the year. In India, guava flowers two to three times a year in the north (rainy, winter, and spring seasons) and distinctly three times in the south (Rathore 1976; Singh and Joon 1984; Samson 1986). In mild tropical or subtropical climates, guava flowers and fruits continuously throughout the year if water and temperatures do not become limiting factors (Rathore 1976). The guava has double sigmoidal curve for fruit growth (Sastry 1965a; Rathore 1976), with three distinct periods: (a) *Rapid* which starts after anthesis and continues for 45 to 60 days depending on fruiting season of the year in areas with multiple cropping seasons, (b) *Relatively slow* which lasts only for 30 to 60 days (after rapid), when seeds attain full maturity and become hard; and (c) *Exponential increase* lasting for 30 to 60 days and ending at fruit maturity. Both fruit height and diameter markedly increase during this period leading the fruit to the 'eating ripeness'. Duration of these growth periods appears inversely proportional to the prevailing temperatures. Depending on guava cultivar and the growing conditions, it takes about 100 to 150 days from bloom to the fruit harvest (Rathore 1976; Samson 1986). During the fruit growth period, the major activities appear to be synthesis of cell wall material and sugars, while pectins rapidly increase during the ripening period (Sastry 1965b).

Adaptation

Guava plant tolerates a wide range of frost-free climatic conditions, although mature trees survive light frosts (Rathore 1976; Campbell 1984; Samson 1986; Martin et al. 1987; Jaiswal and Amin 1992). It thrives well in both humid and dry climates between the sea level and 2,100 m elevation. The optimum temperature required for guava cultivation and high yield of good quality fruit ranges from 20° to 30°C and well distributed rainfall varies from 1,000 to 2,000 mm; however, the fruit quality is poor in areas of high rainfall and high relative humidity (Samson 1986). Guavas develop the best eating quality when they mature during dry period like winter in India (Rathore 1976). Thus, it is beneficial to give guava trees rest (an off season) by withholding irrigation water periodically. It has also been reported that some guava cultivars produce more heavily in areas with distinct winter seasons than in the deep tropics (Morton 1987).

HORTICULTURE

Cultivars

Menzel (1985) and Jaiswal and Amin (1992) have classified guava cultivars into three distinct types: (a) *Dessert* which produce less acidic fruits with mostly white flesh and appealing surface colors and includes 'Allahabad Safeda', 'Lucknow 49 (Sardar)', and 'Tathem White', (b) *Processing* which produce strong acidic fruit usually with red or pink flesh and a high percentage of pulp recovery, for example 'Ka Hua Kula', and (c) *Dual purpose* which yield fruits with a compromise between the processing and the dessert types like 'Etheridge Selection', 'Oakey Pink', and 'Fanretief'. 'Allahabad Safeda', 'Sardar', 'Red Fleshed', 'Chittidar', and 'Nasik' are the best known dessert types of India which produce good yield of high eating quality guava. The best Hawaiian dessert types are 'Beaumont', 'Ruby Supreme', and 'Ka Hua Kula' while 'Supreme' and 'Elisabeth' are famous guavas from the Ivory Coast. 'Centeno Pacific', an eminent Trinidad guava produces more than 50 mt/ha, propagates easily by cuttings, and tolerates flooding. 'Etheridge Selection', 'Oakey Pink', 'Tathem White', and 'Indonesian Seedless', are also important. Nagar and Raja Rao (1983) reported that fruits of the seeded guava 'Allahabad Safeda' grow much larger than those of 'Allahabad Seedless'.

Propagation

Guava is generally propagated from seeds but seedlings are variable in both plant and fruit characteristics. Seedling trees are not as precocious as vegetatively propagated trees which start fruiting in just 2 to 3 years vs 4 to 5 years for seedlings. It is difficult to root guava from mature wood but Prasad et al. (1988) successfully rooted guava hardwood cuttings using bottom heat supplemented with plant growth regulators. Guava propagation by layering, inarching, and budding is achievable. A unique technique of patch budding of guava scions demonstrated at the Horticultural Experiment Center, Basti, India, appears commercially feasible (Morton 1987). Sharma et al. (1992) reported positive influence of an aneuploid dwarfing guava rootstock on tree growth and productivity. Several clonal methods including rooting of succulent green wood, forkert or patch budding, air-layering (marcottage or gooty), layering, and inarching for propagating guava germplasm have been discussed by Chandra (1965) and Samson (1986). Guava trees can be induced to set fruit and produce good yield of marketable fruit during the first season following clonal propagation (Campbell 1984). Micropropagation of guava has been achieved by shoot tip explants from mature trees (Jaiswal and Amin 1987; Amin and Jaiswal 1988; Loh and Rao 1989; Papadatou et al. 1990).

Cultivation

The guava is successfully cultivated in a wide range of growing conditions. It is fairly well adjusted to different rainfall while soil pH ranging from 4.5 to 8.2 is proper for desirable plant performance. While guava trees tolerate poor soils, fruit production is substantially enhanced when grown in rich soils under proper management. As a general rule, guava requires very little attention. Nevertheless, guava trees can be grown as cordons on wire fence. Trees can be planted from 2.5 to 8 m in any combination for rows and tree spacing. The most common tree spacings are 3 × 5 m and 5 × 6 m with trees established along the contours. In order to enhance fruit production and minimize requirements for labor, Mohammed et al. (1984) developed an intensive system for guava growing. They tested plant densities of 27,000 (60 × 60 cm), 37,000 (90 × 30 cm), and 73,000 (45 × 30 cm) plants/ha in combination with the use of growth regulators for plant size control. Highest yield of good quality fruits was obtained at density of 27,000 plants/ha.

Pruning and Fruit Thinning

Regular pruning of bearing tree is essential. Most guava trees, whether propagated from seed or grafts, produce an abundance of suckers which should be removed from trunk up to 50 cm above ground. A framework of four branches representing four quarters of the tree should be established. The crotch angles between the branches and the main stem should be wide enough to facilitate adequate light penetration and provide physical strength to support fruit load at maturity. Desirable tree shape should be maintained by regular but light pruning of unwanted wood. Since guava bears fruit on new growth from mature wood, it is important to keep a balance between the amount of vegetative growth and mature wood to ensure production the next season, and maintain fruiting regularity.

Since healthy guava trees grow and fruit abundantly, there is always a chance for breakage of branches supporting the heavy loads of fruits. Thinning in the early stages of fruit growth increases size of remaining fruits, reduces trunk breakage, and promotes regular bearing (Biswas et al. 1989). Hand thinning is the most common procedure for guava since chemical thinning agents have not yet been worked out.

Harvesting and Storage

Immature guavas do not ripen off the trees; fruits may soften, but never develop abundant color, and typical flavor associated with good eating enjoyment (Menzel 1985; Yadava 1994). Over-ripe fruits drop. There are no visible physical appearances or chemical indices of fruits that consistently reflect the appropriate stage of fruit maturity for harvest. Fruit harvesting should be carried out when the fruit is fully developed, matured, and began to show signs of color change from green to yellowish. Harvested fruits should be wrapped individually in paper towels and packed in padded layers before shipping or refrigeration. Wrapping of guava fruits suppresses weight loss and preserves glossiness (Morton 1987). Storage of wrapped guavas at cool temperatures extends post-harvest life up to 5 weeks with retention of healthy fruit quality and no significant reduction in nutrient contents; however, fruits stored unwrapped loose moisture and shininess (Yadava 1994, 1996). Desiccation, browning of fruit skin tissue with concomitant loss of flesh firmness, and the high rate of physiological loss of weight limit the storage life of guava fruit (Singh et al. 1990). Treatment of harvested guavas with 100 mg L^{-1} morphactin increases storage life of fruit by restricting fungal decay and curtailing loss of color, weight, sugars, ascorbic acid, and non-volatile organic acids (Prasad and Shukla 1979; Morton 1987). Furthermore, guava fruits should be packed in the natural posture (with the pedicel end of the fruit kept upward) in order to retain better quality for longer periods of time (Siddiqui et al. 1991).

Diseases and Pests

Scale insects, mites, thrips, and white flies cause occasional leaf damage on most guava cultivars (Campbell 1984). There are no control measures for sucking moth, birds or fruit bats (Menzel 1985). The guava is a prime host of many fruit flies like Caribbean (*Anastrepha suspensa*), Mediterranean (*Ceratitis capitata*), Mexican (*Anastrepha ludens*), Oriental (*Dacus dorsalis*), and melon fruit fly (*Dacus cucurbitae*). Root-knot nematodes (*Meloidogyne* sp.) damage tree root in lighter soils where some nematicides may provide protection. Fungal organisms causing guava diseases include: oak root-rot (*Clitocybe tabescens*) which causes crown and root rots often resulting in tree death, anthracnose (*Colletotrichum gloeosporioides)* is responsible for the rotting of ripe fruit mainly in the rainy season, and algal spotting (*Cephaleuros virescens*) produces superficial spotting on the fruit and leaves frequently resulting in defoliation. Fruit rots (*Phytophthora parasitica*, *Botryodiplodia* sp., and *Dothiorella* sp.) and fruit canker (*Pestalotia psidii*) are very serious on rainy season guava crops in the humid areas. Fortunately, thus far none of these diseases and pests have been observed at the Fort Valley, Georgia, location.

Utilization

The guava fruit is an excellent source of vitamin C (Rathore 1976; Yadava 1994), up to 2,000 mg/100 g fruit (Campbell 1984; Menzel 1985; Martin et al. 1987). Guava fruit is abundant in dietary fiber (from 5% to 7%), vitamin A, pectin, phosphorus, calcium, and potassium (Wilson 1980; Yadava 1994). Red-fleshed guava fruit contains 3 mg of carotene/100 g of fruit. The predominant nonvolatile organic compounds of guava fruit include citric, malic, lactic, ascorbic, and galacturonic acids (Chan et al. 1971). The strong aroma of guava fruit is attributed to carbonyl compounds (Morton 1987). Comparison of nutritional quality of guava fruits produced at FVSC Agricultural Research Station, Fort Valley, Georgia, indicated that our fruits had nutrient quantities as high as or better than those reported in the literature (Yadava 1996).

Analysis of Fort Valley grown fruits indicated that guava has high levels of dietary fiber, vitamin C, protein, and potassium, but low level of fat and sodium. Guava harvested at Fort Valley at maturity, kept satisfactorily for a period of five weeks under refrigeration at 50°C. Except for a modest loss of moisture from the fruit skin, the storage period showed no significant influence on nutrient contents in proximates, vitamins, or mineral analyses indicating the importance of cold storage for guava fruit to extend its post-harvest life.

Guava fruit has been a staple food of some countries (Menzel 1985) and is still an ideal dessert in Latin America and the Spanish-speaking West Indies. The fruit is also consumed as stewed guava shells (*cascos de guayaba*), that is, guava halves with central seed pulp removed (Rathore 1976). Thick bars of rich guava paste and guava cheese are staple sweets, while guava jelly is almost universally marketed in these countries. Recipes are abundant for using guava fruit in pies, cakes, puddings, sauces, butter, sherbet, marmalade, jam, some dairy products (cheese and ice cream), chutney, relish, catsup, bakery items, and other products (Rathore 1976; Menzel 1985). Guava pulp is eaten fresh or stewed, and used as juice or nectar, jelly, paste, preserves, and many beverages (Jain and Barker 1966; Rathore 1976). In South Africa, guava pulp mixed with cornmeal makes food flakes for breakfast. Guava roots, bark, leaves, and green fruits are medicinally used in tropics to halt gastroenteritis, diarrhea, and dysentery while grinded leaves are applied on wounds, ulcers, and rheumatic areas, whereas chewing whole leaf relieves toothache (Rathore 1976). The larger, improved dessert guava cultivars with thick flesh are utilized for fresh market or as dessert or salad fruits; thus, the fresh market fruit which is gaining importance (Rathore 1976; Campbell 1984; Menzel 1985) appears to hold considerate potential for future commercial production.

CULTIVATION IN PROTECTIVE STRUCTURE

Seedlings of a 'Red-fleshed' line of guava (*Psidium guajava* L.) of unknown origin from India, were raised in the greenhouse. Seeds received no exceptional treatment prior to or during germination. Preliminary investigations with guava seedlings planted in the field indicated its susceptibility to severe cold, however, the root system remained alive to sprout next spring. Thus, two-year-old seedlings of guava were established in protective structure at the Fort Valley State College Agricultural Research Station, Fort Valley, Georgia. No special plant management practices were exercised since all trees and fruits remained free of insect pets and diseases. Irrigation and fertilizers were applied as needed. Weed control was achieved by hand pulling of unwanted plants. During winter, guava trees were protected from cold temperatures using a 6-mil clear polyethylene cover and portable electric heaters. Field grown tree protected by polyethylene, flowered and fruited during Feb. through Aug., while fruit development to maturity and fruit ripening on trees took another 150 days after fruit set.

The four-year-old guava trees established in the field during 1993 and protected from winter temperatures by a 6-mil polyethylene-covered structure (Fig. 1) equipped with two electric heaters, bloomed from Feb. to Aug. 1994. Fruit maturity and tree-ripening followed a similar pattern, and thus, an average of 30 fruits per tree were harvested during Aug. 1994 through Feb. 1995. Fresh fruit weight ranged between 133 and 862 g with an average weight of 390 g/fruit. Monthly yield of guava fruit from 26 trees peaked in Nov. and Dec. (Table 1). Good quality guava fruit was produced during the winter. There have been no incidences of any disease or insect problems. Thus, at full bearing age of 5 to 7 years, 600 guava trees planted 5.2×3.2 m apart, have a yield potential of 25 to 40 t/ha with an annual gross return of $40,000 to $65,000/ha based on price of $1.60/kg. An economic analysis indicates a guava operation could recover costs of protective structure in less than five years.

There is an increasing demand for fresh guava in the United States particularly by the ethnic populations consisting of immigrants from Asia, Latin America, and other warm countries. Therefore, due to its exceptional appeal for the fascinated nutrition-conscious Americans, guava has good potential for commercial production for fresh market (Campbell 1984). We conclude that guava can be successfully produced in Georgia when trees are protected from cold. Unlike India, Georgia experiences only one fruiting season with fruit harvested from Aug. to Feb. Guava trees grown at Fort Valley, have been highly productive and precocious but cold protection in winter in Georgia conditions is required.

Fig. 1. Two-year-old greenhouse-raised guava trees established in a protective structure.

Table 1. Guava fruit harvested from 26 four-year-old trees at Fort Valley, Georgia, during the period from Aug. 1994 through Feb. 1995.

Month	Fruit wt (g)	Fruit yield (kg)
Aug.	359	8
Sept.	418	54
Oct.	462	41
Nov.	406	79
Dec.	357	77
Jan.	347	28
Feb.	323	8

REFERENCES

Amin, M.N. and V.S. Jaiswal. 1988. Micropropagation as an aid to rapid cloning of a guava cultivar. Scientia Hort. 36:89–95.

Biswas, M., A.K. Azad, A. Ahmed, and A.K.M.A. Hossain. 1989. Effect of fruit thinning on fruit size, yield and quality of Kazi Piara guava. Bangladesh Hort. 17:1–4.

Campbell, C.W. 1984. Guava: Tropical fruits and nuts. p. 254–256. In: F.W. Martin (ed.), CRC handbook of tropical food crops. CRC Press, Boca Raton, FL.

Chan, H.T., J.E. Brekke, and T. Chan. 1971. Nonvolatile organic acids in guava. J. Food Sci. 36:237–239.

Chandra, D. 1965. Recent advances in clonal propagation of guava. Allahabad Farmer 39:137–139.

Jain, N.L. and D.H. Barker. 1966. Preparing beverages from guava fruit. Indian Hort. 11:5–7.

Jaiswal, V.S. and M.N. Amin. 1987. In vitro propagation of guava from shoot cultures of mature tree. J. Plant Physiol. 130:7–12.

Jaiswal, V.S. and M.N. Amin. 1992. Guava and jackfruit. p. 421–431. In: F.A. Hammerschlag and R.E. Litz (eds.), Biotechnology of perennial fruit crops, Biotechnology in agriculture 8. C.A.B. Int., Wallingford, UK.

Loh, C.S. and A.N. Rao. 1989. Clonal propagation of guava (*Psidium guajava* L.) From seedling and grafted plants and adventitious shoot formation in vitro. Scientia Hort. 39:31–39.

Martin, F.W., C.W. Campbell, and R.M. Roberte. 1987. Perennial edible fruits of the tropics: An inventory, U.S. Dept. Agr., Agr. Res. Ser., Agr. Handb. 642.

Menzel, C.M. 1985. Guava: An exotic fruit with potential in Queensland. Queensland Agr. J. 111(2):93–98.

Mohammed, S., L. Wilson, and N. Prendergast. 1984. Trop. Agr. (Trinidad) 61:297–301.

Morton, J.F. 1987. Fruits of warm climates, Julia F. Morton Publ., Miami, FL.

Nagar, P.K. and T. Raja Rao. 1983. Endogenous auxins in seeded and seedless fruits of guava. Scientia Hort. 18:323–331.

Papadatou, P., C. Pontikis, E. Ephtimiadou, and M. Lydaki. 1990. Rapid multiplication of guava seedlings by in vitro shoot tip culture. Scientia Hort. 45:99–103.

Prasad, A. and J.P. Shukla. 1979. Studies on the ripening and storage behaviour of guava fruits (*Psidium guajava*). Indian J. Agr. Res. 13:39–42.

Prasad, J., A. Rabbani, and R.A. Ram. 1988. Rooting of hardwood cuttings of guava (*Psidium guajava*) through bottom heat. Prog. Hort. 20:20–23.

Purseglove, J.W. 1968. Tropical crops: Dicotyledonous. Longman, London, UK.
Rathore, D.S. 1976. Effect of season on the growth and chemical composition of guava (*Psidium guajava* L.) fruits. J. Hort. Sci. 51:41–47.
Samson, J.A. 1986. Tropical fruits, 2nd ed., Tropical Agriculture Series, Longman Scientific & Technical, Longman Inc., New York.
Sastry, M.V. 1965a. Biochemical studies in the physiology of guava. I. Physical changes. Indian Food Packer 19:1–4.
Sastry, M.V. 1965b. Biochemical studies in the physiology of guava. II. Major chemical changes. Indian Food Packer 19:5–10.
Sharma, Y.K., A.M. Goswami, and R.R. Sharma. 1992. Effect of dwarfing aneuploid guava rootstock in high density orcharding. Indian J. Hort. 49:31–36.
Siddiqui, S., R.K. Sharma, and O.P. Gupta. 1991. Physiological and quality response of guava fruits to posture during storage. HortScience 26:1295–1297.
Singh, B.P., S.K. Kalra, and D.K. Tandon. 1990. Behavior of guava cultivars during ripening and storage. Haryana J. Hort. Sci. 19:1–6.
Singh, R.B., S.S. Rastogi, R. Singh, S. Ghosh, and M.A. Niaz. 1992. Effects of guava intake on serum total and high-density lipoprotein cholesterol levels and on systemic blood pressure. Am. J. Cardiology 70:1287–1291.
Singh, R.R. and M.S. Joon. 1984. Seasonal influence on physico-chemical composition of guava fruit. South Indian Hort. 32:88–89.
Thonte, G.T. and V.R. Chakrawar. 1982. Physico-chemical characters of the certain types/strains of guava (*Psidium guajava* L.). Progr. Hort. 14:269–272.
Wilson, C.W. 1980. Guava. p. 279–299. In: S. Nagy and P.E. Shaw (eds.), Tropical and sub-tropical fruits: Composition, properties, and uses. AVI, Westport, CT.
Yadava, U.L. 1992. Feasibility of growing guava under the conditions of Middle Georgia. HortScience 27:604.
Yadava, U.L. 1994. Physicochemical properties of guava produced in Georgia. HortScience 29:536–537.
Yadava, U.L. 1996. Guava: An exotic tree fruit with potential in Southeastern USA. HortScience 30:(in press).

Macadamia Production in Southern California

Lawrence T. McHargue

The genus *Macadamia*, a member of the Proteaceae, is native to eastern Australia. Two *Macadamia* species produce edible seeds (*Macadamia integrifolia* Maiden & Betche and *M. tetraphylla* L.A.S. Johnson). *Macadamia integrifolia* is native to the rainforests of southeastern Queensland, and *M. tetraphylla* occurs farther south in the rainforests of New South Wales. *Macadamia integrifolia* is more tolerant of temperature extremes than *M. tetraphylla*. The flowers are borne in pendulous racemes. Flower color ranges from white to cream in *M. integrifolia* and cream to beige in *M. tetraphylla*. Hybrid flower colors range from cream to pink. Both species are insect pollinated. Bee pollination is normal in California.

William Purvis introduced the first macadamias into Hawaii in 1881, though the major development of the Hawaiian macadamia industry occurred after 1930 (Cooper 1988). At one time it produced 90% of the world's supply. Annual macadamia nut production ranged from 21,400 to 22,550 t during the five year period of 1989 to 1994. Approximately 7,100–7,500 ha of macadamia plantings were harvested during the same period (USDA 1993; Rowley et al. 1994). Most Hawaiian macadamia production is from *Macadamia integrifolia* derived cultivars. The University of Hawaii and Hawaiian macadamia growers have actively engaged in research in macadamia cultivation and in the selection of improved macadamia cultivars. The Hawaiian industry has been diligent and aggressive in marketing and sales.

Hawaiian dominance of world production is currently being challenged. Australian plantings and production have expanded dramatically during the past 30 years. There are currently 20,000 ha devoted to macadamia plantings. One third of the area planted to macadamias is mature, another one third is in the early bearing stage, and another third is not yet bearing. Approximately 30% of world production now comes from Australia, and the Australian Macadamia Society projects that Australia will double its percentage of world production within five years (Stapleton 1995). Australian production increased from 4,400 t in 1987 to 21,000 t in 1995. Commercial plantings in Australia were initially based on *Macadamia integrifolia* cultivars developed in Hawaii, but newer selections from Australia are now coming into production. Australia currently has the world's largest macadamia research and selection program, and their industry is actively seeking new and expanded markets.

Macadamia orchards have also been established in Central America, Brazil, Kenya, Indonesia, Taiwan, and southern Africa (Arpaia 1994). Processed nuts grown in Guatemala are being marketed in the United States by one of the larger Hawaiian macadamia producers and processors. However the total world production of macadamia nuts is quite small, and it constitutes less than 0.5% of the total world nut sales (Stapleton 1995).

Macadamia seeds were first brought to California from Australia in 1879 (Ferguson and Arpaia 1990). A number of seedlings were successfully grown in southern California subsequent to that introduction, though Wickson (1910) made no reference to macadamias. There was some interest in Pasadena and Whittier in the 1920s, and a few relic trees from that time remained for decades. I examined one of the early seedling trees and its nuts in the 1950s. However, little interest was shown in commercial macadamia cultivation until the 1950s when several nurserymen began to promote macadamia growing and production. One of these was Cliff Tanner who propagated a number of tropical and subtropical fruits and nuts in addition to the macadamia. Another was Colonel Wells Miller. Macadamia enthusiasts formed the California Macadamia Society in 1953. The Gold Crown Macadamia Association has marketed California grown macadamias for its member growers for over two decades. Macadamia production has not attained the status of an economically important crop in California, despite efforts made by individuals to develop production. Most of the population of the region appears to be unaware that there is any commercial macadamia production in Southern California. Hawaiian and Central American produced nuts dominate the southern California market despite the fact that substantial macadamia production is possible there.

Macadamia trees grow successfully at a number of points in California. Macadamia climatic requirements are similar to Mexican or Guatemalan races of the avocado. Macadamia trees are intolerant of heavy frost. Frost damage begins to occur at –2.8°C (27°F), especially if the period of low temperatures lasts for several hours. Macadamia trees are also intolerant of prolonged periods of high temperatures and low humidity, and this precludes commercial macadamia production in the thermal belts of the San Joaquin Valley or in relatively frost-areas of the Sonoran Desert. Commercial production of macadamias is therefore located in San Diego, Ventura, and Santa Barbara Counties. Climatic conditions are also suitable in Orange and Los Angeles Counties, but those counties are now so intensively urbanized that there is little or no potential land area left for macadamia cultivation. Neither production volumes nor the land area devoted to macadamia production in California are available. There were an estimated 600–800 ha of macadamia plantings in 1988 (Cooper 1988). Long (1988) estimated that there were about 400 ha in San Diego County. It appears that the area devoted to macadamia production has probably declined, but there are no accurate figures to indicate how much. Many of the producing trees have been planted to replace avocado trees that have been infected by avocado root rot (*Phytophthora cinnamomi*). Accurate estimates on the land area occupied by macadamia trees that are scattered through avocado groves are not available. Other producing trees have been planted on small plots of land around homes or on the edges of avocado or citrus groves. No one has determined the extent of area planted in macadamias in this manner, yet there is commercial production from these plantings.

ADVANTAGES

Most cultivars grown in California are derived from *Macadamia tetraphylla* or *M. tetraphylla* x *integrifolia* hybrids (Ferguson and Arpaia 1990). *Macadamia tetraphylla* is less tolerant of temperature extremes than *M. integrifolia*. Its requirement for very even temperatures (7° to 26°C) limits its productivity and geographic range. Hybrid cultivars have generally been more productive in California than those derived from *M. integrifolia*, though the 'James' is an exception. *Macadamia integrifolia* cultivars usually bloom in mid-winter in California. Flower production and pollination are low in cold years resulting in poor yields. Hybrid cultivars usually bloom during the spring when the weather is more moderate and pollination is more certain. They are more tolerant of temperature extremes than *Macadamia tetraphylla,* and are well suited to the coastal climate of southern California (Cooper 1988).

The flavor of cultivars derived from the two *Macadamia* species differs perceptibly. *Macadamia tetraphylla* cultivars produce nuts with a high carbohydrate content, sweeter flavor, and a lower oil content than those from *M. integrifolia*, though they are more difficult to process and stabilize for shipping and storage. The high oil content of *M. integrifolia* gives it an advantage over *M. tetraphylla* for roasting and salting (Stapleton 1995). Subtle but real differences in flavor between the two groups of cultivars offer a particular niche to California growers, for macadamia production in Hawaii and Australia is primarily from *M. integrifolia* cultivars. The sweetness of fresh macadamia nuts from the *M. tetraphylla* x *integrifolia* hybrids usually grown in California makes it unnecessary to salt them, though a few growers roast and salt nuts in the same manner that Hawaiian or Central American producers do. The flavor of California produced macadamias is excellent, and it is a major advantage for growers. The sweetness of California grown macadamias is well suited for various confections. The flavors of candies and other confections made from California nuts are somewhat different from similar products from Hawaii. Frieda's Inc. of Los Alamitos, California buys all of the macadamia nuts included in their products from California sources. Most of the sales volume of California macadamias is repeat business.

Present and potential macadamia production areas in California are both scenic and readily accessible to the large urban populations of Southern California. Direct marketing can be done by growers. People can easily visit macadamia orchards and purchase nuts directly from a grower as they do with a number of other fruits and nuts. Macadamias are popularly perceived as an exotic specialty crop. They therefore have a certain built-in appeal to some segments of the southern California population.

The proximity of production and markets obviate the need to package and preserve California grown nut meats in the same manner as Hawaiian and Central American producers do for their sales to distant markets.

High temperatures or prolonged roasting alters the flavor of the nuts. There is a market for both fresh and packaged nuts, and California growers have an advantage in marketing the former, especially in California and the Southwest. The proximity of production and marketing also permits California growers to market unshelled macadamia nuts to the public more economically than growers located in distant regions. The hardness of macadamia shells requires specially designed and built nutcrackers. There is a certain risk involved in unshelled nut sales, for some nuts are always defective. The purchase of shelled nuts permits the processor to eliminate defective nuts and the customer to observe the quality of the nuts being purchased. The unit price of the nut meats is thus similar in shelled and unshelled nuts.

Macadamia trees in California suffer from fewer diseases and pests than most other crops. They are not particularly susceptible to the oak wilt fungus that has devastated many avocado orchards in southern California.

PROBLEMS

Land is expensive and will become increasingly so as urbanization occurs in those areas best suited for macadamia cultivation. Agricultural land area is being lost due to rapid urbanization, and the average size of all land holdings, including agricultural acreage, is declining. The cost of the labor needed to produce macadamia nuts is higher in California than in other areas. Productivity in yields and in return must therefore be quite high to compete successfully with production in other areas of the world. The cost of irrigation water has risen steadily, and it can be expected to increase even more with increasing competition between urban and rural areas of California for finite supplies.

The high cost of macadamia production dictates high retail prices. Current retail macadamia prices are $6.60 to 8.80/kg ($3 to 4/lb) for nuts in the shell, and $30.80/kg ($14.00/lb) for shelled nuts.

INDUSTRY REQUIREMENTS

There are four principal areas of need if macadamia production is to expand and be profitable in California: improved cultivars, continued progress in cultivation practices, improvements in nut harvesting and handling, and major improvements in publicity and marketing. It is my opinion that there is a potential for increased output and profitability for macadamia production in California. However, that potential will not be realized without a major expenditure of time and effort on the part of those involved in macadamia production and marketing.

Cultivars

Growers and nurserymen must make efforts to improve the quality of the cultivars being grown in California (Cooper 1992). Efforts to select for better and higher yielding cultivars have not been adequate, and there has been no comprehensive program to develop new cultivars (Ferguson and Arpaia 1990). California growers have strong and divergent opinions regarding the desirability of different cultivars. Plant breeders and growers must explore as much genetic diversity as possible from which to select and recombine favorable traits through plant breeding. New and existing cultivars from other areas of the world should be tested in our climate. Macadamia growers and existing nurseries need to make efforts to improve macadamias and cooperate with a state or federal program. The experiment stations of the University of California should initiate a program of macadamia improvement because this is a long term program that requires public support. Louise Ferguson, of the University of California, Davis and the experiment station at Parlier, stated in a telephone conversation on 24 Jan. 1996, that financial support from the macadamia industry is essential for the establishment of such a program.

More information is needed regarding rootstocks because soils in California are highly variable. Clay soils slow the time needed to reach maturity, though the slower growth of the root system results in trees having firmer anchorage. Selections should also be made for cold hardiness. Recently 'Cynthia' has been used as a rootstock for cold resistance (Cooper 1992).

Cultivation Practices

Continued work needs to be done in cultivation practices and methods of propagation. The Mediterranean climate and soils of southern California differ significantly from those of other macadamia producing areas. These differences are reflected in the dominance of *Macadamia tetraphylla* x *integrifolia* hybrid cultivars in California. A research program is needed to ascertain optimum and the most economical cultivation practices.

Harvesting and Handling

Opinions differ regarding the best methods of harvesting. Some growers wait until the nuts drop from the tree while others harvest the nuts from the tree itself. Studies should be undertaken to ascertain differences between harvesting methods in nut quality and efficiency.

Technology used in other crops has been applied to macadamia handling and processing, and some of the machinery being used has been developed by individual growers. Mechanized processing equipment is available from firms, such as the R.P. Barton Co. of Escalon, California; Shaw Macadamia of Gooding, Idaho; or Keiser Manufacturing Co. of Sequin, Texas. However most their equipment is designed to handle large quantities of product, and the equipment is costly. There is a largely unfilled need for small scale harvesting and handling equipment. More and innovative thought is needed to increase efficiency in harvesting, processing, and handling the product.

Publicity and Marketing

There is an acute need for more effective marketing and publicity regarding macadamia production in California. California growers should not simply sell their macadamia crop to brokers who sell the nuts in a world market dominated by cheaper nuts from areas having lower production costs. California produced macadamia nuts will inevitably be unprofitable, even if nut wholesalers would be willing pay a somewhat higher price for them. Wholesaler brokers alone will not and cannot pay a sufficiently high price to bring prosperity to the macadamia industry. Teeter (1993) presented a gloomy analysis of the costs of macadamia production in southern California.

California macadamia nuts cannot compete against other, cheaper nuts if they are marketed and consumed in exactly the same manner as nuts from Central America, Australia, and Hawaii. California growers and marketers must emphasize the different flavors and texture of their product as a niche market if they are to build a successful industry. Growers should emphasize the different flavor of freshly roasted, unsalted nuts. Their product must be perceived as being equal to and somewhat different from the best macadamia nuts grown in other areas, and that perception must be based on reality. The economics of production demand that California grown nuts sell at higher prices than nuts grown in other areas.

Macadamia growers must play the dominant role in marketing and publicizing their crop. They produce a specialty crop that is rightly considered to be a gourmet food. Growers should sell as many nuts and macadamia products as possible themselves. More direct marketing and dealing with fewer handlers between grower and producer will bring higher prices to the grower and more affordable prices for the consumer. Several growers sell nuts and confections directly to the public. More direct marketing efforts are needed. Growers should cater to tourism as the date industry has done in the Coachella Valley or the apple growers have done in Oak Glen. More publicity is needed to acquaint the public with the existence of macadamia growing and macadamia products in California.

Individual outlets for nuts and nut products will probably not be sufficient to sell all the crop, especially if production increases substantially. Growers should deal individually or collectively with specialty food stores such as Farmers Market, Bristol Farms, Trader Joe's, and others who already sell processed nuts from other areas of the world to an affluent and specialty oriented clientele. These firms sell specialty candies and baked goods, and these would be possible outlets for nuts and nut parts. The Gold Crown cooperative currently sells to Frieda's Inc. of Los Alamitos California. Similar marketing arrangements should be sought out by growers. Those involved in macadamia production and marketing should also make efforts to have macadamia nuts or products incorporated into the style of cooking known as "California cuisine."

CONCLUSION

California macadamia growers have both major advantages and disadvantages that confront their industry. They must engage in a major, collective effort if the industry is to expand and be profitable. Decisive leadership and consensus regarding goals for the industry are needed. Those in the macadamia industry must expand their efforts in promotion and marketing. More studies and quantitative data regarding production and marketing are necessary. Thompson Cooper (Internet: 74561.261@CompuServe.com) is currently working to establish an on-line data base for the industry. A long term University of California research program should be undertaken. More knowledge of cultivation practices, better yielding cultivars, and increased marketing efforts are all prerequisites for increased macadamia production and profitability in California.

REFERENCES

Arpaia, M.L. 1994. Macadamia situation for selected countries. California Macadamia Soc. Yearb. 40:27–30.

Cooper, T.H. 1988. Macadamia: a basic understanding. California Macadamia Soc. Yearb. 34:100–104.

Cooper, T.H. 1992. California macadamia tree varieties report. The Macadamia Nut Grower Quart. 7(3):1–7.

Ferguson, L. and M.L. Arpaia. 1990. New subtropical tree crops in California. p. 331–337 In: J. Janick and J.E. Simon (eds.), Advances in new crops. Timber Press, Portland, OR.

Rowley, H.K., D.A. Martin, and R. Wong. 1994. Hawaii macadamia nuts. Final season estimates. California Macadamia Soc. Yearb. 40:114–121.

Stapleton, A. 1995. Australian macadamia industry. Home page of the Australian Macadamia Society. (andrew@nor.com.au) Lismore, New South Wales.

Teeter, J. 1993. Is it worth it? California Macadamia Soc. Yearb. 39:98–105.

U.S. Department of Agriculture. 1993. Agricultural statistics. Government Printing Office, Washington, DC.

Wickson, E.J. 1910. The California fruits and how to grow them. Pacific Rural Press, San Francisco.

VEGETABLES

Specialty Mushrooms*

Daniel J. Royse

Total mushroom production world-wide has increased more than 10-fold in the last 25 years from about 350,000 t in 1965 to about 4,300,000 t in 1991. The bulk of this increase has occurred during the last 10 years. A considerable shift has occurred in the composite of genera that constitute the mushroom supply. During the 1979 production year, the button mushroom, *Agaricus bisporus*, accounted for over 70% of the world's supply. By 1991, only 37% of world production was *A. bisporus*. Mainland China is the major producer (2,200,000 t—or about 50% of the total) of edible mushrooms. In 1993 to 94, the United States produced 346,188 t (or about 8% of the total world supply) of mushrooms. *Agaricus bisporus* accounted for over 90% of total mushroom production value while *Lentinula, Flammulina, Pleurotus, Hypsizygus, Hericium, Morchella,* and *Grifola* were the main specialty genera cultivated. The value of the 1993 to 94 specialty mushroom crop in the U.S. amounted to $28.7 million—a 75% increase over the previous season (USDA 1994). Based on recent and historical trends, it is expected that diversification of the mushroom industry will continue in the United States and many other western countries. The development of improved technology to cultivate each species more efficiently, will allow consumer prices to decline.

PRODUCTION TECHNOLOGY

A list of cultivated species and their common English and Japanese names is given in Table 1. Twelve genera comprise the bulk of cultivated mushrooms as outlined below.

Auricularia spp.

Commonly known as wood ear, *Auricularia auricula* is the first recorded cultivated mushroom (Chang 1993). Total production of *Auricularia* spp. in 1991 exceeded 465,000 t (fresh weight; Table 2). This value is an increase of 346,000 t or 290% over 1986 levels (Chang 1993). *Auricularia* spp. production now represents about 11% of the total cultivated mushroom supply world-wide.

Auricularia auricula and *A. polytricha* commonly are produced on a synthetic medium consisting of sawdust, cotton seed hulls, bran, and other cereal grains or on natural logs of broad-leaf trees (Quimio 1982; Chang and Quimio 1982; Oei 1991). For cultivation on natural logs, members of the oak family (Fagaceae) are preferred, but many other species of both hard and softwoods may be used.

For synthetic medium production of Auricularias, the substrate may be composted for up to 5 days or used directly after mixing. In either case, the mixed substrate (about 2.5 kg wet wt) is filled into heat resistant polypropylene bags and sterilized (substrate temperature 121°C) for 60 min. Composted substrate is prepared by mixing and watering ingredients [sawdust (78%) : bran (20%) : $CaCO_3$ (1%) : sucrose (1%)] in a large pile. The pile then is covered with plastic and turned (remixed) twice at two-day intervals. For direct use of substrate, a mixture of cotton seed hulls (93%), wheat bran (5%), sucrose (1%), and $CaCO_3$ (1%) is moistened to about 60% moisture and then filled into polypropylene bags.

After the substrate has cooled, it is inoculated with either grain or sawdust spawn. The spawn then is mixed into the substrate either mechanically or by hand, and the mycelium is allowed to colonize the substrate (spawn run). Temperatures for spawn run are maintained at about 25°C±2°C for about 28 to 30 days. Light intensity of more than 500 lux during the spawn run may result in premature formation of primordia. Temperature, light intensity and relative humidity all interact to influence the nature and quality of the basidocarps.

In Fujian, China a system has been worked out to intercrop *Auricularia* spp. with sugar cane. Bags containing colonized substrate are suspended in mid air on a rope stretched between rows of sugar cane. The bags then are covered with a thin layer of plastic to help regulate relative humidity. Carbon dioxide generated from the growing mycelium apparently stimulates the growth of the sugar cane (Oei 1991).

---

*Contribution from the Mushroom Research Center, Department of Plant Pathology, 316 Buckhout Laboratory, Pennsylvania State University, University Park, PA 16802.

Flammulina velutipes

Worldwide production of *F. velutipes* (enokitake) has increased from about 100,000 tonnes in 1986 to about 187,000 t in 1991 (an 87% increase). Japan is the main producer of enokitake (Furukawa 1987). In 1986, Japan produced 74,387 t; by 1991, production had risen to 95,123 t. In 1993, Japan produced 103,357 t—an increase of about 8%. From these data, it is evident that a faster growth rate, in terms of total production, is being enjoyed by other countries. In the United States, for example, enokitake production has increased at an estimated rate of 25% or more per year for the last four years.

Production of most enokitake in Japan is based on synthetic substrate contained in polypropylene bottles (Fig. 1). Substrates (primarily sawdust and rice bran; 4 : 1 ratio) are mechanically mixed and filled into heat resistant bottles with a capacity of 800 to 1,000 ml. Sawdust consisting primarily of *Cryptomeria japonica*,

Table 1. Scientific and English and Japanese names for some cultivated specialty mushrooms.

Scientific Name	English name	Japanese name
Agrocybe cylindracea (DC.: Fr.) Maire	South Popular	Yangimatusutake
Armillaria mellea (Vahl:Far.) Kummer	Chiodini, Honey	Naratake
Auricularia auricula (Hook) Underw.	Black ear, Wood ear	Kikurage
Auricularia polyticha (Mont.) Sacc.	Cloud ear, Tree ear, Wood ear	Angekikurage
Coprinus comatus (Mull.:Fr.) S.F.Gray	Shaggy ink cap, Lawyer's wig	
Cordyceps sinensis (Berk.) Sacc.	Chinese catepillar fungus	
Creolophus pergamenus Karsten	Bear's head	Banshariake
Dictyophora indusiata (Vent.:Pers.) Fisch.	Bamboo sprouts, Collared stinkhorn	
Flammulina velutipes (Curt.:Fr) Karst.	Winter, Velvet stem, Golden, snow puff	Enokitake
Ganoderma lucidum (Leyss.:Fr.) Karst.	Ling-Zhi, Reishi	Reishi
Grifola frondosa (Dicks.:Fr.) S.F.Gray	Hen of the woods	Maitake
Hericium erinaceus (Bull.:Fr.) Pers.	Monkeyhead, Bear's head	
Hypsizygus marmoreus (Peck) Bigelow	Shimeji	Bunashimeji
Lentinula edodes (Berk.) Pegler	Black forest, Black, Oak	Shiitake
Lepista nuda (Bull.:Fr.) Cook	Blewit	
Lyophyllum decastes (Fr.:Fr.) Sing	Fried chicken	Hatakeshimeji
Morchella esculenta Pers. ex St. Amans	Morel	
Naematoloma sublateritivum Karsten	Bricktop, Chestnut	Kuritake
Pleurotus abalonus Han *et al.*	Abalone	Kuroawabitake
Pleurotus cornucopiae (Paul.) Roll.	Golden oyster, Horn of plenty	Tamogitake
Pleurotus cystidiosis O.K. Miller	Ohritake	
Pleurotus djmour (Fr.) Boedijn	Rose, Pink oyster	
Pleurotus ostreatus (Jacq.:Fr.) Kumm	Oyster, White oyster, Gray oyster	Hiratake
Pleurotus pulmonarius (Fr.) Quel.	Phoenix-tail	
Panellus serotinus (Fr.) Kuh.	Green oyster, Late fall oyster	Mukitake
Pholiota nameko (T.Ito) S.Ito *et* Imai	Viscid, Nameko	Nameko
Pholiota adiposa (Fr.) Quel	Fat pholiota	Numerisugtake
Tremella fuciformis Berk.	Snow fungus, Silver ear, White jelly	Shirokikurage
Tricholoma matsutake (Ito *et* Iman) Sing.	Pine	Matsutake
Tuber aestivum Vitt.	Summer truffle	
Tuber magnatum Pico ex Vitt.	Piedmont white truffle	
Tuber melanosporum Vitt.	Perigord black truffle	
Volvariella diplasia (Berk & Br.) Sing.	Banana, Straw	
Volvariella volvacea (Bull.:Fr.) Sing.	Straw, Paddy straw	Fukurotake

Table 2. World production of cultivated edible mushrooms in 1986 and 1991 (Chang 1993).

Species	Fresh wt (× 1,000 t) 1986		1991		Increase (%)
Agaricus bisporus	1,215	(55.8%)	1,590	(37.2%)	30.9
Pleurotus spp.	169	(7.8%)	917	(21.5%)	442.6
Lentinula edodes	320	(14.7%)	526	(12.3%)	64.4
Auricularia spp.	119	(5.5%)	465	(10.9%)	290.8
Volvariella volvacea	178	(8.2%)	253	(5.9%)	42.1
Flammulina velutipes	100	(4.6%)	187	(4.4%)	87.0
Tremella fuciformis	40	(1.8%)	140	(3.3%)	250.0
Hericium erinaceus	--	--	66	(1.5%)	--
Pholiota nameko	25	(1.1%)	40	(0.9%)	60.0
Hypsizygus marmoreus	--	--	32	(0.7%)	--
Grifola frondosa	--	--	8	(0.2%)	--
Others	--	--	49	(1.2%)	--
Total	2,176	(100.0%)	2,176	(100.0%)	96.4

Chamaecyparis obtusa or aged (9 to 12 months) *Pinus* spp. appears to offer the best yields. In the United States, a bran-supplemented medium, consisting primarily of corn cobs, serves as the primary medium. After filling into bottles, the substrate is sterilized (4 h at 95°C and 1 h at 120°C), mechanically inoculated and incubated at 18° to 20°C for 20 to 25 days. When the substrate is fully colonized, the original inoculum is removed mechanically from the surface of the substrate and the bottles may be placed upside down for a few days. At the time of original inoculum removal, the air temperature is lowered to 10° to 12°C for 10 to 14 days.

To further improve quality during fruiting, temperatures are lowered to 3° to 8°C until harvest. As the mushrooms begin to elongate above the lip of the bottle, a plastic collar is placed around the neck and secured with a Velcro® strip. This collar serves to hold the mushrooms in place so that they are long and straight. When the mushrooms are 13 to 14 cm long, the collars are removed and the mushrooms are pulled as a bunch from the substrate. The mushrooms then are vacuum packed and placed into boxes for shipment to market.

Ganoderma lucidum

Known as reishi or mannentake to the Japanese and Ling Zhi to the Chinese, *G. lucidum* is renown for its medicinal properties (for an extensive review see Willard 1990). Reishi often is associated with health and recuperation, longevity, wisdom, and happiness (Stamets 1990, 1993). It is believed that certain triterpenes and polysaccharides may account for the multiple activities of Reshi. Thus, considerable time and effort has gone into the isolation and characterization of these compounds. About 100 triterpenes have been isolated from either the fruitbodies or mycelium but only a few have been tested for bio-activity (Mizuno et al. 1995).

Most cultivation of *G. lucidum* is on supplemented sawdust contained in heat-resistant polypropylene bottles (Fig. 2) or bags. Sawdust of hardwoods is supplemented with rice bran (10%) and $CaCO_3$ (3%), moistened with water and filled (700 g) into plastic bags. A plastic collar then is fitted onto each bag and stoppered with a cotton plug. After heat treatment (95° to 100°C for 5 h) the substrate is allowed to cool overnight and then inoculated with grain or sawdust spawn. The inoculated substrate is then incubated for 3 to 4 weeks or until the spawn has fully colonized the substrate.

Mushroom production is initiated by maintaining air temperature at about 28°C with relative humidity in the range of 85% to 90%. Basidiocarps begin to appear in about 1 to 2 weeks after initiation. Approximately

2 to 3 months after the appearance of primordia, mushrooms are ready to harvest. A mushroom is considered mature when the whitish margin around the edge of the basidiocarp has turned red. The substrate may yield another harvest of mushrooms after removal of the first flush.

Grifola frondosa

Japan is the major producer and consumer of *G. frondosa* (Maitake). Commercial production of maitake in Japan (325 t) began in 1981 (Takama et al. 1981). By 1986, production was 2,203 t and, by 1991, production reached 7,950 t (a 261% increase). Japanese production of maitake reached 9,617 t in 1993 (Table 3) and was produced primarily in the provinces of Niigata, Nagano, Gunnma, and Shizuoka.

Commercial production of most *G. frondosa* is on synthetic substrate contained in polypropylene bottles or bags (Fig. 3). A common substrate used for production is composed of sawdust supplemented with rice bran or wheat bran in a 5 : 1 ratio, respectively (Takama et al. 1981). For bottle production, the containers are filled with moistened substrate and sterilized or pasteurized prior to inoculation. Most growers use automated inoculation equipment thereby saving on labor costs. For production in bags, the moistened substrate (2.5 kg) is filled into microfiltered polypropylene bags and sterilized to kill unwanted competitive microorganisms. After cooling (16 to 20 h), the substrate is inoculated and the bags are heat sealed and shaken to uniformly distribute the spawn throughout the substrate. Spawn run lasts about 30 to 60 days depending on strain and substrate formulation. Temperatures then are lowered from about 22° to 14°C to induce fruiting and fruitbody maturation.

Most Maitake is marketed as food. However, Maitake has been shown to have both anti-tumor and anti-viral properties (Jong and Birmingham 1990; Jong et al. 1991; Mizuno and Zhuang 1995). Powdered fruitbodies are used in the production of many health foods such as Maitake tea, whole powder, granules, drinks, and tablets.

Hericium erinaceus

In the wild, *H. erinaceus* occurs on old logs and stumps and on wounds of living trees, especially, maple, beech, oak, and hickory. The fruiting body is formed as a large white mass (5 to 30 cm across) that is toothed in many small tufts. The Chinese were the first to domesticate the fungus and, in 1991, production reached 66,000 t (Table 1). In Japan, the mushroom is cultivated on synthetic substrate in bags and bottles and on logs. It is sold for 1,000 to 1,500 yen per kg ($5.25 to $7.87 per pound).

Polysaccharides in *Hericium* spp. are believed to inhibit a variety of cancers by enhancing the hosts' immune functions (Mizuno 1995b). It also has been suggested that the phenol-analogous compounds hericenone-C, -D, -E, and Y-A-8-c, which induce the synthesis of nerve growth factor, might be effective in treating patients suffering from Alzheimer's disease (Mizuno 1995b).

Fig. 1. Production of enokitake (*Flammulina velutipes*) on synthetic substrate contained in polypropylene bottles; collar removed to show maturing mushrooms.

Fig. 2. Production of reishi (*Ganoderma lucidum*) in bottles.

Hypsizygus marmoreus

Japanese are the main producers and consumers of *H. marmoreus*. Bunashimeji production has increased steadily over the last few years although not as fast as some other types of mushrooms (Royse 1995). In 1986, production of Bunashimeji was 11,439 t in Japan; by 1991 production reached 36,623 t—an increase of 220% (Table 3). Production of *H. marmoreus* increased 38% in the two year period 1991 to 1993.

Bunashimeji usually is produced in polypropylene bottles contained in plastic trays. After the completion of vegetative mycelial growth, bottle lids are removed and the colonized substrate subjected to environmental conditions known to stimulate fruiting. When the mushrooms are mature, the entire cluster of fruiting bodies are removed from the bottles. The mushrooms are packaged by placing an entire cluster (or multiple clusters) into each over-wrapped package. Only one flush of mushrooms is harvested prior to mechanical removal of the "spent" substrate from the bottles. The bottles then are refilled with fresh substrate and the process is repeated.

The antitumor polysaccharide, ß-(1-3)-D-glucan, isolated from *H. marmoreus* showed very high activity (Ikekawa 1995). The water solubility of the polysaccharide was much higher than the same polysaccharide isolated from other fungi. Dried mushroom powder from this mushroom is believed to stimulate the radical-trapping activity of blood (Ikekawa 1995). Excessive free radicals in the blood stream are believed to hasten the aging process.

Lentinula edodes

The cultivation of *L. edodes* (shiitake) first began in China about AD 1100 (Nakamura 1983; Royse et al. 1985; Chang and Miles 1987, 1989). It is believed that shiitake cultivation techniques developed in China were introduced to the Japanese by Chinese growers (Ito 1978).

Cultivation on natural logs. Various species of trees have been used for the cultivation of shiitake (San Antonio 1981). One of the primary species used in one area of Japan in past years was the shii tree—thus the derivation of the name shii-take (Singer 1961). Most production today, however, is on various species of oak (Harris 1986; Stamets and Chilton 1982).

Natural logs usually are cut in the fall (after leaf drop) and may be inoculated within 15 to 30 days of felling. Trees that are cut in the fall also may be left intact through winter and, just before inoculation, cut into lengths of about one meter. Trees that are cut in the summer tend to have bark that is more loosely bound and sugar contents usually are lowest during this time. If trees are cut during the summer, the bark may strip off more easily, increasing the chances of contamination of the wood by competitive organisms. The most effi-

Table 3. Japanese production of *Grifola frondosa* (Maitake), *Hypsizygus marmoreus* (Bunashimeji), *Flammulina velutipes* (Enokitake) and *Pholiota nameko* (Nameko) from 1981 through 1993 (Ohmasa 1994).

	Production (t fresh wt)			
Year	*H. marmoreus*	*G. frondosa*	*F. velutipes*	*P. nameko*
1981	1,885	325	53,282	16,348
1983	4,666	699	55,769	18,141
1985	9,157	1,501	69,530	19,793
1986	11,439	2,203	74,378	20,079
1987	13,688	3,015	78,129	21,054
1989	22,349	6,167	83,200	21,125
1991	36,623	7,950	95,123	21,738
1993	48,479	9,617	103,357	22,613

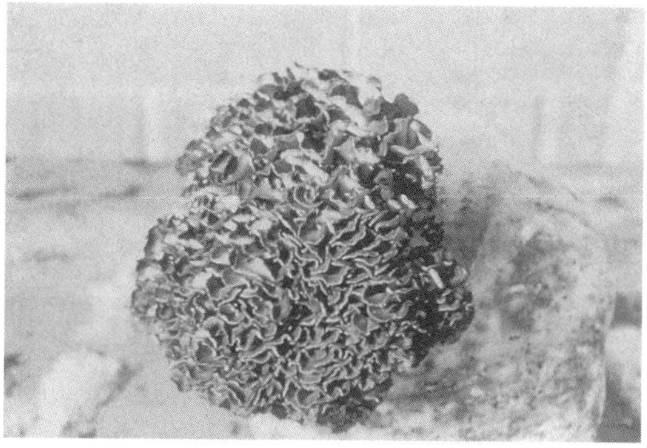

Fig. 3. Maitake (*Grifola frondosa*) fruiting on substrate contained in plastic bags.

cient log diameter appears to be in the 7- to 15-cm range (Ito 1978). Logs greater than 25 cm in diameter often are cut in half prior to inoculation (Royse et al. 1985).

Growers who inoculate the logs with wood-piece spawn drill holes in the logs with high speed drills to correspond to the diameter and length of the wood-piece spawn. Enough holes are drilled in the log to provide spacing of about one hole per 500 cm^2. The wood spawn then is driven into the holes with a hammer and then usually covered with hot wax to prevent excessive drying of the spawn. Sawdust spawn sometimes is used instead of wood-piece spawn.

Spawn run may last from 6 to 9 months, depending on the tree species, log size, spawn cultivar, moisture, temperature, and other variables (Leatham 1982). After the spawn run period the logs often are transferred to a "raising" yard. Raising yards usually are cooler and more moist than the spawn run area. The change in conditions provides an optimum environment for the growth and development of mushrooms. In the raising yard, the logs are arranged to provide for convenient harvesting of the mushrooms. Most production occurs in the spring and fall when conditions are most favorable. However, prices received by the growers usually are lowest during these periods.

Growers may use greenhouses for winter production of mushrooms (Przybylowicz and Donoghue 1988). More overall production is possible, and prices for fresh mushrooms are considerably higher, in winter than during the rest of the year. In the greenhouse method, logs usually are soaked in water (usually less than 48 h) and vibrated mechanically for various periods prior to placement in the greenhouse. After the mushrooms are harvested, the logs are incubated further (up to three months) and the process is repeated (up to five times).

Synthetic log production. Sawdust (Fig. 4a) is the most popular basal ingredient used in synthetic formulations of substrate used to produce shiitake (Miller and Jong 1987). Other basal ingredients that may be used include straw and corn cobs or mixtures thereof. Regardless of the main ingredient used, starch-based supplements such as wheat bran, rice bran, millet, rye, corn, etc. are added to the mix in a 10% to 40% ratio (dry wt) to the main ingredient. These supplements serve as nutrients to provide an optimum growing medium (Royse et al. 1990).

Once the proper ratio of ingredients are selected, they are combined in a mixer and water is added to raise the moisture content of the mix to around 60%. On large farms, the mix then is augured to a machine that fills and weighs the substrate so that a uniform amount is filled into each bag (Fig. 4b). The filled bags are stacked on racks, loaded into a industrial-sized autoclave, sterilized for 2 h at 121°C, cooled and inoculated with shiitake spawn.

After a 20- to 25-day spawn run (Fig. 4c) the bags are removed and the substrate blocks are exposed to an environment conducive for browning of the exterior log surfaces. As the browning process nears completion (4 weeks), primordia begin to form about 2 mm under the surface of the log indicating that the log is ready to produce mushrooms.

Primordium maturation is stimulated by soaking the substrate in water (12°C) for 3 to 4 h. Soaking allows water rapidly to displace carbon dioxide contained in air spaces, providing enough moisture for one flush of mushrooms. Approximately 9 to 11 days after soaking, mushrooms are ready to harvest (Fig. 4d).

The main advantages of using synthetic medium over natural logs is time and efficiency. The cycle for synthetic medium cultivation lasts approximately 4 months from time of inoculation to cleanout. Biological efficiencies for this method may average from 75% to 125%. In contrast, the natural log cultivation cycle usually lasts about 6 years with maximum efficiencies around 33%. The time required on synthetic substrate, therefore, only is about 1/15th that of the natural system with about 3 times the yield efficiency. As a result of these developments, shiitake production in the United States has increased dramatically in the last nine years (Fig. 5).

Shiitake is one of the best known and best characterized mushrooms used for medicinal purposes. Several medicinal properties have been attributed to shiitake in recent years. These properties include antitumor polysaccharides activity (Breene 1990; Mizuno 1995a) and glycoproteins, antiviral nucleic acids, platelet agglutination inhibitive substances, and anti-cholesterol active substances (Tokuda et al. 1974; Fujii et al. 1978; Suzuki et al. 1979; Tokuda and Kaneda 1978; Mizuno 1995a).

Morchella esculenta

Morels (Fig. 6) are some of the most highly prized mushrooms found in the wild. Researchers have long sought to consistently cultivate the morel; until recently this was not possible. In 1982, a report describing the successful production of ascocarps of *Morchella esculenta* under laboratory conditions appeared in the literature (Ower 1982). Since that first report, several patents (Ower et al. 1986, 1988) have issued describing a process for the commercial cultivation of these fungi. While patents have revealed some of the processes involved in predictable production of sporocarps, attempts to practice the invention have met only with limited success.

At present, one company in the United States is producing morels on a commercial scale. Commercial cultivation involves the production of sclerotia, an early overwintering stage of the mushroom. "Nutrient primed" sclerotia are produced in soil placed on a layer of sterilized wheat or rye grain. The production of nutrient primed sclerotia requires about 18 to 21 days under optimum conditions. The sclerotia are harvested, soaked in clean water for 24 h and distributed into a thin layer of pasteurized bark/soil mix. The sclerotia germinate via the production of mycelium. After the mycelium has spread throughout the soil mix, a continuous (12 to 36 h) fine mist of clean water is provided to stimulate the formation of ascocarps.

Several problems have yet to be solved in the commercial production of morels. Consistent fruiting, control of competitive weed molds, poor yields, and small mushroom size are just a few of the problems facing successful cultivation. A better understanding of the many factors contributing to increased yields and quality should lower the cost of commercially produced morels to consumers.

Pleurotus spp.

Oyster mushroom production has increased at rapid rate world-wide during the last few years (Table 2). From 1986 to 1991, oyster mushroom production increased from 169,000 t to 917,000 t (442% increase). China was responsible for most of the production increase. In the United States, production of oyster mushrooms was 882 t in 1994, up 94% from the previous year (USDA 1994). *Pleurotus* spp. (*P. ostreatus* and *P. cornucopiae*) production in Japan peaked in 1989 at about 36,000 t. Production was 24,000 t in 1993, a decrease of 33% in four years.

Fig. 4. Shiitake production on synthetic substrate: a) loading sawdust for use as an ingredient, b) filling polypropylene bags with nutrient supplemented sawdust, c) spawn run in plastic bags, and d) shiitake fruiting from synthetic logs.

In the United States, the primary ingredients used for *Pleurotus* spp. production is chopped wheat straw or cottonseed hulls or mixtures thereof. For production on wheat straw, the material is milled to a length of about 2- to 6-cm. The pH of the material is adjusted with limestone to about 7.5 or higher to provide selectivity against Trichoderma green mold (Stolzer and Grabbe 1991).

After completion of pasteurization (60°C for 1 to 2 h) the substrate is cooled and spawned with the desired strain. At time of spawning, a delayed release supplement (rates of 3% to 10% of dry substrate wt) may be added to increase yield and size of the mushroom (Royse and Schisler 1987; Royse et al. 1991; Royse and Zaki 1991). Use of supplements, however, may cause overheating of the substrate if growers are not able to anticipate and control air temperatures to maintain a steady substrate temperature.

Production of *Pleurotus* spp. on cotton seed hulls has some advantages over straw-based production systems in that chopping of the hulls is not required (Royse 1995). The pasteurized, supplemented hulls are spawned and filled (12 to 15 kg) into clear or black perforated polyethylene bags and then incubated at 23° to 25°C for 12 to 14 days.

In Japan, bottle production of oyster mushrooms is most common (Fig. 7). Substrate is filled into bottles, sterilized and inoculated with *Pleurotus* spawn. Upon completion of spawn run, bottle lids are removed and mushroom emerge from the surface of the substrate. After the mushrooms are harvested they are weighed and packaged for shipment to market.

Pholiota nameko

Japan produced 21,738 t of *P. nameko* in 1991—an increase of only 1,700 t (8% increase) from 1986 levels (Table 2). World-wide production increases averaged 60% over the same time period. In 1991, Japan produced about 54% of the total world production of nameko compared to 80% of total production in 1986. Thus, production of nameko rapidly is gaining popularity in other Asian countries.

Nameko (Fig. 8) means "viscid mushroom" in Japanese. This mushroom is prized for its gelatinous viscosity and for its flavor and is generally used in miso soup, cooked fresh with grated radish, and steamed in pipkin.

Preparation of the medium for nameko production is similar to that for enokitake except that a higher moisture content of the substrate is desirable. A substrate of broad leaf tree sawdust is preferred but research has shown that sawdusts from conifers such *Pinus* spp. and *Cryptomeria japonica* are suitable for growth.

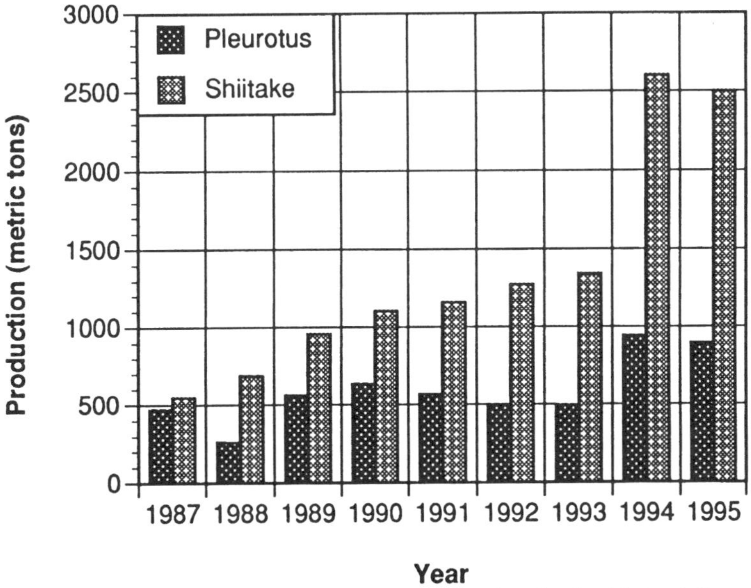

Fig. 5. Shiitake and *Pleurotus* spp. production in the United States from 1987–1994.

Fig. 6. *Morchella esculenta* (morel) fruiting in the wild.

Rice bran usually is added as a supplement in the ratio of 15% for conifer sawdust and 10% for broad-leaf sawdust.

Mushrooms are harvested from the substrate by cutting the stems near the base with scissors. The harvested mushrooms are washed and packed for shipment to market.

Tremella fuciformis

Known as the white jelly fungus or silver ear, *T. fuciformis* has been used as a delicacy food in China for many years. This mushroom can be cultivated on natural logs or on synthetic medium (Quimio et al. 1990). Cultivation techniques used to produce the mushroom on natural logs is similar to that used for shiitake production. In recent years, most production of *T. fuciformis* has been on synthetic substrate using a mixed culture inoculum technique first developed in Fujian, China (Huang 1982).

The mixed culture technique involves the use of "helper" mycelium of *Hypoxylon archeri*, an ascomycete commonly associated in nature with decaying wood. *Hypoxylon archeri* increases the ability of *T. fuciformis* to digest the substrate thereby increasing mushroom yields. Exploitation of this mycelial association is accomplished through use of dual cultures to make mother spawn (Quimio et al. 1990).

Substrate used for mushroom production is the same as that used for spawn production. The supplemented substrate is packed into plastic bags (50 cm long; 9 cm diameter) and ends of the bags are tied with cotton string. Six holes (1 cm diam) then are punched in the filled bags and covered with a breathable fabric. The substrate is sterilized for 6 to 8 h, cooled and inoculated with the mother culture.

After about 30 days of vegetative mycelial growth, the hole covers are removed and the exposed substrate is exposed to conditions favorable for primordia formation (Huang 1982). If optimum conditions are maintained in the growing houses, clusters of jelly fungus should be ready for harvest within 12 to 15 days. Yield for each bag of substrate is in the range of 350 to 500 g fresh weight (35 to 50 g dry weight).

***Volvariella* spp.**

The straw mushroom derives its name from the substrate on which it originally was grown (San Antonio and Fordyce 1972). Cultivation of *Volvariella* was believed to have begun in China as early as 1822 (Chang 1977). In the 1930s, straw mushroom cultivation began in the Philippines, Malaysia, and other Southeast Asian countries (Chang 1982). Production of the straw mushroom increased from 178,000 t in 1986 to about 253,000 t in 1991—a 42% increase. *Volvariella* accounts for approximately 6% of the total world-wide production of edible mushrooms (Table 2).

Many agricultural by-products and waste materials have been used to produce the straw mushroom. These include paddy straw, water hyacinth, oil palm bunch, oil palm pericarp waste, banana leaves and sawdust, cotton waste, and sugarcane waste (Chang 1982; Ho 1985). *Volvariella* is well suited for cultivation in the tropics because of its requirement for higher production temperatures. In addition, the mushroom can be grown on nonpasterized substrate—more desirable for low input agricultural practices.

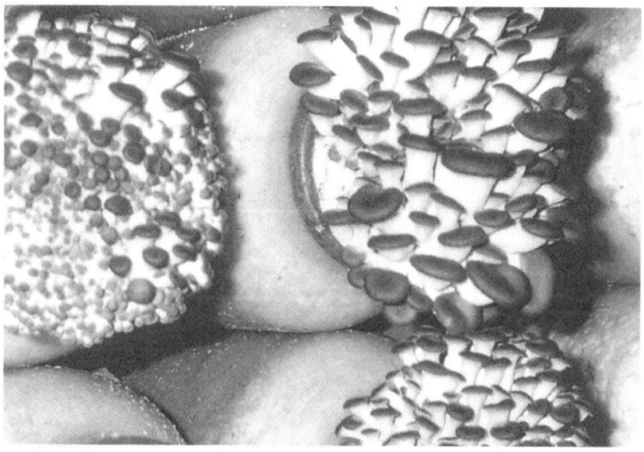

Fig. 7. *Pleurotus* spp. production from polypropylene bottles.

Fig. 8. Nameko (*Pholiota nameko*) production on substrate contained in polypropylene bottles.

In recent years, cotton wastes (discarded after sorting in textile mills) have become popular as substrates for straw mushroom production (Chang 1982). Cotton waste give higher and more stable biological efficiencies (30% to 45%), earlier fructification (four days after spawning) and harvesting (first nine days after spawning) than that obtained using straw as a substratum. Semi-industrialization of paddy straw cultivation on cotton-wastes has occurred in Hong Kong, Taiwan, and Indonesia as a result of the introduction of this method (Chang 1979).

MARKETING

Marketing of specialty mushrooms in the United States is a relatively new enterprise. Since 1984, some farms have seen their production rise as prices have fallen. For example, Donovan (1991) indicates that production of shiitake on their farm has increased from slightly less than 1 t per week in 1984 to over 7 t per week in 1990. At the same time, the price has decreased from US $12.50/kg ($5.50 per pound) to about US $ 8.80/kg ($4.00 per pound). In the 1993 to 94 growing season, the price growers received for shiitake was about $8.14/kg ($3.70 per pound; USDA 1994). Over the past seven years (1987 to 1994) the price of shiitake has declined an average of $0.19/kg ($0.09 per pound) per year (USDA 1994).

In recent years, the trend for specialty mushroom sales has been toward the retail market (Gunn 1992; Sorenson 1992). This trend is driven partly by an increased interest in specialty mushrooms and by the convenience packaged products offer to the consumer. In some retail markets, only 10% of the customers buy 90% of the specialty types (Gunn 1992).

Some merchandisers have projected a steady growth in consumption of specialty mushrooms. As consumers become more aware of specialty mushrooms, demand is expected to increase. Aggressive marketing will help to find new markets for these relatively new products. Therefore, specialty mushroom producers seeking new outlets for their mushrooms may want to check sources listing reputable produce industry firms (Anon. 1995a, b).

Specialty mushrooms are sold fresh, dried, or processed in Japan and China. Most fresh shiitake is collected and shipped to central wholesale markets where brokers and other participants buy the mushrooms through a bidding process in Japan (Hara 1988). Mushrooms then are distributed to retailers for consumer purchase. Other mushrooms, such as *Pleurotus*, may be packaged at the farm and shipped directly to brokers or to retailers.

Dried shiitake is distributed through traders specializing in this mushroom (Hara 1988). These traders (about 400 in Japan in 1988; data not available for China) buy shiitake at special bidding markets and then distribute the product to retailers for in country consumption or to trading firms for overseas export. In recent years, however, exports of shiitake from Japan have declined as the number of shiitake producers have declined and shiitake production has decreased (Anon. 1992; Royse 1995). On the other hand, Chinese production of shiitake and exportation of the product to Japan have increased dramatically in the last five years.

FUTURE OUTLOOK

Production and consumption of specialty mushrooms in the United States and other western countries is expected to increase at an accelerated rate in the years to come (Farr 1983; Royse 1995). As production technology is improved through interdisciplinary efforts, the retail price for specialty mushrooms should decrease. As economies improve in Latin America, production of specialty mushrooms could increase at an even faster rate than in the United States. The culinary advantages offered by specialty mushrooms bode well for the continued growth and development of the specialty mushroom industry worldwide.

REFERENCES

Anon. 1995a. The red book, Vol. 107. Vance Publishing Corp., Overland Park, KS.

Anon. 1995b. The red book produce industry handbook. Vance Publishing Corp., Overland Park, KS.

Anon. 1992. Discussion of fresh Shiitake imports (in Japanese), special produce information. Vol. 13(10):16–27.

Breene, W.M. 1990. Nutritional and medicinal value of specialty mushrooms. J. Food Protection 53:883–894.

Chang, S.T. 1977. The origin and early development of straw mushroom cultivation. Econ. Bot. 31:374–76.

Chang, S.T. 1979. Cultivation of *Volvariella volvacea* from cotton-waste composts. Mushroom Sci. 10(2):609–618.

Chang, S.T. 1982. Cultivation of *Volvariella* mushrooms in Southeast Asia. In: S.T. Chang and T.H. Quimio (eds.), Tropical mushrooms: Biological nature and cultivation methods. The Chinese Univ. Press, Hong Kong.

Chang, S.T. 1993. Mushroom biology: the impact on mushroom production and mushroom products. In: S.T. Chang et al. (eds.), Mushroom biology and mushroom products. The Chinese Univ. of Hong Kong.

Chang, S.T. and P.G. Miles. 1987. Historical record of the early cultivation of *Lentinus* in China. Mushroom J. Tropics 7:31–37.

Chang, S.T. and P.G. Miles. 1989. Edible mushrooms and their cultivation. CRC Press, Boca Raton, FL.

Chang, S.T. and T.H. Quimio (eds.). 1982. Tropical mushrooms: Biological nature and cultivation methods. The Chinese Univ. Press, Hong Kong.

Donovan, K. 1991. Marketing specialty mushrooms. Mushroom News 39(8):9–10.

Farr, D.F. 1983. Mushroom industry: Diversification with additional species in the United States. Mycologia 75:351–360.

Fujii, T., H. Maeda, F. Suzuki, and N. Ishida. 1978. Isolation and characterization of a new antitumor polysaccharide, KS-2, extracted from culture mycelia of *Lentinus edodes*. J. Antibiot. 31:1079–1090.

Furukawa, H. 1987. Mushroom production in Japan. Farming Japan 22(6):12–23.

Gunn, J. 1992. From specialty to mainstream. Mushroom News 40(5):18–22.

Hara, Y. 1988. Trends of the mushroom market. Farming Japan 22(6):28–37.

Harris, B. 1986 Growing shiitake commercially. Science Tech Publishers, Madison, WI.

Ho, K.Y. 1985. Indoor cultivation of straw mushroom in Hong Kong. Mushroom Newsl. Trop. 6(2):4–9.

Huang, N.L. 1982. Cultivation of *Tremella fuciformis* in Fujian, China. Mushroom Newsl. Trop. 2(3):2–5.

Ikekawa, T. 1995. Bunashimeji, *Hypsizigus marmoreus* antitumor activity of extracts and polysaccharides. Food Rev. Int. 11(1):207–209.

Ito, T. 1978. Cultivation of *Lentinus edodes*. In: S.T. Chang and W.A. Hayes (eds.), The biology and cultivation of edible mushrooms. Academic Press, NY.

Jong, S.C., J.M. Birmingham, and S.H. Pai. 1991. Immunomodulatory substances of fungal origin. EOS - J. Immunol. Immunopharmacol. 11:115–122.

Jong, S.C. and J.M. Birmingham. 1990. The medicinal value of the mushroom *Grifola*. World J. Microbiol. and Biotech. 6:227–235.

Leatham, G.F. 1982. Cultivation of shiitake, the Japanese forest mushroom, on logs: A potential industry for the United States. Forest Prod. J. 32:29–35.

Miller, M. and S.C. Jong. 1987. Commercial cultivation of shiitake in sawdust filled plastic bags. Developments in Crop Science, Cultivating Edible Fungi Vol. 10:421–426.

Mizuno, T. 1995a. Shiitake, *Lentinus edodes*: functional properties for medicinal and food purposes. Food Rev. Int. 11(1):111–128.

Mizuno, T. 1995b. Yamabushitake, *Hericium erinaceum*: bioactive substances and medicinal utilization. Food Rev. Int. 11(1):173–178.

Mizuno, T. and C. Zhuang. 1995. Maitake, *Grifola frondosa*: pharmacological effects. Food Rev. Int. 11(1):135–149.

Mizuno, T., G. Wang, J. Zhang, H. Kawagishi, T. Nishitoba, and J. Li. 1995. Reishi, *Ganoderma lucidum* and *Ganoderma tsugae*: bioactive substances and medicinal effects. Food Rev. Int. 11(1):151–166.

Nakamura, N. 1983. An historical study in shiitake (mushroom) culture. Tosen Shuppon, Tokyo.

Oei, P. 1991. Manual on mushroom cultivation: techniques, species and opportunities for commercial application in developing countries. Tool Publ., Amsterdam, The Netherlands.

Ower, R. 1982. Notes on the development of the morel ascocarp: *Morchella esculenta*. Mycologia 74:142–144.

Ower, R.D., G.L. Mills, and J.A. Malachowski. 1986. Cultivation of *Morchella*. US Patent No. 4,594,809.

Ower, R.D., G.L. Mills, and J.A. Malachowski. 1988. Cultivation of *Morchella*. US Patent No. 4,757,640.

Przybylowicz, P. and J. Donoghue. 1988. Shiitake growers handbook. Kendall/Hunt Publ. Co., Dubuque, IA.
Quimio, T.H. 1982. Physiological considerations of *Auricularia* spp. p. 397–408. In: S.T. Chang and T.H. Quimio (eds.), Tropical mushrooms: Biological nature and cultivation methods. Chinese Univ. Press, Hong Kong.
Quimio, T.H., S.T. Chang, and D.J. Royse. 1990. Technical guidelines for mushroom growing in the tropics. FAO Plant Production and Protection Paper 106. FAO, United Nations, Rome.
Royse, D.J. 1995. Specialty mushrooms: cultivation on synthetic substrate in the USA and Japan. Interdisciplin. Sci. Rev. 20:205–214.
Royse, D.J., L.C. Schisler, and D.A. Diehle. 1985. Shiitake mushrooms: consumption, production and cultivation. Interdisciplin. Sci. Rev. 10:329–335.
Royse, D.J. and L.C. Schisler. 1987. Yield and size of *Pleurotus ostreatus* and *Pleurotus sajor-caju* as effected by delayed-release nutrient. Appl. Microbiol. Biotechnol. 26:191–194.
Royse, D.J., S.L. Fales, and K. Karunanandaa. 1991. Influence of formaldehyde-treated soybean and commercial nutrient supplementation on mushroom (*Pleurotus sajor-caju*) yield and in-vitro dry matter digestibility of spent substrate. Appl. Microbiol. Biotechnol. 36:425–429.
Royse, D.J. and S.A. Zaki. 1991. Yield stimulation of *Pleurotus flabellatus* by dual nutrient supplementation of pasteurized wheat straw. Mushroom Sci. 13(2):545–547.
Royse, D.J., B.D. Bahler, and C.C. Bahler. 1990. Enhanced yield of shiitake by saccharide amendment of the synthetic substrate. Appl. Environ. Microbiol. 56:479–482.
San Antonio, J.P. 1981. Cultivation of the shiitake mushroom [*Lentinus edodes* (Berk.) Sing.]. HortScience 16:151–156.
San Antonio, J.P. and C. Fordyce. 1972. Cultivation of the paddy straw mushroom *Volvariella volvacea* (Bul. ex. Fr.) Sing. HortScience 7:461–464.
Singer, R. 1961. Mushrooms and truffles. Leonard Hill Ltd., London.
Sorenson, R. 1992. Marketing of specialty mushrooms in California. Mushroom News 40(10):14–15.
Stamets, P. 1990. A discussion on the cultivation of *Ganoderma lucidum* (Curtis:Fr.) Kar. The reishi of ling zhi mushroom of immortality. McIlvainea 9(2):40–50.
Stamets, P. 1993. Growing gourmet and medicinal mushrooms. Ten Speed Press, Berkeley, CA.
Stamets, P. and J.S. Chilton. 1982. The mushroom cultivator: A practical guide to growing mushrooms at home. Agarikon Press, Olympia, WA.
Stolzer, S. and K. Grabbe. 1991. Mechanisms of substrate selectivity in the cultivation of edible fungi. Mushroom Sci. 13(1):141–146.
Suzuki, F., C. Suzuki, E. Shimomura, H. Maeda, T. Fujii, and N. Ishida. 1979. Antiviral and interferon-inducing activities of a new peptidomannan, KS-2, extracted from culture mycelia of *Lentinus edodes*. J. Antibiot. 32:1336–1345.
Takama, F., S. Ninomiya, R. Yoda, H. Ishii, and S. Muraki. 1981. Parenchyma cells, chemical components of maitake mushroom (*Grifola frondosa* S.F. Gray) cultured artificially, and their changes by storage and boiling. Mushroom Sci. 11(2):767–779.
Tokuda, S. and T. Kaneda. 1978. Effect of shiitake mushroom on plasma cholesterol levels in rates. Mushroom Sci. 10(2):793–796.
Tokuda, S., A. Tagiri, E. Kano, Y. Sugawara, S. Suzuki, H. Sato, and T. Kaneda. 1974. Reducing mechanism of plasma cholesterol by shiitake. Mushroom Sci. 9(1):445–462.
United States Department of Agriculture. 1994. Mushrooms. Agricultural Statistics Board. Washington, DC.
Willard, T. 1990. Reishi mushroom: Herb of spiritual potency and medical wonder. Sylvan Press, Issaquah, WA.

Response of Australian Strains of the Mushroom *Lepista nuda* to Temperature and Substrate

Karen Stott* and Andrew Broderick

Taxonomic classification of macrofungi is constantly changing as new species are discovered in North America, South Africa, and Australia. In Australia there are probably 3000 to 5000 species, most of which have not been properly described (Wood 1992). Improved techniques have enabled microscopic details to be better researched and this has resulted in changes in taxonomic classification.

The recent increase in consumption of wild mushrooms in many countries has provided the impetus to explore new sources of edible mushrooms. In addition, the wild harvesting of edible mushrooms must be curbed to ensure that the environment and ecosystems where these mushrooms grow are not destroyed. Mycologists need to explore new mushroom sources (Purkayastha and Chandra 1985) and develop cultivation techniques.

Worldwide there are 200 edible fungi of which only 25 species are widely accepted as human food and are cultivated (Hashioka and Arita 1978; Chang 1981; Pathak 1986). Fungi other than *Agaricus bisporus* represent about 30% of world production (Olivier 1991). The influx of migrants with a European heritage into Australia has created a demand for edible mushrooms produced locally. Amongst the gilled fungi, there are a number of edible species in the Agaricaceae, Bolbitiaceae, Lepiotaceae, and Tricholomataceae.

The fungus *Lepista nuda*, Tricholomataceae (syn: *Tricholoma nudum, Rhodopaxillus nudus, Clitocybe nuda*) is found in Europe, the Americas, and Australia. It has an international reputation as an excellent edible species and the combination of its lilac color, solid fleshy structure, good shelf life, flavor, and aroma makes the development of commercial cultivation techniques highly desirable. Developing techniques for commercial cultivation requires a detailed understanding of environmental and nutritional parameters which optimise vegetative growth and induce fruiting bodies.

The development of an appropriate substrate requires the chemical, physical, and biological conditioning of composted matter. This creates an environment selective for a particular species that is critical for fruit body production. Factors to be considered in the substrate are microbial activity, physical characteristics, pH, chemical components, aeration, water content, substrate composition, and extent of composting undergone by the substrate. Some species require the application of a casing layer of peat or soil to enhance yield and quality. Casing assists the induction of fruit bodies, but how this occurs is largely unknown. It is thought that casing provides a high water holding layer for hyphae, entraps volatiles released by compost, has a different mycoflora to the substrate, and physically supports the growing fruit body. The physical and chemical characteristics, optimal depth and number of applications for a given casing material can be different for each species. Temperature, light, O_2, CO_2, watering and care must be managed to encourage fruit body formation. Previous studies on substrate requirements of European *Lepista* (Vaandrager and Visscher 1981) indicate that fruit body production is enhanced by the addition of 10% uncomposted straw to commercial *Agaricus* compost. Guinberteau et al. (1989) and Brian et al. (1979) reported that a cold shock is essential for the formation of fruit bodies of *L. nuda* and temperatures of 8° to 15°C have been found to be effective. However Australian isolates (Young 1994) have been found under more variable environmental and substrate conditions than European isolates (Moser 1978, Breitenbach and Kranzlin 1991).

The objective of the present study was to provide the Australian mushroom industry with viable commercial cultivation techniques for Australian species of *Lepista*. This paper describes Australian forms of *L. nuda* isolated from the wild and examines the effect of temperature and substrate on hyphal growth and fructification.

---

*Thanks to the Rural Industries Research and Development Corporation for scholarship funding; INRA-Bordeaux for cultures; N.G. Nair and R. Tolson for supervision.

METHODOLOGY

Isolates

Under Australian conditions *Lepista* appears in the wild from April to July. Isolates of *Lepista* were collected from various sites and voucher documentation, identification, and isolation of strains was carried out. Most isolates were confirmed as *L. nuda* and comparisons were undertaken with French isolates of *Lepista,* provided by Dr. Guinberteau of Station de Recherches sur les Champignons at Institut National de la Recherche Agronomique, France (Table 1).

Temperature

Optimum temperature for hyphal growth is a critical factor in obtaining rapid colonisation of substrate and casing. Isolates were selected from locations with different temperature and environmental characteristics to enable comparison with French isolates. Growth response of four Australian (A1, A2, A3, A4) and four French (F1, F2, F7, F8) isolates of *L. nuda* from warm (A1, A2, F1, F2) and cool (A2, A3, F7, F8) climates were compared at different temperatures to determine maximum and minimum temperatures for hyphal growth.

Isolates were grown on malt extract agar plus 2% yeast (MEAY) at 5°, 12°, 15°, 20°, 25° and 30°C. Radial growth of hyphae was measured on day 8 in two directions and the average taken. Data was analysed by analysis of variance.

Substrate and Cold Shock

Agaricus compost alone or suplemented with 10% uncomposted cereal straw (w/w) was inoculated with *Lepista* at 2%–3% w/w. Trays were sealed and placed at 25°C in a cabinet with temperature and light control. After 12 days substrate was cased with a 75% moisture content 50/50 mix of blond/dark peat. After hyphae had grown through casing, lids were loosened to allow air movement over substrate and cold shock of 12° or 15°C was applied to replicates of both substrates.

RESULTS

Isolates

The location and environment of selected isolates are shown in Table 1. Cool climate isolates (4° to 25°C during growing season) were found in leaf litter under *Rhododendron, Cedrus deodara, Quercus suber* or grass (*Poa pratensis*). Basidiocarp very robust and fleshy, lilac to lilaceous brown; pileus 75–132mm, lilaceous brown, shiny, convex to shallow convex with age; stipe 50–78 mm × 25–33 mm.

Table 1. Location and environment of *Lepista* isolates at time of fruiting.

Country	Strain	Location	Environment
Australia	A1, A4	New South Wales	Warm[z]
	A2, A3	New South Wales	Cool[y]
France	F1	Landes	Warm[x]
	F2	Gironde	Warm[x]
	F7	Pyrenees	Cool[w]
	F8	Haute Alps	Cool[w]

[z]Maximum temperature 24°C, minimum temperature 8°C
[y]Maximum temperature 20°C, minimum temperature 4°C
[x]Average day temperature 14°C
[w]Average day temperature 11°C

Table 2. Response of Australian and French *Lepista* isolates temperature.

Isolate	Maximum radial growth (mm^2)^z	Optimal temperature (°C)
A1	32.6	25–30
A2	25.7	25
A3	24.7	30
A4	24.7	25
F1	12.7	22–25
F2	14.0	24
F7	13.0	22–24
F8	12.0	24–25

^{z}Australian isolates almost double hyphal growth of French isolates.

Warm climate isolates (8° to 35°C during growing season) were found in Kikuyu (*Pennisetum*) and Couch (*Cynodon dactylon*) grass in groups or rings. Basidiocarp strong lilac to lilaceous brown; pileus 30–75 mm, strong lilac, shallow convex with umbo at all stages; stipe 65mm, bright lilac to deep lilac or lilaceous brown.

Temperature

The optimum temperature for Australian isolates was found to be higher than for French isolates as was the minimum growth temperature. The growth rate of Australian isolates is more than double the rate of French isolates at all temperatures except 5°C (Table 2).

Substrate and Cold Shock

The addition of 10% uncomposted cereal straw encouraged hyphal growth, but discouraged the initiation of fruiting bodies and was not beneficial to Australian isolates. A cold shock of 12°C encouraged more hyphal aggregations than 15°C. No fruit bodies were produced.

CONCLUSION

Australian isolates differ from overseas isolates morphologically in shape, form, and size. They are found at different temperatures and in different environments than European species. Whether they are new species or varieties resulting from the Australian environment and habitat has yet to be determined.

The response of Australian isolates to temperature indicates that these isolates have the capacity to colonise substrate more rapidly than French isolates. This would be advantageous to commercial growers as cropping time could be reduced.

The addition of 10% uncomposted straw to *Agaricus* compost had no beneficial effect on the initiation of fruit bodies. Whether it would effect quality or yield is unknown and further research into this area is recommended.

Information from this study can provide the basis for the development of viable commercial cultivation techniques for the exotic mushroom industry in Australia. The differences in cultivation requirements and growth rates of Australian and French strains indicates that a shorter production cycle than is currently achieved with French isolates will be possible.

REFERENCES

Brian, C., J. Guinberteau, and L. Priobe. 1979. Obtention de fructification du pied bleau, *Rhodopaxilus nudus* (Fr. ex Bul.) en conditions artificielles. Academie d'Agriculture de France—extrait du proces-verbal de la Seance du 14 Nov. 1979. 1365–1369.

Brietenbach, J. and F. Kranzlin (eds.). 1991. Fungi of Switzerland, Vol. 3. Boletes and Agarics 1st part. Edition Mykologia, Lucerne, Switzerland. p. 214–216.

Chang, S.T. 1981. Mushrooms and mushroom science. Mushroom Newsletter for the Tropics. 1:3.

Guinberteau, J., J.M. Olivier, and M.R. Bordaberry. 1989. Données récentes sur la culture des "pieds bleus" (*Lepista* sp.) PHM Revue Horticole. 298 (June-July) 17–22.

Hashioka, Y. and I. Arita. 1978. Mushroom Science X (part II). Naturalization of several saprophytic mushrooms under rice-straw-culture. p. 127–135. In: J. Delmas (ed.), Proc. Tenth Int. Congress on the Science and Cultivation of Edible Fungi, France 1978.

Moser, M. 1978. In: G. Kibby (ed.), Keys to Agarics and Boleti (Polyporales, Boletales, Agaricales, Russulales. Roger Phillips. London. p.118.

Olivier, J.M. 1991. Champignon comestibles et diversification. Mushroom Info. 4:20–21.

Pathak, N.C. 1986. Utilization of natural mushrooms flora. p. 43–57. In: M.C. Nair and S. Balakrishnan (eds.), Beneficial fungi and their utilisation. Symposium of the Mycological Society of India, held at the College of Agriculture, Vellayani, Trivandum. Feb. 1984. Scientific Publishers, Jodhpur, India.

Purkayastha, R.P. and A. Chandra. 1985. Manual of Indian edible mushrooms. Today and Tomorrow's Printers and Publishers, New Delhi.

Vaandrager, M. and H.R. Visscher. 1981. Experiments on the cultivation of *Lepista nuda*, the wood blewit. p. 749–760. In: N.G. Nair and A.D. Clift (eds.), Proc. Eleventh Int. Scientific Cong. on the Cultivation of Edible Fungi, Australia, 1981. Sydney. vol. 11.

Wood, A. 1992. Another look at Australian mushrooms. Alumni Papers, Univ. New South Wales. Mar./Apr./May. p. 14–16.

Young, T. 1994. Common Australian fungi—A naturalist's guide. Univ. New South Wales Press. Sydney.

Capsicums: Innovative Uses of an Ancient Crop

Paul W. Bosland

The *Capsicum* genus, chile, represents a diverse plant group, from the well known sweet green bell pepper to the fiery hot, recently exploited habanero (Fig. 1). When Columbus tasted the small red berries he found on his voyage, he believed he had reached India and called them red pepper. Red pepper, *Capsicum*, has been domesticated for 7,000 years. How then, can the most consumed spice in the world be considered a new crop? One reason is that *Capsicum*, chile, is an essential ingredient in the fastest growing food sector in the United States, "Mexican or Southwestern food." In addition, many of the new uses of chile are hidden within manufactured products. Chile is being used as a food flavoring, a coloring agent, a pharmaceutical ingredient, and in other innovative ways. The use, and uses, of the numerous cultivars within the five domesticated species has grown exponentially.

HISTORY

Capsicum has been known since the beginning of civilization in the Western Hemisphere. It has been a part of the human diet since about 7500 BC (MacNeish 1964). It was the ancient ancestors of the native peoples who took the wild chile piquin and selected for the many various types known today. Heiser (1976) states that apparently between 5200 and 3400 BC, the Native Americans were growing chile plants. This places chiles among the oldest cultivated crops of the Americas. As opposed to most domesticated crops, the wild ancestral chiles are not looked upon as worthless or inferior by farming people who cultivate their domestic decedents. The wild *Capsicum annuum* var. *aviculare* is harvested and sold in the marketplace along side the larger-fruited domesticated chiles. *Capsicum* was domesticated at least five times by prehistoric peoples in different parts of South and Middle America. The five domesticated species are *C. annuum* L., *C. baccatum* L., *C. chinense* Jacq., *C. frutescens* L., and *C. pubescens* R. & P. (IBPGR 1983).

Chile is historically associated with the voyage of Columbus (Heiser 1976). Columbus is given credit for introducing chile to Europe, and subsequently to Africa and to Asia. On his first voyage, he encountered a plant whose fruit mimicked the pungency of the black pepper, *Piper nigrum* L. Columbus called it red pepper because the pods were red. The plant was not the black pepper, but a heretofore unknown plant that was later classified as *Capsicum*. *Capsicum* is not related to the *Piper* genus. In 1493, Peter Martyr (Anghiera 1493) wrote that Columbus brought home "pepper more pungent than that from the Caucasus." Chile spread rapidly across Europe into India, China, and Japan. The new spice, unlike most of the solanums from the Western Hemisphere, was incorporated into the cuisines instantaneously. Probably for the first time, pepper was no longer a luxury spice only the rich could afford. Since its discovery by Columbus, chile has been incorporated into most of the world's cuisines. It has been commercially grown in the United States since at least 1600, when Spanish colonists planted seeds and grew chile using irrigation from the Rio Chama in northern New Mexico (DeWitt and Gerlach 1990).

Capsicum terminology is confusing. Pepper, chili, chile, chilli, aji, paprika, and *Capsicum* are used interchangeably for plants in the genus *Capsicum*. *Capsicum* investigators use chile, pepper, or aji, as vernacular terms. *Capsicum* is reserved for taxonomic discussion. The word "chile" is a variation of "chil" derived from the Nahuatl (Aztec) dialect which referred to plants now known as *Capsicum*, whereas "aji" is a variation of "axi" from the extinct Arawak dialect of the Caribbean. This brings up the point of the correct way to spell "chile" (Domenici 1983). The "*e*" ending in chile is the authentic Hispanic spelling of the word, whereas English linguists have changed the *e* to an *i*. Chile pepper has come to mean pungent chile cultivars.

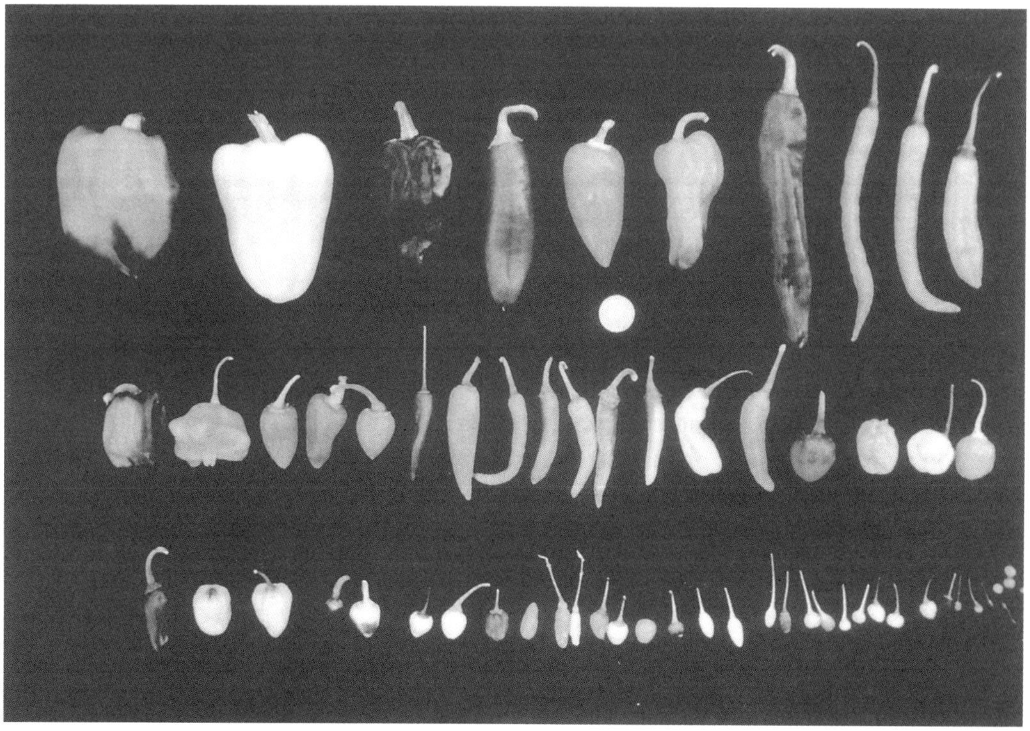

Fig. 1. Diversity in *Capsicum*.

However, chile means pepper (*Capsicum*) whether the fruits are pungent or not. Generally, chili is used to identify the state dish of Texas, which is a combination of pungent chile cultivars and meat (Domenici 1983). Bell pepper generally refers to non-pungent blocky chile types. Additional confusion is present within species designation, because *C. annuum* was sometimes called *C. frutescens* in the scientific literature (Heiser and Smith 1953).

BOTANY

The genus *Capsicum* is a member of the Solanaceae family that includes tomato, potato, tobacco, and petunia. The genus *Capsicum* consists of approximately 22 wild species and five domesticated species (Bosland 1994): *C. annuum*, *C. baccatum*, *C. chinense*, *C. frutescens*, and *C. pubescens*. *Capsicum* is endemic to the western hemisphere and the pre-Columbian distribution extended from the southernmost border of the United States to the temperate zone of southern South America (Heiser 1976). It is a perennial small shrub in suitable climatic conditions, living for a decade or more in tropical South and Central America. *Capsicum* probably evolved from an ancestral form in the Bolivia/Peru area (Heiser 1976). Chile fruits are considered vegetables, but are berries botanically. Chile types usually are classified by fruit characteristics, i.e. pungency, color, shape, flavor, size, and their use (Smith et al. 1987; Bosland 1992). Despite their vast trait differences most chile cultivars commercially cultivated in the world belong to the species, *C. annuum*. The tabasco (*C. frutescens*) and habanero (*C. chinense*), are the best-known exceptions.

Pod-types and Cultivars

Several hundred chile pod-types are grown worldwide. Chile pod-types are a subspecific category that allows for distinguishing among the specific horticultural varieties. Types such as ancho, bell, jalapeño, pasilla, New Mexican, yellow wax, are distinct pod-types that have specific traits for processing and fresh use, flavor and pungency. The most economically important species in the world is *C. annuum* (Greenleaf 1986; Bosland et al. 1988). Many pod-types are central ingredients in ethnic dishes such as ancho and New Mexican for chile rellenos, pasilla for mole sauce, and serrano for pico de gallo. Like wine-tasting, one can discern between the subtle flavors of chiles after a few years of experience; ancho is sweetish, mulatto is chocolaty, mirasol is fruity, and chilpotle is smokey. Grinding the pods produces one flavor, toasting before grinding produces another, and soaking the chiles in water produces yet another. For a more detailed description and use of the various pod-types the reader is referred to Bosland et al. (1988); Bosland (1992); and DeWitt and Bosland (1993).

By the time the Spanish arrived in Mexico, Aztec plant breeders had already developed dozens of pod-types. According to historian Bernardino de Sahagún, who lived in Mexico in 1529, "hot green chiles, smoked chiles, water chiles, tree chiles, beetle chiles, and sharp-pointed red chiles" existed (DeWitt and Gerlach 1990). Undoubtedly, these chiles were the precursors to the large number of pod-types found in Mexico today. Even though many pod-types exist, improvement continues today because of specific needs in the industry. Improvement in quality and yields are occurring, along with refinement of the pod-types to fit special niches of the industry.

Fresh market types such as bell peppers, cuban, and squash, are usually non-pungent, and eaten raw, stir-fried, or prepared in some fashion to season a culinary dish. Bell apparently refers to the fruit's blocky shape, with four lobes preferred in the United States. Bell types were first mentioned in 1699 by the English pirate, Wafer, who found them growing in Panama (Wafer 1699). The conventional green bell pepper has given way to innovative colored pods that can ripen to shades of red, orange, yellow, and brown. A survey in the 1992 issue of *Fresh Trends* stated that 95% of the people who tried a colored bell liked them and 5 out of 6 preferred them over the green bell. More people are trying colored bell peppers than any other produce category. There has been an increase of 75% in the consumption of colored bells in the United States. In 1992, colored bells were tried by as many consumers as leaf lettuce, fresh asparagus, peas, squash, and red potatoes. Consumers bought more colored bells than yellow beans, artichokes, or romaine lettuce. The innovative fruit colors have brought increased consumption and sales.

Chile processing pod-types can be grouped into two main categories. The first consists of a fresh product that is frozen, canned, or pickled Some of the pod-types included in this group are the pimento, jalapeño, serrano, pepperoncini, yellow wax, and the New Mexican. The second category consists of those used as a dehydrated product. Dehydrated pod-types are New Mexican, cayenne, ancho, pasilla, mirasol, piquin, and de arbol (Bosland 1992).

A relatively recent pod-type is the New Mexican, also called long green or 'Anaheim' type. Actually, the pod-type is New Mexican, and 'NuMex Big Jim' and 'Anaheim' are cultivars within this type. The New Mexican pod-type was developed in 1894 when Fabian Garcia at New Mexico State University began improving the local chiles grown by the Hispanic gardeners around Las Cruces, New Mexico. This type, characterized by long green pods that turn red, is the chile of choice for Mexican-style cooked sauces in the United States. Green and red chile represent two developmental states of the same fruit. 'Anaheim' seed originated in New Mexico and was brought to Anaheim, California, where it was widely cultivated. Thus, until Dr. Garcia released the New Mexican pod-type, it did not exist. If pods are left on the plant to be harvested at the red stage, they are usually dried and ground into chile powder (paprika if non-pungent). A green, New Mexican chile pod contains three times the vitamin C of a 'Valencia' orange and provides the minimum daily requirement. As green pods turn red, pro-vitamin A content increases until they contain twice the pro-vitamin A of a carrot (Lantz 1943). A one half tablespoon of red chile powder furnishes the minimum daily requirement of vitamin A.

A peculiar chile category is paprika. It is not a pod-type in the United States, but it is a product. In Europe, there are chile pod-types that are paprikas. This is because in the Hungarian language "paprika" means *Capsicum* (Somos 1984). Paprika is defined in the United States as a sweet, dried, red powder. This mild powder can be made from any type of *C. annuum* that is non-pungent and has brilliant red color. Paprika may be pungent in Hungary, but paprika is always non-pungent in international trade.

A small group of chiles can be classified as ornamental. Although edible, ornamentals are grown primarily for their unusual pod shapes or for their dense foliage and colorful fruits. Ornamental chiles can have all the colors of the rainbow, often displaying pods in four or five colors on the same plant at the same time (Bosland et al. 1994). In the past, they have been called Christmas peppers because of the bright red fruits during the holiday season. Decorations, such as wreaths, made with chile that can be dehydrated are popular in the southwestern United States, and are a major tourist product. A tradition in New Mexico is to harvest mature red chiles and string them into colorful strings (ristras). The ristra is hung near the entrance of the house as a symbol of hospitality. Ornamental chiles have become an innovative way for small farmers to produce a high-value alternative crop.

Exotic Germplasm

Exotic germplasm provides a bountiful source of extraordinary genetic diversity with which to improve the commercial chiles. For the past decade, it has played a significant role in the New Mexico State University chile breeding and genetics program. Because exotic germplasm inhabits a vast array of ecological zones, it can serve as a reservoir of useful and needed genes. However, the potential value of exotic germplasm for improvement of commercial chile is under exploited. Such genetic resources clearly deserve more intensive investigation. The five domesticated species have close wild relatives with which they cross readily, producing viable and fertile hybrids. These relatives have not been evaluated extensively, and may contain useful sources of resistance to viruses, bacteria, and fungi.

The *C. chinense* species, like all *Capsicum* species, originated in the New World. However, the French taxonomist who named this species in 1776 got his seed from China (Smith and Heiser 1957). Habanero, or Scotch Bonnet, is the best known chile of this species. One accession of Habanero has the distinction of being the world's hottest chile (in excess of 200,000 scoville heat units). Fruit shape can vary from long and slender to short and obtuse. Fruit can be extremely pungent and aromatic, with persistent pungency when eaten. *C. chinense* is popular in all tropical regions.

The *C. frutescens* species is represented by two cultivars, tabasco and malagueta. Tabasco is the most common cultivar of *C. frutescens*. The red fruit is the ingredient in Tabasco sauce. The malagueta is a popular cultivar in Brazil. It is not related to *Aframomum melegueta*, the melegueta or Guinea pepper, from Africa which is related to ginger.

In South America, *C. baccatum* is the most commonly grown species, where it is called aji, not chile. Three botanical varieties of *C. baccatum* are recognized: *C. baccatum* var. *baccatum*, *C. baccatum* var. *pendulum*, and *C. baccatum* var. *microcarpum*. *C. baccatum* flowers have yellow, brown, or dark green spots on the corolla. As many different pod-types of chiles (in relation to shape, color, and size) exist within *C. baccatum* as in *C. annuum*. Fruits vary in pungency from non-pungent to very hot. They embody unique aromatics and flavors that can be overpowering to some people. *C. baccatum* is the chile of choice when making ceviche (marinated fish).

Another of the five species, *C. pubescens*, is a relatively unknown chile. It is found from Mexico to Peru, growing in the Andean South America and the Central American highlands. Common names include *rocoto* or *locoto* in South America. Other common names are *manzano* and *peron* because the fruits can be apple- or pear-shaped. Instead of white flowers, it has purple flowers with large nectaries. The presence of conspicuous leaf pubescence and black seeds readily distinguish this chile from any of the other species. This chile is adapted to cooler temperatures, 4.4° to 21.1°C (40° to 70°F), but does not tolerate frost.

The other approximately 20 *Capsicum* species lack extensive study on their biology. It is interesting to note that all wild chiles have small fruits which are eaten with ease by birds, the natural dispersal agent for *Capsicum*. Many of the known wild species have restricted distribution. These species may contain genes for adaptation to unusual environmental conditions as well as disease resistance.

Exotic germplasm will continue to be an important asset for breeding improved commercial chiles. Not only will it be useful in breeding for disease resistance, but it will be used to increase the nutritional quality, yield, and efficiency for machine harvesting of the crop. The enhancement of commercial cultivars by exotic germplasm is dependent on the availability of living material. This may be impossible because the natural habitat for these species is in dire straits. Collection may never be accomplished before their extinction. Tropical deforestation is among the most massive and urgent environmental problems facing *Capsicum* germplasm resources.

The United States National Plant Germplasm System houses an extensive *Capsicum* germplasm collection at the Southern Plant Introduction Station located in Experiment, Georgia. This collection contains approximately 3000 *Capsicum* accessions that includes lines from all over the world. Passport data are recorded upon arrival of the seed at the facility and a USDA Plant Introduction (PI#) Number is assigned. Evaluation data is subsequently entered in the Germplasm Resources Information Network (GRIN) after it has been received. The material held in the USDA collection is the source of germplasm for breeding and research programs throughout the world. Many have been evaluated for descriptors. Preserving genetic diversity in germplasm is of paramount importance.

The genetic diversity of *Capsicum* can be saved only through the use of several strategies. The genetic resources of *Capsicum* need to be enlarged and conserved in base and active genebanks. The improvement of appropriate storage facilities for germplasm and the financial support of those operations is very urgent. It is especially imperative to aid the active collections of Latin America, because *Capsicum* is native to this region. When possible it is desirable to set up *Capsicum* genetic resources reserves in conjunction with relevant biosphere resources and other protected areas.

BREEDING

In breeding chiles, the choice of breeding method depends on the breeding objective and the plant material being used as parents (Greenleaf 1986). The strategy of the chile breeder is to assemble into a cultivar the superior genetic potential for yield, protection against production hazards, and improved quality. Chile cultivars have been developed using selection within introductions and hybridization followed by selection. Hybridization is usually always within a species, but interspecific crosses, especially *C. annuum* by *C. chinense* have been accomplished successfully. Selection methods have included mass, single plant, backcross, and pedigree. Single seed descent and haploid breeding have also been applied to chiles. The only breeding technique so far not applicable to chiles is genetic transformation. This is because the technique to regenerate whole chile plants from single cells has not been ascertained. With the success of other solanaceous crops, e.g. tomato and petunia, the parameters necessary for successful regeneration of chile will be forth coming.

Chile plants are considered a self-pollinating crop (Allard 1960). However, the rates of out-crossing (7% to 91%) recorded by several investigators argue that *Capsicum* should be considered to be facultative cross-pollinating species in field research (Odland and Porter 1941; Franceschetti 1971; Tanksley 1984). The out-crossing is associated with natural insect pollinators. The amount of cross-pollination has an effect not only on the precautions needed for seed production, but also on the breeding methodologies used by the plant breeder (Bosland 1993). Natural pollinators such as insects must be excluded to insure self-pollination.

The commercial production of hybrid chiles has been successful accomplished using hand-emasculation, genic male-sterility, and cytoplasmic male-sterility. Male-sterile chile plants are found as spontaneous mutants in commercial fields. Cytoplasmic male sterility (CMS) in chiles was first discovered by Peterson (1958) in *C. annuum* PI 164835 from India. Unfortunately, Peterson's CMS line is unstable in fluctuating environments, producing pollen at cool temperatures. Some commercial seed companies use the CMS system to produce F_1 hybrids.

Other genetic systems that may assist the chile breeder are the use of trisomic and chromosome mapping. Most *Capsicum* species have $2n = 24$ chromosomes. Meiosis is surprisingly regular within interspecific crosses. The similarity in karyotype of the species was illustrated by Ohta (1962). However, cryptic structural hybridity as defined by Stebbins (1971) appears to be functioning in *Capsicum*.

Pochard (1970) identified primary trisomics that could be distinguished from secondary and tertiary trisomics by their phenotype, frequency, and fertility and by their chromosome configurations in meiosis. Eleven of the 12 possible trisomics could be identified by foliage color, anther/stamen color, and fruit characteristics. Cytologically, only three of the 12 chromosomes in *C. annuum* differ significantly in length. The other nine are metacentric and have lengths too close to permit identification.

The recent development of molecular marker-assisted selection techniques may provide new tools in the breeding of chiles. Isozyme and molecular markers have been applied to chiles. Usefulness of isozyme markers is restricted to the insufficient polymorphic bands and a limited number of detectable loci. Molecular based DNA markers such as Restriction Fragment Length Polymorphism (RFLP) and Random Amplified Polymorphic DNA (RAPD) are unlimited in numbers. A saturated isozyme and RFLP map of chile has been reported (Prince et al. 1993). The map contains 192 chile and tomato genomic cDNA clones with 19 linkage groups with a total coverage of 720 cM. However, specific map positions of 26 RFLP markers in seven linkage groups were not determined and vast regions of the chile genome remain unmapped. This area of research is rapidly adding to our knowledge of chiles and will undoubtedly be important in the future.

A peculiar aspect of chiles is the inability to be regenerated from protoplasts. This limits the technique of genetic transformation. Within other solonaceous crops such as tomato, tobacco, and petunia, excellent progress has been made regenerating plants using genetic transformation to introduce novel genes into the genome. For unexplained reasons, chile has been recalcitrant to being regenerated but many laboratories are addressing this issue. Once this perplexing problem is solved, genetic transformation will be available to introduce novel genes into chiles.

PUNGENCY

The one attribute most typical of chiles is pungency and must be considered one of its most important traits. Some have argued that pungency should be one of the five main taste sensories, along with bitter, sweet, sour, and salty. Chile pungency is a desirable attribute in many foods. In most parts of the world, pungency increases the acceptance of the insipid basic nutrient foods. Many innovative uses of pungency are being studied. Besides new medicinal applications, it has been tried as a barnacle repellent, to repel mice from gnawing on underground electrical cables, and to keep squirrels from eating bird seed. Anti-mugger aerosols with chile pungency as the active ingredient have replaced mace and tear gas in more than a thousand police departments in the United States. The spray will cause attackers to gasp and twitch helplessly for 20 minutes.

Pungency is produced by the capsaicinoids, alkaloid compounds, that are found only in the plant genus, *Capsicum*. The nature of the pungency has been established as a mixture of seven homologous branded-chain alkyl vanillylamides (Hoffman et al. 1983). They often are called capsaicin after the most prevalent one. Dihydrocapsaicin is usually the second most prevalent capsaicinoid, while the other five compounds, norcapsaicin, nordihydrocapsaicin, nornordihydrocapsaicin, homocapsaicin, homodihydrocapsaicin, are considered minor capsaicinoids because of their relative low abundance in most natural products. Capsaicin is a powerful and stable alkaloid that can be detected by human taste buds in solutions of ten parts per million. Capsaicin's composition ($C_{18}H_{27}NO_3$) is similar to peperin ($C_{17}H_{19}NO_3$), that gives black pepper its bite.

The capsaicinoids are produced in glands on the placenta of the fruit. While seeds are not the source of pungency, they occasionally absorb capsaicin because of their proximity to the placenta. No other plant part produces capsaicinoids.

Chile pungency is expressed in Scoville Heat Units (Scoville 1912). The Scoville Organoleptic Test was the first reliable measurement of the pungency of chiles. This test used a panel of five human representatives, who tasted a chile sample and then recorded the heat level. A sample was diluted until pungency could no longer be detected. The organoleptic method or taste test has been the standard method for pungency analysis. Although this method is widely used, it has limitations. Tasters must be trained and their ability to test many samples is restricted by the heat of the test solution. Taster fatigue is a real phenomenon and tasters are also not able to distinguish between the different capsaicinoids. Therefore, the Scoville Organoleptic Test has been replaced with instrumental methods.

The most common instrumental method is high-performance liquid chromatography (HPLC). It provides accurate and efficient analysis of content and type of capsaicinoids present in a chile sample (Woodbury 1980; Collins et al. 1995). HPLC analysis has become the standard method for routine analysis by the processing industry. The method is rapid and can handle a larger number of samples. A common practice today is to multiply capsaicinoid ppm by 15 to convert to SHU.

The chile pungency level has genetic and environmental components. The capsaicinoid content is affected by the genetic make-up of the cultivar, weather conditions, growing conditions, and fruit age. Plant breeders can selectively develop cultivars with varying degrees of pungency. Also, growers can somewhat control pungency by the amount of stress to which they subject their plants. Pungency is increased with increased environmental stress (Lindsay and Bosland 1995). More specifically, any stress to the chile plant will increase the amount of capsaicinoid level in the pods. A few hot days can increase the capsaicinoid content significantly. In New Mexico, it has been observed that even after a furrow irrigation, the heat level will increase in the pods. Anthropopathically, the plant has sensed the flooding of its root zone as a stress, and has increased the capsaicinoid level in its pods. If the same cultivar was grown in both a hot semi-arid region and a cool coastal region, the fruit harvested from the hot semi-arid region would be higher in capsaicinoids than that the fruits harvested in the cool coastal climate.

The medicinal applications of capsaicinoids have brought innovative ideas for their use. Medicinal use of *Capsicums* has a long history, dating back to the Mayas who used them to treat asthma, coughs, and sore throats. The Aztecs used chile pungency to relieve toothaches. The pharmaceutical industry uses capsaicin as a counter-irritant balm for external application (Carmichael 1991). It is the active ingredient in Heet and Sloan's Liniment, two rubdown liniments used for sore muscles. The capsaicin is being used to alleviate pain. Its mode of action is thought to be from nerve endings releasing a neurotransmitter called substance P. Substance P informs the brain that something painful is occurring. Capsaicin causes an increase in the amount of substance P released. Eventually, the substance P is depleted and further releases from the nerve ending are reduced. Creams containing capsaicin have reduced pain associated with post-operative pain for mastectomy patients and for amputees suffering from phantom limb pain. Prolonged use of the cream has also been found to help reduce the itching of dialysis patients, the pain from shingles (Herpes zoster), and cluster headaches. Further research has indicated that capsaicin cream reduces pain associated with arthritis. The repeated use of the cream apparently counters the production of substance P in the joint, hence less pain. Reducing substance P also helps by reducing long-term inflammation, which can cause cartilage break down.

PIGMENTS

Color is very important in paprika and chile powder. Paprika and paprika oleoresin are currently used in a wide assortment of foods, drugs, and cosmetics, as well as for improving the feather color of flamingoes in zoos or koi in aquariums. Carotenoids control pod color with approximately 20 carotenoids contributing to the color of the powder. Carotenoid compounds are yellow to red pigments of aliphatic or alicyclic structures composed of isoprene units, which are normally fat-soluble colors (Bunnell and Bauernfeind 1962). The keto-carotenoids, capsanthin, capsorubin, and cryptocapsin are unique *Capsicum* carotenoids. The major red color in chile comes from the carotenoids capsanthin and capsorubin, while the yellow-orange color is from beta-carotene and violaxanthin. Capsanthin, the major carotenoid in ripe fruits, contributes up to 60% of the total carotenoids. Capsanthin and capsorubin increase proportionally with advanced stages of ripeness; with capsanthin being the more stable of the two (Harkay-Vinkler 1974; Kanner et al. 1977). The amount of carotenoids in fruit tissue depends on factors such as cultivar, maturity stage, and growing conditions (Reeves 1987). The reader is referred to Wall and Bosland (1993) for further information on chile pigments.

CONCLUSION

Innovative uses are still being found for chile. For example, the Black Mountain Brewing Co. in Arizona developed a chile beer. The idea was to produce a spicy beer for a local Mexican restaurant. The idea worked, and now, the beer company is one of the top 20 selling micro-breweries in the United States.

The chile industry has grown from a regional food for tourists to an industry competing on the international market. At present, there is an expanding chile market for dietary consumption. At the same time, chile is also becoming more important in manufacturing other commodities. Chile will become more important as a food coloring agent and this may be the leading use of chile in the future. The most innovative use of chile by humankind may yet be discovered.

The College of Agriculture and Home Economics at New Mexico State University has founded the Chile Institute. The Chile Institute is a non-profit, international organization devoted to the study of *Capsicum*. Some of the goals of The Chile Institute are to assist in preserving *Capsicum* germplasm, provide an international information clearinghouse and be an archive of chile publications. It has already published a bibliography containing more than 6,000 references (Bosland 1995). For further information on The Chile Institute write: The Chile Institute, Box 30003, Dept. 3Q, NMSU, Las Cruces, NM 88003 or phone: (505) 646-3028.

REFERENCES

Allard, R.W. 1960. Principles of plant breeding. Wiley, New York.

Anghiera, P.M. 1493. Decadas de nuevo mundo. (Cited in Andrews 1984).

Andrews, J. 1984. Peppers: the domesticated *Capsicum*s. Univ. of Texas Press, Austin, TX.

Bosland, P.W. 1992. Chiles: a diverse crop. HortTechnology 2:6–10.

Bosland, P.W. 1993. An effective plant field-cage to increase the production of genetically pure chile (*Capsicum* spp.) seed. HortScience 28:1053.

Bosland, P.W. 1994. Chiles: history, cultivation, and uses. p. 347–366. In: G. Charalambous (ed.), Spices, herbs, and edible fungi. Elsevier Publ., New York.

Bosland, P.W. 1995. *Capsicum*: a comprehensive bibliography. 3rd ed. The Chile Institute, New Mexico State Univ., Las Cruces.

Bosland, P.W., A.L. Bailey, and J. Iglesias-Olivas. 1988. *Capsicum* pepper varieties and classification. New Mexico State Univ. Ext. Cir. 530.

Bosland, P.W., J. Iglesias, and M.M. Gonzalez. 1994. 'NuMex Centennial' and 'NuMex Twilight' ornamental chiles. HortScience 29:1090.

Bunnell, R.H. and J.C. Bauernfeind 1962. Chemistry, uses, and properties of carotenoids in foods. Food Technol. 16:36–43.

Carmichael, J.K. 1991. Treatment of herpes zoster and postherpetic neuralgia. Am. Family Physician 44:203–210.

Collins, M.D., L. Mayer-Wasmund, and P.W. Bosland. 1995. Improved method for quantifying capsaicinoids in *Capsicum* using high-performance liquid chromatography. HortScience 30:137–139.

DeWitt, D. and P.W. Bosland. 1993. The pepper garden. Ten Speed Press, Berkeley.

DeWitt, D. and N. Gerlach. 1990. The whole chile pepper book. Little, Brown, and Co., Boston.

Domenici, P. 1983. The correct way to spell chile. Congressional Record 129 (149) (Nov. 3).

Franceschetti, U. 1971. Natural cross pollination in pepper (*Capsicum annuum* L.) Proc. Eucarpia Meeting on Genetic and Breeding of *Capsicum*. Turin, Italy. p. 346–353.

Greenleaf, W.H. 1986. Pepper breeding. p. 67–134. In: Mark J. Bassett (ed.), Breeding vegetable crops. AVI, Westport, CT.

Harkay-Vinkler, M. 1974. Storage experiments with raw material of seasoning paprika with particular reference to the red color pigment components. Acta. Alim. Acad. Sci. Hung. 3:239–249.

Heiser, C.B. 1976. Peppers *Capsicum* (Solanaceae). p. 265–268. In: N.W. Simmonds (ed.), The evolution of crops plants. Longman Press, London.

Heiser, C.B. and P.G. Smith. 1953. The cultivated *Capsicum* peppers. Econ. Bot. 7:214–227.

IBPGR. 1983. Genetic resources of *Capsicum*. Int. Board for Plant Genetic Resources, Rome.

Hoffman, P.G., M.C. Lego, and W.G. Galetto. 1983. Separation and quantitation of red pepper major heat principles by reverse-phase high pressure liquid chromatography. J. Agr. Food Chem. 31:1326–1330.

Kanner, J., S. Harel, D. Palevitch, and I. Ben-gera. 1977. Color retention in sweet paprika powder as affected by moisture contents and ripening stage. J. Food Technol. 12:59–64.

Lantz, E.M. 1943. The carotene and absorbic acid content of peppers. N. Mex. Agr. Expt. Sta. Bul. 306.

Lindsay, K. and P.W. Bosland. 1995. A field study of environmental interaction on pungency. *Capsicum* Eggplant Newsl. 14:36–38.

MacNeish, R.S. 1964. Ancient Mesoamerican civilization. Science 143:531–537.

Odland, M.L. and A.M. Porter. 1941. A study of natural crossing in pepper (*Capsicum frutescens* L.) J. Am. Soc. Hort. Sci. 38:585–588.

Ohta, Y. 1962. Genetic studies in the genus *Capsicum* (in Japanese, English summary). Kihara Inst. Biol. Res, Yokohama, Japan.

Peterson, P.A. 1958. Cytoplasmically inherited male-sterility in *Capsicum*. Am. Nat. 92:111–119.

Pochard, E. 1970. Description of pepper trisomics (*Capsicum annuum* L.) in the progeny of a haploid plant (in French, English summary). Ann. Amerlior. Plant. 20:233–256.

Prince, J.P., E. Pochard, and S.D. Tanksley. 1993. Conservation of a molecular linkage map of pepper and comparison of synteny with tomato. Genome 36:404–417.

Reeves, M.J. 1987. Re-evaluation of *Capsicum* color data. J. Food. Sci. 52:1047–1049.

Scoville, W.L. 1912. Note on *Capsicum*. J. Am. Pharm. Assoc. 1:453.

Smith, P.G. and C.B. Heiser. 1957. Taxonomy of *Capsicum sinense* Jacq. and the geographic distribution of the cultivated *Capsicum* species. Bul. Torrey Bot. Club 84:413–420.

Smith, P.G., B. Villalon, and P.L. Villa. 1987. Horticultural classification of pepper grown in the United States. HortScience 22:11–13.

Somos, A. 1984. The paprika. Akademiai Kiado, Budapest.

Stebbins, G.L. 1971. Processes of organic evolution. Prentice-Hall, Englewood Cliffs, NJ.

Tanksley, S.D. 1984. High rates of cross-pollination in chile pepper. HortScience 19:580–582.

Wafer, L. 1699. A new voyage and description of the isthmus of america. p. 107. (Reprinted from the original edition. Franklin, NY).

Wall, M.M. and P.W. Bosland. 1993. The shelf-life of chiles and chile containing products. p. 487–500. In: G. Charalambous (ed.), Shelf life studies of foods and beverages, chemical, biological, and nutritional aspects. Elsevier Publishing, New York.

Woodbury, J.E. 1980. Determination of *Capsicum* pungency by high pressure liquid chromatography and spectrofluorometric detection. J. Assn. Offic. Anal. Chem. 63:556–558.

Asian Vegetables: Selected Fruit and Leafy Types

Marita Cantwell, Xunli Nie, Ru Jing Zong, and Mas Yamaguchi

The demand for Asian vegetables has expanded in the U.S. marketplace (Morr 1989; Lamberts 1992) due to increased ethnic diversity in the population, increased emphasis on healthy foods, and the desire to have variety in the diet. In 1993 and 1994 production statistics gathered by County Agricultural Commissioners in California, seven counties (Fresno, Monterey, Riverside, San Bernadino, San Luis Obispo, Santa Clara, and Ventura) reported Oriental Vegetables as a production category. In these counties, Asian vegetable production accounted for 1.0% to 5.0% of the farm-gate value of all fresh market vegetables.

In a previous review of Asian vegetables, Yamaguchi (1990) considered several vegetables including water bamboo (*Zizania latifolia*), lotus root (*Nelumbo nucifera*), water spinach (*Ipomoea aquatica*), yard-long beans (*Vigna sesquipedalis*), mitsuba (*Cryptotaenia japonica*), and winter radishes (*Raphanus sativuus*). Our intention here is to review the production and marketing of several other specialty Asian vegetables. In this paper we focus on examples from the extensive groupings of warm-season fruit vegetables from the Cucurbitaceae (Yang and Walters 1992; Ng 1993), including bitter melon (*Momordica charantia*), fuzzy melon (*Benincasa hispida*), angled luffa (*Luffa acutangula*), as well as several cool-season leafy vegetables including Chinese broccoli or gai-lan (*Brassica oleracea* var. *alboglabra*), Chinese flowering cabbages or choy sums (*Brassica campestris* var. *parachinensis*), and the leaves and flower buds of Chinese chives (*Allium tuberosum*). The cucurbits that are included are considered more heat tolerant than other more commonly grown members such as cucumbers and summer squashes. The cool season vegetables included are much appreciated for their quality (Willoughby 1995), but are not yet widely grown in the U.S. An additional emphasis in this paper will be on the postharvest handling characteristics of these vegetables.

ASIAN CUCURBITS

Bitter Melon (*Momordica charantia* L.)

Bitter melon is a cucurbit vine native to Asia with eastern India and southern China proposed as the centers of domestication (Yang and Walters 1992). It is now widely cultivated throughout the world for the immature fruits, and sometimes for the tender leafy shoots or the ripe fruits. Other species used for their immature fruits in a similar way are *M. cochinchinensis* and *M. dioica*. The immature fruits are stuffed, pickled, and sliced into various dishes. The immature fruits, called bitter melon, bitter gourd, or balsam pear, are harvested at developmental stages up to seed hardening; the fruits are warty in appearance and vary in size from 12 to 30 cm long.

The bitter melon vine is a rapidly growing monoecious herbaceous vine (Fig. 1) which needs to be trellised. Bitter melon is best adapted to hot to moderate climates and grows well in a variety of soils. Flowering begins about 1 month after seed planting. A fruit fly (*Strumetia cucurbitae*) attacks the fruit in some production areas, and the fruit are covered with paper protective bags. Fruits can be harvested at any stage of development, but are typically harvested fully sized but green, about 2 weeks after anthesis (Fig. 2).

There are three horticultural groups or types of bitter melons (Yang and Walters 1992): (1) small fruit type, 10–20 cm long, 0.1–0.3 kg in weight, usually dark green, fruit are very bitter; (2) long fruit type, type most commonly grown commercially in China, 30–60 cm long, 0.2–0.6 kg in weight, light green in color with medium size protuberances, and only slightly bitter; and (3) triangular fruit type, cone-shaped, 9–12 cm long, 0.3–0.6 kg in weight, light to dark green with prominent tubercles, moderately to strongly bitter.

The bitter principle, for which the fruit is named, is due to the alkaloid momordicine, not to cucurbitacins as in other members of the Cucurbitaceae. Immature fruits are less bitter than the mature but unripe fruits. Among the different types of bitter melons, smaller darker green types are very bitter, and the lighter green-colored fruit are slightly bitter. Bitter melon is also important for various medicinal properties (Morton 1967), with more recent attention focused on it as a hypoglycemic agent (Perl 1988).

The spongy white interior pulp and seeds of unpeeled immature bitter melon are sliced for use as a vegetable in various Asian dishes. The fruit are parboiled or soaked in salted water to remove excessive bitter principle. Proximate composition is similar to that of other immature cucurbit fruits (Table 1).

Good quality bitter melon should have a fresh appearance and the peel should be of uniform green color and free from visual defects. The developing fruit should be firm without excessive seed development, and free of defects such as decay and splitting, both associated with fruit ripening. When the fruit begins to ripen, the exterior color changes from green to yellow and the pulp becomes gelatinous and orange-red (Fig. 3). Contary to other cucurbits, the seed arils also change color from white-cream to a bright red. Among the more than 14 carotenoids identified during the ripening process, the principal carotenoid is cryptoxanthin (Rodriguez et al. 1976). Coincident with color changes, the fruit pulp loses bitterness and becomes sweet.

The bitter melon is harvested, selected for size and uniformity of fruit surface characteristics and commonly packed in carton or wood boxes containing 5, 10, or 20 kg. It is a chilling sensitive vegetable, and may be air-cooled to 10° to 12°C (Zong et al. 1995). Less commonly, the bitter melon is harvested, hydrocooled and treated with a postharvest fungicide. Bitter melon are intermediate in perishability among vegetable crops (Tables 2–4).

Fuzzy melon (*Benincasa hispida* var. *chieh-gua* How.)

Fuzzy or hairy melon is also called Chinese squash or moqua. Mature fruits of some *Benincasa hispida* [syn. *B. cerifera* (Thunb) Cogn.] are commonly called wax gourd, winter gourd or Chinese preserving melon. These can be stored for many months and are prepared stuffed and baked and in soups.

Fig. 1. Trellised bitter melon plant in commercial production in the San Joaquin Valley, California.

Fig. 2. Different types of bitter melon grown in California.

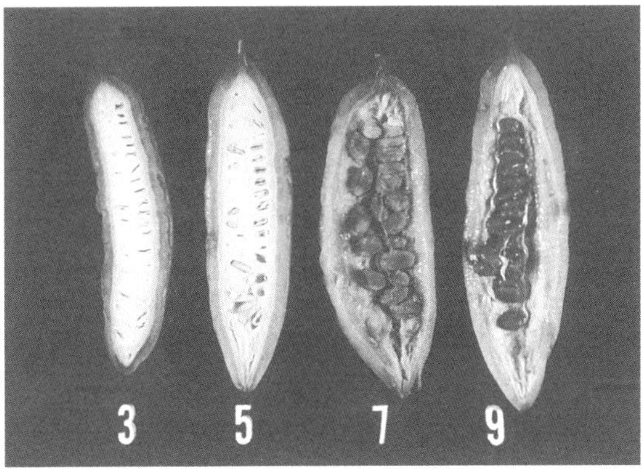

Fig. 3. Various stages of development of bitter melon: immature fruits on the left, ripe fruit with red arils on the right.

Table 1. Nutritive composition per 100 g edible portion of selected Asian vegetables (Source: Wills et al 1984).

	Fruit vegetables			Leafy vegetables	
Component	*Benincasa hispida*	*Luffa acutangula*	*Momordica charantia*	*Allium tuberosum*	*Brassica campestris* var. *parachinensis*
Edible portion (%)	98	54	84	97	94
Water (%)	93.8	94.9	93.8	90.6	94.2
Energy (KJ)	44	20	20	83	49
Protein (g)	0.7	1.2	0.9	2.7	1.3
Fat (g)	0.1	0.1	0.1	0.5	0.3
Carbohydrate (g)	2.0	0.2	0.2	1.0	0.9
Dietary fiber (g)	2.1	3.3	3.3	3.2	2.8
Organic acids (g)	0.04	0.11	0.11	0.19	0.07
Ash (g)	0.7	0.6	0.6	0.9	1.1
Minerals					
Ca (mg)	12	14	22	52	130
K (mg	250	160	260	400	450
Mg (mg)	15	14	16	20	11
Fe (mg)	0.3	0.3	0.9	2.4	0.7
Na (mg)	2	0	3	3	3
Zn (mg)	0.2	0.2	0.1	0.4	0.1
Vitamins					
A (mg)	0.02	0.06	0.04	3.30	1.60
Thiamin (mg)	0.07	0.05	0.05	0.01	0.01
Riboflavin (mg)	0.05	0.01	0.03	0.20	0.10
Niacin (mg)	0.20	0.20	0.40	0.90	0.80
C (mg)	69	18	50	55	46

The immature fuzzy melon fruits have a delicious flavor, stronger and more distinctive than than of immature summer squash, although composition is similar to that of summer squash (Table 1). Flavor can change during storage, with the fuzzy melon taking on an acidic, less agreeable flavor (Table 4).

The diversity and antiquity of cultivars in China suggest that this crop may be indigenous to southern China (Yang and Walters 1992). Cultivars of *B. hispida* are classified on the basis of wax formation on the mature fruit, fruit size, and shape and pubescence of the immature fruits, from which its common name is derived. Generally, separate cultivars are used for immature vegetable production and for mature fruit production (Yang and Walters 1992). One vegetable type has fruits which are cylindrical and roundish with many bristle-like trichomes on the epidermis. The other (jointed gourd) has immature fruits which are narrowed in the center (dumbbell shaped) with their length 2–3 times their width. Small solid green fruit are the best; the younger the fruit, the firmer the flesh.

Fuzzy melon is a vigorous annual vine which is usually allowed to spread over the ground like a pumpkin or winter squash. It can also be grown upright, trellised. The immature fruits can have skin color of various shades of green, depending on variety; all have white flesh. They may be picked within 1 week of flowering, but should not be harvested too mature. Once harvested, the fuzzy melon is susceptible to physical damage but is moderately perishable (Tables 2–4); it is commonly picked and packed in wooden containers of 10 kg or more for commercial handling (Fig. 4).

Table 2. Quality and postharvest defects of selected Asian vegetables.

Product	Harvest indices	Quality criteria	Common postharvest defects
Fruit vegetables			
Benincasa hispida	Size (10–12 cm long)	Fresh appearance with uniform color and size; small size with firm flesh and little seed development	Overmature; seed development; sponginess due to dehydration; pitting due to damage from bristle-like trichomes; flavor changes (increased sourness); chill injury (pitting, decay, sunken areas, exudate)
Luffa acutangula	Size, (10–15 cm long; 2–5 cm diam.)	Fresh appearance; tender; no damage to ridges	Overmature; seed development; dehydration and toughening of peel; loss of green color; damage to ribs causes dehydration and decay; chill injury (discoloration, sunken areas, decay)
Momordica charantia	Size, (10–15 cm long)	Fresh appearance with uniform coloration; firm, without excessive seed development	Overmature; seed development; softening and ripening with internal/external color changes; scuffing of peel; chill injury (pitting, discolored areas, decay)
Leafy vegetables			
Allium tuberosum			
Green leaves	Size (15–30 cm)	Bright green leaves; no defects	Decay; yellowing
Blanched leaves	Size (15–30 cm)	Creamy-yellow leaves	Decay
Flower buds	Size, flower bud development	Well formed but still closed flower buds	Dehydration; toughening of flower stem
Brassica oleracea var. *alboglabra*	Size, flower bud development	15–20 cm tall green stalk; white flower buds buds; developed, not open	Opening and deterioration of flower dehydration, yellowing or decay on leaves
Brassica campestris var. *parachinensis*	Size, flower bud development	15–20 cm tall white stalk; yellow flower buds developed, not open	Opening and deterioration of flower buds; dehydration, yellowing or decay on leaves

Angled luffa *Luffa acutangula* (L.) Roxb.

Angled luffa is another cucurbit native to Asia cultivated since ancient times. The young immature fruit is also called Chinese okra, because of the okra-like shape and external appearance of the tender ridges. Luffas can only be eaten when young as mature fruits become very bitter due to the development of purgative chemicals (Heiser 1979; Yang and Walters 1992). The species used for vegetable production is usually a separate species from the more fibrous species routinely used for sponge production (*L. aegyptiaca*) (Davis and McCourley 1993). Immature fruits of *Luffa acutangula* are considered to be superior in flavor to immature fruits of *L. aegyptiaca* (Ng 1993).

High temperatures (30°–35°C) are required for reasonable (50%) seed germination and early growth in *L. acutangula* (Huyskens et al. 1993). Long days and high temperatures favored the development of more female flowers than short days and lower growing temperatures. Luffa are grown usually on trellises since trained plants are higher yielding. Each plant will yield from 10–23 fruits (Huyskens et al. 1993). Yields are about 3500 to 5000 kg/ha with a density of 10,000 plants/ha.

Table 3. Respiration and ethylene production rates of selected Asian vegetables (modified from Zong et al. 1992, 1995).

Product	Respiration rate (μl CO_2/g-h)				Ethylene production rate (nl/g-h)			
	0°C	5°C	10°C	15°C	0°C	5°C	10°C	15°C
Fruit vegetables								
Benincasa hispida	-.-	8.7	11.4	21.1	-.-	<0.1	<0.1	<0.2
Luffa acutangula	-.-	13.7	19.0	34.4	-.-	<0.1	<0.1	<0.2
Momordica charantia	-.-	7.8	15.0	26.6	-.-	<0.1	0.1–0.2	0.1–0.3
Leafy vegetables								
Brassica oleracea var. *alboglabra*	8.1	10.2	29.2	-.-	0.1	0.1	0.3	-.-
Brassica campestris var. *parachinensis*	4.6	8.2	17.3	-.-	0.1	0.2	0.3	-.-

Table 4. Postharvest handling recommendations for selected Asian vegetables (modified from Zong et al. 1995; Zong and Cantwell unpubl. data).

Product	Susceptibility to water loss	Recommended temperature & relative humidity	Estimated shelf-life (days)
Fruit vegetables			
Benincasa hispida	Low	10–12.5°C, 85–90%	10–14
Luffa acutangula	High	10–12.5°C, 85–90%	5–10
Momordica charantia	Low	10–12.5°C, 85–90%	7–14
Leafy vegetables			
Allium tuberosum,			
Green leaves	Moderate	0–5°C, 90–95%	14
Blanched leaves	High	0–5°C, 90–95%	2–5
Flower buds	Moderate	0–5°C, 90–95%	7
Brassica oleracea var. *alboglabra*	High	0–5°C, 90–95%	7–14
Brassica campestris var. *parachinensis*	High	0–5°C, 90–95%	7–14

After harvest, angled luffa are very perishable. Water loss leads to toughening of the peel, but the fruits are chilling sensitive and the lowest recommended temperature for 1 to 2 weeks of storage is 10°C. Huyskens et al. (1993) showed that luffa rapidly lost weight, sugars, and acids during storage at 4°, 8° and 12°C, with quality losses minimized at 4°C. However, storage at temperatures below 10°C probably induced symptoms of chilling injury (Fig. 5) as reported by Zong et al. (1992). Therefore storage conditions at 10°C (50°F) with high humidity are recommended to reduce weight and quality loss.

ASIAN COOL-SEASON LEAFY VEGETABLES

Chinese chives (*Allium tuberosum* Rottler ex Sprengel, syn. *A. odorum*).

Chinese chives have a long history of cultivation for culinary and medicinal use. Three horticultural products are derived from *Allium tuberosum*: the green (Fig. 6) and blanched leaves and the closed flower buds. In addition, the bulbs may be harvested and used like garlic. The common name Chinese chives is most often used for *A. tuberosum*, but it is also called oriental garlic, Chinese leek, garlic chives, or flowering leek. All parts of the plant have a mild garlic flavor, and like common chives, Chinese chives are mainly used as a fresh culinary herb (Larkcom 1991).

Chinese chives is a perennial plant 30 to 45 cm in height that grows in a clump of 4 to 10 bulbs, each producing 4 to 5 thin flat leaves. Chinese chives is a cool season vegetable which flowers in the hot summer months. The plants live for many years, developing large spreading rhizomes. Chinese chives are usually divided into leafy and flowering types, although most current cultivars are dual purpose. They tolerate a wide range of climates and soils.

As a commercial crop they are grown in beds as a semi-permanent crop for 4 to 5 years, and are intercropped with quick growing leafy greens. Limited harvesting is done during the first year's growth so that reserves build up in the roots. Blanching, or growing the chives in the dark produces a higher value product with soft, yellow leaves and a more subtle garlic flavor. Typically, a green chive production cycle is alternated with a blanched chive production cycle under favorable growing conditions. Under slower growing conditions, only one or two cuttings of blanched chives per year will be taken as it weakens the plant. After harvesting green leaves, resprouting plants are covered with clay pots, or commercially, with plastic films, straw mulches or earth.

Green chives are be stored at least two weeks at low temperatures (Zong and Cantwell unpubl.). Blanched chinese chives are higher respiring and more perishable than the green leaves, and shelf-life is <5 days. Both green and blanched chives are packed in waxed boxes containing 5 to 10 kilos of product. They may be vacuum-cooled, air-cooled or ice-cooled. A paper liner used inside the boxes may result in heat build up, expecially if the product has not been precooled and the box has little venting. Flower buds are bunched and banded and shipped vertically in single layer trays or packed horizontally in boxes used for the leaves.

Chinese broccoli or gai-lan (*Brassica oleracea* L. var. *alboglabra* Bailey)

This excellently flavored broccoli relative is also called Chinese kale or kailan. The edible vegetable consists of a tender green flower stem with buds of what will become white flowers. The leaves and stems are light to medium green in color and are covered with a white haze due to cuticle and wax development (Larkcom 1991).

Fig. 4. Fuzzy melon in wooden containers in the Los Angeles Wholesale market.

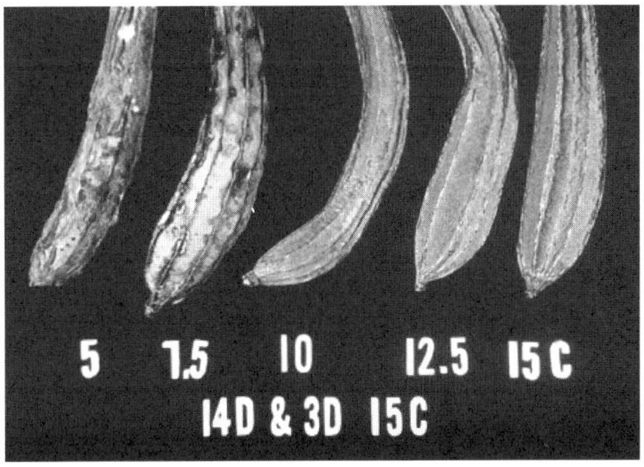

Fig. 5. Effects of different storage temperatures on the visual quality of angled luffa. Severe chilling occurs at temperatures below 10°C. At 12.5°C and above, internal seed development continues. Similar responses to storage temperatures are shown by bitter melon and fuzzy melon.

Fig. 6. Chinese chives packed in a waxed container. The green chives are typically iced or vacuum-cooled to remove field heat and maintain postharvest quality.

Fig. 7. Gai-lan trimmed, bunches and packed in wax containers. Flower buds must be closed to command top price.

Different varieties of gai-lan vary in stem length and color from light to medium green. About 55–70 days are required from a fall seeding to harvest maturity. It grows best during cooler weather. Gai-lan is planted with 2 to 4 rows per bed, 8 to 12 cm between plants, with a planting density of 108,000 to 220,000 plants/ha. Average seasonal yield, with 2 to 3 harvests, is 320–600 18 kg-crates/ha (Shuler 1995).

At harvest the flower buds should be closed (Fig. 7). After harvest the stem becomes tougher than does that of Chinese kale and it may be peeled before use much as broccoli stems are. Gai-lan, like broccoli, is very perishable. Postharvest defects and recommended conditions are described in Tables 2–4.

Chinese flowering cabbages or choy sums (*Brassica campestris* L. ssp. *chinensis* var. *utilis* Tsen et Lee = *B. campestris* ssp. *chinensis* var. *parachinensis* L.H. Bailey)

Choy sums are also referred to as mock pak choy. This vegetable is grown and harvested similar to gai-lan. The white to light green stems are cooked without peeling and have a pleasant, mild flavor. Important quality characteristics are a tender stalk and closed yellow flower buds. For optimum quality, cooler growing weather is required. Under Florida conditions, average days from fall seeding to harvest maturity are 60–70. Choy sum is planted with 3–4 rows per bed with plants spaced 20–30 cm apart; planting density varies from 72,000 to 108,000 plants/ha. Average seasonal yield, with 2 to 3 harvests, is about 620–980 18 kg-cartons/ha (Shuler 1995). Postharvest requirements are similar to those of other cool season leafy vegetables (Tables 3 and 4).

REFERENCES

Allen, P. and C. Wyman. 1986. Glossary of oriental vegetables. Family Farm Series, Small Farm Center, Univ. of California, Davis.

Davis, J. and C.D. DeCourley. 1993. Luffa sponge gourds: a potential crop for small farms. p. 560–561. In: J. Janick and J.E. Simon (eds.), New crops. Wiley, New York.

Dahlen, M. and K. Phillips. 1983. A popular guide to chinese vegetables. Crown Publ., Inc., New York.

Heiser, Jr., C.B. 1979. The gourd book. Univ. Oklahoma Press, Norman, OK.

Hung, L., H. Huang, and H.-F. Yen (eds.). 1992. The nomenclature of vegetable crops. National Taiwan Univ.
Huyskens, S., S. Mendlinger, A. Benzioni, and M. Ventura. 1993. Optimization of agrotechniques in the cultivation of *Luffa acutangula*. J. Hort. Sci. 68:989–994.
Kays, S.J. and J.C. Silva Dias. 1995. Common names of commercially cultivated vegetables of the world in 15 languages. Econ. Bot. 49:115–152.
Lamberts, M. 1992. Production trends for specialty Asian vegetables in Dade County, Florida. Acta Hort. 318:79–85.
Larkcom, J. 1991. Oriental vegetables: The complete guide for kitchen and garden. Kodansha Intl. Ltd., Tokyo.
Lin, C.H. 1989. Diversification and marketing status of vegetable crops in Oriental tropics. Acta Hort. 242:45–58.
Morr, D. 1989. Asian produce enters the American mainstream. Produce Business 5(8):15–19.
Morton, J.F. 1967. The balsam pear-an edible, medicinal and toxic plant. Econ. Bot. 21:57–68.
Myers, C. (ed.). 1991. Specialty and minor crops handbook. The Small Farm Center, Univ. California Publ., Davis. Crops Sheets 001-037.
Ng, T. 1993. New opportunities in the Cucurbitaceae. p. 538–546. In: J. Janick and J.E. Simon (eds.), New crops. Wiley, New York.
Perl, M. 1988. The biochemical basis of the hypoglycemic effects of some plant extracts. vol. 3, p. 49–70. In: L.E. Craker and J.E. Simon (eds.), Herbs, spices and medicinal plants: Recent advances in botany, horticulture, and pharmacology. Oryx Press, Phoenix, AZ.
Rodriguez, D.B., L.C. Raymundo, T.-C. Lee, K.L. Simpson, and C.O. Chichester. 1976. Carotenoid pigment changes in ripening *Momordica charantia* fruits. Ann. Bot. 40:615–624.
Shuler, K.D. 1995. Guidelines for Chinese leafy and root crop vegetable production in South Florida. p. 53–57. In: G.J. Hochmuth and D.N. Maynard (eds.), FACTS 1995 Vegetable Crop Proc.
Walters, T.W. and D.S. Decker-Walters. 1988. Notes on economic plants: Balsam-pear (*Momordica charantia*, Cucurbitaceae). Econ. Bot. 42:286–292.
Willoughby, J. 1995. Asian greenery. Saveur 5:66–76.
Wills, R.B.H., A.W.K. Wong, F.M. Scriven, and H. Greenfield. 1984. Nutrient composition of Chinese vegetables. J. Agr. Food Chem. 32:413–416.
Wittwer, S., Y. Youtai, S. Han, and W. Lianzheng. 1987. Feeding a billion: Frontiers of Chinese agriculture. Michigan State Univ. Press, East Lansing.
Yamaguchi, M. 1973. Production of Oriental vegetables in the United States. HortScience 8:362–370.
Yamaguchi, M. 1983. World vegetables. Principles, production and nutritive values. AVI, New York.
Yamaguchi, M. 1990. Asian vegetables. p. 387–390. In: J. Janick and J.E. Simon (eds.), Advances in new crops. Timber Press, Portland, OR.
Yang, S-L. and T.W. Walters. 1992. Ethnobotany and the economic role of the Cucurbitaceae of China. Econ. Bot. 46:349–367.
Zong, R.J., M. Cantwell, L. Morris, and V. Rubatzky. 1992. Postharvest studies on four fruit-type Chinese vegetables. Acta Hort. 318:345–354.
Zong, R.J., L. Morris, and M. Cantwell. 1995. Postharvest physiology and quality of bitter melon (*Momordica charantia* L.). Postharvest Biol. Tech. 6:65–72.

Growing and Marketing Chinese Vegetables in Central Kentucky*

Wenwei Jia, Mary Witt, and John Strang

In the last 30 years, Asian immigrants entering the United States increased rapidly. Chinese immigrants alone, including persons from Taiwan and Hong Kong, numbered more than 680,000 from 1960 to 1992. The 1994 *Statistical Abstract of the United States* reported total resident population of Asian and Pacific Islanders was 3.8 million in 1980 (1.56% of total U.S. population), 8.4 million in 1992 (3.2%), and is projected to be 11.0 million (4.4%) in 2000 (U.S. Bureau of Statistics 1994). Geographic distribution of Asians is 18.4% in the northeast, 10.6% in the midwest, 15.4% in the south, and 55.7% in western United States (Larckom 1991). Population is almost evenly distributed in the central city, urban, and rural areas. Asian population in Kentucky was 18,712 people in 1995 (Bureau of the Census 1990a). Lexington Kentucky has 1,149 Asian households (3,713 people) or 1.6% of the total Lexington population was Asian (Bureau of the Census 1990b). Lexington ranked sixth in the top ten U.S. cities for increased rate of international immigrants during 1994.

In China, the average vegetable consumption is 300 to 400 g per person per day (Li 1990) which is higher than any western European country or the U.S. Vegetables are basic to Chinese cooking and are found in some form in most dishes. Traditional Chinese cooking methods, such as steaming and stir-frying, are now widely adopted in the west. The use of Chinese vegetables is not limited to Chinese cuisine. Many are used for Oriental and European cooking and may be used as substitutes in, or tasty additions to American dishes.

In China, written record of cultivated vegetables were found 2500 years ago (Li 1990). For thousands of years, China has developed its vegetable industry and has become rich in vegetable resources. More than 170 species, including water, perennial, and wild vegetables are used in China today. Vegetables were also brought to China from all over the world during the last 2000 years. Breeding programs resulted in new vegetable forms and cultivars of vegetables grown for centuries, such as celtuce, Chinese broccoli, Chinese cucumbers, and Chinese melons. (Li 1990). Chinese cabbage, mustard greens, giant oriental radishes, and pak choi are now familiar to the U.S. market, but China has a treasure of lesser-known, delectable, and highly adaptable vegetables which may be easily grown all over the world (Li 1990; Larkcom 1991).

Chinese and Oriental cuisine are of great interest to Americans despite the fact that high quality, authentic fresh ingredients demanded by Chinese cooking are often unavailable. The market demand for specialty and gourmet vegetables continues to expand as more and more attention is given to vegetables by nutritionists, food editors, food suppliers, and restauranteurs. Many vegetables are rich in fiber, minerals, iron, calcium, and vitamins. Vegetables have always been used in China for their curing effects and to keep healthy (Gu, Chen 1989). It is trendy to eat vegetables, especially specialty vegetables, and people are ready for a change.

Data gathered for this experiment of growing Chinese and Oriental vegetables helped determine the feasibility of growing Oriental vegetables for people with interest in diverse vegetables. The research could help farmers in Kentucky grow a diversity of crops and supply new niche markets.

METHODOLOGY

From 1990 to 1995, more than 69 cultivars (42 species) of Chinese vegetables (Table 1) were grown at three University of Kentucky and private farms around Lexington, a private farm in Waddy, Kentucky and in greenhouses of the University of Kentucky. Yield, quality, and growing methods were evaluated. Problems such as insects, diseases, and adaptation of these crops to the climate of Kentucky were evaluated. Taste tests of Chinese vegetables were done in the Department of Horticulture and Landscape Architecture at the Univer-

*Appreciation is extended to D. Slone, J. Pfeiffer, and D. Lowry, U.K. Horticultural Research Farm, for their professional assistance with field and greenhouse research and Drs. J. Buxton and R.G. Anderson for their support. Thanks to Stokes Seed Co., Johnny's Selected Seeds, Burpee Seed Co., Abbott & Cobb, Inc. and Institute of Vegetables & Flowers, CAAS, P.R. China for donations of seed to make this study possible.

Table 1. Chinese Vegetable & Herbs Grown in Central Kentucky, 1991to1995.

Common name	Mandarin name	Scientific name	Parts used
Vegetables			
Adzuki bean (red beans)	Hong xiao dou	*Phaseolus angularis*	Seeds
Amaranthus	Xian cai	*Amaranthus gangeticus*	Greens
Basella	Luo Kui	*Basella rubra*	Greens
Bitter gourd	Ku gua	*Momordica charantia*	Fruit
Bottle gourd	Hu lu gua	*Lagenaria siceraria*	Fruit
Boxthorn	Gou qi	*Lycium arbarum*	Fruit; young shoots
Broccoli (Chinese kale)	Gai lan	*B. rapa* var. *alboglabra*	Greens; flower buds
Burdock	Niu pang	*Arctium lappa*	Root
Celtuce (stem lettuce)	Wo sun	*Lactuca sativa* var. *augustana*	Stem; greens
Chives (garlic chives)	Jiu cai	*Allium tuberosum*	Greens; flowers; flower stems
Choy sum	Cai xin	*B. chinensis* var. *parachinensis*	flower parts; greens
Choy sum (purple flowering pak choi)	Hong cai tai	*B. rapa* var. *purpurea*	Flowers; greens
Chrysanthemum greens	Tung hao	*Chrysanthemum coronarium*	Greens; flowers
Garlic	Suan	*Allium sativum*	Cloves; flower stems; greens
Hairy melon (fuzzy gourd)	Mao gua	*Benincasa hispida* var. *chieh-gua*	Fruit
Hyacinth beans (pig ears)	Bian dou	*Lablab niger*	Fruit
Luffa, angled (Chinese okra)	Si gua	*Luffa acutangula*	Fruit; dry fruit
Luffa, smooth	Si gua	*Luffa cylindrica*	Fruit; dry fruit
Mibuna greens	Ren shen cai	*B. rapa* var. *nipposinica*	Greens
Mizuna greens	Shui cai	*B. rapa* var. *nipposinica*	Greens
Mung bean	Lu dou	*Phaseolus aureus*	Seeds; sprouts
Mustard	Jie cai	*B. juncea*	Greens; stem; root
Napa (headed Chinese cabbage)	Da bai cai	*B. rapa* var. *pekinensis*	greens
Onion, oriental bunching	Da cong	*Allium fistulosum*	Greens
Pak choi	Bai cai	*B. rapa* var. *chinensis*	Greens
Pea shoots	Dou miao	*Pisum sativum*	Sprouts
Pickling melon, oriental	Yue gua	*Cucumis melo* var. *conomon*	Fruit
Pumpkin, Japanese	Nan gua	*Cucurbita moschata*	Fruit
Radish (daikon)	Luo bo	*Raphanus sativus*	Greens; root; sprouts
Soya bean sprouts	Da dou	*Glycine max*	Young pods; dry seeds;
Tai cai (rosette pak choi)	Wu ta cai	*B. chinensis* var. *rosularis*	Greens
Taro	Yu	*Colocasia esculenta*	Tuber
Turnip	Wu jing	*B. rapa* var. *rapifera*	Roots
Watercress	Xi yang cai	*Nasturtium officinale*	Greens
Water spinach	Kong xin cai	*Ipomoea aquatica*	Greens
Winter gourd Wax gourd	Dong gwa	*Benincasa hispida*	Fruit
Yam, Chinese	Shu yu	*Dioscorea batatas*	Tuber
Yard long bean	Chang jiang dou	*Vigna sesquipedalis*	Fruit
Herbs and wild plants			
Ginger	Jiang	*Zingiber officinale*	Root
Lemon grass	Xiang mao cao	*Cymbopogan citratus*	Greens
Sesame	Zhi ma	*Sesamuum indicum*	Seeds
Shepherd's purse	Ji cai	*Capsella bursa-pastoris*	Greens

sity of Kentucky to evaluate preferences of non-Oriental people to Chinese vegetables. Samples of the products were sent to grocery stores, Lexington's farmer's market, and restaurants. Other retail methods included direct sales at three sites of large Chinese communities to investigate market opportunities and acceptable pricing. A survey of Chinese and Oriental vegetable prices was done in Lexington, New York, and Cincinnati.

RESULTS

Five years of growing and marketing experiments showed that napa, yard long beans, Chinese chives, cilantro, winter gourd, bitter gourd, hair gourd, Chinese eggplant, celtuce, and various greens were most popular among Chinese and Oriental people and gave good returns to the grower. Prices for Chinese vegetables sold in central Kentucky were higher than in Cincinnati or New York. Chinese eggplant, Chinese chives, yard long beans, pig ears, and Chinese radishes (daikon) were the most profitable crops on a per meter basis. Typical prices per kilo ranged from $4.00 to $6.00 for Chinese chives or $3.10 to $6.60 for yard long beans. If the produce was fresh and the same produce that was available in their homeland, orientals were willing to pay high prices.

Compared to Beijing, central Kentucky is cooler in spring, hotter in summer and warm longer in fall, making it suitable for growing most Chinese vegetables and especially good for cole crops and all kinds of Chinese radishes. Many vegetables in the cabbage family (*Brassica chinensis*, *B. rapa* var., *B. oleracea*)

Table 2. Schedule for sowing (s) and harvesting (h) Chinese cole crops and daikon radishes in Central Kentucky.

	Date to sow and harvest									
	Spring			Summer			Fall		Winter	
Cultivars	Mar	Apr	May	Jun	Jul	Aug	Sept	Oct	Nov	Dec
Daikon radishes										
April Cross	s[z]	s	h	h	h s	s	s[y] h	h	h	
Minowase										
Summer Cross	s	s	h	h	h s	s	s[x] h	h	h	
New Jersey Cross					s	s	s[x] h	h	h	
White Rat	s	s	s h	h	h s	s	s h	h	h	
Luo Bo	s	s	h	h	h s	s	h	h	h	
Semi-long	s	s	h	h	h s	s	s h	h	h	
Beautiful Heart					s	s	h	h	h	
***Brassica* greens**										
Napa 50 days	s	s	s	h	h	s		h	h	h
Bok choi	s	s	s h	h		s	h	h	h	h
Lei choi	s	s	s h	h	s	s	h	h	h	h
Mei qing choi	s	s	s h	h	s	s	h	h	h	h
Shang hai choi	s	s	s h	h	s	s	s h	h	h	h
Tai cai	s	s	s h	s h	s h	s h	s h	h	h	h
Mustard greens	s	s	s h	h		s h	s h	s h	h	h
Chinese broccoli	s	s	s h	s h	s h	s h	s h	s h	h	h
Flowering pak choi	s	s	s h	s h	s h	s h	s h	s h	h	h

[z]s = sow seeds; h = harvest crop

[y]All napa and daikon radishes can be stored in winter for a continuous supply.

[x]Sow in early September.

Table 3. Planting plan for successive cropping and interplanting in central Kentucky.

Spring	Summer	Fall	Potential income ($\$/m^2$)
Pak choi	Yard long beans, bitter gourd	Pak choi, tai cai or 'White Rat' daikon	35–55
Radishes (daikon)	Bitter gourd or yard long beans		35–55
Cilantro or dill	Cilantro	Cilantro, pak choi, celtuce or dill	30–40
Chinese cabbage or Chinese broccoli	Pig ears		50
Greens, radishes or garlic shoots	Tomato, Amaranthus or chrysanthemum	Chinese cabage, napa or radish (daikon)	30–40

(Larkcom 1991) are staples in the Chinese and Oriental diet. In Beijing, these crops bolt in early summer and grow well only in early fall. Cold weather eliminates them by early Oct. most years. A short window for sowing seeds is typical in Beijing: 10 days around July 24th for daikon and 10 days around Aug. 8th for napa. However, in central Kentucky, *Brassica* crops and radishes can be grown and harvested almost the entire growing season except Aug. (Table 2). The continuous market supply is a decided advantage. Growing methods for the crops are very similar to cole crops presently grown in Kentucky.

Successive plantings during three growing seasons (spring, summer and fall) can bring good returns for farmers. Less land is required with successive plantings and many crops can be produced annually (Table 3). Several planting plans can be used, but rotation of crops in the same family is strongly recommended to reduce pest pressure.

As an example, pak choi could be direct seeded into the field on Mar. 10 or transplanted into the field on Mar. 21. By June 1, pak choi is harvested, and yard long beans and/or bitter gourd are planted in the same location. By Sept. 1, harvest of these two crops is completed, and pak choi, tai cai, and/or 'White Rat' daikon transplants are planted into the same location. Three complete cropping sequences can be realized on the same land area.

Taste tests of 15 cooked Chinese vegetables were conducted in 1991. Americans preferred five vegetables for color: beautiful heart radish, Chinese broccoli, taro, snow peas, and straw mushrooms. Americans preferred straw mushroom, snow peas, radish and Chinese chives for their flavor. Daikon radishes were popular in salads, cooking, and pickling recipes. Their crisp, tender, sweet, and juicy characteristics were among high marks given by the taste panel and were an added bonus for this crop.

SUMMARY

A relatively small volume of each vegetable type is in demand at any one time or place, so this market can easily be saturated. To guard against market saturation, growers should develop a special niche—supply fresh, tender crops harvested at peak, package vegetables attractively, and choose a variety of clients and market mixes. Remember, Chinese vegetables are widely used by all Asians and many Americans.

Specialty vegetables can potentially yield very high returns per square meter. Therefore, they justify more intensive production methods and often require more exacting management.

Since they are often highly perishable, competition is still minimal from distant markets. Produce from California is a key to central Kentucky grocery store prices. Due to the flood in California in 1995, the prices of napa and pak choi were 50% higher for half a year. Kentucky climate gives growers an advantage to grow cole crops and radishes in late spring and early fall when there are no quality products from California.

Freshness of produce is the key to gaining the Chinese market. Without sophisticated post-harvest handling or packaging methods, limit your market radius to easy traveling distance (120 km, 200 miles) to help ensure the freshest specialty produce.

Most importantly, growers must identify specific markets, even before ordering seed. Whether the market is direct retail to consumers, a restaurant, or direct sales to neighborhoods dictates which crops will be grown and what special cultural or post-harvest practices will be required.

REFERENCES

Bureau of the Census. 1990a. Census of population, general population characteristics. Kentucky. U.S. Department of Commerce, Economics and Statistics Administration. Kentucky.

Bureau of the Census. 1990b. Census of population and housing. Summary social, economic, and housing characteristics CPH-5-1. United States.

Gu, Z. and Y. Chen. 1989. Health benefit of vegetables. Chinese Agricultural Science and Technology Publisher.

Larkcom, J. 1991. Oriental vegetables: The complete guide for kitchen and garden. Kodansha Intl. Ltd., Tokyo.

Li, Shuxuan et. al. 1990. Chinese agricultural encyclopedia, Volume of vegetables. Agricultural publisher. Beijing, China.

Lumpkin, T.A., J. Konovsky, K.J. Larsen, and D. McClary. 1993. Potential new specialty crops from Asia: azuki bean, edamame soybean, and astragalus. p. 45–51. In: J. Janick and J.E. Simon (eds.), New crops. Wiley, New York.

U.S. Bureau of Statistics. 1994. Statistical abstract of United States, 114th edition.

Pejibaye Heart-of-Palm in Hawaii: From Introduction to Market

Charles R. Clement, Richard M. Manshardt, Catherine G. Cavaletto, Joseph DeFrank, John Mood, Jr., Natalie Y. Nagai, Kent Fleming, and Francis Zee

Pejibaye or peach palm (*Bactris gasipaes* Kunth) was introduced into Hawaii in 1990 to provide fresh hearts-of-palm for the gourmet restaurant market (Clement et al. 1993). Although heart-of-palm is widely available as a canned product (Tabora et al. 1993), it is little known as a fresh vegetable. The species had been introduced into Hawaii earlier in this century, but no local information existed to guide potential commercial producers. The general objective of the introduction program reported here was to generate the information necessary to guide producers and consumers interested in this new crop.

There are three high-quality edible parts to a juvenile palm stem (Fig. 1). The heart-of-palm is composed of tender developing leaves enclosed within the spear leaf sheath immediately above the apical meristem. The heart accounts for 99% of world trade in vegetable pejibaye (1990 value: U.S.$40 million) and is produced mostly in Latin America, at present (Mora Urpí et al. 1991). The developing stem tissue, extending 10–20 cm

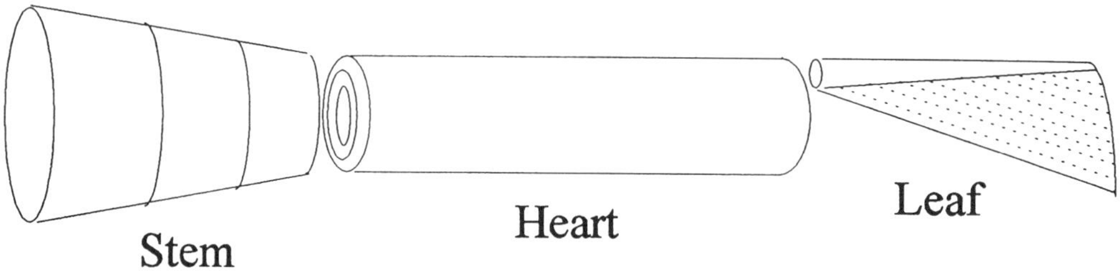

Fig. 1. The three potential palm shoot products.

below the apical meristem, is also edible. Little edible stem is currently utilized; only 1% of trade in fresh pejibaye consists of stem, mostly in the form of chunks in marinade. The edible leaf is similar to the heart but is not wrapped within the leaf sheath. Consequently, it is difficult to manage during processing, but is of very high quality. All edible leaf produced in Latin America is currently discarded. Chefs in Hawaii are enthusiastic about all three products.

Agronomic research done in Latin America provided a useful guide for the initial formulation of crop management practices in Hawaii, especially regarding plantation density, fertilization, plant management, harvesting, and post-harvest handling (Mora Urpí et al. 1984). Hawaiian agriculture operates under different constraints, however. The most important difference is the cost of labor, including social security, liability insurance, and other indirect costs. Consequently, the Latin American experience had to be modified to fit the limitations of the Hawaiian agroeconomic environment.

As part of the introduction program, growth and yield of half-sib progenies from the Benjamin Constant population of the Putumayo landrace (Brazil) were evaluated at three locations in Hawaii. The effect of planting density was also evaluated in Benjamin Constant progenies and progenies from the Yurimaguas population of the Pampa Hermosa landrace (Peru), using three densities at one location on the Island of Hawaii. Other project objectives were to describe juvenile palm morphology, design a planting scheme suitable for Hawaiian agriculture, compare conventional and organic weed control practices, characterize genetic relationships within the introduced germplasm, evaluate product quality, develop a cost-of-production model, and create a market for the fresh product.

PLANT MORPHOLOGY

Juvenile pejibaye plants are essentially all leaves, since no stem develops until just before harvest. There were fewer leaves (6–8) than expected (8–10) at all experimental sites, with the density trial site lowest of all (6–7) (Clement 1995). This may have been due to micronutrient deficiencies, since soils at two locations had been severely degraded by poor management. Adequate nitrogen fertilization masked the leaf deficiency until harvest, when no density effects were observed (see below).

Good sucker (off-shoot) production is essential for long-term survival of the plantation. Sucker number averaged 6–7 (range 0–18) at first harvest, but was substantially lower at second harvest (mean 2, range 0–10) for Benjamin Constant germplasm (Clement 1995). This may have been due to: (1) differences in sucker management, where intense competition among the full sucker complement up to first harvest may have caused abortion of second cycle sucker buds, or (2) different partitioning of resources between shoots and their off-shoots during the second cycle, when shoot growth was much more vigorous.

PLANTATION DENSITY

There were no significant differences in individual plant dimensions or growth rates at 3333, 5000 (the commercial density in Costa Rica), or 6666 plants/ha in either Benjamin Constant or Yurimaguas germplasm at the Island of Hawaii site. There are four possible reasons for the lack of density effects on individual plant dimensions: (1) the 9-plant plots were too small, allowing sufficient light to penetrate the plot from the row spaces; (2) palm growth was sufficiently irregular that most plants received sufficient light from some side; (3) the reduced leaf number (noted above) reduced leaf area index (LAI) and therefore competition; or (4) the densities studied do not significantly affect vegetative growth of pejibaye (Clement 1995).

The first possibility was recognized as a potential limitation before field planting. A shortage of germplasm, because of poor germination, did not permit larger plots. The second possibility is probably important too, since the growth rate of each plant is affected by its unique micro-environment. Also, as harvest progressed, the canopies became more irregular, since some plants were removed each month. The third possibility certainly existed during the experimental period, although the magnitude of its effect on LAI is unknown. The older leaves, which may have senesced prematurely due to micronutrient deficiency, would have a relatively small impact on LAI because they were small. The fourth reason is theoretically possible since Corley (1973) observed that vegetative growth in African oil palm (*Elaeis guineensis* Jacq.) was not reduced at planting densities higher than the optimum for fruiting, although fruit yield dropped dramatically.

PLANTATION DESIGN

Hawaiian farmers require mechanical access to all plants in the field to reduce labor costs. The lack of planting density effects suggested modifications to Latin American practices that would allow easy access without reducing the number of plants per hectare.

Palms are planted in offset double rows with 1-m spacing between plants. Service rows, planted with a legume or grass groundcover, alternate with double rows of palms and permit easy access. This design allows 5000 or more plants/ha. Rows should run along the contours, so that the service rows act to contain erosion. Because all plants are in border rows, the high within-row density should not reduce growth. This design is already being used commercially on the Island of Hawaii.

WEED MANAGEMENT

Weeds in pejibaye plantations in Hawaii are a major consideration. Consequently, the degree of control afforded at two levels each of oxyfluorofen, oryzalin, and paraquat were compared with the suppression provided by woven black polypropylene mat (as the control). Oxyfluorofen and polypropylene mat were equally good conventional weed controls (DeFrank and Clement 1995).

Groundcovers can also act to control weeds, and they fit well into pesticide-free organic production methods. Three vegetative groundcovers [two legumes (*Arachis pintoi*, *Desmodium ovalifolium*) and a grass (*Chloris gayana*)] and polypropylene mat were compared for efficacy in weed control. All of the vegetative groundcovers competed significantly with the palms, as evidenced by delayed harvest. Consequently, polypropylene mat or organic mulch in the tree rows is an essential complement to the groundcovers to avoid competition. The palms must also receive more fertilizer to compensate for that consumed by the groundcovers.

The plantation design in use on the Island of Hawaii integrates the polypropylene mat with *Arachis pintoi*. Two-meter-wide polypropylene mat is used to control weeds in the immediate vicinity of the palms in the double rows. *A. pintoi* is thereby restricted to the service row, where it does not compete with the palms.

Table 1. Benjamin Constant progeny means for earliness (days to harvest), relative growth rate (RGR), unit leaf rate (E_A), and leaf area ratio (LAR) for the period from nursery transplant to first harvest at the best rainfed site, and progeny mean correlation coefficients with earliness.

	Progeny means			
Progeny	Earliness (days to harvest)	RGR	E_A	LAR
B-0	898	11.71	2.66	10.71
B-1	932	11.33	2.59	10.69
B-2	982	10.82	2.45	10.71
B-3	972	10.87	2.51	10.67
B-5	1027	10.31	2.33	10.71
B-8	954	11.01	2.47	10.72
B-9	899	11.85	2.61	10.77
Correlation coefficient[z]		–0.99***	–0.96***	–0.41

[z]Correlation coefficiant of progeny means with progeny means for earliness. Critical values of r are: 0.75, 0.87, 0.95, for p = 0.05, 0.01, 0.001, respectively.

Table 2. Total variation and heritability for relative growth rates of Benjamin Constant progenies determined over different time intervals at the best rainfed site.

Time interval	V_P	CV_P	h^2
292 to 464 days in field	31.9	55.0	0.78[z]
Seedling to 292 days in field	10.7	24.5	1.44[z]
Seedling to first harvest	1.2	9.8	0.74[z]

[z]Inbreeding suspected because of very high h^2
V_P = Total variance, CV_P = coefficient of phenotypic variation, h^2 = heritability (narrow sense)

After full establishment of *A. pintoi,* it is an acceptable weed-suppressing groundcover, requiring only occasional mowing of weeds that penetrate above its canopy.

GROWTH ANALYSIS

Relative growth rate (RGR) in the Benjamin Constant germplasm was highly correlated with earliness (r= –0.99) (Clement 1995), measured as days from nursery transplant to harvest (Table 1). An analysis of Unit Leaf Rate (E_A) and Leaf Area Ratio (LAR), the product of which equals RGR, showed that most of the variation for earliness among Benjamin Constant progenies was due to variation in E_A. Because individual plant heart-of-palm weights were not different among Benjamin Constant progenies (Clement 1995), earliness (as measured by either RGR or days to harvest) is the major component of annual yield in this germplasm.

Benjamin Constant progenies differed in partitioning to sucker number and sucker growth also (Clement 1995). Some partitioned early to sucker number, but not significantly to sucker growth. Others partitioned early to both number and growth. Others partitioned little to sucker number, but significantly to sucker growth. The latter partitioning pattern may be preferable in Hawaii because it will reduce labor costs for sucker pruning, while providing more rapid growth to the next harvest. However, the marked reduction in number of suckers produced by plants in the field between the first and second harvests may impose limitations on the optimum sucker number that is possible.

YIELDS

Mean annual yields of heart-of-palm were 500, 900, and 1000 kg/ha at 5000 plants/ha at the three Benjamin Constant experimental sites (Clement 1995). Yields at the two better sites are comparable to good tropical American figures. The lowest yield was obtained on a Histosol, essentially unweathered lava rock, with rainfall of 4800 mm per year, but where a week with no rain caused severe drought. The intermediate yield was obtained on an Andisol with 3500 mm of rain, degraded by nearly a century of sugarcane production. The highest yield was obtained on an Oxisol degraded by poor management, with 1000 mm of rain annually and supplemental irrigation. The high yield was probably due to high light intensity under cloudless conditions, and yield might have been higher with better irrigation technology. There was considerable variation in yield among Benjamin Constant progenies (Fig. 2).

With two stems per plant after first harvest, cumulative 18-month yields of hearts-of-palm were 800, 1400, and 1500 kg/ha. When both edible stem and edible leaf were added, 18-month yields were 2.5, 4.5, and 4.8 mt/ha, respectively. Yurimaguas yields were consistently 10% higher than Benjamin Constant yields at the 4.5 mt/ha site, apparently because of lower fiber content.

QUANTITATIVE GENETICS

There were strong indications of past inbreeding in the Benjamin Constant germplasm, in that narrow-sense heritabilities for both vegetative traits and growth rates, characters that usually have moderate to low

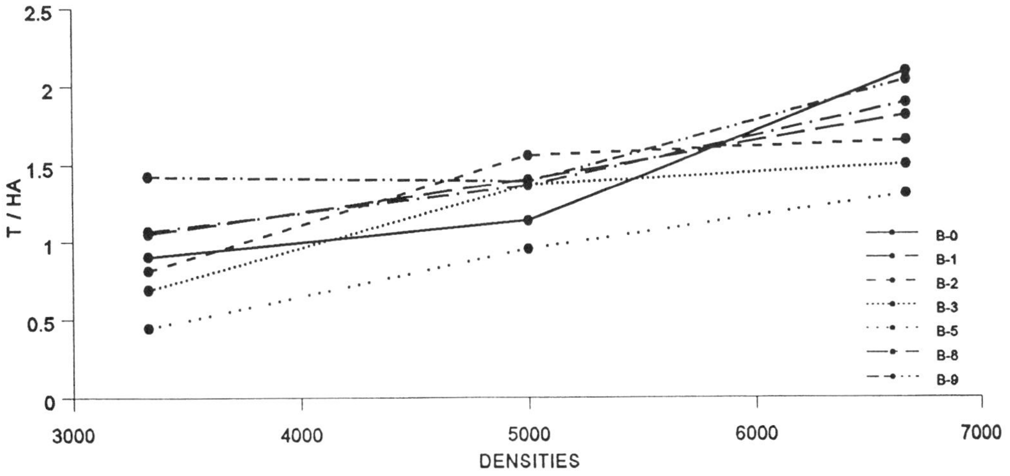

Fig. 2. Mean annual heart-of-palm yields for seven Benjamin Constant progenies planted at three densities at the best rainfed site in Hawaii.

heritabilities, were in excess of 1.0 (Clement 1995). When inbreeding was arbitrarily adjusted to 0.5, RGR, E_A and earliness had moderate heritabilities.

Response to selection for RGR, E_A or earliness depends upon the amount of variation in the population. This, in turn, depends upon the period over which the trait is measured. Three different intervals were examined: a uniform time period (e.g. 292 to 464 days in the field); a uniform developmental period (seedling to harvest); and a mixed category (seedling to 292 days in the field). RGR over the entire seedling-to-harvest period had the lowest variation, but is probably the correct evaluation period (Table 2), since it minimizes variation due to differing developmental morphologies (Clement 1995).

Conventional breeding methods are unlikely to yield rapid genetic gains with this germplasm, because of low variability. Consequently, vegetative propagation is being pursued through tissue culture to capture the best phenotypes for immediate use by farmers.

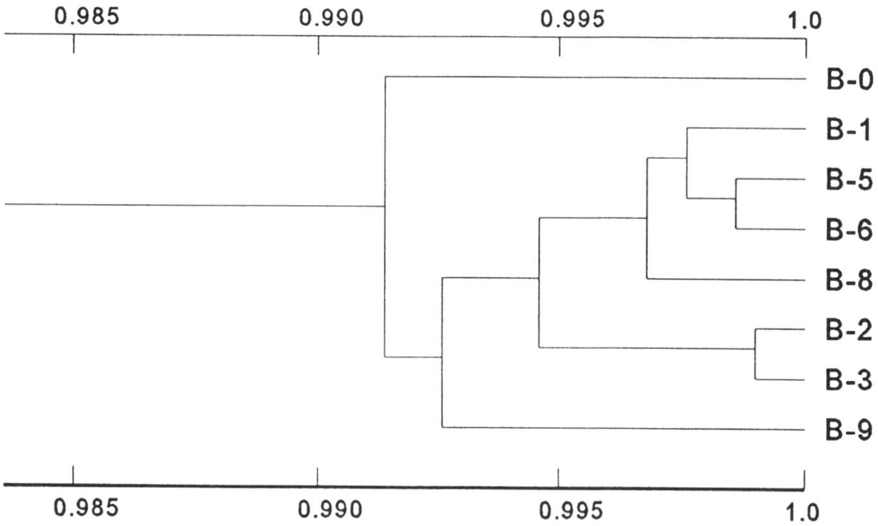

Fig. 3. Dendrogram based on Nei's (1978) unbiased genetic identities among progenies of the Benjamin Constant population in Hawaii. Cophenetic correlation = 0.69.

Table 3. Variation in mean sensory evaluation scores for six quality characteristics of fresh heart-of-palm, comparing values for sib plants within a single Benjamin Constant progeny and for different sections of the same heart (the 1st section includes the apical meristem).

Plants or sections	Evaluation score (1–5 scale)[z]					
	Sweetness	Tenderness	Crispness	Moistness	Astringency	Acridity
Sib plants						
1	2.1	2.7	1.8a[y]	2.8	1.5	1.8ab
2	2.4	2.4	2.2ab	3.0	1.4	1.7a
3	2.4	2.3	2.3b	2.9	1.6	2.3b
Heart sections						
1st	2.6	2.8	1.9	2.7	1.8b	2.1b
2nd	2.1	2.2	2.1	2.9	1.3a	2.0ab
3rd	2.1	2.4	2.2	3.1	1.4a	1.6a

[z]Characteristics were scored by a panel of 11 people, who rated each on a scale of 1 (low expression) to 5 (high expression).

[y]Mean separation in columns by Duncan's new multiple range test, 5% level.

ALLOZYME VARIATION

Averaging across 17 isozyme loci, mean allozyme heterozygosity in Benjamin Constant and Yurimaguas progenies was 0.074 (range 0.038–0.099) and 0.124 (range 0.098–0.153), respectively. In an allogamous crop, such low heterozygosity strongly suggests significant past inbreeding, substantiating results obtained in the quanitative genetic analysis. All Benjamin Constant progenies were very closely related genetically, as shown in the dendrogram based on Nei's (1978) genetic identities (Fig. 3). Allozyme heterozygosity was not correlated with growth rates or vegetative traits in the Benjamin Constant germplasm, probably because of the relative uniformity of this inbred germplasm (Clement 1995).

Table 4. Major components of a cost-of-production model for a 1-ha pejibaye heart-of-palm plantation on the Island of Hawaii.

Variable	$/plant	$/ha
Establishment and first year of management	7.907	50,370
Annualized ownership expense	1.091	6,974
Operating costs (5th year)		
Growing costs	1.035	6,615
Harvesting, processing, shipping costs	2.357	15,067
Total	3.392	21,682
Gross income (5th year)		
Heart[z]	2.880	18,410
Edible stem[y]	3.168	20,251
Edible leaf[x]	?	?
Total gross income	6.048	38,662
Gross margin	2.656	16,982
Ownership costs	1.626	10,454
Total cost of production	5.018	32,135
Economic profit (5th year)[w]	1.030	6,527

[z]0.22 kg/plant, $13.00/kg
[y]0.45 kg/plant, $7.00/kg
[x]0.05 kg/plant, $/kg unknown
[w]Total gross income – Total cost of production

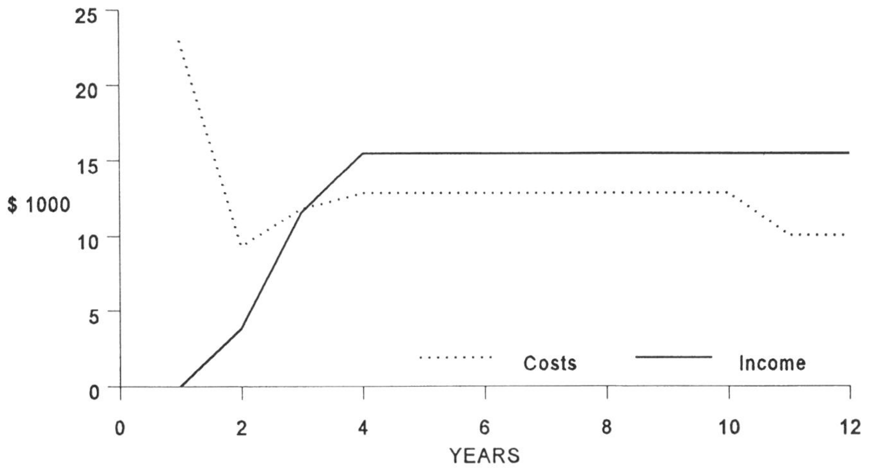

Fig. 4. Hypothetical cash flow for a pejibaye heart-of-palm plantation on the Island of Hawaii, assuming a planting density of 5000 plants/ha and a stem value of $6.00.

QUALITY

There is considerable within-progeny and within-plant variation for heart-of-palm quality characteristics, such as appearance, sweetness, tenderness, crispness, and moistness (Table 3). Total soluble solids varied from 3%–13%, with more variation between harvest dates than between progenies. Yurimaguas germplasm was always slightly sweeter than Benjamin Constant.

Negative sensory characteristics found in some samples were astringency and acridity (Table 3). Acridity was more common, occurring in 5% and 8% of Yurimaguas and Benjamin Constant germplasm, respectively. Human sensitivity to acridity ranges from no reaction to an extremely sensitive, perhaps even allergic reaction. Contrary to the finding of Tabora et al. (1993), most consumers in Hawaii do not like the "bite" (acridity) in fresh heart-of-palm. No biochemical assay is currently available to quantify acridity, which limits our ability to evaluate it or determine the relative importance of genetic and environmental factors. Nonetheless, these negative traits are receiving priority attention because of possible health risks which they may pose for a small minority of the population.

Hearts retain fresh flavor and appearance for two weeks when wrapped in plastic film and refrigerated immediately after harvesting. This shelf-life will allow shipping to and marketing in major U.S. markets. Further research is planned to extend shelf-life.

COST-OF-PRODUCTION MODEL

Establishment of a pejibaye heart-of-palm plantation in Hawaii is expensive ($51,000/ha, amortized over 10 yr), and yearly operating, harvesting, processing and marketing costs totalled $33,000/ha when in full production (3rd year) at a good rainfed site on old sugarcane land on the Island of Hawaii (Table 4). Each shoot (heart = 150–225g, stem = 340–500g, leaves = 45–135g) must be sold for $5.00 to break even. Each dollar received above that price produces a 15% return on investment. Current market price in Hawaii is $12/stem, but is expected to drop as more farmers start to produce. A hypothetical cash flow diagram, assuming $6 per shoot, is presented in Fig. 4. The full cost-of-production model is available upon request from Kent Fleming.

MARKET CREATION

Harvests from the experimental plots were used to introduce the fresh heart-of-palm directly to chefs in Hawaii's finer restaurants and to consumers through participation in gourmet product expositions and food fairs. Interested farmers were urged to negotiate directly with chefs to develop strong relationships and favorable profit margins.

Requirements of individual chefs ranged from 4 to 18 kg/week of palm (including heart, stem, and leaf), so a farmer will need 0.10–0.50 ha to supply one chef. The local restaurant and resort market is estimated to require 80–160 ha of high density plantings. As more area comes into production and supply increases, prices will drop to a level that should allow the demand for hearts-of-palm to expand beyond its present gourmet niche.

California buyers are interested in expanding the area in cultivation in Hawaii to meet mainland and international export market demands. Preliminary research by the USDA/ARS Fruitfly Ecology Lab, Hilo, HI, indicated no fruitfly infestation problem in pejibaye palm hearts. This information is needed by USDA/APHIS to permit importation of heart-of-palm into the continental U.S. Once pejibaye heart-of-palm is cleared for importation, production area may expand beyond the above projections.

REFERENCES

Clement, C.R. 1995. Growth and genetic analysis of pejibaye (*Bactris gasipaes* Kunth, Palmae) in Hawaii. PhD Dissertation, Univ. Hawaii, Honolulu.

Clement, C.R., R.M. Manshardt, J. DeFrank, F. Zee, and P. Ito. 1993. Introduction and evaluation of pejibaye (*Bactris gasipaes*) for palm heart production in Hawaii. p. 465–472. In: J. Janick and J. Simon (eds.), New crops. Wiley, New York.

Corley, R.H.V. 1973. Effects of plant density on growth and yield of oil palm. Expt. Agr. 9:169–180.

DeFrank, J. and C.R. Clement. 1995. Weed control in pejibaye heart of palm plantations in Hawaii. HortScience 30:1215–1216.
Mora Urpí, J., E. Vargas, C.A. López, M. Villaplana, G. Allón, and C. Blanco. 1984. The pejibaye palm (*Bactris gasipaes* H.B.K.). Food and Agriculture Organization of the United Nations (FAO), Universidad de Costa Rica, Banco Nacional de Costa Rica, San José, Costa Rica.
Mora Urpí, J., A. Bonilla, C.R. Clement, and D.V. Johnson. 1991. Mercado internacional de palmito y futuro de la explotación salvaje vs cultivado. Boletin Pejibaye (Guilielma) 3(1–2):6–27.
Nei, M. 1978. Estimation of average heterozygosity and genetic distance from a small number of individuals. Genetics 89:583–590.
Tabora, P.C., M.J. Balick, M.L.A. Bovi, and M.P. Guerra. 1993. Hearts of palm (*Bactris*, *Euterpe* and others). p. 193–218. In: J.T. Williams (ed.), Pulses and vegetables. Chapman & Hall, London.

Starch Noodles from Edible Canna

Michael Hermann

Edible canna (*Canna edulis* Ker-Gawler, Cannaceae) is a starchy root crop with historic importance but is little used today in its native Andean range. Although of pantropical distribution, it is infrequently grown and mostly used as an emergency food. Direct consumption of the fibrous rhizome is limited by its poor eating quality and its long cooking time (2 to 5 hours).

Canna is easy to grow. It has no significant pests or diseases. The crop thrives on residual nutrients and can be grown continuously. Typically yielding 20–40 t per ha, canna is often grown without irrigation on marginal soils or on slopes where its long crop duration (10–12 months) helps prevent erosion.

Nowhere else is canna as important as in north Vietnam where, over the last 30 years, the area in canna has expanded to an estimated 20,000–30,000 ha. Canna is grown for the sole purpose of extracting starch. Wooden drum graters and simple sedimentation tanks are used to extract the starch. Canna starch has the largest known granules (30–100 mm), resulting in fast settling. The rhizomes contain 22% to 25% of dry matter and starch yields range from 12% to 16%.

The overwhelming portion of canna starch production in Vietnam is processed into transparent starch noodles ("cellophane noodles"), a luxury food of south-east Asia and traditionally made of costly mungbean starch. Good cellophane noodles are about 1 mm thick; they display high tensile strength and good transparency. Dry matter loss during prolonged cooking is less than 10%.

Starch noodles of non-canna origin are usually produced through extrusion cooking, which requires the extruded noodles to pass through a cooling water bath. By contrast, canna noodles are manufactured by a different, and previously undescribed, process involving the steam-sheeting of a starch/water dough. The resulting gel sheets are stretched and semi-dried on bamboo frames. The gel sheets are then folded and cut into straight noodles. They are finally dried to a moisture content of about 18% to 21%.

Canna noodles in Vietnam have excellent eating quality, much superior to extrusion noodles made experimentally from sweet potato and cassava starches which are widely available in southeast Asia. Special but as yet poorly understood functional properties of canna starch make it a substitute which has totally replaced expensive mungbean starch as the raw material for cellophane noodles in Vietnam. The high amylose content (25% to 30%) of canna starch as compared with other root starches has been proposed to explain the high peak viscosity observed during gelatinization, which permits the sheets to be easily handled. Canna starch also displays high gel retrogradation (recrystallization) and transparency which is critical to noodle quality.

Canna processing in Vietnam provides employment to many thousands of people in rural communities with as little as 500 m^2 of arable land per capita. Canna use in Vietnam shows how product development can provide new perspectives for crop utilization and stimulate demand for otherwise obsolete crops.

Fig 1. Canna production and noodle processing in Vietnam. A: Muong girl cleaning canna rhizomes (Tu Ly, Hoa Binh Province, 400 m altitude); B: Canna market in Duong Lieu (Ha Tay Province, Red River Delta); C: Canna starch extraction plant (Duong Lieu, Ha Tay Province); D: Drying of canna starch (Duong Lieu, Ha Tay Province); E: Steam-sheeting of canna starch dough (Duong Lieu, Ha Tay Province); F: Stretching canna starch gels (Duong Lieu, Ha Tay Province); G: Canna noodles on sale (Central Market, Hanoi). (Photographs taken November 1992, all north Vietnam).

New Technology for Making Tempeh: A Cultured Soyfood

Gunter Pfaff and Betsy Shipley

Tempeh, a soyfood which originated in Indonesia several hundred years ago, is a valuable source of plant protein. Partially cooked dehulled soybeans are inoculated with the fungus *Rhizopus oligosporus* and left to incubate for about 22 h at 30°C. The white mycelium which develops ties the soybeans together and improves the nutritional profile, resulting in a semi-firm cake which can be served in a large variety of dishes.

When tempeh began to be produced in commercial quantities in the United States, the most prevalent production method was that developed by Martinelli and Hesseltine (1964). The partially cooked, inoculated soybeans were incubated in a perforated plastic bag and frozen after incubation was completed.

There are however numerous advantages to incubating tempeh on trays rather than in plastic bags. The finished tempeh can be pasteurized, cut into various shapes, and premarinated with different flavors before being frozen. In this paper we describe a new method for producing tempeh on trays.

DEVELOPMENT

The first book on tempeh production (Shurtleff and Aoyagi 1980) reported problems utilizing stainless steel trays for tempeh incubation because the heat generated in the heat producing phase of incubation was insufficiently dissipated. The original experiments by Martinelli and Hesseltine were even more negative: "Metal trays without perforations, no matter what kind of covering was used, did not produce a good tempeh, since aeration of the whole mass of soybeans was insufficient for mold development." We first overcame this problem by incubating on a specially designed rack which circulated air under the trays. However this necessitated multiple handling in order to end up with a pasteurized product. We concluded therefore that it was necessary to abandon the traditional incubation room concept.

Tests had shown that circulating water was excellent for heat dissipation. A waterbath incubator would allow pasteurization at the end of the incubation period by raising the waterbath temperature to 80°C for 15 min (Fig. 1).

The substrate is on a stainless steel tray. This tray sits on a support and is partially submerged in the waterbath. An insulated lid covers this tray and seals it from the environment. Air is injected under the lid and drifts over the incubating tempeh to exit at the opposite end. The temperature of the circulating waterbath is closely regulated for both heating and cooling.

The utilization of stainless steel trays for this process was important since food-approved stainless steel is the standard in the commercial food business. Aluminum trays were not considered since the relation of aluminum to Alzheimer's disease is not clear at this time. Other metal trays are available which have a food-approved coating, but this would not stand up to daily use. Fiberglass trays with a food-approved coating would not hold up to the high temperatures reached during pasteurization.

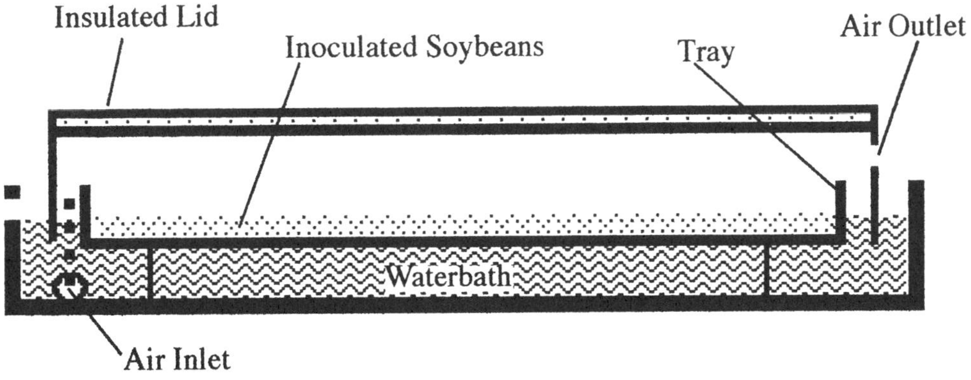

Fig. 1. Simplified diagram of waterbath incubator.

We subsequently constructed an incubator/pasteurizer with 14 large stainless steel trays which has now been operating flawlessly for three years (Fig. 2). This apparatus and method has been patented (Pfaff 1993, 1994).

Our 14 tray unit has consistently produced high quality tempeh without overheating or producing premature sporulation (Fig. 3).

COMMERCIALIZATION

A production incubator/pasteurizer for a larger facility might consist of two waterbath units (one above the other) using a total of 28 trays and producing 70 kg of tempeh every 24 h. Because stainless steel trays are used, the product can easily be premarinated with different flavors before being packaged and frozen. For example, premarinating with a low salt tamari—a wheat free soy sauce—is suggested as an additional product line.

During our market testing in the regional market, we have sold the product as unflavored patties and in grated form. We have produced this tempeh on a part time basis and marketed about 3200 kg/yr. Acceptance by health conscious individuals and chefs indicate this product has a very promising future as a healthy alternative food.

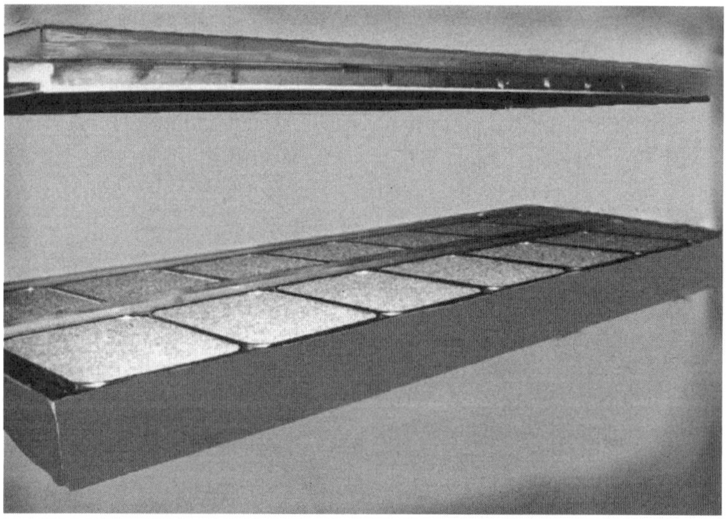

Fig. 2. A 14 tray waterbath incubator/pasteurizer for tempeh production; lid raised, trays filled with substrate; each tray will produce approx. 2 kg of tempeh every 22 h.

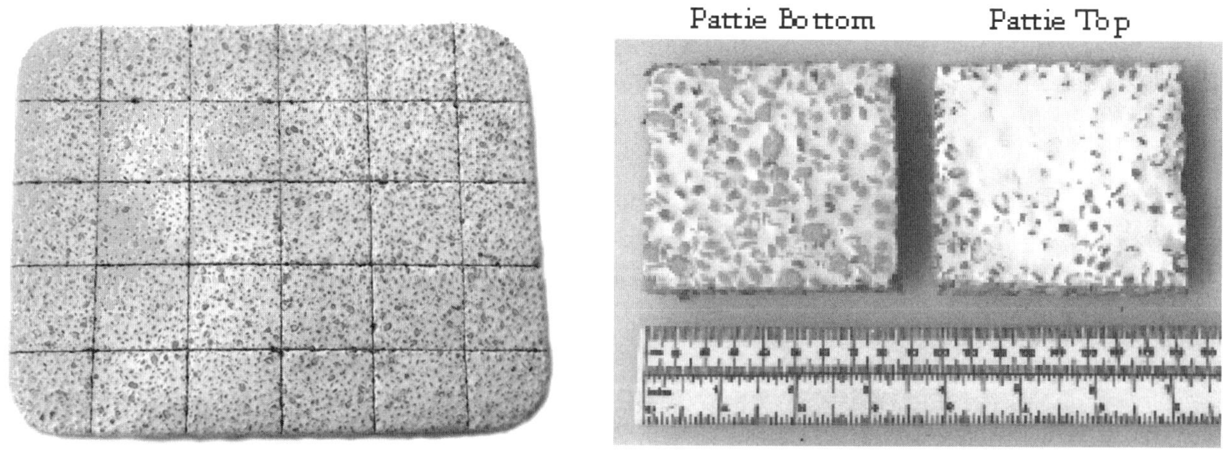

Fig. 3. (Left) Incubated and pasteurized tempeh cake from one tray cut into 26 patties; the four corners will be made into grated tempeh. (Right) The mycelium is fully developed throughout the tempeh because of unrestricted aeration and accurate temperature regulation.

FUTURE PROJECTS

The tempeh we produce (made with organically grown Vinton soybeans and some barley) has an excellent nutritional profile but has 26% calories from fat. This might be improved by creating a low fat soybean. Modification of the *Rhizopus oligosporus* culture (which produces a pure white mycelium) is suggested. Tempeh with a light brown mycelium would be more acceptable to the public since it would look more like the customary meat protein.

REFERENCES

Martinelli, A.F. and C.W. Hesseltine. 1964. Tempeh fermentation: Package and tray fermentations; Food Technol. 18(5):167–171.

Pfaff, G. 1993 and 1994. Apparatus and method for culturing plant materials as foods. U.S Patents 5,228,396 and 5,312,630.

Shurtleff W. and A. Aoyagi. 1980. Tempeh production; The book of tempeh, Vol. II. Center, PO Box 234, Lafayette CA.

Management Methods for Producing Vegetable Amaranth

Bharat P. Singh and Wayne F. Whitehead

Amaranth (*Amaranthus* spp.) is a new crop with ancient history. Members of *Amaranthus* spp. have been grown for centuries for vegetable and grain in different parts of the world (NRC 1984). Amaranth is consumed as vegetable in Africa, Caribbean, China, Greece, India, Italy, Nepal, and South Pacific Islands (Stallknecht and Schulz-Schaeffer 1993). *Amaranthus* species most commonly utilized as vegetable have short plants with wide leaves and small inflorescence (Huang 1980). However, it is common to find same plant type used both for leaf and grain (Saunders and Becker 1984; Tucker 1986).

The awareness in the United States of the potentiality of amaranth, mainly as a grain and to a lesser extent as a vegetable crop, was generated by the Rodale Research Center, Kutztown, Pennsylvania in the mid-1970s. Immigrants from the countries where vegetable amaranth is consumed widely, however, were the real impetus behind creating a demand for this vegetable in the United States. Since then, it has been realized that amaranth can also fill a void for fresh leafy vegetables during the summer months (Makus and Davis 1984; Singh and Whitehead 1993). Amaranth leaves are comparable to spinach (*Spinacia oleracea* L.) in taste (Abbott and Campbell 1982). They are also a good source of dietary fiber and contain high amounts of protein, vitamins, and minerals (Makus and Davis 1984; Teutonico and Knorr 1985; Willis et al. 1984). However, unlike spinach, an ideal season for producing amaranth in the temperate climate is during the hot months of the summer season. In addition, research has shown that green yields from amaranth produced at different locations in the United States are high enough to make commercial exploitation feasible (Berberich 1980; Campbell and Abbott 1982; Makus 1984; Sealy et al. 1990; Singh and Whitehead 1991). In this paper, we present results of cultural studies (planting date, cultivar selection, and N-fertilization) of vegetable amaranth.

METHODOLOGY

Planting Date Study

Field experiments to determine the suitable time period for planting amaranth were carried out during 1992 and 1993. The experiments were conducted on a Dothan sandy loam soil (fine loamy, siliceous, thermic, Plinthic Paleuult). *Amaranthus tricolor* genotype RRC 241 was used for planting. Different planting dates comprised the treatments, which were arranged in a randomized complete block design with four replications. Each plot consisted of four 6.1 m long rows spaced 90 cm apart. Seeds were planted on six dates starting mid-Apr. to mid-Sept. Weeds were controlled mechanically. Plots were irrigated as needed. Crop was harvested 40–44 days from planting. Before harvest, height of five random plants per plot was measured. Two middle rows were used in the yield estimations. Plants were harvested below first leaf nodes and their fresh weight was taken. Dry weight was determined after drying the plant material at 70°C in a forced air oven.

Cultivar Selection Study

Field studies to identify culivars with maximum yield potential were carried out during 1994 and 1995 summer seasons. Soil and other conditions of production for these experiments were similar to the planting date experiments. A total of nine accessions, eight supplied by Rodale Research Center as potential vegetable type and one by the Plant Introduction Station, Ames, Iowa were evaluated. The lay-out of experiments was randomized complete block design with four replications. The plots were 6.1 m long with four rows spaced 90 cm apart. The accessions were planted in mid-June and harvested approximately 40 days after germination. Five randomly selected plants were collected at the time of harvest to measure growth parameters consisting of branch number, leaf number, leaf area, stem and leaf fresh and dry weight. Leaf area was measured with an area meter (LI-COR Model 3100, Lincoln, Nebraska). Height of five random plants per plot was measured. Two middle rows were used in the yield estimations. Fresh weight of the yield rows was taken immediately after harvest. Dry weight was recorded after drying the plant material at 70°C in a forced air oven. Leaf to stem ratios was calculated.

Nitrogen Fertilization Study

Field studies from 1992–1994 were conducted to determine the effect of different rates of N on the vegetative growth of amaranth. The soil and other conditions of production were similar to other experiments. The genotype RRC 241 was used for seeding. The individual plots were 6.1 m long and four rows wide. The experimental design was randomized complete block with four replications. The N levels comprising the treatments were zero (control), 45, 90, and 135 kg/ha N. The N amount assigned to a plot was applied in split application, half at germination and the other half two weeks later. Harvesting and data collections for this study were similar to the cultivar selection study.

RESULTS

Planting Date

Seeds planted in mid-Apr. failed to germinate (Table 1). In all plantings from mid-May onwards, satisfactory amaranth germination was achieved. Mid-June planting produced tallest plants with highest green and

Table 1. Plant height, green yield, and dry matter yield of amaranth at six planting dates[z].

Planting dates	Plant height (cm)	Green yield (Mg/ha)	Dry matter yield (Mg/ha)
Apr. 12	0d[y]	0.0d	0.00e
May 12	18c	1.9bcd	0.33cd
June 15	49a	12.3a	1.24a
July 14	35b	4.1b	0.62b
Aug. 15	30b	3.6bc	0.59bc
Sept. 14	7d	0.7cd	0.15de

[z]1992 and 1993 experiments combined.
[y]Mean separation within column by Duncan's multiple range test, $P = 0.05$.

dry matter yield, while these parameters were lowest for mid-Sept. planted seeds. The range for green yield was 0.70–12.28 Mg/ha and the dry matter yield varied from 0.15–1.24 Mg/ha. The tallest and shortest plants measured 6.9 cm and 49.1 cm, respectively.

Weather data for Apr. to Oct. during 1992 and 1993 indicate that amaranth seeds needed soil temperature about 25°C to germinate and air temperature above 25°C for optimum growth. According to Wagoner (1983), most amaranth species and cultivars germinate when the soil temperature reaches 18°C or above. Webb et al. (1984) found that temperature around 25°C was optimal for germination. Number of growing degree days during the growing season is a major determinant of amaranth plant growth.

Cultivar Selection

Green yield and dry matter yield of the nine plant accessions are presented in Table 2. Out of nine accessions, six were *A. tricolor*, and one each from *A. hybridus, A. cruentus*, and *A. dubius*. *A. tricolor* is the choice vegetable amaranth type in Asia, while *A. hybridus, A. cruentus*, and *A. dubius* are grown as vegetables in Greece, Africa, and Caribbean and South America, respectively.

A. hybridus and *A. cruentus* accessions were taller than other genotypes. *A. dubius* and *A. tricolor* accessions except PI 349553 were of similar height. *A. tricolor* accessions, RRC 389 and RRC 241 had maximum number of leaves and leaf area, respectively. RRC 241 also had the highest leaf fresh and dry weights. *A. hybridus* and *A. cruentus* accessions had the highest stem fresh and dry weights, and green and dry matter yields. RRC 241 produced maximum green yields among *A. tricolor* accessions.

In comparing the green yield among *A. tricolor*, *A. hybridus*, *A. cruentus*, and *A. dubius* species in Mississippi, Igbokwe et al. (1988) also found *A. hybridus* to be highest yielding. Our finding of *A. cruentus* producing higher green yields than *A. tricolor* does not agree with Daloz (1981). Among *A. tricolor* accessions, Kauffman and Gilbert (1981) and Makus and Davis (1984) have also reported RRC 241 to be top performer. *A. tricolor* accessions, PI 349553 and RRC 241 had the highest and *A. cruentus* and *A. hybridus* accessions, RRC 1034 and RRC 843 had the lowest leaf : stem ratios. While both leaf and stem are consumed in some parts of the world, plants with higher leaf : stem ratios are more in demand. Campbell and Abbott (1982) have also reported higher leaf : stem ratio in *A. tricolor* as compared to *A. dubius* and *A. cruentus*.

Table 2. Vegetative growth parameters and yield of nine amaranth accessions[z].

Accession number	Species	Plant height (cm)	Leaf number (no./plant)	Leaf area (cm^2/plant)	Plant weight (g/plant): Stem Fresh	Stem Dry	Leaf Fresh	Leaf Dry	Leaf: stem ratio	Green yield (Mg/ha)	Dry matter yield (Mg/ha)
Hinchoy GL	*A. tricolor*	40b[y]	89bc	1336bc	65b	6.3bc	58bc	10.4b	0.90b	7.3bc	0.91de
RRC 389	*A. tricolor*	48b	133a	1362bc	65b	7.4b	52bcd	9.2bc	0.82b	7.4bc	0.96de
RRC 701	*A. tricolor*	46b	75cd	1069cd	56bc	5.7bc	46cd	9.0bc	0.84b	6.9c	1.00cd
RRC 843	*A. hybridus*	72a	91bc	1351bc	112a	10.2a	67ab	11.7ab	0.61c	11.9a	1.37ab
RRC 1034	*A. cruentu*	67a	105b	1308bc	104a	10.3a	69ab	12.2ab	0.68c	11.1a	1.40a
RRC 1186	*A. dubius*	43b	76cd	1350bc	51bc	4.8bc	56bcd	9.6bc	1.12a	8.4bc	1.10bcd
Hinchoy VL	*A. tricolor*	45b	80cd	1503ab	72b	7.3b	64ab	12.0ab	0.93b	6.9c	0.96de
PI 349553	*A. tricolor*	30c	64d	910d	34c	3.7e	40d	7.2c	1.16a	4.3d	0.69e
RRC 241	*A. tricolor*	43b	93bc	1819a	70b	6.6b	78a	13.8a	1.14a	9.2b	1.25abc

[z]1994 and 1995 experiments combined.
[y]Mean separation within column by Duncan's multiple range test, P = 0.05.

Table 3. Vegetative growth parameters and yield of amaranth at four N-fertilization rates[z].

N-rate (kg N/ha)	Plant height (cm)	Leaf number (no./plant)	Leaf area (cm^2/plant)	Plant weight (g/plant) Stem Fresh	Stem Dry	Leaf Fresh	Leaf Dry	Leaf: stem ratio	Green yield (Mg/ha)	Dry matter yield (Mg/ha)
0	30	57	1013	37	3.2	53	7.6	1.50	5.5	0.83
45	33	71	1484	47	4.5	64	9.9	1.45	8.8	1.21
90	36	71	1766	66	5.4	82	10.2	1.33	12.0	1.57
135	39	82	1705	74	6.3	85	11.1	1.18	12.6	1.53
Significance	L**	L**	Q*	L**	L**	L**	L**	L**	Q**	Q*

[z]1992, 1993, and 1994 experiments combined.
*, ** Significant at P = 0.05 (*) or 0.01 (**) level.

Nitrogen Fertilization

There was a linear increase in plant height from N-fertilization (Table 3). Leaf area increased with N-fertilization until 90 kg/ha. Stem and leaf fresh and dry weights increased linearly with N-fertilization. Quadratic equations provided the best fit for the green and dry matter yield. An R^2 for green yield of 0.70 as compared to 0.51 for the dry matter yield suggested that a higher percentage of the increases in green yield as compared to the dry matter yield could be attributed to N-fertilization probably as a result of an increase in succulence.

The N needed for the growth of a crop will vary depending on the N status of the soil and potential for mineralization. Therefore, optimum N amount reported for maximum amaranth growth by different researchers are substantially different. The reported range varies from 50–200 kg N/ha (Keskar et al. 1981; Subhan 1989; Ramachandra and Thimmaraju 1983). All studies agree that supplemental N is required for optimum amaranth yield.

SUMMARY

The results suggested that for highest yields, amaranth should be seeded in June. *A. hybridus* and *A. cruentus* produced significantly higher green yield than *A. dubius* and *A. tricolor*. RRC 241 yielded the highest among *A. tricolor* accessions. Nitrogen fertilization applied at the rate of 90 kg/ha produced highest vegetable amaranth yields.

REFERENCES

Abbott, J.A. and T.A. Campbell. 1982. Sensory evaluation of vegetable amaranth (*Amaranthus* spp.). HortScience 17:409–410.

Berberich, S. 1980. Amaranth—a hot weather spinach substitute. Agr. Res. 29:15.

Campbell, T.A. and J.A. Abbott. 1982. Field evaluation of vegetable amaranth (*Amaranthus* spp.). HortScience 17:407–409.

Daloz, C.R. 1981. Horticultural aspects of the vegetable amaranths. M.S. thesis, Cornell Univ., Ithaca, NY.

Huang, P.C. 1980. A study of the taxonomy of edible amaranth: an investigation of amaranth both of botanical and horticultural characteristics. p. 142–150. Proc. Second Amaranth Conf. Rodale Press, Emmaus, PA.

Igbokwe, P.E., S.C. Tewari, J.B. Collins, J.B. Tartt, and L.C. Russell. 1988. Amaranth—a potential crop for southern Mississippi. Res. Rep. Mississippi Agr. Forestry Expt. Sta. 13(10):4.

Kauffman, C.S. and L. Gilbert. 1981. Vegetable amaranth summary. Rodale Press, Emmaus, PA.

Keskar, B.G., D.P. Bhore, A.V. Patil, H.N. Sonone, and S.R. Maslekar. 1981. Comparative efficacy of soil and foliar application of nitrogen through urea on yield of leafy vegetable—Chaulai (*Amaranthus blitum* L.) at various seed rates. J. Maharashtra Agr. Univ. 6:68–69.

Makus, D.J. 1984. Evaluation of amaranth as a potential green crop in the mid-south. HortScience 19:881–883.

Makus, D.J. and D.R. Davis. 1984. A mid-summer crop for fresh greens or canning: vegetable amaranth. Ark. Farm Res. 33:10.

National Research Council (NRC). 1984. Amaranth: modern prospects for an ancient crop. National Academy Press, Washington, D.C.

Ramachandra, H.A. and K.R. Thimmaraju. 1983. Effect of different levels of nitrogen and phosphorous on growth components and yield of *Amaranthus* (*Amaranthus gangeticus* L.) cv. 'A-25'. Mysore J. Agr. Sci. 17:158–164.

Saunders, R.M. and R. Becker. 1984. *Amaranthus*: a potential food and feed source. p. 357–396. In: Y. Pomeranz (ed.), Advances in cereal science and technology. Vol. 6. Am. Assoc. Cereal Chem., St. Paul, MN.

Sealy, K.L., E.L. Mcwilliams, J. Novak, F. Fong, and C.M. Kenerly. 1990. Vegetable amaranth: cultivar selection for summer production in the south. p. 396–398. In: J. Janick and J.E. Simon (eds.), Advances in new crops. Timber Press, Portland, OR.

Singh, B.P. and W.F. Whitehead. 1991. Performance of amaranth at six in-row plant densities. HortScience 26:708(Abstr.).

Singh, B.P. and W.F. Whitehead. 1993. Population density and soil pH effects on vegetable amaranth production. p. 562–564. In: J. Janick and J.E. Simon (eds.), New crops. Wiley, New York.

Stallknecht, G.F. and J.R. Schulz-Schaeffer. 1993. Amaranth rediscovered. p. 211–218. In: J. Janick and J.E. Simon (eds.), New crops. Wiley, New York.

Subhan. 1989. Effect of dosage and application time of nitrogen fertilizer on growth and yield of amaranth (*Amaranthus tricolor* L.) (Indonesian, English summary). Penelitian Hortikultura 17:31–40.

Teutonico, R.A. and D. Knorr. 1985. Amaranth: composition, properties and applications of a rediscovered food crop. Food Tech. 39:49–60.

Tucker, J.B. 1986. Amaranth: the once and future crop. BioScience 36:9–13, 59–60.

Wagoner, P. 1983. Summary of the grain amaranth time of planting trials. Report NC-84/30. Rodale Press, Emmaus, PA.

Webb, D.M., J.R. Schaeffer, and C.W. Smith. 1984. Screening of grain amaranth for adaptation to Montana, USA. p. 201–211. Proc. Third Amaranth Conf. Rodale Press, Emmaus, PA.

Willis, R.B.H., A.W.K. Wong, F.M. Scriven, and H. Greenfield. 1984. Nutrient composition of Chinese vegetables. J. Agr. Food Chem. 32:413–416.

Potential Nutritional and Health Benefits of Tree Spinach*

Joseph O. Kuti and Eliseo S. Torres

The tree spinach (*Cnidoscolus chayamansa* McVaughn, Euphorbiaceae), called "chaya" in south Texas, is popular in Mexico and Central America and has been introduced into the United States (mainly South Texas and Florida) for potential uses as a leafy vegetable and/or as a medicinal plant. The plant is an attractive shrub, 3 to 5 m tall (Breckon 1979). The leaves are broad and may consist of 3 or more lobes with fleshy petioles (Fig. 1). The white-colored flowers, which are usually borne on cyme-branched inflorescences, may contain 3-forked arrangements in which the pistillate flowers are located on the basal fork. The staminate flowers are expanded distally from the base of the lobes. Mature seeds and fruit are rare and unknown (McVaugh 1944).

The young shoots and tender leaves of chaya are cooked and eaten like spinach. They comprise part of the staple diet and are the main dietary source of leafy vegetable for the indigenous people of Yucatan peninsula of Mexico and Kekchi people of Alta Verapaz in Guatemala (Harris and Munsell 1950; Booth et al. 1992). There are many underexploited native leafy plants with potential as a traditional food source (NAS 1975). With current renewal of interest in household gardens, attention is being focused on promoting some of these plants as leafy green vegetables among populations in the developing countries (FAO 1987). The edible parts of chaya plant, which taste like spinach when cooked, provide important nutritional sources for protein, vitamins (A and C), minerals (calcium, iron, phosphorus), niacin, riboflavin, and thiamine among populations that cannot afford expensive foods rich in these nutrients (Yang 1979). The plant may also constitute a potentially valuable leafy green vegetable here in the United States and elsewhere.

Chaya traditionally has been recommended for a number of ailments including diabetes, obesity, kidney stones, hemorrhoids, acne, and eye problems (Diaz-Bolio 1975). Chaya shoots and leaves have been taken as a laxative, diuretic, circulation stimulant, to improve digestion, to stimulate lactation, and to harden the fingernails (Rowe 1994). Like most food plants such as lima beans, cassava, and many leafy vegetables, the leaves contain hydrocyanic glycosides, a toxic compound easily destroyed by cooking. Even though some people tend to eat raw chaya leaves, it is unwise to do so.

While the nutritional value of chaya has been demonstrated (Martin and Ruberte 1978; Booth et al. 1992), none of the purported therapeutic values of chaya leaves has been substantiated with scientific experimentation. Therefore, the present study reports on nutritional composition of raw and cooked chaya leaves and the results compared with the nutritional composition of spinach leaves. Also a possible antidiabetic effect of the aqueous leaf extracts or chaya tea, administered through drinking water to streptozotocin-induced diabetic rabbits, was evaluated.

Fig. 1. A potted chaya plant. The young leaves and shoots are edible after being boiled in water.

*Sincere appreciation is extended to Dr. Steven Lukefahr, Associate Professor, Department of Animal and Wildlife Sciences, Texas A&M University-Kingsville for supplying the rabbits and to Dr. Mo Enigbokan, Associate Professor of Pharmacology, College of Pharmacy and Health Sciences, Texas Southern University, Houston, Texas for verifying and giving helpful advice on the animal experimentation.

METHODOLOGY

Nutritional Composition

Young leaves and shoots of *C. chayamansa* were collected from greenhouse-grown plants. Raw and cooked (in microwave oven for 5 min) samples of the leaves and shoot were analyzed for their moisture content, crude fiber, fat, and β-carotene using the AOAC standard methods (1984), for the protein content (N_2 content multiplied by 6.25) using modified semi micro-kjeldahl method of Searle (1974), for mineral contents using an atomic absorption spectrohotometer and for total carbohydrate using gas chromatography. All samples were analyzed in triplicate. Nutritional components and average nutritive value (ANV) of chaya leaves were compared to spinach leaves. The ANV was calculated using the empirical formula proposed by Grubben (1978): ANV/100g = g protein/5 + g fiber + mg Ca^{++}/100 + mg Fe^{++}/2 + mg carotene + mg vit C/40

Possible Antidiabetic Effect

The experimental animals (rabbits) for this study were supplied by Dr. Steven Lukefahr of the Department of Animal and Wildlife Sciences, Texas A&M University–Kingsville. All animals were housed and maintained in compliance with Texas A&M University–Kingsville IACUC policy on animal care and use. The rabbits were fed with standard rabbit chow and given water *ad libitum*. Diabetes was induced by a single subcuteanous injection of 60 mg/kg streptozotocin (STZ), after fasting for 18 h, according to the method described by Bonner-Weir et al. (1981). The rabbits exhibited post-STZ blood glucose levels that were at least double that of the pre-STZ levels one week after diabetes had been induced.

The leaves of *C. chayamansa* were collected from plants grown in the greenhouse. About 10 g of the leaves was extracted with boiling water (1000 mL) for 30 min until the volume of the water had been reduced to 90% of the original. The tea (900 mL) was filtered and used in the subsequent experiments. Two groups of

Table 1. Comparisons of nutritional compositions of leaves of "chaya" (*Cnidoscolus chayamansa* McVaughn) and spinach (*Spinacia oleraceae* L.) per 100 g fresh weight.

Component	chaya	spinach[z]
Water (%)	85.3	90.7
Protein (%)	5.7	3.2
Fat (%)	0.4	0.3
Crude fiber (%)	1.9	0.9
Total CHO (%)	4.2	3.8
Ash (%)	2.2	1.8
Calcium (mg/100g)	199.4	101.3
Phosphorus (mg/100g)	39.0	30.0
Potassium (mg/100g)	217.2	146.5
Iiron (mg/100g)	11.4	5.7
Ascorbic acid (mg/100g)	164.7	48.1
Carotenoids (mg/100g)	0.085	0.014
Average nutritive value[y]	14.94	6.38

[z]Data for spinach were obtained from the USDA (1984).

[y]Average nutritive value according to Grubben emprical formula (1978).

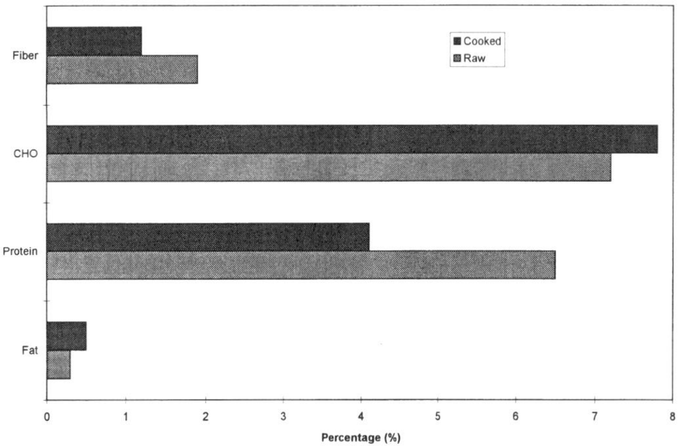

Fig. 2. Proximate fat, protein, carbohydrate and crude fiber compositions of raw and cooked chaya leaves.

8 rabbits each were used. The first group of 8 rabbits were normoglycemic (non-diabetic). Four of the normoglycemic rabbits recieved water (control) only and the remaining 4 received chaya tea treatment only. The second group of 8 rabbits were hyperglycemic (diabetic). Four of the diabetic rabbits received water only and the remaining 4 received chaya tea only.

Before administering the tea or water (control), blood samples were obtained from the ears of 18 h fasted nondiabetic and diabetic rabbits using a capillary tube. Then the tea or water was administered orally through drinking water bottles *ad libitum*. Blood sampling was repeated at hourly intervals for 6 h after the oral administration. Blood glucose was determined using a blood glucometer (Miles Inc., Diagnostic Division, Elkhart, IN, U.S.). The mean blood glucose values ±SE were determined and the significance of the difference between the means of treated and control groups was established by Student's *t*-test.

RESULTS

Nutritional Composition

The nutritional analysis of chaya (*C. chayamansa*) leaves and spinach (*Spinacia oleracea* L.) are presented in Table 1 for comparison. Chaya leaves were found to contain substantially greater amounts of nutrients than the spinach leaves. The chaya leaf is especially high in protein (5.7%), crude fiber (1.9%), calcium (199.4 mg/100 g), potassium (217.2 mg/100 g), iron (11.4 mg/100 g), vitamin C (164.7 mg/100 g), and carotene (0.085 mg/100 g). The levels of chaya leaf nutrients, in this study, agree with published reports (Martin and Ruberte 1978; Munsell et al. 1949; Booth et al. 1992) and are two to threefold greater than most edible leafy green vegetables. In terms of the average nutritive value, chaya leaves [14.9] is by far superior to other leafy green vegetables such as spinach [6.4], amaranth [11.3], Chinese cabbage [7.0], and lettuce [5.4] (Grubben 1978). While some edible leafy green vegetables are usually good sources of mineral macronutrients (Levander 1990), chaya leaf furnishes appreciable quantities of several of the essential mineral macronutrients necessary for human health maintenance. For example, potassium has been shown to be an important mineral nutrient in the control of hypertension and in the reduction of risks of stroke (NRC 1989), calcium is important for ossification and iron is necessary for normal hematopoiesis (Hodges et al. 1978). Brise and Hallberg (1962) reported that vegetables, such as chaya, with high vitamin C content may enhance absorption of nonheme iron.

Analysis of raw and cooked samples of chaya leaves revealed that cooking may increase the relative composition of carbohydrate and fat and decrease relative composition of crude fiber and protein (Fig. 2). On the other hand, cooked samples of chaya leaves were considerably higher in calcium, phosphorus and iron while the potassium content was relatively lower than in the raw samples (Fig. 3). The increase in some of the mineral nutrients may be due to the cooking process, which allows extraction of the nutrients from the tissues, therefore increasing the percentage of mineral elements while decreasing moisture content (Booth et al. 1992).

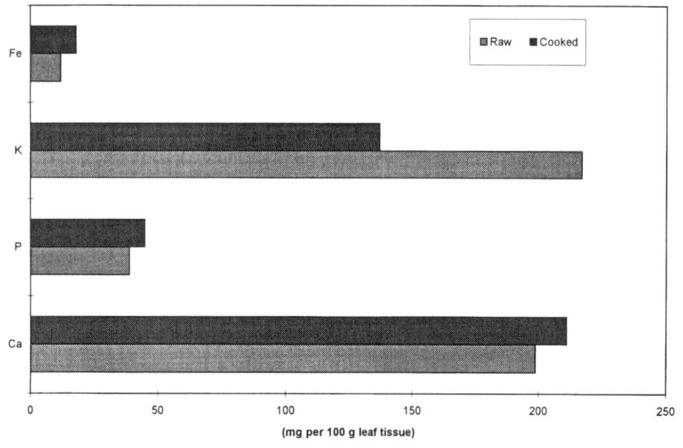

Fig. 3. Proximate mineral macronutrient (calcium, phosphorus, potassium and iron) compositions of raw and cooked chaya leaves.

Possible Antidiabetic Effect

Following the oral administration of chaya tea, the blood glucose levels of the diabetic rabbits were gradually lowered from a high of 118 (baseline at 0.0 h) to 87 six hours after administration. The blood glucose level of 87 is similar to blood glucose levels of normoglycemic rabbits on drinking water (Table 2). The blood glucose levels of non-diabetic control rabbits that were given chaya tea showed a slight increase (i.e. hyperglycemia) above the baseline 85 at 1 to 2 h after administration, but rapidly stabilized thereafter (Table 2). The reason for this transient hyperglycemia is unknown and needs to be investigated. The results obtained in this study suggest that in STZ-induced diabetic rabbits, aqueous leaf extracts of *C. chayamansa* may be effective for treatment of non-insulin dependent diabete mellitus (NIDDM) symptomatology. This is a first report on hypoglycemic effect of chaya plants. The present report is preliminary in nature and additional studies will be needed to properly characterize the antidiabetic potential of chaya in diabetic animals. Also further studies will be necessary to determine the effective dosage, mechanism of the hypoglycemic activity and the active hypoglycemic principle present in the leaves of *C. chayamansa*.

CONCLUSION

The potential of *C. chayamansa* for human food and health has a significant implication for the plant as a horticultural crop. Although demand for chaya, as a medicinal plant, has recently increased among the Hispanic population in the United States, the plant has the potential to make a significant nuritional contribution to the vegetable diet as well, because of its high nutrient content. The development of chaya as a new horticultural crop would transcend the ethnic popularity and create a worldwide market for the plant and its products, whether as a leafy green vegetable and/or as a therapeutic herbal tea.

It is noteworthy that the chaya plant is drought resistant, which is of a particular value in areas with short seasonal rainfall and shortage of green vegetables (Peregrine 1983). Growth of the plant is rapid and edible leaves and shoots could be produced within a short period (8 to 10 weeks). Propagation by cutting is easy and the woody stem sections readily root. Few pests and diseases are known to be of any significance in the cultivation of chaya plants. One disadvantage is the presence of toxic hydrocyanic glucosides in the leaves. However, cooking, which is essential, inactivates the toxic compound. Other *Cnidoscolus* (chaya) species are being examined in our laboratory at Texas A&M University–Kingsville to genetically select species with high leaf and shoot biomass yield and lower hydrocyanic glycoside content. Additionally, we are conducting re-

Table 2. Effect of *Cnidoscolus chayamansa* leaf extract ("chaya" tea) on blood glucose levels of non-diabetic and streptozoctocin-induced diabetic rabbits.

	Blood glucose level (mg/dL)[z]			
	Non-diabetic		Diabetic	
Time (h)	water	"chaya"	water	"chaya"
0.0	87±3.1	85±2.5	112±8.3	118±13.2
1.0	86±2.7	91±3.9	138±4.6	114±7.3
2.0	87±2.6	99±4.3	143±6.4	103±8.7
3.0	87±3.1	82±1.6	139±8.0	96±9.3
4.0	88±3.0	85±2.1	153±6.3	92±5.8
5.0	87±4.7	84±4.2	158±7.4	89±3.6
6.0	87±3.1	82±2.7	162±9.0	87±2.7

[z]Mean±SE.

search on genetic improvement, propagation, field production, potential for processing and marketing of chaya and its products in south Texas.

REFERENCES

AOAC. 1984. Official methods of analysis, 14th ed. Assoc. Official Anal. Chem., Arlington, VA.

Booth, S., R. Bressani, and T. Johns. 1992. Nutrient content of selected indigenous leafy vegetable consumed by Kekchi people of Alta Verapaz, Guatamela. J. Food Compos. Anal. 5:25–34.

Bonner-Weir, S., D.F. Trent, R.N. Honey, and G.C. Weir. 1981. Responses of neonatal rat islets to streptozotocin-limited ß-cell regeneration and hyperglycemia. Diabetes 30:64–69.

Breckon, G.J. 1979. Studies in *Cnidoscolus* (Euphorbiaceae). Brittonia 31:125–148.

Brise, H. and L. Hallberg. 1962. Effect of ascorbic acid on iron absorption. Acta Med. Scand. Suppl. 171:51–58.

Diaz-Bolio, J. 1975. Chaya (*Cnidoscolus chayamansa*, Euphorbiaceae), a marvellous food (in spanish). Tierra 30:407–408, 427–428.

FAO. 1987. Promoting under-exploited food plants in Africa.: A brief for policy markers. Food and Agriculture Organization, Food Policy & Nutrition Div., Rome.

Grubben, G.J.H. 1978. Tropical vegetables and their genetic resources. Int. Board Plant Genetic Resource, FAO-UN, Rome Italy.

Harris, R.S. and H.E. Munsell. 1950. Edible plants of Central America. J. Home Econ. 42:629–631.

Hodges, R.E., H.E. Sauberlich, J.E. Canham, D.L. Wallace, R.B. Rucker, L.A. Mejia, and M. Mohanram. 1978. Hematopoietic studies in vitamin A deficiency. Am. J. Clin. Nutr. 31:876–885.

Levander, O.A. 1990. Fruit and vegetable contribution to dietary mineral intake in human health and disease. HortScience 25:1486–1488.

Martin, F.W., and R. Ruberte. 1978. Chaya, *Cnidoscolus chayamansa* includes composition and nutritional value, culture in Puerto Rico. In: Vegetables of hot humid tropics. USDA, ARS. New Orleans, LA.

McVaugh, R. 1944. The genus *Cnidoscolus*: generic limits and intrageneric groups. Bul. Torrey Bot. Club 71:457–474.

Munsell, H.E., L.O. Williams, L.P. Guild, C.B. Troescher, G. Nightingale, and R.S. Harris. 1949. Composition of food plants of Central America. Food. Res. 14:144–164.

NAS. 1975. Chaya. p. 45–48. In: Underexploited tropical plants with promising economic value. National Academy of Science, Washington, DC.

National Research Council. 1989. Diet and health. National Academy Press, Washington, DC.

Peregrine, W.T.H. 1983. Chaya (*Cnidoscolus aconitifolius*): A potential new vegetable crop for Brunei. Tropical Pest Manag. 29:39–41.

Rowe, L. 1994. Plant guards secret of good health. Valley Morning Star. Sept. 4, p. A1, A12.

Searle, P.L. 1974. Automated colorimetric determination of ammonium in soil extracts with "Technicon Autoanalyzer II" equipment. New Zealand J. Agr. Res. 18:183–187.

USDA. 1984. Agricultural handbook. p. 8–11. USDA, Washington, DC.

Yang, Y.H. 1979. Tropical home gardens as a nutritional intervention. p. 417–436. In: G.E. Inglett and G. Charalambous (eds.), Tropical food chemistry and nutrition, Academic Press, New York.

Precooling: Key Factor for Assuring Quality in New Fresh Market Vegetable Crops*

Glenn H. Sullivan, Lonni R. Davenport, and James W. Julian

Total domestic and export sales of fresh fruits and vegetables produced in the United States reached $65 billion in 1994–95. Marketplace demands for greater selection and consistent high quality, combined with consumers' rising nutritional expectations, have placed increasingly greater performance pressures on all segments the fresh produce marketing chain (How 1991). Producer success in the marketplace for fresh fruits and vegetables is highly correlated with maintaining postharvest quality and shelf-life consistent with current consumer demands (Sullivan et al. 1991). Precooling technologies that quickly remove the field heat and reduce internal respiratory processes were found to be critical in meeting these consumer demands for high quality fresh produce (USDA/ARS 1986). Forced-air and hydrocooling have been found to be the most efficient and economical in enhancing product quality for new muskmelon cultivars and sucrose enhanced bi-color sweet corn. Shelf-life was extended 7 to 14 days, and desired quality characteristics were maintained at near optimum levels.

VALUE-ADDED MARKETING CHAIN

In the past, the principal market for most fresh produce has been the traditional family. Today this traditional family unit accounts for only about 13% of all households in the United States. Individuals living alone now comprise almost 24% of all households, and single parent families account for nearly 22% of all households with children under 18 years of age. Today 40% of all shopping is done by men! Point of purchase selections in the fresh produce departments of retail grocery outlets increased from an average of 67 in 1985 to over 312 in 1995 (Hammel 1995). These trends signal major changes in the way consumers spend their incomes, and how they make their purchases.

Retail supermarkets are still the principal point of purchase for most fresh produce consumers, accounting for over two-thirds of all domestic sales (Fig. 1). But how that produce gets to the point of retail purchase has changed dramatically over the last few decades. Currently, almost 40% of all fresh produce sold through retail supermarkets comes directly from first-handlers in the major producing regions (grower-shippers, independent shippers, and value-added processors) through the retailers' own integrated wholesale distribution facilities (VanSickle 1985). Historically, central terminal markets in the major cities throughout the United States were the dominant assembly points for fresh produce. Today, only about 25% of the volume sold at retail comes through market wholesale institutions who buy fresh produce both from the terminal markets and shipping point markets (Hammel 1995). This shift in the way retail supermarkets source their fresh produce has resulted from the need to achieve greater control over the postharvest handling functions, and thereby better meet the changing consumer demands for higher quality and longer shelf-life at a competitive price. These trends place increasing pressure on the marketing chain to adopt postharvest technologies that help assure greater quality control (Sullivan et al. 1991).

PRECOOLING TECHNOLOGIES

When it comes to produce quality, every minute counts. Research confirms that lowering the respiration rate of fresh vegetables is essential to preserving market quality. The most important technology for lowering respiration rates remains proper precooling of produce within hours of harvest (Jones 1996). Proper precooling preserves product quality by: (1) inhibiting the growth of decay producing microorganisms, (2) restricting enzymatic and respiratory activity, (3) inhibiting water loss, and (4) reducing ethylene production (Hardenburg et al. 1986).

---

*Journal Paper No. 15010 Purdue Univ. Agr. Expt. St., West Lafayette, IN 47907-1165. This research was supported in part by grants from the Indiana Business Modernization and Technology Corporation, Indianapolis, Indiana.

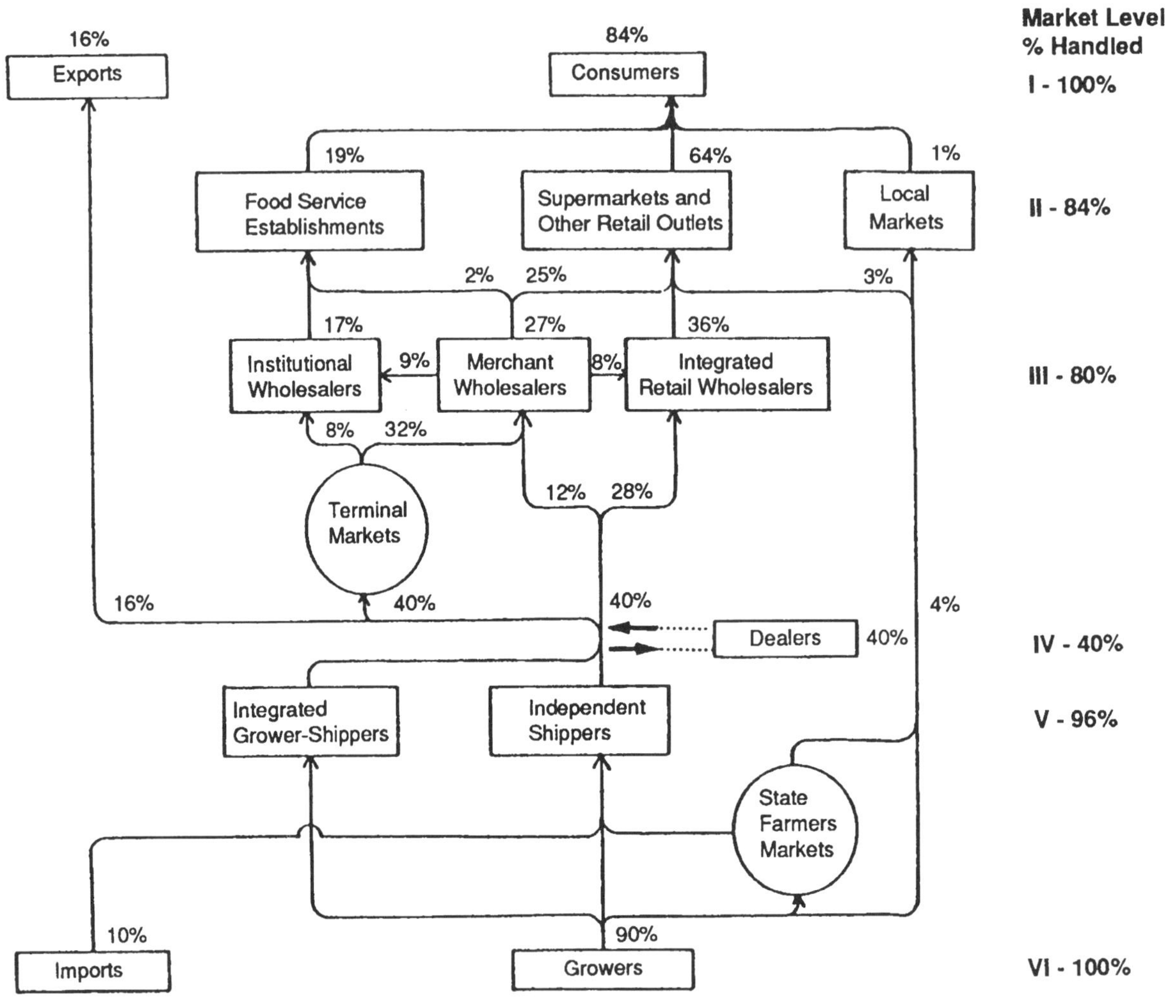

Fig. 1. United States fresh produce marketing system.

The most efficient methods of postharvest cooling fresh produce were found to include forced-air cooling, hydrocooling, and vacuum cooling (Table 1) (Junge et al. 1986). In new muskmelon cultivars and sucrose enhanced bi-color sweet corn, forced-air cooling and hydrocooling respectively proved to be the most effective and economical, preserving optimum quality and increasing market life 7 to 14 days depending on the product. Muskmelon cultivars were preserved at the highest levels of market quality when the internal pulp temperatures (average 95°F/35°C) were reduced to 42°F/5.5°C within 4 h of harvest using forced-air cooling technologies. Sweet corn quality was preserved at optimum levels when harvested at field temperatures below 75°F/24°C and cooled to 32°F/0°C pulp temperatures using hydrocooling technologies (Sullivan and Davenport 1995).

COMMERCIAL MARKET SUCCESS

Research that focused on the market factors influencing commercial success clearly indicated that vegetable producers must increasingly commit themselves to meeting the consumers' demand for higher produce quality. Retail buyers in today's competitive marketplace are increasing their purchases through direct-from-shipping point deliveries, but only from grower-shippers who provide the value-added services demanded by increasingly more sophisticated consumers (Sullivan and Davenport 1991). Product precooling was found to rank as the most essential of these value-added marketing services.

Table 1. Postharvest cooling methods and suitable commodities.

Cooling method	Commodities	Comments
Hydrocooling	Most leafy vegetables, fruits and fruit-type vegetables, sweet corn, snap beans	Very fast cooling; uniform cooling in bulk if properly used, but may vary extensively in packed shipping containers; daily cleaning and sanitation measures essential; product must tolerate wetting; need water-tolerant shipping containers
Forced-air cooling (pressure cooling)	Most fruits, berries, fruit-type vegetables, tubers, and vegetables not susceptible to chilling injury	Much faster than room cooling; cooling rates very uniform if properly used. Container venting and stacking requirements are critical to effective cooling. Economical and efficient.
Package-icing	Most vegetables	Fast cooling; limited to commodities that can tolerate water-ice contact; water-tolerant shipping containers are essential. Economical and efficient.
Room cooling	All commodities	Too slow for many perishable commodities. Cooling rates vary extensively within loads, pallets, and containers.
Vacuum cooling	Leafy vegetables, iceberg lettuce	Commodities must have a favorable surface-to-mass ratio for effective cooling. Causes about 1% weight loss for each 6°C cooled. A procedure that adds water during cooling prevents this weight loss, but equipment is more expensive, and water-tolerant shipping containers are needed.
Transit cooling		
Mechanical refrigeration	All commodities	Cooling in most available equipment is too slow and variable; generally not effective for field heat removal.
Top-icing and channel-icing	Most vegetables	Slow and irregular, top-ice weight reduced net pay load; water-tolerant shipping containers needed.

Given current consumer lifestyles and demand for variety selection in the produce section of retail supermarkets, new vegetable cultivars represent a unique opportunity for grower entry into the marketplace. However, research findings clearly confirm that maintaining an economically viable market position at retail requires new crop producers to focus increasingly greater attention on the product quality demands of an increasingly more dietary diverse consumer population. Precooling is among the most cost-effective and efficient quality enhancements available to commercial new crop producers.

REFERENCES

Hammel, F. 1995. Produce. Supermarket Business 50(9):87+.

Hardenburg, R.E., et al. 1986. The commercial storage of fruits, vegetables and florist and nursery stocks. USDA/ARS, Agr. Handbook 66.
How, R.B. 1991. Marketing fresh fruits and vegetables. VanNostrand Reinhold Publ., New York.
Jones, S. 1996. When time is of the essence. Progressive Grocer 75(2):105.
Junge, K., M. Weimar, D. Blanton, M. Hayenga, and R. Gladon. 1986. Precooling methods for commercial vegetable producers, Dept. of Econ. & Hort., Iowa State Univ., Paper No. 160.
Sullivan, G.H., et al. 1991. Economic and market feasibility for fruit and vegetable industry expansion in Southwest Indiana. USDA/VACDC Study Report.
Sullivan, G.H. and L.R. Davenport. 1991. Fresh vegetable marketing opportunities in Indiana. Purdue Univ. AES, SB-624.
Sullivan, G.H. and L.R. Davenport. 1995. Sweet corn: Marketing and postharvest handling. Purdue Univ. Bul. 707.
VanSickle, J.J. 1985. Marketing—a requirement for profit with fresh produce. Univ. of FL, Food & Res. Econ., Staff paper 273.
USDA, Agricultural Research Service. 1986. The commercial storage of fruits, vegetables, and florist and nursery stock. Agr. Handb. No. 66.

FLORAL AND LANDSCAPE CROPS

New Floral Crops in the United States*

Mark S. Roh and Roger H. Lawson

New crops research and development in the United States has been influenced by floriculture in Europe during the past 10 to 20 years. In the early 1980s several native American plants were imported as new crops to the United States from Europe. For example, *Liatris spicata* Willard, a native American bulbous plant is grown as cut flower in Europe and is being exported to the U.S. Seeds of all commercially important cultivars of *Eustoma grandiflorum* (Raf.) Shinn., a native American herbaceous plant have been imported from Japan after more than 60 years of improvement by Japanese breeders.

The success of new crops programs in Denmark and the Netherlands can be attributed to development of market orientated production systems for export. European producers have developed technology for a year-round production of a single or several crops in highly mechanized greenhouses or outdoors where environments are suitable for certain crops. In contrast, U.S. growers often produce many different crops in a small quantity in a same greenhouse to supply local flower markets.

Development of a new crop includes a combination of superior plant material, production technology, and marketing strategy. If any of these three factors is not properly developed and fully implemented, the chance of success is greatly diminished. The purpose of this presentation is to summarize our observations and conclusions on the process of new crops development after 10 years of research in a cooperative program that included the American Floral Endowment (AFE), the Society of American Florists (SAF) and the U.S. Department of Agriculture (USDA).

THE CHANGING U.S. FLORICULTURE INDUSTRY

The floral industry continues to make a significant contribution to the American economy, despite a decrease in the number of growing establishments. The wholesale value of floriculture crops in the U.S. was estimated to be $3.23 billion in 1994, based on the survey from 36 states of 10,137 growers with gross sales over $10,000 in 1994, up from $1.27 in 1984 (USDA Floriculture Crops 1984, 1994). The commodity sales and percent market shares (%) between 1981 and 1994 (Table 1) indicate that bedding and garden plants sales represent the fastest growing segment of the industry. Bedding and garden plants, that include impatiens, cutting geraniums, grandiflora petunias, pansies, and marigolds, continue to provide the greatest value to operations over $100,000. Of the 4,570 operations, 3,181 produced some type of bedding and garden items, the value totaling $1.28 billion, 9% above the $1.17 billion value in 1993. Potted geraniums from cuttings and seeds returned the highest wholesale value, at $129 million. Of plants sold in flats, bedding impatiens returned the highest wholesale value, at a total of $81.9 million. The average price sold at wholesale was $6.15 per flat. The interest on perennials has increased recently. The top-selling genera of perennials are *Hosta*, *Hemerocallis*, *Coreopsis*, *Astilbe*, grasses, *Phlox*, and *Chrysanthemum*.

Potted flowering plants, that include African violets, chrysanthemums, florist azaleas, Easter lily, and poinsettias were produced by 2,756 operations and were the second largest commodity group, providing 22% of the total wholesale value of crops valued at $654 million. Poinsettias continue to contribute the most value to the group at a total of $206 million, 3% above the prior year.

Cut flowers were produced by 20% of the growers surveyed while market share of fresh cut flowers decreased from 24.4% in 1981 to 17.7% in 1991 (Table 1). According to the national gardening survey (National Gardening Association 1994), container and herb gardening has increased 10% and 15%, respectively, annually from 1989 to 1993. For flower gardening, the increase was almost 7% annually during the same period. Landscaping, tree service, and fruit tree gardening was up 10%, 19%, and 22% annually, respectively, from 1989 to 1993.

---

*We thank the American Floral Endowment (AFE), the Society of American Florist (SAF), and the Fred C. Gloeckner Foundation, and the AFE/SAF/USDA New Crops Committee for their cooperation and the financial support.

Domestic producers of major cut flower commodities have been greatly affected by imports of cut flowers. Favorable climatic conditions, cost of labor and efficient use of transportation systems have increased offshore flower imports to U.S. markets. In 1971, domestically produced pompon chrysanthemum and rose accounted for 95% and 100% market share, respectively. In 1991, however, domestic production of these cut flowers was only 18% and 52%, respectively (Table 2).

Over one-half of American households (57%) purchased a floral product over the period of July 1992 to 1993 (American Floral Endowment Research Report 1993). Each buying household spent an average annual total of $43.69, making 3.7 purchases averaging $11.85 each. The largest dollar volume outlets include supermarkets at 14% of sales, garden centers at 18% of sales, and most particularly florists, accounting for the market share level of 34% of all flowers purchased.

The largest share of consumer dollars, 38% of total sales, was spent on non-calendar occasions. The largest was contributed by sympathy/funeral occasions (8%), by birthday (7%), and home decoration (7%). The calendar events represented on 18% of sales, the largest of which Christmas was responsible for 6% of the sales.

Cut flowers held the largest share of the consumer dollar, accounting for 41% of spending over the year. Cut flowers were most likely to be mixed flowers at 60% of sales. Outdoor bedding and garden plants accounted for another third, flowering house plants represented about 15%, and green plants finished up with an average of about 10%. Poinsettias were the largest selling fresh flowering house plant accounting for 54% of sales in the Christmas season, but equaling out to 21% of all yearly sales. The remainder that includes chry-

Table 1. U.S. floriculture commodity sales and percent market shares.

	Market share (%)					
Year	Cut flowers	Potted flowering	Foliage plants	Bedding plants	Cut foliage	Total sales ($billion)
1981	24.4	17.8	32.4	25.3	NA	1.02
1985	22.0	17.3	27.8	28.9	4.0	1.69
1988	19.8	22.1	21.0	32.9	4.0	2.29
1990	18.6	25.2	18.9	33.0	4.2	2.51
1994	15	22	16	43	4	3.23

Table 2. Domestic and imported pompon chrysanthemums and roses, 1971–1991. USDA Floriculture Crops Report.

	Domestic (% of supply)		Imports (% of supply)		Total supply (1,000)	
Year	Pompon mum	Roses	Pompon mum	Roses	Pompon mum (bunch)	Roses (stems)
1971	94.4	99.8	5.6	0.2	36,509	429,853
1975	73.9	99.0	26.1	1.0	48,340	422,692
1980	47.6	90.6	52.4	9.4	73,133	472,275
1986	30.1	68.1	69.9	31.9	108,633	679,139
1991	18.3	52.0	81.7	48.8	112,742	1,048,417

santhemums, lilies, azaleas, and African violets all came in at less than 10%. "Other" fresh flowering house plants represented 54% of consumer spending from July 1992 to June 1993.

NEW CROPS PROGRAM, FLORAL AND NURSERY PLANTS RESEARCH UNIT

In 1982 the Florist and Nursery Crops Laboratory initiated a program in new crop development to introduce new flowering pot plants to U.S. consumers (Roh and Lawson 1987a). New crops research in the USDA has involved scientists, growers, and the development of marketing channels for newly developed genera or species. Additional new crops programs have been conducted by growers and university researchers working independently or in cooperation.

A new crop in our program was defined as "a plant introduced into commercial culture with a knowledge of production technology and market potential" (Roh and Lawson 1987a). In this regard, any crop can be regarded as a new crop, if there is limited information on culture on controlled flowering (*Aeschynanthus*), if a crop which is very popular in one country, but not known in another country (*Eustoma*), or forgotten, but reintroduced (*Shortia galacifolia* Torr. & Gray). New crops may not effectively compete with well established crops, even when new plants are introduced to the market with complete cultural information. Successful commercialization of a new crop depends not only on the interest of the growers, but results of trial marketing and consumer education. In the final analysis, however, profitability based on the turnover period of a crop in relation to the cost of production and market evaluation will be the basis for a successful introduction. A decision to finally produce and market a new product is the responsibility of the individual grower.

Since the assessment of the market potential of a crop when introduced to commercial culture is time consuming and very difficult, and the outcome is unpredictable, the definition of a new crop could now be stated as "a plant successfully grown commercially based on a knowledge of production technology and available to a consumer with a proven market potential." The term "new" does not necessarily mean that a newly released cultivar is new simply because, for example, flowering time is earlier than the other previously introduced cultivars.

The first step in our program was the acquisition of plant material from both foreign and domestic sources. The plants were initially evaluated to determine general characteristics of propagation, vegetative growth and flowering as influenced by photoperiod and temperature (Roh and Lawson 1992). This production information was provided to grower cooperators and published in trade journal articles. The importance of wide diversity when collecting plant material should be stressed at the beginning of the acquisition. For example, when *Columnea* hybrids were obtained from Austria through the Netherlands, all hybrids required 6 to 8 weeks of 10°C to induce flowering. However, *Columnea* hybrids, 'Bold Adventure' and *C. erythrophae* did not require 10°C for flowering, thus the total production time was reduced by 6 or more weeks as compared to hybrids from Austria.

During the past ten years, several plants have been introduced to the U.S. market through the cooperative program between the USDA and industry. Some of the results of this program including production technology of several new crops is presented.

NEW CROPS ACQUISITION

When we started collecting plants in 1982, we were primarily interested in flowering pot plants that were commercially produced in Europe, but were little known in the U.S. There were two reasons for this decision. First was that flowering pot plants represented the fastest growing segment of the floral business with increased demand for a wider assortment of new products. The second reason was that the production of cut flowers consumed in the U.S. was continuing to move off-shore. U.S. producers were searching for new crops that could replace more traditional cut flower crops.

Europe (Denmark, Netherlands, and Austria)

In the Netherlands more than 400 different kinds of pot plants are sold. These include foliage plants (*Yucca*) and flowering pot plants (*Hebe*) which are not sold widely here in the U.S.

Several *Coprosma* species and hybrids ('Beatson's Gold', 'Kirkii Variegata', and 'Coppershine') first caught our attention in Denmark. Stem tip cuttings rooted very easily and rooted cuttings grew very rapidly. Production experiments on nutritional evaluations revealed that leaves would persist for a longer period of time at the home environments without acclimatization period before placed in home, if plants were fertilized with elevated levels of nitrogen.

There was a limited attempt made to introduce *Coprosma* and introduction was not generally successful. Because it is a foliage, and not a flowering plant, it was not considered sufficiently unique by producers to warrant a major marketing effort through floral outlets. Further, the introduction was not properly timed in early 1980. Now that the cut and pot foliage sales have increased recently ($487 million and 16.8% increase from 1993), and market trends require a small pot size (15 to 25 cm) with value added to increase the profit (Bateman 1995), *Coprosma* could be reintroduced. Perhaps a more successful marketing strategy would have been to sell this plant through garden centers for landscape uses.

Cupressus, 'Goldcrest' is another plant that we collected. A true conifer, the plant is widely grown in Europe as an indoor green plant. 'Goldcrest' has lemon-scented light green foliage. There are more than 2 million plants auctioned in the Netherlands each year. Following introduction of *Cupressus* into the U.S. market, the plant failed to reach its potential. This was partially due to a poor rooting results of cooperating growers and difficulties in growing quality plants outdoors. Plants grow rapidly after propagation. However, rooting of *Cupressus* stem tip cuttings is less than 35% in conventional mist propagation. Rooting of cuttings is significantly increased to more than 85% by controlling watering/misting cycle. Dipping the cuttings in a 2,000 ppm water soluble indole butyric acid (Potassium form of IBA, IBA-K) for a few seconds enhances rooting. Because the cuttings form only one or two roots, they are propagated in Oasis wedge foam cubes to protect tender roots which could be easily damaged during transplanting. For the first 15 to 30 days, cuttings are placed under an intermittent mist, 6 sec of mist every 12 to 15 min over 12 to 16 h. For the next 30 days, misting frequency is decreased, misting every 30 min. Cuttings are finally placed in a greenhouse and watered twice a day. Because the potted plant requires a lot of water proper watering to prevent drying out of the growing medium was essential for successful commercial production.

Aeschynanthus also received our attention in 1984 in Denmark. In the past, *A. specious* Hook., was marketed as green foliage plant due to a fact that controlled flowering was very difficult. Although flowering can be manipulated by controlling temperatures, control is too imprecise to enable growers to force this plant year-round. Although mature plants respond to temperature, flowering from newly rooted cuttings is unpredictable. Cuttings from stock plants grown at 18°/17°C, day/night, flowered earlier as compared to plants grown at higher temperatures. (Whitton et al. 1990).

Achimenes longiflora, D.C., and *Columnea* hybrids were also evaluated as potted plants for forcing during winter. Due to a dormancy that lasts more than 3 months, rhizomes of certain gesneriads, such as *Achimenantha* could not be used. Therefore, tissue culture propagation was employed to maintain stock plants of *Achimenantha*, 'Inferno' *in vitro*. Flower buds were the best source for forming 20 to 33 adventitious shoots per explant that are generally weak. However, from each rhizome scale, three strong shoots were formed in 8 weeks and flowered in less than 70 days, when explants were taken in February. Shoots were transplanted to soil in May. Although stock plants can be maintained in vitro, flower petals are easily damaged, which created difficulties in shipping. Therefore, *Achimenes* and *Achimenantha* was not officially released to growers.

Japan

In the 1980s, Japanese seed companies introduced *Gerbera jamesonii* Hook., as a pot plant. The first of these introductions was the HappiPot from the Sakata Seed Cooperation (Yokohama, Japan). The most outstanding selections of the dwarf forms of seed produced gerbera were tissue cultured by U.S. companies in the late 1980s. Efforts were made in our laboratory to develop procedures for uniform flowering with an increased flowering percentage (Roh and Lawson 1984). *Gerbera* has become an important pot plant in some markets, but the plant did not reach the popularity that was predicted. Flowering of some hybrids from seeds or tissue culture selections was erratic and even 20% to 40% of plants did not flower, particularly during the summer (Roh and Lawson 1984). Gerbera required 8 to 12 h of photoperiod at 21°/15° to 18°C for 40 days to promote flowering. Although new dwarf bicolored forms and other hybrids with more prolific flower production at the first flowering are continuously introduced to the market, the popularity of this pot plant was not dramatically increased.

Another plant that came to our attention in mid 1984 was *Eustoma*. All *Eustoma* cultivars were originally bred for cut flowers in Japan (Roh et al. 1990). In the early 1980s, there were no dwarf genotypes and pot plants were produced from long stemmed cut flower cultivars with the use of growth regulators. We initiated a project on the selection of superior forms from tissue culture. This research resulted in the introduction of the first genetic dwarf *Eustoma* selected as a somaclonal variant (Griesbach et al. 1988). Cultural information of this first dwarf *Eustoma* was made available to commercial growers through trade magazine articles (Lawson 1990; Roh 1988).

Dwarf *Eustoma*, 'Little Belle Blue' was successfully tested and was selected as a winner of the Florastar competition in 1990. However, no official announcement was made since there was not a sufficient number of seeds (a minimum of 500,000 seeds) available for distribution at the time of the planned introduction. Since dwarf cultivars later became available from the commercial companies, there was not sufficient interest by companies to produce 'Little Belle Blue' seeds. New cultivars developed from publically funded research thus has a limitation, when product is not developed commercially.

Dwarf *Clematis* hybrids is an interesting product that we have obtained from Japan. The Sakata Seed Inc., has hybridized genetically dwarf forms. The plants can be grown in winter and sold through floral outlets when they are forced into flower in spring time. Flowers last in the home environment about 10 days. After flowering, the plant can be grown in the garden.

The major disadvantage of dwarf *Clematis* is the slow production time of 2 to 3 years. Because of their genetic dwarf characteristic, stock plants produce a limited number of cuttings. Also, cuttings root slowly. Plants large enough to market as 10 cm pots (one plant) or 15 cm pots (2 plants) will require 2 years. In some markets the plants can be sold at a premium price. Pot *Clematis* will predictably be a successful crop based on a test marketing trial we conducted.

Some of the plants from Japan that we evaluated generally require a minimum of 7 months (*Eustoma*) to 2 to 3 years (*Gentiana* F_1 hybrid, 'Iwate Otome' and *Clematis*) to produce a salable plant. Although these crops are popular in Japan, they will probably be successfully marketed in the U.S. only by specialty growers because of the long cropping time. Several hybrids of dwarf gentian, *Gentiana triflora* var. *japonica* Hara are easily propagated by stem cutting and flower around July/Aug. in Beltsville. However, after flowering, dried flowers must be removed and this could be a limiting factor to promote *Gentiana* as a potted plant. Since the popularity of perennials is increasing now, *Gentiana* and *Clematis* could be reevaluated and introduced as a flowering potted plant for early spring season followed by planting in the garden to flower in subsequent years.

Thailand

Curcuma is a genus of herbaceous plants grown largely in localized forest habitats in Southern Asia (Burma, Thailand, and India). *Curcuma* produces spikes of large concave or hooded bracts of white, pink, or scarlet-orange color. The bracts are the showiest part of the plants. *Curcuma alismatifolia,* Gagnepain can be grown as a cut flower and *C. parviflora*, Will., 'White Angel' is for pot plant production. It is propagated from a rhizome. Rhizomes are stored dry for two to three months at 20° to 25°C before forcing. After forcing at 20°/18°C, day/night, in about 3 months, depending on the duration of rhizome storage, bracts will show color. The post-production longevity is very excellent.

Australia and New Zealand

Some of the most interesting plant material we have evaluated at Beltsville are the plants collected in Australia. *Anigozanthos*, commonly known as kangaroo paw, was the first genus that was intensively investigated (Roh and Lawson 1987b; Roh and Motum 1990). This native Australian plant has been grown as a cut flower and was also bred for landscape uses. Because of the ease of tissue culture, large numbers of uniform plants were available for flowering studies. Research showed that uniform flowering is induced by a low temperature treatment.

Plants less than 11 to 12 weeks old from tissue culture propagation do not respond to an inductive treatment. After this period of vegetative growth, flower buds can be formed by a 4 to 6 weeks of 12.8°C at night. After flower induction, temperatures can be maintained at 15° to 18°C to control the time of flowering.

Although a 12.8°C treatment is necessary for floral induction, some dwarf cultivars can flower even when temperatures are maintained between 15° and 20°C.

Although a considerable effort was made to popularize kangaroo paw as a pot plant, it was not accepted by the U.S. public. Perhaps one reason was the lack of consumer education about the product. Also, plants produced in California and shipped to the East Coast were priced too high, which can be attributed to the high cost of tissue cultured plantlets associated with the image of a new product. This did not compete with potted chrysanthemums or cyclamen that U.S. consumers easily recognized.

The Australian wax flower, *Chamelaucium uncinatum* Schau, has been produced in southern California as a cut flower and filler in flower arrangements. Experiments at Beltsville growing a dwarf form of *Chamelaucium* 'Purple Pride' have demonstrated the potential of this plant for pot production. Cultural information on propagation and controlled flowering has been made available to U.S. producers. Floral induction is accelerated by 8 to 10 h of short day photoperiod at 20°/15° to 18°C and the post-production life of potted plants is very long. Although the plant has not been widely grown, partially due to a long cropping time of about 6 months, it has good potential as a new pot plant. *Chamelaucium* has been recently introduced to Japan and South Africa.

Other woody plant material from Australia with a potential as a new pot plant was *Correa* hybrid, 'Mannii', *C. baeuerlenii* F. Muell, and *C. reflexa* (Labill.) Vent. Propagation and culture of these plants is not difficult. One of the limiting factors to produce these Australian plants is that flower buds are formed only in the middle and lower portion of stem, thus showing vegetative growth at the tip of the stem. Plants *Eriostemon myoporoides* DC that showed shattering of flowers and poor shipping characteristics were also evaluated.

Leptospermum scoparium nanum, 'Kea' from New Zealand produces large dark red flowers. This plant showed some promise as a potted plant. However, shattering of flowers and short post-production keeping quality did not warrant introduction. This plant could be tested and evaluated as a large container plant for outdoor use.

South America

Among many interesting plants from South America, such as the genus of *Leucocoryne* and *Rhodophiala*, we have evaluated *Eucrosia bicolor* var. *bicolor* Ker-Gawler a brightly colored member of the Amaryllis family as a cut flower (Roh and Meerow 1992; Roh et al. 1992, 1993). The red petals and sepals and long yellow stamens are produced on a scape with 6 to 8 florets. Cut stems can be shipped at the tight bud stage. Each flower lasts 5 to 6 days. The multiplication rate is very high and at least 30 small bulblets are produced when one large flowering size bulb is planted. Although only one scape is produced per bulb about Mar. in the greenhouse, by manipulating bulb storage temperature, cut flowers can be harvested from Dec. to May. Initiation of an inflorescence is accelerated by reduced watering before harvesting bulbs followed by a 35°C dry heat treatment for one week, and finally storing bulbs at 20°C for 2 months. When bulbs are stored at 12.5°C, initiation of the inflorescence is inhibited. Therefore, for Dec. cut flower production, bulbs should be stored at 20°C and for May flower production, bulbs are stored at 12.5°C followed by storage at 20°C. Once bulbs have been programmed to initiate flowers, flowers can be harvested in 30 to 45 days.

South Africa

Our laboratory has been cooperating with scientists in the Agricultural Research Council, Republic of South Africa on production of *Ornithogalum*, *Lachenalia*, *Cyrtanthus, Oxalis,* and *Watsonia* as potted plants. *Lachenalia* will be the first crop that will be introduced to the U.S. market in 1996.

Lachenalia is a bulbous geophyte belonging to the Liliaceae. It is native to the south-western Cape and Namaqualand. Inflorescence initiation occurs at high temperature (20° to 25°C) and is inhibited at low temperatures (10° to 15°C) (Louw 1991; Roh et al. 1995). Since bulbs grown in South Africa are harvested in Oct.–Nov., bulbs must be stored for more than 6 to 8 months so that *Lachenalia* can be forced during winter in the northern hemisphere. Bulbs stored at 15°C for 195 days initiated an inflorescence when transferred to 25°C in less than 115 days and flowered without any problems associated with a long term storage (Roh et al. 1995). Storing bulbs at 10° to 15°C for 45 days when 6 to 8 florets were developed accelerated flowering and reduced the leaf length significantly as compared to bulbs stored at 20° to 25°C. Flowering was also accelerated by forcing at 17°/15°C. However, the number of florets were reduced when flowering was accelerated.

Ornithogalum commonly grown as a cut flower, also has been evaluated as a pot plant. Interspecific hybrids between *O. thyrsoides* Jacq., and *O. dubium* were obtained through embryo rescue and a short stemmed *Ornithogalum* with yellow flower color was created (Griesbach et al. 1993). Leaf cuttings can be used for propagation although tissue culture multiplication is a very efficient method used to rapidly increase selected genotypes. *Lachenalia* and *Ornithogalum* are both susceptible to at least one aphid transmitted potyvirus, *Ornithogalum* mosaic virus. Control of this virus will be essential for successful culture of these crops.

SPECIALTY FIELD GROWN CUT FLOWER

Specialty cut flowers produced outdoors are often referred to as summer cut flowers (Armitage 1992, 1993). During the past 10 years, there has been a major expansion of specialty cut flower production in the U.S. due to market demand in the major population centers. American growers discovered that they could grow a diverse range of cut flowers outdoors to satisfy an increase in consumer interest and greater sales. In the U.S., only a few crops such as gladiolus, gypsophila, and peonies have been traditionally grown as fillers in the summer to use in mixed cut flower arrangements.

Many of the new specialty crops are grown as annuals. They can be grown from seed less expensively than perennial or bulb crops. Annuals also have the advantage that they can be timed for harvest by controlling the date of sowing the seed. Several commonly grown summer cut flowers are listed in Table 3. In addition to their use as fresh flowers, some of the genera are also used as dried flowers.

A major problem in the production and marketing of specialty cut flowers is the possible instability of the market. There is a danger that high demand of a particular product may soon result in over-production, a rapid drop in price, and a loss of crop value.

Table 3. List of selected specialty cut flowers.

Annuals	Perennials	Bulbs	Woody
Ageratum	*Achillea*	*Alstroemeria*	*Callicarpa*
Amaranthus	*Aster*	*Dahlia*	*Cornus*
Antirrhinum	*Campanula*	*Gladiolus*	*Hydrangea*
Caryoteris	*Dianthus*	*Iris*	*Ilex*
Celosia	*Delphinium*	*Lilium*	*Salix*
Consolida	*Gypsophila*	*Narcissus*	
Eustoma	*Paeonia*	*Ornithogalum*	
Godetia	*Physostegia*	*Ranunculus*	
Gomphrena	*Tulipa*		
Helianthus	*Zantedeschia*		
Helichrysum			
Limonium			
Matthiola			
Salvia			
Tracholium			
Zinnia			

ANNUALS AND PERENNIALS

The production of bedding plants and perennials has been increasing recently. In our program a little research has been carried out with annuals, since we felt that breeding of new cultivars for an earliness in flowering, new flower color, and other characteristics is a responsibility of commercial companies.

Annuals that include sunflower (*Helianthus annuus* L.) have been evaluated. Sunflower seeds germinate in less than a week at 21° to 23°C and will flower in two to three months at forcing temperatures of 21° to 26°C/15° to 18°C, day/night. An ideal cultivar for potted plant use is 'Big Smile' which grows to about 30 cm (Roh 1993a). A balloon flower, *Platycodon grandiflorus* (Jacq.) A. DC., has been tested from seeds to produce potted plants. Seeds germinate in 15 days at 21°C. Plants from seedlings that germinated in Dec. result in flowering after forcing at 21° to 23°C/18°C under a natural photoperiod condition with supplemental lighting from a high pressure sodium lamps. Among several cultivars evaluated, 'Sentimental Blue' seems to be the ideal one for a potted plant and also bedding plant production. Flowering was unaffected by photoperiod, however, the number of flowers produced per plant was increased by a long day photoperiod (Song et al. 1993).

We also tested *Sinningia cardinalis* and other hybrids starting from seeds to produce plants expressing uniform flower color and flower tube length in seven months. Growth and flowering depend primarily on temperature. Photoperiod does not have a significant effect on flowering. Seeds germinate in about 15 days at 21° to 23°C and seedlings are transplanted in 5 × 5 cm cell packs and grown under a full sun light at 23° to 26°C/18° to 21°C until flower buds become visible. Light intensity is lowered by about 30 percent during the summer for 30 to 40 days to increase the number of flowers. During the summer, supplemental lighting from a high pressure sodium lamps is recommended (Roh 1993b).

WOODY PLANTS FOR EVALUATION

The Florist and Nursery Crops Laboratory at Beltsville was merged with the Research Unit at the US National Arboretum in Washington, DC., in 1993. Since then, woody plants were considered for evaluation and germplasm for greenhouse production, particularly small trees and shrubs that produce colorful berries were collected. Some of genera under evaluation are *Aridsia crenata* and Crape Myrtle (*Lagerstroemia indica* L). as potted plants. Rooting of the *Ardisia* stem cuttings to achieve 100 percent was investigated with the objective of shortening the production time as compared to seed propagation. Many hybrids of Crape Myrtle have been bred and evaluated as indoor flowering plants for an early spring, which will then be planted in the garden as a garden bedding plants to reflower about July. *Callicarpa japonica* and *C. americana* will be investigated.

NEW TECHNOLOGY

Research to find new production methods to shorten the total cropping time or the development of new ways to grow new and old crops is also considered an important aspect of new crops research. Mechanization and modernization of greenhouse systems and facilities which can be controlled automatically by computer is a major advance that allows growers to more accurately manipulate the environment and reduce cropping time.

Recent advances have been made at the University of Kentucky in the production of single stemmed roses (R.G. Anderson 1994, pers. commun.). 'Lady Diana' and 'Samantha' roses grown from cuttings can be harvested in 7 to 8 weeks after rooted cuttings are potted. Cuttings are planted into individual pots so they can be moved by pot handling robots. The plants are grown on a palletized ebb and flood bench system at a density of approximately 100 stems per square meter. It may be possible that cut flowers can be shipped with roots and graded to maintain freshness using this system.

Similarly, *Lilium longiflorum* Thunb., from seeds, *L.* ×*elegans* Thunb., from bulbils, and their interspecific hybrids (LAIH hybrids) from tissue cultured propagules can be produced in 11 months as potted plants or cut flowers (Roh 1992). For seed propagated lilies, attempts have been made to eliminate traits of *L. formosanum* since this is very susceptible to virus infection. *L.* ×*elegans* can be produced from stem bulbils and LA hybrids from tissue cultured propagules. With this method, the bulb production phase in the field can be eliminated.

Method for producing *Lilium* species (Roh 1992) is briefly summarized. It takes 47 weeks from flowering or 40 weeks after the harvest of bulbils to produce commercially acceptable quality cut flowers.

1. 7 weeks: 400 to 500 mg bulbils are harvested 7 weeks after flowering.
2. 8 weeks: Bulbils are packed in peatmoss and treated in a sequence of 5°C for 3 weeks, 10° to 13°C for 14 days, and 5°C for 3 weeks.
3. 12 weeks: Bulbils are potted and grown until termination of scaly leaf formation at 26° to 32°C/21° to 26°C, day/night, under natural long day conditions.
4. 4 weeks: Greenhouse is maintained at 16°C/10°C under a natural photoperiod.
5. 6 weeks: Greenhouse is maintained at 16° to 21°C/10° to 13°C under a 16 h of long day to induce shoot emergence.
6. 10 weeks: Temperature is gradually increased or adjusted from 10°C at night to 18°C to force and control flowering.

This method does not require bulb harvest during the culture and forcing period. In locations where night temperatures can not be maintained at 10° to 13°C for 10 weeks, plug production technology should be employed. Plants grown in 5 to 6 cm pot during the first 6 months can be shipped before commencing a temperature treatment, followed by forcing to flower. Bulbs in 6 to 10 cm pots can be stored frozen at –2°C after 4 weeks of low temperature under a natural photoperiod (step 4). In this case a long day treatment is not required. The methods can be applied to *L. longiflorum* from seeds, or LAIH hybrids from tissue culture. However, this method may not be used when *L.* ×*elegans* hybrids, which does not form bulbils, are propagated by tissue culture.

THE FUTURE OF NEW FLORAL CROPS IN THE UNITED STATES

Gardening in the U.S. is the number one pastime. More people enjoy gardening as a hobby than in other leisure activity. It is the growth in consumer interest that has led to the rapid expansion of purchases of many different types of flowering plants and greens. Although many products purchased represent improved cultivars and a wider selection of traditional crops, others represent "new crops." In our own experience we have conducted limited test marketing of *Cupressus* and *Clematis*. There was good consumer acceptance of both crops, but we were initially unable to identify growers willing to produce sufficient quantities of each crop to conduct a broader regional or national market trial.

Acceptance of a new crop depends on growers, consumers and marketing skills. There is a need to more thoroughly assess the potential of a new crop by applying several plant related criteria such as short turnover production time, good keeping quality, a good display with multiple flowers or buds at the time sale and other characteristics. Consumer preferences, such as consumer expectations, must also be evaluated. (Lawson and Roh 1995).

We know U.S. growers want new plants to produce and the public wants more variety and assortment of new crops. In many fields of business and industry consumer trends and buying habits are closely monitored as forecasts of future product innovations are developed. This type of data gathering and analysis for the floral industry is becoming more important as the industry grows. Future expansion of the floriculture industry in the U.S. is predictable. The challenge is to develop products and marketing plans based on a refined knowledge of consumer trends and buying habits.

REFERENCES

American Floral Endowment. 1993. Annual Report 1993. Edwardsville, IL.

Armitage, A.M. 1992. Field studies of *Achillea* as a cut flower: longevity, spacing, and cultivar response. J. Am. Soc. Hort. Sci. 117:65–67.

Armitage, A.M. 1993. Specialty cut flowers. Timber Press, Portland, OR.

Bateman, C.L. 1995. Foliage production, sales up—what about profits? GrowerTalks. 59(5):50, 52, 54.

Griesbach, R.J., F. Meyer, and H. Koopowitz. 1993. Creation of new flower colors in *Ornithogalum* via interspecific hybridization. J. Am. Soc. Hort. Sci. 118:409–414.
Griesbach, R.J., P. Semeniuk, M. Roh, and R.H. Lawson. 1988. Tissue culture in the improvement of *Eustoma*. HortScience 23:658, 791.
Louw, E. 1991. The effect of temperature on inflorescence initiation and differentiation and development of *Lachenalia* cv. 'Romelia'. M.S. Thesis. Univ. of Pretoria, Pretoria, South Africa.
Lawson, R.H. 1990. Eustoma. Growers' notebook. Greenhouse Manager December, 11–12.
Lawson, R.H. and M.S. Roh. 1995. New crops in the U.S.A. Acta Hort. 397:31–42.
Roh, M.S. 1988. A new dwarf lisianthus. Greenhouse Manager 7(6):98–100.
Roh, M.S. 1992. A method for producing *Lilium elegans*. U.S. Patent 5,138794.
Roh, M.S. 1993a. Growers' notebook—A step-by-step guide to success. Sunflower. Greenhouse Manager 11(12):11.
Roh, M.S. 1993b. Growers' notebook—A step-by-step guide to success. *Sinningia*. Greenhouse Manager 12(6):8.
Roh, M.S., A.H. Halevy, and H.F. Wilkins. 1990. *Eustoma grandiflorum*. p. 322–327. In: Handbook of flowering. Vol. 6. CRC Press, Boca Raton, FL.
Roh, M.S. and R.H. Lawson. 1984. The graces of gerberas. Greenhouse Manager 3(6):79, 82, 86–88, 90, 92, 94–95, 98, 100.
Roh, M.S. and R.H. Lawson. 1987a. Research and development on new crops in the United States Department of Agriculture. Acta Hort. 205:39–48.
Roh, M.S. and R.H. Lawson. 1987b. Kangaroo-paws, a classy new floral experience. Greenhouse Manger 5(9):86–87, 90, 92, 96–97, 97–100, 103, 105, 106–108.
Roh, M.S. and R.H. Lawson. 1992. Propagation and transplant production technology of new floral crops. p. 1–20. In: K. Kurata and T. Kozai (eds.), Transplant production systems. Kluwer, Netherlands.
Roh, M.S., R.H. Lawson, K.C. Gross, and A.W. Meerow. 1992. Flower bud initiation and development of *Eucrosia* as influenced by bulb storage temperatures. Acta Hort. 325:105–111.
Roh, M.S., R.H. Lawson, C.Y. Song, and E. Louw. 1995. Forcing *Lachenalia* as a potted plant. Acta Hort. 397:147–153.
Roh, M.S. and A.W. Meerow. 1992. Flowering of *Eucrosia* influenced by bulb size and watering frequency. HortScience 27:1227.
Roh, M.S., A.W. Meerow, S.H. Kim, and K.B. Lim. 1993. Bulb production, controlled flowering, and postharvest physiology of *Eucrosia*. Acta Hort. 337:59–64.
Roh, M.S. and G.J. Motum. 1990. *Anigozanthos*. p. 37–46. In: Handbook of flowering. Vol. 6. CRC Press, Boca Raton, FL.
Society of American Florists. 1993. The changing floriculture industry. Alexandria, VA.
Song, C.Y., M.S. Roh, S.K. Chung, and R.H. Lawson. 1993. Effect of temperature and light on growth and flowering of potted plant production of *Platycodon*. J. Korean. Soc. Hort. Sci. 34:446–453.
USDA, National Agricultural Statistics Service. Floriculture Crops. 1984 Summary. Sp Cr 6-1 (85). Washington, DC.
USDA, National Agricultural Statistics Service. Floriculture Crops. 1994 Summary. Sp Cr 6-1 (95). Washington, DC.
Whitton, B., W. Healy, and M. Roh. 1990. Flowering of *Aeschynanthus*, 'Koral' at fluctuating and constant temperatures. J. Am. Soc. Hort. Sci. 115:906–909.

Xeriscaping: Sources of New Native Ornamental Plants

Bart C. O'Brien

The use of the word xeriscape as a landscaping term has not received widespread public acceptance, even though its precepts are sound and widely followed. "Xeriscape" has often been interpreted by the general public to mean "zero-scape" and was equated with severe minimalist drought tolerant plantings. The terms "sustainable landscapes" and "appropriate horticulture" better convey the full intent of "xeriscape." Native plant materials are a vital and viable element of the future of xeriscape landscapes and gardens across America and are the primary focus of this paper. As more gardens and landscapes are designed and installed true to their local conditions, opportunities will continue to expand for the use and development of new native plants for ornamental horticulture.

THE XERISCAPE CONCEPT

The word xeriscape was coined in 1981, to capture the idea of water conserving landscapes but conceptually xeriscape involves much more. There are seven essential components of xeriscaping: appropriate planning and design; appropriate/minimal turf areas; appropriate soil preparation and analysis; appropriate plant selection; appropriate watering methods; appropriate use of mulches; and appropriate landscape maintenance. In essence, the underlying concept behind xeriscape is the doctrine of regionally appropriate horticulture.

In practice, however, xeriscape often has been reduced to the landscaping idea of maximizing the efficient use of water in gardens and landscapes by creating areas which group plants by their water requirements. Typically this has been interpreted as growing the plants with the greatest water need nearest the home and transitioning to very low to no irrigation at the zone furthest from the home. This zoning of the landscape according to water use may also have additional benefits pertaining to fire breaks, wildlife value, and in enhancing the diversity of plants that may be grown in the garden.

HISTORY

Due to the dominance of our primarily temperate European horticultural heritage, there has always been a predisposition toward water-loving exotic plants in our landscapes and gardens. The westward expansion and settling of the country brought these same traditions and plants into increasingly dry and warmer regions that are prone to long periods of drought. As long as water was cheap and abundant there wasn't a problem and a more or less standard set of coast-to-coast plants thrived in spacious green havens. Over much of the country, water is the limiting resource governing the use of ornamental plants in gardens. The huge increase in population in the U.S. Southwest has in many areas outstripped local water supplies. In some areas of the country, water for landscape or garden use is seasonally rationed or restricted. The cost of water, even during years of abundant rainfall, is always increasing. Water quality issues, including salinity and contamination by nitrates and bacteria, lead to increased costs due to necessary water treatment programs. As time passed, our use of gardens and landscapes changed from seasonal retreats and visual frames for our homes and buildings to heavily used outside multi-purpose spaces. These changes resulted in the establishment of the xeriscape concept.

CURRENT TRENDS

There are a number of trends in home landscapes and gardens which will influence the ornamental plant industry. Those which I consider to be among the most important are: (1) the decrease in the amount of time, money and expertise that most households have to invest in properly planting and maintaining gardens; (2) the shrinking in average size of new gardens; (3) the increasing use of gardens as living spaces and outdoor rooms; (4) the increasing cost of water, labor, fertilizers, and chemicals; and (5) the restriction or limiting of water use for garden and landscape purposes.

Although there is always a market for colorful annuals, biennials, and short-lived perennials, these are generally more difficult groups to depend upon in landscapes and gardens. Longer lived perennials, subshrubs, shrubs and trees are the backbone of landscapes and gardens and will constitute the primary nursery market.

Under current conditions, I am convinced that a greater diversity of plants which flourish under existing climatic conditions across the country are needed. These changes should be viewed as potential opportunities and require the search for, and development of, perennials, subshrubs, shrubs and trees in the following categories: dwarf or compact forms of existing plants; new plants of small stature; plants that are all season performers; low maintenance plants; long, or continuous, seasons of color, be it flowers, foliage, fruits, stems or bark; good-looking, water-thrifty plants; and plants that attract wildlife.

Native plants which fall into any or all of these categories will be particularly desirable as they have the additional benefit of fitting into the local and regional conditions both visually and ecologically. This opportunity creates a new, relatively unexploited regional niche market for the nursery industry, and should eventually lead to a more striking regionalism in the American landscape vernacular.

THE SEARCH FOR NEW NATIVE ORNAMENTAL PLANTS

There are a number of people and institutions throughout the country looking at our native plants as a direct or secondary source of new introductions for the nursery industry, but anyone can become involved in this process. The methods employed to create new native ornamentals can be as informal as selection from the wild or as sophisticated as plant breeding including even genetic engineering.

A good place to begin the investigation of new native plants for drier portions of our landscapes is to look at the large plant families and genera in the region you are serving, and to concentrate the search on those plants inhabiting drier habitats. In California and the southwest, these would likely include some of the following: Agavaceae (*Agave*, *Nolina*, *Yucca*), Alliaceae (*Allium*), Amaryllidaceae (*Brodiaea*, *Dichelostemma*, *Triteleia*), Asclepidaceae (*Asclepias*), Asteraceae (*Artemisia*, *Erigeron*, *Senecio*, *Haplopappus—Ericameria*, *Hazardia*, *Isocoma*), Cactaceae (*Opuntia*), Crassulaceae (*Dudleya*, *Sedum*), Cyperaceae (*Carex*), Ericaceae (*Arctostaphylos*), Fabaceae (*Astragalus*, *Lupinus*), Fagaceae (*Quercus*), Grossulariaceae (*Ribes*), Iridaceae (*Iris*, *Sisyrinchium*), Hydrophyllaceae (*Phacelia*), Lamiaceae (*Lepechinia*, *Monardella*, *Salvia*), Liliaceae (*Calochortus*, *Lilium*), Malvaceae (*Malacothamnus*, *Sidalcea*, *Sphaeralcea*), Poaceae (*Festuca*), Polygonaceae (*Eriogonum*), Ranunculaceae (*Aquilegia*, *Delphinium*, *Ranunculus*), Rhamnaceae (*Ceanothus*, *Rhamnus*), Rosaceae (*Adenostoma*, *Cercocarpus*, *Heteromeles*, *Holodiscus*, *Prunus*, *Purshia*, *Rosa*, *Spiraea*), Saxifragaceae (*Heuchera*), and Scrophulariaceae (*Penstemon*).

The majority of these large genera and families have yet to be tested or evaluated for horticultural purposes in a serious systematic fashion including most perennial members of *Phacelia*, *Monardella*, *Eriogonum*, and *Astragalus*. A number of these large genera have well known propagation problems (such as commercially viable method for asexual propagation of *Quercus* and *Calochortus*) or are known to be especially difficult to grow and/or maintain under nursery conditions (like *Calochortus* and many native *Delphinium* and *Lupinus* species).

The selection of an individual plant for possible introduction is generally accomplished by finding a desirable feature or set of features that are deemed desirable. These individuals can be found in the wild or in the garden and often represent extremes of the natural variation: compact growth habit, different foliage, albinos, and unusual color forms. Additional factors such as mutations including witch's brooms, variegated foliage, and natural hybrids also contribute to the possible pool of plants to choose from. Traditionally these matters of chance have been the most frequently reported sources of many excellent cultivars: *Heterotheca* (*Chrysopsis*) *villosa* 'San Bruno Mountain', a sterile, free-flowering dwarf selection; *Artemisia pycnocephala* 'David's Choice', a heat tolerant compact selection; *Erigeron* 'W.R.', a compact, free-flowering, heat and cold tolerant chance wild hybrid between *Erigeron glaucus* and an unknown species; and *Acer macrophylla* 'Seattle Sentinel', a fastigiate selection.

Another especially rich area for horticultural selection can be found in nearly all taxonomically confusing groups at either the generic [example: *Arctostaphylos* and *Zauschneria* (*Epilobium*)] or specific [example:

Mimulus aurantiacus or *Mahonia* (*Berberis*) *aquifolium*] level. Taxonomic chaos is often indicative of the extreme plasticity found naturally in the group and by inference the potential horticultural availability of variation to select from. Similarly, there are a number of extremely variable species that are nearly as ripe for selection as the taxonomically complicated groups listed above. These species include *Rhamnus californica*, *Erigeron glaucus*, *Ceanothus maritimus*, *Quercus chrysolepis*, and *Juniperus communis*.

Peripheral populations of a desirable plant are another source of native plants for landscapes and gardens. Plants from these populations may be more tolerant of heat or cold, higher or lower elevations, drought or wetness, may exhibit resistance to diseases or pests, or may be adaptable to a different soil type or condition. An excellent example of this phenomena is *Arctostaphylos uva-ursi* 'Point Reyes', a selection from the fog-bound headlands of the Point Reyes peninsula in Marin County that is extremely tolerant of heat and drought in comparison to typical members of the species. The small relictual San Bernardino Mountains population of *Populus tremuloides* is remarkably well adapted to growing conditions at low elevations in southern California. When plants from these peripheral populations are not themselves horticulturally desirable, they may carry genes which may be useful in a breeding program.

Another particularly rich source of new plants to investigate are the genera that have been dependably used in landscapes and gardens before, and look for a different species with desirable characteristics. Cultivars of the "new" species can be selected directly or the "new" species may be used to create hybrids with the "established" species. The named *Heuchera maxima* x *Heuchera sanguinea* hybrids: 'Genevieve', 'Opal', 'Santa Ana Cardinal', 'Susanna', and 'Wendy', created by Dr. Lee W. Lenz are excellent examples of this approach.

Breeding programs involving controlled crosses and hybridization between species are relatively rarely encountered in the native flora. There are, however, a few exemplary programs which have yielded many fine plants: the pacific coast hybrid irises, the *Mimulus* hybrids (of the section *Diplacus*), the *Lewisia cotyledon* complex, and *Lilium* hybrids.

All new cultivars should be named, described, published and registered with the proper registration authority. A list of all currently accepted registration authorities for ornamental plants is available from the American Association of Botanical Gardens and Arboreta (AABGA).

TESTING, EVALUATION, AND MARKETING

All too frequently, a promising native plant species or cultivar does not live up to its potential. Almost without exception, these plants are victims of poor testing and evaluation.

Establishing a set of desirable traits and characteristics of plants which the public actively wants or desires is a critical first step in the development of a new plant introduction strategy. Consulting with focus groups from the nursery industry, the gardening public, native plant enthusiasts, realtors, horticultural groups and others should provide the researcher with plenty of input on a variety of desirable plant traits. As progress is made in the selection process, the new plants should be planted out in a number of test sites. Test plantings should then be evaluated by a number of outside reviewers. Plants surviving through the testing and evaluation process are then ready to be introduced and marketed to potential user groups. The most recent successful example of a western native plant to go through such a thorough program is *Arctostaphylos uva-ursi* 'Vancouver Jade', an introduction out of the University of British Columbia's well known plant introduction program.

CONSERVATION OF GENETIC RESOURCES

Everyone involved with the selection, use and promotion of new native plant species and cultivars and/or any of their close relatives must be fully aware of the potential negative consequences of their use in gardens and landscapes. Most of these concerns center around the issue of conservation of genetic resources and the genetic pollution of native plant populations in the wild. Three examples from the California flora will serve to illuminate this issue.

Nevin's Barberry [*Mahonia* (*Berberis*) *nevinii*], an extremely rare plant in nature, is found in widely distributed and ecologically varied small populations in southern California. It is, however, a relatively commonly seen landscape plant throughout the state. A molecular level study of these plants showed that there is little to no variation present in the gene pool. Therefore, the seed source of plants to be planted in the vicinity of the remaining individuals in the wild is not of concern—there will be no adverse consequences to the gene pool.

The California dandelion (*Taraxacum californicum*) from the wet meadows in the San Bernardino Mountains is a rare plant whose continued existence is threatened by ongoing hybridization with the common European dandelion (*Taraxacum officinale*), a weedy pest plant that is common throughout the range of the rare species.

The Monterey pine (*Pinus radiata*) is known from five geographically isolated populations (three are from central coastal California and two are on islands off Baja California). This tree is a significant forest tree in the southern hemisphere, particularly in New Zealand. Material from all five populations has been grown together on a massive scale in New Zealand, such that the primary seed source of Monterey pine (even in California) is from New Zealand. The issue of primary concern here is that the "mongrel" Monterey pines of New Zealand mixed origin will genetically pollute the "pure" native stands of these trees, such that the unique genetic character of the California populations will be lost due to homogenization of the gene pool over time (the Baja California populations are not threatened in this way at this time).

CONCLUSION

The development of new native plant crops for use in xeriscape landscapes and gardens is still wide open with innumerable opportunities waiting for the interested or inspired investigator. As more work is done to select and develop the native flora for use in gardens, the more likely it will be that these deserving plants will be used and appreciated by the general public. When the public is aware and appreciative of the beauty and utility of native plants they will be much more open to the conservation and preservation of this essential component of our natural heritage. The single most important caveat regarding the horticultural use of these native plants is that their use should not be allowed to adversely impact native plant genetic resources.

REFERENCES

Beck, Beatrice M. 1990. Drought tolerant planting bibliography. Rancho Santa Ana Botanic Garden, Claremont, CA.

Ellefson, Connie L., Thomas L. Stephens, and Douglas Welsh. 1992. Xeriscape gardening: Water conservation for the American landscape. Macmillan, New York.

Libby, William J. 1990. Genetic conservation of Monterey [pine] and coast redwood. Fremontia. 18(2):15–21.

Ottesen, Carole. 1995. Native plant primer: Trees, shrubs, and wildflowers for natural gardens. Harmony Books, New York.

Perry, Robert C. 1992. Landscape plants for western regions: An illustrated guide to plants for water conservation. Land Design Publishing, Claremont, CA.

Skinner, Mark W. and Bruce M. Pavlik (eds.). 1994. California Native Plant Society's inventory of rare and endangered vascular plants of California. 5th ed. California Native Plant Society, Sacramento.

Wasowski, Sally, with Andy Wasowski. 1995. Native gardens for dry climates. Clarkson Potter / Publ., New York.

Hyacinth Bean: Stems for the Cut Flower Market

Robert G. Anderson, Sharon Bale, and Wenwei Jia

Hyacinth bean *(Dolichos lablab* L., Fabaceae) has been a novelty garden plant in the U.S. for generations (Piper 1915; Bailey 1930). This vigorous, twining vine is characterized by large ternate purple-green leaves and purple petioles. The vines produce hundreds of spikes of lavender flowers in late summer followed by long-lasting deep lavender-purple pods. Hyacinth bean, lablab bean, field bean, sem (India) or pig-ears (Chinese, from the leaf shape) is primarily an ornamental annual vine in the U.S., but it has been used for centuries in India and China as an edible pod and animal forage (Singh and Pandita 1980; Saraswat 1986). The distinctive long-lasting pods are suitable for cut stems for the cut flower industry. In addition, the pods are so unique that they could be used for decoration or harvested for Chinese food wholesalers.

Preliminary studies in 1992, demonstrated high yields, simple production practices and relatively long vase-life for cut pod stems. The objective of this study was to determine the number of cut stems of hyacinth bean pods that could be produced, the seasonal variation in production, the impact of plant density on cut stem production, and the number of pods produced per stem.

METHODOLOGY

Hyacinth beans were planted on 60 lineal meters of fence, 1.5 m tall, for yield trials in 1993 and 1994 at the University of Kentucky Horticulture Research Farm in Lexington. In 1993, three randomized replicates (each 10 m long) were planted with 1 or 3 plants/m. In 1994, plant densities of 0.6 or 1 plant/m were used on the same randomized fence sections. All stems with fruit were harvested when all fruit on a stem were mature enough that the pod was swollen from seed development. This was determined to be best in preliminary vase-life studies. Length of each cut stem and fruit number were determined for each harvest date.

Hyacinth bean seeds were planted May 1 of both years into MetroMix 510 in 10 cm pots and grown in a greenhouse (17° night/25°C day) for 4 to 5 weeks. Plants were fertilized once per week with 100 ppm nitrogen from 20–4–16 fertilizer. Plants were transplanted to the field in the first week of June. Plants were irrigated when necessary during the summer and fertilized monthly, through August. Pesticide applications for grasshoppers were necessary in mid summer both years, but no other insect pests were a problem even though many were present in adjacent fields.

RESULTS

Hyacinth bean is a highly productive plant. Nearly 3800 commercial quality cut stems were harvested from Aug. 20 to Oct. 29 in 1993 with warm Sept. weather (Fig. 1). Yields were reduced to 1800 cut stems in harvests from Aug. 20 to Sept. 10 due to cool Sept. weather in 1994.

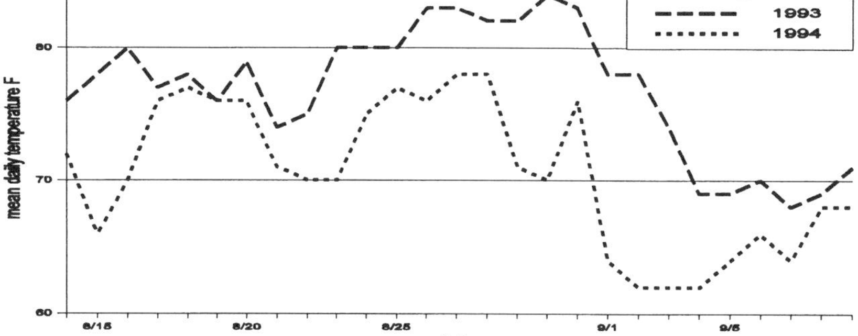

Fig. 1. Contrast in mean daily temperature in 1993 and 1994 during hyacinth bean harvest.

Although there was no significant effect of density on stem production per linear meter of row in both years, the highest production was achieved with the denser planting (Table 1). This relationship between plant density and yield was also demonstrated by Singh and Pandita (1980) at higher plant densities.

As plant density increased, cut stem production per plant decreased proportionately (Table 1). Hyacinth bean is a very vigorous vine and individual plants grew to fill the space on the fence. The mean number of commercial quality cut stems harvested per plant was as high as 55 cut stems per plant in 1993. The relatively low plant densities used in this study still did not identify the maximum yield that could occur on a per plant basis. The lowest plant density used by Singh and Pandita (1980) was approximately 1.5 plants/m and had a mean yield of 19 inflorescences per plant.

Cut stem lengths varied from 25 to 90 cm. The percentage of stems in standard cut flower grades were nearly identical in both plant densities in both years (Table 2). In 1993, approximately 60% of the stems were in medium stem length grades (46–55 cm and 56–65 cm) with just over 10% in the long stem length grades. Longer stem length grades were dominant in 1994 because the harvest period was short due to cool Sept. weather. Longer stems were more common in the early harvests than in later harvests in both years so the 1993 harvests included many shorter stems from the later harvest periods.

Over 80% of the total harvest of cut hyacinth bean stems occurred in the first two weeks of harvest in late Aug. in 1993 and 1994 (Table 3). Yields were so high that harvest occurred on 2–3 day intervals for 12–15 days. However, the number of stems harvested dropped quickly after the first two weeks. It was hoped that harvest would occur over 6–8 weeks, but this was not the case.

The number of purple fruit ranged from one to 30 per cut stem. The mean fruit number was nine fruit per stem regardless of plant density within the rows. Singh and Pandita (1980) measured a mean of 13 fruit per stem on supported plants. Approximately 90% of the cut stems had 4 to 12 fruit on each stem (Fig. 2). This number of fruit per stem was appropriate for use as a cut stem in flower arrangements.

Table 1. Mean yield of commercial quality cut stems of hyacinth bean on a per plant and per meter basis.

Year	Plant density (plants/m)	Mean number of commercial quality cut stems: Per plant	Per m
1993	1	55.6	55.6
	3	23.1	69.3
1994	0.6	47.2	28.3
	1	31.1	31.1

Table 2. Yield and distribution of hyacinth bean stems based on density in 1993 and 1994.

Year	Plant density (plants/m)	Distribution of cut flowers (%)						Total yield (60 linear m)
		36–45 cm	46–55 cm	56–65 cm	66–75 cm	76–85 cm	86–95 cm	
1993	1	24	33	28	11	3		1668
	3	27	34	26	10	2		2080
1994	0.6	4	43	31	17	6	1	849
	1	3	43	32	17	5	1	932

Table 3. Distribution of harvest of cut hyacinth bean stems based on density in 1993 and 1994.

Year	Plant density (plants/m)	Distribution of cut stem harvest (%)							
		Aug. 20	Aug. 30	Sept. 10	Sept. 17	Sept. 24	Oct. 1	Oct. 7	Oct. 29
1993	1	53	29	9	3	2	1	1	1
	3	56	31	8	6	1	1	0.5	1
1994	0.6	54	37	9					
	1	58	34	8					

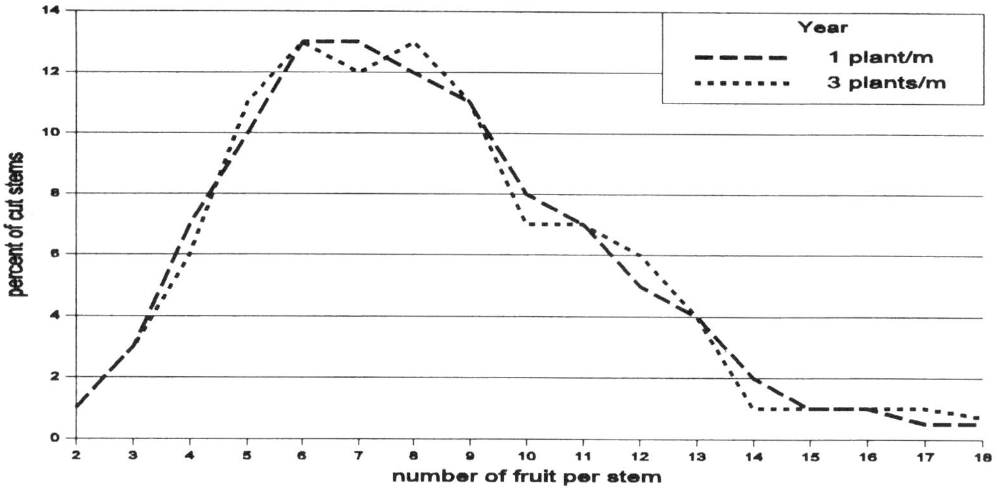

Fig. 2. The distribution of fruit number per stem of all cut hyacinth bean stems in 1993.

SUMMARY

Hyacinth bean cut stems are a unique item for cut flower arrangements. Some local and regional florists enjoyed the special color and texture that hyacinth beans offered their arrangements while others had little interest in use of these stems. With yields as high as 70 cut stems per lineal meter of fence, the potential returns could be quite high [7 bunches/m × $2.50/bunch (a common minimum wholesale price) = $17.50/m] for short term harvest and marketing of this summer/fall crop. However, there is no established market for hyacinth bean cut stems and guarantees of profitability cannot be made.

REFERENCES

Bailey, L.H. 1930. The standard cyclopedia of horticulture. Macmillan, New York.

Piper, C.V. 1915. The bonavist, lablab, or hyacinth bean. USDA. Bul. 318.

Saraswat, K.S. 1986. Ancient crop economy of Harappans from Rohira, Punjab (c. 2,000–1,700 BC). Paleobotanist. 35(1):32–38.

Singh, P. and M.L. Pandita. 1980. Effect of spacing, plants per hill and training on growth and yield of sem (*Dolichos lablab* L.) hyacinth beans. Indian J. Hort. 37:388–391.

New Basil Selections with Compact Inflorescences for the Ornamental Market*

Mario R. Morales and James E. Simon

Basil (*Ocimum basilicum* L., Lamiaceae) has long been prized as a culinary herb used in foods and as a source of essential oil for flavors and fragrances (Simon et al. 1984; 1990). There are many cultivars of basil which vary in their leaf size and color (green to dark purple), flower color (white, red, lavender, purple), growth characteristics (shape, height, flowering time), and aroma, making this plant an increasingly popular culinary and ornamental herb (Darrah 1984; Morales et al. 1993). Primarily a cross pollinating species, many cultivars of basil easily intercross (Krishnan 1981; Nation et al. 1992), and several species of *Ocimum* can form interspecific hybrids (Sobti and Pushpangadan 1982). With the increased interest in the use of basil as ornamental herbs, we initiated a selection and breeding program to identify promising new ornamental basils.

Our first released cultivar, originally identified in an extensive intermated germplasm collection of *Ocimum* spp. in 1989, was developed by selfing and selection from a tall upright basil with an intense lemon aroma (Morales et al. 1993) and released as 'Sweet Dani' in 1996 (Morales and Simon 1996). In 1993, three other new and unusual basil phenotypes with intense purple, maroon and white colored flowers, and very short and compact inflorescences were identified in our basil nursery and selfed. Selections were grown in central Indiana in 1994 and selfed to form S_2 progenies for further selection. Plants are compact, early flowering, and have a "flat-top" appearance which is covered by a mass of flowers at full bloom (Fig. 1). The inflorescences are located mainly at the top of the plant and each has from 10 to 13 whorls of six flowers tightly arranged along the rachis. The plants will produce new inflorescences if the old ones are removed. The growth habits of the purple, maroon and white flowering basils are very similar. These selections are also highly aromatic, and may be used for culinary purposes, as well as being suitable as pot or bedding plants.

EVALUATION OF S_2 LINES

A total of 41 S_2 basil lines (15 purple, 8 maroon, and 18 white) were sown in flats in the greenhouse on May 13, 1995, and transplanted into single rows, 5 m long, with plants spaced 0.5 m apart, at Lafayette, Indiana on June 9, 1995. Days to flower for each line was determined at the beginning of open bloom. Plant growth characteristics including plant height, plant width, leaf width and length, inflorescence length, and overall uniformity and appearance were assessed at full bloom. At harvest (Aug. 11), five plants

Fig. 1. Basil with compact inflorescences: (left) purple, (right) white.

*Journal Paper No. 14,985, Purdue University Agricultural Research Program, West Lafayette, IN 47907-1165.

Table 1. Growth, yield and essential oil characteristics of second generation ornamental basil lines with compact purple, maroon and white inflorescences selected among 41 lines in 1995.

Flower color	Line	Plant Height (cm)	Plant Width (cm)	Leaf Length (cm)	Leaf Width (cm)	Inflorescence length (cm)	Uniformity[z] (rating)	Appearance[y] (rating)	Ess. oil content (%, v/dw)	Days to flower	Main components (% of total essential oil) Linalool	Methyl chavicol	1,8-Cineole
Purple	15	32	42	4.8	2.4	1.7	1	1	1.96	79	48	24	--
	4	32	42	4.4	2.3	1.7	2	1	1.32	66	48	22	--
	7	30	40	4.4	2.2	1.8	4	3	1.64	69	50	23	--
	12	29	40	4.8	2.5	1.7	2	2	2.14	76	60	20	--
Maroon	14	45	60	4.9	2.9	3.9	2	2	2.09	72	80	--	2.5
	7	40	50	4.7	2.7	3.5	2	2	2.07	76	78	--	3.1
White	13	26	30	4.7	2.5	3.2	1	2	0.52	69	58	--	14.7
	18	25	28	4.7	2.7	2.9	2	3	0.56	73	60	--	12.1
	3	25	30	4.2	2.3	3.5	3	3	0.55	66	60	--	10.6
	2	23	29	4.1	2.3	3.2	3	3	0.47	69	59	--	8.4

[z]Scale of 1 (uniform) to 5 (nonuniform).
[y]Scale of 1 (attractive) to 5 (unacceptable).

from each line were randomly cut and dried at 37°–39°C in a forced-air dryer for 18–20 days. Essential oils were extracted via hydrodistillation and oil composition determined by gas chromatography as reported previously (Charles and Simon 1990).

Each of the lines within each of the three basil phenotypes were significantly different in most characters measured suggesting that further selection of targeted traits is possible. Selection at the present time is based on a subjective evaluation of appearance and esthetic quality, including plant shape, structure, flower color intensity, and floral quantity (Table 1).

The purple compact flowering phenotypes appeared most uniform and attractive, had a very compact inflorescence (1.7 to 1.8 cm long), and received positive response from visitors. Lines 4, 7, 12, and 15 were selected for further development. The maroon compact flowering phenotypes were not as intense in flower color as the purple lines. Lines 7 and 14 were selected for further development. The white phenotypes were planted in a compacted soil area that led to reduced growth and lower essential oil yields (0.47% to 0.56% of 100 g of dry material) than the purple and maroon phenotypes. Significant differences between the 18 white lines allowed us to identify and select lines 2, 3, 13, and 18 as the most attractive and best performers.

All phenotypes were highly aromatic. The essential oils from the purple compact flowered lines were rich in linalool (48% to 60%) and methyl chavicol (20% to 24%); while the oils from the maroon compact flowered lines were rich in linalool (78% and 80%) and low in 1,8-cineole (2.5% to 3.1%). The essential oils from the white compact flowered lines were lower in linalool (58% to 60%) and higher in 1,8-cineole (8.4% to 14.7%) than the maroon types.

FUTURE PROSPECTS

After two cycles of selfing and selection, we identified several near-uniform lines of compact-flowering basil with white, maroon, or purple flower. The basil plants are relatively moderate to short (23–47 cm), with a rounded shape and flat-top appearance, and exhibit very short and compact inflorescences (2–6 cm long) which cover the otherwise dark green foliage for several weeks. As days to flowering varied greatly among the lines, it would be possible to select for early or late maturity types. These basils will reflower if the original first blooms are removed, a desirable trait for ornamental purposes. The very short inflorescence length (10 to 13 whorls of six flowers tightly arranged along its rachis) coupled with an intense flower color appear most promising for further development.

Herb collections, which are well-known for their diverse genetic characteristics have principally been screened for materials most suitable for culinary purposes, for processing and essential oils. This study suggests that such herb collections, may also offer a unique source of genetic materials for potential use as ornamentals.

REFERENCES

Charles, D.J. and J.E. Simon. 1990. Comparison of extraction methods for the rapid determination of essential oil content and composition of basil (*Ocimum* spp.). J. Am. Soc. Hort. Sci. 115:458–462.

Darrah, H.H. 1984. The cultivated basils. Buckeye Printing Company, Independence, MO.

Krishnan, R. 1981. Natural outcrossing in sweet basil *Ocimum basilicum* L. Indian Perfumer 23(3/4):74–77.

Morales, M.R. and J.E. Simon. 1996. 'Sweet Dani': a new culinary and ornamental lemon basil. HortScience: (in press).

Morales, M.R., D.J. Charles, and J.E. Simon. 1993. New aromatic lemon basil germplasm. p. 632–635. In: J. Janick and J.E. Simon (eds.), New crops. Wiley, New York.

Nation, R.G., J. Janick, and J.E. Simon. 1992. Estimation of outcrossing in basil. HortScience 27:1221–1222.

Simon, J.E. , A.F. Chadwick, and L.E. Craker. 1984. Herbs: An indexed bibliography 1971–1980; the scientific literature on selected herbs, and aromatic and medicinal plants of the temperate zone. Archon Books, Hamden, CT.

Simon, J.E., J. Quinn, and R.G. Murray. 1990. Basil: a source of essential oils. p. 484–489. In: J. Janick and

J.E. Simon (eds.), Advances in new crops. Timber Press, Portland, OR.
Sobti, S. and P. Pushpangadan. 1982. Studies in the genus *Ocimum*: Cytogenetics, breeding and production of new strains of economic importance, p. 457-472. In: C.K. Atal and B.M. Kapur (eds). Cultivation and utilization of aromatic plants. Jammu-Tawi, India.

Novel Annual and Perennial *Tagetes**

Jacqueline A. Soule

The genus *Tagetes* (Asteraceae) contains 56 species (Soule 1993a) and includes well known, popular, annual bedding plants known as marigolds. There are also three lesser known, but regionally popular, perennial species currently in cultivation.

The four annual species of *Tagetes* commonly cultivated throughout the world as ornamentals are *T. erecta* L., *T. patula* L., *T. lunulata* Ort. and *T. tenuifolia* Cav. These four species were brought into cultivation over two millennia ago in the region that is now western Mexico (Soule 1993a). These early marigolds were used as ornamentals, medicinal plants, and as ritual plants (Nuttall 1920). Shortly after their introduction to the Old World, *Tagetes* were included in well-known herbals (Fuchs 1542; Gerard 1596), in part due to their potential medicinal uses.

They were to be used with care, since they had a strong, "rank odor" (Gerard 1596). Over the last four centuries, much of the strongly scented secondary compounds have been bred out of these annuals, while at the same time, larger and more ornate heads were selected.

Two other annual species of *Tagetes* are currently used in horticulture, *T. filifolia* Lag. and *T. minuta* L. *Tagetes filifolia* is a very short plant, with lacy, filiform leaves and two or three tiny white ray florets and is occasionally found in nursery catalogs under the name "Irish Lace." *Tagetes minuta* is grown in Brazil and other nations for essential oil extraction (Lawrence 1985; Soule 1993b, c) but has escaped cultivation and is considered a noxious weed in parts of southern Africa (Wells et al. 1986).

Since there are 27 species of annual *Tagetes* (Soule 1993a), the six thus far cultivated represent just a beginning of the potential horticultural usefulness of the genus. The current paper discusses four novel annual *Tagetes* which can potentially increase the popularity of the genus with the gardening public.

Of the 56 species in the genus, 29 are perennial (Soule 1993a). Only three of the perennial species are currently cultivated in the horticulture industry: *Tagetes lucida* Cav., *T. lemmonii* A. Gray, and *T. palmeri* A. Gray. The history of the cultivated perennial taxa is tangled and has led to occasional publication of erroneous information on nomenclature, origins, and culture.

Tagetes lemmonii was originally collected in southeastern Arizona in the late 1800s by the Lemmons, a husband and wife collecting team, after which it is named who supplied seed to Asa Gray at Harvard University. The Lemmons proceeded to Oakland, California where plants were established. The progeny of those plants made their way into the nursery trade in southern California, and, surprisingly, England (Anon. 1900). *Tagetes palmeri* is originally from southern Sonora, Mexico. It is used medicinally by local peoples (Soule 1993a) and appears to have been cultivated by the early Spanish missionaries. Even today, healthy specimens of *T. palmeri* can be found growing in the old mission yards of southern California. *T. lemmonii* and *T. palmeri* are sister species (Soule 1993a), and the two are identical in most respects, except that the latter has larger, showier heads, displays greater heat tolerance, and tends to grow into taller, more robust shrubs. *T. lemmonii* tends to be more cold tolerant.

---

*Research was conducted at the University of Texas, Department of Botany, Austin, TX 78713.

Tagetes lucida is a Mexican species which has been in cultivation for over a millennia (Nuttall 1920). *T. lucida* was used by the Aztec as medicinal plants, and in ritual (Ortiz de Montellano 1990). When *T. lucida* was brought to the Old World, it was used as a condiment in France and England (Sweet 1817). Herbarium sheets from the 1800s show that the Spanish and French re-introduced the plant to the New World, calling it "false tarragon" or "falso hypericon." The species is now offered generally as "winter tarragon," but sold in Florida as "Florida tarragon," and in Texas as "Texas tarragon." These last two are not to be confused by the connoisseur with the "Mexican mint-marigold" which is the exact same species, but has been imported from across the border in recent times (Soule 1993b).

Since there are 29 species of perennial marigolds, the three species thus far cultivated represent a bare beginning of the potential horticultural usefulness of the genus. This study was conducted to determine which species of perennial marigolds are the most adaptable to cultivation, and which might potentially prove to be marketable.

METHODOLOGY

During field research for a monograph on the genus (Soule 1993a), native habitats in North, Central, and South America were visited. *Tagetes* populations from Arizona to Argentina were sampled, and both live plants and seed were collected. Plants and seedlings were established in both greenhouse and common garden conditions, and the plants evaluated for their potential use as a new crop for the nursery trade.

To determine overall suitability as an ornamental crop, plants were evaluated based on floral, vegetative, habit, and environmental characters. Floral characters included duration of flowering, color, size, and number of flowering heads. Vegetative characters evaluated were foliage form, color, and density. Habit characters include plant height, overall form, and general plant attractiveness. Environmental characters tested the general acceptability as an ornamental plant, and included cold hardiness, heat tolerance, water requirements, and adaptability to container cultivation. Habitat information is included, and covers habitat preferences and soil requirements.

Four perennial and four annual species of *Tagetes* were selected as the best potential new introductions. The four perennial species are suffrutescent and can be grown to marketable (1 gallon) container size under greenhouse conditions within six months. They can be sown directly after the last frost, or transplanted from greenhouse seedlings. All transplant well, and flowered in the first year. The four annual species can be sown directly after the last frost, or transplanted from greenhouse grown seedlings. All transplant well.

A brief review of Asteraceae terminology is needed for discussion of the floral characteristics. Many small florets cluster together in a head. The heads are surrounded by an involucre, a series of highly modified leaves or bracts. In the case of *Tagetes*, the involucre consists of a single series of fused bracts, and the involucre grades smoothly into the peduncle, two key characters to the genus. At the base of each floret, the ovary develops into a single seeded fruit called an achene. The outer ring of florets in a *Tagetes* head tend to have showy corollas and are called ray corollas or rays. The less showy central florets are disk florets.

PERENNIAL TAGETES

Tagetes campanulata Griseb.

Corollas are brilliant yellow in this species. The heads are 4 to 6 cm across, with generally 5 to 30 heads appearing simultaneously. Each head lasts 10 to 18 days. Bloom begins on mature plants in early summer, and continues through light frost. Peduncles elevate the heads slightly above the level of the foliage.

This mounding shrub is 0.3 to 0.5 m tall. Leaves are medium green and slightly glossy on the upper surface, often silvery on the lower surface. The leaves are narrowly pinnately dissected, but the foliage is very dense, thus a full appearance is maintained.

The species is found in remote canyons of the highlands of northern Argentina. It grows on cliff faces and rocky canyon bottoms, generally in full sun to afternoon shade conditions. The plants tolerate acid to neutral soils, and will bloom throughout the summer if temperatures remain below 37°C. They are normally subject to mild freezing conditions in winter, and in harder freeze will die back to the underground shoots.

Adaptability to cultivation is excellent. With its delicate foliage and brilliant yellow corollas, *T. campanulata*, the canyon marigold, would make a lovely rock wall plant.

***Tagetes mulleri* S.F. Blake**

The corollas are a bright yellow-orange, 4 to 6 cm across, and borne on long peduncles which exceed the foliage. One to 7 heads appear simultaneously, each lasting 8 to 16 days. Mature plants may bloom in early spring, and again in fall after the summer heat, until frost.

It is a low erect shrub 0.2 to 0.3 m tall, with deep green leaves which are moderately pinnately dissected (similar to *T. patula*). In full shade the foliage is more diffuse, but in part shade it can be very dense.

This species is an exception to the rule of thumb that *Tagetes* needs full sun, as they occur in full shade under pines in the Sierra Madre Oriental near Monterrey, Nuevo Leon, Mexico. Soils are limestone, but not highly alkaline due to the pine duff. Plants responded well in acidic to neutral soils. They tolerate heat to around 37°C, after which they cease flowering, but respond as the days cool. Plants withstand mild freezing, but generally die back to perrinating rootstocks. *Tagetes mulleri*, the shade-loving marigold, adapts to cultivation readily, and prefers full shade for at least the afternoon, if not all day.

***Tagetes nelsonii* Greenm.**

The corollas are yellow-orange; individual heads are small, 10 to 15 mm across, but form corymbiform clusters of 30 to 60 heads, producing a showy display, 15 to 20 cm across. Peduncles do not extend beyond the foliage. Flowering begins in late summer and will continue until killing frost.

The plant is an erect shrub that can reach 2 m, but plants can be readily pruned to 1 m. The leaves are very broadly pinnately dissected, and are a deep green color. In some populations, the leaves are covered above and below with a velvety pubescence. Foliage tends to be very dense.

The species is from the Mayan Highlands of Chiapas, Mexico and Guatemala, where it grows in mesic to hydric conditions. In Austin, Texas, plants froze to the ground each year, but came back from underground rootstock three years in a row. Generally, loamy, acidic soils are best, but clay is tolerated. Full sun is preferred, and partial shade is acceptable.

Natives stated that their goats will not eat this species (Soule 1993a). This unpalatability may extend to other ungulates, notably deer which are detrimental to home landscapes in some areas. *Tagetes nelsonii*, the Mayan marigold, displays good adaptability to cultivation.

***Tagetes parryi* A. Gray**

Corollas are orange-yellow, the heads 4 to 5 cm across, borne singlely on short peduncles which do not extend the heads much beyond the foliage. One to 7 heads appear simultaneously, lasting 7 to 15 days each. Flowering on mature plants is virtually year-round. In stressful heat conditions (over 40°C) the plants may cease flowering until cooler weather returns.

This low mounding shrub is 0.2 to 0.3 m tall. The leaves are moderately pinnately dissected, and are a medium green, not glossy or notably pubescent. With full sun and mild pruning a very dense compact form can be maintained.

The species is a limestone endemic and would make a wonderful addition to the nursery trade, especially in limestone rich areas of the southwest. Plants grow in full sun, mesic to xeric conditions and also grow well in rocky, acid to alkaline, porous to clay, soils. They tolerate light freezes, and produce showy flowers virtually year round.

This species has been collected in a pasture with cattle, horses, and goats, and was notably not grazed by these herbivores (Soule 1993a), therefore, *T. parryi* may be a potential "deer-proof" plant for landscape sites with deer problems.

Adaptability to cultivation is excellent in *Tagetes parryi*, the limestone-loving marigold. Peduncles do not extend the heads much beyond the foliage, a preferred characteristic for a border plant, thus in the landscape, *T. parryi* would best be used to create a low border of continuous color.

ANNUAL TAGETES

Tagetes argentina Cabrera

The floral characters of *Tagetes argentina* make it one of the best potential new annuals. The corollas are pure white, and, although each head is only 5 to 6 mm across, 5 to 25 heads together form showy corymbiform clusters. These clusters are on short peduncles which barely exceed the foliage, and since many branches often bloom at once, the entire plant is generally covered in a blaze of white. Achenes directly sown after the last frost grow to flowering in 8 to 12 weeks, and continue until killing frost. The mounding form is encouraged by early pinching of the young plants.

The plants tend to naturally form low mounding shrubs to 0.3 m tall. The foliage is generally dense, formed of pale green leaves which are narrowly pinnately dissected.

Found in central Argentina on granitic and rhyolitic soils, this species responds well to garden soils from acid to neutral. Sandy soils are preferred, but plants did tolerably well in the clay soils of central Texas. In conditions of full sunlight, rich soil, and ample moisture, it forms a compact, rounded plant. Flowering tapered off in the mid-summer as temperatures rose above 37°C, but resumed with cooler weather.

The overall adaptability to cultivation of *Tagetes argentina*, the Argentinian marigold, is excellent, and this beautiful low annual should quickly find its place in home landscapes.

Tagetes minuta L.

The floral characters of *Tagetes minuta* can best be described as non-showy (this plant is included for its nematocidal qualities). The tiny cream to pale yellow corollas are borne in corymbiform clusters of 10 to 50 heads, on peduncles which tend to barely exceed the foliage. Pinching or pruning of young plants results in numerous flowering branches and a more compact form. Directly sown after the last frost, plants begin to flower in late summer to early autumn. In less sunny or far north regions, flowering may not occur.

This species can easily reach 2 m, but responds readily to pinching or pruning and can be kept below 1 m in height. The leaves are narrowly pinnately dissected, with notable glands on them. The scent of the crushed foliage is a musky, anise-like one (Soule 1993b). Foliage is green to yellowish-green, and can become dense with pruning and full sun.

From the seasonally dry grasslands of southern Brazil, *T. minuta* has the potential to be an excellent companion plant in gardens since it releases secondary compounds called thiophenes from its roots (Grainge and Ahmed 1988). Thiophenes have been proven to kill root knot nematodes (Winoto-Suatmadji 1969), a serious garden pest. In good soil and moisture conditions, the plants can quickly reach 2 m, thus it also provides an excellent backdrop in the landscape. Due to the bactericidal secondary compounds released from the roots, one should not plant this near legume vegetables or legume trees. The plants do well in acid to mildly alkaline, sandy to clay, soils. Full sun in mesic conditions are best, although the plants tolerate some drying.

Adaptability to cultivation is generally good, the lack of flowering in some areas is not detrimental in this species which is best planted for its nematocidal properties. Seed is currently offered by one company as the "nematocidal marigold."

Tagetes remotiflora Kunze

The showy, brilliant orange flowers of this marigold are borne on long peduncles which extend beyond the foliage, creating a striking display. The flowering heads can be 5 to 8 cm across, and once flowering begins, the plants may have as many as 50 heads flowering at once. In good conditions a single head will remain showy for 10 to 14 days. When sown directly in the soil after frost, flowering begins in 6 to 10 weeks, and continues until killing frost. In a mild winter in Austin Texas, plants overwintered, and flowered through the next summer.

The vegetative character of *T. remotiflora* is an additional reason for its recommendation here. Most wild populations are high in anthocyanins, resulting in a deep maroon coloration to stems and leaves, especially along the veins. Several populations have such high levels of anthocyanins that the plants are maroon throughout. This character persisted in greenhouse and common garden conditions. The leaves are moderately pinnately dissected (similar to *T. patula*), and range from green to maroon. Depending on the population, the plants range from 0.5 to 2 m in height. They respond very well to pinching, producing a lower, more compact plant with moderately dense foliage.

This species originates from the wild and remote canyons of Guerrero in western Mexico. This may be one of the progenitors of the cultivated *T. erecta* and *T. patula*, as it is native to the region where maize was first cultivated, over five millennia ago (Soule 1993a). Adaptability to cultivation of *T. remotiflora*, the Guerrero marigold, is excellent.

Tagetes subulata **Cerv. ex La Llave et Lex.**

The white ray corollas have orange centers and orange disk corollas, so that this marigold appears to have a golden eye surrounded by white. The heads are 1.5 to 2.5 cm across, and are borne singlely on long peduncles which greatly exceed the foliage. Plants have 5 to 15 heads at once, each head lasts 10 to 17 days on the average. Flowering occurs 6 to 10 weeks after direct sowing, and will continue with adequate moisture until killing frost.

A low plant, not over 0.3 m tall. The leaves are fimbrilate, or very narrowly pinnately dissected, and are a pale green, becoming almost blue-green in intense light situations. Foliage tends to be dense, forming a full backdrop for the floral display.

The species occurs from central Mexico into Colombia, however the white rayed variety is found only in central Mexico. The plants generally occur in grasslands and seasonally dry forests, in alkaline to acid soils. They tolerate rocky clay soils very well, and grow in full sun to part shade conditions.

Adaptability to cultivation is good. *Tagetes subulata*, the lace-leaf marigold, could quickly replace *T. filifolia* in the trade, since the plants are taller, with showier flowering heads, and a denser yet more delicate foliage.

FUTURE PROSPECTS

Tagetes is a popular and well known genus due to the cultivated annual marigolds, planted around the world. Perennial marigolds are not well known, but have some regional popularity. Despite the attraction of novelty, many home owners tend to be wary of change. Therefore, for landscapers and homeowners in search of novel plants, a genus with name recognition may be more quickly accepted than a wholly new genus.

Based on an evaluation of floral, vegetative, habit, and environmental characters, four species of perennial *Tagetes* (*Tagetes campanulata*, *Tagetes mulleri*, *Tagetes nelsonii*, and *Tagetes parryi*) appear highly suitable as novel new horticultural taxa and include resistance to deer predation, limestone tolerance, and adaptability to shade. These novel *Tagetes* could potentially form the basis for new trade in perennial marigolds as landscaping plants.

In recent years, annual marigolds have lost their lead as the single most popular bedding plants in the United States, in part due to the attraction of novel new species in other genera. Four species of annual *Tagetes* (*Tagetes argentina*, *Tagetes minuta*, *Tagetes remotiflora*, and *Tagetes subulata*) including two white-flowered species, appear highly suitable as novel new horticultural taxa. These four novel *Tagetes* could potentially renovate the trade in annual marigolds as bedding plants.

REFERENCES

Anon. 1900. *Tagetes lemmonii*. Gardener's Chron. 2nd Series 27:18.

Fuchs, L. 1542. Historia stirpium. Officina Isingriniana, Brasileae.

Grainge, M. and S. Ahmed. 1988. Handbook of plants with pest control properties. Wiley, New York.

Gerard, J. 1596. The herball or general historie of plantes. (Republished 1963) Gerald Howe, London.

Lawrence, B.M. 1985. A review of the world production of essential oils (for 1984). Perfumer and Flavorist 10(5):1–16.

Nuttall, Z. 1920. Los jardines del antiguo Mexico. Memoires de la Sociedad Cientifica "Antonio Alazate" 37:193–213.

Ortiz de Montellano, B.R. 1990. Aztec medicine, health, and nutrition. Rutgers Univ. Press, New Brunswick.

Soule, J.A. 1993a. Systematics of *Tagetes* (Asteraceae—Tageteae). PhD Diss. Univ. Texas at Austin.

Soule, J.A. 1993b. Medicinal and beverage uses of *Tagetes* (Tageteae: Compositae). Am. J. Bot. 80(6):177 (Abstr.).

Soule, J.A. 1993c. A potential new herbal product from a South American species of *Tagetes*. p. 649–654. In: J. Janick, and J. Simon (eds.), New crops. Wiley, New York.

Sweet, R. 1817. *Tagetes florida*. Brit. Fl. Gard., Ser. I. 5. t. 35.

Wells, J.M., A.A. Balsinhas, H. Joffe, V.M. Engelbrecht, G. Harding, and C.H. Stirton. 1986. A catalogue of problem plants in Southern Africa. Mem. Bot. Survey South Africa 53.

Winoto-Suatmadji, R. 1969. Studies on the effect of *Tagetes* species on plant parasitic nematodes. H.W. Wageningen, Amsterdam.

MEDICINAL, AROMATIC, SPICE, AND BIOACTIVE CROPS

Drug Discovery and Development at the National Cancer Institute: Potential for New Pharmaceutical Crops

Gordon M. Cragg, James E. Simon, Johnson G. Jato, and Kenneth M. Snader

Plants have formed the basis for the treatment of diseases in traditional medicine systems for thousands of years, and continue to play a major role in the primary health care of about 80% of the world's inhabitants (World Health Organization statistic, Farnsworth et al. 1985). In the area of cancer treatment, many claims have been made for the beneficial effects of plants (Hartwell 1982), though many of these claims may be viewed with some skepticism since cancer, as a specific disease entity, is likely to be poorly defined in terms of folklore and traditional medicine. Nevertheless, the discovery and development of efficacious anticancer agents, such as vinblastine and vincristine isolated from the Madagascar periwinkle, *Catharanthus roseus* (L.) G. Don, provided convincing evidence that plants could be a source of novel cancer chemotherapeutic agents. While the natural product isolated as the active compound might not be suitable for development as an effective drug, it can provide a suitable lead for conversion into a clinically useful agent. This approach is well illustrated by the development of the anticancer drugs, etoposide, and teniposide, as semisynthetic derivatives of epipodophyllotoxin, isolated from *Podophyllum peltatum* L. and *P. emodii* Wall. (Cragg et al. 1993a).

PLANT ACQUISITION AND DRUG DISCOVERY

In 1960, the National Cancer Institute (NCI) initiated a plant collection program in collaboration with the United States Department of Agriculture (USDA) (Perdue 1976). During the next twenty one years, over 35,000 plant samples representing some 12,000 to 13,000 species were collected by the USDA, mainly from temperate regions; over 114,000 extracts were tested for antitumor activity, primarily in the in vivo L1210 and P388 mouse leukemia systems. While many active agents belonging to a wide variety of chemical classes were isolated and characterized (Cragg et al. 1993a), few satisfied the stringent requirements for preclinical and clinical development. The major clinically active agents to emerge from this program were paclitaxel isolated from *Taxus brevifolia* Nutt. and other *Taxus* species (Suffness 1995), and hycamptamine (topotecan), CPT-11, and 9-aminocamptothecin, all semisynthetic derivatives of camptothecin isolated from *Camptotheca acuminata* Decne (Wall and Wani 1993). The structures of these compounds are shown in Fig. 1.

In 1986, NCI expanded its program to the collection of plants from tropical and subtropical regions. Contracts were awarded to Missouri Botanical Garden (Africa and Madagascar), New York Botanical Garden (Central and South America), and the Univ. of Illinois at Chicago, assisted by the Arnold Arboretum and the Bishop Museum in Honolulu (Southeast Asia), and were renewed, after open competition, in 1991. Collections have been performed in over 25 countries, with the contractors working in close collaboration with qualified organizations in each source country. The collaboration of the source country organizations and scientists has been indispensable in the procurement of the necessary collection and export permits, in the successful performance of in-field collecting activities and taxonomic identifications, as well as the provision of facilities for the preparation, packaging, and shipment of the samples to the NCI natural products repository in Frederick, Maryland. In turn, the NCI program has provided support for expanded research activities by source country scientists, and the expansion of source country holdings of their flora through deposition of a voucher specimen of each species collected in the national herbarium. Through its Letter of Collection (LOC), the NCI has committed itself to policies of collaboration with source countries in the drug discovery and development process, and fair and equitable compensation in the event of commercialization of a drug developed from a plant collected within their borders (Cragg et al. 1994). A key provision of the LOC is the commitment to utilize source country resources, either through sustainable harvest or cultivation, in the large-scale production of an agent for preclinical and clinical development. In the event that a drug is licensed to a pharmaceutical company for advanced development and commercial production, the successful licensee will be required to seek as its first source of supply the natural resources available from the source country, provided a mutually agreeable fair price can be determined.

To date, over 45,000 plant samples have been collected by the NCI contractors, and over 40,000 have been extracted to yield more than 87,000 organic solvent and aqueous extracts. These extracts are tested in vitro for selective cytotoxicity against panels of human cancer cell lines representing major disease types, including leukemias, breast, central nervous system, colon, lung, ovarian, prostate, and renal cancers (Boyd and Paull 1995), as well as for anti-AIDS activity in a screen comprising human lymphoblastoid cells infected with the live AIDS virus (Bader 1992). Extracts showing significant activity in either screen are subjected to bioassay-guided fractionation aimed at the isolation of the pure, active agents. Of the more than 44,000 extracts tested so far in the in vitro human cancer cell line screen, less than 1% have shown some level of selective cytotoxicity. In some instances, the patterns of differential cytotoxicity have been associated with known classes of compounds such as cardenolides, cucurbitacins, lignans, and quassinoids, but others appear to be new leads which are being investigated further. Over 36,000 extracts have been tested in the anti-AIDS screen, and approximately 10% have exhibited some in vitro activity; however, most of the active extracts are aqueous and, in the majority of cases, the activity has been attributed to the presence of ubiquitous chemotypes, such as polysaccharides and tannins. Such compounds are not a current NCI focus for drug development and are typically eliminated early in the drug discovery process (Cardellina et al. 1993). A number of in vitro active anti-AIDS agents have been isolated and selected for preclinical development. Of these, michellamine B isolated from the leaves of the Cameroon liana, *Ancistrocladus korupensis*, and the calanolides isolated from *Calophyllum* species collected

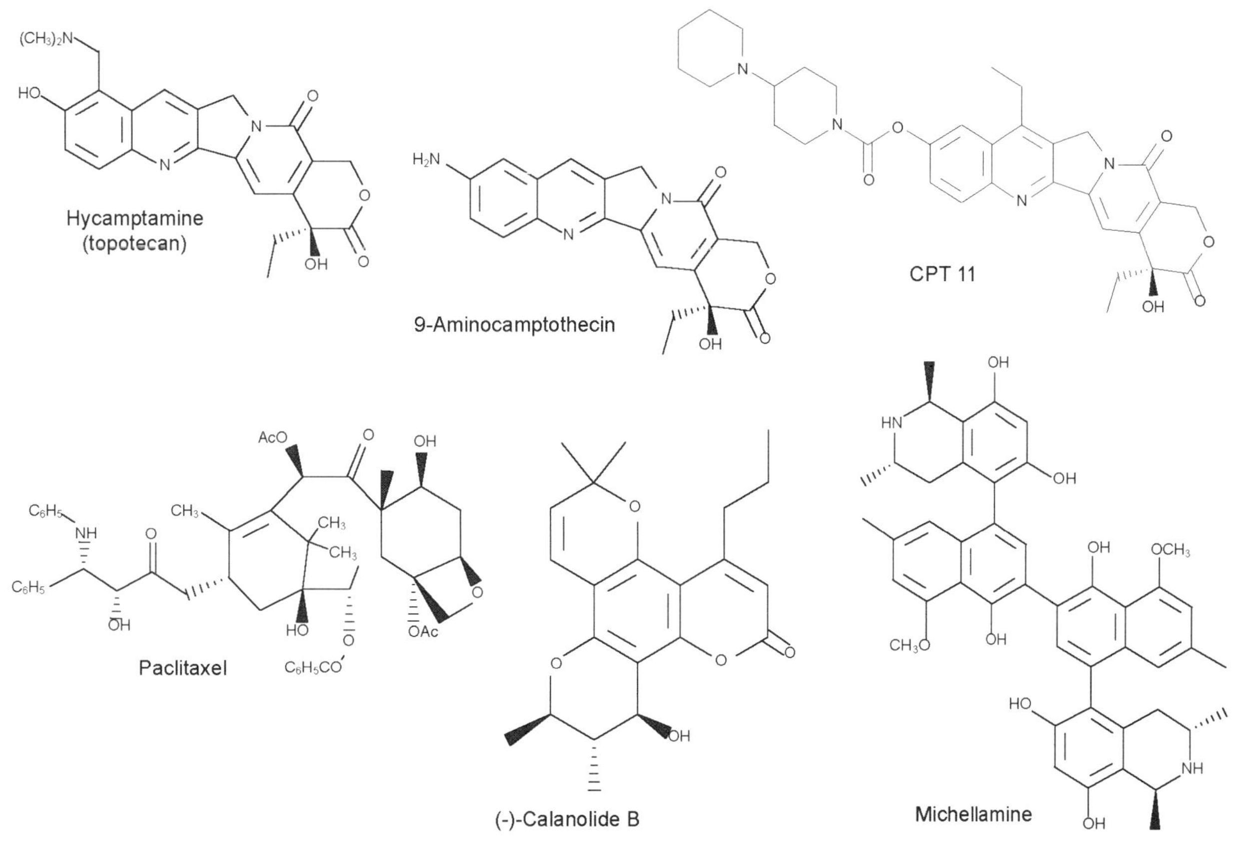

Fig. 1. Clinically active agents identified during plant collection programs conducted by the National Cancer Institute.

in Sarawak, Malaysia, are in advanced preclinical development, and their production is discussed later in this paper.

DRUG DEVELOPMENT: SUPPLY ISSUES AND THE POTENTIAL FOR NEW PHARMACEUTICAL CROPS

Alternative Methods of Production

The isolation of an active compound is the first stage in the development of a new agent which might be developed as a drug for advancement to clinical trials and possibly to commercial use (Grever et al. 1992). While the initial plant sample (0.3–1.0 kg) collected generally yields enough extract (10–40 g) to permit the isolation and structural elucidation of the pure, active constituent, subsequent secondary testing and preclinical development might require gram or even kilogram quantities. Approval of an agent for clinical development could require multi-kilogram quantities.

In order to isolate sufficient quantities of an active agent for preclinical development, re-collections of 5 to 200 kg of the dried plant material might be necessary, preferably from the original collection site. Such large re-collections necessitate surveys to determine the abundance and distribution of the plant, as well as the variation in drug content with the season of harvesting. The feasibility of propagation and the potential for mass cultivation of high-yielding phenotypes of the plant would also need to be assessed. If problems are encountered due to the scarcity of the wild plant or inability to adapt it to cultivation, alternative sources need to be sought. Other species of the same genus or closely related genera may be analyzed for drug content, and other biomass production techniques, such as plant tissue and cell culture can be investigated. Another potential route for bulk production of the active agent is total synthesis, but experience has shown that the complex structures of most bioactive natural products require the development of multi-step bench-scale syntheses which often are not readily adapted to economically feasible large-scale production. Thus, despite over 30 years of extensive research into the synthesis and tissue culture production of the commercial anticancer drugs, vinblastine and vincristine, isolation from the source plant, *Catharanthus roseus*, grown in various regions of the world, remains the most economically viable method of large-scale production.

The Large-Scale Production of Paclitaxel

The development of paclitaxel as an effective drug for the treatment of breast and ovarian cancers illustrates how escalating demands necessitated the development of various methods of biomass production. While the original wild source, the bark of the Pacific yew (*Taxus brevifolia*), provided adequate supplies of the drug for preclinical and early clinical studies, it was soon apparent that the destructive method of harvesting the bark would not meet the large demands resulting from the observation of clinical efficacy against ovarian cancer. An extensive program was initiated to develop alternative, renewable resources , and the current needs are now being met through the harvesting of needles from wild and cultivated *Taxus* species; in addition, scale-up production through tissue culture has been developed by the company, Phyton Inc., but whether such in vitro systems will be commercially viable as a source for paclitaxel or as a source of high taxane-yielding plants is as yet unproven. The evolution and solution of the paclitaxel supply problems has been reviewed in detail (Cragg et al. 1993b).

NCI Policies for Biomass Production

In response to the experience gained in the production of paclitaxel, the NCI has implemented policies which permit the study of various methods of biomass production at an early stage of development of a new anticancer or anti-AIDS agent (Cragg et al. 1993b). Through a Master Agreement (MA) mechanism, pools of qualified organizations have been established with expertise in: the large-scale re-collection of source plant materials; the cultivation of source plants; and source plant tissue culture. Allowance has been made for two phases in the cultivation and tissue culture projects, one involving the initiation of pilot-scale studies aimed at exploring the feasibility of techniques for biomass production, and the second involving the application of methods developed in the feasibility studies to large-scale production. When a plant-derived agent is approved for preclinical development, a Master Agreement Order (MAO) Request for Proposals (RFP) for projects in one or more of the above areas may be issued to the relevant pools of MA Holders who then submit technical and cost proposals addressing the particular RFP specifications. An award is made to the MA Holder whose proposal is judged to be best suited to the Government requirements.

Sustainable Harvesting of *Calophyllum* species. Production of (-)-Calanolide B

Several calanolides have been isolated as potential anti-AIDS drugs from *Calophyllum* species collected in the rainforest regions of Sarawak, Malaysia. Calanolide A is a novel coumarin isolated from leaves and twigs of *Calophyllum lanigerum* Miq. var. *austrocariaceum* (T.C.Whitmore) P.F.Stevens, collected in 1987 as part of the NCI contract with the Univ. of Illinois at Chicago for collections in Southeast Asia (Kashman et al. 1992). Recollections of plant material identified as *C. lanigerum* from the same general location as the original collection failed to yield substantial quantities of calanolide A required for preclinical development. This isomer has not been detected in any related species analyzed thus far, but an extensive analytical survey has shown that the latex of *C. teysmanii* var. *inophylloide* P.F. Stevens collected in the same region yields a related compound, (-)-calanolide B (costatolide) which also shows significant anti-HIV activity (Fuller et al. 1994). Latex collections are made by making small slash wounds in the bark of mature trees. After scraping the latex from the trees, the wounds heal and new collections can be performed. Repeated collections have not affected the health of the trees, and 50 kg has been collected from several hundred trees for the production of sufficient (-)-calanolide B for possible preclinical and clinical development. The latex collections are being performed by the Sarawak State Department of Forests working in collaboration with the NCI collection contractor, the Univ. of Illinois at Chicago. In addition, these two organizations are studying the propagation and cultivation of the source species in Sarawak.

Feasibility Studies of the Cultivation of *Ancistrocladus korupensis*

In 1987, a sample of the leaves of a liana identified as an *Ancistrocladus* species was collected in the Korup region of southwestern Cameroon as part of the NCI contract with Missouri Botanical Garden for collections in Africa and Madagascar. Extracts of the leaves exhibited significant in vitro anti-HIV activity, and the dimeric naphthylisoquinoline alkaloid, michellamine B, was isolated as the active agent (Manfredi et al. 1991; Boyd et al. 1994). The plant was later identified as a new species and named *Ancistrocladus korupensis* D. Thomas & Gereau (Thomas and Gereau 1993). Michellamine B shows in vitro activity against both HIV-1 and HIV-2, and is in advanced preclinical development.

An initial botanical survey indicated that the range of *A. korupensis* was limited to the Korup National Park and that the vines were found in limited abundance with several related species closely resembling *A. korupensis*. Fallen leaves collected from the forest floor were found to contain reasonable quantities of michellamine B, and collections of these leaves should provide sufficient quantities of the drug to complete advanced preclinical development. The collection of fallen leaves thus far, has obviated the necessity for large-scale harvest of fresh leaves, which would be most difficult to collect from a liana, and avoids the possible endangerment of a wild species found endemic primarily in a national park. As larger quantities of leaves would be required if any clinical efficacy is observed, the NCI utilized its Master Agreement mechanism to provide a contract with the Center for New Crops & Plant Products, Purdue Univ., to examine the feasibility of cultivating *A. korupensis* in Cameroon as a potential long-term source of michellamine B. By cultivating high michellamine B yielding plants, the limited wild stands could be preserved and the danger of encroachment into a national park avoided. Purdue Univ. established such a project, working in concert with American scientists such as Roy Gereau and Jim Miller from the Missouri Botanic Garden (the original contracting source which collected this plant), Duncan Thomas from Oregon State Univ., and Cameroon scientists and institutions including the Univ. of Yaounde, the Korup Project, and the World Wide Fund for Nature which is coordinating conservation projects in Korup National Park (Simon et al., 1995). The project is being performed entirely in Cameroon except for the analyses of michellamine B and the entire project staff on site are Cameroonians. This project has the full permission and cooperation of the Cameroon government and is overseen by an Intraministrial Committee for Research on *A. korupensis*. Working in collaboration with the government of the source country is valuable because at present few legal regulations address bioprospecting of plants for non-timber use in Cameroon (Jato et al. 1996).

This project has several objectives. First, to complete a botanical survey of the region so that the total number of wild vines could be determined as well as to identify high michellamine B yielding vines. To accomplish this objective a rapid and quantitative assay was needed. An NCI contractor, Science Applications International Corporation (formerly Program Resources, Inc.), Fort Detrick, MD, developed a very efficient quantitative assay for michellamine B based upon HPLC. Over 1,000 analyses have already been performed and this cooperation enabled the project team to identify high michellamine B yielding phenotypes (Simon et al. 1995).

More than 1,000 kg of dried leaves from the forest floor were also collected in the first year of this project. Leaf collections were made in different regions by a team of locally-hired and trained leaf collectors. Leaves were air-dried, bulked in sacks and stored. The alkaloids appear very stable and not subject to easy degradation. The bulk collection of leaves provides a buffer of raw material for future NCI preclinical studies. As fallen leaves do contain considerable amounts of michellamine B, this technique can be used on a recurrent basis as a sustainable method to collect leaves without any harvesting or cutting of the wild vines. Such a technique can be used for both native stands as well as for cultivated vines (Thomas et al., 1994).

The geographical areas of collection and the averaged michellamine B content from each of the areas was calculated based upon subsampling of the leaf collection bags. Results indicate that the Rengo Camp and the Ekundo–Kundu areas yielded the highest michellamine values averaging 5.5% and 4.5% (w/w basis), respectively. This was followed by Akpasang, Chimpanzee Camp, Mana River, and the lowest, Ikassa. The very high values from the Rengo Camp and the Ekundo–Kundu sites need to be reconfirmed, but suggest that wild vines from these two areas should also be relatively higher in michellamine B than vines from other areas. The completed botanical survey identified several new populations, though all were still in the Korup region. An estimated 10,000 vines are in the wild, and plants are found at an elevational range of 50–160 m in skeletal highly leached soils and with soil pH about 4.0.

An analysis of 791 samples indicated that the average michellamine B content in leaves was 2.11% (dry wt.). This distribution reflects all manner of samples that we collected (single leaf, leaf clusters, and leaves of varying ages). The highest levels of michellamine B were found in mature leaves but young, fully expanded pale green leaves also contained michellamine B. Younger leaves and the older, fallen, brown leaves had significantly less alkaloid (Simon et al. 1995). More than 400 individual vines were sampled for michellamine B and variation between vines was significant. Despite problems in sampling procedures, high michellamine B vines were identified and are being vegetatively propagated. Of interest to note is that some samples appeared devoid of michellamine B.

Many *A. korupensis* seedlings were also collected from the forest by digging the plants up and replanting into polyethylene plastic bags. While lower levels of michellamine B were expected from these young plants, their relative concentrations could be effective markers for high michellamine B. Seedling plants which already exhibit higher michellamine B levels, given the similar age and sampling techniques, may represent a genetic source that will continue to exhibit higher levels of michellamine B over time. Variation in plant growth and in michellamine B content among seedlings was also observed. Plants with the highest michellamine B content (0.6%–0.8% dry wt.) were identified and are also being propagated.

In parallel to the large leaf collection, wild vines were assayed several times for michellamine B content. Samples indicating high values of this dimeric alkaloid were then re-sampled for verification. Vines containing michellamine B contents >3.5% were targeted for propagation. By the end of Year 1, many of the high michellamine B yielding plants were being vegetatively propagated. Methods to improve the vegetative propagation success rate are being investigated. As all work is taking place in the jungle, specialized facilities to propagate and maintain the propagules were needed. Therefore, a medicinal plant nursery was designed and constructed by Purdue and the Korup Project staff at the Korup National Park. This new facility now provides, for the first time, specialized propagation units and a nursery for the *Ancistrocladus* plants which now comprise our germplasm collection.

Field plots have been established including unreplicated demonstrations which provide initial insight into the growth patterns of the plants once introduced into fields of full-sun, shade, and into darkened forest. Seven field studies using more than 5,000 plants are ongoing—all under varying environmental and field conditions which include cleared areas, secondary growth areas, mature forest, and underneath mature palms. In additon, plant population and fertilizer studies in open cleared areas are underway. Plantings in the forest are situated under varying light conditions. Although under natural conditions the seedlings are found only in shaded conditions, plants growing under full-sun are growing rapidly. The plants appear to be responsive to both sunlight and fertilizer, such as nitrogen. Plants growing in full-sunlight look good and have exhibited robust and large leaves. As the plantings mature, trellis systems or trees which the *A. korupensis* plants can eventually use for structural support will be needed. This will provide needed background information and test various shading and trellising systems. Both low and high input production systems within agricultural and agroforestry systems are being explored.

The collection, germplasm preservation, and horticultural studies should permit us to develop strategies to collect the leaves and introduce this plant into cultivation, whether in an open field, through enrichment of the jungle with high yielding clones, or in fallow ground. All this is being done in a manner compatible with the needs of the Cameroon people, the Cameroon government and the preservation of the rainforest to which the plant is endemic.

These studies become integral to drug discovery programs for several reasons. First, we need to ensure the availability of the raw plant material. In general, such a discovery is often focused on a wild plant that has not previously been studied—chemically, botanically, or horticulturally. The procurement of the raw plant material is not always easy. In the case of *A. korupensis*, this plant was not used in local medicine and, thus, is not part of the regional ethnopharmacopia. This suggests that a random screening of flora was successful, since a study only of locally used medicinal plants would never have led to the discovery of this plant. Secondly, as the discovery of new drugs from plants should not infringe upon the natural and undisturbed forest areas to which they may be native, the establishment of partnerships between the countries to which a potential drug plant candidate is native and the country that seeks to develop a drug from the plant is critical. Most tropical countries lack the financial support to initiate and sustain such a preservation and development program.

CONCLUSION

Natural products and plant-derived products continue to be excellent sources of new drug candidates. A program such as the one that has evolved at the National Cancer Institute can achieve three of the most important goals. It can be successful at helping to protect the rights and benefits of the source country which provides the new bioresource. It can anticipate and reduce the problems of providing adequate biomass supplies for drug studies by encouraging early cultivation and plant tissue culture programs. Finally, it can achieve both of these goals in a manner compatible with the needs of the source country, and still be successful in our third and primary goal, that of discovering new pharmacologically useful drug candidates.

REFERENCES

Bader, J.P. 1992. The AIDS antiviral screening programme of the National Cancer Institute, p. 271–275. Biotechnology International. Century, London.

Boyd, M.R., Y.F. Hallock, J.H. Cardellina II, K.P. Manfredi, J.W. Blunt, J.B. McMahon, R.W. Buckheit Jr., G. Bringmann, M. Schaffer, G.M. Cragg, D.W. Thomas, and J.G. Jato. 1994. Anti-HIV michellamines from *Ancistrocladus korupensis*. J. Med. Chem. 37:1740–1745.

Boyd, M.R. and K.D. Paull. 1995. Some practical considerations and applications of the National Cancer Institute in vitro anticancer drug discovery screen. Drug Devel. Res. 34:91–109

Cardellina II, J.H., M.H.G. Munro, R.W. Fuller, K.P. Manfredi, T.C. McKee, M. Tischler, H.R. Bokesch, K.R. Gustafson, J.A. Beutler, and M.R. Boyd. 1993. A chemical screening strategy for the dereplication and prioritization of HIV-inhibitory aqueous natural products extracts. J. Nat. Prod. 56:1123–1129.

Cragg, G.M., M.R. Boyd, J.H. Cardellina II, M.R. Grever, S.A. Schepartz, K.M. Snader, and M. Suffness. 1993a. Role of plants in the National Cancer Institute Drug Discovery and Development Program. p. 80–95. In: D.A. Kinghorn and M.F. Balandrin (eds.), Human medicinal agents from plants. ACS Symposium Series 534. American Chemical Society, Washington DC.

Cragg, G.M., S.A. Schepartz, M. Suffness, and M.G. Grever. 1993b. The taxol supply crisis. New NCI policies for handling the large-scale production of novel natural product anticancer and anti-HIV agents. J. Nat. Prod. 56:1657–1668.

Cragg, G.M., M.R. Boyd, M.R. Grever, T.D. Mays, D.J. Newman, and S.A. Schepartz. 1994. Natural product drug discovery and development at the National Cancer Institute. Policies for international collaboration and compensation. Missouri Botanical Garden Monograph Series. 48:161–167.

Farnsworth, N.R., O. Akerele, A.S. Bingel, D.D. Soejarto, and Z. Guo. 1985. Medicinal plants in therapy. Bul. World Health Org. 63:965–981.

Fuller, R.W., H.R. Bokesch, K.R. Gustafson, J.H. Cardellina II, J.B. McMahon, G.M. Cragg, D.D. Soejarto, and M.R. Boyd. 1994. HIV-inhibitory coumarins from the latex of the tropical rainforest tree *Calophyllum teysmanni* var. *inophylloide*. Biorganic Med. Chem. Letter. 4:1961–1964.

Grever, M.R., S.A. Schepartz, and B.A. Chabner. 1992. The National Cancer Institute: Cancer drug discovery and development program. Seminars in Oncol. 19:622–638.

Hartwell, J.L. 1982. Plants used against cancer. A survey. Quarterman Publications, Lawrence, MA.

Jato, J., J.E. Simon, P. Symonds, A. Koufani, and A.B. Njamnshi. 1996. Rules and regulations on the collection in Cameroon of biological materials for biological testing and drug discovery. J. Ethnopharmacol. (in press).

Kashman, Y., K.R. Gustafson, R.W. Fuller, J.H. Cardellina, II, J.B. McMahon, M.J. Currens, R.W. Buckheit, S.H. Hughes, G.M. Cragg, and M.R. Boyd. 1992. The calanolides, a novel HIV-inhibitory class of coumarin derivatives from the rainforest tree, *Calophyllum lanigerum*. J. Med. Chem. 35:2735–2743.

Manfredi, K.P., J.W. Blunt, J.H. Cardellina, II, J.B. McMahon, L.K. Pannell, G.M. Cragg, and M.R. Boyd. 1991. Novel alkaloids from the tropical plant *Ancistrocladus abbreviatus* inhibit cell-killing by HIV-1 and HIV-2. J. Med. Chem. 34:3402–3405.

Perdue, Jr., R.E. 1976. Procurement of plant materials for antitumor screening. Cancer Treat. Rpt. 60:987–998.

Thomas, D.W., M.R. Boyd, J.H. Cardellina II, R.E. Gereau, J. Jato, and P. Symonds. 1994. Sustainable harvest of *Ancistrocladus korupensis* (Ancistrocladaceae). Leaf litter for research on HIV. Econ. Bot. 48:313–414.

Thomas, D.W. and R.E. Gereau. 1993. *Ancistrocladus korupensis* (*Ancistrocladaceae*): A new species of liana from Cameroon. Novon 3:494–498.

Simon, J.E., J. Jato, P. Symonds, D. Thomas, M.R. Morales, R. Gereau, B. Alkire, E. Jato, and T. McCloud. 1995. Feasibility studies on the cultivation of the Cameroon plant, *Ancistrocladus korupensis*. First Annual Progress Report to the National Cancer Institute, March 10, 1995.

Suffness, M. 1995. Taxol®. Science and applications. CRC Press, Boca Raton, FL.

Wall, M.E. and M.C. Wani. 1993. Camptothecin and analogues: Synthesis, biological in vitro and in vivo activities, and clinical possibilities. p. 149–169. In: D.A. Kinghorn and M.F. Balandrin (eds.), Human medicinal agents from plants. ACS Symposium Series 534, Am. Chem. Soc. Washington, DC.

Ethnobotanical Approaches to the Discovery of Bioactive Compounds

Trish Flaster

The biological diversity of our world is great and we have only begun to investigate her potential. In some areas diversity may be more valuable in its natural state than when used for pasture or timber (Peters et al. 1989). For example, one highly valued marketable commodity of tropical areas is medicinal products. At the present time, half the drugs on the market are derived from natural sources (Reid et al. 1993) and 47 marketed drugs have been derived from 39 tropical forest plants (Farnsworth 1988). Clearly, bioactive molecules from botanicals can continue to be derived from the tropics.

Methods to identify medicinal plant leads from tropical areas include random screening, taxonomic collecting (sampling by botanical family), or ethnobotanical collecting. It has been shown that ethnobotanically-derived compounds have greater activity than compounds derived from random screening and therefore a greater potential for product development (Balick 1990). This paper will concentrate on the ethnobotanical discovery process of plants being used as healing agents as well as environmental and social issues.

Many issues have caused an erosion of the tropical regions of our planet, but for many years only the biological factors associated with this loss have received investigation (MacKerron and Cogan 1993). Recently, it has been recognized that areas of great biological diversity overlap with those of cultural diversity (Manuel Lizzaralde, pers. commun.); when either is disturbed the other is compromised. Clay (1988) and Posey and Balee (1989) have established that the conservation of plants alone was inadequate to maintain an ecosystem because humans are a critical component. Plants are nurtured and manipulated by their human co-inhabitants and only when both are considered as an integral functional group can preservation and conscientious use of species occur.

ETHNOBOTANY

Ethnobotany is a multidisciplinary science defined as the interaction between plants and people. The relationship between plants and human cultures is not limited to the use of plants for food, clothing, and shelter but also includes their use for religious ceremonies, ornamentation, and health care (Schultes 1992). In the past, ethnobotanical research was predominately a survey of the plants used by villagers. A trained botanist identified the plants and recorded their uses. Sometimes an anthropologist was present to translate the disease descriptions, but rarely was a physician available to identify the disease. The results generated a list of plants and their uses which was published in a professional journal, usually in the country of the scientist. Nothing was communicated or returned to the cultural group in exchange for their participation in the survey, nor was any environmental or cultural status or concerns included in the survey.

Basic quantitative and experimental ethnobotany includes basic documentation, quantitative evaluation of use and management, and experimental assessment (Martin 1995). Today, ethnobotanical surveys include applied projects that have the potential to ameliorate poverty levels of these people, allowing them to make more educated decisions about their future directions. These new approaches enhance the quality of the science, provide compensation for the cultural groups, and take into account environmental concerns. This modern approach is based on an interdisciplinary team usually composed of an ethnobotanist, an anthropologist, an ecologist, and a physician. Some of these team members are in-country colleagues who have arranged the details of the expedition as well as the contractual agreements for reciprocal programs of the village or community.

ORGANIZING AN ETHNOBOTANICAL EXPEDITION

The initial steps to this approach are critical. Any errors will jeopardize the entire project. To aggressively enter a community or even a country and take out the information is rarely a successful process. However, abiding by the in-country federal requirements will support the goals of a long-term, mutually productive relationship. The negotiations can be done through a trusted colleague, either a person who has worked within the country and is familiar with these procedures or an in-country person. All permits and permissions must be in order prior to the expedition. This process can taken months or years, but regardless of the time period, must be done properly. Many countries are limiting their genetic resources from scientists because of unhappy past experiences and are slowly devising new guidelines. Some, such as the Philippines, are initiating federal legislation which incorporates reciprocity while most others negotiate on an individual basis.

Prior to the expedition, the in-country collaborator can communicate with the village to find out if they have any immediate needs. For example, a five-member team entering a small village must either provide its own food or shelter or provide funds for the villagers to acquire these things. Sensitivity to the community's needs will lay the groundwork for a trusting relationship. The process should be considered as the normal good manners of invited guests. On a scientific level a sincere relationship is needed to ascertain the real data. A superficial personal relationship will yield a superficial scientific relationship (Joly 1992). Many plants are used in village pharmacopoeias, but only a few powerful ones will make it as pharmaceuticals or powerful dietary supplements.

When a research team arrives at the village a customary welcoming ceremony should occur. This is when permission to work with the healers is requested from the village leader or chief. It is most appropriate at this time to demonstrate your support by offering reciprocity for the information you are about to receive. Reciprocity can be clean water systems, books for their schools, health professionals to visit on a regular basis, or any other program they request. The key here is that it be their request. Most of these requests are not exorbitant, certainly not in excess of $5,000, but this is in addition to payment for the herbs you will be collecting. Be aware of the cultural requirements for payment of the plant collections. Ask your in-country liaison because there are some places where payments are unacceptable and initiate divisiveness between community members.

How the ethnobotanical information is gathered is often a pre-planned technique of the anthropologist or ethnobotanist. The technique depends on whether you want to concentrate on particular disease categories or a general survey, have time for just an interview, or can participate in the process for several months. The key is to elicit objective answers to your questions and to be sure that the diseases you are describing are being understood by both parties. Using signs and symptoms to describe the disease allows for an equal exchange of information. Titles and names can be misleading and unknown. Some native people are often eager to be helpful and may respond positively even if the answer is incorrect. Visual aids are helpful vehicles. This portion of the interview may taken place in the field, in the healer's home, or in a community meeting place. Depending on the culture, it can be with one healer or if the village has several, they may want to offer you a group response. Often because your presence is novel, everyone wants to take part. If the latter is the case, devise plans to work in smaller groups without offending anyone. Often the true healers will stand out from the other general healers.

As the healers identify the plants they use to treat specific diseases, data is recorded, the plants are collected, and voucher specimens are taken. Not all expeditions collect plants, but vouchers are always taken and numbered correlations can be made to the collections. A minimum of one of the collected vouchers needs to be left within the country, another archived at an herbarium for positive identification, and the final one stored under appropriate conditions in the researcher's laboratory. The recording of detailed information during the expedition is crucial for isolating the active molecule and obtaining reliable and reproducible collections for a potential product. No detail is expendable. All the data is entered into a database for future and continual use and returned to all collaborators.

After the plants have been identified to genus, and hopefully species, it is extracted and tested biologically for activity. The critical path is to include the ethnobotanical data into these laboratory processes. The natives are the ethnobotanical technicians and translation of their technical knowledge is required for the discovery of ethnobotanically derived medicines (Schultes 1992). If there is a separation of the field data from the chemistry and biological laboratory assays, the ethnobotanically derived molecule may never be found, or worse, activity could be null. From this point on the process involves chemical and biological assays to isolate the active molecule. This can take from one to nine months or more depending upon the complexity of the molecule and the efficiency and technical competence of the research team. During this time you may need to be acquiring more of the raw botanical material. This allows time to collect more data and compare shipments. Your researchers may know how much raw material would be required from beginning to end of the isolation process. Obtaining this quantity before the process begins may be standard company procedure. Using one lot of material from beginning to end will remove any biological and chemical variation that could exist. However, if collected at the wrong time, activity may only be present in small amounts, if at all.

When isolation and identification of the bioactive molecule is completed and activity is demonstrated, synthesis still may be infeasible or impractical, necessitating further research. Plants have evolved wondrous systems to protect themselves, and it is these "secondary metabolites" that provide flavors and healing substances. However, many plant species have similar metabolic pathways. Running searches to find plants which contain the greatest amount of the same active compound may offer you a choice of species to work with. In some cases, this is very desirable because the plant may be endangered or difficult to import from the country in which it was first found. In these cases the compensation package must reflect these situations.

SUSTAINABLE CROP PRODUCTION

If you are to acquire more material of a plant already researched, then the challenge begins. If the quality of science was high in your original fieldwork then you have all the details you need to expand the horticultural aspects of the project. If not, collect the missing data on the return to the field. This new data may not have been obvious to the healers or yourself before, yet is important at this stage of the project's development. People have different concepts of time and business so these differences need to be taken into consideration in your scientific planning.

At this point, research funds are needed to determine the factors that may enhance the yield of the bioactive molecule. One needs to speak with the healers and understand how, when, and where they harvest, details about drying, and any other particulars about the plant's collection and use. The subtleties are important. We assume or forget that most native peoples in the tropics are excellent managers of their natural resources (Alcorn 1989; Posey 1992) and the best approach may be to observe and integrate their traditional practices into your sustainable program. Additional studies on distribution and density surveys, soil sampling, and multiple plant part sampling are all needed parameters within the design of an ecologically productive program.

A less obvious goal for most agriculturists is a verification of a potential market for this commodity in-country. Many of the plants being investigated may not yield the "pharmaceutical magical bullet" that is desired and having alternative uses such as dietary supplements may offer your collaborators a viable commodity if your goals change. Regardless of the product's end point, the ethnobotanical process can still enhance cultural values and this can have a long-term consequence. This occurs by expanding markets, initiating market economies for the community, and demonstrating to the youth of the community the validity of the elder's plant knowledge.

Several options exist to proceed once the raw data about the plant is compiled. Extractive reserves (Duke 1981) provide forest dwelling peoples an ecologically managed decentralized system of collection. The indigenous groups or communities, if adequate lands are made available, can grow or actively manage economically valuable plant populations within this reserve. Land can also be dedicated as an ecological buffer zone and comparative study plots can be implemented. These plots would demonstrate if the collections were sustainable by comparing the populations in the buffer preserved zone to the actively managed area. The effect of increasing or decreasing a specific plant's density and diversity in secondary forests will be visible and cultural methods that have little impact on the forest can be designed from the data.

Another approach is to work with individual land owners to grow or manage their lands to provide the desired crop. Working with one person who has acquired land allows for direct and responsive communication. Compensation requirements are simplified. They include paying land and community taxes, require participation in community functions, and direct payment to individuals for services or products rendered. The employment of the local community members offers physical protection for the land and often others begin to request participation onto their parcels. The pride and knowledge this creates will assist you in protecting and developing the land, crop, and future product.

Another approach is to work within a community to provide adequate raw botanicals. Once the community has been identified, an all inclusive meeting needs to be convened through the governing body. The purpose of the meeting is to outline your goals and coalesce them with the community's immediate and future needs. For example, what are their political and compensation needs? Many are required to pay imposed federal costs that are beyond their means. This forces them to sell their timber which eliminates agricultural land or hunting grounds. We have the ability to offer them a choice and these meetings can express these potentials.

The decision of who or where to work can be determined by the original field expedition, but if the community has declined participation, another community within the plant's growth range may be chosen. There are advantages to working within communities that are familiar with the cultural practices and use of this plant, but working in a community that has no past history may allow you to readily introduce new ideas (R. Bye, pers. commun. 1985).

When the community agrees with your project their inclusion in the decision making processes and management of the project is necessary for a successful program (Parione 1995). Without their participation there is always a lack of conceptual understanding which had lead to many failures in the past. How you work within the community is determined by their infrastructure. It may be that you contract each member to produce a little or that there is communal land in which the crop can be grown and profits shared. From my experiences, I strongly urge that one person be appointed or voted as leader of the project. This person helps to complete the work on time by communicating and directing the needs of the group and the goals of the project. Training or supplies may be needed. Typical concerns are how to obtain specified yields, coordinate research and test designs with local or foreign laboratories, acquire files to sharpen tools, hire additional staff, supply more food, purchase or make materials to establish seed beds or seedling containers, implement integrated pest management systems, or identify pollinators or predators of the desired plants. The investments required are not philanthropic expenditures, but the true cost of doing business. Your project will take time and needs must be met during this time.

FUTURE PROSPECTS

Culturally sensitive and environmentally sensitive programs are required to successfully acquire data from primitive tropical areas. Valid approaches will become more evident as more collaborative attempts are completed. However, at present, there are not enough data to know which programs are environmentally and socially appropriate. Different programs that are species specific may be required for sustainable production. Very little is known at present on how to even define sustainability. However, we now know that there is a potential for many natural products and that there are many communities who are eager to open their lives to market economies to provide more opportunities for themselves.

Ethnobotanical expeditions can be conducted anywhere that contacts are available and participants agree to cooperate. However, ethnobotanical collecting is a multidisciplinary effort that requires sensitivity to both land and people. The discovery and exploitation of natural products must be a shared program for maximum success. Agricultural programs have other needs and there are many complex decisions that must be analyzed before ascertaining the most appropriate and rewarding approach.

REFERENCES

Alcorn, J. 1989. Bora and Huastec resource management. In: Resource management in amazonia: Indigenous and folk strategies. Advances in Economic Botany. Vol. 7, New York Botanical Garden, Bronx, New York.

Balick, M.J. 1990. Ethnobotany and the identification of therapeutic agents from the rain forest. In: Bioactive compounds from plants. Ciba Foundation Symp. 154.

Clay, J. 1988. Indigenous peoples and tropical forests. Cultural Survival, Cambridge, MA.

Duke, J. 1981. Magic Mountain, 2000 AD. Background Paper for Innovative Biological Technologies for Lesser Developed Countries. OTA Workshop, Nov. 24–25, 1980.

Farnsworth, N. 1988. Screening plants for new medicines. In: E.O. Wilson (ed.), Biodiversity. National Academy Press, Washington, DC.

Joly, J.G. 1992. The secrecy of Indian ethnobotanical data banks. In: Mark Plotkin and Famolare (eds.), Sustainable harvest and marketing of rain forest products. Island Press, CA.

MacKerron, C. and D. Cogan (eds.), 1993, Business in the rainforests. Investor Responsibility Research Center, Washington, DC.

Martin, G. 1995. Ethnobotany. Chapman and Hall, London, New York.

Posey, D.A. and W. Balee (eds.), 1989. Resource management in Amazonia: Indigenous and folk strategies. Advances in Economic Botany Vol. 7, New York Botanical Garden, Bronx, NY.

Posey, D.A. 1992, Traditional knowledge, conservation, and the rain forest harvest. In: Mark Plotkin and Famolare (eds.), Sustainable harvest and marketing of rain forest products. Island Press, CA.

Parione, M. February, 1995. Presentation of the case study on Yanesha Forestry Cooperative at Forestry in the Americas. Community-Based Management and Sustainability. Madison, WI.

Peters, C. et al., 1989. Valuation of an Amazonian rainforest. Nature June 29:655.

Reid, W. et al. 1993. Biodiversity prospecting: Using genetic resources for sustainable development. World Resources Inst., Washington, DC.

Schultes, R.E. 1992. Ethnobotany and technology in the Northwest Amazon: A partnership. In: Mark Plotkin and Famolare (eds.), Sustainable harvest and marketing of rain forest products. Island Press, CA.

Ginseng: Old Crop, New Directions

John T.A. Proctor*

Ginseng (*Panax* spp., Araliaceae) is an herbaceous perennial cultivated for its highly valued root in northeastern China, Korea, and the northern United States, and Canada. The Chinese have used ginseng for thousands of years because of its reputed anti-stress, anti-fatigue, and anti-aging properties (But et al. 1995a). Ginseng use in Western medicine is recent, and expanding, probably because of a growing interest in alternative treatments using natural medicines such as herbs.

North American ginseng production has been reviewed from both Western and Oriental publications on the most important aspects of the industry, botany, and culture of ginseng (Proctor and Bailey 1987). An international ginseng conference organized in 1994 brought the world ginseng community together for the first time (Bailey et al. 1995). The focus of this paper is to review the present industry with special attention to the challenges that confront the industry in the next decade.

*I am indebted to Dean Louttit, Heather Proctor, Jan Schooley, and Gretchen Bozak for their assistance in preparing this paper.

World ginseng production and marketing is at a crossroads. In the marketplace there will be increased competition and traditional methods of marketing will be challenged. For instance, North American producers have sold most of their dried root at the farm to buyers, who export and trade in the Hong Kong market and subsequently distribute and retail. With increased production new approaches are needed to secure guaranteed markets for the North American ginseng industry to remain viable and profitable. Marketing approaches that must be tested include vertical integration, value-added production, and Asian joint ventures.

All phases of ginseng cultivation practices are being questioned. In North America the traditional wooden lath shade is being replaced by black polypropylene shade and every aspect of production from seeding to harvesting and drying is being evaluated for increasing efficiency. In Korea traditional labor sources are vanishing so new shade structures and mechanization are being assessed. As production methods change increasing attention is being given to environmental concerns about fertilizers and pesticides. The ginseng production system in the next decade will include an integration of horticultural, biotechnological, and agrochemical methods in an acceptable economic package that optimizes root quantity and quality and safeguards the environment.

PHARMACOLOGY

There is an enormous literature on the medicinal chemicals in ginseng and their testing to determine biological activity. There were at least 1191 citations that appeared between 1687 and 1975 (Bae 1975). Since that time there has been a dramatic increase in ginseng publications. Books (Lee 1992; Fulder 1980; Duke 1989), and proceedings of conferences (Bailey et al. 1995; Ha et al. 1987) have been published. The Korean Journal of Ginseng Science has been published since 1976. However, despite this vast literature there have been few Western publications.

Medicinal Chemicals

Proctor and Bailey (1987) traced the history and identity of the biologically active components in ginseng. A range of chemical components with biological activity including ginsenosides, alkaloids, polypeptides, and polysaccharides have been isolated (Duke 1989; Liu and Xiao 1992; But et al. 1995a). Shabita et al. (1963) laid the foundation for contemporary ginseng pharmacology by isolation and identification of individual ginsenosides. By 1992, Liu and Xiao reported that 28 ginsenosides had been found in different parts of the ginseng plant. But et al. (1995a) summarized the range of ginsenoside effects and included cardio-protective, immunomodulatory, anti-fatigue, and hepato-protective. Lee (1992) has provided a summary of Korean ginseng (*Panax ginseng* C.A. Meyer) effects as follows:

- Reduces fatigue, increases stamina, and strengthens the general physical condition.
- Exerts a haematopoetic action (formation of red blood cells), protecting against anaemia, hypotension, and heart ailments.
- Improves mental conditions, preventing neurosis and nervous breakdowns.
- Increases the secretion of body fluids and quenches thirst, preventing diabetes.
- Normalizes pulmonary functions, preventing coughing, tuberculosis, and asthma.
- Strengthens the gastrointestinal system, preventing gastroenteritis, diarrhea, and constipation.
- Detoxifies poisons, preventing skin irritation, inflammation, and skin diseases.

But et al. (1995a) claimed that using ginsenosides to evaluate the quality of ginseng products appears to yield results which match some of the traditional empirical preferences.

For American ginseng (*P. quinquefolium* L.) to receive wider acceptance, documented scientific research will be necessary to confirm findings. Work with both American and Korean ginseng will need to be carried out in Western Laboratories. A small start has been made. For instance, Wiklund et al. (1994) showed that *P. ginseng* prepared as standardized product (Pharmaton SA, Lugano, Switzerland) improved the quality of life in a middle aged Swedish population. In Canada, a Ginseng Research Foundation has been formed by ginseng producers and University researchers to support scientific research on the production, properties, uses and clinical effects of ginseng grown in Canada (Francis 1995). This is a small effort but it indicates recognition that for wider acceptance of ginseng and its products supporting clinical testing is essential. The Swiss pharmaceutical company, Pharmaton SA, a successful marketer of ginseng in Europe since the 1960s, agrees (Faleni 1995).

Authenticity

Ginseng is an expensive crop to produce so adulteration or substitution with other cheaper products occurs. To police the marketplace various chemical methods have been developed to ensure that roots (e.g. Oriental vs. American) and value-added products are true to name as indicated on the label.

Oriental ginseng can be distinguished from American using ginsenoside profiles because Oriental contains ginsenoside Rf and American does not (Tanaka et al. 1986; Lang et al. 1993). Protein analysis can be used to distinguish between ginseng species and any included adulterants (see But et al. 1995a for references). Using ginsenoside profiles (Cui et al. 1994), Liberti and Der Marderosian (1978), and Rückert (1980) showed that some commercial value-added products did not contain ginsenosides. In the Liberti and Der Marderosian (1978) study 8 of 23 products did not contain ginseng. In some cases the label indicated that the products contained Siberian ginseng (*Eleutherococcus senticosus*) or Brazilian ginseng (*Pfaffia paniculata*) neither of which contain ginsenosides. The inclusion of ephedrine in a product led to an accusation of doping of a Swedish athlete (Cui et al. 1994).

A comprehensive ginseng evaluation program has been initiated by the American Botanical Council to determine if adulteration has occurred in 250–300 commercial ginseng products (Blumenthal et al. 1995). This program should increase consumer confidence and correct problems in manufacturing and labelling. Manufacturing and retailing reports indicate that ginseng has increased its position and profile in the marketplace. For this trend to continue into the next century, all sectors of the global ginseng community will benefit by ensuring that the consumer has confidence in all products on the market.

PRODUCTION AND TRADE

Estimates of ginseng production in the major centres around the world are given in Table 1. Production statistics are difficult to locate and even if found, reconciliation between sources is often difficult.

World ginseng production increased from 1978 to 1985 (Table 1). Since then South Korean production increased to about 5000 t (Table 1) but has remained around this figure (Park and Kim 1995). Annual production figures often reflect growing conditions for the crop, particularly the incidence of disease (see below) rather than change in crop area. The variation in Wisconsin production from 1985 to 1994 (range 519 to 736

Table 1. Estimated world ginseng production in selected years. Sources: Polczinski 1982; Proctor and Bailey 1987; Anon. 1990; Proctor et al. 1990; Liu and Xiao 1992; Park 1994; Statistics Canada (Ginseng, Canadian Exports 1994); Wisconsin Dept. of Agr. Trade and Consumer Protection (Wisconsin Cultivated Ginseng Export Program 1994); and Park and Kim 1996.

	Production in dry wt(t)			
Country	1978	1985	1990	1993
South Korea	2,300	3,140	4,590	5,000
China	600	1,560	3,350	5,000 est
United States	200	519	522	692
Japan	125	100	65	38
Canada	40	102	232	347
Total	3,265	5,421	8,759	11,077 est

t; mean 623 t) likely reflects high incidence of disease rather than grower changes in crop area (Table 2).

In Canada, ginseng production has increased steadily in the last 10 years with British Columbia's first crop sold in 1986 (Table 2). My projection is that Canada's production will not increase beyond present levels.

The major unknown in world ginseng is Chinese production and its marketing. One development in China has been to grow American Ginseng, known in China as Wulong ginseng for local markets to replace costly imports, and also possibly export (Wang et al. 1995). The project was initiated in 1975 and large scale planting is now taking place after much research and development.

Another problem associated with Chinese ginseng root is that of adulteration of American root and value-added products in the Hong Kong market (But et al. 1995b). American Ginseng is 5 to 10 times more expensive than comparable Chinese ginseng in the Hong Kong market, hence the attraction of adulteration. To counter the problem chemical analyses have been developed to allow identity of ginseng constituents of value-added products (Cheung et al. 1994). To protect its interests in the Hong Kong market the Ginseng Board of Wisconsin has organized a labelling system for genuine American ginseng products and has been fairly successful.

Hong Kong continues to be a major importer, distributor, processor, and retailer of ginseng (But et al. 1995b; Evans 1995). Over 80% of American ginseng grown in North America is shipped to the Hong Kong market as is much of the Oriental ginseng from China and Korea. Redistribution of ginseng from Hong Kong is world-wide with major destinations being Taiwan, Japan, Malaysia, Singapore, and the U.S. Hong Kong has become a major center for ginseng trade because of a strong herb industry, a large population of Chinese origin, its strategic location, and a strong finance and communication system. In addition, Hong Kong has become a center for scientific research into Chinese traditional medicine and cures (Evans 1995). The Chinese University of Hong Kong is helping to integrate Chinese and Western techniques and approaches to testing traditional medicines and cures. Ginseng is included in these studies. The prominence of Hong Kong as a ginseng world center may change in the next decade because of its return to China in 1997, changes in the herb industry, and increasing use of Western medications. However, Evans (1995) claims that these changes will be minimal. He points to the major transformation of Southern China that has taken place and the amount of trade and investment between Hong Kong and the mainland.

There has been a strong European market for ginseng since its introduction by Pharmaton SA in the 1960s. Well established markets are in Scandinavia, Poland, Germany, Spain, Holland, Belgium, the UK, France, and Italy. The Eastern European countries, which already have a strong herb and medicinal plant industry, are projected to be receptive to ginseng.

Table 2. Estimated North American ginseng production from 1985 to 1994 based on export figures. (Sources: Statistics Canada (Ginseng, Canadian Exports 1994) and Wisconsin Dept. of Agr. Trade and Consumer Protection (Wisconsin Cultivated Ginseng Export Program 1994). Adding 10% to the production figures provides an estimate of total production since North America exports 90% of its total ginseng production.

Country State/Province	Production in dry wt (t)									
	1985	1986	1987	1988	1989	1990	1991	1992	1993	1994
U.S.										
Wisconsin	519	687	594	704	667	522	533	577	692	736
Canada										
Ontario	102	76	104	126	103	161	93	217	255	324
British Columbia	0	7	10	15	19	71	35	117	92	164
Total	621	770	708	845	789	754	661	911	1,039	1,224

Cultivation is also being attempted in Australia (Baxter et al. 1995) and New Zealand (Follett et al. 1995). Ginseng use in oceania is likely to rise because of increasing immigration from South East Asian countries and demand for natural medicines. Since imported ginseng is used to meet market demands, domestic production, and a move to agricultural diversification are stimulating interest in ginseng.

In the past, North American ginseng growers have sold their crops to buyers who exported the roots to Hong Kong (Chan 1984). These growers have become uneasy about these traditional buyers and are considering developing their own joint ventures with established marketing systems in the Orient. This is a daunting task given the distances, regulations, and cultural complexities involved. Some corporate ginseng growing and marketing has occurred in the last decade e.g. Chai-Na-Ta Corp. (Gill 1995).

CULTIVATION

The two major ginsengs in trade are *Panax ginseng* C.A. Meyer (Oriental ginseng), and *Panax quinquefolium* (American ginseng). There are several minor Oriental ginseng species but not of economic significance. The reader is referred to the authoritative guide by Hu (1976) for further details.

The deciduous aerial stem of ginseng has 3 leaflets in seedlings, 5 leaflets per leaf in mature plants with leaf number correlating with age of the plant. Ginseng plants start to flower in the second year. Hu (1976) described the flowers of Oriental ginseng as small, 2–3 mm across; sepals 5, green; petals 5, cream yellow, ovate, apex obtuse; stamens 5, filaments, short; pistil 1, ovary inferior, 2 locular; styles 2, united at base; disk cup-shaped. A typical flowering 4-year old American ginseng plant may carry 30–40 berries in an umbel inflorescence with an average of two cream white seeds, 5–6 mm long and 4–5 mm wide, in each berry.

The ginseng plant has a fleshy taproot which often has two to five laterals and is light yellowish white in color. Dry matter content of the root is about 30%.

Ginseng is cultivated under artificial shade or simulated forest conditions to meet its requirements of about 30% of full sunlight (Proctor 1980; Stathers and Bailey 1986; Proctor and Bailey 1987; Proctor et al. 1988; Proctor et al. 1990; Persons 1995). Artificial shade growing is the more common cultivation method worldwide. Woods growing accounts for about 4% of U.S. production (see Table 2) and provides a high quality root for an expanding niche market (Persons 1995).

Propagation

Seeding is still the principal method of propagating ginseng yet little is known about seed after-ripening, stratification, and germination particularly for American ginseng (Proctor and Bailey 1987). In North America ginseng seeds are after-ripened in stratification beds outdoors after they are collected in Aug./Sept. (Polczinski 1982). They will not grow until the second spring following harvest (18–22 months). Oriental ginseng can be germinated and induced to grow in 8 to 10 months (Proctor et al. 1990). If American ginseng seed could be manipulated like Oriental then spring planting would be feasible. The establishment of ginseng plantings in the spring, instead of the traditional fall period, could offer advantages particularly in relation to reduction in damping-off and root rot diseases (Gotlieb 1981). Disease is the limiting factor in ginseng production (Parke and Shotwell 1989; Proctor and Bailey 1987) because it reduces the stand of ginseng seedlings, the productive life of the ginseng planting, and the number, size, and quality of marketable roots.

The establishment of ginseng gardens in the spring has been difficult because of the inability to store stratified seed from fall to spring. In preliminary work we have been able to stratify and germinate "green" (Aug. harvested) seed in the following May very successfully in the greenhouse (80%–100% germination) but with limited success in the field (30%–40% germination). We have used a combination of seed treatment regimes and growth regulators, particularly gibberellic acid, to achieve this early stratification and germination. We propose to refine our preliminary work so that spring planting is a viable and profitable production system and an alternative to fall planting (Hovius et al. 1995).

There are many aspects of ginseng seed stratification that have not been investigated but which probably influence germination, seedling emergence and crop stand. For instance, there have been no studies on ginseng seed development in the umbel in relation to germination. In other plants with an umbellate inflorescence such as carrot, variability in seedling weight at emergence is associated with variation in embryo length, which is influenced by umbel order and the date of harvest (Gray and Steckel 1983).

Tissue Culture

An efficient tissue culture regeneration system may be an alternative to seeding. Such a system would allow asexual propagation of the crop, the selection and propagation of uniform germplasm, and manipulation of the genetic make-up of the plant.

Early attempts to tissue culture ginseng from various tissues were reviewed by Proctor and Bailey (1987). Since then callus, shoots, embryoids, and plantlets have been cultured from roots (Jiu 1992), seed (Lee et al. 1990; Arya et al. 1993), protoplast (Arya et al. 1991), and young flower buds (Shoyama et al. 1987) of Oriental ginseng. The two reports for American ginseng were of root cultures producing callus, embryonic calli, and plantlets (Wang 1990) and leaf and root explants giving somatic embryos (Tirojah and Punja 1995). Although considerable progress has been made much more research and development is needed before tissue culture regeneration systems can replace seeding.

The in vitro culture of Oriental ginseng cells or tissue for the production of ginsenosides has been successful (e.g. Furuya et al. 1983; Choi et al. 1990; Inomata et al. 1993; Choi et al. 1995). Similar systems for American ginseng have not been reported. For ginsenoside production from ginseng cell cultures to be commercially successful large amounts of ginsenosides must be produced rapidly. Choi et al. (1995) selected two ginseng cell lines, KGC 13 and KGC 15, for ginsenoside production and of these KGC 13 produced the highest level reported to date with 80 mg/15 ml culture medium in 34 days. Their work also suggested that some cultured lines may produce more ginsenosides than intact roots. Further work is needed to check the stability of selected lines and details of bioreactor culture.

Plant cell culture for ginsenosides may have a place in production of these metabolites for inclusion in value-added products. It will complement field culture of roots rather than replace it since consumer preference is for whole roots.

Pest Control

Disease control remains the central problem in commercial cultivation of ginseng globally. As early as 1930 Whetzel et al. reported that the average yield of ginseng in the United States was one-sixth to one-third of expected because of the numerous diseases that attack the crop. Ohh (1986) has quantified losses from diseases in Korea as follows: anthracnose (20%–47%), damping off (5%–50%), root rot (1%–60%), phytophthora (2%–30%), and alternaria blight (10%–20%).

A plea by Proctor and Bailey (1987) for authoritative guides or bulletins on ginseng diseases, particularly for North America, has been answered e.g. Parke and Shotwell 1989; Brammall and Fisher 1993; Anon. 1995; Reeleder and Fisher 1995a, b). An illustrated compendium, *Diseases and Pests of Vegetable Crops in Canada* (Howard et al. 1994) includes discussion of 8 fungal diseases, 3 insect pests, 3 nutritional and other disorders, and nematodes and slugs in ginseng. A more extensive list for Korean ginseng is presented in Table 3.

Research on diseases in ginseng in North America has received increasing attention in the last decade. Darmono and Parke (1990) characterized the chlamydospores of *Phytophthora cactorum* and Darmono et al. (1991) isolated and determined the pathogenicity of *Phytophthora* from forest soils and ginseng gardens. Reeleder and Brammall (1994), working on damping-off and root rot diseases, demonstrated that isolates of *Pythium* spp., *Cylindrocarpon destructans* and *Rhizoctonia solani* reproduced symptoms observed in the field. However, new pest problems are also being reported. Hudelson and Parke (1995) reported an extreme problem in establishing ginseng seedling stands. Healthy-looking seedlings develop a red discoloration of the foliage and the roots rot and the disorder has been called "mystery seedling disease" (MSD). Pathogenicity trials suggest the involvement of *Cylindrocarpon*, *Pythium*, and a fungus putatively identified as *Septonema*. Control of MSD has been difficult with soil fumigation and fungicides having only limited success.

Table 3. Some common pest problems in Korean ginseng production and chemicals used for control. (Adapted from Proctor et al. 1990.)

Pest[z] Common name	Scientific name	Chemical applied for control[y]
Diseases		
Alternaria blight	*Alternaria panax* Whetzel	Bordeaux mixture, Captafol
Anthracnose	*Colletotrichum panaciocola* Uyeda et Takimoto	Iprodione, Mancozeb, Polyoxin
Grey mold	*Botrytis cinerea* Persoon	
Phytophthora blight	*Phytophthora cactorum* (Cohn.et Lebt.) Schroter	Benomyl Metalaxyl
Rhizoctonia stem rot	*Rhizoctonia solani* Kuhn	
Root rot complex	*Fusarium solani* (Mart.) Sacc.	Rizolex Benomyl, Captafol, Gyron
Seedling rot	*Phythium ultimum* Trow	
Sclerotinia white	*Sclerotinia sclerotiorum* (Lib.) DeBary *Sclerotinia inor* Jagger	Captafol, Metalaxyl
Phoma stem blight	*Phoma panacis* Nakata et Takimoto	Rizolex
Damping-off	*Rhizoctonia solani* Kuhn	
Insects		
African mole cricket	*Gryllotalpa africana* Palisot de Beauvois	Deltamethrin
Black cutworm	*Agrotis ipsilon* Hufnagel	
Common cutworm	*Agrotis fucosa* Butler	
Dark gray cutworm	*Agrotis tokionis* Butler	
Japanese wheat wireworm	*Agriotes sericeus* Candaze	
Korean black chafer	*Holotrichia diomphalia* Bates	Diazinon, Etoc
Larger black chafer	*Holotrichia morosa* Waterhouse	
Small velvety chafer	*Serica orientalis* Motschylsky	
European corn borer	*Pyrausta nubilalis* Hubner	
Rodents and other pests		
Striped field mouse	*Apodemus agrarius* Thomas	Brodifacoum
Snail	*Bradybaena sieboldiana* Pfeifer	Bordeaux mixture
Slug	*Limacella agrestisuarians*	Adams
Nematodes		
Potato rot nematode	*Ditylenchus destructor* Thomas	
Northern root-knot nematode	*Miloidogyne halpa* Chitwood	Mocap

[z]Pests listed are taken from Korean Monopoly Corp. et al., (1984).
[y]Although specific chemicals are mentioned, they should be regarded only as examples and in no way are to be considered as recommendations. The primary source for this listing from, Korea Monopoly Corp. (1987).

Insect problems in ginseng are minor compared to diseases (Table 3). A new insect problem is pit-making pittosporum scale, *Asterolecanium arabidis* (Signoret) which causes depressions in the stems at the feeding sites and distorted shoot development (Schooley pers. commun.). It is thought that the scale has been carried by birds into the ginseng gardens from adjacent woodlots containing host plants.

Globally, the commercial cultivation of ginseng requires the application of pesticides. Registered pesticides for Korean pests are listed in Table 3. Pesticide evaluation and registration in other countries is ongoing. In Canada some pesticides are registered as in Korea (iprodione and mancozeb), some await full registration (metalaxyl), whereas others (chlorothalonil and pentachloronitrobenzene) are registered in Canada but not in Korea (Anon. 1995).

Ginseng growers are concerned about pesticide use and residues and are continually looking for new strategies for pest control. For example, Yu and Ohh (1995) reported that Korean ginseng growers use soil amendments of antagonistic microbes plus organic matter to reduce root rot. In North America, Joy and Parke (1995) are investigating the biological control potential of *Burkholderia cepacia*, strain AMMD, for *Alternaria* leaf blight. This disease is a major problem worldwide (Anon. 1995 and Table 3) and the preferred control chemical is mancozeb. A biological control agent as an alternative to mancozeb is needed.

Another approach to disease control in ginseng would be to breed resistant types. This is very long-term because of breeding difficulties such as the 20 months required for seed germination and yield assessment in a perennial plant. Biotechnology may have a place in the breeding of ginseng. *Agrobacterium*-mediated transformation has already been achieved in ginseng (Lee et al. 1995; Tirajoh and Punja 1995).

A major problem in ginseng cultivation is that a second planting made in the same ground will fail, a phenomenon known as "replant disease." Growers recognize the problem and do not replant the same land. This poses problems where there is a limited land base for cultivation.

Ginseng replant disease is very complex (Li 1995; Yu and Ohh 1995) with no single factor being responsible, even though Ohh et al. (1992) confirmed that *Cylindrocarpon destructans* was the causal organism leading to replant problems.

A concerted, international effort is needed to solve the replant disease problem. Such a solution will likely include biological, ecological, and chemical approaches. The characteristic features of ginseng replant disease must be understood. For example, does the disease induce leaf symptoms and, if it does, is it in year 1, 2, or 3? Does the disease inhibit root growth leading to weak, necrotic, and poorly branched roots with altered shoot, root ratio? Does it cause mineral nutrient imbalances? How long do the pathogens persist in the soil?

The root-knot nematode (*Miloidogyne* sp.) reduced the growth of naturally infested 6-year old Korean ginseng roots by reducing fresh and dry weight and root length and diameter (Ahn et al. 1983). Vrain (1993) reported similar results for American ginseng. Soil fumigants for nematode control are recommended (Anon. 1995) and may also reduce ginseng replant disease. However, it is expected that use of soil fumigants will be prohibited in the near future. Hence, new approaches that are more environmentally acceptable are needed. For instance, pre-planting with plants having nematocidal properties e.g. members of the *Compositae* (Gommers 1973), and steam pasteurization need to be evaluated. Availability of self-propelled field bed steamers makes this approach attractive.

Mechanization

Ginseng production is extremely labor intensive. Also, labor supply and cost are major factors in the efficiency and optimization of ginseng production. Therefore, increased mechanization in ginseng production is receiving international attention (Mok et al. 1993; van Dalfsen et al. 1993).

In the establishment of ginseng gardens bed plows and bed groomers are used extensively and are now fairly standard throughout the North American industry. The sowing of seed has progressed rapidly in the last decade. The one-row, hand-held Planet Junior which gave both a crude-sizing capability and a spreading device that produced scattering of the seed to reduce bunching has been replaced with tractor mounted multiple-row seeders. Although these units allow seeding of 2 ha or more per day there is still an opportunity to improve plant stand and final root quality using sized seed and precision seeding.

Straw spreaders range from the single spreader of small rectangular bales to large spreaders using large rectangular bales ($1.2 \times 1.2 \times 2.4$ m). The latter totally mechanize the straw spreading system, including the handling of the bales.

The traditional wood lathe shade system is rapidly being replaced by plastic shade systems. The plastic system requires fewer shade support posts thus allowing fewer and wider passes with the sprayer and fewer posts to purchase and install. Some growers are evaluating retractable shade. Light sensors or time clocks are being used to withdraw shade. Shade removal may allow drying out of the foliage in gardens, thus reducing disease incidence. Possibilities exist for improving yield by allowing more light to reach the plants under cloudy conditions, or earlier or later in the day. Possibly all artificial shade for ginseng growing will be plastic by the turn of the century.

Many different types of sprayers are used. The most common type uses a hydraulic boom; sophisticated sprayers have large spray tanks and complete hydraulic drive systems in self-propelled units.

The objective in root harvest and post-harvest is to mechanize the complete process. The most advanced diggers convey the root into bulk bins. These bulk bins can be moved with fork lifts to cold storage or processing lines. At the processing lines the bins are mechanically dumped for root washing, inspection, and loading onto drier trays. These drier trays can be moved mechanically into dryers. Most ginseng driers in North America are modified tobacco kilns with the necessary technology being taken from the tobacco industry. After drying, the roots are inspected again and loaded into drums with automatic weight recording. The system described above is the one that all growers should reach within the next decade.

FUTURE PROSPECTS

Ginseng production has increased in the major growing areas in the last decade and may increase further in these areas and in new geographical locations. The tradition that Oriental ginseng be grown in the Orient only, and American ginseng in North America only, was broken in the 1970s by China. The impact of this change may be felt in the next few years as increasing volume of this Chinese-grown American root makes its way into the marketplace. There may be opportunities for North Americans to grow Oriental ginseng and export it to Europe. Presently European markets use mostly Oriental ginseng. Some substitution for high quality ginseng root and adulteration of products in the marketplace by unscrupulous dealers have caused concern and reduced consumer confidence. Better labelling and policing supported by sophisticated scientific root and product analyses is overcoming this.

The marketplace demands high quality roots. A large quantity of low quality, poorly dried and inappropriately shipped roots destroyed the Canadian ginseng business for about 100 years from 1750 to 1850 (Hellyer 1984). Most problems leading to low quality roots start with the grower and can be avoided. These include inadequate soil drainage, untimely and poorly applied pesticides, and neglect of good sanitary practices.

A range of chemicals have been isolated from ginseng with the ginsenosides receiving most attention. Oriental testing has shown that the biological activity of the ginsenosides appears to match traditional and empirical claims and uses. Scepticism still remains, particularly in the West, although sales of ginseng products are rising steadily. Extensive clinical testing is needed to sustain and expand ginseng use.

Seeding is still the major method of propagation of ginseng in spite of some success in culturing different parts of the plants. A major remaining challenge in tissue culture is to produce flowering plantlets which give viable seed.

The production of ginsenosides in large-scale plant cell culture has been successful. More remains to be done in selecting cell lines that produce large amounts of ginsenosides quickly in bioreactor culture.

Opportunities exist for shortening the stratification period of American ginseng seed to allow spring planting. This may reduce disease incidence. Since only one-third of ginseng seed sown ultimately produces plants harvested after 3 years any approach that reduces disease incidence and improves seed germination, seedling emergence and crop stand must be pursued.

Disease is the major problem in ginseng cultivation from seed stratification, soil preparation prior to planting, right through to drying of the roots. Replant disease remains as an unresolved problem and needs full characterization and new approaches for control. Much progress has been made in research and related extension activities in disease control. Continued progress is expected with the new, concerted global attack on disease research and investigation of new strategies. Work on biocontrol of diseases shows promise. The application of *Agrobacterium*-mediated transformation for disease resistance is interesting and exciting.

Pest control studies must continue to investigate timing and amount of pesticide applied, to include integrated pest management, and to consider new methods of soil fumigation. Use of soil fumigants will be prohibited. Steam pasteurization is more environmentally acceptable than chemical fumigation and may be a replacement for it.

Decreased labor populations and increased associated costs for ginseng production are causing rapid mechanization in every aspect of the ginseng industry. Engineers, machinery dealers, and fabricators, and growers are being challenged to increase efficiency by mechanization.

At present ginseng production is an established part of North American medicinal crops industry. Trade could expand if ginseng finds an increasing market here and abroad.

REFERENCES

Ahn, Y.J., S.H. Ohh, J.H. Kim, and S.K. Lee. 1983. Effect of root-knot nematode infection on ginseng growth and inorganic nutrients in ginseng roots. Korean J. Ginseng Sci. 7:35–43.

Anon. 1990. An industry development plan for British Columbia ginseng producers. B.C. Ministry Agr. Fisheries and Food, Victoria.

Anon. 1995. Ginseng pest control recommendations. Ontario Ministry of Agriculture, Food and Rural Affairs, Simcoe.

Arya, S., J.R. Liu, and T. Eriksson. 1991. Plant regeneration from protoplasts of *Panax ginseng* (C.A. Meyer) through somatic embryogenesis. Plant Cell Rep. 10:277–281.

Arya, S., I.D. Arya, and T. Eriksson. 1993. Rapid multiplication of adventitious somatic embryos of *Panax ginseng*. Plant Cell, Tissue Organ Cult. 34:157–162.

Bae, H.W. 1975. Abstracts of Korean ginseng studies (1687–1975). The Res. Inst., Office of Monopoly, Seoul.

Bailey, W.G., C. Whitehead, J.T.A. Proctor, and J.T. Kyle (eds.). 1995. Proc. Int. Ginseng Conf. Vancouver 1994, Canada.

Baxter, L.B., S.D. Martin, K.J. Saltmarsh, A.J. Monks, and M. Barber. 1995. The potential for commercial ginseng production in Southern Australia. p. 67–71. In: W.G. Bailey, C. Whitehead, J.T.A. Proctor, and J.T. Kyle (eds.), Proc. Int. Ginseng Conf. Vancouver 1994, Canada.

Blumenthal, M., D.V.C. Awang, H.H. Fong, J.F. Fitzloff, J.T. Arnason, I. Ibraham, and T. Hall. 1995. Analysis of North American commercial ginseng: American Botanical Council's ginseng evaluation program. p. 109–111. In: W.G. Bailey, C. Whitehead, J.T.A. Proctor, and J.T. Kyle (eds.), Proc. Int. Ginseng Conf. Vancouver 1994, Canada.

Brammall, R. and P. Fisher. 1993. Botrytis blight of ginseng. Ont. Min. Agr. and Food Factsheet, Toronto, Canada. Agdex 260/631.

But, P.P.H., S.Y. Hu, and H. Cao. 1995a. The ginseng plant: products, and quality. p. 24–34. In: W.G. Bailey, C. Whitehead, J.T.A. Proctor, and J.T. Kyle (eds.), Proc. Int. Ginseng Conf. Vancouver 1994, Canada.

But, P.P.H., T.Y.S. Li, and A.Y.K. But. 1995b. Ginseng in Hong Kong. p. 44–49. In: W.G. Bailey, C. Whitehead, J.T.A. Proctor, and J.T. Kyle (eds.), Proc. Int. Ginseng Conf. Vancouver 1994, Canada.

Chan, D. 1984. Operations of a Hong Kong ginseng importer. p. 112–115. In: J.T.A. Proctor, (ed.), Proc. 6th North Am. Ginseng Conf., Univ. of Guelph, Ontario, Canada.

Cheung, K.S., H.S. Kwan, P.P.H. But, and P.C. Shaw. 1994. Pharmacognostical identification of American and Oriental ginseng roots by genomic finger-printing using arbitrarily primed polymerase chain reaction (AP-PCR). J. Ethnopharm. 42:67–69.

Choi, K.T., J.C. Park, and I.O. Ahn. 1990. Saponin production in tissue culture of ginseng (*Panax ginseng* C.A. Meyer). Korean J. Ginseng Sci. 14:107–111.

Choi, K.T.,C.H. Lee, I.O. Ahn, J.H. Lee, and J.C. Park. 1995. Characteristics of the growth and ginsenosides in the suspension–cultured cells of Korean ginseng (*Panax ginseng* C.A. Meyer), p. 259–269. In: W.G. Bailey, C. Whitehead, J.T.A. Proctor, and J.T. Kyle (eds.), Proc. Int. Ginseng Conf. Vancouver 1994, Canada.

Cui, J., M. Gale, P. Eneroth, and I. Bjorkhem. 1994. What do commercial ginseng preparations contain? Lancet 344:134.

Darmono, T.W. and J.L. Parke. 1990. Chlamydospores of *Phytophthora cactorum*: their production, structure and infectivity. Can. J. Bot. 68:640–645.

Darmono, T.W., M.L. Owen, and J.L. Parke. 1991. Isolation and pathogenicity of *Phytophthora cactorum* from forest and ginseng garden soils in Wisconsin. Plant Dis. 75:610–612.

Duke, J.A. 1989. Ginseng:a concise handbook. Reference Publications, Inc., Algonac MI.

Evans, B. 1995. Ginseng and the new world order. p. 309–312. In: W.G. Bailey, C. Whitehead, J.T.A. Proctor, and J.T. Kyle (eds.), Proc. Int. Ginseng Conf. Vancouver 1994, Canada.

Faleni, R. 1995. Clinical safety and efficacy of G115. p. 215–218. In: W.G. Bailey, C. Whitehead, J.T.A. Proctor, and J.T. Kyle (eds.), Proc. Int. Ginseng Conf. Vancouver 1994, Canada.

Follett, J.M., B.M. Smallfield, and M.H. Douglas. 1995. Development of ginseng as a new crop for New Zealand. p. 72–77. In: W.G. Bailey, C. Whitehead, J.T.A. Proctor, and J.T. Kyle (eds.), Proc. Int. Ginseng Conf. Vancouver 1994, Canada.

Francis, T. 1995. Opening ceremony on behalf of the Canadian Ginseng Research Foundation. p. 12. In: W.G. Bailey, C. Whitehead, J.T.A. Proctor, and J.T. Kyle (eds.), Proc. Int. Ginseng Conf. Vancouver 1994, Canada.

Fulder, S. 1980. The root of being. Hutchinson & Co., London.

Furuya, Y., T. Yoshikawa, Y. Ohihara, and H. Oda. 1983. Saponin production in cell suspension cultures of *Panax ginseng*. Planta Med. 48:83–87.

Gill, G. 1995. Raising the global standard of ginseng. p. 303–304. In: W.G. Bailey, C. Whitehead, J.T.A. Proctor, and J.T. Kyle (eds.), Proc. Int. Ginseng Conf. Vancouver 1994, Canada.

Ginseng, Canadian Exports. 1994. Statistics Canada, Int. Trade Division, Ottawa, Canada.

Gommers, F.J. 1973. Nematicidal principles in Compositae. Mededeelingen Landbouwhogeschool Wageningen 17:71–73.

Gotlieb, A.R. 1981. Spring planting to control root rot during ginseng establishment. p. 39–49. In: N.A. Lapp (ed.), Proc. 3rd Nat. Ginseng Conf. Asheville, NC.

Gray, D. and J.R.A. Steckel. 1983. Some effects of umbel order and harvest date on carrot seed variability and seedling performance. J. Hort. Sci. 58:73–82.

Ha, Hongchien, Yamamoto, Masahiro, Joo, Chung No, and Kuo, Sheng-Chu. 1987. Proc. Int. Symp. on Ginseng Res., R.O.C. Taiwan.

Hellyer, W.C. 1984. Canadian ginseng production—past and present. p. 3–6. In: J.T.A. Proctor, (ed.), Proc. 6th North Am. Ginseng Conf. Univ. Guelph, Ontario, Canada.

Hovius, M.H.Y., J.T.A. Proctor, and R. Reeleder. 1995. Shortening the dormancy period and improving germination in American ginseng seed. HortScience 30:867. (Abstr.)

Howard, R.J., J.A. Garland, and W.L. Seaman. 1994. Diseases and pests of vegetable crops in Canada, an illustrated compendium. Entomological Society of Canada, Ottawa.

Hu, S.Y. 1976. The genus *Panax* (ginseng) in Chinese medicine. Econ. Bot. 30:11–28.

Hudelson, B.D. and J.L. Parke. 1995. Etiology of a seedling root rot of American ginseng in Wisconsin. p. 133–140. In: W.G. Bailey, C. Whitehead, J.T.A. Proctor, and J.T. Kyle (eds.), Proc. Int. Ginseng Conf. Vancouver 1994, Canada.

Inomata, S., M. Yokoyama, Y Gozu, T. Shimizu, and M. Yanagi. 1993. Growth pattern and ginsenoside production of *Agrobacterium*-transformed *Panax ginseng* roots. Plant Cell Rep. 12:681–686.

Jiu, S.Y. 1992. Plant regeneration from adventitious root-derived calli of ginseng (*Panax ginseng* C.A. Meyer). J. Agr. Assoc. of China (N.S.) 41:48.
Joy, A.E. and J.L. Parke. 1995. Biocontrol of Alternaria leaf blight on American ginseng by *Burkholderia cepacia* AMMD. p. 93–100. In: W.G. Bailey, C. Whitehead, J.T.A. Proctor, and J.T. Kyle (eds.), Proc. Int. Ginseng Conf. Vancouver 1994, Canada.
Korean Monopoly Corp., Korea Ginseng and Tobacco Research Inst., and Ginseng Growers Assoc. of Korea. 1984. Ginseng: Diseases, insects and color pictures. Korean Monopoly Corp., Shintanjin, Korea.
Korean Monopoly Corp. 1987. Standard methods of ginseng cultivation, 3rd ed. Shintanjin, Korea.
Lang, W.S., Z.C. Lou, and P.P.H. But. 1993. High-performance liquid chromatographical analysis of ginsenosides in *Panax ginseng, P. quinquefolium* and *P. notoginseng*. J. Chinese Pharm. Sci. 3:133–143.
Lee, F.C. 1992. Facts about ginseng, the elixir of life. Hollym Intern. Corp., Elizabeth, NJ.
Lee, H.S., S.W. Kim, K.W. Lee, T. Eriksson, and J.R. Liu. 1995. *Agrobacterium*-mediated transformation of ginseng (*Panax ginseng*) and mitotic stability of the inserted ß-glucuronidase gene in regenerants from isolated protoplasts. Plant Cell Rep. 14:545–549.
Lee, H.S., J.R. Liu, S.G. Yang, Y. H. Lee, and K.W. Lee. 1990. In vitro flowering of plantlets regenerated from zygotic embryo-derived somatic embryos of ginseng. HortScience 25:1652–1654.
Li, T.S.C. 1995. The effects of chemical and organic treatment on ginseng seedlings planted in old ginseng soil. p. 201–204. In: W.G. Bailey, C. Whitehead, J.T.A. Proctor, and J.T. Kyle (eds.), Proc. Int. Ginseng Conf. Vancouver 1994, Canada.
Liberti, L. and A. Der Marderosian. 1978. Evaluation of commercial ginseng products. J. Pharm. Sci. 67:1487–1489.
Liu, Chang-Xiao and Pie-Gen Xiao. 1992. Recent advances on ginseng research in China. J. Ethnopharm. 36:27–38.
Mok, S.K., D.Y. Shin, S.K. Cheon, T.S. Lee, J. Paeng, and K.S. Kim. 1993. Development of agricultural machinery in ginseng cultivation. p. 111–112. In: Program and Abstracts, 6th Int. Ginseng Symp., Korea Ginseng and Tobacco Research Inst., Seoul.
Ohh, S.H. 1986. Diseases of ginseng:environmental and host effect on disease outbreak and growth of pathogens. Korean J. Ginseng Sci. 5:73–84.
Ohh, S.H., Y.H. Yu, K.H. Kim, and D.H. Cho. 1992. Studies on control of soil-borne diseases and insects of ginseng and development of antifungal compound. p. 121–184. In: Korea Ginseng and Tobacco Research Inst., Ginseng Cultivation Bul.
Park, H. 1994. Recent trend of import and export of *P. ginseng* in Japan. Korean J. Ginseng Sci. 18:200–203.
Park, M.K. and S.S. Kim. 1995. Ginseng cultivation, production, products and use in Korea. p. 35–40. In: W.G. Bailey, C. Whitehead, J.T.A. Proctor, and J.T. Kyle (eds.), Proc. Int. Ginseng Conf. Vancouver 1994, Canada.
Parke, J.L. and K.M. Shotwell. 1989. Diseases of cultivated ginseng. Univ. of Wisconsin: Ext. Res. Bul. 3465.
Persons, W.S. 1995. American ginseng farming in its woodland habitat. p. 78–83. In: W.G. Bailey, C. Whitehead, J.T.A. Proctor, and J.T. Kyle (eds.), Proc. Int. Ginseng Conf. Vancouver 1994, Canada.
Polczinski, L.C. 1982. Ginseng (*P. quinquefolius* L.) culture in Marathon County, Wisconsin: historical growth, distribution and soils inventory. MS Thesis, Univ. of Wisconsin, Stevens Point.
Proctor, J.T.A. 1980. Some aspects of the Canadian culture of ginseng (*P. quinquefolius* L.), particularly the growing environment. p. 39–47. In: Proc. 3rd Int. Ginseng Symp., Korea Ginseng Research Inst., Seoul.
Proctor, J.T.A. and W.G. Bailey. 1987. Ginseng: industry, botany, and culture. Hort. Rev. 9:188–236.
Proctor, J.T.A., T.S. Wang, and W.G. Bailey. 1988. East meets west; cultivation of American ginseng in China. HortScience 23:968–973.
Proctor, J.T.A., J.C. Lee, and S.S. Lee. 1990. Ginseng production in Korea. HortScience 25:746–750.

Reeleder, R.D. and R.A. Brammall. 1994. Pathogenicity of *Cylindrocarpon destructans*, and *Rhizoctonia solani* to ginseng seedlings in Ontario. Can. J. Plant Pathol. 16:311–316.

Reeleder, R.D. and P. Fisher. 1995a. Diseases of ginseng. Ont Min. Agr. Food and Rural Affairs Factsheet, Toronto, Canada. Agdex 260/630.

Reeleder, R.D. and P. Fisher. 1995b. Rhizoctonia disease of ginseng. Ont Min. Agr. Food and Rural Affairs Factsheet, Toronto, Canada. Agdex 260/638.

Rückert, K.H. 1980. Quality control of ginseng preparations by means of HPLC: a safe method for the pharmaceutical industry. p. 217–222. In: Proc. 3rd Int. Ginseng Symp., Korea Ginseng Research Inst., Seoul.

Shibata, S., O. Tanaka, M. Sado, and S. Tsuchima. 1963. The genuine saponin of ginseng. Tetrahedron Lett. 12:795–800.

Shoyama, Y., K. Kamura, and I. Nishioka. 1987. Somatic embryogenesis and clonal multiplication of *Panax ginseng*. Planta Med. 54:155–156.

Stathers, R.J. and W.G. Bailey. 1986. Energy receipt and partitionning in a ginseng shade canopy and mulch environment. Agr. For. Meteorol. 37:1–14.

Tanaka, O., R. Kasai, and T. Morita. 1986. Chemistry of ginseng and related plants: recent advances. Abstracts of Chinese Medicines 1:130–152.

Tirajoh, A. and Z.K. Punja. 1995. Tissue culture and *Agrobacterium*-mediated transformation of American ginseng (*P. quinquefolium* L.). p. 144–155. In: W.G. Bailey, C. Whitehead, J.T.A. Proctor, and J.T. Kyle (eds.), Proc. Int. Ginseng Conf. Vancouver 1994, Canada.

Van Dalfsen, K.B., W.G. Bailey, and Y. Guo. 1993. Mechanization: a key to survival in the global ginseng industry. p. 113–114. In: Program and Abstracts, 6th Int. Ginseng Symp., Korea. Ginseng and Tobacco Research Inst., Seoul.

Vrain, T.C. 1993. Pathogenicity of *Pratylenchus penetrans* to American ginseng (*Panax quinquefolium*) and to kiwi (*Actinidia chinensis*). Can. J. Plant Sci. 73:907–912.

Wang, A.S. 1990. Callus induction and plant regeneration of American ginseng. HortScience 25:571–572.

Wang, Z., S. Zhang, Y. Wang, F Zhao, G. Wang, and Y. Du. 1995. A study on the introduction, acclimatization and application of Wulong ginseng. p. 84–88. In: W.G. Bailey, C. Whitehead, J.T.A. Proctor, and J.T. Kyle (eds.), Proc. Int. Ginseng Conf. Vancouver 1994, Canada.

Whetzel, H.H., J. Rosenbaum, J. W. Brann, and J.A. McClintock. 1930. Ginseng diseases and their control. U.S. Dept. Agr. Farmers Bul. 736.

Wiklund, I., J. Karberg, and B. Lund. 1994. A double-blind comparison of the effect on quality of life of a combination of vital substances including standardized ginseng G115 and placebo. Current Therapeutic Res. 55:32–42.

Wisconsin Ginseng Export Program. 1994. Wisconsin Dept. of Agr. Trade and Consumer Protection, Madison, WI.

Yu, Y.H. and S.H. Ohh. 1995. Problems and present status of research on ginseng diseases in Korea. p. 120–130. In: W.G. Bailey, C. Whitehead, J.T.A. Proctor, and J.T. Kyle (eds.), Proc. Int. Ginseng Conf. Vancouver 1994, Canada.

Distribution of Artemisinin in *Artemisia annua**

Jorge F.S. Ferreira and Jules Janick

Artemisia (*Artemisia annua* L., Asteraceae), known in the United States as sweet Annie or annual wormwood, is an annual herb native to Asia, most probably China, where it is known as *qinghao*. The plant has become naturalized in many countries including Argentina, Bulgaria, France, Hungary, Romania, Italy, Spain, the United States, and the former Yugoslavia (Gray 1884; Bailey and Bailey 1976; Klayman 1989, 1993). Artemisia is mentioned in the *Chinese Handbook of Prescriptions for Emergency Treatments* of 340 AD for treatment of fevers. In 1971, extraction of aerial parts of *A. annua* with low-boiling solvents, such as diethylether, produced a compound mixture with antimalarial properties on infected mice and monkeys. The main active principle, artemisinin (formerly referred to as arteannuin and as *qinghaosu* in Chinese), was isolated and had its structure correctly defined in 1972 in China as a sesquiterpene lactone with an endoperoxide bridge. Artemisinin is now available commercially in China and Vietnam as an antimalarial drug efficacious against drug-resistant strains of *Plasmodium*, the malarial parasite. A semisynthetic drug based on artemisinin (artemether) has been recently registered in Africa as Paluther. Artemisinin also has phytotoxic activity, even on *A. annua,* and is a candidate as a natural herbicide (Duke et al. 1987; Chen et al. 1991).

Artemisinin production by *A. annua* is usually in the range of 0.01% to 0.4% but some clones produce over 1% (Delabays et al. 1993). Artemisinin can also be obtained from artemisinic acid which occurs at concentrations as much as 10-fold higher than artemisinin (Acton et al. 1985). Recently, Vonwiller et al. (1993) reported an extraction method which makes possible the extraction of both compounds from the same plant material, thus increasing the final production of artemisinin.

BOTANY

Artemisia annua, a vigorous weedy annual (Hall and Clements 1923), is a short day plant with a critical photoperiod of 13.5 hr (Ferreira et al. 1995a). The chromosome number is $2n = 36$ (Benn et al. 1982). The plant is usually single-stemmed reaching about 2m in height with alternate branches and alternate, deeply dissected, aromatic leaves ranging from 2.5 to 5.0 cm in length. Tiny yellow nodding flowers (capitula) only 2 or 3 mm across are displayed in lose panicles containing numerous, greenish or yellowish, bisexual central (disc) florets containing little nectar and pistillate marginal (ray) florets (Fig. 1). The involucre is imbricated with several rows of bracts. The central flowers are perfect and can be either fertile or sterile. Ovaries are inferior and unilocular and each generates one achene, ca. 1 mm in length and faintly nerved. The pistillate marginal florets in the capitulum produce numerous achenes without pappus. The pollen is tricolpate and smooth, typical of anemophilous species, and has vestigial or no spines (Stix 1960). It has an internal, complex, columellae-tecta configuration in the exine, which is common to all taxa of the tribe Anthemideae and varies from two to three layers in *A. annua* (Skvarla and Larson 1965). The plant is naturally cross-pollinated by insects and wind action, which is unusual in the Asteraceae (McVaugh 1984).

Non-glandular T-shaped trichomes and 10-celled biseriate glandular trichomes occur on leaves, stems, and inflorescences. The morphology and origin of the glandular trichomes has been described for leaves (Duke and Paul 1993) and capitula (Ferreira and Janick 1995) using light and/or scanning electron microscopy. The essential oils of *A. annua* contains at least 40 volatile compounds and several nonvolatile sesquiterpenes, of which artemisinin and related compounds are the ones of most interest due to their antimalarial properties (Charles et al. 1991; Woerdenbag et al. 1994).

---

*We acknowledge the Southern Weed Science Laboratory (USDA/ARS), for the use of their electron microsopy facilities, and Rex N. Paul for taking the pictures.

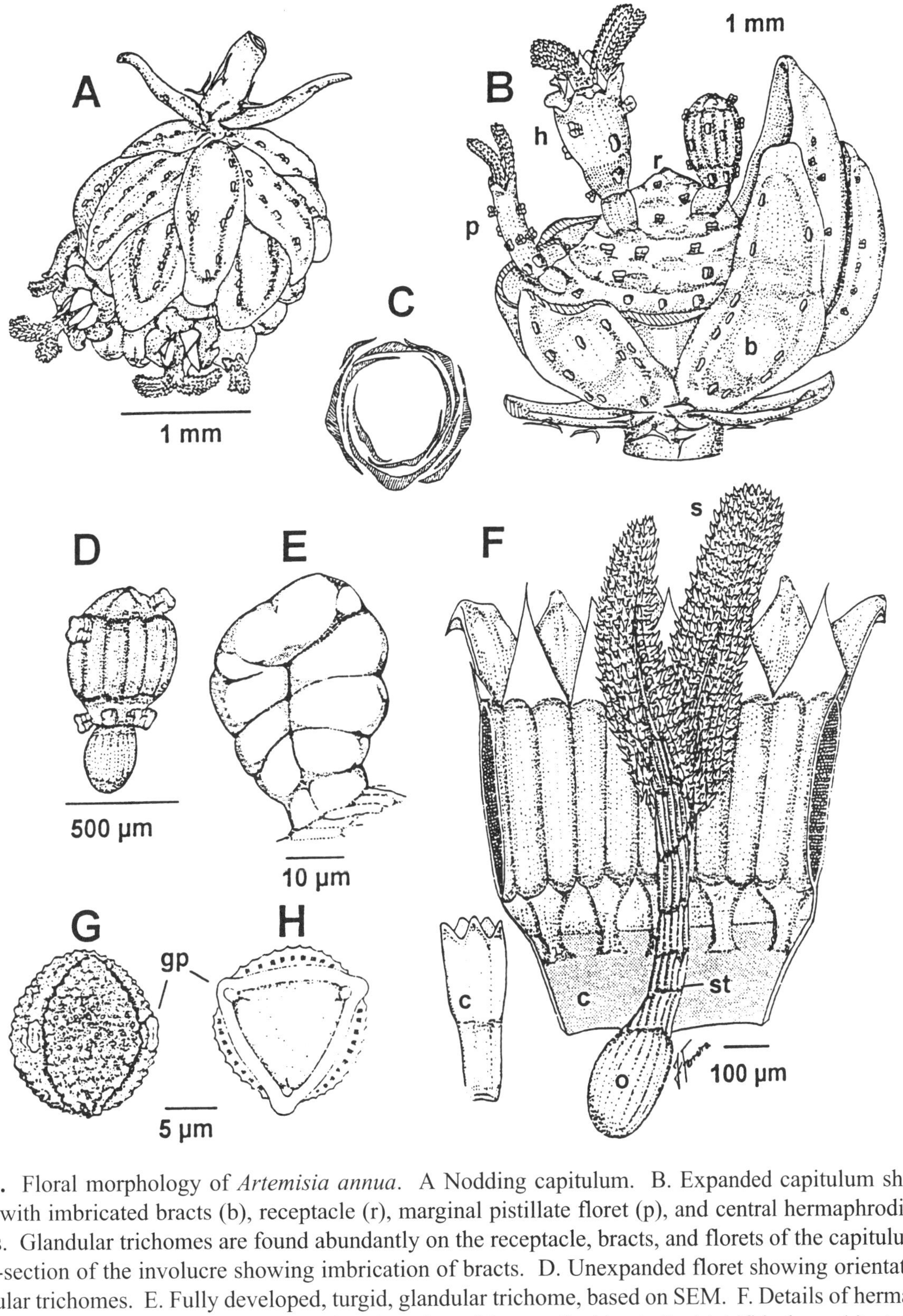

Fig. 1. Floral morphology of *Artemisia annua*. A Nodding capitulum. B. Expanded capitulum showing calyx with imbricated bracts (b), receptacle (r), marginal pistillate floret (p), and central hermaphroditic (h) florets. Glandular trichomes are found abundantly on the receptacle, bracts, and florets of the capitulum. C. Cross-section of the involucre showing imbrication of bracts. D. Unexpanded floret showing orientation of glandular trichomes. E. Fully developed, turgid, glandular trichome, based on SEM. F. Details of hermaphroditic floret with lobed anthers attached to basal portion of the corolla (c), pistil with bifid stigma (s), style (st), and ovary (o). Note that in a hermaphroditic floret, the stigma reaches this state of development only after pollen shed. G. Tricolpate pollen grain with vestigial spines, characteristic of wind-pollinated species, and germination pores (gp) bulging from the furrows. H. Pollen cross section based on light microscopy shows details of bulging germination pores. (Source: Ferreira and Janick 1995).

ARTEMISININ DISTRIBUTION

In Vivo

Artemisinin can be quantified by various analytical procedures including thin layer chromatography, gas chromatography, high-performance liquid chromatography with ultraviolet or electrochemical detection (see Ferreira et al. 1994), radioimmunoassay, and enzyme-linked immunosorbant assay (see Ferreira and Janick 1996a). Artemisinin has been detected from aerial parts of the plant, mostly in leaves and inflorescences with low levels in stems and none in pollen or roots (Table 1). The occurrence of artemisinin in the achene (seed) is due to the presence of floral remnants.

Although some authors reported artemisinin being highest during preflowering stages (Acton et al. 1985; Liersch et al. 1986; Woerdenbag et al. 1991; El-Sohly 1990; Woerdenbag et al. 1994), others reported artemisinin reaching its peak during flowering (Singh et al. 1988; Pras et al. 1991; Morales et al. 1993; Ferreira et al. 1995a; Laughlin 1995). Artemisinin reached its peak during full flowering in a Chinese clone for both greenhouse and field conditions (Ferreira et al. 1995b).

Reports on the distribution of artemisinin throughout the plant have been inconsistent. Artemisinin has been reported to be higher at the top of the plant in some clones (Charles et al. 1990; Laughlin 1995) and equally distributed in others (Laughlin 1995). We analyzed six clones derived from Chinese material during both vegetative and flowering stages with samples taken from the bottom, middle, and top parts of the plant, and found a relatively even distribution of artemisinin along the main stem (Table 2). Five of six clones showed the same or higher levels of artemisinin at the flowering stage. Although plants from all clones were harvested at the same date, they were in different stages of development. Branches of one clone (clone 1) were collected sequentially from the bottom to the top of the plant. In this clone artemisinin showed a slight increase in artemisinin toward the top of the plant (Table 3).

Table 1. Artemisinin content of different organs and structures of greenhouse- and field-grown *Artemisia annua,* determined by HPLC-EC (Source: Ferreira et al. 1995a).

Organ/ structure	Artemisinin (% DW × 1000)	
	Greenhouse	Field
Leaves	3–30	6–60
Main stems	0–3	0.4–7
Side stems	0	0.4–14
Roots	0	0
Flowers	12–42	104–264
Pollen	0	ND[z]
Seed husks	nd	116
Seeds[y]	36	81

[z]Not determined.
[y]Containing floral debris.

Table 2. Artemisin content of six clones of *Artemisia annua* by location on the plant and date.

Clone	Date[z]	Artemisinin (% DW)			
		Base	Middle	Top	Mean
1	Sept 7	0.022	0.014	0.019	0.018
	Oct 8	0.044	0.055	0.054	0.051
2	Sept 7	0.020	0.023	0.019	0.021
	Oct 8	0.079	0.070	0.094	0.081
3	Sept 7	0.043	0.043	0.054	0.046
	Oct 8	0.034	0.026	0.023	0.027
4	Sept 7	0.080	0.078	0.080	0.079
	Oct 8	0.081	0.146	0.143	0.123
5	Sept 7	0.100	0.127	0.107	0.111
	Oct 8	0.086	0.123	0.127	0.112
6	Sept 7	0.163	0.153	0.146	0.154
	Oct 8	0.245	0.181	0.234	0.220
7	Sept 7	0.071	0.073	0.071	0.072
	Oct 8	0.095	0.100	0.112	0.102
Mean	Sept 7	0.071 a[y]	0.073 a	0.071 a	0.072a
	Oct 8	0.095 b	0.100 b	0.112 b	0.102b

[z]All plants vegetative Sept 7; all plants flowering Oct 8.
[y]Mean separation in rows and columns by Duncan's Multiple Range Test (5% level).

In Vitro

Artemisinin is produced by differentiated (shoots + roots) shoot cultures (Martinez and Staba 1988; Fulzele et al. 1991; Whipkey et al. 1992; Ferreira and Janick 1996b) but occur only in trace levels, if at all, in shoots without roots (Martinez and Staba 1988; Jha et al. 1988; Fulzele et al. 1991; Woerdenbag et al. 1993; Paniego and Giuliette 1994). Brown (1994) reported 0.0038% of artemisinin being produced by callus cultures bearing shoots but did not specify whether or not they had roots. Most workers (Martinez and Staba 1988; Tawfiq et al. 1989; Fulzele et al. 1991; Kim et al. 1992) did not detect artemisinin in roots, although Nair et al. (1986) and Jha et al. (1988) reported trace amounts. Weathers et al. (1994) reported high levels (0.4%) of artemisinin in hairy root cultures of *A. annua* transformed with *Agrobacterium rhyzogenes*, but this was not confirmed by Jaziri et al. (1995). With the exception of Jha et al. (1988), most authors (He et al. 1983; Nair et al. 1986; Tawfiq et al. 1989; Fulzele et al. 1991; Kim et al. 1992; Woerdenbag et al. 1993; Brown 1994; Paniego and Giuliette 1994; Ferreira and Janick 1996b) either reported no artemisinin or only trace amounts produced from callus, cell, or the spent liquid media from these cultures. Ferreira and Janick (1996b) presented evidence that artemisinin production in shoots is enhanced by the presence of roots. The highest levels of artemisinin (0.287% DW) were obtained in hormone-free medium when the root production was maximized.

GLANDULAR TRICHOMES AS SITES OF ARTEMISININ ACCUMULATION

Although artemisinin immunolocalization has not been achieved, there is strong circumstantial evidence that the compound is sequestered in the glandular trichomes (Figs. 1D, E and 2A). Duke and Paul

Table 3. Artemisinin content of *Artemisia annua* branches (base to apex) of clone 1 harvested Oct. 8, 1994.

Position[z]	Artemisinin content (% DW)
1 (apex)	0.040
2	0.046
3	0.043
4	0.041
5	0.033
6	0.037
7 (base)	0.023

[z]Five branches were composited at each position.

Fig. 2. Florets of *Artemisia annua* pictured by JEOL JSM840 SEM, at 5KV. Top, hermaphroditic florets of normal biotype showing glandular trichomes. Bottom, hermaphroditic florets of glandless biotype. Bar size = 100 µm.

(1993) and Duke et al. (1994) described the development of such glands in leaves of *A. annua* and reported that neither artemisinin or artemisitene were detected from a glandless biotype and that virtually all artemisinin could be extracted by a 5-sec leaf dip in chloroform, without visible damage to other leaf epidermal cells, from the biotype with glands. Fig. 2B shows florets of the glandless biotype photographed under the scanning electron microscope.

Artemisinin content (% DW) was shown to be 4 to 11 times higher in the inflorescences as compared to leaves (Ferreira et al. 1995a) and the presence and development of glandular trichomes in the inflorescences was associated with artemisinin production based on extraction studies (Ferreira and Janick 1995). The glandular trichomes are more prominent in the corolla and receptacles florets than in leaves, stems, or bracts. Although these glands are present since the early stage of development on both leaves and inflorescences, artemisinin increases at anthesis, suggesting that it accumulates as the glands reach physiological maturity, a stage which coincides with the end of cell expansion in floret development. As glands approach maturity, there appears to be a cellular discharge into the subcuticular space around the apical cells and the contents are spread over the epidermis when the glands burst. After anthesis, artemisinin decreases and so does the number of intact glands. The association of artemisinin with glandular trichomes sequestration explains why artemisinin was not detected in parts of the plant that do not bear glands, such as pollen or roots (Ferreira et al. 1995a) or in a glandless biotype (Duke et al. 1994). Glandular trichomes are observed in leaves and stems of differentiated shoot cultures and artemisinin content of shoot cultures in vitro was similar to artemisinin content in vegetative clones grown in the greenhouse (Ferreira et al. 1995b).

CONCLUSION

Artemisinin is a sesquiterpene lactone which is produced both in vivo and differentiated in vitro cultures, by *Artemisia annua* and is equally distributed throughout the plant. Artemisinin appears to be sequestered in glandular trichomes which occur in stems, leaves, and inflorescences. The association of peak artemisinin with flowering is related to the abundance of glandular trichomes in the inflorescence, particularly florets and receptacle. In vitro studies indicate that the biosynthesis of artemisinin is enhanced by the presence of roots. *Artemisia annua* is unlikely to be produced economically by chemical synthesis or by in vitro production, thus *A. annua* is a potential new antimalarial crop for temperate areas.

REFERENCES

Acton, N., D.L. Klayman, and I.J. Rollman. 1985. Reductive electrochemical HPLC assay for artemisinin (qinghaosu). Planta Med. 51:445–446.

Bailey, L.H. and E.Z. Bailey. 1976. Hortus third. MacMillan Publ. Co., New York.

Bennett, M.D., J.B. Smith, and J.S. Heslop-Harrison. 1982. Nuclear DNA amounts in angiosperms. Proc. Royal Soc. London B 216:179–199.

Brown, G.D. 1994. Secondary metabolism in tissue culture of *Artemisia annua*. J. Nat. Prod. 57(7):975–977.

Charles, D.J., E. Cebert, and J.E. Simon. 1991. Characterization of the essential oils of *Artemisia annua* L. J. Ess. Oil Res. 3:33–39.

Charles, D.J., J.E. Simon, K.V. Wood, and P. Heinstein. 1990. Germplasm variation in artemisinin content of *Artemisia annua* using an alternative method of artemisinin analysis from crude plant extracts. J. Nat. Prod. 53:157–160.

Chen, P.K., G. Leather, and M. Polatnick. 1991. Comparative study on artemisinin, 2,4-D, and glyphosate. J. Agr. Food Chem. 39:991–994.

Delabays, N., A. Benakis, and G. Collet. 1993. Selection and breeding for high artemisinin (qinghaosu) yielding strains of *Artemsia annua*. Acta Hort. 330:203–206.

Duke, S.O., K.C. Vaughn, E.M. Croom Jr., and H.N. Elsohly. 1987. Artemisinin, a constituent of annual wormwood (*Artemisia annua*), is a selective phytotoxin. Weed Sci. 35:499–505.

Duke, S.O. and R.N. Paul. 1993. Development and fine structure of glandular trichomes of *Artemisia annua* L. Int. J. Plant Sci. 154(1):107–118.

Duke, M.V., R.N. Paul, H.N. Elsohly, G. Sturtz, and S.O. Duke. 1994. Localization of artemisinin and artemisitene in foliar tissues of glanded and glandless biotypes of *Artemisia annua* L. Int. J. Plant Sci. 155:365–372.

El-Sohly, H.N. 1990. A large scale extraction technique of artemisinin from *Artemisia annua*. J. Nat. Prod. 53:1560–1564.

Ferreira, J.F.S., D.J. Charles, K. Wood, J.E. Simon, and J. Janick. 1994. A comparison of gas chromatography and high performance liquid chromatography for artemisinin analyses. Phytochem. Anal. 5:116–120.

Ferreira, J.F.S. and J. Janick. 1995. Floral morphology of *Artemisia annua* with special reference to trichomes. Int. J. Plant Sci. 156:807–815.

Ferreira, J.F.S. and J. Janick. 1996a. Immunoquantitative analysis of artemisinin from *Artemisia annua* using polyclonal antibodies. Phytochemistry 41:97–104.

Ferreira, J.F.S. and J. Janick. 1996b. Roots as an enhancing factor for the production of artemisinin in shoot cultures of *Artemisia annua*. Plant Cell Tissue Organ Cult. 44:211–217.

Ferreira, J.F.S., J.E. Simon, and J. Janick. 1995a. Developmental studies of *Artemisia annua*: Flowering and artemisinin production under greenhouse and field conditions. Planta Med. 61:167–170.

Ferreira, J.F.S., J.E. Simon, and J. Janick. 1995b. Relationship of artemisinin content of tissue-cultured, greenhouse-grown, and field-grown plants of *Artemisia annua*. Planta Med. 61:351–355.

Fulzele, D.P., A.T. Sipahimalani, and M.R. Heble. 1991. Tissue cultures of *Artemisia annua*: organogenesis and artemisinin production. Phytother. Res. 5:149–153.

Gray, A. 1884. Synoptical flora of North America. Vol. 1, Part II. Smithsonian Institution, Washington, DC. Univ. Press, John Wilson and Son, Cambridge.

Hall, H.M. and F.E. Clements. 1923. The phylogenetic method in taxonomy. The North American species of *Artemisia, Chrysothamnus* and *Atriplex*. Carnegie Inst. Wash., Washington.

He, X-C., M-Y. Zeng, G-F. Li, and Z. Liang. 1983. Callus induction and regeneration of plantlets from *Artemisia annua* and changes of qinghaosu contents. Acta Bot. Sin. 25:87–90.

Jaziri, M., K. Shimomura, K. Yoshimatsu, M.-L. Fauconnier, M. Marlier, and J. Homes. 1995. Establishment of normal and transformed root cultures of *Artemisia annua* L. for artemisinin production. J. Plant Physiol. 145:175–177.

Jha, J., T.B. Jha, and S.B. Mahato. 1988. Tissue culture of *Artemisia annua* L.: A potential source of an antimalarial drug. Curr. Sci. 57:344–346.

Kim, N-C., J-G. Kim, H-J. Lim, T-R. Hahn, and S-U. Kim. 1992. Production of secondary metabolites by tissue culture of *Artemisia annua* L. J. Korean Agr. Chem. Soc. 35:99–105.

Klayman, D.L. 1989. Weeding out malaria. Nat. Hist. Oct.:18–26.

Klayman, D.L. 1993 *Artemisia annua*: From weed to respectable antimalarial plant. p. 242–255. In: A.D. Kinghorn and M.F. Balandrin (eds.), Human Medicinal Agents from Plants. Am. Chem. Soc. Symp. Series. Washington, DC.

Laughlin, J.C. 1995. The influence of distribution of antimalarial constituents in *Artemisia annua* L. on time and method of harvest. Acta Hort. 390:67–73.

Liersch, R., H. Soicke, C. Stehr, and H-U. Tüllner. 1986. Formation of artemisinin in *Artemisia annua* during one vegetation period. Planta Med. 52:387–390.

Martinez, B.C. and J. Staba. 1988. The production of artemisinin in *Artemisia annua* L. tissue cultures. Adv. Cell Cult. 6:69–87.

McVaugh, R. 1984. Compositae. Vol. 12. Flora Novo-Galiciana: a descriptive account of the vascular plants of Western Mexico, ed. W.R. Anderson. University of Michigan Press, Ann Arbor.

Morales, M.R., D.J. Charles, and J.E. Simon. 1993. Seasonal accumulation of artemisinin in *Artemisia annua* L. Acta Hort. 344:416–420.

Nair, M.S.R., N. Acton, D.L. Klayman, K. Kendrick, D.V. Basile, and S. Mante. 1986. Production of artemisinin in tissue cultures of *Artemisia annua*. J. Nat. Prod. 49:504–507.

Paniego, N.B. and A.M. Giuliette. 1994. *Artemisia annua* L.: dedifferentiated and differentiated cultures. Plant Cell Tiss. Organ Cult. 36:163–168.
Pras, N., J.F. Visser, S. Batterman, H.J. Woerdenbag, T.M. Malingré, and C.B. Lugt. 1991. Laboratory selection of *Artemisia annua* L. for high artemisinin yielding types. Phytochem. Anal. 2:80–83.
Singh, A., R.A. Vishwakarma, and A. Husain. 1988. Evaluation of *Artemisia annua* strains for higher artemisinin production. Planta Med. 54:475–476.
Skvarla, J.J. and D.A. Larson. 1965. An electron microscopy study of pollen morphology in the Compositae with special reference to the Ambrosiinae. Grana Palinol. 6:210–269.
Stix, E. 1960. Pollenmorphologische untersuchungen an Compositen. Grana Palynol. 2:41–104.
Tawfiq, N.K., L.A. Anderson, M.F. Roberts, J.D. Phillipson, D.H. Bray, and D.C. Warhurst. 1989. Antiplasmodial activity of *Artemisia annua* plant cell cultures. Plant Cell Rep. 8:425–428.
Vonwiller, S.C., R.K. Haynes, G. King, and H.J. Wang. 1993. An improved method for the isolation of qinghao (artemisinic) acid from *Artemisia annua*. Planta Med. Lett. 59:562–563.
Weathers, P.J., R.D. Cheetham, E. Follansbee, and K. Tesh. 1994. Artemisinin production by transformed roots of *Artemisia annua*. Biotech. Lett. 16:1281–1286.
Whipkey, A., J.E. Simon, D.J. Charles, and J. Janick. 1992. In vitro production of artemisinin from *Artemisia annua* L. J. Herbs Spices Med. Plants. 1:15–25.
Woerdenbag, H.J., N. Pras, R. Bos, J.F. Visser, H. Hendriks, T.M. Malingré. 1991. Analysis of artemisinin and related sesquiterpenoids from *Artemisia annua* L. by combined gas chromatography/mass spectrometry. Phytochem. Anal. 2:215–219.
Woerdenbag, H.J., J.F.J. Lüers, W. van Uden, N. Pras, T.M. Malingré, and A.W. Alfermann. 1993. Production of the new antimalarial drug artemisinin in shoot cultures of *Artemisia annua* L. Plant Cell Tiss. Organ Cult. 32:247–257.
Woerdenbag, H.J., N. Pras, N.G. Chan, B.T. Bang, R. Bos, W. van Uden, P. Van Y., N.V. Boi, S. Batterman, and C.B. Lugt. 1994. Artemisinin, related sesquiterpenes, and essential oil in *Artemisia annua* during a vegetation period in Vietnam. Planta Med. 60:272–275.

Wild Rose Germplasm Evaluation in Chile*

Jean Paul Joublan, Marisol Berti, Humberto Serri, Rosemarie Wilckens, Felicitas Hevia, and Inés Figueroa

A number of *Rosa* species (Rosaceae) native to southern Europe and North Africa were introduced to Chile by Spanish colonizers. These species are now widely distributed in Chile from Santiago (33°S) to Aisen (45ºS) and from sea level to 2000 m (Navas 1976; Hoffmann et al. 1992). Chile annually exports between 3600 to 4500 t of dehydrated rosehips to Europe.

Rose hips are the enlarged floral cups (receptacles) which surround numerous small, hard dry fruits (achenes) commonly called seeds. Rose hips are bright orange and oval and become fleshy but are not true fruits (Fig. 1). They are collected from the wild by peasants and their families. Rose hips are brought to the processing plant where they are dehydrated and seeds are removed; dry pulp is exported, mainly, to European countries. The dry pulp is used in herbal teas and marmelades and has been used as a pigment for laying hens and broiler chickens (Burgos 1976; Cortés 1976; Larraín 1978). It contains large amounts of vitamin C (1000–2000 mg/100g), riboflavin, pectins, nicotinic acid, and malic acid (Israel and Benado

*Contribution from the Plant Production Department, Universidad de Concepción, Chillán, Chile. This project has been financed by Soc. Agrícola y Forestal Casino Ltda. and FONTEC-CORFO project No. 95-0491.

1977). The vitamin C content varies with the site of collection, harvest date, and dehydration methods (Galeb 1976; Ziegler et al. 1986).

The achenes are a by-product of pulp extraction and have been used to feed pigs, hens, chickens, and chinchilla rabbits (Villagrán 1976; Moraga 1978; Peña 1978; Fuente et al. 1979; Seitz 1979; Barbet 1987; Voullieme and Hiriart 1978, 1980a, b). Achenes contain 8% oil composed by oleic, linoleic, linolenic, and transretinoic acids (Soto 1978). This oil is used in the cosmetic industry and is reported to accelerate the regeneration of damaged tissue, and promote wound healing (Valladares et al. 1985, 1986).

Information about the *Rosa* species that exist in Chile is scarce. "Rosa mosqueta" is the common name used for at least three species [*Rosa moschata*, *R. rubiginosa* (= *eglanteria*) and *R. canina*] which are collected from the wild. Morphological differences are evident in the wild material indicating that more than one species and probably several subspecies and ecotypes have developed since introduction. The objective of this study was to evaluate differences in characteristics, size, pulp thickness, and vitamin C of rose hips from wild material existing between latitudes 36° and 38° S where most processing plants exist.

METHODOLOGY

Rose hips and shoots were collected at 30 locations between Cauquenes (36°S) and Perquenco (38°S), Chile, and brought to the University of Concepción, Chillán. Locations sampled were at altitudes from 0 to about 2000 m. Forty hips were collected from each plant sampled and 20 hips were used to measured diameter, length, pulp thickness and the other 20 were used to determine weight, and industrial pulp yield. Vitamin C content was obtained from a sample of 20 hips, homogeneous in color. Samples were frozen and ascorbic acid content was measured according to the AOAC method (1969).

RESULTS

Preliminary results determined that rose hips were very variable among and within locations. All fruits collected from locations in the northern part of the sampled areas had thorns on fruit pedicels and some sparcely distributed thorns in the fruits. Clearly most plants South of Angol (42°S) did not have thorns in the fruit pedicel. Thorns in fruit pedicels have been used to classified species. According to Navas (1976), *R. canina* pedicels are glabrous and *R. rubiginosa* has thorns on the pedicels, and *R. moschata* has pubescent pedicels. Flowers are needed for a definitive classification.

Average hip length and diameter vary between 13 to 24 mm and 9 to 14 mm, respectively. Pulp thickness vary between 2.3 to 1.1 mm and mean fruit weight vary between 2.5 to 2.7 g (Table 1). Industrial pulp

Fig. 1. Rosehips fruits at harvest time in Chile.

Table 1. Variability in rose hips collected at 60 different locations in Chile.

Location	Ascorbic acid (mg/100 g)	Length (mm)	Diameter (mm)	Fruit weight (mg)	Pulp thickness (mm)
Angol	1181	22	13	2170	1.8
Antuco	5289	20	14	1730	1.5
Antuco	2085	20	13	1378	1.5
Cabrero	1474	21	14	1642	1.6
Cabrero	2082	21	13	1342	1.5
Cauquenes	2534	16	11	1160	1.1
Cauquenes	4469	17	11	1065	1.4
Cerro Negro	1910	18	13	1553	1.7
Cerro Negro	3493	22	13	1613	1.7
Co Colorado	3916	19	14	1423	1.4
Colliguay	2129	20	13	1630	1.7
Collipulli	4254	21	13	1163	1.5
Collipulli	2752	19	14	1520	1.6
Florida	3447	20	13	1540	1.6
Florida	2157	21	13	1660	1.6
Hualqui	1825	20	13	1440	1.5
Huepil	3675	20	12	1448	1.5
Laja	2318	20	13	1395	1.4
Larqui	4151	18	13	1588	1.5
Las Rosas	3492	19	11	1388	1.8
Las Rosas	2064	24	13	1817	2.1
Las Rosas	3976	21	13	1875	1.9
Las Rosas	1300	20	15	2555	2.4
Lolenco	2397	20	14	1993	1.6
Lumaco	2503	17	10	793	1.4
Lumaco	4437	16	11	928	1.4
Lumaco	3013	20	12	1753	1.7
Lumaco	2666	21	11	1545	1.9
Millantu	3226	19	12	1268	1.5
Mininco	2566	19	12	1472	1.5
Mulchen	3624	20	12	1355	1.9
Nacimiento	4656	19	13	1470	1.5
Nereo	6694	18	14	1585	1.7
Nereo	3435	19	13	1510	1.6
Ninhue	2175	20	13	1630	1.5
Ninhue	2542	13	9	555	1.2
Paso Hondo	3110	21	13	1453	1.5
Paso Hondo	4156	21	13	1538	1.6
Pemuco	3835	18	12	1248	1.5
Perquenco	4097	23	14	1538	1.6
Portezuelo	1095	21	14	1698	1.6
Portezuelo	1389	19	13	1598	1.7
Puente Perales	3075	19	12	1253	1.4
Quilaco	2923	20	12	1438	1.5

Table 1. Continued.

Location	Ascorbic acid (mg/100 g)	Length (mm)	Diameter (mm)	Fruit weight (mg)	Pulp thickness (mm)
Quilleco	2281	18	11	815	1.3
Quirihue	4221	21	14	1598	1.5
Quirihue	3091	19	14	1715	1.5
Rere	2677	22	13	1523	1.6
Rere	2699	22	14	1615	1.6
San Nicolas	2382	17	11	1065	1.5
San Nicolas	1968	19	13	1445	1.6
San Rosendo	2901	21	13	1623	1.5
Sn. Carlos Pur	3480	23	14	1660	1.8
Sta. Barbara	3391	22	11	1063	1.4
Tomeco	2680	18	13	1365	1.5
Traiguen	2681	21	13	1598	1.6
Trintre	2250	16	10	835	1.4
Yumbel	4438	20	14	1523	1.6
Yumbel	2537	21	13	1440	1.6
Yungay	--	19	14	--	1.5
Yungay	4072	20	13	1498	1.6
LSD (0.05)		1	1	33	0.1

yield (kg of fresh fruit needed to produce 1 kg of dry pulp) vary between 2.7 to 5.3 kg. Neighboring plants differ significantly in their fruit length, diameter, pulp thickness, weight, and industrial pulp yield. Ascorbic acid content varied among locations, but this was related directly with the moisture content of the fruits (Table 1). Variability in data collected is influenced by the environment and the genetic background. Further evaluations will be performed to select plants with the desired characteristics for commercial planting, management, and industrial processing. Selection is an important goal for processing companies because they would receive a more uniform product that could have a better industrial yield.

REFERENCES

AOAC. 1969. Ascorbic acid official method. p. 838. In: Official Methods of Analysis of AOAC.

Barbet, S. 1987. Uso de aquenio de rosa mosqueta (*Rosa eglanteria* L.) en la alimentación de conejos chinchilla mediano. Tesis Ing. Agr. Univ. Austral Chile.

Burgos, G.A. 1976. Uso de la rosa mosqueta (*Rosa* aff. *rubiginosa* L.) como pigmentante en raciones para ponedoras en jaula. Tesis Ing. Agr. Univ. Concepción.

Cortés, J.C. 1976. Uso de la rosa mosqueta (*Rosa* aff. *rubiginosa* L.) como pigmentante en raciones para broilers. Tesis Ing. Agr. Univ. Concepción.

Fuente, F., H. Rodríguez, G. Burgos, R. Villagrán, J. Larraín, and C. Ebert. 1979. Utilización del fruto deshidratado de mosqueta (*Rosa rubiginosa* L.) como pigmentante de la yema de huevo. Simiente 49 (1):20.

Galeb, P. 1976. Industrialización de la rosa mosqueta (*Rosa* Aff. *rubiginosa* L.). Investigación Agrícola. (Chile) 2(1):39–41.

Hoffmann, A., C. Farga, J. Lastra, and E. Veghazi. 1992. Plantas medicinales de uso común en Chile. Ed. Fundación Claudio Gay, Santiago, Chile

Israel, J.M. and T.S. Benado. 1977. Aspectos preliminares del aprovechamiento de la rosa mosqueta (*Fructus cynosbati*) en Chile. Alimentos 2(1):5–8.

Larraín, J.F.J. 1978. Uso de la cascarilla de mosqueta *Rosa* aff. *rubiginosa* como fuente pigmentante en raciones para ponedoras en jaulas. Tesis Ing. Agr. Univ. Concepción.

Moraga, C. 1978. Determinación de la energía metabolizable del aquenio de rosa mosqueta harneada (*Rosa eglanteria* L.) y su utilización en raciones para broilers. Tesis Ing. Agr. Univ. Austral Chile.

Navas, L.E. 1976. Flora de la Cuenca de Santiago de Chile. Tomo II. Ediciones Univ. Chile.

Peña, J.E. 1978. Utilización de la pepa de mosqueta (*Rosa* aff. *rubiginosa* L.) en la alimentación de aves de postura durante el período de recría. Tesis. Ing. Agr. Univ. Concepción.

Seitz, C.G. 1979. Determinación in vivo de la digestibilidad del aquenio de rosa mosqueta (*Rosa eglanteria* L.) en cerdos. Tesis Ing. Agr. Univ. Austral Chile.

Soto, G. 1978. Caracterización del aceite crudo de semilla de mosqueta (*Rosa* aff.*rubiginosa*). Tesis Fac. Bioquímica Univ. Concepción.

Valladares, J., M. Palma, C. Sandoval, and F. Carvajal. 1985. Crema de aceite de mosqueta (*Rosa* aff. *rubiginosa* L.) I. Parte: Formulación, preparación y aplicación sistemática en clínica. Anal. Real Academia de Farmaceutica 51:327–332.

Valladares, J., M. Palma, C. Sandoval, and Carvajal. 1986. Crema de aceite de mosqueta (*Rosa* aff. *rubiginosa* L.) II. Parte: Estudio de las propiedades físico-químicas de estabilidad, eficacia, cosmética y aplicación sistemática en clínica. Anal. Real Academia Farmaceutica 51:597–612.

Villagrán, R. 1976. Utilización del concho de mosqueta (*Rosa* Aff. *rubiginosa* L.) como fuente de pigmentación en raciones para ponedoras. Tesis Ing. Agr. Univ. Concepción.

Voullieme, A. and M. Hiriart. 1978. Aquenio de rosa mosqueta (*Rosa eglanteria* L.) Alternativas alimentarias a nivel pecuario. 1. Características morfológicas, composición química y valor energético. Agro-Sur 6(1):30-32.

Voullieme, A. and M. Hiriart. 1980a. Aquenio de rosa mosqueta (*Rosa eglanteria* L.) Alternativas alimentarias a nivel pecuario. 2. Material soluble y pared celular.: Mejoramiento de la calidad nutricional. Agro-Sur 8(1):18–20.

Voullieme, A. and M. Hiriart. 1980b. Aquenio de rosa mosqueta (*Rosa eglanteria* L.) Alternativas alimentarias a nivel pecuario. 3. Digestibilidad in vitro y efecto del tamaño de partícula sobre su medición. Agro-Sur 8(2):104–105

Ziegler, S.J., B. Meier, and O. Sticker. 1986. Fast and selective assay of L-ascorbic in rose hips by RP-HPLC coupled with electrochemical and/or spectrophotometric detection. Planta Med. 5:383–387.

Domestic Production of Herbs and Spices

Dario Cisneros

Strict rules and regulations regarding agricultural practices in the growing of gourmet herbs forced Burns Philp Food Inc., to close its farming operations in the state of California. This led to the consolidation of the largest herb farm in the country, located in the Mesilla Valley in south central New Mexico. Annual and perennial herbs are grown, harvested, and dehydrated on the company's 850 acre farm. Extensive management from ground preparation through dehydration practices ensures the availability of high quality gourmet spices from year to year.

The reason for producing domestic herbs is to produce the highest quality spices in the world through total control of the crop from sowing the seeds to dehydrating and processing the finished goods. We control the cost of quality herbs and guarantee consistent high quality year after year. This gives us the flexibility to meet demand and control the availability of as many products as possible. High quality herbs are perceived to be consistent with domestic labeling, thus giving Burns Philp Food Inc. a distinct selling advantage over the competition.

HISTORY

Burns Philp Food Inc. owned two herb farms in California. The first located in San Jacinto/Hemet in the southern part of the state consisted of 200 acres (80.9 ha) where basil, summer savory, and tarragon where grown. The northern farm was located in Dixon and consisted of 180 acres (72.8 ha). Dill weed, thyme, rosemary, and cilantro where grown in this farm. Both of these farms operated since the early 1940s.

The consolidation of the herb farming operations required extensive research for an area that would have well drained productive soils, good reliable and cost efficient source of water, suited climate for herb production, sufficient acreage for expansion, affordable land prices, good labor source, major highway system, and available energy. After a long period of researching the southwest and Mexico, the company decided on the Mesilla Valley in south central New Mexico.

The Berino farm consists of 850 acres (343 ha) located within a five mile (8 km) radius of the plant. The facility consists of a 82,000 ft^2 (7618 m^2) building. Tunnel dryers, processing room, warehouse, maintenance shop, laboratory, and offices make up this building. The winter annuals grown are cilantro, dill weed, and chervil. The summer annuals grown are basil and summer savory. Perennials grown are rosemary, tarragon, thyme, and marjoram.

CULTURAL PRACTICES

The cultural practices used for ground preparation are numerous and costly but necessary because no pesticides are labeled for the use on herbs and spices. This forces the farm to use additional operations to germinate weeds before planting is performed. It is not uncommon for 20 field operations to occur before the beds are prepared for planting purposes. All herbs are grown on 40" (101.6 cm) centers and are mechanically harvested using forage harvesters modified to cut and transport the crop to our harvest wagons. The harvest operation occurs throughout the day, the herbs are harvested and introduced into the dehydration tunnels within 1 h from cutting. This ensures the highest quality finished good and reduces the chances of spoilage.

All incoming product is weighed fresh for wet to dry ratios. Depending on what herb is being processed, dry weight of finished goods varies from 7.7% to 4.5% of fresh weight. All product is unloaded and conveyed through a leaf reel to remove any dirt or insects that might be on the product. From there the fresh product is loaded on trays that are manually introduced into the dehydration tunnel. Fresh product takes anywhere from 4 to 6 h to dry at temperatures of 90° to 140°F (32° to 60°C). Once the product is dried it is removed from the tunnel and unloaded onto strippers that separate leaves from stems. The leaves are conveyed onto vibratory screens and introduced into the processing lines. In the processing lines the leaves are sized and any remaining sticks or stems are removed. This is achieved using seed clippers, indents, swecos, gravity air separators, and lots of vibrating screens. Every processing line contains different magnets to remove metal and installation of metal detectors is underway. finished goods are packed in plastic liners and boxed in corrugated boxes. Composite samples of every box are sent to the laboratory for final analysis. The laboratory test for bulk index, bulk density, water activity, sieves, volatile oils, moistures, micro analysis, and any other test that a certain customer requires. Once the laboratory results are obtained the product becomes a finished good or is held back for further processing. The product is then moved to the warehouse and awaits shipment to our bottling plants or customers.

Coriander Under Irrigation in Argentina

Gustavo Luayza, Roberto Brevedan, and Rosana Palomo

Coriander (*Coriandrum sativum* L., Umbelliferae), shows considerable potential as an aromatic crop plant in Argentina. It is an annual herb with a slender, erect, hollow branching stem growing to a height of 0.30 to 0.70 m. It has compound leaves and light pink or white flowers in a compound umbel. The fruit is globular a diachene and contains two seeds, one in each achene. Roots are slender and highly branched.

Coriander requires a temperate climate and though it can tolerate a warm temperate climate there is a noticeable reduction in yield in this latter case. The concentration of essential oil in the fruits decreases at temperatures above 21°C; the ideal temperature for swelling the grain is a constant 15° to 18°C from the time of seed formation to swelling (Hornok 1986). Though it typically grows in arid regions, it produces well under irrigation.

Very few nutrients are required by this crop. It requires a soil of medium fertility but the best yields are obtained in light and fertile soils. It does not do well in impervious clay soils. Though it is the practice in other countries to fertilize coriander crops, in Argentina this is not customary. Studies have indicated that yield in Argentina responds to the application of up to 90 kg N/ha (Barreyro et al. 1993). No response to P has been observed, though this fact is attributed to the good availability of this element already in the soil (Barreyro et al. 1993). A diminished availability of P in the soil however gives rise to significant increases in yield (Brevedan et al. 1994).

In the cereal growing region coriander is sown between the end of May and the end of July. When sown after this date the plants are less developed and yield decreases. The distance between rows varies between 0.18 and 0.35 m. Sowing is carried out by means of a seed drill, using a seed density of 20 kg fruit/ha. The sowing depth ranges from 2 to 6 cm, depending on the type and humidity of the soil, the quality of the seeds, etc.

During winter the plants remain in the form of rosette plants; with the onset of spring the stems elongate and branch, bearing floral buds. Full bloom is reached after 20 or 30 days and after a further 25 or 30 days the plants begin to mature. Full maturity is reached in the final 25 to 30 days. The total cultivation cycle lasts from 145 to 190 days.

Coriander plants require a total of approximately 400 mm of water during the development cycle, some 100 mm for germination and to last through the winter period, during which they are in the form of rosette plants, and the remaining 300 mm for stem elongation, flowering, and fructification (Lafourcade et al. 1985).

Harvesting can commence when approximately half the fruits on the plant have turned gray. There are a number of inconveniences involved in coriander harvesting, such as the facts that the fruit ripens progressively and the mature seeds are highly susceptible to shattering, exposing them to particular risk under unfavorable meteorological conditions such as rains and strong winds during harvesting time. Harvesting should be carried out in the early morning or late afternoon to prevent seed shattering.

Coriander fruits can be harvested by two methods (Arizio et al. 1994). One is to leave the fruits to completely mature and dry whilst still on the plants, taking 3 to 6 days, harvest them with a windrow and subsequently collect and thresh the cut and dried material. Alternatively, the plants can be harvested directly with or without the application of a desiccant using a wheat combine harvester.

Diseases or pests affecting the cultivation of coriander in Argentina are minimal, and the economic incidence of such effects on production is low. Nevertheless, weeds present a serious problem and can lead to a considerable reduction in yield. The most commonly used herbicide is linuron, which eliminates some grasses and most broadleaf weeds. Trifluralin is also used.

Coriander is cultivated throughout an estimated total of 2840 ha in Argentina, most of this area situated in the humid Pampa (Buenos Aires and southern Santa Fe provinces). Production is around 2500 t/ha, 1500 t of this being sold on the domestic market and the rest being exported mainly to Brasil, the United States, and the United Kingdom (Arizio et al. 1994).

Coriander fruit and the oil obtained therefrom are used in the medicinal, food, and perfume industries. Steam distillation of the coarsely crushed fruit is the most common method used to obtain the volatile oil, with a yield of 0.5% to 1%. Coriander is one of the principle ingredients of curry and of other spice mixtures. In some countries young leaves are used to flavor salads and soups. The green dehydrated parts are used to prepare stock cubes.

There are extensive areas in Argentina where the cultivation of coriander is feasible, one of the most promising being the Lower Valley of the Colorado River in the extreme south of the Province of Buenos Aires. This region, which is under the jurisdiction of CORFO (Corporación de Fomento del Valle del Río Colorado) covers a total of 535,000 ha, 90,000 ha of which are under effective irrigation. According to an extensive report published by SERCOPLAN (Technical Cooperation and Planning Service) (1987), the zone is a productive mosaic characterized by intensive agricultural activities such as horticulture, fruit growing, seed production and the cultivation of cereals, as well as beef production. About 57% of the area under irrigation is dedicated to forage crops, 35% to grain, fruit and forestry, and 8% to horticultural production.

The zone has a semiarid climate with a negative balance, where the annual water deficit is 300 to 500 mm throughout all seasons of the year. The mean annual temperature is 15°C. The first frost is likely to occur in Apr. and the last in Nov., with an average of 43 frosts per year. The soils are not difficult to cultivate and are widely suitable for the traditional crops of the zone. Drainage problems caused by the low permeability of some soils with a clay-texture horizon or with compacted materials which impede the downward seepage of water, together with the presence of brackish water and continuous irrigation, have given rise to a process of salinization over large areas. In an attempt to deal with this problem of a permanent layer of saline water close to the surface, an extensive network of drainage channels has been constructed.

Of the 95% of utilizable land in the area, 90% is actually exploited. The productive structure of the region is characterized by high dispersion in land ownership, inadequate use of soils, and serious problems in commercialization of the produce.

Despite the variety of agricultural and livestock production in the region, the sandy soils and variable salinity levels mean that land use in the area is marginal, irrigation being the only element serving to compensate for the other defects. The irrigation infrastructure comprises a network covering a total of 5,441 km and a drainage network covering 3,738 km. The system as a whole has a catchment capacity of 43 m^3/s. The main deficiencies lie in channelling the water at plot level and in the systematization of the soils. A total of 25,000 ha are currently under a systematization project and 2,600 ha are being levelled with laser beams.

Despite the promising potential of the zone in terms of its natural resources and infrastructure, a census carried out in the area shows a low level of demographic growth (0.2% between 1960 and 1980) reflecting stagnating development: low capacity of the region's economy to absorb permanent and temporary labor, and a low level of investment, owing largely to vagaries in commercialization possibilities.

In an attempt to find a solution to this growth problem the above-cited report published by SERCOPLAN (1987) recommends the development of new crops for which there is a more stable demand on both the domestic and foreign markets and whose production process involves a minimum of risk. It is within the framework of this search for solutions to the mentioned problems and recognition of the potential of the region that the idea was put forward of introducing the cultivation of coriander, a crop well suited to the agro-climatic conditions of the area.

The objectives of the present work are to evaluate the relative advantages of different sowing seasons and different row spacings in coriander cultivation in the irrigation zone of the province of Buenos Aires.

METHODOLOGY

Field studies were conducted in Igarzábal and Mayor Buratovich in the Lower Valley of the Colorado River. Soil texture in both locations is loamy sand, with pH 7.15 and 7.00, respectively. The plots were irrigated to supplement rainfall during the growing season, with a average of 6 irrigations. Weeds were controlled manually. Sowing density was 20 kg/ha.

Plant height was measured periodically on five plants chosen at random from each plot. At maturity, 10 plants were randomly selected from each plot and the number of umbels per plant counted. The weight of 1000 fruits was determined and the diameter measured of 200 fruits. Lodging was rated visually at maturity on each plot on a scale of 1 (all plants erect) to 5 (≥80% of plants severely lodged). Five lineal meters were harvested from the three center rows of each plot in order to calculate yield. Threshing was carried out manually and cleaning done by means of a winnowing machine.

There were 6 planting dates: the first (1) and the second (2) fortnight of July, the first (3) and the second (4) fortnight of Aug., and the first (5) and the second (6) fortnight of Sept. Treatments were replicated five times in a completely randomized design. Individual plots consisted of 17 rows, 6 m long.

There were four treatments, consisting in rows set 0.15, 0.30, 0.45, or 0.70 m apart. Treatments were replicated four times in a randomized complete block design. Individual plots were 6 m long and the number of rows varied according to the spacing, keeping the total plot area constant for all treatments, approximately 17 m^2.

RESULTS

It was found that the later the sowing date, the shorter the total growth cycle of coriander plants, decreasing from 170 days for those sown on the first date to 100 days for those sown on the last date, a 41% reduction with respect to the first sowing date (Table 1). The sowing period lasted for 81 days whereas maturation lasted only 25 days. As reported by Alvarez and Curioni (1993), later sowing also reduced the duration of the vegetative period. Luayza et al. (1994) previously reported a high correlation between the length of time from emergence to initial bloom and yield.

Late sowing also caused a reduction in plant height (Table 1). There was no lodging. Fruit size was larger in plants sown during the first three dates than in those sown during the last three, and early sowing produced a larger number of umbels per plant and a higher yield.

An analysis of the results of different row spacings showed that the height and number of plants per unit area was similar (Table 2). With greater row spacing, 0.70 m, the lodging rate was high; at 0.45 m the percentage of lodged plants was low and at 0.30 m and 0.15 m it was zero.

The number of umbels per plant was significantly higher with a row spacing of 0.70 m than with spacings of 0.15 and 0.30 m (Table 2). There was no significant difference in the diameter and weight of the fruit nor in the yields obtained with the three closest row spacings, though there was a reduction in yield at distances greater than 35% with respect to the average spacing. These results for the most part coincide with those reported by Rubió et al. (1986), who also found no difference between the yields of the three closest spacings. However, Rubió et al. (1987, 1992) reported higher yield with narrower spacing.

Table 1. Agronomic characteristics of coriander as affected by planting date. Data averaged over years.

Planting date	Initial bloom (days)	Maturity (days)	Height (cm)	Umbels per plant	1000 fruit wt (g)	Yield (kg/ha)
1 (July 1 to 15)	100	170	0.68a[z]	12.5a	12.1a	1731a
2 (July 16 to 31)	87	153	0.64a	12.1a	12.1a	1415b
3 (Aug. 1 to 15)	77	142	0.66a	11.7a	12.3a	1495b
4 (Aug. 16 to 31)	59	122	0.59a	10.0b	11.7b	1032c
5 (Sept. 1 to 15)	55	118	0.63a	9.7b	11.8b	853cd
6 (Sept. 16 to 30)	42	100	0.53a	8.8c	11.8b	805d

[z] Values within a column followed by the same letter do not differ according to Duncan's Multiple Range Test (5% level).

Table 2. Agronomic characteristics of coriander as affected by row width. Data averaged over years.

Row width (m)	Plants/m^2	Height (m)	Lodging score	Umbels per plant	1000 fruit wt (g)	Yield (kg/ha)
0.15	21.0a[z]	0.55a	1.0c	12.8b	12.0a	2435a
0.30	20.5a	0.54a	1.0c	12.5b	12.3a	2605a
0.45	21.3a	0.57a	1.5b	13.4ab	12.5a	2504a
0.70	18.6a	0.62a	4.0a	13.8a	12.0a	1862b

[z]Values within a column followed by the same letter do not differ according to Duncan's Multiple Range Test (5% level).

The results obtained indicate the benefits of early sowing for obtaining higher yield, showing this increase in yield to be due mainly to an increase in the number of fruits per plant. The best yields are obtained with narrowly spaced rows.

SUMMARY

Coriander (*Coriandrum sativum* L.) has shown considerable potential as an aromatic crop plant in Argentina. It is an annual herb originating in Mediterranean countries. It has an erect branching stem growing to a height of about 30 to 70 cm and has compound leaves. The mature dried fruit is characterized by a pleasant aroma and is used for flavoring a variety of food products as well as some alcoholic liqueurs. The essential oil of the fruit, from which it derives its aroma, can be isolated by steam distillation. The plant does not require rich soil and grows well in deep loamy soils.

Current economic conditions favor the introduction of new crops into the irrigated areas of Argentina. Among such new crops, coriander has shown particular promise. The Lower Colorado River Valley (39° 23' S, 62°37' W) covers an area of more than 500,000 ha, 90,000 ha of which are under effective irrigation.

The sowing of coriander in the Lower Colorado River Valley can be carried out from autumn to the end of winter and even up to spring. In this area the greatest yields are obtained from seeds sown at the beginning of winter. Late sowing leads to a progressive and significant decline in yield. The later sowing takes place, the shorter the total growth cycle, which can vary from 170 to 100 days. The number of umbels per plant and the weight of 1000 fruits were found to be higher in early-sown plants. Seeds sown in narrowly spaced rows (up to 45 cm) gave higher yields (35%) than those sown in widely spaced rows.

REFERENCES

Alvarez, A. and A. Curioni. 1993. Efecto de dos épocas de siembra sobre el rendimiento y la producción de biomasa aérea en coriandro (*Coriandrum sativum* L.). Anales SAIPA 11:211–215.

Arizio, O., A. Curioni, and M. García. 1994. Coriandro (*Coriandrum sativum* L.): Importancia económica, cultivo y comercialización. 2[das] Jornadas de Actualización en Cultivos no Convencionales Aromáticos Medicinales (Luján; Argentina):33–59.

Barreyro, R.A., G.E. Sánchez Vallduvi, and R. Bezus. 1993. Respuesta del coriandro (*Coriandrum sativum* L.) a la fertilización nitrogenada y fosforada. Anales SAIPA 11:195–200.

Brevedan, R.E., G.G. Luayza, and R.I. Palomo. 1994. Respuesta a la fertilización de un cultivo de coriandro bajo riego. Anales SAIPA 12:337–343.

Hornock, L. 1986. Effect of environmental factors on growth, yield and on the active principles of some spices plants. Acta Hort. 188:169–176.

Lafourcade, H.W., M.S. Rubió, and M.A. Elechosa. 1985. Coriandro (*Coriandrum sativum* L.). Acintacnia 2(15):30–32.

Luayza, G.G., I.R. Palomo, and R.E. Brevedan. 1994. Efecto de la época de siembra sobre el crecimiento y desarrollo de coriandro cultivado bajo riego. Anales SAIPA 12:353–362.
Rubió, M.S., M.A. Elechosa, and H.W. Lafourcade. 1986. Ensayo de espaciamiento entre líneas en coriandro (*Coriandrum sativum* L.). SAIPA 7:94–98.
Rubió, M.S., M.A. Elechosa, and H.W. Lafourcade. 1987. Ensayo de espaciamiento entre líneas en coriandro (*Coriandrum sativum* L.). SAIPA 8:53–58.
Rubió, M.S., M.A. Elechosa, and H.W. Lafourcade. 1992. Ensayo de espaciamiento entre líneas en coriandro (*Coriandrum sativum* L.). Resultados del 2º año. SAIPA 9–10:22–26.
SERCOPLAN 1987. Development Plan for Patagones and Villarino, Province of Buenos Aires. Prepared by SERCOPLAN, the Technical Cooperation and Planning Service.

Greenhouse Production of Garlic Chives and Cilantro

Robert G. Anderson and Wenwei Jia

Fresh herbs continue to increase in consumption in the U.S. Fresh cilantro was the third most common herb used in southern California restaurants (Anon. 1991). Field grown fresh cilantro is commonly available at supermarkets but the quality is decreased by shipment from California. Chinese or garlic chives are a less common herb, but commonly used in the Orient for cooking (Brewster 1994; Larkcom 1991). Greenhouse production of cilantro and garlic chives could tap the ever expanding market for ethnic vegetables and herbs.

Trials were initiated to determine if cilantro and garlic chives could be grown in typical greenhouse bedding plant/vegetable transplant trays. Cell size and plant density were varied to determine the yield of the plants in a greenhouse situation.

METHODOLOGY

Cilantro (*Coriandrum sativum*) was grown as a short term transplant. Seed germination required 5–7 days, seedlings were transplanted 10–12 days later and foliage was harvested 25–30 days later. Garlic chives (*Allium tuberosum*) were grown as a perennial, leaves were harvested every 4–8 weeks depending on the time of year.

Subirrigation techniques, capillary mat or "float" system, were used for both plants in this study. Capillary mat subirrigation utilized a constant water table system to maintain uniform water availablility in the cells (Buxton 1994). The "float" system utilized polystyrene trays that are placed (floated) on the surface of water, 15 cm deep, held inside a wooden frame with a plastic liner.

Garlic chives and cilantro were transplanted and grown from Oct. 1993 to May 1994 in polystyrene and polypropylene trays (Table 1). Individual experimental units were 0.1 square meter. Garlic chives and cilantro were grown in greenhouses with ambient light levels and set points of 17°C night temperature and 23°C day temperature. Plants were fertilized with water soluble fertilizer, 20–4–16, at levels of 50–70 ppm N. Fresh weights of the foliage and plant number were noted at the completion of the experiments.

RESULTS

Garlic Chives

Garlic chive plants grew well on both types of subirrigation systems. Because the "float" system was so easy to maintain, this irrigation system was used in all experiments. Initial trials demonstrated an effect of cell size on the harvested fresh weight of garlic chive foliage. Subsequent experiments demonstrated that the mean fresh weight (g/plant) was statistically similar in each trial (Table 2). Thus, a limiting plant density was not reached in these experiments because individual Chinese chive plants are quite small with 3–8 narrow leaves per plant.

Differences in the harvested fresh weight on a square meter basis were apparent in these experiments but this was due to changes in plant numbers and variable initial plant densities in the experiments (Table 2). There were no differences in harvested fresh weight of chive foliage with increased levels (80 ppm N com-

Table 1. Specifications of polystyrene and polypropylene trays used in this study.

Cell dimensions width × depth (cm)	Cell volume (cm^3)	Cell number/m^2
	Polystyrene trays	
2.4 × 5.7	17	1044
2.8 × 7.6	23	861
3.8 × 11.5	61	549
5.1 × 7.6	75	310
7.6 × 7.6	160	138
	Polypropylene trays	
2.1 × 2 cm	6	1076
2.1 × 4 cm	11	1076

Table 2. The effect of plant density and cell volume on the yield of garlic chive plants.

Initial plants/m^2	Cell volume (cm^3)	No. plants per cell	Foliage fresh weight		Plant number increase (%)
			g/plant	g/m^2	
Transplanted Oct. 7, harvested Dec. 15, 1993					
828	160	6	1.49	1741	145
930	75	3	1.47	1732	128
1098	61	2	1.49	2255	125
Transplanted Oct. 7, harvested Dec. 15, 1993					
1550	61		1.45	3757	171
1550	75		1.30	2751	141
1550	160		1.20	2399	129
Transplanted Nov. 5, harvested Feb. and Apr. 5, 1994					
828	160	6	1.49	2867	233
930	75	3	1.47	2792	205
1098	61	2	1.49	3600	220
Transplanted Oct. 5, harvested Dec., 1993, Feb., Apr., and May 10, 1994					
138	160	1	2.72	1356	361
272	160	2	3.12	3056	360
549	61	1	2.88	5154	326
1098	61	2	2.84	8321	267

Table 3. The effect of plant density and cell volume on the yield of cilantro foliage harvested Nov. 11, 1993 from seed sown Oct. 5 and transplanted Oct. 25.

Plant number/m²	Cell volume (cm³)	Fresh weight g/plant	Fresh weight g/m²
689	61	1.52	984.5
839	75	1.30	1090.0
968	23	1.40	1353.6
1076	17	1.36	1466.6
1377	61[z]	1.35	1859.3

[z]2 plants/cell

Table 4. The effect of cell volume and plant density on the yield of cilantro foliage.

Cell volume (cm³)	Plants/m²	Foliage fresh weight g/plant	Foliage fresh weight g/m²
Seed sown Oct. 5, transplanted Oct. 15, harvested Nov. 15, 1993			
61	1550	2.04	3155
75	1550	2.05	2926
160	1550	1.98	3065
Seed sown Sept. 27, transplanted Oct. 15, harvested Nov. 11, 1993			
6	1550	1.66	2564
11	1550	1.61	2491
75	1550	1.51	2336
Seed sown Feb. 27, transplanted Mar. 15, harvested Apr. 12, 1994			
61	775	4.44	3440
75	775	5.64	4373
160	775	4.67	3617

pared to 40 ppm N) of fertilizer. Garlic chives grown for six and eight months had increased yields in fresh weight and plant numbers compared to those grown for two months (Table 2). Per plant fresh weight in the six-month-old plants was not larger than the two-month-old plants due to plant damage from thrips. Thrips can be a significant problem on garlic chives in the greenhouse. No pesticides are cleared for thrips on fresh herbs in the greenhouse so thrips screening and excellent santitation must be used to manage these insects.

Cilantro

Cilantro was a successful short term crop requiring only 50 to 60 days for production. This fast production time led to high greenhouse space utilization due to the rapid turnover. Cilantro did not perform well in the "float" system but did well on a constant water table capillary mat subirrigation system.

Table 5. Potential returns for garlic chives and cilantro.

Assumptions for winter greenhouse crop of garlic chive foliage

1. Perennial crop, harvest every 8 weeks from Oct. 1 to June 1.
2. Each crop requires 8 weeks and plants are grown at 1550 plants m^{-2}
3. Utilize 50 m^2 of bench space for each crop.
4. Receive $.70 per 2 oz (30g) package.

Gross returns calculations

2.5 kg foliage $m^{-2} \times 200$ $m^{-2} \times 4$ crops = 2,000 kg × 5% shrink = 1,900 kg
1,900 kg/30 g per 2 oz. package = 63,333 packages
63,333 packages × $0.70 = $44,333.00
$44,333/200 m^2 bench space = $222 m^{-2} gross returns for winter crop

Assumptions for winter greenhouse crop of cilantro foliage

1. Thirteen crops, sow seed every 2 weeks from Oct. 1 to Apr. 1. Use greenhouse from Oct. 1 to June 1.
2. Each crop requires 8 weeks and plants are grown at 775 plants m^{-2}
3. Utilize 50 m2 of bench space for each crop.
4. Receive $.70 per 2 oz (30g) package.

Gross returns calculations

3 kg foliage $m^{-2} \times 50^{-2}$ m × 13 crops = 1,950 kg × 5% shrink = 1,852,500 g
1,852.5 kg/30 g per 2 oz. package = 61,750 packages
61,750 packages × $0.70 = $43,225.00
$43,225/200 m^2 bench space = $216 m^{-2} gross returns for winter crop

---

The effects of cell size and plant density on yield of cilantro were variable. Initial trials demonstrated that lower plant densities may increase the harvested fresh weight of cilantro foliage even though the means were not statistically different (Table 3). Subsequent experiments demonstrated no statistical differences in the mean fresh weight on per plant basis in each trial (Table 4).

Differences in the harvested fresh weight on a square meter basis were statistically similar in these experiments due to the variability within the treatments (Table 4). There were no differences in harvested fresh weight of cilantro foliage with increased levels of fertilizer (100 ppm N compared to 50 ppm N). The yield of cilantro foliage was increased in the experiment completed in the spring. This was probably due to increased light levels in the greenhouse in the spring where these plants received a total of 992 mol m^{-2} PAR during the crop while plants in the fall received 435 mol m^{-2} PAR.

SUMMARY

Garlic chives and cilantro were easy to grow and produced high yields in bedding plant/vegetable transplant trays. The potential returns for greenhouse production are based on numbers from these trials and typical wholesale prices for these products (Table 5). Unfortunately, the market for these herbs is highly variable and any grower should determine market opportunities before production is initiated.

REFERENCES

Anonymous. 1991. Rising culinary herb use offers opportunity for growers. GrowerTalks March:86.

Brewster, J.L. 1994. Onions and other vegetable alliums. CAB Int., Wallingford, UK.

Buxton, J.W., W. Jia, and G. Hou. 1994. Providing a constant, optimum moisture/air ration in plug trays during seed germination and seedling growth. HortScience 29:502.

Larkcom, J. 1991. Oriental vegetables: The complete guide for garden and kitchen. Kodansha Int., New York.

Basil Seed Oils*

Paul Angers, Mario R. Morales, and James E. Simon

The genus *Ocimum* (Lamiaceae), which includes sweet basil, offers a wide diversity among its more than 50 species (Bailey 1924; Darrah 1980), particularly regarding plant growth, morphology, physical appearance and essential oil content and composition (Morales et al. 1993; Simon et al. 1984). Seed oil composition constitutes another characteristic which contributes to the rich diversity of the *Ocimum* genus.

Previous studies on basil seed oils include the drying properties of *O. basilicum* (Earle et al. 1960), *O. kilimandscharicum* (Barker et al. 1950; Henry and Grindley 1944), and *O. sanctum* (Nadkarni and Patwardhan 1952); the isolation of the compounds in seed oil that might be responsible for medicinal properties (Metha and Metha 1943); and the suitability of *O. basilicum* seed oil for cosmetic uses (Domokos and Perédi 1993). In this paper, we analyzed the chemical composition and characterized the physical properties of the fixed oil from seeds of four *Ocimum* species: *O. basilicum*, *O. canum*, *O. gratissimum*, and *O. sanctum*.

METHODOLOGY

Mature dry seeds of seven basil chemotypes: citral, linalool, methyl chavicol, and methyl cinnamate (*O. basilicum*); camphor (*O. canum*); eugenol (*O. sanctum*); and geraniol (*O. gratissimum*), from our breeding program at Purdue University, were washed and finely ground in a mortar. Classification of *O. canum* should be considered temporary due to possible interspecific hybridization. The oil from the flour was extracted in a Butt-type apparatus following A.O.C.S. recommendations (Link, 1973). The extract was concentrated under a stream of dry nitrogen or argon for quantitative determination of the oil content. Transesterification of acylated fatty acids was carried out with 2.0 N sodium hydroxide in dry methanol (Pelick and Mahadvan 1975). Iodine values (Hanus method) and saponification values were both measured according to AOAC (Cunniff 1995). Refractive indices were measured at 20°C with a refractometer (Bausch and Lomb, Rochester, NY), coupled with a controlled temperature bath and circulator (Forma Scientific model 2067, Marietta, Ohio). Methyl esters and glycerides were analyzed by gas chromatography, as described in Angers et al. (1996).

RESULTS

Oil content and characteristics of basil seeds evaluated in this study compared to linseed are shown in Table 1 and values reported in the literature are presented in Table 2. Oil content in the seeds averaged 22%, with a minimum of 18% in the camphor chemotype (*O. canum*) and a maximum of 26% in the linalool chemotype (*O. basilicum*). While the oil content of these basils is lower than that found in many commercial oil plants such as rape (*Brassica napus* L., 38%–44%) and linseed (flax, *Linum usitatissimum* L., 32%–43%), the actual physico-chemical and fatty acid profiles approached that of linseed oil. The refractive indices (1.460–1.481), saponification (191–200) and iodine values (172–200, Hanus) were all characteristic of highly unsaturated oils. The oils were very high (96% among all basils) in triacylglycerols (TAG), while the total for both mono- and diacylglycerols reached only 2% each. TAGs with carbon number (CN) 54 constituted 70%–84% of the total TAGs; with CN 52, 15%–27%; and with CN 50, 1%–3%. TAGs with CN 48 and 56 constituted less than 1% each.

Unsaturated fatty acids averaged 89%, including α-linolenic (43.8% to 64.8%), linoleic (17.8% to 31.3%), and oleic (8.5% to 13.3%). The most abundant saturated fatty acids were palmitic (6.1% to 11.0%) and stearic (2.0% to 4.0%). Values from the literature (Table 2) range from 5.3% to 15.4% for oleic acid, 14.0% to 66.1% for linoleic acid, and 15.7% to 65.0% for linolenic acid (Patwardhan 1930; Henry and Grindley

*Journal Paper No. 14,984, Purdue University Agricultural Research Program, Purdue University, West Lafayette, IN. This work was supported in part by the Center for New Crops and Plant Products and a postdoctoral fellowship (NSERC of Canada) to the senior author, which we gratefully acknowledge. Thanks are due to E. Tousignant (Food Science Department, Laval University, Canada) for help in glyceride analysis.

Table 1. Analytical values, glyceride content and fatty acid composition of oil from seeds of different *Ocimum* chemotypes compared to linseed[z].

	Species and chemotype							
	O. basilicum				*O. canum*	*O. gratissimum*	*O. sanctum*	*Linum usitatissimum*
Variable	Citral	Linalool	Methyl chavicol	Methyl cinnamate	Camphor	Geraniol	Eugenol	Linseed[w]
Oil content (% w/w)[y]	20	26	21	24	18	20	22	32–43
Refractive index (nD^{20})	1.481	1.480	1.479	1.479	1.472	1.460	1.477	1.477–1.482
Saponification value	199	200	200	200	--	194	191	192
Iodine value	198	198	184	190	200	178	172	180
Class (%)								
Monoacyglycerols	2	1	1	3	1	1	3	
Diacylglycerols	2	1	1	2	2	1	3	
Triacylglycerols	96	98	98	95	97	98	94	
Carbon number (%)[x]								
50	2	1	1	1	1	3	3	
52	16	18	20	20	15	23	27	
54	82	81	78	78	84	74	70	
Composition (mol%)								
Palmitic (16:0)	6.8	7.4	8.8	7.8	6.1	10.0	11.0	6
Stearic (18:0)	2.2	2.0	2.8	2.4	2.3	2.1	4.0	4
Palmitoletic (16:1)	0.3	0.2	0.2	0.2	0.2	0.3	0.2	--
Oleic (18:1)	9.7	8.7	9.5	11.6	8.5	8.6	13.3	22
Linoleic (18:2)	18.3	21.7	21.3	20.6	17.8	31.3	26.8	16
α-Linolenic (α-18:3)	62.5	60.0	57.4	57.4	64.8	47.4	43.8	52
γ-Linolenic (γ-18:3)	0.3	0.1	0.2	0.3	0.3	0.2	trace	--
Arachidic (20:0)	0.2	trace	trace	trace	0.2	0.2	0.2	--
Eicosenoic (20:1)	trace[v]	trace	trace	trace	trace	trace	trace	--

[z]Modified from Angers et al. 1996. Data based on four replications.
[y]Based on seed dry weight.
[x]Carbon number is relative to triacylglycerols.
[w]Analytical and glyceride value from Bernardini (1985); fatty acid composition from Patterson (1989).
[v]Trace values are lower than 0.1%.

Table 2. Analytical values and fatty acid composition of basil seed oil reported in the literature.

Variable	Species						
	O. basilicum	*O. canum*	*O. kilimand-scharicum*	*O. kilimand-scharicum*	*O. pilosum*	*O. sanctum*	*O. viride*
Oil content (% w/w)[z]	24	10	13	16	17	18	10
Refractive index (nD)	1.474^{40}	1.471^{40}	1.485^{25}	1.477^{40}	1.475^{32}	1.479^{30}	1.481^{20}
Saponification value	188	195	--	193	187	182	--
Iodine value	191	180	193	196	173	173	169
Composition (mol%)							
Palmitic (16:0)	}8[y]	7.0	8.2	}8[y]	3.3	6.9	8.7
Stearic (18:0)		0.2	--		6.4	2.1	2.7
Oleic (18:1)	15	11.1	5.3	17	15.4	9.0	14.3
Linoleic (18:2)	22	60.4	16.2	14	56.3	66.1	32.5
α-Linolenic (α-18:3)	50	21.3	65.0	61	18.5	15.7	39.2
γ-Linolenic (γ-18:3)	--	--	--	--	--		--
Arachidic (20:0)	--	--	5.3	--	--	--	2.6
Eicosenoic (20:1)	--	--	--	--	--	--	--
References	Earle et al. 1960	Patwardan 1930	Barker et al. 1950	Henry and Grindley 1944	Khan et al. 1961	Nadkarni and Patwardhan 1952	Barker et al. 1950

[z]Based on seed dry weight.
[y]Values for saturated fatty acids.

1944; Barker et al. 1950; Nadkarni and Patwardhan 1952; Earle et al. 1960; Khan et al. 1961). The total of linoleic and linolenic acids appears to be fairly constant for all species, ranging from 78% to 82%, except in *O. sanctum* where it reached only 71%.

The composition of basil seed oil suggests that the oil would be suitable for industrial purposes, much in the same way as linseed oil is used. An estimated basil seed yield in Indiana of 1400 kg/ha at 22% seed oil, would yield 300 kg/ha of oil, or 180 kg/ha of linolenic acid. In contrast, seed yields of flax averaged 765 kg/ha (FAO 1995) and at a 40% oil content, the estimated yield of linolenic acid from flax (linseed) is 150 kg/ha.

Techniques and conditions for basil cultivation are well known, as the plant has been grown as a culinary herb and source of essential oil for many years. Seed and oil content yields could be increased by plant selection. Seeds of basil do not readily dehisce and can be harvested using a combine. A high linolenic acid oil, such as that found in *Ocimum basilicum* and *O. canum*, could be used in the paint, varnish and ink industries, and as a source of linolenic acid, while oils with lower linolenic acid content, such as those of *O. gratissimum* and *O. sanctum*, might be used by the food industry.

REFERENCES

Angers, P., M.R. Morales, and J.E. Simon. 1996. Fatty acid variation in seed oil among *Ocimum* species. J. Am. Oil Chem. Soc. 73:393–395.

Bailey, L.H. 1924. Manual of cultivated plants. MacMillan, New York.

Barker, C., H.C. Dunn, and T.P. Hilditch. 1950. African drying oils. V. Some Nigerian and Sudanese drying oils. J. Soc. Chem. Ind. 69:71–75.

Bernardini, E. 1985. Vegetable oils and fats processing, BE. Oil, Rome, p.16.

Cunniff, P. (ed.). 1995. Official methods of analysis of the association of official analytical chemists, 16th ed., Association of Official Analytical Chemists, Methods 920.158 and 920.160.

Darrah, H.H. 1980. The cultivated basils. Buckeye Printing, Independence, MO.

Domokos, J. and J. Perédi. 1993. Studies on the seed oils of basil (*Ocimum basilicum* L.) and summer savory (*Satureja hortensis* L.). Acta Hort. 344:312–314.

Earle, F.R., R.A. McGuire, J. Mallan, M.O. Bagby, and I.A. Wolff. 1960. Search for new industrial oils. II. Oils with high iodine values. J. Am. Oil Chem. Soc. 37:48–50.

FAO 1995. Production yearbook 1994, 48:114: Rome.

Henry, A.J. and D.N. Grindley. 1944. The oils of the seeds of *Ocimum kilimandscharicum*, *Euphorbia calycina*, *E. erythraeae*, *Sterculia tomentosa*, and *Trichilia emetica*. J. Soc. Chem. Ind. 63:188–190.

Khan, S.A., M.I. Qureshi, M.K. Bhatty, and Karimulah. 1961. Composition of the seed oils of *Salvia spinosa* and *Ocimum pilosum* (*O. Basilicum*). Pakistan J. Sci. Res. 13:41–43.

Link, W.E. (ed.). 1973. Method Aa. p. 4–38. In: Official methods and recommended practices of the Am. Oil Chem. Soc., 3rd ed., Am. Oil Chem. Soc., Champaign, IL.

Mehta, C.R. and T.P. Metha. 1943. Chemical examination of *Ocimum canum* Sims. Current Sci. 12:300–301.

Morales, M.R., D.J Charles, and J.E. Simon. 1993. New aromatic lemon basil germplasm. p. 632–639. In: J. Janick and J.E. Simon (eds.), New crops. Wiley, New York.

Nadkarni, G.B. and V.A. Patwardhan. 1952. Fatty oil from the seeds of *Ocimum sanctum* Linn. (Tulsi). Current Sci. 21:68–69.

Patterson, H.B.W. 1989. Handling and storage of oilseeds, oils, fats and meal. p. 112–113. Elsevier Applied Science, London.

Pelick, N. and V. Mahadvan. 1975. Lipid derivatives and gas liquid chromatography. p. 27–29. In: E.G. Perkins (ed.), Analysis of lipids and lipoproteins. Am. Oil Chem. Soc., Champaign, IL.

Simon, J.E., A.F. Chadwick, and L.E. Craker. 1984. Herbs: An indexed bibliography 1971–1980; the Scientific literature on selected herbs, and aromatic and medicinal plants of the temperate zone. Archon Books, Hamden, CT.

Selection of *Boronia* for Essential Oils and Cut Flowers*

Julie A. Plummer, Joanne M. Wann, John A. Considine, and Z.E. (Ted) Spadek

Several species of *Boronia* (Rutaceae) are endemic to the southwest of Western Australia. They grow in wet or seasonally wet low-lying areas and are usually associated with jarrah (*Eucalyptus marginata* Sm.) forests, paperbark (*Melaleuca parviflora* Lindl.) flats, and creeks. Two species are commercially exploited; *B. megastigma* Nees. (brown boronia) for its essential oils and *B. heterophylla* F. Muell. (red boronia, kalgan) for cut flowering stems (Plummer 1995). In the past total production was from plant material harvested from natural stands. With increasing demand and the need to improve quality these industries have shifted to harvesting cultivated plants.

B. megastigma plants bear a profusion of strongly scented bell-like flowers (Fig. 1). Petals are yellow on the inside, while outer petal color varies from yellow (cv Lutea) through red to dark brown. Outer petal color may be solid or in vertical red (cv Harlequin) or brown stripes (Plummer 1995). Essential oils are extracted from the flowers with hexane to yield about 0.4%–0.8% concrete. Concrete is washed with alcohol or distilled to yield 60% boronia absolute (Guenther 1949; Penfold and Willis 1954). Boronia absolute is valued at between US$4,000 to $10,000/kg, depending on quality and purity. It can be used in perfumes, cosmetics, and as a food flavoring where it enhances most fruit essences (Davies and Menary 1984; Weyerstahl et al. 1994). There are over 150 compounds in boronia concrete. β-Ionone is by far the most important, but other compounds such as dodecyl acetate enhance the perfume. The monoterpene hydrocarbons, α-pinene, β-pinene, and limonene detract from the aroma. Three cultivated genotypes have been shown to have a high degree of variability in their essential oil components (Davies and Menary 1984) but the diversity within the species has not been examined.

Cut flowering stems of *B. heterophylla* have been exported from Australia to Japan for 7 years and markets are growing in southeast Asia, Europe, and Canada. *B. heterophylla* is an upright shrub which produces a spectacular display of vibrant pink bell-shaped flowers each about 1 cm long in early spring. Cultivated plants can yield 8–12 stems 12–15 months from planting, 20–30 stems in the third year, and 30–60 stems in the fourth year with a commercial life of 5–6 years. Stems are usually 60 cm long and are picked with >50% of the flowers open (Plummer 1995). Although demand currently exceeds supply, the limited flowering season (1–2 weeks) and lack of flower color options are obstacles to future expansion.

Demand for both boronia absolute from *B. megastigma* and cut flowering stems of *B. heterophylla* is

Fig. 1. Young (2-year-old) *Boronia megastigma* plant in cultivation in Nannup, Western Australia.

*Joint contribution of Plant Sciences, Faculty of Agriculture, The University of Western Australia, Nedlands, WA 6907 and the Agricultural Chemistry Laboratory, Chemistry Centre of Western Australia, 125 Hay Street, East Perth, WA 6004, Australia. Research supported in part by the Rural Industries Research and Development Corporation, Sunglow Flowers Pty. Ltd. and Plantex Australia Pty. Ltd.

increasing. The export market potential is substantial and is likely to continue its rapid growth. Exports from Western Australia alone are expected to increase by 20% per annum. Boronia exports from Western Australia were valued at US$46,000 FOB (freight on board) in 1992/93 while the total value of boronia exports from Australia was US$180,000 FOB in the same year (Anon. 1994). Current wholesale prices for *B. heterophylla* range from US$2.00 to $3.50 per bunch, depending on supply. Wholesale prices for *B. megastigma* blossom for essential oil production range from US$10 to $15/kg.

Commercial production of both species is dependent on a few cultivars and selection of superior genotypes would be beneficial. The aim of this study was to investigate variation in the content of essential oil components of *B. megastigma* within and between populations, and over different seasons, and to extend the color range and production period of *B. heterophylla*.

METHODOLOGY

Boronia megastigma

For plant sampling in Aug. to Sept. 1993 *B. megastigma* flowers were collected from plants in 13 natural populations and from a yellow flowering selection growing in cultivation. In Aug. to Sept. 1994 flowers were collected from 16 natural populations. Flowers from 25 randomly selected plants were collected from the majority of populations, but where numbers of flowering plants were limited, fewer plants were sampled (n = 24 for the Walpole-2 and Palgarup-1 populations; n = 20 for the Yallingup population; n = 22 for the Barlee population).

Details of flower color, plant age, and plant vigor were collected for all plants. Plant age was determined using the number of years since the last fire or major soil disturbance as an approximate age. Seedlings were determined as less than 1 year old, young plants 1–2 years, mature plants 2 to 6 years and old plants greater than 6 years. Plant vigor was determined by the density of leaves and flowers on each plant. Weak plants had few flowers and little foliage, average plants had moderate amounts of foliage and flowers, and vigorous plants had abundant flowers and lush green foliage. Oil component contents within each category were compared using analysis of variance and means were separated using Fisher's PLSD. The amount of shading by surrounding vegetation was determined for 273 of the plants sampled in 1993 using a Model-A forest densiometer (Lemmon 1957).

Seven to 10 plants from 12 natural populations initially sampled in 1993 were re-sampled in 1994 in order to investigate the quantity of oil components produced by individual plants over successive years. Comparisons were made using paired 2-tailed t-test.

For oil analysis 12 flowers collected from each plant were extracted with absolute ethanol (10 ml). Collection vials were weighed before and after flower collection and the difference used to determine flower fresh weight. Flowers were extracted for at least 24 h before oil analysis. An aliquot (4 ml) of the extract was analysed without further purification or concentration using a gas liquid chromatograph (Hewlett Packard 5890A) fitted with dual columns (50 m length, 0.25 mm internal diameter BPX70 from SGE International and 50 m length, 0.2 mm internal diameter ULTRA1 from Hewlett Packard) connected to a single injection port and twin ionization detectors. The injector and detector temperatures were set at 275°C. An initial oven temperature of 60°C was held for 5 minutes and then increased at a rate of 8°C min^{-1} to a final temperature of 260°C which was held for 10 min. Ethyl un-decanoate (1 $\mu g\ ml^{-1}$) was used as an internal standard. The quantity ($\mu g\ g^{-1}$ fresh weight of flowers) of α-pinene, β-pinene, limonene, α-ionone, β-ionone, and dodecyl acetate was determined using the areas produced by the ULTRA 1 column. The BPX70 column was used to confirm the presence or absence of α-ionone, β-ionone and dodecyl acetate. α-Pinene, β-pinene, and limonene co-eluted with the solvent peak on this column.

Principal components analyses were performed on the oil data. The first analysis used the oil data from all individuals sampled to plot the orthogonal factor scores and determine the relationship between the different oil components. The second analysis used the mean data for each oil component from each population to determine the relationship between different populations in terms of the 5 oil components.

Boronia heterophylla

New populations of *B. heterophylla* were sought in 1994. Plants were tagged and their flowering period, flower density, flower color, and plant vigor were determined. Cutting material was taken from plants to assess their ease of propagation and subsequent growth and flowering in cultivation.

RESULTS

Boronia megastigma

The 29 *B. megastigma* populations identified cover most of the remaining natural range of the species however, much of this region is now agricultural land (Fig. 2). Flowers from all plants contained β-ionone. α-Ionone was either absent or present in undetectable amounts and was therefore excluded from further analysis. β-Pinene was absent from half of the plants, α-pinene from about a quarter of plants and limonene and dodecyl acetate were absent from only a few plants. Oil contents were used to rank each oil component (Table 1). Ranking was determined by a fixed percentage of plants whose oil component quantity fell within the stated range. The extremely high β-ionone producers (those in the top 5%) contained 803 to 1787 $\mu g\ g^{-1}$ of β–ionone. β-Ionone was the major essential oil component, followed by dodecyl acetate. α-Pinene and limonene were often present in similar quantities, and β-pinene was usually the least abundant of the oil components examined.

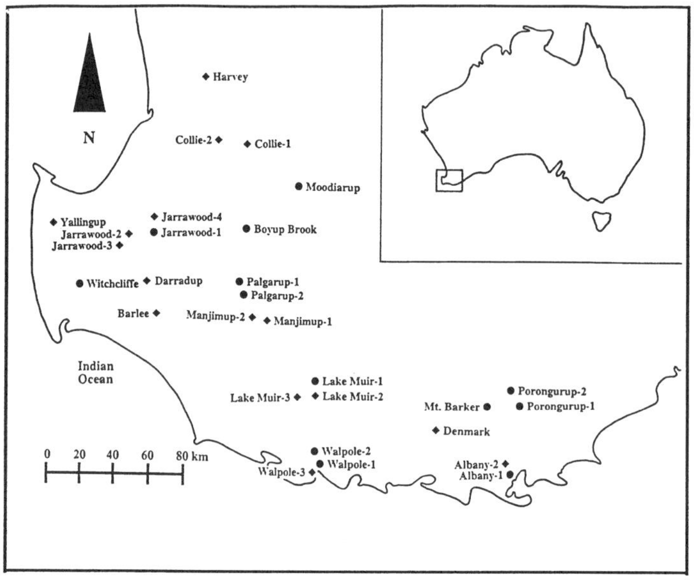

Fig. 2. *Boronia megastigma* populations in the south west of Western Australia. ● denotes populations sampled in 1993, and ◆ denotes populations sampled in 1994.

Table 1. Ranking of the 5 oil components determined using the gas liquid chromatography results from the 718 plants sampled in 1993 and 1994.

Rank	% of total plants	Oil content ($\mu g\ g^{-1}$ FW)				
		α-Pinene	β-Pinene	Limonene	β-Ionone	Dodecyl acetate
Extremely high	5	420–1001	298–940	264–893	803–1787	577–1291
Very high	5	303–419	207–297	184–263	646–802	518–576
High	10	193–302	140–206	118–183	548–645	427–517
Medium	60	0–192	0–139	33–117	260–547	178–426
Low	10	0	0	25–32	199–259	115–177
Very low	5	0	0	15–24	162–198	84–114
Extremely low	5	0	0	0–14	70–161	0–83

The mean quantity of oil components varied considerably between populations. α-Pinene, β-pinene and limonene were commonly absent from individual plants, but each of these compounds was present in at least one plant in each population. Location had some influence on the content of the monoterpene hydrocarbons. Southeastern populations had quite high concentrations of α-pinene, β-pinene, and limonene, while northern populations all had low concentrations. Location was not always a good indicator of the content of oil components. In the Jarrahwood area, Jarrahwood-3 produced more than twice the amount of α-pinene and β-pinene as Jarrahwood-2 and Jarrahwood-4, even though the three populations are only a short distance apart. Location did not influence β-ionone and dodecyl acetate. The highest mean β-ionone content for a population was 736 $\mu g\ g^{-1}$ at Albany-1 and the lowest was 227 $\mu g\ g^{-1}$ at Porongurup-1.

The quantity of oil components from plants within populations varied considerably. The Walpole-1 population sampled in 1993 was typical of the variation in β-ionone content within populations. β-Ionone concentration ranged from very low (175 $\mu g\ g^{-1}$) to extremely high (1342 $\mu g\ g^{-1}$).

The relationship between oil components based on their quantitative presence in plants was investigated using principal components analysis. Results indicated that β-ionone and dodecyl acetate were closely related and quite distinct from the monoterpene hydrocarbons. The was also a close relationship among the monoterpene hydrocarbons. α-Pinene and β-pinene, and to a lesser extent, limonene were closely related.

Principal components analysis was also used to investigate the relationship between *B. megastigma* populations based on the mean content of oil components at each site. Results failed to show any distinct relationships between populations. However, among the populations sampled in each year there were certain populations which were closely related in terms of the mean content of the 5 oil components. Some populations in close proximity to each other were similar in oil composition, but others were distinct. Boyup Brook, Palgarup-1 and Palgarup-2 were within 33 km of each other and were closely related (1993 data). Jarrahwood-2 and Jarrahwood-4 were closely related, but Jarrahwood-3 was quite distinct (1994 data).

Canopy cover above *B. megastigma* plants affected concentrations of the monoterpene hydrocarbons but not β-ionone ($y = 0.6x + 365$; $r^2 = 0.005$) and dodecyl acetate ($y = 0.8x + 273$; $r^2 = 0.009$). Plants shaded by higher densities of canopy cover produced lower concentrations of α-pinene ($y = -3x + 267$; $r^2 = 0.09$), β-pinene ($y = -2.2x + 156$; $r^2 = 0.09$), and limonene ($y = -2.1x + 160$; $r^2 = 0.09$).

Plant age, plant vigor, and flower color each affected the concentration of at least one of the five essential oil components investigated. Similar trends occurred in both years but only 1993 data is presented. Age affected four essential oil components (Table 2). Young plants contained less α-pinene and limonene than mature plants and less β-pinene than seedlings, mature and old plants. Old plants had the most dodecyl acetate. Seedlings had more β-pinene and limonene, and less β-ionone and dodecyl acetate than older plants but, due to the low sample size of seedlings, the differences were not significant. Vigorous plants tended to produce more β-pinene and limonene than weak plants (Table 3). Red flowers contained the lowest concentrations of β-ionone and dodecyl acetate (Table 4). There was a tendency for red/brown flowers to have less β-ionone and dodecyl acetate than brown flowers. Yellow flowers contained no β-pinene.

Table 2. The influence of plant age on concentration ($\mu g\ g^{-1}$ fresh flower weight) of oil components of *Boronia megastigma*. Mean separation in columns by Fisher's PLSD, 5% level.

	Component ($\mu g\ g^{-1}$ FW)				
Plant age	α-Pinene	β-Pinene	Limonene	β-Ionone	Dodecyl acetate
Seedling	156ab	231b	125ab	234a	169ab
Young	90a	29a	44a	395a	274a
Mature	159b	93b	99b	382a	297a
Old	144ab	87b	60ab	391a	364b

In plants sampled in both years β-ionone and dodecyl acetate content differed between the 1993 and 1994 sampling. The mean quantity of β-ionone and dodecyl acetate was higher in 1994 than 1993, while α-pinene, β-pinene and limonene content did not differ (Table 5). The mean quantity of the majority of oil components for each population sampled in 1994 was the same as for 1993. Where differences did occur, they could be attributed to either large differences in an individual plant, or numerous small differences in several plants within the population.

Twenty six plants were found to have no α-pinene in at least 1 of the years sampled. Of the plants found to have no α-pinene present in either the 1993 or 1994 analysis, 69% were found to have no α-pinene in both years. No β-pinene was found in 53 plants in at least 1 of the years sampled. Of these plants, 85% had no β-pinene in both years. Of the plants lacking limonene, only 21% had no limonene in both years. Many plants consistently had two or more of the hydrocarbon monoterpenes absent, and several were lacking all three.

Table 3. The influence of plant vigour on concentration (μg g^{-1} fresh flower weight) of oil components of *Boronia megastigma*. Mean separation in columns by Fisher's PLSD, 5% level.

	Component (μg g^{-1} FW)				
Plant vigor	α-Pinene	β-Pinene	Limonene	β-Ionone	Dodecyl acetate
Weak	120a	76ab	48a	411a	347a
Average	145a	80a	84a	383a	303a
Vigorous	207a	161b	157b	379a	265a

Table 4. The influence of flower colour on concentration (μg g^{-1} fresh flower weight) of oil components of *Boronia megastigma*. Mean separation in columns by Fisher's PLSD, 5% level.

	Component (μg g^{-1} FW)				
Flower color	α-Pinene	β-Pinene	Limonene	β-Ionone	Dodecyl acetate
Yellow	287a	0	27a	380ab	273abc
Red	121a	100a	86a	263a	172a
Red/brown	140a	79a	84a	378b	280b
Brown	157a	88a	90a	407b	338c

Table 5. Mean ± SE quantity of oil components from flowers of *Boronia megastigma* sampled in 1993 and 1994 (n = 109).

	Component (μg g^{-1} FW)				
Year	α-Pinene	β-Pinene	Limonene	β-Ionone	Dodecyl acetate
1993	157±17	86±11	80±7	415±16	323±16
1994	142±15	81±10	84±7	452±16	398±17

Boronia heterophylla

Seven populations of *B. heterophylla* were located. A number of known sites no longer contained boronias or they had been recently burnt out by fires. Plants bearing flowers with a wide range of pink tones were identified but white flowers proved elusive. Most petals had solid color but some presented either vertical or horizontal bands of deeper color. Some populations had uniform plants with consistent pink colored flowers whilst others showed considerable variation in both plant form and petal color. Flowering period was very uniform with nearly all plants flowering in early Sept. (early spring, southern hemisphere). The exceptions were one plant which flowered two weeks earlier and others which flowered up to several weeks later than the general population. Some plants proved easier than others to propagate from cuttings. The resulting plants will be assessed in cultivation during their first flowering season in Sept. 1996.

DISCUSSION

Considerable diversity was identified both within and between populations of *B. megastigma* and *B. heterophylla*. In *B. megastigma* flowers, β-ionone was the major oil component followed by dodecyl acetate. These two compounds impart much of the desirable perfume of boronia absolute and their close association following principle components analysis is promising for the selection of genotypes with very high yields of both compounds. The monoterpene hydrocarbons, α-pinene, β-pinene, and limonene are undesirable in boronia absolute. These components were also linked but were quite distinct from β-ionone and dodecyl acetate. These relationships indicate that selection of genotypes which produce very high quality oil with abundant β-ionone and dodecyl acetate and little or no monoterpenes is possible.

The relationships between β-ionone and dodecyl acetate and α-pinene, β-pinene, and limonene is supported by the work of Bussell et al. (1995) on volatile oil ontogeny in *B. megastigma.* The stigma, stamen, and staminode tissue contained dodecanol, β-ionone, dodecyl acetate, and heptadecene among the principal components. β-Ionone and dodecyl acetate concentrations were highest in the stigma extracts. Extracts of the combined ovary, nectary, and receptacle tissue contained primarily α-pinene, β-pinene, and limonene. The petal tissue contained relatively low concentrations of all of the above-mentioned volatile compounds.

The recorded characteristics of *B. megastigma* and *B. heterophylla* varied considerably between populations. Although current populations were quite isolated from one another, this is likely to be a recent event due to a number of factors, such as over-picking, frequent fires, and land clearing for agriculture. Boronia flowers are pollinated by an unidentified small moth (Bussell et al. 1995). It is possible that as populations have become more isolated, the movement of pollinators could be restricted to individual populations or a few close populations. This would result in distant populations becoming more genetically isolated over time. For example *B. megastigma* populations in the southeastern area of their range had quite high concentrations of α-pinene, β-pinene, and limonene while northern populations all had low concentrations.

Populations in close proximity may remain genetically similar not only due to pollen transfer but also water flow which could wash seeds down river from one population to another. Populations within the same water catchment area, e.g. (*B. megastigma*) Jarrahwood-2 and Jarrahwood-4, appeared closely related, while nearby populations located within different catchment areas e.g. Jarrahwood-3 were occasionally quite distinct. However, location did not always explain the differences between populations. There are many environmental factors which may affect flower color, oil content, and plant vigor.

Canopy cover was used as a crude measure of shading levels. In open conditions with little canopy cover and shade, *B. megastigma* produced high amounts of α-pinene, β-pinene, and limonene. In dense forest where canopy cover and shade levels were high, monoterpene hydrocarbons were produced in lower quantities. A number of essential oil species have decreased yields when they are grown under shade. In peppermint (*Mentha piperita*) high oil yield, including the production of limonene, results from high light intensity (Clark and Menary 1980). Reduced oil production occurs under low radiation in Japanese mint (*Mentha arvensis*) (Duriyaprapan and Britten 1982). In contrast to our results, experiments by Bussell (1996) showed that 70% shade cloth (30% of ambient light) had no effect on α-pinene and limonene concentrations (β-pinene was not tested) in three clones of *B. megastigma* and β-ionone and dodecyl acetate concentrations increased. Although the genetic material used by Bussel (1996) was different, the shade levels much higher and the plants grown under cultivation, these differences highlight the need to further investigate the influence of shading. Light may also affect flower color in *B. heterophylla*. Anthocyanins contribute to red flower color and anthocyanin production in many species is modified by light quality and intensity. Flowers of *B. heterophylla* are known to fade during senescence (Plummer 1995) but the influence of light on pigment intensity remains unknown.

Plant age, vigor, and flower color affected the concentration of oil components in *B. megastigma*. Unfortunately concentrations of oil components produced by seedlings was not a good indication of production in the adult plant. This reduces the possibility of screening seedling material for high quality oil in a breeding program. Red-flowering plants are unlikely to be used in commercial production of oil as they produced low amounts of the favorable compounds, β-ionone and dodecyl acetate. Plants with yellow flowers produced very little limonene and no β-pinene but relatively high levels of β-ionone and dodecyl acetate. Bussell (1996) also found β-pinene to be absent from yellow flowering *B. megastigma*. Apart from these trends flower color does not appear to be a useful tool in selecting plants with high oil quality.

Some differences in oil content were detected between 1993 and 1994. The concentrations of the undesirable oil components, α-pinene, β-pinene, and limonene were fairly consistent which permits genetic selection for low or no production. The change in β-ionone and dodecyl acetate content was only marginal but offers the possibility of environmental manipulation, such as site selection, to increase yield.

Boronia heterophylla plants with a range of solid and striped flower colors, and plants which flower outside the usual two-week flowering period were identified. Most of these plants could be propagated successfully. Plants will need to be examined in cultivation to determine if these characteristics are genetically controlled. Plants will also need to be assessed for their ability to produce numerous high quality stems and withstand the substantial pruning at harvest. Selected plants with these attributes will increase the color range and harvest period.

Although many of the original populations of *B. megastigma* and *B. heterophylla* no longer exist much variation exists in the remaining plants. Natural pollination of boronias requires the presence of a pollinator moth and her movements tend to ensure out-crossing. Thus the remaining plants probably contain most of the original genetic diversity. Boronias can be manually self and cross pollinated (Plummer and Considine 1995) and a breeding program has commenced. *B. megastigma* plants with high yields of β-ionone and dodecyl acetate and low or no monoterpene production were identified. Collection from natural stands is inadequate to meet demand and expansion of cultivation is underway. Currently boronia absolute is only used in the food flavoring sector and development of new uses will require expanded production. *B. heterophylla* plants with a range of pink flower tones and different flowering periods were identified. Although these characteristics will need to be confirmed under conditions of commercial cultivation they show promise for the improvement of product quality and the expansion of production of these new crops.

REFERENCES

Anon. 1994. The Australian wildflower industry: A review. Paper 94/9. Rural Industries Research and Development Corportation Research Papers. Australian Publ. Service, Canberra.

Bussell, B.M., J.A. Considine, and Z.E. Spadek. 1995. Flower and volatile oil ontogeny in *Boronia megastigma*. Ann. Bot. 76:457–463.

Bussell, B.M. 1996. Studies on the ecology and biology of *Boronia megastigma* Nees. M. Hort. Sci. Thesis, Univ. Western Australia, Nedlands.

Clark, R.J. and R.C. Menary. 1980. Environmental effects on peppermint (*Mentha piperita* L.) I. Effect of daylength, photon flux density, night temperature and day temperature on the yield and composition of peppermint oil. Austral. J. Plant Physiol. 7:685–692.

Davies, N.W. and R.C. Menary. 1984. Volatile constituents of *Boronia megastigma* flowers. Perfumer Flavorist 8:3–8.

Duriyaprapan, S. and Britten, E.J. 1982. The effects of solar radiation on plant growth, oil yield and oil quality of Japanese mint. J. Expt. Bot. 33:1319–1324.

Guenther, E. 1949. Oil of *Boronia megastigma* (boronia flower oil). p. 368–367. In: The essential oils. Van Nostrand, New York.
Lemmon, P.E. 1957. A new instrument for measuring forest overstory density. J. Forestry. 55:667–668.
Penfold, A.R. and J.L. Willis. 1954. The essential oil industry of Australia. Econ. Bot. 8:316.
Plummer, J.A. 1996. Boronia. p. 191–214. In: M. Burchett and K.A. Johnson (eds.), Australian native plants—Horticulture and uses. Univ. New South Wales Press, Sydney, Australia.
Plummer, J.A. and J.A. Considine. 1995. New boronias, more colours, better oils. Austral. Hort. 93 (5):35–36.
Weyerstahl, P., H. Marschall, W.-R. Bork, and R. Rilk. 1994. Megastigmanes and other constituents of the absolute from *Boronia megastigma* from Tasmania. Liegigs Ann. Chem. 1994:1043–1047.

Monthly Variations in Biological Activity of *Asimina triloba*

Holly A. Johnson, John Gordon, and Jerry L. McLaughlin*

The acetogenins are a class of compounds endogenous to the Annonaceae. Uvaricin, the first acetogenin discovered, was isolated from *Uvaria accuminata* in 1982 (Jolad et al. 1982). Since then, over 230 annonaceous acetogenins have been identified. The genera studied thus far which produce the acetogenins are *Annona*, *Asimina*, *Goniothalamus*, *Rollinia*, *Uvaria*, and *Xylopia*.

Asimina triloba, the North American paw paw tree or Indiana banana, is native to eastern and central North America. The tree bears an exotic flavored fruit while the bark, roots, twigs, and seeds contain the majority of the potent acetogenins (Ratnayake et al. 1992). The tree is usually pest and disease resistant, and acetogenins present inhibit the feeding of insects and herbivores.

Most acetogenins are potent biologically active molecules. They are inhibitors of Complex I (NADH: ubiquinone oxidoreductase) in mitochondria and, thus, cause a decrease in cellular ATP levels (Londerhausen et al. 1991; Ahammadsahib et al. 1993; Lewis et al. 1993). The acetogenins have shown patented antitumor, pesticidal, and antifeedant activities (Mikolajczak et al. 1988, 1989; Ikekawa et al. 1991; McLaughlin and Hui 1993). The chemical structures of the acetogenins consist of a long aliphatic chain, usually a variable number of tetrahydrofuran rings, followed by another aliphatic chain, ending in a gamma lactone ring. Anywhere from two to six hydroxyls can be found in each acetogenin, but molecules containing three hydroxyls have the highest biological activity (Landolt et al. 1995). Acetogenins are polyketide derived molecules and current research is being conducted to demonstrate that fatty acids are their precursors.

---

*This work was supported by RO1 grant no. CA30909 from the National Cancer Institute, National Institutes of Health. R. Neal Peterson of the Paw Paw Foundation is greatfully acknowledged.

Of all the acetogenins, the adjacent bis-THF ring compounds are the most potent biologically (Hui et al. 1989; Zhao et al. 1994a, b; Landolt et al. 1995). Bullatacin (Fig. 1) inhibits mammalian solid tumor cells at concentrations often a billion times lower than Adriamycin®, as shown in Table 1 (Zhao, et al. 1994b).

Asimicin, another adjacent bis-THF compound is the epimer of bullatacin at position 24 (Rupprecht et al. 1986). Asimicin has been shown to be effective and the whole class of compounds have been patented as natural pesticides as shown in Table 2 (Mikolajczak et al. 1988). Asimicin is a more effective insecticide than pyrethrins in three out of four assays performed. The F005 (crude extract) of *Asimina triloba* is also active in the pesticide tests. The F005 extract is 30% more effective in the mosquito larvae assay when compared with rotenone, a classic Complex I mitochondrial electron transport inhibitor (Ernster et al. 1963; Singer 1979). For the nematode assay, pyrethrins killed no nematodes at 10 ppm, whereas asimicin (0.1 ppm) and the F005 extract (10 ppm) showed 100% lethality after 72 hours.

The F005 bark extract has been field tested for pesticidal activity. Figure 2A displays a tree with Eastern tent catepillars (*Malacosoma americanum*) nested in a *Prunus serotina* tree. Tween 80, a non-ionic surfactant, was used to solubilize 0.5% F005 extract in water. The concentration of Tween 80 used was 1.0%. The aqueous surfactant extract solution was sprayed on the catepillars. After a thirty minute period, the majority of the insects were killed and fell from the tree (Fig. 2B).

In another experiment, the same 0.5% F005 bark extract solubilized in Tween 80 and water was sprayed on *Phlox* plants infested with powdery mildew fungus. The *Phlox* plants were sprayed with the extract solution and observed 10 days later. Fig. 3 shows the *Phlox* without F005 (left-hand side) and with F005 sprayed (right-hand side). The F005 extract solution inhibited the powdery mildew fungus from growing on the *Phlox*.

Asimicin and bullatacin along with over 30 other bioactive acetogenins and other compounds (Rupprecht et al. 1986; Zhao et al. 1992, 1993, 1994a, b, 1995; Woo et al. 1995a, b) are present in the crude extract of *Asimina triloba*. For pesticidal purposes crude extracts rather than pure compounds can be used. Using the mixture saves much time and is more economical for the preparation of the pesticide. The crude extract produces a potent acetogenin mixture; hence, this mixture presents a host of different compounds to target different insects. The crude extract could also help to decrease insect adaptation and development of resistance to the acetogenins.

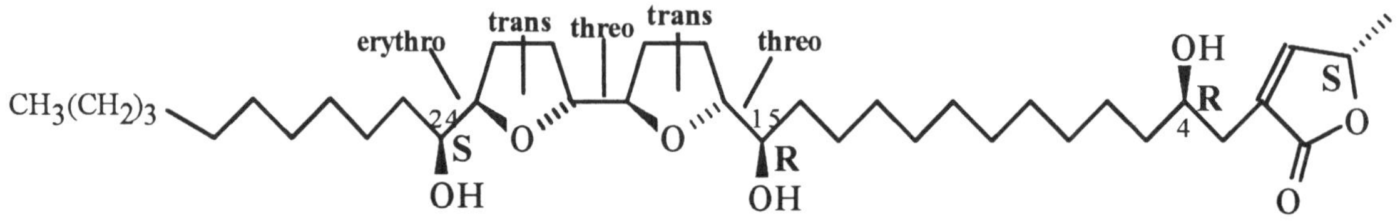

Fig. 1. Structure of bullatacin.

Table 1. The bioactivities of bullatacin in the brine shrimp test and human solid tumor cell lines.

Compound	BST LC_{50} (μg/ml)	A-549 ED_{50} (μg/ml)	MCF-7 ED_{50} (μg/ml)	HT-29 ED_{50} (μg/ml)
Bullatacin	1.59×10^{-3}	1.25×10^{-13}	>10	1.0×10^{-12}
Adriamycin	2.57×10^{-1}	1.04×10^{-4}	1.76×10^{-2}	1.53×10^{-4}

BST = brine shrimp lethality test
Adriamycin® = standard anticarcinoma drug
ED_{50} = effective dose resulting in 50% cell death
LC_{50} = lethal concentration resulting in 50% deaths

A-549 = human lung carcinoma
MCF-7 = human breast carcinoma
HT-29 = human colon carcinoma

METHODOLOGY

A. triloba twig samples from two trees (labelled east and west) were collected by John Gordon at an orchard in Amherst, New York. The twigs were dried and pulverized in a Wiley Mill to a fine powder. Originally, *A. triloba* was extracted with 95% ethanol followed by two sets of partitions (Fig. 4) before reaching F005 (Ratnayake et al. 1992). The bioactivity of the extracts was monitored by the brine shrimp lethality test (BST) (Meyer et al. 1982).

A new, more rapid, extraction procedure was developed to monitor the biological activity of multiple biomass samples of paw paw. This new method requires fewer steps and uses even less solvent. The twig samples from each month were extracted in triplicate. The residues of the three separate extractions for

Table 2. Comparison of pesticidal activity of asimicin and paw paw extract vs. standard insecticides.

Compound	Test organism	Time of exposure (h)	Mortality Rate (%)						
			0.1 ppm	1 ppm	10 ppm	50 ppm	100 ppm	500 ppm	5000 ppm
Asimicin, purified	MBB	72			70	100			
F005 (extract)	MBB	72					60		
Pyrethrins, (57% pure)	MBB	72			0	100			
Asimicin	MA	24					20	100	
F005 (extract)	MA	24							80
Pyrethrins	MA	24			20	100			
Asimicin	ML	24		100					
F005 (extract)	ML	24		10	80				
Pyrethrins	ML	24			100				
Rotenone, (97% pure)	ML	24			50		100		
Asimicin	NE	72	100						
F005 (extract)	NE	72		0	100				
Pyrethrins	NE	72			0				

MBB = mexican bean beetle
MA = melon aphid
ML = mosquito larvae
NE = nematode (*Caenorhabditis elegans*)

Fig. 2. Left, *Malacosoma americanum* on *Prunus serotina* tree. Right, pesticidal activity of paw paw extract: 30 min. after 0.5% F005 bark extract and 1.0% Tween 80 were applied.

each tree each month were screened for activity using the BST. For each extraction, 2.5 g of dried twigs was placed in a scintillation vial and extracted with 20 ml of dichloromethane for 24 h. The dichloromethane eluate was filtered, and the solvent was left to evaporate in a scintillation vial in the hood overnight. The next day the dilutions for the brine shrimp assay were made starting from the residue (F003) left in the vial. The new modified assay is summarized in Fig. 5.

RESULTS

The LC_{50} value for each assay was calculated using a Finney Probit analysis program on an IBM personal computer (McLaughlin et al. 1991). The mean LC_{50} value was calculated for each tree each month. Graphical results are shown in Fig. 6. The east tree was sampled all 12 months from Sept. 1994 through Aug. 1995. The west tree was not sampled June and July 1995.

Fig. 3. Antifungal activity of paw paw extract. *Phlox* with powdery mildew. The plants on the right were sprayed with 0.5% F005 paw paw bark extract + 1.0% Tween 80. Photograph taken after treatment (10 d).

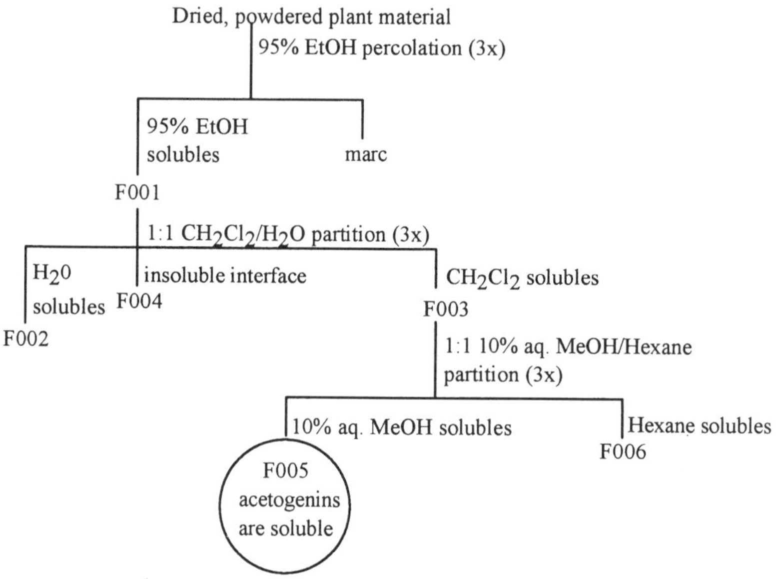

Fig. 4. Original extraction procedure for *Asimina triloba*.

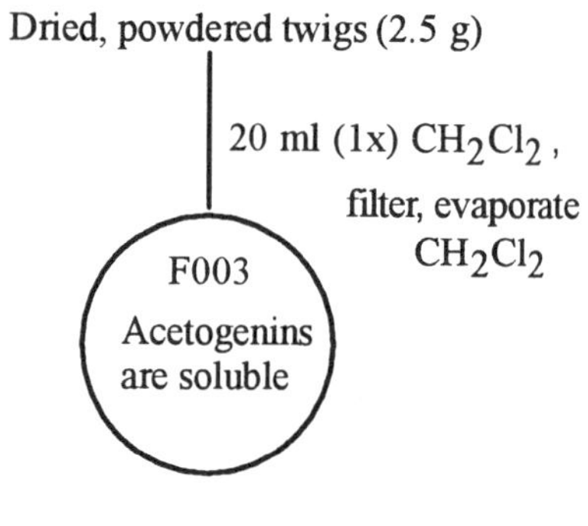

Fig. 5. Modified extraction scheme for *Asimina triloba*.

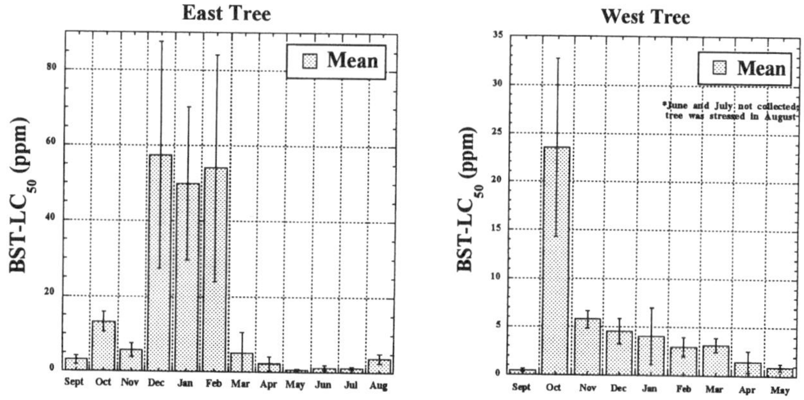

Fig. 6. Monthly biological activity of *Asimina triloba*.

The east tree data shows that the twigs were more biologically active during the summer months of 1995. The month of May showed the highest biological activity with a mean BST of 0.7922 ± 0.250 ppm SD. The months of July and June had activities of 1.2644 ± 0.430 and 1.2972 ± 0.754 ppm, respectively. The winter months had a significant decrease in activity compared to the summer months. There was a one to two order of magnitude decrease in activity in late fall to winter.

The west tree was not collected June and July due to the tree under-going the stress of massive sampling and the drought. The tree was stressed by drought in the month of Aug., and these results were discarded. The top three months of biological activity for this tree were Sept. (0.4680 ± 0.181 ppm), May (0.9073 ± 0.341 ppm), and Apr. (1.4447 ± 1.0864 ppm), respectively. Also, the content of acetogenins was higher in the west tree (high in spring and summer; low in fall and winter) suggesting that genotypic variations might be significant. The data gathered from these two trees, from Sept. 1994 to Aug. 1995, indicates that biomass content of paw paw twigs would be optimized if collections were made in the months of May through July. Slight variations might be expected in other geographical regions; but the production of acetogenins by the trees shows definite monthly variations and seems to be maximized when protection from insects would be most beneficial to the plant species.

This new, rapid, assay will be used to monitor the biological activity of twig samples from a plantation of over 670 *A. triloba* trees at the Western Maryland Resource and Education Center in Keddysville, Maryland. It is hoped that the new extraction method and the brine shrimp lethality assay will then provide data to help find the most biologically active *A. triloba* genotypes.

REFERENCES

Ahammadsahib, K.I., R.M. Hollingworth, P.J. McGovern, Y.-H. Hui, and J.L. McLaughlin. 1993. Inhibition of NADH:ubiquinone reductase (mitochondrial complex I) by bullatacin, a potent antitumor and pesticidal Annonaceous acetogenin. Life Sci. 53:1113–1120.

Ernster, L., G. Dallner, and G.F. Azzone. 1963. Differential effects of rotenone and amytal on mitochondrial electron transfer J. Biol. Chem. 238:1124–1131.

Hui, Y.-H., J.K. Rupprecht, Y.-M. Liu, J.E. Anderson, D.L. Smith, C.-J. Chang, and J.L. McLaughlin. 1989. Bullatacin and bullatacinone:two highly potent acetogenins from *Annona bullata*. J. Nat. Prod. 52:463–477.

Ikekawa, N., Y. Fujimoto, and T. Ikekawa, Sumura Limited Co. 1991. Novel bis-tetrahydrofuran compounds and their preparations used as anticancer agents. Japan Patent No. He 3-41076.

Jolad, S.D., J.J. Hoffman, K.H. Schram, J.R. Cole, M.S. Tempesta, G.R. Kriek, and R.B. Bates. 1982. Uvaricin, a new antitumor agent from *Uvaria accuminata* (Annonaceae). J. Org. Chem. 47:3151–3153.

Landolt, J.L., K.I. Ahammadsahib, R.M. Hollingworth, R. Barr, F.L. Crane, N.L. Buerck, G.P. McCabe, and J.L. McLaughlin. 1995. Determination of structure activity relationships of Annonaceous acetogenins by inhibition of oxygen uptake in rat liver mitochondria. Chemico-Biol. Interact. 98:1–13.

Lewis, M.A., J.T. Arnason, J.R. Philogene, J.K. Rupprecht and J.L. McLaughlin. 1993. Inhibition of respiration at site I by asimicin, an insecticidal acetogenin of the paw paw, *Asimina triloba* (Annonaceae). Pestic. Biochem. Physiol. 45: 15–23.

Londerhausen, M., W. Liecht, F. Leib., and H. Moeschler. 1991. Molecular mode of action of annonins. Pestic. Sci. 33: 427–438.

McLaughlin, J.L. and Y.-H. Hui. 1993. Chemotherapeutically active acetogenins. U.S. Patent No. 5,229,419.

McLaughlin, J.L., C.J. Chang, and D.L. Smith. 1991. "Bench-top" bioassays for the discovery of bioactive natural products: an update. Stud. Nat. Prod. Chem. 9:383–409.

Meyer, B.N., N.R. Ferrigni, J.B. Putnam, L.B. Jacobsen, D.E Nichols, and J.L. McLaughlin. 1982. Brine shrimp: a convenient general bioassay for active plant constituents. Planta Med. 45:31–34.

Mikolajczak, K.L., J.L. McLaughlin, and J.K. Rupprecht. 1988. Control of pests with Annonaceous acetogenins. U.S. Patent No. 4,721,727.

Mikolajczak, K.L., J.L. McLaughlin, and J.K. Rupprecht. 1989. Control of pests with Annonaceous acetogenins. U.S. Patent No. 4,855,319.

Ratnayake, S., J.K. Rupprecht, W.M. Potter, and J.L. McLaughlin. 1992. Evaluation of various parts of the paw paw tree, *Asimina triloba* (Annonaceae) as commercial sources of the pesticidal Annonaceous acetogenins. J. Econ. Entom. 85:2353–2356.

Rupprecht, J.K., C.J. Chang, J.M. Cassady, J.L. McLaughlin, K.L. Mikolajczak, and D. Weisleder. 1986. Asimicin, a new cytotoxic and pesticidal acetogenin from the paw paw, *Asimina triloba*. Heterocycles 24:1197–1201.

Singer, I.P. 1979. Biomembranes. Part F, Bioenergetics, oxidative phosphorylation. Methods Enzymol. 55:454–462.

Woo, M.H., L. Zeng, Q. Ye, Z.M. Gu, G. Zhao, and J.L. McLaughlin. 1995. 16,19-Cis Murisolin and murisolin A, two novel bioactive mono tetrahydrofuran Annonaceous acetogenins from *Asimina triloba* seeds. Bioorg. Med. Chem. Lett. 5:1135–1140.

Woo, M.H., L. Zeng, and J.L. McLaughlin. 1995. Asitribin and asiminenins A and B, novel bioactive annonaceous acetogenins from the seeds of *Asimina triloba*. Heterocycles 41:1731–1742.

Zhao, G.-X., Y.-H. Hui, J.K. Rupprecht, K.V. Wood, and J.L. McLaughlin. 1992. Additional bioactive compounds and trilobacin: a novel highly cytotoxic acetogenin from the bark of *Asimina triloba* (Annonaceae). J. Nat. Prod. 55:347–356.

Zhao, G.X., M.J. Rieser, Y.-H. Hui, L.R. Miesbauer, D.L. Smith, and J.L. McLaughlin. 1993. Biologically active acetogenins from the stem bark of *Asimina triloba*. Phytochemistry 33:1065–1073.

Zhao, G.-X., L.R. Miesbauer, D.L. Smith, and J.L. McLaughlin. 1994a. Asimin, asiminacin, and asiminecin: novel highly cytotoxic asimicin isomers from *Asimina triloba*. J. Med. Chem. 37:1971–1976.

Zhao, G., J.H. Ng, J.F. Koslowski, D.L. Smith, and J.L. McLaughlin. 1994b. Bullatin and bullanin: two novel, highly cytotoxic acetogenins from *Asimina triloba*. Heterocycles 38:1897–1908.

Zhao, G.-X., Z.-M. Gu, L. Zeng, J.-F. Chao, J.F. Koslowski, K.V. Wood, and J.L. McLaughlin. 1995. The absolute stereochemistry of trilobacin and trilobin: a novel highly potent acetogenin from the stem bark of *Asimina triloba* (Annonaceae). Tetrahedron 51:7149–7160.

Rapeseed Meal as a Natural Pesticide*

Harbans L. Bhardwaj, Anwar A. Hamama, D. Morris Porter, and Paul F. Reese, Jr.

The current interest in glucosinolates and the associated degradation products stems from the possibility of using them as pesticides. Glucosinolates are organic anions possessing a β-D-thioglucose moeity (Brown et al. 1991) decomposing to form isothiocyanates, the same class of chemical compounds produced by the decomposition of metam-sodium (Vapam).

Cruciferous plant tissues or tissue extracts, which contain high amounts of glucosinolates, have been reported to be phytotoxic (Jimenez-Osornio and Gliessman 1987; Fenwick et al. 1983) and insecticidal (Lichtenstein et al. 1964). The most thoroughly described control of a plant pathogen is for *Aphanomyces* root rot of peas, in which a variety of cruciferous plant amendments applied to infested soil reduced either the level of *Aphanomyces euteiches* (f. sp. *pisi* W.F. Pfender & D.J. Hagedorn) or the severity of the disease symptoms (Papavizas 1966; Papavizas and Lewis 1971; Lewis and Papavizas 1971, Chan and Close 1987). Glucosinolates from rapeseed (*Brassica napus* L.) meal were shown to have anti-fungal activity against *Cylindrocladium parasiticum* (Crous, Wingefield and Alfenas), casual agent of Cylindrocladium black rot (CBR) of peanuts, by Adamsen et al. (1991).

Considerable demand exists in the United States for industrial feedstocks and products from high erucic acid oil. Existence of glucosinolates restricts the use of protein-rich, defatted meal, from industrial rapeseed, as livestock feed. Development of defatted rapeseed meal as a natural annually-renewable biocide can not only help the environment but also has the potential to support industrial base of U.S. industry by providing a boost to development of rapeseed as a domestic source of high erucic acid oil.

The main objective of these studies was to ascertain efficacy of glucosinolate-rich rapeseed meal to control CBR in peanut and SCN (soybean cyst nematode, *Heterodera glycines* Ichinohe) in soybean. Our intent in conducting these experiments was not to develop rapeseed as a new crop to provide meal for use as a natural pesticide but to find a use of the meal so that production of industrial rapeseed as an oilseed can be facilitated. Accessions of *Brassica napus* and *B. rapa* were evaluated for winter hardiness in a separate study.

METHODOLOGY

CBR and SCN incidence was determined in replicated field experiments following three rates of soil-treatment with rapeseed meal (1, 2, or 3 t/ha), a commercial pesticide (Vapam for peanut and Aldicarb for soybean), and an untreated control. Both Vapam and Aldicarb were applied at rates recommended in the production area. The rapeseed meal was obtained from the Department of Agricultural Engineering, University of Idaho.

Two peanut cultivars, 'NC6' (susceptible) and 'NC10C' (resistant), were used for CBR studies. Six soybean lines/cultivars (susceptible/tolerant, G88-20092 and 'Hutcheson'; susceptible/intolerant, 'Essex' and 'Toano'; and resistant 'Asgrow 5403' and 'Deltapine 415') were used for SCN studies. The layout of field experiments was a split-plot design with soil-treatments in main-plots and cultivars in sub-plots. Each sub-plot consisted of four rows with 1 m and 0.45 m spacing between rows, respectively for peanuts and soybean. The rapeseed meal and Vapam treatments were applied to peanut experiments with a tractor-driven spreader/sprayer. In the soybean experiments, rapeseed meal was spread manually in each main-plot. The rapeseed meal in soybean experiments was spread on May 2, soybean were planted on May 23, and Aldicarb was applied on May 27, 1994. Rapeseed meal was spread in peanut experiments on May 2 and peanuts were planted on May 16, 1994. The peanut sub-plots were 15 m long whereas soybean plots were 6 m long. A

*This material is based upon work supported by the Cooperative State Research, Education, and Extension Service, U.S. Department of Agriculture, under Cooperative Agreement No. 93-COOP-1-9523. Any opinions, findings, conclusions, or recommendations expressed in this publication are those of the authors and do not necessarily reflect the view of the U.S. Department of Agriculture.

constant length was harvested from two middle rows of each plot at maturity. The peanut plots were scored for CBR incidence in Sept., 1994 with a disease severity index of 1 (all plants healthy in 15m row) to 10 (all plants in a 15m row length exhibiting typical CBR symptoms). Each row of each sub-plot was scored and averaged for CBR incidence.

Soil samples were taken on May 2, June 29, Oct. 7, 1994 in soybean experiments. These samples were kept under refrigeration until determination of SCN cysts and larvae by Department of Plant Pathology, Virginia Tech, Blacksburg.

PESTICIDAL EVALUATION

The rapeseed meal was effective as a pesticide to control CBR of peanut. The rapeseed meal reduced the disease incidence by 7%, 25%, and 70% over control in 'NC6', a susceptible cultivar, when 1, 2, or 3 t/ha, respectively of rapeseed meal was used as soil treatment. The lowest disease incidence on 'NC6' occurred with Vapam (Table 1). However, a clear reduction in disease incidence was observed with increasing rate of rapeseed meal. Similar results were also obtained from 'NC10C', a resistant cultivar. All three rapeseed meal treatments resulted in significantly lower disease incidence as compared to control. However, the lowest incidence still occurred when Vapam was used, indicating that resistance alone may not be adequate to control CBR incidence in peanut. The yield of 'NC6', following soil treatment with 3 t/ha of rapeseed meal, was similar to that following treatment with Vapam and the yield following soil treatment with 1 or 2 t/ha was significantly higher than that from the control treatment. In the case of 'NC10C', Vapam treatment resulted in significantly higher yield as compared to the other four treatments. The yield differences among rapeseed meal treatments and control were not significant.

The results with soybean were unclear. The number of cysts was significantly lower when 2 t/ha rapeseed meal was used with resistant cultivars but increased with susceptible-tolerant ones (Table 2). The larval population of soybean cyst nematodes, generally, increased over time when 2 or 3 t/ha of rapeseed meal was used. The seed yields (Table 3) of soybean cultivars were similar following five soil treatments.

GERMPLASM EVALUATIONS

In a separate experiment, 938 accessions of *Brassica napus* and *B. rapa*, supplied by USDA-ARS, Plant Introduction Station, Ames, Iowa, were evaluated in single row observation plots during 1993-94 for winter

Table 1. Cylindrocladium black rot (CBR) disease scores on peanut cultivars at Suffolk, Virginia following rapeseed meal and other soil treatments during 1994.

Soil treatment	CBR severity scores[z]		Yield (kg/ha)	
	NC10C[x]	NC6[y]	NC10C	NC6
Untreated	4.9a[w]	8.9a	3612b	2643c
Vapam	1.2c	1.8c	4364a	4359a
Meal (1 t/ha)	3.4b	8.3a	3594b	3221b
(2 t/ha)	2.6b	6.7b	3897b	3194b
(3 t/ha)	2.7b	2.7b	3959b	4027a

[z]Disease index from 1 (all plants healthy) to 10 (all plants exhibiting typical CBR symptoms).
[y]Resistant to CBR.
[x]Susceptible to CBR.
[w]Mean separation in columns by Duncan's Multiple Range Test (5% level).

Table 2. Soybean cyst nematode populations following soil treatments with rapeseed meal or Aldicarb during 1994.

Treatment	Cultivar	Sample date[z]		
		T1	T2	T3
			No. cysts	
Meal (1 t/ha)	Resistant	63	33	56
	Susceptible-Intolerant	63	41	45
	Susceptible-Tolerant	63	48	88
(2 t/ha)	Resistant	71a[y]	62ab	31b
	Susceptible-Intolerant	71	85	51
	Susceptible-Tolerant	71	61	69
(3 t/ha)	Resistant	92	73	61
	Susceptible-Intolerant	92	64	67
	Susceptible-Tolerant	92	60	78
Aldicarb	Resistant	41	45	35
	Susceptible-Intolerant	41	45	65
	Susceptible-Tolerant	41b	37b	102a
Untreated	Resistant	50	45	42
	Susceptible-Intolerant	50	64	55
	Susceptible-Tolerant	50	66	75
			No. larvae	
Meal (1 t/ha)	Resistant	170	70	227
	Susceptible-Intolerant	170	88	178
	Susceptible-Tolerant	170	110	260
(2 t/ha)	Resistant	133	205	137
	Susceptible-Intolerant	133	87	165
	Susceptible-Tolerant	133b	102b	450a
(3 t/ha)	Resistant	43b	133a	150a
	Susceptible-Intolerant	43	155	153
	Susceptible-Tolerant	43b	88b	282a
Aldicarb	Resistant	110	162	103
	Susceptible-Intolerant	110	122	193
	Susceptible-Tolerant	110	197	258
Untreated	Resistant	63	97	110
	Susceptible-Intolerant	63	62	106
	Susceptible-Tolerant	63b	148ab	297a

[z]T1 = Sampled before planting, May 2, 1994
T2 = Sampled after planting, June 29, 1994
T3 = Sampled before harvest, Oct. 7, 1994
[y]Mean separation between T1, T2, or T3 within treatments and cultivars by Duncan's Multiple Range Test (5% level)

Table 3. Seed yield of soybean cultivars following soil treatments with rapeseed meal or Aldicarb during 1994.

	Seed yield (kg/ha)		
Soil treatment	Resistant[z]	Susceptible/ intolerant	Susceptible/ tolerant
Untreated	2368[y]	1336[y]	1800[y]
Aldicarb	2688	1566	1997
Meal (1 t/ha)	2700	1510	2103
(2 t/ha)	2401	1179	2004
(3 t/ha)	2488	1046	1972

[z]Classification of soybean cultivars based on reaction to infestation by soybean cyst nematode.
[y]No significant differences within columns (Duncan's Multiple Range Test, 5% level).

hardiness at Petersburg, Virginia. Five hundred accessions from this collection were identified to be cold-tolerant. The glucosinolate content in seeds of these accessions, when determined using TRUBLUGLU meter (Truscott et al. 1991), varied from 38 to 77 µmol/g. The glucosinolate content in 456 *napus* accessions as a group was 49 µmol/g which was significantly greater than the mean glucosinolate content of 44 µmol/g in 44 *rapa* accessions. The oil content was determined in 159 accessions. The 122 *napus* accessions (38.4%) had significantly higher oil content than that of 37 *rapa* accessions (36.5%). A significant positive correlation (0.63** for *napus* and 0.79** for *rapa* accessions) existed between oil content and glucosinolate content.

The results from one year's field studies indicate that rapeseed meal has pesticidal properties against *Cylindrocladium parasiticum*, casual agent of Cylindrocladium black rot (CBR) of peanuts. Further evaluations are needed to confirm these results and to determine if rapeseed meal has potential to control soybean cyst nematode.

REFERENCES

Adamsen, F.J., D.M. Porter, and D.L. Auld. 1991. Rapeseed meal as a potential biological control of CBR of peanut. Am. Peanut Res. Educ. Soc. Abstr. 23:38.

Brown, P.D., M.J. Morra, J.P. McCaffrey, D.L. Auld, and L. Williams, III. 1991. Allelochemicals produced during glucosinolate degradation in soil. J. Chem. Ecol. 17:2021–2034.

Chan, M.K.Y. and R.C. Close. 1987. *Aphanomyces* root rot of peas. 3. Control by the use of cruciferous amendments. New Zealand J. Agr. Res. 30:225–233.

Fenwick, G.R., R.K. Heaney, and W.J. Mullin. 1983. Glucosinolates and their breakdown products in food and food plants. Crit. Rev. Food Sci. Nutr. 18:123–201.

Jimenz-Osornio, J.J. and S.R. Gliessman. 1987. Allelopathic interference in a wild mustard (*Brassica campestris* L. var. *italica*) intercrop agroecosystem. p. 262–288. In: G.R. Waller (ed.), Allelochemicals: Role in agriculture and forestry.

Lewis, J.A. and G.C. Papavizas. 1971. Effect of sulfur-containing volatile compounds and vapors from cabbage decomposition on *Aphanomyces euteiches*. Phytopathology 61:208–214.

Lichtenstein, E.P., D.G. Morgan, and C.H. Mueller. 1964. Naturally occurring insecticides in cruciferous crops. J. Agr. Food Chem.. 12:158–161.

Papavizas, G.C. 1966. Suppression of *Aphanomyces* root rot of peas by cruciferous amendments. Phytopathology 56:1071–1975.

Papavizas, G.C. and J.A. Lewis. 1971. Effect of amendments and fungicides on *Aphanomyces* root rot of peas. Phytopathology 61:215–220.
Truscott, R.J.W., J. Tholen, G. Buzza, and D.I. McGregor. 1991. Glucosinolate measurement in rapeseed using reflectance. The TRUBLUGLU meter. p. 1425–1429. In: D.I. McGregor (ed.), Proc. 8th Int. Rapeseed Congress, Groupe Consultatif International de Recherche sur le Colza (GCIRC) and Canola Council of Canada. Vol. 5 of 6.

Potential of Sugar Beet Nematode-Resistant Radishes and Mustard for Use in Sugar Beet Rotations

James M. Krall, David W. Koch, Fred A. Gray, and Li Mei Yun

Current production of sugar beet (*Beta vulgaris* L.) relies on extensive use of pesticides. Annual losses from pests in the Western United States are greater than 25% (Roberts and Thomason 1981). In Wyoming, 65% of fields in long-term sugar beet production are infested with the sugar beet (cyst) nematode (SBN), *Hederodera schachtii* Schmidt (F.A. Gray pers. commun.). In Wyoming, aldicarb is applied to 30% of sugar beet hectarage and an increasing hectarage is treated with the pre-plant soil fumigant 1,3-dichloropropene (Legg et al. 1992). Continued use of nematicides threatens the sustained production of sugar beets because of their high cost and environmental risks.

Agronomically-desirable nematode-resistant sugar beet cultivars are unavailable. In Wyoming and other areas sugar beets are grown in two- or three-year rotations. Longer rotations, which would help in the control of SBN, are not economically practical due to lack of adapted and profitable crops. SBN-resistant "trap crop" varieties are available and their use is currently a major control practice in Europe (Peterson 1992). Seed of nematode-resistant radish (*Raphanus sativus* L.) and mustard (*Sinapsis alba* L.) has recently become available in the United States. Technology for using these cultivars in United States sugar beet rotations has not been adequately developed. In western Nebraska, trap crop cultivars grew best when planted between mid-May and mid-Aug., when soil temperatures at 15 cm were above 20°C (Wilson et al. 1993). A major limitation of this method is the limited amount of time and growing degree days following harvest of the main crop. Full-season production is probably not justified economically as a routine practice. The objectives of this study were to determine the potential for biological control of SBN with SBN-resistant cultivars and the potential for utilizing these crops for fall grazing.

METHODOLOGY

A cooperating farmer's (Hefenieder) field on which the SBN had been a problem was selected for initiating a test in 1992. Following malt barley (*Hordeum vulgare* L.) harvest in 1992, straw was baled and removed and the field was tilled twice with a tandem disk. 'Pegletta' radish was planted with a disk drill at 24 kg ha^{-1} on July 28. The experimental design was a split-plot with six replications. Main plots (1992) were radish-seeded as a second crop or unseeded (normal fallow). Subplots (at sugar beet planting in 1993) were aldicarb rates of 0, 2.5 or 5.0 kg ha^{-1} a.i. The latter rates represent one-half and full-label rates of aldicarb, applied as Temik® 15G. Main plots were 12 by 12 m and subplots were 4 by 12 m. Ammonium nitrate was surface-applied at 56 kg ha^{-1} at time of radish planting. Plots were furrow-irrigated twice during radish growth.

Another cooperator's field (Mosegard) was used to study the effect of grazing the SBN-resistant radish cultivar, 'Adagio', on SBN control. Malt barley was harvested Aug. 2, 1994. The field was burned and on Aug. 4, seed (22 kg ha^{-1}) was mixed with fertilizer (67 kg ha^{-1} N as ammonium nitrate) and applied with a spinner-broadcast type applicator. Replicated strips were left unseeded. The field was roller-harrowed to cover the seed and firm the seedbed. The field was furrow-irrigated twice. Sethoxydim at 0.31 kg ha^{-1} a.i. and oil concentrate at 2.3 l ha^{-1} was applied for the control of volunteer barley on Aug. 27, when barley was about 10 cm in height. The experimental design was a randomized complete block with five replications. Treatments were: (1) unseeded and ungrazed; (2) radish-seeded and fall-grazed; and (3) radish-seeded and fall growth plowed down, rather than grazed. Following evaluation of radish stand and growth, 150 lambs averaging 33.1 kg grazed on 1.8 ha from Oct. 13 until Nov. 12. Lambs were weighed to determine daily weight gain and gain ha^{-1}. Following grazing the entire field was moldboard plowed. Sugar beets were planted on the entire field in 1995 with no nematicides applied to the study area and sugar beet yields and quality components were evaluated.

On the above two test sites, populations of SBN were determined before and after the radish crop, at sugar beet planting, at mid-season and at sugar beet harvesting. Two hundred cm^3 of soil was processed through an elutriator and cysts, eggs, and juveniles counted. SBN counts were adjusted for efficiency of extraction and nematode population data were subjected to covariance analysis.

Trap crop mustard was evaluated for grazing potential at the Torrington Research and Extension Center, where there are no SBN. In 1993, following harvest of feed barley, loose straw was removed and 'Metex' mustard was planted on Aug. 16 at 21 kg ha^{-1} with a disk drill into stubble averaging 20 cm in height. Soil analysis showed low phosphorus, therefore, 72 kg N and 93 kg P ha^{-1} were applied. The field was divided into three seeded blocks (0.15 ha each) and three unseeded blocks (0.05 ha each). Lambs grazed 0.11 ha of each seeded block. Eight lambs grazed each block from Oct. 29 until Nov. 12 and weight gain was determined. The entire field was spring-plowed and uniformly planted to sugar beets in 1994.

In 1994 the same grazing study was repeated on a different field of the Torrington R & E following winter wheat (*Triticum aestivum* L. emend. Thell) harvest. The methods were the same, except that phosporus fertility was higher and no phosphorus fertilizer was applied. Planting date was on July 26 and plots were grazed Nov. 4 to 14.

RESULTS

SBN-resistant radish cultivars, 'Pegletta' and 'Adagio', planted as a second crop following malt barley reduced soil populations of SBN 54% and 71% over a $2\frac{1}{2}$-month period (Table 1). The greater reduction on the Mosegard field, compared with the Hefenieder field, may have been related to better soil moisture and the greater amount of plant growth. Radishes at Hefenieders appeared to have been overwatered.

Table 1. Top growth and sugar beet nematode (SBN) reduction with SBN-resistant radish and mustard cultivars.

Location/year	Preceeding crop	Trap crop planted	Planting date	Top growth (Mg/ha)	Growing degree days (base 4.4°C)	SBN reduction[z] (%)
Hefenieder/1992	Malt barley	Pegletta radish	July 28	1.56	1758	54
Mosegard/1994	Malt barley	Adagio radish	Aug. 4	2.52	1658	71
Torr. R&E[y]/1993	Feed barley	Metex mustard	Aug. 12	1.90	1279	--
Torr. R&E/1994	Winter wheat	Metex mustard	July 26	8.56	2168	--

[z]From the time when trap crop was planted until plow down in the fall.
[y]Fields at the Torrington Research and Extension Center are not infested with SBN.

On the Hefenieder field SBN populations at sugar beet planting in April 1993 were below the estimated damage threshold (EDT = 2.8 eggs and/or juveniles cm^{-3}) (Griffin 1981) (Table 2). SBN population remained lower than with the unseeded fallow treatment throughout the sugar beet season. At mid-season (July) the SBN population was 26% lower for radish-seeded, compared to conventional fallow plots. At sugar beet harvest in Sept. the SBN population was significantly (25%) lower for radish-seeded compared to unseeded plots. Aldicarb, band-applied along the seed row at sugar beet planting, did not significantly reduce SBN population, even at the full-label rate. While aldicarb application did not increase sugar beet yield, radish inclusion in the rotation the previous year increased sugar beet yield 8.78 Mg ha^{-1}. Estimated cost of growing SBN-resistant mustard and radish is \$192 and \$206 ha^{-1}. The yield increase with radishes on this field was well above the estimated 4.9 Mg ha^{-1} needed to offset the cost of growing radishes, based on the average price of sugar beets, \$45.83 Mg^{-1}, over the past five years.

Table 2. Effect of 'Pegletta' radish (grown as a second crop in 1992) and aldicarb on SBN population and sugar beet yields in 1993 (Hefenieder field).

	Sugar beet nematode population (No. eggs and/or juveniles cm^{-3} of soil)					
Treatment	Aug. 1992	Oct. 1992	Apr. 1993	July 1993	Sept. 1993	Sugar beet yield (Mg/ha)
Radish effect						
Unseeded (fallow)	2.04 a[z]	1.93 a	2.55 a	4.66 a	6.83 a	31.72 b
Radish seeded	2.90 a	1.33 a	2.48 a	3.43 a	5.10 b	40.50 a
Nematicide effect[y]						
Untreated (no aldicarb)	--	--	2.33 a	4.33 a	6.19 a	37.78 a
Half rate of aldicarb	--	--	2.78 a	4.30 a	5.39 a	35.16 a
Full rate of aldicarb	--	--	2.44 a	3.52 a	6.66 a	35.38 a
Interaction	--	--	n.s.	n.s.	n.s.	n.s.

[z]Means followed by different letter within columns are significantly different at the $P < 0.05$ level.
[y]Half-label and full-label rate of aldicarb, 2.5 and 5.0 kg ha^{-1}, applied at sugar beet planting in Apr. 1993.

Table 3. Forage production, quality and performance of lambs grazing SBN-resistant radishes and mustard.

Cooperator/ Location	Crop	Forage yield (Mg/ha) Total[z]	Forage yield (Mg/ha) Trap crop	Crude protein (g/kg)	Utili-zation (g/kg)	ADG[y] (kg)	Lamb gain/ha (kg)	Value of lamb[x] (\$/ha)
Mosegard	Radish	3.00	2.52	154	740	0.169	251	319
Torr. R&E								
1993	Mustard	2.63	1.90	129	870	0.309	308	393
1994	Mustard	8.74	8.56	126	460	0.336	1230	709

[z]Includes volunteer grain.
[y]Average daily gain.
[x]Based on average Oct. lamb price (\$1.25/kg) for the past five years.

Both trap-crop cultivars are high in forage quality and are readily consumed by lambs (Table 3). 'Metex' mustard is capable of high yields of topgrowth, as evidenced by the 1994 planting at the Torrington Research and Extension Center. Performance (average daily gain) was greater with mustard than with radish. Gains on mustard were nearly comparable to gains on hay and grain diets. With all three grazing studies the value of lamb produced ha^{-1} was greater than the cost of growing the crops. On the Mosegard field there was no difference in SBN populations with grazing vs plowdown of radishes. Radishes were not grazed until growth had stopped and soil temperatures were too low for SBN activity.

REFERENCES

Griffin, G.D. 1981. Relationship of plant age, soil temperature and population density of *Heterodera schachtii* on the growth of sugar beet. J. Nematol. 13:184–190.

Legg, D.E., M. Ferrell, D.T. Taylor, and D.L. Kellogg. 1992. Pesticide use in Wyoming, 1990. Coop. Ext. Serv. Bul. RJ-211.

Peterson, P.H. 1992. Biological nematode control. Saaten Union, Hannover, Germany,

Roberts, P.A. and I.J. Thomason. 1981. Sugarbeet pest management: Nematodes. Univ. Cal., Div. of Agr. Sci. Special Publ. 3272.

Spittel, A. 1986. Biological control of beet nematodes with resistant oil radish and resistant yellow mustard. Publ. Saten-Union GMBH, Glucherstrabve 7, 3000 Hannover, Germany.

Wilson, R.G., E.D. Kerr, and P. Provance. 1993. Growth and development of oil-radish and yellow mustard in Nebraska. J. Sugar Beet Res. 3:159–167.

CROPS IN SPACE

Plant Nutrient Composition Altered With Controlled Environments for Future Space Life-Support Systems*

S. Suzanne Nielsen, Martha A. Belury, Kwangok P. Nickel, and Cary A. Mitchell

Future manned outposts on the Moon and Mars will rely on self-sustaining life-support systems that utilize plants to revitalize atmosphere, recycle water, and provide edible biomass for vegetarian diets without resupply from Earth (Mitchell 1993). The crops selected for such Controlled Ecological Life-Support Systems (CELSS) most likely will be grown hydroponically under controlled environment conditions (Mitchell 1994).

There is increasing evidence that the nutritional content of food crops is modified by controlled environments (CE) relative to soil culture in the field (McKeehen et al. 1996 a,b). This finding raises questions regarding diet composition, food safety, bioavailability, and palatability for limited vegetarian diets.

Calcium metabolism during long-term space flight is of great concern due to bone demineralization caused by a weightless environment (Schneider et al. 1989). Since the first generation of CELSS diets likely will be composed of plant-derived sources, it is important that such diets provide sufficient levels of available calcium. Calcium bioavailability can be affected by dietary levels of phytate and oxalate, which form insoluble complexes of calcium salts. Pure vegetarian diets may contain high amounts of these antinutrients (Havala and Dwyer 1993). Because the risk of renal stone formation is high due to increased urinary calcium excretion during space flight (Lane et al. 1993), calcium oxalate stone formation could be minimized by contolling the level of oxalate in the diet.

The present study compared the macronutrient content of the edible portion for nine CELSS candidate crop species grown in controlled environments or in the field, investigated the mineral content and the partitioning of N into components of nutritional significance in plant biomass as a function of growth environment, and investigated the calcium bioavailability of CELSS diets using a rat model.

METHODOLOGY

CELSS Candidate Crops

Select cultivars of nine CELSS candidate crops [soybean (*Glycine max* (L.) 'Hoyt'), cowpea (*Vigna unguiculata* (L.) 'IT87D-941-1'), peanut (*Arachis hypogaea* (L.) 'Ga Red'), wheat (*Triticum aestivum* (L.) 'Yecora Rojo'), rice (*Oryza sativa* (L.) 'Ai-Nan-Tsao'), canola (*Brassica napus* (L.) 'Westar'), potato (*Solanum tuberosum* (L.) 'Norland'), sweet potato (*Ipomoea batatus* (L.) Lam 'Tu-82-155'), and lettuce (*Lactuca sativa* (L.) 'Waldmann's Green')] were grown under controlled environment (CE) conditions, and the same cultivars of all but cowpea and peanut were grown under field conditions (see Table 1). The typically edible part of each plant was dried and ground prior to analysis or incorporated into animal diets.

Analysis of Plant Material

Fat, ash, carbohydrate, total N, protein N, total nonprotein N (NPN), nitrate N, and nonnitrate NPN were determined as described by McKeehen (1994) for seeds or grain of CE and field-grown soybeans, wheat, rice, and canola, for tubers of CE and field potato, for leaves of CE and field lettuce, and for seeds of CE cowpea. For all other CE and field samples (values taken from the literature or by personal communication) and for all handbook (Haytowitz and Matthews 1986) values, protein content presumably was determined by the standard Kjeldahl procedure, fat and ash contents presumably were determined by standard AOAC (Association of Official Analytical Chemists International) methods (AOAC 1990), and carbohydrate content was calculated by difference [Carbohydrate = 100% - (% protein + % fat + % ash)]. Mineral analysis [cal-

*Paper No. 15006 of Purdue University Agricultural Research Programs. Research supported in part by NASA grant NAGW-2329

cium (Ca) and phosphorus (P)] were performed by Inductively Coupled Plasma-Atomic Emission Spectroscopy (ICP-AES), as described by McKeehen (1994). Energy content for all crops was calculated as: kcal/100g dry matter = (g protein/100g × 4 kcal/g) + (g carbohydrate/100g × 4 kcal/g) + (g fat/100g × 9 kcal/g).

Calcium Bioavailability Study Using Rats

Diets formulated with CELSS candidate crops were used to study calcium bioavailability in a rat model system. Calcium bioavailability in CELSS diets was evaluated by the femur ^{45}Ca uptake method (Weaver et al. 1987), which has been validated for the absorption of calcium in many previous studies (Buchowski et al. 1989; Koo et al. 1993).

Forty male Sprague-Dawley rats (100–125 g) (Harlan Sprague-Dawley, Indianapolis, IN) were housed in stainless steel cages with a reverse 12 h light-dark cycle. Animals were randomly assigned to groups of 8 rats each.

All the crops except kale, broccoli, mustard greens, bokchoy, and canola oil were grown in the semi-controlled environment of the Purdue University Horticulture greenhouses. Vegetables were freeze-dried and diets were prepared in a powdered form. The composition of CELSS diets is shown in Table 3. Diets 1 and 2 contained lower levels of calcium compared with the control diet. Diet 3 was formulated with calcium-rich vegetables (kale, bokchoy, broccoli, and mustard greens) (Tables 3 and 4). Because calcium is most efficiently absorbed at low levels (Heaney at al. 1975), all diets were adjusted to contain similar levels of calcium. Therefore, calcium as $CaCO_3$ was supplemented in Diets 1 and 2. Rats were given free access to the semi-purified control diet [(American Institute of Nutrition (AIN)-76A] and deionized water for one week. They were meal-trained by fasting for 12 h each day for 3 d before receiving test meals labeled with ^{45}Ca. Four g of three CELSS diets and the control diet were extrinsically labeled with 7 $\mu Ci\,^{45}Ca$ as a tracer and test meals were fed to rats. One group of rats was injected intraperitoneally (IP) with 7 $\mu Ci\,^{45}Ca$ in 300 µl 0.9% saline 2 h after receiving an unlabeled 4 g of the control diet. It was assumed that the radioactivity of an IP dose that reached the femur represented 100% calcium absorption. Rats were allowed to consume test meals for a 4-h period and the remaining food was collected for counting of radioactivity in order to calculate the exact amount of labeled food intake. All rats were sacrificed with CO_2 24 h after administration of ^{45}Ca and the femurs were removed. Femurs were dried in a vacuum oven at 70°C overnight, dissolved with 3 ml concentrated HNO_3 in 25 ml volumetric flasks, and diluted with double deionized water. A 1 ml aliquot was counted for radioactivity by liquid scintillation spectroscopy. Absorption of ^{45}Ca from the test meal was calculated from the ratio of the percent ^{45}Ca dose accumulated in the femur compared to that in the femur of rats receiving the IP injection.

RESULTS

Proximate Composition of CELSS Candidate Crops

Nitrogen determination presumably was the only difference between assays of samples according to methods reported by McKeehen (1994) used for most CE and field samples of CELSS crops, compared to the assays used to obtain all other CE, field, and handbook values. The standard Kjeldahl procedure (AOAC 1990) is commonly used for protein analysis, but it measures both protein N and some, but not all, of the nitrate N. This method alone is inappropriate to determine the true protein content of plant materials that have a significant fraction of N as nitrate (Goyal and Hafez 1990). McKeehen (1994) used a total N procedure by Goyal and Hafez (1990), with a predigestion procedure to include NO_3^- N in the Kjeldahl digestion. To differentiate protein N from total N, McKeehen (1994) determined protein N content after precipitating protein in 6% trichloroacetic acid (TCA) (Bensadoun and Weinstein 1976), and then measuring N content of the washed pellet by the Kjeldahl method (AOAC 1990).

Protein content values obtained under CE were higher than field and handbook values of the typical edible portion of CELSS candidate crops (Table 1). Some handbook values are considerably lower than CE or field values, likely in part because of differences in the N analysis methods described herein. The higher total N content in CE conditions compared to field conditions is attributed to luxuriant uptake of N

Table 1. Proximate composition and energy content (calculated) of edible portion of some CELSS candidate species, as de-determined for edible portions of plants grown under controlled environments (CE), in the field, or as reported in USDA Handbook No. 8 (Hdbk).

Crop Condition	Specific content (% dry wt) Protein[z]	Fat	Ash	Carbohydrate[y]	Energy content (Kcal /100g dry wt)	Reference
Soybean						
CE[x]	53.4	19.6	5.9	20.7	472.8	Nielsen et al. 1996
Field	46.6	18.1	5.6	29.7	468.1	Nielsen et al. 1996
Hdbk[w]	43.7	21.8	5.3	29.2	487.8	Haytowitz and Matthews 1986
Cowpea						
CE[v]	28.3	1.3	4.7	65.7	387.7	--[o]
Field[u]	28.6	2.0	3.3	66.1	396.8	Nielsen et al. 1993
Hdbk[w]	26.7	1.4	3.7	68.2	392.2	Haytowitz and Matthews 1986
Peanut						
CE	29.3	49.5	3.0	18.5	636.7	--[n]
Field	24.8	47.9	2.7	24.6	628.7	Belitz and Grosch, 1987
Hdbk[w]	31.6	52.7	2.5	13.2	653.5	Haytowitz and Matthews 1986
Wheat						
CE[t]	24.4	1.4	2.0	72.1	398.6	McKeehen et al. 1996a
Field	23.2	1.5	1.9	73.4	399.9	McKeehen et al. 1996a
Hdbk[w]	18.9	2.2	2.2	76.7	402.2	Haytowitz and Matthews 1986
Rice						
CE	17.0	3.1	1.9	78.0	407.9	McKeehen et al. 1996a
Field	10.7	2.4	1.7	85.2	405.2	McKeehen et al. 1996a
Hdbk[w]	9.0	3.1	1.5	86.4	409.5	Haytowitz and Matthews 1986
Canola						
CE	29.7	33.8	4.5	32.0	551.0	Stephens 1995
Field[s]	27.1	37.0	8.2	27.7	552.2	Stephens 1995
Hdbk	22.8	46.0	8.0	23.2	598.0	--[m]
Potato						
CE[r]	17.9	1.8	6.5	73.8	383.0	McKeehen et al. 1996a
Field	12.0	0.5	6.1	81.4	378.0	McKeehen et al. 1996a
Hdbk[w]	9.8	0.5	4.2	85.5	385.7	Haytowitz and Matthews 1986
Sweet potato						
CE[q]	3.3	2.0	1.3	93.4	404.8	--[l]
Field	2.4	1.1	3.6	92.9	391.1	--[l]
Hdbk[w]	5.8	1.4	3.4	89.5	393.8	Haytowitz and Matthews 1986
Lettuce						
CE[p]	28.7	3.4	17.7	50.3	346.6	McKeehnen et al. 1966b
Field	19.3	4.2	15.6	60.9	358.6	McKeenhen et al. 1996b
Hdbk[w]	21.7	5.0	15.0	58.3	365.0	Haytowitz and Matthews 1986

[z]Protein content determined by total N × 6.25 for CE and field plants of soybean, wheat, rice, canola, potato and lettuce and for CE plants for cowpea. Note that not all N is protein for some species, especially under CE conditions. Method for true total N is a modified Kjeldahl procedure, as described by McKeehen (1994). Protein content of all other samples presumably determined by standard Kjeldahl procedure. Handbook protein values are expressed using N conversion factor of 6.25.

[y]Carbohydrate values are calculated by difference: % Carbohydrate = 100% - (% Protein + % Fat + % Ash)
[x] CE values are means of 350 and 1000 ppm CO_2 conditions.
[w]Soybeans: mature seeds, dry, raw
Cowpeas: mature seeds, dry, raw
Peanuts: raw, without skins
Wheat: whole grain, hard red spring
Rice: brown, raw
White potatoes: raw
Sweet potatoes: raw, all commerical varaieties
Lettuce: raw, loose-leaf
[v]CE values are means of values for 4 planting densitites at 2 levels of CO_2, for 'IT87D-941-1' cowpeas, described by Ohler (1994).
[u]Values reported are means of values for 100 improved cowpea lines from the International Institute of Tropical Agriculture (IITA) in Ibadan, Nigeria. The 'IT87D-941-1' cowpea line tested in the CE was also an improved line from IITA.
[t]CE values are means of three samples, with 2 from growth chamber and 1 from biomass production chamber.
[s]Field values are means of samples grown at 2 locations.
[r]CE values are means of 2 samples from biomass production chamber.
[q]CE conditions ulitized nutrient film technique.
[p]CE values are means of 400 and 1000 ppm CO_2 conditions.
[o]Nielsen, S.S. 1995 unpubl. data. Purdue Univ., West Lafayette, IN.
[n]Lee, J.Y. 1995 pers. commun. Tuskegee Univ., Tuskegee, AL.
[m]DeClercq, D. 1995 pers. commun. Canadian Grain Commission, Winnipig, Manitoba, Canada.
[l]Loretan, P. 1995 pers. commun. Tuskegee Univ., Tuskegee, AL.

by hydroponically grown plants. Fat and ash contents were relatively unaffected by environment. For most species, carbohydrate content values tended to be lower under CE than field or handbook values. Specific energy content was relatively unchanged by different growth environments.

Nitrogen Allocation within CELSS Candidate Crops

Protein content calculated from true total N for the typical edible portion of CELSS candidates was higher under CE than under field conditions (Table 2). This total N included a significant amount of NPN. Total NPN was highest for soybean seeds (CE and field) and potato tubers (CE). This NPN did not include nitrate in seeds or tubers, but nitrate was significant in CE-grown leaf lettuce. Nitrate is known to accumulate in the vegetative material of rice, wheat, white potato, soybeans, and cowpeas when grown under CE conditions (McKeehen et al. 1996a, Nielsen 1995).

Calcium and Phosphorus Content

The Ca/P ratios of edible plant parts differed for CE compared to field and handbook values (Table 5). This ratio ideally is maintained at a ratio of 1:1 in the diet for proper calcium absorption and retention in humans (Stare and Williams, 1984). The Ca/P ratios of all seeds, grains, and tubers were quite low due to low Ca levels in the typical edible parts of most plants. Especially for potato under CE, Ca content was reduced and P content increased, to reduce the Ca/P ratio. Although the Ca content is high in green, leafy vegetables such as lettuce, the higher P content of lettuce leaves under CE compared to field conditions led to a Ca/P ratio under CE half that of under field conditions.

Calcium Bioavailability

The average Ca absorption by rats fed CELSS diets was slightly higher than that of the control diet (Table 6). However, there was no significant difference ($p = 0.34$; ANOVA) in Ca absorption between control and CELSS diets in spite of different levels of phytate, oxalate, and total dietary fiber (Nickel et al.

Table 2. Percent protein (% nitrogen × 6.25) of edible portion of some CELSS candidate species, as derived from total N, trichloroacetic acid (TCA)-precipitated N, total nonprotein N (NPN), nitrate N, and nonnitrate NPN, for edible portion of plants grown under controlled environments (CE) and in the field.[z]

	Protein (%)					
Crop Condition	From total N	From TCA-precipitated N	From total NPN[y]	From nitrate N	From nonnitrate NPN[x]	Reference
Soybean						
CE	53.4	42.5	10.9	0.0	10.9	Nielsen et al. 1996
Field	46.6	35.1	11.5	0.0	11.5	Nielsen et al. 1996
Cowpea						
CE	28.3	24.4	3.9	0.0	3.9	--[v]
Field	NA[w]	NA	NA	NA	NA	
Wheat						
CE	24.4	19.3	5.1	0.0	5.1	McKeehen et al. 1996a
Field	23.2	16.7	6.5	0.0	6.5	McKeehen et al. 1996a
Rice						
CE	17.0	16.6	0.4	0.0	0.4	McKeehen et al. 1996a
Field	10.7	9.5	1.2	0.0	1.2	McKeehen et al. 1996a
Canola						
CE	29.7	24.7	5.0	NA	NA	Stephens 1995
Field	27.1	22.6	4.5	NA	NA	Stephens 1995
Potato						
CE	17.9	6.5	11.4	0.1	11.3	McKeehen et al. 1996a
Field	12.0	6.5	5.5	0.1	5.4	McKeehen et al. 1996a
Lettuce						
CE	28.7	22.9	5.8	4.1	1.7	McKeehen et al. 1996b
Field	15.7	17.7	1.6	0.2	1.4	McKeehen et al. 1996b

[z]Percent N values for total, protein, nonprotein, and nitrate were all multiplied by 6.25 to express values as percent protein, to allow for comparisons.
[y]Calculated by difference between % protein total N and % protein from TCA-precipitated N. Methods for total N and TCA-precipitated N are as described by McKeehen (1994).
[x]Calculated by difference between % protein from total NPN and % protein from nitrate N. Method for nitrate N is as described by McKeehen (1994).
[w]NA = data not available.
[v]S.S. Nielsen 1995 unpubl. data. Purdue Univ., West Lafayette, IN.

1996). Calcium supplementation of Diets 1 and 2 showed Ca absorption to exceed that of the control diet. These data support earlier work suggesting that Ca supplementation of CELSS diets may be necessary for optimal Ca absorption in rats (Saha et al. 1996). It is notable that Diet 3 containing Ca-rich vegetables showed high bioavailability of Ca.

SUMMARY

Hydroponic growth of food crops under CE conditions represents an opportunity to manipulate and control the composition of edible plant parts. The predictability of biomass composition for crops grown under standardized, modified conditions represents a form of "nutritional value added" in preparing balanced veg-

Table 3. Composition of experimental diets for rats.[z,y]

Ingredients	Composition (g/100g) Diet 1	Diet 2	Diet 3
Wheat	10	20	15
Brown rice	20	20	12
Soybeans	10	10	10
Cowpea	10	10	5
Potato	15	15	15
Sweet potato	10	10	10
Peanut	3	3	2
Tomato	4	4	2
Lettuce	5	5	2
Canola oil	3	3	3
Wheat straw	10	--	--
Kale	--	--	6
Bokchoy	--	--	6
Broccoli	--	--	4
Mustard greens	--	--	8

[z]Control diet was provided as a semi-purified AIN-76A diet.
[y]All protocols involving animals were approved by the Purdue University Animal Care and Use Committee.

Table 4. Compositional analysis of experimental diets.[z]

Diet	Moisture (%)	Fat (g)	Protein (g)	Ash (g)	CHO (g)	Ca[w] (mg)	Ca[v] (mg)	P (mg)	Ca/P (ratio)
Control[y]	8.5	5.2	25.4	2.8	58.1	449.3	--	504.6	0.89
Diet 1[x]	8.9	7.0	23.5	6.4	54.2	179.5	489.0	524.8	0.93
Diet 2[x]	9.7	7.0	27.8	5.4	50.1	167.6	464.6	510.8	0.91
Diet 3[x]	8.9	7.0	26.6	6.7	50.8	508.8	--	480.4	1.06

[z]Units are based on 100 g wet sample.
[y]Control diet was provided as semi-purified AIN-76A diet.
[x]Diets 1, 2 and 3 are described in Table 3.
[w]Calcium values before supplementation with $CaCO_3$.
[v]Calcium values after supplementation with $CaCO_3$.

etarian diets for crews of long duration space missions. Hydroponic culture conditions combined with optimizing environments enhanced the protein content of the edible portions of soybean, wheat, rice, potato, and lettuce. Since legumes and cereals constitute the staple basis of vegetarian diets, amounts of both types of crops grown in CE would have to be reduced when balancing dietary protein for composition and amount. Because protein and carbohydrate vary reciprocally but fat is relatively unaffected by growth environment, caloric density is similar across growth environments for each crop. CE conditions enhanced the true protein content of soybeans and rice in particular. For the starch crops rice, wheat, and potato, total nitrogen (N) and non-protein N (NPN) contents of plant biomass generally increased under CE/hydroponic conditions relative

Table 5. Calcium (Ca) and phosphorus (P) contents (ppm, dry weight basis), and Ca/P ratio for edible portion of plants grown under controlled environments (CE) or in the field, and as reported in USDA Handbook No. 8 (Hdbk).[z]

Crop	Condition	Ca (ppm)	P (ppm)	Ca/P ratio	Reference
Soybean	CE	2,560	7,610	0.34	Nielsen et al. 1996
	Field	1,383	7,200	0.19	Nielsen et al. 1996
	Hdbk	3,030	7,697	0.39	Haytowitz and Matthews 1986
Cowpea	CE	707	8,071	0.09	--[x]
	Field	NA[y]	NA	NA	
	Hdbk	1,249	4,815	0.26	Haytowitz and Matthews 1986
Wheat	CE	567	4,069	0.14	McKeehen et al. 1996a
	Field	562	3,451	0.16	McKeehen et al. 1996a
	Hdbk	287	3,806	0.08	Haytowitz and Matthews 1986
Rice	CE	122	4,198	0.03	McKeehen et al. 1996a
	Field	103	3,408	0.03	McKeehen et al. 1996a
	Hdbk	378	3,013	0.13	Haytowitz and Matthews 1986
Canola	CE	2,150	7,441	0.29	Stephens, 1995
	Field	4,571	8,204	0.56	Stephens, 1995
	Hdbk	NA	NA	NA	NA
Potato	CE	214	4,091	0.05	McKeehen et al. 1996a
	Field	470	3,267	0.14	McKeehen et al. 1996a
	Hdbk	333	2,186	0.15	Haytowitz and Matthews 1986
Lettuce	CE	6,758	6,529	1.04	McKeehen et al. 1996a
	Field	6,349	3,074	2.07	McKeehen et al. 1996a
	Hdbk	11,333	4,167	2.72	Haytowitz and Matthews 1986

[z]Refer to footnotes of Table 1 for details on CE and field conditions for some crops.
[y]NA = Data not available.
[x]S.S. Nielsen 1995 unpubl. data. Purdue Univ., West Lafayette, IN.

Table 6. Mean ^{45}Ca uptake (±SD) by femur and ^{45}Ca absorption by rats[z,y]

Diets	^{45}Ca uptake (%)	^{45}Ca absorption (% of IP dose)
Control[x]	2.78±0.17	87.85±5.25
Diet 1[w]	2.93±0.17	92.59±5.49
Diet 2[w]	2.89±0.15	91.18±4.63
Diet 3[w]	2.79±0.17	89.86±5.15
IP[v]	3.16±0.16	

[z]n = 6–8 rats

[y] $$\% \, ^{45}\text{Ca absorption} = \frac{\% \, ^{45}\text{Ca of oral dose in femur}}{\% \, ^{45}\text{Ca of IP dose in femur}} \times 100$$

[x]Control diet was provided us a semi-purified American Institute of Nutrition (AIN) -76 diet.

[w]Diets 1, 2, and 3 are described in Table 3.

to the field, especially for leafy plant parts and roots. The nature of the NPN in seeds and tubers is yet uncharacterized in terms of diet safety and nutrition. Vegetative plant material is known to accumulate nitrate and other NPN (Aldrich 1980). Potentially toxic NO_3^- accumulated in vegetative tissues of CE-grown plants (e.g., lettuce) but was excluded from seeds and tubers. For lettuce leaves and radish roots, CE/hydroponic culture generally increased total N, protein N, and NO_3^-. While nitrogen-containing compounds are vital components for optimal crop yield, excess dietary nitrogen may be toxic to humans. Controlled environments enhanced phosphorous content of leaf lettuce, potato tuber, and rice grain. Controlled environments decreased the calcium/phosphorus ratio of lettuce leaves and white potato tubers. A calcium bioavailability study in rats suggested that calcium supplementation of diets from CE plants may be necessary for optimal calcium absorption. Bioavailability of minerals and macronutrients are issues deserving attention in CELSS vegan diets, and the degree to which controlled environments enhance or retard content and bioavailability require further research.

REFERENCES

Aldrich, S.R. 1980. Protein, nitrates, and nitrites in foods and feeds, p. 97–110. In: Nitrogen in relation to food, environment, and energy. Agr. Expt. Sta., Univ. Illinois, Urbana–Champaign.

AOAC. 1990. Official methods of analysis. 15th ed. Association of Official Analytical Chemists, Arlington, VA.

Belitz, H.D., and W. Grosch (ed.). 1987. Food chemistry. Translation from 2nd German ed. Springer Verlag, Berlin, Germany.

Bensadoun, A., and D. Weinstein. 1976. Assay of protein in the presence of interfering materials. Anal. Biochem. 70:241–250.

Buchowski, M.S., K.C. Sowizral, F.W. Lengemann, D.W Campen, and D.D. Miller. 1989. A comparison of extrinsic tracer methods for estimating calcium bioavailability to rats from dairy foods. J. Nutr. 119:228–234.

Goyal, S.S., and A.A. Hafez. 1990. Quantitative reduction and inclusion of plant tissue nitrate in Kjeldahl digestion. Agron. J. 82:571–576.

Havala, S. and J. Dwyer. 1993. Positon of the American Dietetic Association: Vegetarian diets. J. Am. Diet. Assoc. 93:1317–1319.

Haytowitz, D.B. and R.H. Matthews. 1986. Composition of foods. Ag. Handb. 8. USDA, Washington DC.

Heaney, P.R., P.D. Sarille, and R.R Recker. 1975. Calcium absorption as a function of calcium intake. J. Lab. Clin. Med. 85:881–890.

Koo, J., C. M. Weaver, M.J. Neylan, and G.D. Miller. 1993. Isotopic tracer techniques for assessing calcium absorption in rats. J. Nutr. Biochem. 4:72–76.

Lane, H.W., A. D. LeBlanc, L. Putcha, and P.A. Whitson. 1993. Nutrition and human physiological adaptations to space flight. Am. J. Clin. Nutr. 58:583–588.

McKeehen, J.D. 1994. Nutrient content of select controlled ecological life-support system candidate species grown under field and controlled environment conditions. M.S. Thesis. Purdue Univ., West Lafayette, IN.

McKeehen, J.D., C.A. Mitchell, R. M. Wheeler, B. Bugbee, and S.S. Nielsen. 1996a. Excess nutrients in hydroponic solutions alter nutrient content of rice, wheat, and potato. Adv. Space. Res. 18(4/5):73–83.

McKeehen, J.D., D.J. Smart, C.L. Mackowiak, R.M. Wheeler, and S.S. Nielsen. 1996b. Effect of CO_2 levels on nutrient content of lettuce and radish. Adv. Space Res. 18(4/5):85–92.

Mitchell, C. 1993. The role of bioregenerative life-support systems in a manned future in space. Trans. Kans. Acad. Sci. 96(1–2):87–92.

Mitchell, C. 1994. Bioregenerative life-support systems. Am. J. Clin. Nutr. 60:820S–824S.

Nickel, K.P., Nielsen, S.S., Mitchell, C.A., and Belury, M.A. 1996. Calcium bioavailability of vegan diets for a controlled ecological life-support system. FASEB J. 10:A232.

Nielsen, S.S., W.E. Brandt, and B.B. Singh. 1993. Genetic variability for nutritional composition and cooking time of improved cowpea lines. Crop Sci. 33:469–472.

Nielsen, S.S., L.J. Jurgonski, D.J. Smart, and B. Burgbee. 1996. Controlled environments alter nutrient content of soybeans. Paper presented at Council on Space Research (COSPAR) Meeting, Birmingham, England, July 15, 1996.

Ohler, T.A. 1994. Evaluation of cowpea (*Vigna unguiculata* (L.) *Walp*) as a candidate species for inclusion in controlled ecological llife-support systems. M.S. Thesis, Purdue Univ., West Lafayette, IN.

Saha, P.R., and P.R. Trumbo. 1996. The nutritional adequacy of limited vegan diet for a controlled ecological life support system. Adv. Space Res. 18(4/5):63–72.

Schneider, V.S., A. LeBlanc, and R.C. Rambaut. 1989. Bone and mineral metabolism, p. 214–221 In: Space Physiology and Medicine, A.E. Nicogossian, C.L. Huntoon, and S.L. Pool (eds.), Lea & Fibiger, Philadelphia, PA.

Stephens, S.D. 1995. Effect of growing conditions on the proximate and fatty acid compositions of three *Brassica napus* varieties and determination of the storage stability of screw-pressed extracted oils and meals from soybean, peanut and canola. M.S. Thesis, Purdue Univ., West Lafayette, IN.

Weaver, C. M., B.R. Martin, J.S. Ebner, and C.A. Kruger. 1987. Oxalic acid decreases calcium absorption in rats. J. Nutr. 117:1903–1906.

Quinoa: Candidate Crop for NASA's Controlled Ecological Life Support Systems

Greg Schlick and David L. Bubenheim

Quinoa (*Chenopodium quinoa* Willd.) is being considered as a potential "new" crop for NASA's Controlled Ecological Life Support System (CELSS). The CELSS concept will utilize plants to remove carbon dioxide from the atmosphere and generate food, oxygen, and water for the crew of long-term human space missions. Criteria for selection of potential crops include nutritional composition, harvest index, canopy stature, and life cycle duration.

Quinoa is a potential crop for CELSS due to it's high productivity and desirable nutritional characteristics. Of special interest is quinoa's high protein and mineral concentrations, in addition, the amino acid is suitable for a balanced human diet. Typically, CELSS has had to combine the nutritional values of several crops to obtain a suitable concentrations of amino acids for a human diet; quinoa may serve as a suitable complement to a balanced human diet.

Quinoa seeds have an exceptionally well balanced amino acid profile. For example, lysine, an essential amino acid deficient in most grain crops exceeds the concentrations necessary for proper amino acid nutrition in humans. In addition, the sulfur containing amino acids, cystine, and methionine are found in concentrations that are unusually high compared to other plants. Many minerals are found at concentrations greater than that reported for most grain crops; providing they are found in bioavailable forms, calcium, magnesium, and potassium are found in sufficient quantities for a balanced human diet. Most of the other minerals necessary for a balanced diet are found in greater concentrations than in most other crops as well. The pericarp of the quinoa seed contains saponins which may useful in the human diet and other applications for long-term space travel.

Conventionally, the seeds of quinoa are considered to be the edible form of the plant while the leaves have generally been overlooked. The leaves and sprouts can be eaten raw or cooked and provide a substantial amount of nutritive value. Utilizing quinoa as a leafy vegetable and a grain crop may provide nutritional and harvest versatility greater than that of a purely vegetative crop or that of a crop grown solely for it's seeds.

Significant increases in yield potential and nutrient content is attainable from crop plants grown in controlled environments (Bubenheim 1991). Ultimately, for a fully regenerative CELSS to be achievable we must maximize the productivity and quality of a each crop while minimizing the volume, mass, time and energy inputs. Because of the significant food value associated with quinoa the feasibility of controlled environment culture has been investigated. This manuscript summarizes efforts to select appropriate cultivars of quinoa for controlled environment production and efforts to determine optimal growing conditions.

METHODOLOGY

Cultivar Selection in the Greenhouse

Ten different accessions of quinoa were examined for potential use in controlled environment hydroponic production systems. The genotypes evaluated in the initial screening and their origins were: QL3 and QH3 from the Univ. of Cambridge, England; CO407XISLUGA and CO407 Heat Tolerant Population 1 from Colorado State University, Fort Collins, Colorado; 'Real', 'Sajama', and 'Cochabamba 250' from Bolivia; 'Blanca de Junin' from Peru; and 'Faro', and 'Temuco' from Chile. The primary selection criteria for advancement of the genotypes in the selection process were canopy height, biomass production, seed yield, harvest index, and life cycle duration.

Because of the general lack of experience anywhere in the hydroponic culture of quinoa, the preliminary trials focused on providing an appropriate nutritional regime. Following several seed-to-seed trials the candidate accession list was trimmed to four: 'Real', CO407, CO407 Heat Tolerant Population 1, and QL3. The other cultivars exhibited extreme growth habits under all conditions and were eliminated from the screening early in the process. A detailed characterization of the crop physiology of quinoa grown in controlled environment, hydroponic production was pursued using the four cultivars and results are presented only for those genotypes. A ranking was developed through a series of production studies while general responses to environmental factors such as photoperiod, nutrient composition, planting density, and temperature were determined. The top ranking cultivar was the first to be advanced to growth chamber trials.

Results of greenhouse studies, presented in Fig. 1 and Table 2, were all from plants grown in a greenhouse using a recirculating, deep-film hydroponic system maintained at 21°(± 3.0°C) with elevated carbon dioxide levels 1000 mmol mol^{-1} (Schlick and Bubenheim 1993). The plants were grown under both long and short day conditions (daily photon flux of 49.7 and 15.3 mol m^{-2} d^{-1}, respectively). The nutrient solution (Table 1) was maintained at a pH of 5.8 and a electrical conductivity of 0.50 mS (seedling solution) during the first 20 days and then increased to an electrical conductivity of 1.50 mS for the duration of the experiment. The nutrient solution was delivered to the hydroponic trays at a flow rate of 14 liters min^{-1}.

Growth Chamber Yield Studies

The initial efforts to define the yield potential for quinoa in highly controlled environments were focused on a series of experiments designed to determine photoperiod and irradiance responses. Three nearly identical growth chambers (EGC Model M-12) each with approximately 0.8 m^2 of growing area were modified to increase the number of environmental variables controlled and to better maintain those conditions in the growing environment. The standard C-3 microprocessor (standard EGC control system) was removed and replaced with a Optomux analog and digital brain board (Opto 22) which supplied and received the input/output signal to and from a computer running Paragon 500 Control Software (Intec Controls Corp.). The control package allowed precise and complete computer control for all the environmental parameters.

The photoperiod response of quinoa was quantified by presenting three photoperiod treatments (16, 12, and 8 h) while maintaining the daily photosynthetic photon flux consistent among all photoperiods at 51.8 mol

$m^{-2} d^{-1}$. The instantaneous photosynthetic photon flux levels maintained in the chambers were 800, 1200, and 1800 μmol $m^{-2} s^{-1}$, respectively. In a subsequent trial, designed to quantify the irradiance response of quinoa, photoperiod was maintained at 8 h with irradiance treatments of 800, 1200, and 1800 μmol $m^{-2} s^{-1}$. The daily photosynthetic photon flux was 25.9, 34.6 and 51.8 mol $m^{-2} d^{-1}$, respectively.

In all of the growth chamber studies, the relative humidity was maintained at 80% (±0.1% RH), temperature was maintained at 21°(±0.5°C), and the carbon dioxide level was maintained at 1000 mmol mol^{-1} (±10 mmol mol^{-1}). The root zone environment consisted of a continually recirculating deep-film solution with a delivery rate of 14 liters min^{-1}. The nutrient solution (Table 1) was maintained at a pH of 5.8 and the electrical conductivity was maintained at 0.50 mS during the first 20 days and then increased to 1.50 mS for the remainder of the experiment.

Assessing Growth and Yield

For both the greenhouse and growth chamber experiments growth, development, and morphology were characterized over time. Individual plants in the communities were measured weekly for height, branch number, internode length and leaf area index. At the same time, the days to floral initiation, days to flowering, duration of seed fill, and days to maturity were determined. At harvest, the plants were oven dried at 60°C to a uniform weight before dry mass was determined for seeds, leaves, stems, and roots, and yield analysis performed.

Table 1. Hydroponic nutrient solution for quinoa in controlled environments.

Nutrient Source	Seedling solution concn.	Standard solution concn.
$Ca(NO_3)_2 \cdot 4H_20$	0.5 mM	2 mM
KNO_3	0.5 mM	2 mM
KH_2PO_4	0.15 mM	0.6 mM
$MgSO_4 \cdot 7H_2O$	0.13 mM	0.5 mM
K_2SO_4	0.13 mM	0.5 mM
$FeNO_3 \cdot 9H_2O$	2.5 μM	10 μM
$FeCl_3$+HEDTA	11.3 μM	45 μM
$MnCl_2 \cdot 4H_2O$	0.75 μM	3 μM
$ZnSO_4 \cdot 7H_2O$	0.75 μM	3 μM
$CuSO_4 \cdot 5H_2O$	0.05 μM	0.18 μM
$Na_2Mo_4 \cdot 2H_2O$	0.02 μM	0.09 μM
H_3BO_3	0.5 μM	2 μM

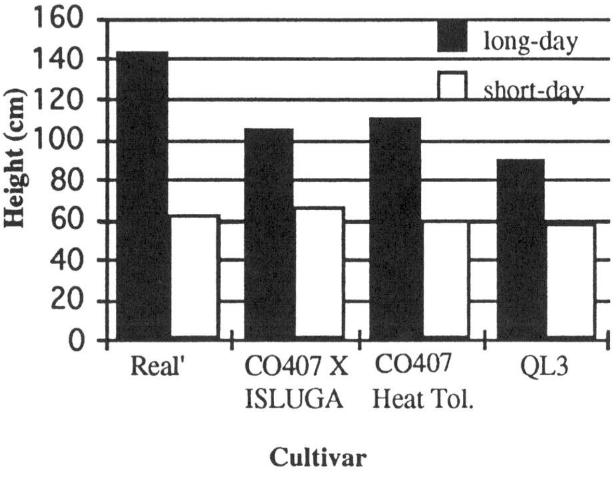

Fig. 1. Effect of daylength on the height characteristics of four selected *C. quinoa* cultivars grown in hydroponic greenhouse conditions.

Table 2. Growth and yield characteristics of four selected quinoa cultivars grown in hydroponic greenhouse conditions.

Cultivar	Days to flower	Days to harvest	Harvest index	Total biomass (g m^{-2})
CO407 x ISLUGA	24	103	0.37	542
CO407 heat tol.	27	103	0.17	453
Real	30	89	0.28	184
QL3	27	89	0.49	503

Analysis of Nutritional Value

Total proximate analysis, mineral composition, and amino acid profile were determined for each of the separated tissues: roots, leaves, stems, and seeds. Fatty acid profile, carbohydrate profile, total starch content, ascorbic acid (vitamin C), thiamin (vitamin B1), riboflavin (vitamin B2), niacin (vitamin B3), vitamin A, alpha-tocopherol (vitamin E), and beta-carotene analysis was, at this time, only conducted on the seed and leaf tissue.

RESULTS AND DISCUSSION

Cultivar Determination

It is critical for a candidate CELSS crop to meet certain criteria which include small canopy stature, high harvest index and a favorable nutritional composition. While most of the cultivars examined exhibited a positive response to controlled environment production in reference to harvest index and nutritional composition, many grew exceedingly tall, some growing as tall as 2.5 m in height. Fig. 1 shows the height characteristics of four of the examined accessions. The size limitations of controlled environment chambers and eventually the size constraints of extra-terrestrial crop production systems force volume/height considerations in cultivar selection. Of the examined cultivars, 'Real' and QL3 showed the greatest promise in canopy stature, biomass partitioning, and seed yields (Table 2). 'Real' exhibited short height (60 cm) and maintained relative seed yield under short day conditions. The height of the QL3 canopy was similar at 58 cm under short days but was also the shortest under long day conditions. Additionally, QL3 exhibited little branching and consistently had the highest seed yield and harvest index under both long and short day conditions, compared with the other cultivars evaluated. The performance of QL3 in controlled environment production presented here is equivalent to or surpasses the performance of other crops selected as primary candidates for CELSS. The canopy height, yield, biomass accumulation rate, life cycle duration, and harvest index are all appropriate for continued consideration of quinoa as a CELSS crop and greater definition of yield potential is justified.

Planting Density

Previous crop production studies using controlled environment hydroponic systems (Bubenheim 1991) demonstrated that dramatic increases in yield and harvest index resulted in part from a higher planting densities when compared to average field seeding rates. In controlled environments, higher planting densities result in improved light interception and increased yields. Optimal field planting density reported for quinoa is approximately 32 plants m^{-2} (Johnson 1985). From our yield per unit area, harvest index, and biomass accumulation data, it is apparent that a planting density of approximately 140 plants m^{-2} appears to be optimal for hydroponic controlled environment production (data not shown).

Photoperiod and Irradiance Effects on Canopy Height, Biomass Accumulation and Allocation, and Yield

It appears that QL3 responds favorably to a combination of high irradiance (1800 μmol m^{-2} s^{-1}) and short day lengths (8 h). Maintaining a daily photon flux of 51.8 mol d^{-1} and increasing the photoperiod results in an increase in canopy height. Maintaining the same photoperiod and altering the instantaneous flux results in a similar height response. This suggests that the optimal photo conditions for maintaining a short stature, high seed yield, and maximum harvest index for QL3 is short days with high instantaneous flux.

Biomass production and allocation was also affected by photoperiod and irradiance (Tables 3 and 4). Long days (12 and 16 h) and low instantaneous photon flux levels promoted biomass production. However, the allocation of biomass differed greatly depending on the photoperiod and irradiance environment. While both shoot mass (including seeds) and root mass increased with increased biomass accumulation, the proportion of the shoot mass partitioned to the seed was greater in the short photoperiod and high irradiance environments. Plants that grew under a short photoperiod had a relatively constant shoot/root ratio. The plants that grew under long day conditions exhibited smaller ratios. The harvest index (seed only) in the 8 h/1800 μmol m^{-2} s^{-1} treatment was greater than that of the other treatments.

Most of the attention to this crop has been focused around the seed, but it is plausible that the leaves may provide an additional source of food and nutrition. If the edible leaves are added into the harvest index calculation, the harvest index can exceed 80%. Alternate harvesting techniques need to be examined, such as leaf harvest during the vegetative growth phase, to determine the effects, if any, this has on seed fill and yield.

Nutritional Composition

The protein concentrations (Table 5) found in the seed (20.9%) of quinoa grown in controlled environments is consistently greater than the average values obtained from field studies (16.5%). The increase in seed protein values is likely due to increased nutrient availability characteristic of recirculating hydroponic culture systems and the decreased stress of controlled environments. The values shown in Table 5 have been corrected to show zero moisture content. Due to the lack of a specific nitrogen:protein conversion ratio, a factor of 6.25 was used to calculate protein content.

Lipid concentrations in the seed and leaves are consistent with that reported for field data. For the seed, 64% of the lipids are in the form of polyunsaturates, 17% are in the form of mono-unsaturates and the remainder are saturated (19%).

Table 3. Total biomass allocation in quinoa cv. QL3 as influenced by photoperiod and irradiance.

Treatment						
Photoperiod (h)	Instantaneous flux (μmol $m^{-2}s^{-1}$)	Daily flux (mol $m^{-2}s^{-1}$)	Total biomass (g m^{-2})	Shoot mass (g m^{-2})	Root mass (g m^{-2})	Shoot/root ratio
8	1800	51.8	922±49.1	851±27.4	71±3.8	11.9
12	1200	51.8	1450±81.0	1314±32.4	136±6.4	8.9
16	900	51.8	1593±93.2	1444±49.5	149±5.2	9.7
8	1800	51.8	1021±102.3	944±51.5	76±11.2	12.4
8	1200	34.6	1336±87.9	1241±47.9	95±6.2	13.1
8	900	25.9	1546±108.0	1435±57.6	112±6.6	12.8

Table 4. Yield characteristics of quinoa cv. QL3 as influenced by photoperiod and irradiance.

Treatment						
Photoperiod (h)	Instantaneous flux (μmol $m^{-2}s^{-1}$)	Daily flux (mol $m^{-2}s^{-1}$)	Seed yield (g m^{-2})	Leaf yield (g m^{-2})	Harvest index seed only	Harvest index seed&leaf
8	1800	51.8	507±11.2	250±12.2	0.55	0.82
12	1200	51.8	542±19.3	526±17.1	0.34	0.67
16	900	51.8	406±17.1	521±17.4	0.28	0.64
8	1800	51.8	592±17.7	128±21.0	0.58	0.71
8	1200	34.6	494±10.2	432±14.6	0.37	0.69
8	900	25.9	557±22.6	499±8.2	0.36	0.68

Table 5. Proximate analysis of quinoa cv. QL3 grown in controlled environment and comparison with quinoa grown in a field environment (corrected for zero moisture).

Growing condition Organ	Component (%)				
	Protein[z]	Fat	Ash	Carbohydrate	Crude fiber
Hydroponically grown					
Seeds	20.9	5.5	4.9	78.5	2.8
Stems	11.2	0.7	7.2	57.8	27.2
Roots	13.3	0.4	15.0	74.8	26.3
Leaves	18.7	1.0	24.7	52.8	8.4
Field grown[y]					
Seeds	16.6	6.4	3.9	69.3	3.3
Leaves	21.9	3.8	19.1	35.0	13.7

[z]Total Kjeldahl Nitrogen
[y]Koziol 1992

The ash content in the seeds of controlled environment produced plants is consistent with that of field data (5.5% and 6.4%, respectively), yet the ash content in leaves of controlled environment plants is higher than that observed in field trials (24.7% and 19.1%, respectively).

Crude fiber concentrations found in seed is comparable to field results (2.8% and 3.3%, respectively) presented by Koziol (1992). The content of fiber found in the leaves of quinoa grown in controlled environments (8.4%) is considerably lower than any of the previously reported fiber concentrations (13.7%) from field studies. The reason for the low concentrations in the leaves of the hydroponically grown quinoa is unknown at this time, but may be related to the decreased stress in the controlled environment.

Carbohydrate concentration is greater in the controlled environment grown quinoa (78.5% seeds; 52.8% leaves) than in quinoa grown under field conditions (69.3% seeds; 35.0% leaves) (Koziol 1992). The quality of the seed carbohydrate is not affected by controlled environment production; approximately 47.5% of the carbohydrates is in the form of starch, similar to the field results.

Table 6 summarizes the amino acid composition of quinoa from multiple field trials (Risi and Galwey 1984) and of the controlled environment studies completed at NASA/Ames Research Center. Of primary interest to CELSS is the high lysine and sulfur containing amino acid (methionine and cystine) concentrations. Currently, CELSS has found it necessary to combine nutritional values of crops such as soybeans (*Glycine max*) and wheat (*Triticum aestivum*) to obtain a suitable amino acid pattern that meets the needs of humans during long-term space travel. Amino acid analysis from field trials indicate quinoa as a good source of lysine, methionine, cystine as well as all the other essential amino acids and meets or exceeds the recommendations for a proper amino acid nutrition. In hydroponic production, all amino acid concentrations have remained similar or increased in concentration with the exception of the sulfur-containing amino acids. The relative decrease in cystine and methionine may be due to a number of factors, such as a sub-optimal sulfur content within the nutrient solution or a response to controlled environment production (lack of stress) which may suppress the formation of the sulfur metabolites. Sulfur concentrations in the nutrient solution and specific stress response is currently being studied to better understand the factors influencing sulfur containing amino acid synthesis in quinoa.

Quinoa has a very high mineral value and may prove to be beneficial for long-term human space travel. The mineral concentrations reported for quinoa seem to vary dramatically. This may due to the soil type and mineral composition of the region and/or fertilizer application. In a controlled environment, the nutrient composition can easily be modified and maintained to meet the demands required for optimal plant growth or human nutritional requirements. It may also be possible to modify the nutrient solution in order to manipulate the mineral concentrations found in the tissue of quinoa. Table 7 provides a comparison of quinoa grown hydroponically to those of field grown plants reported by Koziol (1992) and Souci (1994). Table 8 compares the mineral concentrations of quinoa with other major grains and leafy vegetable crops. Quinoa is a good source of calcium, phosphorus, iron, sodium, and potassium. Modifications made to the nutrient solution have a direct effect on the mineral concentrations in the tissues. This may prove to be an excellent way to provide the proper mineral requirements for human nutrition to the crew of a space mission.

Quinoa appears to be a good source of vitamin E (10.04 mg/100 g), which is substantially higher than other grain crops and a adequate thiamin (0.68 mg/100 g), which is found in similar concentrations to other grain crops and niacin (1.53 mg/100 g), which is lower than reported from other grain crops, but still in biologically significant amounts. While not presented here, many of the vitamin concentrations from quinoa grown in controlled environments increased from the values presented from field grown studies. Quinoa may prove to a valuable source for some of the vitamins required for a human nutrition during space travel.

The pericarp of the quinoa seed contains saponins, a plant glycoside. Saponins are a structurally diverse group of naturally occurring compounds found mainly in plants (Price et al. 1987). These saponins impart a

Table 6. Amino acid composition of quinoa compared with wheat, soybean, skimmed milk, and amino acid requirements of humans.

	Content (g/100g protein)						
Amino acid	Hydroponic quinoa leaves	Hydroponic quinoa seeds	Field quinoa[z] seeds	Wheat[z]	Soybean[z]	Skimmed milk[z]	Amino acid requirements[y]
Isoleucine	3.2	3.9	5.2	3.8	4.9	6.3	4.0
Leucine	5.6	6.4	6.7	6.8	7.6	9.7	6.7
Lysine	3.5	5.9	6.2	2.9	6.4	7.7	5.0
Phenylalanine	3.9	4.1	3.8	4.5	4.9	4.9	3.2
Arginine	4.0	9.4	7.9	4.8	7.2	3.7	2.0
Histidine	1.2	3.0	2.7	2.2	2.5	2.6	1.7
Alanine	3.9	4.0	4.4	3.8	4.3	4.0	-.-
Aspartic acid	8.0	9.0	8.1	5.3	12.0	8.3	-.-
Glutamic acid	14.0	15.0	14.0	27.0	18.0	23.0	-.-
Glycine	4.3	5.3	5.7	4.0	4.2	2.2	-.-
Proline	3.6	3.5	4.0	10.0	5.5	11.0	-.-
Serine	3.6	4.4	4.6	5.0	5.6	6.0	-.-
Tyrosine	2.6	3.2	3.1	3.1	3.5	5.0	3.2
Phenylalanine plus Tyrosine	6.5	7.3	6.9	7.6	8.4	9.9	6.4
Cysteine	0.6	1.0	1.4	2.3	1.5	0.9	1.3
Methionine	0.8	1.0	1.4	1.7	1.4	2.5	1.9
Threonine	3.5	3.5	4.1	3.1	4.2	4.6	3.4
Tryptophan	1.6	1.1	1.2	1.1	1.3	1.4	1.1
Valine	4.0	4.5	4.6	4.7	5.0	6.9	4.6
Cysteine plus Methionine	1.4	2.0	2.8	4.0	2.9	3.4	3.2

[z]Souci et al. 1994
[y]Scott 1986

bitter taste and tend to foam in aqueous solutions. Koziol (1992) reports concentrations ranging from 0.01%–4.65% for saponins (dry weight) in quinoa. Until recently, saponins have been considered to be highly toxic. While, extremely toxic to cold-blooded animals, the oral toxicity to mammals is low. Saponins present in common foods, including quinoa, seem to free from significant oral toxicity (Oser 1966; Ishaaya et al. 1969; Phillips et.al. 1979; Malinow 1982 and 1984; Oakenfull and Sidhu 1990) in humans. Saponins naturally present in foodstuffs are non-toxic and suggests they may even be beneficial in a human diet. Ruales (1992) showed that the digestibility of quinoa protein is comparable to that of other high quality food proteins and that the saponins do not exert any negative effect on the nutritive quality of the protein.

However, because saponins may impart a bitter taste, the separation of saponins from quinoa seed is easily accomplished by rinsing the seed in cold alkaline water or by mechanical abrasion. Saponins are useful in producing soaps, detergents, shampoos, cosmetics, and pharmaceuticals (Johnson and Ward 1993). The broad spectrum of useful products made with saponins may prove to be beneficial for long-term space habitation.

CONCLUSIONS

Quinoa is clearly a candidate for inclusion among the suite of crops to be grown as part of a Controlled Ecological Life Support System for long-term space exploration. Quinoa responded well to controlled envi-

Table 7. Tissue analysis of all plant parts of quinoa grown in hydroponic solution and seed analysis in field conditions.

Treatment	P (%)	Mg (%)	K (%)	Ca (%)	S (%)	Na (ppm)	Zn (ppm)	Cu (ppm)	Fe (ppm)	Mn (ppm)
Hydroponic										
Roots	0.85	0.84	3.89	1.67	-.-	1232	80	14	2076	159
Stems	0.18	0.11	3.26	0.99	-.-	49	27	0	89	27
Leaves	0.37	0.94	11.81	2.92	0.69	159	88	4	293	353
Seeds	0.27	0.92	2.17	0.16	0.24	36	94	2	186	48
Field										
Seed[z]	0.33	0.28	0.80	0.08	-.-	96	25	0	79	28
Leaves[y]	0.04	0.08	0.38	0.15	-.-	2890	5	1	9	--

[z]Souci et al. 1994
[y]Koziol 1992

Table 8. Comparative mineral nutrition values of quinoa seeds and leaves with other major seed and leafy vegetable crops.

	Content (mg/100g)				
Crop	Calcium	Phosphorus	Iron	Sodium	Potassium
Quinoa seed	160	270	19	4	2170
Wheat	38	341	3	8	381
Soybean	201	550	7	5	1800
Quinoa leaves	2920	370	29	16	1181
Spinach	126	55	4	65	633
Cabbage	35	30	1	4	266

ronment production practices with large increases in seed production, maintenance of short canopy stature, and increased harvest index, as compared with the field. Quinoa productivity is similar to that of other current candidate species. In addition, the very desirable nutritional composition, leading to the original interest in quinoa, seems to be enhanced in controlled environment production. The protein values increase and the relative proportions of amino acids meet or exceed the requirements of humans, except for cysteine and methionine. The mineral concentration is exceptional and has the potential of being easily manipulated to meet the dietary needs for human nutrition. The high protein levels, unique amino acid pattern, vitamins, minerals and desirable growth and yield features make quinoa a likely candidate crop for CELSS applications.

REFERENCES

Bubenheim, D. 1991. Plants for water recycling, oxygen regeneration, and food production. Waste Management Res. 9:435–443.

Ishaaya, I., Y. Birk, A. Bondi, and Y. Tencer. 1969. Soyabean saponins IX. Studies of their effect on birds, mammals and cold blooded organisms. J. Sci. Food Agr. 20:433–436.

Johnson, D.L. and R. Croissant. 1985. Quinoa production in Colorado. Service In Action no.112. Colorado State Univ., Cooperative Extension, Fort Collins.

Johnson, D.L. and S. Ward. 1993. Quinoa. p. 222–227. In: J. Janick and J.E. Simon (eds.), New crops. Wiley, New York.

Koziol, M.J. 1992. Chemical composition and nutritional evaluation of quinoa. J. Food Comp. Anal. 5:35–68.

Malinow, M., W. McNulty, D. Houghton, S. Kessler, P. Stenzel, S. Goodnight, E. Bardana, J. Polatay, P. McLaughlin, and A. Livingston. 1982. Lack of toxicity of alfalfa saponins in monkeys. J. Med. Primatol. 11:106–118.

Malinow, M., P. McLaughlin, E. Bardana, S. Craig. 1984. Elimination of toxicity from diets containing alfalfa seeds. Food Cosmet. Toxicol. 22:583–587.

Oakenfull, D. and G. Sidhu. 1990. Could saponins be a useful treatment for hypercholesterolaemia? Eur. J. Clin. Nut. 44:79–88.

Oser, B. 1966. An evaluation of *Yucca mohavensis* as a source of food grade saponin. Food. Cosmet. Toxicol. 4:57–61.

Phillips, J. K., Butterworth, I. Gaunt, J. Evans, and P. Grasso. 1979. Long-term toxicity study of quillaja extract in mice. Food Cosmet. Toxicol. 17:23–27.

Risi, J. and N.W. Galwey. 1984. The Chenopodium grains of the Andes: Inca crops for modern agriculture. Adv. Appl. Biol. 10:145–217.

Ruales, J. and B.M. Nair. 1992. Nutritional quality of the protein in quinoa (*Chenopodium quinoa* Willd) seeds. Plant Foods Hum. Nutr. 42(1):1–12.

Schlick, G. and D.L. Bubenheim. 1993. Quinoa: An emerging "new" crop with potential for CELSS. NASA Tech. Paper #3422.

Scott, M.L. 1986. Nutrition of humans and selected animal species. Wiley, New York.

Souci, S.W., W. Fachmnn, and H. Kraut. 1994. Food composition and nutrition tables. CRC Press, Boca Raton, FL.

Afterword

Jules Janick

With the onset of the development of agriculture 10,000 years ago, during what is now known as the Neolithic Revolution, humans and crop plants have undergone a convergence of evolutionary development. It is certainly clear that the evolution of our crop plants has been markedly altered as a result of their domestication by humans. In many cases their development has been so changed as to almost obliterate their obvious origins, making many of our crop plants taxonomic nightmares. In turn, the development of crop plants and increases in agricultural technology have permitting explosive growth of populations that no doubt have had a tremendous effect on human evolution. Traits such as swiftness of foot and keenness of eye, once essential to the hunter, may no longer be as relevant to our destiny as traits affecting social behavior and abstract reasoning. The explosive growth of population so far has been paralleled by an explosive growth in food production in most of the world, so far delaying the gloomy predictions of the dour Thomas Malthus.

However, recent advances in agricultural technology have led to a reduction in diversity of our crop plants. Technology and politics have interacted in an unplanned way to exacerbate this trend, and subsidization of surpluses in some crops have retarded the development of alternate crop strategies and reduced diversification. The reliance of human civilization on a handful of crops seems on the face of it perilous, and many, such as Noel Vietmeyer, have expressed their deep concern. Biodiversity has come to be regarded as a talisman to provide insurance against crop disasters from biotic and abiotic factors, as a hedge against depletion of nonrenewable resources such as petroleum, as a way to provide a more sustainable agriculture through their benefits to our ecology, and as a means for improving human welfare by increasing diet diversity and adding valuable pharmacological resources. The dangers of reduced diversity and monoculture become apparent in the United States through natural disasters such as the outbreak of southern corn blight in the 1970s, gray mold in 1995, and drought in the great plains in the winter of 1995/96.

However, the success of soybean in the United States and canola in Canada and the emergence of new crop candidates in the last decade, such as cactus fruit, crambe, meadowfoam, Oriental vegetables, pearl millet, and *Taxus* (for Taxol), underscore the possibility of Biodiversity Engineering as an emerging agricultural field. Could it be that we new croppers are the vanguard of a new agricultural movement? We have competition from proponents of new uses for old crops and biotechnologists, both proposing approaches that might reduce rather than expand diversity. Yet both of these concepts can be shown to be compatible with the new crops movement, because both new uses technology and biotechnology are necessary techniques to expand the economic usefulness of new crops and to adapt them quickly. As Abraham Lincoln said in another context: "We are not enemies, but friends."

With the publication of this third volume on new crops, a case can be made that the progress and problems and rational for new crops has been well articulated. Where do we go from here? The position paper *Diversifying U.S. Crop Production* (see page 98) has made the case for increased federal involvement in new crop research and development and proposes the creation of a "Jeffersonian Initiative," in honor of the personal commitment of Thomas Jefferson to improve American agriculture through the development of new crops. It proposes a Jefferson Institute of Crop Diversification be established comprised of a national research and development center and a series of cooperative regional centers to address the interlocking problems of breeding, production, utilization and marketing of new and underexploited crops. Peter Felker has independently made the suggestion of a grants program for new crops research.

Exploitation of new crops is not a new and untried technology but rather the continuation of the American tradition of exploration and evaluation of new crop solutions. A very small part of the billions spent for crop subsidies is needed to promote new crops research and development, and at present there is no voice but ours. The Jefferson Initiative proposes the kind of substantial, long-term and coordinated framework necessary for the creation of a successful national strategic program in new crops development. At present the European Economic Community is in the lead in this endeavor as they search for new industrial crops, emphasizing biodiesels. It is clear that a new attitude is needed. As of this moment, the decline in feed grains surpluses, poor yields in 1995, and increased demand from China have produced increases in commodity prices. Thus, in the short term, there will be increased demand for traditional crops. The investment of a small part of the expected windfall into a more rational policy to exploit the tremendous diversity that is our heritage and which we have chosen to ignore at our peril would appear to be a wise and prudent investment in the future.

Index of Species, Crops, and Products

B

C

D

G

H

M

N

T

X

Y

Z

Index of Authors